HANDBOOK OF WEATHER, CLIMATE, AND WATER

HANDBOOK OF WEATHER, CLIMATE, AND WATER

Dynamics, Climate, Physical Meteorology, Weather Systems, and Measurements

Edited by

THOMAS D. POTTER

BRADLEY R. COLMAN

A JOHN WILEY & SONS, INC., PUBLICATION

Published by John Wiley & Sons, Inc., Hoboken, New Jersey.
Published simultaneously in Canada.

Library of Congress Cataloging-in-Publication Data:

The handbook of weather, climate, and water : dynamics, climate physical meteorology, weather systems, and measurements / co-chief editors, Thomas D. Potter and Bradley R. Colman.
p. cm.
Includes bibliographical references.
ISBN 0-471-21490-6 (acid-free paper)
1. Dynamic meteorology. 2. Atmospheric physics. 3. Weather.
I. Potter, Thomas D., 1955- II. Colman, Bradley Roy.
QC880.H34 2003
551.51′5--dc21

2003003584

Printed in the United States of America.

10 9 8 7 6 5 4 3 2 1

CONTENTS

PREFACE

The *Handbook of Weather, Climate, and Water* provides an authoritative report at the start of the 21st Century on the state of scientific knowledge in these exciting and important earth sciences. Weather, climate, and water affect every person on earth every day in some way. These effects range from disasters like killer storms and floods, to large economic effects on energy or agriculture, to health effects such as asthma or heat stress, to daily weather changes that affect air travel, construction, fishing fleets, farmers, and mothers selecting the clothes their children will wear that day, to countless other subjects.

During the past two decades a series of environmental events involving weather, climate, and water around the globe have been highly publicized in the press: the Ozone Hole, Acid Rain, Global Climate Change, El Niños, major floods in Bangladesh, droughts in the Sahara, and severe storms such as hurricane Andrew in Florida and the F5 tornado in Oklahoma. These events have generated much public interest and controversy regarding the appropriate public policies to deal with them. Such decisions depend critically upon scientific knowledge in the fields of weather, climate, and water.

One of two major purposes of the Handbook is to provide an up-to-date accounting of the sciences that underlie these important societal issues, so that both citizens and decision makers can understand the scientific foundation critical to the process of making informed decisions. To achieve this goal, we commissioned overview chapters on the eight major topics that comprise the Handbook: Dynamics, Climate, Physical Meteorology, Weather Systems, Measurements, Atmospheric Chemistry, Hydrology, and Societal Impacts. These overview chapters present, in terms understandable to everyone, the basic scientific information needed to appreciate the major environmental issues listed above.

The second major purpose of the Handbook is to provide a comprehensive reference volume for scientists who are specialists in the atmospheric and hydrologic areas. In addition, scientists from closely related disciplines and others who wish to get an authoritative scientific accounting of these fields should find this work to be of great value. The 95 professional-level chapters are the first comprehensive and integrated survey of these sciences in over 50 years, the last being completed in 1951 when the American Meteorological Society published the Compendium of Meteorology.

The *Handbook of Weather, Climate, and Water* is organized into two volumes containing eight major sections that encompass the fundamentals and critical topic areas across the atmospheric and hydrologic sciences. This volume contains sections on the highly important topics of Dynamics, Climate, Physical Meteorology, Weather Systems, and Measurements. Dynamics describes the nature and causes of atmospheric motions mostly through the laws of classical Newtonian physics supplemented by chaos theory. Climate consists of the atmosphere, oceans, cryosphere and land surfaces interacting through physical, chemical, and biological processes. Physical Meteorology presents the laws and processes governing physical changes in the atmosphere through chapters on atmospheric thermodynamics, atmospheric radiation, cloud physics, atmospheric electricity and optics, and other physical topics. The section on Weather Systems describes remarkable advances in weather forecasting gained by an increased understanding of weather systems at both large and small scales of motion and at all latitudes. The section on Measurements describes the many advances in the sensing of atmospheric conditions through constantly improving instrumentation and data processing.

To better protect against weather, climate, and water hazards, as well as to promote the positive benefits of utilizing more accurate information about these natural events, society needs improved predictions of them. To achieve this, scientists must have a better understanding of the entire atmospheric and hydrologic system. Major advances have been made during the past 50 years to better understand the complex sciences involved. These scientific advances, together with vastly improved technologies such as Doppler radar, new satellite capabilities, numerical methods, and computing, have resulted in greatly improved prediction capabilities over the past decade. Major storms are rarely missed nowadays because of the capability of numerical weather-prediction models to more effectively use the data from satellites, radars, and surface observations, and weather forecasters improved understanding of threatening weather systems. Improvements in predictions are ongoing. The public can now rely on the accuracy of forecasts out to about five days, when only a decade or so ago forecasts were accurate to only a day or two. Similarly, large advances have been made in understanding the climate system during the past 20 years. Climate forecasts out beyond a year are now made routinely and users in many fields find economic advantages in these climate outlooks even with the current marginal accuracies, which no doubt will improve as advances in our understanding of the climate system occur in future years.

Tom Potter
Brad Colman

Color images from this volume are available at ftp://ftp.wiley.com/public/sci_tech_med/weather/.

DEDICATION AND ACKNOWLEDGMENTS

Many people have assisted in the production of this Handbook—the Contributing Editors, the Authors, our editors at Wiley, friends too numerous to mention, and our families who supported us during the long process of completing this work. Professor Peter Shaffer, University of Washington, is owed deep appreciation for his untiring generosity in sharing his experience and talent to solve many problems associated with this large project. They all deserve much credit for their contributions and we want to express our deep thanks to all of them.

Finally, we want to dedicate this work to Tom Lockhart, the Contributing Editor of the Measurements part of the Handbook. Tom passed away in early 2001 and we regret that he will not be able to see the results of his efforts and those of his colleagues in final form.

Tom Potter
Brad Colman

CONTRIBUTORS

Steven A. Ackermann, NOAA-CIRES Climate Diagnostic Center, 1225 West Dayton Street, Madison, WI 53706-1695

Amanda Adams, University of Wisconsin-Madison, Department of Atmospheric and Ocean Sciences, 1225 West Dayton Street, Madison, WI, 53706-1695

W. D. Bach, Jr., Army Research Program, AMXRL-RO-EN, P.O. Box 12211, Research Triangle Park, NC 27709-2211

R. A. Baxter, 26106 Alejandro Road, Valencia, CA 91355

Craig F. Bohren, Pennsylvania State University, Department of Meteorology, College of Earth and Mineral Sciences, 503 Walker Building, University Park, PA 16802-5013

Gordon Bonan, NCAR, Box 3000, Boulder, CO 80307-3000

Fred V. Brock, Oklahoma Climate Survey, 100 East Zboyd Street, Room 1210, Norman, OK 73019

H. Brooks, NOAA/NSSL, 1313 Halley Circle, Norman, OK 73069

John R. Christy, University of Alabama in Huntsville, Earth Sciences Center, Huntsville, AL 35899

Robert E. Dickinson, Georgia Tech, School of Earth and Atmospheric Sciences, Atlanta, GA 30332-0340

NOLAN J. DOESKEN, Colorado State University, Department of Atmospheric Sciences, Colorado Climate Center, Fort Collins, CO 80523

C. DOSWELL III, NOAA/NSSL, 1313 Halley Circle, Norman, OK 73069

D. DOWELL, NOAA/NSSL, 1313 Halley Circle, Norman, OK 73069

LAWRENCE B. DUNN, National Weather Service, 2242 West North Temple, Salt Lake City, UT 84116-2919

DAVID R. EASTERLING, NCDC, 151 Patton Avenue, Room 120, Asheville, NC 28801-5001

PAUL M. FRANSOLI, 1112 Pagosa Way, Las Vegas, NV 8912

GRANT W. GOODGE, P.O. Box 2178, Fairview, NC 28730

JOHN GYAKUM, McGill University, Department of Atmospheric and Oceanic Sciences, 805 Sherbrooke Street West, Montreal, Quebec H3A 2K6

W. H. HAGGARD, 150 Shope Creek Road, Asheville, NC 28805

JOHN HALLETT, Desert Research Center, Atmospheric Sciences Center, 2215 Raggio Parkway, Reno, NV 89512-1095

JACKSON R. HERRING, NCAR, Box 3000, Boulder, CO 80307-3000

ALAN L. HINCKLEY, Campbell Scientific Inc., 815 West 1800 North, Logan, UT 84321-1784

R. HOLLE, NOAA/NSSL, 1313 Halley Circle, Norman, OK 73069

JOHN HOREL, University of Utah, Meteorology Department, 135 South 1460 East, Salt Lake City, UT 84112-0110

B. JOHNS, NOAA/NSSL, 1313 Halley Circle, Norman, OK 73069

D. JORGENSON, NOAA/NSSL, 1313 Halley Circle, Norman, OK 73069

EUGENIA KALNAY, University of Maryland, Department of Meteorology, 3433 Computer and Space Sciences Building, College Park, MD 20742-2425

THOMAS R. KARL, NCDC, 151 Patton Avenue, Room 120, Asheville, NC 28801-5001

KRISTINA KATSAROS, NOAA/AOML, 4301 Rickenbacker Causeway, Miami, FL 33149

PAUL KUSHNER, GFDL, Princeton Forrestal Route, P.O. Box 308, Princeton, NJ 08542-0308

THOMAS J. LOCKHART, deceased

WALTER A. LYONS, FMA Research Inc., 46050 Weld County Road 13, Fort Collins, CO 80524

FRANK D. MARKS, Jr., NOAA Hurricane Research Division, 43001 Rickenbacker CSWY, Miami, FL 33149

THOMAS B. MCKEE, Colorado State University, CIRA, Foothills Campus, Fort Collins, CO 80523-1375

GERALD MEEHL, NCAR, Box 3000, Boulder, CO 80307-3000

JOHN MITCHELL, Meteorology Office, Hadley Center, London Road, Bracknell, Berkshire RG12 2SY

JOHN W. NIELSEN-GAMMON, Texas A&M University, Department of Atmospheric Sciences, 3150 TAMU, College Station, TX 77843-3150

HAROLD D. ORVILLE, South Dakota School of Mines and Technology, 501 East St. Joseph Street, Rapid City, SD 57701-3995

ROBERT PINCUS, NOAA-CIRES Climate Diagnostic Center, 1225 West Dayton Street, Madison, WI 53706-1695

ARTHUR L. RANGNO, Department of Atmospheric Sciences, University of Washington, Box 351640, Seattle, WA 98195

ROBERT M. RAUBER, University of Illinois at Urbana-Champaign, Department of Atmospheric Sciences, 105 South Gregory Street, Urbana, IL 61801-3070

SCOTT J. RICHARDSON, Oklahoma Climate Survey, 100 East Zboyd Street, Room 1210, Norman, OK 73019

MURRY SALBY, University of Colorado, Box 311, Boulder, CO 80309-0311

EDWARD S. SARACHIK, University of Washington, JISAO, 4909 25th Avenue NW, Box 354235, Seattle, WA 98195-4235

JOSEPH W. SCHIESL, 9428 Winterberry Lane, Fairfax, VA 22032

D. SCHULTZ, NOAA/NSSL, 1313 Halley Circle, Norman, OK 73069

D. STENSRUD, NOAA/NSSL, 1313 Halley Circle, Norman, OK 73069

KYLE SWANSON, Geosciences, University of Wisconsin-Milwaukee, Milwaukee, WI 53201

ROBERT N. SWANSON, 1216 Babel Lane, Concord, CA 94518-1650

KEVIN TRENBERTH, NCAR, Climate Analysis Section, Box 3000, Boulder, CO 80307-3000

JOSEPH TRIBBIA, NCAR, P.O. Box 3000, Boulder, CO 80307-3000

GREG TRIPOLI, University of Wisconsin-Madison, Department of Atmospheric and Oceanic Sciences, 1225 W. Dayton Street, Madison, WI, 53706-1695

JEFFREY B. WEISS, NCAR, Box 3000, Boulder, CO 80307-3000

S. Weiss, NOAA/NSSL, 1313 Halley Circle, Norman, OK 73069

L. Wicker, NOAA/NSSL, 1313 Halley Circle, Norman, OK 73069

Earle R. Williams, FMA Research Inc., 46050 Weld County Road 13, Fort Collins, CO 80524

Josh Wurman, 1945 Vassar Circle, Boulder, CO 80305

Robert Young, RM Young Co., 2801 Aeropark Drive, Traverse City, MI 49686

D. Zaras, NOAA/NSSL, 1313 Halley Circle, Norman, OK 73069

Edward J. Zipser, University of Utah, Department of Meteorology, 135 South 1460 East, Salt Lake City, UT 84112-0110

SECTION 1

DYNAMIC METEOROLOGY

Contributing Editor: Joseph Tribbia

CHAPTER 1

OVERVIEW—ATMOSPHERIC DYNAMICS

JOSEPH TRIBBIA

The scientific study of the dynamics of the atmosphere can broadly be defined as the attempt to elucidate the nature and causes of atmospheric motions through the laws of classical physics. The relevant principles are Newton's second law of motion applied to a fluid (the atmosphere), the equation of mass continuity, the ideal gas law, and the first law of thermodynamics. These principles are developed in detail in the contribution by Murry Salby. Since the empirical discovery and mathematical statement of these laws was not completed until the middle of the nineteenth century, as defined above, atmospheric dynamics was nonexistent before 1875. Nonetheless, attempts at applying dynamical reasoning using principles of dynamics appeared as early as 1735, in a work discussing the cause of the trade winds. Hadley's contribution and a complete history of theories of the atmospheric general circulation can be found in the monograph by Lorenz (1967).

The recognition that the laws enumerated above were sufficient to describe and even predict atmospheric motions is generally attributed to Vilhelm Bjerknes. He noted this fact in a study (1904) detailing both the statement of the central problem of meteorology (as seen by Bjerknes), weather prediction, and the system of equations necessary and sufficient to carry out the solution of the central problem. The chapter by Eugenia Kalnay describes the progress toward the solution of the central problem made since 1904 and the current state-of-the-art methods that marry the dynamical principles spelled out by Bjerknes and the computational technology brought to applicability by John von Neumann, who recognized in weather prediction a problem ideally suited for the electronic computer.

If Bjerknes' central problem and its solution were the sole goal of dynamical meteorology, then the chapters by Salby and Kalnay would be sufficient to describe

Handbook of Weather, Climate, and Water: Dynamics, Climate, Physical Meteorology, Weather Systems, and Measurements, Edited by Thomas D. Potter and Bradley R. Colman.
ISBN 0-471-21490-6

both the scientific content of the field and its progress to date. However, as noted above, atmospheric dynamics also includes the search for dynamical explanations of meteorological phenomena and a more satisfying explanation of why weather patterns exist as they do, rather than simply Force = (mass)(acceleration). The remaining chapters in the part demonstrate the expansion of thought required for this in three ways. The first method, exemplified by Paul Kushner's chapter, is to expand the quantities studied so that important aspects of atmospheric circulation systems may be more fully elucidated. The second method, exemplified by the chapters of Gerry Meehl and Kyle Swanson, develops dynamical depth by focusing on particular regions of Earth and the understanding that can be gained through the constraints imposed by Earth's geometry. The third method of expanding the reach of understanding in atmospheric dynamics is through the incorporation of techniques and ideas from other related scientific disciplines such as fluid turbulence and dynamical systems. These perspectives are brought to bear in the chapters of Jackson Herring and Jeffrey Weiss, respectively.

The focus of the chapter by Kushner is vorticity and potential vorticity. Anyone familiar with the nature of storms, e.g., both tropical and extratropical cyclones, will note the strong rotation commonly associated with these circulations. As Kushner shows, the local measure of rotation in a fluid can be quantified by either the vorticity or the related circulation. The recognition of the importance of circulation and vorticity in atmospheric systems can be traced at least as far back as von Helmholtz (1888). However, the most celebrated accomplishment in the first half of the twentieth century within atmospheric dynamics was the recognition by Carl G. Rossby (1939) that the most ubiquitous aspects of large-scale atmospheric circulations in middle latitudes could be succinctly described through a straightforward analysis of the equation governing vorticity. Rossby was also one of the first to see the value of the dynamical quantity, denoted by Ertel as potential vorticity, which, in the absence of heating and friction, is conserved by fluid parcels as they move through the atmosphere. The diagnostic analysis and tracking of this quantity forms the basis of many current studies in atmospheric dynamics, of both a theoretical and observational nature, and Kushner's chapter gives a concise introduction to these notions.

The chapters by Meehl and Swanson review the nature of motions in the tropics and extratropics, respectively. These geographic areas, distinguished from each other by their climatic regimes, have distinctive circulation systems and weather patterns that necessitate a separate treatment of the dynamics of each region. The dominant balance of forces in the tropics, as discussed by Meehl, is a thermodynamic balance between the net heating/cooling of the atmosphere by small-scale convection and radiation and the forced ascent/descent of air parcels that leads to adiabatic cooling/heating in response. This thermodynamic balance is sufficient to explain the mean circulations in the equatorial region, the north–south Hadley circulation and east–west Walker cell, the transient circulations associated with the El Niño–Southern Oscillation (ENSO) phenomenon, the monsoon circulations of Australia and Asia, and the intraseasonal Madden–Julian Oscillation. Meehl also explains the interactions among these circulations.

In contrast to the tropics, the main focus in the extra-tropics are the traveling cyclones and anticyclones, which are the dominant cause of the weather fluctuations seen at midlatitudes in all seasons except summer. These variations, which are symbolized on weather maps with the familiar high- and low pressure centers and delimiting warm and cold fronts, are dynamically dissected by Swanson and explained in terms of inherent instabilities of the stationary features that arise due to the uneven distribution of net radiative heating, topography, and land mass over Earth's surface. In the process of dynamically explaining these systems, Swanson makes use of the quasi-geostrophic equations, which are a simplification of the governing equations derived by Salby. This quasi-geostrophic system is a staple of dynamical meteorology and can be formally derived as an approximation of the full system using scale analysis (cf. Charney, 1948, or Phillips, 1963). The advantage of such reduced equations is twofold: the reduction frequently leads to analytically tractable equations as shown by Swanson's examples and, with fewer variables and degrees of freedom in the system, it is almost always easier to directly follow the causal dynamical mechanisms.

The chapters by Herring and Weiss bring in paradigms and tools from the physics of fluid turbulence and the mathematics of dynamical systems theory. The entire field of atmospheric dynamics is but a subtopic within the physics of fluid dynamics. The study of fluid motions in the atmosphere, ocean, and within the fluid earth is frequently referred to as geophysical fluid dynamics (GFD), so it is not surprising that ideas from fluid turbulence would be used in atmospheric dynamics, as well as in the rest of GFD. What is different in the application in the large-scale dynamics of the atmosphere is the notion of viewing the atmosphere as a turbulent (nearly) two-dimensional flow. The perspective given by Herring was conceived in the late 1960s by George Batchelor and Robert Kraichnan, and further developed by C. Leith, Douglas Lilly, and Herring. Prior to this time it was thought that two-dimensional turbulence was an oxymoron since turbulence studied in the laboratory and observed in nature is inherently three dimensional. As Herring shows, the two-dimensional turbulence picture of the atmosphere has enabled a dynamical explanation of the spectrum of atmospheric motions and elucidated the growth in time of forecast errors, which initiate in small scales and propagate up the spectrum to contaminate even planetary scales of motion. This notion of forecast errors contaminating the accuracy of forecasts was first investigated by Philip Thompson (1957) using the methodology of Batchelor's statistical theory of homogeneous turbulence. Herring's chapter is a summary of subsequent developments using this methodology.

A seminal study by Edward Lorenz (1963) is the predecessor of the review given by Weiss, detailing the use of a dynamical system's perspective and deterministic chaos in quantifying the predictability of the atmosphere. Lorenz' study was the starting point for two research fields: the application of dynamical systems theory to atmospheric predictions and the mathematical topic of deterministic chaos. Weiss' chapter summarizes the scientific developments relevant to atmospheric dynamics and climate and weather prediction since 1963.

In any brief summarization of an active and growing field of research as much, or more, will be left out as will be reviewed. The chapters presented in this part are to be viewed more as a sampler than an exhaustive treatise on the dynamics of atmospheric motions. For those intrigued by works presented here and wishing to further learn about the area, the following texts are recommended in addition to those texts and publications cited by the individual authors: *An Introduction to Dynamical Meteorology* (1992) by J. R. Holton, Academic Press; *Atmosphere-Ocean Dynamics* (1982) by A. E. Gill, Academic Press; and *Geophysical Fluid Dynamics* (1979) by J. Pedlosky, Springer.

REFERENCES

Bjerknes, V. (1904). Das problem von der wettervorhersage, betrachtet vom standpunkt der mechanik un der physik, *Meteor. Z.* **21**, 1–7.

Charney, J. G. (1948). On the Scale of Atmospheric Motions, *Geofys. Publik.* **17**, 1–17.

Hadley, G. (1735). Concerning the cause of the general trade-winds, *Phil. Trans.* **29**, 58–62.

Lorenz, E. N. (1963). Deterministic Nonperiodic Flow, *J. Atmos. Sci.* **20**, 130–141.

Lorenz, E. N. (1967). The Nature and Theory of the General Circulation of the Atmosphere, World Meteorological Organization/Geneva.

Phillips, N. A. (1963). Geostrophic motion, *Rev. Geophys.* **1**, 123–176.

Thompson, P. D. (1957). Uncertainty of the initial state as a factor in the predictability of large scale atmospheric flow patterns, *Tellus* **9**, 257–272.

von Helmholtz, H. (1888). On Atmospheric motions, *Sitz.-Ber. Akad. Wiss. Berlin* 647–663.

CHAPTER 2

FUNDAMENTAL FORCES AND GOVERNING EQUATIONS

MURRY SALBY

1 DESCRIPTION OF ATMOSPHERIC BEHAVIOR

The atmosphere is a fluid and therefore capable of redistributing mass and constituents into a variety of complex configurations. Like any fluid system, the atmosphere is governed by the laws of continuum mechanics. These can be derived from the laws of mechanics and thermodynamics governing a discrete fluid body by generalizing those laws to a continuum of such systems. The discrete system to which these laws apply is an infinitesimal fluid element or *air parcel*, defined by a fixed collection of matter.

Two frameworks are used to describe atmospheric behavior. The *Eulerian description* represents atmospheric behavior in terms of field properties, such as the instantaneous distributions of temperature and motion. It lends itself to numerical solution, forming the basis for most numerical models. The *Lagrangian description* represents atmospheric behavior in terms of the properties of individual air parcels, the positions of which must then be tracked. Despite this complication, the laws governing atmospheric behavior follow naturally from the Lagrangian description because it is related directly to transformations of properties within an air parcel and to interactions with its environment.

The Eulerian and Lagrangian descriptions are related through a kinematic constraint: The field property at a given position and time must equal the property possessed by the air parcel occupying that position at that instant. Consider the property ψ of an individual air parcel, which has instantaneous position $(x, y, z, t) =$

Handbook of Weather, Climate, and Water: Dynamics, Climate, Physical Meteorology, Weather Systems, and Measurements, Edited by Thomas D. Potter and Bradley R. Colman.
ISBN 0-471-21490-6

$(\mathbf{x}, t)$. The incremental change of $\psi(\mathbf{x}, t)$ during the parcel's displacement $(dx, dy, dz) = d\mathbf{x}$ follows from the total differential:

$$\begin{aligned} d\psi &= \frac{\partial \psi}{\partial t} dt + \frac{\partial \psi}{\partial x} dx + \frac{\partial \psi}{\partial y} dy + \frac{\partial \psi}{\partial z} dz \\ &= \frac{\partial \psi}{\partial t} dt + \nabla \psi \cdot d\mathbf{x} \end{aligned} \tag{1}$$

The property's rate of change following the parcel is then

$$\begin{aligned} \frac{d\psi}{dt} &= \frac{\partial \psi}{\partial t} + u \frac{\partial \psi}{\partial x} + v \frac{\partial \psi}{\partial y} + w \frac{\partial \psi}{\partial z} \\ &= \frac{\partial \psi}{\partial t} + \mathbf{v} \cdot \nabla \psi \end{aligned} \tag{2}$$

where $\mathbf{v} = dx/dt$ is the three-dimensional velocity. Equation (2) defines the *Lagrangian derivative* of the field variable $\psi(\mathbf{x}, t)$, which corresponds to its rate of change following an air parcel. The Lagrangian derivative contains two contributions: $\partial \psi / \partial t$ represents the rate of change introduced by transience of the field property ψ, and $\mathbf{v} \cdot \nabla \psi$ represents the rate of change introduced by the parcel's motion to positions of different field values.

Consider now a finite body of air having instantaneous volume $V(t)$. The integral property

$$\int_{V(t)} \psi(x, y, z, t)\, dV'$$

changes through two mechanisms (Fig. 1), analogous to the two contributions to $d\psi/dt$: (1) Values of $\psi(\mathbf{x}, t)$ within the material volume change due to unsteadiness of the field. (2) The material volume moves to regions of different field values. Relative to a frame moving with $V(t)$, this motion introduces a flux of property ψ across the material volume's surface $S(t)$. Collecting these contributions and applying Gauss' theorem [see, e.g., Salby (1996)] leads to the identity

$$\begin{aligned} \frac{d}{dt} \int_{V(t)} \psi\, dV' &= \int_{V(t)} \left\{ \frac{\partial \psi}{\partial t} + \nabla \cdot (\mathbf{v}\psi) \right\} dV' \\ &= \int_{V(t)} \left\{ \frac{d\psi}{dt} + \psi \nabla \cdot \mathbf{v} \right\} dV' \end{aligned} \tag{3}$$

Known as *Reynolds' transport theorem*, (3) constitutes a transformation between the Lagrangian and Eulerian descriptions of fluid motion, relating the time rate of change of some property of a finite body of air to the corresponding field variable and the motion field $\mathbf{v}(\mathbf{x}, t)$. Applying Reynolds' theorem to particular properties

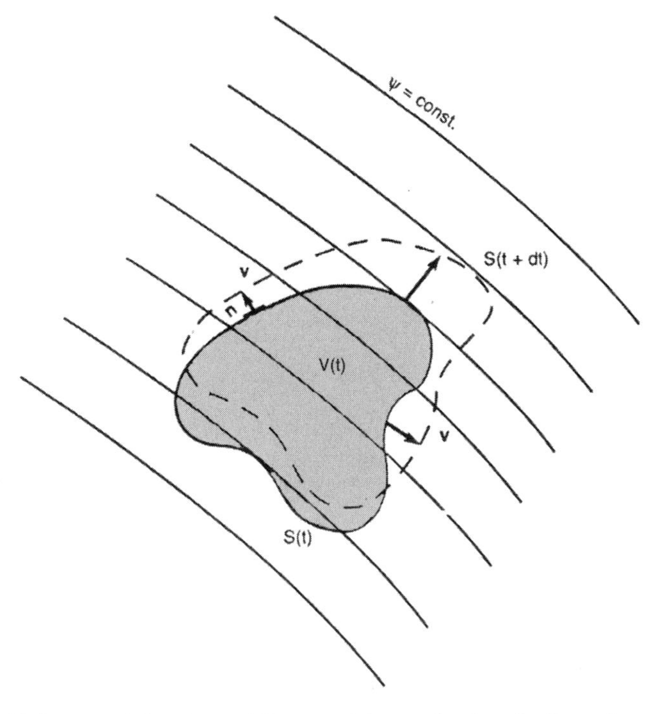

Figure 1 Finite material volume $V(t)$, containing a fixed collection of matter, that is displaced across a field property ψ. (Reproduced from Salby, 1996.)

then yields the field equations governing atmospheric behavior, which are known collectively as the *equations of motion*.

2 MASS CONTINUITY

A material volume $V(t)$ has mass

$$\int_{V(t)} \rho(\mathbf{x}, t)\, dV'$$

where ρ is density. Since $V(t)$ is comprised of a fixed collection of matter, the time rate of change of its mass must vanish

$$\frac{d}{dt}\int_{V(t)} \rho(\mathbf{x}, t)\, dV' = 0 \tag{4}$$

Applying Reynolds' theorem transforms (4) into

$$\int_{V(t)} \left\{ \frac{d\rho}{dt} + \rho \nabla \cdot \mathbf{v} \right\} dV' = 0 \tag{5}$$

where the Lagrangian derivative is given by (2). This relation must hold for "arbitrary" material volume $V(t)$. Consequently, the quantity in brackets must vanish identically. Conservation of mass for individual bodies of air therefore requires

$$\frac{d\rho}{dt} + \rho \nabla \cdot \mathbf{v} = 0 \tag{6}$$

which is known as the *continuity equation.*

Budget of Constituents

For a specific property f (i.e., one referenced to a unit mass), Reynolds' transport theorem simplifies. With (6), (3) reduces to

$$\frac{d}{dt}\int_{V(t)} \rho f \, dV' = \int_{V(t)} \rho \frac{df}{dt} dV' \tag{7}$$

where ρf represents the concentration of property f (i.e., referenced to a unit volume).

Consider now a constituent of air, such as water vapor in the troposphere or ozone in the stratosphere. The specific abundance of this species is described by the mixing ratio r, which measures its local mass referenced to a unit mass of dry air (i.e., of fixed composition). The corresponding concentration is then ρr.

Conservation of the species is then expressed by

$$\frac{d}{dt}\int_{V(t)} \rho r \, dV' = \int_{V(t)} \rho \dot{P} \, dV' \tag{8}$$

which equates the rate that the species changes within the material volume to its net rate of production per unit mass $\dot{P}$, collected over the material volume. Applying (7) and requiring the result to be satisfied for arbitrary $V(t)$ then reduces the constituent's budget to

$$\frac{dr}{dt} = \dot{P} \tag{9}$$

Like the equation of mass continuity, (9) is a partial differential equation that must be satisfied continuously throughout the atmosphere. For a long-lived chemical species or for water vapor away from regions of condensation, $\dot{P} \cong 0$, so r is approximately conserved for individual air parcels. Otherwise, production and destruction of the species are balanced by changes in a parcel's mixing ratio.

3 MOMENTUM BUDGET

In an inertial reference frame, Newton's second law of motion applied to the material volume $V(t)$ may be expressed

$$\frac{d}{dt}\int_{V(t)} \rho \mathbf{v}\, dV' = \int_{V(t)} \rho \mathbf{f}\, dV' + \int_{S(t)} \boldsymbol{\tau} \cdot \mathbf{n}\, dS' \tag{10}$$

where $\rho\mathbf{v}$ is the concentration of momentum, $\mathbf{f}$ is the body force per unit mass acting internal to the material volume, and $\boldsymbol{\tau}$ is the *stress tensor* acting on its surface (Fig. 2). The stress tensor $\boldsymbol{\tau}$ represents the vector force per unit area exerted on surfaces normal to the three coordinate directions. Then $\boldsymbol{\tau} \cdot \mathbf{n}$ is the vector force per unit area exerted on the section of material surface with unit normal $\mathbf{n}$, representing the flux of vector momentum across that surface.

Incorporating Reynolds' theorem for a concentration (7), along with Gauss' theorem, casts the material volume's momentum budget into

$$\int_{V(t)} \left\{ \rho \frac{d\mathbf{v}}{dt} - \rho \mathbf{f} - \nabla \cdot \boldsymbol{\tau} \right\} dV' = 0 \tag{11}$$

As before, (11) must hold for arbitrary material volume, so the quantity in brackets must vanish identically. Newton's second law for individual bodies of air thus requires

$$\rho \frac{d\mathbf{v}}{dt} = \rho \mathbf{f} + \nabla \cdot \boldsymbol{\tau} \tag{12}$$

to be satisfied continuously.

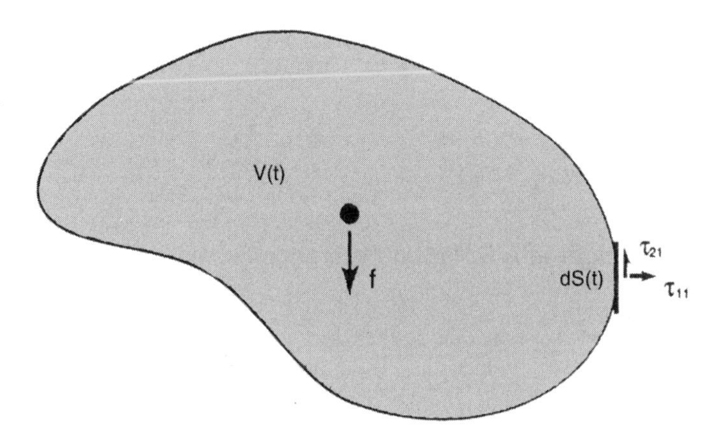

Figure 2 Finite material volume $V(t)$ that experiences a specific body force $\mathbf{f}$ internally and a stress tensor τ on its surface. (Reproduced from Salby, 1996.)

The body force relevant to the atmosphere is gravity

$$\mathbf{f} = \mathbf{g} \tag{13}$$

which is specified. The stress term, on the other hand, is determined autonomously by the motion. It represents the divergence of a momentum flux, or force per unit volume, and has two components: (i) a normal force associated with the pressure gradient ∇p and (ii) a tangential force or drag $\mathbf{D}$ introduced by friction. For a *Newtonian fluid* such as air, tangential components of the stress tensor depend linearly on the shear. Frictional drag then reduces to (e.g., Aris, 1962)

$$\begin{aligned} \mathbf{D} &= -\frac{1}{\rho}\nabla \cdot \boldsymbol{\tau} \\ &= -\frac{1}{\rho}\nabla \cdot (\mu \nabla \mathbf{v}) \end{aligned} \tag{14}$$

where μ is the coefficient of viscosity. The right-hand side of (14) accounts for diffusion of momentum, which is dominated in the atmosphere by turbulent diffusion (e.g., turbulent mixing of an air parcel's momentum with that of its surroundings). For many applications, frictional drag is expressed in terms of a turbulent diffusivity v as

$$\mathbf{D} = -v \frac{\partial^2 \mathbf{v}}{\partial z^2} \tag{15}$$

in which horizontal components of $\mathbf{v}$ and their vertical shear prevail.

With these restrictions, the momentum budget reduces to

$$\frac{d\mathbf{v}}{dt} = \mathbf{g} - \frac{1}{\rho}\nabla p - \mathbf{D} \tag{16}$$

which comprise the so-called *Navier–Stokes equations*. Also called the *momentum equations*, (16) assert that an air parcel's momentum changes according to the resultant force exerted on it by gravity, pressure gradient, and frictional drag.

Momentum Budget in a Rotating Reference Frame

The momentum equations are a statement of Newton's second law, so they are valid in an inertial reference frame. The reference frame of Earth, on the other hand, is rotating and consequently noninertial. The momentum equations must therefore be modified to account for acceleration of the frame in which atmospheric motion is observed.

Consider a reference frame that rotates with angular velocity $\mathbf{\Omega}$. A vector $\mathbf{A}$ that appears constant in the rotating frame rotates when viewed from an inertial frame.

During an interval dt, $\mathbf{A}$ changes by a vector increment $d\mathbf{A}$ that is perpendicular to the plane of $\mathbf{A}$ and $\mathbf{\Omega}$ (Fig. 3) and has magnitude

$$|d\mathbf{A}| = A \sin\theta \cdot \Omega\, dt$$

where θ is the angle between $\mathbf{A}$ and $\mathbf{\Omega}$. The time rate of change of $\mathbf{A}$ apparent in an inertial frame is then described by

$$\left(\frac{d\mathbf{A}}{dt}\right)_i = \mathbf{\Omega} \times \mathbf{A} \tag{17}$$

More generally, a vector $\mathbf{A}$ that has the time rate of change $d\mathbf{A}/dt$ in the rotating reference frame has the time rate of change

$$\left(\frac{d\mathbf{A}}{dt}\right)_i = \frac{d\mathbf{A}}{dt} + \mathbf{\Omega} \times \mathbf{A} \tag{18}$$

in an inertial reference frame.

Consider now the position $\mathbf{x}$ of an air parcel. The parcel's velocity $\mathbf{v} = d\mathbf{x}/dt$ apparent in an inertial reference frame is then

$$\mathbf{v}_i = \mathbf{v} + \mathbf{\Omega} \times \mathbf{x} \tag{19}$$

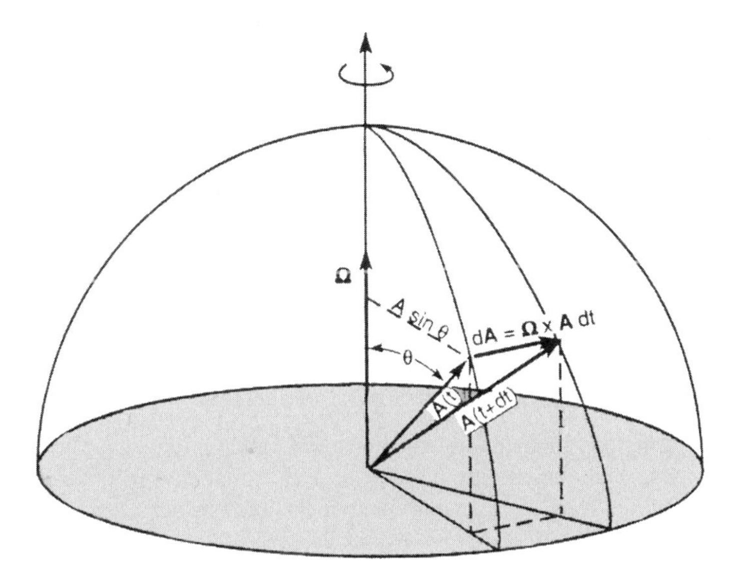

Figure 3 A vector $\mathbf{A}$ fixed in a rotating reference frame changes in an inertial reference frame during an interval dt by the increment $|d\mathbf{A}| = A \sin\theta \cdot \Omega\, dt$ in a direction orthogonal to the plane of $\mathbf{A}$ and Ω, or by the vector increment $d\mathbf{A} = \mathbf{\Omega} \times \mathbf{A}\, dt$. (Reproduced from Salby, 1996.)

Likewise, the acceleration apparent in the inertial frame is given by

$$\left(\frac{d\mathbf{v}_i}{dt}\right)_i = \frac{d\mathbf{v}_i}{dt} + \mathbf{\Omega} \times \mathbf{v}_i$$

which upon consolidation yields for the acceleration apparent in the inertial frame:

$$\left(\frac{d\mathbf{v}_i}{dt}\right)_i = \frac{d\mathbf{v}}{dt} + 2\mathbf{\Omega} \times \mathbf{v} + \mathbf{\Omega} \times (\mathbf{\Omega} \times \mathbf{x}) \tag{20}$$

Earth's rotation introduces two corrections to the parcel's acceleration: (i) $2\mathbf{\Omega} \times \mathbf{v}$ is the *Coriolis acceleration*. It acts perpendicular to the parcel's motion and the planetary vorticity $2\mathbf{\Omega}$. (ii) $\mathbf{\Omega} \times (\mathbf{\Omega} \times \mathbf{x})$ is the *centrifugal acceleration* of the air parcel associated with Earth's rotation. When geopotential coordinates are used (e.g., Iribarne and Godson, 1981), this correction is absorbed into the *effective gravity*: $\mathbf{g} = -\nabla\Phi$, which is defined from the geopotential Φ.

Incorporating (20) transforms the momentum equations into a form valid in the rotating frame of Earth:

$$\frac{d\mathbf{v}}{dt} + 2\mathbf{\Omega} \times \mathbf{v} = -\frac{1}{\rho}\nabla p - g\mathbf{k} - \mathbf{D} \tag{21}$$

where g is understood to denote effective gravity and $\mathbf{k}$ the upward normal in the direction of increasing geopotential. The correction $2\mathbf{\Omega} \times \mathbf{v}$ is important for motions with time scales comparable to that of Earth's rotation. When moved to the right-hand side, it enters as the Coriolis force: $-2\mathbf{\Omega} \times \mathbf{v}$, a fictitious force that acts on an air parcel in the rotating frame of Earth. Because it acts orthogonal to the parcel's displacement, the Coriolis force performs no work.

Component Equations in Spherical Coordinates

In vector form, the momentum equations are valid in any coordinate system. However, those equations do not lend themselves to standard methods of solution. To be useful, they must be cast into component form, which then depend on the coordinate system in which they are expressed.

Consider the rectangular Cartesian coordinates $\mathbf{x} = (x_1, x_2, x_3)$ having origin at the center of Earth (Fig. 4) and the spherical coordinates

$$\mathbf{x} = (\lambda, \phi, r)$$

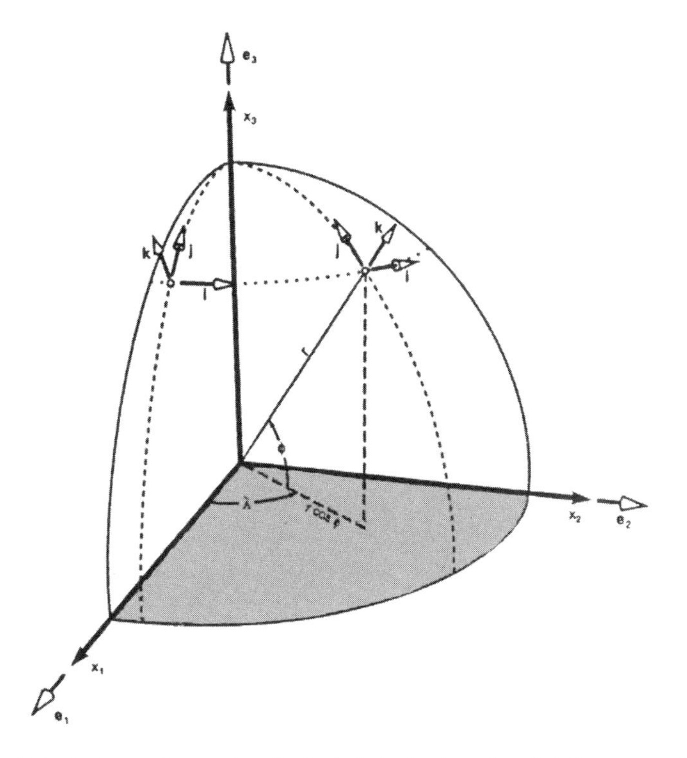

Figure 4 Spherical coordinates: longitude λ, latitude ϕ, and radial distance r. Coordinate vectors $\mathbf{e}_\lambda = \mathbf{i}$, $\mathbf{e}_\phi = \mathbf{j}$, and $\mathbf{e}_\gamma = \mathbf{k}$ change with position (e.g., relative to fixed coordinate vectors $\mathbf{e}_1$, $\mathbf{e}_2$, and $\mathbf{e}_3$ of rectangular Cartesian coordinates). (Reproduced from Salby, 1996.)

both fixed with respect to Earth. The corresponding unit vectors

$$\begin{aligned} \mathbf{e}_\lambda &= \mathbf{i} \\ \mathbf{e}_\phi &= \mathbf{j} \\ \mathbf{e}_r &= \mathbf{k} \end{aligned} \tag{22}$$

point in the directions of increasing λ, ϕ, and r and are mutually perpendicular.

The rectangular Cartesian coordinates can be expressed in terms of the spherical coordinates as

$$\begin{aligned} x_1 &= r \cos\phi \cos\lambda \\ x_2 &= r \cos\phi \sin\lambda \\ x_3 &= r \sin\phi \end{aligned} \tag{23a}$$

which may be inverted for the spherical coordinates:

$$\lambda = \tan^{-1}\left(\frac{x_2}{x_1}\right)$$
$$\phi = \tan^{-1}\left(\frac{x_3}{x_1^2 + x_2^2}\right) \tag{23b}$$
$$r = \sqrt{x_1^2 + x_2^2 + x_3^2}$$

Parcel displacements in the directions of increasing longitude, latitude, and radial distance are then described by

$$dx = r\cos\phi\, d\lambda$$
$$dy = r\, d\phi \tag{24a}$$
$$dz = dr$$

in which vertical distance is measured by height

$$z = r - a \tag{24b}$$

where a denotes the mean radius of Earth. Velocity components in the spherical coordinate system are then expressed by

$$u = \frac{dx}{dt} = r\cos\phi\frac{d\lambda}{dt}$$
$$v = \frac{dy}{dt} = r\frac{d\phi}{dt} \tag{25}$$
$$w = \frac{dz}{dt} = \frac{dr}{dt}$$

The vector momentum equations may now be expressed in terms of the spherical coordinates. Lagrangian derivatives of vector quantities are complicated by the dependence on position of the coordinate vectors **i**, **j**, and **k**. Each rotates in physical space under a displacement of longitude or latitude. For example, an air parcel moving along a latitude circle at a constant speed u has a velocity $\mathbf{v} = u\mathbf{i}$, which appears constant in the spherical coordinate representation, but which actually rotates in physical space (Fig. 4). Consequently, the parcel experiences an acceleration that must be accounted for in the equations of motion.

Consider the velocity

$$\mathbf{v} = u\mathbf{i} + v\mathbf{j} + w\mathbf{k}$$

Since the spherical coordinate vectors **i**, **j**, and **k** are functions of position $\hat{\mathbf{x}}$, a parcel's acceleration is actually

$$\begin{aligned}\frac{d\mathbf{v}}{dt} &= \frac{du}{dt}\mathbf{i} + \frac{dv}{dt}\mathbf{j} + \frac{dw}{dt}\mathbf{k} + u\frac{d\mathbf{i}}{dt} + v\frac{d\mathbf{j}}{dt} + w\frac{d\mathbf{k}}{dt} \\ &= \left(\frac{d\mathbf{v}}{dt}\right)_C + u\frac{d\mathbf{i}}{dt} + v\frac{d\mathbf{j}}{dt} + w\frac{d\mathbf{k}}{dt}\end{aligned} \tag{26a}$$

where the subscript refers to the basic form of the Lagrangian derivative in Cartesian geometry

$$\left(\frac{d}{dt}\right)_C = \frac{\partial}{\partial t} + \mathbf{v}\cdot\nabla \tag{26b}$$

Corrections appearing on the right-hand side of (26) can be evaluated by expressing the spherical coordinate vectors in terms of the fixed rectangular Cartesian coordinate vectors $\mathbf{e}_1$, $\mathbf{e}_2$, $\mathbf{e}_3$. Lagrangian derivatives of the spherical coordinate vectors then follow as

$$\begin{aligned}\frac{d\mathbf{i}}{dt} &= u\left(\frac{\tan\phi}{r}\mathbf{j} - \frac{1}{r}\mathbf{k}\right) \\ \frac{d\mathbf{j}}{dt} &= -u\frac{\tan\phi}{r}\mathbf{i} - \frac{v}{r}\mathbf{k} \\ \frac{d\mathbf{k}}{dt} &= \frac{u}{r}\mathbf{i} + \frac{v}{r}\mathbf{j}\end{aligned} \tag{27}$$

With these, material accelerations in the spherical coordinate directions become

$$\frac{du}{dt} = \left(\frac{du}{dt}\right)_C - \frac{uv\tan\phi}{r} + \frac{uw}{r} \tag{28a}$$

$$\frac{dv}{dt} = \left(\frac{dv}{dt}\right)_C + \frac{u^2\tan\phi}{r} + \frac{vw}{r} \tag{28b}$$

$$\frac{dw}{dt} = \left(\frac{dw}{dt}\right)_C - \left(\frac{u^2+v^2}{r}\right) \tag{28c}$$

Introducing (28), making use of the atmosphere's shallowness ($z \ll a$), and formally adopting height as the vertical coordinate then casts the momentum equations into component form:

$$\frac{du}{dt} - 2\Omega(v \sin\phi - w \cos\phi) = -\frac{1}{\rho a \cos\phi}\frac{\partial p}{\partial \lambda} + uv\frac{\tan\phi}{a} - \frac{uw}{a} - D_\lambda \tag{29a}$$

$$\frac{dv}{dt} + 2\Omega u \sin\phi = -\frac{1}{\rho a}\frac{\partial p}{\partial \phi} - \frac{u^2 \tan\phi}{a} - \frac{uw}{a} - D_\phi \tag{29b}$$

$$\frac{dw}{dt} - 2\Omega u \cos\phi = -\frac{1}{\rho}\frac{\partial p}{\partial z} - g + \frac{u^2 + v^2}{a} - D_z \tag{29c}$$

4 FIRST LAW OF THERMODYNAMICS

For the material volume $V(t)$, the first law of thermodynamics is expressed by

$$\frac{d}{dt}\int_{V(t)} \rho c_v T\, dV' = -\int_{S(t)} \mathbf{q}\cdot\mathbf{n}\, dS' - \int_{V(t)} \rho p \frac{d\alpha}{dt} dV' + \int_{V(t)} \rho \dot{q}\, dV' \tag{30}$$

where c_v is the specific heat at constant volume, so $c_v T$ represents the internal energy per unit mass, $\mathbf{q}$ is the local heat flux so $-\mathbf{q}\cdot\mathbf{n}$ represents the heat flux "into" the material volume, $\alpha = 1/\rho$ is the specific volume so $p d\alpha/dt$ represents the specific work rate, and $\dot{q}$ denotes the specific rate of internal heating (e.g., associated with the latent heat release and frictional dissipation of motion).

In terms of specific volume, the continuity equation (6) becomes

$$\frac{1}{\alpha}\frac{d\alpha}{dt} = \nabla\cdot\mathbf{v} \tag{31}$$

Incorporating (31), along with Reynolds' transport theorem (3) and Gauss' theorem, transforms (30) into

$$\int_{V(t)} \left\{ \rho c_v \frac{dT}{dt} + \nabla\cdot\mathbf{q} + p\nabla\cdot v - \rho\dot{q} \right\} dV' = 0$$

Since $V(t)$ is arbitrary, the quantity in brackets must again vanish identically. Therefore, the first law applied to individual bodies of air requires

$$\rho c_v \frac{dT}{dt} = -\nabla\cdot\mathbf{q} - p\nabla\cdot\mathbf{v} + \rho\dot{q} \tag{32}$$

to be satisfied continuously.

The heat flux can be separated into radiative and diffusive components:

$$\begin{aligned} \mathbf{q} &= \mathbf{q}_R + \mathbf{q}_T \\ &= \mathbf{F} - k\nabla T \end{aligned} \tag{33}$$

where $\mathbf{F}$ is the net radiative flux and k is the thermal conductivity in Fourier's law of heat conduction. The first law then becomes

$$\rho c_v \frac{dT}{dt} + p\nabla \cdot \mathbf{v} = -\nabla \cdot \mathbf{F} + \nabla \cdot (k\nabla T) + \rho\dot{q} \tag{34}$$

Known as the *thermodynamic equation*, (34) expresses the rate that a material element's internal energy changes in terms of the rate that it performs work and the rate that it absorbs heat through convergence of radiative and diffusive energy fluxes.

The thermodynamic equation is expressed more compactly in terms of another thermodynamic property, one that accounts collectively for a change of internal energy and expansion work. The *potential temperature* θ is defined as

$$\frac{\theta}{T} = \left(\frac{p_0}{p}\right)^{\kappa} \tag{35}$$

where $\kappa = R/c_p$, R is the specific gas constant for air, and temperature and pressure are related through the ideal gas law

$$p = \rho RT \tag{36}$$

and θ is related to a parcel's entropy. It is conserved during an adiabatic process, as characterizes a parcel's motion away from Earth's surface and cloud. Incorporating θ into the second law of thermodynamics (e.g., Salby, 1996) then leads to the fundamental relation

$$c_p T \frac{d\ln\theta}{dt} = \frac{du}{dt} + p\frac{d\alpha}{dt} \tag{37}$$

where d/dt represents the Lagrangian derivative. Making use of the continuity equation (31) transforms this into

$$\frac{\rho c_p T}{\theta}\frac{d\theta}{dt} = \rho c_v \frac{dT}{dt} + p\nabla \cdot \mathbf{v} \tag{38}$$

Then incorporating (38) into (34) absorbs the expansion work into the time rate of change of θ to yield the thermodynamic equation

$$\rho \frac{c_p T}{\theta} \frac{d\theta}{dt} = -\nabla \cdot \mathbf{F} + \nabla \cdot (k\nabla T) + \rho\dot{q} \qquad (39)$$

Equation (39) relates the change in a parcel's potential temperature to net heat transfer with its surroundings. In the absence of such heat transfer, θ is conserved for individual air parcels.

REFERENCES

Aris, R. (1962). *Vectors, Tensors, and the Basic Equations of Fluid Mechanics*, Englewood Cliffs, NJ, Prentice Hall.

Iribarne, J., and W. Godson (1981). *Atmospheric Thermodynamics*, Reidel, Dordrecht.

Salby, M. (1996). *Fundamentals of Atmospheric Physics*, San Diego, Academic.

CHAPTER 3

CIRCULATION, VORTICITY, AND POTENTIAL VORTICITY

PAUL J. KUSHNER

1 INTRODUCTION

Vorticity and circulation are closely related quantities that describe rotational motion in fluids. Vorticity describes the rotation at each point; circulation describes rotation over a region. Both quantities are of fundamental importance to the field of fluid dynamics. The distribution and statistical properties of the vorticity field provide a succinct characterization of fluid flow, particularly for weakly compressible or incompressible fluids. In addition, stresses acting on the fluid are often interpreted in terms of the generation, transport, and dissipation of vorticity and the resulting impact on the circulation.

Potential vorticity (PV) is a generalized vorticity that combines information about both rotational motion and density stratification. PV is of central importance to the field of geophysical fluid dynamics (GFD) and its subfields of dynamical meteorology and physical oceanography. Work in GFD often focuses on flows that are strongly influenced by planetary rotation and stratification. Such flows can often be fully described by their distribution of PV. Similarly to vorticity, the generation, transport, and dissipation of PV is closely associated with stresses on the fluid.

Vorticity, circulation, and PV are described extensively in several textbooks (e.g., Holton, 1992; Gill, 1982; Kundu, 1990; Salmon, 1998). This review is a tutorial, with illustrative examples, that is meant to acquaint the lay reader with these concepts and the scope of their application. An appendix provides a mathematical summary.

Handbook of Weather, Climate, and Water: Dynamics, Climate, Physical Meteorology, Weather Systems, and Measurements, Edited by Thomas D. Potter and Bradley R. Colman.
ISBN 0-471-21490-6

2 CIRCULATION AND VORTICITY: DEFINITIONS AND EXAMPLES

Circulation is a physical quantity that describes the net movement of fluid along a chosen *circuit*, that is, a path, taken in a specified direction, that starts at some point and returns to that point.

> **Statement 1.** The *circulation* at a given time is the average, over a circuit, of the component of the flow velocity tangential to the circuit, multiplied by the length of the circuit. Circulation therefore has dimensions of length squared per unit time.

Consider the circuit drawn as a heavy curve in Figure 1 for a fluid whose velocity **u** is indicated by the arrows. At each point we may split the velocity into components tangential to and perpendicular to the local direction of the circuit. The tangential component is defined as the dot product $\mathbf{u} \cdot \hat{\mathbf{l}} = |\mathbf{u}| \cos\theta$, where $\hat{\mathbf{l}}$ is a unit vector that points in the direction tangential to the circuit and θ is the angle between **u** and $\hat{\mathbf{l}}$. Note that $\hat{\mathbf{l}}$ points in the specified direction chosen for the circuit. Where θ is acute, the tangential component is positive, where oblique, negative, and where a right angle, zero. To calculate the circulation, we average over "all" points along the circuit. (Although we cannot actually average over all points along the circuit, we can approximate such an average by summing over the tangential component $\mathbf{u} \cdot \hat{\mathbf{l}} = |\mathbf{u}| \cos\theta$ at evenly and closely spaced points around the circuit and by dividing by the number of points.) The circulation would then be this average, multiplied by the length of the circuit. (In the Appendix, the circulation is defined in terms of a line integral.)

The circulation is not defined with reference to a particular point in space but to a circuit that must be chosen. For example, to measure the strength of the primary near-surface flow around a hurricane in the Northern Hemisphere, a natural circuit is a circle, oriented horizontally, centered about the hurricane's eye, of radius somewhat larger than the eye, and directed counterclockwise when viewed from above. We thus have specified that the circuit run in the "cyclonic" direction, which is the direction of the primary flow around tropical storms and other circulations with low-pressure centers in the Northern Hemisphere (Fig. 2).

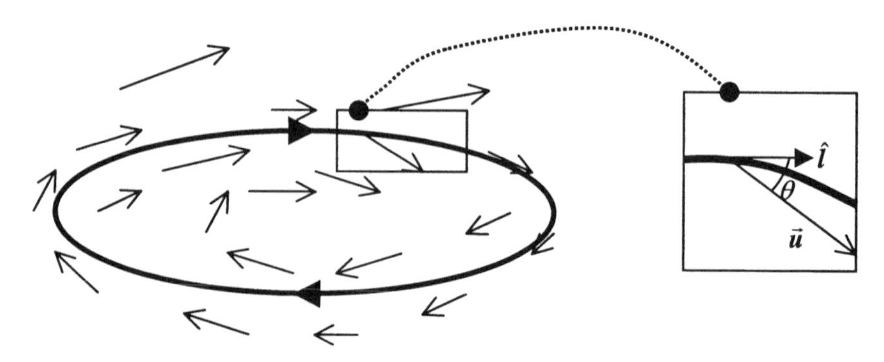

Figure 1 Heavy curve, dark arrows: circuit. Light arrows: flow velocity.

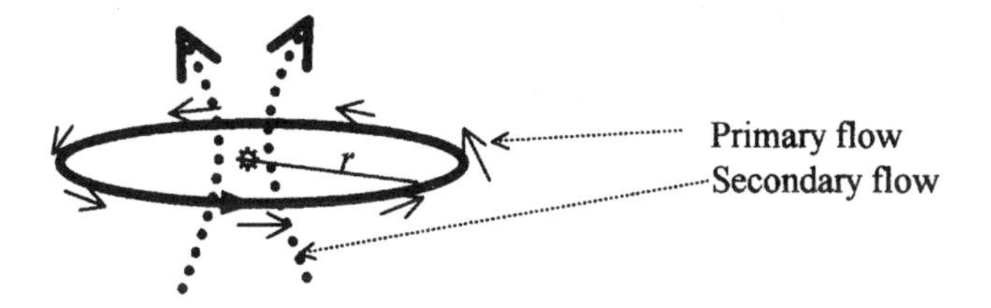

Figure 2 Circuit shown with bold curve. ✿ symbol represents a paddle wheel with its axis in the vertical and located at the center of the circle.

Suppose that the average flow tangential to the circuit is approximately $u = 40\,\mathrm{m/s}$ and that the radius of the circle is $r = 100\,\mathrm{km}$. The circulation, Γ, using the definition above, is then the average tangential flow times the circumference of the circle: $\Gamma = 2\pi r u \approx 2.5 \times 10^7\,\mathrm{m^2/s}$. If the circuit were chosen to run clockwise, i.e., opposite to the primary flow, instead of counterclockwise, the circulation would be of the same strength but of opposite sign.

Circulation is a selective measure of the strength of the flow: It tells us nothing about components of motion perpendicular to the circuit or components of the flow that average to zero around the circuit. For example, suppose the hurricane is blown to the west with a constant easterly wind. This constant wind will not contribute to the circulation: Since $\cos(\theta + \pi) = -\cos\theta$, every point on the north side of the circle that contributes some amount to the circulation has a counterpart on the south side that contributes an equal and opposite amount. As another example, the circuit in Figure 2 would not measure the strength of the hurricane's secondary flow, which runs toward the eye near the surface and away from the eye aloft (dotted arrows in Fig. 2). The reader might try to imagine a circuit that would measure this secondary flow.

The hurricane example might suggest that the fluid must flow around the chosen circuit for the circulation to be nonzero; however, this need not be true. For example, consider the circuit shown in Figure 3, which shows a vertically oriented 50- × 200-m rectangular circuit in a west-to-east flow whose strength increases linearly with height, according to the formula $u(z) = az$, where z is the height in meters and $a = 0.01/\mathrm{s} = (10\,\mathrm{m/s})\,\mathrm{km}$ is the "vertical shear," that is, the rate of change of the wind with respect to height. This is a value of vertical shear that might be encountered in the atmospheric boundary layer. If the direction of the circuit is chosen, as in the figure, to be clockwise, the circulation in this example is the average of the along-circuit component of the flow weighted by the length of each side,

$$\frac{50 \times u(\mathrm{top}) - 50 \times u(\mathrm{bottom}) + 200 \times 0 + 200 \times 0}{50 + 50 + 200 + 200} = \frac{50 \times 200a}{500} = 0.2\,\mathrm{m/s}$$

times the perimeter of the rectangle, giving $\Gamma = 100\,\mathrm{m^2/s}$. Note that the vertical sides make no contribution to the circulation since the flow is perpendicular to them.

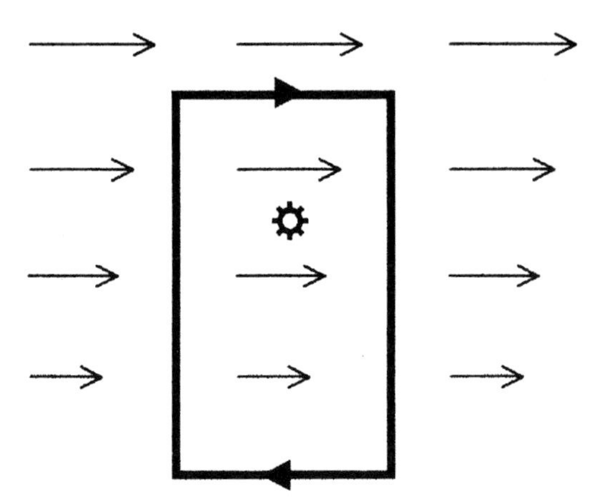

Figure 3 Circuit shown with bold line. ✿ symbol represents paddle wheel with rotation axis coming out of page.

Thus, shear alone can make the circulation nonzero even though the flow is not actually moving around the circuit.

What, then, does the circulation represent in these examples, if not the strength of the flow around the circuit? In general, circulation represents the ability of a given flow to rotate mass elements in the fluid and objects placed in the flow. Imagine a paddle wheel placed within the circuits in Figures 2 and 3 [see, e.g., Holton (1992), Fig. 4.6], with axis perpendicular to the plane of the circuit. In the hurricane illustration, the paddle wheel rotates counterclockwise; in the vertical shear-flow, clockwise. In both examples, when the paddle wheel turns in the same direction as the circuit in the illustration, the sign of the circulation is positive.

Circulation depends on many details about the circuit: its size, shape, orientation, direction, and location. For example, in Figure 3, we could make the circulation arbitrarily large by increasing the length of the sides of the circuit. It is useful to define a quantity that measures rotation but that does not refer to the details of any particular circuit. It is also useful to define a "microscopic" quantity that measures rotation at individual points instead of a "macroscopic" quantity that measures rotation over a region. This leads us to vorticity.

Statement 2. The *vorticity* in the right-hand-oriented direction perpendicular to the plane of a circuit is equal to the circulation per unit area of the circuit in the small-area limit. Vorticity is, therefore, a quantity with dimensions of inverse time. In order to define the x, y, and z components of the vorticity, we consider circuits whose perpendicular lies in each of these directions.

Statement 2 requires some explanation. Consider the circulation for a circuit which, for simplicity, we assume to be flat. This defines a unique direction that is perpendicular to the plane of the circuit if we use the "right-hand rule." This rule works as follows: Curl the fingers of the right hand in the direction of the circuit, e.g., clockwise in Figure 3. Then point the right-hand thumb in a direction perpendicular to the circuit: into the page in Figure 3. In statement 2, we refer to this direction as the "right-hand-oriented direction normal to the plane of the circuit." Now, imagine reducing the size of the circuit until it becomes vanishingly small. In Figure 3, for example, we could imagine reducing the loop to $5 \times 20\,\mathrm{m}$, then $5 \times 20\,\mathrm{mm}$, and so on. In statement 2, we refer to this as the "small-area limit." Therefore, vorticity describes rotation as a microscopic quantity.

We can calculate vorticity in the previous examples. For the shear-flow example (Fig. 3), it is easy to show that the circulation around the loop is aA, where A is the area of the rectangular loop. With more effort, it can be shown that the circulation is aA for any loop if A is taken to be the cross-sectional area of the loop in the plane of the flow. Therefore, the vorticity in the plane of the flow is into the page and equal to $a = 0.01/\mathrm{s}$ everywhere. The reader should try to verify that the other two components of the vorticity are zero because the associated circulation is zero. In hurricanes, as is typical of other vortices, the vorticity varies strongly as a function of distance from the eye. The average vertical component of vorticity in our example is upward and equal to the circulation divided by the total area of the circle: $\Gamma/\mathrm{A} = 2\pi ru/(\pi r^2) = 2u/r \approx 10^{-3}/\mathrm{s}$.

In GFD, there are three types of vorticity: absolute, planetary, and relative. The *absolute vorticity* is the vorticity measured from an inertial frame of reference, typically thought of as outer space. The *planetary vorticity* is the vorticity associated with planetary rotation. Earth rotates with a period of $T \approx 24\,\mathrm{h}$ from west to east around the north pole–south pole rotation axis (Fig. 4). Consider a fluid with zero ground speed, meaning that it is at rest when measured from the frame of reference of Earth. This fluid will also be rotating with period T from west to east. The fluid is said to be "in solid-body rotation" because all the fluid elements maintain the same distance from one another over time, just as the mass elements do in a rotating solid. The distance from the axis of rotation is $r\cos\theta$, the component of the velocity tangential to this rotation is $2\pi r\cos\theta/T$, the circumference of the latitude circle is $2\pi r\cos\theta$, and the circulation, by statement 1, is the product of the two last

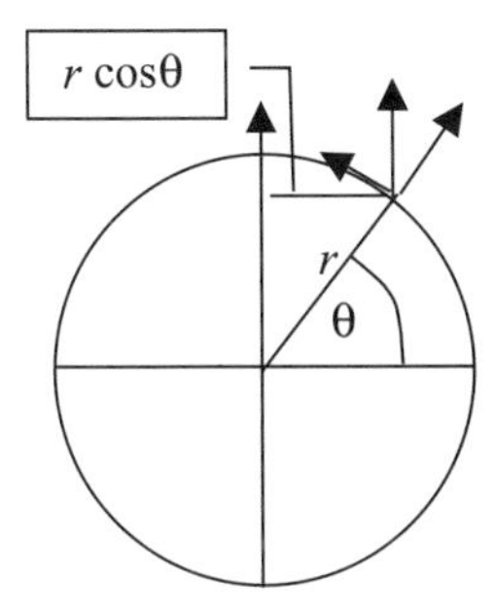

Figure 4 Geometry of Earth. Latitude is θ The vertical vector is the vorticity of solid-body rotation. The distance from the surface to the axis of rotation is $r\cos\theta$. The component of the planetary vorticity perpendicular to the surface is $4\pi\sin\theta/T$. The component tangential to the surface is $4\pi\cos\theta/T$.

quantities $(2\pi r \cos\theta)^2/T$. The vorticity points from the south pole to the north pole and has magnitude $4\pi/T = 1.4 \times 10^{-4}$ per second, where we have divided the circulation by the area of the latitude circle. This is the planetary vorticity. Finally, the *relative vorticity* is the vorticity of the velocity measured with respect to the ground, i.e. in the rotating frame of reference of Earth. The absolute vorticity is the sum of the relative vorticity and the planetary vorticity (since the velocities are additive).

The comparative strengths of planetary vorticity and relative vorticity determine, in large part, the different dynamical regimes of GFD. The planetary vorticity has a component perpendicular to the planet's surface with value $4\pi \sin\theta/T$. This component points radially away from the center of the planet in the Northern Hemisphere where $\sin\theta > 0$, and toward the center of the planet in the Southern Hemisphere where $\sin\theta < 0$. From the viewpoint of an observer on the surface in the Northern Hemisphere, this component points vertically up toward outer space and, in the Southern Hemisphere, vertically down toward the ground. For motions that are characterized by scales of 1000 km or greater in the atmospheric midlatitudes, or 100 km or larger in the oceanic midlatitudes, the radial, "up/down" component of the planetary vorticity is an order of magnitude larger than the relative vorticity in this direction. This is the dynamical regime for midlatitude cyclones and oceanic mesoscale eddies, for which geostrophic balance and quasi-geostrophy hold (see chapter by Salby). At scales smaller than this, the relative vorticity can be comparable to or larger than the planetary vorticity. For instance, in the hurricane example, the vertical component of the vorticity was determined to be 10^{-3}/s in a region 100 km across. Often even larger are the values of the horizontal components of vorticity associated with vertical shear—recall the vertical shear example, with vorticity of magnitude 10^{-2}/s. Although it is large values of the vertical component of vorticity that are associated with strong horizontal surface winds, the availability of large values of horizontal vorticity associated with vertical shear can have a potentially devastating impact. For example, the tilting into the vertical of horizontal shear vorticity characterizes the development of thunderstorms and tornadoes (e.g., Cotton and Anthes, 1989).

Having defined vorticity in terms of the circulation, it is also useful to go from the microscopic to the macroscopic and define circulation in terms of the vorticity. We first introduce the idea of a vector *flux*: the flux of any vector field through a surface is the average of the component of the vector perpendicular to the surface, multiplied by the area of the surface. For example, the flux of a 10-m/s flow passing through a pipe of cross-sectional area 0.5 m^2 is a volume flux of 5 m^3/s.

Statement 3. The circulation for a given circuit is equal to the flux of the vorticity through any surface bounded by the circuit.

To illustrate statement 3, we consider Figure 5, which shows adjacent rectangular circuits with circulation values C_1, C_2 and areas A_1, A_2 that are small enough to have unique values of the vorticity $Z_1 = C_1/A_1$, $Z_2 = C_2/A_2$ pointing into the page. The total circulation for the larger rectangular region formed by joining the two smaller

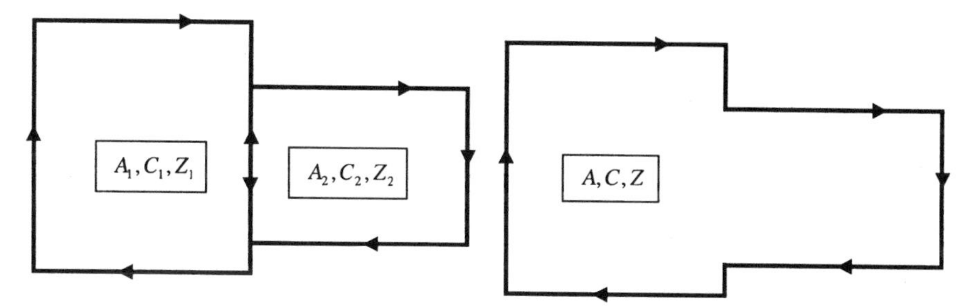

Figure 5 Calculation of net circulation.

rectangles is $C = C_1 + C_2$. This is because, along the shared side (marked in the figure with circulation arrows pointing in both directions), the tangential flow component for circuit 1 is equal and opposite to that for circuit 2. The average component of the vorticity for this region is $(Z_1A_1 + Z_2A_2)/(A_1 + A_2) = (C_1 + C_2)/(A_1 + A_2) = C/A$, where A is the total area. The flux of vorticity is then $(C/A) \times A = C$, which is the total circulation, consistent with statement 3. We can repeat this calculation to include more rectangles in order to determine the average component of the vorticity over a large region. Although we have illustrated the simple case in which the surface bounded by the circuit is flat, statement 3 holds for any surface, flat or curved, bounded by the circuit. This is because statement 3 is a statement of the Stokes's theorem of vector calculus (see Appendix).

Statement 3 shows that vorticity distributed over a small region can be associated with a circulation well away from that region. Typical vortices in geophysical flows tend to have a core of strong vorticity surrounded by a region of relatively weak or zero vorticity. Suppose the vortex covers an area A and has a perpendicular component of vorticity of average value Z. The circulation induced by this flux, for any circuit enclosing the vortex, is AZ. Consider a circuit that spans an area larger than A. The perimeter of such a circuit is proportional to its average distance l from the center of the vortex. (Recall the hurricane example for which the radius is l and the perimeter is $2\pi l$.) Then the induced tangential flow speed associated with the vortex, from statement 1, is proportional to AZ/l. That is, for typical localized vortex distributions, the flow around the vortex varies as the reciprocal of distance from the center. The ability of vorticity to induce circulation "at a distance" and the variation with the reciprocal of distance of the flow strength are key to understanding many problems in GFD.

Statement 3 also implies that mechanisms that change the vorticity in some region can change the circulation of any circuit that encloses that region. For instance, in the hurricane example of Figure 2, suppose that drag effects near the surface reduce the vertical component of the vorticity in some location (see next section). By statement 3, this would reduce the average circulation around the circuit. In other words, the reduction in vorticity would decelerate the flow around the circuit. In this way, we see that vorticity transport, generation, and dissipation are associated with stresses that accelerate or decelerate the flow.

In the final example of this section, we will illustrate, with an idealized model, how vorticity transport gives rise to flow accelerations on a planetary scale (I. Held, personal communication). A thin layer of fluid of constant density and depth surrounds a solid, featureless, "billiard ball" planet of radius r. By "thin," we mean that the depth of the fluid is much less than r. Both planet and fluid layer are rotating, with period T, from west to east. As we have seen, the vorticity of the solid-body rotation is the planetary vorticity, and the component of this vorticity perpendicular to the surface is $4\pi \sin\theta/T$. We will learn, shortly, that this normal component, in the absence of applied forcing or friction, acts like a label on fluid parcels and stays with them as they move around. In other words, the component of the absolute vorticity normal to the surface of the fluid is a tracer.

Suppose, now, that a wave maker near the equator generates a wave disturbance, and that this wave disturbance propagates away from the region of the wave maker. Away from the wave maker, the parcels will be displaced by the propagating disturbance. Given that the component of the absolute vorticity normal to the surface is a tracer, fluid elements so displaced will carry their initial value of this quantity with them. For example, a fluid element at 45N latitude will preserve its value of vorticity, $4\pi \sin(\pi/4)/T = 1.0 \times 10^{-4}$ per second as it moves north and south. Now, $4\pi \sin\theta/T$ is an increasing function of latitude: there is a gradient in this quantity, from south to north. Therefore, particles from low-vorticity regions to the south will move to regions of high vorticity to the north, and vice versa. There will be, therefore, a net southward transport of vorticity, and a reduction in the total vorticity poleward of that latitude. Notice that this transport is down-gradient. Thus, the circulation around that latitude, for a west-to-east circuit, will be reduced and the fluid at that latitude will begin to stream westward as a result of the disturbance. The transport of vorticity by the propagating disturbance gives rise to a stress that induces acceleration on the flow. This acceleration is in the direction perpendicular to the vorticity transport.

We can estimate the size of the disturbance-induced acceleration. Suppose, after a few days, that particles are displaced, on average, by 10 degrees latitude, which corresponds to a distance of about 1000 km. Since $4\pi \sin\theta/T$ has values between -1.4×10^{-4} and 1.4×10^{-4} per second for $T = 24$ h, a reasonable estimate for the difference between a displaced particle's vorticity and the background vorticity is 10^{-5} per second for a 10 degrees latitude displacement. The average estimated southward transport of the vorticity is then the displacement times the perturbation vorticity: $1000\,\text{km} \times 10^{-5}$ per second $= 10$ m/s. This is an estimate of the westward flow velocity induced by the displacement over a few days and corresponds to reasonable values of the observed eddy-induced stresses on the large-scale flow.

3 POTENTIAL VORTICITY: DEFINITION AND EXAMPLES

The previous example describes the effect on the horizontal circulation of redistributing the vorticity of solid-body rotation. Although the example seems highly idealized, the wave-induced stress mechanism it illustrates is fundamental to Earth's large-scale atmospheric circulation. The physical model of a thin, fixed-density, and constant-depth fluid, which is known as the barotropic vorticity model, is deceptively simple but has been used as the starting point for an extensive body of work in GFD. This work ranges from studies of the large-scale ocean circulation (Pedlosky 1996), to the analysis of the impact of tropical disturbances such as El Niño on the midlatitude circulation, and to early efforts in numerical weather forecasting. Such applications are possible because the idealizations in the example are not as drastic as they initially appear. For example, Earth's atmosphere and ocean are quite thin compared to the radius of Earth: Most of the atmosphere and world ocean are contained within a layer about 20 km thick, which is small compared to Earth's radius (about 6300 km). In addition, large-scale atmospheric speeds are characteristically 10 to 20 m/s, which is small compared to the speed of Earth's rotation ($2\pi r \cos\theta/T = 460\cos\theta\,\text{m/s} \approx 300\,\text{m/s}$ in the midlatitudes)—the atmosphere is not so far from solid-body rotation. This is consistent with the idea that at large scales, atmospheric and oceanic flows have vertical relative vorticity components that are much smaller than the planetary vorticity. Perhaps the most drastic simplifications in the example are that the layer of fluid has constant depth and constant density. In this section, we consider variations of fluid layer depth and of fluid density; this will lead us to potential vorticity (PV).

Since large-scale atmospheric circulations typically occur in thin layers, these flows are typically *hydrostatic*. This means that there is a balance between the vertical pressure force and the force of gravity on each parcel, that vertical accelerations are weak, and that the pressure at any point is equal to the weight per unit area of the fluid column above that point. Consider a thin hydrostatic fluid of constant density but of variable depth. It can be shown (e.g., Salmon 1998) that the state of the fluid may be completely specified by three variables: the two horizontal components of the velocity and the depth of the fluid. The system formed by these three variables and the equations that govern them is known as the *shallow-water model*. These three variables are independent of depth, which implies that there is only a single component of vorticity: the vertical component of vorticity associated with the north–south and east–west components of motion.

The shallow-water model provides a relatively simple context to begin to think about PV:

Statement 4. The *shallow-water PV* is equal to the absolute vorticity divided by the depth of the fluid; it has dimensions of inverse time–length.

By this definition, the PV can be changed by either changing the vorticity or by changing the depth of the fluid column. The depth of the fluid column can be changed, in turn, by the fluid column encountering variations in depth of the topography below the fluid or in the height of the fluid's surface. For example, if the PV of a fluid column is held constant and the depth of the fluid decreased, by, for example, shoaling (i.e., moving the column up a topographic slope), the result is a reduction in the column's vorticity.

The generalization of PV to fluids in which the density is not constant can be approached in stages. To start with, consider, instead of a single layer of fluid, a system consisting of two or more thin, constant-density, hydrostatic layers in which each layer lies under another layer of lighter fluid. For this system, the PV in each layer is the ratio of the vertical vorticity to the layer depth. If density is instead taken to vary continuously, extra complications are added. First, at least six variables are required to specify the state of the fluid: the three components of the velocity, the pressure, the density, and another thermodynamic variable such as the temperature or the specific entropy. The specific entropy is the entropy per unit mass (see Section 4). Additional variables are needed to account for salinity in the ocean and moisture in the atmosphere; we neglect these additional factors. The second complication is that, since the velocity is three dimensional, so now is the vorticity vector. The generalization of the shallow-water PV to fluids with variable density is known as Ertel's PV.

Statement 5. *Ertel's PV* is equal to the projection of the absolute vorticity vector onto the spatial gradient of the specific entropy, divided by density. Its dimensions depend on the physical dimensions of the entropy measure used.

To explain statement 5 in more detail: Recall that the projection of vector **a** onto vector **b** is $\mathbf{a} \cdot \mathbf{b}$. The gradient of the specific entropy is a vector, pointing perpendicular to a surface of constant specific entropy, in the direction of increasing specific entropy. Its value is equal to the rate of change of specific entropy per unit along-gradient distance (Fig. 6). Despite the additional complexity, there are analogies between the shallow-water PV (statement 4) and Ertel's PV (statement 5). For example, for fixed Ertel's PV, we can decrease a column's vorticity by decreasing the thickness between two surfaces of specific entropy, since this would increase the gradient. This reduction in thickness is analogous to compressing the fluid column, as in the shallow-water example. The connection between statements 4 and 5 will be discussed in more detail below.

We have defined circulation, vorticity, and PV but have made little explicit reference to fluid dynamics, that is, to the interactions and balances within a fluid that allow us to predict the behavior of these quantities. Dynamical considerations justify our interest in defining PV as we have here, and link the PV back to our original discussion of circulation. Another important loose end is to look at the impact of planetary rotation in more detail, since rotation dominates large-scale motions on Earth. These topics will be taken up in the next section.

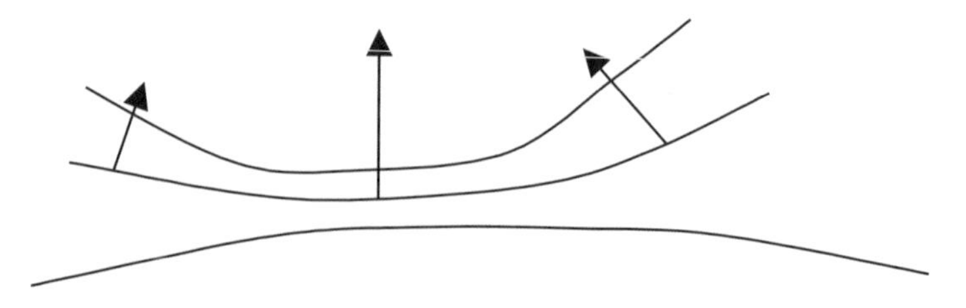

Figure 6 Contours: surfaces of constant specific entropy. Arrows: specific-entropy gradient, pointing in direction of increasing specific entropy.

4 DYNAMICAL CONSIDERATIONS: KELVIN'S CIRCULATION THEOREM, ROTATION, AND PV DYNAMICS

We begin our discussion of the dynamical aspects of circulation, vorticity, and PV by returning to the shallow-water system, which consists of a thin hydrostatic layer of fluid of constant density. In this section, we will need to introduce several new concepts. First, we introduce the concept of a *material circuit.* Points on a material circuit do not stay fixed in space, as in the circuits in Figures 2 and 3, but instead follow the motion of the fluid. In the presence of shear, mixing, and turbulence, material circuits become considerably distorted over time.

Statement 6. In the absence of friction and applied stresses, the absolute circulation in the shallow-water model is constant for a material circuit.

Statement 6 asserts that the circulation measured along a material circuit, as it follows the flow, will not change. This is a special case of Kelvin's circulation theorem, which applies to a fluid with variable density, and which will be discussed shortly.

Statement 6 follows from applying, to a material circuit, the fluid momentum equations that express Newton's second law of motion. Newton's second law expresses the balance between the acceleration of the flow and the sum of the forces per unit mass acting on the fluid. For the shallow-water fluid, these forces include the pressure force, friction near solid boundaries, and applied stresses (such as the stress exerted by the wind on the ocean). The pressure force is not mentioned in statement 6 because, when averaged around a circuit, it cannot generate circulation on any circuit in the shallow-water system. To understand this, we need to remind the reader of the concept of *torque.* Torque is a force that acts on a body through a point other than the body's center of mass. Forces that act through a body's center of mass cause acceleration in the direction of the force. Torque, on the other hand, causes rotation. For example, it is a torque imparted by the vorticity in the fluid that sets spinning the paddle wheels in Figures 2 and 3. Torque is a vector, which, by convention, points in the direction of the imparted rotation using the right-hand rule.

The pressure force in the shallow-water system acts through the center of fluid *elements*, that is, fluid columns in the small-area limit. Therefore, the pressure force cannot directly impart a rotational torque to fluid elements. This can be used to show that the pressure force cannot generate circulation on a circuit of finite size. Nevertheless, pressure forces can compress or expand the circuit horizontally. Variations in the area of the circuit then induce rotation indirectly because of statement 6. To see this, suppose a material circuit that surrounds a fluid column is compressed horizontally. We note that the mass of the column is another quantity that is constant following the motion—in the absence of mass sources and sinks, no mass will leak into or out of the vertical column that the circuit encloses. Since the density is a constant throughout the fluid, and since the mass is constant following the motion, the volume of the column must be conserved following the motion of the fluid. Therefore, horizontal compression increases proportionally the height of the column.

At the same time, if the area of the material circuit is reduced, statement 6 shows that the vorticity within the circuit must increase to maintain a constant circulation. The increase in vorticity associated with horizontal compression and vertical stretching reflects local angular momentum conservation [see, e.g., Salmon (1998)].

As pointed out in the previous section, the depth of the fluid column may also change if the column passes over topographic features of the solid underlying surface. Columns passing over ridges obtain negative vorticity, and over valleys, positive vorticity, relative to their initial vorticity.

The interpretation of statement 6 in the presence of planetary rotation is fundamental to the study of large-scale GFD. Statement 6 holds for the absolute circulation, that is, the circulation measured from an inertial, nonrotating frame. An important mechanism for generating relative vorticity involves the constraint of having to conserve the absolute circulation as fluid moves north and south. Consider a ring of fluid, at rest in the rotating frame, of small surface area A, and located at latitude θ. From our earlier discussion of the rotating fluid on the sphere, the vorticity for this ring is the planetary vorticity $4\pi \sin\theta/T$ and the circulation for this ring is the planetary circulation $4\pi A \sin\theta/T$ for the right-hand oriented path around the circuit. If the fluid is free of friction and external applied stresses, then statement 6 tells us that the sum of the planetary and relative circulations will be constant following the motion of the ring. Thus, if the ring maintains its original area and is displaced northward, it will acquire a positive (cyclonic, counterclockwise in the Northern Hemisphere) relative circulation.

The final application of statement 6 we will consider concerns the effect of applied stresses and friction on the circulation. The proof of statement 6, which we have not detailed, shows how circulation may be generated under conditions that depart from the statement's assumptions. In particular, frictional and applied stresses, such as those found in atmospheric and oceanic boundary layers, can impart vorticity to a fluid by a mechanism known as *Ekman pumping*. Ekman pumping describes the way a frictional fluid boundary layer responds to interior circulation. For example, a cyclonic-relative circulation, i.e. one with positive relative vorticity, can cause upwelling in the boundary layer. This tends to decrease the depth of the fluid column, to reduce the relative vorticity, and therefore to counteract the positive circulation. On the other hand, applied stresses (such as atmospheric wind stress on the ocean) play a major role in generating large-scale oceanic circulation. These effects can be modeled in a simple way within the shallow-water system.

Just as the macroscopic circulation has a microscopic counterpart, namely the vorticity, the macroscopic statement of conservation of circulation (statement 6) has a microscopic counterpart.

Statement 7. In the absence of friction and applied stresses, the shallow-water PV (statement 4) is constant following the motion of the fluid.

To demonstrate statement 7, we apply the following argument, using statement 6 in the small-area limit (see Fig. 7):

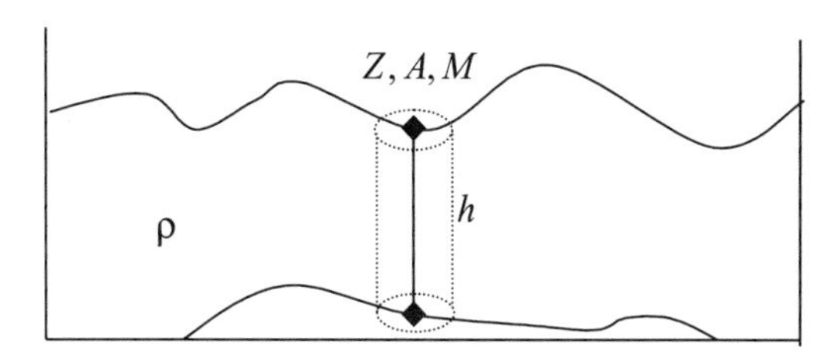

Figure 7 Demonstration of statement 7.

1. In the small-area limit, from statement 6, the circulation ZA is a constant of the motion, where Z is the vorticity and A is the area of the material circuit.
2. $ZA = (ZM)/(\rho h)$, where M is the mass of the fluid enclosed by the circuit, ρ the density, and h the depth.
3. Recall that M is constant following the motion and that ρ is a simple constant.
4. Thus, Z/h, which is the shallow-water PV (statement 4), is also a constant of the motion.

That PV is a constant following the motion is consistent with the example in which the vorticity and the height increased proportionally to the horizontal compression of the column. All the important mechanisms discussed in the context of definition 6—the roles of fluid column stretching and compression by the pressure force and topography, and the ability of planetary rotation to induce relative circulation as systems move north and south—can be interpreted in terms of the conservation of shallow-water PV (statement 7). The PV can be thought of as the planetary vorticity a fluid column would have if it were brought to a reference latitude and depth. This is the origin of the word *potential* in the name *potential vorticity*.

At the end of Section 2, we discussed an example that used the barotropic vorticity model, that is, a thin fluid of constant density and depth. For fixed depth, statement 6 is the same, but the PV is now, simply, the absolute vorticity. That is, in the absence of column stretching, the absolute vorticity is constant following the motion. This property was used to argue that particles disturbed by the propagating wave in the example would retain their initial values of absolute vorticity.

The final topic of this section concerns the dynamics of fluids with variable density. If density is not constant, it is possible for the pressure force to induce a net rotational torque. This will result in an important modification to the circulation theorem. To start with, we define a *barotropic* fluid to be one for which surfaces of constant density are aligned with surfaces of constant pressure. Put in another way, in a barotropic fluid, the density is a function only of the pressure, and not of the temperature or other thermodynamic variables.

Statement 8. In a barotropic fluid without friction or applied stresses, the circulation is constant following the motion of the fluid (Kelvin's circulation theorem).

Similarly to statement 6, statement 8 is proved by applying the momentum equations to a circuit. In a barotropic fluid, the net pressure force around a fluid element points through its center of mass, as in the shallow-water system, and the pressure force cannot exert a net rotational torque. A fluid that is not barotropic is said to be *baroclinic*. In baroclinic fluids, the net pressure force on a fluid element does not pass through the center of mass of the element. This results in a torque on the element that generates rotation, that is, vorticity, and, therefore, circulation. The impact of baroclinicity can be felt at small and large scales in the atmosphere and ocean. On scales of a few kilometers, the baroclinicity associated with the thermal contrast between land and sea generates sea breeze circulations. On planetary scales, it is the baroclinicity associated with the large-scale equator-to-pole temperature gradient that provides the source of energy for midlatitude atmospheric and oceanic eddies.

As for the shallow-water case, Kelvin's circulation theorem (statement 8) has a microscopic counterpart.

Statement 9. In the absence of friction, applied stresses, and applied heating, Ertel's PV, (statement 5), is constant following the motion of the fluid.

In order to justify statement 9, we need to discuss the concept of specific entropy in more detail. In thermodynamics, entropy is a thermodynamic variable that can be expressed as a function of other thermodynamic variables, such as temperature, pressure, and density. When heat is applied to a thermodynamic system, the change in entropy is related to the amount of heat transferred to the system. In the absence of heating, the entropy of the system does not change: the system is said to be *adiabatic*. In an adiabatic fluid, this implies that the entropy per unit mass, i.e., the specific entropy, is a constant of the motion. Therefore, surfaces of constant entropy (isentropic surfaces) are material surfaces, that is, surfaces that move with the fluid. In other words, the motion in an adiabatic fluid is along isentropic surfaces. Adiabatic fluids are relevant to GFD because large-scale atmospheric and oceanic flows are often approximately adiabatic.

Now, let us consider a simple fluid such as an ideal gas, for which entropy is a function of pressure and density. Suppose the fluid is adiabatic, so that the flow is along isentropic surfaces. On these surfaces, the pressure is a function of density. That is, the flow along isentropic surfaces in an adiabatic fluid is barotropic. This implies that statement 8 holds: The circulation is constant following the flow on an isentropic surface. Now, consider statement 8 applied in the small-area limit to two closely spaced isentropic surfaces (Fig. 8). The demonstration of statement 9 proceeds as follows:

1. In the small-area limit, from statement 8, the circulation ZA is a constant of the motion, where Z is the component of the vorticity normal to the isentropic surface and A is the area of each circuit in the small-area limit.

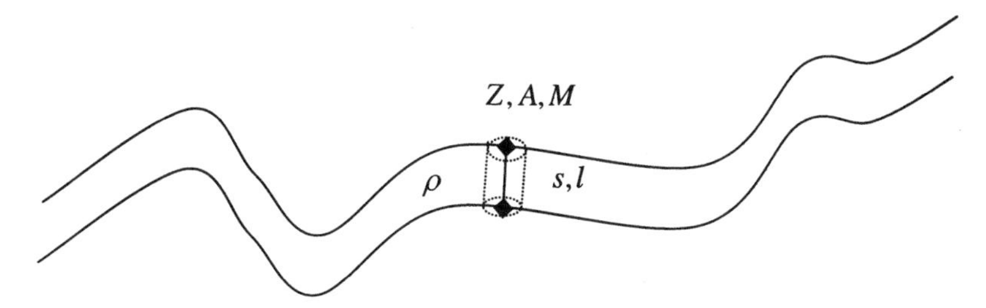

Figure 8 Demonstration of statement 8.

2. The entropy difference between the two isentropic surfaces, s, is a constant of the motion, since the entropy value of each surface is a constant of the motion.
3. Therefore, the product AZs is a constant of the motion.
4. The area $A = V/l$, where V is the volume enclosed by the material column and l is the perpendicular distance between the isentropic surfaces.
5. The volume $V = M/\rho$, where ρ is the density, and M is the mass of the column enclosed by the circuit and the two isentropic surfaces.
6. The product $MZs/(l\rho)$ is a constant of the motion, as is M. Thus, $(Z/\rho)(s/l)$ is a constant of the motion.
7. The quantity s/l represents the change of entropy per unit distance between the isentropic surfaces, and therefore represents the specific entropy gradient.
8. Therefore, the quantity $(Z/\rho)(s/l)$ is Ertel's PV (statement 5) and is a constant of the motion. This demonstrates statement 9.

The dynamics of Ertel's PV incorporate similar mechanisms to the one discussed for shallow-water PV dynamics, with the fluid depth replaced by the distance between isentropic surfaces. It is very useful to have simpler model systems, such as the shallow-water model, with which to build up our intuition concerning PV dynamics. For example, there is a strong analogy between changes in thickness of the fluid column brought about by mass sources and sinks in the shallow-water model and thermally induced changes in the distribution of isentropic surfaces.

5 CONCLUSION

Potential vorticity is a powerful unifying tool in the atmospheric and oceanic sciences because it combines apparently distinct factors, such as topography, stratification, relative vorticity, and planetary vorticity, into a single dynamical quantity that is conserved, following the motion, like a chemical tracer. Indeed, much current work in the field is characterized by "PV thinking:" how PV is generated, maintained, converted from one form to another, transported, and dissipated. To conclude, there follows a list of a few important topics of current interest in which PV dynamics is central. Hoskins et al. (1985), Salmon (1998), and Andrews et al. (1987) are excellent starting points for further investigation.

- *Rossby-Wave Dynamics* The primary large-scale oscillations in the extra-tropical atmosphere and oceans are Rossby waves, which are supported by the PV gradients largely associated with planetary rotation. In the "billiard-ball planet" example discussed in Section 2, it is Rossby waves that carry the signal of the wave maker away from the equator. Rossby-wave dynamics underlies our understanding of the midlatitude remote response to El Niño events in the tropical Pacific, of the propagation of upper tropospheric disturbances, and of the dynamics of large-scale waves in the stratosphere (see Chapter 4).
- *Baroclinic Instability* The development of atmospheric cyclones and of oceanic eddies can be understood in terms of the interactions of regions of opposite-signed PV gradients (see Chapter 4).
- *Geostrophic Turbulence* Our understanding of atmosphere/ocean large-scale turbulence is framed by PV, which behaves as an "active" tracer that can be mixed and diffused downgradient (see Chapter 6).
- *Balance Models* Balance models, such as the quasi-geostrophic model (Chapter 4, Swanson), are filtered equations that model the slow, large-scale motions associated with Rossby waves, baroclinic instability, and geostrophic turbulence. These models are based on some of the dynamical regimes mentioned above; for example, the assumption that the vertical component of relative vorticity is small compared to the vertical component of planetary vorticity. In these models, the PV is the sole dynamical variable, and all other variables can be obtained from the PV by an inversion procedure. This inversion procedure is analogous to the one involved in obtaining the circulation from the vorticity distribution (see Section 2).
- *Rossby-Wave Activity Propagation and Wave-Induced Circulations* Recall that the southward flux of vorticity in the barotropic vorticity model example at the end of Section 2 gives rise to a westward acceleration. More generally, the flux of PV in both the shallow-water model and fluids with variable density can be associated with a wave-induced torque, generally in the direction transverse to the flux. In this way, the existence of persistent extratropical jets, such as the atmospheric jet stream and the jets in the Antarctic Circumpolar Current, can be understood in terms of the flux, often downgradient, of PV. These fluxes are also the starting point for a theory of "wave activity" propagation that is the foundation of models of the overturning circulation in the stratosphere.
- *Symmetries, Conservation Laws, and Hamiltonian Structure* The existence of PV as a constant of the motion has a profound connection to the mathematical structure of the equations of motion. All conserved quantities for a physical system—spatially distributed quantities such as energy and momentum, and local quantities such as PV and wave activity, are connected to symmetries in the system through Noether's theorem. This idea is the starting point for Hamiltonian fluid dynamics, which is reviewed in Salmon (1998). The symmetry connected to the PV is the so-called particle relabeling

symmetry, which reflects the fact that the dynamics of an adiabatic fluid is unaffected by changing the particle labels on isentropic surfaces. Such considerations provide a theoretical basis for wave activity invariants and generalized nonlinear stability theorems for GFD, as well as for the systematic derivation of balance models with desirable symmetry.

6 APPENDIX: MATHEMATICAL SUMMARY

In this section, we provide a summary, using the notation of vector calculus, of the main results. First, the circulation, C, in statement 1, may be represented by a line integral

$$C = \oint_{\text{circuit}} |\mathbf{u}| \cos \theta \, ds$$

where $\mathbf{u}$ is the velocity measured from the inertial frame. In this notation, the length of the circuit is $L = \oint_{\text{circuit}} ds$ and the average of the tangential flow component is C/L. This is consistent with statement 1. By the Stokes theorem, the circulation also satisfies

$$C = \iint_{\text{region}} \text{curl } \mathbf{u} \cdot \hat{\mathbf{n}} \, dA$$

where the double-integral sign and the word *region* indicate that a flux integral is being taken over the region bounded by the circuit, and $\hat{\mathbf{n}}$ is a unit vector perpendicular to the surface. The vector curl of the velocity is, in Cartesian coordinates,

$$\text{curl } \mathbf{u} = \left(\frac{\partial w}{\partial y} - \frac{\partial v}{\partial z}, \frac{\partial u}{\partial z} - \frac{\partial w}{\partial x}, \frac{\partial v}{\partial x} - \frac{\partial u}{\partial y} \right)$$

Statement 2 in the small-area limit implies that the vorticity, $\boldsymbol{\zeta}$, is equal to the curl of the velocity:

$$\boldsymbol{\zeta} = \text{curl } \mathbf{u}$$

Statement 3 is therefore equivalent to the Stokes's theorem for finite-area surfaces.

For the shallow-water system, the vorticity is vertical:

$$\boldsymbol{\zeta} = \left(\frac{\partial v}{\partial x} - \frac{\partial u}{\partial y} \right) \mathbf{z}$$

and the potential vorticity, statement 4, is

$$q = \frac{(\partial v/\partial x) - (\partial u/\partial y)}{h - h_b}$$

where h is the height of the fluid surface and h_b is the height of the solid surface beneath the fluid. Ertel's PV, statement 5, is

$$q = \boldsymbol{\zeta} \cdot \frac{\nabla s}{\rho}$$

where s is the specific entropy and ρ is the density.

The circulation theorem for the shallow-water model, statement 6, and Kelvin's circulation theorem, statement 8, are written

$$\frac{DC}{Dt} = 0$$

where D/Dt is the material derivative following the motion, defined by

$$\frac{D}{Dt} = \frac{\partial}{\partial t} + \mathbf{u} \cdot \nabla$$

and C is the circulation of a material circuit. The conservation of PV following the motion for shallow-water (statement 7) or three-dimensional stratified flow (statement 9) is

$$\frac{Dq}{Dt} = 0$$

REFERENCES

Andrews, D. G., J. R. Holton, and C. B. Leovy (1987). *Middle Atmosphere Dynamics*, Orlando, Academic.

Cotton, W. R., and R. A. Anthes (1989). *Storm and cloud dynamics*, San Diego, Academic.

Gill, A. E., (1982). *Atmosphere-Ocean Dynamics*, San Diego, Academic.

Holton, J. R. (1992). *An Introduction to Dynamic Meteorology*, 3rd Edition, San Diego, Academic.

Hoskins, B. J., M. E. McIntyre, and A. W. Robertson (1985). On the use and significance of isentropic potential vorticity maps, *Q. J. Roy. Met. Soc.* **111**, 877–946.

Kundu, P. K. (1990). *Fluid Mechanics*, San Diego, Academic.

Pedlosky, J. (1996). *Ocean Circulation Theory*, Berlin, Springer.

Salmon, (1998) *Lectures on Geophysical Fluid Dynamics*, New York, Oxford University Press.

CHAPTER 4

EXTRATROPICAL ATMOSPHERIC CIRCULATIONS

KYLE SWANSON

1 INTRODUCTION

As one moves poleward in Earth's atmosphere, the character of the circulation changes markedly. At large scales and low frequencies, tropical circulations are fundamentally ageostrophic, epitomized by the thermally direct Hadley, Walker, and monsoon circulations. In contrast, at large scales and low frequencies in the extratropics, the effect of Coriolis accelerations due to Earth's rotation become important, and circulations are approximately in geostrophic balance. As such, away from the surface, the extratropical large-scale wind satisfies

$$(u, v, w) \approx (u_g, v_g, 0) \equiv \lfloor -a^{-1}\partial_\phi \psi, (a \cos\phi)^{-1}\partial_\lambda \psi, 0 \rfloor \tag{1}$$

where the geostrophic stream function is given by:

$$\psi \equiv f_0^{-1}(\Phi - \Phi_0) \tag{2}$$

Here $\Phi_0(z)$ is a suitable reference geopotential profile and $f_0 = 2\Omega \sin\phi_0$ is the Coriolis parameter evaluated at some extratropical latitude ϕ_0. The extent to which geostrophic balance adequately describes the extratropical circulation is measured by the smallness of the Rossby number, $\mathrm{Ro} = U/(f_0 L)$, which is approximately 0.1 for velocity and length scales $U \sim 10\,\mathrm{m/s}$ and $L \sim 1000\,\mathrm{km}$ that, respectively, characterize large-scale extratropical dynamical motions.

Handbook of Weather, Climate, and Water: Dynamics, Climate, Physical Meteorology, Weather Systems, and Measurements, Edited by Thomas D. Potter and Bradley R. Colman.
ISBN 0-471-21490-6

In addition to geostrophic balance, the extratropical atmospheric circulation is also in hydrostatic balance. Hydrostatic balance provides a relation between the potential temperature and mass fields of the form

$$\theta_e \equiv \theta - \theta_0(z) = HR^{-1} f_0 e^{\kappa z/H} \partial_z \psi \tag{3}$$

where $\theta_0(z) = HR^{-1} f_0 e^{\kappa z/H} \partial_z \Phi_0$ is a reference potential temperature. The combination of hydrostatic and geostrophic balance, through cross differentiation of (1) and (3), yields the thermal wind relations that link the vertical shear of the geostrophic wind to horizontal potential temperature gradients. Therefore, the jets that dominate the zonally averaged tropospheric flow have their origin in the density contrasts of the equator–pole temperature gradient and are referred to as baroclinic.

Approximate geostrophic balance, along with the existence of a robust zonally averaged flow provide the basis for theoretical inquiries into the dynamics of observed large-scale motions in the extratropical atmosphere. The former allows for a simpler description of the underlying dynamics, while the latter allows for meaningful insights into dynamical phenomena to be gained through the study of linear fluctuations about a zonal basic state flow. In this chapter, we apply these simplifications to examine the dynamics of several important classes of large-scale motions in the extratropical atmospheric circulation.

2 QUASI-GEOSTROPHIC THEORY

Geostrophy by itself does not provide an adequate description of large-scale dynamical motions, as it is a time-independent diagnostic relation between the mass and velocity fields. To understand how flows in approximate geostrophic balance evolve in time, it is necessary to include the important effects of $O(\text{Ro})$ ageostrophic winds, which includes the vertical wind. The quasi-geostrophic (QG) approximation includes the effects of these ageostrophic winds on the $O(1)$ geostrophic fields in a simplified but consistent manner and provides a concise, idealized system with which to study large-scale extratropical dynamical motions.

The QG approximation has its roots in the geometrical simplification of replacing the horizontal spherical coordinates (λ, ϕ) by the eastward and northward Cartesian coordinates (x, y) in the full primitive equations and restricting the flow domain to some neighborhood of the latitude ϕ_0. Here x is eastward distance and y northward distance from some origin (λ_0, ϕ_0). This geometric simplification allows the QG dynamical approximations to be made with full rigor (Pedlosky 1987, Section 6.2-5) and once made provides the important conceptual simplification that the *only* effect of Earth's sphericity is the variation of the Coriolis parameter f with latitude,

$$f = f_0 + \beta y \tag{4}$$

where $\beta = 2\Omega a^{-1} \cos\phi_0$. This so-called *β-plane approximation* captures the most important dynamical effect of the variation of f with latitude. Once this simplification is made, the QG dynamical equations naturally emerge from an expansion of the equations of motion in the small parameter Ro.

In the absence of forcing and dissipation,* QG dynamics are described by the material conservation of quasi-geostrophic potential vorticity (QGPV) by the horizontal geostrophic flow

$$(\partial_t + u_g \partial_x + v_g \partial_y) q = 0 \tag{5}$$

where

$$\begin{aligned} q &= \zeta_g + f_0 \rho_0^{-1} \partial_z \lfloor \rho_0 \theta_e (\partial\theta_0/\partial z)^{-1} \rfloor + f_0 + \beta y \\ &= \partial_{xx}\psi + \partial_{yy}\psi + \rho_0^{-1}\partial_z(\rho_0 \varepsilon \partial_z \psi) + f_0 + \beta y \end{aligned} \tag{6}$$

is the QGPV, ζ_g is the vertical component of the relative geostrophic vorticity, and $\varepsilon(z) \equiv f_0^2/N^2(z)$. QG dynamics are layer-wise equivalent to a shallow water system insofar as variations in the vertical only appear implicitly in the elliptic nature of the q-ψ relation. Although ageostrophic circulations do not explicitly occur in the QGPV evolution equation, these circulations are vital to QG dynamics and may be obtained diagnostically from the geostrophic mass field and its time evolution (Holton, 1992, Section 6.4).

This system is completed by the specification of upper and lower boundary conditions. The lower boundary condition over a rigid surface is

$$(\partial_t + u_g \partial_x + v_g \partial_y)(\partial_z \psi + N^2 h_T / f_0) = 0 \quad \text{at} \quad z = 0 \tag{7}$$

where $h_T(x, y)$ is the bottom topography. Recalling the connection between the vertical derivative of the geostrophic stream function and the perturbation potential temperature (3), condition (7) simply describes the material conservation of potential temperature in the geostrophic wind at the lower boundary. If there is a rigid upper boundary, another condition of the form (7) is applied there. In the atmosphere, this condition is replaced by the requirement that either the perturbation energy density decay as $z \to \infty$, or alternatively, that the energy flux be directed upward as $z \to \infty$.

The robust extratropical zonally averaged flow suggests studying small-amplitude fluctuations about that flow. Dividing the stream function into a zonally symmetric background and a perturbation,

$$\psi(x, y, z, t) = \psi_0(y, z) + \psi'(x, y, z, t) \tag{8}$$

*Forcing and dissipation are neglected not because they are unimportant to large-scale atmospheric dynamical motions, but because their inclusion is not necessary to understand the dynamical features examined herein. However, in general they will have important modifying effects in any given circumstances and will be crucial in the study of any steady-state motions.

the evolution of the perturbation QGPV is described by

$$(\partial_t + u_{g0}\partial_x)q' + v'_g q_{0y} + \mathbf{u}'_g \cdot \nabla q' = 0 \tag{9}$$

Here $q' = \partial_{xx}\psi' + \partial_{yy}\psi' + \rho_0^{-1}\partial_z(\rho_0 \varepsilon \partial_z \psi')$; $\mathbf{u}'_g = (u'_g, v'_g) = (-\partial_y\psi', \partial_x\psi')$; $u_{g0} = -\partial_y\psi_0$; and

$$q_{0y} = \beta - \partial_{yy}u_{g0} - \rho^{-1}\partial_z(\rho\varepsilon\partial_z u_{g0}) \tag{10}$$

is the background QGPV gradient. The perturbation (rigid surface) boundary condition is

$$(\partial_t + u_{g0}\partial_x)(\partial_z\psi' + N^2 h'/f_0) + v'_g \eta_{0y} + \mathbf{u}'_g \cdot \nabla(\partial_z\psi') = 0 \quad \text{at} \quad z = 0 \tag{11}$$

where

$$N^2 f_0^{-1}\partial_y[h_T] - \partial_z u_{g0} \tag{12}$$

is proportional to the basic state potential temperature gradient measured along the sloping lower boundary, and the brackets indicate a zonal average.

We are free to seek wavelike solutions to Eqs. (8) and (9) of the form,

$$\psi'(x, y, z, t) = \Psi(y, z)e^{ik(x-ct)} \tag{13}$$

where k is the (non-negative) wavenumber in the x (zonal) direction and c is the complex phase speed, related to the frequency $\omega = kc$. Substituting in solutions of this form and neglecting terms that are quadratic in perturbation quantities yields the equation

$$(u_{g0} - c)(\rho^{-1}\partial_z\rho\varepsilon\partial_z + \partial_{yy} - k^2)\Psi + q_{0y}\Psi = 0 \tag{14}$$

with the lower boundary condition

$$(u_{g0} - c)\partial_z\Psi + \eta_{0y}\Psi = 0 \quad \text{at} \quad z = 0 \tag{15}$$

along with the appropriate upper boundary condition for the problem at hand. Jumps in N or the shear, common at the tropopause, may be included in (14) by applying analytically derived jump conditions.

The linear, homogeneous system [Eqs. (14) and (15)] defines an eigenvalue problem for c. A particularly illuminating case is where u_{g0} and N are constant, $\rho = \rho_0 \exp(-z/H)$, and the topography h_T vanishes. In this situation, solutions to (14) may be sought of the form

$$\Psi = A\exp(imz)\exp\left(\frac{z}{2H}\right)\exp(ily) \tag{16}$$

where the complex constant A may be adjusted to meet the boundary condition (15). For this particular solution, the zonal phase speed is given by

$$c = \frac{\omega}{k} = u_{g0} - \frac{\beta}{k^2 + l^2 + \varepsilon(m^2 + \frac{1}{4}H^2)} \tag{17}$$

Waves of this type are called *Rossby waves* and provide an archetype for low-frequency ($\omega < f_0$) motions in the extratropics. Rossby waves differ from other, more familiar wave types such as gravity waves, since it is the planetary vorticity gradient β that acts as a restoring mechanism for the wave. An illuminating discussion of the vorticity gradient as a restoring mechanism is found in Pedlosky (1987, Section 3.16).

Rossby waves have a number of remarkable properties. First, the zonal phase speed is everywhere westerly relative to the basic state zonal flow u_{g0}. However, as the total wavenumber $K^2 = k^2 + l^2 + \varepsilon(m^2 + 1/4H^2)$ gets larger, Rossby waves move more and more nearly with the basic flow u_{g0}. These properties are entirely consistent with observed large-scale atmospheric waves. Rossby waves are also dispersive, i.e., the group velocity $\mathbf{c}_g \equiv (\partial\omega/\partial k, \partial\omega/\partial l, \partial\omega/\partial m)$ that describes how a packet of Rossby waves moves with time differs from the phase velocity of the waves comprising that packet in both magnitude and direction. Since information travels with the group velocity, rather than the phase velocity, the group velocity also describes the propagation of the energy of the packet. The difference between the Rossby wave phase and group velocities is most apparent for vertical propagation, since the vertical group velocity and phase velocity are oppositely directed. If the lines of constant phase propagate upwards (downwards), the energy in the wave propagates downwards (upwards).

While Rossby waves provide a nontrivial example of nearly geostrophic motions in the extratropics, the actual extratropical flow situation is significantly more complicated than that considered above. Specifically, vertical shear and associated meridional temperature gradients allow for the existence of instabilities that can amplify with time as well as propagate. The characteristics of such instabilities, along with their importance in the extratropical atmospheric circulation form the topic of the next section.

3 BAROCLINIC INSTABILITY AND FRONTOGENESIS

The most striking observational feature of the extratropical atmosphere is the spectrum of vigorous synoptic-scale transient eddies that provide much of the day-to-day variability in the middle latitude weather. These eddies arise from the release of potential energy stored in the equator–pole temperature gradient. The release of this potential temperature is marked by a down gradient flux of heat from equator to pole, leading to an equator–pole temperature gradient that is significantly smaller than radiative processes would provide acting alone. Curiously, this heat flux can be linked to the growth of transient eddies through the budget for the perturbation eddy energy

$$E' = \int\int\int \frac{1}{2}\rho[u_g'^2 + v_g'^2 + \varepsilon(\partial_z\psi')^2]\,dx\,dy\,dz \tag{18}$$

The equation governing E' is obtained by multiplying (9) by ψ', integrating over all space, and employing the boundary condition (11). This equation can be written

$$\frac{dE'}{dt} = \int\int \mathbf{Q}\cdot\nabla u_{g0}\,dy\,dz \tag{19}$$

where the $\mathbf{Q}$ vector is given by

$$\mathbf{Q} = (Q_y, Q_z) \equiv (-\rho\overline{u_g' v_g'}, \rho\varepsilon\overline{v_g'\partial_z\psi'}), \tag{20}$$

and the overbar denotes an integral over x. It has been assumed that the vertical flux terms vanish by a radiation condition as $z \to \infty$, and the analogous lateral flux terms vanish by virtue of meridional confinement in a channel, or equivalently, by decay of perturbations at large y.

The integral in (19) represents the conversion of energy between the background flow and perturbations. In the integrand, $Q_z\partial_z u_{g0}$ is the baroclinic conversion term, and represents energy exchange with the potential energy of the basic state. By thermal wind balance, positive vertical shear implies colder air toward the pole; and, in this situation, perturbations with a poleward heat flux tap the baroclinic energy and grow. This term releases potential energy by reducing the tilt of the mean isentropes.

The term $Q_y\partial_y u_{h0}$ is the barotropic conversion term and represents exchanges with the basic state kinetic energy. In the extratropics, it typically acts as an energy sink. Synoptic eddies thus act as an intermediary, transforming potential energy due to differential heating into kinetic energy of barotropic jets.

For any basic state flow with vertical or horizontal shear, an initial perturbation can be crafted that will extract energy and begin to grow. However, as time passes, the structure of the perturbation will generally change, and growth may cease or even turn to decay. Stability theory seeks to identify energy-extracting structures that can persist long enough to amplify by a significant amount and to understand what basic state flow circumstances allow for such growth. This is elegantly discussed with reference to the wave action budget of the perturbation. If we linearize (9), multiply by q', and average over x, we obtain

$$\partial_t A + \nabla\cdot\mathbf{Q} = 0 \tag{21}$$

where the wave action (or more properly the pseudomomentum) A is

$$A = \frac{1}{2}\rho\overline{q'^2}q_{0y}^{-1} \tag{22}$$

An analogous budget for the variance of temperature at the ground can be obtained by multiplying (11) by $\rho\partial_z\psi'$ and averaging in x. Doing so yields

$$\partial_t B + Q_z(0) = 0 \quad \text{at} \quad z = 0 \tag{22}$$

where

$$B = \frac{1}{2}\rho\varepsilon\overline{(\partial_z\psi')^2}\eta_{0y}^{-1} \tag{23}$$

Integrating (21) over space, assuming boundary conditions that make the horizontal flux terms vanish, and using (22) to reduce the bottom flux contribution yields the following relation for the total pseudomomentum $\prod$:

$$\frac{d}{dt}\prod = 0 \qquad \prod = \int\int A\,dy\,dz + \int B\,dy \tag{24}$$

This conservation relation yields the desired stability criterion. Specifically, if q_{0y} and η_{0y} each have uniform sign throughout the domain, have the same sign as each other, and the magnitudes of q_{0y} and η_{0y} are finite and bounded away from zero, then the system is stable in the sense that the perturbation enstrophy

$$Z' = \int\int\frac{1}{2}\rho\overline{q'^2}\,dy\,dz \tag{25}$$

remains small for all times if it is initially small. If this criterion is not met, sustained exponential disturbance growth is possible. However, such growth is not inevitable, as (24) provides a necessary rather than a sufficient condition for disturbance growth.

The archetypal scenario permitting baroclinic instability in the extratropics is q_{0y} greater than zero everywhere, positive vertical shear at the ground, and a sufficiently flat lower boundary to keep η_{0y} negative. This condition is almost ubiquitously satisfied throughout the extratropical troposphere in both hemispheres. In this situation, a growing disturbance can be regarded as a coupled QGPV/surface temperature motion. The QGPV anomaly aloft induces a motion that stirs up a surface temperature anomaly, and the surface temperature anomaly in turn further stirs up the QGPV aloft. The influence depth of a QGPV or surface temperature perturbation of length L is the "deformation depth" $f_0 L/N$ provided this is not much in excess of the density scale height H. If a QGPV anomaly is at altitude D, motions with

$L \ll ND/f_0$ will have QGPV dynamics uncoupled from the surface dynamics, and cannot be expected to cohere and grow. Thus, short-wave instabilities must also be shallow, and deep modes filling out a density scale height must have horizontal scales comparable to the radius of deformation $L_d = NH/f_0$, if not longer. These arguments are quite general and carry over to strongly nonlinear motions.

The simplest possible model that satisfies the necessary conditions for instability in a continuous atmosphere is the Eady model. The Eady model has $\rho = \text{constant}$, $\beta = 0$, $\partial_z u_{g0} = \Lambda = \text{constant}$, and rigid lids at $z = 0$ and H_T. While this scenario provides only a crude model of the atmosphere, it allows for insights into the dependence of a growing disturbance on horizontal scale and stability. The primary simplification is the vanishing potential vorticity gradient in the interior of the fluid. Since the QGPV is zero initially in the interior, it will remain so for all time. Therefore, all the dynamics involve the time-dependent boundary condition (11) applied at each boundary. Despite the vanishing of the QGPV gradient, the Eady model satisfies the necessary conditions for instability because vertical shear of the basic state flow at the upper boundary provides an additional term in (24) that is equal and opposite to the lower boundary integral.

For this situation, (11) and the condition that q' vanish in the interior lead to wave solutions on each boundary (considered in isolation), whose effective restoring force is the meridional temperature gradient. Such waves are called Rossby edge waves, where edge waves are boundary-trapped disturbances. The wave at the lower boundary propagates eastward (relative to the surface flow), and the wave at the lid propagates westward. Localized warm surface air is associated with low pressure and cyclonic flow, and is effectively a positive QGPV perturbation. Cold air at the surface features high pressure and anticyclonic flow, analogous to a negative QGPV perturbation.

Baroclinic instability occurs in this model when the waves at each boundary amplify the temperature at the opposite boundary. For this to happen, the "deformation depth" defined above must be at least comparable to the distance between the boundaries. If the disturbance horizontal length scale is too short, or if the boundaries are too far apart, unstable modes do not exist.

Figure 1 shows the disturbance structure for a wave with equal zonal and meridional wavenumber ($k = l$). For this situation, the wavelength of the perturbation that grows fastest is

$$L_m = \frac{2\sqrt{2}\pi L_d}{(H_T \alpha_m)} \approx 4000 \text{ km} \tag{26}$$

similar to the length scale of observed synoptic-scale transients. The growth rate of the instability is approximately given by

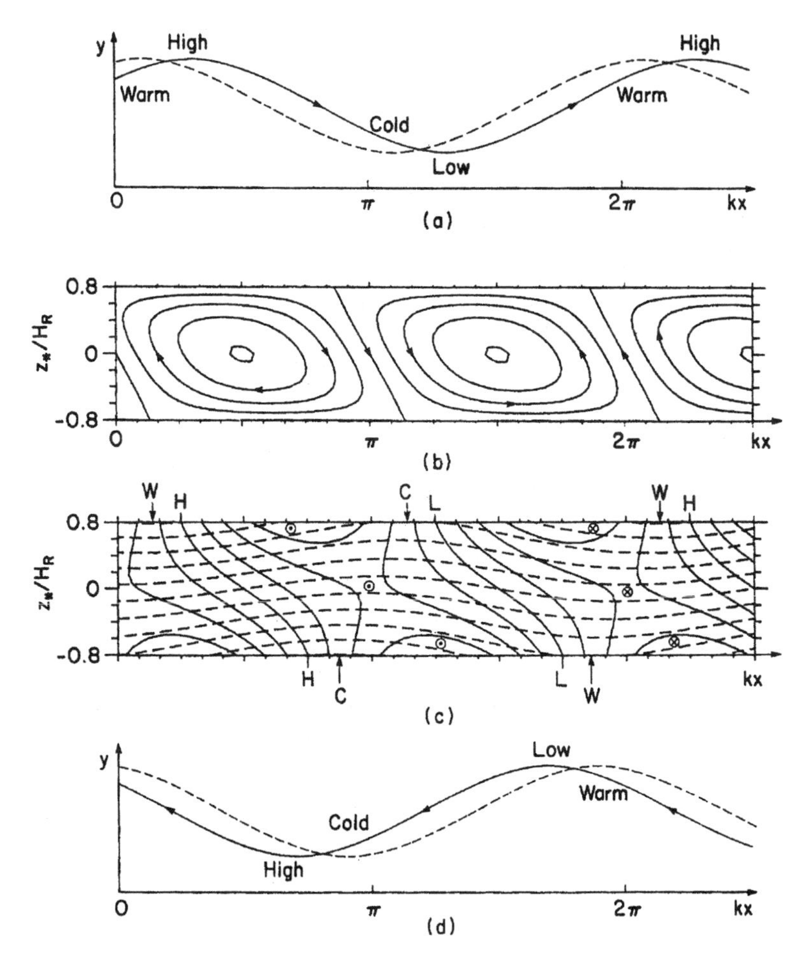

Figure 1 Properties of the most unstable Eady wave. (*a* and *d*) Geopotential height and temperature patterns on the upper and lower boundaries, respectively. (*b*) The stream function for the ageostrophic flow, showing the overturning in the *x*-*z* plane. Ascent is just to the east of the surface trough and descent is just to the west of the surface ridge. (*c*) Contours of meridional geostrophic wind v_g' (solid) and isentropes (dashed) in the *x*-*z* plane. *H* and *L* indicate the ridge and trough axes; *W* and *C* denote the phase of the warmest and coldest air. Circles with enclosed crosses indicate the axis of maximum poleward flow, and circles with enclosed dots indicate the axis of maximum equatorward flow (from Gill, 1982).

$$\mathrm{Im}(\omega) \approx \frac{0.25 f_0 \partial_z u_{g0}}{N} \tag{27}$$

This leads to an *e*-folding time of 2 or 3 days for typical extratropical values of the shear and *N*, again consistent with observed synoptic eddy growth rates. Figure 1*b* shows that for this particular wavelength, as for all unstable baroclinic waves, the trough and ridge axes slope westward with height. This phase displacement allows the velocity and temperature perturbations to be partly in phase at each boundary, thus leading to a poleward heat flux. The axes of the warmest and coldest air, however, tilt eastward with height. East of the trough, where the perturbation meridional velocity is poleward, Figure 1*c* shows that the ageostrophic vertical motion is upward. Thus, parcel motion is poleward and upward in the region where temperature perturbations are positive, and vice versa in regions where the temperature perturbations are negative. Through condensation in ascending, adiabatically cooling trajectories, this vertical velocity pattern provides a link to the hydrological cycle in the extratropical troposphere and, in particular, provides the "comma cloud" signature of surface troughs.

The scaling of the growth rate with N in (27) suggests that baroclinic instability would produce a statically stable extratropical atmosphere on Earth even in the absence of moisture. Local radiative convective equilibrium would drive N to zero but would have a strong equator–pole temperature gradient. Such a flow would be violently unstable to baroclinic instability. It is presumed that baroclinic instability increases N so as to adjust toward a more stable state. Eddies could do this by taking cold air manufactured in polar regions and sliding it under warmer air, and vice versa for warm air produced near the tropics. Either by vertical heat fluxes that increase N or by horizontal heat fluxes that reduce the equator–pole temperature gradient (and hence the vertical shear), baroclinic instability acts to adjust the extratropical flow toward a state that is neutrally stable to baroclinic instability.

In addition to their role in the tropospheric general circulation, synoptic-scale eddies also drive weather-related events at smaller scales. It has long been understood that extratropical cyclones have cold frontal regions in which the temperature typically changes rapidly, often accompanied by shifts in the wind and by precipitation. In general, these fronts are considered to form in a growing nonlinear baroclinic wave, though this secondary role in no way diminishes their importance in practical meteorology. Rather, the theoretical emphasis shifts toward the mechanism for generating fronts, namely frontogenesis.

The basis of frontogenesis lies in the advection of surface temperature (7). The kinematics of frontogenesis can be understood by considering a certain class of horizontal velocity fields called *deformation fields*. Such deformation fields, which are *local* features of large-scale horizontal wave motions, contain confluent regions that tend to concentrate a large-scale preexisting temperature gradient, squeezing isotherms together. A prototypical deformation field is given by $u = -\gamma x$, $v = \gamma y$, where γ is a constant with dimension $(\text{time})^{-1}$. Suppose the potential temperature field is initially oriented so the isotherms are parallel to the y axis. Then at the ground, where w vanishes, the time evolution of the potential temperature must satisfy

$$\partial_t \theta = -u \partial_x \theta = \gamma x \partial_x \theta \tag{28}$$

since θ is independent of y. The solution to (28) is easily verified to be

$$\theta(t) = \theta_0(x e^{\gamma t}) \tag{29}$$

where $\theta_0(x)$ is the surface distribution of potential temperature at the initial instant. The temperature gradient at the surface is

$$\partial_x \theta = e^{\gamma t} \dot{\theta}(x e^{\gamma t}) \tag{30}$$

where the dot indicates derivative with respect to the argument. Therefore, the surface temperature gradient increases exponentially with time in regions of confluence in the deformation field.

Naturally, as time goes by and the temperature field changes as a result of this confluence, the changing thermal wind in the y direction will produce, by Coriolis accelerations, flow in the x direction that will alter the initial deformation velocity field. Thus, the above solution will be valid only initially and even then describes the structure of the temperature gradient at the ground. However, this solution does show that large-scale flow features can generate strong density gradients.

A more complete theoretical explanation of frontogenesis lies outside of the class of motions described by quasigeostrophic theory. This is because fronts are inherently *anisotropic*, with a very narrow frontal zone compared to the spatial extent along the front. A different scaling of the primitive equations, namely the semi-geostrophic theory originally developed by Hoskins and Bretherton, explicitly recognizes this anisotropy and provides a more complete dynamical theory for the formation of fronts [see Hoskins (1982) for a review]. While such a theory lies beyond the scope of our treatment, it does present a number of features absent in the deformation model of frontogenesis examined above. Foremost among these features is that the frontogenetic mechanism is so powerful in semigeostrophic theory that infinite temperature gradients can be generated in a finite amount of time. Of course, in reality instabilities can be expected to occur in regions of such gradients, and accompanying mixing of surface temperature will tend to limit the sharpness of the frontal zone. However, the prediction of a discontinuity indicates the vigor of the frontogenetic process.

4 STATIONARY PLANETARY WAVES

Another striking component of the observed extratropical atmospheric circulation are the large-scale stationary planetary (Rossby) waves that dominate the zonally asymmetric flow during the Northern Hemisphere winter. Flow over larger-scale surface features (e.g., the Tibetan plateau), along with longitudinally varying diabatic heating force these waves, which subsequently propagate zonally and meridionally throughout the troposphere, and vertically into the upper atmosphere. In the troposphere, these waves influence the regional climates of Earth, as they affect the position and strength of the midlatitude storm tracks. In the stratosphere, breaking vertically propagating Rossby waves instigate strong changes in the stratospheric flow, called sudden warmings (Andrews et al., 1987, Section 6). More generally, the breaking of these vertically propagating waves plays an important role in the overall exchange of air between the stratosphere and the troposphere, as this breaking forces planetary-scale vertical ascent and descent of air (Holton et al., 1995).

First, let us consider the zonal and meridional propagation of stationary planetary waves. This problem is most easily treated using a barotropic model on the sphere, linearized about a zonally symmetric but latitudinally varying mean flow. In this model, stationary waves are solutions of the equation

$$u_{g0}(a\cos\phi)^{-1}\partial_\lambda\zeta_g' + v_g'(a^{-1}\partial_\phi\zeta_g + \beta) = -u_{gs}f(h_0 a\cos\phi)^{-1}\partial_\lambda h_T - r\zeta' \tag{31}$$

where $\zeta_g \equiv -a^{-1}\partial_\phi u_g + (a\cos\phi)^{-1}\partial_\lambda v_g$ is the relative perturbation vorticity, primes indicate deviation from zonal mean background states, and h_0 is the resting depth of the barotropic fluid. The damping coefficient r is taken to be independent of latitude.

The response of this system with h_T equal to the Northern Hemisphere topography is shown in Figure 2, where u_{g0} and u_{gs} equal the zonally averaged Northern Hemisphere wintertime 300 mb and surface winds, respectively. The response field appears to be comprised of two wavetrains propagating equatorwards, one produced by the Rockies and the other by the Tibetan plateau.

These structures can be explained by Rossby wave ray-tracing theory. If we transform the linearized, unforced inviscid vorticity equation into Mercator coordinates $x = a\lambda$, $a^{-1}dy = (\cos\phi)^{-1}d\phi$, then (30) takes the form

$$\hat{u}\partial_x(\partial_{xx} + \partial_{yy})\psi' = -\hat{\beta}\partial_x\psi' \tag{32}$$

where $\hat{u} \equiv u_{g0}(\cos\phi)^{-1}$ and $\hat{\beta} \equiv \cos\phi(\beta + a^{-1}\partial_\phi\zeta_{g0})$. For a wave of the form $\psi' = \mathrm{Re}\,\tilde{\psi}\exp(ikx)$, one has

$$\partial_{yy}\tilde{\psi} = (k^2 - \hat{\beta}/\hat{u})\tilde{\psi} \tag{33}$$

Given a source localized in latitude, the structure of (32) requires that zonal wavenumbers $k > k_s \equiv (\hat{\beta}/\hat{u})^{1/2}$ remain meridionally trapped in the vicinity of the source, while wavenumbers $k < k_s$ are free to propagate away from the source.

A source localized in longitude as well as latitude can be regarded as producing two rays for each $n < n_s$, where $n = ak$ is the zonal wavenumber on the sphere, and $n_s = ak_s$. These rays correspond to the two possible meridional wavenumbers, $al = \pm(n_s^2 - n^2)^{1/2}$. Since the meridional group velocity of nondivergent Rossby waves $c_{gy} = \partial\omega/\partial l = 2\hat{\beta}kl(k^2 + l^2)^{-2}$ has the same sign as l, the positive (negative) sign corresponds to poleward (equatorward) propagation. The zonal group velocity $c_{gx} = \partial\omega/\partial k = u_{g0} + \hat{\beta}(k^2 - l^2)(k^2 + l^2)^{-2}$ is eastward for stationary planetary waves with $k \sim l$. Therefore, the rays can be traced downstream from their source, noting that the zonal wavenumber remains unchanged because the mean flow is independent of x, whereas the meridional wavenumber adjusts to satisfy the local dispersion relation. The ray path is tangent to the vector group velocity (c_{gx}, c_{gy}) for a particular k and l, and is always refracted toward the larger "refraction index" $(n_s^2 - n^2)^{1/2}$, i.e., rays turn toward larger n_s in accordance with Snell's law for

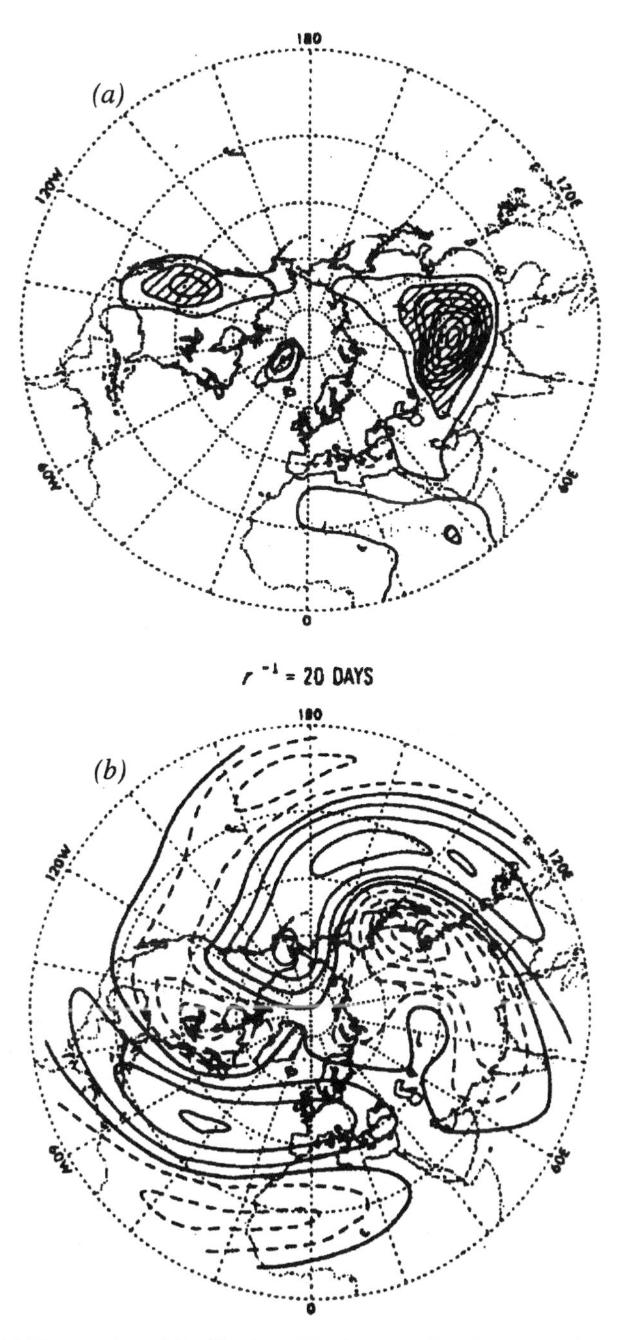

Figure 2 (*a*) Topography of the Northern Hemisphere. (*b*) The stream function response of the Northern Hemisphere of a barotropic model, forced using the zonal mean winds and Northern Hemisphere topography. Solid contours are positive or zero; dashed contours are negative (from Held, 1982).

optics. The atmosphere tends on average to exhibit low values of the index of refraction toward the polar regions, with high values toward the subtropical regions. Therefore, a wave excited somewhere in extratropics will refract away from the polar latitudes and propagate in the direction of the tropics.

When the zonal mean flow varies with latitude, two important complications occur. The first is the existence of a turning latitude where $n = n_s$. At this latitude, both the meridional wavenumber and meridional group velocity go to zero; thus, for a wave propagating poleward and eastward out of the tropics, it would continue to be refracted until at some point its troughs and ridges are aligned entirely north/south. After this point, the wave continues to turn and begins to propagate back toward the tropics.

The second complication is posed by the existence of a critical latitude, where $u_{g0} = 0$. Critical latitudes are generally marked by complicated dynamics because the implicit linear assumption that particle position deviations from rest are small does not hold. For the case of a Rossby wave train, as it approaches its critical latitude, linear theory predicts that the meridional wavenumber $l \rightarrow \infty$, while the meridional group velocity vanishes. The development of small scales while propagation slows leads linear theory to predict the absorption of the Rossby wave train at critical latitudes. However, the inclusion of nonlinearities opens a whole set of possibilities, including the reflection of the equatorward propagating wave train back into the extratropics. More generally, down gradient fluxes of potential vorticity associated with stationary wave breaking (or for that matter, with synoptic transient wave breaking) at critical latitudes located on the flanks of the upper tropospheric or stratospheric jets plays an important role in the momentum budget of the extratropical atmospheric general circulation. Specifically, this wave breaking results in significant wave-induced forcing of the zonal basic state flow.

Stationary Rossby waves not only propagate zonally and meridionally but also vertically. To understand this vertical propagation, let us return to the continuously stratified QG equations of Section 2. We are free to seek stationary solutions to Eqs. (14) and (15) with $c = 0$. For the case where u_{g0} and N are constant, but including topography $h_T = \exp(ily)\exp(ikx)$, the dispersion relation (17) leads to a condition on the vertical wavenumber m. Provided that $0 < u_{g0} < \beta\,(k^2 + l^2 + \varepsilon(2H)^{-2})^{-1}$, m will be real and given by

$$m = \pm N f_0^{-1}(\beta u_{g0}{}^{-1} - (k^2 + l^2) - \varepsilon(2H)^{-2})^{1/2} \tag{34}$$

Examination of the vertical group velocity $(\partial\omega/\partial m)$ reveals that it has the same sign as m, so choosing the positive branch of (34) corresponds to a wave propagating upward from a source at the ground and is an acceptable solution. For winds outside this range, m will be imaginary, and the waves will be evanescent rather than propagate with height. The most important fact highlighted by this simple system is that only sufficiently weak westerlies permit vertical stationary wave propagation. For parameters relevant to Earth's atmosphere, stationary wavenumbers three and greater will not readily propagate into the stratosphere, thus accounting for the predominance of wavenumbers one and two in the winter stratosphere. Additionally, the summer stratospheric easterlies effectively block the propagation of all stationary waves, accounting for the observed zonal character of the summer stratospheric circulation.

Of particular importance is the external Rossby wave, defined as the evanescent mode with the gravest vertical structure, which tends to dominate the response downstream of a localized forcing. For realistic flows such as the one shown in Figure 3*a*, the external mode is equivalent barotropic, with geopotential amplifying with height in the troposphere and decaying with height in the stratosphere. The stationary Green's function response to a "spike" imposed topographic forcing $h_T = \sin(ly)\delta(x)$, where $\delta(x)$ is the Dirac delta function, is shown in Figure 3*c*. The emergence of the external mode in the far-field is apparent, due to the rapid vertical propagation of longer wavelength stationary Rossby waves and destructive interference among shorter wavelength stationary Rossby waves as they propagate upward and downward, confined by their midtropospheric turning levels.

5 ISENTROPIC POTENTIAL VORTICITY

In any fluid dynamical problem, the existence of conserved quantities provides a foundation on which to build firm theoretical descriptions of dynamical phenomena. The usefulness of having such a conserved quantity has been apparent throughout the above discussion, as, for example, the development of the necessary conditions for stability in Section 3 explicitly used the conservation of QGPV to develop the global pseudomomentum conservation relation (24). In the atmosphere, the Rossby–Ertel isentropic potential vorticity (IPV) goes beyond QGPV in that it is conserved by the full dynamics of the atmosphere in the absence of forcing and dissipation, rather than by just the simplified QG subset of those dynamics.

The conservation of IPV in the full dynamical equations of motion can be shown as follows. In a frictionless, adiabatic atmosphere a close material contour on an isentropic surface remains on that surface and its absolute circulation

$$C_a = \oint \mathbf{u} \cdot d\mathbf{l} = \int \zeta \cdot \mathbf{n}\, dS \tag{35}$$

is conserved. For an elemental cylinder of cross-sectional area S between the isentropic surfaces θ and $\theta + \delta\theta$, $C_a = \zeta \cdot \mathbf{n}S$, and the mass $m = \rho S\,\delta h$ and $\delta\theta$ are also conserved. Here $\mathbf{n}$ is the unit normal to the isentropic surface and h is the distance in this direction. Therefore

$$\frac{C_a}{m}\delta\theta = \rho^{-1}\zeta \cdot \mathbf{n}\frac{\delta\theta}{\delta h} \tag{36}$$

is also conserved. In the infinitesimal limit this gives material conservation of the IPV

$$P = \rho^{-1}\zeta \cdot \nabla\theta \tag{37}$$

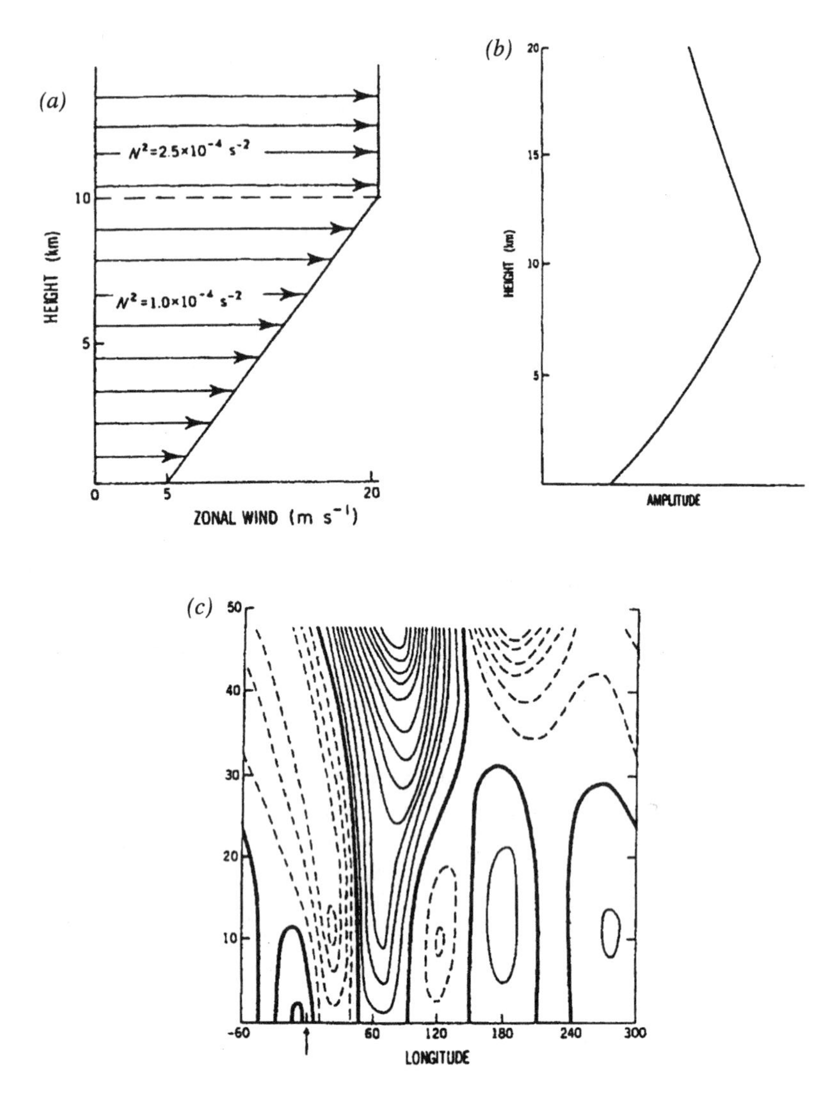

Figure 3 (*a*) Idealized zonal wind and static stability profile. (*b*) Vertical structure of the external Rossby wave for this flow profile. (*c*) Stationary response to "spike" topography for this flow profile, where the arrow marks the location of the source. Solid lines are positive contours and dashed lines are negative (from Held, 1982).

At the level of the hydrostatic approximation, $\zeta_a \cdot \mathbf{n} = \zeta_{a\theta}$, a quantity resembling the vertical component of absolute vorticity but with horizontal derivatives evaluated on isentropic surfaces. Then (37) is similarily approximated by

$$P = \sigma^{-1}\zeta_{a\theta} \tag{38}$$

where $\sigma = -g^{-1}\partial p/\partial\theta$ plays the role of density in isentropic coordinates.

In a frictionless, adiabatic atmosphere IPV is conserved by the two-dimensional (2D) motion on an isentropic surface; and similarly θ is conserved by the 2D motion on an iso-IPV surface. Most importantly, if the interior IPV and boundary θ distributions are known, then *balanced* motions may be determined by inversion of an elliptic operator. The simplest balance assumption is geostrophy; under this assumption, given the IPV/surface θ distributions one could deduce the structure of the waves and instabilities that comprise the large-scale extratropical atmospheric circulation. However, much more general balance assumptions can be made; for example, the assumption of some form of semigeostrophic balance would allow the study of IPV-dynamical interactions in the vicinity of fronts.

Provided the elliptic operator used for the IPV inversion is linear, a straightforward examination of the relative roles of various IPV disturbances in a given dynamical situation is possible. For example, the use of a QG inversion operator to study observed cyclogenesis leads to a similar dynamical picture to that outlined for the Eady model above, namely the intensification of the surface temperature anomalies by winds that stem from the upper tropospheric IPV anomalies, and vice versa. However, IPV allows more general investigations within this framework than does QGPV; for example, the importance of diabatic production of IPV by latent heat release and surface friction during cyclogenesis can be using such a decomposition. Hoskins et al. (1985) review a number of applications of IPV inversion, along with an in-depth review of the associated concepts underlying IPV conservation and atmospheric dynamics.

A cross section of the time-averaged IPV (Fig. 4) shows that the strongest gradients of IPV along isentropic surfaces occur at the tropopause. The other region of

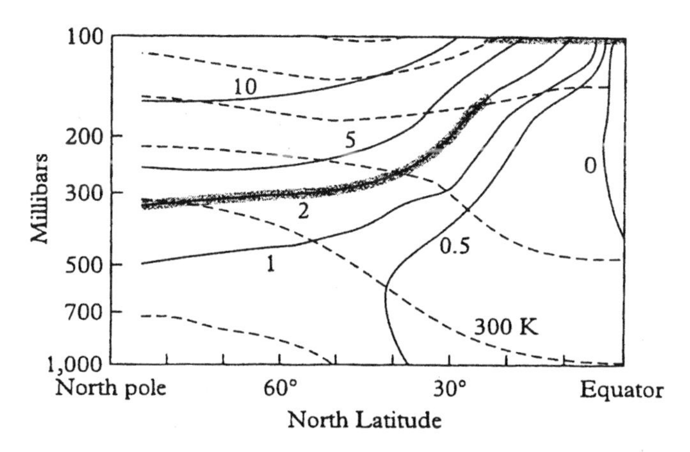

Figure 4 Climatology of isentropic potential vorticity (IPV) and potential temperature θ for the Northern Hemisphere winter. Isentropes are dashed and drawn in intervals of 30 K; IPV has units of $10^{-6}\,m^2\,K\,(kg\,s)^{-1} = 1$ PV unit. Contours of IPV are 0, 0.5, 1, 2, 5, and 10 PV units, with the contour at 2 PV units stippled to indicate the approximate position of the dynamical tropopause.

importance is the large horizontal temperature gradient at the surface in midlatitudes. The natural decomposition of the general circulation into those isentropic surfaces that are everywhere above the tropopause (the "overworld"), those that intersect the tropopause ("middleworld"), and those that intersect the surface ("underworld") provides an alternative framework in which to view global-scale dynamical processes in the extratropical atmosphere. The qualitative link between IPV and the mass field in (38) makes IPV an ideal tool with which to study planetary-scale dynamical motions; examples of such motions include the structure of the mean meridional mass transport in the troposphere, the interaction between thermal and mechanical sources of IPV at Earth's surface and their role in the overall general circulation, and the exchange of air between the stratosphere and troposphere. Coupled with the invertibility principle, this link makes IPV a powerful diagnostic tool for studying the extratropical atmospheric circulation.

REFERENCES

Andrews, D. G., J. R. Holton, and C. B. Leovvy (1987). *Middle Atmosphere Dynamics*, International Geophysics Series, Vol. 40, Academic, New York.

Gill, A. E. (1982). *Atmosphere-Ocean Dynamics*, International Geophysics Series, Vol. 30, Academic, New York.

Held, I. M. (1982). Stationary and quasi-stationary eddies in the extratropical tropopshere: Theory, In *Large-Scale Dynamical Processes in the Atmosphere*, B. J. Hoskins and R. P. Pearce (Eds.), Academic, New York, pp. 127–168.

Held, I. M., and B. J. Hoskins (1985). Large-scale eddies and the general circulation of the troposphere, *Adv. Geophys.* **28A**, 3–31.

Holton, J. R. (1992). *An Introduction to Dynamical Meteorology*, 3rd ed., International Geophysics Series, Vol. 48, Academic, New York.

Holton, J. R., P. H. Haynes, M. E. McIntyre, A. R. Douglass, R. B. Rood, and L. Pfister (1995). Stratosphere–troposphere exchange, *Rev. Geophys.* **33**, 403–439.

Hoskins, B. J. (1982). The mathematical theory of frontogenesis, *Ann. Rev. Fluid Mech.* **14**, 131–151.

Hoskins, B. J., M. E. McIntyre, and A. W. Robertson (1985). On the use and significance of isentropic potential vorticity maps, *Quart. J. Roy. Met. Soc.* **111**, 877–946.

Pedlosky, J. P. (1987). *Geophysical Fluid Dynamics*, 2nd ed. Springer, New York.

Pierrehumbert, R. T., and K. L. Swanson (1995). Baroclinic instability, *Ann. Rev. Fluid Mech.* **27**, 419–167.

CHAPTER 5

DYNAMICS OF THE TROPICAL ATMOSPHERE

GERALD MEEHL

1 THERMALLY DIRECT CIRCULATIONS AND MONSOONS

The heating of Earth's surface that follows the annual cycle of solar forcing produces fundamentally ageostrophic thermally direct circulations in the tropical atmosphere. Hadley postulated that two thermally driven north–south cells, roughly straddling the equator, must be necessary to transport heat and energy surpluses from the tropics toward the poles. These meridional cells now bear his name. Subsequently, Sir Gilbert Walker recognized that, in addition to the north–south Hadley cells, important components of large-scale tropical circulations are associated with tremendous east–west exchanges of mass. The term *Walker circulation* was first used by Bjerknes (1969) to describe these exchanges of mass in the equatorial Pacific. Subsequently, Krishnamurti (1971) showed that the Walker circulation in the equatorial Pacific is just a component of large-scale east–west circulations emanating from centers of organized tropical convective activity associated with regional monsoon regimes.

For example, applications of the Hadley and Walker circulation concepts have linked the monsoon regimes of Asia and northern Australia to weather and climate in other parts of the tropics and subtropics. Figure 1 shows one such interpretation of the east–west Walker circulation connecting the Asia–Australia region to the eastern Pacific (Webster et al., 1998). Another branch of east–west circulation, termed the *transverse monsoon* here, connects those monsoon regions to eastern Africa. The north–south Hadley components in Figure 1 are termed *lateral monsoon* and involve large-scale mass overturning from the respective summer monsoon regimes to

Handbook of Weather, Climate, and Water: Dynamics, Climate, Physical Meteorology, Weather Systems, and Measurements, Edited by Thomas D. Potter and Bradley R. Colman.
ISBN 0-471-21490-6

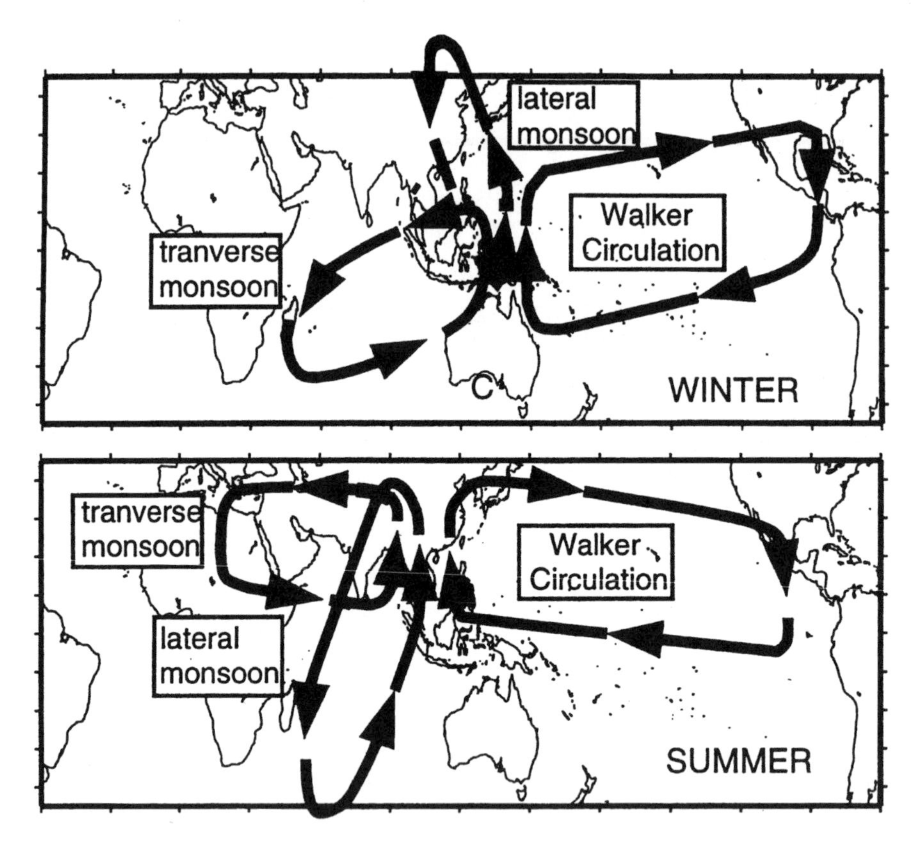

Figure 1 Schematic of the (*a*) south Asian and (*b*) Australian monsoon divergent wind circulations. Three major components identified are the transverse monsoon (Walker-type east–west circulation west of the monsoons), lateral monsoon (Hadley-type north–south circulations), and Walker circulation (east–west circulation east of the monsoons) (Webster et al., 1998).

the winter hemisphere subtropics, thus depicting the classic Hadley cell type of circulation.

The latitude where the sun is most directly overhead in its seasonal march is where the surface is most strongly heated in the tropics. The air directly above that warm surface rises and draws nearby air in to take its place. The warm air rises, cools and condenses, and forms convective clouds. The low-level convergence at that latitude forms a zonally oriented line of convection called the intertropical convergence zone (ITCZ). The arrangement of land and ocean in the tropics dictates that the ITCZ does not simply follow the sun's motion from northern to southern hemisphere with the seasonal cycle. Off-equatorial land masses in the tropics tend to heat more than the adjacent oceans, thus giving rise to seasonal regional convective systems called monsoons. Though the word *monsoon* has a number of interpretations, monsoon regimes exist wherever there is a regional organization of convection associated with an off-equatorial land mass and an equatorward ocean moisture source. This particular distribution of land and ocean disrupts the normal zonal orientation of the ITCZ and contributes to the east–west Walker-type circulations (e.g., Fig. 1). These regional monsoon regimes are recognized to exist in tropical regions of south Asia, northern Australia, southwestern North America, west Africa, and, in some definitions, South America.

All these monsoon regimes share certain characteristics. With the approach of local summer, intensifying solar radiation heats the land surface forming a heat low. Since the land surface has heated faster than the adjacent ocean, the heat low begins to draw in moist low-level air from the ocean. This moist warm air rises and forms convective clouds and rainfall. The latent heating from the convective systems powers a positive feedback, with even stronger low-level moist inflow from the ocean intensifying convection and precipitation over land. An upper level high forms overhead with outflow in all directions from the regional convective center associated with the monsoon circulation. This feeds into the thermally direct mass circulations mentioned above and connects the monsoon regimes with other areas in the tropics and subtropics. The massive amounts of latent heating associated with the monsoon regimes provide a major energy source for the general circulation of the global atmosphere.

2 EL NIÑO–SOUTHERN OSCILLATION

It was the efforts of Walker and others to try to understand the factors that influence the Indian monsoon in particular that led to the discovery of the east–west circulations evidenced by large-scale exchanges of mass between the Indian and Pacific Oceans described above. This phenomenon became known as the Southern Oscillation. These investigators associated the fluctuations of mass between the Indian and Pacific regions with temperature and precipitation variations over much of the globe. Bjerknes (1969) later linked sea surface temperature (SST) anomalies in the equatorial Pacific (which came to be known as El Niño and La Niña), and changes in the east–west atmospheric circulation near the equator, to the Southern Oscillation. The entire coupled ocean–atmosphere phenomenon is called *El Niño–Southern Oscillation* (ENSO).

ENSO and its connections to the large-scale east–west circulation can best be illustrated by correlations of sea level pressure between one node, representing the Indian–western Pacific region (in this case Darwin on the north coast of Australia), with all other points (Fig. 2). Positive values cover most of the south Asian and Australian regions, with negative values east of the dateline in the Pacific. Also note positive values over northeastern South America and the Atlantic. Thus, when convection and precipitation in the south Asian or Australian monsoons tends to be below normal, there is high pressure in those regions with a corresponding decrease of pressure and enhanced rainfall over the tropical Pacific, with higher pressure and lower rainfall over northeastern South America and tropical Atlantic.

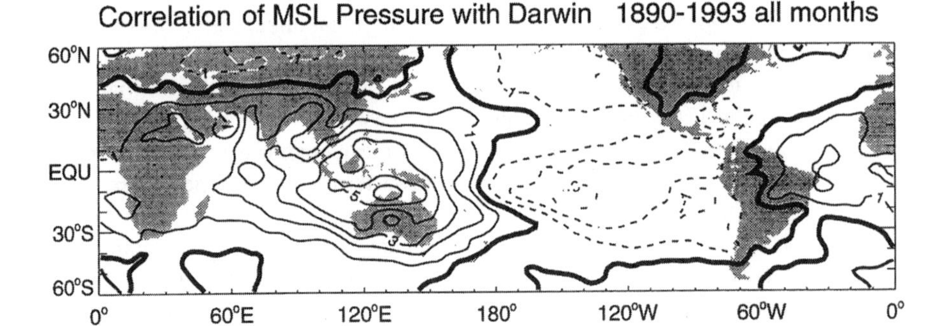

Figure 2 Distribution of the zero-lag correlation of surface pressure variations with the Darwin surface pressure and all other points for monthly data for the period 1890–1993 (Webster et al., 1998).

Since the atmosphere and ocean are dynamically coupled in the tropics, ENSO and its associated fluctuations of atmospheric mass is associated with SST anomalies as a consequence of the mostly ageostrophic surface winds involved with sea level pressure anomalies such as those implied in Figure 2. This dynamic coupling is reflected with the association between El Niño and La Niña events and the Southern Oscillation noted above. For example, anomalously low pressure and enhanced precipitation over the tropical Pacific are usually associated with anomalously warm SSTs there. In the extreme, this is known as an El Niño event. In the opposite extreme, anomalously cold SSTs east of the dateline have come to be known as a La Niña event. These are often associated with low pressure and strong Asian–Australian monsoons west of the dateline, with high pressure and suppressed rainfall east of the dateline. The conditions that produce El Niño and La Niña events, with a frequency of roughly 3 to 7 years, involve dynamically coupled interactions of ocean and atmosphere over a large area of the tropical Indian and Pacific Oceans (Rasmusson and Carpenter, 1982). Figure 3 shows low-level wind, precipitation, and SST anomalies for the northern summer season during the 1997–1998 El Niño event. Westerly surface wind anomalies occur to the west of the largest positive SST anomalies in the central and eastern equatorial Pacific. These wind anomalies reduce upwelling of cool water and contribute to the positive SST anomalies. The warmer water is associated with greater evaporation and increased precipitation. As a result of the anomalous Walker circulation, there is suppressed precipitation over much of Southeast Asia.

El Niño and La Niña events are extremes in the system that tends to oscillate every other year to produce the tropospheric biennial oscillation, or TBO (Meehl, 1997). Thus there are other years, in addition to the Pacific SST extremes, that have similar coupled processes but lower amplitude anomalies also involving the Asian–Australian monsoons and tropical Indian and Pacific SST anomalies (Fig. 4). The TBO encompasses all the ENSO processes of dynamically coupled air–sea interaction in the tropical Indian and Pacific Oceans, large-scale east–west cir-

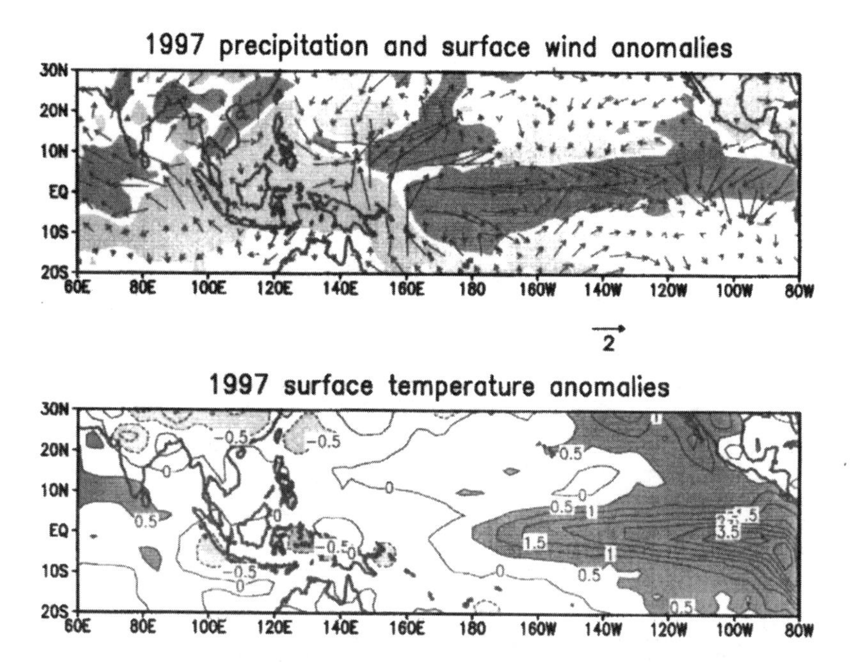

Figure 3 (*a*) Precipitation and surface wind anomalies for the 1997 June–July–August–September seasonal average during the 1997–1998 El Niño event; dark gray shading indicates precipitation anomalies greater than 0.5 mm/day, light gray less than −0.5 mm/day, scaling arrow below part (*a*) indicates surface wind anomalies of 2 m/s; (*b*) as in (*a*) except for SST anomalies with dark gray greater than 0.5°C, light gray less than −0.5°C.

culations in the atmosphere, the Asian–Australian monsoons, and coupled SST anomalies in Indian and Pacific involving internal ocean wave dynamics.

Figure 4 illustrates these coupled interactions for a sequence from a weak Australian monsoon during anomalously warm conditions in the tropical Pacific that, in the extreme, is an El Niño event, through to the onset of anomalously cold SSTs in the Pacific (in the extreme, a La Niña event) associated with a strong Indian monsoon, and on to a strong Australian monsoon the following southern summer. In the December–January–February (DJF) season prior to a strong Indian monsoon, there is a relatively weak Australian monsoon (Fig. 4*a*), with warm SST anomalies to the west in the Indian Ocean, to the east in the central and eastern Pacific, and relatively cool SSTs north of Australia. Relatively strong convection over the western Indian Ocean and eastern African areas, along with strong convection over the central equatorial Pacific, are associated with an atmospheric Rossby wave response over Asia such that there is an anomalous ridge of positive 500 hPa heights and warm land temperatures. Anomalous easterly winds in the western Pacific induce upwelling Kelvin waves in the ocean that propagate eastward. These begin to raise the thermocline in the central and eastern Pacific over the course of the next few months, as indicated in Figure 4*b*. Westerly anom-

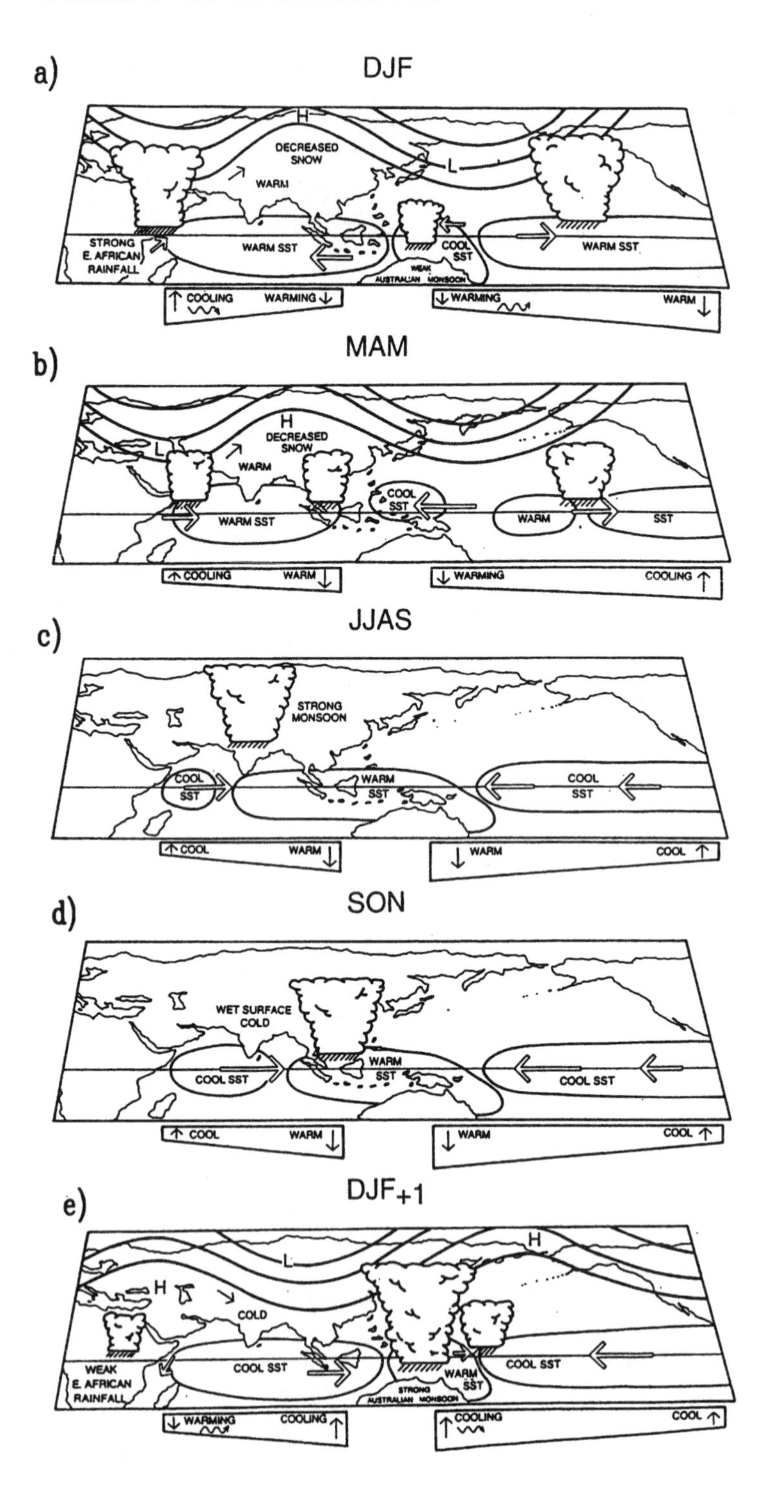
a)
DJF
H
DECREASED
SNOW
WARM
L
STRONG
E. AFRICAN
RAINFALL
WARM SST
COOL
SST
WEAK
AUSTRALIAN MONSOON
WARM SST
COOLING
WARMING
WARMING
WARM
b)
MAM
H
DECREASED
SNOW
L
WARM
COOL
SST
WARM SST
WARM
SST
COOLING
WARM
WARMING
COOLING
c)
JJAS
STRONG
MONSOON
COOL
SST
WARM
SST
COOL
SST
COOL
WARM
WARM
COOL
d)
SON
WET SURFACE
COLD
WARM
SST
COOL SST
COOL SST
COOL
WARM
WARM
COOL
e)
DJF+1
H
L
H
COLD
WEAK
E. AFRICAN
RAINFALL
COOL SST
WARM
SST
STRONG
AUSTRALIAN MONSOON
COOL SST
WARMING
COOLING
COOLING
COOL

alous surface winds in the far western Indian Ocean set off downwelling equatorial Kelvin waves in the ocean and contribute to a deepening of the thermocline in the eastern equatorial Indian Ocean in March–April–May (MAM, Fig. 4*b*).

In the MAM season prior to a strong Indian monsoon (Fig. 4*b*), the upwelling Kelvin waves from the easterly anomalous winds in the western Pacific during DJF continuing into MAM act to raise the thermocline in the central and eastern Pacific.

During June–July–August–September (JJAS) (Fig. 4*c*), convection and precipitation over the Indian monsoon region is strong. Anomalous winds over the equatorial Indian ocean remain westerly, and the shallow thermocline in the west is evidenced by anomalously cool SSTs appearing in the western equatorial Indian Ocean. There are strong trades, and a shallow thermocline and cool SSTs in the central and eastern equatorial Pacific, thus completing the SST transition there.

The September–October–November (SON) season after the strong Indian monsoon (Fig. 4*d*) has anomalously strong convection traversing with the seasonal cycle to the southeast, encountering warm SSTs set up by the dynamical ocean response to the westerly winds in the equatorial Indian Ocean in JJAS with a dipole of SST anomalies in the Indian Ocean.

Finally, in Figure 4*e* in the DJF following a strong Indian monsoon, there is a strong Australian monsoon, and strong easterlies in the equatorial Pacific as part of the strengthened Walker circulation associated with cool SSTs there. Anomalous westerlies in the far western equatorial Pacific associated with the strong Australian monsoon convection start to set off downwelling equatorial oceanic Kelvin waves, which begin to deepen the thermocline to the east and set up the next transition to warm SSTs in the central and eastern equatorial Pacific the following MAM–JJAS.

3 SUBSEASONAL TROPICAL ATMOSPHERIC WAVES

The westerly winds in the far western equatorial Pacific involved with transitions of SST anomalies in the Pacific have been associated with a subseasonal phenomenon called the *Madden–Julian Oscillation* or MJO (Madden and Julian, 1972). The MJO is a dominant mode of subseasonal tropical convection. It is associated with an eastward-moving envelope of convection that progresses around the global tropics with a period of about 30 to 70 days and is most evident in clouds and convection in

---

Figure 4 Schematic representation of coupled processes for (*a*) the DJF season after a weak Indian monsoon with a weak Australian monsoon and anomalously warm SSTs in the tropical Pacific, (*b*) MAM, (*c*) a strong Indian monsoon associated with anomalously cold SSTs in the Pacific, (*d*) SON and (*e*) DJF with a strong Australian monsoon. Note that conditions in DJF preceding the strong monsoon in (*a*) are opposite to those in the DJF following the strong monsoon in (*e*) indicative of the TBO in the coupled system. The TBO encompasses El Niño and La Niña events along with other years with similar but lower amplitude anomalies. Thus the coupled processes depicted here apply to TBO and ENSO (Meehl and Arblaster, 2002).

the Indian and Pacific sectors. Particularly active MJO events, as they move across the western Pacific, are associated with bursts of westerly anomalous winds that force eastward propagating Kelvin waves in the ocean. These oceanic Kelvin waves affect the depth of the thermocline to the east and can contribute to a transition of SST anomalies in the eastern Pacific associated with the onset of some El Niño events (McPhaden, 1999). For example, Kessler et al. (1995) show how successive MJO westerly wind events in the western Pacific can lead to the forcing of oceanic Kelvin waves and successive or stepwise advection of warm SSTs from the western Pacific toward the east to contribute to El Niño onset.

The MJO is just one of a number of convectively coupled tropical waves (i.e., atmospheric wave dynamics coupled to convection such that their signatures are seen in analysis of tropical clouds) that exist in the equatorial atmosphere. Subseasonal tropical convection is organized on larger scales by certain modes that are unique to the equatorial region. An entire class of equatorially trapped modes organizes atmospheric motions and convection at low latitudes. The theory for these "shallow water" modes was developed by Matsuno (1966).

Formally, equatorial wave theory begins with a separation of the primitive equations, linearized about a basic state with no vertical shear, governing small motions in a three-dimensional stratified atmosphere on an equatorial Beta plane, into the "vertical structure" equation and "shallow-water" equations. Four parameters characterize the equatorial wave modes that are the zonally (and vertically) propagating equatorially trapped solutions of the shallow-water equations. These include meridional mode number n, frequency ν, "equivalent depth" h of the "shallow" layer of fluid, and zonal wavenumber s. The internal gravity wave speed c is related to the equivalent depth as

$$c = \sqrt{gh} \tag{1}$$

and links the vertical structure equation and shallow-water equations as a separation constant. The equatorial Rossby radius R, is related to the vertical wavelength of free waves and to the meridional scaling by

$$R_e = \left(\frac{\sqrt{gh}}{\beta} \right)^{1/2} \tag{2}$$

where β is the latitudinal gradient of the Coriolis parameter.

The theoretical dispersion relationship will fully characterize the wave, given the meridional mode number and wave type, provided two out of h, ν, and s are specified. It is assumed that tropical waves associated with convective heating are internal modes with wavelike vertical structures, and the resulting solution of the shallow-water equations are either antisymmetric or symmetric about the equator. Convection is usually related to the divergence or temperature field, and modes of even meridional mode number n are antisymmetric, and odd n are symmetric. Figure 5 shows examples of circulation and divergence associated with some of these symmetric

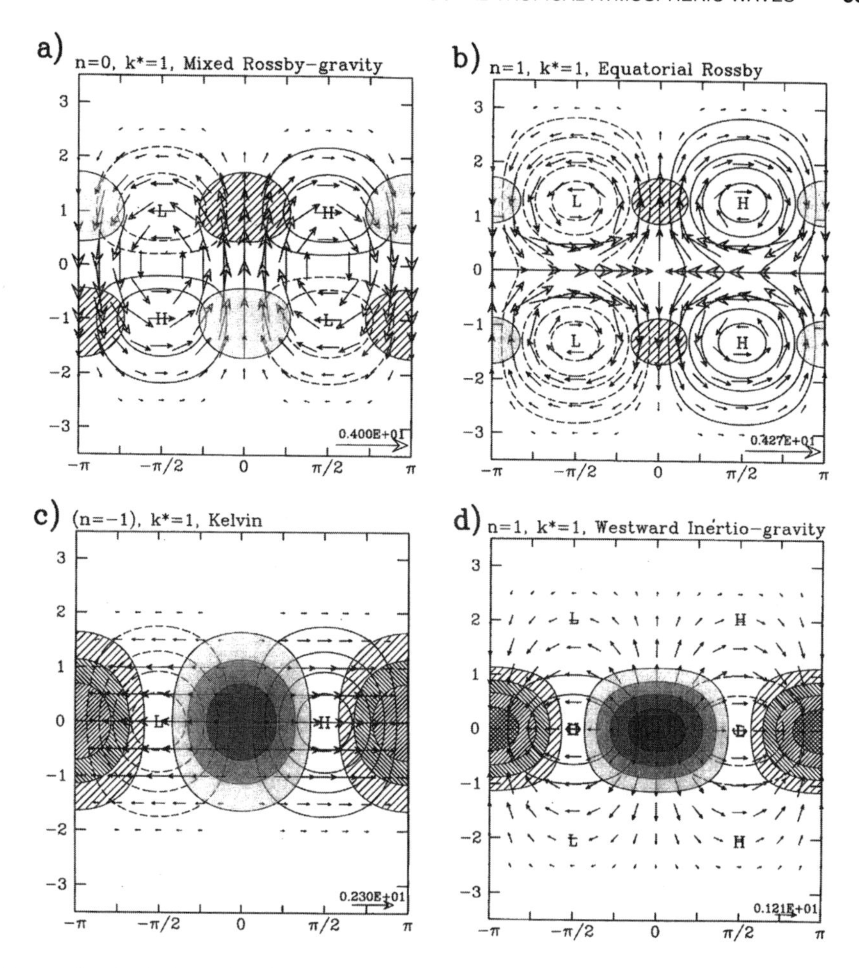

Figure 5 Depiction of various atmospheric tropical waves from shallow water theory: (*a*) $n=0$ mixed Rossby gravity, (*b*) $n=1$ equatorial Rossby, (*c*) $n=-1$ Kelvin wave, and (*d*) $n=1$ westward inertiogravity wave. Hatching indicates convergence, shading is divergence.

modes, the $n=0$ mixed Rossby gravity, $n=1$ equatorial Rossby, $n=-1$ K, and $n=1$ westward inertiogravity waves.

A space–time analysis of satellite-observed outgoing longwave radiation (OLR), a proxy for tropical convection, is shown in Figure 6. By separating the OLR into its antisymmetric and symmetric components about the equator, as well as removing a red background spectrum, the spectral peaks of these convectively coupled equatorial waves are shown in Figure 6*a* for the antisymmetric components and Figure 6*b* for the symmetric (Wheeler and Kiladis, 1999). The modes occur for a wide range of space and time scales, from synoptic to planetary, and submonthly to intraseasonal. Waves depicted in Figure 5 appear in Figure 6 as Kelvin waves, equatorial Rossby

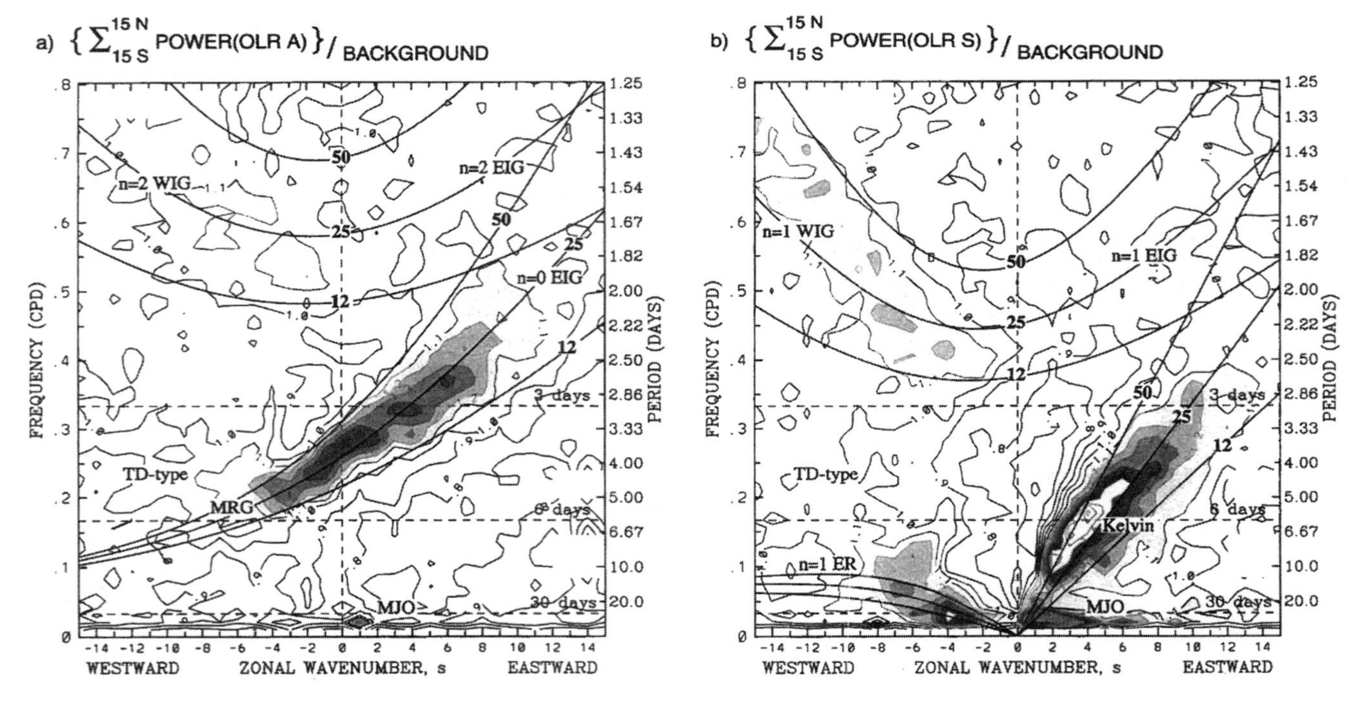

Figure 6 (*a*) Antisymmetric (with respect to equator) observed OLR power divided by the background power. The background power is calculated as a smoothed average of the power of the antisymmetric and symmetric components. The power is calculated for the 1979 to 1996 period and summed between 15 S and 15 N. Contour interval is 0.1, and shading begins at a value of 1.1 for which the spectral signatures are statistically significant above the background at the 95% level (based on 500 degrees of freedom). Contours less than 1.0 are dashed. Superimposed are the dispersion curves of the even meridional mode-numbered equatorial waves for the three equivalent depths of $h = 12$, 25, and 50 m. (*b*) Same as (*a*) except for the symmetric component of OLR, and the corresponding odd meridional mode-numbered equatorial waves. Frequency spectral bandwidth is 1/96 cpd (Wheeler and Kiladis, 1999).

(ER) waves, mixed Rossby gravity (MRG) waves, and westward inertiogravity (WIG) waves. Eastward inertiogravity (EIG) waves, along with tropical depression (TD) type and the MJO are also depicted. The mode with the most power is the MJO, which is unique in these spectra since it does not correspond to any particular shallow-water mode. It has most of its power in the eastward-propagating symmetric zonal wavenumbers 1 through 6 (Fig. 6*b*), with a substantial contribution to the zonal mean as well. There is also some MJO power in the same region of the antisymmetric spectrum in Figure 6*a*.

The periodic and apparently linear appearance of these waves would imply that they are predictable for a lead time of about half their period. Such relationships can be applied to predict tropical rainfall variability on the submonthly time scale using knowledge of the base state provided by the MJO (Lo and Hendon, 2000). The MJO itself can provide forecast information concerning the onset of ENSO or swings in the TBO, while the interannual time scale base states affect MJO and submonthly convection in the tropics in a continuum of upscale and downscale interactions involving tropical dynamics (Meehl et al., 2001).

REFERENCES

Bjerknes, J. (1969). Atmospheric teleconnections from the equatorial Pacific, *Mon. Wea. Rev.* **97**, 163–172.

Kessler, W. S., M. J. McPhaden, and K. M. Weickmann (1995). Forcing of intraseasonal Kelvin waves in the equatorial Pacific, *J. Geophys. Res.* **100**, 10,613–10,631.

Krishnamurti, T. N. (1971). Tropical east-west circulations during the northern summer, *J. Atmos. Sci.* **28**, 1342–1347.

Lo, F., and H. H. Hendon (2000). Empirical extended-range prediction of the Madden-Julian Oscillation, *Mon. Wea. Rev.* **128**, 2528–2543.

Madden, R. A., and P. R. Julian (1972). Description of global-scale circulation cells in the tropics with a 40–50 day period, *J. Atmos. Sci.* **29**, 1109–1123.

Matsuno, T. (1966). Quasi-geostrophic motions in the equatorial area, *J. Meteorol. Soc. Jpn.* **44**, 25–43.

McPhaden, M. (1999). Genesis and evolution of the 1997–98 El Niño, *Science* **283**, 950–954.

Meehl, G. A. (1997). The South Asian monsoon and the tropospheric biennial oscillation, *J. Climate* **10**, 1921–1943.

Meehl, G. A., and J. M. Arblaster (2002). The tropospheric biennial oscillation and Asian-Australian monsoon rainfall, *J. Climate* **15**, 722–744.

Meehl, G. A., R. Lukas, G. N. Kiladis, M. Wheeler, A. Matthews, and K. M. Weickmann (2001). A conceptual framework for time and space scale interactions in the climate system, *Clim. Dyn.* **17**, 753–775.

Rasmusson, E. M., and T. H. Carpenter (1982). Variations in tropical sea surface temperature and surface wind fields associated with the southern oscillation/EL Niño, *Mon. Wea. Rev.* **110**, 354–384.

Webster, P. J., V. O. Magana, T. N. Palmer, J. Shukla, R. A. Tomas, M. Yanai, and T. Yasunari (1998). Monsoons: Processes, predictability, and the prospects for prediction, *J. Geophys. Res.* **103**, 14,451–14,510.

Wheeler M., and G. N. Kiladis (1999). Convectively-coupled equatorial waves: Analysis of clouds and temperature in the wavenumber-frequency domain, *J. Atmos. Sci.* **56**, 374–399.

CHAPTER 6

TURBULENCE

JACKSON R. HERRING

1 TWO-DIMENSIONAL AND QUASI-GEOSTROPHIC TURBULENCE

Large-scale motions of Earth's atmosphere are perforce nearly two dimensional since the atmosphere's thickness is thin compared to Earth's radius. However, more important dynamical considerations also dictate near two dimensionality: Flows that are both stable against convection, and rapidly rotating, exhibit strong two-dimensional motion regardless of their vertical dimensions. The atmosphere, on average, meets these conditions and it is this latter constraint that determines the dynamic nature of its approximate two dimensionality. To illustrate this point, we consider a simple model consisting of a thin layer of incompressible fluid, subjected to the gravitational force **g** and rotation $\boldsymbol{\Omega}$. The equations of motion for such a fluid are the Boussinesq approximation to the Navier–Stokes equations:

$$\{\partial_t + \mathbf{u} \cdot \nabla\}\mathbf{u} = \hat{\mathbf{g}}\alpha T - 2\boldsymbol{\Omega} \times \mathbf{u} - \nabla p + \nu\nabla^2\mathbf{u} \tag{1}$$

$$\{\partial_t + \mathbf{u} \cdot \nabla\}\theta = \hat{\beta} w + \nabla \cdot \overline{\mathbf{u}\theta} + \kappa\nabla^2\theta \tag{2}$$

$$\nabla \cdot \mathbf{u} = 0 \tag{3}$$

In Eq. (1), $\mathbf{u} = (u, v, w)$ denotes the velocity, with w the velocity component parallel to Earth's radius, u the eastward component, and v the northern. The temperature T is split into its horizontal average $\bar{T}$, and a fluctuation about that average θ, with $\hat{\beta} \equiv -d\bar{T}/dz$; α is the coefficient of thermal expansion of air $(= 1/T)$. We consider these equations in a Cartesian coordinate frame tangential to Earth's sphere at a latitude $\pi/2 - \vartheta$. Here, δp is the deviation of the pressure field

Handbook of Weather, Climate, and Water: Dynamics, Climate, Physical Meteorology, Weather Systems, and Measurements, Edited by Thomas D. Potter and Bradley R. Colman.
ISBN 0-471-21490-6

from that necessary to provide a balance with gravitational and centrifugal forces. Its role in Eq. (1) is to preserve Eq. (3).

Consider now Eqs. (1) to (3) in the limit of large $\mathbf{g}$ and $\mathbf{\Omega}$. Then under suitable conditions [so as to neglect breaking gravity waves, and hence $\nabla \cdot \overline{\mathbf{u}\theta}$ in (2)] the flow separates into two nearly noninteracting components. The first is comprised of rapidly fluctuating gravity and inertial waves, and the second is a slowly evolving component in which terms in (1) $\sim\hat{\mathbf{g}}$ and $\sim\Omega$ nearly cancel. This separation is much analogous to the separation of acoustics and incompressible flow for low Mach numbers. It is the second "balanced" part that constitutes the focus of this chapter. For it, Eqs. (1) to (3) may be replaced by a simpler set, the *quasi-geostrophic* system (hereafter indicated by QG), whose derivation we now sketch. The QG equations do not contain gravity waves or other rapid fluctuations, as do Eqs. (1) to (3), hence their simplicity. Since the right-hand side of (1) contains individual terms (each of which is large), we may expect, in the balanced state, a cancellation among the largest of these. The equation for vorticity, $\omega = \nabla \times \mathbf{u}$ follows from (1) as

$$\partial_t \omega = -\mathbf{u} \cdot \nabla\omega + \omega \cdot \nabla\mathbf{u} + 2(\mathbf{\Omega} \cdot \nabla)\mathbf{u} - \mathbf{g} \times \nabla\theta - \beta v + \nu\nabla^2\omega \tag{4}$$

An independent constraint balancing the effects of rotating and gravity may be formed from the evolution equation for $\mathbf{g}\cdot(\nabla \times \omega)$:

$$+2(\mathbf{\Omega} \cdot \nabla)\omega - \alpha\mathbf{g}\nabla_{\perp}^2\theta = 0 + \cdots \tag{5}$$

Here $\beta = 2\mathbf{\Omega}\sin\theta/R$, with R the earth's radius, and θ the polar angle.

In Eq. (5), $\nabla_{\perp}^2 \equiv \partial_x^2 + \partial_y^2$, and $+\ \cdots$ indicates that only leading-order terms are recorded. Anticipating that $\mathbf{u}$ is nearly two dimensional, but with a possible z dependency, we represented it by a stream function, $\psi(x, y, z)$: $(u, v) = (-\partial_y, \partial_x)\psi(x, y, z)$. In terms of ψ, Eq. (5) simplifies to

$$\alpha g\theta = 2\Omega\partial_z\psi + \cdots \tag{6}$$

The evolution equations for QG may now be obtained by eliminating $(\partial_z w)$ between (2) and the z component of (4):

$$(\partial_t + \mathbf{u} \cdot \nabla)\left\{\omega - \frac{4\Omega^2}{\alpha\hat{\beta}g}\frac{\partial^2}{\partial z^2}\psi\right\} = -\beta v \tag{7}$$

In Eq. (7), we omit the viscous contribution ($\sim\nu$) from (4). In deriving (7), we neglect $(\Omega_x O_x + \Omega_y\partial_y)\omega$ as compared to $\Omega_3\partial_3\omega$ since the vertical extent of the atmosphere is much smaller than its horizontal extent. The divergence of Eq. (1) and Eq. (6) may be used to relate the stream function to the pressure δp,

$$-\delta p = 2\Omega\psi + \cdots \tag{8}$$

For Eq. (7) to be a reasonable approximation, it is clear [from Eq. (4)] that the term proportional to Ω on the right-hand-side of (4) should dominate the term: $\omega \cdot \nabla \mathbf{u}$. Such dominance is indicated by the *smallness* of the Rossby number:

$$R_o = |\omega|/\Omega \sim |u|/(\Omega \ell) \tag{9}$$

where ℓ estimates the length scale of typical horizontal gradients of the flow. For a more comprehensive statement of conditions under which the quasi-geostrophic equations are valid; see Pedlosky (1987).

We next make a few comments concerning the physics of the quasi-geostrophic approximation, confining our remarks to the case of inviscid flows, with $\beta = 0$.* First note that (7) simply states that the quantity

$$Q \equiv \left\{ \omega - \frac{4\Omega^2}{\alpha \beta g} \frac{\partial^2}{\partial z^2} \psi + \beta y \right\}, \qquad \omega = -\nabla_\perp^2 \psi \tag{10}$$

is conserved along streamlines. Q is the potential vorticity, and for flows confined to a given spatial volume, both it and an associated kinetic energy,

$$E = \int_{\mathcal{U}} d\mathbf{r} \left\{ (\nabla_\perp \psi)^2 + \frac{4\Omega^2}{\alpha \hat{\beta} g} \left\{ \frac{\partial}{\partial z} \psi \right\}^2 \right\}, \quad \text{and} \quad \mathcal{E}(z) \equiv \int dxdy\, Q^2 \tag{11a,b}$$

are conserved. $\mathcal{U}$ denotes the volume to which the flow is confined, which, for simplicity, we take as a rectangular box, L_x, L_y, L_z. $\mathcal{E}$ is called the total enstrophy, and $Q(x, y, z)^2$ the enstrophy density. In Eq. (11a,b), $\nabla_\perp = \partial_x + \partial_y$.

Flows described by (7) are known to be subject to a variety of instabilities, which lead to intrinsic temporal fluctuations in Q. We shall not discuss these instabilities here, except to remark that they lead, through the effects of nonlinearities, to the existence in the flow of eddies with a broad distribution of sizes. A convenient description of this distribution stems from a Fourier analysis of $Q(\mathbf{r})$:

$$Q(r) = \sum_{\mathbf{k}} \exp(i\mathbf{k} \cdot \mathbf{r}) Q(\mathbf{k}) \tag{12}$$

where $\mathbf{k} = 2\pi\ (\pm n/L_x, \pm m/L_y, \pm l/L_z)$, and (n, m, l) range over all positive integers.

A basic question in the study of turbulent flows is the distribution of Q among available modes $\mathbf{k}$. The simplest indicators of this distribution are moments

$$\langle E(\mathbf{k}) \rangle \equiv (k_\perp^2 + (4\Omega^2/(\alpha \hat{\beta} g)) k_z^2) \langle |\psi(\mathbf{k})|^2 \rangle$$
$$\mathcal{E}(\mathbf{k}) \equiv \langle |Q(\mathbf{k})|^2 \rangle$$

*Two-dimensional turbulence with $\beta \neq 0$ (Rossby wave turbulence) is important in the study on large-scale planetary flows. Statistical studies of the sort described here were initiated by Holloway and Hendershott (1977). For recent interesting advances, see Galperin et al. (2001).

where the angular brackets denote ensemble averages.

We now explore briefly the implications of the inviscid constraints E and $\mathcal{E}$ constants of motion for QG flow, confining our attention to the case $\beta = 0$ in (10). For the vertical variability, we take a simple form

$$\psi(\rho, z, t) = \frac{1}{2}(\psi_2 + \psi_1) + \frac{1}{2}(\psi_1 - \psi_2)\cos(\pi z/L) \tag{13}$$

Here $\rho = (x,y)$. Evaluating (7) at $z = (0,L)$ with (13) gives

$$(\partial_t + \mathbf{u_i} \cdot \nabla)\{-\nabla_\perp^2 \psi_i - \varepsilon(-1)^i(\psi_1 - \psi_2)\} = 0, \qquad i = (1, 2) \tag{14a,b}$$

Here, $\mathbf{u_i} = (-\partial_y, \partial_x)\psi_i$, etc., and

$$\varepsilon = 2\Omega^2\pi^2/(\alpha\hat{\beta}gL^2) \equiv 1/L_R^2 \tag{14c}$$

In (14c), L_R is the Rossby radius of deformation, which for the atmosphere is ~400 km. Note that $\psi_1 \equiv \psi_2$ is a solution to (14). Such flows are barotropic, and strictly two dimensional. Flows for which $\psi_1 - \psi_2 \neq 0$ are designated as baroclinic since in that case the pressure gradient is inclined to $\mathbf{g}$, according to (8).† Our discussion now focuses on the qualitative behavior of the flow at horizontal scales above and below L_R. For Eqs. (14a) and (14b), conservation laws (11a) and (11b) specialize to

$$\frac{d}{dt}\mathcal{J}_i = 0, \qquad \mathcal{J}_i = \frac{1}{2}\sum_{\mathbf{k}} |Q_i(\mathbf{k})|^2 = 0, \qquad (i = 1, 2) \tag{15a}$$

$$\frac{d}{dt}\mathcal{I} = 0, \qquad \mathcal{I} = \frac{1}{2}\sum_{\mathbf{k}}\{u_1^2 + u_2^2 + \varepsilon(\psi_1 - \psi_2)^2\} \tag{15b}$$

where

$$Q_i(\mathbf{k}) = k^2\psi_i - \varepsilon(-1)^i(\psi_1 - \psi_2) \quad i = 1, 2, \quad u_i^2 = k^2|\psi_i|^2 \tag{15c}$$

It is of interest to ask what distribution in $\mathbf{k}$ emerges as the most probable having given values of total energy, $\mathcal{I}$, and total enstrophy, $\mathcal{J}_1 + \mathcal{J}_2$. Such a distribution fixes $\langle|\psi_i(\mathbf{k})|^2\rangle$, as well as $\langle\psi_1(\mathbf{k})\psi_2(\mathbf{k})\rangle$. If $\mathcal{I}(\psi_1, \psi_2)$ and $\mathcal{J}_1(\psi_1, \psi_2) + \mathcal{J}_2(\psi_1, \psi_2)$ are, separately, constants of motion, then so is their linear combination,

†Equations (14a) and (14b) have more respectability than just a simple vertical collocation of (7). They describe exactly the behavior of two immiscible fluids (the lighter above) under QG conditions.

$\bar{\alpha}\mathcal{I} + \bar{\beta}(\mathcal{J}_1 + \mathcal{J}_2)$, where $\bar{\alpha}$, and $\bar{\beta}$ are arbitrary constants. We may diagonalize the quadratic form $\bar{\alpha}\mathcal{I} + \bar{\beta}(\mathcal{J}_1 + \mathcal{J}_2)$ by a change of variables:

$$\psi_1 = \frac{1}{\sqrt{2}}(\phi_1 + \phi_2), \quad \psi_2 = \frac{1}{\sqrt{2}}(\phi_1 - \phi_2) \tag{16}$$

In terms of ϕ_1, ϕ_2,

$$2(\bar{\alpha}\mathcal{I} + \bar{\beta}(\mathcal{J}_1 + \mathcal{J}_2)) = \left\langle \sum_{\mathbf{k}} a(k)|\phi_1|^2 + (a(k) + \delta(k))|\phi_2|^2 \right\rangle \tag{17}$$

where

$$\alpha(k) = \bar{\alpha}k^2 + \bar{\beta}k^4, \quad \delta(k) = 2\varepsilon(\bar{\alpha} + 2\varepsilon(k^2 + \varepsilon)) \tag{18}$$

Here, ϕ_1 is the barotropic part of the flow, and ϕ_2, the baroclinic. According to standard thermodynamics, the most probable distribution of $\psi_1(\mathbf{k}), \psi_2(\mathbf{k})$ should be a negative exponential of $\bar{\alpha}\mathcal{I} + \bar{\beta}(\mathcal{J}_1 + \mathcal{J}_2)$, with $\bar{\alpha}, \bar{\beta}$ adjusted so that the invariants $\mathcal{I}$ and $\mathcal{J}_1 + \mathcal{J}_2$ have prescribed values. This yields;

$$\langle|\phi_1|^2\rangle(\mathbf{k}) \sim \frac{1}{a(k)}, \quad \langle|\phi_2|^2\rangle(\mathbf{k}) \sim \frac{1}{a(k) + \delta(k)} \tag{19}$$

$$\langle|\psi_1(\mathbf{k})|^2\rangle = \langle|\psi_2(\mathbf{k})|^2\rangle \sim \frac{1}{a(k)} + \frac{1}{a(k) + \delta(k)} \tag{20}$$

and,

$$R_{12} \equiv \frac{\langle\psi_1\psi_2\rangle}{\sqrt{\langle|\psi_1|^2\rangle\langle|\psi_2|^2\rangle}} = \frac{\delta(k)}{2a(k) + \delta(k)} \tag{21}$$

Thus, the flow approaches maximum two dimensionality at large scales $[R_{12}(k) \to 1, k \to 0]$. As k increases beyond $\sqrt{\varepsilon}$, levels 1 and 2 becoming increasingly decorrelated so that $R_{12}(k) \to 4\varepsilon^2/k^2$, as $k/\sqrt{\varepsilon} \to \infty$.[‡]

Distributions (19) to (21) hold for inviscid flows confined to a finite wavenumber domain and have no energy or enstrophy transfer among scales of differing sizes. They represent an end state toward which nonlinear interactions impel the system, but which molecular dissipation prevents.

[‡]The analysis sketched here has been given more substantial form by Salmon et al. (1978). For an application of equipartitioning ideas to ocean basin circulation, see Holloway (1992) and Griffa and Salmon (1989).

2 ENERGY AND ENSTROPHY CASCADE

We turn next to dissipative flows and the scale-size distribution of two-dimensional flows. For simplicity, we focus on barotropic flows, however much of the discussion may be made for baroclinic flows as well. We again ignore the variation of Coriolis force with latitude.

Although equipartitioning arguments of the last section are qualitatively instructive, they are unable to incorporate dissipative effects, which are essential in determining the scale-size distribution of the flow. In atmospheric flows molecular dissipation (viscous and diffusive effects) are centered at very small scales, well below the range of most meteorological interest. However, even if dissipation is negligible in a given spectral range of $\mathbf{k}$, its presence at very small scales induces a flux of energy (or enstrophy, for QG flow) which is vital in determining scale-size distributions, such as that for $E(\mathbf{k})$, or $Q(\mathbf{k})$. Thus, for the viscous form of (4), the principal range of interest is that for which inertial forces $(\mathbf{u} \cdot \nabla\omega)$ far exceed dissipation $(\nu\nabla^2\omega)$, so that

$$R_e = |\mathbf{u} \cdot \nabla\omega|/(\nu|\nabla^2\omega|) \gg 1 \tag{21}$$

This equation defines a Reynolds number. It is clear that its value depends on specifying a length scale to estimate the gradient operator.

In three-dimensional flows, it is well known that as $R_e \to \infty$ dissipation acts to significantly reduce the total energy:

$$E(t) = \int_0^\infty dkE(k, t)$$

on a time scale $\sim(\sqrt{E}/\mathcal{L})^{-1}$, where $\mathcal{L}$ pertains to the energy containing range (that region in wavenumber that contributes most significantly to E). Thus, the decrease of total energy is independent of ν. The underlying physics is that of a cascade. If we discretize wavenumber space in bins Δk_i, in which Δk_1 contains most of the total energy, and $(\Delta k_{i+1}/\Delta k_i) = 2$, then energy is passed from a given bin to its right-hand neighbor at a rate independent of i, until a value of k is reached for which $R_e(1/k) \sim 1$. The range over which this constant flux is maintained is called the inertial range, and the range for which $R_e(k) \leq 1$, the dissipation range. The wave-number distribution of energy in the inertial range is wholly calculable from the constant flux of energy, F, and the decay time scale noted above. Thus, the energy within $\Delta k_i \sim k_i E(k_i)$, and its residency time is $\sim 1/k_i\sqrt{k_i E(k_i)}$, so that $F \sim [k_i E(k_i)(k_i\sqrt{E(k_i)k_i}]$, or

$$E(k) = CF^{2/3}k^{-5/3} \tag{22}$$

A central question in turbulence theory is to determine the equation of motion of $E(k, t)$. We write this evolution equation in the form

$$\dot{E}(k, t) = T(k, t) - 2\nu k^2 E(k, t) + \mathcal{F}(k) \tag{23}$$

where, for simplicity, we assume isotropy. In (23) $\mathcal{F}(k)$ is a possible external forcing function. The energy transfer function, $T(k, t)$ represents the effects of the nonlinearity and is a functional of the energy spectrum, during the entire history of the flow. The idea of a constant of energy flux noted above for the inertial range may be more formally put in terms of (22) as

$$\mathcal{F} = -\int_0^k T(p, t)dp \tag{24}$$

where k is within the inertial range. Energy conservation for inviscid flows is expressed by

$$\int_0^\infty dk\, T(k, t) = 0 \tag{25}$$

We now discuss how ideas of cascade apply to two-dimensional and QG flows, for which both the energy and enstrophy are inviscid constants of motion. In this case, we must have in addition to (25),

$$\int dp\, p^2 T(p, t) = 0 \tag{26}$$

We illustrate the physics of energy and enstrophy transfer by means of a simple model of $T(k, t)$, originally proposed by Leith (1967) and modified by Pouquet et al. (1975). Leith proposed that T be modeled as a diffusion in wavenumber space, with a diffusion rate fixed by the large-scale strain field. It seems a plausible suggestion, and if we assume the straining field acts rapidly, such approximation follows in a nearly exact manner, if we assume the large-scale strain is independent of the vorticity upon which it acts (Kraichnan, 1968). The form for $T(k)$ is

$$T(k) = \frac{\partial}{\partial k}\left\{\frac{1}{k}\frac{\partial}{\partial k}\left\{\sqrt{\int_0^k p^2 dp\, E(p)} k^3 E(k)\right\}\right\} \tag{27}$$

We note that both $\int_0^\infty dk\, T(k)$, and $\int_0^\infty k^2 dk\, T(k)$ vanish, so that for $\nu = 0$ Eq. (27) conserves total energy and total enstrophy. In our discussion of three-dimensional turbulence, we saw that at large R_e, a cascade, together with a constant flux, was associated with a constant of motion (i.e., energy). The question we now address is how does the existence of two constants of motion translate into possible inertial ranges (with associated constant fluxes) for energy and enstrophy? The use of

Eq. (27) in an initial value problem shows that T spreads energy to both small and large k, but primarily to small k. The converse is true for $k^2E(k)$. However, for decaying turbulence, the only consistent constant flux inertial range is a forward fluxed enstrophy inertial range. Then defining

$$\eta = -d\mathcal{E}/dt \tag{28}$$

Eq. (28) gives

$$E(k) \sim \eta^{2/3}k^{-3}/O(\ln^{1/3}(k)) \tag{29}$$

On the other hand, we may verify by direct substitution that for (27), $T(k^{-5/3}) = 0$. These remarks apply to freely decaying turbulence, and the situation changes if we consider turbulence forced at some intermediate wavenumber $k_f[\mathcal{F}(\mathbf{k}) \sim \delta(k - k_f)$, for example]. Then we may solve Eqs. (23) to (27) outside the dissipation range to find:

$$E(k) \sim \eta^{2/3}k_f^{-3}(k_f/k)^{-5/3}, k \le k_f, \sim \eta^{2/3}k^{-3}/(1 + 2\ln(k/k_f))^{1/3}, k > k_f \tag{30}$$

Thus, for stationary two-dimensional turbulence, energy is fluxed to small k, and enstrophy to large k. However, for a steady state to exist, the energy fluxed toward $k \to 0$ must somehow be dissipated by friction or some other large-scale dissipation.

The ideas of a dual cascade carries over to QG flow in which the continuous degrees of freedom are permitted in the vertical direction. First, we may note that the equipartitioning formalism implies in this case an isotropic form for $\psi(\mathbf{k}) = \psi(|\mathbf{k}|)$. This, combined with the form of the enstrophy inertial range sketched above comprises the basis for Charney's (1971) discussion of QG turbulence in the atmosphere. The problem has subsequently been examined *via* the statistical theory of turbulence (Herring, 1980) and high-resolution direct numerical simulation (McWilliams et al., 1994). Both studies examined decaying QG flows. In general, scales larger than the energy peak developed into strong two-dimensional flow with scales smaller than the energy peak mixed barotropic and baroclinic. Above k_{peak}, the statistical theory's prediction for $Q(\mathbf{k})$ displayed weak baroclinic anisotropy just beyond the energy peak, which weakens into isotropy as $k \to \infty$. The simulation, on the other hand, had a much more pronounced baroclinicity beyond the energy peak that persisted over the entire range $k\phi > k_{\text{peak}}$.

3 COHERENT STRUCTURES

Our considerations of spectra have said nothing of the structures imbedded in the turbulence, and in fact two-point covariances cannot distinguish between strain and vorticity, except to say that their root-mean-square (rms) values are equal. Thus, flows with intense, isolated vortices surrounded by expansive strain fields would have the same spectra as flows in which the vortex regions are spatially merged, inseparable from the strain regions. The latter situation could be obtained by picking the amplitudes, $Q(\mathbf{k})$ according to a Gaussian distribution. Such a field would have $T(k) = 0$, thereby excluding energy transfer, which gives rise to inertial ranges. This suggests that some simple measure of non-Gaussianity would be an indicator of the presence of structures, although the converse is not the case. The simplest such measure is the kurtosis,

$$\mathcal{K} \equiv \langle Q^4(\rho)\rangle / \langle Q^2(\rho)\rangle^2 \tag{31}$$

Here, the angular brackets indicate an average over the spatial domain of the flow. For the case in which $Q(\rho)$ consists of isolated pulses of width w separated by empty areas, of area A, $\mathcal{K} = A/w$, while for Gaussian fields, $\mathcal{K} = 3$.

Numerical simulations of large R_e two-dimensional decaying flows (McWilliams, 1984) indicate that flows that start with Gaussian initial conditions develop intense, well-separated vortices, and that $\mathcal{K}(t)$ systematically increases during the flows development with values of $\mathcal{K} = 60$ typical at late stages. The same comment may be made of homogeneous quasi-geostrophic flows. For comparison, similar studies of three-dimensional flows have values of $\mathcal{K} \sim 6$. As the flow develops, the vortices become more isolated and more circular, and the vortex cores probably do not participate in the cascade. Thus regions contributing to enstrophy cascade are confined to thin regions surrounding vortex cores, so that the turbulence lives on a subspace available to it, with transfer to small scales that is progressively reduced during the decay.

One simple consequence derived from Eqs. (23) to (27) is that the decay of enstrophy is $\mathcal{E}(t) \sim t^{-p}, p = 2$, first proposed by Batchelor (1969). Such conclusions are also reached by more complete statistical theories, which are reviewed by Lesieur (1990, p. 269 et seq.). But high-resolution numerical simulations show a much slower decay. Thus Carnavale et al. (1991) find $p = 0.5$, while Chasnov (1997) finds from his 4096 resolution simulation that $p = 0.8$. However, these results for decaying turbulence appear not to bring into question the nature of the inverse cascade discussed elsewhere [see, e.g., Gotoh (1998) or Foias and Chae (1993)].

4 STRATIFIED THREE-DIMENSIONAL TURBULENCE AND WAVES

Stably stratified turbulence shares certain features of two-dimensional turbulence; indeed it has been proposed that motion in the larger scales of the mesoscale are explained as inverse cascading quasi two-dimensional turbulence (Gage, 1978; Lilly, 1983). To explore this issue, we write the Boussinesq equations in a convenient nondimensional form:

$$(\partial_t - \nabla^2)\mathbf{u} = -\nabla p - \mathbf{u} \cdot \nabla \mathbf{u} + \hat{g} N \theta - 2\mathbf{\Omega} \times \mathbf{u} \tag{32}$$

$$(\partial_t - \sigma\nabla^2)\theta = -Nw - \mathbf{u} \cdot \nabla\theta \tag{33}$$

$$\nabla \cdot u = 0 \tag{34}$$

Here, $N = \sqrt{g\alpha\hat{\beta}T_0}$. We take the stratification to be constant and stable $N > 0$. Now introduce, in Fourier space, the representation

$$\mathbf{u} = \mathbf{e}_1\phi_1 + \mathbf{e}_2\phi_2, \mathbf{e}_1(\mathbf{k}) = \mathbf{k} \times \hat{\mathbf{g}}/|\mathbf{k} \times \hat{\mathbf{g}}|, \mathbf{e}_2(\mathbf{k}) = \mathbf{k} \times \mathbf{e}_1/|\mathbf{k} \times \mathbf{e}_1| \tag{35}$$

where ϕ_1 describes horizontal motion (and hence its name "vortical mode"), while ϕ_2 describes gravity waves (for the case $\Omega = 0$). In terms of $(\phi_1\phi_2, \theta)$ (33) may be written:

$$\partial_t \begin{pmatrix} \phi_1 \\ \phi_2 \\ \theta \end{pmatrix} = M \begin{pmatrix} \phi_1 \\ \phi_2 \\ \theta \end{pmatrix} + \mathrm{NL}\{\phi, \theta\} \tag{36}$$

Nonlinearities are indicated symbolically here by 'NL', and

$$M = \begin{pmatrix} 0 & 2\Omega\cos\varphi & 0 \\ -2\Omega\cos\varphi & 0 & -N\sin\varphi \\ 0 & N\sin\varphi & 0 \end{pmatrix} \tag{37}$$

Here, $\hat{\mathbf{g}} \cdot \hat{\mathbf{k}} = \cos\varphi$. For simplicity, we take $\hat{\mathbf{g}} \| \mathbf{\Omega}$. The eigenvalues of M are

$$\lambda = (0, \pm i\mathcal{R}), \mathcal{R} = \sqrt{4\Omega^2\cos^2\varphi + N^2\sin^2\varphi} \tag{38}$$

Some insight may be gained from supposing that the nonlinear terms are weak and act as random agitators of (ϕ_1, ϕ_2, θ). Such a linearization of (36) is called rapid-distortion theory. Then for a given N, Ω the dominant structures activated would be those that minimize the frequency, λ. For $N/\Omega \gg 1$, gravity waves having $\vartheta \sim 0$ dominate, and for $N/\Omega \ll 1$, inertial waves with $\vartheta \sim \pi/2$ dominating. The former condition is satisfied by shear layers ("pancakes"), while the latter by vertically oriented columns. Of course, simple linear reasoning cannot resolve structural issues, for example, whether the shear layers are organized into circular pancakes or the vertical spacing between "pancakes."

Turbulence and waves coexist in stably stratified flows, but their time scales may be disparate: The time scale for waves is $\sim 1/N$, while the eddy turn over time scale is $1/\sqrt{k^3E(k)}$. Thus, if the scale-size distribution is less steep than k^{-3}, waves will dominate the large scales, and turbulence the small. If small scales are isotropic [with spectra (22)], the scale at which these two time scales are equal is $L_O = \sqrt{\varepsilon/N^3}$, known as the Ozmidov scale (Ozmidov 1965). Eddies whose $k > k_O$ are unstable to vertical overturning; and vice versa: at scales smaller than k_O, an rms Richardson number $R_i \sim N^2/|\partial_z u|^2 \leq 1$. We should mention here that this picture may be a gross oversimplification of the physics: There has yet to emerge a direct verification of it from either numerical simulations or experiments.

Stable stratification severely suppresses vertical eddy diffusion of scalar fields. Neglecting molecular processes, a vertically moving parcel of fluid exhausts its kinetic energy and stops. In reality the parcel's kinetic energy is refreshed through thermal exchanges with neighboring parcels and can continue its vertical migration. In terms of turbulence concepts, we may argue that the gravity wave part of the spectrum ($k \leq k_O$) contributes little to eddy diffusion, with only scales with $k > k_O$ active in vertical diffusion. Eddy diffusion may be estimated by treating each Fourier mode of the vertical velocity as a random component contributing to the parcel's velocity,

$$dZ_k(t)/dt = w(k, t), \; \kappa_{\text{eddy}} = \sum_{\mathbf{k}} \langle Z_k dZ_k/dt \rangle \sim \int dk E(k) \tau(k) \tag{39}$$

where $\tau(k)$ is the correlation time associated with $\langle w(k, t) w(-k, t') \rangle \sim 1/\sqrt{k^3 E(k)}$. Here, $w(k, t)$ is the kth component of the Lagrangian vertical velocity field. The equation for κ_{eddy} may be found in Lesieur (1990, p. 285). If in the sum in (39) we suppress scales larger than $1/k_O$,

$$\kappa_{\text{eddy}} \sim \varepsilon/N^2 \tag{40}$$

This represents a reduction by a factor $(L_O/L)^{4/3}$ of the unstratified value of eddy diffusivity, $u_{\text{rms}}\, L \sim \varepsilon^{1/3}\, L^{4/3}$. Although the above estimate has been given a more quantitative form by the theory of Weinstock (1978), numerical simulations (Kimura and Herring, 1996) as well as observational studies (Britter et al., 1985) suggest much smaller eddy diffusivity than (40). It is interesting to note that rapid distortion theory is able to account qualitatively for the suppression of eddy diffusivity (Kaneda and Ishida, 2000).

The search for the "balanced" dynamics of (32) and (34) in the limit $\boldsymbol{\Omega} \to 0$, $N \to \infty$ is of vital interest. As noted above, it was thought early on that the "reduced system" would sufficiently resemble two-dimensional turbulence so that the latter paradigm would be useful in understanding the large horizontal scales of mesoscale variability [for which $E(k) \sim k^{-5/3}$ is observed]. Early analysis suggested two-dimensional turbulence in horizontal, independent layers (Riley et al., 1982; Lilly, 1983). Current mathematical reformulation of the reduced system (Majda and Grote, 1997) now seems able to explain how "pancakes" form, as well as their vertical spacing. Of course, with rotation present, quasi-geostrophy may be invoked as an explanation, but the difficulty is that as the smaller scales of mesoscale variability are approached, R_o [see (9)] becomes $\sim$1 with the approximate $-\frac{5}{3}$ range still present. The issue has been examined numerically by Herring and Métais (1989), who concluded that forced flow did indeed become two dimensionally layered with horizontal motion that varies from layer to layer. But layer edges are rough, and frictional effects associated with roughness tends to seriously suppress inverse cascade. A recent study of the mesoscale variability by Lindborg (1999) indicates that the dynamics of the $\sim k^{-5/3}$ range (thought by Gage and Lilly to be inverse cascading two-dimensional turbulence) does not conform to that of inverse cascading two-dimensional turbulence. His analysis compared theoretical estimates of the third-order structure function [which gives a measure of $T(k, t)$ as in (23)] to that gathered from an ensemble of aircraft observations. It remains to be seen if the theory of Majda and Grote will be able to give insight into the issue of inverse cascade.

REFERENCES

Batchelor, G. K. (1969). Computation of the energy spectrum in homogeneous two-dimensional turbulence, *Phys. Fluids Suppl.* **12**(II), 233–239.

Britter, R. E., J. C. R. Hunt, G. L. Marsh, and W. H. Snyder (1983). The effects of stable stratification on the turbulent diffusion and the decay of grid turbulence, *J. Fluid Mech.* **127**, 27–44.

Carnavale, G. F., J. C. McWilliams, Y. Pomeau, J. B. Weiss, and W. R. Young (1991). Evolution of vortex statistics in two-dimensional turbulence, *Phys. Rev. Lett.* **66**, 2735–2737.

Charney, J. G. (1971). Quasigeostrophic turbulence, *J. Atmos. Sci.* **28**, 1087–1095.

Chasnov, J. R. (1997). On the decay of two-dimensional homogeneous turbulence, *Phys. Fluids* **9**, 171–180.

Foias, C., and D. Chae (1993). A probability measure representing 2-D homogeneous turbulence, in S. Kida (Ed.), *Unstable and Turbulent Motion of Fluid*, World Scientific, Singapore, pp. 131–140.

Gage, K. S. (1979). Evidence for a $k^{-5/3}$ law inertial range in mesoscale two-dimensional turbulence, *J. Atmos. Sci.* **36**, 1950–1954.

Galperin, B., S. Sukoriansky, and H.-P. Huang (2001). Universal n^{-5} spectrum of zonal flows on giant planets, *Phys. Fluids* **13**, 1545–1548.

Gotoh, T. (1998). Energy spectrum in the inertial and dissipation ranges of the two-dimensional steady turbulence, *Phys. Rev. E* **57**, 2984–2991.

Griffa, A., and R. Salmon (1989). Wind-driven ocean circulation and equilibrium statistical mechanics, *J. Marine Res.* **47**, 457–492.

Herring, J. R. (1980). Statistical theory of quasi-geostrophic turbulence, *J. Atmos. Sci.* **37**, 969–977.

Herring, J. R. and O. Métais (1989). Numerical experiments in forced stably stratified turbulence. *J. Fluid Mech.* **202**, 97–115.

Holloway, G. (1992). Representing topographic stress for large-scale ocean models, *J. Phys. Ocean* **22**, 1033–1046.

Holloway, G., and M. C. Hendershott (1977). Stochastic modeling for non-linear Rossby waves, *J. Fluid Mech.* **82**, 747–765.

Kaneda, Y., and T. Ishida (2000). Suppression of vertical diffusion in strongly stratified turbulence, *J. Fluid Mech.* **402**, 311–327.

Kimura, Y., and J. R. Herring (1996). Diffusion in stably stratified turbulence, *J. Fluid Mech.* **328**, 253–269.

Kraichnan, R. H. (1968). Small-scale structure convected by turbulence, *Phys. Fluids* **11**, 945–953.

Kraichnan, R. H. (1975). Statistical dynamics of two dimensional turbulence, *J. Fluid Mech.* **67**, 155–175.

Leith, C. E. (1971). Atmospheric predictability and two-dimensional turbulence, *J. Atmos. Sci.* **28**, 145–161.

Lesieur, M. (1990). *Turbulence in Fluids*, 2nd ed., Dordrecht, Kluwer Academic.

Lilly, D. G. (1983). Stratified turbulence and the mesoscale variability of the atmosphere, *J. Atmos. Sci.* **40**, 749–761.

Lindborg, E. (1999). Can the atmospheric energy spectrum be explained by two-dimensional turbulence? *J. Fluid Mech.* **388**, 259–288.

Majda, A. J., and M. J. Grote (1997). Model dynamics and vertical collapse in decaying strongly stratified flows, *Phys. Fluids* **9**(10), 2932–2940.

McWilliams, J. C. (1984). The emergence of isolated vortices in turbulent flows, *J. Fluid Mech.* **146**, 21–43.

McWilliams, J. C., J. B. Weiss, and I. Yavneh (1994). Anisotropy and coherent vortex structures in planetary turbulence, *Science* **264**, 410–413.

Ozmidov, R. V. (1965). On the turbulent exchange in a stably stratified ocean. *Izu. Accad. Sci. USSR Atmos. Oceanic Phys.* **1**, 493–497.

Pedlosky, J. (1987). *Geophysical Fluid Dynamics*, 2nd ed., New York, Springer.

Pouquet, A., M. Lesieur, J. C. Andre, and C. Basedevant (1975). Evolution of high Reynolds number two-dimensional turbulence, *J. Fluid Mech.* **72**, 305–319.

Riley, J. J., R. W. Metcalfe, and M. A. Weissman (1982). Direct numerical simulations of homogeneous turbulence in density stratified fluids, Proc. AIP Conf. On Nonlinear Properties of Internal Waves, (AIP, Woodbury, 1981) pp. 79–112.

Salmon, R., G. Holloway, and M. C. Hendershott (1978). The equilibrium statistical mechanics of simple quasi-geostrophic models, *J. Fluid Mech.* **75**, 691–703.

Weinstock, J. (1978). Vertical turbulent diffusion in a stably stratified fluid, *J. Atmos. Sci.* **35**, 1022–1027.

CHAPTER 7

PREDICTABILITY AND CHAOS

JEFFREY B. WEISS

Attempts to predict the weather are probably as old as humanity itself. One can imagine our earliest ancestors anxiously examining a stormy sky trying to decide whether to stay in their caves or venture out to hunt and gather. By the early nineteenth century, science viewed the universe as a deterministic and predictable clockworks, a view concisely expressed by Laplace (1825):

> An intelligence that, at a given instant, could comprehend all the forces by which nature is animated and the respective situation of the beings that make it up, if moreover it were vast enough to submit these data to analysis, would encompass in the same formula the movements of the greatest bodies of the universe and those of the lightest atoms. For such an intelligence nothing would be uncertain, and the future, like the past, would be open to its eyes.

Laplace's viewpoint reduces the problem of weather prediction to that of finding the right equations of motion, Laplace's forces, and the exact initial state of the appropriate system, Laplace's respective positions of the beings. Apart from the inherent randomness arising from the quantum nature of the universe, Laplace's view is still correct, but of limited utility. The problem is one of approximations. We will never know the exact equations for the entire universe and the exact positions and velocities of all its particles. For linear systems, such as an archetypical clock, approximate knowledge of only a small portion of the universe, the current location of the clock's hands to within some uncertainty, allows prediction of its entire future to within a similar uncertainty. However, for nonlinear dynamical systems, such as the atmosphere, the question is: How accurately do we need to know the equations and the initial state to make a prediction within a desired error?

Handbook of Weather, Climate, and Water: Dynamics, Climate, Physical Meteorology, Weather Systems, and Measurements, Edited by Thomas D. Potter and Bradley R. Colman.
ISBN 0-471-21490-6

In what follows, the focus will be on the atmosphere, but the issues are similar for any other component of the climate system and for its entirety.

Consider the one-dimensional linear dynamical system

$$\frac{dx}{dt} = ax \tag{1}$$

which has a solution

$$x(t) = e^{at}x(0) \tag{2}$$

If $a > 0$, the system is unstable and $x(t) \rightarrow \infty$, while if $a < 0$ the system is stable, $x(t) \rightarrow 0$. When $a < 0$, the point $x = 0$ is said to be the attractor of the dynamical system in that all initial conditions are attracted to $x = 0$ as $t \rightarrow \infty$.

Suppose we know the initial condition $x(0)$ and the parameter a to within (positive) uncertainties $\delta x(0)$ and δa, respectively. Then, if those uncertainties are small, the uncertainty, or error, $\delta x(t)$ in the prediction $x(t)$ for $t > 0$ is

$$\delta x(t) = e^{at}|x(0)|\delta a\ t + e^{at}\delta x(0) \tag{3}$$

When the system is unstable, the error, as well as the system itself, grows exponentially to infinity; but, when the system is stable, the error eventually decays to zero. Assume now that we know the dynamics exactly, $\delta a = 0$. Since the error growth is exponential, it is convenient to define the mean rate of exponential error growth λ as

$$\lambda = \frac{1}{t}\ln\frac{\delta x(t)}{\delta x(0)} = a \tag{4}$$

So stability corresponds with errors shrinking and negative λ, while instability causes error growth and positive λ.

Most natural systems are, however, not linear. The atmosphere, ocean, and climate system are presumed to be extremely high-dimensional deterministic dissipative dynamical systems. They are dynamical systems in that their state at a future time is a function of their state at a previous time. They are dissipative in that energy is not conserved but is input through forcing, e.g., shortwave solar radiation, and is drained away through damping, e.g., viscosity and outgoing longwave radiation. The dimensionality of a dynamical system refers to the number of variables, or degrees of freedom, needed to fully describe its state. The atmosphere is a continuous fluid described by partial differential equations and its dimensionality is infinite. Discretization of the continuum, necessary for numerical modeling, renders the dimensionality finite, but any reasonable discretization results in a high dimensionality. Current numerical weather prediction models have $O(10^6)$ degrees of freedom.

Lorenz (1975) classified predictability into the first and second kinds. Predictions of the first kind are deterministic predictions, predictions of the specific state of a system evolving from an initial condition. This is the kind of predictability addressed by Laplace: Given an initial state with some uncertainty how well can we predict the future? Predictability of the second kind takes over after predictability of the first kind is no longer possible. Once the errors in an initial value problem have grown so large that a prediction of the first kind is useless, all that can be predicted are the statistics of the system. Those statistics will depend on the properties of the external forces. Some of the external forces for the atmosphere are incoming solar radiation, sea surface temperature (SST), and CO_2 concentration. Prediction of the second kind, statistical prediction, attempts to predict how the structure of the attractor and its resulting statistics respond to changes in the forcing.

1 NONLINEAR DYNAMICS AND CHAOS

Dynamical systems theory discusses the time evolution of a physical system in terms of its phase space, which is defined by the collection of all the variables needed to determine the state of the system at a given time (Lichtenberg and Lieberman, 1992). A single state is then a point in phase space. For a simple pendulum, the state is defined by its position and velocity, and its phase space is two dimensional. The state of the atmosphere is defined by the temperature, wind velocity, humidity, etc. at every point in space. Determining the appropriate state variables of a complex system is often extremely difficult. As a continuous dynamical system evolves in time, the state changes continuously, and the point representing the system's state moves through phase space tracing out a one-dimensional curve called the phase space trajectory.

If one starts a dissipative dynamical system at an arbitrary initial condition, it will typically display some transient behavior and then eventually settle down to its long-term behavior. This long-term behavior is the system's attractor and can take several forms. If the system sits at an equilibrium state, such as the pendulum hanging straight down, then the attractor is a single point in phase space, called a stable fixed point. If the system undergoes regular oscillations, the attractor is a closed curve. If the system undergoes quasiperiodic oscillations, i.e., the superposition of oscillations with irrationally related frequencies, the attractor covers a torus. Motion on all of these attractors is regular and predictable.

Most nonlinear dynamical systems, however, do not display any of these behaviors. First, the attractors are typically fractals, i.e., complex sets with fractional dimension. Attractors with fractal dimension are said to be strange attractors. Second, even though the dynamics is restricted to evolve on an attractor, two nearby states will rapidly separate. Dynamics with this sensitive dependence on initial conditions is said to be chaotic. While there are examples of chaotic nonstrange attractors, as well as nonchaotic strange attractors, it seems that dissipative nonlinear systems typically have chaotic strange attractors and the two words are often used interchangeably. Sensitive dependence on initial conditions refers to predictions of the first kind, while the properties of the attractor as a whole are predictions of the second kind.

Sensitive dependence was first discovered by Lorenz using the so-called Lorenz equations:

$$\begin{aligned}\frac{dx}{dt} &= \sigma(y - x)\\ \frac{dy}{dt} &= -xz + rx - y\\ \frac{dz}{dt} &= xy - bz\end{aligned} \tag{5}$$

Despite being only three dimensional, the Lorenz system and its variants continue to be used in predictability studies. The Lorenz equations display sensitive dependence on initial conditions (Fig. 1) and have a strange attractor (Fig. 2).

Sensitive dependence obviously has a significant impact on predictability and is quantified in terms of Lyapunov exponents, which are a generalization of the error growth rate defined in Eq. (4). Consider a nonlinear dynamical system

$$\frac{d\boldsymbol{y}}{dt} = \boldsymbol{F}(\boldsymbol{y}) \tag{6}$$

with a solution $\boldsymbol{X}(t)$. Consider an initially small perturbation to this trajectory, $\boldsymbol{x}(0)$, which gives a new trajectory $\boldsymbol{X}'(t) = \boldsymbol{X}(t) + \boldsymbol{x}(t)$. The difference, or error, $\boldsymbol{x}(t)$ evolves via the linearization of the dynamical equations:

$$\begin{aligned}\frac{d\boldsymbol{x}}{dt} &= \frac{d\boldsymbol{X}'}{dt} - \frac{d\boldsymbol{X}}{dt}\\ &= \boldsymbol{F}(\boldsymbol{X} + \boldsymbol{x}) - \boldsymbol{F}(\boldsymbol{X})\\ &= \mathbf{M}(\boldsymbol{X})\boldsymbol{x}\end{aligned} \tag{7}$$

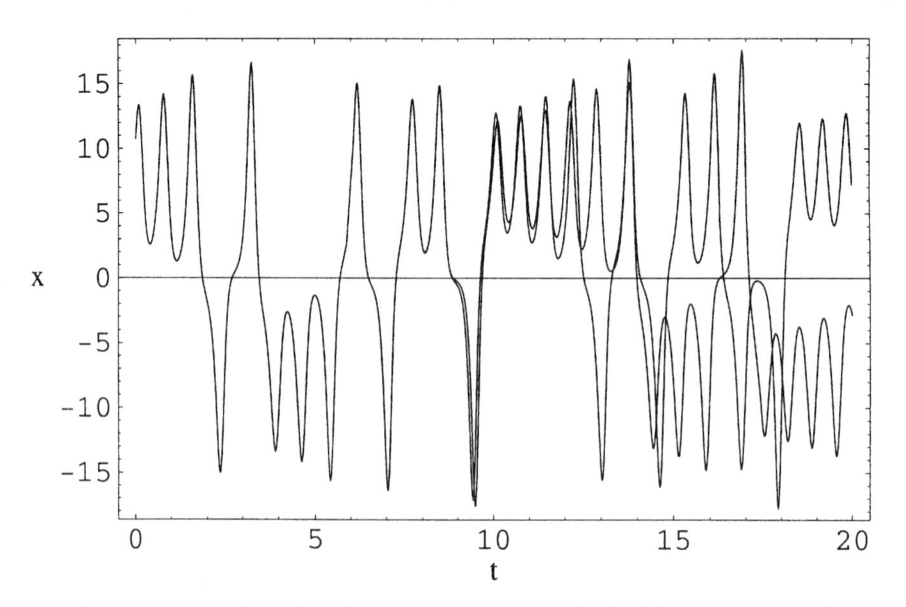

Figure 1 Two trajectories of the Lorenz equations with initial separation 0.00001.

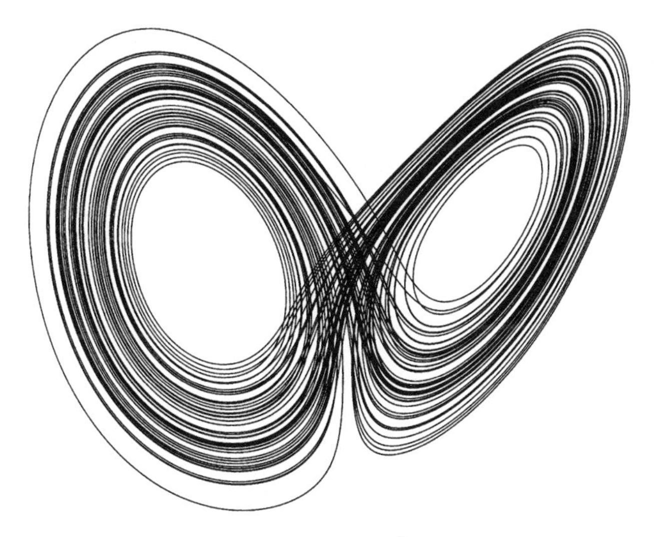

Figure 2 Lorenz attractor at $b = \frac{8}{3}$, $\sigma = 10$, and $r = 28$.

where the linearization of the dynamics at $\boldsymbol{X}$ is given by the matrix

$$M_{ij}(\boldsymbol{X}) = \left.\frac{\partial F_i(\boldsymbol{y})}{\partial y_j}\right|_{\boldsymbol{y}=\boldsymbol{X}} \tag{8}$$

and terms of $O(x^2)$ are ignored because the perturbation is assumed to be small.

The distance, or error, between the two trajectories is given by $d = \|\boldsymbol{x}\|$ where the norm $\|\cdot\|$ must be specified. The choice of norm is highly subjective and the appropriate choice depends on the situation. Some simple common norms are quadratic functions of the state variables, such as the total energy or total enstrophy. However, if your goal is to plan a picnic, then a local precipitation norm would be more appropriate.

Since Eq. (7) shows that the error growth is linear in the error we expect the error to grow exponentially as in (3). However, due to the nonlinear nature of the dynamics, when d becomes large enough that $O(d^2) \sim O(d)$, the linearized equation (7) breaks down. To follow error growth for long times requires either including nonlinear terms in the error growth equation or shrinking the initial d as the prediction time grows so that the final error remains small. Including nonlinear effects causes difficulty over long times since eventually the error grows to the size of the attractor and then remains roughly constant. Averaging error growth over these long times results in zero averaged error growth. A measure of the error growth over the whole attractor thus requires making the initial error smaller as the time gets longer. The result is the Lyapunov exponent λ:

$$\lambda = \lim_{\substack{t\to\infty \\ d(0)\to 0}} \frac{1}{t} \ln \frac{d(t)}{d(0)} \tag{9}$$

The Lyapunov exponent is thus the mean exponential error growth rate linearized about the true trajectory averaged over the entire attractor. If the Lyapunov exponent is positive, errors grow, while if it is negative, errors decay.

For an n-dimensional dynamical system there are actually n Lyapunov exponents, typically ordered from largest to smallest. Over long times, the largest Lyapunov exponent will dominate error growth and is given by Eq. (9). The spectrum of Lyapunov exponents can be understood in terms of the propagator matrix **G** of Eq. (7), defined by:

$$\boldsymbol{x}(t) = \mathbf{G}(t, 0)\boldsymbol{x}(0) \tag{10}$$

For a quadratic norm, the error is given by:

$$\begin{aligned} \|\boldsymbol{x}\|^2 &= \langle \mathbf{G}(t, 0)\boldsymbol{x}(0), \mathbf{G}(t, 0)\boldsymbol{x}(0) \rangle \\ &= \langle \boldsymbol{x}(0), \mathbf{G}(t, 0)^{\dagger}\mathbf{G}(t, 0)\boldsymbol{x} \rangle \end{aligned} \tag{11}$$

where $\mathbf{G}^{\dagger}$ is the adjoint of **G** with respect to the norm. The matrix $\mathbf{G}^{\dagger}\mathbf{G}$ is sometimes called the Oseledec matrix. The n eigenvalues σ_i and the n eigenvectors of $\mathbf{G}^{\dagger}\mathbf{G}$ are called the singular values and singular vectors, respectively, of **G**. The Lyapunov exponents λ_i are related to the singular values by:

$$\lambda_i = \lim_{t \to \infty} \frac{1}{t} \ln \sigma_i \tag{12}$$

Since chaotic systems have sensitive dependence on initial conditions, their largest Lyapunov exponent must be positive. Positive largest Lyapunov exponents impose a practical limit on predictability. While the Laplacian view is still valid in that knowledge of the exact initial condition, $\boldsymbol{x}(0) = 0$, allows perfect prediction, any small errors in the initial condition grow rapidly. Increasing the length of time a forecast error is below some threshold requires decreasing the errors in the initial condition. Since errors grow exponentially, the gain in prediction time is much smaller than the associated decrease in initial error. In particular, decreasing the initial error by a factor of e results in an increase in prediction time of only $1/\lambda$. Current estimates for the atmosphere indicate that the maximum practical prediction time is about 2 weeks.

2 STOCHASTIC DYNAMICS

Although the atmosphere and other components of the climate system are fundamentally deterministic chaotic dynamical systems, they can often be considered as random systems. Due to the high dimensionality of the atmosphere, there is a broad range of space and time scales. If one is interested in the slower motion of the large scales, then the faster variation of the small scales can be parameterized as random noise. Loss of predictability in the resulting stochastic system is not due to the

Lyapunov exponent of the deterministic part of the system, but rather to the lack of predictability in the noise itself. Physically, however, since the noise is merely a parameterization of fast processes with large Lyapunov exponents, it is the deterministic chaos of the parameterized processes that ultimately is the cause of the loss of predictability.

The simplest such system is a deterministic stable fixed point perturbed by white noise,

$$\frac{d\boldsymbol{x}(t)}{dt} = \mathbf{A}\boldsymbol{x}(t) + \vec{\xi}(t), \tag{13}$$

where $\boldsymbol{x}$ represent the state of the system in terms of its deviation from a stable fixed point, $\mathbf{A}$ is the linear operator describing the deterministic dynamics, and $\vec{\xi}$ is multivariate Gaussian white noise. A stable fixed point requires that all of the eigenvalues of $\mathbf{A}$ be negative. In climatic applications, the fixed point is usually taken to be the mean state of the system and $\boldsymbol{x}$ represents the anomaly. This kind of stochastic system has been used to study a number of climatic systems including the El Niño–Souther Oscillation (ENSO) and extratropical atmospheric dynamics.

One of the most interesting features of these systems is that they can amplify the noise. As for the calculation of Lyapunov exponents, the growth of anomalies $d = \|\boldsymbol{x}\|$, is governed by Eq. (11). Now, however, the propagator $\mathbf{G}$ has both deterministic and stochastic components. In most climatic applications the deterministic dynamics is non-normal, i.e., $\mathbf{A}\mathbf{A}^{\dagger} - \mathbf{A}^{\dagger}\mathbf{A} \neq 0$. In this case the eigenvectors of $\mathbf{A}$ are not orthogonal. For normal systems, the stability of the fixed point ensures that all perturbations decay to zero monotonically. Non-normality, however, allows the nonorthogonal eigenvectors to interfere and transient perturbation growth is possible. For a given time interval t, the maximum growth ratio is the largest eigenvalue of $\mathbf{G}^{\dagger}(t, 0)\mathbf{G}(t, 0)$, which is the largest singular value of $\mathbf{G}$. The perturbation that grows the most is the eigenvector of $\mathbf{G}^{\dagger}(t, 0)\mathbf{G}(t, 0)$ corresponding to this eigenvalue, which is the corresponding singular vector of $\mathbf{G}$.

One can treat a time-evolving state perturbed by noise in a similar fashion. The deterministic dynamics is described by its linearization about the time-evolving state, Eq. (7), which is perturbed by adding noise. The propagator now depends on the time-evolving state, but the error growth is still determined by its singular values and singular vectors.

3 ENSEMBLE FORECASTING

When producing a forecast, it is becoming more common to provide an estimate of how correct the forecast is likely to be (Ehrendorfer, 1997). For a perfect model, forecast errors are due only to initial condition errors. The Lyapunov exponents are measures of error growth averaged over the entire attractor. However, the short-time error growth can vary significantly from one location on the attractor to another. This variation is described by so-called local Lyapunov exponents. The distribution of local Lyapunov exponents on the Lorenz attractor shows that the dynamics is least predictable near the unstable fixed point that divides the two lobes (Fig. 3). The variation in local Lyapunov exponents indicates that some atmospheric states are inherently more predictable than others.

An estimate of the believability of a forecast can be made by producing a large number, or ensemble, of forecasts, each starting from a different initial condition, where the ensemble of initial conditions is chosen to reflect the estimated distribution of initial condition errors. A forecast is more likely to be correct when the resulting distribution of forecasts is narrow than when it is large. Furthermore, any detailed structure in the distribution of forecasts indicates the type of forecast errors likely to occur.

Due to the cost of making forecasts, and the high dimensionality of atmospheric dynamics, it is impractical to use an ensemble large enough to adequately sample all possible initial condition errors. Since many directions in phase space correspond to stable directions, errors in these directions will decay and not affect the forecast quality. Thus, a good estimate of forecast error can be obtained with a small ensemble, provided the ensemble of initial conditions is distributed among the directions responsible for the largest error growth.

Currently, there are two main techniques for producing such an ensemble in an operational forecast: breeding methods and singular vectors. The European Center for Medium-Range Weather Forecasting (ECMWF) ensemble prediction system uses singular vectors, discussed above. The breeding method, used by National Center for Environmental Prediction (NCEP), starts two initial conditions some small distance apart and integrates them both using the fully nonlinear model resulting in a control solution and a perturbed solution. As the model evolves the two

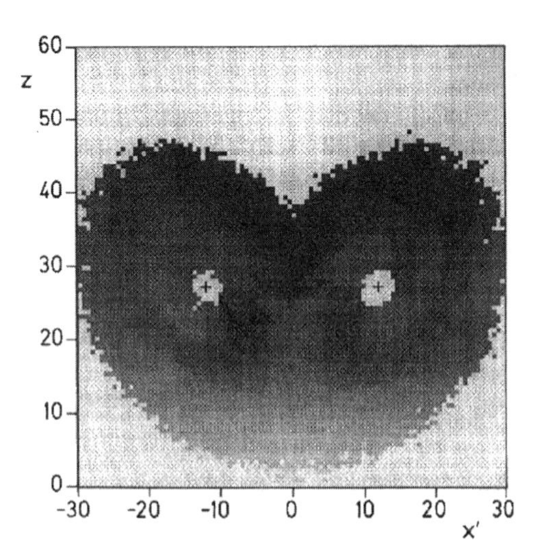

Figure 3 Distribution of local Lyapunov exponents on the Lorenz attractor, with darker (lighter) shadings indicating more (less) predictable regions (Eckhardt and Yao, 1993).

solutions separate. After some time interval the difference between the control and perturbation is rescaled back to its original amplitude and a new perturbed solution is begun. After this cycle has been repeated several times, components of the perturbation in stable and less unstable directions have been removed due to the rescalings. The resulting bred modes are then used to initialize the ensemble.

Ensemble forecasts accurately represent forecast error only if the model itself is accurate enough to approximately follow the true trajectory of the system. If the model error is large, then the entire ensemble will evolve into a completely different region in phase space than the true system. In this case, the ensemble spread is irrelevant to the forecast error.

4 SEASONAL TO CLIMATE PREDICTABILITY

Predicting the atmosphere beyond the roughly 2-week limit of deterministic predictability requires predicting its statistics rather than its actual state. Statistical prediction is possible if there is some slow external forcing that significantly affects the structure of the attractor. Note that the division between internal and external variability is somewhat subjective. For example, in predicting the effect of ENSO on the extratropics, the variation in tropical SST is considered an external forcing, while when considering the climate response to increasing CO_2, the behavior of tropical SST is internal variability.

The land and ocean surface provide boundary conditions for the atmosphere and, to the extent that they evolve slowly compared to the rapid dynamics of the weather, they allow longer term prediction. For example, the variation of tropical SSTs associated with ENSO evolves on a seasonal time scale. In addition, the relatively weak nonlinearity in the tropical atmosphere–ocean system allows successful predictions of ENSO well beyond the 2-week limit of extratropical weather. Thus, ENSO prediction provides some degree of climate prediction. The degree to which the state of the ENSO system affects the structure of the seasonal attractor can be seen in Figure 4, which shows the effects of an El Niño episode on December–February climate.

One factor that can significantly impact predictability is regime behavior, which occurs when a system spends significant amounts of time in localized regions of the attractor, followed by relatively rapid transitions to other localized regions of the attractor. For example, the two lobes of the Lorenz attractor in Figure 2 are regimes. There is very little understanding of the generic regime behavior of high dimensional chaotic attractors, and the role of regimes in intraseasonal to climate variability is still a subject of debate despite decades of research.

Intraseasonal regimes have been described in a number of ways including variations in the Rossby wave amplitude, blocking events, and teleconnection patterns. It has been suggested that the warming of the equatorial Pacific in 1976, which affected subsequent ENSO cycles, is an example of an interdecadal regime shift. Climate events such as the Little Ice Age may indicate centennial-scale regimes and glacial–interglacial transitions indicate regimes on even longer time scales.

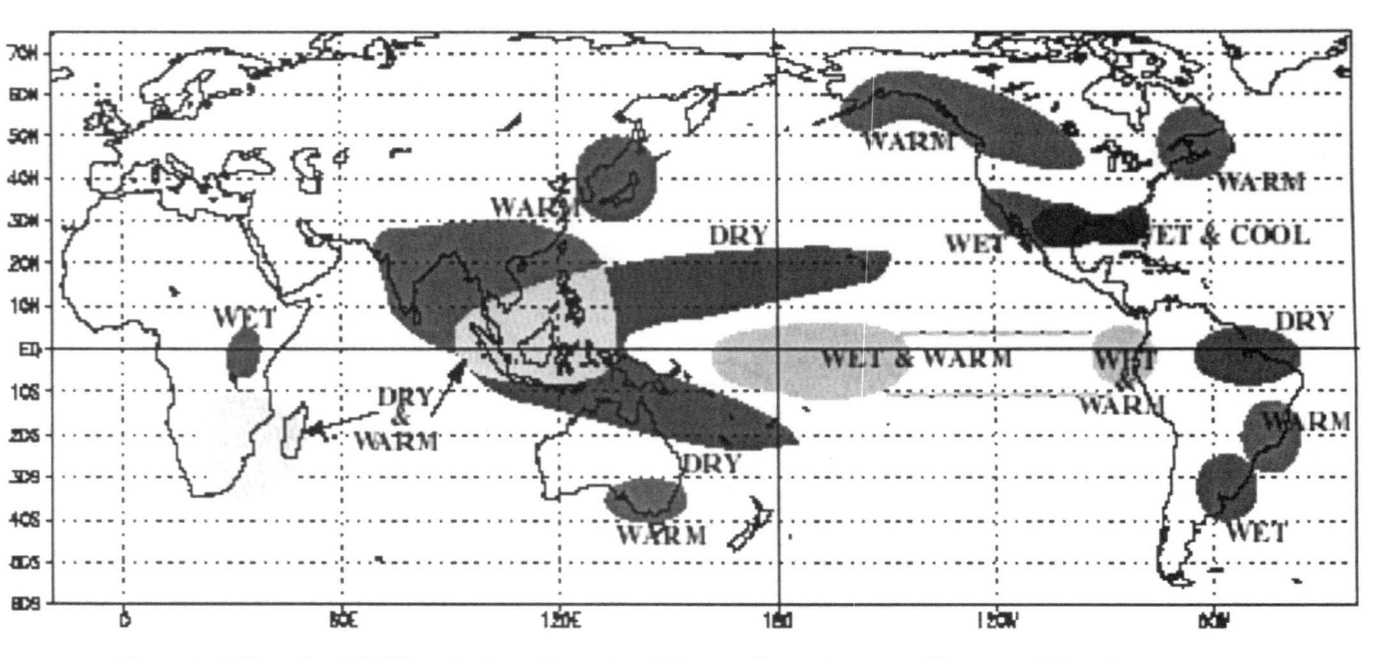

Figure 4 Effect of an El Niño episode on December–February climate (courtesy Climate Prediction Center, NOAA).

Since weather statistics depend on the regime of the atmosphere, one avenue for longer-term prediction is to focus on regime dynamics. Rather than predict the specific evolution of the system trajectory, the goal is to predict the sequence of regimes and their duration. Unfortunately, attempts at predicting the timing and outcome of regime transitions have not been particularly successful.

Recently, Palmer (1999) has analyzed the predictability of climate under the assumption that the climate attractor is composed of a number of distinct quasi-stationary regimes. Regimes are regions of the attractor that are relatively stable. For example, the regimes in the Lorenz attractor are localized around unstable fixed points, which, despite being unstable, have stable subspaces that attract trajectories. In the context of stochastic systems, the regimes may be modeled as truly stable attractors and the transitions are then solely due to the random noise. The areas of phase space between the regimes, the "saddles," are more unstable, and the system passes through these areas relatively quickly. An important feature of such saddles is that they show sensitive dependence: The relationship between the specific location where the system crosses the saddle and the regime it next visits can have fractal structure. Palmer noted that small-amplitude forcing will affect this fine-scale structure of the saddle much more than the relatively stable regime structure. The stability of the behavior of the attractor to external forcing is called structural stability. Thus, small-amplitude forcing will not change the structure of the regimes, i.e., they are structurally relatively stable, but will affect the relative probability of the system visiting a regime, which is structurally unstable. In the context of anthropogenic CO_2 increase, the resulting climate change will not show new climate regimes but, rather, a shift in the occurrence of existing regimes. Thus, the fact that recent climate changes are similar to naturally occurring variations does not disprove the possibility that the changes have an anthropogenic cause.

REFERENCES

Eckhardt, B., and D. Yao (1993). Local Lyapunov exponents in chaotic systems, *Physica D* **65**, 100–108.

ECMWF (1996). *Predictability*, Proceedings of a seminar held at ECMWF on predictability: 4–8 September 1995 Reading, European Centre for Medium-Range Weather Forecasts.

Ehrendorfer, M. (1997). Predicting the uncertainty of numerical weather forecasts: a review, *Meteorol. Z.* **6**, 147–183.

Laplace, P. S. (1825). *Philosophical Essay on Probabilities*, translated from the fifth French edition of 1825 by A. I. Dale, New York, Springer, 1995.

Lichtenberg, A. J., and M. A. Lieberman (1992). *Regular and Chaotic Dynamics*, New York, Springer.

Lorenz, E. N. (1975). Climatic predictability, in *The Physical Basis of Climate and Climate Modelling*, GARP Publication Series, (ICSU/WMO, Geneva), Vol. **16**, pp. 132–136.

NATO (1996). *Decadal Climate Variability: Dynamics and Predictability*, Proceedings of the NATO Advanced Study Institute held at Les Houches, France, February 13–24, 1995, L. T. Anderson and J. Willebrand (eds.), Berlin, Springer.

Palmer, T. N. (1999). A nonlinear dynamical perspective on climate prediction, *J. Climate* **12**, 575–591.

CHAPTER 8

HISTORICAL OVERVIEW OF NUMERICAL WEATHER PREDICTION

EUGENIA KALNAY

1 INTRODUCTION

In general, the public is not aware that our daily weather forecasts start out as initial-value problems on the major National Weather Service supercomputers. Numerical weather prediction provides the basic guidance for weather forecasting beyond the first few hours. For example, in the United States, computer weather forecasts issued by the National Centers for Environmental Prediction (NCEP) in Washington, DC, guide forecasts from the U.S. National Weather Service (NWS). NCEP forecasts are performed running (integrating in time) computer models of the atmosphere that can simulate, given today's weather observations, the evolution of the atmosphere in the next few days. Because the time integration of an atmospheric model *is an initial-value problem*, the ability to make a skillful forecast requires both that *the computer model be a realistic representation of the atmosphere and that the initial conditions be accurate*. In what follows we will give examples of the evolution of numerical weather prediction at NCEP, but they are representative of what has taken place in all major international operational weather centers such as the European Centre for Medium Range Weather Forecasts (ECMWF), the United Kingdom Meteorological Office (UKMO), and the weather services of Japan, Canada, and Australia.

Formerly the National Meteorological Center (NMC), the NCEP has performed operational computer weather forecasts since the 1950s. From 1955 to 1973, the forecasts included only the Northern Hemisphere; they have been global since 1973. Over the years, the quality of the models and methods for using atmospheric observations has improved continuously, resulting in major forecast improvements.

Handbook of Weather, Climate, and Water: Dynamics, Climate, Physical Meteorology, Weather Systems, and Measurements, Edited by Thomas D. Potter and Bradley R. Colman.
ISBN 0-471-21490-6

Figure 1 shows the longest available record of the skill of numerical weather prediction. The S_1 score (Teweles and Wobus, 1954) measures the relative error in the horizontal gradient of the height of the constant-pressure surface of 500 hPa (in the middle of the atmosphere, since the surface pressure is about 1000 hPa) for 36-h forecasts over North America. Empirical experience at NMC indicated that a value of this score of 70% or more corresponds to a useless forecast, and a score of 20% or less corresponds to an essentially perfect forecast. Twenty percent was the average S_1 score obtained when comparing analyses hand-made by several experienced forecasters fitting the same observations over the data-rich North America region.

Figure 1 shows that current 36-h 500-hPa forecasts over North America are close to what was considered "perfect" 30 years ago: The forecasts are able to locate generally very well the position and intensity of the large-scale atmospheric waves, major centers of high and low pressure that determine the general evolution of the weather in the 36-h forecast. Smaller-scale atmospheric structures, such as fronts, mesoscale convective systems that dominate summer precipitation, etc., are still difficult to forecast in detail, although their prediction has also improved very significantly over the years. Figure 1 also shows that the 72-h forecasts of today are as accurate as the 36-h forecasts were 10 to 20 years ago. Similarly, 5-day forecasts, which had no useful skill 15 years ago, are now moderately skillful,

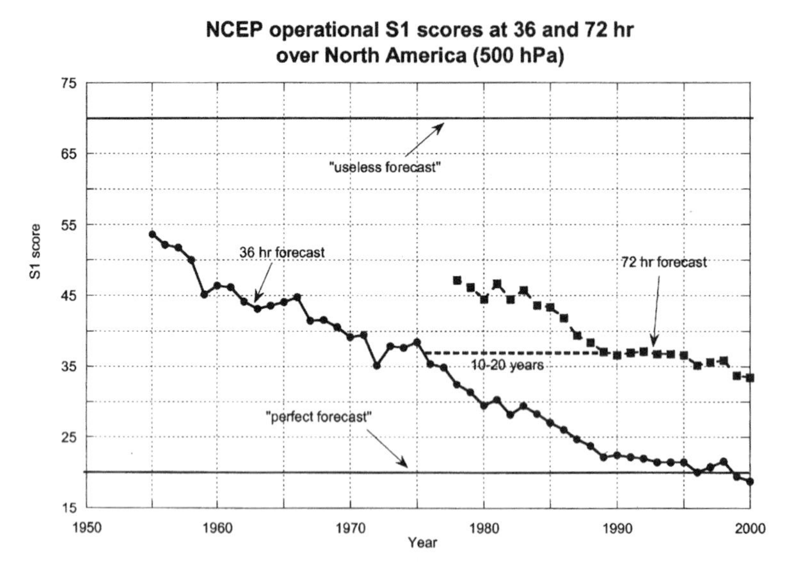

Figure 1 Historic evolution of the operational forecast skill of the NCEP (formerly NMC) models over North America. The S_1 score measures the relative error in the horizontal pressure gradient, averaged over the region of interest. The values $S_1 = 70\%$ and $S_1 = 20\%$ were empirically determined to correspond, respectively, to a "useless" and a "perfect" forecast when the score was designed. Note that the 72-h forecasts are currently as skillful as the 36-h forecasts were 10 to 20 years ago. (Data courtesy C. Vlcek, NCEP.)

and during the winter of 1997–1998, ensemble forecasts for the second week average showed useful skill (defined as anomaly correlation close to 60% or higher).

The improvement in skill over the last 40 years of numerical weather prediction apparent in Figure 1 is due to four factors:

- Increased power of supercomputers, allowing much finer numerical resolution and fewer approximations in the operational atmospheric models.
- Improved representation of small-scale physical processes (clouds, precipitation, turbulent transfers of heat, moisture, momentum, and radiation) within the models.
- Use of more accurate methods of data assimilation, which result in improved initial conditions for the models.
- Increased availability of data, especially satellite and aircraft data over the oceans and the Southern Hemisphere.

In the United States, research on numerical weather prediction takes place in national laboratories of the National Oceanic and Atmospheric Administration (NOAA), the National Aeronautics and Space Administration (NASA), the National Center for Atmospheric Research (NCAR), and in universities and centers such as the Center for Prediction of Storms (CAPS). Internationally, major research takes place in large operational national and international centers (such as the European Center for Medium Range Weather Forecasts, NCEP, and the weather services of the United Kingdom, France, Germany, Scandinavian, and other European countries, Canada, Japan, Australia, and others). In meteorology there has been a long tradition of sharing both data and research improvements, with the result that progress in the science of forecasting has taken place in many fronts, and all countries have benefited from this progress.

2 PRIMITIVE EQUATIONS, GLOBAL AND REGIONAL MODELS, AND NONHYDROSTATIC MODELS

As envisioned by Charney (1951, 1960), the filtered (quasi-geostrophic) equations, introduced by Charney et al. (1950), were found not accurate enough to allow continued progress in Numerical Weather Prediction (NWP), and were eventually replaced by primitive equations models. These equations are conservation laws applied to individual parcels of air: conservation of the three-dimensional momentum (equations of motion), conservation of energy (first law of thermodynamics), conservation of dry air mass (continuity equation), and equations for the conservation of moisture in all its phases, as well as the equation of state for perfect gases. They include in their solution fast gravity and sound waves, and therefore in their space and time discretization they require the use of a smaller time step. For models with a horizontal grid size larger than 10 km, it is customary to replace the vertical component of the equation of motion with its hydrostatic approximation, in which the vertical acceleration is neglected compared with gravitational acceleration (buoyancy). With this approximation, it is convenient to use atmospheric pressure, instead of height, as a vertical coordinate.

The continuous equations of motions are solved by discretization in space and in time using, for example, finite differences. It has been found that the accuracy of a model is very strongly influenced by the spatial resolution: In general, the higher the resolution, the more accurate the model. Increasing resolution, however, is extremely costly. For example, doubling the resolution in the 3-space dimensions also requires halving the time step in order to satisfy conditions for computational stability. Therefore, the computational cost of a doubling of the resolution is a factor of 2^4 (3-space and one time dimension). Modern methods of discretization attempt to make the increase in accuracy less onerous by the use of implicit and semi-Lagrangian time schemes (Robert, 1981), which have less stringent stability conditions on the time step, and by the use of more accurate space discretization. Nevertheless, there is a constant need for higher resolution in order to improve forecasts, and as a result running atmospheric models has always been a major application of the fastest supercomputers available.

When the "conservation" equations are discretized over a given grid size (typically a few kilometers to several hundred kilometers) it is necessary to add "sources and sinks," terms due to small-scale physical processes that occur at scales that cannot be explicitly resolved by the models. As an example, the equation for water vapor conservation on pressure coordinates is typically written as:

$$\frac{\partial \bar{q}}{\partial t} + \bar{u}\frac{\partial \bar{q}}{\partial x} + \bar{v}\frac{\partial \bar{q}}{\partial y} + \bar{\omega}\frac{\partial \bar{q}}{\partial p} = \bar{E} - \bar{C} + \frac{\partial \overline{\omega' q'}}{\partial x}$$

where q is the ratio between water vapor and dry air, x and y are horizontal coordinates with appropriate map projections, p pressure, t time, u and v are the horizontal air velocity (wind) components, $\omega = dp/dt$ is the vertical velocity in pressure coordinates, and the primed product represents turbulent transports of moisture on scales unresolved by the grid used in the discretization, with the overbar indicating a spatial average over the grid of the model. It is customary to call the left-hand side of the equation, the "dynamics" of the model, which are computed explicitly.

The right-hand side represents the so-called physics of the model, i.e., for this equation, the effects of physical processes such as evaporation, condensation, and turbulent transfers of moisture, which take place at small scales that cannot be explicitly resolved by the dynamics. These subgrid-scale processes, which are sources and sinks for the equations, are then "parameterized" in terms of the variables explicitly represented in the atmospheric dynamics.

Two types of models are in use for numerical weather prediction: global and regional models. Global models are generally used for guidance in medium-range forecasts (more than 2 days), and for climate simulations. At NCEP, for example, the global models are run through 16 days every day. Because the horizontal domain of global models is the whole Earth, they usually cannot be run at high resolution. For more detailed forecasts it is necessary to increase the resolution, and this can only be done over limited regions of interest, because of computer limitations.

Regional models are used for shorter-range forecasts (typically 1 to 3 days) and are run with resolutions several times higher than global models. In 1997, the NCEP global model was run with 28 vertical levels, and a horizontal resolution of 100 km for the first week, and 200 km for the second week. The regional (Eta) model was run with a horizontal resolution of 48 km and 38 levels, and later in the day with 29 km and 50 levels. Because of their higher resolution, regional models have the advantage of higher accuracy and ability to reproduce smaller scale phenomena such as fronts, squall lines, and orographic forcing much better than global models. On the other hand, regional models have the disadvantage that, unlike global models, they are not "self-contained" because they require lateral boundary conditions at the borders of the horizontal domain. These boundary conditions must be as accurate as possible because otherwise the interior solution of the regional models quickly deteriorates. Therefore, it is customary to "nest" the regional models within another model with coarser resolution, whose forecast provides the boundary conditions. For this reason, regional models are used only for short-range forecasts. After a certain period, proportional to the size of the model, the information contained in the high-resolution initial conditions is "swept away" by the influence of the boundary conditions, and the regional model becomes merely a "magnifying glass" for the coarser model forecast in the regional domain. This can still be useful, for example, in climate simulations performed for long periods (seasons to multiyears), and which therefore tend to be run at coarser resolution. A "regional climate model" can provide a more detailed version of the coarse climate simulation in a region of interest.

More recently the resolution of regional models has been increased to just a few kilometers in order to resolve better mesoscale phenomena. Such storm-resolving models as the Advanced Regional Prediction System (ARPS) cannot be hydrostatic since the hydrostatic approximation ceases to be accurate for horizontal scales of the order of 10 km or smaller. Several major nonhydrostatic models have been developed and are routinely used for mesoscale forecasting. In the United States the most widely used are the ARPS, the MM5, the RSM, and the U.S. Navy model. There is a tendency toward the use of nonhydrostatic models that can be used globally as well.

3 DATA ASSIMILATION: DETERMINATION OF INITIAL CONDITIONS FOR NWP PROBLEM

As indicated previously, NWP is an initial value problem: Given an estimate of the present state of the atmosphere, the model simulates (forecasts) its evolution. The problem of determination of the initial conditions for a forecast model is very important and complex, and has become a science in itself (Daley, 1991). In this brief section we introduce the main methods used for this purpose [successive corrections method (SCM), optimal interpolation (OI), variational methods in three and four dimensions, 3D-Var and 4D-Var, and Kalman filtering (KF)]. More detail is available in Chapter 5 of Kalnay (2001) and Daley (1991).

In the early experiments, Richardson (1922) and Charney et al. (1950) performed hand interpolations of the available observations, and these fields of initial conditions were manually digitized, which was a very time-consuming procedure. The need for an automatic "objective analysis" became quickly apparent (Charney, 1951), and interpolation methods fitting data were developed (e.g., Panofsky, 1949; Gilchrist and Cressman, 1954; Barnes, 1964).

There is an even more important problem than spatial interpolation of observations: There is not enough data to initialize current models. Modern primitive equations models have a number of degrees of freedom on the order of 10^7. For example, a latitude–longitude model with typical resolution of one degree and 20 vertical levels would have $360 \times 180 \times 20 = 1.3 \times 10^6$ grid points. At each grid point we have to carry the values of at least 4 prognostic variables (two horizontal wind components, temperature, moisture) and surface pressure for each column, giving over 5 million variables that need to be given an initial value. For any given time window of ± 3 h, there are typically 10,000 to 100,000 observations of the atmosphere, two orders of magnitude fewer than the number of degrees of freedom of the model. Moreover, their distribution in space and time is very nonuniform (Fig. 2), with regions like North America and Eurasia, which are relatively data rich, and others much more poorly observed.

For this reason, it became obvious rather early that it was necessary to use additional information (denoted background, first guess, or prior information) to prepare initial conditions for the forecasts (Bergthorsson and Döös, 1955). Initially climatology was used as a first guess (e.g., Gandin, 1963), but as forecasts became

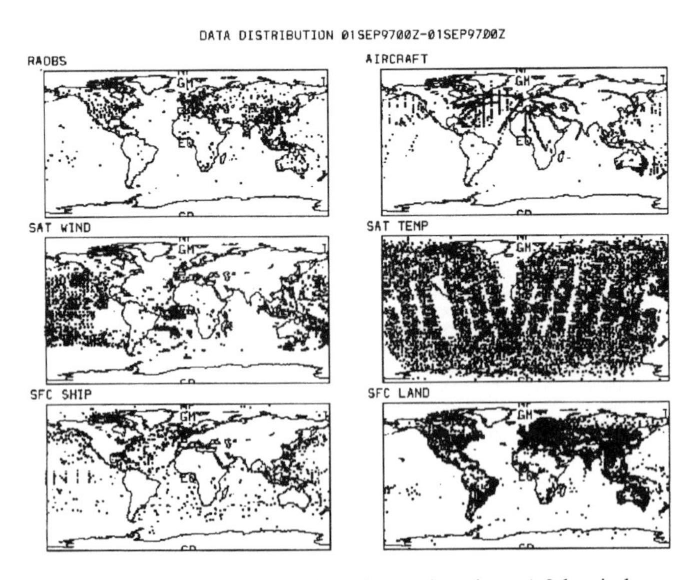

Figure 2 Typical distribution observations in a ± 3-h window.

better, a short-range forecast was chosen as first guess in the operational data assimilation systems or "analysis cycles." The intermittent data assimilation cycle shown schematically in Figure 3 is continued in present-day operational systems, which typically use a 6-h cycle performed 4 times a day.

In the 6-h data assimilation cycle for a global model, the model 6-h forecast x^b (a three-dimensional array) is interpolated to the observation location and, if needed, converted from model variables to observed variables y^o (such as satellite radiances or radar reflectivities). The first guess of the observations is therefore $H(x^b)$, where H is the observation operator that performs the necessary interpolation and transformation. The difference between the observations and the model first guess $y^o - H(x^b)$ is denoted "observational increments" or "innovations." The analysis x^a is obtained by adding the innovations to the model forecast (first guess) with weights that are determined based on the estimated statistical error covariances of the forecast and the observations:

$$x^a = x^b + W[y^o - H(x^b)] \tag{1}$$

Different analysis schemes (SCM, OI, variational methods, and Kalman filtering) differ by the approach taken to combine the background and the observations to

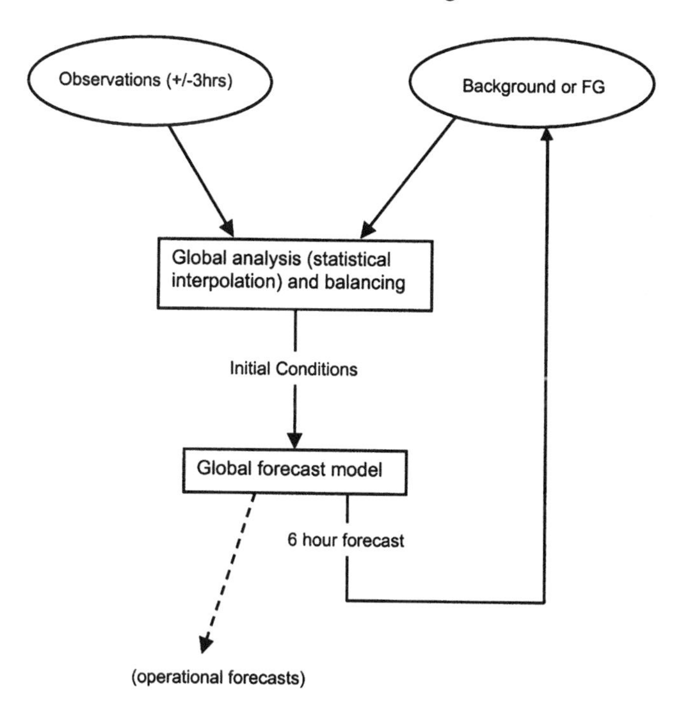

Figure 3 Flow diagram of a typical intermittent (6 h) data assimilation cycle. The observations gathered within a window of about ± 3 h are statistically combined with a 6-h forecast denoted background or first guess. This combination is called "analysis" and constitutes the initial condition for the global model in the next 6-h cycle.

produce the analysis. In optimal interpolation (OI; Gandin, 1963) the matrix of weights W is determined from the minimization of the analysis errors. Lorenc (1986) showed that there is an equivalency of OI to the variational approach (Sasaki, 1970) in which the analysis is obtained by minimizing a cost function:

$$J = \frac{1}{2}\{[y^o - H(x)]^T R^{-1}[y^o - H(x)] + (x - x^b)^T B^{-1}(x - x^b)\} \tag{2}$$

This cost function J in (2) measures the distance of a field x to the observations (first term in the cost function) and the distance to the first guess or background x^b (second term in the cost function). The distances are scaled by the observation error covariance R and by the background error covariance B, respectively. The minimum of the cost function is obtained for $x = x^a$, which is the analysis. The analysis obtained in (1) and (2) is the same if the weight matrix in (1) is given by

$$W = BH^T(HBH^T + R^{-1})^{-1}[y^o - H(x)^b] \tag{3}$$

In (3) **H** is the linearization of the transformation H.

The difference between optimal interpolation (1) and the three-dimensional variational (3D-Var) approach (2) is in the method of solution: In OI, the weights W are obtained using suitable simplifications. In 3D-Var, the minimization of (2) is performed directly, and therefore allows for additional flexibility.

Earlier methods such as the successive corrections method, (SCM; Bergthorsson and Döös, 1955; Cressman, 1959; Barnes, 1964) were of a form similar to (1), but the weights were determined empirically, and the analysis corrections were computed iteratively. Bratseth (1986) showed that with a suitable choice of weights, the simpler SCM solution will converge to the OI solution. More recently, the variational approach has been extended to four dimensions, by including in the cost function the distance to observations over a time interval (assimilation window). A first version of this expensive method was implemented at ECMWF at the end of 1997 (Bouttier and Rabier, 1998; Andersson et al., 1998). Research on the even more advanced and computationally expensive Kalman filtering [e.g., Ghil et al. (1981) and ensemble Kalman filtering, Evensen and Van Leeuwen (1996), Houtekamer and Mitchell (1998)] is included in Chapter 5 of Kalnay (2001). That chapter also discusses in some detail the problem of enforcing a balance in the analysis in such a way that the presence of gravity waves does not mask the meteorological signal, as it happened to Richardson (1922). This "initialization" problem was approached for many years through "nonlinear normal mode initialization" (Machenauer, 1976; Baer and Tribbia, 1976), but more recently balance has been included as a constraint in the cost function (Parrish and Derber, 1992), or a digital filter has been used (Lynch and Huang, 1994).

In the analysis cycle, no matter which analysis scheme is used, the use of the model forecast is essential in achieving "four-dimensional data assimilation" (4DDA). This means that the data assimilation cycle is like a long model integration in which the model is "nudged" by the data increments in such a way that it remains

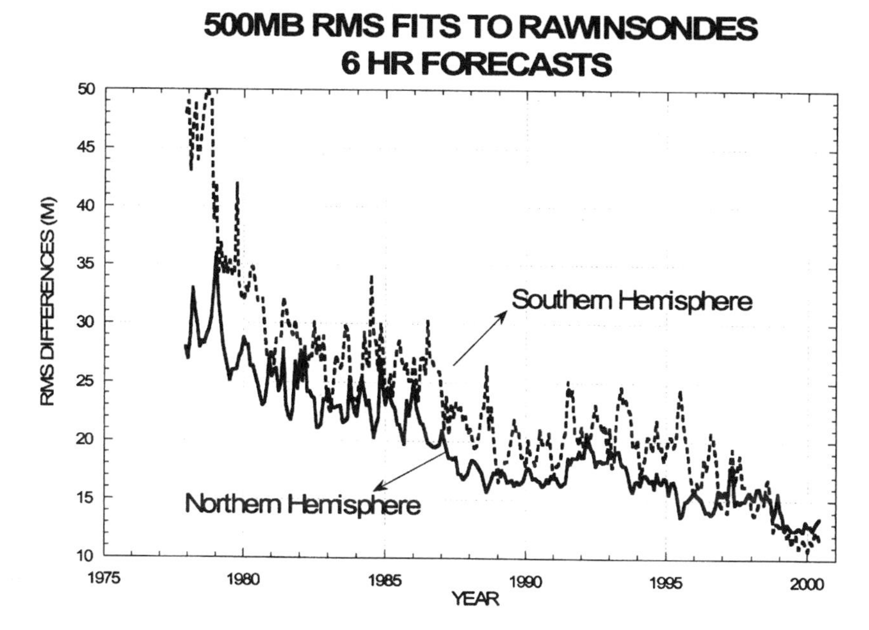

Figure 4 RMS observational increments (differences between 6-h forecast and rawinsonde observations) for 500-hPa heights. (Data courtesy of Steve Lilly, NCEP.)

close to the real atmosphere. The importance of the model cannot be overemphasized: It transports information from data-rich to data-poor regions, and it provides a complete estimation of the four-dimensional state of the atmosphere. Figure 4 presents the root-mean-square (rms) difference between the 6-h forecast (used as a first guess) and the rawinsonde observations from 1978 to the present (in other words, the rms of the observational increments for 500-hPa heights). It should be noted that the rms differences are not necessarily forecast errors since the observations also contain errors. In the Northern Hemisphere the rms differences have been halved from about 30 m in the late 1970s to about 15 m in the present, equivalent to a mean temperature error of about 0.75 K, not much larger than rawinsonde observational errors. In the Southern Hemisphere the improvements are even larger, with the differences decreasing from about 47 m to about 17 m, close to the present forecast error in the Northern Hemisphere. The improvements in these short-range forecasts are a reflection of improvements in the model, the analysis scheme used in the data assimilation, and, to a lesser extent, the quality and quality control of the data.

4 OPERATIONAL NUMERICAL PREDICTION

Here we focus on the history of operational numerical weather prediction at the NMC (now NCEP), as reviewed by Shuman (1989) and Kalnay et al. (1998), as an example of how major operational centers have evolved. In the United States operational numerical weather prediction started with the organization of the Joint Numerical Weather Prediction Unit (JNWPU) on July 1, 1954, staffed by members of the U.S. Weather Bureau (later National Weather Service, NWS), the Air Weather Service of the U.S. Air Force, and the Naval Weather Service.* Shuman (1989) pointed out that in the first few years, numerical predictions could *not* compete with those produced manually. They had several serious flaws, among them overprediction of cyclone development. Far too many cyclones were predicted to deepen into storms. With time, and with the joint work of modelers and practicing synopticians, major sources of model errors were identified, and operational NWP became the central guidance for operational weather forecasts.

Shuman (1989) included a chart with the evolution of the S_1 score (Teweles and Wobus, 1954), the first measure of error in a forecast weather chart that, according to Shuman (1989), was designed, tested, and modified to correlate well with expert forecasters' opinions on the quality of a forecast. The S_1 score measures the average relative error in the pressure gradient (compared to a verifying analysis chart). Experiments comparing two independent subjective analyses of the same data-rich North American region made by two experienced analysts suggested that a "perfect" forecast would have an S_1 score of about 20%. It was also found empirically that forecasts with an S_1 score of 70% or more were useless as synoptic guidance.

Shuman (1989) pointed out some of the major system improvements that enabled NWP forecasts to overtake and surpass subjective forecasts. The first major improvement took place in 1958 with the implementation of a barotropic (one-level) model, which was actually a reduction from the three-level model first tried, but which included better finite differences and initial conditions derived from an objective analysis scheme (Bergthorsson and Döös, 1954; Cressman, 1959). It also extended the domain of the model to an octagonal grid covering the Northern Hemisphere down to 9 to 15°N. These changes resulted in numerical forecasts that for the first time were competitive with subjective forecasts, but in order to implement them JNWPU (later NMC) had to wait for the acquisition of a more powerful supercomputer, an IBM 704, replacing the previous IBM 701. This pattern of forecast improvements, which depend on a combination of better use of the data and better models, and would require more powerful supercomputers in order to be executed in a timely manner has been repeated throughout the history of operational NWP. Table 1 (adapted from Shuman, 1989) summarizes the major improvements in the first 30 years of operational numerical forecasts at the NWS. The first primitive equations model (Shuman and Hovermale, 1968) was implemented in 1966. The first regional system (Limited Fine Mesh or LFM model; Howcroft, 1971) was implemented in 1971. It was remarkable because it remained in use for over 20 years, and it was the basis for Model Output Statistics (MOS). Its development was frozen in 1986. A more advanced model and data assimilation system, the Regional Analysis and Forecasting System (RAFS) was implemented as the main guidance for North

*In 1960 the JNWPU divided into three organizations: the National Meteorological Center (National Weather Service), the Global Weather Central (U.S. Air Force), and the Fleet Numerical Oceanography Center (U.S. Navy).

TABLE 1 Major Operational Implementations and Computer Acquisitions at NMC between 1955 and 1985

Year	Operational Model	Computer
1955	Princeton three-level quasi-geostrophic model (Charney, 1954). Not used by the forecasters	IBM 701
1958	Barotropic model with improved numerics, objective analysis initial conditions, and octagonal domain	IBM 704
1962	Three-level quasi-geostrophic model with improved numerics	IBM 7090 (1960) IBM 7094 (1963)
1966	Six-layer primitive equations model (Shuman and Hovermale, 1968)	CDC 6600
1971	Limited-area fine mesh (LFM) model (Howcroft, 1971) (first regional model at NMC)	
1974	Hough functions analysis (Flattery, 1971)	IBM 360/195
1978	Seven-layer primitive equation model (hemispheric)	
1978	Optimal interpolation (Bergman, 1979)	Cyber 205
Aug 1980	Global spectral model, R30/12 layers (Sela, 1982)	
March 1985	Regional Analysis and Forecast System based on the Nested Grid Model (NGM; Phillips, 1979) and optimal interpolation (DiMego, 1988)	

Adapted from Shuman (1989).

America in 1982. The RAFS was based on the multiple Nested Grid Model (NGM; Phillips, 1979) and on a regional optimal interpolation (OI) scheme (DiMego, 1988). The global spectral model (Sela, 1982) was implemented in 1980.

Table 2 (from Kalnay et al., 1998) summarizes the major improvements implemented in the global system starting in 1985 with the implementation of the first comprehensive package of physical parameterizations from GFDL. Other major improvements in the physical parameterizations were made in 1991, 1993, and 1995. The most important changes in the data assimilation were an improved OI formulation in 1986, the first operational three-dimensional variational data assimilation (3D-VAR) in 1991, the replacement of the satellite retrievals of temperature with the direct assimilation of cloud-cleared radiances in 1995, and the use of "raw" (not cloud-cleared) radiances in 1998. The model resolution was increased in 1987, 1991, and 1998. The first operational ensemble system was implemented in 1992 and enlarged in 1994.

Table 3 contains a summary for the regional systems used for short-range forecasts (to 48 h). The RAFS (triple-nested NGM and OI) were implemented in 1985. The Eta model, designed with advanced finite differences, step-mountain coordi-

TABLE 2 Major Operational Implementations and Computer Acquisitions at NMC between 1955 and 1985

Year	Operational Model	Computer Acquisition
April 1985	GFDL physics implemented on the global spectral model with silhouette orography, R40/18 layers	
Dec. 1986	New optimal interpolation code with new statistics	
1987		2nd Cyber 205
Aug. 1987	Increased resolution to T80/18 layers, Penman-Montieth evapotranspiration and other improved physics (Caplan and White, 1989; Pan, 1989)	
Dec. 1988	Implementation of hydrostatic complex quality control (Gandin, 1988)	
1990		Cray YMP/8cpu/ 32 megawords
Mar. 1991	Increased resolution to T126 L18 and improved physics, mean orography (Kanamitsu et al., 1991)	
June 1991	New 3D variational data assimilation (Parrish and Derber, 1992; Derber et al., 1991)	
Nov. 1991	Addition of increments, horizontal, and vertical OI checks to the CQC (Collins and Gandin, 1990)	
Dec. 7, 1992	First ensemble system: one pair of bred forecasts at 00Z to 10 days, extension of AVN to 10 days (Toth and Kalnay, 1993; Tracton and Kalnay, 1993)	
Aug. 1993	Simplified Arakawa-Schubert cumulus convection (Pan and Wu, 1995). Resolution T126/28 layers	
Jan. 1994		Cray C90/16cpu/ 128 megawords
March 1994	Second ensemble system: 5 pairs of bred forecasts at 00Z, 2 pairs at 12Z, extension of AVN, a total of 17 global forecasts every day to 16 days	
Jan. 10, 1995	New soil hydrology (Pan and Mahrt, 1987), radiation, clouds, improved data assimilation; reanalysis model	

(*continued*)

TABLE 2 (*Continued*)

Year	Operational Model	Computer Acquisition
Oct. 25, 1995	Direct assimilation of TOVS cloud-cleared radiances (Derber and Wu, 1997). New PBL based on nonlocal diffusion (Hong and Pan, 1996); improved CQC	Cray C90/16cpu/ 256 megawords
Nov. 5, 1997	New observational error statistics; changes to assimilation of TOVS radiances and addition of other data sources	
Jan. 13, 1998	Assimilation of non cloud-cleared radiances (Derber et al., pers. comm.); improved physics	
June 1998	Resolution increased to T170/40 layers (to 3.5 days); improved physics; 3D ozone data assimilation and forecast; nonlinear increments in 3D; VAR	

Adapted from Shuman (1989).

TABLE 3 Major Changes in the NMC/NCEP Regional Modeling and Data Assimilation Since 1985

Year	Operational Model	Computer
March 1985	Regional Analysis and Forecast System (RAFS) based on the triply Nested Grid Model (NGM; Phillips, 1979) and optimal interpolation (OI; DiMego, 1988); resolution: 80 km/16 layers	Cyber 205
August 1991	RAFS upgraded for the last time: NGM run with only two grids with inner grid domain doubled in size; Implemented Regional Data Assimilation System (RDAS) with 3 hourly updates using an improved OI analysis using all off-time data including Profiler and ACARS wind reports (DiMego et al., 1992) and complex quality control procedures (Gandin et al., 1993)	Cray YMP 8 processors 32 megawords
June 1993	First operational implementation of the Eta model in the 00Z & 12Z early run for North America at 80 km and 38 layer resolution (Mesinger et al., 1988; Janjic, 1994; Black et al., 1994)	

(*continued*)

TABLE 3 (*Continued*)

Year	Operational Model	Computer
September 1994	The Rapid Update Cycle (RUC; Benjamin et al., 1994) was implemented for CONUS domain with 3 hourly OI updates at 60-km resolution on 25 hybrid (sigma-theta) vertical levels	Cray C-90 16 processors 128 megawords
September 1994	Early Eta analysis upgrades (Rogers et al., 1995)	
August 1995	Mesoscale version of the Eta model (Black, 1994) was implemented at 03Z and 15Z for an extended CONUS domain, with 29-km and 50-layer resolution and with NMC's first predictive cloud scheme (Zhao and Black, 1994) and new coupled land–surface–atmosphere package (2 layer soil)	Cray C-90 16 processors 256 megawords
October 1995	Major upgrade of early Eta runs: 48-km resolution, cloud scheme and Eta Data Assimilation System (EDAS) using 3 hourly OI updates (Rogers et al., 1996)	
January 1996	New coupled land–surface–atmosphere scheme put into early Eta runs (Chen et al., 1997; Mesinger, 1997)	
July–August 1996	Nested capability demonstrated with twice daily support runs for Atlanta Olympic Games with 10-km 60-layer version of Meso Eta	
February 1997	Upgrade package implemented in the early and Meso Eta runs	
February 1998	Early Eta runs upgraded to 32 km and 45 levels with 4 soil layers; OI analysis replaced by 3-dimensional variational (3D-VAR) with new data sources; EDAS now partially cycled (soil moisture, soil temperature, cloud water/ice, and turbulent kinetic energy)	
April 1998	RUC (3 hourly) replaced by hourly RUC II system with extended CONUS domain, 40-km and 40-level resolution, additional data sources and extensive physics upgrades	

TABLE 3 (*Continued*)

Year	Operational Model	Computer
June 1998	Meso runs connected to early runs as single 4 day system for North American domain at 32-km and 45-level resolution, 15z run moved to 18z, added new snow analysis; all runs connected with EDAS, which is fully cycled for all variables	

From compilations by Fedor Mesinger and Geoffrey DiMego, personal communication, 1998.

nates, and physical parameterizations, was implemented in 1993, with the same 80-km horizontal resolution as the NGM. It was denoted "early" because of a short data cutoff. The resolution was increased to 48 km, and a first "mesoscale" version with 29 km and reduced coverage was implemented in 1995. A cloud prognostic scheme was implemented in 1995, and a new land-surface parameterization in 1996. The OI data assimilation was replaced by a 3D-VAR analysis in 1998, and at this time the early and meso-Eta models were unified into a 32-km/ 45-level version. Many other less significant changes were also introduced into the global and regional operational systems and are not listed here for the sake of brevity. The Rapid Update Cycle (RUC), which provides frequent updates of the analysis and very short-range forecasts over the continental United States (CONUS), developed at NOAA's Forecast System Laboratory, was implemented in 1994 and upgraded in 1998 (Benjamin et al., 1994).

The 36-h S_1 forecast verification scores constitute the longest record of forecast verification available anywhere. They were started in the late 1940s for subjective surface forecasts, before operational computer forecast guidance, and for 500 hPa in 1954, with the first numerical forecasts. Figure 1 includes the forecast scores for 500 hPa from 1954 until the present, as well as the scores for the 72-h forecasts. It is clear that the forecast skill has improved substantially over the years, and that the current 36-h 500-hPa forecasts are close to a level that in the 1950s would have been considered "perfect" (Shuman, 1989). The 72-h forecasts have also improved and are now as accurate as the 36-h forecasts were about 15 years ago.

Figure 5 shows threat scores for precipitation predictions made by expert forecasters from the NCEP Hydrometeorological Prediction Center (HPC, the Meteorological Operations Division of the former NMC). The threat score (TS) is defined as the intersection of the predicted area of precipitation exceeding a particular threshold (P), in this case 0.5 inches in 24 h, and the observed area (O), divided by the union of the two areas: $\mathrm{TS} = (P \cap O)/(P \cup O)$. The bias (not shown) is defined by P/O. The TS, also known as critical success index (CSI) is a particularly useful score for quantities that are relatively rare. Figure 4 indicates that the forecasters' skill in predicting accumulated precipitation has been increasing with time, and that the current average skill in the 2-day forecast is as good as the 1-day forecasts were in the 1970s. Beyond the first 6 to 12 h, the forecasts are based mostly on numerical

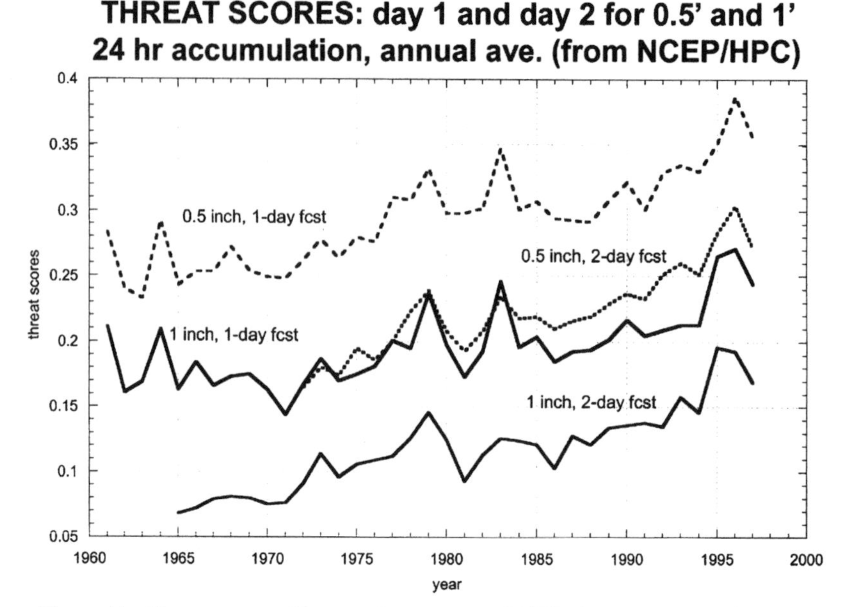

Figure 5 Threat scores of human forecasters at NCEP. (Data courtesy of J. Hoke.)

guidance, so that the improvement reflects, to a large extent, improvements of the numerical forecasts, which the human forecasters in turn improve upon based on their knowledge and expertise. The forecasters also have access to several model forecasts, and they use their judgment in assessing which one is more accurate in each case. This constitutes a major source of the "value-added" by the human forecasters.

The relationship between the evolution of human and numerical forecasts is clearly shown in a record compiled by the late F. Hughes (1987), reproduced in Figure 6. It is the first operational score maintained for the "medium-range" (beyond the first 2 days of the forecasts). The score used by Hughes was a standardized anomaly correlation (SAC), which accounted for the larger variability of sea level pressure at higher latitudes compared to lower latitudes. The fact that until 1976 the 3-day forecast scores from the model were essentially constant is an indication that their rather low skill was more based on synoptic experience than on model guidance. The forecast skill started to improve after 1977 for the 3-day forecast, and after 1980 for the 5-day forecast. Note that the human forecasts are on the average significantly more skillful than the numerical guidance, but it is the improvement in NWP forecasts that drives the improvements in the subjective forecasts.

5 THE FUTURE

The last decades have seen the expectations of Charney (1951) fulfilled and an amazing improvement in the quality of the forecasts based on NWP guidance.

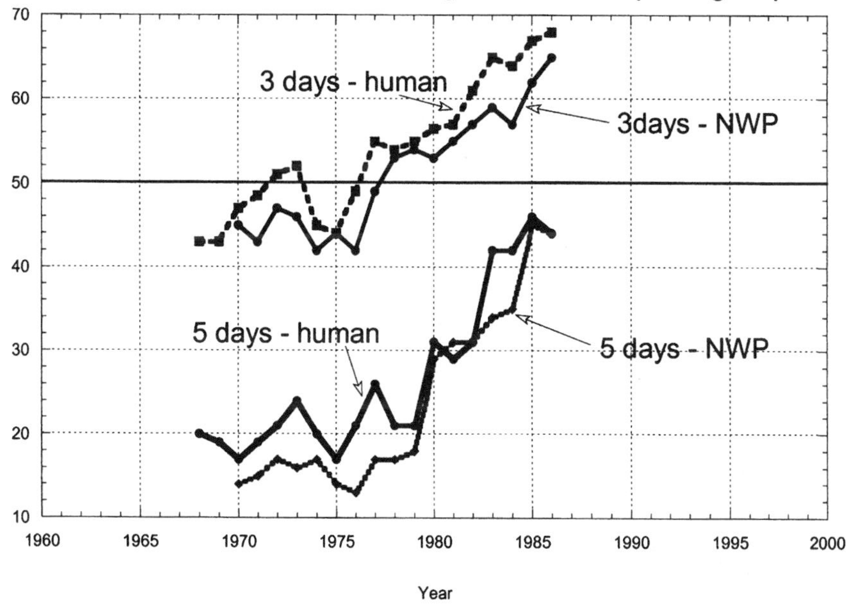

Figure 6 Hughes data: comparison of the forecast skill in the medium-range from NWP guidance and from human forecasters.

The next decade will continue seeing improvements, especially in the following areas:

- Detailed short-range forecasts, using storm-scale models able to provide skillful predictions of severe weather
- More sophisticated methods of data assimilation able to extract the maximum possible information from observing systems, especially remote sensors such as satellites and radars
- Development of adaptive observing systems, where additional observations are placed where ensembles indicate that there is rapid error growth (low predictability)
- Improvement in the usefulness of medium-range forecasts, especially through the use of ensemble forecasting
- Fully coupled atmospheric-hydrological systems, where the atmospheric model precipitation is appropriately scaled down and used to extend the length of river flow prediction
- More use of detailed atmosphere–ocean–land coupled models, where the effect of long-lasting coupled anomalies such as sea surface temperatures (SST) and

soil moisture anomalies leads to more skillful predictions of anomalies in weather patterns beyond the limit of weather predictability (about 2 weeks)

- More guidance to government and the public on areas such as air pollution, ultraviolet (UV) radiation and transport of contaminants, which affect health
- An explosive growth of systems with emphasis on commercial applications of NWP, from guidance on the state of highways to air pollution, flood prediction, guidance to agriculture, construction, etc.

REFERENCES

Andersson, E., J. Haseler, P. Undén, P. Courtier, G. Kelly, D. Vasiljevic, C. Brankovic, C. Cardinali, C. Gaffard, A. Hollingsworth, C. Jakob, P. Janssen, E. Klinker, A. Lanzinger, M. Miller, F. Rabier, A. Simmons, B. Strauss, J.-N. Thepaut, and P. Viterbo (1998). The ECMWF implementation of three-dimensional variational assimilation (3D-Var). III. Experimental results, *Quart. J. Roy. Meteor. Soc.* **124**, 1831–1860.

Baer, F., and J. Tribbia (1977). On complete filtering of gravity modes through non-linear initialization, *Mon. Wea. Rev.* **105**, 1536–1539.

Barnes, S. (1964). A technique for maximizing details in numerical map analysis, *J. Appl. Meteor.* **3**, 395–409.

Bengtsson, L. (1999). From short-range barotropic modelling to extended-range global weather prediction: A 40-year perspective, *Tellus* **51 (A-B)**, 13–32.

Benjamin, S. G., G. A. Grell, J. M. Brown, R. Bleck, K. J. Brundage, T. L. Smith, and P. A. Miller (1994). An operational isentropic/sigma hyrid forecast model and data assimilation system, in *Proceedings, The Life Cycles of Extratropical Cyclones*, Vol. III, Bergen, Norway, June 27–July 1, 1994, S. Gronas and M. A. Shapiro (Eds.), Geophysical Institute, Bergen, Norway, University of Bergen, pp. 268–273.

Bergthorsson, P., and B. Döös (1955). Numerical weather map analysis, *Tellus* **7**, 329–340.

Bergthorsson, P., B. Döös, S. Frykland, O Hang, and R. Linquist (1955). Routine forecasting with the barotropic model, *Tellus* **7**, 329–340.

Black, T. L. (1994). The new NMC mesoscale Eta Model: Description and forecast examples, *Wea. Forecasting* **9**, 265–278.

Bratseth, A. (1986). Statistical interpolation by means of successive corrections, *Tellus* **38A**, 439–447.

Caplan, P., and G. White (1989). Performance of the National Meteorological Center's Medium-Range Model, *Wea. Forecasting* **4**, 391–400.

Charney, J. G. (1951). *Dynamical Forecasting by Numerical Process. Compendium of Meteorology*, Boston, American Meteorological Society.

Charney, J. G. (1962). Integration of the primitive and balance equations, in Proc. of the International Symposium on Numerical Weather prediction, Nov. 1960, Tokyo, Meteorological Society of Japan.

Charney, J. G., R. Fjørtoft, and J. von Neuman (1950). Numerical integration of the barotropic vorticity equation, *Tellus* **2**, 237–254.

Collins, W. G. (1998). Complex quality control of significant level radiosonde temperatures, *J. Atmos. Oceanogra. Tech.* **15**, 69–79.

Collins, W. G., and L. S. Gandin (1990). Comprehensive hydrostatic quality control at the National Meteorological Center, *Mon. Wea. Rev.* **118**, 2752–2767.

Courtier, P., J.-N. Thepaut, and A. Hollingsworth (1994). A strategy for operational implementation of 4d-Var using an incremental approach, *Quart. J. Roy. Meteor. Soc.* **120**, 1367–1387.

Daley, R. (1991). *Atmospheric Data Analysis*, Cambridge University Press.

Derber, J. C., and W.-S. Wu (1998). The use of TOVS cloud-cleared radiances in the NCEP SSI analysis system, *Mon. Wea. Rev.* **126**, 2287–2302.

Derber, J. C., D. F. Parrish, and S. J. Lord (1991). The new global operational analysis system at the National Meteorological Center, *Wea. Forecasting* **6**, 538–547.

DiMego, G. J. (1988). The National Meteorological Center regional analysis system, *Mon. Wea. Rev.* **116**, 977–1000.

DiMego, G. J., K. E. Mitchell, R. A. Petersen, J. E. Hoke, J. P. Gerrity, J. J. Tuccillo, R. L. Wobus, and H. M. H. Juang (1992). Changes to NMC's regional analysis and forecast system, *Wea. Forecasting* **7**, 185–198.

Evensen, G., and P. J. Van Leeuwen (1996). Assimilation of GEOSAT altimeter data for the Aghulas current using the ensemble Kalman filter with a quasigeostrophic model, *Mon. Wea. Rev.* **124**, 85–96.

Gandin, L. S. (1963). Objective analysis of meterological fields, *Gidrometerologicheskoe Izdatelstvo*, Leningrad; English translation by Israeli Program for Scientific Translations, Jerusalem, 1965.

Gandin, L. S. (1988). Complex quality control of meteorological observations, *Mon. Wea. Rev.* **116**, 1137–1156.

Gandin, L. S., L. L. Morone, and W. G. Collins (1993). Two years of operational comprehensive hydrostatic quality control at the NMC, *Wea. Forecasting*, **8**(1), 57–72.

Gilchrist, B., and G. Cressman (1954). An experiment in objective analysis, *Tellus* **6**, 309–318.

Haltiner, G. J., and R. T. Williams (1980). *Numerical Prediction and Dynamic Meteorology*, New York, Wiley.

Hong, S. Y., and H. L. Pan (1996). Nonlocal boundary layer vertical diffusion in a medium-range forecast model, *Mon. Wea. Rev.* **124**, 2322–2339.

Howcroft, J. G. (1971). Local forecast model: Present status and preliminary verification. NMC Office note 50, National Weather Service, NOAA, US Dept of Commerce. Available from the National Centers for Environmental Prediction, 5200 Auth Rd., Rm. 100, Camp Springs, MD 20746.

Hughes, F. D. (1987). Skill of Medium-range Forecast Group, Office Note #326, National Meteorological Center, NWS, NOAA, US Dept of Commerce.

Janjic, Z. I. (1994). The step-mountain eta coordinate model: Further developments of the convection, viscous sublayer, and turbulence closure schemes, *Mon. Wea. Rev.* **122**, 927–945.

Kalnay, E. (2001). *Atmospheric Modeling, Data Assimilation and Predictability*, Cambridge University Press.

Kalnay, E. S., S. J. Lord and R. McPherson (1998). Maturity of operational numerical weather prediction: The medium range, *Bull. Am. Meteor. Soc.* **79**, 2753–2769.

Kanamitsu, M., J. C. Alpert, K. A. Campana, P. M. Caplan, D. G. Deaven, M. Iredell, B. Katz, H.-L. Pan, J. Sela, and G. H. White (1991). Recent changes implemented into the global forecast system at NMC, *Wea. Forecasting* **6**, 425–436.

Lorenc, A. (1981). A global three-dimensional multivariate statistical interpolation scheme, *Mon. Wea. Rev.* **109**, 701–721.

Lorenc, A. (1986). Analysis methods for numerical weather prediction, *Quart. J. Roy. Meteor. Soc.* **112**, 1177–1194.

Lorenz, E. N. (1963). Deterministic nonperiodic flow, *J. Atmos. Sci.* **20**, 130–141.

Lorenz, E. N. (1965). A study of the predictability of a 28-variable atmospheric model, *Tellus* **17**, 321–333.

Lorenz, E. N. (1982). Atmospheric predictability experiments with a large numerical model, *Tellus* **34**, 505–513.

Lorenz, E. N. (1993). *The Essence of Chaos*, University of Washington Press.

Lynch, P., and X.-Y. Huang (1994). Diabatic initialization using recursive filters, *Tellus* **46A**, 583–597.

McPherson, R. D., K. H. Bergman, R. E. Kistler, G. E. Rasch, and D. S. Gordon (1979). The NMC operational global data assimilation system, *Mon. Wea. Rev.* **107**, 1445–1461.

Mesinger, F. (1996). Improvements in quantitative precipitation forecasts with the Eta regional model at the National Centers for Environmental Prediction: The 48-km upgrade, *Am. Meteor. Soc.* **77**, 2637–2649.

Mesinger, F., Z. I. Janjic, S. Nickovic, D. Gavrilov, and D. G. Deaven (1988). The step-mountain coordinate: Model description and performance for cases of Alpine lee cyclogenesis and for a case of an Appalachian redevelopment, *Mon. Wea. Rev.* **116**, 1493–1518.

Pan, H.-L. (1990). A simple parameterization scheme of evapotranspiration over land for the NMC Medium-Range Forecast Model, *Mon. Wea. Rev.* **118**, 2500–2512.

Pan, H.-L., and L. Mahrt (1987). Interaction between soil hydrology and boundary-layer development, *Boundary-Layer Meteor.* **38**, 185–202.

Pan, H.-L., and W.-S. Wu (1995). Implementing a mass flux convection parameterization package for the NMC Medium-Range Forecast Model, Office note 409, National Meteorological Center.

Parrish, D. F., and J. D. Derber (1992). The National Meteorological Center spectral statistical interpolation analysis system, *Mon. Wea. Rev.* **120**, 1747–1763.

Phillips, N. A. (1956). The general circulation of the atmosphere, a numerical experiment, *Q. J. Roy Met. Soc.* **82**, 123–164.

Phillips, N. A. (1979). The nested grid model, NOAA, Tech. Report NWS 22, US Dept of Commerce, Washington DC. Available from the National Centers for Environmental Prediction, 5200 Auth Rd., Rm. 100, Camp Springs, MD 20746.

Richardson, L. F. (1922). *Weather Prediction by Numerical Process*, Cambridge University Press; reprinted by Dover (1965) with a new introduction by Sydney Chapman.

Robert, A. J. (1981). A stable numerical integration scheme for the primitive meteorological equations, *Atmosphere-Ocean* **19**, 35–46.

Rogers, E., T. L. Black, D. G. Deaven, G. J. DiMego, Q. Zhao, M. Baldwin, N. W. Junker, and Y. Lin (1996). Changes to the Operational Early Eta Analysis/Forecast System at the National Centers for Environmental Prediction, *Wea. Forecasting* **11**, 319–413.

Sasaki, Y. (1970). Some basic formalisms in numerical variational analysis, *Mon. Wea. Rev.* **98**, 875–883.

Sela, J. G. (1980). Spectral modeling at the National Meteorological Center, *Mon. Wea. Rev.* **108**, 1279–1292.

Shuman, F. G. (1989). History of numerical weather prediction at the National Meteorological Center, *Wea. Forecasting* **4**, 286–296.

Shuman, F. G., and J. B. Hovermale (1968). An operational six-layer primitive equation model, *J. Appl. Meteor.* **7**, 525–547.

Toth, Z., and E. Kalnay (1993). Ensemble forecasting at NMC: The generation of perturbations, *Bull. Am. Meteor. Soc.* **74**, 2317–2330.

Toth, Z., and E. Kalnay (1997). Ensemble forecasting at NCEP: The breeding method, *Mon. Wea. Rev.* **125**, 3297–3318.

Tracton, M. S., and E. Kalnay (1993). Ensemble forecasting at NMC: Practical aspects, *Wea. Forecasting* **8**, 379–398.

Zhao, Q., T. L. Black, and M. E. Baldwin (1997). Implementation of the cloud prediction scheme in the eta model at NCEP, *Wea. Forecasting* **12**, 697–711.

SECTION 2

THE CLIMATE SYSTEM

Contributing Editor: Robert E. Dickinson

CHAPTER 9

OVERVIEW: THE CLIMATE SYSTEM

ROBERT E. DICKINSON

The climate system consists of the atmosphere, cryosphere, oceans, and land interacting through physical, chemical, and biological processes. Key ingredients are the hydrological and energy exchanges between subsystems through radiative, convective, and fluid dynamical mechanisms. Climate involves changes on seasonal, year-to-year, and decadal or longer periods in contrast to day-to-day weather changes. However, extreme events and other statistical measures are as, or more, important than simple averages. Climate is seen to impact human activities most directly through the occurrence of extremes. The frequency of particular threshold extremes, as, for example, the number of days with maximum temperatures above 100°F, can change substantially with shifts in climate averages.

1 THE ATMOSPHERE

The atmosphere is described by winds, pressures, temperatures, and the distribution of various substances in gaseous, liquid, and solid forms. Water is the most important of these substances. Also important are the various other radiatively active ("greenhouse") gases, including carbon dioxide and liquid or solid aerosol particulates. Most of the mass of the atmosphere is in the troposphere, which is comprised of the layers from the surface to about 12 km (8 km in high latitudes to 16 km at the equator) where the temperature decreases with altitude. The top of the troposphere is called the tropopause. Overlying this is the stratosphere, where temperatures increase with altitude to about 50 km or so (Fig. 1). The tropospheric temperature decreases with altitude are maintained by vertical mixing driven by moist and dry convection.

Handbook of Weather, Climate, and Water: Dynamics, Climate, Physical Meteorology, Weather Systems, and Measurements, Edited by Thomas D. Potter and Bradley R. Colman.
ISBN 0-471-21490-6

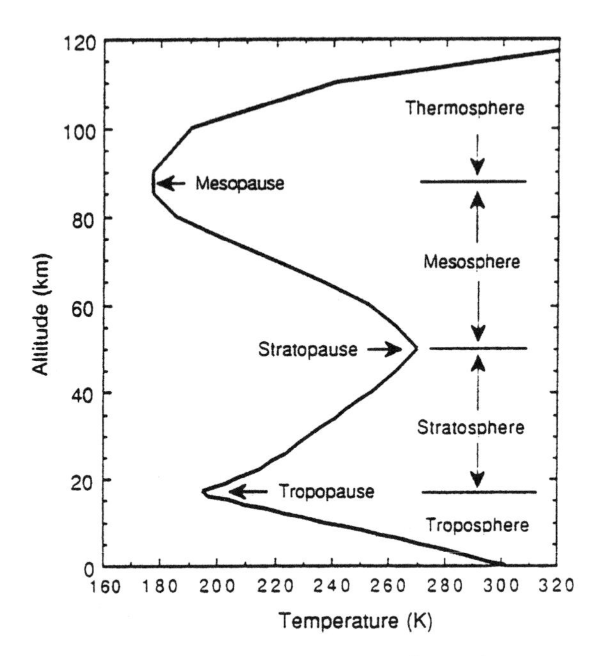

Figure 1 Main zones of the atmosphere defined according to the temperature profile of the standard atmosphere profile at 15°N for annual-mean conditions (Hartmann, 1994).

The temperature increases with altitude in the stratosphere in response to increasing heating per the unit mass by ozone absorption of ultraviolet radiation. The variation of temperature structure with latitude is indicated in Figure 2. The troposphere is deepest in the tropics because most thunderstorms occur there. Because of this depth and stirring by thunderstorms, the coldest part of the atmosphere is the tropical tropopause. In the lower troposphere temperatures generally decrease from the equator to pole, but warmest temperatures shift toward the summer hemisphere, especially in July. Longitudinally averaged winds are shown in Figure 3. Because of the geostrophic balance between wind and pressures, winds increase with altitude where temperature decreases with latitude. Conversely, above about 8 km, where temperatures decrease toward the tropical tropopause, the zonal winds decrease with altitude. The core of maximum winds is referred to as the jet stream. The jet stream undergoes large wavelike oscillations in longitude and so is usually stronger at a given latitude than in its longitudinal average. These waves are especially noticeable in the winter hemisphere as illustrated in Figure 4.

2 GLOBAL AVERAGE ENERGY BALANCE

Solar radiation of about $342\,\mathrm{W/m^{-2}}$ entering Earth's atmosphere is absorbed and scattered by molecules. The major gaseous absorbers of solar radiation are water vapor in the troposphere and ozone in the stratosphere. Clouds and aerosols likewise

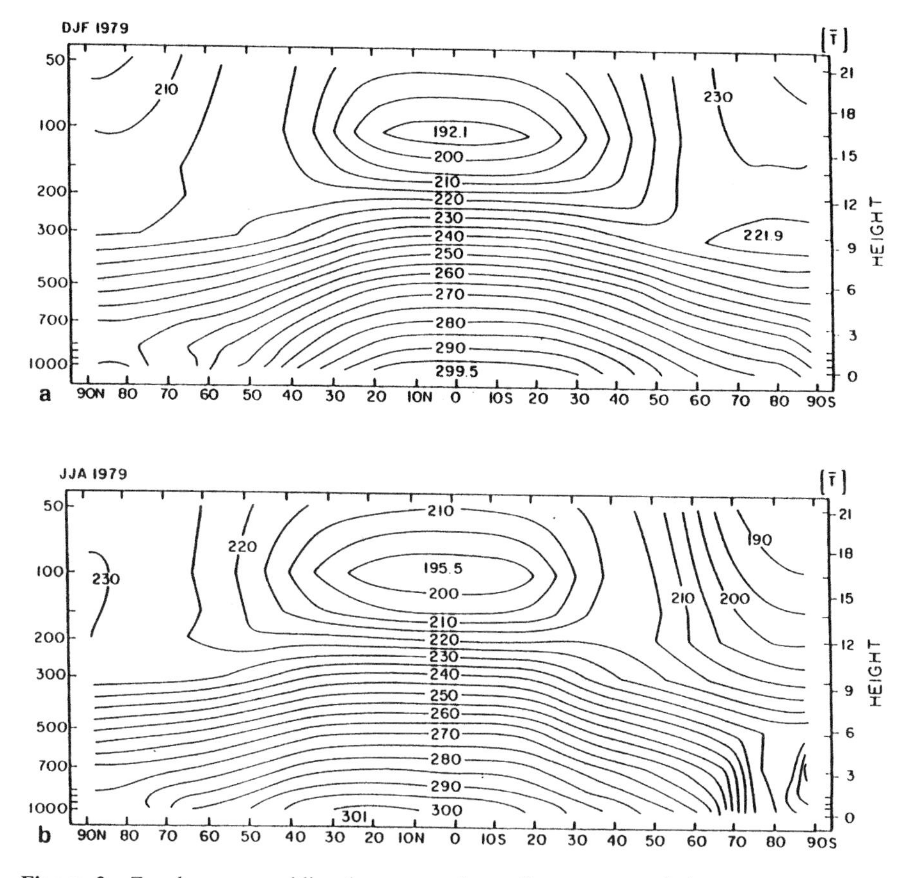

Figure 2 Zonal mean meridional cross sections of temperature during two seasons. (*a*) December 1978–February 1979 and (*b*) June–August 1979. Height scale is approximate and the contour interval is 5 K (Grotjahn, 1993).

scatter and absorb. Clouds are the dominant scatterer and so substantially enhance the overall planetary reflected radiation, whose ratio to incident solar radiation, about 0.31, is referred to as *albedo*. Thermal infrared radiation, referred to as *longwave*, is controlled by clouds, water vapor, and other greenhouse gases. Figure 5 (Kiehl and Trenberth, 1997) illustrates a recent estimate of the various terms contributing to the global energy balance. The latent heat from global average precipitation of about 1.0 m per year is the dominant nonradiative heating term in the atmosphere.

Because of the seasonally varying geometry of Earth relative to the sun, and the differences in cloudiness and surface albedos, there are substantial variations in the distribution of absorbed solar radiation at the surface and in the atmosphere, as likewise in the transfer of latent heat from the surface to the atmosphere. This heterogeneous distribution of atmospheric heating drives atmospheric wind systems, either directly or through the creation of available potential energy, which is utilized

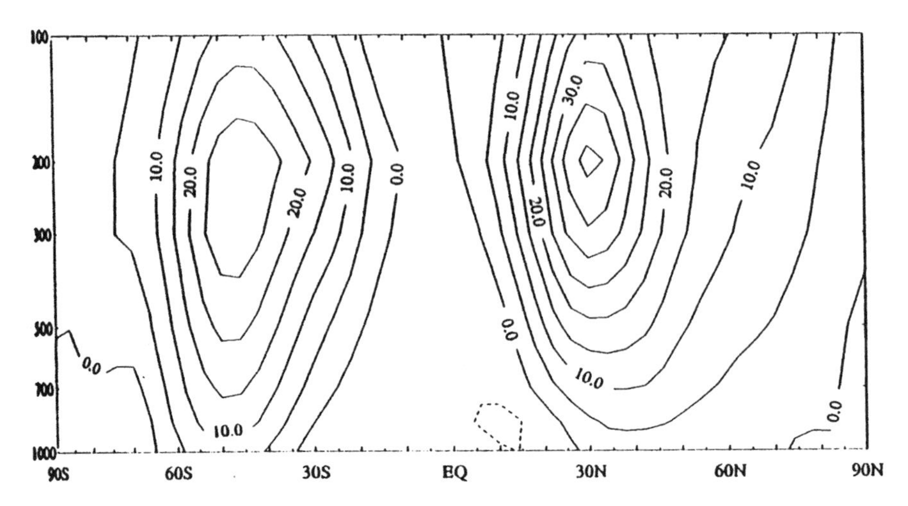

Figure 3 Meridional cross sections of longitudinally averaged zonal wind (top panels, m/s) for DJF (Holton, 1992).

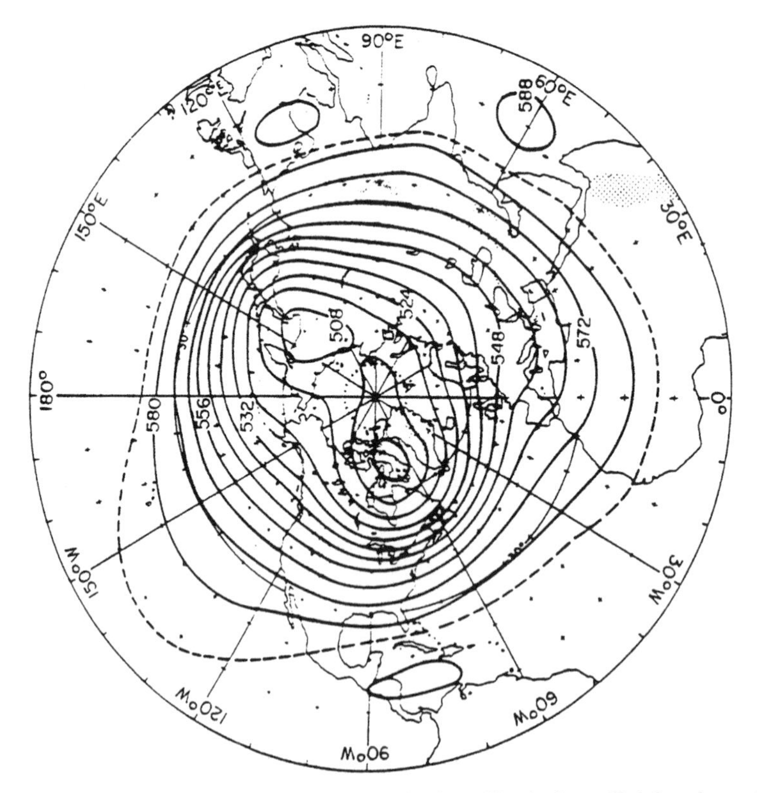

Figure 4 Mean 500-mb contours in January, Northern Hemisphere. Heights shown in tens of meters (Holton, 1992).

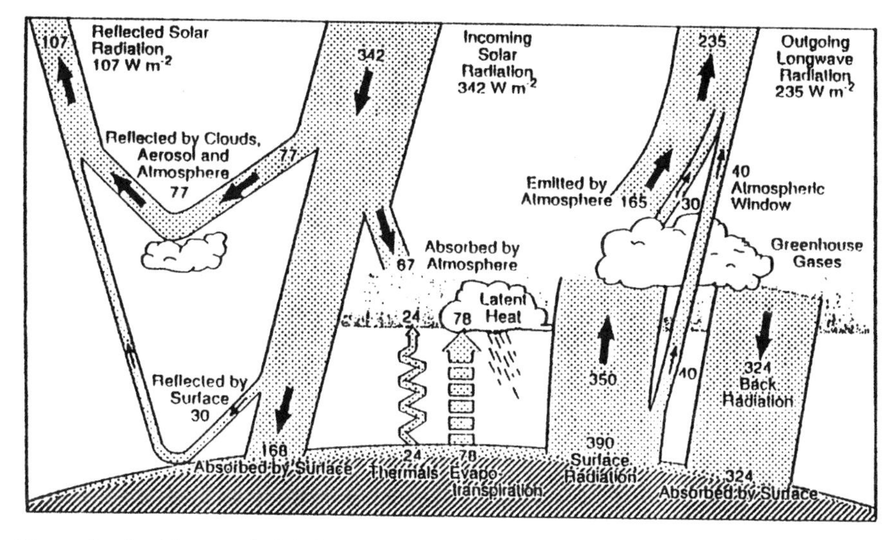

Figure 5 Earth's annual global mean energy budget based on the present study. Units are W/m^2 (Kiehl and Trenberth, 1997).

to maintain random occurrences of various kinds of instabilities, such as thunderstorms and wintertime cyclonic storm systems. These dynamical systems hence act to redistribute energy within the atmosphere and so determine the distributions of temperature and water vapor. Likewise, the balances at the surface between fluxes of radiative, latent, and thermal energies determine surface temperatures and soil moistures. The properties of the near-surface air we live in are determined by a combination of surface and atmospheric properties, according to processes of the atmospheric boundary layer. Thus climatic anomalies in surface air may occur either because of some shift in atmospheric circulation patterns or through some modification of surface properties such as those accompanying deforestation or the development of an urban area.

3 THE ATMOSPHERIC BOUNDARY LAYER

The term *boundary layer* is applied in fluid dynamics to layers of fluid or gas, usually relatively thin, determining the transition between some boundary and the rest of the fluid. The atmospheric boundary layer is the extent of atmosphere that is mixed by convective and mechanical stirring originating at Earth's surface. Such stirring is commonly experienced by airplane travelers as the bumps that occur during takeoff or landing, especially in the afternoon, or as waves at higher levels in flying over mountainous regions. The daytime continental boundary layer, extending up to several kilometers in height, is most developed and vigorously mixed, being the extent to which the daytime heating of the surface drives convective overturning of the atmosphere. The land cools radiatively at night, strongly stabilizing the atmosphere against convection, but a residual boundary layer extends up to about 100 m stirred by the flow of air over the underlying rough surface. This diurnal variation of fluxes over the ocean is much weaker and the boundary layer is of intermediate height. The temperature of the atmosphere, when stirred by dry mixing, decreases at a rate of 9.8 K/km. Above the boundary layer, temperatures decrease less rapidly with height, so that the atmosphere is stable to dry convection. A layer of clouds commonly forms at the top of the daytime and oceanic boundary layers and contributes to the convection creating the boundary layer through its radiative cooling (convection results from either heating at the bottom of a fluid or cooling at its top). Also, at times the clouds forming near the top of the boundary layer can be unstable to moist convection, and so convect upward through a deep column such as in a thunderstorm.

4 ATMOSPHERIC HYDROLOGICAL CYCLE

The storage, transport, and phase changes of water at the surface and in the atmosphere are referred to as the hydrological cycle. As already alluded to, the hydrological cycle is closely linked to and driven by various energy exchange processes at the surface and in the atmosphere. On the scale of continents, water is moved from oceans to land by atmospheric winds, to be carried back to the oceans by streams and rivers as elements of the land hydrological cycle. Most of the water in the atmosphere is in its vapor phase, but water that is near saturation vapor pressure (relative humidity of 100%) converts to droplets or ice crystals depending on temperature and details of cloud physics. These droplets and crystals fall out of the atmosphere as precipitation. The water lost is replenished by evaporation of water at the surface and by vertical and horizontal transport within the atmosphere. Consequently, much of the troposphere has humidities not much below saturation. Saturation vapor pressure increases rapidly with temperature (about 10% per kelvin of change). Hence, as illustrated in Figure 6, the climatological concentrations of water vapor vary from several percent or more when going from near-surface air to a few parts per million near the tropical tropopause. Water vapor concentrations in the stratosphere are close to that of the tropical tropopause, probably because much of the air in the lower stratosphere has been pumped through the tropical tropopause by moist convection.

5 CLIMATE OF THE STRATOSPHERE

The dominant radiative processes in the stratosphere are the heating by absorption of solar ultra violet (UV) radiation and cooling by thermal infrared emission from carbon dioxide and, to a lesser extent, ozone molecules. The stratospheric absorption of UV largely determines how much harmful UV reaches the surface. Ozone in the upper troposphere and lower stratosphere additionally adds heat by absorption of thermal emission from the warmer surface and lower layers. The stratosphere,

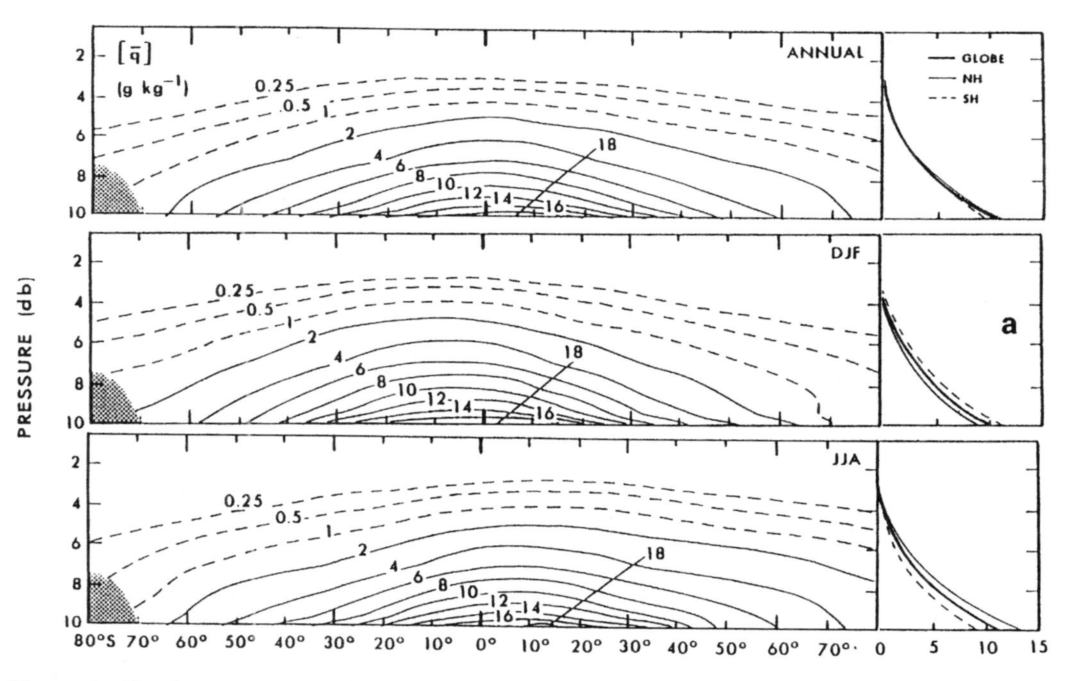

Figure 6 Zonal mean cross sections of the specific humidity in g/kg for annual, DJF, and JJA mean conditions. Vertical profiles of hemispheric and global mean values are shown on the right (Peixoto and Oort, 1992).

furthermore, enhances the greenhouse warming of CO_2 in the troposphere through substantial downward thermal emissions to the troposphere.

How changes of ozone change stratospheric and tropospheric radiative heating depends on the amounts of overlying ozone and, for thermal effect, on pressure and radiative upwelling depending on underlying temperatures.

Besides radiative processes, stratospheric climate is characterized by its temperature and wind patterns and by the chemical composition of its trace gases. At midstratosphere, temperature increases from winter pole to summer pole with an accompanying eastward jet stream in the winter hemisphere extending upward from the tropospheric jet steam. This wind configuration allows planetary wave disturbances to propagate into the stratosphere, contributing significant temporal and longitudinal variabilities. Conversely, the westward jet, found in the summer stratosphere attenuates wave disturbances from below, and so is largely zonally symmetric, changing only with the seasonal heating patterns.

6 THE CRYOSPHERE

The term *cryosphere* refers to the components of the climate system dominated by water in its frozen phase, that is, in high latitudes and extratropical winter conditions. Elements include snow, its distribution and depths, sea ice, its distribution and properties, high-latitude ice caps, and temperate glaciers. The largest volume of frozen water is stored in ice caps, and glaciers. This storage acts to remove water from the oceans. How it changes with climate change is, hence, of interest for determining changing sea levels.

Ice is highly reflective of sunlight, especially in crystal form. The loss of solar heating because of this high albedo acts to substantially reduce high-latitude temperatures especially in spring and early summer where near-maximum solar radiation sees white snow-covered surfaces. This high albedo can be substantially masked by cloud cover and, over land, tall vegetation such as conifer forests.

7 THE OCEAN

Oceans are a major factor in determining surface temperatures and fluxes of water into the atmosphere. They store, release, and transport thermal energy, in particular, warming the atmosphere under wintertime and high-latitude conditions, and cooling it under summer and tropical conditions.

How the oceans carry out these services depends on processes coupling them to the atmosphere. Atmospheric winds push the oceans into wind-driven circulation systems. Net surface heating or cooling, evaporation, and precipitation determine oceanic densities through controlling temperature and salinity, hence oceanic buoyancy. This net distribution of buoyancy forcing drives "thermohaline" overturning of the ocean, which acts to transport heat. Climate of the surface layers of the ocean includes the depth to which waters are stirred by waves and net heating or cooling.

Heating acts to generate shallow warm stable layers, while cooling deepens the surface mixed layers. Under some conditions, convective overturning of cold and/or high-salinity water can penetrate to near the ocean bottom.

REFERENCES

Grotjahn, R. (1993). Zonal average observations, Chapter 3, in *Global Atmospheric Circulations: Observations and Theories*, New York, Oxford University Press.

Hartmann, D. L. (1994). Atmospheric temperature, Chapter 1.2, in *Global Physical Climatology*, San Diego, Academic Press.

Holton, J. R. (1992). The observed structure of extratropical circulations, Chapter 6.1, in *An Introduction to Dynamic Meteorology*, San Diego, Academic.

Kiehl, J. T. and K. E. Trenberth (1997). Earth's annual global mean energy budget, *J. Clim.* **78**, 197–208.

Peixoto, J. P., and A. H. Oort (1992). Observed atmospheric branch of the hydrological cycle, Chapter 12.3, *Physics of Climate*, New York, American Institute of Physics.

CHAPTER 10

THE OCEAN IN CLIMATE

EDWARD S. SARACHIK

1 INTRODUCTION

Earth's present climate is intrinsically affected by the ocean—the climate without the ocean would be different in many essential ways: Without the evaporation of water from the sea surface, the hydrological cycle would be different; without ocean heat transport and uptake, the temperature distribution of the globe would be different; and without the biota in the ocean, the total amount of carbon in the atmosphere would be many time its current value. Yet, while we may appreciate the role of the ocean in climate, the difficulty and expense of making measurements below the ocean's surface has rendered the vast volume of the ocean a sort of *mare incognita.* Why is the ocean so important in Earth's climate, and which of its properties are of special significance for climate? How have we learned about the ocean and its role in climate, and what more do we need to know?

2 PROPERTIES OF THE OCEAN AND PROCESSES IN THE OCEAN

The ocean covers 70% of Earth's surface to an average depth of about 4 km. The mass of the ocean is 30 times and its heat capacity 120 times that of the atmosphere, and the ocean contains 80 times the carbon dioxide stored in the atmosphere.

The density of the ocean is controlled both by its temperature and by its salt content. Ocean density increases with salinity and decreases with temperature. Unlike fresh water, which has a maximum density at 4°C (so that colder water and ice float on 4°C water and the temperature at the bottom of a nonfrozen lake is 4°C), water saltier than 26 parts per thousand of water is continuously denser as

Handbook of Weather, Climate, and Water: Dynamics, Climate, Physical Meteorology, Weather Systems, and Measurements, Edited by Thomas D. Potter and Bradley R. Colman.
ISBN 0-471-21490-6

the temperature is lowered, and the temperature at the bottom of the ocean is closer to 1°C.

Since heat always diffuses from warm to cool temperatures, why does not the temperature of the deep ocean eventually adjust and become the same as its surface temperature? Cold water constantly sinks at high latitudes (both in the Northern and Southern Hemispheres) and fills the deeper parts of the oceans with cold water so that the water at depth is always cold, even when the surface temperature is very warm. This circulation is called the thermohaline circulation.

About 7% of the ocean surface is covered by sea ice. Growth of sea ice radically changes the nature of the ocean surface: Sea ice reflects solar radiation, thereby preventing it from being absorbed by the surface and blocks the transfer of heat and moisture from the surface of the ocean to the atmosphere.

The average salinity in the global oceans is 34.7 parts per thousand salt to water by weight. As the total amount of salt in the ocean is constant, changes in salinity only occur because of additions and subtractions of fresh water. Salinity decreases as rain falls on the ocean or river water enters the ocean, and it increases as water evaporates from the surface of the ocean. As sea ice grows, it rejects salt into the ocean thereby increasing its salinity. Similarly, as ice melts, it dilutes the surrounding ocean and lowers its salinity. A specific parcel of water can either increase or decrease its salinity by mixing with parcels with different salinities.

3 HOW THE OCEAN INTERACTS WITH THE ATMOSPHERE TO AFFECT THE CLIMATE

The ocean interacts with the atmosphere at (or very near) the sea surface where the two media meet. Visible light can penetrate into the ocean several tens of meters, but heat, moisture, and momentum, carbon dioxide, and other gases exchange directly at the surface. Sea ice forms at the surface and helps to determine the local exchanges. The basic problem of the ocean in climate is to explain these interchanges and to determine those characteristics of the ocean that affect these exchanges.

The ocean may be considered to interact with the atmosphere in two distinct ways: passively and actively. It interacts passively when the ocean affects the atmosphere but does not change the essential manner in which the atmosphere is operating. An example of a passive interaction is the oceanic response to adding CO_2 to the atmosphere where the ocean simply delays the greenhouse warming of the atmosphere as heat and CO_2 enters the ocean.

Active interaction with the atmosphere produces results that would not otherwise be there—an example is where the warming of the atmosphere reduces the thermohaline circulation and produces a climate reaction that could not have been obtained without the essential interaction of the ocean. In particular, since the northern branch, say, of the thermohaline circulation brings cold water from high latitudes toward the equator, and since the water must be replaced by warm water that is cooled as it moves northward, the net effect of the thermohaline circulation is to transport heat northward and thereby warm the higher latitudes. As the atmosphere warms, the water becomes less dense both by the effect of temperature and by increased rainfall, a necessary concomitant of global warming. Since the atmosphere sees a reduced north–south temperature gradient at the sea surface, it reacts fundamentally differently than if the thermohaline circulation were at full strength.

Our present models of greenhouse warming have the ocean acting in both the active and passive modes—active when warming leads to a slowed thermohaline circulation and passive when heat and CO_2 simply enter the ocean surface and is therefore lost to the atmosphere. Another example of active interaction is El Niño, a phenomenon that would not exist were it not for the active interaction of the atmosphere and the ocean (see the chapter by Trenberth). The ocean also has been inferred (by examining the composition of ancient ice stored in the Greenland and Antarctic ice sheets) to have, and probably actively take part in causing, climatic variability on time scales anywhere from decades to a few thousand years, a type of variability not seen on Earth since it emerged from the last glacial maximum some 18,000 years ago.

4 MEASURING THE OCEAN

The ocean is remarkably poorly measured. While the global atmosphere is constantly probed and analyzed for the purposes of weather prediction, until recently no such imperative existed for the ocean. Our ability to measure the ocean is severely limited basically by the inability of radiation to penetrate very far into the ocean—this requires direct in situ measurements of the interior of the ocean. The world's oceanographic research fleet is small and incapable of monitoring the breadth and depth of the world's ocean, although valuable research measurements are constantly being taken at selected places. As a result of ocean observations, we know the basic pathways of water in the ocean, we have a good idea of the transports by the ocean, we have some idea of the basic mechanisms for much of the major ocean currents, and we have a good idea of how the exchanges at the surface are accomplished. Yet we cannot measure vertical velocity (it is far too small), and so we remain completely ignorant of the processes by which the interior of the ocean affects the surface. Similarly, we are ignorant of the basic processes of mixing and friction in the ocean, both basic to being able to model the ocean for climate purposes.

Major oceanographic programs have been conducted in the last decade (the World Ocean Circulation Experiment, WOCE) and (the Tropical Ocean–Global Atmosphere, TOGA), and while they have taught us much about the ocean circulation and El Niño, respectively, the basic lesson is that, unless we can make continuous long-term measurements beneath the surface of the ocean, it will forever remain unknown territory.

5 MODELING THE OCEAN

Because the ocean is poorly measured, and because much of what we need to know about the past and predict about the future cannot be directly known, only inferred, models have played a particularly important role in the development of oceanography and, in particular, the role of the ocean in climate.

The basic tool of climate studies is the coupled model, where the various components of the climate system—the atmosphere, ocean, cryosphere, biosphere, and chemosphere—are simultaneously and consistently coupled. The ocean component of such a coupled model exchanges its heat, fresh water, and momentum with the atmosphere at the sea surface. The test of the successful coupling of the atmosphere and ocean is the correct simulation of the time-varying sea surface temperature and surface winds, both of which are relatively easy to measure: Directly by ship or mooring, remotely by satellite, or by a combination of the two.

The development of off-line ocean-only models requires the heat, momentum, and freshwater forcing from the atmosphere to be known. Since precipitation and evaporation, in particular, are so poorly measured over the ocean, it is a continual struggle to know whether errors in the ocean model are due to errors in the model itself or errors in the forcing of the ocean by the atmosphere.

Ocean models themselves are relatively simple in concept: The known equations of water and salt are discretized and time stepped into the future. The discretization process requires a trade-off between fine resolution for accuracy and the need to simulate over long periods of time, which, because of limited computer resources, requires coarser resolution. While the equation of state of seawater relating density to salt, temperature, and pressure cannot be written down simply, it has, over the course of time, become known to high accuracy.

What makes ocean modeling difficult is the specification of those mixing processes that unavoidably cannot be resolved by whatever resolution is chosen. We are beginning to understand that enhanced small-scale mixing occurs near bottom topography and near boundaries: Purposeful release experiments, where a dye is released and then followed in time to see how the dye cloud evolves, has revealed this to us. Larger scale mixing, where parcels are interchanged because of the large-scale circulation (but still unresolved by the ocean models) itself is more problematic, but recent advances in parameterizing these unresolved mixing effects have shown promise.

6 THE FUTURE OF THE OCEAN IN CLIMATE

It is clear that the ocean is a crucial component of the climate system. Since so much of what is not known about the past and future of the climate system depends on active interactions with the ocean, it is clear that we have to learn more about its essential processes. How to go about learning about the ocean is the difficult question.

Direct measurements are very expensive, and satellites, while giving a global look at the entire ocean, see only its surface. Designs are currently underway for a Global Ocean Observing System (GOOS), but the cost of implementing such a system in toto is prohibitive, even if shared among the wealthier countries of the world.

It is likely that a combination of studies, perhaps conducted for entirely different purposes, will advance the field most rapidly. In particular, the advent of the El Niño–Southern Oscillation (ENSO) prediction, which requires subsurface ocean data as initial conditions, has made almost permanent the Tropical Atmosphere-Ocean (TAO) array in the tropical Pacific, giving an unprecedented and continuous view of a significant part of the tropical ocean. We may extend the reasoning to say that, where predictability is indicated and shows societal or economic value, the measurement systems to produce the initial data will almost certainly be implemented. The promise of predicting climate from seasons to a few years will expand the ocean-observing system considerably. Additional expansions will come from resource monitoring, pollution monitoring, and various types of monitoring for national security purposes. While monitoring for security has traditionally meant the data is classified, once taken, data can eventually reach the research arena—the vast amount of Soviet and U.S. data that was declassified after the end of the cold war has shown this.

Observations can be also combined with models to give "value-added" observations. Data at individual points in the ocean exist without reference to neighboring points unless models are used to dynamically interpolate the data to neighboring points using the equation of motion of a fluid. This so-called four-dimensional data assimilation is in the process of development and shows promise as a powerful way of optimally using the ocean data that can be taken.

Models can also be compared with other models. While this might seem sterile, fine-resolution models can be used to develop parameterizations of large-scale mixing for use in coarse-resolution ocean models that can be run the long times needed to participate in coupled model simulations of climate. Advances in computer power will ultimately allow successive refinements in resolution so that finer scale resolution models can be run directly.

We close by reemphasizing the crucial role that the ocean plays in climate and climate variability and the necessity to know more about the ocean for all aspects of the climate problem.

CHAPTER 11

PROCESSES DETERMINING LAND SURFACE CLIMATE

GORDON BONAN

1 INTRODUCTION

Energy is continually flowing through the land–atmosphere system. As the sun's radiation passes through the atmosphere, some of it is absorbed, primarily by water vapor and clouds, and some is reflected back to space by clouds and particles suspended in the air. The remainder reaches Earth's surface, where it is either absorbed or reflected upwards. The solar radiation absorbed by the surface provides the warmth needed to maintain life and is used in biological activities such as photosynthesis. In turn, it provides energy to warm the atmosphere. Its surface emits radiation in the infrared waveband in proportion to its temperature raised to the fourth power. Most of this longwave radiation is absorbed by water vapor, clouds, carbon dioxide, and other gases in the atmosphere, heating the atmosphere. Without this heating, Earth's effective temperature would be 33°C cooler than it is now.

The solar and longwave radiation absorbed by Earth's surface constitute the net radiation at the surface. This energy is either stored or returned to the atmosphere as sensible or latent heat. Objects that absorb radiation become warmer than their surroundings and lose some of that energy by convection. For example, heat will normally be lost from a warm surface to the cooler air surrounding it. The transfer of this heat determines the temperature of air and is called sensible heat because it can be felt. Heat is also lost from the surface by evapotranspiration. Evapotranspiration determines the amount of water in the atmosphere, but it also cools the surface because the change of water from liquid to gas requires a large amount of heat,

Handbook of Weather, Climate, and Water: Dynamics, Climate, Physical Meteorology, Weather Systems, and Measurements, Edited by Thomas D. Potter and Bradley R. Colman.
ISBN 0-471-21490-6

which is transferred from the surface to the atmosphere as latent heat. The net radiation at the surface that is not returned to the atmosphere as sensible or latent heat is stored at the surface. Heat storage is very important for the diurnal cycle of temperature over land. Soils have a much smaller heat capacity than water. Consequently, land heats and cools faster than water.

Surface properties determine these energy fluxes and the resulting surface climate. Forests are "darker", hence absorbing more solar radiation, than grasslands. Forests are also taller—that is "rougher"—than shrubs or grasses and exert more stress on the fluid motion of the atmosphere. Deserts and shrublands, with their dry soils, have less evapotranspiration than well-watered forests or crops. Observational studies and the advent of sophisticated mathematical models of atmospheric physics, atmospheric dynamics, and surface ecological and hydrological processes have allowed scientists to examine how these different surfaces affect climate. Numerous studies all point to the same conclusion: The distribution of vegetation on Earth's surface is an important determinant of regional and global climates; consequently, natural and human-induced changes in Earth's surface can alter the climate.

Tropical deforestation is one example of the way in which human alterations of the natural landscape are changing climate. Climate model experiments show that the replacement of tropical forests with pastures causes a warmer, drier climate. Desertification, in which deserts encroach into forest landscapes, also results in a warmer, drier climate. Conversely, vegetation expansion into deserts, as happened 6000 years ago in North Africa, causes a cooler, wetter climate. By masking the high albedo of snow, the boreal forest creates a warmer climate compared to simulations in which the boreal forest is replaced with tundra vegetation. In Asia, monsoon circulations are created by land–sea temperature contrasts. A high land albedo, such as occurs when there is increased snow cover in Tibet, cools the surface, decreasing the land–sea temperature contrast and causing less rainfall.

2 SURFACE ENERGY FLUXES AND TEMPERATURE

The radiation that impinges on a surface or object must be balanced by the energy re-radiated back to the atmosphere, energy lost or gained as sensible and latent heat, and heat storage. More formally, the energy balance at the surface is

$$(1 - r) \cdot S\downarrow + a \cdot L\downarrow = L\uparrow + H + \lambda \cdot E + G$$

The first term in this equation, $(1 - r) \cdot S\downarrow$, is the solar radiation absorbed by the surface. $S\downarrow$ is the radiation onto the surface and r is the albedo, which is defined as the fraction of $S\downarrow$ that is reflected by the surface. The remainder, $(1 - r)$, is absorbed by the surface. The second term, $a \cdot L\downarrow$, is the atmospheric longwave radiation absorbed by the surface, where a is the fraction of the incoming radiation $L\downarrow$ that is absorbed. Together, the absorbed solar radiation and longwave radiation comprise the radiative forcing, Q_a. This must be balanced by:

1. Longwave radiation emitted by the surface ($L\uparrow$) in proportion to its absolute temperature, in kelvins, raised to the fourth power.
2. Energy transferred to or from the surface by convection (H). This sensible heat flux is directly proportional to the temperature difference between the surface and air and inversely proportional to a transfer resistance.
3. Energy used to evaporate water ($\lambda \cdot E$). The latent heat flux is directly proportional to the vapor pressure difference between the surface and air and is inversely proportional to a transfer resistance.
4. Energy stored in the soil via conduction (G). When the ground beneath the surface is colder than the surface, heat is conducted into the ground. At night, when the surface is colder than the ground, heat is transferred from the ground to warm the surface.

These energy flows at the surface must be balanced. This balance is maintained by changing the surface temperature. For example, the temperature of a surface will rise as more radiation is received on the surface. As a result, more energy is returned to the atmosphere as longwave radiation and as sensible heat. More energy is stored in the ground via conduction. As the latent heat flux increases, the surface temperature cools. During the day, when the surface is warmer than the air, this decreases the sensible heat flux. If the underlying soil is a good conductor of heat, and there is a large soil heat flux, the surface will not be as hot and the sensible heat flux will decrease to compensate for the increased heat storage. Conversely, if the soil is a poor conductor of heat, little heat will be transferred from the surface to the soil and the surface will be hot.

The importance of these energy fluxes in determining surface temperature can be illustrated with a simple example. Suppose a leaf has a radiative forcing of 1000, 700, and 400 W/m^2, which are representative of values for a clear sky at mid-day, a cloudy sky at mid-day, and at night, when solar radiation is zero and the surface receives only longwave radiation. If the only means to dissipate this energy is through re-radiation (i.e., $H=0$ and $\lambda \cdot E=0$) and there is no heat storage ($G=0$), the leaf surface would attain temperatures of 91, 60 and 17°C with the high, moderate, and low radiative forcings (Table 1). When heat loss by convection

TABLE 1 Temperatures of Well-Watered Leaf for Radiative Forcings[a]

	Temperature (°C)						
		$L\uparrow + H$			$L\uparrow + H + \lambda E$		
Q_a (W/m^2)	$L\uparrow$	0.1 m/s	0.9 m/s	4.5 m/s	0.1 m/s	0.9 m/s	4.5 m/s
1000	91	53	39	34	39	33	31
700	60	40	34	32	32	29	29
400	17	26	28	29	23	26	27

[a]Air temperature is 29°C, relative humidity is 50%, and wind speeds are 0.1, 0.9, and 4.5 m/s.

is included, leaf temperature depends on wind speed because the transfer resistance decreases with high wind speeds. Under calm conditions, with a wind speed of 0.1 m/s, leaf temperature decreases by 38°C with the high radiative forcing and by 20°C with the moderate forcing (Table 1). Higher wind speeds lead to even cooler temperatures. At the low radiative forcing, convection warms the leaf because it is colder than the surrounding air and heat is transferred from the air to the leaf. Latent heat exchange is also a powerful means to cool a surface because of the large amount of energy required to evaporate water. For a well-watered leaf, under calm conditions and high radiative forcing, evapotranspiration cools the leaf an additional 14°C to a temperature of 39°C (Table 1). Higher winds result in even lower temperatures.

3 HYDROLOGIC CYCLE

As the preceding example shows, evapotranspiration is an effective means to cool a surface, particularly at high radiative forcing and low wind speed. The rate of latent heat loss depends on the amount of water present. A well-watered site has more water to evaporate than a dry site. Because more energy goes into latent heat rather than sensible heat, the lower atmosphere is likely to be cool and moist. A dry surface, on the other hand, has low latent heat flux, high sensible heat flux, and the air is likely to be warm and dry. Typical values of the Bowen ratio (the ratio of sensible to latent heat) are: 0.1 to 0.3 for tropical rain forests, where high annual rainfall keeps the soil wet; 0.4 to 0.8 for temperate forests and grasslands, where less rainfall causes drier soils; 2.0 to 6.0 for semiarid regions; and greater than 10.0 for deserts.

The water stored on land (ΔW) is the difference between water input as precipitation (P) and water loss via evapotranspiration (E) and runoff (R):

$$\Delta W = P - E - R$$

In many regions, water is stored as snow in winter and not released to the soil until the following spring. Snow is climatologically important because its high albedo reflects a majority of the solar radiation. Snow has a low thermal conductivity, and a thick snow pack insulates the soil. In spring, a large portion of the net radiation at the surface is used for snow melt, preventing the surface from warming. Precipitation is greatly modified by vegetation. Some of the rainfall is intercepted by foliage, branches, and stems. The remainder reaches the ground as throughfall, stemflow, and snowmelt.

Only a portion of the liquid water reaching the ground surface infiltrates into the soil. The actual amount depends on soil wetness, soil type, and the intensity of the water flux. Table 2 shows representative hydraulic conductivity when the soil is saturated. Sandy soils, because of their large pore sizes, can absorb water at fast rates. Loamy soils absorb water at slower rates. Clay soils, because of their small pores, have the lowest hydraulic conductivity. The water that does not infiltrate into

TABLE 2 Hydraulic Conductivity at Saturation and Water Contents

	Volumetric Water Content (mm^3/mm^3)			Hydraulic Conductivity (mm/s)
	Wilting Point	Field Capacity	Saturation	
Sand	0.07	0.23	0.40	0.176
Sandy loam	0.11	0.32	0.44	0.035
Loam	0.15	0.39	0.45	0.007
Silty clay loam	0.22	0.42	0.48	0.002
Clay	0.29	0.45	0.48	0.001

the soil accumulates in small depressions or is lost as surface runoff, which flows overland to streams, rivers, and lakes.

The water balance of the soil is the difference between water input via infiltration and water loss from evapotranspiration and subsurface drainage. Since the latent heat flux decreases as soil becomes drier, the amount of water in the soil is a crucial determinant of the surface climate. Two hydraulic parameters determine the amount of water a soil can hold. Field capacity is the amount of water after gravitational drainage. Sandy soils, because of their large pores, hold less water at field capacity than clay soils, with their small pores (Table 2). Wilting point is the amount of water in the soil when evapotranspiration ceases. Because water is tightly bound to the soil matrix, clay soils have very high wilting points. The difference between field capacity and wilting point is the available water holding capacity. Loamy soils hold the most water; sands and clays hold the least amount of water.

4 VEGETATION

The ecological characteristics of the surface vary greatly among vegetation types. Some plants absorb more solar radiation than others; some plants are taller than others; some have more leaves; some have deeper roots. For example, the albedo of coniferous forests generally ranges from 0.10 to 0.15; deciduous forests have albedos of 0.15 to 0.20; grasslands have albedos of 0.20 to 0.25. As albedo increases, the amount of solar radiation absorbed at the surface decreases and if all other factors were equal the surface temperature would decrease. However, plants also vary in height. Trees are taller than shrubs, which are taller than grasses. Taller surfaces are "rougher" than shorter surfaces and exert more stress on atmospheric motions. This creates more turbulence, increasing the transfer of sensible and latent heat away from the surface. Plants also increase the surface area from which sensible and latent heat can be transferred to the atmosphere. The leaf area of plant communities is several times that of the underlying ground. A typical ratio of leaf area to ground area (called the leaf area index) is 5 to 8. Plant communities differ greatly in leaf area index. A dry, unproductive site supports much less foliage than a moist, nutrient-rich grassland or forest. The rooting depth of plants is important because this determines how deep in the soil water can be obtained for transpiration. Shallow rooted plants have a much smaller volume of water for evapotranspiration than deep rooted plants.

Leaves have microscopic pores, called stomata, through which they absorb carbon dioxide from the atmosphere during photosynthesis. These pores open and close in response to environmental factors such as light, temperature, atmospheric CO_2 concentration, and soil water. When they are open, the plant absorbs CO_2; but water also diffuses out of the leaf to the surrounding air—a process known as transpiration. Plants differ greatly in their stomatal physiology, especially responses to environmental factors. Some plants photosynthesize, and hence have open stomata, at lower light levels than others. Plants differ in the water content at which stomata close. They differ in optimum temperatures for photosynthesis, and they differ in their responses to increasing CO_2 concentration. These different stomatal physiologies contribute to variations in latent heat flux, and hence surface temperature, among vegetation types.

5 COUPLING TO ATMOSPHERIC MODELS

Different surfaces, with different soil and vegetation types, can create vastly different climates. These differences can be examined by coupling models of surface energy fluxes, which depend on the hydrological and ecological state of the land, with atmospheric numerical models. The surface model provides to the atmospheric model albedo and emitted longwave radiation, which determine the net radiative heating of the atmosphere; sensible and latent heat fluxes, which determine atmospheric temperature and humidity; and surface stresses, which determine atmospheric winds. In turn, the atmospheric model provides the atmospheric conditions required to calculate these fluxes: temperature, humidity, winds, precipitation, and incident solar and longwave radiation.

Land surface models account for the ecological effects of different vegetation types and the hydrological and thermal effects of different soil types. Although the equations needed to model these processes at a single point are well understood, the scaling over large, heterogeneous areas comprised of many soils and many plant communities is much less exact. Moreover, when coupled to a global climate model that may be used to simulate the climate of thousand of points on the surface for tens to hundreds of years, there is a need to balance model complexity with computational efficiency.

CHAPTER 12

OBSERVATIONS OF CLIMATE AND GLOBAL CHANGE FROM REAL-TIME MEASUREMENTS

DAVID R. EASTERLING AND THOMAS R. KARL

1 INTRODUCTION

Is the planet getting warmer?
Is the hydrologic cycle changing?
Is the atmospheric/oceanic circulation changing?
Is the weather and climate becoming more extreme or variable?
Is the radiative forcing of the climate changing?

These are the fundamental questions that must be answered to determine if climate change is occurring. However, providing answers is difficult due to an inadequate or nonexistent worldwide climate observing system. Each of these apparently simple questions are quite complex because of the multivariate aspects of each question and because the spatial and temporally sampling required to address adequately each question must be considered on a global scale. A brief review of our ability to answer these questions reveals many successes, but points to some glaring inadequacies that must be addressed in any attempt to understand, predict, or assess issues related to climate and global change.

Handbook of Weather, Climate, and Water: Dynamics, Climate, Physical Meteorology, Weather Systems, and Measurements, Edited by Thomas D. Potter and Bradley R. Colman.
ISBN 0-471-21490-6

2 IS THE PLANET GETTING WARMER?

There is no doubt that measurements show that near-surface air temperatures are increasing. Best estimates suggest that the warming is around 0.6°C (+−0.2°C) since the late nineteenth century (IPCC, 2001). Furthermore, it appears that the decade of the 1990s was the warmest decade since the 1860s, and possibly for the last 1000 years. Although there remain questions regarding the adequacy of this estimate, confidence in the robustness of this warming trend is increasing (IPCC, 2001). Some of the problems that must be accounted for include changes in the method of measuring land and marine surface air temperatures from ships, buoys, land surface stations as well as changes in instrumentation, instrument exposures and sampling times, and urbanization effects. However, recent work evaluating the effectiveness of corrections of sea surface temperatures for time-dependent biases, and further evaluation of urban warming effects on the global temperature record have increased confidence in these results. Furthermore, by consideration of other temperature-sensitive variables, e.g., snow cover, glaciers, sea level and even some proxy non-real-time measurements such as ground temperatures from boreholes, increases our confidence in the conclusion that the planet has indeed warmed. However, one problem that must be addressed is that the measurements we rely upon to calculate global changes of temperature have never been collected for that purpose, but rather to aid in navigation, agriculture, commerce, and in recent decades for weather forecasting. For this reason there remain uncertainties about important details of the past temperature increase and our capabilities for future monitoring of the climate. The IPCC (2001) has summarized latest known changes in the temperature record, which are summarized in Figure 1.

Global-scale measurements of layer averaged atmospheric temperatures and sea surface temperatures from instruments aboard satellites have greatly aided our ability to monitor global temperature change (Spencer and Christy, 1992a,b; Reynolds, 1988), but the situation is far from satisfactory (Hurrell and Trenberth, 1996). Changes in satellite temporal sampling (e.g., orbital drift), changes in atmospheric composition (e.g., volcanic emissions), and technical difficulties related to overcoming surface emissivity variability are some of the problems that must be accounted for, and reduce the confidence that can be placed on these measurements (NRC, 2000). Nonetheless, the space-based measurements have shown, with high confidence, that stratospheric temperatures have decreased over the past two decades. Although perhaps not as much as suggested by the measurements from weather balloons, since it is now known that the data from these balloons high in the atmosphere have an inadvertent temporal bias due to improvements in shielding from direct and reflected solar radiation (Luers and Eskridge, 1995).

3 IS THE HYDROLOGIC CYCLE CHANGING?

The source term for the hydrologic water balance, precipitation, has been measured for over two centuries in some locations, but even today it is acknowledged that in

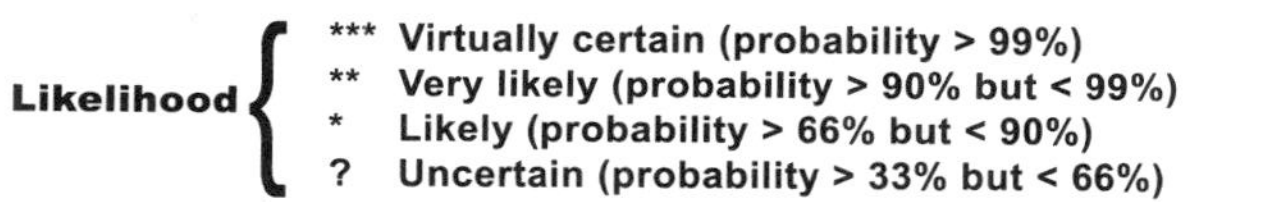

Figure 1 Schematic of observed variations of selected indictors regarding (*a*) temperature and (*b*) the hydrologic cycle (based on IPCC, 2001). See ftp site for color image.

many parts of the world we still cannot reliably measure true precipitation (Sevruk, 1982). For example, annual biases of more than 50% due to rain gauge undercatch are not uncommon in cold climates (Karl et al., 1995), and, even for more moderate climates, precipitation is believed to be underestimated by 10 to 15% (IPCC, 1992). Progressive improvements in instrumentation, such as the introduction of wind shields on rain gauges, have also introduced time-varying biases (Karl et al., 1995). Satellite-derived measurements of precipitation have provided the only large-scale ocean coverage of precipitation. Although they are comprehensive estimates of large-scale spatial precipitation variability over the oceans where few measurements exist, problems inherent in developing precipitation estimates hinder our ability to have much confidence in global-scale decadal changes. For example, even the landmark work of Spencer (1993) in estimating worldwide ocean precipitation using the microwave sounding unit aboard the National Oceanic and Atmospheric Administration (NOAA) polar orbiting satellites has several limitations. The observations are limited to ocean coverage and hindered by the requirement of an unfrozen ocean. They do not adequately measure solid precipitation, have low spatial resolution, and are affected by the diurnal sampling inadequacies associated with polar orbiters, e.g., limited overflight capability. Blended satellite/in situ estimates also show promise (Huffman et al., 1997); however, there are still limitations, including a lack of long-term measurements necessary for climate change studies.

Information about past changes in land surface precipitation, similar to temperature, has been compared with other hydrologic data, such as changes in stream flow, to ascertain the robustness of the documented changes of precipitation. Figure 1 summarizes some of the more important changes of precipitation, such as the increase in the mid to high latitude precipitation and the decrease in subtropical precipitation. Evidence also suggests that much of the increase of precipitation in mid to high latitudes arises from increased autumn and early winter precipitation in much of North America and Europe. Figure 2 depicts the spatial aspects of this change, reflecting rather large-scale coherent patterns of change during the twentieth century.

Other changes related to the hydrologic cycle are summarized in Figure 1. The confidence is low for many of the changes, and it is particularly disconcerting relative to the role of clouds and water vapor in climate feedback effects.* Observations of cloud amount long have been made by surface-based human observations and more recently by satellite. In the United States, however, human observers have been replaced by automated measurements, and neither surface-based or spaced-based data sets have proven to be entirely satisfactory for detecting changes in clouds. Polar orbiting satellites have an enormous difficulty to overcome related to sampling aliasing and satellite drift (Rossow and Cairns, 1995). For human observers changes in observer schedules, observing biases, and incomplete sampling have created major problems in data interpretations, now compounded by a change to new automated measurements at many stations. Nonetheless, there is still some confi-

*An enhancement or diminution of global temperature increases or decreases due to other causes.

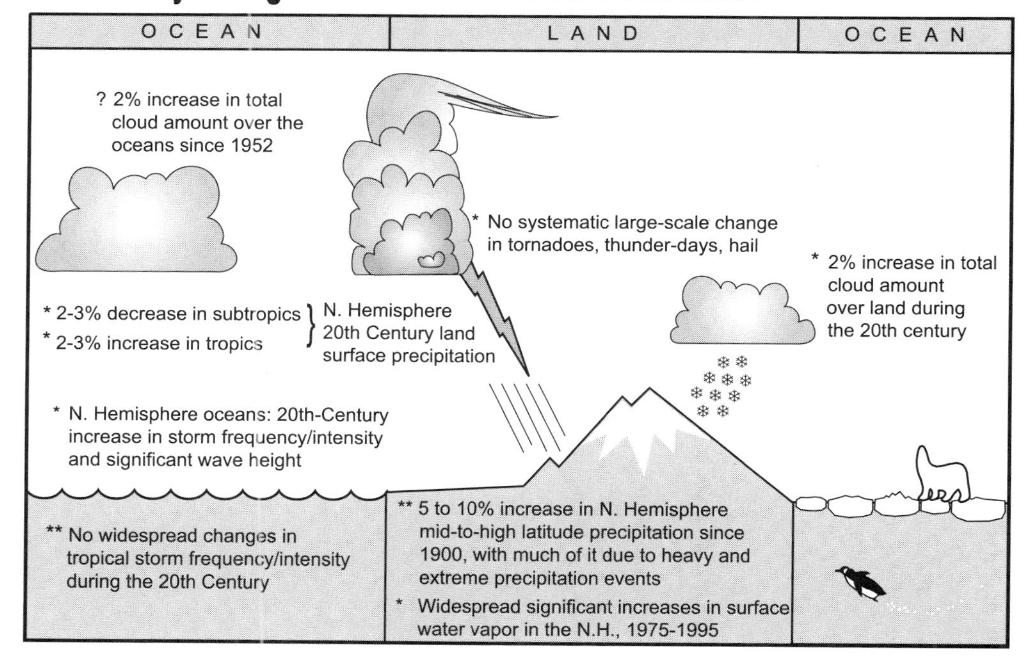

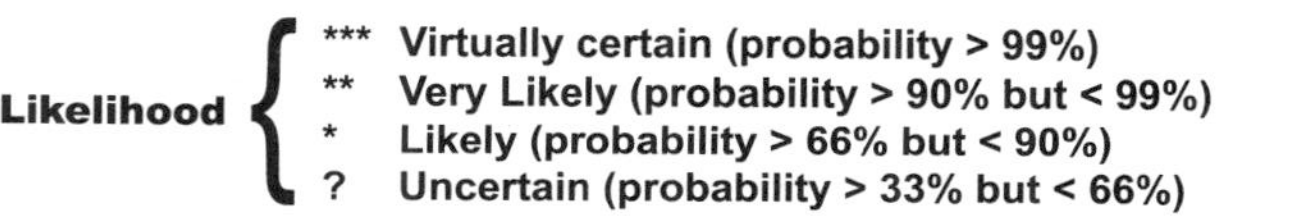

Figure 2 Precipitation trends over land 1900–1999. Trend is expressed in percent per century (relative to the mean precipitation from 1961–1990) and magnitude of trend is represented by area of circle with green reflecting increases and brown decreases of precipitation. See ftp site for color image.

dence (but low) that global cloud amounts have tended to increase. On a regional basis this is supported by reductions in evaporation as measured by pan evaporimeters over the past several decades in Russia and the United States, and a worldwide reduction in the land surface diurnal temperature range. Moreover, an increase in water vapor has been documented over much of North America and in the tropics (IPCC, 2001).

Changes in water vapor are very important for understanding climate change since water vapor is the most important greenhouse gas in the atmosphere. The measurement of changes in atmospheric water vapor is hampered by both data processing and instrumental difficulties for both weather balloon and satellite retrievals. The latter also suffers from discontinuities among successive satellites and errors introduced by changes in orbits and calibrations. Upper tropospheric water vapor is particularly important for climate feedbacks, but, as yet, little can be said about how it has varied over the course of the past few decades.

4 IS THE ATMOSPHERIC/OCEANIC CIRCULATION CHANGING?

Surprisingly, there is a considerable lack of reliable information about changes in atmospheric circulation, even though it is of daily concern to much of the world since it relates to day-to-day weather changes. Analyses of circulation are performed every day on a routine basis, but the analysis schemes have changed over time, making them of limited use for monitoring climate change. Moreover, even the recent reanalysis efforts by the world's major numerical weather prediction centers, whereby the analysis scheme is fixed over the historical record, contains time-varying biases because of the introduction of data with time-dependent biases and a changing mix of data (e.g., introducing satellite data) over the course of the reanalysis (Trenberth and Guillemot, 1997). Even less information is available on measured changes and variations in ocean circulation.

A few major atmospheric circulation features have been reasonably well measured because they can be represented by rather simple indices. This includes the El Niño–Southern Oscillation (ENSO) index, the North Atlantic Oscillation (NAO) index, and the Pacific–North American (PNA) circulation pattern index. There are interesting decadal and multidecadal variation, but it is too early to detect any long-term trends. Evidence exists that ENSO has varied in period, recurrence interval, and strength of impact. A rather abrupt change in ENSO and other aspects of atmospheric circulation seems to have occurred around 1976–1977. More frequent ENSOs with rare excursions into its other extreme (La Niña) became much more prevalent. Anomalous circulation regimes associated with ENSO and large-amplitude PNA patterns persisted in the North Pacific from the late 1970s into the late 1980s, affecting temperature anomalies. Moreover, the NAO has been persistent in its association with strong westerlies into the European continent from the late 1980s until very recently when it abruptly shifted. As a result, temperature anomalies and storminess in Europe have abruptly changed over the past 2 years compared to the past 7 or 8 years.

Increases in the strength of the Southern Hemisphere circumpolar vortex during the 1980s have been documented (van Loon et al., 1993; Hurrell and van Loon, 1994) using station sea level pressure data. This increase was associated with a delayed breakdown in the stratospheric polar vortex and ozone deficit in the Antarctic spring. A near-global sea level pressure data set has been used to identify changes in circulation patterns in the Indian Ocean. Allan et al. (1995) and Salinger et al. (1995) find that circulation patterns in the periods 1870–1900 and 1950–1990 were more meridional than those in the 1900–1950 period, indicating intensified circulation around anticyclones. These changes may be related to changes in the amplitude of longwave troughs to the south and west of Australia and the Tasman Sea/New Zealand area and a subsequent decrease in precipitation in Southwest Australia (Nicholls and Lavery, 1992; Allan and Haylock, 1993).

5 IS THE WEATHER AND CLIMATE BECOMING MORE EXTREME OR VARIABLE?

Climate and weather extremes are of great interest. Due to inadequate monitoring as well as prohibitively expensive access to weather and climate data held by the world's national weather and environmental agencies, only limited reliable information is available about large-scale changes in extreme weather or climate variability. The time-dependent biases that affect climate means are even more difficult to effectively eliminate from the extremes of the distributions of various weather and climate elements. There are a few areas, however, where regional and global changes in weather and climate extremes have been reasonably well documented (Easterling et al., 2000a).

Interannual temperature variability has not changed significantly over the past century. On shorter time scales and higher frequencies, e.g., days to a week, temperature variability may have decreased across much of the Northern Hemisphere (Karl and Knight, 1995). Related to the decrease in high-frequency temperature variability there has been a tendency for fewer low-temperature extremes, but widespread changes in extreme high temperatures have not been noted.

Trends in intense rainfall have been examined for a variety of countries. Some evidence suggests an increase in intense rainfalls (United States, tropical Australia, Japan, and Mexico), but analyses are far from complete and subject to many discontinuities in the record. The strongest increases in extreme precipitation are documented in the United States and Australia (Easterling et al., 2000b)

Intense tropical cyclone activity may have decreased in the North Atlantic, the one basin with reasonably consistent tropical cyclone data over the twentieth century, but even here data prior to World War II is difficult to assess regarding tropical cyclone strength. Elsewhere, tropical cyclone data do not reveal any long-term trends, or if they do they are most likely a result of inconsistent analyses. Changes in meteorological assimilation schemes have complicated the interpretations of changes in extratropical cyclone frequency. In some regions, such as the North Atlantic, a clear trend in activity has been noted, as also in significant wave heights in the northern half of the North Atlantic. In contrast, decreases in storm frequency and wave heights have been noted in the south half of the North Atlantic over the past few decades. These changes are also reflected in the prolonged positive excursions of the NAO since the 1970s.

6 IS THE RADIATIVE FORCING OF THE PLANET CHANGING?

Understanding requires a time history of forcing global change. The atmospheric concentration of CO_2, an important greenhouse gas because of its long atmospheric residence time and relatively high atmospheric concentration, has increased substantively over the past few decades. This is quite certain as revealed by precise measurements made at the South Pole and at Mauna Loa Observatory since the late 1950s, and from a number of stations distributed globally that began operating in subsequent decades. Since atmospheric carbon dioxide is a long-lived atmospheric constituent and it is well mixed in the atmosphere, a moderate number of well-placed stations operating for the primary purpose of monitoring seasonal to decadal changes provides a very robust estimate of global changes in carbon dioxide.

To understand the causes of the increase of atmospheric carbon dioxide, the carbon cycle and the anthropogenic carbon budget must be balanced. Balancing the carbon budget requires estimates of the sources of carbon from anthropogenic emissions from fossil fuel and cement production, as well as the net emission from changes in land use (e.g., deforestation). These estimates are derived from a combination of modeling, sample measurements, and high-resolution satellite imagery. It also requires measurements for the storage in the atmosphere, the ocean uptake, and uptake by forest regrowth, the CO_2 and nitrogen fertilization effect on vegetation, as well as any operating climate feedback effects (e.g., the increase in vegetation due to increased temperatures). Many of these factors are still uncertain because of a paucity of ecosystem measurements over a sustained period of time. Anthropogenic emissions from the burning of fossil fuel and cement production are the primary cause of the atmospheric increase.

Several other radiatively important anthropogenic atmospheric trace constituents have been measured for the past few decades. These measurements have confirmed significant increases in atmospheric concentrations of CH_4, N_2O, and the halocarbons including the stratospheric ozone destructive agent of the chloroflourocarbons and the bromocarbons. Because of their long lifetimes, a few well-placed high-quality in situ stations have provided a good estimate of global change. Stratospheric ozone depletion has been monitored both by satellite and ozonesondes. Both observing systems have been crucial in ascertaining changes of stratospheric ozone that was originally of interest, not because of its role as a radiative forcing agent, but its ability to absorb ultraviolet (UV) radiation prior to reaching Earth's surface. The combination of the surface- and space-based observing systems has enabled much more precise measurements than either system could provide by itself. Over the past few years much of the ozonesonde data and satellite data has been improved using information about past calibration methods, in part because of differences in trends between the two observing systems.

Figure 3 depicts the IPCC (1995) best estimate of the radiative forcing associated with various atmospheric constituents. Unfortunately, measurements of most of the forcings other than those already discussed have low or very low confidence, not only because of our uncertainty about their role in the physical climate system, but because we have not adequately monitored their change. For example, estimates of changes in sulfate aerosol concentrations are derived from model estimates of source emissions, not actual atmospheric concentrations. The problem is complicated because of the spatially varying concentrations of sulfate due to its short atmospheric lifetime. Another example is measurements of solar irradiance, which have been taken by balloons and rockets for several decades, but continuous measurements of top-of-the-atmosphere solar irradiance did not begin until the late 1970s with the *Nimbus 7* and the Solar Maximum Mission satellites. There are significant absolute differences in total irradiance between satellites, emphasizing the critical need for overlaps between satellites and absolute calibration of the irradiance measurements to determine decadal changes. Spectrally resolved measurements will be a key element in our ability to model the effects of solar variability, but at the present time no long-term commitment has been made to take these measurements. Another important forcing that is estimated through measured, modeled, and estimated changes in optical depth relates to the aerosols injected high into the atmosphere by major volcanic eruptions. The aerosols from these volcanoes are sporadic and usually persist in the atmosphere for at most a few years. Improved measurements of

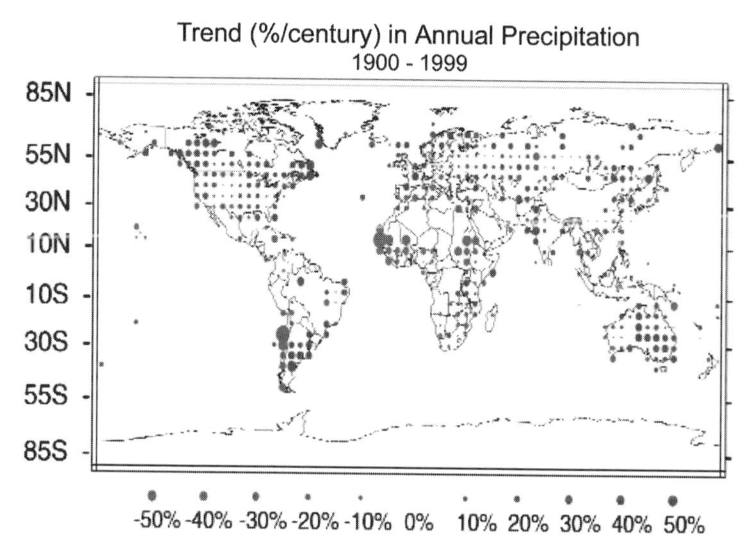

Figure 3 (see color insert) Estimates of globally and annually averaged radiative forcing (in W/m^{-2}) for a number of agents due to changes in concentrations of greenhouse gases and aerosols and natural changes in solar output from 1750 to the present day. Error bars are depicted for all forcings (from IPCC, 2001). See ftp site for color image.

aerosol size distribution and composition will help better understand this agent of climate change.

7 WHAT CAN WE DO TO IMPROVE OUR ABILITY TO DETECT CLIMATE AND GLOBAL CHANGE?

Even after extensive reworking of past data, in many instances we are incapable of resolving important aspects concerning climate and global change. Virtually every monitoring system and data set requires better data quality, continuity, and fewer time-varying biases if we expect to conclusively answer questions about how the planet has changed, because of the need to rely on observations that were never intended to be used to monitor the physical characteristics of the planet of the course of decades. Long-term monitoring, capable of resolving decade-to-century-scale changes, requires different strategies of operation.

In situ measurements are currently in a state of decay, decline, or rapid poorly documented change due to the introduction of automated measurements without adequate precaution to understand the difference between old and new observing

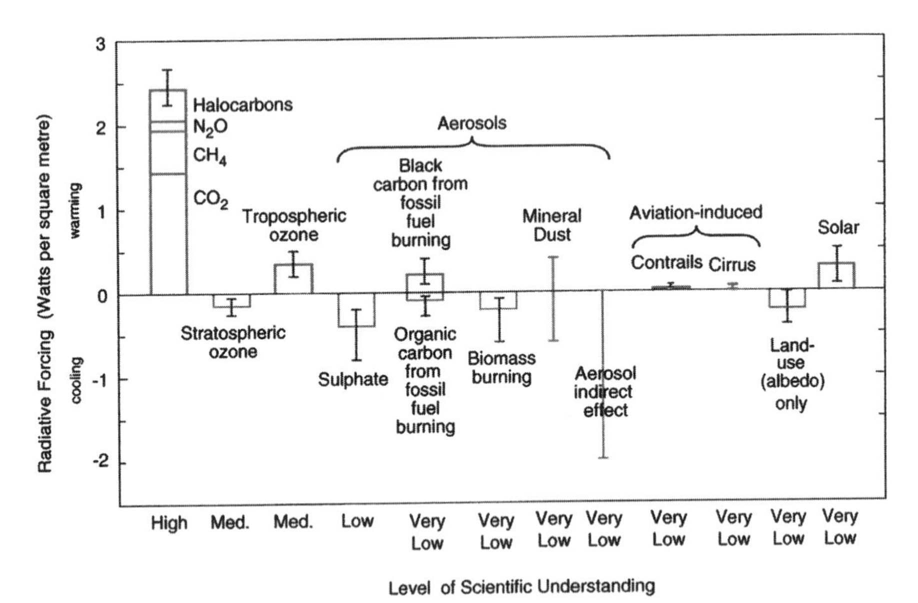

Figure 4 Global, annual-mean radiative forcings (Wm^{-2}) due to a number of agents for the period from pre-industrial (1750) to the present. The height of the vertical bar denotes the central or "best" estimate, no bar indicates that it is not possible to provide a "best" estimate. The vertical line indicates an estimate of the uncertainty range and the level of scientific understanding is a subjective judgement about the reliability of the forcing estimate based on such factors as assumptions, degree of knowledge of the physical/chemical mechanisms, etc. (From IPCC 2001). See ftp site for color image.

systems. Satellite-based systems alone will not and cannot provide all the necessary measurements. Much wiser implementation and monitoring practices must be adopted for both space-based and surface-based observing systems in order to adequately understand global change. The establishment of the Global Climate Observing System (GCOS) is a high priority (Spence and Townsend, 1995), and continued encouragement by the World Meteorological Organization (WMO) of a full implementation of this system in all countries is critical. Furthermore, in the context of the GCOS, a number of steps can be taken to improve our ability to monitor climate and global change.

These include:

1. Prior to implementing changes to existing environmental monitoring systems or introducing new observing systems, standard practice should include an assessment of the impact of these changes on our ability to monitor environmental variations and changes.
2. Overlapping measurements in time and space of both the old and new observing systems should be standard practice for critical environmental variables.
3. Calibration, validation, and knowledge of instrument, station, and/or platform history are essential for data interpretation and use. Changes in instrument sampling time, local environmental conditions, and any other factors pertinent to the interpretation of the observations and measurements should be recorded as a mandatory part of the observing routine and be archived with the original data. The algorithms used to process observations must be well documented and available to the scientific community. Documentation of changes and improvements in the algorithms should be carried along with the data throughout the data archiving process.
4. The capability must be established to routinely assess the quality and homogeneity of the historical database for monitoring environmental variations and change, including long-term high-resolution data capable of resolving important extreme environmental events.
5. Environmental assessments that require knowledge of environmental variations and change should be well integrated into a global observing system strategy.
6. Observations with a long uninterrupted record should be maintained. Every effort should be made to protect the data sets that document long-term homogeneous observations. Long term may be a century or more. A list of prioritized sites or observations based on their contribution to long-term environmental monitoring should be developed for each element.
7. Data-poor regions, variables, regions sensitive to change, and key measurements with inadequate temporal resolution should be given the highest priority in the design and implementation of new environmental observing systems.

8. Network designers, operators, and instrument engineers must be provided environmental monitoring requirements at the outset of network design. This is particularly important because most observing systems have been designed for purposes other than long-term monitoring. Instruments must have adequate accuracy with biases small enough to resolve environmental variations and changes of primary interest.

9. Much of the development of new observation capabilities and much of the evidence supporting the value of these observations stem from research-oriented needs or programs. Stable, long-term commitments to these observations, and a clear transition plan from research to operations, are two requirements in the development of adequate environmental monitoring capabilities.

10. Data management systems that facilitate access, use, and interpretation are essential. Freedom of access, low cost, mechanisms that facilitate use (directories, catalogs, browse capabilities, availability of metadata on station histories, algorithm accessibility and documentation, on-line accessibility to data, etc.), and quality control should guide data management. International cooperation is critical for successful management of data used to monitor long-term environmental change and variability.

REFERENCES

Allan, R. J., and M. R. Haylock (1993). Circulation features associated with the winter rainfall decrease in southwestern Australia, *J. Climate* **6**, 1356–1367.

Allan, R. J., J. A. Lindesay, and C. J. C. Reason (1995). Multidecadal variability in the climate system over the Indian Ocean region during the austral summer, *J. Climate* **8**, 1853–1873.

Easterling, D. R., G. Meehl, C. Parmesan, S. Changnon, T. Karl, and L. Mearns (2000a). Climate extremes: Observations, modeling, and impacts, *Science* **289**, 2068–2074.

Easterling, D. R., J. L. Evans, P. Ya. Groisman, T. R. Karl, K. E. Kunkel, and P. Ambenje (2000b). Observed variability and trends in extreme climate events: A brief review, *Bull. Am. Meteor. Soc.* **81**, 417–426.

Elliott, W. P. (1995). On detecting long-term changes in atmospheric moisture, *Climatic Change* **31**, 219–237.

Groisman, P. Y., and D. R. Legates (1995). Documenting and detecting long-term precipitation trends: Where we are and what should be done, *Climatic Change* **31**, 471–492.

Huffman, G. J., R. F. Adler, P. Arkin, A. Chang, R. Ferraro, A. Gruber, J. Janowiak, A. McNab, B. Rudolf, and U. Schneider (1997). The Global Precipitation Climatology Project (GPCP) Combined Precipitation Dataset, *Bull. Am. Meteor. Soc.* **78**, 5–20.

Hurrell, J. W., and K. E. Trenberth (1996). Satellite versus surface estimates of air temperature since 1979, *J. Climate* **9**, 2222–2232.

Hurrell, J. W., and H. van Loon (1994). A modulation of the atmospheric annual cycle in the Southern Hemisphere, *Tellus* **46A**, 325–338.

IPCC (1992). *Climate Change, 1992, Supplementary Report*, WMO/UNEP, J. T. Houghton, B. A. Callander, and S. K. Varney (Eds.), New York, Cambridge University Press, pp. 62–64.

IPCC (2001). *Climate Change, 2001: The Scientific Basis. Contribution of Working Group I to the Third Assessment Report of the Intergovernmental Panel on Climate Change*, J. T. Houghton, Y. Ding, D. J. Griggs, M. Noguer, P. J. van der Linden, X. Dai, K. Maskell, and C. A. Johnson (Eds.), New York, Cambridge University Press.

Karl, T. R., R. W. Knight, and N. Plummer (1995). Trends in high-frequency climate variability in the twentieth century, *Nature* **377**, 217–220.

Luers, J. K., and R. E. Eskridge (1995). Temperature corrections for the VIZ and Vaisala radiosondes, *Appl. Meteor.* **34**, 1241–1253.

National Research Council (NRC) (2000). *Reconciling Observations of Global Temperature Change, Report of the Panel on Reconciling Temperature Observations*, Washington, DC, National Academy Press.

Nicholls, N., and B. Lavery (1992). Australian rainfall trends during the twentieth century, *Int. J. Climatology* **12**, 153–163.

Reynolds, R. W. (1988). A real-time global sea surface temperature analysis, *J. Climate* **1**, 75–86.

Rossow, W. B., and B. Cairns (1995). Monitoring changes in clouds, *Climatic Change* **31**, 175–217.

Salinger, M. J., R. Allan, N. Bindoff, J. Hannah, B. Lavery, L. Leleu, Z. Lin, J. Lindesay, J. P. MacVeigh, N. Nicholls, N. Plummer, and S. Torok (1995). Observed variability and change in climate and sea level in Oceania, in *Greenhouse: Coping with Climate Change*, W. J. Bouma, G. I. Pearman, and M. R. Manning (eds.), CSIRO, Melbourne, Australia 100–126.

Sevruk, B. (1982). Methods of correcting for systematic error in point precipitation measurements for operational use, *Hydrology Rep.* **21**, World Meteorological Organization, Geneva 589.

Spence, T., and J. Townsend (1995). The Global Climate Observing System (GCOS), *Climatic Change* **31**, 131–134.

Spencer, R. W. (1993). Global oceanic precipitation from the MSU during 1979–1992 and comparison to other climatologies, *J. Climate* **6**, 1301–1326.

Spencer, R. W., and J. R. Christy (1992a). Precision and radiosonde validation of satellite gridpoint temperature anomalies, Part I: MSU channel 2. *J. Climate* **5**, 847–857.

Spencer, R. W., and J. R. Christy (1992b). Precision and radiosonde validation of satellite gridpoint temperature anomalies, Part II: A tropospheric retrieval and trends during 1979–90, *J. Climate* **5**, 858–866.

Trenberth, K. E., and C. J. Guillemot (1997). Evaluation of the atmospheric moisture and hydrologic cycle in the NCEP reanalysis, *Clim. Dyn.* **14**, 213–231.

van Loon, H., J. W. Kidson, and A. B. Mullan (1993). Decadal variation of the annual cycle in the Australian dataset, *J. Climate* **6**, 1227–1231.

CHAPTER 13

WHY SHOULD WE BELIEVE PREDICTIONS OF FUTURE CLIMATE?

JOHN MITCHELL

1 INTRODUCTION

Three-dimensional models of the atmosphere are based on laws of classical physics, including the conservation of momentum (the Navier–Stokes equations), heat (the first law of thermodynamics, the perfect gas law), mass and water vapor, allowing for sources and sinks. The state variables are temperature. the northerly and easterly wind components, and water vapor. For the ocean, salt is included rather than water vapor, winds are replaced by currents, and the equation of state for seawater is used (Fig. 1).

The state variables are held on a three-dimensional grid, which for current atmospheric models is typically 200 to 300 km in the horizontal and on about 20 levels in the vertical. This gives over 60,000 basic variables. The equations are solved to produce the rates of change of these variables, and hence new values of the state variables a small time interval ahead. This process is repeated over and over again to produce the evolution of the system. In practice, a variety of numerical techniques (explicit, implicit, semi-implicit, semi-Lagrangian) and spatial representations (spectral and finite difference) are used. The time step varies typically from 10 min to an hour.

Other variables are diagnosed each time step from the state variables (e.g., cloudiness, precipitation and latent heat release, radiative heating rates, surface evaporation, ground wetness, and snow cover) and, where appropriate, are used for the source and sink terms in the basic equations. Processes that occur on a scale too small to be represented explicitly by the model grid have to be included approxi-

Handbook of Weather, Climate, and Water: Dynamics, Climate, Physical Meteorology, Weather Systems, and Measurements, Edited by Thomas D. Potter and Bradley R. Colman.
ISBN 0-471-21490-6

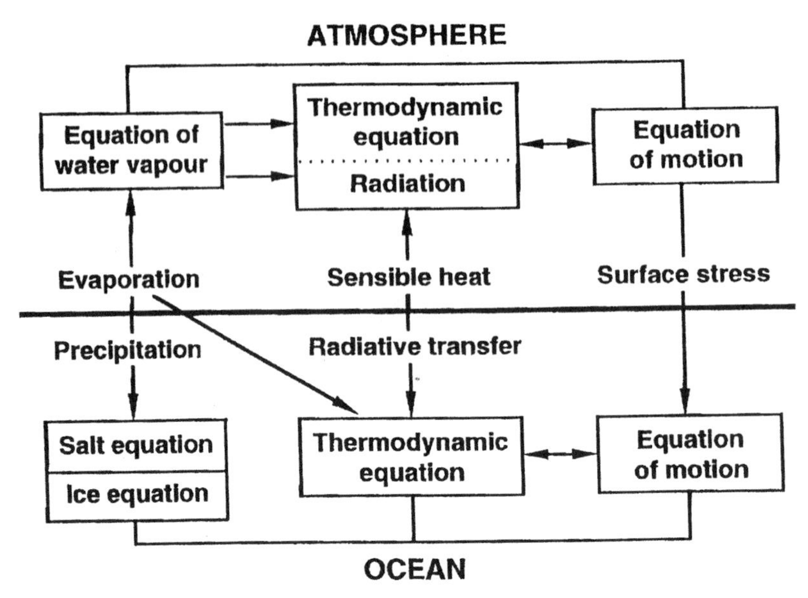

Figure 1 Some factors that affect climate.

mately, by representing them in an idealized way in terms of the grid-scale variables (*parameterization*). The parameterization may be based on one or a combination of the following: well-established theory (e.g., for radiative transfer), field or laboratory observations (e.g., for turbulent mixing of heat, moisture and momentum in the boundary layer), finer scale models (e.g., clouds) and experimentation with the general circulation models (GCM).

The oceanic component is similar to the atmosphere in resolution, although, ideally, higher resolution is required to represent mesoscale eddies, which are the oceanic equivalent of weather, and the parameterizations are generally simpler than in the atmosphere.

Climate models are numerical models of the atmosphere, ocean, and other components of the climate system that are used to understand climate and past climate change and to predict future climate. They range in complexity from simple globally averaged models that attempt to model changes in the energy balance of the climate system to complex three-dimensional GCMs. GCMs were originally developed in parallel with numerical weather prediction models in the 1960s. Indeed, several current GCMs are low-resolution versions of weather prediction models.

Simple climate models are useful in illustrating some aspects of climate change and can be tuned to mimic some of the global mean changes found in GCMs. However, only GCMs can provide information on the detailed geographical distribution of climate change. State-of-the-art GCMs include a full representation of the atmosphere, ocean, sea ice, and land surface (Fig. 2). The exact formulation of a climate model will depend on the use to which it is put. For example, accurate

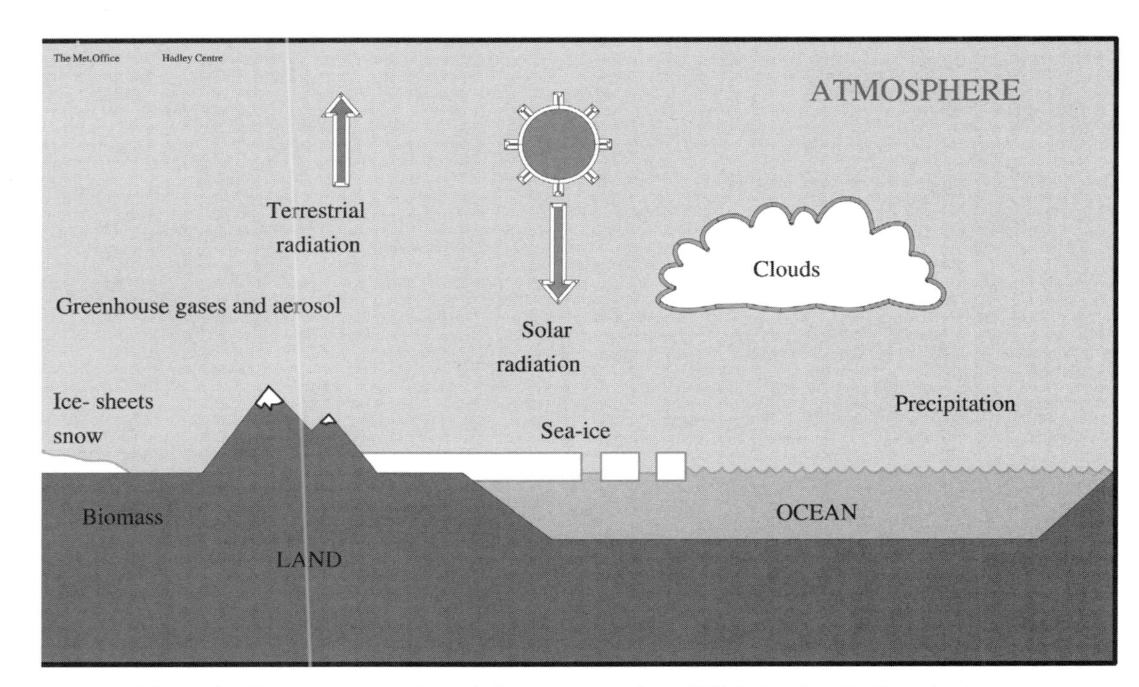

Figure 2 Basic structure of coupled ocean atmosphere GCM. See ftp site for color image.

representation of the radiative effects of greenhouse gases is required if the model is to be used for studies of anthropogenic climate change. Some new GCMs include carbon and sulfur models (to understand and predict changes in atmospheric CO_2 and sulfate aerosol concentrations and their effect on climate) and detailed atmospheric chemistry (to study ozone depletion and changes in tropospheric trace gases).

Climate models are used differently from weather prediction models. Weather prediction models prescribe an initial state of the basic variables from observations, and the equations are stepped forward in time to give the evolution of the atmosphere. For a few days, it is possible to relate individual features in the forecast to features that evolve at the corresponding time in the real world. As the forecast proceeds further, this correspondence disappears as errors due to inexact initial data and model inadequacies grow because of the nonlinear nature of the equations of motion. In other words, the chaotic nature of the equations limits the length of time of deterministic forecasts to about a week or so (depending upon processes involved and the scale of the system being forecast).

A climate model is usually run for long enough that its statistics are independent of the initial conditions. The experiment is then continued over a number of years, decades, or even centuries (depending on the availability of computer time and the application), and the statistics of the simulation over the final period are analyzed. This will include not only the time means, but variability on daily to annual or longer time scales, storm tracks, extreme events, and so on. If the effect of a particular change is being investigated (e.g., doubling atmospheric carbon dioxide concentrations, or changing Earth's orbital parameters), then the experiment is repeated with this change in the model, and the statistics of the two experiments are compared. Both the GCM and observed climate display variations on all time scales due to

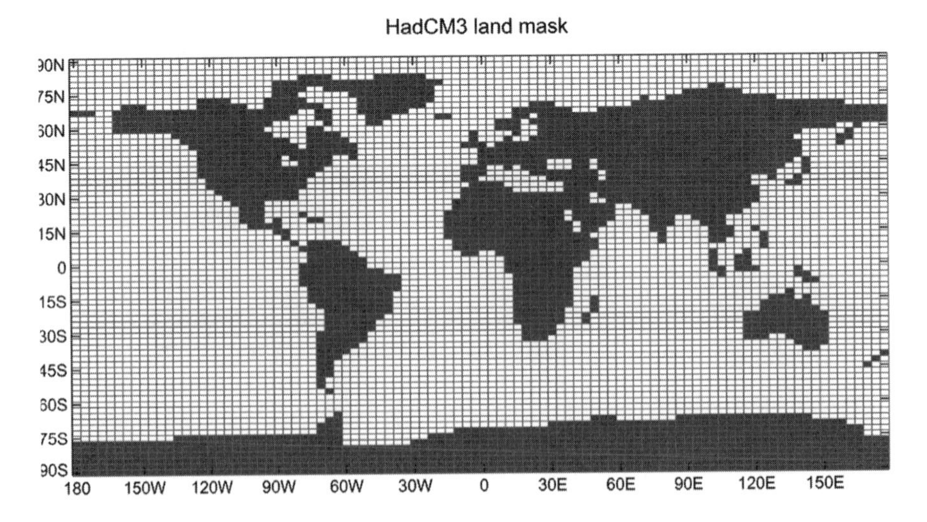

Figure 3 Typical climate model grid.

internal variability; statistical tests may be needed to demonstrate that the differences between the control and anomaly simulations are due to the change in the model, and not to internal variations.

What has been the evolution of climate change in the recent past, and likely changes in the near future? These questions are addressed using coupled ocean atmosphere GCMs. Models have been integrated from preindustrial times to the present using the estimated changes in climate forcing (factors governing climate due to human and natural causes), assuming that the preindustrial climate was in quasi-equilibrium, and then extended assuming some future scenario of greenhouse gas concentrations. As in weather forecasting, errors in initial data and the model will contaminate these "hindcasts" and forecasts. Hence, the most recent studies use an ensemble of simulations started from different initial conditions to give a range of possible future projections taking into account the uncertainty in initial conditions.

How reliable are GCMs for predicting future climate change? With climate, this is argued in several ways. First, they are based on well-established physical laws and principles based on a wealth of observations. Second, they reproduce many of the main features of current climate and its interannual variability, including the seasonal cycle of temperature, the formation and decay of the major monsoons, the seasonal shifts of the major rain belts, the average daily temperature cycle, and the variations in outgoing radiation at high elevations in the atmosphere, as measured by satellites. Similarly many features of the large-scale features in the ocean are reproduced by current climate models.

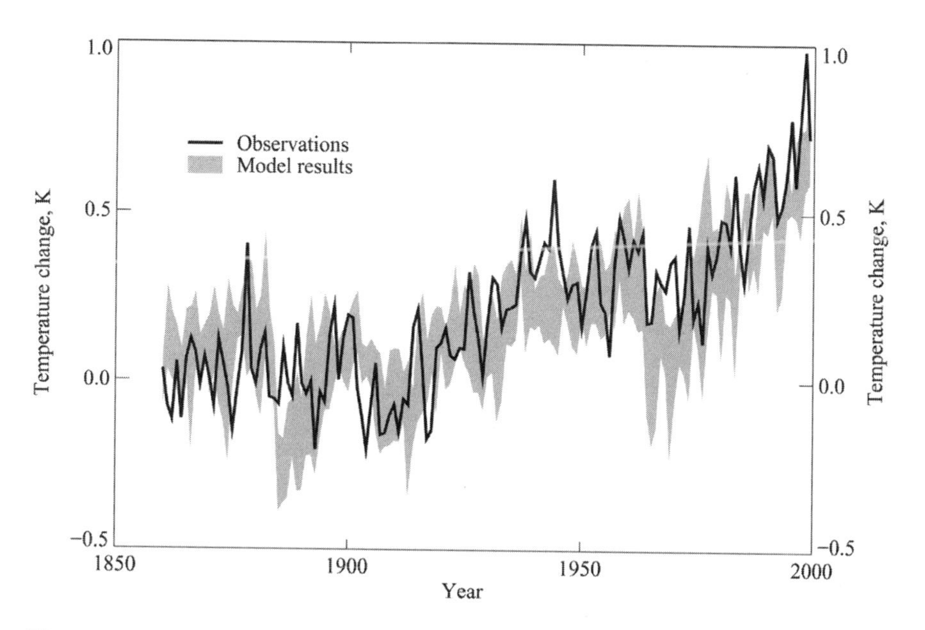

Figure 4 Simulations of recent changes in global mean surface temperature (greyband) compared to observations (black line).

A model may produce a faithful representation of current climate, yet give unreliable predictions of climate change. Hence models have been tested to reproduce both past climates and recent climatic events. Simulations of temperature and precipitation changes 6000 years ago (due to changes in Earth's orbital parameters) and the last glacial maximum (prescribing reconstructed changes in ice sheets, orbital parameters, and greenhouse gases) compare tolerably well with reconstructions from palaeodata. For example, models reproduce the strengthening of the North African monsoon 6000 years ago and the approximate level of cooling estimated for the last ice age. The value of these comparisons is limited by the possibility that other factors not included in the model may have contributed to change in these periods, and the uncertainty in the reconstructions of temperature and precipitation from the palaeodata.

Simulations of the last 130 years have been made driven with the observed increase in greenhouse gases and an estimate of the increase in sulfate particles. The simulated global mean warming agrees tolerably well with observations. Forecasts of the global mean cooling due to the eruption of Mount Pinatubo 1991 were also very successful. However, it is possible in both these instances that errors in the estimate of the factors governing climate in these cases (e.g., changes in ozone and aerosol concentrations) were fortuitously canceled by errors in the model's sensitivity to those changes. Recent patterns of change including cooling of the upper atmosphere, a reduction in diurnal temperature range, and a tentative increase in precipitation in high northern latitudes are in broad qualitative agreement with available observations.

Various factors limit our confidence in models. The sensitivity of global mean temperature change with increasing atmospheric carbon dioxide had different models which varied by a factor of over 2, largely as a result of uncertainty in the treatment of cloud and related processes. Pronounced differences exist in the regional changes in model responses to increasing greenhouse gases, although the broad scale features are quite robust, including enhanced warming in high northern latitudes in all seasons but summer, generally greater warming over the land than ocean, limited warming over the Southern Ocean and northern North Atlantic, and increased annual mean runoff in high latitudes. The validation against recent observed climate change is hampered not only by uncertainty in the factors affecting climate but also in how much of the recent observed changes are due to natural variability and naturally forced events (e.g., due to changes in solar intensity).

In summary, climate models are numerical models used to study climate and climate change. They reproduce many features of observed climate and climate variability, some of the broad features of past climates. Predictions of future global mean temperature change vary by a factor of 2 or so for a given scenario of anthropogenic greenhouse gas emissions and are reasonably consistent on global scales, but not on regional scales. The main areas of current research are reducing model errors by improving the representation of physical process (particularly clouds), reducing uncertainties in the factors affecting climate, particularly in the recent past, and estimating the magnitude of natural climate variability.

BIBLIOGRAPHY

Houghton, J. T., Y. Ding, D. J. Griggs, M. Noguer, P. S. van der Linden, X. Dai, K. Maskell, and C. A. Johnson, (Eds.) (2001). *Climate Change 2001: The Scientific Basis*. The Third Assessment Report of the IPCC: Contribution of Working Group I, Cambridge University Press, Cambridge, MA.

Trenberth, K. E. (1995). *Numerical Modeling of the Atmosphere—Numerical Weather Prediction Haltiner and Martin Climate Modeling—Climate Systems Modeling*, Cambridge University Press, Cambridge, MA.

CHAPTER 14

THE EL NIÑO–SOUTHERN OSCILLATION (ENSO) SYSTEM

KEVIN TRENBERTH

1 ENSO EVENTS

Every 3 to 7 years or so, a pronounced warming occurs of the surface waters of the tropical Pacific Ocean. The warmings take place from the international dateline to the west coast of South America and result in changes in the local and regional ecology and are clearly linked with anomalous global climate patterns. These warmings have come to be known as *El Niño events*. Historically, El Niño referred to the appearance of unusually warm water off the coast of Peru as an enhancement of the normal warming about Christmastime (hence Niño, Spanish for "the boy Christ-child") and only more recently has the term come to be regarded as synonymous with the basinwide phenomenon. The atmospheric component tied to El Niño is termed the *Southern Oscillation* (SO) whose existence was first noted late in the 1800s. Scientists call the interannual variations where the atmosphere and ocean collaborate together in this way El Niño–Southern Oscillation (ENSO).

The ocean and atmospheric conditions in the tropical Pacific are seldom close to average, but instead fluctuate somewhat irregularly between the warm phase of ENSO, the El Niño events, and the cold phase of ENSO consisting of cooling of the central and eastern tropical Pacific, referred to as *La Niña events* (La Niña is "the girl" in Spanish). The most intense phase of each event lasts about a year.

This chapter briefly outlines the current understanding of ENSO and the physical connections between the tropical Pacific and the rest of the world and the issues involved in exploiting skillful but uncertain predictions.

Handbook of Weather, Climate, and Water: Dynamics, Climate, Physical Meteorology, Weather Systems, and Measurements, Edited by Thomas D. Potter and Bradley R. Colman.
ISBN 0-471-21490-6

2 THE TROPICAL PACIFIC OCEAN–ATMOSPHERE SYSTEM

The distinctive pattern of average sea surface temperatures in the Pacific Ocean sets the stage for ENSO events. Key features are the "warm pool" in the tropical western Pacific, where the warmest ocean waters in the world reside and extend to depths of over 150 m, warm waters north of the equator from about 5 to 15° N, much colder waters in the eastern Pacific, and a cold tongue along the equator that is most pronounced about October and weakest in March. The warm pool migrates with the sun back and forth across the equator but the distinctive patterns of sea surface temperature are brought about mainly by the winds.

The existence of the ENSO phenomenon is dependent on the east–west variations in sea surface temperatures in the tropical Pacific and the close links with sea level pressures, and thus surface winds in the tropics, which in turn determine the major areas of rainfall. The temperature of the surface waters is readily conveyed to the overlying atmosphere, and, because warm air is less dense, it tends to rise while cooler air sinks. As air rises into regions where the air is thinner, the air expands, causing cooling and therefore condensing moisture in the air, which produces rain. Low sea level pressures are set up over the warmer waters while higher pressures occur over the cooler regions in the tropics and subtropics, and the moisture-laden winds tend to blow toward low pressure so that the air converges, resulting in organized patterns of heavy rainfall. The rain comes from convective cloud systems, often as thunderstorms, and perhaps as tropical storms or even hurricanes, which preferentially occur in the *convergence zones*. Because the wind is often light or calm right in these zones, they have previously been referred to as the *doldrums*. Of particular note are the Inter-Tropical Convergence Zone (ITCZ) and the South Pacific Convergence Zone (SPCZ), which are separated by the equatorial dry zone. These atmospheric climatological features play a key role in ENSO as they change in character and move when sea surface temperatures change.

There is a strong coincidence between the patterns of sea surface temperatures and tropical convection throughout the year, although there is interference from effects of nearby land and monsoonal circulations. The strongest seasonal migration of rainfall occurs over the tropical continents, Africa, South America and the Australian–Southeast Asian–Indonesian maritime region. Over the Pacific and Atlantic, the ITCZ remains in the Northern Hemisphere year round, with convergence of the trade winds favored by the presence of warmer water. In the subtropical Pacific, the SPCZ also lies over water warmer than about 27°C. The ITCZ is weakest in January in the Northern Hemisphere when the SPCZ is strongest in the Southern Hemisphere.

The surface winds drive surface ocean currents, which determine where the surface waters flow and diverge, and thus where cooler nutrient-rich waters upwell from below. Because of Earth's rotation, easterly winds along the equator deflect currents to the right in the Northern Hemisphere and to the left in the Southern Hemisphere and thus away from the equator, creating upwelling along the equator. The presence of nutrients and sunlight in the cool surface waters along the equator and western coasts of the Americas favors development of tiny plant species called phytoplankton, which are grazed on by microscopic sea animals called zooplankton, which in turn provide food for fish.

The winds largely determine the sea surface temperature distribution along with differential sea levels and the heat content of the upper ocean. The latter is related to the configuration of the thermocline, which denotes a region of sharp temperature gradient within the ocean separating the well-mixed surface layers from the cooler abyssal ocean waters. Normally the thermocline is deep in the western tropical Pacific (on the order of 150 m) and sea level is high as waters driven by the easterly tradewinds pile up. In the eastern Pacific on the equator, the thermocline is very shallow (on the order of 50 m) and sea level is relatively low. The Pacific sea surface slopes up by about 60 cm from east to west along the equator.

The tropical Pacific, therefore, is a region where the atmospheric winds are largely responsible for the tropical sea surface temperature distribution which, in turn, is very much involved in determining the precipitation distribution and the tropical atmospheric circulation. This sets the stage for ENSO to occur.

3 INTERANNUAL VARIATIONS IN CLIMATE

Most of the interannual variability in the tropics and a substantial part of the variability over the Southern Hemisphere and Northern Hemisphere extratropics is related and tied together through ENSO. ENSO is a natural phenomenon arising from coupled interactions between the atmosphere and the ocean in the tropical Pacific Ocean, and there is good evidence from cores of coral and glacial ice in the Andes that it has been going on for millennia. The region of the Pacific Ocean most involved in ENSO is the central equatorial Pacific (Fig. 1), especially the area 5° N to 5° S, 170° E to 120° W, not the traditional El Niño region along the coast of South America. The evolution of sea surface temperature covering ENSO events after 1950 is shown in Figure 2.

Inverse variations in pressure anomalies (departures from average) at Darwin (12.4° S 130.9° E) in northern Australia and Tahiti (17.5° S 149.6° W) in the South Pacific Ocean characterize the SO. Consequently, the difference in pressure anomalies, Tahiti minus Darwin, is often used as a Southern Oscillation Index (SOI), also given in Figure 2. The warm ENSO events clearly identifiable since 1950 occurred in 1951, 1953, 1957–1958, 1963, 1965, 1969, 1972–1973, 1976–1977, 1982–1983, 1986–1987, 1990–1995, and 1997–1998. The 1990–1995 event is sometimes considered as three events as it faltered briefly in late 1992 and early 1994 but reemerged strongly in each case, and the duration is unprecedented in the past 130 years. Worldwide climate anomalies lasting several seasons have been identified with all of these events.

The SO is principally a global-scale seesaw in atmospheric sea level pressure involving exchanges of air between eastern and western hemispheres (Fig. 3) centered in tropical and subtropical latitudes with centers of action located over Indonesia (near Darwin) and the tropical South Pacific Ocean (near Tahiti). Higher than normal pressures are characteristic of more settled and fine weather,

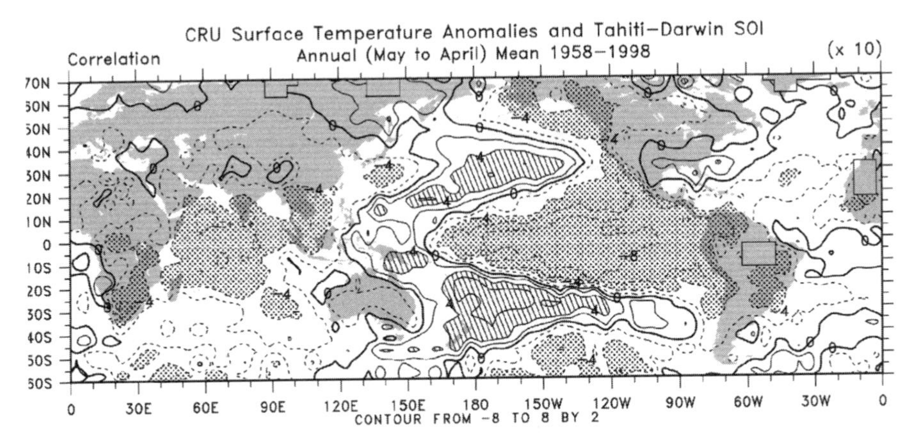

Figure 1 (see color insert) Correlation coefficients of the SOI with sea surface temperature seasonal anomalies for January 1958 to December 1998. It can be interpreted as the sea surface temperature patterns that accompany a La Niña event, or as an El Niño event with signs reversed. Values in the central tropical Pacific correspond to anomalies of about 1.5°C [From Trenberth and Caron, 2000]. See ftp site for color image.

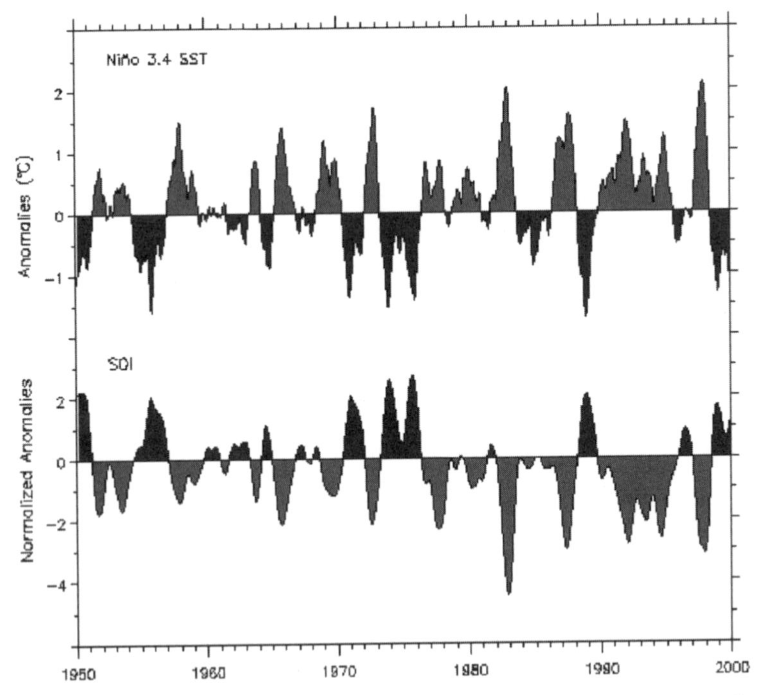

Figure 2 (see color insert) Time series of areas of sea surface temperature anomalies from 1950 through 2000 relative to the means of 1950–79 for the region most involved in ENSO 5° N–5° S, 170° E –120° W (top) and for the Southern Oscillation Index. El Niño events are in grey and La Niña events in black. See ftp site for color image.

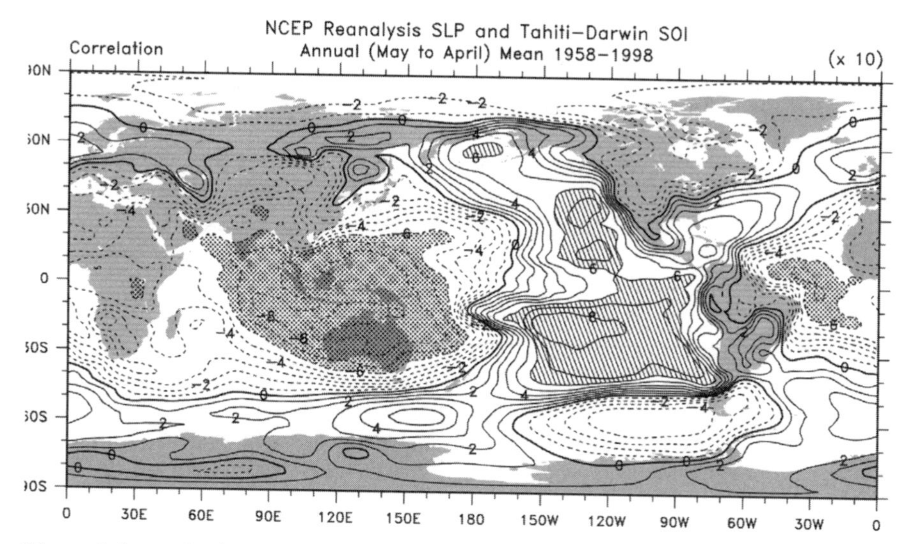

Figure 3 (see color insert) Map of correlation coefficients ($\times 10$) of the annual mean sea level pressures (based on the year May to April) with the SOI showing the Southern Oscillation pattern in the phase corresponding to La Niña events. During El Niño events the sign is reversed [From Trenberth and Caron, 2000]. See ftp site for color image.

with less rainfall, while lower than normal pressures are identified with "bad" weather, more storminess, and rainfall. So it is with the SO. Thus for El Niño conditions, higher than normal pressures over Australia, Indonesia, southeast Asia, and the Philippines signal drier conditions or even droughts. Dry conditions also prevail at Hawaii, parts of Africa, and extend to the northeast part of Brazil and Colombia. On the other end of the seesaw, excessive rains prevail over the central and eastern Pacific, along the west coast of South America, parts of South America near Uruguay, and southern parts of the United States in winter (Fig. 4). In the winters of 1992–1993, 1994–1995, and 1997–1998 this included excessive rains in southern California, but in other years (e.g., 1986–1987 and 1987–1988 winters) California is more at risk for droughts. Because of the enhanced activity in the central Pacific and the changes in atmospheric circulation throughout the tropics, there is a decrease in the number of tropical storms and hurricanes in the tropical Atlantic during El Niño.

The SO has global impacts; however, the connections to higher latitudes (known as teleconnections) tend to be strongest in the winter of each hemisphere and feature alternating sequences of high and low pressures accompanied by distinctive wave patterns in the jet stream and storm tracks in midlatitudes. For California, all of the winter seasons noted above were influenced by El Niños, but their character differed and the teleconnections to higher latitudes were not identical. Although warming is

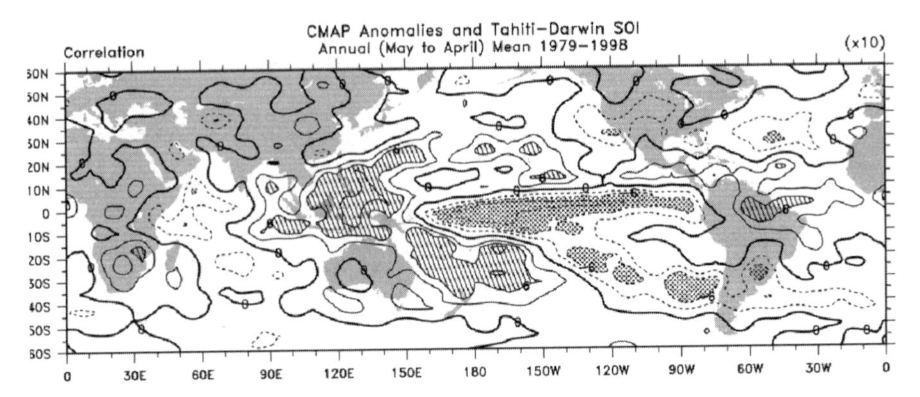

Figure 4 Correlations of annual mean precipitation with the SOI for 1979–1998. The pattern has the phase corresponding to La Niña events, and during El Niño events the sign is reversed, i.e., the stippled areas are wetter and hatched areas drier in El Niño events [From Trenberth and Caron, 2000]. See ftp site for color image.

generally associated with El Niño events in the Pacific and extends, for instance, into western Canada, cool conditions typically prevail over the North and South Pacific Oceans. To a first approximation, reverse patterns occur during the opposite La Niña phase of the phenomenon. However, the latter is more like an extreme case of the normal pattern with a cold tongue of water along the equator.

The prominence of the SO has varied throughout this century. Very slow long-term (decadal) variations are present; for instance, SOI values are mostly below the long-term mean after 1976. This accompanies the generally above normal sea surface temperatures in the western Pacific along the equator (Fig. 2). The prolonged warm ENSO event from 1990–1995 is very unusual and the 1997–1998 event is the biggest on record in terms of sea surface temperature anomalies. The decadal atmospheric and oceanic variations are even more pronounced in the North Pacific and across North America than in the tropics and also present in the South Pacific, with evidence suggesting they are at least in part forced from the tropics. It is possible that climate change associated with increasing greenhouse gases in the atmosphere, which contribute to global warming, may be changing ENSO, perhaps by expanding the west Pacific warm pool.

Changes associated with ENSO produce large variations in weather and climate around the world from year to year, and often these have a profound impact on humanity and society because of droughts, floods, heat waves, and other changes that can severely disrupt agriculture, fisheries, the environment, health, energy demand, and air quality and also change the risks of fire. Changes in oceanic conditions can have disastrous consequences for fish and sea birds and thus for the fishing and guano industries along the South American coast. Other marine creatures may benefit so that unexpected harvests of shrimp or scallops occur in some places. Rainfalls over Peru and Ecuador can transform barren desert into lush growth and benefit some crops but can also be accompanied by swarms of grasshoppers and increases in the populations of toads and insects. Human health is affected by mosquito-borne diseases such as malaria, dengue, and viral encephalitis and by water-borne diseases such as cholera. Economic impacts can be large, with losses typically overshadowing gains. This is also true in La Niña events and arises because of the extremes in flooding, drought, fires, and so on that are outside the normal range.

4 MECHANISMS OF ENSO

During El Niño, the trade winds weaken, causing the thermocline to become shallower in the west and deeper in the eastern tropical Pacific (Fig. 5), while sea level falls in the west and rises in the east by as much as 25 cm as warm waters surge eastward along the equator. Equatorial upwelling decreases or ceases and so the cold tongue weakens or disappears and the nutrients for the food chain are substantially reduced. The resulting increase in sea temperatures warms and moistens the overlying air so that convection breaks out and the convergence zones and associated rainfall move to a new location with a resulting change in the atmospheric circulation. A further weakening of the surface trade winds completes the positive feedback cycle leading to an ENSO event. The shift in the location of the organized rainfall in the tropics and the latent heat released alters the heating patterns of the atmosphere. Somewhat like a rock in a stream of water, the anomalous heating sets up waves in the atmosphere that extend into midlatitudes, altering the winds and changing the jet stream and storm tracks.

Although the El Niños and La Niñas are often referred to as "events" that last a year or so, they are somewhat oscillatory in nature. The ocean is a source of moisture, and its enormous heat capacity acts as the flywheel that drives the system through its memory of the past, resulting in an essentially self-sustained sequence in which the ocean is never in equilibrium with the atmosphere. The amount of warm water in the tropics builds up prior to and is then depleted during ENSO. During the cold phase with relatively clear skies, solar radiation heats up the tropical Pacific Ocean, the heat is redistributed by currents, with most of it being stored in the deep warm pool in the west or off the equator. During El Niño heat is transported out of the tropics within the ocean toward higher latitudes in response to the changing currents, and increased heat is released into the atmosphere mainly in the form of increased evaporation, thereby cooling the ocean. Added rainfall contributes to a general warming of the global atmosphere that peaks a few months after a strong El Niño event. It has therefore been suggested that the time scale of ENSO is determined by the time required for an accumulation of warm water in the tropics to essentially recharge the system, plus the time for the El Niño itself to evolve. Thus a major part of the onset and evolution of the events is determined by the history of what has occurred 1 to 2 years previously. This also means that the future evolution is predictable for several seasons in advance.

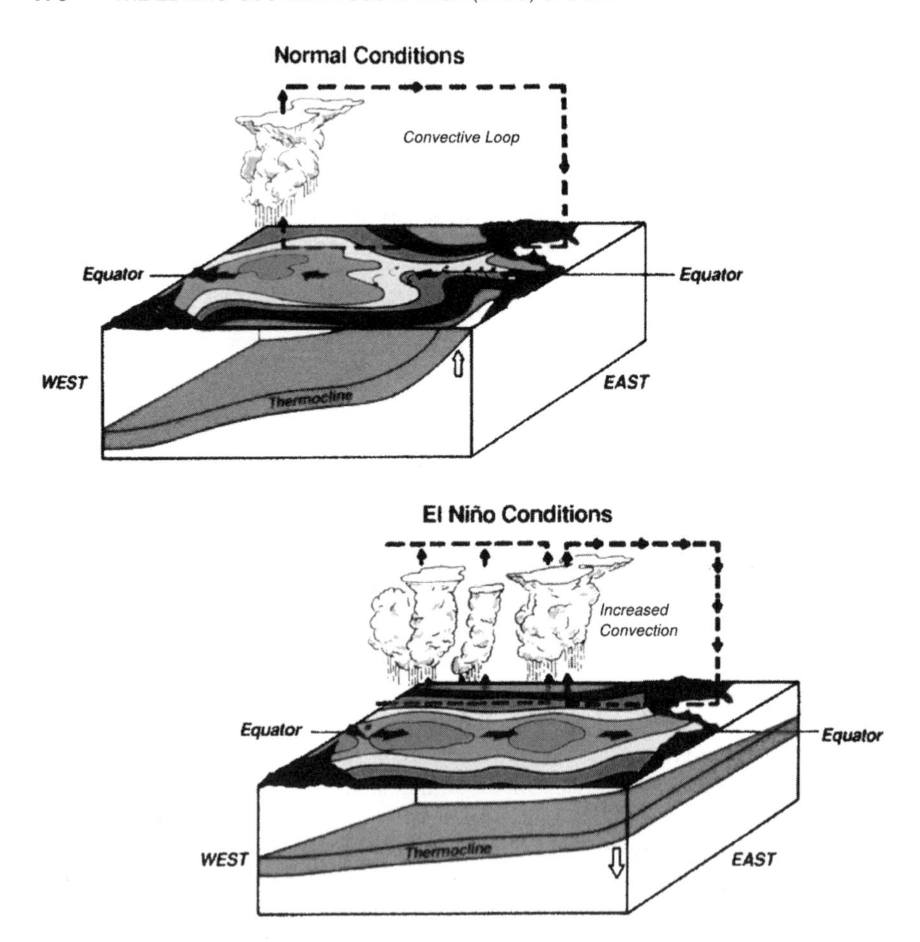

Figure 5 (see color insert) Schematic cross section of the Pacific Basin with Australia at lower left and the Americas at right depicting normal and El Niño conditions. Total sea surface temperatures exceeding 29°C are in gold and the colors change every 1°C. Regions of convection and overturning in the atmosphere are indicated. The thermocline in the ocean is shown in blue. Changes in ocean currents are shown by the black arrows. (Copyright University Corporation for Atmospheric Research. Reprinted by permission). See ftp site for color image.

5 OBSERVING ENSO

In the early 1980s, the observing system for weather and climate revolved around requirements for weather forecasting, and there was little routine monitoring of ocean conditions. Oceanographic observations were mostly made on research cruises, and often these were not available to the community at large until months or years later. Satellite observations of the ocean surface and atmosphere provided some information on sea surface temperatures and clouds, but processing of all the relevant information for ENSO was not routinely available. This meant that in 1982–1983, during the largest El Niño seen to that point, the tropical Pacific was so poorly observed that the El Niño was not detected until it was well underway. A research program called Tropical Oceans–Global Atmosphere (TOGA) was begun in 1985 until 1994 to explore ENSO and to build a system to observe and perhaps predict it. The understanding of ENSO outlined above has developed largely as a result of the TOGA program and an observing system has been put in place.

The ENSO observing system has developed gradually and was fully in place at the end of 1994, so the benefits from it and experience with it are somewhat limited. A centerpiece of this observing system is an array of buoys in the tropical Pacific moored to the ocean bottom known as the TAO (Tropical Atmosphere–Ocean) array. The latter is maintained by a multinational group spearheaded in the United States by the National Oceanic and Atmospheric Administration's (NOAA's) Pacific Marine Environmental Laboratory (PMEL). This array measures the quantities believed to be most important for understanding and predicting ENSO. Each buoy has a series of temperature measurements on a sensor cable on the upper 500 m of the mooring, and on the buoy itself are sensors for surface wind, sea surface temperature (SST), surface air temperature, humidity, and a transmitter to a satellite. Some buoys also measure ocean currents down to 250 m depth. Observations are continually made, averaged into hourly values, and transmitted via satellite to centers around the world for prompt processing. Tapes of data are recovered when the buoys are serviced about every 6 months by ships.

Other key components of the ENSO observing system include surface drifting buoys, which have drogues attached so that the buoy drifts with the currents in the upper ocean and not the surface wind. In this way, displacements of the buoy provide measurements of the upper ocean currents. These buoys are also instrumented to measure sea surface temperatures and, outside of the tropics, surface pressure. These observations are also telemetered by satellite links for immediate use.

Observations are also taken from all kinds of ships, referred to as "volunteer observing ships." As well as making regular observations of all surface meteorological observations, some of these ships are recruited to do expendable bathythermograph (XBT) observations of the temperatures of the upper 400 m of the ocean along regular ship lines. Another valuable part of the observing system is the network of sea level stations. Changes in heat content in the ocean are reflected in changes in sea level. New measurements from satellite-borne altimeters are providing much more comprehensive views of how sea level changes during ENSO.

Considerable prospects exist for satellite-based remote sensing, including observations of SSTs, atmospheric winds, water vapor, precipitation, aerosol and cloud properties, ocean color, sea level, sea ice, snow cover, and land vegetation. Continuity of consistent calibrated observations from space continues to be an issue for climate monitoring because of the limited lifetimes of satellites (a few years); replacement satellites usually have somewhat different orbits and orbits decay with time. All the ship, buoy, and satellite observations are used together to provide analyses, for instance, of the surface and subsurface temperature structure.

6 ENSO AND SEASONAL PREDICTIONS

The main features of ENSO have been captured in models that predict the anomalies in sea surface temperatures. Lead times for predictions in the tropics of up to about a year have been shown to be practicable. It is already apparent that reliable prediction of tropical Pacific SST anomalies can lead to useful skill in forecasting rainfall anomalies in parts of the tropics, notably those areas featured in Figure 4. While there are certain common aspects to all ENSO events in the tropics, the effects at higher latitudes are more variable. One difficulty is the vigor of weather systems in the extratropics, which can override relatively modest ENSO influences. Nevertheless, systematic changes in the jet stream and storm tracks do occur on average, thereby allowing useful predictions to be made in some regions, although with some uncertainty inherent.

Skillful seasonal predictions of temperatures and rainfalls have the potential for huge benefits for society. However, because the predictability is somewhat limited, a major challenge is to utilize the uncertain forecast information in the best way possible throughout different sectors of society (e.g., crop production, forestry resources, fisheries, ecosystems, water resources, transportation, energy use). Considerations in using a forecast include how decisions are made and whether they can be affected by new information. The implication of an incorrect forecast if acted upon must be carefully considered and factored into decisions along with estimates of the value of taking actions. A database of historical analogs (of past societal behavior under similar circumstances) can be used to support decision making. While skillful predictions should positively impact the quality of life in many areas, conflicts among different users of information (such as hydroelectric power versus fish resources in water management) can make possibly useful actions the subject of considerable debate. The utility of a forecast may vary considerably according to whether the user is an individual versus a group or country. An individual may find great benefits if the rest of the community ignores the information, but if the whole community adjusts (e.g., by growing a different crop), then supply and market prices will change, and the strategy for the best crop yield may differ substantially from the strategy for the best monetary return. On the other hand, the individual may be more prone to small-scale vagaries in weather that are not predictable. Vulnerability of individuals will also vary according to the diversity of the operation, whether there is irrigation available, whether the farmer has insurance, and whether he or she can work off the farm to help out in times of adversity.

ACKNOWLEDGMENTS

This research was partially sponsored by NOAA Office of Global Programs grant NA56GP0247 and the joint NOAA/NASA grant NA87GP0105.

BIBLIOGRAPHY

Chen, D., S. E. Zebiak, A. J. Busalacchi, and M. A. Cane (1995). An improved procedure for El Nino forecasting: Implications for predictability, *Science* **269**, 1699–1702.

Glantz, M. H. (1996). *Currents of Change. El Nino's Impact on Climate and Society*, Cambridge University Press, Cambridge, New York.

Glantz, M. H., R. W. Katz, and N. Nicholls (Ed.) (1991). *Teleconnections Linking Worldwide Climate Anomalies*, Cambridge University Press, Cambridge, New York.

Kumar, A., A. Leetmaa, and M. Ji (1994). Simulations of atmospheric variability induced by sea surface temperatures and implications for global warming, *Science* **266**, 632–634.

McPhaden, M. J. (1995). The Tropical Atmosphere–Ocean Array is completed, *Bull. Am. Meteor. Soc.* **76**, 739–741.

Monastersky, R. (1993). The long view of the weather, *Science News* **144**, 328–330.

National Research Council (1996). *Learning to Predict Climate Variations Associated with El Nino and the Southern Oscillation: Accomplishments and Legacies of the TOGA Program*, National Academy Press, Washington D.C.

Patz, J. A., P. R. Epstein, T. A. Burke, and J. M. Balbus (1996). Global climate change and emerging infectious diseases, *J. Am. Med. Assoc.*, **275**, 217–223.

Philander, S. G. H. (1990). *El Nino, La Nina, and the Southern Oscillation*, Academic Press, San Diego, CA.

Suplee, C. (1999). El Nino/La Nina, *National Geographic*, March, pp. 72–95.

Time (1983). Australia's "Great Dry," Cover story, March 28, pp. 6–13.

Trenberth, K. E. (1997a). Short-term climate variations: Recent accomplishments and issues for future progress, *Bull. Am. Met. Soc.* **78**, 1081–1096.

Trenberth, K. E. (1997b). The definition of El Nino, *Bull. Am. Met. Soc.* **78**, 2771–2777.

Trenberth, K. E. (1999). The extreme weather events of 1997 and 1998, *Consequences* **5**(1), 2–15, (http://www.gcrio.org/CONSEQUENCES/vol5no1/extreme.html).

Trenberth, K. E., and J. M. Caron (2000). The Southern Oscillation revisited: Sea level pressures, surface temperatures and precipitation, *J. Climate*, **13**, 4358–4365.

SECTION 3

PHYSICAL METEOROLOGY

Contributing Editor: Gregory Tripoli

CHAPTER 15

PHYSICAL ATMOSPHERIC SCIENCE*

GREGORY TRIPOLI

1 OVERVIEW

Perhaps the oldest of the atmospheric sciences, physical meteorology is the study of laws and processes governing the physical changes of the atmospheric state. The underlying basis of physical meteorology is the study of *atmospheric thermodynamics*, which describes relationships between observed physical properties of air including pressure, temperature, humidity, and water phase. Generally, thermodynamics as well as most of physical meteorology does not seek to develop explicit models of the processes it studies, such as the evolution of individual air molecules or individual water droplets. Instead, it formulates relationships governing the statistics of the microscale state and processes. Temperature and the air density are two examples of such statistics representing the mean kinetic energy of a population of molecules and the mass of molecules of air per unit volume, respectively.

A second major branch of physical meteorology is the study of the growth of water and ice precipitation called *cloud microphysics*. Here theories of how liquid and ice hydrometeors first form and subsequently evolve into rain, snow, and hail are sought. The interaction of water with atmospheric aerosols is an important part of atmospheric microphysics, which leads to the initial formation of hydrometeors. The quantification of these theories leads to governing equations that can be used to simulate and predict the evolution of precipitation as a function of changing atmospheric conditions. By studying the processes that lead to the formation of clouds,

*The derivations, concepts, and examples herein are based partially on course notes from the late Professor Myron Corrin of Colorado State University. Other derivations are taken from the Isbane and Godson text, the Emanuel (1994: *Atmospheric Convection*, Oxford University Press) text, and the Dutton (*The Ceaseless Wind*) book.

Handbook of Weather, Climate, and Water: Dynamics, Climate, Physical Meteorology, Weather Systems, and Measurements, Edited by Thomas D. Potter and Bradley R. Colman.
ISBN 0-471-21490-6

clouds themselves must be categorized by the processes that form them, giving rise to the study of *cloud types*. From the discipline of microphysics emerges the study of *weather modification*, which studies the possibility of purposefully modifying natural precipitation formation processes to alter the precipitation falling to the surface. Another discipline associated with microphysics is *atmospheric electricity*, which studies the transfer of electrical energy by microphysical processes.

A third major branch of physical meteorology is the study of *atmospheric radiation*. This area of study develops theories and laws governing the transfer of energy through the atmosphere by radiative processes. These include the adsorbtion, transmission, and scattering of both solar radiation and terrestrial radiation. A primary goal of atmospheric radiation studies is to determine the net radiative loss or gain at a particular atmospheric location that leads to thermodynamic change described by thermodynamics theory. Since radiative transfer is affected by details of the atmospheric thermal, humidity, and chemical structure, it is possible to recover some details of those structures by observing the radiation being transferred. This has given rise to a major branch of atmospheric radiation called *remote sensing*, the goal of which is to translate observations of radiation to atmospheric structure. The study of *atmospheric optics* is another branch of atmospheric radiation that applies concepts of radiative transfer to study the effect of atmospheric structure on visible radiation passing through the atmosphere. Phenomena such as blue sky, rainbows, sun-dogs, and so on are subjects of this area.

Finally, *boundary layer* meteorology studies the transfer of heat, radiation, moisture, and momentum between the surface and the atmosphere. The transfer of energy occurs on a wide range of scales from the molecular scale all the way to the scales of thermal circulations. The challenge is to quantify these transfers on the basis of observable or modelable atmospheric quantities. This also gives rise to the need to represent the evolution of the soil, vegetation, and water surfaces. This has given rise to branches of atmospheric and other earth sciences aimed at developing models of the soil, water, and vegetation.

In this chapter we will concentrate our efforts on laying down the foundations of basic thermodynamics, microphysics, and radiative transfer and surface layer theory.

2 ATMOSPHERIC THERMODYNAMICS

Atmospheric thermodynamics is the study of the macroscopic physical properties of the atmosphere for which temperature is an important variable. The thermodynamic behavior of a gas, such as air, results from the collective effects and interactions of the many molecules of which it is composed. However, because these explicit microscale processes are too small, rapid, and numerous to be observable and too chaotic to be predictable, their effects are not described by the Newtonian laws governing the molecular processes. Instead, classical thermodynamics consists of laws governing the observed macroscopic statistics of the behavior of the microscopic system of air and the suspended liquids and solids within. As a consequence, thermodynamic laws are only valid on the macroscale where a sufficient population of molecular entities is present to justify the statistical approach to their behavior.

Classical atmospheric thermodynamics seeks to:

- Develop an understanding for a statistically significant mass of air called a *parcel*.
- Classify the interacting forms of energy contained within a parcel.
- Never ask "why" in terms of the first principles of atomic and molecular structure and the energetics of interactions between atoms and molecules.
- Derive all principles from *observables*. For example, the first law of thermodynamics begins with a classification of energy into thermal and nonthermal forms as is observed.
- Create an efficient *book-keeping* system in which the interacting forms of energy and mass are defined in a convenient and nonredundant form.
- Quantify the amount and rate of energy transfer from one form to another.
- Provide a means of calculating a true equilibrium state among these energies as well as defining the conditions for constrained equilibrium.

The atmospheric application of classical thermodynamics differs in that particular attention is paid to the properties of air and forms of the thermodynamic laws and relationships that are most useful for application to the atmosphere. We define *air* to be the gas present in Earth's lower atmosphere where air is sufficiently dense for the empirical laws to apply. This only excludes regions on the order of 50 km and higher above the surface.

This section will begin by defining the basic concepts needed, then move into the thermodynamics of gasses and then specifically into the thermodynamics of air. Finally the thermodynamics of the mixed-phase system, including liquid and ice processes, will be described in Chap. 16.

Basic Concepts, Definitions, and Systems of Units

We begin by defining a *system* to represent a portion of the universe selected for study. It may or may not contain matter or energy. Systems are classified as *open*, *isolated*, or *closed*. In contrast, *surroundings* refer to the remainder of the universe outside the system. An *isolated system* can exchange neither matter nor energy with its surroundings. The universe is defined by the first law as an isolated system! A *closed system* can exchange energy but not matter with its surroundings. An *open system*, on the other hand, can exchange both energy and matter with its surroundings.

Next we define a *property* of the system to be an observable. It results from a physical or chemical measurement. A system is therefore characterized by a set of properties. Typical examples are mass, volume, temperature, pressure, composition, and energy. Properties may be classified as intensive or extensive. An *intensive property* is a characteristic of the system as a whole and not given as the sum of these properties for portions of the system. Pressure, temperature, density, and specific heat are intensive properties. We will adopt a naming convention that uses capital letters for variables describing an "intensive" property. An *extensive property* is dependent on the dimensions of the system and is given by the sum of the extensive properties of each portion of the system. Examples are mass, volume, and energy. *Extensive variables will be named with lowercase letters.*

A homogeneous system is one in which each intensive variable has the same value at every point in the system. Then any extensive property z can be represented by the mass-weighted intensive properties:

$$z = mz \tag{1}$$

where z is the specific property.

We define *phases* to be subsets of a system that are homogeneous in themselves but are physically different from other portions. A *heterogeneous system* is defined to be a system composed of two or more phases. In this case, any extensive property will be the sum of the mass-weighted phases of the system, i.e.,

$$Z = \sum_{\alpha} m_{\alpha} z_{\alpha} \tag{2}$$

where α refers to the particular phase.

An *inhomogeneous system* is one in which intensive properties change in a continuous way, such as how pressure or temperature change from place to place in the atmosphere. Our thermodynamic theory will be applied only locally where the system can be considered homogeneous or heterogeneous as an approximation.

The study of the atmosphere requires the definition of an *air parcel*. This is a small quantity of air whose mass is constant but whose volume may change. The parcel should be considered to be small enough to be considered infinitesimal and homogeneous but large enough for the definitions of thermodynamic and microphysical statistics to be valid. In an atmosphere, where air movements are not confined to a rigid container and therefore expand and contract against local forces of pressure gradient, the parcel quantification is a natural choice for defining an air sample.

Energy is the *ability to perform work* and is a *property* defined in the first law of thermodynamics. Energy is an extensive variable that is normally defined as a product of an intensive and extensive variable. The intensive term defines, by differencing, the direction of the energy transfer. The extensive term is also called the capacity term. An example is mechanical energy in the air, which is written as a product of a pressure (intensive) and volume change (extensive). When other intensive variables are substituted, other forms of energy are represented. For instance, replace pressure in the example above with temperature and the energy is thermal energy; replace it with chemical potential and one defines chemical energy and so on.

The *state* of a system is defined by a set of system properties. A certain minimum number of such properties are required to specify the state. For instance, in an ideal gas, any three of the four properties—pressure, volume, number of moles, and temperature—will define the state of the system. The so-called *equation of state* states an empirical relationship between the state properties that define the system.

If the state of a system remains constant with time if not forced to change by an outside force, the system is said to be in a state of *equilibrium*; otherwise we say the system is in a state of *nonequilibrium*. Independence of the state to time is a necessary but not sufficient condition for equilibrium. For instance, the air within an evaporating convective downdraft may have a constant cool temperature due to the evaporation rate equaling a rate of warming by convective transport. The system is not in thermodynamic equilibrium but is still not varying in time. In that case we say the system is *steady state*.

When there is an equilibrium that if slightly perturbed in any way will accelerate away from equilibrium, it is referred to an *unstable equilibrium*. *Metastable equilibrium* is similar to unstable equilibrium except with respect to perhaps only a single process. For instance, when the atmosphere becomes supersaturated over a plain surface of ice, no ice will grow unless there is a crystal on which to grow. Introduce that crystal and growth will begin, but without it the system remains time independent even to other perturbations such as a temperature or pressure perturbation.

A *process* is the description of the manner by which a change of state occurs. Here are some process definitions:

- *Change to the System* Any difference produced in a system's state. Therefore any change is defined only by the initial and final states.
- *Isothermal Process* A change in state occurring at constant temperature.
- *Adiabatic Process* A change of state occurring without the transfer of thermal energy or mass between the system and its surroundings.
- *Diabatic Process* A process resulting from the exchange of mass or energy with the surroundings. An example is radiative cooling.
- *Cyclic Process* A change occurring when the system (although not necessarily its surroundings) is returned to its initial state.
- *Reversible or Quasi-Static Process* A special, idealized thermodynamic process that can be reversed without change to the universe. It will be shown that this is possible only when the conditions differ infinitesimally from equilibrium. Hence at any given point during a reversible process the system is nearly in equilibrium. One can think of a reversible process as being composed of small irreversible steps, each of which have only a small departure from equilibrium (see Fig. 1).

In association with these concepts of processes, we also define the *internal derivative* (d_i) to be the change in a system due to a purely adiabatic process. Conversely, the *external derivative* (d_e) is defined to be the change in a system due to a purely diabatic process.

We can say that the total change of a system is the sum of its internal and external derivatives, i.e.,

$$d = d_i + d_e \tag{3}$$

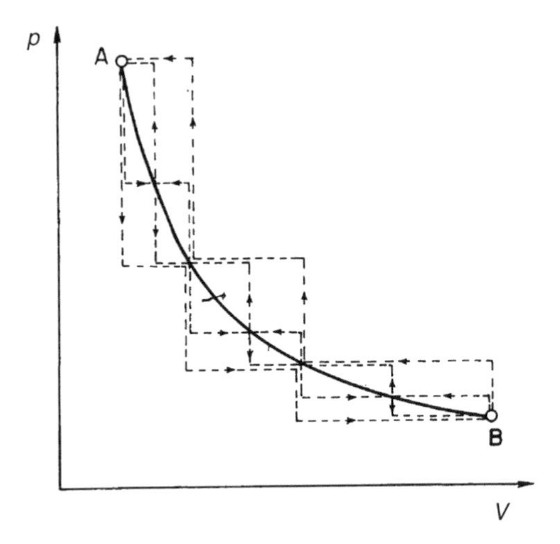

Figure 1 Reversible process as a limit of irreversible processes (from Iribarne and Godson, 1973).

where d is the total derivative. Now, it follows from the definition of adiabatic that

$$\oint d = \oint d_e$$

$$\oint d_i = 0$$

Zeroth Principle of Thermodynamics: Definition of Temperature (*T*)

We define a *diathermic wall* as one that allows thermal interaction of a system with surroundings but not mass exchange. The *zeroth principle of thermodynamics* states: *If some system A is in thermal equilibrium with another system B separated by a diathermic wall, and if A is in thermal equilibrium with a third system C, then systems B and C are also in thermal equilibrium.* It follows that all bodies in equilibrium with some reference body will have a common property that describes this thermal equilibrium and that property is defined to be *temperature*. The so-called *reference body* defines this property and is called a *thermometer*. A number is assigned to the property by creating a temperature scale. This is accomplished by finding a substance that has a property that changes in correspondence with different thermal states. For instance, Table 1 defines five thermometric substances and corresponding properties.

A thermometer is calibrated to changes in its observed thermometric property, obeying the zeroth principle. The thermometer must be much smaller than the substance measured so that the changes brought to the substance by exchanges from the thermometer can be neglected.

A linear temperature scale may be defined generally as:

$$T_X = aX + b \tag{4}$$

where T_X is an arbitrary temperature scale, a represents the slope, b is a constant representing the intercept of the linear scale, and X is the thermometric property of the substance. Use of this scale requires two well-defined thermal states and an approximate linear relationship of T to changes in the substance's thermometric property in order to determine a and b. The Celsius, or Centigrade, temperature scale is defined in this way by the freezing point of pure water at one atmosphere to be 0°C and the boiling point of pure water at one atmosphere to be 100°C. The Fahrenheit temperature scale, used primarily in the United States for nonscientific measurements and unofficially in the United Kingdom, is calibrated by assigning the boiling point of pure water at one atmosphere to be 212°F and the freezing point of pure water at one atmosphere to be 32°F.

A more general scale can be derived using the thermometric properties of gasses listed as the first two examples in Table 1. Using gas as the thermometric substance and pressure as the thermometric property, it is observed that pressure decreases as the thermal state of the gas falls. We then can define the lowest possible thermodynamic state to be that where pressure decreases to zero (at constant volume). Letting the temperature be zero at that point, defines the intercept to also be zero. Now only a single well-defined thermodynamic state is necessary to define the scale. This is chosen to be the triple point for pure water where all three phases of water can coexist. The actual value of this "absolute" temperature scale at the triple point is arbitrary and its choice defines the value of the slope a. We can define this scale to be such that the thermal increments for one degree of temperature equal that of the Celsius scale. We do this by solving for the intercept of the Celsius scale, using pressure as the thermometric property. This yields $b = -273.15$°C. Then defining the freezing point at 1 atm to be 273.15° and the triple point of water to be 273.16° of the absolute temperature, the scale is set to have equal increments to Celsius temperature but have an absolute scale beginning at absolute zero. This scale is called the *Kelvin* scale. The Kelvin temperature is detonated by K. Similarly, an absolute temperature scale is also defined for Fahrenheit units and is called the

TABLE 1 Empirical Scales of Temperature

Thermometric Substance	Thermometric Property X
Gas, at constant volume	Pressure
Gas, at constant pressure	Specific Volume
Thermocouple, at constant pressure and tension	Electromotive force
Pt wire, at constant pressure and tension	Electrical resistance
Hg, at constant pressure	Specific Volume

Source: Irabarne and Godson (1973).

Rankine scale having units of degrees Rankine (R). Generally, atmospheric scientists work with the Celsius and Kelvin scales.

The relationships between the four major scales are

$$T = T_C + 273.15$$
$$T_R = T_F + 459.67$$
$$T_F = \frac{5}{9} T_C + 32$$

where T, T_C, T_F, and T_R are the Kelvin, Celsius and Fahrenheit and Rankine temperatures, respectively.

Ideal Gas and Equation of State for the Atmosphere

A gas is composed of molecules moving randomly about, spinning, vibrating and occasionally colliding with each other. If held within a container, the gas molecules will also collide with the container wall, exerting a force on the walls. The nature of a molecule's movement and the movement of the entire population of molecules is chaotic and unpredictable on a time scale of only a fraction of a second. We make no attempt to quantify a physical description of this process with Newtonian mechanics.

The statistical properties of a gas that are measured include its volume, its temperature, and its pressure. The temperature is a measure of the average kinetic energy of the molecules, which is proportional to the average of the square of the speed of the molecules. The pressure is proportional to force per unit area exerted on a plane surface resulting from the collisions of gas molecules with that surface. Newtonian mechanics tells us that force will be proportional to the average change in momentum of the molecules as they reflect off the surface, which is then proportional to the mass and normal velocity component of each molecule and the total number of molecules hitting the surface.

The mass of the gas is defined to be

$$m \equiv nM \tag{5}$$

where n is the number of moles and M is the molecular weight (in grams per mole). It is often advantageous to express an extensive thermodynamic quantity as the ratio of that quantity to its total mass. That ratio is referred to as the *specific* value of that quantity. For instance, the *specific volume* is given by:

$$\alpha \equiv \frac{V}{m} \equiv \frac{1}{\rho} \tag{6}$$

where V is the volume and ρ is the *density* K/g^3.

We can now define the *equation of state* for the atmosphere. Observations suggest that under normal tropospheric pressures and temperatures, air gases exhibit nearly identical behavior to an ideal gas. To quantify this observation, we define an *ideal gas* to be one that obeys the empirical relationship; also called the *ideal gas law* or the *equation of state.*

$$pV = nR^*/T = mR^*T/M = mRT \tag{7}$$

where n is the number of moles, M is the molecular weight of the gas, R^* is the universal gas constant, and $R = R^*/M$ is the *specific gas constant* valid only for a particular gas of molecular weight M. Observations show that, $R^* = 8.3143$ J/mol/K and is the same for all gases having pressures and temperatures typical of tropospheric conditions.

Employing these definitions, the equation of state for an ideal gas can be written as:

$$p = \frac{RT}{\alpha} = \rho RT \tag{8}$$

It may be shown, by comparison with measurements, that for essentially all tropospheric conditions, the ideal gas law is valid within 0.1%. Under stratospheric conditions the departure from ideal behavior may be somewhat higher.

We now consider multiple constituent gases such as air. In such a case, the partial pressure (p_j) is defined to be the pressure that the jth individual gas constituent would have if the same gas at the same temperature occupied an identical volume by itself. We can also define the *partial Volume* (V_j) to be the volume that a gas of the same mass and pressure as the air parcel would occupy by itself.

Dalton's law of partial pressure states that the total pressure is equal to the sum of the partial pressures:

$$p = \sum_j p_j \tag{9}$$

where p is the total pressure and the sum is over all components (j) of the mixture. Since each partial pressure independently obeys the ideal gas law, volume can be written

$$V = \sum V_j \tag{10}$$

where V is the total volume. Also, it follows that

$$P_j = n_j R^* T/V \tag{11}$$

and

$$p = (R^*T/V) \sum_j n_j \tag{12}$$

If we divide through by the total mass of the gas (m_j) and note that $n_j = m_j/M_j$, where M_j is the molecular weight of the species i, and total mass is

$$m = \sum m_j \tag{13}$$

The statement of the ideal gas law (equation of state) for a mixture of gases is then

$$p = \rho \frac{R^*}{\bar{M}} T = \rho \bar{R} T \tag{14}$$

where $\bar{M}$ is the mean molecular weight of the mixture, which is given by:

$$\bar{M} \equiv \frac{\sum m_j}{\sum \frac{m_j}{M_j}} = \sum M_j N_j \tag{15}$$

where N_j is the molar fraction of a particular gas given by $N_j = p_j/p = V_j/V$.

Composition of Air The current atmosphere of Earth is composed of "air," which we find to contain:

1. *Dry air*, which is a mixture of gases described below
2. *Water*, which can be in any of the three states of liquid, solid, or vapor
3. *Aerosols*, which are solid or liquid particles of small sizes

The chemical composition of dry air is given in Table 2.

Water vapor and the liquid and solid forms of water vary in its volume fraction from 0% in the upper atmosphere to as high as 3 to 4% near the surface under humid conditions. Because of this variation, dry air is treated separately from vapor in thermodynamic theory.

Among the trace constituents are carbon dioxide, ozone, chlorofluorocarbons (CFCs), and methane, which, despite their small amounts, have a very large impact on the atmosphere because of their interaction with terrestrial radiation passing through the atmosphere, or in the case of CFCs because of their impact on the ozone formation process.

Note from Table 2 that the molecular weighted average gas constant for dry air is $R_d = \bar{R} = 287.05$ J/kg/K, which we also call the *dry air gas constant.*

Work by Expansion

If a system is not in mechanical equilibrium with its surroundings, it will expand or contract. Assume that λ is the surface to the system that expands infinitesimally to λ'

TABLE 2 Composition of Dry Air near the Earth's Surface

Gas	Symbol	Molecular Weight	Molar (or Volume) Fraction	Mass Fraction	Specific Gas Constant (J/kg K)	$m_j R_j/m$ ($\bar{R}$) (J/kg K)
Nitrogen	N_2	28.013	0.7809	0.7552	296.80	224.15
Oxygen	O_2	31.999	0.2095	0.2315	259.83	60.15
Argon	Ar	39.948	0.0093	0.0128	208.13	2.66
Carbon Dioxide	CO_2	44.010	0.0003	0.0005	188.92	0.09
Helium	He		0.0005			
Methane	CH_4		0.00017			
Hydrogen	H		0.00006			
Nitrous Oxide	N_2O		0.00003			
Carbon Monoxide	CO		0.00002			
Neon	Ne		0.000018			
Xenon	X		0.000009			
Ozone	O_3		0.000004			
Krypton	Kr		0.000001			
Sulfur Dioxide	SO_2		0.000001			
Nitrogen Dioxide	NO_2		0.000001			
Chlorofluorocarbons	CFC		0.00000001			
Total			1.0000	1.0000		$\bar{R} = 287.05$

in the direction ds (see Fig. 2). Then the surface element $d\sigma$ has performed work against the external pressure P. The work performed is

$$(dW)_{d\sigma} = P\, d\sigma\, d\lambda \cos\phi = P_{\text{surr}}\, dV \tag{16}$$

where dV is the change in volume.

For a finite expansion, then

$$W = \int_i^f p\, dV \tag{17}$$

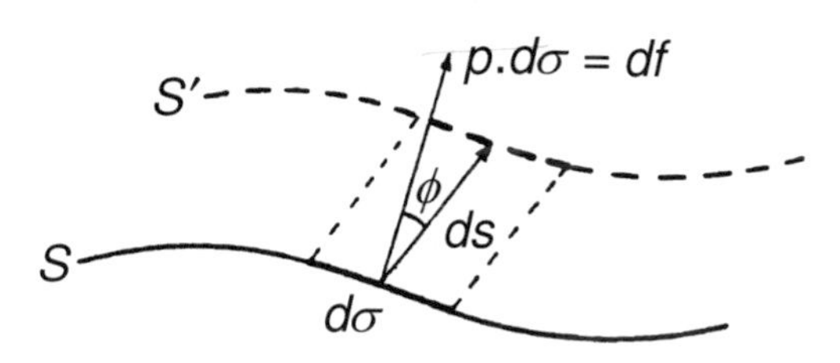

Figure 2 Work of expansion (from Iribarne and Godson, 1973).

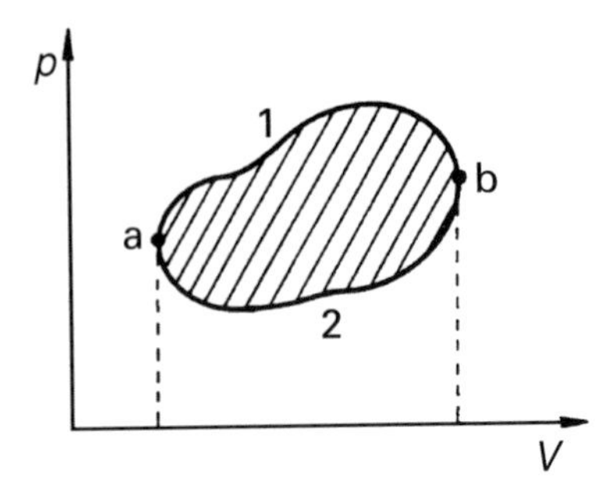

Figure 3 Work of expansion in a cycle (from Iribarne and Godson, 1973).

where i, f stand for the initial and final states. Since the integrand is not a total derivative, total work over a cyclic process can be nonzero. Therefore

$$W = \oint p\, dV = \left(\int_a^b P\, dV\right)_1 - \left(\int_a^b P\, dV\right)_2 \tag{18}$$

which is defined to be positive if integrated in the clockwise sense. This is the area enclosed by the trajectory in the graph given in Figure 3.

The work by expansion is the only kind of work that we shall consider in our atmospheric systems. Assuming the pressure of the system is homogeneous and in equilibrium with the surroundings, we can adopt the notation

$$dW = P\, dV \tag{19}$$

where W is the work performed on the surroundings by the system.

First Law (or Principle) of Thermodynamics

The *first law of thermodynamics* can be simply stated: *The energy of the universe is constant*. Let us consider the general concept of just what is meant by *energy*. We know that if we apply force to an object over some distance and as a result we accelerate the object, we will have performed *work* on the object, measured in the energy units of *force* × *distance* = *joules*. That work will precisely equal the increase in kinetic energy of the object, the total of which is measured by half its mass multiplied by the square of its new velocity. Moreover, we understand that our effort expends energy from the *working substance*. Depending on the method used to accelerate the object, the energy was previously stored in some other form; perhaps the "potential energy" of a compressed spring, "chemical energy" stored in the muscle cells of one's arm, or perhaps the "kinetic energy" of another object that strikes the object that was accelerated. There are obviously many possibilities for working substances featuring different types of stored energy. Energy, in a general sense, quantifies a potential to apply a force and so perform work.

The first law requires that for any thermodynamic system an energy budget must exist that requires any net energy flowing into a system to be accounted for by changes in the *external energy* of the system or the *internal energy* stored within the system. The external energy of the system includes its kinetic energy of motion and its energy of position, relative to forces outside the system such as gravitational, electrical, magnetic, chemical, and many other forms of potential energy between the system and its surroundings. The same energies can also exist within the system between the molecules, atoms, and subatomic particles composing the system and are termed internal energies. Kinetic energy of molecular motions relative to the movement of the center of mass of the system is measured by the system's temperature and represent the *thermal internal energy*. The other internal energies are *potential internal energies* and may include a potential energy against the intermolecular forces of attraction (*latent heat*), chemical potential energy (e.g., a gas composed of oxygen and hydrogen can potentially react), and so on.

Because we probably are not even aware of all of the forms of internal forces that exist in a system, we make no attempt to evaluate the total internal energy. Instead, for the purposes of classical thermodynamics, we need only consider changes in internal energy occurring during *allowable processes*. For instance, for atmospheric studies we choose to ignore nuclear reactions and even most chemical reactions. But we cannot neglect the changes in internal potential energy due to intermolecular forces of liquid and ice water phases in the atmosphere. The intermolecular forces are substantial, and the energy needed to overcome them is the latent heat of vaporization and melting. Thermodynamic energy transfers between thermal and these internal potential energies drive the general circulation of Earth's atmosphere! Our limited thermodynamic discussion will ignore the treatment of some internal energies such as surface energy of droplets, chemical energy in photochemical processes, electrical energy in thunderstorms and the upper atmosphere, chemical processes involving CFCs, and other processes known to have important secondary impacts on the evolution of the atmospheric thermodynamic system. These processes can be added to the system when needed, following the methods described in this chapter.

In most texts of classical thermodynamics, *closed* thermodynamic systems are assumed. A *closed* system allows no mass exchange between the parcel (system) and the surroundings. This approximation greatly simplifies the thermodynamic formulation. Later, we can still account for molecular transfer by representing it as a noninteracting and externally specified source of mass or energy rather than as an explicitly interacting component of the formulation.

One such open mass flux that must be accounted for with an open system is the mass flux of precipitation. We affix our parcel coordinate relative to the center of mass of the dry air, which is assumed to move identically to vapor. The liquid and ice components of the system, however, may attain a terminal velocity that allows it to flow into and out of the parcel diabatically. As a result, we will derive thermodynamic relationships for an *open system*, or one where external fluxes of mass into and out of the system are allowed. To retain some simplicity, however, we will make the assumption that external mass fluxes into and out of the system will be of constituents having the same state as those internal to the system. Hence we will allow rain to fall into our parcel, but it will be assumed to have the same temperature as the system. Although the effect of this assumption can be scaled to be small, there are instances where it can be important. One such instance is the case of frontal fog, where warm droplets falling into a cool air mass result in the formation of fog. These moist processes will be described in detail in the next chapter.

As demanded by the first law, we form an energy budget, first for a simple one-component system of an ideal gas:

$$\text{Energy exchanged with surroundings} = \text{change in energy stored}$$
$$Q - W = \Delta U - \sum_k \Delta A_k \tag{20}$$

where Q represents the flow of thermal energy into the system including radiative energy and energy transferred by conduction, W is the work performed on the surroundings by the system, U is the thermal internal energy, and A_k is the kth component of potential internal energy, summed over all of the many possible sources of internal energy. The work, as described in the previous section, is the work of expansion performed by the parcel on the surroundings. It applies only to our gas system and does not exist in the same form for our liquid and ice systems. In those systems, it should be reflected as a gravitational potential term; however, it is typically neglected.

In general, we treat the air parcel as a *closed* system, whereby there are no mass exchanges with the surroundings. This is generally a good approximation as molecular transfers across the boundaries can usually be neglected or included in other ways. One exception occurs and that is when we are considering liquid and ice hydrometeors. If we fix our parcel to the center of mass of the dry air, then there can be a substantial movement of liquid or ice mass into and out of the parcel. Clearly this process must be considered with an *open* thermodynamic system, where at least fluxes of liquid and ice are allowed with respect to the center of mass of the dry air parcel. Hence we will consider the possibility that A_k may change in part because of exchanges of mass between the parcel and the environment, particularly present with falling precipitation. Combining Eq. (20) with (19) for an infinitesimal process, we obtain the form of the first law:

$$\delta Q = dU + P\,dV - \sum_k dA_k \tag{21}$$

where it should be noted that P is the pressure of the surroundings, as $P\,dV$ defines work performed on the surroundings and $-dV$ is the change in volume of the surroundings. Note that for the adiabatic case, $\delta Q = 0$ by definition.

Because we require conservation of energy, it is evident that the change in both internal thermal energy (dU) and internal potential energy (dA_k) must be exact differentials, only dependent on the initial and final states of the process and not the path that the process takes. Hence *U and A_k must also be state variables.*

Heat Capacities *Heat capacity* is a property of a substance and is defined to be the rate at which the substance absorbs (loses) thermal energy (Q) compared to the rate at which its temperature (T) rises (falls) as a result. It is a property usually defined in units of dQ/dT (energy per temperature change) or in units of dq/dT (energy per mass per temperature change) in which case it is called specific heat capacity. If heat is passed into a system, it may be used to either increase the internal thermal or potential energy of the system or can be used to perform work. Hence there are an infinite number of possible values for heat capacity depending on the processes allowable. It is useful to determine the heat capacity for special cases where the allowable processes are restricted to only one. Hence we neglect the storage of potential internal energy and restrict the system to constant volume or constant pressure processes. Hence,

$$
\begin{aligned}
C_v &= \left(\frac{\delta Q}{dT}\right)_V \qquad c_v = \left(\frac{\delta q}{dT}\right)_\alpha \\
C_p &= \left(\frac{\delta Q}{dT}\right)_p \qquad c_p = \left(\frac{\delta q}{dT}\right)_p
\end{aligned}
\tag{22}
$$

where C_v and C_p are the heat capacities at constant volume and pressure, respectively, and where c_v and c_p are the specific heat capacities at constant volume and pressure, respectively. It can be expected that $C_p > C_v$, since in the case of constant pressure, the parcel can use a portion of the energy to expand, and hence perform work on the surroundings, reducing the rise in temperature for a given addition of heat.

Since the first law requires that the change in internal energy, U, be an exact differential, U must also be a state variable; i.e., its value is not dependent upon path. Hence $U = u(P, \alpha, T)$. If we accept that air can be treated as an ideal gas, then the equation of state eliminates one of the variables, since the third becomes a function of the other two. Employing Euler's rule,

$$
dU = \left(\frac{\partial U}{\partial V}\right)_T dV + \left(\frac{\partial U}{\partial T}\right)_V dT \tag{23}
$$

Substituting Eq. (23) into the first law [Eq. (20)], we get

$$
\delta Q = \left(\frac{\partial U}{\partial T}\right)_V dT + \left(\frac{\partial U}{\partial V} + P\right)_T dV - \sum_k dA_k. \tag{24}
$$

For a *constant volume process*,

$$
C_v = \left(\frac{\delta Q}{dT}\right)_V = \left(\frac{\partial U}{\partial T}\right)_V \tag{25}
$$

Hence, returning to Eq. (23),

$$dU = C_v\,dT + \left(\frac{\partial U}{\partial V}\right)_T dV. \tag{26}$$

It has been shown experimentally that $(\partial U/\partial V)_T = 0$, indicating that air behaves much as an ideal gas. Hence the internal energy of a gas is a function of temperature only provided that C_v is a function of temperature only. If we assume that possibility, $\partial C_v/\partial V = 0$. Experiments show that for an ideal gas the variation of C_v with temperature is small, and since we hold air to be approximately ideal, it follows that C_v is a constant for air. The first law [Eq. (24)] is therefore written as:

$$\delta q = C_v\,dT + P\,dV - \sum_k dA_k. \tag{27}$$

One can also obtain an alternate form of the first law by replacing dV with the ideal gas law and then combining q^{ith} Eq. (27) to yield

$$\delta Q = (C_v + R^*)\,dT - V\,dP - \sum_k dA_k. \tag{28}$$

For an isobaric process, $(dP = 0)$, one finds

$$C_p = \left(\frac{\delta q}{\partial T}\right)_p = C_v + R^* \tag{29}$$

or

$$C_p - C_v = R^*. \tag{30}$$

Similarly for dry air, the specific heat capacities are

$$c_{pd} - c_{vd} = R_d \tag{31}$$

Equation (30) demonstrates that $C_p > C_v$. Statistical mechanics show that specific relationships between specific heats at constant volume and at constant pressure can be found for a gas depending on how many atoms form the gas molecule. Table 3

TABLE 3 Relationship between Specific Heats and Molecular Structure of Gas

Monotonic gas	$C_v = \frac{3}{2}R$	$C_p = C_v + R = \frac{5}{2}R^*$
Diatomic gas	$C_v = \frac{5}{2}R$	$C_p = \frac{7}{2}R$

shows these relationships for diatomic and monoatomic gases. Since dry air, composed mostly of molecular oxygen and nitrogen is nearly a diatomic gas,

$$c_{pd} = 1004 \text{ J/kg/K}, \qquad c_{vd} = 717 \text{ J/kg/K}. \tag{32}$$

Substituting Eq. (30) into Eq. (28) we obtain

$$\delta Q = C_p \, dT - V \, dP - \sum_k dA_k \tag{33}$$

This form of the first law is especially useful because changes of pressure and temperature are most commonly measured in atmospheric science applications. Note that $C_p \, dT$ is not purely a change in energy, nor is $V \, dp$ purely the work.

Enthalpy For convenience, we introduce a new state variable called *enthalpy*, defined as:

$$H \equiv U + PV. \tag{34}$$

Differentiating across we obtain

$$dH = dU + p \, dV + V \, dP, \tag{35}$$

and substituting Eq. (21), to eliminate dU:

$$\delta Q = dH - V \, dP - \sum_k dA_k. \tag{36}$$

Using Euler's rule with Eq. (33), we obtain

$$\left(\frac{\partial H}{\partial T} \right)_p = C_p. \tag{37}$$

Enthalpy differs only slightly from energy but arises as the preferred energy variable when the work term is expressed with a pressure change rather than a volume change. The enthalpy variable exists purely for convenience (since pressure is more easily measured than volume) and is not demanded or defined by any thermodynamic law.

Reversible Adiabatic Processes in the Atmosphere: Definition of Potential Temperature (θ) These types of processes are of great importance to the atmospheric scientist. Ignoring radiative effects, or heating effects of condensation, the dry adiabatic process represents the thermal tendencies that air will experience as it rises and the pressure lowers or as it sinks and the pressure rises. In the absence of condensation, this represents the bulk of the temperature change a parcel will experience if its rise rate is fast enough to ignore radiative effects. We defined an adiabatic process as one for which $Q = 0$. Then for an adiabatic process, combining Eq. (33), the equation of state [Eq. (7)], and then integrating we obtain

$$T = kP^{\kappa} \tag{38}$$

where $\kappa = R^*/C_p = 0.29$ and k is a constant of integration that can be determined from a solution point. Equation (38) is known to atmospheric scientists as the *Poisson equation*, not to be confused with the second-order partial differential equation also known as "Poisson's equation." The Poisson equation states that, given a relationship between temperature and pressure at some reference state, the value of temperature can be calculated for all other pressures the system might obtain through reversible dry adiabatic processes.

It is often convenient to define *potential temperature (θ)* to be the temperature that a parcel would have at 1000 kPa pressure. It follows directly from Eq. (38) that

$$\theta \equiv T\left(\frac{p_{00}}{P}\right)^{R/c_p}, \tag{39}$$

where $p_{00} = 1000\,\text{kPa}$ is the reference pressure at which potential temperature (θ) equals temperature (T). The potential temperature proves extremely useful to meteorologists. For all reversible dry adiabatic atmospheric thermodynamic processes, θ remains constant and represents a conserved property of the flow. Later we will show θ also represents the dry entropy of the flow and that it is one in a hierarchy of entropy variables that are conserved in flow under various permitted dry and moist processes.

Second Law (or Principle) of Thermodynamics

The zeroth principle of thermodynamics defined temperature to be a quantity that is determined from two bodies in thermal equilibrium. The first law stated the principle of energy conservation during any thermodynamic process. The second law deals with the direction of energy transfer during a thermodynamic process occurring when two bodies are not in thermodynamic equilibrium. It is again an empirical statement, not derived in any way from first principles but rather from observations of our perceived universe. The second law can be stated in two ways that can be shown to be equivalent:

1. *Thermal energy will not spontaneously* flow from a colder to a warmer object.
2. A thermodynamic process always acts down gradient in the universe to reduce differentiation overall and hence mix things up and reduce order overall. If we define entropy to be a measure of the degree of disorder, then the second law states: *The entropy of the universe increases or remains the same as a result of any process. Hence, the entropy of the universe is always increasing.*

Just as the first law demanded that we define the thermal variable temperature, the second law requires a new thermodynamic variable *entropy*. To form a mathematical

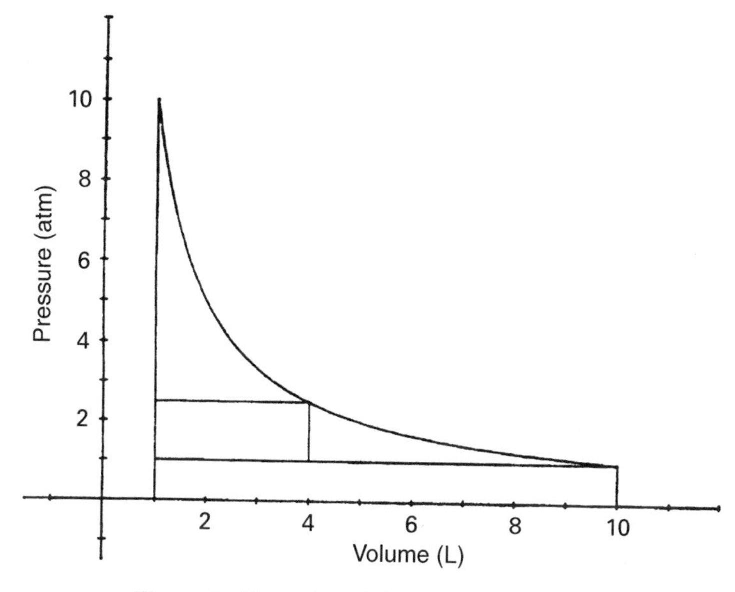

Figure 4 Illustration of three isothermal processes.

treatment for entropy, we consider several special thermodynamic processes that reveal the nature of entropy behavior and provide guidance to its form.

Mathematical Statement of the Second Law It is illustrative to examine the amount of work performed by a parcel on its surroundings under *isothermal* conditions, implying that the internal energy of the system is held constant. This will demonstrate an interesting behavior addressed by the second law.

Consider a fixed mass of an ideal gas confined in a cylinder fitted with a movable piston of variable weight. The weight of the piston and its cross-sectional area determine the pressure on the gas. Let the entire assembly be placed in a constant temperature bath to maintain isothermal conditions. Let the initial pressure of the gas be 10 atm and the initial volume be 1 liter. Now consider three isothermal processes (Fig. 4) (Sears, 1953):

Process I The weight on the piston is changed so as to produce an effective pressure of 1 atm and since $PV = nR^*T$, its volume becomes 10 liters. The work of expansion is given by:

$$W_{\text{exp}} = P\Delta Va$$
$$(1 \text{ atm} = 1013 \times 10^2 \text{ Pa}) \tag{40}$$

Since P_{surr} is constant at 1 atm, the work is $1(10 - 1) = 9$ Latm. This work is performed on the surroundings. With respect to the system, the energy transfer (δQ) is therefore -9 Latm.

Process II This will occur in two stages: (1) Decrease the pressure exerted by the piston to 2.5 atm so that the specific volume will be 4 liters and then (2) reduce the pressure to 1 atm with a specific volume of 10 liters. The work of expansion becomes the sum:

$$
\begin{aligned}
w_{\text{exp}} &= P_1\,\Delta V_1 + P_2\,\Delta V_2 \\
&= 7.5 + 6.0 \\
&= 13.5\ \text{Latm}
\end{aligned}
$$

The mechanical energy transfer from the system (δQ) is thus -13.5 Latm.

Process III Let the pressure exerted by the piston be continuously reduced such that the pressure of the gas is infinitesimally greater than that exerted by the piston (otherwise no expansion would occur). Then the pressure of the gas is essentially equal to that of the surroundings and since $dw = P\,dV$,

$$
\begin{aligned}
W_{\text{exp}} &= \int_{V_1}^{V_2} P\,d\alpha = R^*T \ln\left(\frac{V_2}{V_1}\right) \\
&- -23.03\ \text{Latm}
\end{aligned}
$$

In each of the above three cases, the system responded to the removal of external weight to the piston by changing state and performing a net work on the surroundings while maintaining isothermal conditions within the system by absorbing heat from the attached heat reservoir, the reservoir being part of the surroundings. In each case, the total change in the system state was identical, although the energy absorbed by the system from the surroundings, equal also to the work performed on the surroundings by the process varied greatly.

Note that process III differed from the first two processes in that it was an *equilibrium* process, or one where the difference between the intensive driving variable (pressure in this case) of the system differed only infinitesimally from the surroundings. The maximum work possible for the given change in system state occurs in such an equilibrium process. All other processes, not in equilibrium, produce less net work on the surroundings and thus also absorb less energy from the heat reservoir!

It is important to note that, unlike the nonequilibrium processes, *the net work for the equilibrium process is dependent only on the initial and final states of the system*. Hence if the equilibrium process is run in reverse, the same amount of work is performed on the system as the system performed on the surroundings in the forward direction. In the case of the nonequilibrium process, this is not the case and so the amount of energy released to the surroundings for the reverse process is unequal to that absorbed in the forward direction. Hence the system is *irreversible*. Hence we can equate an irreversible process to one which is not in equilibrium.

For an isothermal reversible process involving an ideal gas, $dU = 0$ and hence $\delta Q = p\,dV = dW_{max}$. Thus δQ is dependent only on the initial and final states of the system. *Hence, for a reversible process, Q and W behave like state functions*. As a consequence, since for either the irreversible or reversible processes, $\delta Q = \delta W$, it follows that:

$$\delta Q_{rev} = \delta W_{max} \tag{42}$$

Since the δQ_{rev} is related to a change in state, then the isothermal reversible result described by process III must apply for the general isothermal case, i.e.,

$$\frac{\delta Q_{rev}}{T} = R^* d \ln V \tag{43}$$

where the right-hand side of Eq. (43) is an exact differential. By this reasoning it is concluded that the left-hand side of Eq. (43) must also be an exact differential of some variable that we will define to be the *entropy* (*S*) of the system (joules per kelvin). Since the change in entropy of the system is always an exact differential, its value after any thermodynamic process is independent of the path of that process and so *entropy, itself, is also a state function*. We can write:

$$dS = \frac{\delta Q_{rev}}{T} \tag{44}$$

From Eq. (42) we can also show that

$$\delta W_{max} = T\,dS \tag{45}$$

which states that the maximum work that can be performed by a change in state is equal to the temperature multiplied by the total change in entropy from the initial to the final state of the system. Since we showed that the net work performed on the surroundings as a result of the isothermal process must be less than or equal to dW_{max}, then it is implied that

$$dS > \frac{\delta Q}{T} \tag{46}$$

for an *irreversible* process. It is also reasoned that by virtue of the second law, and obvious from the example described above, that a process where $dS < \delta Q/T$ would be a *forbidden* process. Since the entropy of the surroundings must also satisfy these constraints, the entropy of the universe can either remain constant for a reversible process or increase overall for an irreversible process. This is the mathematical statement of the second law for the case of isothermal processes.

We must now expand this concept to a system not constrained to isothermal conditions. To build a theory applicable to a general process, we consider the following additional particular processes:

1. *Adiabatic Reversible Expansion of an Ideal Gas* For a reversible adiabatic expansion, $\delta Q = 0$ and so $dS = 0$. This is an *isentropic process*. Since a

reversible dry adiabatic process is one where θ is conserved, a process with constant θ is also called an isentropic process.

2. *Heating of a Gas at Constant Volume:* For a reversible process at constant volume, the work term vanishes:

$$dS = \frac{dQ_{\text{rev}}}{T} = C_v \frac{dT}{T} = C_v d \ln T \tag{47}$$

3. *Heating of an Ideal Gas at Constant Pressure:* For a reversible process:

$$dS = \frac{dQ_{\text{rev}}}{T} = C_p \frac{dT}{T} = C_p d \ln T \tag{48}$$

The Carnot Cycle We demonstrated earlier that a net work can also be performed on the surroundings as a system goes through a cyclic process. That work, we showed, was equal to the area traced out on a P–V diagram by that cyclic process. The maximum amount of work possible by any process was that performed by a *reversible process*. We can call our *system* that performs work on the surroundings as the *working substance*.

Now we ask, *Where does the energy for the work performed originate?* For a cyclic process the initial and final states of the working substance are the same so any work performed over that process must come from the *net* heat absorbed by the system from the surroundings during that process. We say *net* because the system typically *absorbs* heat and also *rejects* heat to the surroundings. If the system converted all of the heat absorbed to work without rejecting any we would say that the system is 100% efficient. That cannot happen typically. For instance, the engine in a car burns gasoline to heat air within the cylinder that expands to do work. But much of the heat given off from the gas actually ends up warming the engine and ultimately the surroundings rather than making the engine turn. The degree to which the energy warms the surroundings compared to how much is used to turn the engine and ultimately move the car (more energy is lost to friction in the engine and transmission and so on) is a measure of how efficient the car is.

For any engine, we can quantify its efficiency as:

$$\eta \equiv \frac{\text{Mechanical work performed by engine}}{\text{Heat absorbed by engine}} \equiv \frac{Q_1 - Q_2}{Q_1} = \frac{q_1 - q_2}{q_1} \tag{49}$$

where Q_1 is the heat absorbed by the engine and Q_2 is the heat rejected by the engine to the surroundings. In terms of specific heat:

$$\eta = \frac{q_1 - q_2}{q_1} \tag{50}$$

The maximum efficiency that any engine can ever achieve would be that of an engine powered by a reversible process. Processes such as friction in the engine or trans-

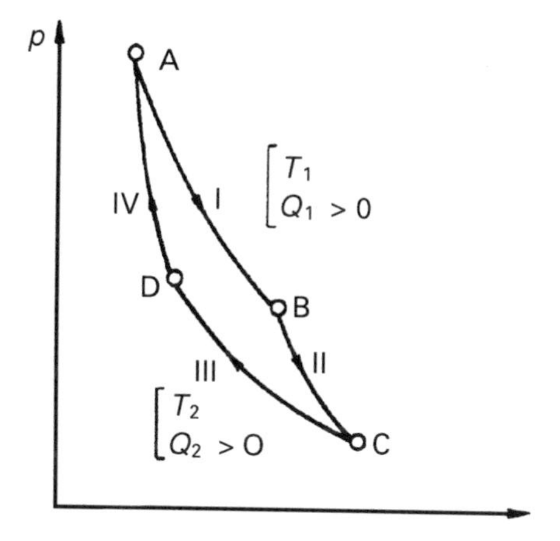

Figure 5 Carnot cycle (from Iribarne and Godson, 1973).

mission, not perfectly insulated walls of the pistons, and so on would all be irreversible parts of the cyclic process in a gasoline engine. Now let's imagine an engine where the thermodynamic cyclic process is carried out perfectly reversibly. One simple reversible thermodynamic process that fits this bill is the *Carnot cycle*, which powers an imaginary engine called a *Carnot engine*. No one can build such a perfect frictionless engine, but we study it because it tells us what the *maximum efficiency possible* is for cyclic process. Yes, even the Carnot engine is not 100% efficient and below we will see why.

First let's describe the Carnot cycle. Simply put, the Carnot cycle is formed by two adiabatic legs and two isothermal legs so that the net energy transfers can be easily evaluated with simple formulations for adiabatic or isothermal processes. Figure 5 shows such a process. The four legs of the process are labeled I to IV beginning and ending at point A. To visualize this physically, we can imagine some working substance contained within a cylinder having insulated walls and a conducting base, and a frictionless insulated beginning piston. The cycle is explained as:

I The cylinder is placed on a heat reservoir held at $T = T_1$. Beginning at $T = T_1$, and $V = V_A$, the working substance is allowed to slowly expand maintaining isothermal conditions.

II The cylinder is placed on an insulated stand and allowed to expand further to volume V_C and temperature T_2. Since the working substance is totally insulated for stage II, the process is *adiabatic*.

III The cylinder is taken from the stand and placed on a heat reservoir where the working substance is held at $T = T_2$ and slowly compressed (isothermally) to specific volume V_D.

IV The cylinder is replaced on the insulated stand and the working substance is slowly compressed (perhaps adding weight to piston slowly) causing the substance to warm adiabatically back to temperature T_1 and specific volume V_A

This process is illustrated graphically in Figure 5. The cycle is reversible and consists of two isotherms at temperatures T_1 and T_2, where $T_2 < T_1$ and two adiabats, i.e., θ_1 and θ_2.

The net work performed by the system is the area formed by the intersection of the four curves. We can compute the work A and the heat Q for the four steps of the process (direction indicated by arrows) in the following table:

$$
\begin{array}{lll}
\text{(I)} & \Delta U_{\mathrm{I}} = 0 & \delta Q_{\mathrm{I}} = -A_{\mathrm{I}} = \int_A^B P\,dV = R^* T_1 \ln V_B/V_A \\
\text{(II)} & \delta Q_{\mathrm{II}} = 0 & -A_{\mathrm{II}} = -\Delta U_{\mathrm{II}} = C_V(T_1 - T_2) \\
\text{(III)} & \Delta U_{\mathrm{III}} = 0 & \delta Q_2 = -A_{\mathrm{III}} = -R^* T_2 \ln V_C/V_D \\
\text{(IV)} & \delta Q_{\mathrm{IV}} = 0 & -A_{\mathrm{IV}} = -\Delta U_{\mathrm{IV}} = -C_V(T_1 - T_2)
\end{array}
$$

Since this is a cyclic process, the sum of the four terms Δu is zero. The work terms on the two adiabats cancel each other. The gas effectively absorbs the quantity of heat $Q_1 > 0$ from the warmer reservoir and rejects it $Q_2 < 0$ in the colder reservoir. The total work is then $W = A_{\mathrm{I}} + A_{\mathrm{III}} = -(Q_1 + Q_2)$.

We can now relate q_1 and q_2 to the temperatures of the two heat reservoirs. From Poisson's equation,

$$\frac{T_1}{T_2} = \left(\frac{V_C}{V_B}\right)^{R^*/C_p} = \left(\frac{V_D}{V_A}\right)^{R^*/C_p}, \tag{51}$$

and so:

$$\frac{V_B}{V_A} = \frac{V_C}{V_D}, \tag{52}$$

which, substituting into the expressions for Q_1 and Q_2 gives

$$\frac{Q_1}{T_1} + \frac{Q_2}{T_2} = 0, \tag{53}$$

or, alternatively

$$\left| \frac{|Q_2|}{Q_1} \right| = \frac{T_2}{T_1}. \tag{54}$$

Now we can calculate the thermodynamic efficiency of the Carnot engine:

$$\eta = \frac{T_1 - T_2}{T_1}. \tag{55}$$

Hence, *the Carnot cycle achieves its highest efficiencies when the temperature difference between the two heat reservoirs is the greatest!*

It is useful to generalize this result to the case of irreversible processes. To do so, we consider the second law, and its implication that heat can only flow from a warmer toward a cooler temperature. As a first step, it is noted that Eq. (55) is valid for any reversible cycle performed between two heat sources T_1 and T_2, independent of the nature of the cycle and of the systems. This is a statement of *Carnot's theorem*.

Just as Eq. (55) holds for a reversible cycle, it can be shown that for any *irreversible* cycle, the second law requires that $(Q_1/T_1) + (Q_2/T_2) < 0$, i.e., heat *must* flow from warm to cold and not vice versa. It can also be shown that any reversible cycle can be decomposed into a number of Carnot cycles (reversible cycles between two adiabats and two isotherms). Therefore, for any irreversible process it can be shown that

$$\oint \frac{\delta Q}{T} \leq 0 \tag{56}$$

which implies Eq. (46).

The Carnot cycle can also be viewed in reverse, in which case it is a refrigerating machine. In that case a quantity Q_2 of heat is taken from a cold body (the cold reservoir) and heat Q_1 is given to a hot reservoir. For this to happen, mechanical work must be performed on the system by the surroundings. In the case of a refrigerator, an electric motor supplies the work.

In recent years, atmospheric scientists have discovered that certain types of weather systems derive their energy from a Carnot cycle. In particular, the tropical cyclone* is powered by a Carnot-like cycle created by the radial circulation inward to the storm center at the surface, up within the eye-wall clouds, outward at the storm top, and then back downward far away from the storm center. State I, or the Carnot cycle, is equivalent to the air moving inward toward the storm center along the ocean surface at relatively constant temperature and toward low pressure in the storm center. The heat reservoir is the ocean. In this case it heats not only through the transfer of sensible heat but also through the transfer of latent heat. Stage II is found in the eye wall, where the heated air rises moist adiabatically under falling temperatures. Stage III occurs in the outflow at high levels away from the storm center. The outflowing air maintains a constant temperature while slowly subsiding and pressure

*Tropical cyclones are known as "hurricanes" in the Atlantic and eastern Pacific oceans, while in the western Pacific they are known as "typhoons."

rising. The heat is lost to space by transfer through the emission of long-wave radiation. Hence the cold plate is the radiation sink in space. Finally, stage IV occurs as the air finally sinks back to the ground approximately moist adiabatically. This occurs mainly within the downdrafts of convective clouds in the periphery of the storm. The result is a net gain in energy of the entire system, which is manifested in the cyclone winds. The storm loses energy primarily to surface friction, which is proportional to the speed of the winds. Therefore, the more energy released by the Carnot cycle, the stronger the surface winds that come into equilibrium with the Carnot cycle release. As with the discussion above, the efficiency of the cycle is proportional to the temperature difference between storm top and the sea surface. As a result, the strength attainable by a tropical cyclone is closely related to the sea surface temperature.

Energy harnessed by the Carnot cycle is likely the basis for other types of weather for which thermally driven convection is important. These include a host of convective systems and possibly some aspects of Earth's general circulation that derive their energy from thermally direct circulations.

Restatement of First Law with Entropy The second law is a statement of inequality regarding the limits of entropy behavior. When combined with the first law, the resulting relationships become inequalities:

$$T\,ds \geq dU + P\,dV - \sum_k dA_k \tag{57}$$

$$T\,dS \geq dH - V\,dP - \sum_k dA_k \tag{58}$$

for the internal energy and enthalpy forms, respectively. Alternatively, we may write for the special case of a reversible process:

$$dU = T\,dS - P\,dV + \sum_k dA_k \quad \text{and} \tag{59}$$

$$dH = T\,dS + V\,dP + \sum_k dA_k. \tag{60}$$

Note this differs from previous forms of the first law in that we have an equation for the parcel (or system) in terms of state function variables only. The difference between the general Eqs. (57) and (58) and the special case Eqs. (59) and (60) is a positive source term for entropy resulting from a nonequilibrium reaction that has not yet been determined. In the world of dry atmospheric thermodynamics, we can live with assuming the dry system is in equilibrium and so the second form is sufficient. However, when we begin to consider moist processes, the system is not always in equilibrium, and so the sources for entropy from nonequilibrium processes will have to be evaluated. The framework for the evaluation of these effects involves the creation of free energy relations described in the next chapter.

Free Energy Functions In their present form, Eqs. (57) and (58) relate thermodynamic functions that involve dependent variables that include the extensive variable S. It is again convenient to make a variable transformation to convert the dependence to variations in T instead of S. The transformation is made by defining the so-called *free energy functions* called the *Helmholtz free energy* (F) and the *Gibbs free energy* (G) are defined as:

$$F \equiv U - TS$$

$$G \equiv H - TS$$

In derivative form, they are written:

$$dF = dU - T\,dS - S\,dT$$

$$dG = dH - T\,dS - S\,dT$$

For our application to atmospheric problems, we will work primarily with the Gibbs free energy because of its reference to a constant pressure process.

Combining Eq. (61) with Eq. (60) we obtain

$$dG = -S\,dT + V\,dP + \sum_k dA_k + \sum_k dG_k \tag{61}$$

where G_k refers to Gibbs energies for each constituent of the system (k) necessary to form an equality of Eq. (58). The restrictions imposed by the second law and reflected by those inequalities result in the requirement that $\sum_k dG_k \leq 0$. Because $G = G(T, V, n_j)$ is an exact differential, then

$$dG = \left(\frac{\partial G}{\partial T}\right)_{P,n_k} dT + \left(\frac{\partial G}{\partial P}\right)_{T,n_k} dP + \left(\frac{\partial G}{\partial n_k}\right)_{P,T,n_j} dn_k \tag{62}$$

We can evaluate the potential internal energy and Gibbs free energy terms of Eq. (61) as:

$$\sum_k dA_k = \left(\frac{\partial U}{\partial n_k}\right)_{T,P} dn_k$$

$$\sum_k dG_k = \left(\frac{\partial G_\mu}{\partial n_k}\right)_{T,P} dn_k$$

where G_μ is the Gibbs free energy resulting from chemical potential. The terms of Eq. (62) are then evaluated:

$$\left(\frac{\partial G}{\partial n_k}\right)_{T,P} = \left(\frac{\partial U}{\partial n_k}\right)_{T,P} dn_k + \left(\frac{\partial G_\mu}{\partial n_k}\right)_{T,P} dn_k,$$

$$\left(\frac{\partial G}{\partial T}\right)_{V,n_k} = -S, \text{ and}$$

$$\left(\frac{\partial G}{\partial V}\right)_{T,n_k} = -P.$$

Concept of Equilibrium Equilibrium is not an absolute but is defined in terms of permitted processes. In statics, equilibrium is described as the state when the sum of all forces is zero. This may be stated as a principle of work: *If a system is displaced minutely from an equilibrium state by a change in one of the system properties, the sum of all the energy changes is zero*. Neglecting internal potential energies (A_k), this leads to the statement that for a single component system of ideal gas, undergoing a constant temperature, constant volume process, $dF = 0$ while for a constant temperature process and constant pressure process $dG = 0$ at equilibrium. Hence if one were to plot G (or F) as a function of system properties, equilibrium will appear as a minimum along the G (or F) curve.

In general, we define a spontaneous process as one in which the system begins out of equilibrium and moves toward equilibrium, resulting in an increase in entropy. The rate of the process is undetermined thermodynamically. Consider the Gibbs free energy for a constant temperature and constant pressure process. Since equilibrium occurs at a minimum in Gibbs free energy, we can define:

A *spontaneous* process	$\Delta G < 0$
An *equilibrium* process	$\Delta G = 0$
A *forbidden* process	$\Delta G > 0$

The *Molar chemical potential* of species k (μ_k) is defined as:

$$\mu_j \equiv \left(\frac{\partial G_\mu}{\partial n_k}\right)_{T,p,n_j}. \tag{63}$$

For a system of only one component of a set on noninteracting components:

$$\mu_k \, dn_k = G_k(m) = H_k - TS_k \tag{64}$$

in which $G_k(m)$ is the molar free energy of species k. We now write:

$$dG = -S\,dT + V\,dP + \sum_k dA_k + \sum_k \mu_k\,dn_k. \tag{65}$$

Hence the third term on the right-hand side (RHS) would represent free energies resulting from the potential for interaction between all components in the system, while the last term on the RHS represents the noninteracting free energy of each component, for example, surface free energy.

Consider a system composed of a single ideal gas and held in a state of equilibrium where $dG = 0$. Ignoring internal potential energies, Eq. (65) becomes

$$d\mu = S\,dT - V\,dP. \tag{66}$$

Using the equation of state and integrating, we obtain

$$\mu = \mu_0(T) + nR^*T \ln P, \tag{67}$$

where $\mu_0(T)$ is the *standard chemical potential* at unit pressure (usually taken to be 1 atm). Hence, for an ideal gas with two constituents:

$$\Delta G = -W_{\max} = (\mu_2 - \mu_1)(n_2 - n_1) = nRT\ \ln \frac{p_2}{p_1}. \tag{68}$$

We can also look at the chemical potential of a condensed phase, using the concept of equilibrium. Consider liquid water in equilibrium with water vapor at a temperature T and a partial pressure of e_s. Applying the principle of virtual work, we transfer dn moles of water from one phase to the other under the condition:

$$\mu_l\,dn = \mu_v\,dn,$$
$$\mu_l = \mu_v = \mu_l^0 + R_vT\ \ln e_s,$$

where μ_l and μ_v are the channel potentials of liquid and vapor respectively.

This is a general statement and demonstrates that the chemical potential of any species is constant throughout the system (if the equilibrium constraints permit transfer of the species). We can also see that for a nonequilibrium phase change:

$$\Delta G = \mu_v - \mu_l = R_vT\ \ln \frac{e_v}{e_s}, \tag{69}$$

where e_v is the partial pressure of vapour not at equilibrium.

Hence if $e_v < e_s$, then $\Delta G < 0$ for a process of condensation and hence that is a forbidden process. By the same token, evaporation becomes a spontaneous process. When $\mu_v = \mu_l$, the system is in equilibrium and at *saturation*. These concepts can be extended to include other free energies affecting condensation or sublimation such as solution effects, curvature effects, to process of chemical equilibrium and so on.

REFERENCES

Dutton, J. (1976). *The Ceaseless Wind*, McGraw-Hill.

Emanuel, K. A. (1994). *Atmospheric Convection*, Oxford University Press.

Hess, S. L. (1959). *Introduction to Theoretical Meteorology*, Holt, Reinhardt, and Winston.

Iribarne, J. V., and W. L. Godson (1973). *Atmospheric Thermodynamics*, D. Reidel Publishing Co.

Sears, F. W. (1953). *Thermodynamics*, Addison-Wesley.

Wallace, J. M., and P. V. Hobbs, (1977). *Atmospheric Science: An Introductory Survey*, Academic Press.

CHAPTER 16

MOIST THERMODYNAMICS*

AMANDA S. ADAMS

1 LATENT HEATS—KIRCHOFF'S EQUATION

Consider the effect resulting from mass fluxes between constituents of the parcel within the system. In particular, mass fluxes between the three possible phases of water are considered. We can express these phase transformations by:

$$dH = \delta Q + V\,dP + \sum_k \left(\frac{\partial H_k}{\partial n_k}\right) dn_k \tag{1}$$

where $\partial H_k/\partial n_k$ represents the energy per mole of the n_k constituent of the system. We let the system of moist air be composed of n_d moles of dry air, n_v moles of vapor gas, n_l moles of liquid water and n_i moles of ice. Then

$$dH = \delta Q + V\,dP + \left(\frac{\partial H_v}{\partial n_v} - \frac{\partial H_l}{\partial n_l}\right) dn_v + \left(\frac{\partial H_i}{\partial n_i} - \frac{\partial H_l}{\partial n_l}\right) dn_i \tag{2}$$

*The derivations, concepts, and examples herein are based partially on course notes from the late Professor Myron Corrin of Colorado State University. Other derivations are taken from the Isbane and Godson text, the Emanuel (1994: *Atmospheric Convection*, Oxford University Press) text, and the Dutton (*The Ceaseless Wind*) book.

Handbook of Weather, Climate, and Water: Dynamics, Climate, Physical Meteorology, Weather Systems, and Measurements, Edited by Thomas D. Potter and Bradley R. Colman.
ISBN 0-471-21490-6

We define latent heats to be the differences in enthalpy per mole between two phases at the same temperature and pressure and is expressed as:

$$\begin{aligned} L_{lv} &\equiv \left(\frac{\partial H_l}{\partial n_l} - \frac{\partial H_v}{\partial n_v} \right) \\ L_{il} &\equiv +\left(\frac{\partial H_i}{\partial n_i} - \frac{\partial H_l}{\partial n_l} \right) \\ L_{iv} &\equiv L_{il} + L_{lv} \end{aligned} \tag{3}$$

L_{lv}, L_{il}, and L_{iv} are the latent heats of condensation, melting, and sublimation defined as positive quantities with units of joules per mole. Employing these definitions, Eq. (2) can be written:

$$dH = \delta Q + V\, dP - L_{iv}\, dn_v + L_{il}\, dn_i \tag{4}$$

In order to find Kirchoff's equation, we write Eq. (4) for three adiabatic homogenous systems consisting of vapor, liquid, and ice held at constant pressure:

$$\begin{aligned} C_{pv} T &= \frac{\partial H_v}{\partial n_v}, \\ C_l T &= \frac{\partial H_l}{\partial n_l}, \text{ and} \\ C_i T &= \frac{\partial H_i}{\partial n_i}. \end{aligned} \tag{5}$$

where C_{pv}, C_l, and C_i are the heat capacity of vapor at constant pressure, the heat capacity of liquid, and the heat capacity of ice. Applying (5) to (3) and differencing with respect to temperature we obtain Kirchoff's equation:

$$\begin{aligned} \frac{\partial L_{lv}}{\partial T} &= C_{pv} - C_l, \\ \frac{\partial L_{il}}{\partial T} &= C_l - C_i, \text{ and} \\ \frac{\partial L_{iv}}{\partial T} &= C_{pv} - C_i. \end{aligned} \tag{6}$$

In the "enthalpy per mass" form, Kirchoff's equation is:

$$\frac{\partial l_{lv}}{\partial T} = c_{pv} - c_l,$$
$$\frac{\partial l_{il}}{\partial T} = c_l - c_i, \text{ and} \tag{7}$$
$$\frac{\partial l_{iv}}{\partial T} = c_{pv} - c_i.$$

Given heat capacities determined observationally as a function of temperature, we can determine the variation of latent heat with temperature. Kirchoff's law can also be used to study reaction heats of chemical changes.

2 GIBBS PHASE RULE

For the case of a simple homogenous gas, we found that there were three independent variables, being the pressure, temperature, and volume assuming the number of moles was specified. Once we require that the gas behave as an ideal gas, the imposition of the equation of state (and the implicit assumption of equilibrium) reduces the number of independent variables by one to two.

If we now look at a system consisting entirely of water, but allow for both liquid and gas forms, and again assume equilibrium, then not only must the vapor obey the ideal gas law, but the chemical potential of the liquid must equal that of the vapor. This means that if we know the pressure of the vapor, only one water temperature can be in true equilibrium with that vapor, i.e., the temperature that makes the liquid exert a vapor pressure equal to the vapor pressure in the air. Hence by requiring equilibrium with two phases, the number of degrees of freedom is reduced to one.

If we allow for all three phases, i.e., vapor, liquid, and ice, then there is only one temperature and pressure where all three states can exist simultaneously, called the *triple point*.

Now consider a two-component system such as one where we have water and air mixed. Consider a system allowing only liquid but not the ice phase of water. Now we have to consider the equilibrium as is applied to the dry air by the ideal gas law in addition to the equilibrium between the liquid and ice water. For this case, the water alone had one independent variable and the dry air had two, for instance, its partial pressure and temperature. If the vapor pressure is specified, then the temperature of the water is specified and, for equilibrium, so is the air temperature. Hence adding the additional component of air to the two-phase water system increased the number of independent variables to two.

A general statement concerning the number of independent variables of a heterogeneous system is made by *Gibbs phase rule*, which states:

$$v = c - \phi + 2, \tag{8}$$

where v is the number of independent variables (or degrees of freedom), c is the number of independent species, and ϕ is the number of phases total. Hence for a water–air mixture, allowing two phases of water and one phase of dry air, there is 1 degree of freedom. So if we specify vapor pressure, then liquid temperature is known, air temperature is known, and so air pressure is fixed assuming the volume of the system is specified.

3 PHASE EQUILIBRIUM FOR WATER

Figure 1 shows where equilibrium exists between phases for water. Note that equilibrium between any two phases is depicted by a line, suggesting one degree of freedom, while all three phases are only possible at the triple point. Note that the curve for liquid–ice equilibrium is nearly of constant temperature but weakly slopes to cooler temperatures at higher pressures. This is attributable to the fact that water increases in volume slightly as it freezes so that freezing can be inhibited some by applying pressure against the expansion.

Note also that at temperatures below the triple point there are multiple equilibria, i.e., one for ice–vapor and another (dashed line) for liquid–vapor. This occurs because the free energy necessary to initiate a freezing process is high, and local equilibrium between the vapor and liquid phase may exist.

Note that at high temperatures, the vapor pressure curve for water abruptly ends. This is the critical point (p_c) where there is no longer a discontinuity between the liquid and gaseous phase. As we showed earlier, the latent heat of evaporation l_{lv} decreases with increasing temperature. It becomes zero at the critical point.

Since the triple point is a well-defined singularity, we define thermodynamic constants for water at that point, and these values are given in Table 1. At the critical point, liquid and vapor become indistinguishable. Critical point statistics are given in

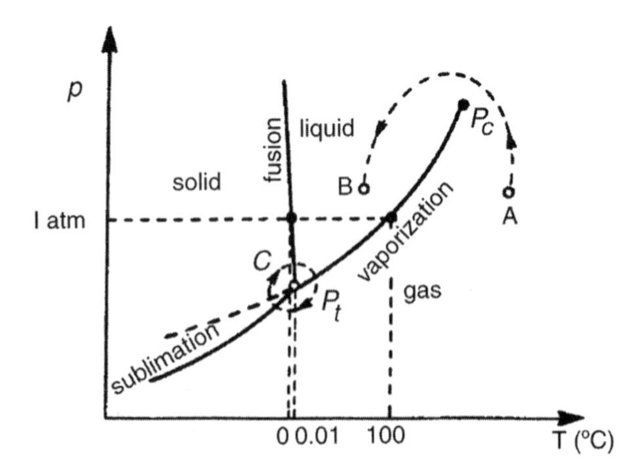

Figure 1 Phase-transition equilibria (from Iribarne and Godson, 1973).

TABLE 1 Triple Point Values for Water

Variable	Symbol	Value
Temperature	T_t	273.16 K
Pressure	p_t	610.7 Pa, 6.107 mb
Ice density	$\rho_{i,t}$	$917\,kg/m^{-3}$
Liquid density	$\rho_{l,t}$	$1000\,kg/m^{-3}$
Vapor density	$\rho_{v,t}$	$0.005\,kg/m^{-3}$
Specific volume ice	$\alpha_{i,t}$	$1.091 \times 10^{-3}\ m^3/kg$
Specific volume liquid	$\alpha_{l,t}$	$1.000 \times 10^{-3}\ m^3/kg$
Specific volume vapor	$\alpha_{v,t}$	$2.060 \times 10^{2}\ m^3/kg$
Latent heat of condensation	$l_{vl,t}$	$2.5008 \times 10^{6}\ J/kg$
Latent heat of sublimation	$l_{vi,t}$	$2.8345 \times 10^{6}\ J/kg$
Latent heat of melting	$l_{il,t}$	$0.3337 \times 10^{6}\ J/kg$

Table 2. These phase relationships can also be seen schematically with an Amagat–Andrews diagram (Fig. 2). Note how the isotherms follow a hyperbola-like pattern at very high temperatures as would be expected by the equation of state for an ideal gas. However, once the temperature decreases to values less than T_C, an isothermal process goes through a transition to liquid as the system is compressed isothermally. The latent heat release and the loss of volume of the system keep the pressure constant while the phase change occurs. After the zone of phase transition is crossed, only liquid (or ice at temperatures below T_t) remains, which has very low compressibility and so a dramatic increase in pressure with further compression.

Phase equilibrium surfaces can be displayed also in three dimensions as a *p-V-T* surface and seen in Figure 3. Here we see that at very high temperatures, the region where phases coexist is lost. Note how the fact that water expands upon freezing results in the kink backward of the liquid surface. Note that we can attain the saturation curves shown in Figure 1 by cutting cross sections through the *p-V-T* diagram and constant volume or temperature (Fig. 3).

We can use these figures to view how the phase of a substance will vary under differing conditions, since the phases of the substance must lie on these surfaces. Figure 4 shows one such system evolution beginning at point *a*. Note the liquid system exists at point *a* under pressure p_1. The pressure presumably is exerted by an atmosphere of total pressure p_1 above the surface of the liquid. Holding this pressure constant and heating the liquid, we see that the system warms with only a small volume increase to point *b* where the system begins to evolve to a vapor phase at

TABLE 2 Critical Point Values for Water

Variable	Symbol	Value
Temperature	T_c	647 K
Pressure	p_c	2.22×10^{7} Pa (218.8 atm)
Specific volume vapor	$\alpha_{c,t}$	$3.07 \times 10^{-3}\ m^3/kg$

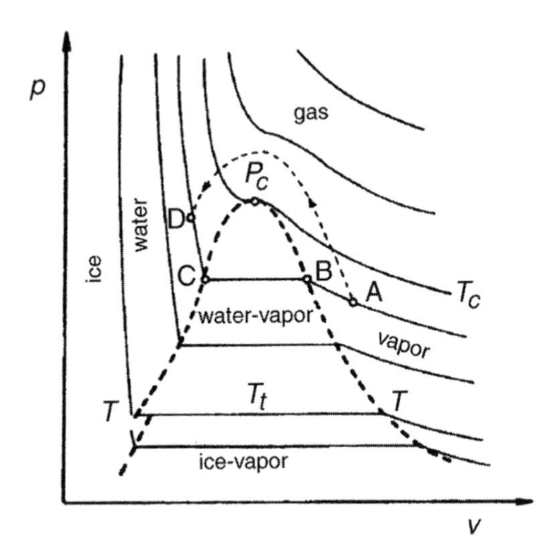

Figure 2 Amagat–Andrews diagram (from Iribarne and Godson, 1973).

much higher volume as the temperature holds constant. The evolution from b to c is commonly called *boiling* and occurs when the vapor pressure along the vapor–liquid interface becomes equal to the atmospheric pressure. Hence we can also view the saturation curves as boiling point curves for various atmospheric pressures.

Alternatively, if the system cools from point a, the volume very slowly decreases until freezing commences at point d. There we see the volume more rapidly decrease as freezing occurs. Of course, for water the volume change is reversed to expansion and is not depicted for the substance displayed on this diagram.

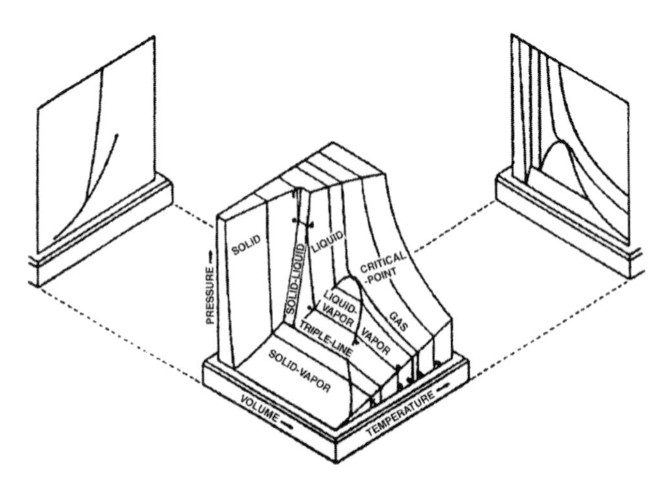

Figure 3 Projection of the p-V-T surface on the p-T and p-V planes. (Sears, 1959).

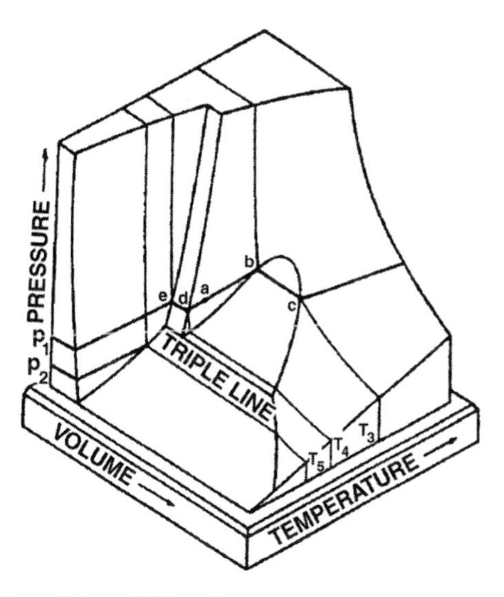

Figure 4 Projection of the p-V-T surface on the p-T and p-V planes. (Sears, 1959).

4 CLAUSIUS–CLAPEYRON EQUATION

We will now express mathematically the relation between the changes in pressure and temperature along an equilibrium curve separating two phases. The assumption of equilibrium between two systems a and b, requires that at the interface between the systems:

$$\begin{aligned} g_a &= g_b, \\ \mu_a &= \mu_b, \\ T_a &= T_b, \text{ and} \\ u_a &= u_b. \end{aligned}$$

where we will assume that a and b are of different phases. For equilibrium, we also require infinitesimal changes in conditions along the interface to preserve equilibrium. Hence,

$$\begin{aligned} dg_a &= dg_b, \\ d\mu_a &= d\mu_b, \\ dT_a &= dT_b, \text{ and} \\ du_a &= du_b. \end{aligned} \tag{9}$$

The Gibbs free energy was defined to be

$$g = u + p\alpha - Ts, \tag{10}$$
$$= h - Ts. \tag{11}$$

By virtue of Eq. (11),

$$\begin{aligned} dg_a &= (du_a) - s_a dT + (-T ds_a + p d\alpha_a) + \alpha_a dp \\ &= (du_b) - s_b dT + (-T ds_b + p d\alpha_b) + \alpha_b dp = dg_b \end{aligned}$$

where the terms in parenthesis drop out [because of Eq. (9) and the first law] to give:

$$(s_b - s_a)\, dT = (\alpha_b - \alpha_a)\, dp, \text{ and}$$

$$\frac{s_b - s_a}{\alpha_b - \alpha_a} = \frac{dp}{dT} \tag{12}$$

From Eq. (11) it follows that along the interface,

$$g_b - g_a = h_b - h_a - T(s_b - s_a) = 0, \tag{13}$$

and since, $l_{ab} = h_b - h_a$, we can write

$$l_{ab} = T(s_b - s_a). \tag{14}$$

Hence, we can rewrite Eq. (12) as:

$$\frac{dp}{dT} = \frac{l_{ab}}{T(\alpha_b - \alpha_a)} \tag{15}$$

which is the general form of the Clapeyron equation.

The physical meaning of this equation can be illustrated by the process depicted in Figure 5. Consider the four-step process shown. Beginning at T, P in the lower left point of the cycle we can move to the point $T + dT$ and $p + dp$ in either of the two paths shown. Since g is a state function, the path does not matter for the change in g and hence the change in g along both paths can be equated, yielding the equation. Hence the equation defines how the equilibrium vapor pressure must vary with temperature based on the value of latent heat.

The Clapeyron equation can be applied between any two systems having differing phases to produce expressions for the variations of equilibrium vapor pressure with temperatures. Its integrated solution leads directly to expressions for the value of

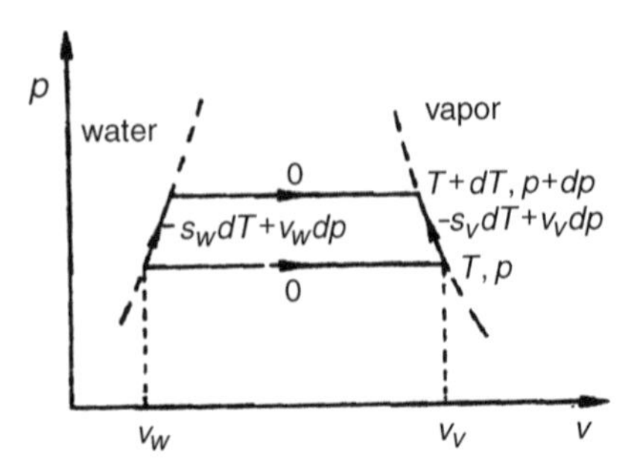

Figure 5 Cycle related to Clapeyron equation (Hess, 1959).

equilibrium vapor pressures with respect to a particular phase change for any temperature.

Equilibrium Between Liquid and Solid Clapeyron Equation

Applying the phase equilibrium to a phase transition between liquid and ice yields

$$\frac{de}{dT} = -\frac{l_{il}}{T(\alpha_i - \alpha_l)} \tag{16}$$

where we are using the symbol e for pressure of the liquid and ice, and we will always take the latent heat l_{il} as the positive definite difference in enthalpy between a liquid and frozen phase where the liquid phase has the greater enthalpy. For a freezing process considered here, the enthalpy change passing from liquid to ice will be negative. Hence, the negative sign appears on the right-hand side (RHS) of Eq. (16). Since the volume of the ice is greater than that of the liquid, the change in vapor pressure with temperature is negative. This is evidenced in Figure 1 as a negative slope to the liquid–ice phase equilibrium curve. Note also that the slope of the ice–liquid curve is very large resulting from the relatively small volume change between liquid and ice phases [denominator of Eq. (16)].

Equilibrium Between Liquid and Vapor: Clausius–Clapeyron Equation

Applying the phase equilibrium to a phase transition between liquid and vapor under equilibrium conditions yields

$$\frac{de_s}{dT} = -\frac{l_{lv}}{T(\alpha_l - \alpha_v)} \tag{17}$$

where e_s is the vapor pressure of the equilibrium state, and the latent heat of vaporization l_{lv} is defined as the absolute value of the difference in enthalpies between the liquid and ice phase. Since the enthalpy of the vapor state is higher than that of the liquid, the negative sign appears in front of Eq. (17). Since $\alpha_v \gg \alpha_l$, we can drop the specific volume of liquid water compared to that of vapor. Hence, applying the equation of state to the vapor specific volume:

$$\frac{d \ln e_s}{dT} = \frac{l_{lv}}{R_v T^2}. \tag{18}$$

This is the Clausius–Clapeyron equation and we will use it for a number of applications. The equation tells us how the saturation vapor pressure varies with temperature. Hence given an observation point, such as the triple point, we can determine e_s at all other points on the phase equilibrium diagram.

Equilibrium Between Ice and Vapor

Similarly, for equilibrium between ice and vapor, we can write

$$\frac{d\ \ln\ e_{si}}{dT} = \frac{l_{iv}}{R_v T^2}, \tag{19}$$

where we again take l_{il} to be the absolute value of the difference in enthalpy between the vapor and solid phase of water.

Computation of Saturation Vapor Pressure (e_s and e_{si})

For a precise integration of the Clausius–Clapeyron equation, one must consider the latent heat variation with temperature. To a good first approximation, let temperature be independent of latent heat and then integrate, yielding

$$\ln\ e_s = -\frac{l_{lv}}{R_v T} + \text{const.} \tag{20}$$

This formulation has assumed latent heat to be independent of temperature. This is a good approximation for sublimation, but not as good for freezing or condensation. We can improve on this by using a more precise formulation for latent heat variation and specific heat variation with temperature with a series expansion:

$$\ln\ e_s = \frac{1}{R_v}\left[-\frac{l_0}{T} + \Delta\alpha\ \ln\ T + \frac{\Delta\beta}{2}T + \frac{\Delta\gamma}{6}T^2 + \cdots\right] + \text{const.} \tag{21}$$

where the integration constant is determined empirically.

We need not achieve this much accuracy for most meteorological considerations, since we cannot measure vapor pressure that precisely anyway. Hence we can make approximations that allow us to calculate e_s or e_{si} with sufficient accuracy. Here are three approximations:

1. Solving the constant at the triple point temperature (T_t), and assuming latent heat constant, we obtain

$$e_s = 10^{(9.4051-2354/T)}$$
$$e_{si} = 10^{(10.5553-2667/T)}$$

2. Assuming heat capacity constant and retaining two terms in a series, we get *Magnus's formula (liquid–vapor only)*:

$$e_s = 10^{((-2937.4/T)-4.9283\log T+23.5518)} \tag{22}$$

TABLE 3 Constants for Teten's Formula

	Water	Ice
a	17.2693882	21.8745584
b	35.86	7.66

This has been simplified by Tetens (1930) and later by Murray (1966) to be easily computed from the relationship (for pressure in millibars):

$$e_s = 6.1078 \exp\left[\frac{a(T - 273.16)}{T - b}\right] \tag{23}$$

where the constants are defined in Table 3.

3. Many atmospheric scientists use the Goff–Gratch (1946) formulation given by:

$$\begin{aligned} e_s = 7.95357242x10^{10} \exp\Bigg\{ & -18.1972839\left(\frac{T_s}{T}\right) + 5.02808 \ln\left(\frac{T_s}{T}\right) \\ & - 70242.1852 \exp\left(\frac{-26.1205253}{T_s/T}\right) \\ & + 58.0691913 \exp\left[-8.03945282\left(\frac{T_s}{T}\right)\right]\Bigg\} \end{aligned} \tag{24}$$

for saturation over liquid where $T_s = 373.16\,\text{K}$. For saturation over ice:

$$\begin{aligned} e_s = 5.57185606x10^{10} \exp\Bigg\{ & -20.947031\left(\frac{T_0}{T}\right) - 3.56654 \ln\left(\frac{T_0}{T}\right) \\ & - \frac{2.01889049}{T_0/T}\Bigg\} \end{aligned} \tag{25}$$

where $T_0 = 273.16\,\text{K}$.

Tables 4 and 5 show how the accuracy of Teten's formula compares to Goff–Gratch as reported by Murray (1966). Both of these forms of saturation vapor pressure are commonly used as a basis to compute saturation vapor pressure.

5 GENERAL THEORY FOR MIXED-PHASE PROCESSES WITHIN OPEN SYSTEMS

Often thermodynamic theory is applied to precipitating cloud systems containing liquid and ice that often are not in equilibrium. In many thermodynamic texts, the simplifying assumption of equilibrium is made that can invalidate the result.

TABLE 4 Saturation Vapor Pressure over Liquid

t (°C)	Goff–Gratch (mb)	Tetens (mb)	Difference (%)
−50	0.06356	0.06078	-1.6×10^{-2}
−45	0.1111	0.1074	-1.6×10^{-2}
−40	0.1891	0.1842	-1.6×10^{-2}
−35	0.3139	0.3078	-1.7×10^{-2}
−30	0.5088	0.5018	-2.1×10^{-2}
−25	0.8070	0.7993	-4.5×10^{-2}
−20	1.2540	1.2462	2.8×10^{-2}
−15	1.9118	1.9046	5.8×10^{-2}
−10	2.8627	2.8571	1.9×10^{-3}
−5	4.2149	4.2117	5.2×10^{-4}
0	6.1078	6.1078	-1.8×10^{-7}
5	8.7192	8.7227	-1.9×10^{-4}
10	12.272	12.2789	-2.2×10^{-4}
15	17.044	17.0523	-1.8×10^{-4}
20	23.373	23.3809	-1.1×10^{-4}
25	31.671	31.6749	-3.7×10^{-5}
30	42.430	42.426	2.5×10^{-5}
35	56.237	56.221	7.1×10^{-5}
40	73.777	73.747	9.5×10^{-5}
45	95.855	95.812	9.7×10^{-5}
50	123.40	123.35	7.5×10^{-5}

Moreover, ice processes are often neglected, despite the existence of ice processes within the vast majority of precipitating clouds. We will adopt the generalized approach to the moist thermodynamics problem, developed by Dutton (1973) and solve for the governing equations for a nonequilibrium three-phase system first and then find the equilibrium solution as a special case.

TABLE 5 Saturation Vapor Pressure over Ice

t (°C)	Goff–Gratch (mb)	Tetens (mb)	Difference (%)
−50	0.03935	0.03817	-9.4×10^{-3}
−45	0.07198	0.07032	-8.8×10^{-3}
−40	0.1283	0.1261	-8.5×10^{-3}
−35	0.2233	0.2205	-8.5×10^{-3}
−30	0.3798	0.3764	-9.2×10^{-3}
−25	0.6323	0.6286	-1.3×10^{-2}
−20	1.032	1.028	1.2×10^{-1}
−15	1.652	1.648	4.1×10^{-3}
−10	2.597	2.595	9.9×10^{-4}
−5	4.015	4.014	1.7×10^{-4}
0	6.107	6.108	-6.3×10^{-5}

Besides the nonequilibrium phase changes, the formation of liquid and ice hydrometeors create a heterogeneous system having components of gas, solids, and liquids, each of which may be moving at different velocities. We must therefore forego our traditional assumption of an adiabatic system and instead consider an *open system*. In doing so, we fix our coordinate system relative to the center of the *dry air parcel*. Although we can assume that vapor will remain stationary relative to the dry air parcel, we must consider movements of the liquid and solid components of the system relative to the parcel, hence implying diabatic effects. We therefore generalize the derivation of Eq. (4) to include these effects. It will be convenient to work in the system of specific energies (energy per mass).

We will assume that our system is composed of several constituents, each of mass m_j. Then the total mass is

$$m = \sum_j m_j. \tag{26}$$

We account for both internal changes (d_i), which result from exchanges between phases internal to the parcel, and external changes (d_e), which result from fluxes of mass into and out of the parcel. The total change in mass is thus

$$dm_j = d_i m_j + d_e m_j. \tag{27}$$

By definition and for mass conservation, we require

$$d_i m = \sum_j d_i m_j = 0, \text{ and}$$

$$d_e m = \sum_j d_e m_j = d_e m_l + d_e m_i.$$

As discussed above, external mass changes are restricted to liquid or ice constituents, while gas constituents are assumed not to move relative to the system. An exception can be made, although not considered here, for small-scale turbulent fluxes relative to the parcel. From our original discussion of the first law, we can write the generalized form of Eq. (4) in a relative mass form as:

$$dH = \delta Q + V\, dP + \sum_j h_j\, dm_j \tag{28}$$

where $h_j = (\partial H_j / \partial m_j)$. We found earlier that the terms involving h_j eventually result in the latent heat term.

We can split the heating function, $\delta Q = Q_i + Q_e$. Here Q_i is the diabatic change due to energy flowing into or out of the system that does not involve a mass flow. Q_e accounts for diabatic heat fluxes resulting from mass fluxes into or out of the system. The diabatic heat source of conduction between the parcel and an outside source would be accounted for in Q_i if no mass back and forth flow into and out of the parcel is not explicitly represented. Even the case of turbulent heat transfer into and out of the parcel would not affect Q_i unless the explicit fluxes of mass in and out were represented. The heat flux associated with the movement of precipitation featuring a different temperature than the parcel, into and out of the parcel, would also be represented by Q_e, as it would appear as a different enthalpy for the precipitation constituent, and be accounted for by the $d_e m_j$ term. As is conventional, we will assume $Q_e = 0$. Now;

$$Q_i = d_i H - V\,dp - \sum_j h_j\,d_i m_j, \text{ and} \tag{29}$$

$$Q_e = d_e H - \sum_j (h_j - h)\,d_e m_j. \tag{30}$$

We can similarly split the enthalpy tendency into its internal and external parts:

$$dH_i = Q_i + V\,dp - \sum_j h_j\,d_i m_j, \text{ and} \tag{31}$$

$$dH_e = -\sum_j (h_j - h)\,d_e m_j. \tag{32}$$

Ignoring the noninteracting free energies, Gibbs' relation for the mixed-phase system is written in "per mass" form as:

$$T\,dS = dH - V\,dp - \sum_j h_j\,dm_j - \sum_j \mu_j\,dm_j, \tag{33}$$

where μ_j is the *chemical potential per mass* rather than per mole as we first defined it. We can now divide the entropy change between internal and external changes, and employing Eq. (29)

$$\begin{aligned} T\,d_i S &= Q_i - \sum_j h_j\,d_i m_j - \sum_j \mu_j\,d_i m_j, \text{ and} \\ T\,d_e S &= -\sum_j \mu_j\,d_e m_j, \end{aligned} \tag{34}$$

for the external change. Since, $\mu_j = g_j = h_j - Ts_j$, and Eq. (30), then

$$T\,d_e S = -\sum_j h_j\,d_e m_j - \sum_j Ts_j\,d_e m_j. \tag{35}$$

Now we apply our results to the particular water–air system. Consider a mixture composed of m_d grams of dry air, m_v grams of vapor, m_l grams of liquid water, and m_i grams of ice. We require

$$\begin{aligned} d_i m_d = d_e m_d &= 0, \\ d_i m_v + d_i m_l + d_i m_i &= 0, \text{ and} \\ d_e m_v &= 0. \end{aligned} \tag{36}$$

The internal enthalpy equation [Eq. (31) and internal entropy Eq. (36)] can then be written for this mixed-phase system:

$$d_i H = Q_i + V\, dp + \sum_j h_j\, d_i m_j, \text{ and} \tag{37}$$

$$d_i S = \frac{Q_i}{T} - \frac{\sum_j h_j\, d_i m_j}{T} - \frac{\sum_j \mu_j\, d_i m_j}{T}. \tag{38}$$

For the air–water system Eq. (38) becomes

$$d_i S = -\frac{l_{iv}}{d_i} m_v + \frac{l_{il}}{T} d_i m_i - \frac{a_{iv}}{T} d_i m_v + \frac{a_{il}}{T} d_i m_i + \frac{Q_i}{T}, \tag{39}$$

where the a_{iv} and a_{il} are the specific affinities of sublimation and melting. They are defined as:

$$a_{iv} \equiv \mu_l - \mu_i,$$
$$a_{il} \equiv \mu_l - \mu_i.$$

The affinity terms take into account the entropy change resulting if the two interacting phases are out of equilibrium. They would be expected to be nonzero, for instance, if rain drops are evaporating in an environment where the air is subsaturated. Under equilibrium conditions, they would vanish. An example is cloud drops growing by condensation in an environment where the vapor pressure is equal to the saturation vapor pressure over liquid. The affinity is the Gibbs free energy available to drive a process and unavailable to perform work. Notice the symmetry between the affinity declarations and the latent heat declarations.

If we apply Eq. (39) to adiabatic melting, condensation, and sublimation, we obtain

$$\begin{aligned} T(s_l - s_i) &= l_{il} + a_{il}, \\ T(s_v - s_l) &= l_{lv} + a_{lv}, \text{ and} \\ T(s_v - s_i) &= l_{iv} + a_{iv}. \end{aligned} \tag{40}$$

We can state that the total entropy is equal to the sum of the entropy in each system component:

$$S = m_d s_d + m_v s_v + m_l s_l + m_i s_i, \tag{41}$$

where the dry air entropy is

$$s_d \equiv c_{pd} \ln T - R_d \ln p_d. \tag{42}$$

We have required that the only external fluxes are those of precipitation. Hence, we find

$$\begin{aligned} d_iS &= dS - d_eS, \\ &= dS - s_l d_e m_l - s_i d_e m_i, \\ &= d(m_d s_d + m_v s_v) + s_i d_i m_l \\ &\quad + s_i d_i m_i + m_l ds_l + m_i ds_i = \frac{A_{lv}}{T} d_i m_v - \frac{A_{il}}{T} d_i m_i + \frac{Q_i}{T}. \end{aligned} \tag{43}$$

Combining Eqs. (36) and (40) with (4) we obtain:

$$d\left(m_d s_d + \frac{m_v l_{lv}}{T}\right) - d_i\left[\frac{w_i l_{il}}{T} + m_v d\left(\frac{A_{lv}}{T}\right)\right] - m_i d\left(\frac{A_{il}}{T}\right) + (m_v + m_l + m_i)\, ds_l = \frac{Q_i}{T}. \tag{44}$$

If we now assume that the liquid is approximately incompressible, then $ds_l = c_l(dT/T)$. Defining mixing ratio of the jth constituent as:

$$r_j \equiv \frac{m_j}{m_d}, \tag{45}$$

and divide Eq. (44) by m_d. Then we obtain the following general relation:

$$d\left\{c_{pd} \ln T - R_d \ln p_d + \frac{r_v l_{lv}}{T}\right\} - d_i\left(\frac{r_i l_{il}}{T}\right) + r_v d\left(\frac{A_{lv}}{T}\right) - r_i d\left(\frac{A_{il}}{T}\right) + (r_v + r_l + r_i) c_l \frac{dT}{T} = \frac{q_i}{T}. \tag{46}$$

The differentials account for the interrelationship between temperature change, pressure change, and liquid phase change. Additional diabatic heating tendencies by radiative transfer, molecular diffusion, and eddy diffusion can be accounted for with the internal heating term q_i. Neglected are the effects of chemical reaction on entropy, although the heating effect can be included in the heating term. Generally, these effects are small and can be neglected.

What is included are not only the latent heating effects of the equilibrium reaction but the effects on heating resulting from the entropy change in nonequilibrium reactions. Hence, since entropy change removes energy from that available for work, nonequilibrium heating from phase change modifies the enthalpy change resulting from phase change. The affinity terms account for these effects.

Notice that both affinity terms are inexact differentials, implying that their effect is irreversible. The heat storage term [last term on the left-hand side (LHS)] is also an exact differential if we neglect the variation of c_l and we assume the total water, defined as:

$$r_T = r_v + r_l + r_i, \tag{47}$$

is a constant. If there is a change, i.e., a diabatic loss or gain of moisture by precipitation, then the differential is inexact and the process is irreversible. This term also contains all of the net effects of diabatic fluxes of moisture for systems that are otherwise in equilibrium between phases. The heating term, q_i, is purely diabatic and defines a diabatic thermal forcing on the system such as by radiative transfer or molecular diffusion and turbulence.

Note that, although the first term on the LHS is written as a total derivative, the external derivative of the quantity in brackets is in fact zero. Hence all reversible terms appear as internal derivatives while the reversible terms contain both internal derivatives for moist adiabatic process and external derivatives for diabatic flux terms.

6 ENTHALPY FORM OF THE FIRST–SECOND LAW

We now derive what is perhaps a more commonly used form of Eq. (46). Whereas Eq. (46) is in a form representing entropy change, it can be rewritten as an enthalpy change. To do so we make some manipulations.

First, we make the following assumptions:

1. Neglect the curvature effects of droplets.
2. Neglect solution effects of droplets.

These assumptions are really quite unimportant for the macrosystem of fluid parcels. However, the assumptions would be critical when discussing the microsystem of the droplet itself, since these effects strongly influence nucleation and the early growth of very small droplets. With these assumptions, the chemical potential of the vapor is defined:

$$\mu_l = \mu_o + R_v T \ \ln \ e_v, \tag{48}$$

where e_v is the atmospheric partial pressure of the vapor. The chemical potential of the liquid and ice are defined as the chemical potential of vapor, which would be in equilibrium with a plane pure surface of the liquid or ice and are given by:

$$\mu_l = \mu_o + R_v T \ \ln \ e_s, \tag{49}$$

$$\mu_i = \mu_o + R_v T \ \ln \ e_{si}, \tag{50}$$

where e_s and e_{si} are the saturation vapor pressures of the ice and liquid defined by the temperature of the liquid or ice particle. This temperature need not be the same as the vapor temperature T for this equation.

Combining the equation of state applied to vapor and to dry air, it can be shown that:

$$d \ln p(R_d + r_v R_v) = R_d d \ln p_d + r_v R_v d \ln e_v. \tag{51}$$

Combining Kirchoff's equation (5), with Eqs. (46) to (51) and (18) and (19),

$$c_{pm} d \ln T - R_m d \ln p + \frac{l_{lv}}{T} dr_v - \frac{l_{il}}{T} d_i r_i = \frac{q_i}{T} \tag{52}$$

where $c_{pm} = c_p + r_v c_{vp} + r_i c_i + r_l c_l$ is the effective heat capacity of moist air and $R_m = R_d + r_v R_v$ is the moist gas constant (not to be confused with the gas constant of moist air).

Although Eq. (52) appears different from Eq. (46), it contains no additional approximations other than the neglect of the curvature and solution effects implicit in the assumed form of chemical potential. It is simpler and easier to solve than the other form because the affinity terms and latent heat storage terms are gone. Note, however, that there are some subtle inconveniences. In particular, each term is an *inexact differential* that means that they will not vanish for a cyclic process. This makes it more difficult to integrate Eq. (52) analytically. Nevertheless, it is a convenient form for applications such as a numerical integration of the temperature change during a thermodynamic process.

Some of the effects of precipitation falling into or out of the system are included in Eq. (52) implicitly. To see this look at the change in vapor. It is a total derivative because only internal changes are allowed. The ice change, on the other hand, is strictly written as an internal change. Hence it is the internal change that implies a phase change, and knowing the ice phase change and liquid phase change, the liquid phase change is implicitly determined since the total of all internal phase changes are zero. Since, by virtue of the assumption that a heterogenous system is composed of multiple homogenous systems, we assumed that hydrometeors falling into or out of the system all have the same enthalpy as those in the system itself, there is no explicit effect on temperature.

This assumption can have important implications. For instance, frontal fog forms when warm rain droplets fall into a cold parcel, hence providing an external flux of heat and moisture through the diabatic movement of the rain droplet relative to the parcel. We neglect this effect implicitly with eq. (52). By requiring $d_e H = \sum_j d_e(m_j h_j) = \sum_j h_j d_e m_j$, we only considered the external changes due to an external flux of water with the same enthalpy of the parcel. The neglect of these effects is consistent with the pseudo-adiabatic assumption that is often made. That assumption assumes condensed water immediately disappears from the system, and so the heat storage effects within the system and for parcels falling into or out of the system can be neglected. So far, the pseudo-adiabatic assumption has only been partially made because we still retain the heat storage terms of the liquid and ice phases within the system. It is unclear whether there is an advantage to retaining them only partially.

7 HUMIDITY VARIABLES

Water vapor, unlike the "dry" gases in the atmosphere exists in varying percentages of the total air mass. Obviously, defining the amount of vapor is critical to understanding the thermodynamics of the water–air system. We have developed a number of variables to define the vapor, liquid, and ice contents of the atmosphere. Below is a list of the variables used to define water content.

Vapor Pressure (e_v)

Vapor pressure represents the partial pressure of the vapor and is measured in pascals. The saturation vapor pressure over a plane surface of pure liquid water is defined to be e_s while the vapor pressure exerted by a plane surface of pure ice is e_{si}.

Mixing Ratio and Specific Humidity

We have already introduced *mixing ratio* to be $r_v = \rho_v/\rho_d$, and employing the equation of state, we can relate mixing ratio to vapor pressure:

$$r_v = \frac{\epsilon e_v}{p - e_v} \tag{53}$$

where $\epsilon \equiv M_d/M_v$. We define *specific humidity* to be the ratio of $q_v = \rho_v/\rho$. It follows that

$$q_v = \frac{r_v}{1 + r_v}$$

Then, similar to Eq. (53) we can write for specific humidity:

$$q_v = \frac{\epsilon e_v}{p} \frac{1}{(1 + (r_v/\epsilon))} \tag{54}$$

To a reasonable approximation, one can show that $e_v \ll p$ and hence:

$$q_v \sim r_v \sim \epsilon \frac{e_v}{p} \tag{55}$$

Relative Humidity

We define relative humidity to be the ratio of the vapor pressure to the vapor pressure exerted by a plain surface of pure water. There is a relative humidity for liquid (H_l) and a relative humidity over ice (H_i):

$$H_l = \frac{e_v}{e_s}$$

$$H_i = \frac{e_v}{e_{si}}$$

which is approximately equal to

$$H_{l'} \sim \frac{r_v}{r_s} \tag{56}$$

$$H_{i'} \sim \frac{r_v}{r_{si}} \tag{57}$$

where r_s and r_{si} are the saturation mixing ratios over liquid and ice.

8 TEMPERATURE VARIABLES

Virtual Temperature (T_v)

We now apply the ideal gas law to the mixture of air and vapor. Applying the total pressure is given by Dalton's law, $p = p_d + e_v$, to the equation of state separately to the vapor and dry air components, we obtain the modified equation of state:

$$p = \rho R_d T \frac{\left(1 + \frac{M_d}{M_v} r_v\right)}{1 + r_v} = \rho R_d T_v \tag{58}$$

where $T_v \equiv T(1.0 + 0.61 r_v)$ is the virtual temperature. Note the effect of adding moisture is to increase the virtual temperature over the temperature. Since the total density is a function of the pressure and virtual temperature, the addition of moisture actually lowers the air density. This tends to be opposite to the common perception that humid air is "heavy."

Dew Point Temperature (T_d)

This is the temperature to which moist air must be cooled at constant pressure and vapor mixing ratio in order to become saturated over a plane surface of pure water.

Dew point temperature can be calculated with the following algorithm:

$$T_d = \frac{35.86 \ln e_s - 4947.2325}{\ln e_s - 23.6837} \tag{59}$$

where e_s is in millibars and T_d is in Celsius.

Wet-Bulb Temperature (T_w)

The wet bulb temperature is the temperature that a ventillated thermometer wrapped in a wet cloth will have due to evaporation from the cloth. This will be colder than the air temperature.

The wet-bulb temperature (T_w) may be defined by the isobaric or the adiabatic process.

Isobaric Process T_w is the temperature to which air will cool by evaporating water into it at constant pressure until it is saturated. The latent heat is assumed to be supplied by the air. Note that r_v is not kept constant and so T_w differs from T_d.

When measured by a *psychrometer*, the air is caused to move rapidly past two thermometer bulbs, one dry and the other shrouded by a water-soaked cloth. When thermal equilibrium is reached on the wet bulb, the loss of heat by air flowing past the wet bulb must equal the sensible heat, which is transformed to latent heat. Hence,

$$(T - T_w)(c_p + r_v c_{pv}) - [r_s(T_w, p) - r_v]l_{lv}, \tag{60}$$

where T is the temperature of the air approaching the wet bulb.

Given temperature and mixing ratio, and a suitable relation for obtaining $r_s(T_w)$ and l_{lv}, one can solve for T_w. Alternatively, one can measure T and T_w directly with a psychrometer, and knowing pressure solve for r_v.

Adiabatic Process One can find T_w graphically with the aid of a thermodynamic diagram using the following steps:

1. Begin with pressure and mixing ratio.
2. Reduce pressure dry-adiabatically until saturation is reached to find temperature and pressure of lifting condensation level (LCL).
3. Increase pressure moist-adiabatically from LCL to original pressure.
4. The temperature at the original pressure is T_w.

This is sometimes called the *wet-adiabatic wet-bulb temperature*. It differs at most a few tenths of a degree from the other wet-bulb temperature.

9 ENTROPY VARIABLES FOR MOIST AIR

Potential Temperature for Moist Air (θ)

We earlier derived the Poisson's equation for θ relative to a dry air parcel. Later we pointed out that the conservation of θ in a dry parcel, was equivalent to the conservation of *entropy* for a dry adiabatic, reversible process. Hence, we could show that

$$c_{pd} d \ln \theta = ds. \tag{61}$$

We now extend our definition of θ to a system containing vapor gas as well as dry air. To derive this, we return to Eq. (52) and assume an adiabatic reversible system with no condensation or sublimation. We then can define θ_m for an adiabatic process:

$$c_{pm} d \ln \theta_m = d(c_{pm} \ln T) - d(R_m \ln p) = \frac{q_i}{T}. \tag{62}$$

Assuming that c_{pm} and R_m are constant for an adiabatic process, we now define θ_m to be

$$\theta_m \equiv T\left(\frac{p_{oo}}{p}\right)^{R_m/c_{pm}}. \tag{63}$$

Now, Eq. (63) can be written as:

$$d\theta_m = \frac{\theta_m}{T}\frac{q_i}{c_{pm}} \tag{64}$$

The term θ_m differs only about 1% from the θ defined earlier. As a result, distinction between θ and θ_m is usually neglected in meteorological applications and the simpler θ is used as the standard form or potential temperature.

The term θ, unlike T, takes into account the natural cooling of air as it rises from pressure change so that one can compare air parcels at different elevations to determine which parcel is warmer or colder when brought to a common elevation. Our intuitive concepts such as "cold air sinks" or "warm air rises" do not work over deep atmospheric depths because of this. Hence it normally gets colder with height, but the low-level air does not start to rise. Our intuitive concepts do work, however, when we use θ as our temperature variable. If warmer θ occurs below colder θ air, the underlying warm air will, in fact, rise spontaneously. Features such as a cold air mass do appear as cold dense flowing masses when viewed with θ instead of temperature. Figure 6 depicts a frontal system flowing southward over the central United States. Note that the cold air dams up against the Rockies to the west. One can see the wavelike feature on the eastern side of the cold air mass representing warm front and cold frontal features.

Using θ in atmospheric science naturally makes the air conceptually easier to understand as a fluid. It also makes formulations depicting the dynamics and evolution of the flow easier and more straightforward. Since air will tend to conserve its potential temperature unless diabatic effects are occurring, air naturally tends to move along θ surfaces. As a consequence, there is a better relationship between

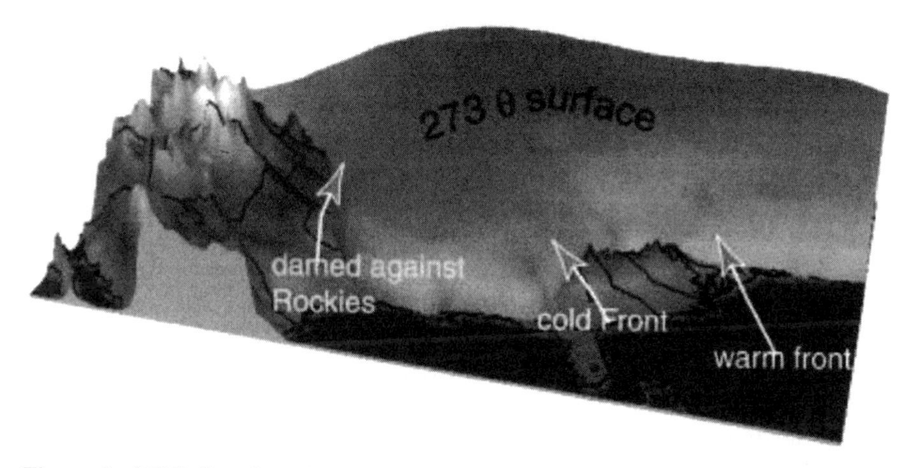

Figure 6 273 K θ surface the March, 1993 Storm of the Century. Note that the cold θ surface, flows along the ground like a heavy fluid and even dams up against the mountains. The wave like feature in the southeastern quadrant are the warm front and cold front created by the cyclone moving up the east coast at the time of this drawing.

air flows along a θ surface than along say a horizontal surface, or even a topographical surface.

Equivalent Potential Temperature (θ_e)

We will now find the *equivalent potential temperature* for an air parcel undergoing phase transition. We again assume an adiabatic, reversible system with multiple phases. Strictly, this is possible only at the triple point temperature, otherwise the ice and liquid cannot both be in equilibrium with vapor at the same time. Hence, we will find diabatic sources to any equivalent potential temperature that we define when both liquid and ice are present at other than the triple point. Equation (46) depicts entropy change for such an irreversible system. Note that there is not a general condition of equilibrium at any temperature to reduce this equation to an exact differential. Since entropy is a state variable, the entropy of the final state is determined by the state parameters of the final state, which themselves are dependent on path. Hence we *cannot* integrate Eq. (46) as we did with (62) to find a moist entropy similar to Eq. (64).

Nevertheless, as previously stated, entropy is an absolute, and according to Eq. (41), it is a sum of the entropies of each component, which for specific entropy is written

$$s = s_d + r_v s_v + r_l s_l + r_i s_i. \tag{65}$$

Employing the definition of specific entropy we can write

$$s_d = c_{pv} \ln T - R_d \ln p_d, \tag{66}$$

$$s_v = c_{pv} \ln T - R_v \ln e_v, \tag{67}$$

$$s_{lv} = c_{pv} \ln T - R_v \ln e_s, \tag{68}$$

$$s_{iv} = c_{pv} \ln T - R_v \ln e_{si}. \tag{69}$$

where s_{lv} is the equilibrium entropy of the liquid surface and s_{iv} is the equilibrium entropy over the ice surface, defined by the entropy of vapor at saturation vapor pressure as given by the Clausius–Clapeyron equation. The entropies for pure liquid and ice are, respectively,

$$s_l = c_l \ln T \tag{70}$$

$$s_i = c_i \ln T \tag{71}$$

Now substituting Eqs. (40), (47), and (66) to (71) into Eq. (65), we obtain an equation describing the total entropy of a mixed-phase system:

$$s = (c_{pd} + r_T c_l) \ln T - R_d \ln p_d + r_v \frac{l_{lv}}{T} + \frac{a_{lv}}{T} - r_v \frac{l_{il}}{T} + \frac{a_{il}}{T}. \tag{72}$$

Although the effects of curvature, solution, chemical changes, and other fairly minor considerations have been neglected, this equation accurately defines the total specific entropy of a parcel. Under equilibrium conditions, and except for the minor omissions described, we would expect s to be invariant for moist adiabatic processes. We can readily identify the differential of Eq. (72) within Eq. (46) and so determine the source of entropy to be

$$ds = d_e\left(\frac{l_{il}}{T} r_i\right) - \frac{A_{lv}}{T} dr_v + \frac{A_{il}}{T} dr_i - c_l \frac{dT}{T} d_e r_T + \frac{q_i}{T}. \tag{73}$$

As expected, the irreversible (diabatic) effect of nonequilibrium is proportional to the amount of phase transition occurring under nonequilibrium conditions in addition to external mass fluxes of water and diabatic heat sources. Hence Eq. (73) describes the change in moist entropy resulting from irreversible processes.

We can now follow the same procedure as we did with Eqs. (61) and (62) to define an equivalent potential temperature (θ_e) for a moist adiabatic process to be

$$c_{pl} d \ln \theta_e \equiv ds \tag{74}$$

where $c_{pl} = c_{pd} + r_T c_l$. Integrating Eq. (74) and defining the arbitrary constant we get

$$c_{pl} \ln \theta_e \equiv s + R_d \ln P_{oo} \tag{75}$$

Substituting Eq. (72) into Eq. (75), we obtain the expression:

$$\theta_e = T\left(\frac{P_{oo}}{p_d}\right)^{R_d/c_{pl}} (H_l)^{[(r_v+r_i)R_v)]/c_{pl}} (H_i)^{-r_v R_v/c_{pl}} e^{[l_{lv}r_v - l_{il}r_i]/c_{pl}T}. \tag{76}$$

Similar to θ_m for dry adiabatic reversible processes in moist air, θ_e is conserved for all moist adiabatic processes carried out at equilibrium. We can relate the changes of θ_e to the irreversible changes in entropy by rewriting Eq. (73) as:

$$d\theta_e = \frac{\theta_e}{c_{pl}T}[l_{il}d_e r_i + R_v T \ln H_l(dr_v - d_i r_i) - R_v T \ln H_i d_i r_i$$

$$-c_l T \ln T d_e(r_l + r_i) + q_i] \tag{77}$$

Note the inclusion of the external derivative for ice. This demonstrates that for the case of an adiabatic system in equilibrium, and neglect of heat capacity of liquid, θ_e is conserved.

Note that the more vapor in the air the greater the θ_e. It is proportional to the amount of potential energy in the air by the effects of temperature, latent heat, and even geopotential energy combined. Hence θ_e tends to be high on humid warm days and low on cool and/or dry days. It also tends to be greater at high elevations than low elevations for the same temperature because of the lower pressure present at higher elevations. The statement: "The higher the θ_e is at low levels, the greater the potential for strong moist convection," is analogous to the statement: "The higher the θ at low levels, the greater the potential for dry convection." Whereas, we look at a tank of relatively incompressible water and say warm (cold) water rises (sinks), we look at the dry atmosphere and say warm (cold) θ rises (sinks) and look at the moist saturated atmosphere and we state warm (cold) θ_e rises (sinks).

If we ignore diabatic heating effects such as radiation, friction, and heat conduction at the surface, θ_e is nearly perfectly conserved in the atmosphere! θ_e acts like a tracer such that one can identify an air parcel from its θ_e and trace where it came from, even in the midst of moist processes. Figure 7 depicts the 345 K θ_e surface for a supercell thunderstorm viewed from the northeast. A thunderstorm occurs when warm θ_e builds up under a region of colder θ_e, creating the condition where the warm θ_e air will rise if a cloud forms and the moisture within the warm θ_e air condenses. Also, the condition evolves, where the colder θ_e air at middle levels will sink if evaporation takes place. Both of these processes are occurring in Figure 7. The rising currents of warm θ_e air form a "tree trunk"-like structure of the 345 K θ_e, very analogous to rising plume in a lava lamp. To the west (right in Fig. 7) rain is falling into the cold θ_e air at middle levels where it is evaporating causing the cold θ_e

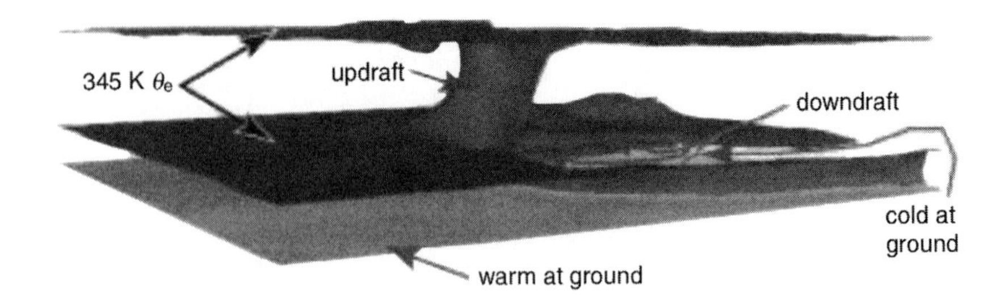

Figure 7 345 K θ_e surface of a model simulation of supercell thunderstorm observed over Montana on 2 August, 1981. The view is from the northeast. Generally θ_e increases with height, i.e., the warmest air is on top. But summertime and tropical conditions allow warm θ_e to be produced near the surface creating a situation where plumes of warm θ_e rise and cold plumes sink as can be seen in the figure. The surface plane is colored with surface temperature where cold air from evaporational cooling is found in the western sector of the storm resulting from cold evaporating downdrafts carrying cold θ_e from middle levels to the surface.

air to sink to the surface. The surface, colored according to its temperature, is cold in the middle of the pool of cold θ_e air sinking to the ground. The power of using equivalent potential temperature to understand the inner structures of complex moist convective weather systems is demonstrated in this figure. Many other weather systems evolve from energy released by phase change including tropical cyclones and some middle latitude cyclones.

10 PSEUDO-ADIABATIC PROCESS (θ_{ep})

Because entropy is a function of pressure, temperature, and liquid and ice water, a multiphase process cannot be represented on a two-dimensional thermodynamic diagram. It is convenient to define a *pseudo-adiabatic* process, where the heat capacity of liquid and ice is neglected.

Omitting the liquid water term (and references to ice) from Eq. (72), and differentiating we can write

$$ds_p = (c_{pd} + r_v c_l) d \ln T - R_d d \ln p_d + d\left(\frac{r_v l_{lv}}{T}\right) - d(r_v R_v \ln H_l) \tag{78}$$

Since the first term on the RHS now contains r_v instead of r_T, an exact differential is not possible. Bolton integrated numerically and derived the following expression for the *pseudo-equivalent potential temperature* θ_{ep}, as:

$$\theta_{ep} = T\left(\frac{p_{oo}}{p}\right)^{0.2854(1-0.28r_v)} \exp\left[r_v(1 + 0.81r_v)\left(\frac{3376}{T_{sat}} - 2.54\right)\right] \tag{79}$$

where T_{sat} is a saturation temperature defined as one of:

$$T_{sat} = \frac{2840}{3.5 \ln T - \ln e - 4.805} + 55, \tag{80}$$

or

$$T_{sat} = \frac{1}{\dfrac{1}{T - 55} - \dfrac{\ln(H_l)}{2840}} + 55. \tag{81}$$

T_{sat} can also be determined graphically to be the temperature of the lowest condensation level (LCL).

The pseudo-adiabatic (θ_{ep}) isentropes parallel the *adiabatic wet-bulb potential temperature* (θ_w). The θ_{ep} is related to θ_w by:

$$\theta_{ep} = \theta_w \exp\left[r_{v'}(1 + 0.81 r_{v'})\left(\frac{3376}{\theta_w} - 2.54\right)\right], \tag{82}$$

where

$$r_v \equiv r_{sl}(p_{oo}, \theta_w). \tag{83}$$

The pseudo-equivalent potential temperature is interpreted as a two-step process being pseudo-adiabatic ascent to zero pressure followed by dry adiabatic descent to $p_{oo} = 1000\,\text{mb}$. This is how we label the moist adiabats on a thermodynamic diagram.

11 NEGLECTING HEAT STORAGE

A common approximation made to the first–second law is to neglect the heat storage by water. This greatly simplifies the heat capacity and gas constant terms such that $c_{pm} \simeq c_{pd}$ and $R_m \simeq R_d$. This effectively reduces Eq. (52) to:

$$c_{pd} d \ln T - R_d d \ln p + \frac{l_{lv}}{T} dr_v - \frac{l_{il}}{T} d_i r_i = \frac{q_i}{T} \tag{84}$$

Generally the effects of this approximation are negligible for meteorological applications. As a result of this approximation, $\theta_m \simeq \theta$ and

$$\theta_e \simeq \theta \exp\frac{l_{lv} r_v - l_{il} r_i}{c_{pd} T_{LCL}} \tag{85}$$

where T_{LCL} is the temperature at the lowest condensation level at which the relative humidity (H_l) is 1. It also follows that:

$$d \ln \theta - \frac{l_{iv}}{c_{pd}T} dr_v + \frac{l_{il}}{c_{pd}} d_i r_i = q_i \tag{86}$$

These approximations are standard in most applications.

12 HYDROSTATIC BALANCE AND HYPSOMETRIC EQUATION

It is sometimes useful to consider the force balance responsible for suspending an air parcel above the surface. The force upward results from the pressure change across the parcel per height change, which gives T net force per parcel volume. The downward force per parcel volume is gravity multiplied by parcel density. Setting these forces equal we obtain the hydrostatic balance:

$$dp = -\rho g \, dz \tag{87}$$

where z is the height coordinate.

Strict hydrostatic balance is not adhered to in the atmosphere, otherwise there would not be vertical acceleration. But the balance is usually very close, and nearly exact over a time average. Assuming the balance, variations in pressure can be converted to variations in height.

We can substitute the equation of state for density to obtain the hypsometric equation:

$$\frac{d \ln p}{dz} = -\frac{g}{R_d T_v} \tag{88}$$

The hypsometric can be integrated to give

$$p = p_0 \exp \frac{-g\Delta z}{R_d \overline{T_v}} \tag{89}$$

where the subscript 0 refers to an initial value, the Δz is the height change from the initial value, and the bar represents an average value during the change from the initial state. This equation is commonly used to find the height change between two pressures or the pressure change between two heights. A common use of this formula is to find the pressure the atmosphere would have at sea level given a pressure measured at a surface location above sea level.

13 DRY AND MOIST ADIABATIC LAPSE RATES (Γ_d AND Γ_m)

A by-product of assuming hydrostatic balance is that we can determine how temperature would change with height for different thermodynamic processes. For an adiabatic process, where $d\theta = 0$, it is easily shown that Eq. (88) requires that the *dry lapse rate* of temperature (Γ_d) be

$$\Gamma_d = -\frac{dT}{dz} = \frac{g}{c_{pd}} = 9.8°\text{C/K}. \tag{90}$$

Similar equations can be derived for a *moist lapse rate* Γ_m occurring in a saturated (w.r.t. liquid) atmosphere. Then applying Eq. (84) for a no-ice system:

$$\frac{dT}{dz} = \frac{\dfrac{g}{c_{pd}}}{1 + \dfrac{L_{lv}}{c_{pd}T}\dfrac{dr_{vs}}{dT}}, \tag{91}$$

where r_{vs} is the saturation mixing ratio, and where we have assumed that the air parcel remains at 0% supersaturation. Notice that the moist adiabatic lapse rate is reduced from the dry adiabatic lapse rate because the latent heating of condensation counters a portion of the expansional cooling. This effect weakens with decreasing temperature as the rate of change of saturation mixing ratio with height decreases.

Dry and Moist Static Energy (h and h_d)

It is sometimes convenient to work with a thermodynamic variable that is directly related to enthalpy. Recall that we looked at this in the first section where we defined gravitational potential energy. It is useful to do the same here, further subdividing the contributing energies. In most applications of static energy, direct measurement of condensate is not possible and so we cannot consider the energies stored in the liquid or ice phases.

The partial work term $-R_d d \ln P$ found in the first law is conceptually difficult. We learned that the term comes from the work performed by expansion that depletes the parcel of total enthalpy. This can either reduce its temperature or slow it down, i.e., reduce its kinetic energy. For large-scale flows, we can generally neglect the effect on the parcel's kinetic energy and assume a nonaccelerating, hydrostatic balance. Then we can employ the hydrostatic Eq. (87) and transform $R_d d \ln P$ into $g\, dz$. Under the hydrostatic assumption, the partial work term is thus shown to be equivalent to a conversion of enthalpy to geopotential energy.

Ignoring the effects of ice, and diabatic processes, Eq. (87) then is written:

$$d(c_{pd}\, dT - g\, dz + l_{lv}\, dr_v) = q. \tag{92}$$

We now define the dry and moist static energies to be

$$h_m \equiv c_{pd}T - g\,dz + lr_v \text{ and}$$
$$h \equiv c_{pd}T - g\,dz,$$

respectively. Notice that the dry static energy is closely related to θ and the moist static energy is closely related to θ_e. The three energy storage terms involved are the sensible heat ($c_{pd}T$), the geopotential energy (gz), and the latent heat ($L_{vl}r_v$). It is now convenient to study the total energy budget of a large-scale atmospheric system to and assess conversions between kinetic energy and static energy.

REFERENCES

Dutton, J. (1976). *The Ceaseless Wind*, McGraw-Hill.

Emanuel, K. A. (1994). *Atmospheric Convection*, Oxford University Press.

Hess, S. L. (1959). *Introduction to Theoretical Meteorology*, Holt, Reinhardt, and Winston.

Iribarne, J. V., and W. L. Godson (1973). *Atmospheric Thermodynamics*, D. Reidel Publishing Co.

Sears, F. W. (1953). *Thermodynamics*, Addison-Wesley.

Wallace, J. M., and P. V. Hobbs, (1977). *Atmospheric Science: An Introductory Survey*, Academic Press.

CHAPTER 17

THERMODYNAMIC ANALYSIS IN THE ATMOSPHERE

AMANDA S. ADAMS

1 ATMOSPHERIC THERMODYNAMIC DIAGRAMS

Before computations were made on computers, atmospheric scientists regularly employed graphical techniques for evaluating atmospheric processes of all kinds including radiation processes, dynamical processes, and thermodynamic processes. Over the last few decades, the use of graphical techniques has almost completely disappeared with the exception of thermodynamic diagrams and conserved variable thermodynamic diagrams.

Classical Thermodynamic Diagrams

The thermodynamic diagram is used to graphically display thermodynamic processes that occur in the atmosphere. The diagram's abscissa and ordinate are designed to represent two of the three state variables, usually a pressure function on one and a thermodynamic function on another. Any dry atmospheric state may be plotted. Unfortunately, any moist state cannot be plotted as a unique point since that must depend on the values of r_v, r_l, and r_i. However, vapor content can be inferred by plotting dewpoint temperature, and moist processes can be accounted for by assuming certain characteristics of the moist process such as if the process is pseudo-adiabatic.

Handbook of Weather, Climate, and Water: Dynamics, Climate, Physical Meteorology, Weather Systems, and Measurements, Edited by Thomas D. Potter and Bradley R. Colman.
ISBN 0-471-21490-6

There are three characteristics of a thermodynamic diagram that are of paramount importance. They are:

1. *Area is proportional to the energy of a process or the work done by the process.* An important function of a thermodynamic diagram is to find the energy involved in a process. If the diagram is constructed properly, the energy can be implied, by the area under a curve or the area between two curves representing a process.
2. *Fundamental lines are straight.* Keeping the fundamental lines straight aids in use of the thermodynamic diagram.
3. *Angle between isotherms (T) and isentropes (θ) are to be as large as possible.* One of the major functions of the thermodynamic diagram is to plot an observed environmental sounding and then compare its lapse rate to the dry adiabatic lapse rate. Since small differences are important, the larger the angle between an isotherm and an isentrope, the more these small differences stand out.

Over the years there have been several different thermodynamic diagrams designed. While they all have the same basic function of representing thermodynamic processes, they have a few fundamental differences. Before these differences can be discussed, it is important to have an understanding of the variables and lines found on a thermodynamic diagram.

Dry Adiabatic Lapse Rate and Dry Adiabats

One of the lines on a thermodynamic diagram, known as the dry adiabat, is representative of a constant potential temperature (θ). Following along a dry adiabat the potential temperature will remain constant, while the temperature will cool at the dry adiabatic lapse rate. The dry adiabatic lapse rate represents the rate at which an unsaturated parcel will cool as it rises in the atmosphere, assuming the parcel does not exchange heat or mass with the air around it, hence behaving adiabatically. The dry adiabatic lapse rate can be derived from a combination of the first law of thermodynamics, the ideal gas law, and the hydrostatic approximation (see chapters 15 and 16), given by:

$$-\frac{dT}{dz} = \frac{g}{cp}$$

Thus the dry adiabatic lapse rate for an ascending parcel is simply g/c_p and is approximately equal to 9.8°C/km or 9.8 K/km. The dry adiabat lines on a thermodynamic diagram are beneficial in that the user may ascend an unsaturated parcel along the dry adiabats and simply read off the new temperature.

Moist Adiabatic Lapse Rate and Pseudo-Adiabats

When a parcel is saturated, it does not suffice to use the dry adiabatic lapse rate. A parcel that is saturated will experience condensation of the water vapor in the parcel. As water vapor condenses the process releases latent heat. This release of latent heat keeps the parcel from cooling at the dry adiabatic lapse rate. The parcel cools at a slower rate, which is dependent on temperature. This reduced rate of temperature decrease has a slope approximately parallel to the moist adiabat and exactly parallel to a line of constant equivalent potential temperature. The term pseudo-adiabat is often used to describe the moist adiabat on a thermodynamic diagram. The moist adiabat on a thermodynamic diagram ignores the contribution by ice, which is more difficult to represent because freezing and melting do not typically occur in equilibrium as does condensation. In a pseudo-adiabatic process the liquid water that condenses is assumed to be immediately removed by idealized instantaneous precipitation. Thus the word *pseudo-adiabat* is used since the line on a thermodynamic diagram is not truly moist adiabatic. If ice is neglected, an expression for the moist adiabatic lapse rate can be derived in a manner similar to the dry adiabatic lapse rate, by using the first law for moist adiabatic hydrostatic ascent. The moist adiabatic lapse rate can never be greater than the dry adiabatic lapse rate. At very cold temperatures, around $-40°C$, the moist adiabatic lapse rate will approach the dry adiabatic lapse rate. In the lower troposphere the moist adiabatic lapse rate may be as small as $5.5°C/km$. The equation for the moist adiabatic lapse rate is given as:

$$-\frac{dT}{dz} = \frac{g}{cp}\left[\frac{1 + \dfrac{L_{vl} r_{sl}}{R_d T}}{1 + \dfrac{L_{vl}}{R_d}\dfrac{r_{sl}}{T}\dfrac{\varepsilon L_{vl}}{c_p T}}\right]$$

Types of Thermodynamic Diagrams

There are several different thermodynamic diagrams. All the diagrams serve the same basic function; however, there are differences among these diagrams that should be understood before choosing a diagram. The *emagram* was named by Refsdal as an abbreviation for "energy per unit mass diagram" (Hess, 1959). The emagram uses temperature along the abscissa and the log of pressure along the ordinate. The isobars (lines of constant pressure) and the isotherms (lines of constant temperature) are straight and at a right angle to each other. With pressure plotted on a logarithmic scale, the diagram is terminated at 400 mb, well before the tropopause.

On the *tephigram* one coordinate is the natural log of potential temperature (θ) and the other is temperature. Thus, on a tephigram an isobar is a logarithmic curve sloping upward to the right. The slope decreases with increasing temperature. In the range of meteorological observations, the isobars are only gently sloped. When using the tephigram, the diagram is usually rotated so that the isobars are horizontal with decreasing pressure upward. The pseudo-adiabats are quite curved, but lines of

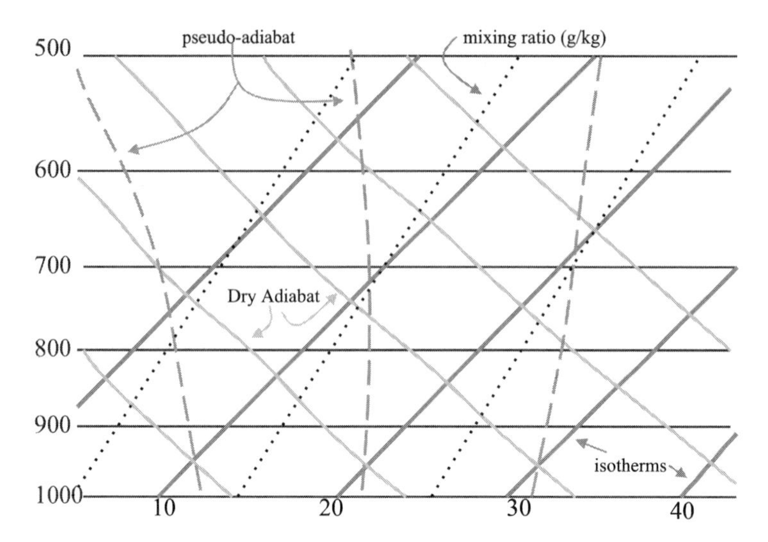

Figure 1 Skew T–log P diagram. See ftp site for color image.

constant saturation mixing ratio tend to be straight. The angle between the isotherms and isentropes is 90°, making this type of thermodynamic diagram particularly good for looking at variations in stability of an environmental sounding.

The skew T–log P diagram is now the most commonly used thermodynamic diagram in the meteorological community (Fig. 1). This diagram first suggested by Herlofson in 1947 was intended as a way to increase the angle between the isotherms and isentropes on an emagram (Hess, 1959). The isobars are plotted on a log scale, similar to the emagram. The isotherms slope upward to the right at 45° to the isobars. This angle of the isotherms is what gives this diagram its name, since the temperature is skewed to the right. The isentropes are skewed to the left, at approximately a 45° angle to the isobars. However, the isentropes are not perfectly straight, although they remain nearly straight in the meteorological range typically used. With both the isotherms and isentropes (dry adiabats) at approximately 45° to the isobars, the angle between the isotherms and isentropes is close to 90°. The pseudo-adiabats are distinctly curved on this diagram. In order for the pseudo-adiabats to be straight, the energy–area proportionality of the diagram would need to be sacrificed.

The Stüve diagram is another type of thermodynamic diagram. The Stüve diagram has pressure in mb along the x axis and temperature along the y axis. This coordinate system allows the dry adiabats (isentropes) to be straight lines. However, this diagram does not have area proportional to energy. The uses of the Stüve diagram are limited when compared to the uses of energy conservation thermodynamic diagrams. This diagram is still occasionally used in the meteorological community, but because of familiarity rather than usefulness.

The characteristics of the thermodynamic diagrams discussed is summarized in the following table:

Attribute	Emagram	Tephigram	Skew T–log P	Stüve
Area α energy	Yes	Yes	Yes	No
T vs. θ angle	45°	90°	Almost 90°	45°
p	Straight	Gently curved	Straight	Straight
T	Straight	Straight	Straight	Straight
θ	Gently curved	Straight	Gently curved	Straight
θ_{ep}	Curved	Curved	Curved	Curved
r_s	Gently curved	Straight	Straight	Straight

With multiple thermodynamic diagrams available it may seem overwhelming to a new user to determine the appropriate diagram for the task. The skew T–log P diagram and tephigram are the most commonly used thermodynamic diagrams. Their strength lies in the angle between the isentropes and isotherms. The tephigram is most widely used by tropical meteorologists because of the ability of the tephigram to show small changes in stability. In the tropics convective potential may be modified by only small changes in the vertical stability of the environment, and the tephigram best captures small changes in the environmental lapse rate. The skew T–log P diagram is the most commonly used thermodynamic diagram in the midlatitudes. The straight isobars and the straight up vertical temperature profile make the diagram intuitively easy to comprehend. The ultimate decision of which diagram to use relies on the familiarity of the diagram to the user. While a summer air mass in the middle latitudes may best show instability on a tephigram, a meteorologist may prefer to use the familiar skew T–log P diagram.

2 ATMOSPHERIC STATIC STABILITY AND APPLICATIONS OF THERMODYNAMIC DIAGRAMS TO THE ATMOSPHERE

Early atmospheric scientists sent up kites with thermometers attached to acquire an understanding of the vertical profile (atmospheric sounding) of the atmosphere. Today, atmospheric scientists use balloons rather than kites and radiosondes rather than thermometers in an effort to understand the vertical structure of the atmosphere. Radiosondes collect temperature, moisture, wind speed, and wind direction at various pressure levels in the atmosphere. The data gained from radiosondes, when displayed on a thermodynamic diagram, provide forecasters and scientific investigators with information that can be used to diagnose not only the current state of the atmosphere but also the recent history of the local atmosphere and the likelihood of future evolution. An air mass originating over a land mass will have inherently different characteristics to its thermodynamic profile than an air mass originating over the oceans. An air mass influenced by motions associated with an approaching weather system will acquire characteristics associated with those circulations. A trained meteorologist can read an atmospheric sounding plotted on a thermodynamic diagram like a book, revealing the detailed recent history of the air mass.

Radiosondes are launched twice daily around the world, at 0000 Universal Time, Coordinated (UTC), also known as Greenwich Mean Time (GMT) or Zulu time (Z), and 1200 UTC. While the data from radiosondes are available only twice daily, when the data is plotted on a thermodynamic diagram, they can be used to study potential changes in the atmosphere. Thermodynamic diagrams can be used to examine the potential warmth of the atmosphere near the surface during the day, the potential cooling of temperature at night, and the potential for fog. The potential for cloudiness can be found due to the movements induced by approaching or withdrawing weather systems or by the development of unstable rising air parcels. When the vertical thermodynamic structure is combined with the vertical wind structure, information on the structure, and severity of possible cumulus clouds can be ascertained. The sounding can also describe the potential for local circulations such as sea breezes, lake effect snow, downslope windstorms, and upslope clouds and precipitation.

An atmospheric sounding plotted on a thermodynamic diagram is one of the most powerful diagnostic tools available to meteorologists. The applications of the thermodynamic diagrams are far too many to be adequately covered in this discussion. This discussion will focus on some of the basic interpretive concepts that can be applied generally to atmospheric soundings plotted on thermodynamic diagrams.

Environmental Structure of the Atmosphere

The sounding plotted on a thermodynamic diagram represents a particular atmospheric state. The state may be observed by radiosonde, satellite derived, or forecasted by a computer atmospheric numerical model. The data plotted on the diagram represent the temperature and dew-point temperature at various pressure levels throughout the atmosphere. Individual observations are plotted, and then the points are connected forming the dew point and temperature curves. The dew-point temperature is always less than or equal to the temperature for any given pressure level. The plotted curves represent the *environmental* temperature and dew-point temperature, as a function of pressure.

The environmental temperature and dew-point temperature curves will divulge much information to a well-trained observer who can deduce where the sounding is from, as well as identify air masses and the processes involved in their formation. Air masses originate from specific source regions and carry the characteristics of that source region.

Arctic Air Mass Air masses that originate near one of the poles have a nearly isothermal temperature profile. In the polar regions, the main process driving the temperature profile is radiational cooling. Light winds allow very little vertical mixing of the air. When coupled with the lack of solar radiation for much of the year, radiational cooling of the surface is the remaining process. Initially the air near Earth's surface cools faster. The rate at which energy is emitted is dependent on temperature to the fourth power (Stefan–Boltzmann equation), and therefore the warmer air above will then cool faster than the colder air at the surface. This results in the atmosphere moving toward a constant temperature, and this result is

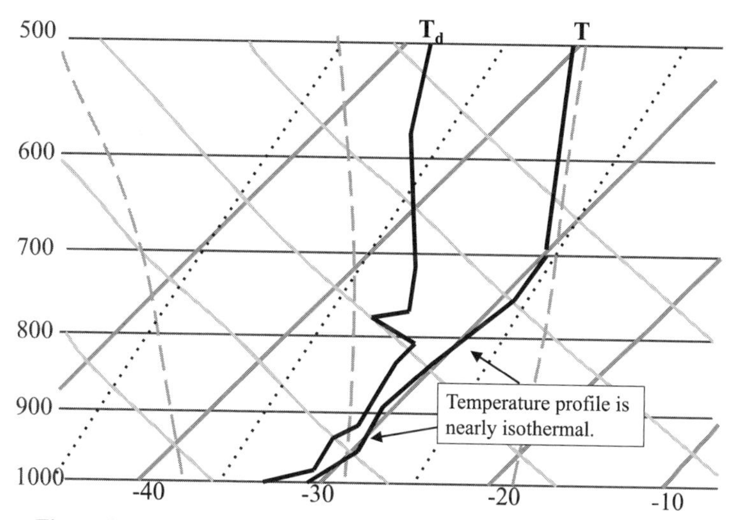

Figure 2 Characteristic arctic air mass. See ftp site for color image.

observable on a skew-T diagram by the isothermal profile that begins at the surface (Fig. 2). The tropopause is found at a much lower height near the poles. An arctic sounding will illustrate the low tropopause. Both maritime and continental arctic air masses will have an isothermal temperature profile; the difference is found in the degree of saturation of the air.

Marine Tropical Air Mass In a marine tropical environment (Fig. 3) the temperature profile is nearly moist adiabatic (parallels the pseudo-adiabats). The air in this region is close to saturation near the surface and thus very humid. Above the planetary boundary layer, the air is dominated by sinking motion between cumulus clouds and will tend to be dry except for some moisture mixed from the cumulus clouds. Since saturated air parcels in cumulus updrafts force much of the sinking motion outside clouds, the temperature profile closely parallels the pseudo-adiabat. A dry air maximum resulting from the sinking around 700 mb characterizes the dew-point profile of a marine tropical environment. Below 700 mb surface moisture mixes upward driven by solar heating and solar-powered surface evaporation. The dry air maximum is the feature that truly distinguishes the maritime tropical air mass from other air masses.

Well-Mixed Air Mass Perhaps one of the easiest air masses to identify is one that is well mixed (Fig. 4). If a layer of the atmosphere has sufficient turbulence, the layer will become well mixed with the moisture and potential temperature evenly distributed.

The temperature profile will take on the slope of the adiabatic lapse rate and the dew-point temperature will follow a constant mixing ratio value. Mixing ratio is a measure of the actual amount of water vapor in the air, and in an environment with a

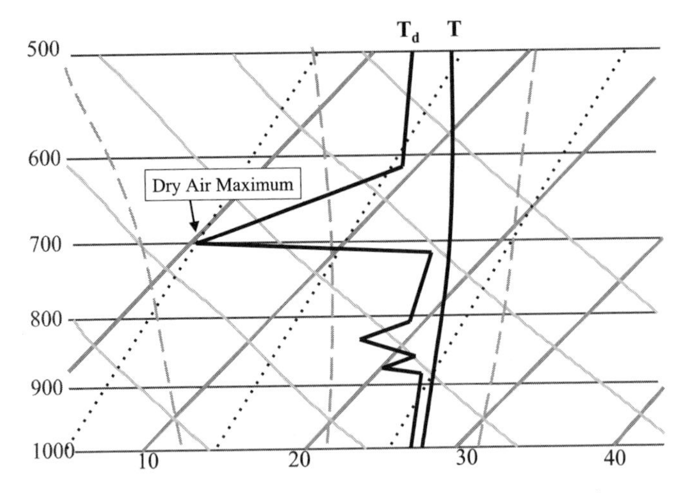

Figure 3 Marine tropical air mass. See ftp site for color image.

lot of turbulent mixing the water vapor will become evenly distributed. Well-mixed layers generally form in a capped boundary layer. However, a well-mixed layer may form in the boundary layer of one region and then be advected into another region. If the region that the well-mixed layer is advected into is at a lower elevation, the well-mixed layer may ride over the existing boundary layer, resulting in an elevated mixed layer. This is commonly observed in the Great Plains of the United States, where a well-mixed layer is advected off the Mexican Plateau or Rocky Mountains, over the boundary layer of the Plains, which is very moist due to flow from the Gulf of Mexico.

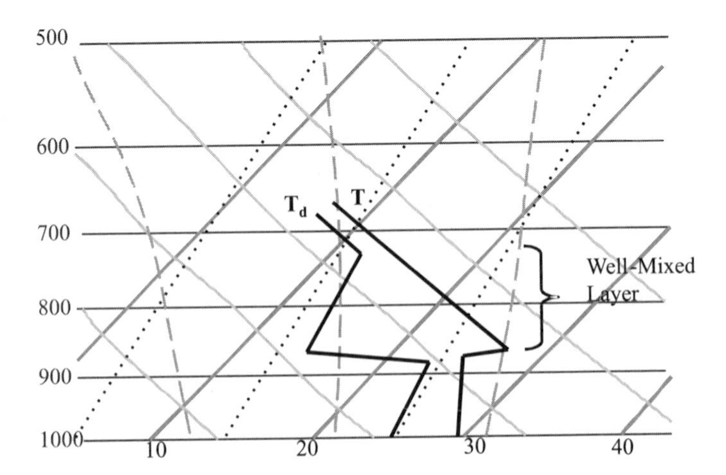

Figure 4 Well-mixed layer in the atmosphere. See ftp site for color image.

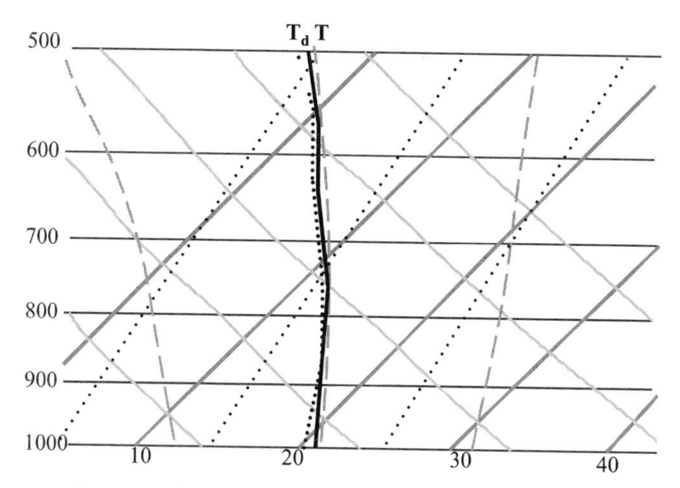

Figure 5 Deep convection. See ftp site for color image.

Deep Convection Layers of the atmosphere that include clouds are identifiable because the temperature and dew-point temperatures will be close together indicating the air is saturated. In the case of deep convection (Fig. 5), the dew-point temperature and temperature, within the updraft, are so close together that they are almost indistinguishable from each other. They are close together throughout a deep layer of the atmosphere, typically the surface up to the tropopause. The temperature profile of the environment will closely parallel the moist adiabat in the case of deep convection. In deep convection, where the atmosphere is basically saturated throughout, all rising parcels will cool at the moist adiabatic lapse rate.

Critical Levels on a Thermodynamic Diagram

The information on a thermodynamic diagram may be used not only to determine where air masses originated from but to also determine what processes the atmosphere may undergo. There are many different critical levels that can be diagnosed from a sounding plotted on a thermodynamic diagram. A thermodynamic diagram may be used to determine where a parcel will become either saturated and/or buoyant. The level at which the parcel becomes saturated depends on not only the moisture content but the processes the atmosphere is undergoing, for this reason there are multiple condensation levels dependent on the process occurring.

Lifting Condensation Level (LCL) This is the pressure level at which a parcel lifted dry adiabatically will become saturated. This level is important for two reasons. First, since the parcel becomes saturated at this point, it represents the height of cloud base. Second, since the parcel is saturated, if it continues to rise above this level, the parcel will cool at the moist adiabatic lapse rate. In a conditionally unstable atmosphere, a parcel is unstable only if saturated. A conditionally unstable parcel lifted to its LCL will typically be cooler than the environment at the LCL. Further lifting will cause the parcel to cool at a slower rate than the environment and eventually will become warmer than the environment. To find the LCL on a thermodynamic diagram first determine from which level you wish to raise a parcel. (Typically the LCL is found for a parcel taken from the surface, but in certain atmospheric conditions, it may be more pertinent for a meteorologist to raise a parcel from a level other than the surface, especially if there is strong convergence at another level.) Initially the parcel is assumed to have the same temperature and moisture content (dew-point temperature) as the environment. Starting from the temperature follow the dry adiabat (constant potential temperature), since the parcel temperature will change at the dry adiabatic lapse rate until it becomes saturated. From the dew-point temperature, follow the intersecting line of constant mixing ratio. The actual grams per kilogram of water vapor in the air is assumed to not change, so long as the parcel rises the mixing ratio remains constant. Note that while the actual amount of water vapor remains constant, the relative humidity will increase. The pressure level at which the mixing ratio value of the parcel intersects the potential temperature of the parcel is the level at which the parcel is saturated. This level is the lifting condensation level (Fig. 6).

Level of Free Convection (LFC) In the atmosphere there are many mechanisms that may force a parcel to rise, and thus initiate convection. The level of free convection is the level above which a parcel becomes buoyant and thus will continue to rise without any additional lifting. Typically, the LFC is found at a height above the LCL. To determine the LFC of a parcel, one need only follow the parcel path until it crosses the environmental temperature profile (Fig. 6). The path the parcel will follow is along the dry adiabat until the LCL is reached, and along the pseudo-adiabat above the LCL. Due to the existence of layers of increased stability (low

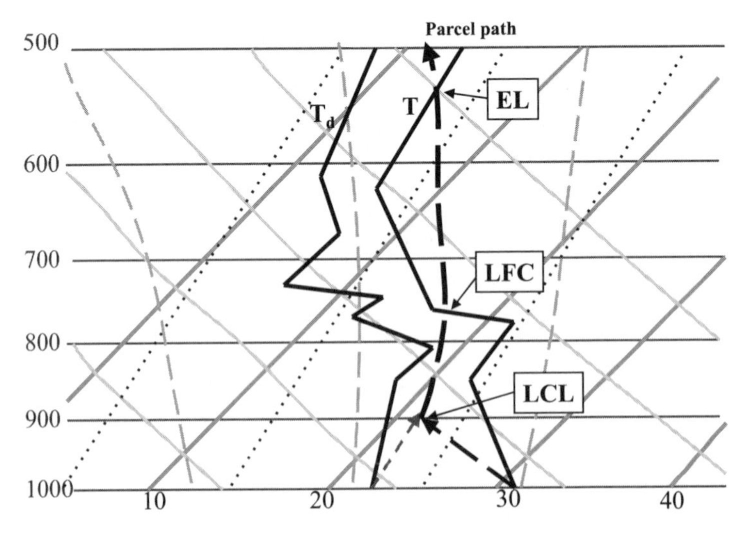

Figure 6 Determining the LCL, LFC, and EL of a parcel. See ftp site for color image.

lapse rates) in the atmosphere, a parcel may have multiple levels of free convection. When using the LFC to determine how much lifting or vertical motion is needed to produce free convection, the LFC at the lowest pressure level (highest above the ground) is typically used.

Equilibrium Level (EL) The equilibrium level is the level at which a parcel buoyant relative to the environment is no longer buoyant. At the equilibrium level, the parcel and the environment have the same temperature. If a parcel rises above the equilibrium level, it will become colder than the environment and thus negatively buoyant. To determine the EL, follow the parcel path from the LFC until it intersects the environmental temperature (Fig. 6). Often the equilibrium level will be found near the beginning of the tropopause. The tropopause is isothermal in nature and thus very stable. On a sounding with multiple levels of free convection, there will also be multiple equilibrium levels. In an extremely stable environment, with no level of free convection, there will also be no equilibrium level. It is noteworthy to mention that while the equilibrium level represents where a parcel is no longer positively buoyant, it does not mean the parcel cannot continue rising. Above the equilibrium level, while the parcel is negatively buoyant, a parcel that reaches this level will have a certain amount of kinetic energy due to its vertical motion. This kinetic energy will allow the parcel to continue rising until the energy associated with the negative buoyancy balances the kinetic energy.

Convective Condensation Level (CCL) The convective condensation level can be used as a proxy for the level at which the base of convective clouds will begin. The CCL differs from the LCL in that the LCL assumes some sort of lifting, and the CCL assumes the parcel is rising due to convection alone. To determine the convective condensation level on a thermodynamic diagram use the dew-point temperature and follow the mixing ratio line that intersects that dew-point temperature until it crosses the environmental temperature (Fig. 7). This represents the level at which a parcel rising solely due to convection will become saturated and form the base of a cloud.

Convective Temperature (CT) The convective temperature is the temperature that must be reached at the surface to form purely convective clouds. If daytime heating warms the surface to the convective temperature, thermals will begin to rise and, upon reaching the CCL, be not only saturated but also positively buoyant. The convective temperature is determined by first locating the CCL. The dry adiabat, which intersects the convective condensation level, is followed down to the surface (Fig. 7). The temperature at the surface represents the convective temperature. If the surface is able to warm to the convective temperature, then the LCL and CCL are the same.

Mixing Condensation Level (MCL) Clouds can form as a result of turbulent mixing rather than lifting or convection. The height at which a cloud will form due to mixing can be determined on a thermodynamic diagram and is referred to as the

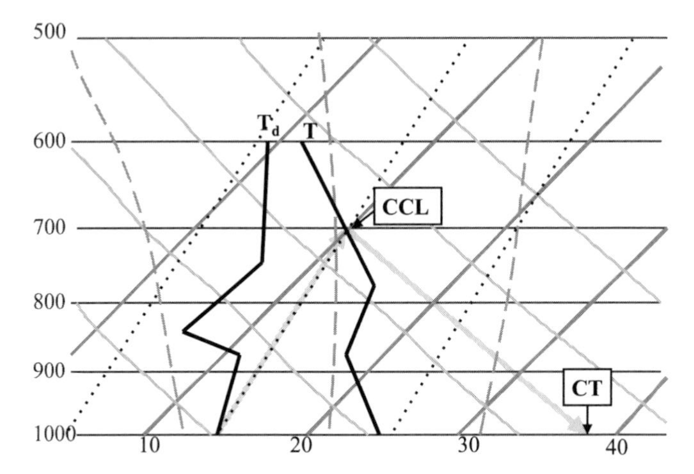

Figure 7 Determining the CCL and CT. See ftp site for color image.

mixing condensation level (Fig. 8). The MCL is determined on a thermodynamic diagram by estimating the average water content (mixing ratio) and the average potential temperature (θ) within the layer. The pressure level at which the average mixing ratio and average potential temperature intersect is the mixing condensation level. The idea is that in a well-mixed layer the amount of moisture will be constant with height, as will the potential temperature. This process often occurs in a shallow strongly capped boundary layer with wind speeds over 10 m/s. The mixing process that forms the cloud is the same process that allows one to "see their breath" on a cold day. While the boundary layer may initially be unsaturated throughout, turbulent mixing can redistribute the moisture and result in saturation near the top of the boundary layer.

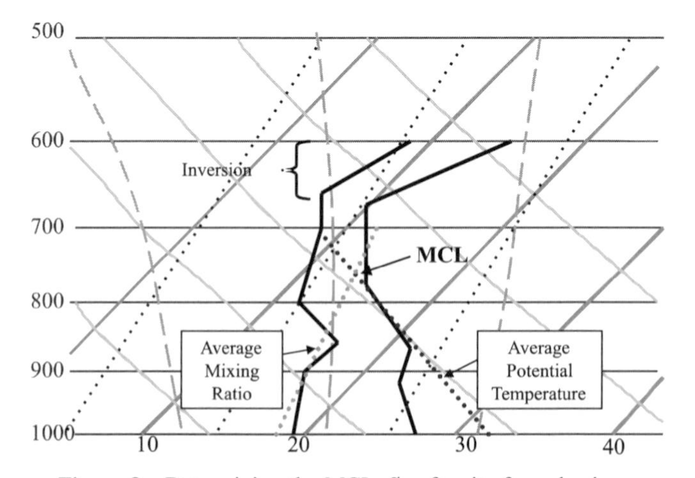

Figure 8 Determining the MCL. See ftp site for color image.

Diagnosing Stability and Parcel Path

The thermodynamic profile of the atmosphere is indicative of the stability of the atmosphere. Understanding the stability of various layers of the troposphere is important in forecasting what processes may occur. To discuss stability, the environment is compared to a parcel. A parcel can be thought of as an entity of the atmosphere with specific temperature and moisture content, which is assumed to not interact with the air around it, and thus undergoes purely adiabatic processes. With this assumption about a parcel, it is easy to ascertain on a thermodynamic diagram the temperature a parcel would have as a result of being either raised or lowered in the atmosphere. This temperature of the parcel at various pressure levels can be drawn on a thermodynamic diagram and is referred to as the parcel path. In general, it is assumed that an unsaturated parcel will change temperature at the dry adiabatic lapse rate, while a saturated parcel will change temperature at the pseudo-adiabatic (moist) lapse rate, as it moves up or down from its initial location.

Stability is determined by comparing the temperature of a parcel to the temperature of the environment to which the parcel rises or sinks (Fig. 9). A parcel that is warmer than the environment is unstable and will rise. A parcel colder than the environment is stable and will sink. However, it is important to remember that a parcel will cool as it rises. Therefore, the stability at a single point is not as important as stability throughout the layer. By comparing the environmental lapse rate to the lapse rate of the parcel, stability can be determined. A layer is stable if the environmental lapse rate is greater than the parcel's lapse rate. A layer is unstable if the environmental lapse rate is less than the parcel's lapse rate. The lapse rate of the parcel is dependent on whether the parcel is unsaturated or saturated. An unsaturated parcel will cool at the dry adiabatic lapse rate (DALR), of approximately 9.8°C/km in the troposphere. A saturated parcel will cool at the moist adiabatic lapse rate. The moist adiabatic lapse rate (MALR) is not constant (hence the use of a thermody-

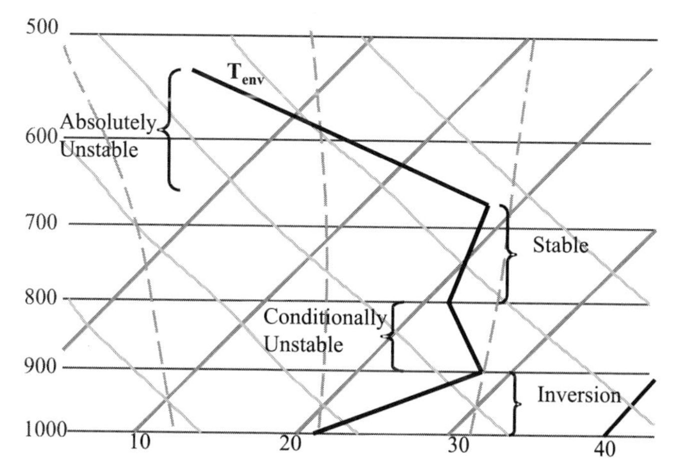

Figure 9 Using the environmental temperature lapse rate to determine stability. See ftp site for color image.

namic diagram for ease), but averages about 6°C/km in the troposphere. The moist adiabatic lapse rate is always less than the dry adiabatic lapse rate. Therefore, if the environmental lapse rate (ELR) is greater than both the moist and dry adiabatic lapse rates, the parcel is *absolutely stable* in that its stability is not dependent on whether the parcel is unsaturated or saturated. A layer is *absolutely unstable* when the environmental lapse rate is less than both the dry and moist adiabatic lapse rates. In the case of an absolutely unstable layer, both dry and saturated parcels will be unstable relative to the environment and will rise. An absolutely unstable layer is rarely observed in a sounding because when a layer is absolutely unstable it will instantly overturn and mix the air in the layer. If an absolutely unstable layer is plotted on a thermodynamic diagram, it is likely that the observation is an error, with the possible exception being when it is plotted in the lowest 50 m near the surface. When the environmental lapse rate is less than the dry adiabatic lapse rate, but greater than the moist adiabatic lapse rate, the atmosphere in that layer is *conditionally unstable*. In a conditionally unstable environment a saturated parcel will be unstable, while an unsaturated parcel is stable. The average environmental lapse rate of the troposphere is conditionally unstable:

DALR > MALR > ELR	Absolutely stable
DALR > ELR > MALR	Conditionally unstable
ELR > DALR > MALR	Absolutely unstable

As mentioned previously in this chapter, one of the benefits of a thermodynamic diagram is that area is proportional to energy. By comparing the parcel path to the environmental temperature profile on one of these diagrams, the *convective available potential energy* (CAPE) and *convective inhibition* (CIN) can be ascertained (Fig. 10). The equation for CAPE is found by integrating from the LFC to the EL to determine the area where the parcel is positively buoyant:

$$\text{CAPE} = g \int_{LFC}^{EL} \frac{\theta_{parcel} - \theta_{env}}{\theta_{\text{env}}} dz$$

If a parcel reaches the LFC, the maximum updraft speed that could develop can be estimated by converting the convective available potential energy into kinetic energy. This gives an equation for the updraft speed of

$$w = \sqrt{2 * \text{CAPE}}$$

The CIN represents the area from the surface to the LFC where the parcel is colder than the environment. Whereas the CAPE provides an idea of how much energy is available once the LFC is reached, the CIN gives an approximation of how much energy must be expended to lift a parcel to the LFC. CAPE and CIN play an important role in determining the possibility for severe weather to occur. Convective inhibition is important for severe weather because it allows CAPE above the bound-

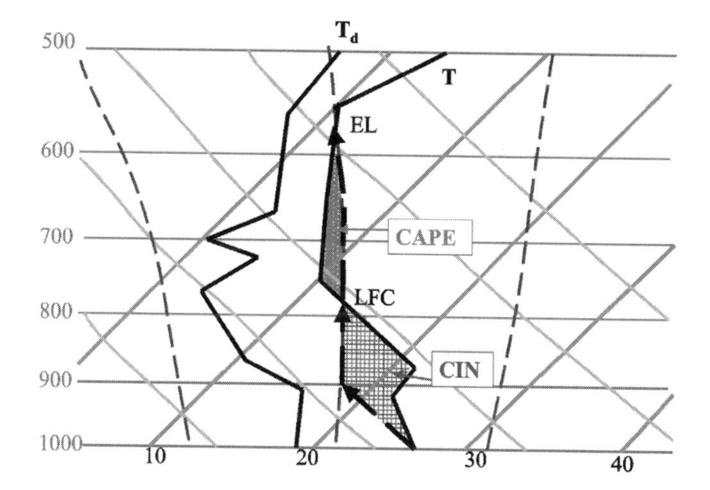

Figure 10 Path of parcel rising from the surface is denoted by the arrows. See ftp site for color image.

ary layer to build. Convective inhibition also keeps parcels from freely rising everywhere, and thus the convection will be more focused along a specific forcing mechanism when significant CIN is present. Large values of CAPE are conducive to severe weather because the more CAPE the more energy a potential storm will have.

Inversions

When the environmental lapse rate is such that the temperature actually increases with height, the layer in which this occurs is called an inversion. An inversion is a very stable layer. The high stability of an inversion can act to "cap" a layer of the atmosphere, preventing parcels from rising above the inversion. Inversions can form as a result of several different processes. While the environmental temperature profile is the same for different types of inversions, the environmental dew-point profile will appear different depending on the processes involved.

Subsidence Inversion The most distinct characteristic of a subsidence inversion (Fig. 11) is the dryness of the inversion. A subsidence inversion occurs as the result of widespread sinking air. The air aloft is initially cooler, and thus has less water vapor. As the air sinks, compressional heating warms it, but no additional water vapor is added. The dew-point temperature will decrease throughout the inversion, with the driest air at the top of the inversion. A subsidence inversion will commonly occur due to the position of the jet streak, causing wide-scale subsidence.

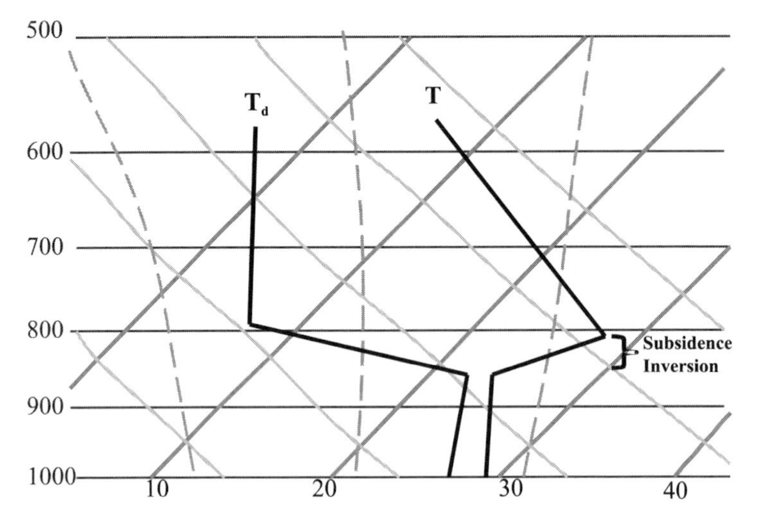

Figure 11 Subsidence inversion. See ftp site for color image.

Radiational Inversion Strong radiational cooling can act to form an inversion. The dew-point temperature in this inversion will typically follow the slope of the temperature profile. However, the dew-point temperature profile does not have to follow the temperature profile slope in the case of a radiational inversion. See Figure 12. An inversion formed by radiational cooling is typically deduced by where in the vertical the inversion occurs. Nighttime cooling of Earth's surface will form an inversion near the surface visible on morning soundings. A radiational inversion can also be observed at the top of a cloud layer due to the cloud top cooling.

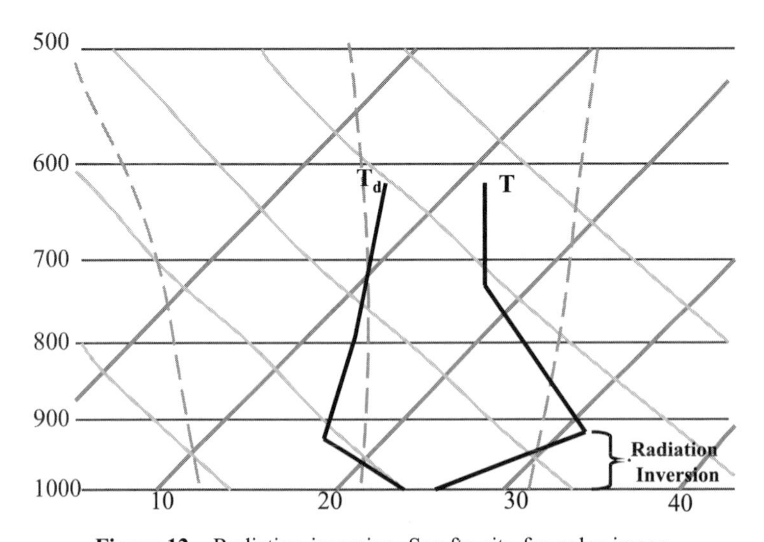

Figure 12 Radiation inversion. See ftp site for color image.

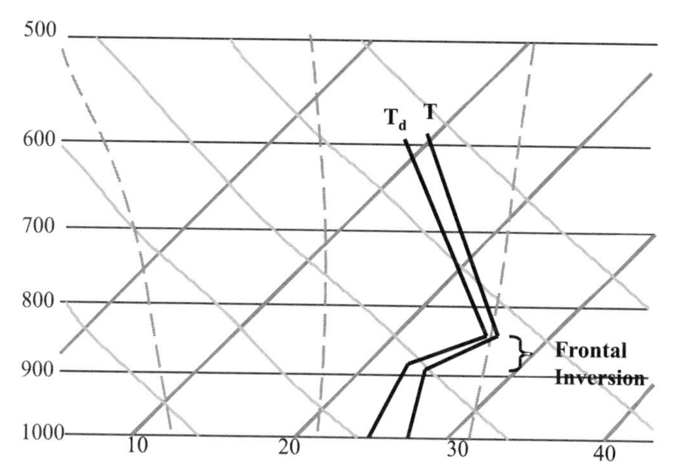

Figure 13 Frontal inversion. See ftp site for color image.

Frontal Inversion In the case of a frontal inversion the dew-point temperature will generally increase with height through the inversion (Djuric, 1994) (Fig. 13). Along a frontal boundary warm, less dense, air will override the colder denser air at the surface. This results in the air aloft being warmer than the surface, and this is depicted in the environmental temperature profile as an inversion. A frontal inversion is differentiated from a radiational inversion in that it is typically much deeper than a radiational inversion. Also, meteorologists with many tools at their disposal will be able to determine the location of the front with surface and upper air analyses.

Inversions, even shallow ones, can play an important role in thermodynamic processes. The height of the LFC, the amount of CAPE, and the amount of convective inhibition a sounding possesses is very dependent on the strength and height of any inversions. The processes involved in the creation of an inversion are also important to the stability of the atmosphere. While a frontal inversion is representative of a stable layer, a trained meteorologist may infer that there is lifting in advance of the front, which could force convection.

BIBLIOGRAPHY

Bohren, C. F., and Albrecht, B. A. (1998). *Atmospheric Thermodynamics*, Oxford University Press, New York, NY.

Djuric', D. (1994). *Weather Analysis*, Prentice Hall Inc., Englewood Cliffs, New Jersey.

Emanuel, Kerry A. (1994) Atmospheric Convection, Oxford University Press, New York, NY.

Hess, Seymour L., (1959) Introduction to Theoretical Meteorology, Krieger Publishing Company, Malabar, FL.

Glossary of Meteorology, (2nd ed.) (2000). American Meteorological Society, Boston, MA.

Wallace, J. M., and P. V. Hobbs (1977). *Atmospheric Science An Introductory Survey*, Academic, San Diego, CA.

CHAPTER 18

MICROPHYSICAL PROCESSES IN THE ATMOSPHERE

ROBERT M. RAUBER

1 INTRODUCTION

Viewed from space, the most distinct features of Earth are its clouds. Most of the world's clouds form within rising air currents forced by atmospheric circulations that can be local or extend over thousands of kilometers. Clouds and fogs also form when air cools to saturation either from radiative cooling or from conduction of heat from atmosphere to Earth's surface. Clouds can also develop as air masses with different thermal and moisture properties mix together. In all cases, clouds have direct influence on atmospheric motions through latent heat exchange, absorption and scattering of solar and terrestrial radiative energy, and redistribution of water vapor. They modulate Earth's climate, reducing the amount of solar radiation reaching Earth and trapping terrestrial radiation. Particle scavenging, chemical reactions, and precipitation processes within clouds continuously alter the trace gas and aerosol composition of the atmosphere. Clouds are an important component of Earth's hydrological cycle.

Clouds are broadly classified according to their altitude, the strength, orientation, and extent of their updrafts, their visual shape, and whether or not they are precipitating. The internationally accepted cloud classification first proposed by Pharmacist Luke Howard in 1803 identifies four broad categories: cumulus (clouds with vertical development), stratus (layer clouds), cirrus (high, fibrous clouds), and nimbus (precipitating clouds), and numerous secondary categories. For example, clouds forming at the crest of waves generated by airflow over mountains are called altocumulus lenticularis because they are often lens shaped. Excellent

Handbook of Weather, Climate, and Water: Dynamics, Climate, Physical Meteorology, Weather Systems, and Measurements, Edited by Thomas D. Potter and Bradley R. Colman.
ISBN 0-471-21490-6

photographs of clouds with their classifications can be found in the International Cloud Atlas (World Meteorological Organization, 1987).

Clouds are composed of ensembles of water and ice particles, most of which form on aerosol particles that range in diameter from about 10^{-8} to 10^{-5} m. Raindrops typically grow to diameters between 10^{-4} and 5×10^{-3} m, while large hailstones reach diameters of about 5×10^{-2} m. Clouds typically have dimensions between 10^5 and 10^7 m, and individual particle paths through clouds may extend 10^4 to 10^5 m. As water drops and ice particles follow these paths, they can encounter temperatures that can range from 30 to -50°C, pressures from 1050 to 250 mb, and humidity conditions from supersaturation with respect to water to nearly dry air. The complexities of the ever-changing cloud environment and this enormous range of scales must all be considered when investigating the physics governing cloud and precipitation processes.

This chapter explores the fundamental principles and key issues within the discipline of *cloud microphysics*, a discipline specifically concerned with the formation, growth, and fallout of cloud and precipitation particles. An extensive body of cloud microphysics literature has been published over the last century. Limited references are provided in this chapter. Readers who wish to explore individual topics in greater depth, or quickly refer to the scientific literature summarized in this chapter, should consult Pruppacher and Klett (1997), the most authoritative treatise on cloud microphysics currently available.

2 ATMOSPHERIC AEROSOL

The atmosphere consists of a mixture of gases that support a suspension of solid and liquid particles called the atmospheric aerosol. Chemical reactions between gases, aerosol, and cloud particles continually modify the chemical structure, concentration, and distribution of aerosol in the atmosphere. Cloud droplets and ice crystals form on specific aerosol called cloud condensation nuclei and ice nuclei. Once formed, cloud droplets and ice crystals scavenge atmospheric gases and aerosol and provide an environment for additional chemical reactions.

Sources, Sinks, and Formation Mechanisms

Aerosol particles (AP) form during chemical reactions between gases (gas-to-particle conversion), as droplets containing dissolved or solid material evaporate (drop-to-particle conversion), and through mechanical or chemical interactions between the earth or ocean surface and the atmosphere (bulk-to-particle conversion). Gas-to-particle conversion primarily involves the transformation of sulfur oxides, nitrogen oxides, and gaseous hydrocarbons to sulfates, nitrates, and solid hydrocarbon particles. These reactions typically involve water vapor, often require solar radiation, and usually include intermediate states involving sulfuric, nitric, or other acids. Drop-to-particle conversion occurs when cloud drops containing dissolved or suspended material evaporate. Tiny droplets originating at the sea surface when air bubbles break also produce AP when they evaporate. Bulk-to-particle conversion involves wind erosion of rocks and soils and decay of biological material.

Mineral dust from Earth's land surfaces, salt particles from the ocean, organic material and gas emissions from aquatic and terrestrial plants, combustion products from human activities and natural fires, volcanoes, and even meteor bombardment all are sources of atmospheric aerosol. The continents are a much larger source than the oceans, and urbanized areas are a larger source than rural areas. Consequently, the highest concentration of aerosol in the lower atmosphere can normally be found over cities and the lowest concentrations over the open ocean. The concentration of AP decreases rapidly with height, with approximately 80% of AP contained in the lowest kilometer of the atmosphere. Aerosol concentrations are reduced through self-coagulation, precipitation processes, and gravitational settling. The residence time of aerosol in the atmosphere depends on the size and composition of the particles and the elevation where they reside (Fig. 1). Smaller particles ($<0.01\ \mu m$ radius) collide rather quickly due to thermal diffusion and coagulate, while large particles ($>10\ \mu m$ radius) fall out quickly due to their increased fall velocity. Particles with radii between 0.01 and $10\ \mu m$ have the longest residence times, typically from 1 to 10 days in the lower troposphere, weeks in the upper troposphere, and months to years at altitudes above the tropopause.

Number Concentration, Mass, and Size Distribution

Aerosol particles range in size from molecular clusters consisting of a few molecules to about $100\ \mu m$. Aerosol spectra fall into four size groups: Aitken particles (dry radii $r < 0.1\ \mu m$), large particles ($0.1 < r < 1\ \mu m$), giant particles ($1 < r < 10\ \mu m$), and ultragiant particles ($r > 10\ \mu m$). Whitby (1978) showed that the aerosol size distribution is comprised of three modes, each related to different physical processes (Fig. 2). The *nuclei mode*, which consists of the smallest particles, develops during chemical reactions associated with gas-to-particle conversion. This mode is large in polluted regions and essentially absent in pristine environments. The larger *accumulation mode* forms through coagulation of smaller particles and continued growth of existing particles during vapor condensation and chemical reactions. The accumulation mode develops as a result of aging of the aerosol population. The largest mode, the *coarse particle mode*, is comprised of particles that originate at Earth's surface, either through mechanical disintegration of the solid earth or through evaporation of tiny droplets produced during bubble breakup at the ocean surface. These aerosols differ in chemical composition from those comprising the accumulation mode.

The number concentration of aerosol in the troposphere varies substantially from location to location. In very polluted air over cities, the number concentration may approach $10^6\ cm^{-3}$. Over land, average concentrations range from 10^3 to $10^5\ cm^{-3}$ while over the oceans, average concentrations typically range from a few hundred to $10^3\ cm^{-3}$. The mass concentration varies in a similar way. Over cities, the mass concentration of aerosol typically ranges from about 100 to $200\ \mu g/m^{-3}$, while over

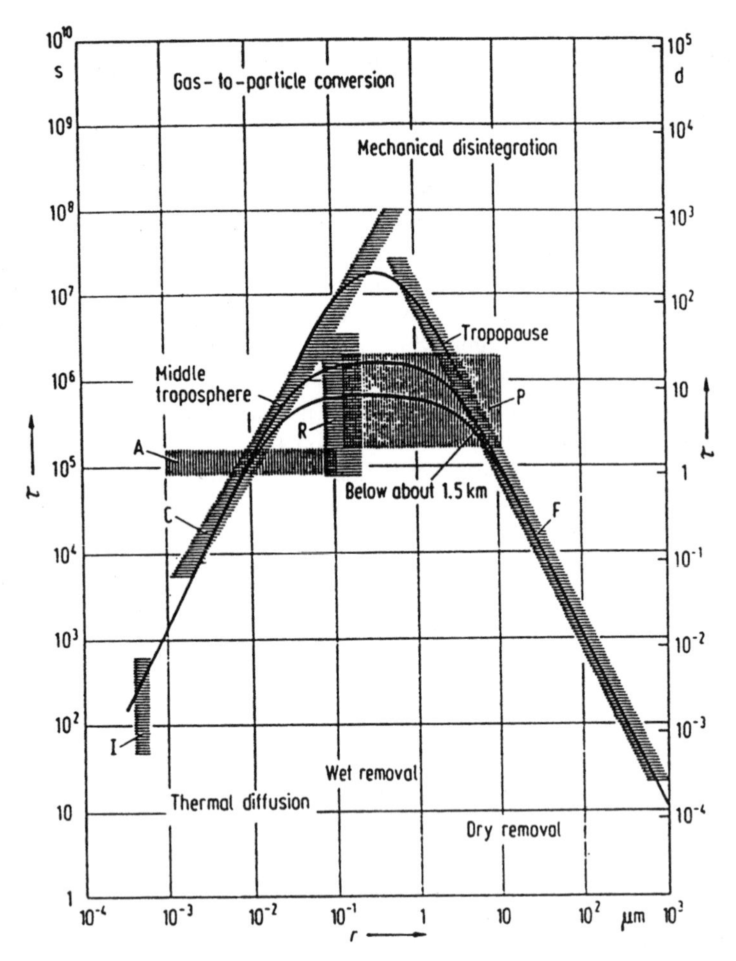

Figure 1 Residence time of aerosol particles as a function of their radius and altitude. I, small ions; A, Aitken particles; C, from thermal diffusion of aerosol particles; R, based on radioactivity data; P, removal by precipitation; F, removal by sedimentation. Solid lines are empirical fits for three atmospheric levels (Jaenicke, 1988).

the oceans, mass concentrations are nearly an order of magnitude smaller (15 to 30 $\mu g/m^{-3}$). The number concentration of Aitken particles decreases exponentially with height between the surface and ~5 km with a scale height of about 1 km (Fig. 3). Between 5 km and the tropopause, the Aitken particle concentration is nearly constant. In the lower stratosphere, to a height of about 20 km, the Aitken particle concentration decreases slowly with a scale height of about 7 km. The number concentration of "large" particles decreases by about 3 orders of magnitude over the 4 lowest kilometers of the atmosphere, and reaches a minimum near the tropopause. Unlike Aitken particles, the concentration of large particles increases

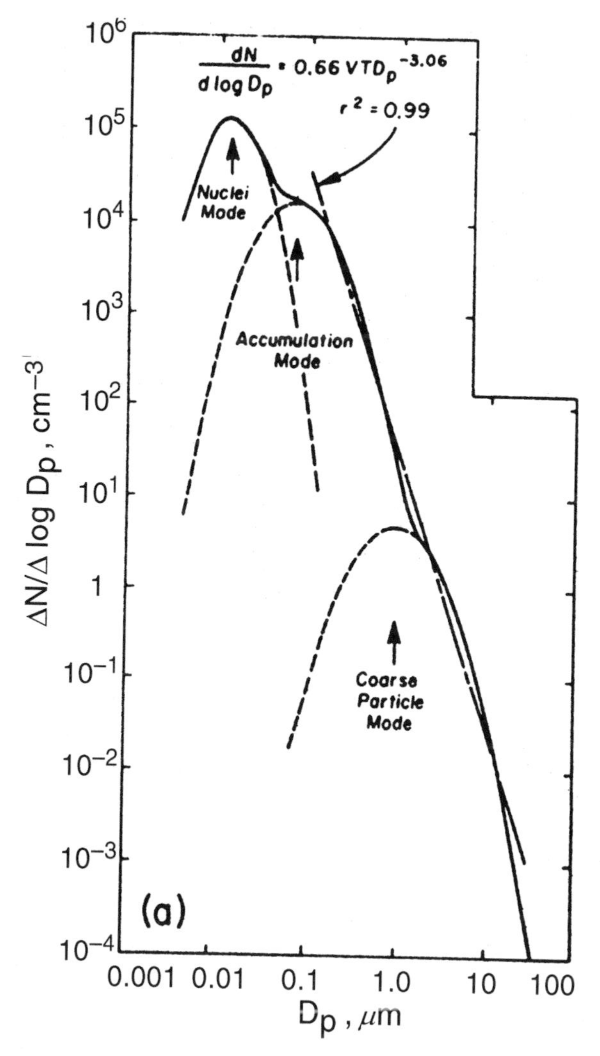

Figure 2 Number size distribution of aerosol particles in urban conditions. N, number of particles; D_p, diameter of particles; V_T, total volume concentration; r, correlation coefficient between the power law in the figure and experimental data (Whitby, 1978).

with height in the stratosphere, with maximum concentrations occurring at an altitude of about 20 km. This stratospheric aerosol layer, called the *Junge layer* after its discoverer, consists primarily of sulfate particles. The number concentration of giant particles, such as sea salt particles, decreases rapidly with height above the surface. The effects of gravitational settling and cloud scavenging reduce the concentration of sea salt aerosol to near insignificance above about 3 km.

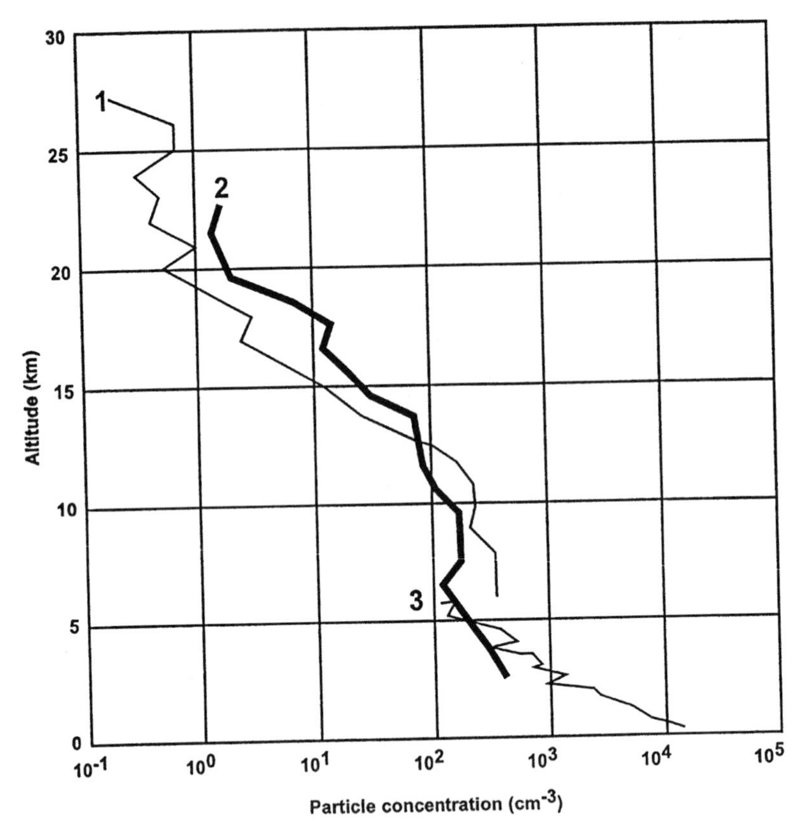

Figure 3 Vertical profiles of Aitken particles: (1) Average concentration per reciprocal cubic centimeter from seven flights over Sioux Falls, South Dakota (44°N); (2) average concentration per cubic centimeter from four flights over Hyderabad, India (17°N); (3) average concentration per cubic centimeter based on flights over the Northeast United States. (Adapted from Junge, 1963.)

Because of the wide range of sizes of aerosol particles, it has been customary to use logarithmic size intervals to characterize the aerosol particle size distribution. If we express the total concentration of aerosol particles with sizes greater than r as

$$N(r) = \int_{\log(r)}^{\infty} n(r) d \log(r) \tag{1}$$

then the number of particles in the size range r, $r + d \log r$ is given by

$$n(r) = -\frac{dN}{d \log(r)} \tag{2}$$

The corresponding volume, surface, and mass distributions, expressed in log radius, are given respectively by:

$$\frac{dV(r)}{d\log(r)} = \frac{4\pi r^3}{3}\frac{dN}{d\log(r)} \tag{3}$$

$$\frac{dS(r)}{d\log(r)} = 4\pi r^2 \frac{dN}{d\log(r)} \tag{4}$$

$$\frac{dM(r)}{d\log(r)} = \frac{4\pi r^3 \rho(r)}{3}\frac{dN}{d\log(r)} \tag{5}$$

where $\rho(r)$ is the particle density. Figure 4 shows examples of aerosol number, surface, and volume distributions from different environments. The accumulation and course particle modes discussed by Whitby (1978) appear in many of the volume distributions. The nuclei mode appears only in the number distributions since these particles contribute negligibly to the volume because of their small

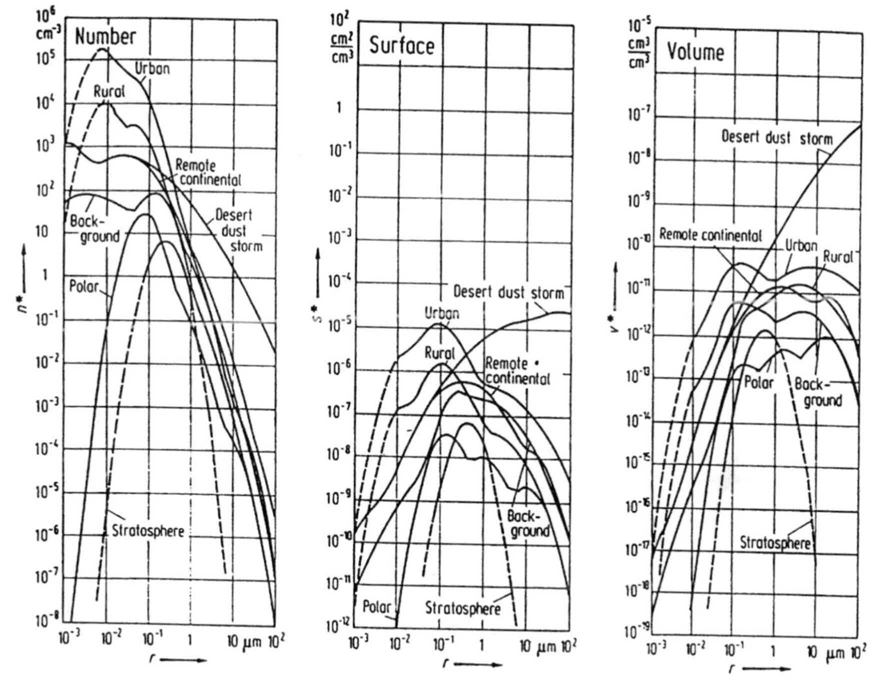

Figure 4 Number $[n^*(r) = dN/d\log(r)]$, surface $[s^*(r) = dS/d\log(r)]$, and volume $[v^*(r) = dV/d\log(r)]$ size distributions for various aerosols (Jaenicke, 1988).

radius. Note that the concentration of aerosol with $r > 0.1$ μm decreases with increasing size such that the aerosol number distribution can be approximated by:

$$\frac{dN}{d\,\log(r)} = Cr^{-\beta} \tag{6}$$

where β generally falls between a value of 3 and 4.

3 FORMATION OF INDIVIDUAL CLOUD DROPLETS AND ICE CRYSTALS IN THE ATMOSPHERE

Cloud droplets and ice crystals form in the atmosphere through a spontaneous process called *nucleation*. *Homogeneous nucleation* is said to occur when a droplet or ice crystal forms from water vapor in the absence of any foreign substance. Homogeneous nucleation of ice crystals also occurs when water droplets freeze without the action of an ice nucleus. The water droplets may be very highly diluted solutions (cloud droplets) or concentrated solutions (haze droplets). *Heterogeneous nucleation* is said to occur when a condensation nucleus is involved in the formation of a water droplet or an ice nucleus is involved in the formation of an ice particle.

Homogeneous Nucleation of Water Droplets in Humid Air

The change in free energy during the homogeneous nucleation of a water droplet in humid air has two contributions: a negative change per unit volume associated with the creation of the new phase and a positive change associated with the creation of the surface interface between the phases. The former is proportional to the volume, while the latter is proportional to the surface area. Because the surface-to-volume ratio is large for small droplets, the surface term dominates until the embryo becomes sufficiently large. The thin surface layer of the drop has unique properties. Molecules within this layer find themselves in an asymmetric force field, attracted only to neighboring molecules located in the interior. This net attractive force causes the surface of the drop to be in a state of tension described by the surface tension, $\sigma_{w,v}$, which has units of energy/unit area. When humid air is cooled beyond saturation, a metastable state develops where the water vapor becomes supersaturated without condensing to form the new phase. The metastable state exists because the path to the lower energy state (a water droplet) initially involves an increase in free energy associated with the droplet surface. Homogeneous nucleation requires the spontaneous collision and aggregation of a sufficient number of water molecules so that the embryonic droplet will be large enough to overcome this "energy barrier." Although statistical mechanics are required to strictly describe this process, classical thermodynamic approaches that assume the embryos are distributed according to the Boltzmann law and are water spheres that have macroscopic densities and surface tensions provide an adequate description. With these assumptions, classical thermodynamics predicts the energy barrier to be

$$\Delta F = -\left(\frac{4\pi r^3 \rho_w}{3M_w}\right) R^* T \ln\left(\frac{e_r}{e_{s,\infty}}\right) + 4\pi r^2 \sigma_{w,v} \tag{7}$$

where ΔF is the Helmholtz free energy change, e_r is the water vapor pressure over a spherically curved surface of radius r, $e_{s,\infty}$ is the saturation vapor pressure over a plane water surface, T is temperature, R^* is the universal gas constant, M_w is the molecular weight of water, ρ_w is the density of water, r is the radius of the embryo, and $\sigma_{w,v}$ is the surface tension at the water–vapor interface. The ratio $e_r/e_{s,\infty}$ is the saturation ratio, $S_{w,v}$, over the droplet surface. The first term on the right is the decrease in free energy associated with the creation of the droplet, and the second term is the increase in free energy associated with the creation of the droplet's surface. Taking the limit $\Delta F/\Delta r = 0$ to determine the equilibrium radius r_{eq} gives the Kelvin equation:

$$S_{w,v} = \exp\left(\frac{2M_w \sigma_{w,v}}{R^* T \rho_w r_{\text{eq}}}\right) \tag{8}$$

Substituting r_{eq} into (7) gives the height of the energy barrier:

$$\Delta F_{\max} = \frac{16\pi M_w^2 \sigma_{w,v}^3}{3[R^* T \rho_w \ln(S_{w,v})]^2} \tag{9}$$

The Kelvin equation relates the equilibrium radius of a droplet to the environmental saturation ratio. A droplet must grow by chance collisions of water molecules to a radius equal to r_{eq} to be stable. Numerical evaluation of the Kelvin equation shows that an embryo consisting of 2.8×10^5 molecules would be in equilibrium at an environmental supersaturation of 10% ($S_{v,w} = 1.10$). A supersaturation of 500% would be required for an embryo consisting of 58 molecules to be stable (Rogers and Yau, 1989). The rate at which embryos of a critical radius form at a given supersaturation, the nucleation rate, J, has been determined using statistical mechanics to be

$$\begin{aligned} J &= \frac{\beta}{\rho_w}\left(\frac{2N_a^3 M_w \sigma_{w,v}}{\pi}\right)^{1/2}\left(\frac{e_{s,\infty}}{R^* T}\right)^2 S_{w,v} \exp\left(\frac{-\Delta F_{\max}}{kT}\right) \\ &\approx 10^{25}\ \text{cm}^{-3}/\text{s}\ \exp\left(\frac{-\Delta F_{\max}}{kT}\right) \end{aligned} \tag{10}$$

where β is the condensation coefficient, N_a is Avogadro's number, and k is the Boltzmann constant. Numerical evaluations of J for different $S_{w,v}$ show that J varies over 115 orders of magnitude as the cloud supersaturation increases from

200 to 600%. The threshold for nucleation is nominally considered to be $J = 1$ droplet cm^{-3}/s. Supersaturations exceeding 550% are required before $J = 0.1$ droplet cm^{-3}/s. Since supersaturations in natural clouds rarely exceed a few percent, we conclude from consideration of J and the Kelvin equation that homogeneous nucleation of water droplets does not occur in Earth's atmosphere.

Heterogeneous Nucleation of Water Droplets

Homogeneous nucleation of cloud droplets directly from vapor cannot occur in natural clouds because supersaturations in clouds rarely exceed a few percent. Cloud droplets form through a heterogeneous nucleation process that involves aerosol particles. In cloud chambers, where supersaturations of several hundred percent can be achieved, nearly all aerosol will initiate droplets. Aerosol that nucleate droplets under these high supersaturation conditions are called *condensation nuclei*. Aerosol that nucleate droplets in the low supersaturation conditions of natural clouds are called *cloud condensation nuclei* (CCN). The number of CCN is a function of cloud supersaturation (s_w) as well as the mass, composition, and concentration of the aerosol population in the region where the clouds are forming. The supersaturation required to activate a cloud droplet is a strong inverse function of particle mass, so only larger aerosol act as CCN. Figure 5 shows the concentration of CCN as a function of supersaturation in maritime and continental environments worldwide. The concentration of CCN typically varies from about 20 to 300 cm^{-3} in maritime air and from about 200 to 1000 cm^{-3} in continental air. Typically about 50% of maritime aerosol and 1% of continental aerosol act as CCN at $s_w = 1\%$. The relationship between CCN concentrations and supersaturation (%) in Figure 5 can be approximated as a power law of the form

$$N_{\text{CCN}} = C(S_w)^k \tag{11}$$

Measurements in unpolluted maritime environments have found values of C and k ranging from $25 \leq C \leq 250\ cm^{-3}$ and $0.3 \leq k \leq 1.4$. Values of C and k measured in continental environments are $600 \leq C \leq 5000$ and $0.4 \leq k \leq 0.9$.

Aerosols that act as CCN are normally hygroscopic and totally or partially water soluble. The chemical composition of CCN depends on the proximity to sources. The primary components of marine CCN are non-sea-salt sulfates (NSS). These aerosols are produced from organic gases such as dimethylsulfide (DMS) and methanesulfonate during gas-to-particle reactions. Measurements suggest that DMS, which is produced primarily by marine algae, is emitted into the atmosphere from the ocean at a global rate of 34 to 56 Tg/yr. The second component of the marine CCN spectra is sea salt. Sea salt is injected into the atmosphere by bubble bursting and spray. Studies suggest that submicron CCN in the marine atmosphere are primarily NSS, while supermicron particles are sea salt, and that sea salt particles contribute only about 1% to the total CCN concentration over the ocean.

The higher CCN concentrations found in continental air masses arise primarily from natural sources rather than anthropogenic activities. The production rate of

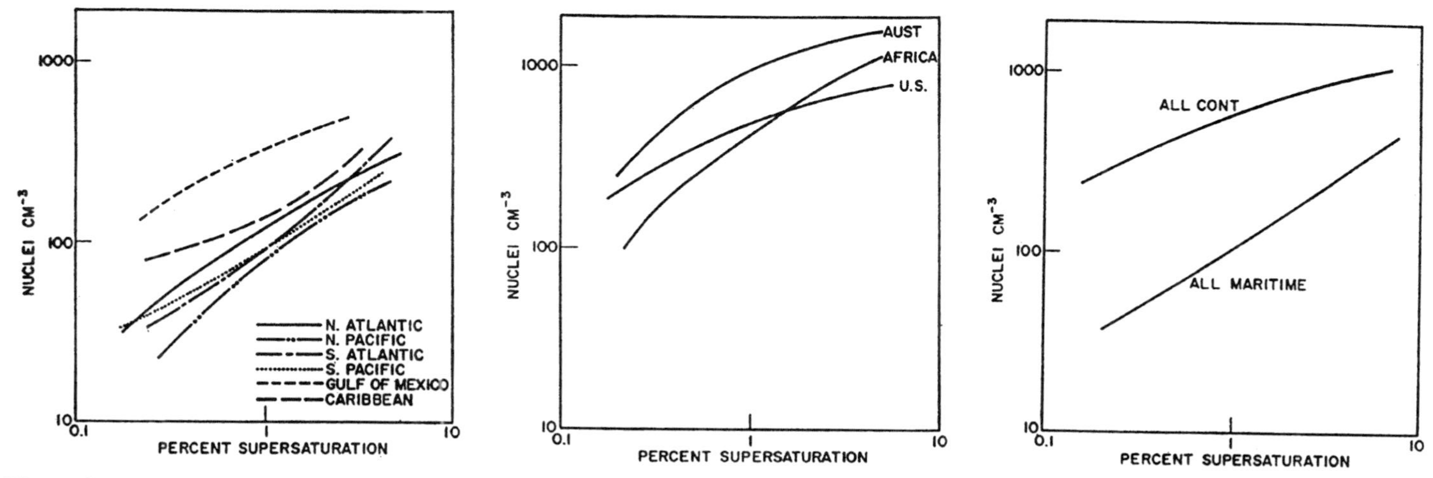

Figure 5 Median concentrations of CCN for (left) different oceanic regions, (center) different continents, and (right) all continents and oceans (Twomey and Wojciechowski, 1969).

CCN due to human activities in the United States has been estimated at about 14% of that from natural sources. Cloud and fog droplets sampled over land frequently contain residue of combustion products and organic material, often of biogenic origin. Forest fires and sugar cane burns, for example, produce large concentrations of CCN. Studies of subequatorial and Saharan air over West Africa suggest that CCN are produced from bush fire smoke, bacterial decomposition of plants and associated sulfur-containing gases, and emission of droplets rich in soluble substances by plants.

The surface of a pure water droplet consists of a layer of water molecules, each with the potential to evaporate and enter the humid air above the surface. Vapor molecules within the humid air also collide with and enter the drop. When a non-volatile solute is present in the droplet, solute molecules or ions occupy some sites on the droplet surface. Thus, fewer water molecules are available to evaporate into the air. However, vapor molecules can enter the solution at the same rate as before. Equilibrium can only be established when the vapor pressure decreases over the solution so that the rate that water molecules leave the solution equals the rate at which they enter the solution from the air. Raoult's law,

$$\frac{e_r'}{e_r} = \frac{n_0}{n_s + n_0} = \frac{\left(\frac{4}{3}\pi r_{\text{eq}}^3 \rho_w - m_s\right)/M_w}{\left(\frac{4}{3}\pi r_{\text{eq}}^3 \rho_w - m_s\right)/M_w + i m_s / M_s} = \left[1 + \frac{i m_s M_w}{M_s\left(\frac{4}{3}\pi r_{\text{eq}}^3 \rho_w - m_s\right)}\right]^{-1} \tag{12}$$

which can be derived from principles of equilibrium thermodynamics, describes this process formally. Raoult's law states that the vapor pressure over a solution droplet (e_r') is reduced from that over a pure water droplet (e_r) by an amount equal to the mole fraction of the solvent. In (12), n_0 is the number of molecules of water in the droplet, n_s the number of molecules of solute, m_s is the mass of the solute, and M_s is the molecular weight of the solute. In nature, soluble particles are typically salts or other chemicals that dissociate into ions when they dissolve. For solutions in which the dissolved molecules dissociate, the number of moles of solute, n_s, must be multiplied by the factor i, the degree of ionic dissociation. Unfortunately, the factor i, called the van't Hoff factor, has not been determined for many substances. Other quantities such as the rational activity coefficient, the mean activity coefficient, and the molal or practical osmotic coefficient have been used as alternate expressions to the van't Hoff factor and are tabulated in many chemical reference books (Pruppacher and Klett, 1997).

The Kelvin equation (8) describes the equilibrium vapor pressure over a curved water surface, e_r, and Raoult's law (12) describes the reduction in vapor pressure e_r'/e_r over a solution droplet. Multiplying (8) by (12) to eliminate e_r, gives

$$S_{s,v} = \frac{e_r'}{e_{s,\infty}} = \left[1 + \frac{i m_s M_w}{M_s\left(\frac{4}{3}\pi r_{\text{eq}}^3 \rho_w\right) - m_s}\right]^{-1} \exp\left(\frac{2 M_w \sigma_{w,v}}{R^* T \rho_w r_{\text{eq}}}\right) \tag{13}$$

which describes the equilibrium saturation ratio over a solution droplet. For a sufficiently dilute solution, this equation can be simplified by approximating $e^x \approx 1 + x$ and $(1 + y)^{-1} \approx 1 - y$. Ignoring the small product $x \times y$, and m_s compared to the mass of water, one obtains

$$S_{s,v} = 1 + \frac{a}{r_{\rm eq}} - \frac{b}{r_{\rm eq}^3} \tag{14}$$

where $a = 2M_{\rm w}\sigma_{w,v}/R^*T\rho_w = 3.3 \times 10^{-5}/T$, $b = 3im_sM_w/4\pi M_s\rho_w = 4.3im_s/M_s$ in cgs units with T in kelvins. Equations (13) and (14) are forms of the Köhler equation, first derived by the Swedish meteorologist H. Köhler in the 1920s. Curves 2 to 6 in Figure 6 show solutions of the Köhler equation for droplets containing fixed masses of NaCl and NH_4SO_4, common components of CCN. Curve 1 shows the solution for a pure droplet, the Kelvin equation. For a given $S_{s,v}$, droplets on the left side of the maximum in the Köhler curves are in stable equilibrium. If a droplet in equilibrium on the left side of the curve experiences a small increase in radius due to chance collection of vapor molecules, the droplet would find the vapor pressure around itself less than that required for equilibrium at its new radius. Physically, less water molecules would be striking the droplet from the vapor field than would be evaporating from the droplet. As a result, the droplet would shrink, returning to its position on the equilibrium curve. The opposite would happen if the droplet lost water molecules—it would grow back to its equilibrium size. Note that when $S_{s,v} \leq 1$, all droplets growing from CCN remain on the left side of the curves. The small droplets in stable equilibrium on the left side of the curve are called *haze droplets*. Over some cities, where soluble particles are abundant and relative humidities high, haze droplets can severely restrict visibility.

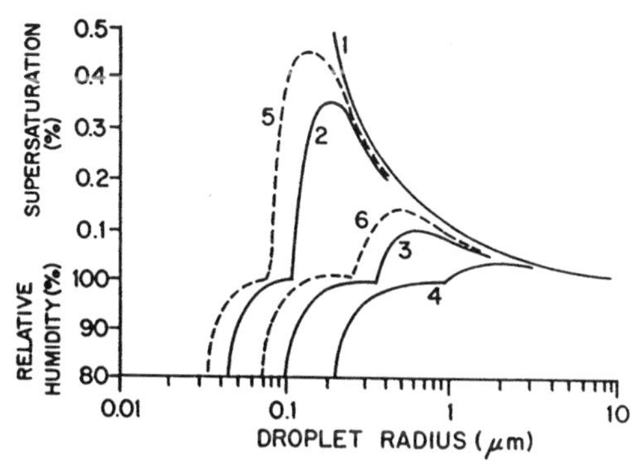

Figure 6 Variations of the relative humidity and supersaturation of air adjacent to droplets of (1) pure water and solution droplets containing the following fixed masses of salt: (2) 10^{-19} kg of NaCl, (3) 10^{-18} kg of NaCl, (4) 10^{-17} kg of NaCl, (5) 10^{-19} kg of $(NH_4)_2SO_4$, and (6) 10^{-18} kg of $(NH_4)_2SO_4$ (Wallace and Hobbs, 1977).

A droplet at the peak of a curve, gaining a few molecules by random collisions, would find the vapor pressure around it greater than that required for equilibrium. More molecules would strike the droplet than evaporate. Vapor would rapidly deposit on the droplet and it would quickly grow into cloud droplet. At the peak, and on the right side of the curves, droplets are in unstable equilibrium. Droplets that reach the peak in their Köhler curves are said to be *activated*. Such droplets are called *cloud droplets*. The peak of a curve (the critical radius) is the transition point between the effect of the solute and drop curvature. When the supersaturation ($S_{s,v} - 1$) in the atmosphere exceeds a critical value, droplets containing solute of a critical mass will rapidly grow from haze droplets into cloud droplets. As indicated in Figure 6, larger nuclei, which produce stronger solution droplets, are much more likely to become cloud droplets because they require lower supersaturation to activate.

Homogeneous Nucleation of Ice Crystals in Humid Air

Equations analogous to (8) and (10) can be derived for the homogeneous nucleation of ice in humid air by assuming that the embryonic ice particle is spherical. In (8), for example, $S_{w,v}$, $\sigma_{w,v}$, and ρ_w are replaced by $S_{i,v} = e_{r,i}/e_{si,\infty}$, the saturation ratio with respect to ice, $\sigma_{i,v}$, the surface tension at the ice surface, and ρ_i, the density of ice. Numerical evaluation of the Kelvin equation for ice embryos shows that the required supersaturations exceed those for water droplet nucleation. Calculations of J for ice nucleation and the Kelvin equation both show that this process does not occur in the atmosphere.

Homogeneous Nucleation of Ice Particles in Supercooled Water Droplets

Homogeneous nucleation of ice in a water droplet requires that a stable icelike molecular structure form within the droplet through statistical fluctuations in the arrangement of the water molecules. The development of such an ice embryo is favored compared to homogeneous nucleation of ice in humid air because the water molecules in the droplet will be in direct contact with any icelike molecular structures created through statistical fluctuations. Unlike homogeneous nucleation of ice in moist air, the process of homogeneous nucleation in water involves two energy barriers. The first, ΔF_i, is associated with the increase in free energy at the ice crystal–liquid surface interface. The second, $\Delta F'$, exists because energy is required to break the bonds between individual water molecules before they can realign themselves to join the ice embryo. The nucleation rate of ice in water therefore depends on the number of liquid molecules per unit volume per unit time that will contact an ice structure of critical size, the probability that an icelike structure will exist in a droplet, and the probability that one of these liquid molecules will overcome the energy barrier and become free to attach to the ice structure. The rate equation becomes

$$J = 2N_c\left(\frac{\rho_w kT}{\rho_i h}\right)\left(\frac{\sigma_{w,i}}{kT}\right)^{1/2}\exp\left(-\frac{\Delta F'}{R^*T}+\frac{\Delta F_i}{kT}\right) \tag{15}$$

where h is Planck's constant, $\sigma_{w,i}$ is the interface energy per unit area between the ice surface and the water, and N_c is the number of water molecules in contact with a unit surface area of ice. Figure 7 shows measurements of J from several experiments. J increases from about 1 to $1020\,\mathrm{cm}^{-3}/\mathrm{s}$ as the temperature decreases from -32 to $-40°\mathrm{C}$. Until recently, insufficient information was available on the properties of supercooled water and on $\Delta F'$. Pruppacher (1995) showed that incorrect extrapolations of data for these properties to large supercoolings, and a lack of understanding of the behavior of water molecules at large supercoolings led to the disagreement between J calculated from classical theory and experimental data (Fig. 7).

Pruppacher (1995) resolved the discrepancies between the classical nucleation equation, laboratory data, and the results of an earlier molecular theory. Pruppacher noted that because water molecules become increasingly bonded at colder temperatures, $\Delta F'$ might be expected to increase with decreasing temperature. However, earlier cloud chamber experiments found that $\Delta F'$ decreases sharply at temperatures

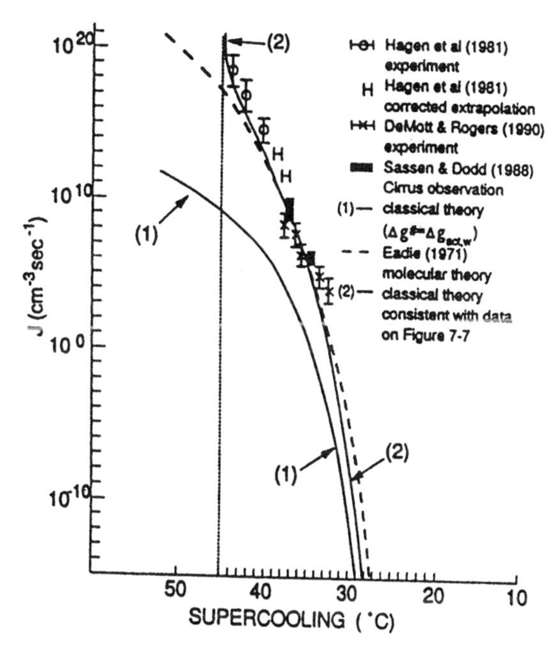

Figure 7 Variation of the rate of homogeneous nucleation in supercooled water. Data are from different experiments. The solid line (1) is from early classical theory. The solid line (2) is from the revision of classical theory discussed by Pruppacher (1995). The dashed line is from molecular theory (Pruppacher, 1995).

colder than −32°C. Pruppacher used nucleation rates measured in laboratory experiments and inferred from field observations, and recent measurements of the physical properties of supercooled water, to calculate $\Delta F'$ and showed using the classical nucleation equation that $\Delta F'$ indeed decreases at temperatures colder than −32°C. This behavior was attributed to the transfer of clusters of water molecules, rather than individual water molecules, across the ice–water interface. With clusters, the only hydrogen bonds that must be broken are those at the periphery of the cluster.

Pruppacher (1995) summarized the experiments of many investigators to determine the homogeneous nucleation temperature threshold for ice in water droplets. He reasoned that (1) the larger the volume of a droplet, the larger is the probability of a density fluctuation in the droplet and the larger the probability that an ice embryo will be produced and (2) the probability for ice formation in a given sized droplet increases with increasing time of exposure of the droplet to a given range of temperatures. Reexamining the available experimental data, he showed that the lowest temperature at which virtually all pure water droplets froze was a function of droplet diameter, with the spread in the data attributable to the different techniques used in the experiments to support the drops (Fig. 8). Figure 8 shows that freezing occurs at −40°C for 1-μm droplets and −35°C for 100-μm droplets. Pruppacher's results apply to cloud droplets, which are nearly pure water droplets, their solute concentrations typically less than 10^{-3} mol/L.

Unactivated haze droplets consist of much stronger solutions whose chemical consistencies depend on the parent CCN. Recent interest in understanding the importance of polar stratospheric clouds in ozone depletion and the role of cirrus in climate change has fueled investigations of the homogeneous nucleation of haze droplets. The homogeneous nucleation temperature in strong solution haze droplets

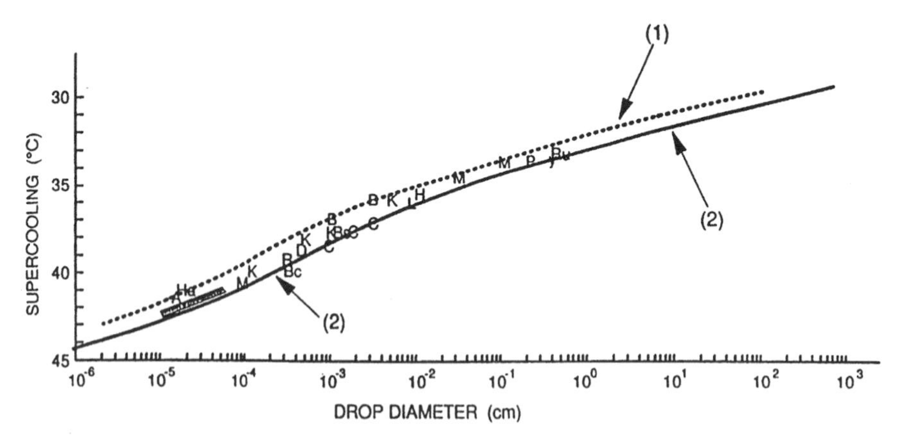

Figure 8 Lowest temperature to which extra pure water drops of a given size and exposed to cooling rates between 1°C/min and 1°C/s have been cooled in various laboratory experiments, indicated by different letters. Lines 1 and 2: Temperature at which 99.99% of a population of uniform-sized water drops freezes when exposed to cooling rates of 1°C/min and 1°C/s, respectively (Pruppacher, 1995).

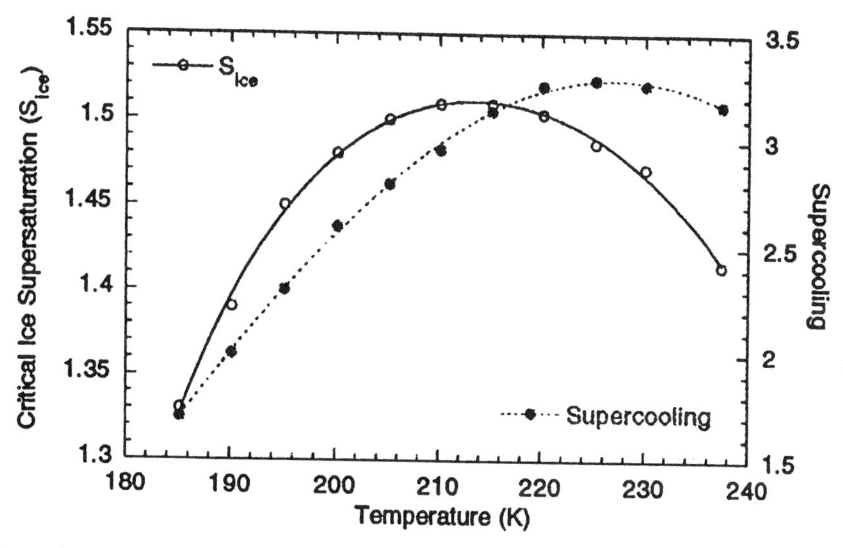

Figure 9 Critical ice supersaturation and the extent of supercooling required to achieve a nucleation rate $J = 1\ cm^{-3}/s$ for strong H_2SO_4 solution droplets. The ice supersaturation is defined as the ratio of the ambient water vapor pressure over the ice saturation vapor pressure when $J = 1\ cm^{-3}/s$. Supercooling is defined as the critical nucleation temperature minus the temperature of ice that has a vapor pressure equal to the ambient water vapor pressure (Tabazadeh and Jensen, 1997).

is depressed in proportion (nonlinearly) with increased solution molality. The equilibrium size of a solution haze droplet increases and solution molality decreases with increasing ambient relative humidity. As a result, the homogeneous nucleation temperature is depressed most at low relative humidities. Two factors of interest are the ice supersaturation where homogeneous nucleation occurs and the extent to which a sulfate solution can be supercooled above ice saturation over the solution drop before ice nucleation is possible. Figure 9 shows calculations of homogeneous nucleation of droplets containing sulfate aerosols made by Tabazadeh and Jensen (1997). Their results, which correspond to $J = 1\ cm^{-3}/s$, show that supercoolings of about 3 K (below the equilibrium freezing temperature of a strong H_2SO_4 solution) and ice supersaturation between 40 and 50% are required for homogeneous nucleation at ambient temperatures between −33 and −63°C (240 and 210 K). These conditions can exist in upper tropospheric clouds, suggesting that the high ice particle concentrations observed in some cirrus clouds are due to homogeneous nucleation of haze droplets.

Heterogeneous Nucleation of Ice Crystals

Ice particles form in the atmosphere through homogeneous nucleation of supercooled water droplets, heterogeneous nucleation processes that involves aerosol particles called *ice nuclei* (IN), and shattering of existing ice particles during collisions or droplet freezing events. Homogeneous nucleation of ice particles in supercooled water is limited to temperatures colder than about -33°C. At warmer temperatures, primary ice particle formation requires ice nuclei.

Most ice nuclei are composed of clay particles such as vermiculite, kaolinite, and illite and enter the atmosphere during wind erosion of soils. Combustion, volcanic eruptions, and airborne microorganisms are also sources of IN. Ice nuclei function in four modes: (1) *deposition* or *sorption-nuclei* adsorb water vapor directly on their surfaces to form ice; (2) *condensation-freezing nuclei* act first as CCN to form drops and then as IN to freeze the drops; (3) *immersion nuclei* become incorporated into a drop at $T > 0^\circ$C and act to initiate freezing after the drop has been transported into a colder region; and (4) *contact nuclei* initiate freezing of a supercooled drop on contact with the drop surface. The fact that IN can function in these different modes has made their measurement difficult. Instruments designed to measure IN, which include rapid and slow expansion chambers, mixing chambers, thermal precipitation devices, membrane filters, and other devices, typically create conditions favoring one mode, and often only measure the dependence of IN concentration on one variable, such as temperature [see reviews by Vali (1985) and Beard (1992)]. In addition, the time scales over which ice nucleation can occur in natural clouds often differs substantially from those characterizing the measurements. Measurements made with different instruments at workshops in controlled conditions with air samples drawn from the same source have shown considerable scatter. Absolute values of concentrations of ice nuclei should therefore be viewed with some caution.

The concentration of ice nuclei in the atmosphere is highly variable (Fig. 10). In general, ice nuclei concentrations increase by an order of magnitude with each 4°C decrease in temperature but can vary by nearly an order of magnitude at any temperature. The equation

$$N_{\mathrm{IN}} = A \exp(\beta \, \Delta T) \tag{16}$$

where $A = 10^{-5}$ per liter, $\beta = 0.6/^\circ$C, N_{IN} the number of active ice nuclei per liter active at temperatures warmer than T and $\Delta T = T_0 - T$ is the supercooling in degrees centigrade, reasonably approximates this behavior. Ice nucleation can occur at relative humidities below water saturation, provided that the air is supersaturated with respect to ice. Experiments have shown that at a given temperature, N_{IN} increases with increasing supersaturation with respect to ice (s_i) according to the relationship

$$N_{\mathrm{IN}} = C^k s_i \tag{17}$$

where the values of C and k depend on the air mass. The concentration of IN can undergo orders of magnitude variations at a single location from day to day. Explanations forwarded for these rapid changes include advection of dust from desert windstorms, downward transport of stratospheric IN created from meteor bombardment, IN production from evaporation of cloud and precipitation particles, and preactivation of IN in the cold upper troposphere followed by transport to the surface. Studies of the vertical profile of IN concentrations have found that

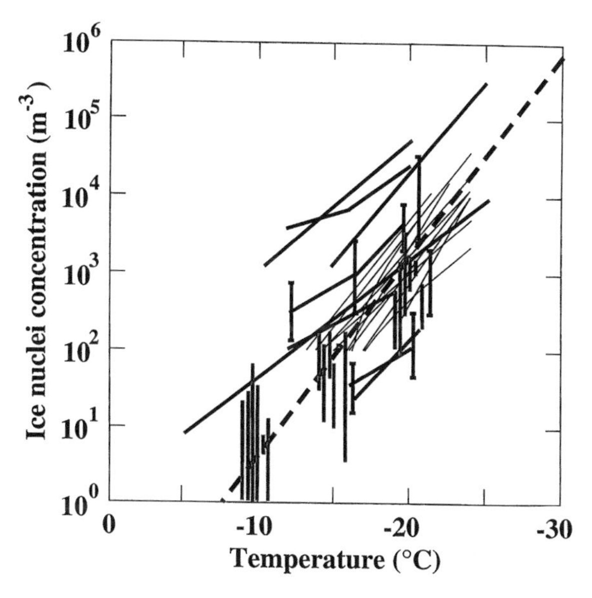

Figure 10 Mean or median worldwide measurements of ice nuclei concentrations from Bigg and Stevenson (1970) (vertical gray bars), compilations of 11 studies by various authors in Götz et al. (1992) (thick lines), and compilations of 10 additional studies by Pruppacher and Klett (1997) (thin lines). Bigg and Stevenson's data at -10, -15 and -20°C are spread over a small temperature range for clarity. The heavy dashed line is $N_{IN} = 10^{-5}\ \exp(0.6\Delta T)$.

concentrations generally decrease with height in the troposphere above the surface mixed layer, although evidence exists for higher concentrations of IN in the vicinity of the jet stream.

Because ice nuclei must provide a stable solid substrate during the growth of ice embryos, they are normally water insoluble. Nucleation occurs when an ice crystal with a specific lattice structure first forms on a substrate that has a different lattice structure. The probability of an ice nucleation event increases when the lattices of atoms composing the ice crystal and ice nuclei closely align. For the IN atoms and ice atoms to bond, a strain in the bonds must be accommodated between the out-of-place atoms in the IN and ice lattices. The surface free energy of the interface increases in response to increased elastic deformation. The greater the lattice mismatch, the higher the surface free energy and the less likely the ice crystal will form. Particle surfaces normally contain contaminants, cracks, crevices, and particular growth patterns and may possess specific electrical properties due to ions or polar molecules. Certain of these locations are much more effective at adsorbing water molecules onto the surface of the aerosol and enhance the nucleating capability of the substance. Since the water molecule is polar, and the ice lattice is held together by hydrogen bonding, substances that exhibit hydrogen bonds on their surfaces act as more effective nucleants. For this reason, some organics exhibit strong ice nucleating behavior. Ice nuclei are large aerosols, typically having radii larger than 0.1 μm.

The theory governing the nucleation rate, *J*, of an ice embryo in a saturated vapor, or of an ice embryo in a supercooled water drop, proceeds in a similar way to the theory concerning heterogeneous nucleation of a drop. The theory assumes an ice embryo forms a spherical cap with a contact angle with the surface substrate. Hydrophobic substances have large contact angles and act as poor ice nucleants. The equations are analogous to those used for nucleation of a water droplet, except that all terms applying to water and vapor now apply to ice and vapor, or ice and water. A discussion of this theory, and more complicated extensions of the classical theory, is presented by Pruppacher and Klett (1997). The theory predicts that at $-5^{\circ}C$, particles with radii smaller than 0.035 μm and a contact angle of 0 will not be effective as ice nuclei. The threshold is 0.0092 μm at $-20^{\circ}C$. Few if any particles will exhibit contact angles of zero, so actual ice nuclei will have to be somewhat larger than these values.

Inside a water droplet, nuclei sizes can be somewhat smaller, with the threshold at least 0.010 and 0.0024 μm at -5 and $-20^{\circ}C$, respectively. Experiments have shown that the exact value of the cutoff is also dependent on the chemical composition of the particle and on its mode of action (deposition, freezing, or contact). In the case of deposition nuclei, it also depends on the level of supersaturation with respect to ice. Some organic chemicals have been found to have somewhat smaller sizes and still act as ice nuclei.

There is considerable experimental evidence showing that atmospheric ice nuclei can be preactivated. Preactivation describes a process where an ice nucleus initiates the growth of an ice crystal, is subjected to an environment where complete sublimation occurs, and then is involved in another nucleation event. The particle is said to be preactivated if the second nucleation event occurs at a significantly warmer temperature or lower supersaturation. Ice nuclei can also be deactivated, that is lose their ice nucleating ability. This effect is due to adsorption of certain gases on to the surface of the nucleus. Pollutants such as NO_2, SO_2, and NH_3 have been found to decrease the nucleation ability of certain aerosols. There has also been evidence from laboratory and field experiments that the nucleating ability of silver iodide particles decreases when the aerosols are exposed to sunlight.

The very poor correspondence between ice nucleus measurements and ice particle concentrations in clouds has yet to be adequately explained. Hypotheses forwarded to explain these observations generally focus on more effective contact nucleation, particularly in mixed regions of clouds where evaporation can lead to the formation of giant ice nuclei, and secondary ice particle production, which involves shattering of existing ice particles during collisions or droplet freezing events. The relative importance of each of these mechanisms in real clouds is still uncertain.

4 FORMATION OF RAIN IN WARM CLOUDS

Raindrops form through one of two microphysical paths. The first occurs when ice particles from high, cold regions of clouds fall through the melting level and become raindrops. In cloud physics literature, clouds that support this process are called "cold" clouds. "Warm" clouds are clouds that develop rain in the absence of ice processes. In the tropics, shallow clouds such as trade wind cumulus and stratocumulus produce rain entirely through warm rain processes. Deep tropical convective clouds and summertime midlatitude convection also support active warm rain processes, although melting ice also contributes to rainfall.

The warm rain process occurs in three steps: (1) activation of droplets on CCN, (2) growth by condensation, and (3) growth by collision and coalescence of droplets. The rapid production of warm rain in both maritime and continental clouds remains one of the major unsolved problems in cloud physics. The central problem lies in the transition from step 2 to 3. Theory predicts that growth by condensation will create narrow drop spectra in clouds. Growth by coalescence, on the other hand, requires large and small cloud droplets within droplet spectra. Determining how broad droplet spectra develop in clouds has been a central theme of cloud physics research.

Growth of a Single Droplet by Condensation

Cloud droplets form when the supersaturation in the atmosphere becomes sufficiently large so that haze droplets reach their critical radii and begin to grow in the unstable regime to the right of the Köhler curves (Fig. 6). The equation for the growth of a pure water droplet, first derived by Maxwell in 1890, is obtained by assuming that (1) heat transfer and vapor concentration in the vicinity of the droplet both satisfy the diffusion equation, (2) an energy balance exists at the surface of the droplet such that the latent heat added during condensation balances the diffusion of heat away from the droplet, and that (3) the concentration of droplets in the vapor field is independent of direction outwards from the droplet. Modifications must be made to account for curvature and solute effects. Additionally, one must account for kinetic effects near the drop surface. These include vapor molecules striking the surface of the droplet and rebounding rather than condensing, and the direct heat transfer that occurs at the droplet–vapor interface as water molecules cross the interface between the liquid and gas phases.

Accounting for these effects, the equation describing the diffusional growth of a droplet is

$$r\frac{dr}{dt} = \frac{S_{w,v} - 1 - a/r + b/r^3}{((L_v/R_vT) - 1)(L_v\rho_w/KT^2f(\alpha)) + (\rho_w R_v T/De_{s,\infty}g(\beta))} \tag{18}$$

where a and b are defined in Eq. (14), L_v is the latent heat of vaporization, R_v is the gas constant for water vapor, D is the diffusion coefficient, and K, the thermal conductivity. The quantities $f(\alpha)$ and $g(\beta)$ are given by:

$$f(\alpha) = \frac{r}{r + [(K/\alpha p)(2\pi R_d T)^{1/2}/(C_v + (R_d/2))]} = \frac{r}{r + l_\alpha} \tag{19}$$

$$g(\beta) = \frac{r}{r + (D/\beta)(2\pi/R_v T)^{1/2}} = \frac{r}{r + l_b} \tag{20}$$

In Eqs. (19) and (20), the accommodation coefficient, α, characterizes the transfer of heat by molecules arriving at and leaving the interface between the liquid and vapor phase, and the condensation coefficient, β, is the fraction of vapor molecules hitting the droplet surface that actually condense. In these equations, C_v is the specific heat at constant volume, p is pressure, and R, the gas constant of dry air. The terms l_α and l_β can be considered length scales. As r becomes large such that $r \gg l_\alpha$, l_β, $f(\alpha)$ and $g(\beta)$ approach unity. This essentially is the case by the time a droplet reaches $r = 5\,\mu\text{m}$, as is apparent from Figure 11, which compares growth by condensation with and without kinetic effects.

Equation (18) relates the radius of a droplet growing by vapor diffusion to the supersaturation in the atmosphere, and the environmental pressure and temperature. It is clear that the rate of growth of the radius of a droplet is inversely proportional to its size. Solution and curvature effects are most important when the droplets are very small, as indicated by the $1/r$ and $1/r^3$ dependence. Table 1 gives calculated growth times in seconds for droplets with different nuclear masses of NaCl at $T = 273\,\text{K}$, $P = 900\,\text{mb}$ and a supersaturation of 0.05%. Note that for a droplet with a large nucleus, growth to $r = 20\,\mu\text{m}$ takes 5900 s or 1.63 h. Growth to a radius of 50 μm takes 41,500 s, or approximately a half-day. A very small drizzle droplet is about 50 μm radius. Individual cloud parcels are unlikely to exist in a supersaturated environment for half a day. Clearly, diffusional growth alone is inadequate to explain the formation of rain in clouds.

Equation (18) is valid for a stationary droplet. As droplets grow, they fall through the air. Air flowing around the droplet causes the field of vapor molecules to interact differently with the droplet surface than for a still droplet. The net result is that the growth rate (or evaporation rate if the relative humidity <100%) of the droplet increases. The effect of ventilation has been studied using both numerical simulations and by making direct measurements in wind tunnels. For droplets with

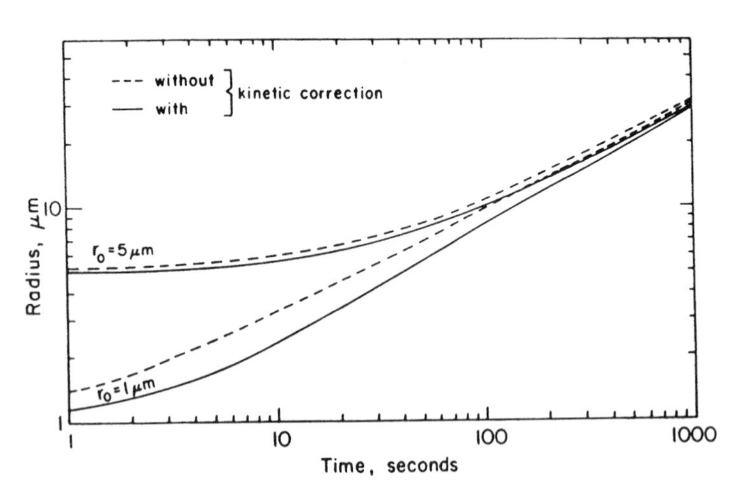

Figure 11 Comparison of condensation growth, with and without kinetic effects, for droplets of initial radii of 1 and 5 μm (Rogers and Yau, 1989).

TABLE 1 Time (s) required for Droplet to Grow from Initial Radius of 0.75 μm to Specified Size at 273 K, Pressure of 900 mb, and Supersaturation $[100(S_{w,v} - 1)] = 0.05$ for Salt Nuclei with 3 Nuclear Masses

Radius (μm)	10^{-14} g	10^{-13} g	10^{-12} g
2	130	7	1
5	1,000	320	62
10	2,700	1,800	870
15	5,200	4,200	2,900
20	8,500	7,400	5,900
25	12,500	11,500	9,700
30	17,500	16,000	14,500
35	23,000	22,000	20,000

From Best (1951).

$r \leq 50$ μm, ventilation can be ignored, but for droplets of greater sizes, the ventilation effect increases with droplet size. In the growth equation, ventilation is accounted for by multiplying the numerator on the right by f_v, the ventilation coefficient.

Growth of Population of Droplets by Condensation

Cloud droplets will activate on CCN when a parcel of air rises through cloud base. The larger, more soluble CCN activate first, followed by the more numerous, but less effective CCN. The new droplets each extract water vapor from the parcel. The maximum supersaturation occurs, and the droplet concentration is determined, when the rate of vapor deposition on the drops equals the rate at which vapor is supplied through cooling of the rising parcel. Above this point in a cloud, the supersaturation remains below its peak value and no additional droplets are activated. The cloud droplet concentration therefore depends on the spectrum of CCN entering the cloud, the updraft strength, and the cloud base temperature.

Equation (18) shows that the growth rate of activated cloud droplets decreases with radius. This equation predicts that the width of a spectrum of different sized droplets narrows with time, since smaller droplets in a rising parcel grow faster than larger droplets. This is illustrated by the measurements and computations reported by Fitzgerald (1972). Fitzgerald found close agreement between cloud droplet spectra measured 200 to 300 m above cloud base with that predicted by condensation theory including kinetic effects (see Fig. 12).

For collisions between cloud droplets to occur, the droplet spectrum must evolve so that droplets possess a range of sizes and fall speeds. Calculations suggest that the collision efficiency of two falling drops only becomes appreciable ($>10\%$) when the radius of the larger collector drop (R) and smaller collected droplet (r) reach $R = 25$ μm and $r = 7.5$ μm. Equation (18) predicts that a droplet with a large nucleus will require 1.6 h to grow to $r = 20$ μm, yet many tropical maritime clouds produce

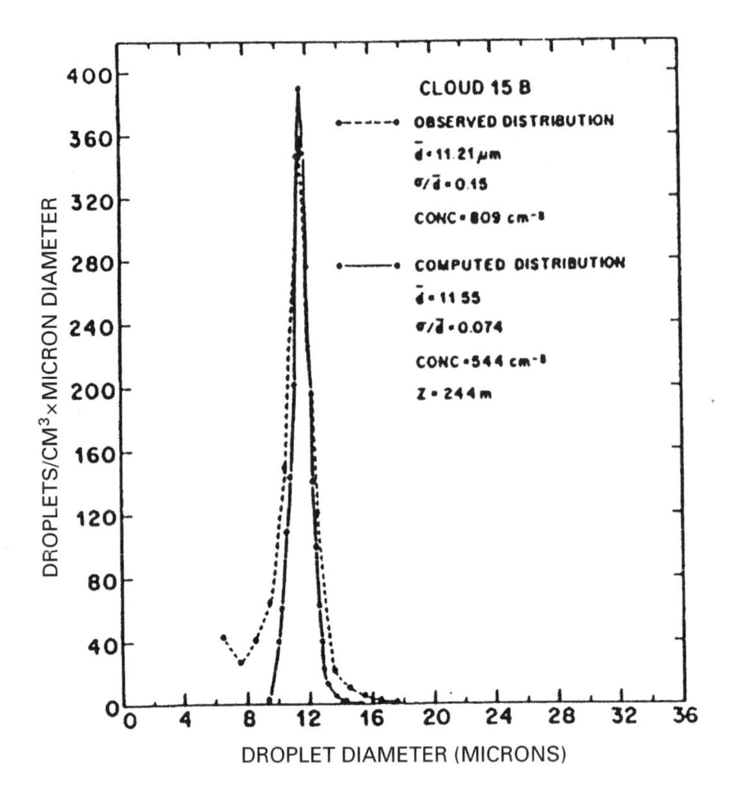

Figure 12 Computed and measured cloud droplet distributions at a height of 244 m above cloud base (Fitzgerald, 1972).

rain in about 20 min. Continental clouds, which have much higher droplet concentrations, also produce rain in short times compared to the prediction of Eq. (18). Although continued condensation within a parcel may eventually generate droplets larger than about 25 μm radius to initiate coalescence, the time for this process to produce rain is generally much too long.

Proposed Mechanisms for Broadening of Droplet Spectra

Several mechanisms have been suggested to explain the production of large cloud droplets capable of initiating coalescence. These include (1) favorable CCN spectra, such as low concentrations of smaller nuclei and high concentrations of larger nuclei; (2) mechanisms that enhance condensation growth of a few larger cloud droplets through mixing of air parcels; (3) stochastic condensation, the mixing of droplets (rather than parcels) that have different supersaturation histories; and (4) fluctuations in supersaturation or collision rates induced by turbulence.

One of the earliest mechanisms hypothesized to explain the onset of precipitation is that very large (>10 μm radius) wettable or soluble nuclei exist below cloud base that, when carried through cloud base, can almost immediately begin to collect cloud droplets. Measurements in marine environments show that giant salt particles can often occur in significant concentrations. Within the moist region below cloud base such salt particles will deliquesce into much larger solution droplets [e.g., a NaCl particle of 1 ng (5 μm) has a 25-μm equilibrium radius at 99% relative humidity]. Thus, there are often significant concentrations of coalescence nuclei that can enter the base of maritime clouds. Numerical parcel models initialized with CCN spectra from observations have been used to evaluate the role of giant CCN in the production of warm rain. Calculations show that raindrops can form on these CCN in time scales comparable to observed cloud lifetimes. Drop trajectory calculations also suggest that accretion of cloud water on giant and ultragiant nuclei can account for the formation of rain in observed time scales when these large particles are present at cloud base. Nevertheless, questions remain as to how important these large salt particles really are to the warm rain process in maritime clouds. Uncertainties about the growth rate of these particles in updrafts remain, since their growth is highly nonequilibrium. Also, the residence time of larger salt particles in the atmosphere is short, with the highest concentrations located close to the ocean surface. Much less is known about the role giant particles play in initiating coalescence in continental clouds. Measurements show that 50-μm-diameter particles exist in concentrations of 100 to 1000 m^{-3} in the boundary layer over the High Plains of the United States. These are typically insoluble soil particles that will not grow as they pass through cloud base. They are large enough, however, to make effective coalescence nuclei without undergoing growth by condensation.

A second mechanism to explain the onset of precipitation involves broadening of the drop size distribution during the mixing process. Early studies of parcels subjected to velocity fluctuations were unable to reproduce the appreciable broadening of the cloud droplet spectra that occurs in warm cumulus. These studies invoked the process of *homogeneous* mixing in which dry air is mixed into a cloudy parcel of air such that all droplets are exposed to the dry air and evaporate until the parcel is again saturated. This process has been shown to produce a constant modal size, but a decrease in mean size because of a broadening toward smaller sizes. Absent in this process is the large drop tail to the spectrum required for coalescence growth.

Laboratory experiments by Latham and Reed (1977) raised questions about the applicability of the homogeneous mixing process. Their experiments suggested that the mixing process occurs in such a way that some portions of a cloudy parcel are completely evaporated by mixing with unsaturated air while other portions remain unaffected. Broadening of the droplet spectrum toward larger sizes will occur by subsequent condensation of the parcel if it continues to ascend, since there is reduced competition for vapor among the remaining droplets. Further studies have showed that an idealized inhomogeneous mixing process will produce appreciable broadening compared to homogeneous mixing. However, other studies have found that special dynamical sequences of vertical mixing can achieve the same spectral broadening without invoking inhomogeneous mixing concepts. Observations in clouds over the High Plains of the United States found that the large drop peak was more consistent with a closed parcel environment than a mixed one and that there was no evidence that mixing or cloud age increased the size or concentration of the largest drops. Considerable uncertainty remains concerning the importance of mixing to the broadening of the droplet spectrum.

The stochastic condensation mechanism invokes the idea that droplets can have different supersaturation histories. This process considers the mixing of droplets, rather than parcels. Cooper (1989) describes the theory of stochastic condensation. Regions of clouds having fine-scale structure, such as mixed regions of clouds outside of adiabatic cores, would best support this type of process. Measurements within warm, orographic clouds in Hawaii have shown little broadening in the laminar clouds, but appreciable broadening in the breaking wave regions of the clouds near cloud top, suggesting the importance of stochastic condensation in producing large droplets.

A fourth hypothesis is that the breadth of the droplet spectra in clouds can be increased by turbulent fluctuations. Conceptually, turbulence may be thought to induce vertical velocity fluctuations that induce fluctuations in supersaturation. These fluctuations, in turn, lead to the production of new droplets at locations throughout the cloud. Turbulent fluctuations can also lead to drop clustering, which can create opportunities for favorable supersaturation histories and/or enhanced collision rates. A general approach has been to examine the evolution of a distribution of droplets exposed to a supersaturation that varied with a known distribution, such as a normal distribution, and to derive analytic expressions for the droplet size distribution as a function of time. This approach predicts continued dispersion of the spectra with time. This approach has been criticized on the basis that updrafts and supersaturation are highly correlated. A droplet that experiences a high supersaturation is likely to be in a strong updraft and will arrive at a given position in a cloud faster, which means there will be less time for growth. Conversely, a droplet that experiences a low supersaturation will grow slower but have a longer time to grow. The net result is that the droplets arrive at the same place with approximately the same size. Supersaturation fluctuations in clouds can arise not only from fluctuations in the vertical velocity, but also from fluctuations in integral radius (mean radius multiplied by the droplet concentration) of the droplet spectra. Realistic droplet spectra can be obtained in model simulations when the fluctuations in mean radius are negatively correlated with vertical motions. Unfortunately, experimental data in cumulus show the opposite behavior—the largest droplets occur in upward moving parcels.

Recent studies have provided conflicting evidence regarding the role of turbulence in spectral broadening. Studies show that turbulence can lead to significant trajectory deviations for smaller droplets in clouds, leading to larger relative vector velocities for droplets and collector drops and enhanced collision rates. Results indicate that cloud turbulence supports spectral broadening and more rapid production of warm rain. Studies have also examined whether turbulence can create regions of preferential concentration in conditions typical of cumulus clouds, and whether nonuniformity in the spatial distribution of droplets and/or variable vertical velocity in a turbulent medium will contribute to the broadening of the drop size distribution.

These studies suggest that turbulence severely *decreases* the broadening of the size distribution compared to numerical experiments performed in the absence of turbulence. The importance of turbulence in drop spectral broadening is still a matter of debate. Villiancourt and Yau (2000) provides a review of recent understanding of the role of turbulence in spectral broadening.

Growth by Coalescence

When sufficiently large (>25 μm radius) cloud droplets exist in a cloud, their growth is accelerated by collision and coalescence with neighboring droplets. The nonlinear behavior of the flow around a drop affects both the drop's terminal velocity and the manner in which a smaller drop is swept around a larger drop as the two approach in free fall.

Single Drop Hydrodynamics The Reynolds number, given by $N_{\text{Re}} = 2V_T r/\nu$, where V_T is the drop terminal velocity, and ν is the kinematic viscosity, is the ratio of inertial to viscous forces and is typically used as a scaling parameter to describe the flow regimes around a drop. The characteristics of flow around a sphere as a function of the Reynolds number are well documented. When N_{Re} is very small (<0.4), the droplet is in the so-called Stokes flow regime where viscous forces dominate and the flow field around a droplet is symmetric and laminar. At about $N_{\text{Re}} = 2$ (about $r = 50$ μm at standard temperature and pressure), a wake appears behind the droplet. At about $N_{\text{Re}} = 20$ (about $r = 150$ μm), an eddy of rotating fluid appears behind the drop. This eddy grows as N_{Re} increases to 130 ($r = 350$ μm), at which point the wake becomes unstable. At $N_{\text{Re}} = 400$ ($r = 650$ μm) eddies are carried downstream by the turbulent wake.

The problem of determining the flow around a drop is complicated by shape deformation due to hydrodynamic and sometimes electrical forces. The nature of the deformation is primarily related to the droplet size. Observations in wind tunnel experiments and model calculations have provided information on the nature of drop deformation. Numerical modeling has further clarified the nature of drop deformation under the influence of external hydrostatic pressure. Chuang and Beard (1990) summarize the effect of hydrodynamic and electric forces on both charged and uncharged drops. Figure 13 from their study shows the shapes drops can assume as they fall in various electric fields characteristic of thunderstorms.

The drag force on the droplet and the force of gravity determine the terminal velocity of a droplet. For Stoke's flow, determining the balance between these forces is relatively straightforward, but in the more complicated flow regimes, a single accurate analytical expression for the terminal velocity is not available. Rogers and Yau (1989) suggest the following general expression:

$$V_t = ar^b \tag{21}$$

for drops that remain spherical, where the terms a and b have different values depending on the Reynold's number. Gunn and Kinzer (1949) provide the most

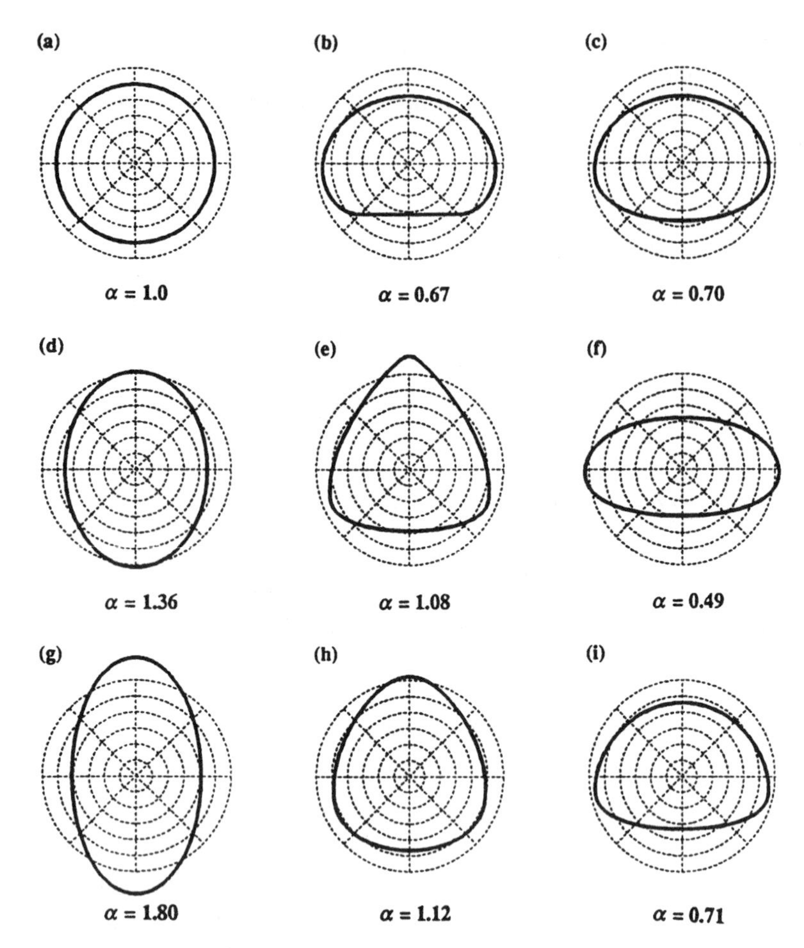

Figure 13 Modeled drop shapes and axis ratios (α = vertical/horizontal dimension) for 5-mm diameter drops with various distortion effects: (*a*) stationary drop (surface tension only), (*b*) drop resting on a flat surface, (*c*) raindrop falling at terminal velocity, (*d*) stationary drop in strong vertical electric field, (*e*) raindrop falling in a strong vertical electric field, (*f*) highly charged drop falling at terminal velocity, (*g*) stationary drop in the maximum vertical electric field before disintegration, (*h*) charged raindrop falling in maximum thunderstorm electrical field with upward directed electric force, and (*j*) same as (*i*), but with downward electric force (Chuang and Beard, 1990).

accurate observational data over a large range of drop sizes. Their measurements, made at 1013 mb and 20°C, are shown in Figure 14.

Collection Efficiency When two droplets interact in free fall, the outcome of the interaction will depend on the droplet's sizes and the offset, x, between the drop centers. Inside some critical offset, x_0 (see Fig. 15), the outcome will be a collision, while outside x_0 the outcome will be a miss. For drops acting as rigid spheres, the

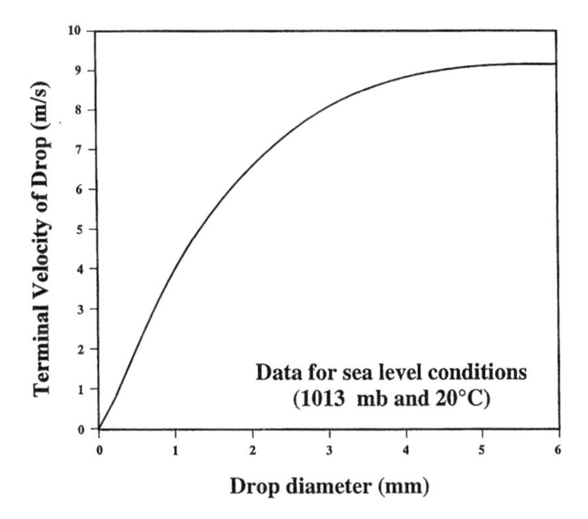

Figure 14 Terminal velocity of a raindrop as a function of its diameter at 1013 mb and 20°C (adapted from Gunn and Kinzer, 1949).

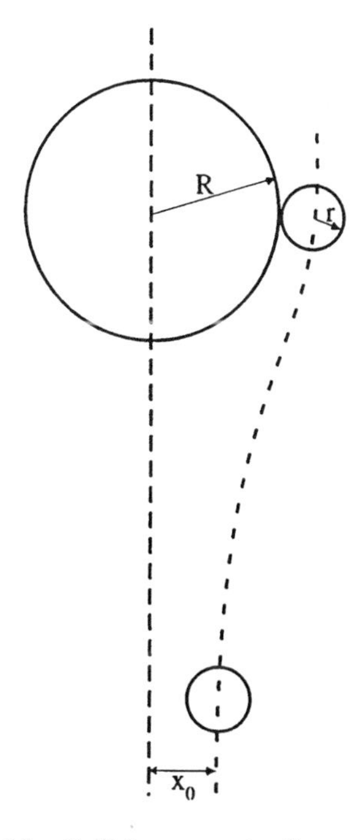

Figure 15 Collision geometry for two droplets.

collision efficiency, E_{col}, is defined as

$$E_{col} = \frac{(x_0)^2}{(R+r)^2} \tag{22}$$

the ratio of the collection cross section to the geometric cross section for the large and small drops. Theoretical calculations for collector drops with $R < 75\,\mu m$ are shown in Figure 16. Typically most newly formed cloud droplets will range in size from about $4 < r < 8\,\mu m$. A 25-μm-radius collector drop will have a collection efficiency of about 10% for these droplets. For this reason, the 25-μm radius is generally considered a threshold size for initiation of coalescence. Theoretical collision efficiencies for larger R and r/R ratios rapidly approach unity and can even exceed unity when r/R approaches 1 and $R > 40\,\mu m$ due to the possibility of capture

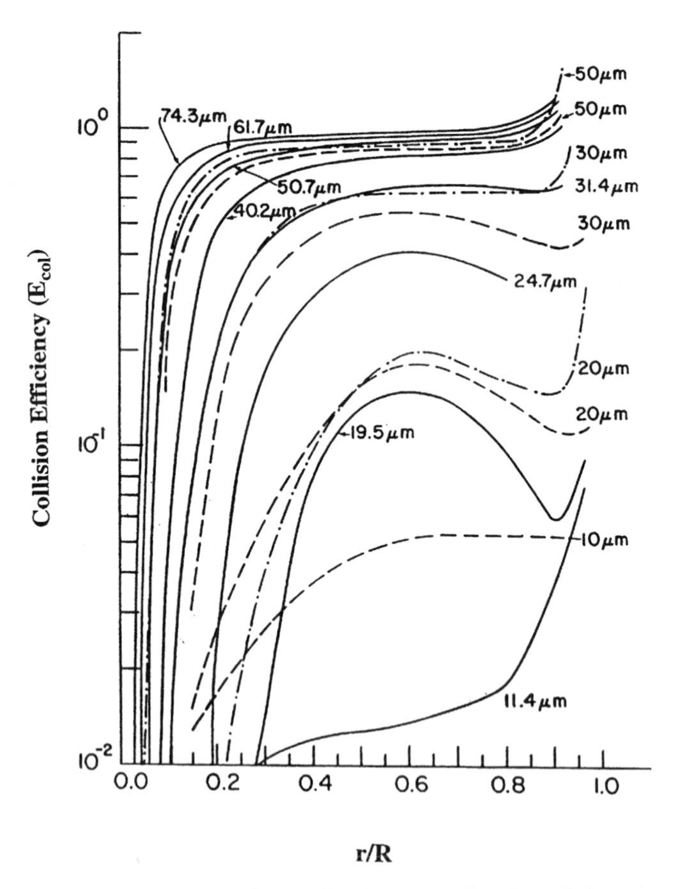

Figure 16 Collision efficiencies for small spheres as a function of the ratio of their radii. Solid lines from Schlamp et al. (1976), dashed lines from Klett and Davis (1973), dashed-dot lines from Lin and Lee (1975). (Adapted from Schlamp et al., 1976.)

of the small droplet in the large droplet wake. Unfortunately, no reliable laboratory studies exist for $R < 40\ \mu m$, so the critical efficiencies for the coalescence threshold remain untested.

The calculations of E_{col} in Figure 16 assume that the drops are rigid spheres. Drizzle-sized drops ($R > 100\ \mu m$) deform upon approach to another droplet as the air film between the drops drains. Studies of small precipitation-sized drops in the laboratory have shown that the outcome of collisions can be complicated, with the possibility for coalescence, temporary coalescence, and satellite production. These outcomes depend on R and r, the drop separation, charge, and the ambient relative humidity. Coalescence depends upon the drop–droplet interaction time and their impact energy. Formulas to calculate coalescence efficiencies, E_{coal}, over a wide range of drop sizes are now available. The collection efficiency, E, is the product of E_{col} and E_{coal}. Beard and Ochs (1984) show that as E_{col} approaches unity, E_{coal} decreases, leading to a maximum E for drop–droplet pairs in the size range $R = 100$, $r = 12\ \mu m$ (see Fig. 17).

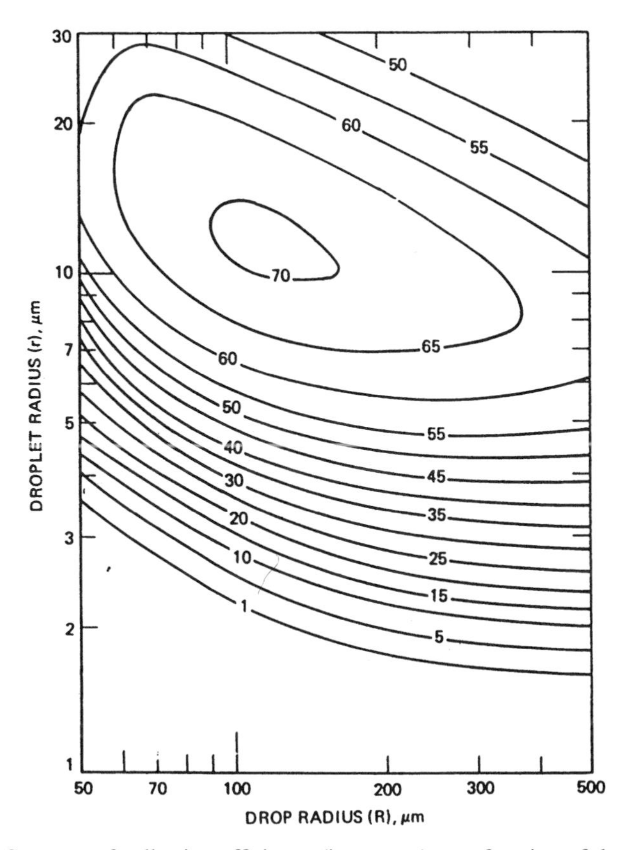

Figure 17 Contours of collection efficiency (in percent) as a function of drop and droplet radius (Beard and Ochs, 1984).

Growth by Coalescence Two methods have been used for calculating growth rates of cloud droplets by coalescence. The first, called *continuous collection*, is used to calculate the growth of single drops that pass through a cloud of smaller droplets. These single drops are assumed not to interact with drops of similar size so as to lose their identity. The continuous collection equation is

$$\frac{dM}{dt} = \sum K(R, r) w_L(r) \Delta r \tag{23}$$

where M is the mass of the drop, K is the collection kernel, and w_L is the liquid water content per unit size interval Δr. The collection kernel, the volume of air swept out per unit time by the larger drop, is given by $K = \pi E(R + r)^2 [V_T(R) - V_T(r)]$, and $w_L = n(r) 4\pi r^3 \rho_w / 3$, where E is the collection efficiency, $V_T(R)$ and $V_T(r)$ are the terminal velocities of the large and small drops, ρ_w is the liquid water content, and n is the small droplet concentration.

Calculations show that small raindrops can form in about 20 to 30 min provided coalescence nuclei ($R > 50\,\mu\text{m}$) are present in the cloud and the smaller cloud droplets have radii of at least 7 μm. Szumowski et al. (1999) applied the continuous growth equation in marine cumulus using a Lagrangian drop-growth trajectory model in wind fields derived from dual-Doppler radar analyses. The evolution of the radar-observed clouds in this study provided a strong constraint on the growth of raindrops. Raindrops were found to grow to sizes ranging from 1 to 5 mm from coalescence nuclei in about 15 to 20 min. Their calculations suggest that coalescence growth is efficient in marine clouds that contain giant and ultragiant sea salt aerosol and low (100 to 300 cm^{-3}) droplet concentrations. In continental clouds, which generally have narrow droplet distributions and high (800 to 1500 cm^{-3}) droplet concentrations, continuous growth may take significantly longer.

In nature, the droplets composing a cloud have a continuous distribution of sizes, and droplets of all sizes have the potential to collide with drops of all other sizes. The drops of most importance to the precipitation process occupy the large end tail of the drop size distribution. To consider rain formation, the statistical aspects of interactions between droplets must be considered. There is an element of probability associated with the collection process, and, occasionally, collisions will occur between larger droplets. A general method for calculating coalescence growth is to assess the probability that droplets of one size will collect droplets of another size for all possible droplet size pairs in the droplet distribution. The term *stochastic collection* is applied when the collection probability is used to determine the number of droplets of a given size, r, that are created by collisions of smaller pairs, $r - \Delta r_1$ and $r - \Delta r_2$, and destroyed by collisions of r with other droplets. Stochastic collection considers rare collisions between larger droplets and consequently predicts accelerated coalescence in warm clouds.

Possible interactions between drops include small droplets collecting small droplets (*autoconversion*), large drops collecting small drops (*accretion*), and large drops collecting large drops (*large hydrometeor self-collection).* Bulk microphysical cloud models often parameterize these three processes. Another outcome is *droplet breakup* where two drops collide with the result being a number of smaller drops.

Collisions between large drops will result in a dramatic broadening of the tail of the distribution and an increase in the size and number of precipitating drops. However, these drops are in such low concentration that collisions of this type must be considered a rare event from a statistical point of view. Simulations of the collection process using stochastic approaches must accurately consider the most rare events.

The continuous form of the stochastic coalescence equation is given by:

$$\begin{aligned}\frac{\partial N(m,t)}{\partial t} &= \frac{1}{2}\int_0^m N(m',t)N(m-m',t)K(m',m-m')dm \\ &\quad - \int_0^\infty N(m,t)N(m',t)K(m,m')dm' \qquad (24)\end{aligned}$$

where N is the concentration of drops of mass m at time t and K is the collection kernel. The first term describes the coalescence of two drops of mass m' and $m - m'$ to form a droplet of mass m. The average concentration of drops of mass m will increase for every coalescence of two smaller drops of masses m' and $m - m'$. For every two drops "destroyed" one is created, thus the factor of $\frac{1}{2}$. The second term describes coalescence of drops of mass m with drops of mass m'. This term represents the "destruction" of drops of mass m, since any mass m drop that combines with any other drop will no longer be mass m.

Care must be taken to avoid numerical diffusion when integrating the stochastic collection equation. Considerable efforts have been made to develop integration schemes that accurately portray the evolution of the droplet spectra. Berry and Reinhardt's (1974) study indicated that the formation of raindrops in warm clouds can be fairly rapid. In their numerical experiments, initial droplet sizes were specified using gamma distributions of their masses, with the liquid water content as $1\ \text{g/m}^3$ and mean drop sizes ranging from 10 to 18 μm radius (see example in Fig. 18). Drizzle drops were produced in 10 to 22 min and raindrops in 20 to 30 min in different experiments. Faster growth occurred for larger mean initial sizes. The growth of drops after they reached drizzle drop size was nearly independent of the drop spectrum. Berry and Reinhardt's studies show that growth of raindrops in warm clouds can occur by stochastic processes within time scales observed in nature, provided the initial spectra has sufficiently large mean radii and the spectra contain droplets with radii of at least 30 μm radius.

5 GROWTH OF ICE PARTICLES IN ATMOSPHERE

Ice crystals form in the atmosphere through either homogeneous or heterogeneous nucleation. Once formed, ice particles grow by three mechanisms: diffusion of water vapor, accretion of supercooled droplets, and aggregation with neighboring crystals.

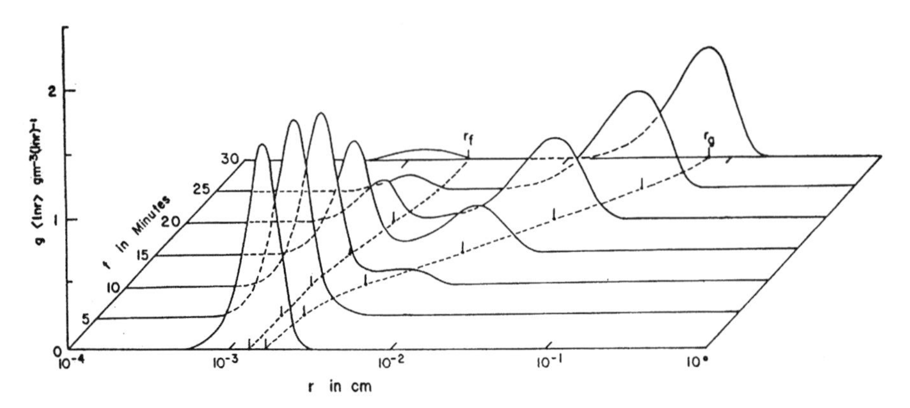

Figure 18 Example of development of a droplet spectrum using stochastic coalescence (from Berry and Reinhardt, 1974). The radius r_f denotes the mean mass, which follows the cloud droplet peak, and r_g denotes the radius of mean-square mass, which follows the raindrop peak.

Accretion leads to the formation of graupel and hail, while aggregation produces snowflakes. Ice particles sometimes shatter during growth or evaporation, increasing particle concentrations. Mechanisms that enhance ice particle concentrations in clouds in this manner are called *secondary* ice particle production mechanisms.

Diffusional Growth of Ice Particles

The Clausius–Clapeyron equation describes the equilibrium condition between two phases. By comparing the solution of the Clausius–Clapeyron equation for the cases of water and vapor, and ice and vapor, it can be easily shown (e.g., Rogers and Yau, 1989; Pruppacher and Klett, 1997) that the saturation vapor pressure over water, $e_{s,\infty}$, will always exceed the saturation vapor pressure over ice, $e_{si,\infty}$, for $T < 0°C$. This is a consequence of the fact that the latent heat of sublimation, L_s, is larger than L_v. This difference has major implications for the growth of ice in clouds. Air saturated with respect to ice will always be subsaturated with respect to water. Conversely, air saturated with respect to water will always be supersaturated with respect to ice. As a consequence, supercooled water droplets and ice crystals can *never exist in a cloud together in equilibrium*. Ice crystals, always subject to greater supersaturations, will rapidly grow, extracting vapor from the surroundings, while coexisting supercooled droplets will evaporate, supplying vapor to the surroundings. The differences in relative humidity with respect to ice and water as a function of temperature can be found by taking the ratios $e/e_{s,\infty}$ and $e/e_{si,\infty}$ and plotting them as a function of temperature. Figures 19 and 20 show that, at cold temperatures, air can be significantly below saturation with respect to water (relative humidity with respect to water, $\text{RH}_w \approx 65$ to 70%) and still be supersaturated with respect to ice. High in the atmosphere, at cirrus cloud levels, ice crystals can grow at moderate relative humidities with respect to water. Even at lower altitudes, particularly in

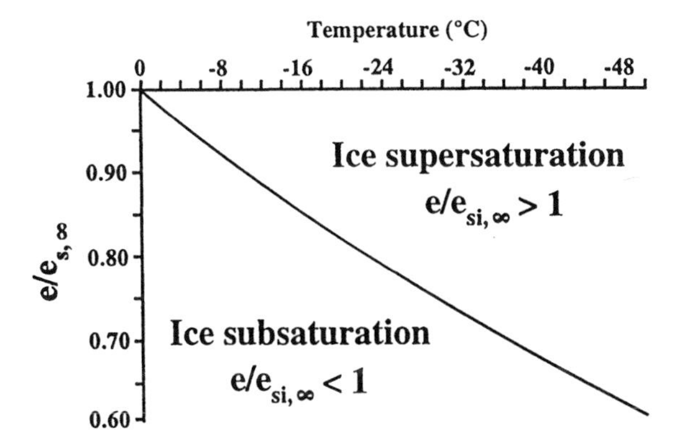

Figure 19 Ice saturation as a function of saturation ratio with respect to water (adapted from Pruppacher and Klett, 1997).

wintertime, ice particles can grow at $RH_w < 100\%$. On the other hand, a cloud that has a relative humidity with respect to water of 100% will be supersaturated with respect to ice by a substantial amount. For example, at -10°C, the relative humidity with respect to ice will be about 112%. At -20°C, it will be 120%. In an environment where $RH_w = 100\%$ exactly, water drops will not grow, yet ice crystals in this same environment will experience enormous supersaturations. As a consequence, they grow very rapidly to become large crystals.

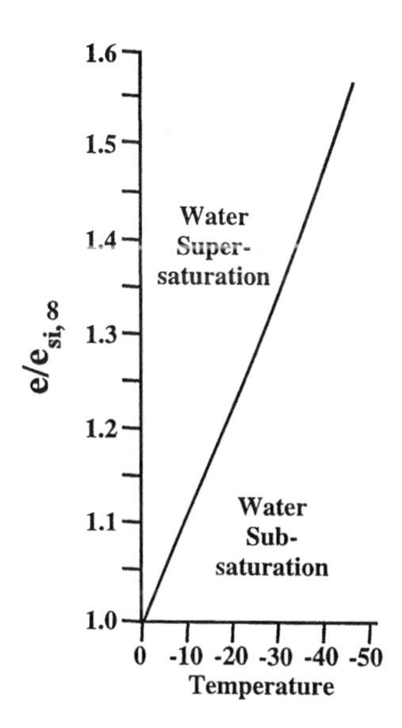

Figure 20 Water saturation as a function of saturation ratio with respect to ice (adapted from Pruppacher and Klett, 1997).

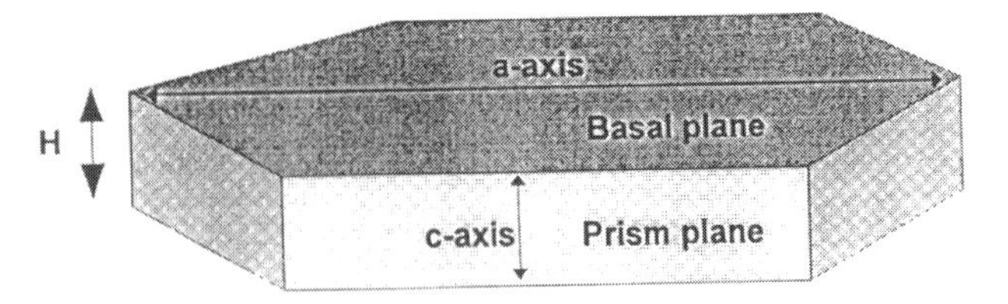

Figure 21 Geometry of a simple ice crystal.

From a crystallographic point of view, ice crystal growth normally occurs on two planes, the *basal* and *prism* planes. The basal plane of an ice crystal refers to the hexagonal plane illustrated in Figure 21. The axis that passes from the outside edge of the crystal through the center along the basal plane is called the a axis. The a axis can be thought of as the diameter of the circle that circumscribes the basal plane. The prism plane refers to any plane along the vertical axis of the crystal, the c axis. Prism planes appear rectangular in Figure 21. The height of a crystal, H, is shown in Figure 21. Growth along the basal plane implies that the c axis will lengthen and H will become larger. Growth along the prism plane implies that the a axis will lengthen and the diameter of the crystal will become larger.

Laboratory experiments have shown that the rate of growth along the c axis relative to growth along the a axis of crystals varies significantly as a function of temperature and supersaturation. These variations occur in a systematic manner, resulting in the characteristic shapes, or "habits," of ice crystals, summarized in Figure 22. As temperature decreases, the preferred growth axis changes from the a axis (platelike crystals, 0 to -4°C) to the c axis (columnar crystals, -4 to -9°C), to the a axis (platelike crystals, -9 to about -22°C), to the c axis (columnar crystals,

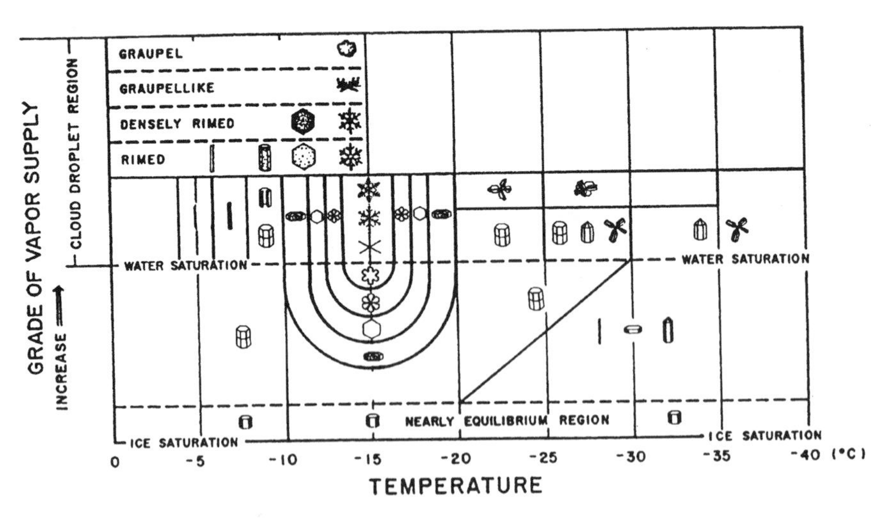

Figure 22 Natural ice crystal habits that form when crystals grow in different temperature and humidity conditions (Magono and Lee, 1966).

−22°C and colder). The first two transitions are well defined, while the latter is diffuse, occurring between about −18 and −22°C. At high supersaturations with respect to ice (water supersaturated conditions), the transition as a function of temperature is, in the same order as above, a plate, a needle, a sheath, a sector, a dendrite, a sector, and a sheath. The temperatures of these transitions are evident from Figure 22. At low supersaturations with respect to ice (below water saturation), the transition in habit as a function of temperature is from a plate, to a hollow (or solid) column, to a thin (or thick) plate to a hollow (or solid) column, depending on the level of saturation. Observations of ice particles in the atmosphere show that the same basic habits appear in nature, although the structure of the particles is often more complicated. This can be understood when one considers that the particles fall through a wide range of temperatures and conditions of saturation.

Ice particles possess complex shapes and, as a result, are more difficult to model than spherical droplets. However, the diffusional growth of simple ice crystals can be treated in a similar way as drops by making use of an analogy between the governing equations and boundary conditions for electrostatic and diffusion problems. The electrostatic analogy is the result of the similarity between the equations that describe the electrostatic potential about a conductor and the vapor field about a droplet. Both the electrostatic potential function outside a charged conducting body and the vapor density field around a growing or evaporating droplet satisfy Laplace's equation. The derivation of the growth equation for ice crystals proceeds exactly as that for a water droplet, except the radius of a droplet, r, is replaced by the electrostatic capacitance, C, the latent heat is the latent heat of sublimation, L_s, and the saturation is that over ice, $S_{i,v}$, rather than water. In the case of a spherical ice crystal $C = r$. The result is

$$\frac{dM}{dt} = \frac{4\pi C(S_{i,v} - 1)}{L_s/KT\big((L_s/R_vT) - 1\big) + R_vT/De_{si,\infty}} \tag{25}$$

The problem in using this equation is determining the capacitance factor C, which varies with the shape of the conductor. For a few shapes, C has a simple form. These shapes are not the same as ice crystals but have been used to approximate some ice crystal geometries. To use Eq. (25) to determine the axial growth rate of ice particles, the term dM/dt must be reduced to a form containing each of the axes. In general, this will depend on the crystal geometry. To solve for a or c, an additional relationship is required between da/dt and dc/dt. This relationship is generally obtained from measurements of the axial relationships determined from measurements of a large number of crystals. Laboratory experiments conducted to determine the importance of surface kinetic effects in controlling ice crystal growth have shown that surface kinetic effects are important in controlling crystal habit since kinetic effects control how vapor molecules are transported across the crystal surface and incorporated into the crystal lattice. Ryan et al. (1976) has measured the growth rates of ice crystals in the laboratory. Figure 23 from their study shows that the a axis

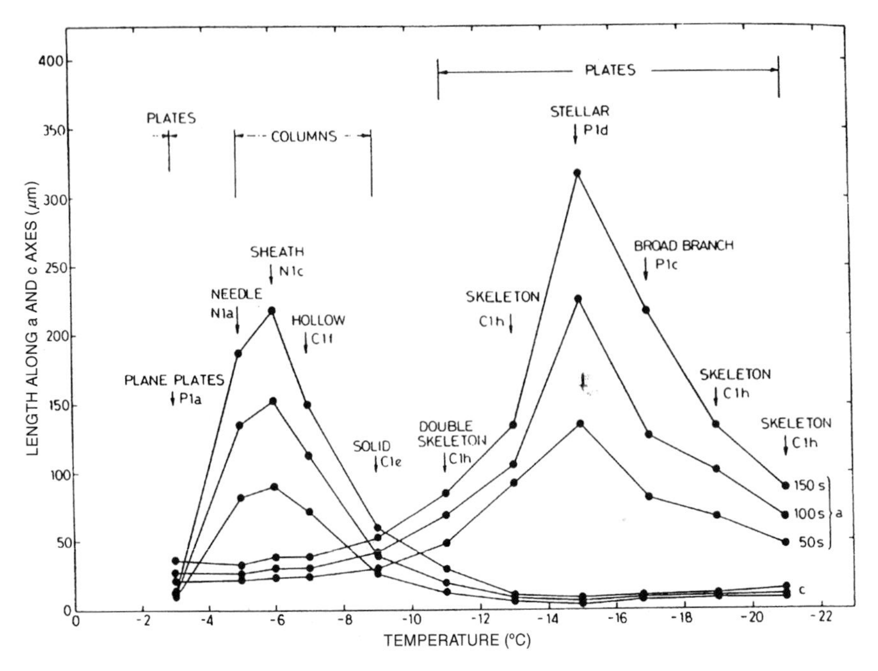

Figure 23 Variation of crystal axial dimensions with temperature for 50, 100, and 150 s after seeding (Ryan et al., 1976).

grows most rapidly at $T > -9°C$, with a peak at $-15°C$, while the c axis grows most rapidly at $T < -9°C$, with a peak at $-6°C$.

Ice Particle Growth by Accretion

Growth by accretion occurs when ice particles collide with supercooled water droplets, causing them to freeze on to the ice surface. For supercooled water to be present in substantial quantities in a mixed-phase cloud, it is necessary for the condensate supply rate to exceed the bulk diffusional growth rate of the ice particles, after accounting for any dry air entrainment that may be occurring. High condensate supply rates require sustained large vertical velocities. For this reason, accretion is an important growth process in clouds with localized regions of sustained high vertical velocities.

The most common type of cloud that meets this requirement is a cumulus cloud, particularly cumulus clouds with large vertical development. Regions of other types of cloud systems also have sustained high vertical velocities. Examples include orographic clouds, particularly near the maximum terrain slope, and convective cores of frontal bands. A second way supercooled water can appear in a cloud is through vertical transport of droplets upward through the 0°C level. This process can occur in any type of cloud with a moderate updraft near the melting level. The supercooled water in this case will generally be confined to the warmer range of subfreezing temperatures. Clouds that generally do not contain much supercooled water are those with weak vertical velocities, such as stratus clouds. In such clouds, the accretion process will not be as important, if it occurs at all.

Studies of accretion have focused primarily on the conditions required for riming onset, and the late stages of riming, when the particles reach the graupel or hail stage of growth. The onset of riming has been studied theoretically and in field and laboratory experiments. These studies generally show that platelike crystals must grow to a threshold size of about 150 μm radius before any droplets accrete to their surfaces. Droplets must be larger than about 5 μm radius before they are collected. Collection efficiencies increase with increasing ice particle and cloud droplet size, reaching about 0.9 when the ice particle radius approaches 400 μm and the droplet radius reaches 35 μm.

Graupel develops in a cloud when small supercooled cloud droplets collect in large numbers on falling ice particles. The parent ice particle, or embryo, of a graupel particle can have various shapes. Certain types of ice particles serve as graupel embryos more effectively than others. Studies have shown that aggregates of ice crystals can provide excellent graupel embryos in hailstorms because of their ability to rapidly collect water droplets. Aggregate embryos can be entrained into the main updraft core of hailstorms from the debris clouds of previously active convective towers. Large drops present in the core updraft can also serve as graupel embryos upon freezing. Graupel characteristics, including density, mass, and terminal velocities have been measured in a number of field studies. In general, the data show a great deal of spread, since the observations come from very different cloud systems.

Hail develops when extreme growth by accretion occurs in thunderstorms. Hail growth occurs in one of two ways, depending on the surface temperature of the hailstone and the rate of supercooled water accumulation. As a hailstone collects supercooled water droplets while falling through a thunderstorm, latent heat is released at the hailstone's surface as the droplets freeze, raising the surface temperature, T_s, of the hailstone. If $T_s < 0°C$, droplets continue to freeze immediately on the hailstone's surface. If, however, T_s reaches 0°C, water impacting on the hailstone will no longer immediately freeze. Water then spreads across the surface of the stone, drains into porous regions, and can shed from the stone's surface. The former situation, when $T_s < 0°C$, is called the *dry growth regime*. The latter situation, when $T_s = 0°C$, is called the *wet growth regime*. For dry growth, the growth rate of the spherical hailstone due to collection of water drops can be approximated by the continuous collection equation:

$$\left(\frac{dM}{dt}\right)_{\text{dry}} = E\pi R_H^2 V_{T,H} w_L \tag{26}$$

where R_H is the radius of the hailstone and $V_{T,H}$ is its terminal velocity. The terminal velocity and radius of cloud drops are neglected because they are small relative to R_H and $V_{T,H}$ [compare Eqs. (23) and (26)]. The same equation applies in the wet growth regime provided the hailstone does not shed water. If the hailstone grows in the wet growth regime, and sheds excess liquid water from its surface, its growth rate will be determined by the rate at which the collected water can be frozen. Calculations show that at $-5^{\circ}C$, the effective liquid water content for wet growth is less than $1\ g/m^3$. At $-10^{\circ}C$ the effective liquid water content is near $1\ g/m^3$. Higher liquid water content is required for wet growth at colder temperatures. The threshold values of liquid water content are well within the range of values found in cumulonimbus, implying that shedding of drops and wet growth are important processes in thunderstorms.

Hailstones exhibit alternating layers of higher and lower density. The density variations are due to variations in the concentration of bubbles trapped in the stone. Generally bubbles are grouped in concentric layers of higher and lower density, which in turn are associated with the different growth regimes. Field studies have shown that hailstone embryos can be conical graupel, large frozen drops, and occasionally large crystals. Hailstones sometimes exhibit irregular structure while others have distinct conical or spherical shapes. The largest known hailstone, collected near Coffeyville, Kansas, weighed 766 g and had a circumference of 44 cm.

Ice Particle Aggregation

Snowflakes form through aggregation of ice crystals. Several mechanisms have been proposed to explain the adhesion of ice particles to one another. Snowflakes of sizes greater than 1 cm are often composed of planar and spatial dendritic crystals, needle-like crystals or, at higher cloud levels, radiating assemblages of bulletlike crystals. All of these crystals have shapes that easily interlock during collisions, suggesting that mechanical interlocking is initially important for crystals to attach to each other. Two mechanisms have been identified that bond ice particles together once in contact. The first, called sintering, occurs because ice particles in point contact form a nonequilibrium system. To minimize the total free surface energy of the system, a requirement for equilibrium, water molecules must diffuse from ice surfaces toward the point of contact, strengthening the bond between the particles. The second bonding mechanism appears to be associated with the presence of a liquid layer on the surface of ice crystals. This layer is thought to enhance a crystal's sticking efficiency. Experiments have shown that the surface electrical conductivity of ice increases significantly at temperatures warmer than $-10^{\circ}C$, suggesting the surface contains a quasi-liquid layer. Studies using optical and magnetic resonance techniques support the existence of the layer. However, the existence of a liquid layer on the surface of ice crystals at subfreezing temperatures is still not universally accepted.

Larger aggregates typically first develop between -15 and $-12^{\circ}C$, the temperature range where dendritic crystals form, although aggregates have been observed at colder temperatures in cirrus. Aggregates form most rapidly near the melting level. Enhanced aggregation near the melting level is apparently due to increased collection efficiency as a liquid layer develops on the surface of the crystals. The rate of aggregation in clouds also depends on ice particle concentrations. Calculations and field observations both suggest that aggregation is enhanced in the presence of high crystal concentrations.

Ice Particle Concentrations and Evidence for Ice Multiplication

Ice particles first appear in clouds when ice nucleation occurs. Measurements of ice nuclei suggest that they should be active primarily at cold temperatures ($<-20^\circ$C) and that the concentration of active nuclei should be an exponential function of temperature. According to measurements of ice nuclei (Fig. 10), few ice particles should present in clouds whose tops are warmer than -10°C.

Measurements of ice particle concentrations in clouds have not shown this expected relationship. Ice particle concentrations often exceed by several orders of magnitude the concentrations expected on the basis of ice nuclei measurements. This is particularly true at warm temperatures ($>-10^\circ$C). Conversely, at temperatures between -20°C and the homogeneous nucleation threshold (-35 to -40°C) ice particle concentrations are often less than expected on the basis of ice nuclei measurements.

Determining the origin of ice particles in clouds remains one of the major unsolved problems in cloud physics. It is possible that ice nucleation occurs in clouds under conditions that are not well understood. It is also possible that secondary mechanisms for ice enhancement occur. Three mechanisms have been studied extensively: (1) fragmentation of colliding ice particles, (2) ice splintering during the riming process, and (3) shattering of drops during freezing.

Ice particles that fragment during collisions are typically more delicate crystals such as dendrites, sectors, side planes, and needles and sheaths. Particles, that seldom fragment are thick and thin plates, solid and hollow columns, and scrolls. Experiments to examine fragmentation show that dendritic crystals readily fragment in conditions expected in clouds. On the other hand, breakup of rime accumulated on crystals during collisions is unlikely in clouds. Overall, data suggest that only clouds containing dendritic particles are likely to have a significant increase in measured ice particle concentrations due to fragmentation during collisions.

The high ice particle concentrations observed in clouds provoked a large number of research groups in the early 1970s to try to find a strong ice multiplication mechanism to explain the observations. A breakthrough on the problem was reported by Hallett and Mossop (1974), who showed that splintering of ice during the riming process led to copious ice production under a limited range of conditions. Subsequent studies determined the conditions required for splinter production and clarified the mechanism for splintering.

These studies together showed that copious numbers of ice splinters could be produced during riming when: (1) the temperature was between -3 and -8°C; (2) large ice particles were present in the cloud to collect cloud droplets, (3) droplets with diameters >24 μm were present, and (4) droplets smaller than 13 μm diameter were present. The small droplets, upon freezing onto a falling ice particle, provide sites for the larger droplets to come into point contact with the ice, attach, and freeze. Laboratory experiments indicate that supercooled drops accreting on to another already frozen droplet in this manner will freeze in distinct modes that depended on temperature. In the temperature range characteristic of ice multiplication, droplets freeze such that the outer shell of the droplet freezes first. The expanding ice builds internal pressure on the liquid in the interior of the drop until the shell ruptures at its weakest point ejecting liquid, which freezes creating a protuberance or a fragment. This process, called the Hallett–Mossop process after its discoverers, is believed to be important in many cloud systems.

The final mechanism that has been studied extensively is droplet fragmentation by shattering or splinter ejection during freezing. Droplet shattering during freezing appears to be limited to droplets $>50\,\mu m$ diameter. Larger drops are more likely to fragment than smaller drops. Even when droplets do not fragment, the nature of the freezing process can cause them to eject an ice splinter as the outer shell ruptures. Because of the large droplet criteria, this process tends to be favored in maritime clouds with larger drop distributions, and much less favored in continental clouds, particularly those with large updrafts, such as cumulus.

Despite the fact that these ice multiplication mechanisms are well understood, they still cannot account for the observed high ice particle concentrations in many clouds. One attractive hypothesis that fits the observations is that rapid nucleation of ice occurs at cloud boundaries, particularly cloud top. It has been shown observationally and theoretically that a narrow layer of supercooled water frequently exists at the top of cold clouds, even at temperatures $<-20°C$. This liquid layer is believed to exist because shear, radiation, and entrainment enhance vertical velocities near cloud top, and because ice crystals are naturally very small near cloud top. Because of these factors, the condensate supply rate frequently exceeds the bulk ice particle growth rate and liquid is produced. Broad droplet spectra are often present in these regions. Entrainment of dry air leads to evaporation of droplets near cloud top. Contact nucleation may be favored in these conditions. Also, evaporation of drops can lead to giant aerosol particles that, in deep cumulus, may carry significant charge. These particles may be more effective as ice nuclei. At present, the importance of each of these ice-forming mechanisms in different clouds is speculative at best. Clearly, more research is required before we understand all the possibilities and processes associated with the formation of the ice phase in clouds.

LIST OF SYMBOLS

A	constant in IN concentration equation
a	ice crystal axis along the basal plane
C	constant in CCN, IN concentration equations, capacitance factor
C_v	specific heat of air at constant volume
c	ice crystal axis along prism plane
D	diffusion coefficient
E_{col}	collision efficiency
E_{coal}	coalescence efficiency
E	collection efficiency

e'_r	vapor pressure over a solution droplet
e_r	vapor pressure over a spherically curved water surface of radius r
$e_{r,i}$	vapor pressure over a spherically curved ice surface of radius r
$e_{s,\infty}$	saturation vapor pressure over a plane water surface
$e_{si,\infty}$	saturation vapor pressure over a plane ice surface
F	Helmholz free energy
ΔF_i	free energy change at ice–liquid interface
$\Delta F'$	free energy change associated with molecular realignment
f_v	ventilation coefficient
h	Planck's constant
i	van't Hoff factor
J	nucleation rate
K	thermal conductivity, collision kernel
k	Boltzmann constant
k	constants in CCN, IN concentration equations
L_v	latent heat of vaporization
L_s	latent heat of sublimation
$n(r)$	number of aerosol particles in the size range r, $r + d \log r$
m	mass of a drop
m_s	mass of solute
$M(r)$	total mass of aerosol particles with radii greater than r
M_w	molecular weight of water
M_s	molecular weight of the solute
N	number concentration
N_a	Avagadro's number
N_c	number of water molecules in contact with ice surface
N_{CCN}	number of CCN active at a specified supersaturation
N_{IN}	number of ice nuclei
$N(r)$	total number of aerosol particles with radii greater than r
N_{Re}	Reynolds number
n_0	number of molecules of water
n_s	number of molecules of solute
p	pressure
R	large drop radius
R^*	universal gas constant
R_d	gas constant for dry air
R_v	gas constant for water vapor
RH_w	relative humidity with respect to water
r	particle radius
r_{eq}	equilibrium radius of a droplet
$S(r)$	total surface area of aerosol particles with radii greater than r
$S_{w,v}$	saturation ratio over a water surface
$S_{s,v}$	saturation ratio over a solution droplet
$S_{i,v}$	saturation ratio over an ice surface
s_w	supersaturation with respect to water

s_i	supersaturation with respect to ice
T	temperature
T_s	surface temperature of a hailstone
t	time
$V(r)$	total volume of aerosol particles with radii greater than r
V_T	terminal velocity
$V_{T,H}$	terminal velocity of a hailstone
x_0	collision cross section
α	accomodation coefficient
β	condensation coefficient, constant in IN concentration equation
ν	kinematic viscosity
ρ	aerosol particle density
ρ_w	density of water
$\sigma_{w,v}$	surface tension at a water–vapor interface
$\sigma_{w,i}$	surface tension at a water–ice interface
$\sigma_{i,v}$	surface tension at an ice–vapor surface
ρ_i	density of ice

REFERENCES

Beard, K. V. (1992). Ice initiation in warm-base convective clouds: An assessment of microphysical mechanisms, *Atmos. Res.* **28**, 125–152.

Beard, K. V., and H. T. Ochs III (1984). Collection and coalescence efficiencies for accretion, *J. Geophys. Res.* **89**, 7165–7169.

Berry, E. X., and R. L. Reinhardt (1974). An analysis of cloud drop growth by collection: Parts I–IV, *J. Atmos. Sci.* **31**, 1814–1831, 2118–2135.

Best, A. C. (1951). The size of cloud droplets in layer-type cloud, *Quart. J. Roy. Met. Soc.* **77**, 241–248.

Bigg, E. K., and C. M. Stevenson (1970). Comparison of concentrations of ice nuclei in different parts of the world, *J. Rech. Atmos.* **4**, 41–58.

Chuang, C. C., and K. V. Beard (1990). A numerical model for the equilibrium shape of raindrops, *J. Atmos. Sci.* **47**, 1374–1389.

Cooper, W. A. (1989). Effects of variable droplet growth histories on droplet size distributions. Part I: Theory, *J. Atmos. Sci.* **46**, 1301–1311.

Fitzgerald, J. W. (1972). A study of the initial phase of cloud droplet growth by condensation: Comparison between theory and observation, Ph.D. Dissertation, Dept. of Geophys. Sci., University of Chicago.

Gunn, R., and G. D. Kinzer (1949). The terminal velocity of fall for water droplets in stagnant air. *J. Meteor.* **6**, 243–248.

Hallett, J., and S. C. Mossop (1974). Production of secondary particles during the riming process, *Nature* **249**, 26–28.

Jaenicke, R. (1988). In G. Fischer (Ed.), *Numerical Data and Functional Relationships in Science and Technology*, Landolt-Börnstein New Series, V: Geophysics and Space

Research, 4: Meteorology, b: Physical and Chemical Properties of Air, Berlin, Springer, pp. 391–457.

Junge, C. E. (1963). Large scale distribution of condensation nuclei in the troposphere, *J. Rech. Atmos.* **1**, 185–189.

Klett, J. D., and M. H. Davis (1973). Theoretical collision efficiencies of cloud droplets at small Reynolds numbers, *J. Atmos. Sci.* **30**, 107–117.

Latham, J., and R. L. Reed (1977). Laboratory studies of the effects of mixing on the evolution of cloud droplet spectra, *Quart. J. Roy. Meteor. Soc.* **103**, 297–306.

Lin, C. L., and S. C. Lee (1975). Collision efficiency of water drops in the atmosphere, *J. Atmos. Sci.* **32**, 1412–1418.

Magono, C., and C. W. Lee (1966). Meteorological classification of natural snow crystals, *J. Fac. Sci.* **7**(2), 321–362.

Pruppacher, H. R. (1995). A new look at homogeneous ice nucleation in supercooled water drops, *J. Atmos. Sci.* **52**, 1924–1933.

Pruppacher, H. R., and J. D. Klett (1997). *Microphysics and Clouds and Precipitation*, 2nd ed., Dordrecht, Kluwer Academic.

Rogers, R. R., and M. K. Yau (1989). *A Short Course in Cloud Physics*, Pergamon Press, Oxford, England.

Ryan, B. F., E. R. Wishart, and D. E. Shaw (1976). The growth rates and densities of ice crystals between −3°C and −21°C, *J. Atmos. Sci.* **33**, 842–850.

Schlamp, R. J., S. N. Grover, H. R. Pruppacher, and A. E. Hamilec (1976). A numerical investigation of the effect of electric charges and vertical external electric fields on the collision efficiency of cloud drops, *J. Atmos. Sci.* **33**, 1747–1750.

Szumowski, M. J., R. M. Rauber, and H. T. Ochs III (1999). The microphysical structure and evolution of Hawaiian rainband clouds. Part III: A test of the ultragiant nuclei hypothesis, *J. Atmos. Sci.* **56**, 1980–2003.

Tabazadeh, A., and E. J. Jensen (1997). A model description for cirrus cloud nucleation from homogeneous freezing of sulfate aerosols, *J. Geophys. Res.* **102**, D20, 23845–23850.

Twomey, S., and T. A. Wojciechowski (1969). Observations of the geographical variation of cloud nuclei, *J. Atmos. Sci.* **26**, 684–688.

Vaillancourt, P. A., and M. K. Yau (2000). Review of particle-turbulence interactions and consequences for cloud physics, *Bull. Amer. Met. Soc.* **81**, 285–298.

Vali, G. (1985). Atmospheric ice nucleation—a review, *J. Rech. Atmos.* **19**, 105–115.

Wallace, J. M., and P. V. Hobbs (1977). *Atmospheric Science, An Introductory Survey*, Academic Press, Orlando, Florida, p. 163.

Whitby, K. T. (1978). The physical characteristics of sulfur aerosols, *Atmos. Environ.* **12**, 135–159.

World Meteorological Organization (1987). *International Cloud Atlas*, Vol. II, Geneva, World Meteorological Organization.

CHAPTER 19

RADIATION IN THE ATMOSPHERE: FOUNDATIONS

ROBERT PINCUS AND STEVEN A. ACKERMANN

1 OVERVIEW

"What is the weather like?" Meteorologists answer this question in thousands of locations every day when we make observations to determine the state of the atmosphere. The answer is summarized in a few carefully chosen quantities, typically wind speed and direction, relative humidity, and temperature. In the language of physics these roughly correspond to momentum, mass (of water), and energy. Weather forecasting is the science and craft of predicting how these interrelated quantities will change with time.

Energy is transferred through the atmosphere via five processes: convection, advection, conduction, phase change, and radiation. The first two processes involve the movement of mass from one place to another; conduction occurs when two bodies at different temperatures are in contact; phase changes release latent heat. Radiation is fundamentally different because it allows energy to be transferred between two locations without any intervening material.

Why Study Radiation?

Because radiation can transport energy even without a medium, it is the only way in which Earth interacts with the rest of the universe. It is radiation, in fact, that determines Earth's climate, since in the long term the planet must shed as much energy as it absorbs. Even within the Earth–atmosphere system radiation can be a powerful player in determining the local energy budget. Radiation is why it is usually

Handbook of Weather, Climate, and Water: Dynamics, Climate, Physical Meteorology, Weather Systems, and Measurements, Edited by Thomas D. Potter and Bradley R. Colman.
ISBN 0-471-21490-6

warmer during the day, when the sun shines, than at night, and why the surface air temperature is higher on cloudy nights than clear ones.

Radiation is also the basis for remote sensing, the ability to measure the state of the atmosphere from a remote location (usually the ground or outer space). Remote sensing includes everything from simple cloud imagery, through radar estimates of precipitation, to ground-based sounding. Understanding the capabilities and limits of remote-sensing measurements requires learning something about radiation.

Earth's atmosphere is primarily made of gases, but it also contains liquids and solids in the form of aerosols and clouds. Radiation interacts in fundamentally different ways with gases and condensed materials. Gases, as we will see, interact with radiation in just a few ways but have complicated spectral structure. Clouds and aerosols, on the other hand, affect the radiation in each spectral region in about the same way but make the mathematics much more complicated.

Nature of Radiation

In the atmospheric sciences, the word *radiation* means "electromagnetic radiation," which is different than the particle radiation emitted from radioactive decay. Visible light is one kind of electromagnetic radiation, as are gamma- and x-rays, ultraviolet, infrared, and microwave radiation, and radio waves.

When radiation is measured using very sensitive instruments at extremely low light levels, it is observed that the energy does not arrive continuously but rather in small, finite amounts. These and other observations described in physics texts, including the photoelectric and Compton effects, suggest that radiation can be thought of as a collection of photons, tiny but discrete packets of energy traveling at the speed of light. This is the particle view of radiation.

We can also describe radiation as an electromagnetic phenomenon. The interactions among electric and magnetic fields, matter, charges, and currents are described by Maxwell's equations. The only nonzero solution to these equations in empty space is a traveling wave. Constants in Maxwell's equations predict the wave velocity, which is exactly the speed of light. Light behaves as a wave in many circumstances, too, diffracting when passed through a slit and reflecting from discontinuities in the medium. These observations are the motivation to describe radiation in purely electromagnetic terms.

How do we reconcile these two views? It is tempting to say that light can be both wave and particle, but this is not quite accurate; rather, there are circumstances in which light behaves like a wave and others in which it behaves like a particle. In this and the following chapter we will primarily use the wave model, which is usually more useful in the context of meteorology.

Radiation is also the single aspect of atmospheric science in which quantum mechanics plays a role. This theory, developed in the first few decades of the twentieth century, is based on the idea that the world is not continuous at very small scales, but is divided up ("quantized") into discrete elements: An electron's angular momentum, for example, can take on only certain values. As we will see, a complete description of radiation requires us to invoke ideas from quantum mechanics several times.

2 FOUNDATIONS

Geometry

The directional aspects of radiative transfer are most naturally expressed in spherical coordinates, illustrated in Figure 1. Location is specified by radius r from the origin, zenith angle θ, and azimuthal angle φ, with differential increments dr, $r\,d\theta$, and $r\sin\theta\,d\varphi$. The amount of radiation depends on direction, which in three dimensions is specified with solid angle Ω, measured in steradians (str). The differential element $d\Omega$ is the product of the polar and azimuthal angle differentials. There are 4π steradians in a sphere:

$$\int_0^{2\pi}\int_0^{\pi} \sin\theta\,d\theta\,d\varphi = \int_{\Omega} d\Omega = 4\pi \tag{1}$$

Polar angle is often replaced with its cosine $\mu = \cos\theta$, $d\mu = \sin\theta\,d\theta$, so that $d\Omega = d\mu\,d\varphi$. Polar angle is measured relative to a beam directed upwards, so $\theta = \pi$ and $\mu = -1$ for a beam pointing straight down, and $\mu > 0$ for radiation traveling upwards.

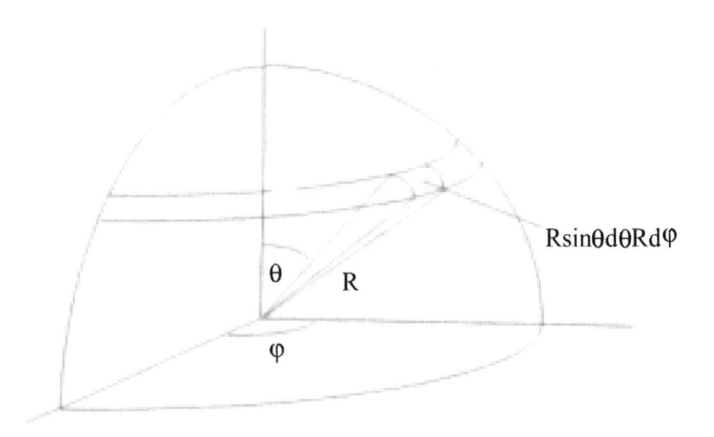

Figure 1 Geometry in polar coordinates. Radius r is measured from the origin, zenith angle θ from the vertical, and azimuthal angle φ from the south.

Describing Electromagnetic Waves

Electromagnetic waves can be characterized in terms of their velocity c, the frequency of oscillation ν, and the wavelength λ. These quantities are related:

$$c = \lambda \nu \tag{2}$$

Wavelength is the distance between successive maxima of field strength and has dimensions of length, while frequency has dimensions of inverse time, so velocity is measured in distance per time. In a vacuum the speed of light $c = 3 \times 10^8$ m/s. This value can change in other materials depending on the index of refraction m, which varies with frequency. In air $m \approx 1$ so c is nearly unchanged; in water at visible wavelengths $m \approx 1.33$, so within cloud drops the speed of light is diminished by about 25%. The inverse of wavelength is the wavenumber $k = 1/\lambda$. In atmospheric applications wavelength is commonly measured in microns ($1\ \mu\text{m} = 10^{-6}$ m), nanometers ($1\ \text{nm} = 10^{-9}$ m), or angstroms ($1\ \text{Å} = 10^{-10}$ m), with frequency in megahertz ($1\ \text{MHz} = 10^6\ \text{s}^{-1} = 0^6$ Hz) or gigahertz ($1\ \text{GHz} = 10^9$ Hz) and wavenumber expressed in inverse centimeters.

Name	*Spectral Region*
X-ray	$\lambda < 10$ nm
Ultraviolet (UV)	$10 < \lambda < 400$ nm
Visible	$0.4 < \lambda < 0.7\ \mu$m
Near-infrared (near-IR)	$0.7 < \lambda < 3.5\ \mu$m
Middle-IR	$3.5 < \lambda < 30\ \mu$m
Far-IR	$30 < \lambda < 100\ \mu$m
Microwave	$1\ \text{mm} < \lambda < 1$ m

The plane in which the electric field oscillates determines the polarization of the radiation. In the atmosphere, though, this plane is rarely constant (i.e., radiation in the atmosphere is usually unpolarized) so we'll ignore this aspect.

Describing Radiation

If we stand in a field on a clear day and look around the sky, we will notice that lots of light comes from the sun and little from the other directions. It is brighter in the open field than in the shadow of a tree, and darker at night than at noon. The amount of radiation, then, depends on space, time, direction, and wavelength. Because light travels so fast, we usually ignore the time dependence.

The fundamental measure of radiation is the amount of energy traveling in a given direction at a certain wavelength. This measure is called spectral intensity I_λ and has dimensions of power per unit area per solid angle per spectral interval, or units of $\text{W/m}^{-2}\ \text{str}\ \mu\text{m}$. Spectral intensity is assumed to be monochromatic, or consisting of exactly one wavelength, and depends on position.

If we want to know the total amount of energy traveling in a given direction (say, the amount of energy entering a camera lens or a satellite detector), we must integrate I_λ across some portion of the spectrum to compute intensity (or broadband intensity) I:

$$I = \int I_\lambda \, d\lambda \tag{3}$$

The limits of integration in (3) depend on the application. If the film in our camera is sensitive only to visible light, for example, the integration is over the visible portion of the spectrum, while if the lens is behind a colored filter we include only those wavelengths the filter passes. Intensity has units of $\mathrm{W/m^2}$ str; we use the term broadband intensity when the integration is over a large part of the spectrum (e.g., all the infrared or all the visible). Unless the radiation interacts with the medium neither I_λ nor I change with distance.

Imagine next a sheet of back plastic placed on the ground in the sunlight. How much energy does the sheet absorb? If we start with just one wavelength, we see that the sheet absorbs energy at a rate F_λ

$$F_\lambda = \int_{-1}^{1} \int_{0}^{2\pi} I_\lambda |\mu| \, d\varphi \, d\mu \tag{4}$$

Weighting by μ accounts for geometry: A ray encountering the surface at an angle is spread over a wider area than a beam coming straight at the surface. F_λ is called the spectral flux and has units of $\mathrm{W/m^2\,\mu m}$. Flux is traditionally divided into upward- and downward-going components:

$$\begin{aligned} F_\lambda^{\downarrow} &= \int_{-1}^{0} \int_{0}^{2\pi} I_\lambda |\mu| \, d\varphi \, d\mu \\ F_\lambda^{\uparrow} &= \int_{0}^{1} \int_{0}^{2\pi} I_\lambda |\mu| \, d\varphi \, d\mu \end{aligned} \tag{5}$$

The black sheet absorbs at all wavelengths, so the total amount of energy absorbed F is computed by integrating over both solid angle and spectral interval:

$$F^{\downarrow} = \int_{-1}^{0} \int_{0}^{2\pi} \int I_\lambda |\mu| \, d\lambda \, d\varphi \, d\mu = \int_{-1}^{0} \int_{0}^{2\pi} I |\mu| \, d\lambda \, d\varphi \, d\mu = \int F_\lambda^{\downarrow} \, d\lambda \tag{6}$$

Flux (or broadband flux) has units of $\mathrm{W/m^2}$. The terms *radiance* and *irradiance* are also used in textbooks and the technical literature; these correspond to our terms *intensity* and *flux*.

Why are so many different quantities used to describe radiation? Because the two main applications of radiation, remote sensing and energy budget computations, require fundamentally different kinds of information. Remote-sensing instruments, for example, usually have a finite field of view and a finite spectral sensitivity; interpreting measurements from these sensors therefore requires calculations of intensity, which may be broadband, narrowband, or essentially monochromatic, depending on the detector.

Imagine, though, wanting to compute the rate at which radiation heats or cools a layer of air in the atmosphere. Here we need to know the net amount of energy remaining in the layer, which we can calculate by considering the decrease in both upwelling and downwelling fluxes as they cross a layer between z_1 and z_2:

$$\begin{aligned} E_{\text{in}} &= F^{\uparrow}(z_1) - F^{\uparrow}(z_2) + F^{\downarrow}(z_2) - F^{\downarrow}(z_1) \\ &= F^{\uparrow}(z_1) - F^{\downarrow}(z_1) - [F^{\uparrow}(z_2) - F^{\downarrow}(z_2)] \\ &= F^{\text{net}}(z_1) - F^{\text{net}}(z_2) \end{aligned} \tag{7}$$

where we have defined the net flux as $F^{\text{net}}(z) = F^{\uparrow}(z) - F^{\downarrow}(z)$. As z_1 and z_2 get very close together, the difference in (7) becomes a differential, and the rate of heating can be related to the divergence of the radiative flux though the air density ρ and heat capacity c_p:

$$\frac{dT(z)}{dt} = -\frac{1}{\rho c_p} \frac{dF^{\text{net}}(z)}{dz} \tag{8}$$

3 SOURCES OF RADIATION

Imagine setting a cast-iron skillet over a really powerful burner and turning on the gas. At first the pan appears unchanged, but as it heats it begins to glow a dull red. As the pan becomes hotter it glows more brightly, and the color of the light changes too, becoming less red and more white.

In the idealized world inhabited by theoretical physicists, the cast-iron skillet is replaced by a block. The block has a cavity inside it and a small hole opening into the cavity. The block can be heated to any temperature. Measurements of the radiation emerging from this cavity show that:

1. The spectral intensity B_λ emerging from the cavity at each wavelength depends only on the temperature T of the block, and not on the material or the shape of the cavity.
2. The total amount of energy emitted by the cavity increases as the temperature increases.
3. The wavelength at which B_λ reaches its maximum value decreases with temperature.

The Planck Function

A theoretical explanation for cavity radiation was the single most compelling unsolved problem in physics at the beginning of the twentieth century. The best fit to observations had been suggested by Wien:

$$B_\lambda(T) = \frac{c_1}{\lambda^5}\frac{1}{e^{c_2/\lambda T}} \tag{9}$$

where c_1 and c_2 were determined experimentally. In October 1900 Max Planck proposed a small modification to this relationship, namely

$$B_\lambda(T) = \frac{c_1}{\lambda^5}\frac{1}{e^{c_2/\lambda T} - 1} \tag{10}$$

Planck's relationship, now called the *Planck function*, was a better fit to the data but was not an explanation for why radiation behaves this way. To develop a theoretical model, Planck imagined that the atoms in the cavity walls behave like tiny oscillators, each with its own characteristic frequency, each emitting radiation into the cavity and absorbing energy from it. But it proved impossible to derive the properties of cavity radiation until he made two radical assumptions:

1. The atomic oscillators cannot take on arbitrary values of energy E. Instead, the oscillators have only values of E that satisfy $E = nh\nu$ where n is an integer called the quantum number, ν the oscillator frequency, and h a constant.
2. Radiation occurs only when an oscillator changes from one of its possible energy levels to another; that is, when n changes value. This implies that the oscillators cannot radiate energy continuously but only in discrete packets, or quanta.

By December of 1900 Planck had worked out a complete theory for cavity radiation, including the values of c_1 and c_2:

$$c_1 = 2hc^2 \qquad c_2 = \frac{hc}{k} \tag{11}$$

In these equations k is Boltzmann's constant, which appears in statistical mechanics and thermodynamics, and h is Planck's constant. Planck used observations of $B_\lambda(T)$ to determine the values $h = 6.63 \times 10^{-34}$ J/s and $k = 1.38 \times 10^{-23}$ J/K. Equations (10) and (11) were the basis for the development of quantum mechanics, which fundamentally changed the way physicists looked at the world.

Blackbody Radiation and Its Implications

In the atmospheric sciences we refer not to cavity radiation but to *blackbody radiation*, a blackbody being anything that radiates according to (10). Blackbody radiation is isotropic, meaning that it is emitted in equal amounts in all directions.

Wavelength of Maximum Emission

The wavelength at which blackbody radiation reaches its maximum intensity can be found by taking the derivative of $B_\lambda(T)$ with respect to wavelength, setting the result to 0, and solving for λ. That is, we solve

$$\frac{\partial B_\lambda(T)}{\partial \lambda} = 0 \tag{12}$$

for the wavelength $\lambda_{\max}$, which yields *Wien's displacement law*

$$\lambda_{\max} = 2898\ \mu\text{m}\ K/T \tag{13}$$

At temperatures typical of Earth's atmosphere (say, $T = 288$ K), the maximum wavelength is about 10 μm in the infrared portion of the spectrum, while the sun (at a temperature about 5777 K) is brightest at about 0.5 μm, where visible light is green.

Total Amount of Energy Emitted

How much energy does a blackbody radiate at a given temperature? This quantity is the blackbody broadband intensity and is computed by integrating over all wavelengths:

$$B(T) = \int_0^\infty B_\lambda(T)\, d\lambda = \frac{\sigma T^4}{\pi} \tag{14}$$

Equation (14) is the *Stefan–Boltzmann relation* and makes use of the Stefan–Boltzmann constant $\sigma = 5.67 \times 10^{-8}\ \text{W/m}^2\,\text{K}^{-4}$. Because blackbody radiation is isotropic, the total amount of radiation lost by an object into each hemisphere is

$$F(T) = \int_0^{2\pi} \int_0^1 B(T)\mu\, d\mu\, d\varphi = \sigma T^4 \tag{15}$$

Figure 2 shows $B_\lambda(\lambda, T)$ plotted for two values of T, roughly corresponding to the average surface temperatures of Earth and sun. The curves are each normalized, since otherwise the much greater intensity produced at solar temperatures would swamp the intensity produced by terrestrial objects. Almost all the sun's energy is produced at wavelengths less than about 4 μm, while almost all the energy emitted

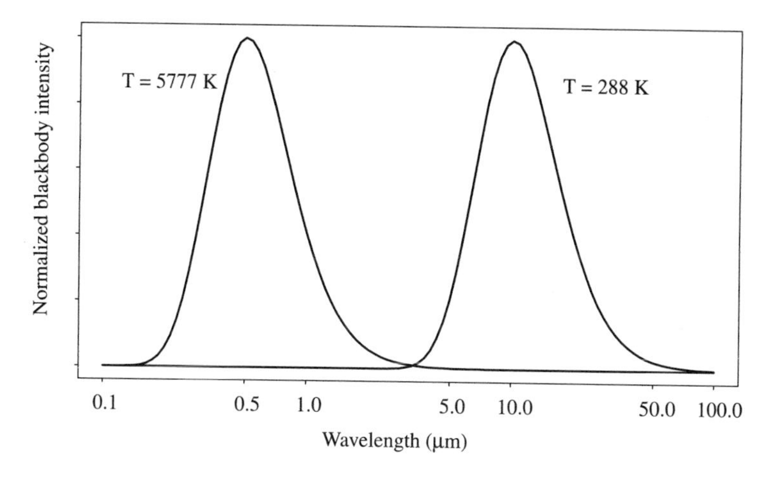

Figure 2 Blackbody intensity B_λ as a function of wavelength for temperatures corresponding to the surfaces of Earth and sun. The curves are arbitrarily normalized.

by Earth is produced at wavelengths longer than 4 μm; this convenient break lets us treat solar radiation and terrestrial radiation as independent.

Blackbody radiation helps us understand the light produced by our cast-iron skillet. The pan emits radiation even at room temperature, although the emission is mostly in the infrared portion of the spectrum. As the skillet is heated, the total amount of radiation emitted increases (as per the Stefan–Boltzmann relation) and the wavelength of maximum emission gets shorter (as per Wien's displacement law) until some of the emission is in the longer (red) part of the visible spectrum. If the pan were made even hotter, it would glow white, like the filament in an incandescent light bulb.

Emissivity, Energy Conservation, Brightness Temperature

The radiation emitted by any object can be related to the blackbody radiation

$$I_\lambda(T) = \varepsilon_\lambda B_\lambda(T) \tag{16}$$

where ε_λ is the emissivity of the object, which varies between 0 and 1. If ε_λ does not depend on wavelength, we say that the object is a gray body; a blackbody has a value of $\varepsilon_\lambda = 1$.

How are emission and absorption related? Imagine an object that absorbs perfectly at one wavelength but not at all at any other wavelength (i.e., an object with $\varepsilon_\lambda = 1$ at one value of $\lambda = \lambda^*$ and $\varepsilon_\lambda = 0$ everywhere else), illuminated by broadband blackbody radiation from a second body. The object absorbs the incident radiation and warms; as it warms the emission (which occurs only at λ^*, remember) increases. Equilibrium is reached when the amount of energy emitted by the particle E_{out} is equal to the amount absorbed E_{in}. At wavelength λ the body acts as a blackbody, so emission depends only on the equilibrium temperature and $E_{out} = I_{\lambda^*}(T) = B_{\lambda^*}(T)$. The body is exposed to broadband blackbody radiation, so $E_{in} = B_{\lambda^*}(T)$. But since equilibrium implies that $E_{in} = E_{out}$, the absorption at every wavelength other than λ^* must be zero. This chain of reasoning, known as *Kirchoff's law*, tells us that the absorptivity and emissivity of objects is the same at every wavelength.

The Planck function finds another application in the computation of brightness temperature. If we make measurements of monochromatic intensity I_m at some wavelength λ, and assume that $\varepsilon_\lambda = 1$, we can invert the Plank function to find the temperature T_b at which a blackbody would have to be in order to produce the measured intensity

$$T_b = \frac{c_2}{\lambda} \frac{1}{\ln(c_1/I_m\lambda^5 + 1)} \tag{17}$$

where T_b is called the equivalent blackbody temperature or, more commonly, *brightness temperature*. It provides a more physically recognizable way to describe intensity.

4 ABSORPTION AND EMISSION IN GASES: SPECTROSCOPY

The theory of blackbody radiation was developed with solids in mind, and in the atmospheric sciences is most applicable to solid and liquid materials such as ocean and land surfaces and cloud particles. The emissivity of surfaces and suspended particles is generally high in the thermal part of the spectrum, and tends to change fairly slowly with wavelength.

In gases, however, emissivity and absorptivity change rapidly with wavelength, so blackbody radiation is not a useful model. Instead, it is helpful to consider how individual molecules of gas interact with radiation; then generalize this understanding by asking about the behavior of the collection of molecules in a volume of gas. We will find that the wavelengths at which gases absorb and emit efficiently are those that correspond to transitions between various states of the molecule; understanding the location and strength of these transitions is the subject of spectroscopy.

Spectral Lines: Wavelengths of Absorption and Emission

Imagine the simplest possible atom, a single electron of mass m_e circling a single proton. In a mechanical view the electrostatic attraction between the unlike charges balances the centripetal acceleration of the electron, so the velocity v of the electron, and therefore its angular momentum, are related to the distance r between the electron and the proton. In 1913 Niels Bohr postulated that the angular momentum of the electron is quantized, and can take on only certain values such that

$$m_e v r = \frac{nh}{2\pi} \tag{18}$$

where n is any positive integer. This implies that the electron can only take on certain values of v and r, so the total energy (kinetic plus potential) stored in the atom is quantized as well. If the energy levels of the atom are discrete, the changes between the levels must also be quantized.

It is these specific changes in energy levels that determine the *absorption and emission lines*, the wavelengths at which the atom absorbs and emits radiation. The energy of each photon is related to its frequency and so its wavelength

$$E = h\nu = \frac{hc}{\lambda} \tag{19}$$

A photon striking an atom may be absorbed if the energy in the photon corresponds to the difference in energy between the current state and another allowed state. In a population of atoms or molecules (i.e., in a volume of gas) collisions between molecules mean that there are molecules in many states, so a volume of gas has many absorption lines.

In the simple Bohr atom the energy of the atom depends only on the state of the electron. Polyatomic molecules can contain energy in their electronic state, as well as in their vibrational and rotational state. The energy in each of these modes is quantized, and photons may be absorbed when their energy matches the difference between two allowable states in any of the modes.

The largest energy differences (highest frequencies and shortest wavelengths, with absorption/emission lines in the visible and ultraviolet) are associated with transitions in the electronic state of the molecules. At the extreme, very energetic photons can completely strip electrons from a molecule. Photodissociation of ozone, for example, is the mechanism for stratospheric absorption of ultraviolet radiation.

Energy is also stored in the vibration of atoms bound together in a stable molecule. Vibrational transitions give rise to lines in near-infrared and infrared, between those associated with electronic and rotational transitions. The amount of energy in each vibrational mode of a molecule depends on the way the individual atoms are arranged within the molecule, on the mass of the atoms, on the strength of the bonds holding them together, and on the way the molecule vibrates. Vibrational motion can be decomposed into normal modes, patterns of motion that are orthogonal to one another. In a linear symmetric molecule such as CO_2, for example, the patterns are symmetric stretch, bending, and antisymmetric stretch, as shown in Figure 3. The state of each normal mode is quantized separately.

The way atoms are arranged within a molecule also plays a role in how energy is stored in rotational modes. Carbon dioxide, for example, contains three atoms arranged in a straight line, and thus has only one distinct mode of rotation about

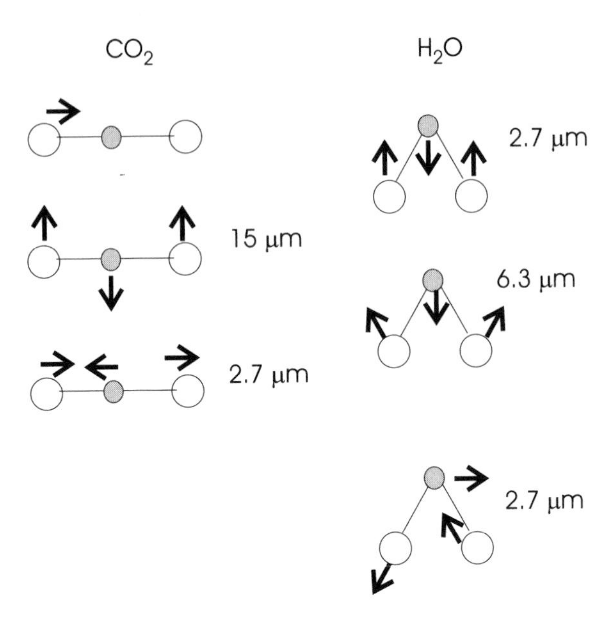

Figure 3 Vibrational normal modes of CO_2 and H_2O molecules. Any pattern of vibration can be projected on to these three modes, which are all orthogonal to one another. Also shown are the wavelengths corresponding to each vibrational mode. See ftp site for color image.

the center molecule and one quantum number associated with rotation. Complicated molecules like ozone or water vapor, though, have three distinct axes of rotation, and so three modes of rotation and three quantum numbers associated with the rotational state, each of which may change independently of the others. For this reason the absorption spectrum of water is much more complicated than that of CO_2.

Changes in vibrational and rotational states may also occur simultaneously, leading to a very rich set of absorption lines. Because the changes in energy associated with rotation are so much smaller than those associated with vibration, the absorption spectrum appears as a central line with many lines (each corresponding to a particular rotational change) clustered around it, as shown in Figure 4. More energetic lines (with higher wavenumbers and shorter wavelengths) are associated with simultaneous increases in the rotational and vibrational state, while the lines to the left of the central line occur when rotational energy is decreased while vibrational energy increases.

What gives rise to the family of absorption lines flanking the central wavenumber in Figure 4? Carbon dioxide has only one axis of rotation, so the different energies are all associated with the same mode, but the family of lines implies that there is a set of rotational transitions, each with a different energy. If J is the quantum number describing the rotational states, and ΔJ the change between two states, there are two possibilities. The different lines might correspond to different values of ΔJ, but this is prohibited by quantum mechanics. In fact, each line corresponds to $\Delta J = 1$, but neighboring states differ in energy by various amounts depending on J.

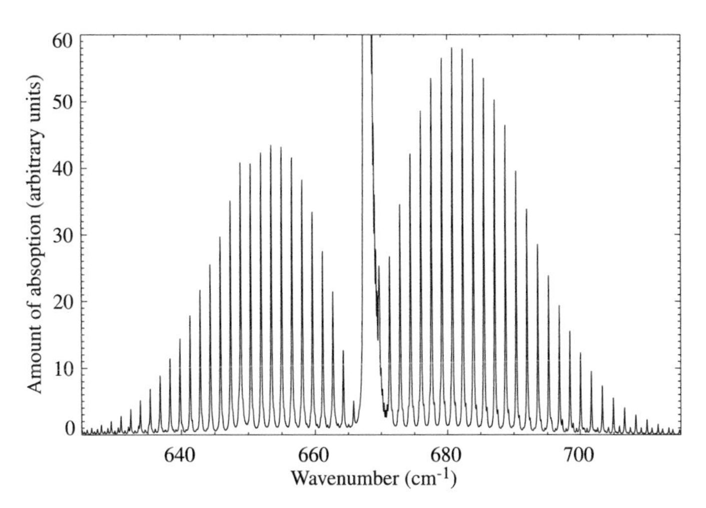

Figure 4 Pure vibration and vibration-rotation lines near the 15 µm (666 cm^{-1}) absorption line of carbon dioxide. The spectrum is computed for gases at 300 K and 1000 mb.

Line Strength

The quantum mechanics of electronic, vibrational, and rotational transitions determines the location of absorption and emission lines in gases. Molecules do not float freely in the atmosphere, however, but rather are members of a large ensemble, or population. The temperature of the gas determines the mean energy of the molecules, and random collisions between them redistribute the energy among the possible states. The amount of absorption due to any particular transition, then, is the product of how likely the transition is (as determined by quantum mechanics) and the fraction of molecules in the originating state. The effectiveness of absorption is called the *line strength* S and has dimensions of area per mass or per molecule.

Line strength depends in part on temperature. The atmosphere is almost everywhere in thermal equilibrium, which means that the average amount of energy per molecule is determined by the temperature, but this energy is constantly being redistributed among molecules by frequent collisions. The collisions also transfer energy between different states, so that a fast-moving molecule may leave a collision moving more slowly but rotating more rapidly. The abundance of molecules in a state with energy E depends on the ratio of E to the average kinetic energy kT. This explains the increase and decrease of line strengths to either side of the central line in Figure 4: The most common rotational state is $J=7$ or $J=8$, and the number of molecules in other states, and so the line strength, decreases the more J differs from the most common value.

Line Shape

Each molecular transition, be it electronic, vibrational, or rotational, corresponds to a particular change in energy, which implies that absorption and emission take place at exactly one frequency. But as Figure 4 shows, absorption and emission occur in a narrow range of wavelengths surrounding the wavelength associated with the transition. We say that absorption lines are affected by natural broadening, pressure (or collision) broadening, and Doppler broadening. We describe these effects in terms of a line shape function $f(\nu - \nu_0)$, which describes the relative amount of absorption at a frequency ν near the central transition frequency ν_0. Line broadening affects the spectral distribution of absorption but not line strength.

Natural Broadening of Absorption Lines

Natural broadening of absorption lines has a quantum mechanical underpinning. Because molecules spend a finite amount of time in each state, the Heisenberg uncertainty principle tells us that the energy of the state must be somewhat uncertain. This blurring of energy levels implies a similar blurring in the transition energies between levels, though the effect is small compared to other broadening mechanisms.

Doppler Broadening of Absorption Lines

If we stand beside a racetrack and listen to an approaching car, the whine of the engine will seem to increase in pitch, then decrease as the car passes us. This change in frequency is known as the Doppler shift and occurs whenever an object and observer move relative to one another. The frequency ν of the emitted sound or light appears to increase or decrease from its unperturbed value ν_0 depending on how fast the object moves v relative to the speed of the wave c:

$$\nu = \nu_0\left(1 \pm \frac{v}{c}\right) \tag{20}$$

Because the atmosphere is almost always in thermal equilibrium, the velocities of a collection of gas molecules follow a Maxwell–Boltzmann distribution. The probability $p(v)$ that a molecule will have a radial velocity v depends on the temperature T, since molecules move faster at higher temperatures, and on the mass m of the molecules

$$p(v) = \sqrt{\frac{m}{2\pi kT}}\exp\left(-\frac{mv^2}{2kT}\right) \tag{21}$$

We can compute the Doppler line shape $f_D(\nu - \nu_0)$ around an absorption line by combining (20) and (21):

$$f_D(\nu - \nu_0) = \frac{1}{\alpha_D\sqrt{\pi}}\exp\left(-\frac{(\nu - \nu_0)^2}{\alpha_D^2}\right) \tag{22}$$

where we have defined the Doppler line width

$$\alpha_D = \frac{\nu_0}{c}\sqrt{\frac{2kT}{m}} \tag{23}$$

Absorption lines associated with heavier molecules are broadened less than those associated with light molecules, since for the same mean thermal energy (as measured by temperature) heavier molecules move more sluggishly than light ones. Doppler broadening also increases in importance at shorter wavelengths. It has the greatest effect high in the atmosphere, where pressure is low.

Pressure Broadening of Absorption Lines

When pressure is high, however, collision or pressure broadening is the most important process determining the shape of absorption and emission lines. The exact mechanisms causing pressure broadening are so complicated that there is no exact theory. What is clear, however, is that the width of the absorption line increases as the frequency of collisions between molecules increases. Pressure-broadened spectral lines are well-characterized by the Lorentz line profile:

$$f_L(\nu - \nu_0) = \frac{\alpha_L}{\pi}\frac{1}{(\nu - \nu_0)^2 + \alpha_L^2} \tag{24}$$

The Lorentz line width α_L depends on the mean time t between collisions:

$$\alpha_L = \frac{1}{2\pi t} \tag{25}$$

In practice the value of α_L is determined for some standard pressure and temperature p_0 and T_0, then scaled to the required pressure and temperature using kinetic theory:

$$\alpha_L = \alpha_{L0}\frac{p}{p_0}\left(\frac{T_0}{T}\right)^n \tag{26}$$

where n varies but is near $\frac{1}{2}$. When molecules collide with others of the same species, we say that the line is self-broadened; when the collisions are primarily with other gases we say the line is subject to foreign broadening. For the same width pressure broadening affects the line wings, those frequencies further from the central frequency, more dramatically than Doppler broadening, as Figure 5 demonstrates.

Both pressure and Doppler broadening can occur simultaneously. The line shape that results, the convolution of $f_L(\nu - \nu_0)$ and $f_D(\nu - \nu_0)$, is called the Voigt line shape, which does not have an exact analytic form. It is instructive, though, to

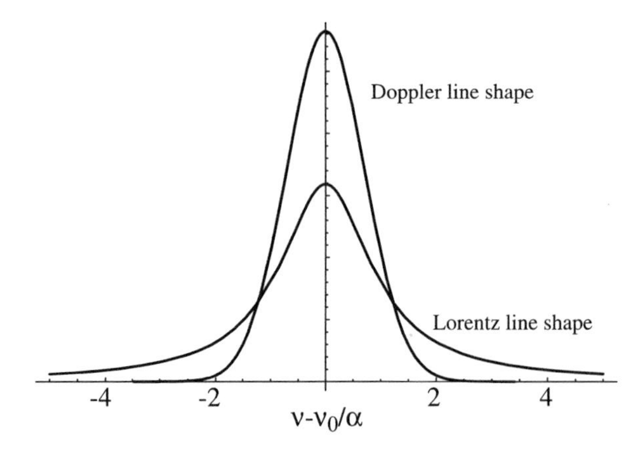

Figure 5 Lorentz and Doppler line shapes for equivalent line strengths and widths. Line wings are more strongly affected by pressure than Doppler broadening.

examine the relative importance of Doppler and pressure broadening before invoking anything more complicated. A useful approximation is

$$\frac{\alpha_L}{\alpha_D} \approx 10^{-12} \frac{\nu_0}{p} \tag{27}$$

when ν_0 is measured in hertz and p in millibars. Throughout most of the spectrum and most of the atmosphere this ratio is much less than one, telling us that the Lorentz line profile is usually a pretty good description of line shape.

Practical Applications

Absorption spectra almost never need to be determined directly from spectroscopy. Spectroscopic databases (HITRAN is popular) have been compiled for all the gases in Earth's atmosphere. These databases contain the line centers and parameters describing line shape as a function of temperature and pressure, and can be accessed with standard programs (e.g., LBLRTM) that provide the amount of absorption at very high spectral resolution.

Radiative Transfer Equation for Absorption

We are at last ready to compute the fate of radiation as it travels through a medium. Radiation may be absorbed or emitted by the medium, and may also be *scattered* or redirected without a change in intensity. We will develop the radiative transfer equation by adding each process in turn in order to keep the mathematics clear.

We will begin by considering a medium that absorbs but does not emit or scatter radiation. This is the framework we might use, for example, to describe the absorption of ultraviolet or visible light by gases in the atmosphere, where relatively low temperatures mean that emission is effectively zero.

Imagine a pencil of monochromatic radiation crossing a small distance ds between two points S_1 and S_2, as illustrated in Figure 6. The amount of radiation absorbed along this path depends linearly on the amount of incident radiation (more incoming photons means a greater likelihood that a photon will strike a molecule), the gas density (higher density increases the number of molecules encountered), and on how effectively the molecules absorb radiation at this wavelength. These are the three factors that appear in the radiative transfer equation for absorption:

$$\frac{dI_\lambda}{ds} = -I_\lambda k_\lambda \rho \tag{28}$$

where ρ is the density of the absorbing gas and k_λ the mass absorption coefficient (m^2/kg). Notice that k_λ has the same units as absorption line strength S. In fact, the absorption coefficient at some wavelength due to one particular line i is the product of the line shape and the line strength:

$$k_i(\nu) = S_i f(\nu - \nu_0) \tag{29}$$

and k_λ is the sum contributions from all lines from all gases.

The radiative transfer equation can be integrated along the path between S_1 and S_2:

$$\int_{S_1}^{S_2} \frac{1}{I_\lambda} dI_\lambda = \ln I_\lambda \Big|_{S_1}^{S_2} = -\int_{S_1}^{S_2} k_\lambda \rho \, ds \tag{30}$$

$$I_\lambda(S_2) = I_\lambda(S_1) \exp\left(-\int_{S_1}^{S_2} k_\lambda \rho \, ds \right) \tag{31}$$

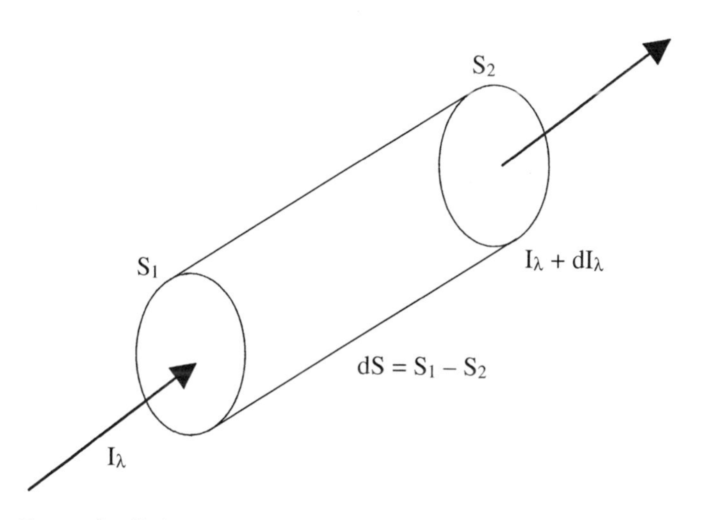

Figure 6 Beam of radiation passing through an absorbing medium.

where we leave the integral over the path unevaluated because ρ and k_λ may change with distance. We insist that the light be monochromatic since the integration does not make sense if k_λ is variable. Equation (31) is called *Beer's law*, or more formally the Beer–Lambert–Bouguer law.

How can we relate a path between arbitrary points S_1 and S_2 to the atmosphere? Two assumptions are usually made: that the atmosphere is much more variable in the vertical than in the horizontal, and that Earth is so large that the surface can be considered flat. These simplifications lead to the plane-parallel coordinate system, in which directions are specified through zenith angle θ and azimuthal angle φ but the medium varies only with vertical position z. In this system the path between S_1 and S_2 is related to the vertical displacement by $1/\mu$, so we may write

$$I_\lambda(z_2) = I_\lambda(z_1)\exp\left(-\frac{1}{\mu}\int_{z_1}^{z_2} k_\lambda \rho \, dz\right) \tag{32}$$

The integral in the exponential defines the *optical thickness* τ, also called the optical depth:

$$\tau = -\int_{z_1}^{z_2} k_\lambda \rho \, dz \tag{33}$$

The minus sign in (33) is an unfortunate hangover from the days when radiative transfer was dominated by astrophysicists. Since they were thinking about other stars, they set up a coordinate system in which $\tau = 0$ at the top of the atmosphere, so that $\mu \, d\tau < 0$. Though it is a little confusing, (33) does lead to much simpler forms of both the radiative transfer equation

$$\mu \frac{dI_\lambda}{d\tau} = I_\lambda \tag{34}$$

and to a more transparent form of Beer's law

$$I_\lambda(\tau_2) = I_\lambda(\tau_1)\exp\left[\frac{(\tau_2 - \tau_1)}{\mu}\right] = I_\lambda(\tau_1)\exp\left[\frac{-(\tau_2 - \tau_1)}{|\mu|}\right] \tag{35}$$

For downwelling radiation $\tau_2 > \tau_1$ and $\mu < 0$ and for upwelling radiation both signs are reversed, so intensity always decreases along the path. According to Beer's law, then, intensity in an absorbing medium falls off exponentially with optical depth.

The transmissivity T_λ of the layer between z_1 and z_2 is defined as

$$T_\lambda = \frac{I_\lambda(z_2)}{I_\lambda(z_1)} \tag{36}$$

which for an absorbing medium T_λ can be computed as

$$T_\lambda = \exp\left(\frac{-|\tau_2 - \tau_1|}{|\mu|}\right) \tag{37}$$

In a strictly absorbing medium the light that is not transmitted must be absorbed; so the absorptivity a_λ of the medium is

$$a_\lambda = 1 - T_\lambda = 1 - \exp\left(\frac{|\tau_2 - \tau_1|}{|\mu|}\right) \tag{38}$$

Computing Optical Depth along Inhomogeneous Paths

Optical depth is formally defined by (33), but computing its value is not always straightforward since both density and the absorption coefficient may change with height in the atmosphere. We can roughly account for the change in density with height by defining the specific gas ratio $q = \rho/\rho_{\text{air}}$ and invoking the hydrostatic equation $dp/dz = -\rho g$ in the definition of optical mass u:

$$u(p_1, p_2) = -\frac{1}{g}\int_{p_2}^{p_1} q\, dp \tag{39}$$

which has dimensions of mass per unit area.

The value of k_λ may also change with height if, for example, the wavelength in question is on the wing of an absorption line, so that the local strength changes as pressure and temperature change. We could adjust k_λ to some average value along the path. It is more common, however, to scale the optical mass to the reference temperature T_0 and pressure p_0 at which the mass absorption coefficient is measured. There are various approaches to scaling, though a common tactic is to compute a scaled absorber amount $\tilde{u}$ as

$$\tilde{u} = u\left(\frac{p_e}{p_0}\right)^m \left(\frac{T_0}{T_e}\right)^n \tag{40}$$

where m and n depend on the gas, and p_e and T_e are the effective pressure and temperature along the path, computed for temperature as

$$T_e = \frac{\int T\, du}{\int du} \tag{41}$$

and similarly for pressure.

Radiative Transfer Equation for Emission and Absorption

Imagine next a material that emits and absorbs at a wavelength λ. This might apply, for example, to the transfer of infrared radiation through Earth's atmosphere: Vibrational transitions in common gases give rise to absorption lines in the infrared, where temperatures through most of the atmosphere give rise to strong emission.

Kirchoff's law tells us that the emission and absorption per amount of material (or per unit of optical depth) must be the same, but emission adds to the intensity while absorption reduces it. Since $\mu\, d\tau < 0$, the radiative transfer equation is

$$\mu \frac{dI_\lambda}{d\tau} = I_\lambda - B_\lambda \tag{42}$$

This relation, known as Schwartzchild's equation, holds strictly for any medium that does not scatter radiation at the wavelength in question; (34) is a useful approximation when temperatures are such that B_λ is very small at the wavelength in question.

We can solve (42) using an integrating factor: We multiply both sides of the equation by $e^{-\tau/\mu}/\mu$ and collect terms in I_λ:

$$\mu e^{-\tau/\mu} \frac{dI_\lambda}{d\tau} - e^{-\tau/\mu} I_\lambda = -e^{-\tau/\mu} B_\lambda \tag{43}$$

or

$$\mu \frac{d}{d\tau} I_\lambda e^{-\tau/\mu} = -B_\lambda e^{-\tau/\mu} \tag{44}$$

To find the intensity propagating in direction μ at some vertical optical depth τ, we integrate (44) starting at one boundary, being careful to account for the sign of μ and for the fact that B_λ may vary with height. The downwelling intensity $I_\lambda^\downarrow$, at optical depth τ, for example, is

$$I_\lambda^\downarrow(\tau) e^{-\tau/\mu} - I_\lambda^\downarrow(0) = \frac{-1}{\mu} \int_0^\tau B_\lambda(\tau') e^{-\tau'/\mu}\, d\tau' \tag{45}$$

or

$$I_\lambda^\downarrow(\tau) = I_\lambda^\downarrow(0) e^{\tau/\mu} - \frac{1}{\mu} \int_0^\tau B_\lambda(\tau') e^{(\tau-\tau')/\mu}\, d\tau' \tag{46}$$

while the upwelling intensity is

$$I_\lambda^\uparrow(\tau) = I_\lambda^\uparrow(\tau^*) e^{-(\tau^*-\tau)/\mu} + \frac{1}{\mu} \int_\tau^{\tau^*} B_\lambda(\tau') e^{(\tau'-\tau^*)/\mu}\, d\tau' \tag{47}$$

where τ^* refers to the bottom of the atmosphere.

Equations (46) and (47) tell us that the downwelling intensity at some height τ consists of the intensity incident at the boundary and attenuated by the intervening medium (the first term on the right-hand side) plus contributions from every other part of the medium (the integral over the blackbody contribution at every height), each attenuated by the medium between the source at τ' and the observation location τ.

As an example, imagine a satellite in orbit at the top of the atmosphere, looking straight down at the ground, which has emissivity 1 and is at temperature 294 K. At 10 μm, blackbody emission from the ground is found with (10) and (11) as $B_\lambda(294\,\mathrm{K}) = 2.85\,\mathrm{W/m^{-2}\,str\,\mu m}$. The clear sky is nearly transparent ($\tau \approx 0$) at 10 μm, so the upwelling nadir-directed ($\mu = 1$) intensity at the top of the atmosphere is essentially the same as the intensity at the ground. But if a thin ($\tau = 1$ at 10 μm), cold ($T = 195$ K) cirrus cloud drifts below the satellite, the outgoing intensity will be reduced. If we assume that the cloud has constant temperature, we can find the upwelling intensity with (47), using $\tau = 0$ at the top of the cloud, $\tau^* = 1$ at the cloud base, and $B_\lambda(\tau') = B_\lambda(195\,\mathrm{K})$ for $0 < \tau' < \tau^*$:

$$I_\lambda^\uparrow(\tau) = B_\lambda(294\,\mathrm{K})e^{-1} + B_\lambda(195\,\mathrm{K})(1 - e^{-1}) = 1.2\,\mathrm{W/m^{-2}\,str\,\mu m} \qquad (48)$$

which corresponds to a brightness temperature of $T_b = 250$ K. This effect is clear in infrared satellite imagery: Thick cirrus clouds appear much colder (i.e., have lower brightness temperatures) than thin ones at the same level.

It seems on the face of things that computing radiative transfer in the infrared is not that hard, since we can now predict the intensity and flux using (46) and (47) if we know the boundary conditions and the state of the atmosphere. Life, alas, is not that simple. Any practical use of radiative transfer involves integration over some spectral interval, and spectral integration in the infrared is where things become difficult. We learned in the section on spectroscopy that the absorption and emission characteristics of gases change very rapidly with wavelength, being large near absorption lines and small elsewhere. The atmosphere is composed of many gases, so the absorption structure as a function of wavelength is extremely rich. Brute force spectral integration, while theoretically possible, is computationally prohibitive in practice. We will address more practical methods for spectral integration, including band models and k distributions, in the next chapter.

5 FULL RADIATIVE TRANSFER EQUATION, INCLUDING ABSORPTION, EMISSION, AND SCATTERING

What is Scattering?

The absorption of radiation by a gas molecule is a two-step process. First, the photon must pass close enough to the molecule for an interaction to occur, and second, the photon's energy must match the difference between the molecule's current state and another allowed state. But what happens to those photons that interact with molecules but whose energies do not match an allowed transition? These photons essentially bounce off the molecule and are redirected; we call this process *scattering*.

Gases do scatter radiation, but the vast majority of scattering in the atmosphere occurs when light interacts with condensed materials, primarily clouds and aerosols. Formally, we say that photons that are absorbed undergo inelastic interactions with the medium, while elastic collisions cause scattering. The likelihood of each kind of interaction need not be the same; that is, the extinction coefficients for scattering and absorption are not identical.

The radiative transfer equation as we have been writing it describes intensity, a quantity associated with a particular direction of propagation. When we discuss scattering, the direction of the beam becomes more important than when considering only absorption and emission because photons can be scattered both out of the beam into other directions and into the beam from radiation traveling in any other direction.

Accounting for Scattering

When we first wrote the radiative transfer equation, we assumed that the medium absorbed but did not emit radiation at the wavelength in question, as occurs during the absorption of solar radiation in Earth's atmosphere. In going from (34) to (42) we included the effects of emission, as when considering the transfer of infrared radiation in the atmosphere. The blackbody emission contribution in (42) is called a source term. Scattering from the beam into other directions is an additional reduction in intensity, while scattering into the beam from other directions adds a second source term.

To write the complete radiative transfer equation, we must distinguish the amount of absorption and emission along the path from the amount of scattering. We do so by introducing a mass scattering coefficient $k_{s\lambda}$ with dimensions of area per mass, as an analog to the mass absorption coefficient $k_{a\lambda}$. Both absorption and scattering diminish the beam, while scattering of radiation traveling in any other direction into the beam can add to the intensity. The full radiative transfer equation is therefore

$$\begin{aligned} \frac{dI_\lambda(\mu,\varphi)}{ds} = {} & -(k_{s\lambda}+k_{a\lambda})\rho I_\lambda(\mu,\varphi) + k_{a\lambda}\rho B_\lambda + \\ & \frac{k_{s\lambda}\rho}{4\pi}\int_0^{2\pi}\int_{-1}^{1} P(\mu',\varphi' \to \mu,\varphi) I(\mu',\varphi')\,d\mu'\,d\varphi' \end{aligned} \tag{49}$$

where we have made explicit the direction of the beam (specified by μ, φ). The last term in (49) accounts for the scattering of radiation into the beam traveling in direction μ,φ from every other direction. The quantity $P(\mu',\varphi' \to \mu,\varphi)$ is called the *single scattering phase function*, or often simply the phase function, and describes how likely it is that radiation traveling in the μ',φ' direction will be scattered into the μ,φ direction. The phase function is reciprocal, so that

$P(\mu',\varphi' \rightarrow \mu,\varphi) = P(\mu,\varphi \rightarrow \mu',\varphi')$, and is defined such that the integral over the entire sphere is 4π.

We divide both sides of the equation by $(k_{s\lambda} + k_{a\lambda})\rho$ and relate path length differential to the vertical displacement to obtain

$$\mu \frac{dI_\lambda(\mu,\varphi)}{d\tau} = I_\lambda(\mu,\varphi) - (1 - \omega_0)B_\lambda - \frac{\omega_0}{4\pi} \int_0^{2\pi} \int_{-1}^{1} P(\mu',\varphi' \rightarrow \mu,\varphi) I(\mu',\varphi')\, d\mu'\, d\varphi' \tag{50}$$

where we have now defined the *single scattering albedo* $\omega_0 = k_{s\lambda}/(k_{s\lambda} + k_{a\lambda})$, which is the likelihood that a photon is scattered rather than absorbed at each interaction. Single scattering albedo varies between zero and one; the lower limit corresponds to complete absorption and the upper to complete scattering.

Equation (50) is the plane parallel, unpolarized, monochromatic radiative transfer equation in full detail. Despite its length, it describes only four processes: extinction by absorption and by scattering out of the beam into other directions, emission into the beam, and scattering into the beam from every other direction. The equation is, unfortunately, quite difficult to solve because it is an integrodifferential equation for intensity; that is, intensity appears both in the differential on the left-hand side and as part of the integral on the right-hand side of the equation.

Before we can even begin to solve this equation, we have to come to grips with the way particles scatter light. When we consider absorption and emission, we need only to determine the mass absorption coefficient $k_{a\lambda}$. When we include scattering in the radiative transfer equation, though, we require three additional pieces of information: the mass scattering coefficient $k_{s\lambda}$, along with the phase function $P(\mu',\varphi' \rightarrow \mu,\varphi)$ and single scattering albedo ω_0.

6 SINGLE SCATTERING

When we are concerned with emission and absorption, it is spectroscopy, a combination of quantum mechanics and statistical mechanics, that lets us determine $k_{a\lambda}$ from knowledge of the temperature, pressure, and chemical composition of the atmosphere. To compute the single scattering parameters within a volume, we begin with knowledge of the way light interacts with individual particles, which comes from solutions to Maxwell's equations; this knowledge is then combined with information about the statistics of different particle types within the volume.

Scattering from a particle is most naturally computed in a frame of reference centered on the particle. In particular, it is easiest to describe the phase function in terms of a *scattering angle* Θ between the incident and scattered radiation. The phase function is often summarized using the *asymmetry parameter* g:

$$g = \frac{1}{2} \int_{-1}^{1} \cos\Theta P(\Theta) d \cos\Theta \tag{51}$$

which is the average cosine of the scattering angle. When $g > 0$, more light is scattered into the forward than backward direction. $P(\mu',\varphi' \to \mu,\varphi)$ is, of course, related to $P(\Theta)$:

$$\cos\Theta = \mu'\mu + \sqrt{(1-\mu'^2)}\sqrt{(1-\mu^2)}\cos(\varphi' - \varphi) \tag{52}$$

Computing Scattering from a Single Particle

The task of computing the intensity scattered from a single particle is conceptually straightforward: Maxwell's equations are solved inside and outside the particle subject to the boundary conditions on the particle's surface. In practice this is such a difficult feat that it is possible only in circumstances when geometric or size considerations are favorable, or when it is possible to make simplifying assumptions of one kind or another.

In the electromagnetic terms of Maxwell's equations, cloud drops and aerosol drops differ from the gaseous atmosphere only in their index of refraction $m = n_r + in_i$. The index of refraction is complex: The real part primarily determines the speed of light within the medium, while the imaginary part determines the amount of absorption per amount of material. The value of m varies with wavelength and can also depend on the state of the material; the indices of refraction of water and ice, for example, can differ dramatically at certain wavelengths, as Figure 7 shows.

Imagine a particle with monochromatic radiation I_{inc} incident upon it. The total surface area projected in the direction of the beam's origin is called the particle's geometric cross section C_{geo}, and the power incident on the particle is $P_{inc} = C_{geo}I_{inc}$. We can generalize this idea to define scattering and absorption cross sections for the particle though the rate at which each process removes energy from the beam: $C_{sca} = P_{sca}/I_{inc}$, $C_{abs} = P_{abs}/I_{inc}$. Extinction includes both scattering and absorption, so the extinction cross section C_{ext} is the sum of C_{sca} and C_{ext}. The radiative cross sections of a particle are related to its geometric cross section but also depend on the particle shape and index of refraction. This makes it useful to unscramble the two influences, defining efficiencies $Q_j = C_j/C_{geo}$, where j denotes one of scattering, absorption, or extinction.

The single scattering parameters of a particle depend on the particle's size, composition, and shape, and on the wavelength of radiation being scattered through the index of refraction. The relative sizes of the particle and the radiation determine the methods that must be used to compute single scattering parameters. The relationship is embodied in the *size parameter* x, the ratio between some characteristic radius r and the wavelength; for spheres, for example, the size parameter is defined as $x = 2\pi r/\lambda$.

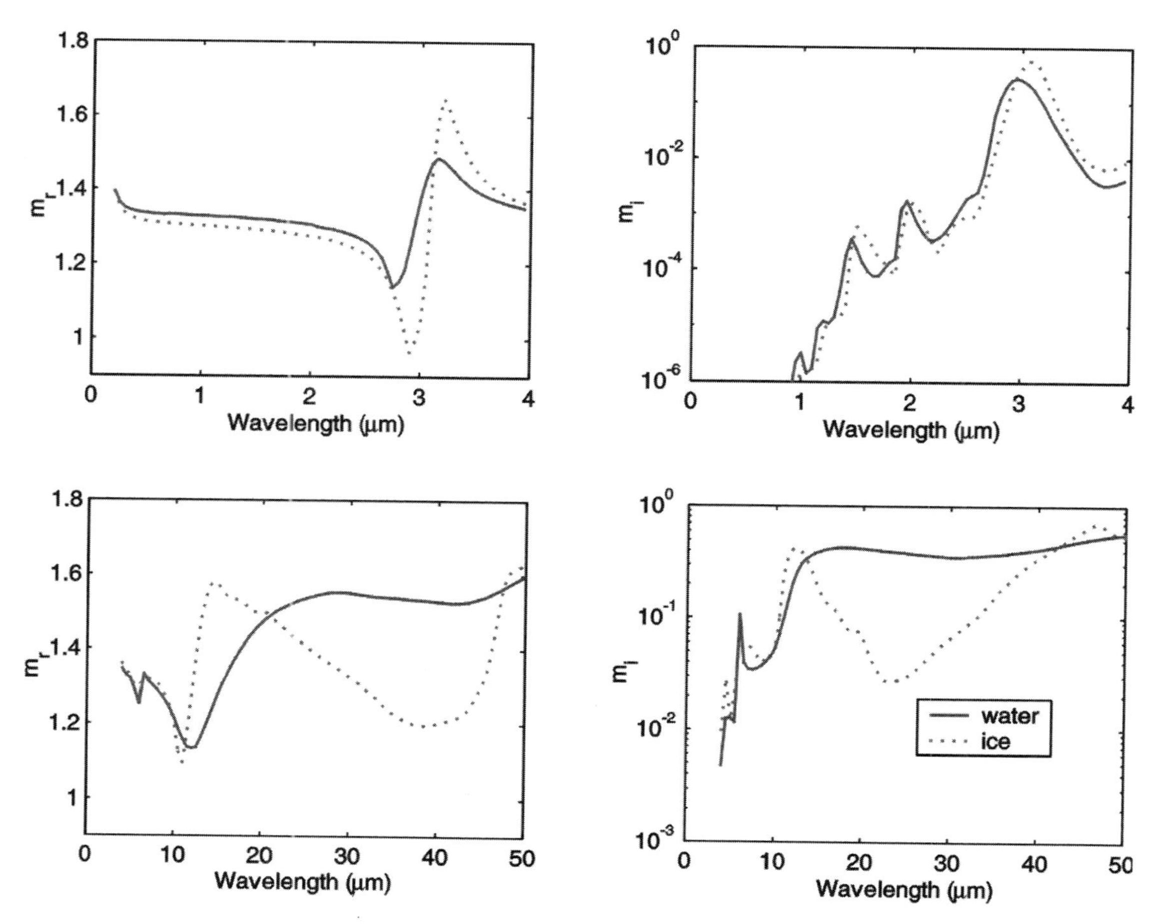

Figure 7 Indices of refraction for liquid water and ice as a function of wavelength. Upper set of panels is a closer look at the left-hand portion of the lower set of panels. See ftp site for color image.

Scattering by Small Particles: Rayleigh Theory

When materials are placed within a constant electric field, the molecular charges can become displaced from one another so that the material is polarized. In a traveling wave the electric field varies in time and space; so, in general, different parts of a particle subject to radiation are polarized differently. But if the particle is very small compared to the wavelength of radiation (i.e., if $x \ll 1$), the entire particle is subject to the same field at any given moment. This allows us to treat the radiation scattered from small particles as if it were emitted from a dipole oscillating at the same frequency as the incident wave.

The theory of scattering from small particles is named after its developer, Lord Rayleigh, who published the work in 1871. Rayleigh used an elegant and succinct dimensional argument to show that the scattering efficiency varies as the size parameter to the fourth power, which for particles of roughly constant size means that scattering depends strongly on wavelength. The full expression for scattering efficiency is

$$Q_{\text{sca}} = \frac{8}{3}x^4\left|\frac{m^2 - 1}{m^2 + 2}\right|^2 \tag{53}$$

which implies that scattering cross section depends on the size of the particle to the sixth power and (if m does not depend too strongly on wavelength) on λ^{-4}. The theory provides a simple phase function for small particles shown in Figure 9.

Microwave radiation encountering cloud drops is subject to Rayleigh scattering, which is why radar beams reflect so strongly from large drops and precipitation. And Rayleigh scattering of visible light by gas molecules is why skies are blue and sunsets red: Gas molecules are all about the same size, but blue light is scattered from the incoming sunbeam into the open sky much more efficiently than red light.

Scattering by Round Particles: Lorenz–Mie Theory

Some of the most dramatic sights in the atmosphere come from the scattering of visible light by clouds. Cloud drops are typically about 10 μm in size, and ice crystals are an order of magnitude larger. For visible light this yields size parameters much larger than 1, so Rayleigh theory is not applicable. In warm clouds, though, surface tension acts to minimize the particle surface and make the drops round. *Lorenz–Mie theory*, developed around the turn of the twentieth century, takes advantage of this symmetry to develop an exact solution for scattering from homogeneous spheres. The technique computes the radiation field by finding a series solution to the wave equation for the scattered wave in spherical coordinates centered on the particle, then expanding the incident radiation in the same coordinates and matching the boundary conditions.

The application of Lorenz–Mie theory is routine, and codes are freely available in several computer languages. The calculation requires the relative index of refraction of the particle and its size parameter and provides the phase function, extinction efficiency, and single scattering albedo; examples are shown in Figures 8, 9, and 10. Because the number of terms used to expand the incoming wave increases with particle size, calculations require more terms as the size parameter increases.

When x is very small, the extinction efficiency increases rapidly (echoing Rayleigh theory) at a rate that depends on the index of refraction. The efficiency, plotted in Figure 8, then oscillates as interference between the scattered and incident radiation changes from constructive to destructive with small changes in particle size. At very large values of x (not shown) Q_{ext} approaches 2, implying that an area twice as large as the particle's cross section is removed from the beam. This is called the extinction paradox and highlights the role of diffraction in particle scattering. We might think that a large particle casts a shadow exactly as large as its cross section, and that this shadow corresponds to the amount of extinction. But every particle has an edge, and the light passing near this edge is diffracted, or diverted very slightly from its initial direction. Half of the extinction is due to diffraction and half to absorption and scattering into other directions. Diffraction contributes to a large forward peak in scattering phase function of moderately sized, weakly absorbing spherical particles such as cloud drops in the visible; this peak can be five or more orders of magnitude larger than other parts of the phase function, as Figure 9 shows.

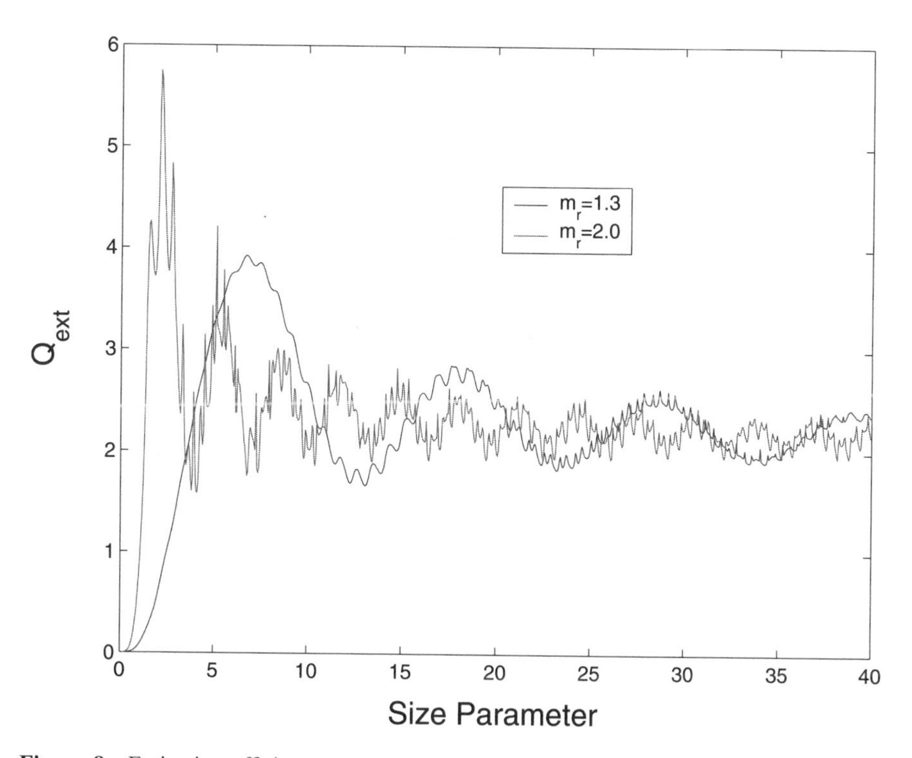

Figure 8 Extinction efficiency, as computed with Mie theory for spherical particles, as a function of size parameter and index of refraction. The value asymptotes to 2 (the "extinction paradox") for large particles as both refraction and diffraction act. See ftp site for color image.

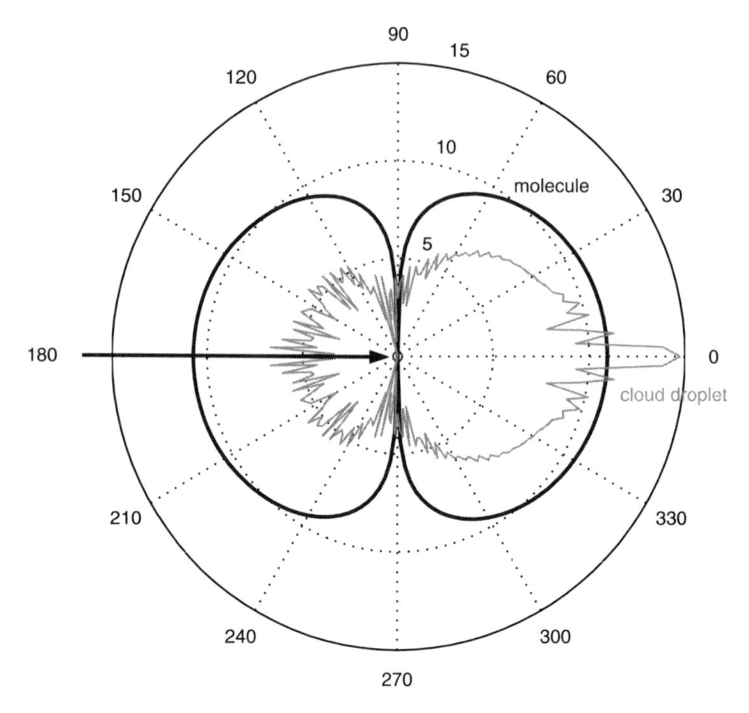

Figure 9 Scattering phase functions for small particles computed using Rayleigh theory (dark line) and a spherical particle of moderate size parameter. The radial axis is on a logarithmic scale. See ftp site for color image.

Regardless of drop size or details of the drop size distribution, in fact, the value of the asymmetry parameter g is always quite near to 0.86 in water clouds.

Scattering by Arbitrary Particles

Mie theory and Rayleigh theory are the two most common methods of computing scattering from particles in the atmosphere. Neither is applicable to the scattering of visible light by either ice crystals or mineral aerosols. Other techniques have been developed for irregular particles, but these tend to be more complicated and difficult to use.

Very large particles (size parameters greater than about 50) are said to be in the "geometric optics limit," and their single scattering properties can be computed by ray tracing. The particle is oriented in space and a series of infinitely thin rays are assumed to illuminate it. The direction and intensity is computed using the Fresnel relations for reflection and transmission each time the ray encounters an interface, and the ray is absorbed as it travels through the medium; the total radiation field is the sum of all the reflected and transmitted components for all the rays. Diffraction is computed separately.

Irregularly shaped particles of intermediate size are the biggest challenge. If the phase function is not required, we can use anomalous diffraction theory (ADT) to find the extinction efficiency and single scattering albedo. ADT applies to large particles with an index of refraction near one. It makes the simplifying assumption that extinction is due primarily to interference between waves slowed by their passage through the medium, and those diffracted around the particle edge. Though it ignores internal reflection and refraction, it is analytically tractable and reasonably accurate.

Relatively small particles can be treated using the discrete dipole approximation (DDA), which breaks the particle up in small volumes relative to the wavelength of radiation, treats each volume as a dipole, and computes the interaction among each dipole pair. This gets very computationally expensive as the particle size increases, since the number of dipoles increases as the volume cubed and the number of interactions as the number of dipoles squared; the current range of applicability is to size parameters less than 5 to 10.

An alternative to DDA is the finite-difference time-domain (FDTD) technique. Here the particle is discretized in space; then electric and magnetic fields that vary in time and space are imposed and the solution to Maxwell's equations is integrated forward in time. The incident wave can be monochromatic, but is more commonly a

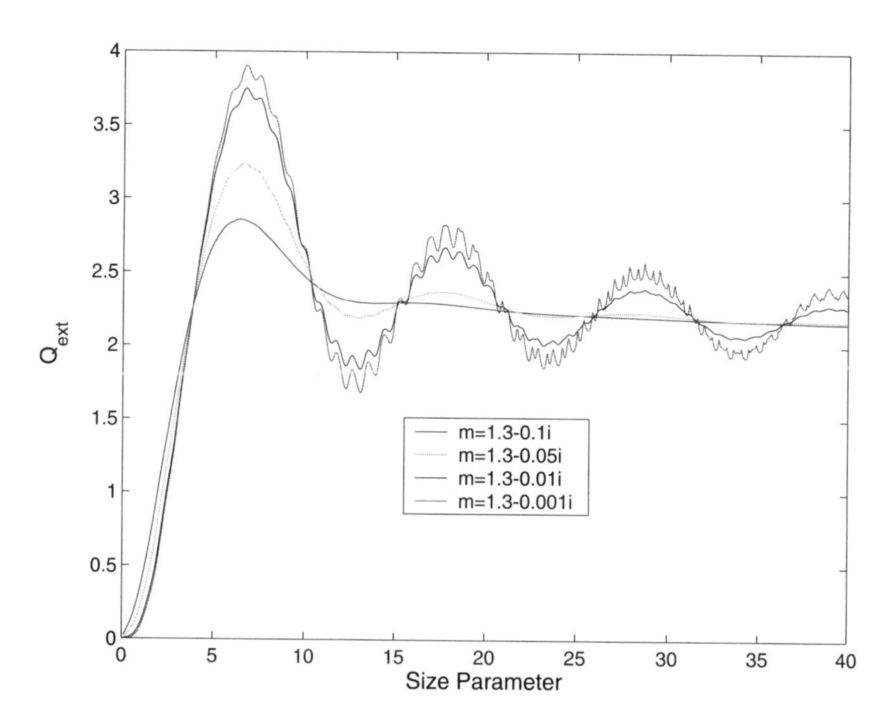

Figure 10 Extinction efficiencies of spherical particles for absorbing particles. In more absorptive materials the variation of Q_{ext} with size parameter is damped. See ftp site for color image.

pulse, since the latter can provide information about many frequencies at once through a Fourier analysis. FDTD can be used for size parameters less than 10 to 20.

Integrating over a Particle Size Distribution

Individual photons are scattered by individual particles, but in treating a beam of radiation we have to consider the many different particles encountered along a differential path $d\tau$, so that the phase function, single scattering albedo, and extinction coefficient represent the entire distribution of drops encountered by the beam. Because cloud drops and aerosols are separated by distances much greater than their characteristic size, we can treat each interaction as independent, and simply add up the contributions from different kinds of particles according to their relative abundance. We have to account for all classes of particles that scatter light uniquely and must weight the effects according to their contributions to the change in intensity.

Let us make this concrete by asking ourselves how light is scattered within warm clouds. The drops within a cloud are all round and have the same index of refraction, but vary in radius r according to the droplet size distribution $n(r)$, which has dimensions of number per volume per radius increment. If we add up all the drops in the distribution, we find the total droplet number concentration

$$\int_0^\infty n(r)\, dr = N \tag{54}$$

where N is measured in number per volume. Drop size distributions within clouds are often represented with the two-parameter gamma distribution:

$$n(r) = \frac{N}{\Gamma(\alpha) r_n} \left(\frac{r}{r_n}\right)^{\alpha-1} \exp\left(-\frac{r}{r_n}\right) \tag{55}$$

in which $\Gamma(\alpha)$ is the Euler gamma function, r_n a characteristic radius, and α is related to the variance of the distribution. The gamma distribution is useful because moments of the drop size distribution can be found analytically.

The scattering properties of the drops depend on the drop radius. Each drop has an extinction cross section $C_{\text{ext}}(r)$, in units of area; if we add up the contributions from each drop size, we obtain the volume extinction coefficient k_{ext} in units of area per volume, or inverse length

$$k_{\text{ext}} = \int_0^\infty C_{\text{ext}}(r) n(r)\, dr = \int_0^\infty \pi r^2 Q_{\text{ext}}(r) n(r)\, dr \tag{56}$$

To compute the average single scattering albedo $\langle \omega_0 \rangle$ we divide the average amount of scattering by the average amount of extinction:

$$\langle \omega_0 \rangle = \frac{\int_0^\infty C_{\text{sca}}(r) n(r)\, dr}{\int_0^\infty C_{\text{ext}}(r) n(r)\, dr} \tag{57}$$

while the average phase function is weighted according to the amount of light scattered by each drop size:

$$\langle P(\Theta)\rangle = \frac{\int_0^\infty P(\Theta)C_{\text{sca}}(r)n(r)\,dr}{\int_0^\infty C_{\text{sca}}(r)n(r)\,dr} \tag{58}$$

Single scattering parameters are not particularly sensitive to the exact drop size distribution, but are controlled primarily by the total surface area of drops within the cloud. The radius associated with the average surface area of the drops is called the *effective radius* r_e:

$$r_e = \frac{\int_0^\infty r^3 n(r)\,dr}{\int_0^\infty r^2 n(r)\,dr} \tag{59}$$

In most water clouds r_e is of order 10 μm. The phase function, single scattering albedo, and extinction coefficient vary smoothly with effective radius, in part because integration over the size distribution smoothes out rapid oscillations like those in Figure 8.

The drops within water clouds vary only in their radius, while a collection of aerosols or ice crystals might contain particles with differing shapes or indices of refraction. If this is the case the sums (integrals) in (56) though (58) must be extended to account for every combination of shape, size, and material, but the idea remains unchanged.

Relating Cloud Optical and Physical Parameters

Imagine a cloud of thickness z made up of drops following a size distribution $n(r)$. We can compute the optical thickness of this cloud by integrating the extinction coefficient, given by (56), to find

$$\tau = \int_0^z k_{\text{ext}}\,dz = \int_0^z \int_0^\infty \pi r^2 Q_{\text{ext}}(r)n(r)\,dr\,dz \approx \int_0^z \int_0^\infty 2\pi r^2 n(r)\,dr\,dz \tag{60}$$

where we can make the last approximation in the visible, where cloud drops have large size parameters.

How can this be related to the cloud physical properties? The liquid water content (LWC) of this cloud is the sum of the water mass in each drop:

$$\text{LWC} = \int_0^\infty \rho_w \frac{4}{3}\pi r^3 n(r)\,dr \tag{61}$$

and the liquid water path (LWP) is the liquid water content integrated through the depth of the cloud:

$$\text{LWP} = \int_0^z \text{LWC}\,dz = \int_0^z \int_0^\infty \rho_w \frac{4}{3}\pi r^3 n(r)\,dr\,dz \tag{62}$$

We multiply (62) by $\frac{3}{2}\rho_w$ and use the definition of effective radius r_e from (59):

$$\frac{3}{2\rho_w}\frac{\text{LWP}}{r_e} = \int_0^z \int_0^\infty 2\pi r^3 n(r)\,dr\,dz \frac{\int_0^\infty r^2 n(r)\,dr}{\int_0^\infty r^3 n(r)\,dr} \approx \tau \tag{63}$$

The last relationship is exact if $n(r)$ is constant with height; the approximation holds to good accuracy for most distributions and is very useful.

7 SIMPLIFYING THE RADIATIVE TRANSFER EQUATION

The full radiative transfer equation is usually applied to solar radiation, for which we know the boundary condition at the top of the atmosphere, namely that the atmosphere is illuminated by the sun from a particular direction denoted μ_0,φ_0:

$$I(\tau = 0, \mu < 0, \varphi) = F_s \delta(\mu_0 - \mu)\delta(\varphi_0 - \varphi) \tag{64}$$

Here F_s is the solar flux at the top of the atmosphere and $\delta(x)$ is the Kroenecker delta function, with is infinite when its argument is zero and zero otherwise, but integrates to one.

If sunlight were not scattered, it would obey Beer's law and intensity would decrease exponentially with optical depth. This implies that energy traveling in any direction other than μ_0, φ_0 has been scattered at least once. We will therefore decompose the total intensity field into direct and diffuse components, depending on whether the photons have or have not been scattered at least once:

$$I(\tau, \mu, \varphi) = I_{\text{dir}}(\tau, \mu, \varphi) + I_{\text{dif}}(\tau, \mu, \varphi) \tag{65}$$

The boundary condition in (64) applies to the direct intensity, which is depleted by any scattering and any absorption, and so follows Beer's law. There is no incoming diffuse intensity at the top of the atmosphere, but the diffuse intensity can be increased by scattering from either itself or from the direct beam. The equation describing diffuse intensity is therefore:

$$\begin{aligned}
\mu\frac{dI_{\text{dif}}(\mu, \varphi)}{d\tau} &= I_{\text{dif}}(\mu, \varphi) - (1 - \omega_0)B - \frac{\omega_0}{4\pi}\int_0^{2\pi}\int_{-1}^{1} P(\mu', \varphi' \to \mu, \varphi) I_{\text{dif}}(\mu', \varphi')\,d\mu'\,d\varphi' \\
&\quad - \frac{\omega_0}{4\pi}\int_0^{2\pi}\int_{-1}^{1} P(\mu', \varphi' \to \mu, \varphi) I_{\text{dir}}(\mu', \varphi')\,d\mu'\,d\varphi' \\
&= I_{\text{dif}}(\mu, \varphi) - (1 - \omega_0)B - \frac{\omega_0}{4\pi}\int_0^{2\pi}\int_{-1}^{1} P(\mu', \varphi' \to \mu, \varphi) I_{\text{dif}}(\mu', \varphi')\,d\mu'\,d\varphi' \\
&\quad - \frac{\omega_0}{4\pi} P(\mu_0, \varphi_0 \to \mu, \varphi) F_s \exp\left(-\frac{\tau}{\mu_0}\right)
\end{aligned} \tag{66}$$

The last term in (66) is called the single scattering source term, and the penultimate contribution is called the multiple scattering source term, which is the redistribution of diffuse intensity from one direction into another.

In many applications the full details of the intensity field are more information than we need. One of the easiest ways we can simplify the radiative transfer equation is to average it over azimuth φ. We will begin by defining the azimuthally averaged phase function P_0:

$$P_0(\mu' \rightarrow \mu) = \frac{1}{2\pi}\int_0^{2\pi} P(\mu', \varphi' \rightarrow \mu, \varphi)\, d\varphi \tag{67}$$

and azimuthally average intensity I_0:

$$I_0(\mu) = \frac{1}{2\pi}\int_0^{2\pi} I(\mu, \varphi)\, d\varphi \tag{68}$$

We then average (66) by integrating both sides over 2π radians in φ and dividing the entire equation by 2π. Since the blackbody source term is isotropic, this results in

$$\begin{aligned}\mu\frac{dI_0}{d\tau} = I_0 - (1-\omega_0)B - \frac{1}{2\pi}\int_0^{2\pi}\frac{\omega_0}{4\pi}\int_0^{2\pi}\int_{-1}^{1} P(\mu', \varphi' \rightarrow \mu, \varphi) I(\mu', \varphi')\, d\mu'\, d\varphi'\, d\varphi \\ - \frac{\omega_0}{4\pi} P(\mu_0, \varphi_0 \rightarrow \mu, \varphi) F_s \exp\left(-\frac{\tau}{\mu_0}\right)\end{aligned} \tag{69}$$

We can simplify the second to last term by recalling that the phase function is reciprocal:

$$\begin{aligned}&\frac{1}{2\pi}\int_0^{2\pi}\frac{\omega_0}{4\pi}\int_0^{2\pi}\int_{-1}^{1} P(\mu', \varphi' \rightarrow \mu, \varphi) I(\mu', \varphi')\, d\mu'\, d\varphi'\, d\varphi \\ &\quad = \frac{\omega_0}{2}\frac{1}{2\pi}\int_0^{2\pi}\int_{-1}^{1} P_0(\mu' \rightarrow \mu) I(\mu', \varphi')\, d\mu'\, d\varphi' \\ &\quad = \frac{\omega_0}{2}\int_{-1}^{1} P_0(\mu' \rightarrow \mu) I_0(\mu')\, d\mu'\end{aligned} \tag{70}$$

which yields the azimuthally averaged radiative transfer equation, from which we have dropped the subscript:

$$\mu\frac{dI}{d\tau} = I - (1-\omega_0)B - \frac{\omega_0}{2}\int_{-1}^{1} P(\mu' \rightarrow \mu) I(\mu')\, d\mu' - \frac{\omega_0}{4\pi} P(\mu_0 \rightarrow \mu) F_s \exp\left(\frac{-\tau}{\mu_0}\right) \tag{71}$$

In the remainder of this chapter we will focus on solving (71) rather than any of the more involved and complete versions of the radiative transfer equation. The choice is a simplification, but it is a very useful one: Most methods for solving the azimuthally dependent version of the radiative transfer equation use a Fourier expansion in azimuth, and (71) is the lowest order moment in such a treatment.

Delta Scaling

When discussing single scattering, we found that diffraction from particle edges has as big an impact on extinction as refraction within the particle itself, but that this light is diverted only slightly from its initial direction. Phase functions as strongly peaked as those in Figure 9 cause numerical headaches in solving the radiative transfer equation. And in fact, the width of the diffraction peak for cloud particles is so narrow that the light may as well not be scattered at all.

In most problems in solar radiative transfer we replace the original phase function, which contains a large, narrow forward peak, with two components: A delta function in the forward direction and a smoother, scaled phase function. We define f as the fraction of light scattered directly forward; then we approximate

$$P(\cos\Theta) \approx 2f\partial(1-\cos\Theta) + (1-f)P'(\cos\Theta) \tag{72}$$

We choose the asymmetry parameter of the scaled phase function so that the asymmetry parameter of the original phase function is unchanged:

$$g = \frac{1}{2}\int_{-1}^{1} P(\cos\Theta)\, d\cos\Theta = f + (1-f)g' \tag{73}$$

The value of f can be chosen in a variety of ways, depending on how much information is available. One approach is to define $f = g$ so that $g' = 0$; the more common delta-Eddington approximation sets $f = g^2$.

We could apply this scaling to any form of the radiative transfer equation that includes scattering. For the purposes of illustration we will substitute (72) into the azimuthally averaged radiative transfer equation (71), omitting the source terms for clarity:

$$\mu\frac{dI}{d\tau} = I - f\omega_0 I - \frac{(1-f)\omega_0}{2}\int_{-1}^{1} P'(\mu' \to \mu)I(\mu')\, d\mu' \tag{74}$$

Dividing both sides of this equation by $(1-\omega_0 f)$ yields

$$\mu\frac{dI}{(1-\omega_0 f)\, d\tau} = I - \frac{1-f}{1-\omega_0 f}\frac{\omega_0}{2}\int_{-1}^{1} P'(\mu' \to \mu)I(\mu')\, d\mu' \tag{75}$$

This is exactly the same form as the original equation if we scale variables:

$$\tau' = (1-\omega_0 f)\tau \qquad \omega_0' = \frac{(1-f)\omega_0}{1-\omega_0 f} \qquad g' = \frac{g-f}{1-f} \tag{76}$$

so that any technique we have for solving the original radiative transfer equation works on the scaled version as well. In the scaled system both optical thickness and the asymmetry parameter are reduced, so less forward scattering combines with less extinction to produce the same reflection, transmission, and absorption as the unscaled system. The systems agree, of course, when the direct and diffuse beams are added together. In practice, almost all calculations of radiative transfer in the solar system are made using scaled versions of the radiative transfer equation.

8 SOLVING THE RADIATIVE TRANSFER EQUATION SIMPLY

How can we approach the solution of even the azimuthally averaged radiative transfer equation? Intensity may vary with both optical depth and polar angle μ, and ignoring the vertical variation would mean we could not compute even such simple quantities as radiative heating rates. We can, however, try to find the intensity at only a few angles. In fact, computing the intensity field at just two angles, one each in the upward and downward hemispheres, is a lot like computing upward and downward fluxes defined in (5).

Two-stream methods are those that describe the radiation field with just two numbers. They have the advantage of being analytically soluble, which makes them very fast and thus suitable for, say, use in a numerical climate model. They are generally good at computing fluxes and therefore useful in heating rate calculations, but they cannot be used when the angular details of the intensity field are important. There are more than a few two-stream methods, and in every one simplifications need to be made about both the intensity and the phase function.

The following examples illustrate the computation of fluxes in the visible part of the spectrum, where there is no emission, but two-stream models are also applicable to calculations in absorbing and emitting atmosphere.

Eddington's Solution

In the Eddington approximation we expand both intensity and phase function to first order in polar angle. That is, we assume that each varies linearly with μ:

$$I(\mu) = I_0 + I_1\mu, \qquad P(\mu \to \mu') = 1 + 3g\mu\mu' \tag{77}$$

This means we can compute the upward and downward fluxes analytically:

$$F^+ = \pi(I_0 - 2I_1/3) \qquad F^- = \pi(I_0 + 2I_1/3) \tag{78}$$

To find the intensity we substitute (77) into (71):

$$\begin{aligned} \mu\frac{d(I_0 + I_1\mu)}{d\tau} &= I_0 + I_1\mu - \frac{\omega_0}{2}\int_{-1}^{1}(1 + 3g\mu\mu')(I_0 + I_1\mu')\,d\mu' \\ &\quad - \frac{\omega_0}{4\pi}(1 - 3g\mu\mu_0)F_s\exp\left(\frac{-\tau}{\mu_0}\right) \end{aligned} \tag{79}$$

Now we can evaluate the scattering integral

$$\mu \frac{d(I_0 + I_1\mu)}{d\tau} = I_0 + I_1\mu - \omega_0(I_0 + I_1 g\mu) - \frac{\omega_0}{4\pi}(1 - 3g\mu\mu_0)F_s \exp\left(\frac{-\tau}{\mu_0}\right) \tag{80}$$

and rearrange terms

$$\mu \frac{dI_0}{d\tau} + \mu^2 \frac{dI_1}{d\tau} = I_0(1 - \omega_0) + I_1\mu(1 - \omega_0 g) - \frac{\omega_0}{4\pi}(1 - 3g\mu\mu_0)F_s \exp\left(\frac{-\tau}{\mu_0}\right) \tag{81}$$

We can break (81) into two pieces by observing that it contains both odd and even powers of μ. What we will do, then, is integrate the equation over μ from -1 to 1, which will leave only terms in even powers of μ. We then multiply (81) by μ and repeat the integration. The two resulting equations are

$$\frac{dI_1}{d\tau} = 3(1 - \omega_0)I_0 - \frac{3\omega_0}{4\pi}F_s \exp\left(\frac{-\tau}{\mu_0}\right) \tag{82}$$

$$\frac{dI_0}{d\tau} = (1 - \omega_0 g)I_1 + \frac{3\omega_0}{4\pi}g\mu_0 F_s \exp\left(\frac{-\tau}{\mu_0}\right) \tag{83}$$

Equations (82) and (83) are a set of two first-order coupled linear differential equations. We can uncouple the equations and find a solution by differentiating one equation with respect to optical depth and substituting the other. This yields, for example

$$\frac{d^2 I_0}{d\tau^2} = k^2 I_0 - \frac{3\omega_0}{4\pi}F_s \exp\left(\frac{-\tau}{\mu_0}\right)(1 + g - \omega_0 g) \tag{84}$$

where we have defined the eigenvalue $k^2 = 3(1-\omega_0)\,(1-\omega_0 g)$. The solutions to (84) and the analogous equation for I_1 are a sum of exponentials in $k\tau$, i.e.,

$$I_0 = Ae^{k\tau} + Be^{-k\tau} + \psi e^{-\tau/\mu_0} \tag{85}$$

where A, B, and ψ are determined from the boundary conditions at the top and bottom of the medium and from the particular solution.

The general solution is very complicated but is tractable in some limits. The simplest case is a single homogeneous layer of total optical depth τ^* over a nonreflective surface. If the layer is optically thin ($\tau^* \ll 1$), the reflected and transmitted fluxes are

$$R = \omega_0\left(\frac{1}{2} - 3g\mu_0/4\right)\frac{\tau^*}{\mu_0} \qquad T = 1 - R - \left(\frac{\tau^*}{\mu_0}\right)(1 - \omega_0) \tag{86}$$

If the layer does not absorb (i.e., if $\omega_0 = 0$), the reflected flux is

$$R = \frac{(1-g)\tau^* + \left(\frac{2}{3} - \mu_0\right)(1 - e^{-\tau^*/\mu_0})}{\frac{4}{3} + (1-g)\tau^*} \tag{87}$$

Reflectance increases with optical thickness, rapidly at first and then more slowly as shown in Figure 11. Reflectance at a given optical depth also increases as μ_0 decreases (this is not shown), since the radiation must change direction less drastically to be reflected into the upward hemisphere. Decreases in the asymmetry parameter increase the amount of reflection, since photons are more likely to be scattered backwards, and since most photons, especially near cloud top, are headed downward. Even relatively optically thick clouds transmit at least 20% of the flux incident on them, which is why it is not pitch dark at the surface even on very cloudy days.

Computing Flux: Two-Stream Model

Eddington's approach to the radiative transfer equation is to expand both the intensity and the phase function to first order in angle; the flux can be computed once the approximate intensity field is known. In the two-stream model we first average the radiative transfer equation and the phase function to get a differential equation for fluxes; then we compute the solution. The solutions are quite similar, as we expect them to be, and the choice of exactly which method to use is a little arbitrary.

Let us begin with the azimuthally averaged radiative transfer equation (71) and integrate it over each hemisphere to find the flux. Ignoring emission, we have for the downward flux

$$\begin{aligned}\int_{-1}^{0} \mu \frac{dI}{d\tau}\, d\mu = \int_{-1}^{0} I\, d\mu - \frac{\omega_0}{2}\int_{-1}^{0}\int_{-1}^{1} P(\mu' \to \mu) I(\mu')\, d\mu'\, d\mu \\ - \frac{\omega_0}{4\pi}\int_{0}^{1} P(\mu_0 \to \mu) F_s \exp(-\tau/\mu_0)\, d\mu\end{aligned} \tag{88}$$

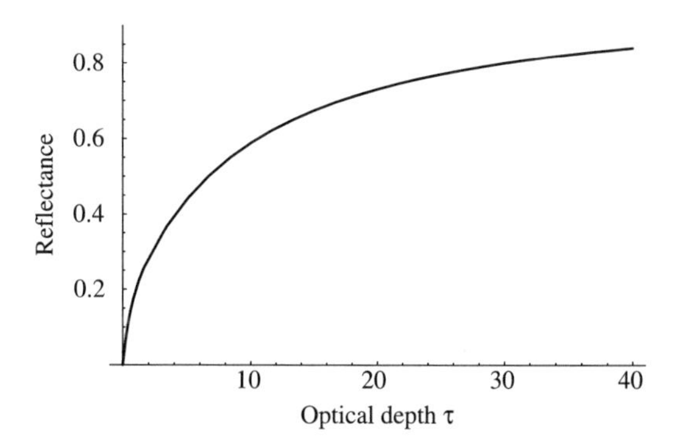

Figure 11 Reflectance of a nonabsorbing layer as computed from the Eddington approximation. The solar zenith angle is 60°.

or

$$\begin{aligned}\bar{\mu}\frac{dF^{\downarrow}}{d\tau} = F^{\downarrow} &- \frac{\omega_0}{2}\int_{-1}^{0}\int_{-1}^{0} P(\mu' \to \mu) I^{\downarrow}(\mu')\, d\mu'\, d\mu \\ &- \frac{\omega_0}{2}\int_{-1}^{0}\int_{0}^{1} P(\mu' \to \mu) I^{\uparrow}(\mu')\, d\mu'\, d\mu \\ &- \frac{\omega_0}{4\pi}\int_{0}^{1} P(\mu_0 \to \mu) F_s \exp\left(\frac{-\tau}{\mu_0}\right) d\mu\end{aligned} \tag{89}$$

Implicit in going from (88) to (89) is a choice about the ratio:

$$\bar{\mu} = \frac{\int_{-1}^{0} \mu \frac{dI}{d\tau}\, d\mu}{\int_{-1}^{0} \frac{dI}{d\tau}\, d\mu} \tag{90}$$

Different two-stream approximations differ in this choice. We then make use of the reciprocity of the phase function and define the *backscattering coefficients*:

$$b(\mu) = \frac{1}{2}\int_{-1}^{0} P(\mu' \to \mu)\, d\mu' \qquad b = \frac{1}{2}\int_{0}^{1} b(\mu)\, d\mu \tag{91}$$

which define the fraction of flux scattered into the opposite hemisphere. Applying these definitions, and doing the analogous calculation for the upward flux, we arrive at the two-stream equations:

$$\bar{\mu}\frac{dF^{\downarrow}}{d\tau} = F^{\downarrow} - \omega_0(1-b)F^{\downarrow} + \omega_0 b F^{\uparrow} - \frac{\omega_0}{2\pi}[1 - b(\mu_0)]F_s \exp\left(\frac{-\tau}{\mu_0}\right) \tag{92}$$

$$\bar{\mu}\frac{dF^{\uparrow}}{d\tau} = F^{\uparrow} - \omega_0(1-b)F^{\uparrow} + \omega_0 b F^{\downarrow} - \frac{\omega_0}{2\pi} b(\mu_0) F_s \exp\left(\frac{-\tau}{\mu_0}\right) \tag{93}$$

These are two first-order, linear, coupled differential equations with constant coefficients. They can be uncoupled by differentiating one with respect to optical thickness, then substituting the other. As with the Eddington approximation, the complete solution for reflected or transmitted flux is the sum of exponentials in optical thickness, with coefficients determined by the boundary conditions.

9 SOLVING RADIATIVE TRANSFER EQUATION COMPLETELY

Analytic approximations like the Eddington and two-stream methods are useful and often accurate enough for flux calculations. If intensity is required, though, we must find numerical ways to solve the radiative transfer equation. In numerical weather prediction, the continuous Navier–Stokes equations of motion are approximated by finite differences between points on a spatial grid. The integrals in the radiative transfer equation are over direction, so the discretization is over angle, turning the continuous equation into one at discrete ordinates or directions. There are two popular approaches to solving this equation, which of course give the same numerical results.

Adding–Doubling Method

Imagine two layers, one of which overlies the other. The upper layer has flux transmittance and reflectance T_1 and R_1, and the lower layer T_2 and R_2. How much total flux R_T is reflected from the combination of layers? Some flux is reflected from the first layer (R_1); some of the flux transmitted through the first layer is reflected from the second layer and transmitted through the first layer ($T_1R_2T_1$); some of this flux reflected from the second layer is reflected back downwards, where some portion is reflected back up ($T_1R_2R_1R_2T_1$), and so on. Because reflection and transmission are both less than one, we can use the summation formula for geometric series to compute the total reflection:

$$\begin{aligned} R_T &= R_1 + T_1R_2T_1 + T_1R_2R_1R_2T_1 + T_1R_2R_1R_2R_1R_2T_1 + \cdots \\ &= R_1 + T_1R_2[1 + R_1R_2 + R_1R_2R_1R_2 + \cdots]T_1 \\ &= R_1 + T_1R_2[1 + R_1R_2]^{-1}T_1 \end{aligned} \tag{94}$$

where the term in square brackets accounts for the multiple reflections between the layers. There is a similar relation for transmission.

Equation (94) and its analog for transmission may be combined with the transmission and reflection results from the Eddington or two-stream solutions to compute the transmittance and reflectance of layered atmospheres, in which the single scattering albedo or asymmetry parameter changes with optical thickness, or to account for a reflecting surface beneath the atmosphere.

The adding–doubling method extends this idea to intensity by replacing the reflection and transmission terms in (94) with matrices that account for the forward and backward scattering from one polar angle into another. Rather than beginning with analytical results for arbitrary layers, we find R and T for a layer of optical depth $d\tau \ll 1$, assuming that photons are scattered no more than once. The reflection and transmission of a layer of $2d\tau$ can be computed using (94) and the reflection and transmission matrices for the original layer. Repeating this process we can find the reflection and transmission of arbitrarily thick layers. Arbitrarily complicated atmospheres can be built up by superimposing layers with different properties and computing the reflection and transmission matrices for the combination. Reflection from surfaces fits naturally into the adding framework.

Eigenvector Solutions

As an alternative to the adding–doubling method we can approximate the continuous radiative transfer equation with a version that is discrete in polar angle

$$\frac{d}{d\tau}\begin{pmatrix} \mathbf{I}^+ \\ \mathbf{I}^- \end{pmatrix} = \begin{pmatrix} -t & -r \\ r & t \end{pmatrix}\begin{pmatrix} \mathbf{I}^+ \\ \mathbf{I}^- \end{pmatrix} - \begin{pmatrix} \mathbf{S}^+ \\ \mathbf{S}^- \end{pmatrix} \tag{95}$$

where $\mathbf{I}^{\pm}$ are vectors containing the intensity $I(\mu_j)$ at each of N angles (each discrete ordinate). The t and r matrices contain the phase function information:

$$t_{j,j'} = \frac{1}{\mu_j}\left[\frac{\omega_0}{2}P(\mu_j \to \mu_{j'}) - 1\right] \qquad r_{j,j'} = \frac{1}{\mu_j}\frac{\omega_0}{2}P(\mu_j \to -\mu_{j'}) \tag{96}$$

and the source vector is

$$\mathbf{S}^{\pm} = \frac{\pm 1}{\mu_j}\frac{\omega_0}{4\pi}P(\mu_0 \to \pm\mu_j)F_s e^{-\tau/\mu} \tag{97}$$

.

To solve the matrix version (95) of the radiative transfer equation, we treat the homogeneous equation as an eigenvector problem. That is, we set the source term to zero and substitute solutions of the form $\mathbf{I}^{\pm} = \mathbf{G}^{\pm}e^{-k\tau}$. This results in

$$\begin{pmatrix} -t & -r \\ r & t \end{pmatrix}\begin{pmatrix} \mathbf{G}^+ \\ \mathbf{G}^- \end{pmatrix} = k^2\begin{pmatrix} \mathbf{G}^+ \\ \mathbf{G}^- \end{pmatrix} \tag{98}$$

The solution for intensity is then a sum of the eigenvectors and the particular solution for the source term, with weights determined from the boundary conditions. The solution can be extended to more than one layer by ensuring that intensities match at the layer boundaries. This technique is usually called "discrete ordinates" in the scientific literature, though many solutions use discrete forms of the radiative transfer equation.

Radiative Transfer in Two and Three Dimensions

In our development of the radiative transfer equation we assumed that the atmosphere varies only in the vertical, which leads to a radiation field that depends only on direction and vertical position. But the real atmosphere varies in the horizontal as well as the vertical. In particular, cloud optical properties can change dramatically over relatively short distances. A full description of a radiation field depends on horizontal position as well as vertical position and direction, and in the three-dimensional radiative transfer equation the derivative becomes a gradient operator to account for transfers in the horizontal as well as the vertical.

Traditionally, radiative transfer in two and three dimensions has been computed using Monte Carlo integration techniques. The domain is divided into discrete volumes, within which the optical properties (extinction coefficient, single scattering albedo, and phase function) are considered constant. The paths of many, many photons are then traced from their source until they leave the domain. In solar radiation calculations, for example, the photons are typically introduced at the top of the domain, then travel a random distance depending on the amount of extinction they encounter along the trajectory. They then undergo a scattering and/or absorption event, and continue in a new direction related to the scattering phase function; this process continues until they exit the domain. Monte Carlo methods are best suited to the computation of domain-averaged quantities, but are very flexible. An alternative is to estimate the radiation field at each grid point, then iterate until the field is self-consistent.

Three-dimensional radiative transfer is much more computationally expensive than one-dimensional calculations, and it is difficult to measure atmospheric properties well enough to provide useful input fields. Three-dimensional effects can be significant, however, especially in remote-sensing applications.

10 FROM THEORY TO APPLICATIONS

In this chapter we have laid out the foundations of radiative transfer. We have discussed the physical mechanisms that underlie absorption and scattering and talked about ways to compute the quantities needed for remote-sensing and

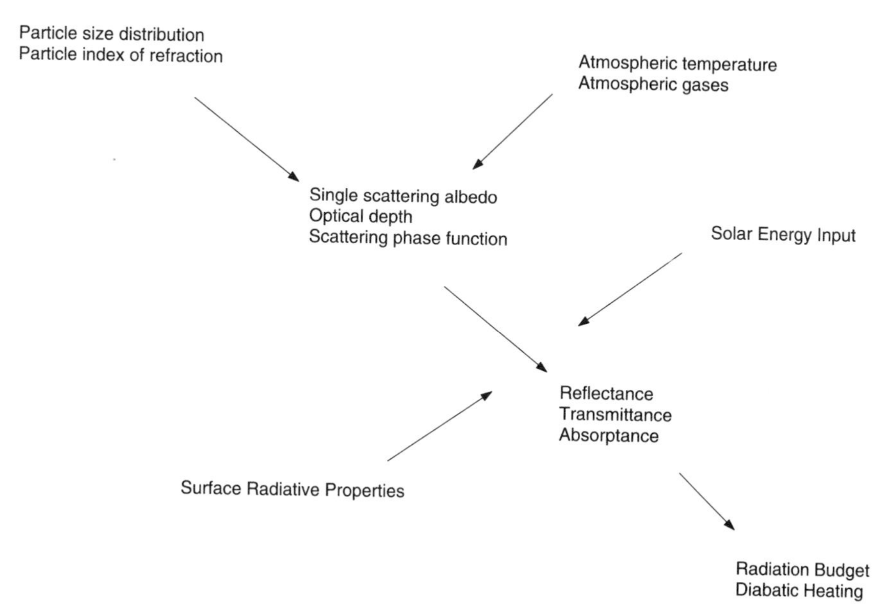

Figure 12 Topics in radiative transfer and remote sensing. This chapter has covered the upper half of the diagram, as well as the computation of reflectance, transmittance, and absorptance; the other topics are discussed in the next chapter.

energy budget applications. The relationship between the various topics, and those to be covered in the following chapter, is shown in Figure 12.

In day-to-day practice much of this work is routine. There are codes available to compute the absorption spectrum of an arbitrary atmosphere, to compute the single scattering parameters of individual particles, and any number of models available for solving the radiative transfer equation, each of which is tailored to a specific application. An understanding of the underlying theory is essential for a critical assessment, but most current research in radiative transfer is in the applications, that we discuss in the next chapter.

CHAPTER 20

RADIATION IN THE ATMOSPHERE: OBSERVATIONS AND APPLICATIONS

STEVEN A. ACKERMANN AND ROBERT PINCUS

1 OVERVIEW

The previous chapter discussed the physical mechanisms that underlie absorption and scattering and presented approaches to compute the quantities needed for remote sensing and energy budget applications. In this chapter we discuss observations of radiative fluxes and apply the equations to remote sensing the atmosphere. The boundary conditions at the top and bottom of the atmosphere are required to compute the transfer of radiation through the atmosphere. Thus, the chapter begins by describing the boundary conditions at the top of atmosphere and at the surface. These conditions are then used to calculate the radiative heating of the atmosphere in the next section. After discussing radiative heating profiles, observations of the energy budget at the top of atmosphere are presented, followed by a discussion of the greenhouse effect.

Satellite observations are routinely used to help answer the question "What is the weather like?" In addition, satellite observations are routinely used to determine the state of Earth's surface as well as cloud properties. Section 6 provides examples of satellite remote sensing the surface and atmospheric conditions.

2 BOUNDARY CONDITIONS

The boundary conditions at the top and bottom of the atmosphere include the properties of the surface and the incoming solar radiation at the top of the

Handbook of Weather, Climate, and Water: Dynamics, Climate, Physical Meteorology, Weather Systems, and Measurements, Edited by Thomas D. Potter and Bradley R. Colman.
ISBN 0-471-21490-6

atmosphere. To determine the amount of incoming solar radiation, we must consider the sun–Earth geometric relationships.

The Sun and Its Relationship to Earth

Earth receives energy from the sun. This energy drives atmospheric and oceanic circulations. In this section we briefly discuss some of the properties of the sun before presenting methods of computing sun–Earth astronomical relationships.

The sun, which primarily consists of hydrogen and helium, is 4.6 billion years old and is approximately 1.5×10^8 km from Earth. The radius of the sun is approximately 700,000 km with a mass of 2×10^{35} g. The temperature of the sun is about 5×10^6 K at center decreasing to about 5800 K at the surface.

As with the Earth's atmosphere, the sun can be categorized into different layers, including the photosphere, corona, and the chromosphere. The photosphere is the visible region, or the *surface* of the sun with a thickness of approximately 500 km within which the temperature decreases from 8000 to 4000 K. The photosphere is often marked by features called sunspots, which are associated with convection of the sun's hot gases. Sunspots do not occur in the polar regions of the sun. The sunspot minima and maxima occur in a cycle that has a period of approximately 11 years. During sunspot maxima the sun is disturbed with particle outbursts and solar flares. During sunspot minima the sun is quiet or less active. Pairs of sunspots often have opposite magnetic polarities. For a given sunspot cycle, the polarity of the leading spot is always the same for a given hemisphere. With each new sunspot cycle the polarities reverse; thus, the sunspot cycle is often thought of as a 22-year period. Sunspots appear as dark spots on the surface because these areas are at lower temperatures than the surrounding surface; however, satellite measurements show that more radiation is emitted during a sunspot maximum than a sunspot minimum.

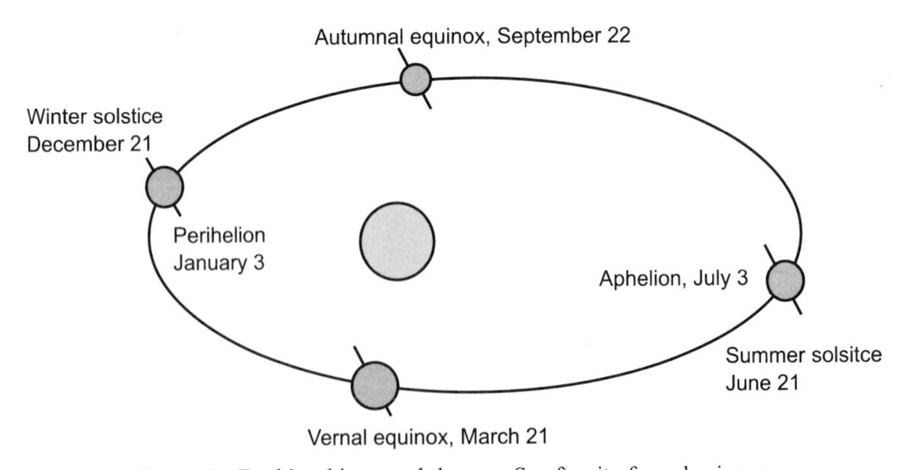

Figure 1 Earth's orbit around the sun. See ftp site for color image.

This may result from increased emission by the faculae, which are bright areas around the sunspots.

The chromosphere extends up to 5000 km above the photosphere. The temperature ranges from 4000 to 6000 K near the photosphere and increases to 10^6 K at 5000 km. The corona, which can be thought of as the sun's atmosphere, extends from the solar disk outward many millions of kilometers. The corona is visible during total solar eclipses. The stream of gas that flows out of the corona and into the solar system is called the solar wind, which provides the energy to produce Earth's aurora borealis and aurora australis.

Sun–Earth Astronomical Relationships

To calculate the solar flux distribution within the atmosphere and ocean, we need to know where the sun is with respect to the geographical region of interest. Such relationships are depicted in Figures 1 through 3 and formulated below.

The mean sun–Earth distance, r_0, is defined to be one astronomical unit (1 AU). The minimum sun–Earth distance (perihelion) is about 0.983 AU and occurs on approximately January 3. The maximum distance (aphelion) is 1.1017 AU and occurs on approximately July 3. The sun–Earth distance, r, is known accurately for every day of the year and is published in the *American Ephemeris and Nautical Almanac*; however, for applications in atmospheric radiative transfer, mathematical approximations are simpler to use. What is required is the reciprocal of the square of

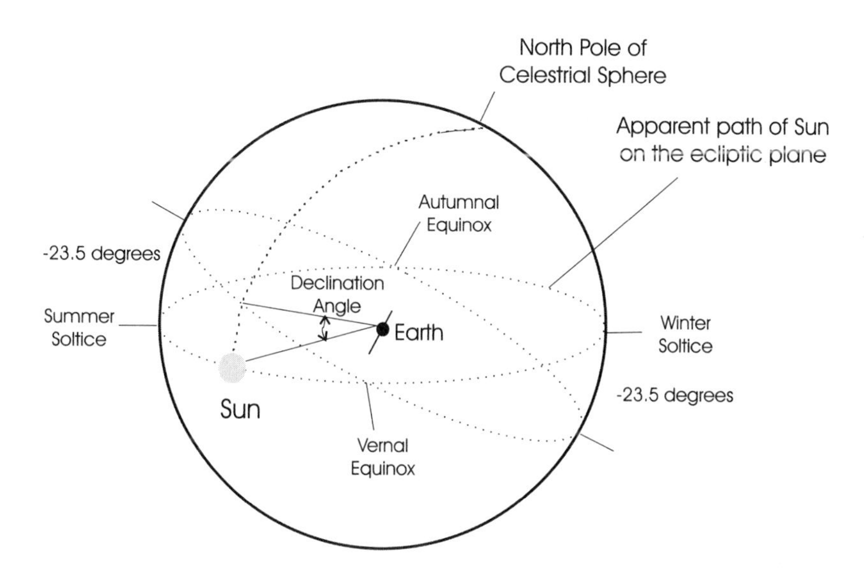

Figure 2 Definition of Sun-Earth astronomical relations. See ftp site for color image.

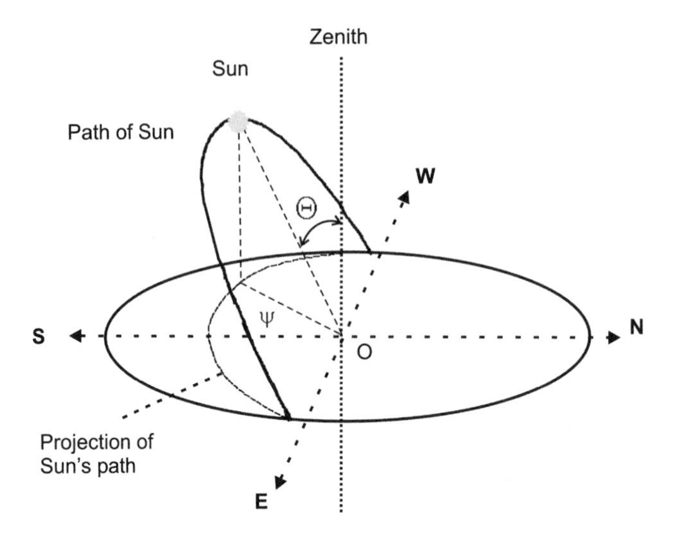

Figure 3 Definition of Earth's azimuth and zenith angles. See ftp site for color image.

the radius vector of Earth, or $(r_0/r)^2$, which is an eccentricity correction factor of Earth's orbit. A common approximation is

$$\left(\frac{r_0}{r}\right)^2 = 1.000110 + 0.034221\cos\Gamma + 0.001280\sin\Gamma + 0.000719\cos 2\Gamma + 0.000077\sin 2\Gamma \tag{1}$$

where $\Gamma = 2\pi(d_n - 1)/365$ and d_n is the day of the year, which ranges from 1 on January 1 to 365 (366 on a leap year) on December 31.

To determine the solar flux within Earth's atmosphere requires defining some fundamental parameters that describe the position of the sun relative to a location on Earth. The ecliptic plane is the plane of revolution of Earth around the sun. Earth rotates around its polar axis, which is inclined at approximately 23.5° from the normal to the ecliptic plane. The angle between the polar axis and the normal to the ecliptic plane remains unchanged as does the angle between Earth's celestial equator plane and the ecliptic plane (Fig. 1). The angle between the line joining the centers of the sun and Earth to the equatorial plane changes and is known as the solar declination, δ. It is zero at the vernal and autumnal equinoxes and approximately 23.5° at the summer solstice and −23.5° at the winter solstice. A useful approximate equation to express the declination in degrees is

$$\delta = 0.006918 - 0.399912\cos\Gamma + 0.070257\sin\Gamma - 0.006758\cos 2\Gamma + 0.000907\sin 2\Gamma \tag{2}$$

The solar zenith angle θ_0 is the angle between the local zenith and the line joining the observer and the sun. The solar zenith angle ranges between 0° and 90° and is determined from

$$\cos\theta_0 = \sin\delta\sin\phi + \cos\delta\cos\phi\cos\omega \tag{3}$$

where ϕ is the geographic latitude. The hour angle ω is measured at the celestial pole between the observer's longitude and the solar longitude. The hour angle is zero at noon and positive in the morning, changing 15° per hour.

The solar azimuth ψ is the angle at the local zenith between the plane of the observer's meridian and the plane of a great circle passing through the zenith and the sun. It is zero at the south and measured positive to the east and varies between $\pm 180°$ and is calculated from

$$\cos\Psi = \frac{(\cos\theta_0\sin\phi - \sin\delta)}{\sin^{-1}[\cos\theta_0]\cos\phi} \tag{4}$$

and $0° \leq \Psi \leq 90°$ when $\cos\Psi \geq 0$ and $90° \leq \Psi \leq 180°$ when $\cos\Psi \leq 0$.

Incoming Solar Radiation at the Top of the Atmosphere

Earth's rotation causes daily changes in the incoming solar radiation while the position of its axis relative to the sun causes the seasonal changes.

The Solar Constant

The flux at the top of Earth's atmosphere on a horizontal surface is

$$F = S_0\left(\frac{r_0}{r}\right)^2\cos\theta_0 \tag{5}$$

where S_0 is the solar constant, the rate of total solar energy at all wavelengths incident on a unit area exposed normally to rays of the sun at 1 AU.

The solar constant can be measured from satellites or derived from the conservation of energy. The power emitted by the sun is approximately 3.9×10^{26} W. The radius of the sun is approximately 7×10^8 m, so the flux at the surface of the sun is $F_{\text{sun}} \approx 6.3 \times 10^7\ \text{W/m}^{-2}$. What is the irradiance reaching Earth (S_0)? Assuming the power from the sun is constant, by conservation of energy the total power crossing a sphere at a radius equivalent to the Earth–sun distance must be equal to the power emitted at the sun's surface, so

$$S_0(4\pi R_{\text{es}}^2) = F_{\text{sun}}(4\pi R_s^2) \tag{6}$$

$$S_0 = F_{\text{sun}}\left[\frac{R_s^2}{R_{\text{es}}^2}\right] \approx 1368\ \text{W/m}^2 \tag{7}$$

The amount of solar energy reaching Earth is not constant, but a function of distance from the sun, and therefore time of year, as indicated by Eq. (5).

The spectral distribution of the solar flux at the top of the atmosphere is shown in Figure 4. While the sun is often considered to have a radiative temperature of approximately 5777 K, Figure 4 demonstrates that the sun is not a perfect blackbody. Also shown in Figure 4 is the percentage of solar energy below a given wavelength (dotted line) confirming that most of the sun's energy resides at wavelengths less than 4 μm.

The daily radiation on a horizontal surface in joules per square meter per day ($\mathrm{J/m^2}$ day) is

$$F_{\text{day}} = \int_{\text{sunrise}}^{\text{sunset}} F\,dt = 2\int_0^{\omega_s} F\,d\omega \frac{24}{2\pi} \tag{8}$$

After converting dt to hour angle,

$$\frac{d\omega}{dt} = \frac{2\pi \text{ radians}}{24 \text{ hour}}$$

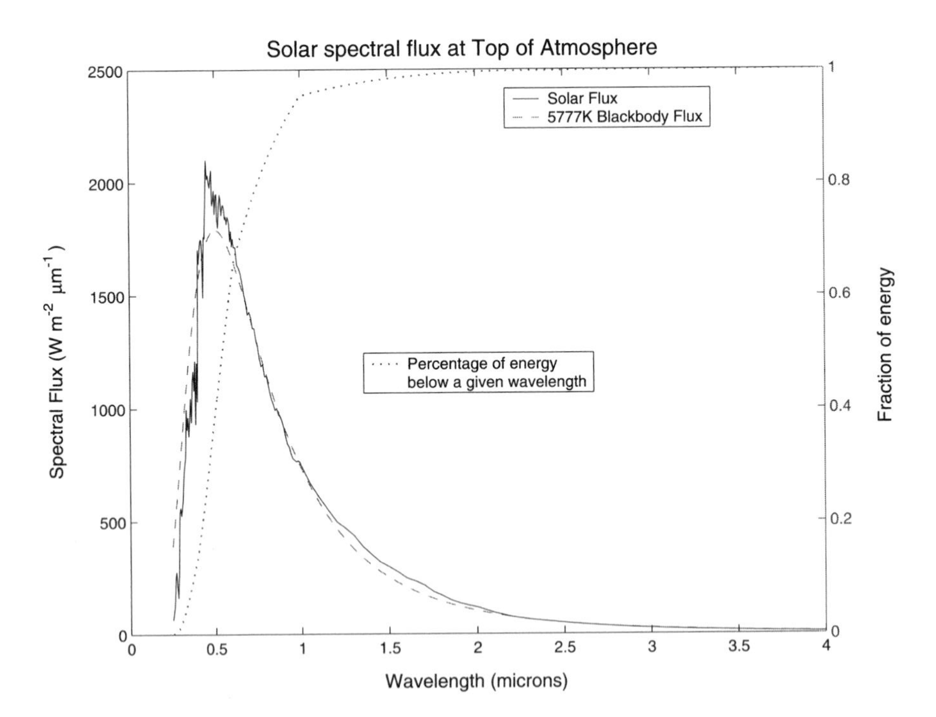

Figure 4 Spectral distribution of solar energy at the top of the atmosphere. Solid line is incoming solar spectral flux, and dashed line is the radiation emitted by a blackbody with a temperature of 5777 K. See ftp site for color image.

we obtain the average daily insolation on a level surface at the top of the atmosphere as

$$F_{\text{day}} = \frac{S_0(r_0/r)^2(\omega_s \sin\delta \sin\phi + \cos\delta \cos\phi \sin\omega_s)}{\pi} \tag{9}$$

In polar regions during summer, when the sun is always above the surface, ω_s equals π and the extraterrestrial daily flux is

$$F_{\text{day}} = S_0\left(\frac{r_0}{r}\right)^2 \sin\delta \sin\phi$$

The daily variation of insolation at the top of the atmosphere as a function of latitude and day of year is depicted in Figure 5. Since Earth is closer to the sun in January, the Southern Hemisphere maximum insolation during summer is about 7% higher than the maximum insolation during Northern Hemisphere summer. The insolation of the polar regions is greater than that near the equator during the summer solstice. This results from the longer days, despite the high solar zenith angles at these high

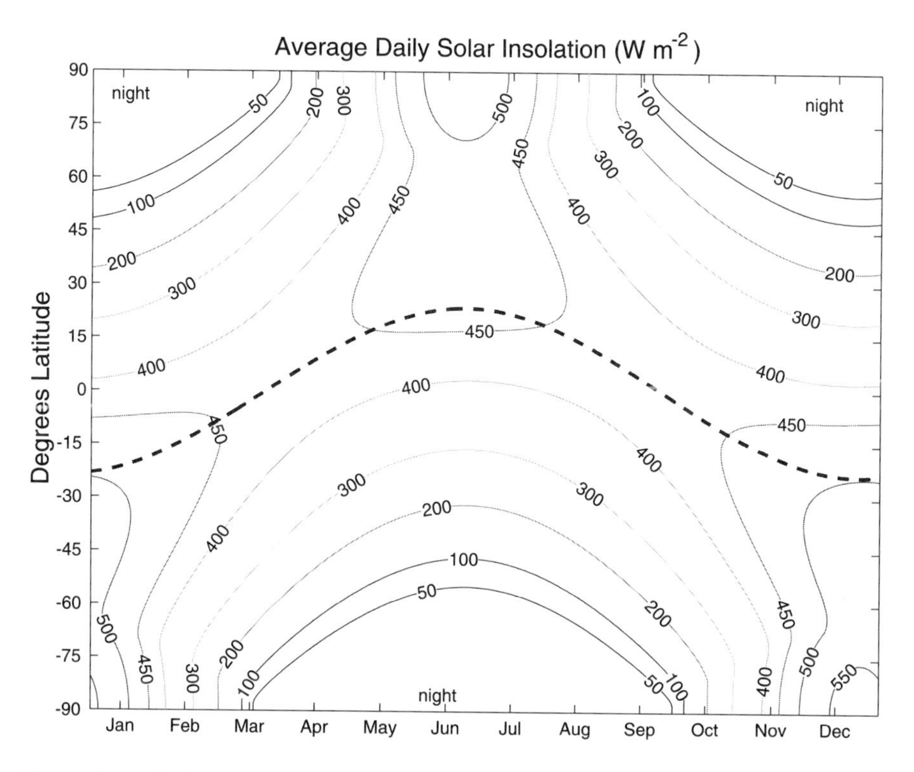

Figure 5 Daily insolation at top of atmosphere as function of latitude and day of year. Also shown is the solar declination angle (thick dashed line). See ftp site for color image.

latitudes. However, the annual average insolation at the top of the atmosphere at the poles is less than half the annual average insolation at the equator.

Surface Radiative Properties

The surface albedo ρ_g is defined as the ratio of the radiation reflected from the surface to the radiation incident on the surface. The surface albedo varies from approximately 5% for calm, deep water to over 90% for fresh snow. The surface albedo over land depends on the type and condition of the vegetation or bare ground. Thus, over land, the surface albedo varies from location to location and with time. Over water, the surface albedo is also a strong function of solar zenith angle.

The surface albedo depends on the wavelength of the incident radiation. Figure 6 is an example of spectral reflection of various surfaces. Snow is very reflective at visible wavelengths (0.4 to 0.7 μm) and less reflective an the near-infrared wavelengths. Plants have higher reflectances in the near-infrared than in the visible. Photosynthesis is effective at absorbing visible energy. When plants dry out, their chlorophyll content decreases and the reflectance at visible wavelength increases.

In the longwave spectral region we generally speak of surface emissivity, or emittance, $\varepsilon_g = 1 - \rho_g$. It is common in models to assume that the surface emittance

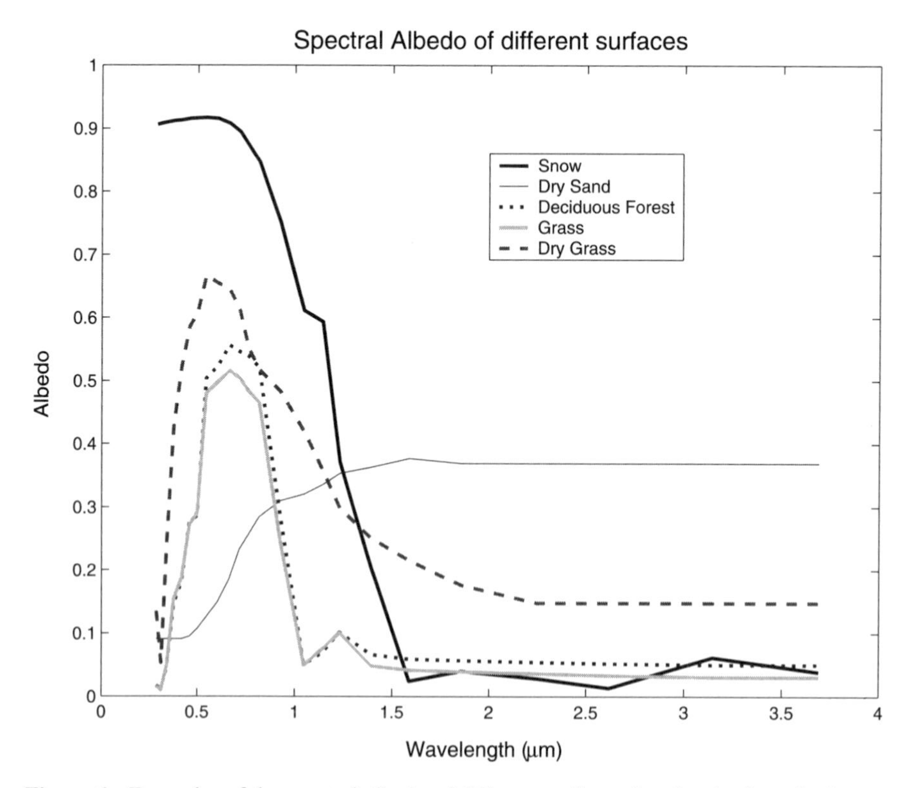

Figure 6 Examples of the spectral albedo of different surfaces. See ftp site for color image.

TABLE 1 Infrared Emissivities of Some Surfaces

Water	0.92–0.96
Ice	0.96
Fresh snow	0.82–0.99
Dry sand	0.89–0.90
Wet sand	0.96
Desert	0.90–0.91
Dry concrete	0.71–0.88
Pine forest	0.90
Grass	0.90

in the infrared (ε_g) is spectrally independent and equal to 1, even though the infrared emissivity of most surfaces is between 0.9 and 0.98 (Table 1). Assuming that the surface emissivity is 1 leads to approximately a 5% error in upward fluxes. The emittance of some surfaces is also wavelength dependent. In particular, sand and desert surface emissivity varies in the 8- to 12-μm window regime between 0.8 and 0.95.

3 RADIATIVE HEATING RATES

This section provides examples of radiative heating profiles for clear and cloud conditions. If more radiative energy enters the layer than leaves the layer, a heating of the layer results. Since the atmosphere is too cold to emit much radiation below 3 μm, atmospheric solar radiative heating rates must all be positive. Since the atmosphere emits and absorbs longwave radiation, longwave radiative heating rates can be positive (net energy gain) or negative (net energy loss).

Radiatively Active Gases in the Atmosphere

The three major absorbing and emitting gases in the stratosphere and troposphere are ozone, carbon dioxide, and water vapor. There are also important minor constituents such as chlorofluorocarbons (CFCs) and methane. While N_2 and O_2 are the most abundant gases in the atmosphere, from an atmospheric energetics point of view they are of small importance. The solar energy below 0.2 μm is absorbed by O, NO, O_2, and N_2 before it reaches the stratosphere.

Ozone (O_3) primarily absorbs in the ultraviolet and in the 9.6-μm region. Solar radiation in the Hartley spectral band (0.2 to 0.3 μm) is absorbed in the upper stratosphere and mesosphere by ozone. Absorption by O_3 in the Huggins band (0.3 to 0.36 μm) is not as strong as in the Hartley bands. Ozone absorbs weakly in the 44- to 0.76-μm region, and strongly around the 9.6-μm region, where radiation is emitted by the surface.

Carbon dioxide is generally a weak absorber in the solar spectrum with very weak absorption in the 2.0-, 1.6-, and 1.4-μm bands. The 2.7-μm band is strong enough that it should be included in calculations of solar absorption, though it overlaps with H_2O. The 4.3-μm band is important more in the infrared region due to the small amount of solar energy in this band. This 4-μm band is important for remote sensing atmospheric temperature profiles. CO_2 absorbs significantly in the 15-μm band from about 12.5 to 16.7 μm (600 to 800 cm^{-1}). It is these differences in the shortwave and infrared properties of CO_2 (and atmospheric water vapor) that lead to the greenhouse effect.

Water vapor absorbs in the vibrational and rotational bands (ground-state transitions). In terms of radiative transfer through the atmosphere, the important water vapor absorption bands in the solar spectrum are centered at 0.94, 1.1, 1.38, 1.87, 2.7, and 3.2 μm. In the infrared, H_2O has a strong vibrational-absorption band at 6.3 μm. The rotational band extends from approximately 13 μm to 1 mm. In the region between these two infrared water vapor bands is the continuum, 8 to 13 μm, known as the atmospheric window. The continuum enhances absorption in the lower regions of the moist tropical atmosphere.

Radiative Heating Rates under Clear Skies

The previous chapter discussed the procedures to calculate radiative fluxes and heating rates in the atmosphere. In this section we discuss examples of radiative heating and cooling under clear-sky conditions.

Atmospheric radiative heating rates due to absorption of solar energy are given in Figure 7 for different solar zenith angles and for tropical conditions. As the solar zenith angle decreases, the total heating of the atmosphere decreases as the solar energy incident on a horizontal surface at the top of atmosphere decreases. Figure 8 demonstrates the absorption by the individual gases if they existed in the atmosphere alone. The large heating in the stratosphere is due to absorption of solar energy by O_3. A minimum in the heating occurs in the upper troposphere. The increased heating in the lower troposphere is due to water vapor. CO_2 contributes little to the solar heating of the atmosphere.

Infrared heating rates are shown in Figure 9 for standard tropical, midlatitude summer, and subarctic summer conditions. Negative values indicate a cooling. The larger cooling rates in the lower troposphere for the tropical conditions arise due to the warmer temperatures and larger amounts of water vapor. The contributions by H_2O, CO_2, and O_3 to the cooling in a tropical atmosphere are shown in Figure 10.

In the tropical moist atmosphere, the water vapor continuum (8- to 13-μm region) makes significant contributions to the cooling to the lowest layers of the atmosphere. CO_2 accounts for the large cooling in the stratosphere. The positive radiative heating rates by O_3 in the stratosphere arise due to the large amounts of radiation in the 9.6-μm band.

Radiative Heating under Cloudy Conditions

Clouds significantly alter the radiative heating and cooling of the atmosphere and at Earth's surface. Clouds also undergo physical changes (e.g., particle size distribution, water content, and cloud top and base altitude) as they form, grow, and dissi-

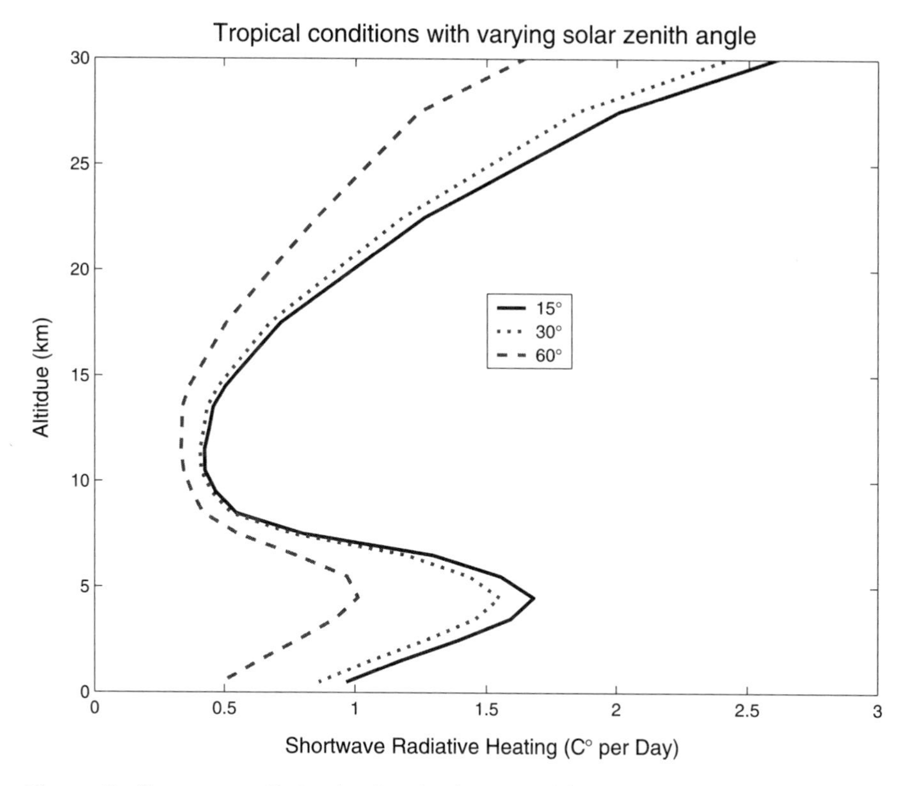

Figure 7 Shortwave radiative heating in degrees celsius per day for standard tropical conditions and three solar zenith angles. See ftp site for color image.

pate. These physical changes are capable of affecting the radiative characteristics of the cloud. Radiative processes have a strong influence on the convective structure and water budget of clouds and affect the particle growth rate.

The previous chapter examined the appropriate equations for calculating the radiative properties of clouds. In this section, we examine the radiative impact of clouds on the atmospheric heating profile. Figure 11 shows the impact of four clouds, each 1 km thick, with differing cloud microphysics on the radiative heating of a midlatitude summer atmosphere with a solar zenith angle of 15°. Three of the clouds have r_{eff} radius of 20 μm with differing ice water content (IWC). Heating rates are expressed in degrees Celsius per day. The larger the IWC the greater the energy convergence, expressed as a heating, within the cloud, as more solar radiation is absorbed. The larger the IWC, the more the heating below cloud base is reduced because of the reduced amount of solar energy below the cloud.

The presence of the cloud affects the heating profiles above the cloud. As the cloud optical depth increases, more energy is reflected upward, enabling absorption in gaseous absorption bands. The particle size also impacts the heating profile. Comparing the two clouds with the same IWC, but different r_{eff} indicates that the cloud with the large particle size absorbs less solar energy.

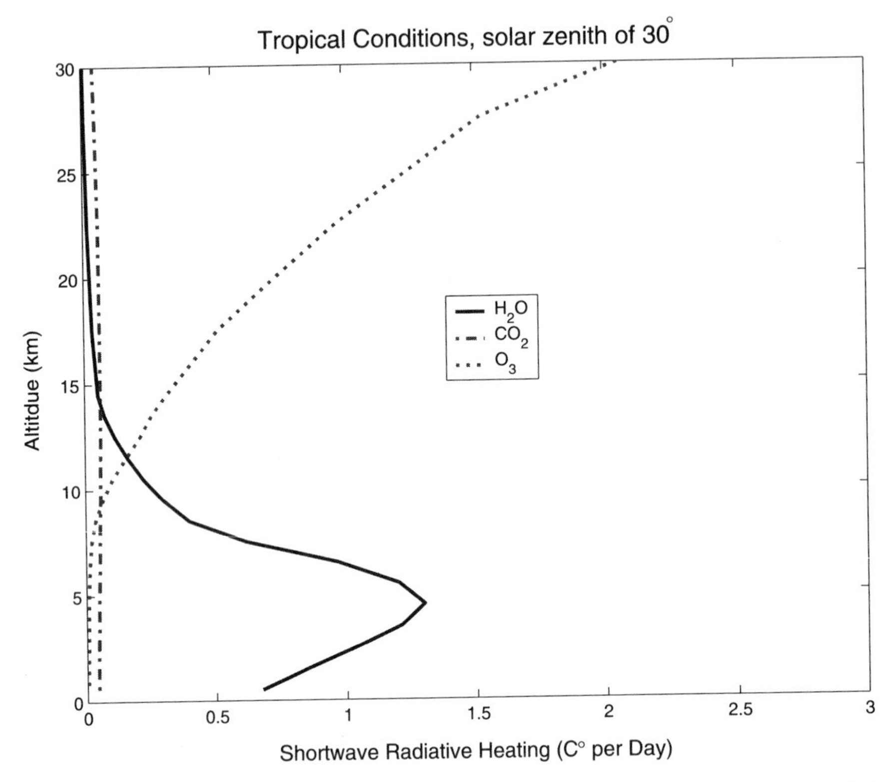

Figure 8 Shortwave radiative heating for standard tropical conditions by H_2O, CO_2, and O_3 and a solar zenith angle of 30°. See ftp site for color image.

Figure 12 depicts the impact of clouds on the longwave heating, for a midlatitude summer atmosphere. The cloud is 2 km thick with a top at 10 km. The top of the cloud has a net energy loss, expressed as a cooling. The cloud base can experience a radiative warming or cooling, depending on the IWC. Increasing the IWC reduces the cooling below cloud base as more energy is emitted by the cloud into this lower layer. Increasing the particle size reduces the cloud radiative heating, if the IWC is fixed.

We will now consider how radiation interacts with a cloudy layer, accounting for multiple scattering within the cloud. The previous chapter discussed how radiation interacts with a particle, or distribution of particles, in terms of the single scattering properties of the particles. There are some useful limits of the two stream model to consider:

$$\lim_{\delta \to 0} R = \frac{\omega_0 \delta}{\mu_0} \gamma_3 \tag{10}$$

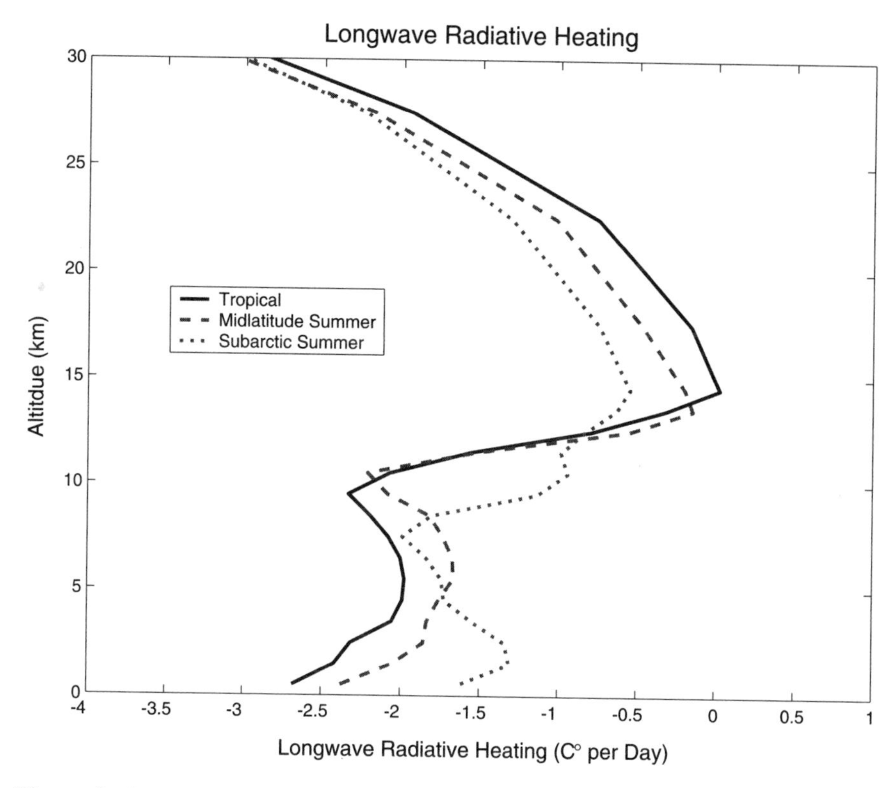

Figure 9 Longwave radiative heating for standard tropical, midlatitude summer, and subarctic summer conditions. See ftp site for color image.

$$\lim_{\delta \to 0} A = \frac{\delta}{\mu_0}(1 - \omega_0) \tag{11}$$

$$\lim_{\delta \to \infty} R = \frac{\omega_0(\alpha_2 + k\gamma_3)}{(1 + k\mu_0)(k + \gamma_1)} \tag{12}$$

These limits become useful in understanding the relationship between a cloud's microphysical properties and its radiative properties. For example, the reflectance is inversely proportional to the cosine of the solar zenith angle. Thus, as the sun gets lower in the sky, the reflectance of a cloud increases. Cloud albedo increases with optical depth, or the water path, though it does approach a limiting value. Reflectance is directly proportional to the single scattering albedo. As the particles composing a cloud get smaller, ω_0 increases, and so does the cloud reflectance.

Cloud absorption decreases with increasing solar zenith angle, the opposite of the case for gases. For optically thin clouds, and thin aerosol layers, the absorption is proportional to $(1 - \omega_0)$. The variation of absorption with increasing r_e is a combination of two competing factors: Droplet absorption $1 - \omega_0$ increases with increasing drop size, while the extinction, and thus optical depth, decreases with increasing

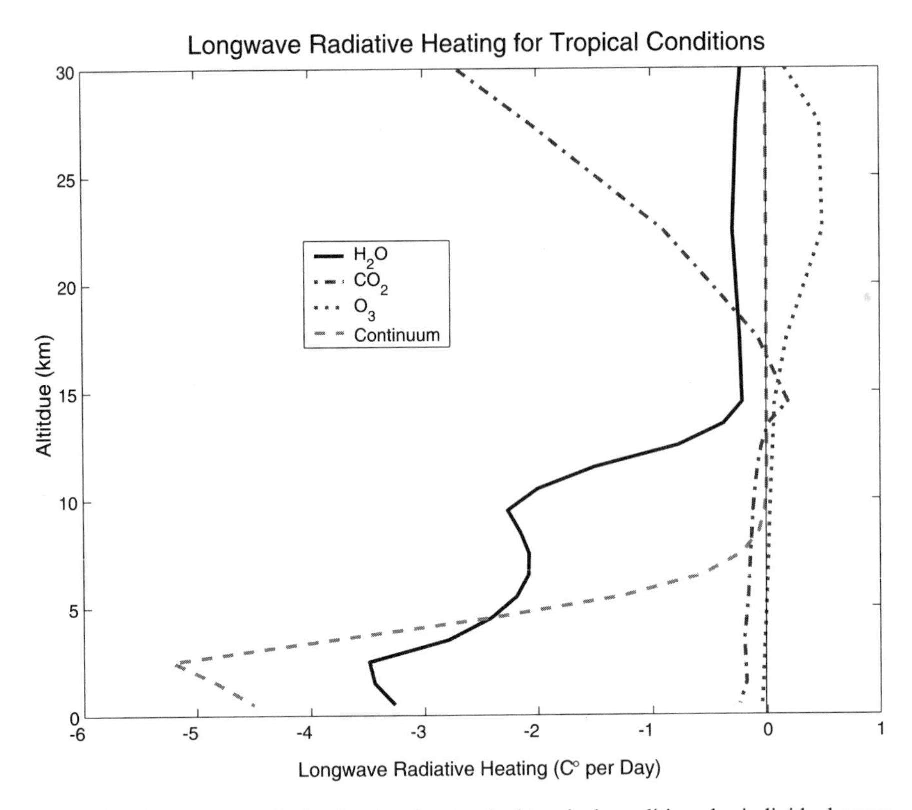

Figure 10 Longwave radiative heating for standard tropical conditions by individual gases. See ftp site for color image.

r_e. For low liquid water path (LWP) clouds and in the weak absorption regime, the combination of these effects renders A approximately independent of r_e. In the strong absorption regime, these two effects combine to produce an absorption that is a decreasing power of r_e. By contrast, reflectance varies primarily as a function of $1/r_e$ through the dependence of extinction on droplet size.

Figures 13 and 14 show the reflectance and transmittance of a water cloud as a function of effective radius for three different liquid water contents (LWC) and three different wavelengths. The red lines are for wavelengths of 0.5 μm and represents the case of no absorption. The blue and black lines are for wavelengths of 2.1 and 2.9 μm, which represent moderate and strong absorption, respectively. The previous chapter demonstrated the inverse dependence of σ_{ext} on r_e. Combining this with the two-stream model, we see that cloud reflectance decreases as an approximate inverse function of r_e for low LWP clouds. The sensitivity between R and r_e decreases with increasing LWP and is most acute at weakly absorbing wavelengths. The cloud transmittance increases with increasing r_e. The limit is a function of the index of refraction.

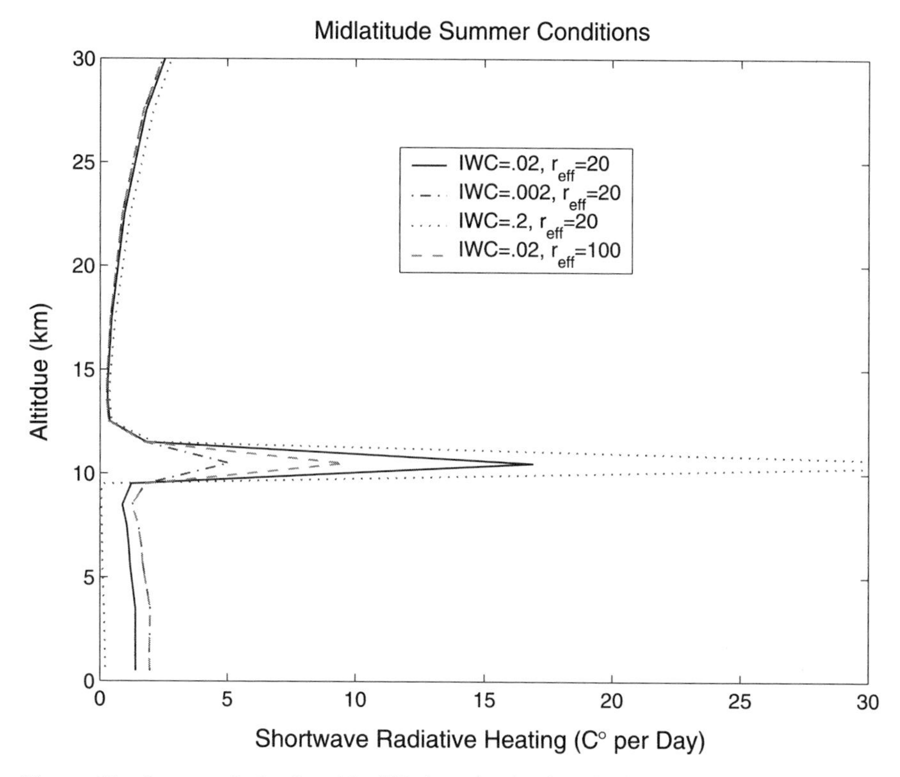

Figure 11 Impact of clouds with differing cloud microphysical properties on radiative heating in a midlatitude summer atmosphere. Clouds are 1 km thick and the solar zenith angle is 15°. See ftp site for color image.

Figure 15 shows the cloud absorptance as a function of effective radius for three different LWPs and two different wavelengths. This figure demonstrates that the absorption of solar radiation by a cloud layer varies with size spectra in a manner dependent on the cloud LWP. For deep clouds, with large LWPs, the absorption increases monotonically with r_e, due to the droplet radius dependence on $1 - \omega_0$. For thin cloud layers and low LWPs, the absorption depends on the combined effects associated with the variation of $1 - \omega_0$ and σ_{ext} with r_e. For small LWP, absorption decreases with increasing r_e.

Volcanic Aerosols

Volcanoes can inject gases and particles into the stratosphere, where the residence times for particles are on the order of 1 year, in contrast to a residence time of 1 week for tropospheric particles. Sulfur-bearing gases such as SO_2 are photochemically converted to sulfuric acid aerosols, which quickly establish themselves as the aerosol species. The dominant visible optical depth of the resulting particulate cloud depends

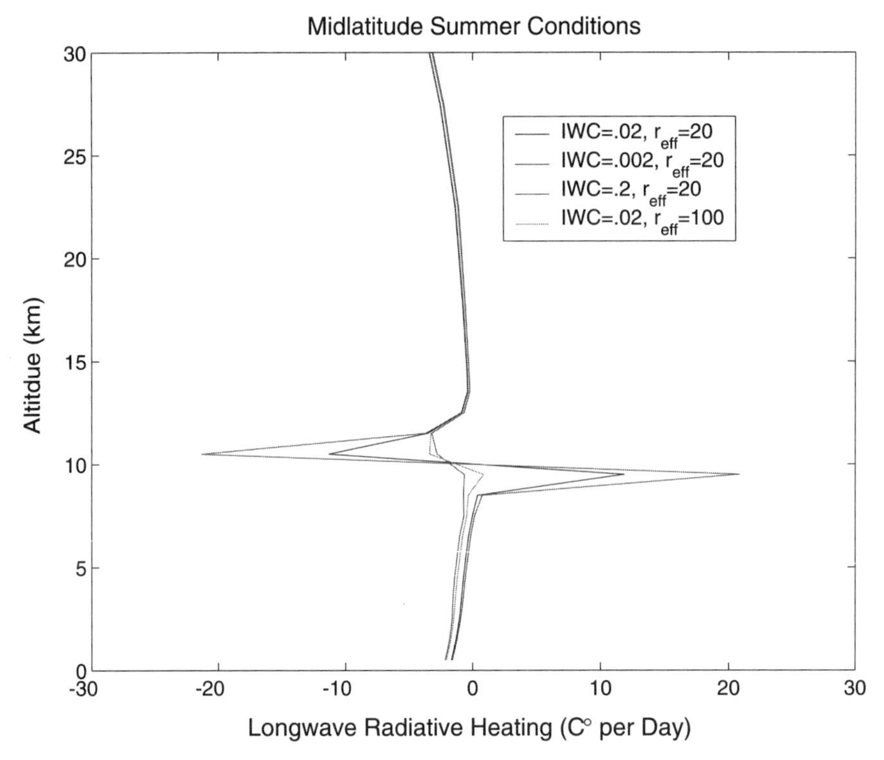

Figure 12 Impact of clouds with differing cloud microphysical properties on the longwave radiative heating in a midlatitude summer atmosphere. Cloud is 2-km thick with a top at 10 km. See ftp site for color image.

on the sulfur content of the volcanic effluents. Thus, although Mt. St. Helens and El Chichon injected similar quantities of ash into the stratosphere, the former caused an optical enhancement of 10^{-2} at 0.55 μm over the Northern Hemisphere during the first year, while the latter caused an enhancement in excess of 10^{-1} because it was a sulfur-rich volcano.

The single scattering albedo of the volcanic aerosols is quite close to unity (0.995) at visible wavelengths. Thus, these aerosols can be expected to cause Earth's albedo to increase, temperatures in the lower stratosphere to increase, and temperature in the troposphere to decrease. Measurements of the El Chicon cloud imply that the thermal effects of the volcanic aerosols are important for the heat balance of both the troposphere and stratosphere, although it dominates in the latter. The stratospheric cloud consisted of sulfuric acid droplets with very small amounts of volcanic ash. The particle size distribution had a mean mode radius of 0.3 μm. Spectral measurements of optical depth show a relatively flat distribution through the visible spectrum with a peak optical depth at 0.55 μm.

The sign and magnitude of thermal infrared heating due to a stratospheric volcanic aerosol of a given optical depth varies significantly with the height and

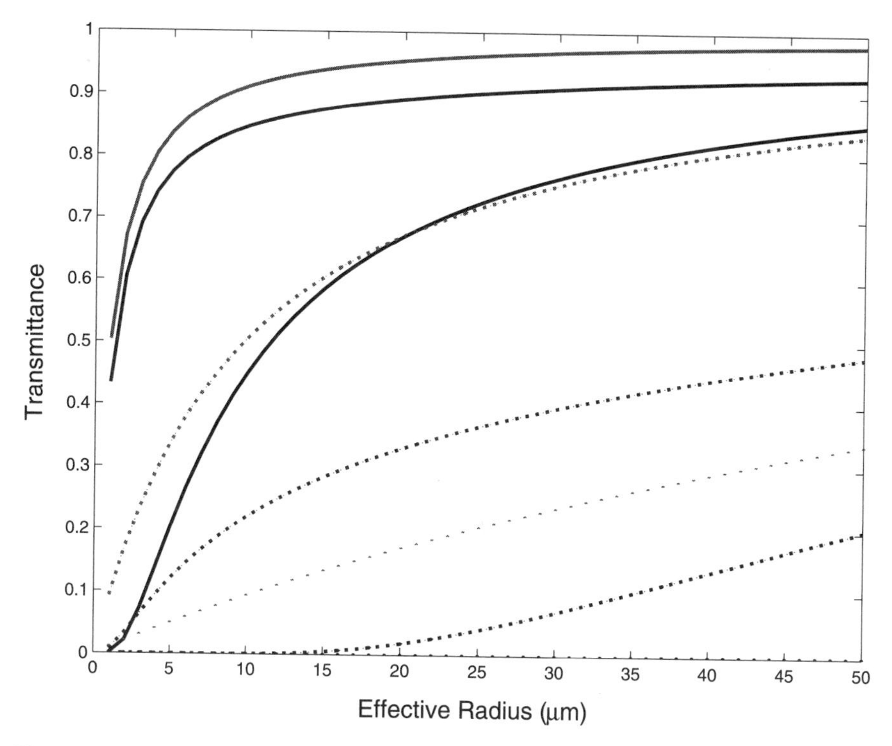

Figure 13 Impact of effective radius on transmittance as function of three wavelengths and three ice water contents. Three wavelengths are 0.5 μm (red), 2.1 μm (blue), and 2.9 μm (black). Liquid water content (LWC) of 0.01, 0.1, and 1 g^{-3} are represented by the solid, dashed, and dotted lines, respectively. See ftp site for color image.

latitude of the aerosols, and also with the height of the tropospheric cloud beneath the aerosol layer and its emissivity. The aerosol of El Chicon was mostly sulfuric acid particles and had significant opacity in the 760- to 1240-cm^{-1} window region. Thus, the stratospheric aerosols can cause a net heating by absorption of upwelling radiation from a warm surface; the aerosols own infrared emission is small at the cold stratospheric temperatures.

Modeling results indicate that these volcanic particles caused a warming by the lower stratosphere of several degrees and a cooling of the troposphere of a few tenths by a degree over their first year.

The largest volcanic eruption in recent history has been that of Mount Pinatubo. Observations of the Earth Radiative Budget Experiment (ERBE) have indicating that the aerosol is causing a net radiative cooling of Earth's atmosphere system. The Mount Pinatubo aerosols resulted in an enhancement of the Earth–atmosphere albedo, with a smaller shift in the lower outgoing longwave radiation.

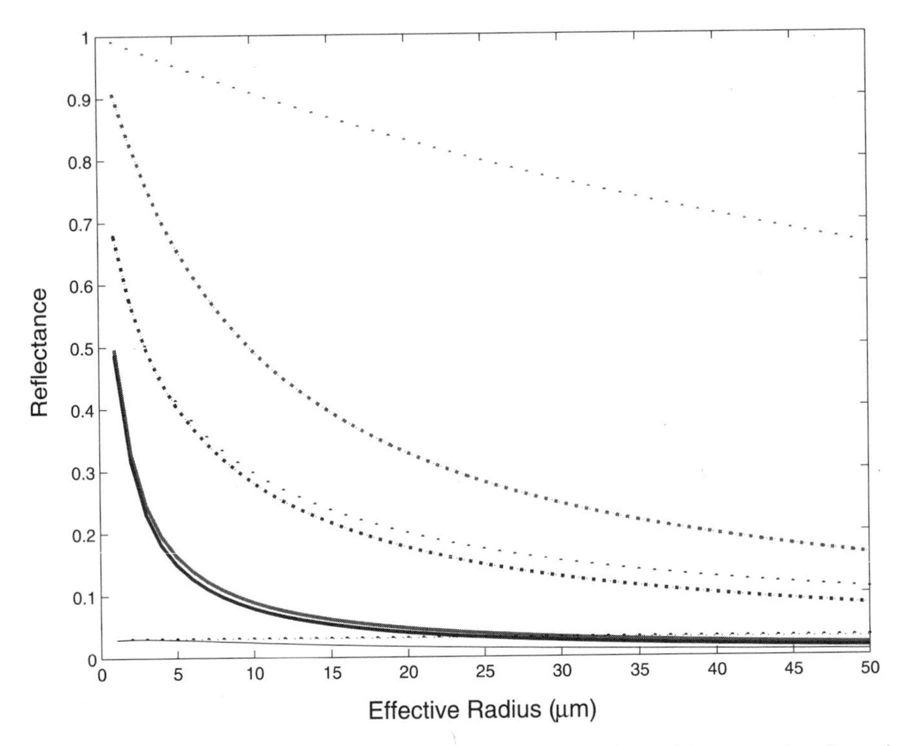

Figure 14 Impact of effective radius on reflectance as function of three wavelengths and three ice water contents. Three wavelengths are 0.5 μm (red), 2.1 μm (blue), and 2.9 μm (black). The LWC and 0.01, 0.1, and 1 g^{-3} are represented by the solid, dashed, and dotted lines, respectively. See ftp site for color image.

4 TOP-OF-ATMOSPHERE RADIATION BUDGETS

If we consider the planet as a system, Earth exchanges energy with its environment (the solar system) via radiation exchanges at the top of the atmosphere. The balance between radiative energy gains and radiative energy losses at the top of the atmosphere is referred to as the Earth radiation budget. The determination of Earth's radiation budget is essential to atmospheric modeling and climate studies as it determines net energy gains and losses of the planet. Radiation budget experiments have used satellites to measure the fundamental radiation parameters:

- Amount of solar energy received by the planet
- Planetary albedo (the portion of incoming solar radiation reflected back to space)
- Emitted terrestrial radiation [also referred to as outgoing longwave radiation (OLR)]
- Net planetary energy balance (difference between the absorbed solar energy and the OLR)

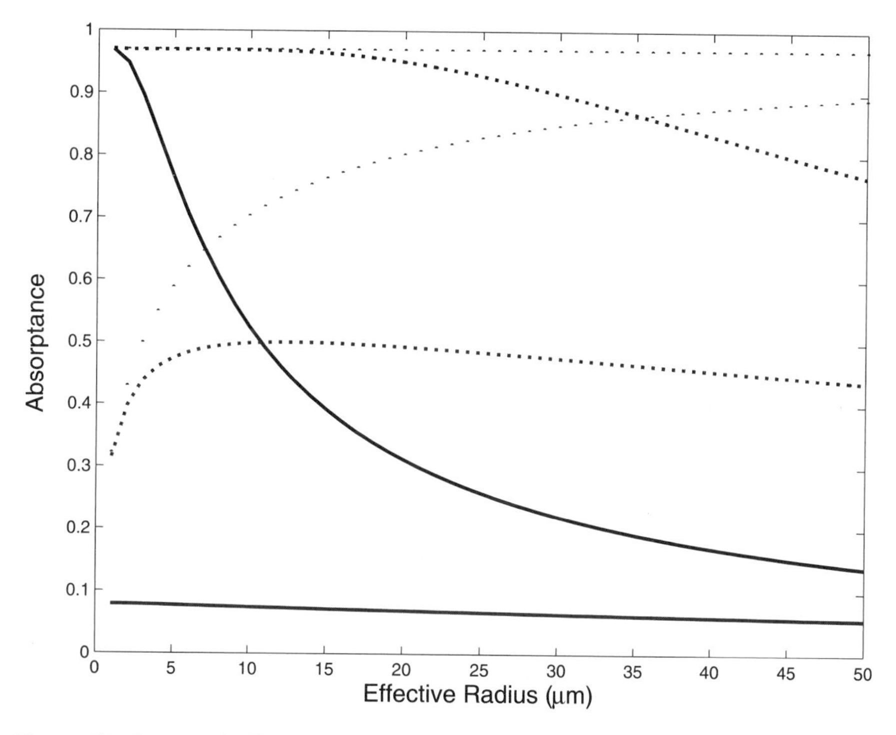

Figure 15 Impact of effective radius on absorption as function of three wavelengths and three ice water contents. Three wavelengths are 0.5 μm (red), 2.1 μm (blue), and 2.9 μm (black). The LWP of 10, 100, and 1000 g^{-2} are represented by the solid, dashed, and dotted lines, respectively. See ftp site for color image.

These elements are described in Figure 16.

Averaged over the globe for a year, the incoming shortwave flux is $(S_0/4)$ where S_0 is the solar constant, 1368 W/m^2. Of this incoming energy, approximately 17%, or 82 W/m^2, is absorbed by the atmosphere, with about 103 W/m^2 (30%) reflected back to space (24% from the atmosphere due to clouds, aerosols, and Rayleigh scattering and 6% from the surface). The net shortwave flux at the surface is 157 W/m^2, or about 53% of that incident at the top of atmosphere. In the longwave, 239 W/m^2 leaves Earth to balance the shortwave gain. Of this 239 W/m^2, approximately 57% is due to emission by the atmosphere and 13% is due to surface emission that is transmitted through the atmosphere. Compared to the top of the atmosphere, the surface loses approximately 51 W/m^2 in the longwave, 88% of which is absorbed by the atmosphere. The surface also gains longwave energy emitted by the atmosphere. Thus, while the net top of atmosphere flux is zero, the surface balance is 21 W/m^2. To retain energy balance the surface must lose or store energy. Neglecting storage, the surface loses 21 W/m^2, which is transferred to the atmosphere to balance the net radiative loss. This is accomplished via latent and sensible heat fluxes. Thus, the atmospheric radiative cooling is balanced by the latent

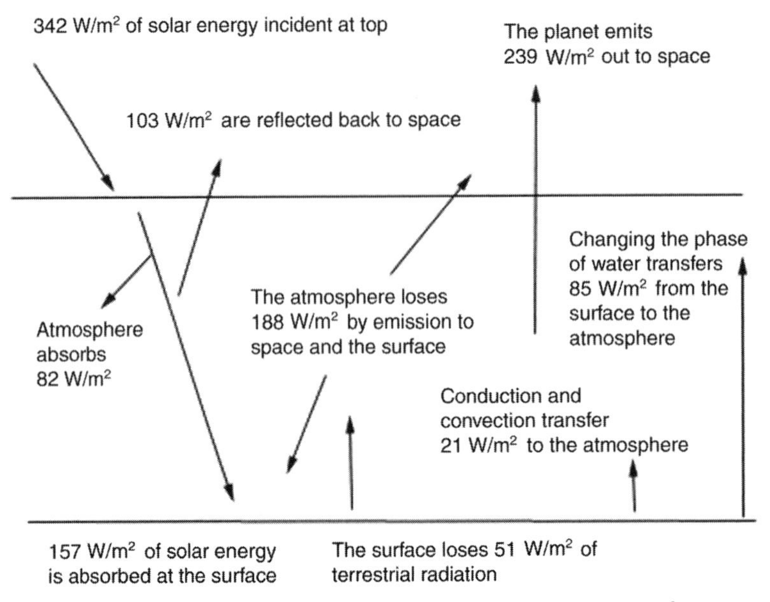

Figure 16 Average annual global energy budget in W/m^2.

heat of condensation, which is released in regions of precipitation, and by conduction of sensible heat from the surface.

Earth's Top of the Atmosphere Radiation Budget

Independent satellite observations of the longwave emitted energy to space have indicated that on a global annual mean basis, the absorbed solar radiation is in balance with the outgoing longwave radiation (to within instrument noise). This balance exists on a hemispherical scale, indicating that there is little net cross-equatorial energy transport.

Seasonal and latitudinal variations in temperature are driven by the variations in the incoming solar radiation. Measurements of the solar insolation at the top of the atmosphere are important, since it is the primary climate external forcing mechanism. Methods of calculating these variations have already been discussed. Measurements from satellites indicate that there are small variations in the solar insolation, on the order of 0.1 to 0.3%.

The global annual averaged albedo is approximately 0.30. The globally and annually averaged planetary albedo is a key climate variable since it, combined with the solar insolation, determines the radiative energy input to the planet. Variations in the global mean albedo result from the eccentricity of Earth's orbit and geographical differences between the Northern and Southern Hemispheres. The annual average albedo of the Northern and Southern Hemispheres is nearly the same, demonstrating the important influence of clouds.

There are significant variations in the month-to-month global mean albedo, longwave, and net flux (Fig. 17). The annual cycle in the global monthly means are due to Earth's orbit about the sun and the geographical differences between the Northern and Southern Hemispheres. The range in the OLR throughout the year is approximately $10\,\mathrm{W/m^2}$, with a maximum in July and August. This maximum results from there being more land in the Northern Hemisphere than the Southern. The annual average global albedo is approximately 30%, with an amplitude of approximately 2%. The planetary albedo reaches a maximum around October and November. This albedo variation reduces the impact of the annual variation in incoming solar radiation on the net radiation budget.

The amplitude of the annual cycle of globally averaged net flux is approximately $26\,\mathrm{W/m^2}$. This is similar to the peak-to-peak amplitude in the external forcing associated with variations in the solar insolation due to Earth–sun geometry. The interannual variation of the hemispherical averaged net flux shows maximum heating during the summer with the largest changes occurring for the transition from solstice to equinox.

The zonal annual average absorbed solar radiation exceeds the outgoing longwave radiation in the tropical and subtropical regions, resulting in a net radiative heating of the surface–atmospheric column; while in the mid to polar latitudes there

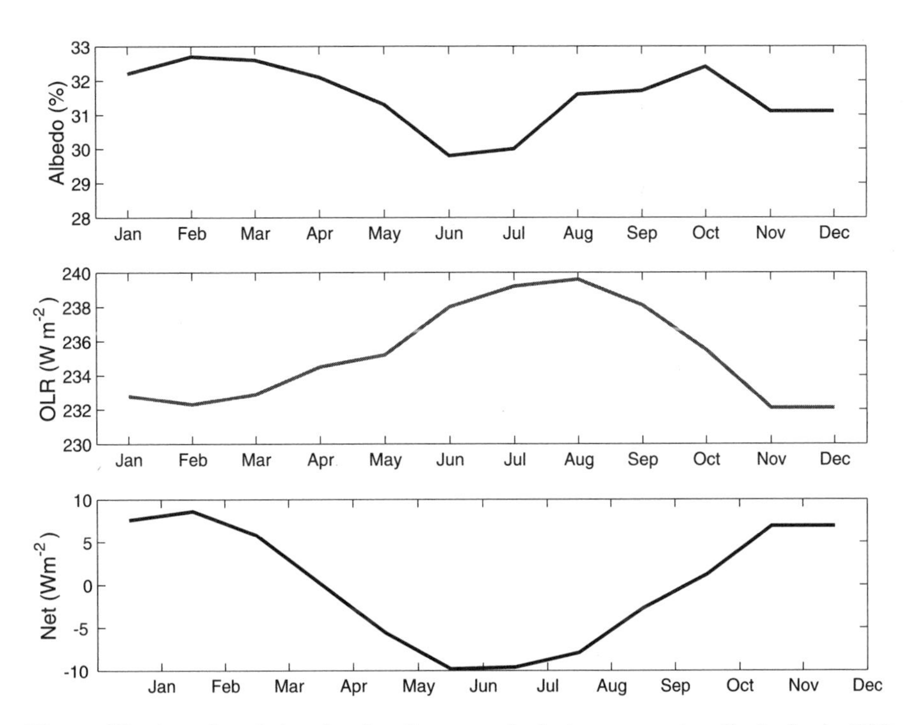

Figure 17 Annual variation in planet's energy budget components: albedo (top), OLR (middle), and net radiation (bottom). See ftp site for color image.

is a net divergence (Fig. 17). This equator-to-pole gradient in radiative energy is the primary mechanism that drives the atmospheric and oceanic circulations. On an annual and long-term basis no energy storage and no change in the global mean temperature occurs, so the zonal mean radiative budget must be balanced by meridional heat transport by the atmosphere and oceans.

The minimum in emitted flux by the planet located near the equator is due to the high cloud tops associated with the Inter-Tropical Convergence Zone (ITCZ) (Fig. 18). This minimum migrates about the equator as seen in the seasonal profiles and is seen as a maximum in albedo. Note also the large emission in the vicinity of the subtropical highs and the corresponding lower albedos. The lowest values of OLR are associated with the Antarctic plateau in winter. This region is very cold and the high altitude means that most of the surface-emitted radiation escapes to space. Maximum OLR occurs in the tropics. Throughout the year the OLR slowly increases toward the summer hemisphere.

The largest albedos are associated with the polar regions, which are snow covered and have high solar zenith angles. The increasing albdeo with latitude is, in general, due to the increasing solar zenith angle (Fig. 19). In the Northern Hemisphere the albedo is larger in summer than winter, due to the increase in cloud cover and optical

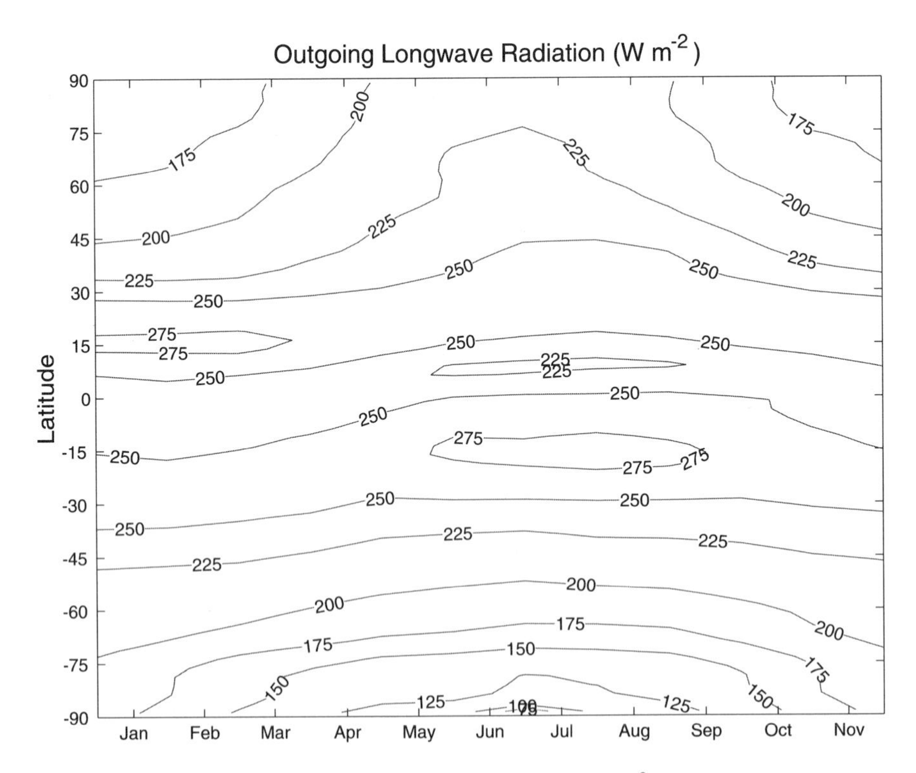

Figure 18 Annual cycle of monthly mean OLR in W/m^{-2} determined from ERBE observations. See ftp site for color image.

thickness. In the tropical regions the albedo variation is influenced primarily by weather disturbances and their associated cloud distributions. In the polar region, variations are due to the distribution of major ice sheets and the decreasing mean solar elevation angle with latitude. The annual variation in the zonal mean absorbed solar radiation follows the variation of the solar declination due to the annual variation of the incoming solar energy being greater than the annual variation of the albedo. The net energy also exhibits this same dependency.

There are net radiation energy gains in the tropics and subtropics with energy losses over the polar regions (Fig. 20). The region of net energy gains tracks the solar declination. Differences between the Northern and Southern Hemispheres result from differences in land distributions. While the minimum OLR occurs over Antarctica, the minimum net losses occur between approximate 60 S and 70 S. Maximum energy gains occur in the Southern Hemisphere subtropical regions during December and January. Differences between land and water are more evident in a regional analysis of Earth's radiation budget.

Figures 21 and 22 are the ERBE measured January and July OLR. Maximum OLR occurs over the deserts and cloud free ocean regions of the subtropical highs. Relative minimums in tropical regions result from high, thick cirrus associated with

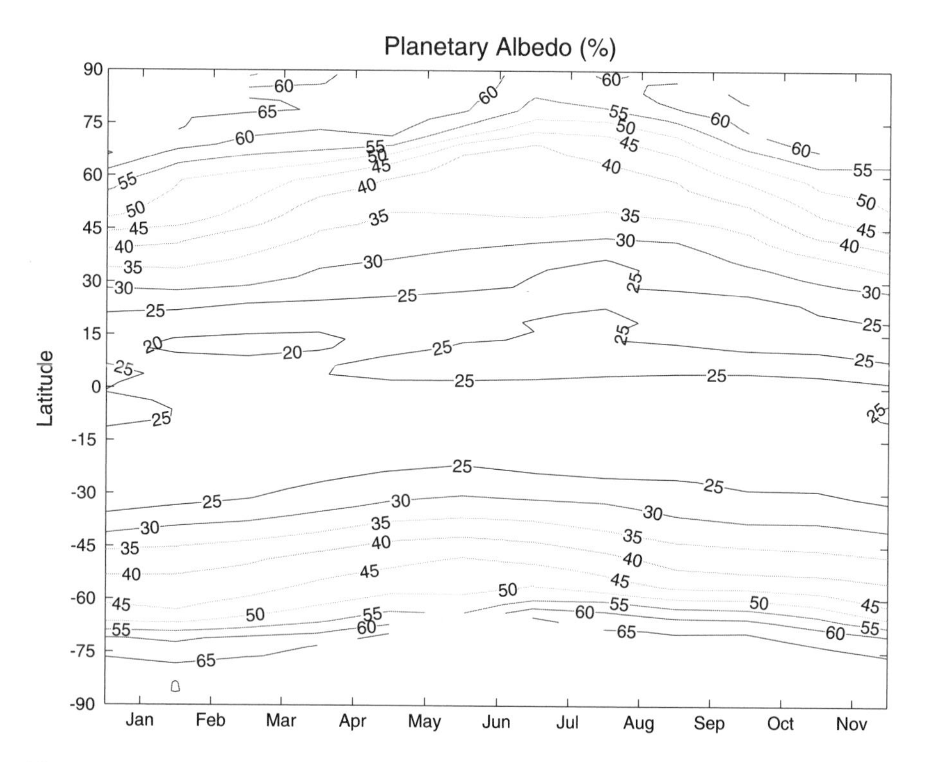

Figure 19 Annual cycle of monthly mean albedo in percent determined from ERBE observations. See ftp site for color image.

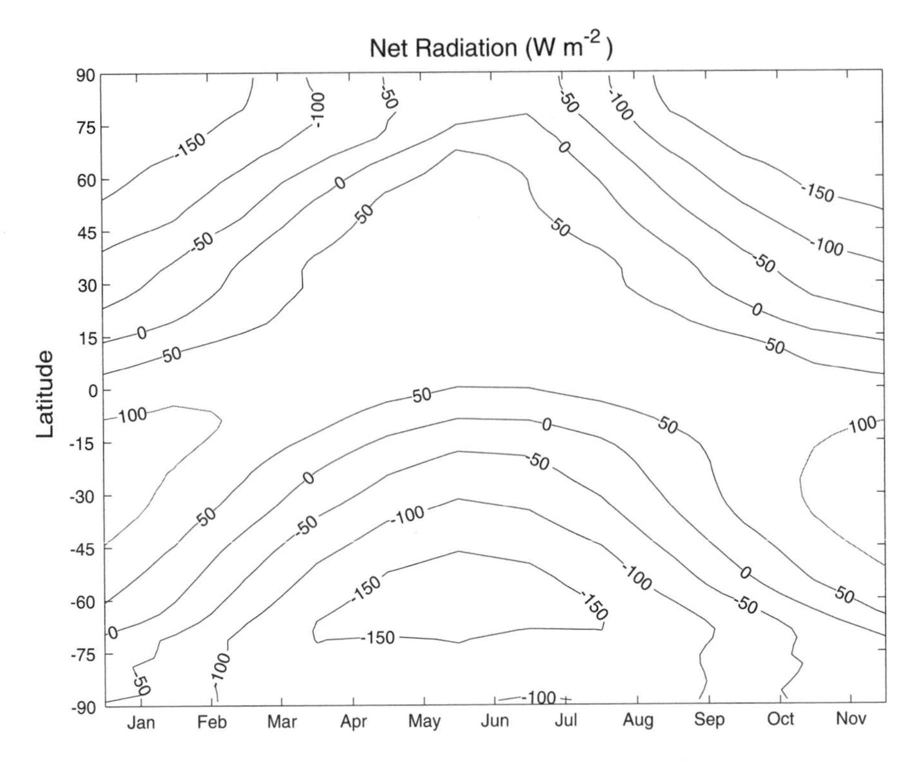

Figure 20 Annual cycle of monthly mean net radiation in W/m^{-2} determined from the ERBE observations. See ftp site for color image.

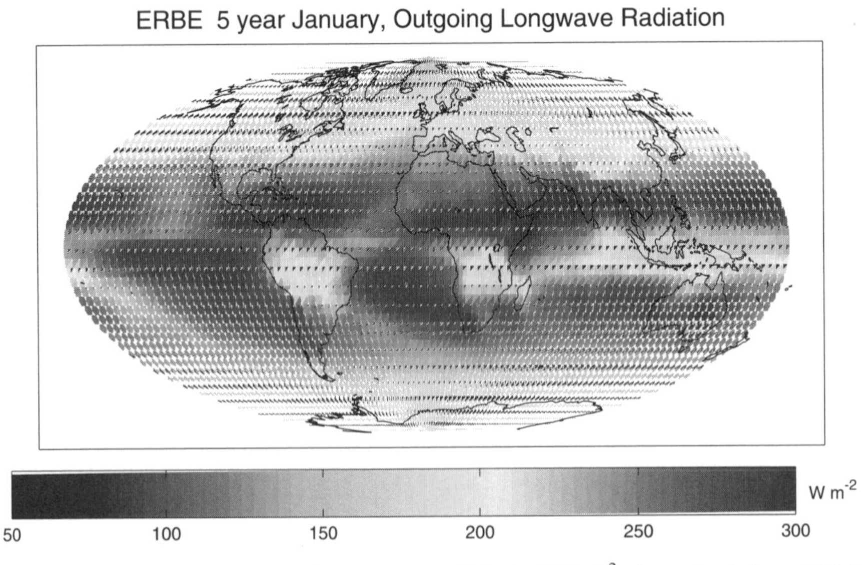

Figure 21 (see color insert) January mean OLR in W/m^{-2} determined from ERBE observations. See ftp site for color image.

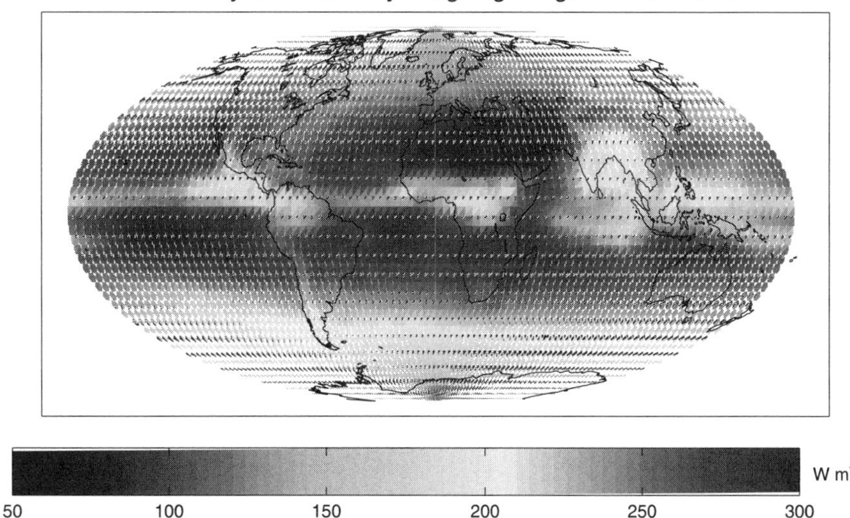

Figure 22 (see color insert) July mean OLR in W/m^{-2} determined from ERBE observations. See ftp site for color image.

tropical convection regions (e.g., Indonesia, Congo Basin, central South America). These thick clouds also yield local minima in the January and July averaged planetary albedos (Figs. 23 and 24). Regional albedo maps for January and July indicate the presence of maritime stratus regions located off the west coast of continents.

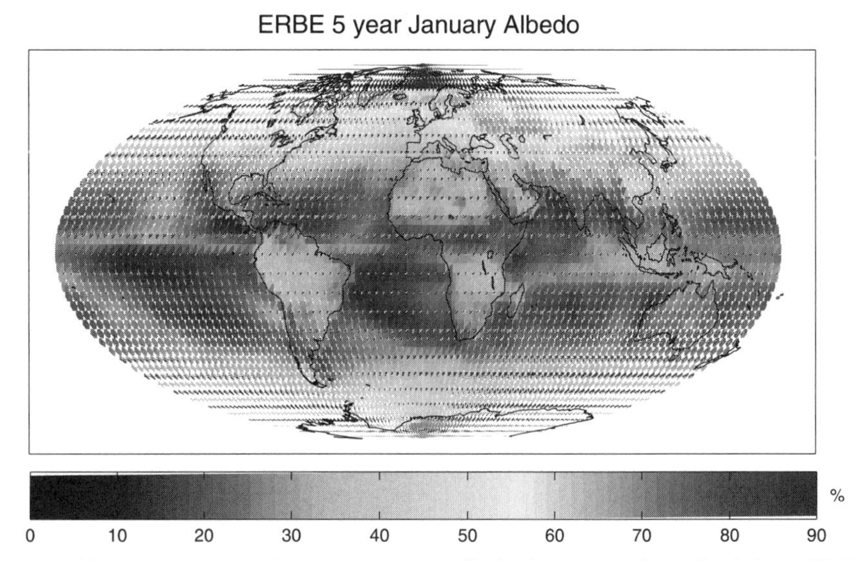

Figure 23 (see color insert) January mean albedo in percent determined from ERBE observations. See ftp site for color image.

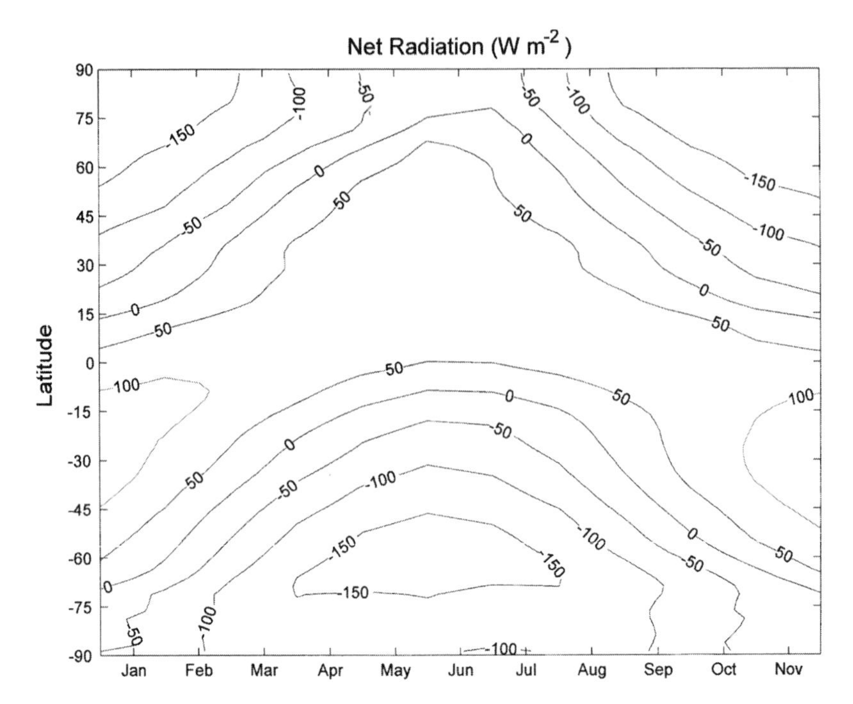

Figure 24 July mean albedo in percent determined from ERBE observations. See ftp site for color image.

These features do not appear on the OLR maps because of the similarity between the sea surface temperature and cloud effective temperature. The ITCZ appears in the albedo maps as a regional enhancement, while in maps of OLR it is a relative minimum.

Analysis of the regional distribution of the net radiative energy gains and losses for January and July is shown in Figures 25 and 26, respectively. These figures clearly indicate that the summer hemisphere gains radiative energy while the winter hemisphere is a net radiative sink. In July, the large desert regions of northern Africa and Saudi Arabia also exhibit a net radiative loss. This arises from the high surface albedos and high surface temperatures and overall cloud-free conditions. The largest energy gains are associated with cloud-free ocean regions in the summer hemisphere, due to the relatively low albedo and high solar input.

The measured outgoing longwave radiation and albedo also indicate regional forcing mechanisms. For example, in the tropics longitudinal variations can be as large as the zonal averages and are associated with east–west circulations. While tropical regions in general display a net radiative heating, the Sahara is radiatively cooling. This is due to the high surface albedo, the warm surface temperatures, and the dry and cloud-free atmosphere. The radiative cooling is maintained by subsidence warming, which also has a drying effect and therefore helps maintain the desert.

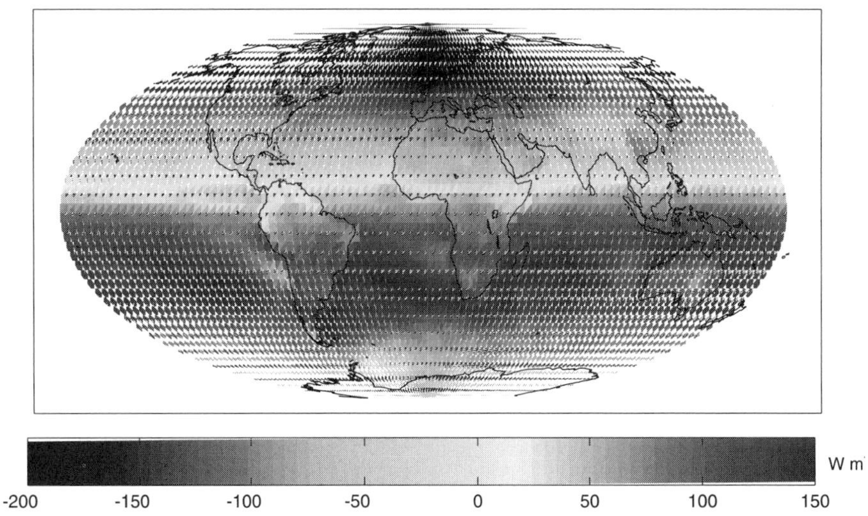

Figure 25 (see color insert) January mean net radiation in W/m^{-2} determined from ERBE observations. See ftp site for color image.

Cloud-Radiative Forcing

One of the major research problems in radiative transfer applications is how clouds affect the energy budget of the planet and thus impact climate. The net radiative heating H within a column is

ERBE 5 year mean July Net Radiation

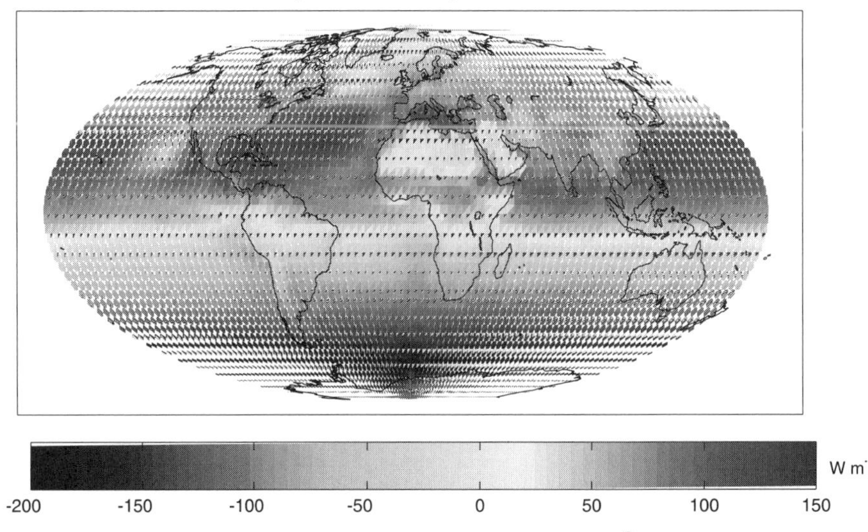

Figure 26 (see color insert) July mean net radiation in W/m^{-2} determined from ERBE observations. See ftp site for color image.

$$H = S_0(1 - \alpha) - \text{OLR} \tag{13}$$

where $S_0(1 - \alpha)$ is the absorbed solar energy and OLR is the outgoing longwave energy. To describe the effects of clouds on H from observations we define the cloud forcing,

$$C = H - H_{\text{clr}} \tag{14}$$

where H_{clr} is the clear-sky net heating. Thus, cloud-radiative forcing is the difference between the all-sky (cloudy and clear regions) and cloud-sky fluxes. We can write

$$C = C_{\text{SW}} + C_{\text{LW}} \tag{15}$$

where

$$C_{\text{SW}} = S(\alpha_{\text{clr}} - \alpha) \tag{16}$$

and

$$C_{\text{LW}} = \text{OLR}_{\text{clr}} - \text{OLR} \tag{17}$$

In the longwave, clouds generally reduce the LW emission to space and thus result in a heating. While in the SW, clouds reduce the absorbed solar radiation, due to a generally higher albedo, and thus result in a cooling of the planet. Results from ERBE indicate that in the global mean, clouds reduce the radiative heating of the planet. This cooling is a function of season and ranges from approximately -13 to $-21\ \text{W/m}^2$, with an uncertainty of approximately $5\ \text{W/m}^2$. On average, clouds reduce the globally absorbed solar radiation by approximately $50\ \text{W/m}^2$, while reducing the OLR by about $30\ \text{W/m}^2$. These values may be compared with the $4\ \text{W/m}^2$ heating predicted by a doubling of CO_2 concentration.

Variations in net cloud forcing are associated with surface type, cloud type, and season. These dependencies can be seen in maps of January and July cloud net radiative forcing (Figs. 27 and 28). Maritime stratus tend toward a negative net cloud forcing as the shortwave effects dominate the longwave. The deserts of North Africa and Saudi Arabia have a positive cloud radiative forcing as the longwave dominates the cloud impact on the albedo.

5 RADIATION AND THE GREENHOUSE EFFECT

The energy that fuels the atmospheric and oceanic circulations originates from the sun. If the total energy content of the Earth–atmosphere system does not vary significantly with time, there must be a close balance between the incoming absorbed solar radiation and the outgoing terrestrial emitted thermal radiation. In

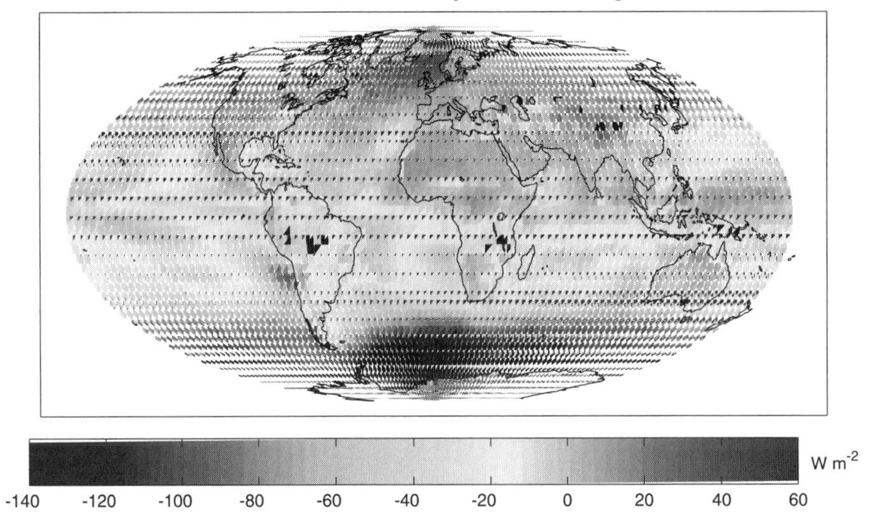

Figure 27 (see color insert) January cloud radiative forcing in W/m^{-2}. See ftp site for color image.

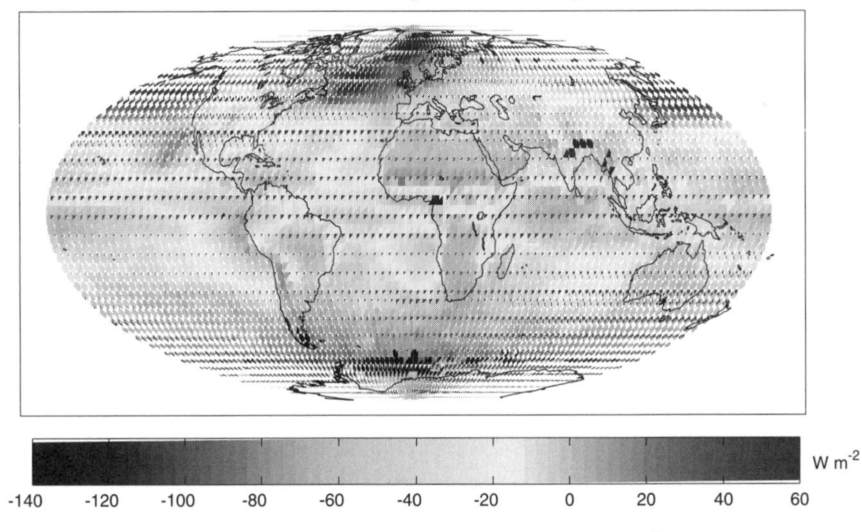

Figure 28 (see color insert) July cloud radiative forcing in W/m^{-2}. See ftp site for color image.

other words, when averaged over the entire globe for the year, the net energy gain equals the net loss

$$\text{Absorbed sunlight} = \text{terrestrial emission} \tag{18}$$

$$S_0 \pi R_e^2 (1 - \alpha_e) = 4\pi R_e^2 \sigma T_e^4 \tag{19}$$

where α_e is the albedo (broadband reflectance) of Earth and T_e is the effective temperature. For an albedo of $\alpha_e = 0.3$, $T_e \approx 255^\circ$, which is considerably cooler than the average surface temperature of 288 K.

The average surface temperature is greater than the effective radiative temperature because of the radiative properties of the atmosphere. The atmosphere is relatively transparent to solar radiation while absorbing and emitting longwave radiation effectively.

The spectral selectivity of absorption by atmospheric gases is the fundamental cause of the greenhouse effect. Much of the shortwave, or solar, energy passes through the atmosphere and warms the surface. While the atmosphere is transparent to shortwave radiation, it is efficient at absorbing terrestrial or longwave radiation emitted upward by the surface. So, while carbon dioxide and water vapor comprise only a very small percentage of the atmospheric gases, they are extremely important because of their radiative properties, i.e., their abilities to absorb this longwave radiation and emit it throughout the atmosphere.

Gases that are transparent to solar energy while absorbing terrestrial energy will warm the atmosphere because they allow solar energy to reach the surface and inhibit longwave radiation from reaching outer space. These radiatively active gases are called greenhouse gases. In addition to carbon dioxide, other important greenhouse gases are water vapor, methane (CH_4), and CFCs. Methane and CFCs are important because they absorb terrestrial radiation in the 10- to 12-μm infrared (IR) atmospheric window. The concentration of these latter two gases has also been increasing. Increases in the greenhouse gases with time can potentially result in a climate change, as the atmosphere becomes more effective at absorbing longwave energy emitted by the surface.

Greenhouse warming or the enhanced greenhouse effect are the terms used to explain the relationship between the observed rise in global temperatures and an increase in atmospheric carbon dioxide. In this chapter we have considered one aspect of the enhanced greenhouse effect, absorption and emission of radiation by certain atmospheric gases. How does this play a role in the enhanced greenhouse effect? Let us use carbon dioxide as an example. Increasing the carbon dioxide concentrations in the atmosphere does not appreciably affect the amount of solar energy that reaches the surface. However, since carbon dioxide absorbs longwave radiation, the amount of longwave energy emitted by the surface and absorbed by the atmosphere increases as the atmospheric concentration of carbon dioxide rises. The increased absorption increases the temperature of the atmosphere. The warmer atmospheric temperatures increase the amount of longwave energy emitted by the atmosphere toward the surface, which increases the energy gain of the surface, warming it. Thus, increased concentration of carbon dioxide may result in a warming of Earth's atmosphere and surface.

Water vapor is the most effective greenhouse gas because of its strong absorption of longwave energy emitted by the surface. A warmer atmosphere can mean more water vapor in the atmosphere and possibly more clouds.

Clouds and the Greenhouse Effect

Increased concentration of greenhouse gases can lead to a warming of the atmosphere. As the air temperature warms, the relative humidity initially decreases. Evaporation depends on relative humidity; so, as the atmosphere warms, more evaporation occurs, which adds more water to the atmosphere and enhances the greenhouse warming. Earth and atmosphere keep heating up until the energy emitted balances the amount of sunlight absorbed. But greenhouse gases are not the whole story of climate change. Clouds have a large impact on the solar and terrestrial energy gains of the atmosphere. Clouds reflect solar energy and reduce the amount of solar radiation reaching the surface, and thus cause a cooling of Earth. The thicker the cloud, the more energy reflected back to space, and the less solar energy available to warm the surface and atmosphere below the cloud. By reflecting solar energy back to space, clouds tend to cool the planet. Clouds are also very good emitters and absorbers of terrestrial radiation.

Clouds block the emission of longwave radiation to space. Thus, in the longwave, clouds act to warm the planet, much as greenhouse gases do. To complicate matters, the altitude of the cloud is important in determining how much they warm the planet. Cirrus are cold clouds. Thick cirrus therefore emit very little to space because of their cold temperature, while at the same time cirrus are effective at absorbing the surface-emitted energy. Thus, with respect to longwave radiation losses to space, cirrus tend to warm the planet. Stratus also warm the planet but not as much as cirrus. This is because stratus are low in the atmosphere and have temperatures that are more similar to the surface than cirrus clouds. Stratus absorb radiation emitted by the surface, but they emit similar amounts of terrestrial radiation to space as the surface. To complicate matters still further, how effective a cloud is at reflecting sunlight is a function of how large the cloud droplets or cloud ice crystals are.

Figure 29 depicts the dependence of cloud radiative forcing as a function of cloud temperature and the change in the planetary albedo due to the cloud. Thin, cold clouds tend to warm the planet while thick, warm clouds tend to cool. The zero line indicates those clouds that have no net effect on the top of the atmosphere energy budget. Cold, thick clouds, such as convective systems, have little impact on the energy budget at the top of the atmosphere.

So clouds can either act to cool or warm the planet, depending on how much of Earth they cover, how thick they are, how high they are, and how big the cloud particles are. Measurements indicate that on average, clouds' reflection of sunlight dominates the clouds' greenhouse warming. Thus, today's distribution of clouds tends to cool the planet. But this may not always be the case. As the atmosphere warms the distribution of cloud amount, cloud altitude, and cloud thickness may all

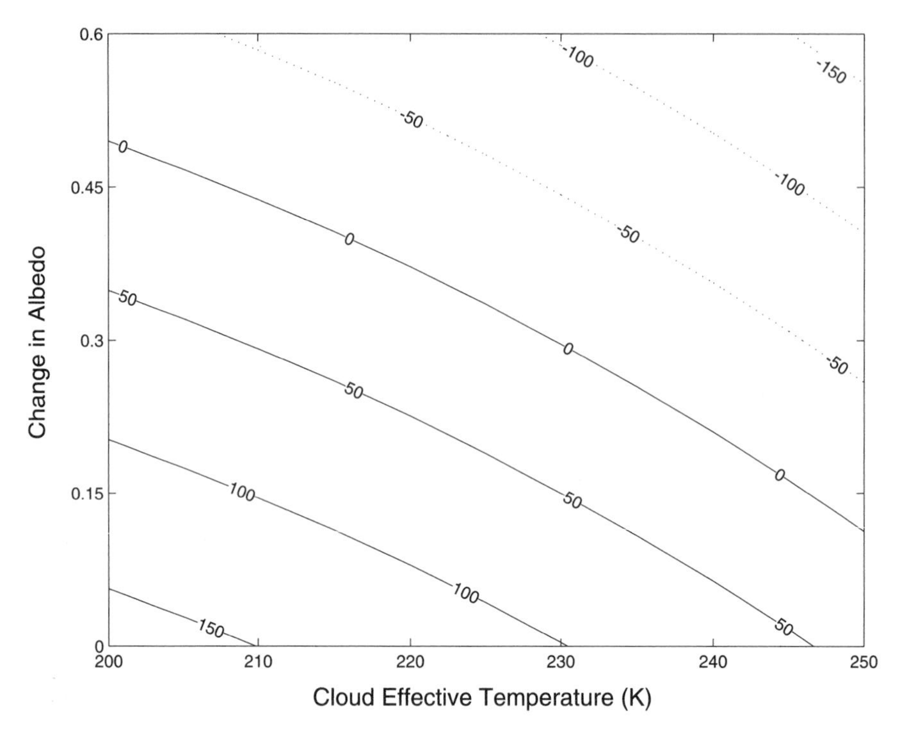

Figure 29 Change in net radiation at top of atmosphere as function of cloud effective temperature and change in planetary albedo due to the cloud. Contours are in W/m^{-2} (after D. Hartmann, (1994) Global Physical Climatology, Academic Press, NY).

change. We do not know what the effect of clouds will be on the surface temperatures if the global climate changes. Clouds could dampen any greenhouse warming by increasing cloud cover or decreasing cloud altitude. On the other hand, clouds could increase a warming if the cloud cover decreases or cloud altitude increases. Climate is so sensitive to how clouds might change that an accurate prediction of climate change hinges on correctly predicting cloud formation.

Radiative Equilibrium

A fundamental challenge of modern science is to predict climate. Recent concerns about global warming and the effect of greenhouse gases added to the atmosphere by humans have heightened the need to understand the processes that cause climate variations. Models are used to gain a better understanding of the atmosphere. We can use a simple model to derive the temperature distribution of an atmosphere in radiative equilibrium by assuming that the only source term is due to thermal emission: $B = \sigma T^4$. For demonstration, we also assume that the atmosphere absorbs at all wavelengths. The upwelling and downwelling fluxes are then

$$\frac{d(F\uparrow - F\downarrow)}{d\tau} = (F\uparrow + F\downarrow) + 2B \tag{20}$$

$$\frac{d(F\uparrow + F\downarrow)}{d\tau} = (F\uparrow - F\downarrow) \tag{21}$$

We can directly solve this set of equations to yield the emission as a function of optical depth:

$$B(\tau) = \frac{F\uparrow - F\downarrow}{2}\tau + B_0 \tag{22}$$

where B_0 is the emission of the first layer of atmosphere ($\tau = 0$).

For thermal equilibrium, the atmosphere and surface outgoing longwave flux must balance the incident solar flux, F_0, so that

$$F\uparrow(0) = F_0 \tag{23}$$

and

$$(F\uparrow - F\downarrow) = F_0 \tag{24}$$

yielding

$$B(\tau) = \frac{F_0}{2}(\tau + 1) \tag{25}$$

At the surface, the upwelling flux emitted by the surface must balance the absorbed solar flux and the downwelling flux emitted by the atmosphere [$F\downarrow(\tau_s)$].

$$B(T_s) = F_0 + F\downarrow(\tau_s) \tag{26}$$

which leads to

$$B(T_s) = B(\tau_s) + \frac{F_0}{2} \tag{27}$$

This simple model demonstrates that the atmospheric temperature profile is a function of its optical depth. It also shows that the surface temperature is warmer than the overlying air. The greater the optical depth the greater the difference between the surface temperature and the effective temperature of the planet.

6 SATELLITE REMOTE SENSING

Meteorologists use two basic methods of observing the atmosphere: in situ and remote-sensing methods. In situ methods, for "in place," measure the properties of the air in contact with an instrument. Remote-sensing methods obtain information without coming into physical contact with the region of the atmosphere being measured. Remote sensing the atmosphere is the emphasis of this section. In remote sensing, radiation measurements are used to infer the state of the atmosphere. The inference of the atmospheric state is often referred to as the retrieval process (see page 39 for discussion of retrieval process).

There are two basic types of remote sensing the atmosphere: active sensors and passive sensors. Active remote-sensing instruments emit energy, such as a radio wave or beam of light, into the atmosphere and then measure the energy that returns. Passive remote-sensing instruments measure radiation emitted by the atmosphere, Earth's surface, or the sun. Much of what we observe about the atmosphere using remote-sensing techniques results from how radiation interacts with molecules or objects, such as water drops, suspended in the atmosphere. The previous chapter discussed the principles on which remote-sensing methods are based, while this section looks at some applications of passive remote sensing from satellites.

Remote Sensing the Surface

In remote sensing the surface of Earth from a satellite, we select spectral regions, or channels, in which the atmosphere is transparent. These are called atmospheric windows. In the solar spectrum, or the shortwave, very little absorption occurs in the visible region; however, Rayleigh scattering is large. Rayleigh scattering is well understood and can be handled via modeling as described in the previous chapter. There are also windows in the near-infrared spectral regions where Rayleigh scattering is smaller.

Analysis of the spectral reflectance of different surfaces generally shows a distinct difference between the visible and near-IR regions. Observations in both the visible (such as 0.58 to 0.68 μm) and near-infrared (such as 0.725 to 1.1 μm) are useful for monitoring surface conditions. Vegetation regions generally have reflectances in the near-infrared (NIR) that range from 20 to 40%, while visible reflectances generally range from 5 to 15%. Soils also have a higher reflectivity in the NIR than in the visible while the opposite is true for snow.

A common method of monitoring surface vegetation is through the normalized difference vegetation index (NDVI):

$$\mathrm{NDVI} = \frac{R_{\mathrm{NIR}} - R_{\mathrm{VIS}}}{R_{\mathrm{NIR}} - R_{\mathrm{VIS}}} \tag{28}$$

This has long been used to monitor the vegetation, and changes in vegetation, of the entire Earth. NDVI for vegetation generally range from 0.3 to 0.8, with the larger values representing "greener" surfaces. Bare soils range from about 0.2 to 0.3.

Identifying snow cover is important for weather and hydrological forecasting. To detect the presence of snow, recent satellite instruments include observations at 0.66 and 1.6 μm (Fig. 30). The atmosphere is transparent at both these wavelengths, while snow is very reflective at 0.66 μm and not reflective at 1.6 μm. The normalized difference snow index (NDSI),

$$\mathrm{NDSI} = \frac{R_{0.66} - R_{1.6}}{R_{0.66} + R_{1.6}} \tag{29}$$

is used to monitor the extent of snow cover. At visible wavelengths (e.g., 0.66 μm), snow cover is just as bright as clouds and is therefore difficult to distinguish from cloud cover. However, at 1.6 μm, snow cover absorbs sunlight and therefore appears much darker than clouds. This allows the effective discrimination between snow cover and clouds. Values of NDSI <0.4 typically indicate the presence of snow. Figure 30 demonstrates the ability to separate clouds from snow using observations at these wavelengths.

Sea surface temperature (SST) is another surface property we are interested in from a meteorological perspective. An IR window in the atmosphere for sensing

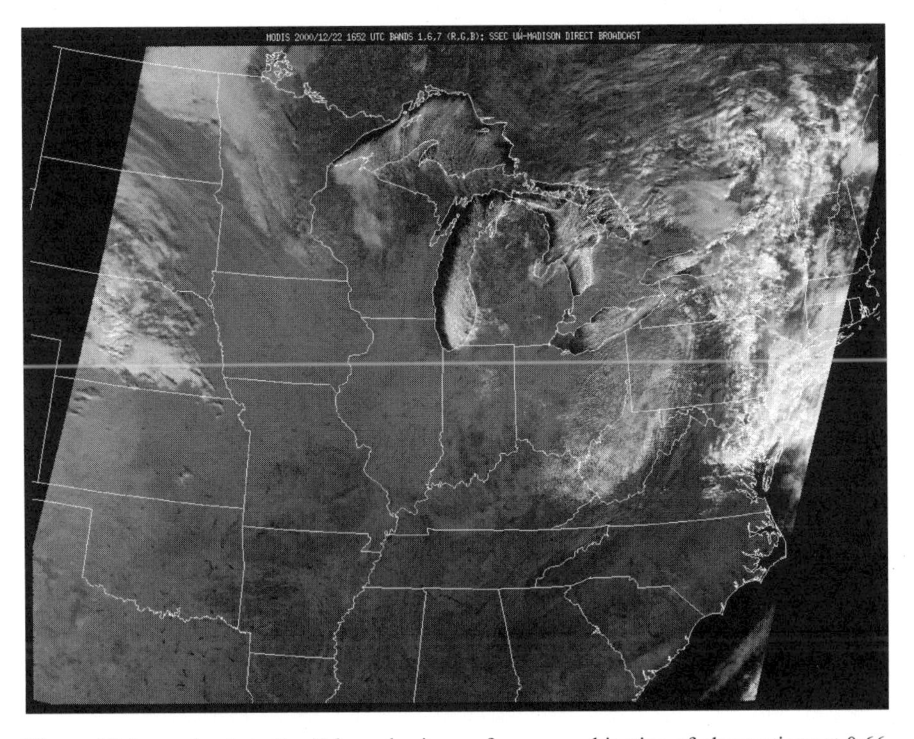

Figure 30 (see color insert) False color image from a combination of observations at 0.66, 1.64, and 2.14 μm are used to detect snow. In this false color images, land surfaces are green, water surfaces are black, snow cover is red, and clouds are white. See ftp site for color image.

surface temperature is the 10- to 12-μm region where absorption by water vapor is weak. Figure 31 is a MODIS (moderate resolution imaging spectrometer) 11-μm image that clearly indicates changes of SST in the vicinity of the Gulf Stream. Most of the radiation in this band is emitted by the surface and transmitted through the atmosphere. In a warm moist atmosphere the difference between the SST and the brightness temperature at 11 μm (BT_{11}) can approach 10°C. This difference is often

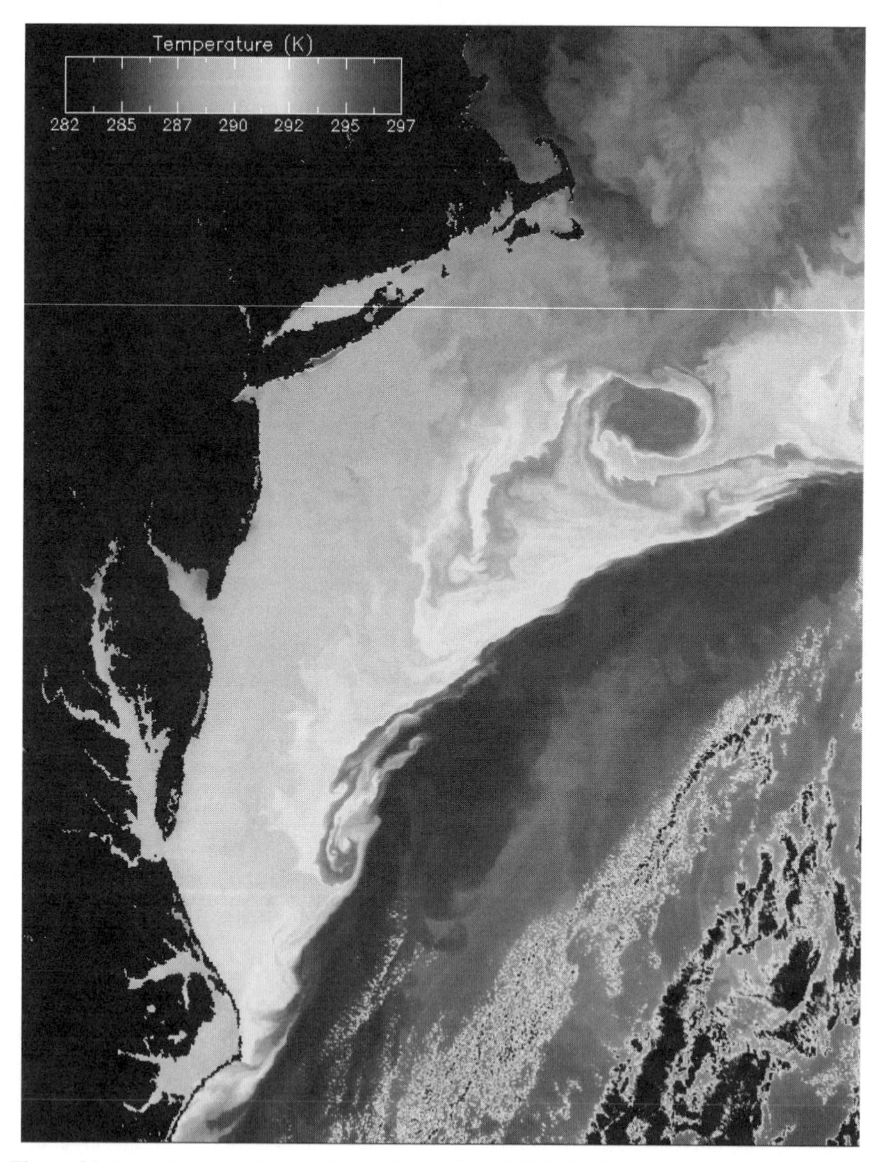

Figure 31 (see color insert) An 11-μm image from MODIS of Atlantic Ocean off east coast of North America. See ftp site for color image.

corrected for by making observations at more than one wavelength, such as 11, 12, 3.7, and 8.5 μm. Differences between these channels represent the total amount of water vapor in the column. For example, the 12-μm channel has more absorption and therefore $(BT_{11} - BT_{12})$ is positive; the greater this difference the larger the water vapor loading of the atmosphere. Observations at these wavelengths are used daily to derive SST. The SST from satellite observations is typically determined from a regression derived empirically using observations from drifting buoys.

Remote Sensing of Clouds

Clouds are generally characterized by higher reflectance and lower temperature than the underlying Earth surface. As such, simple visible and infrared window threshold approaches offer considerable skill in cloud detection. However, there are many surface conditions when this characterization of clouds is inappropriate, most notably over snow and ice. Additionally, some cloud types such as thin cirrus, low stratus at night, and small cumulus are difficult to detect because of insufficient contrast with the surface radiance. Cloud edges cause further difficulty since the instrument field of view will not always be completely cloudy or clear. There are many different methods of detecting clouds. In this section we review some of the more common approaches.

The simplest cloud measurement technique is the threshold method in which an equivalent blackbody temperature or a spectral reflectance threshold is selected that distinguishes between cloud and noncloud in infrared or visible satellite images. Information on cloud top temperature is obtained by comparing the observed brightness temperature with an atmospheric temperature profile—this approach usually underestimates the cloud height. Using a visible or near-infrared reflectance threshold works well for determining clear-sky ocean scenes that are free of sun glint.

Another straightforward approach employs two channels in combination. For example, the split window technique makes use of observations near 11 and 12 μm to detect clouds over oceans. Cloud classification is accomplished by considering the 11-μm blackbody temperature and the difference between 11 and 12 μm. Clear scenes have warm temperatures and brightness temperature differences that are negative, usually less than about $-1°$. Another simple two-channel technique uses visible and infrared observations. In this approach observed visible reflectance and equivalent blackbody temperature are compiled, and observations are then classified based on their relative brightness and temperature. For example, clear sky oceans would be warm and dark while convective clouds would be cold and bright. Classification of the cloud is accomplished by either assigning thresholds or by employing maximum-likelihood statistical techniques.

Once clear pixels can be determined, these observations can be combined with cloudy observations to derive cloud properties. One such approach is the CO_2 slicing technique. In this technique a cloud pressure function $G(\lambda_1, \lambda_2, p)$ is defined as an expression involving a pair of differences of two column radiances and $[I(\lambda^{clr})$ and $I(\lambda^{cld})]$, one clear and one cloud contaminated with a cloud at pressure level p_c. If you express the observed column radiances in terms of differences, then the implied vertical integration need only go from the surface to the cloud top because the above cloud components subtract; then by taking the ratio of two observed radiances, you can remove the coefficient of cloud amount that results from expanding these terms into clear and cloudy portions. It is necessary to make an assumption that the cloud is infinitesimally thin, so that you need only work with a single transmittance function below the cloud. This method enables us to assign a quantitative cloud top pressure to a given cloud element using observed radiances for the CO_2 spectral bands. Defining the radiance from a partly cloudy scene as

$$I_\lambda = F I_\lambda^{\text{cld}} + (1 - F) I_\lambda^{\text{clr}} \tag{30}$$

where F is the fractional cloud cover. The cloud radiance is given by

$$I_\lambda^{\text{cld}} = \varepsilon_\lambda I_\lambda^{\text{bcld}} + (1 - \varepsilon_\lambda) I_\lambda^{\text{clr}} \tag{31}$$

where ε is the cloud emissivity, and $I_\lambda{}^{\text{bcld}}$ is the radiance from an opaque cloud. Thus

$$I_\lambda^{\text{clr}} = B_\lambda(T_{ps}) \tau\lambda(p_s) + \int_{ps}^{0} B_\lambda(T_p)\, d\tau_\lambda \tag{32}$$

$$I_\lambda^{\text{bcld}} = B_\lambda(T_{pc}) \tau\lambda(p_c) + \int_{pc}^{0} B_\lambda(T_p)\, d\tau_\lambda \tag{33}$$

where p_c is the cloud top pressure. Integrating by parts (e.g., $\int u\dot{v}\, dx = uv - \int v\dot{u}\, dx$) and subtracting the two terms yields

$$I_\lambda^{\text{clr}} - I_\lambda^{\text{bcld}} = \int_{p_c}^{p_s} \tau_\lambda(p)\, dB_\lambda(T_p) \tag{34}$$

and

$$I_\lambda - I_\lambda^{\text{clr}} = F\varepsilon_\lambda \int_{p_c}^{p_s} \tau_\lambda(p)\, dB_\lambda(T_p) \tag{35}$$

If two wavelengths are chosen that are close to one another, then $\varepsilon_{\lambda_1} \approx \varepsilon_{\lambda_1}$, which leads to

$$\frac{I_{\lambda_1} - I_{\lambda_1}^{clr}}{I_{\lambda_2} - I_{\lambda_2}^{clr}} = \frac{\int_{p_c}^{p_s} \tau_{\lambda_1}(p)\, dB_{\lambda_1}(T_p)}{\int_{p_c}^{p_s} \tau_{\lambda_2}(p)\, dB_{\lambda_2}(T_p)} \tag{36}$$

In practice, the left-hand side is determined from observations, and the right-hand side is calculated from a known temperature profile and the profiles of atmospheric transmittance for the spectral channels as a function of p_c. The optimum cloud top pressure is determined when the difference between the right-hand and left-hand sides of the equation are a minimum.

Once the cloud height has been determined, the effective cloud amount $\eta = F\varepsilon$ is determined from a window channel observation.

$$F\varepsilon = \frac{I_w - I_w^{\text{clr}}}{B_w(T_{pc}) - I_w^{\text{clr}}} \tag{37}$$

The CO_2 slicing approach is good for detecting clouds in the upper troposphere but fails for clouds in the lower troposphere.

Remote-Sensing Atmospheric Temperature Profiles

The retrieval of the atmospheric temperature and moisture profile is often accomplished using spectral observations in the infrared. The appropriate equation for the transfer of infrared radiation is

$$I_\lambda = B_\lambda(T_{\text{sfc}})T_\lambda(0) + \int_{z=0}^{\infty} B_\lambda(T)\frac{dT_\lambda(z)}{dz}dz \tag{38}$$

where I_λ is the observed radiance at wavelength λ, T_{sfc} is the surface temperature, T_λ is the transmittance, and $B_\lambda(T)$ is the Planck function containing information on the atmospheric temperature. The term

$$\frac{dT_\lambda(z)}{dz} = W(z) \tag{39}$$

is referred to as the weighting function. The intensity measured by a satellite radiometer due to the emission from a layer in the atmosphere at location z, is determined from the layer blackbody emission $B_\lambda(T)$ weighted by the factor $W(z)$. The weighting function is of fundamental importance to vertical sounding the atmosphere from satellite observations. The weighting distribution depends on the strength and distribution of the absorbing gas.

To retrieve atmospheric temperature profiles, satellite radiometers make measurements in the carbon dioxide absorption bands because carbon dioxide is relatively uniformly mixed in the atmosphere, and thus the vertical distribution is known. Observations are made at spectral regions across the carbon dioxide absorption band, including weak and strong absorption regions. Figure 32 shows the weighting functions in the 12- to 15-μm spectral region of the Geostationary Orbiting Environmental Satellite (GOES) sounder radiometer. The 14.7-μm region is a strong absorption region of carbon dioxide and so the weighting function peaks in the stratosphere. The weighting function at 14.7 μm is near zero for pressures greater than 500 mb, indicating that radiance observations at this wavelength receive no contribution from the lower atmosphere and surface. The 13.4-μm spectral region is a weak absorption region and weighting function peaks at the surface, with only small

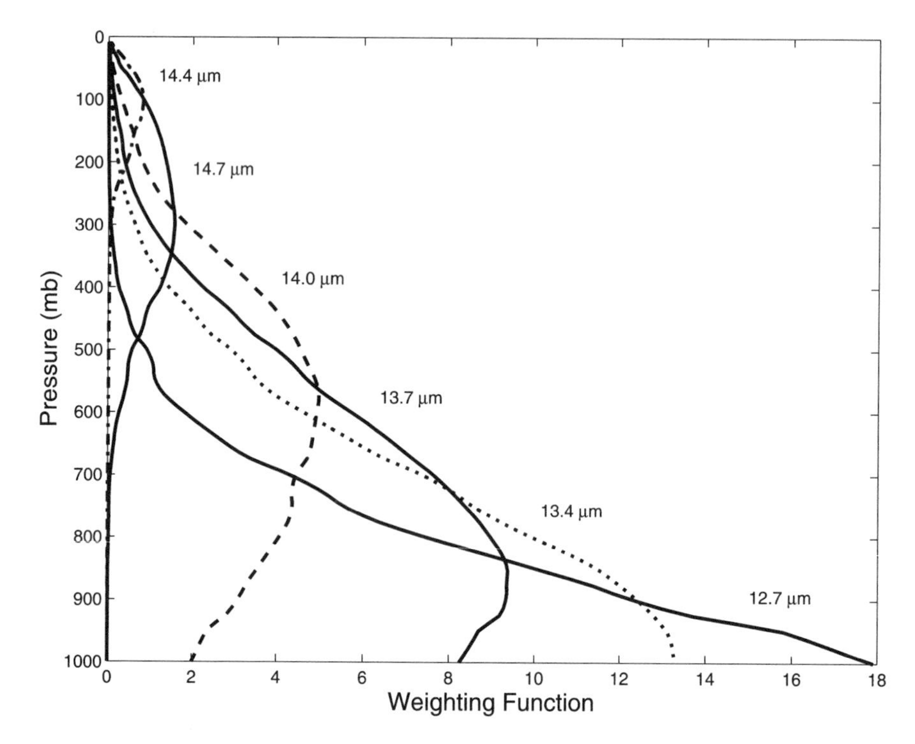

Figure 32 GOES sounder channels used in retrieval of temperature profiles.

contributions in the stratosphere. The location of peak in the weighting function indicates the region of the atmosphere that makes the largest contribution to the radiance being measured. The width of the weighting function characterizes the vertical resolution of the retrieval.

Given different spectral observations, I_λ, we wish to solve for the temperature profile, given the distribution of absorbing gas. This solution is not straightforward as the equation is nonlinear and the problem is underconstrained so that no unique solution exists. To simplify the problem we consider the discrete form of the radiative transfer equation

$$I_i = (B_0)_i T_i(z = 0) + \sum_{j=1}^{m} B_{i,j} K_{i,j} T_{i,j-1} - T_{i,j} \tag{40}$$

where i represents a spectral channel and j is the atmospheric level. Or

$$I_i = (B_0)_{iri}(z = 0) + \sum_{j=1}^{m} B_{i,j} K_{i,j} \tag{41}$$

The vector I_λ represents our spectral channel observations, K_{ij} is the discrete weighting function elements and includes the surface term, and the unknowns are the vector $B_{i,j}$. Our problem is to invert this equation to solve for $B_{i,j}$ from which the temperature profile follows. The problem is underconstrained: Given I_λ we cannot say anything about the individual values of $B_{i,j}$. We require a priori constraints, for example, by reducing the number of layers over which the profile is specified or by specifying the representation of $B_{i,j}$. The numerical methods used to retrieve temperature from the radiance measurements are described elsewhere. Retrievals often start with a first-guess profile, for example, from in situ radiosonde observations or a numerical forecast model. The first-guess profiles are then adjusted until calculated radiances match the observed radiances within some threshold. Because the observed radiances arise from deep and overlapping layers, as indicated in the weighting function, the retrieved temperature profiles are for layers of the atmosphere and do not resolve the sharp temperature gradients sometimes observed in radiosonde measurements. However, the satellite provides better spatial and temporal resolution and can be very useful for forecasting.

An example is monitoring the convective conditions of the atmosphere using the lifted index. The lifted index is the temperature difference found by subtracting the temperature of a parcel of air lifted from the surface to 500 mb from the existing

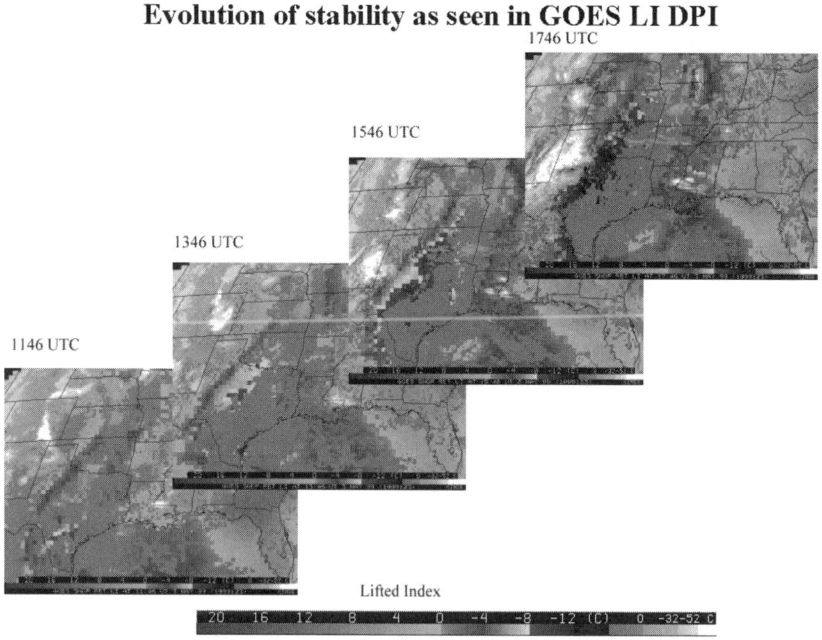

Evolution of stability as seen in GOES LI DPI

Figure 33 (see color insert) Sequence of GOES-derived lifted index on May 3, 1999. Red regions indicate the potential areas of thunderstorm development. Strong tornadoes developed in southwest Oklahoma before 22 UTC. The lifted index versus color code is given in the legend. See ftp site for color image.

GOES Sounder PW pattern on 20 August 1999

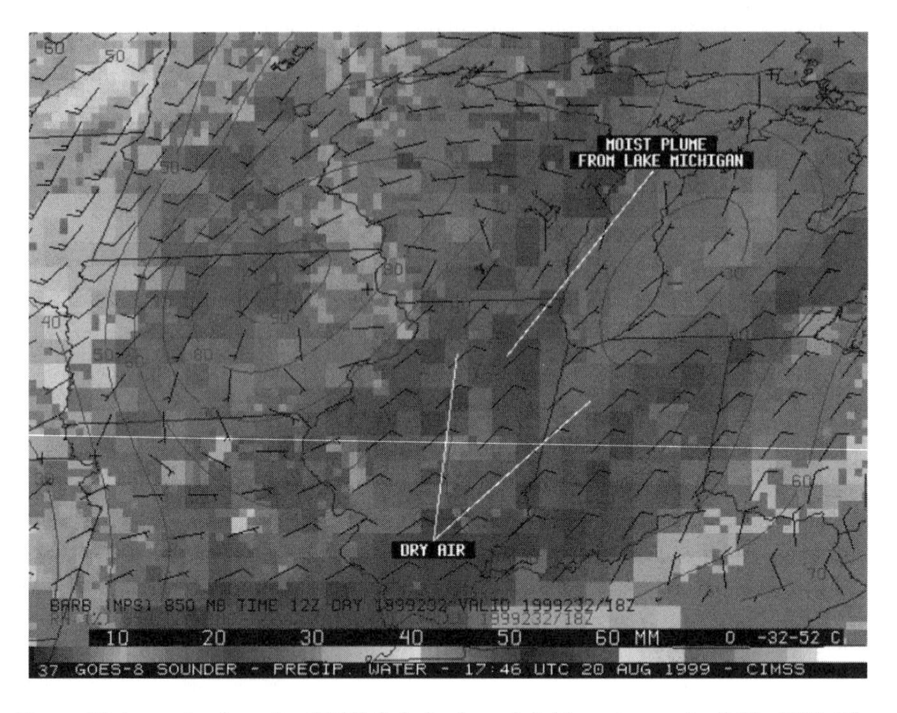

Figure 34 (see color insert) GOES-8-derived precipitable water on April 20, 1999. Blue regions indicate regions of low PW and green relative high amounts of PW. The satellite analysis clearly shows dry region south of Great Lakes. See ftp site for color image.

temperature at 500 mb. The lifted index numbers are related to thunderstorm severity. Values less than or equal to -6 indicates conditions are very favorable for development of thunderstorms with a high likelihood that if they occur, they would be severe with high winds and hail. The satellite-observed lifted index on May 3, 1999, over the south central United States is shown in Figure 33. The time sequence of the satellite-derived lifted index clearly shows (see the red region) the region favorable for the development of severe thunderstorms.

Remote-Sensing Atmospheric Moisture

Once the temperature profile of the atmosphere has been determined, infrared observations in water vapor absorption bands can be used to infer atmospheric water content. GOES observations at water vapor absorption bands are routinely used by the National Oceanic and Atmospheric Administration (NOAA) to derive the vertically integrated water vapor, or precipitable water (PW). An example is shown

Cloud field pattern on 20 August 1999

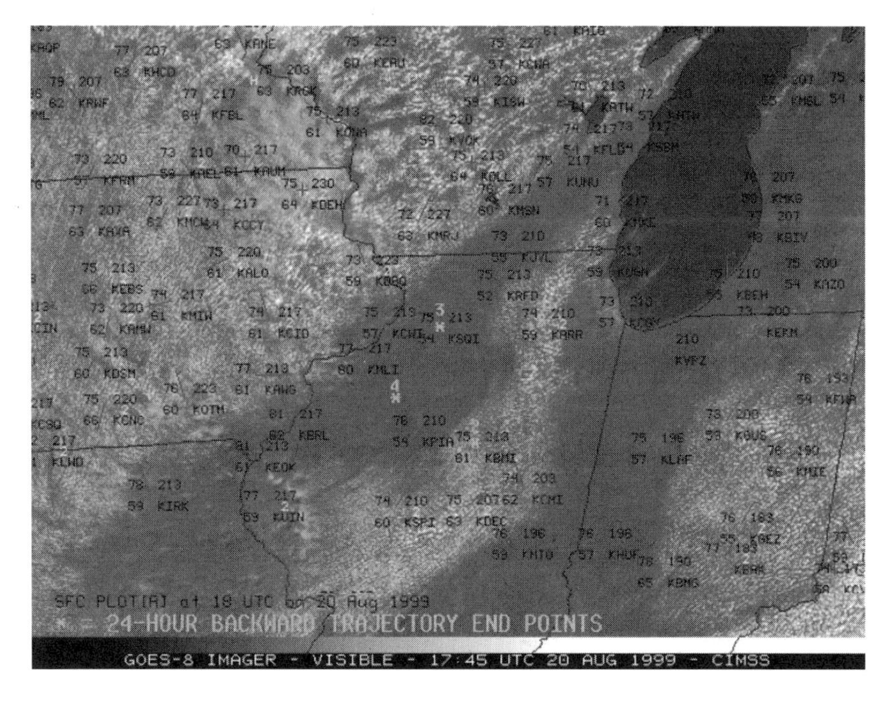

Figure 35 GOES-8-visible image on April 20, 1999 at 1715 UTC. Cumulus clouds are absent in the regions of low PW. See ftp site for color image.

in Figure 34. The GOES sequence of observations captures the large region of relatively dry air with less than 20 mm of PW (blue enhancement) in the western Great Lakes region on August 20, 1999. During the late morning and afternoon hours, the GOES-derived PW shows an elongated moist region with PW greater than 20 mm (green enhancement) across Illinois.

This moisture distribution impacted the cloud formation on this day. Cumulus formation was suppressed on either side of this moist plume, as seen in the GOES visible image. The GOES-derived PW remained over these cloud-free areas during the entire day (see Fig. 35).

CHAPTER 21

THE CLASSIFICATION OF CLOUDS

ARTHUR L. RANGNO

1 INTRODUCTION

Official synoptic weather observations have contained information of the coverage of various types of clouds since 1930. These cloud observations are based on a classification system that was largely in place by the late 1890s (Brooks, 1951). In recent years, these cloud observations have also had increased value. Besides their traditional role in helping to assess the current condition of the atmosphere and what weather may lie ahead, they are also helping to provide a long-term record from which changes in cloud coverage and type associated with climate change might be discerned that might not be detectable in the relatively short record of satellite data (Warren et al., 1991). This article discusses what a cloud is, the origin of the classification system of clouds, and contains photographs of the most commonly observed clouds.

2 WHAT IS A CLOUD?

As defined by the World Meteorological Organization (1969), a cloud is an aggregate of minute suspended particles of water or ice, or both, that are in sufficient concentrations to be visible—a collection of "hydrometeors," a term that also includes in some cases, due to perspective, the precipitation particles that fall from them.

Handbook of Weather, Climate, and Water: Dynamics, Climate, Physical Meteorology, Weather Systems, and Measurements, Edited by Thomas D. Potter and Bradley R. Colman.
ISBN 0-471-21490-6

Clouds are tenuous and transitory. No single cloud element, even within an extensive cloud shield, exists for more than a few hours, and most small clouds in the lower atmosphere exist for only a few minutes. In precise numbers, the demarcation between a cloud and clear air is hard to define. How many cloud drops per liter constitute a cloud? When are ice crystals and snow termed "clouds" rather than precipitation? When are drops or ice crystals too large to be considered "cloud" particles, but rather "precipitation" particles?

These questions are difficult for scientists to answer in unanimity because the difference between cloud particles and precipitation particles, for example, is not black and white; rather they represent a continuum of fallspeeds. For some scientists, a 50-μm diameter drop represents a "drizzle" drop because it likely has formed from collisions with other drops, but for others it may be termed a "cloud" drop because it falls too slowly to produce noticeable precipitation, and evaporates almost immediately after exiting the bottom of the cloud. Also, the farther an observer is from falling precipitation, the more it appears to be a "cloud" as a result of perspective. For example, many of the higher "clouds" above us, such as cirrus and altostratus clouds, are composed mainly of ice crystals and even snowflakes that are settling toward the Earth; they would not be considered a "cloud" by an observer inside them on Mt. Everest, for example, but rather a very light snowfall. Some of the ambiguities and problems associated with cloud classification by ground observers were discussed by Howell (1951).

3 ORIGIN OF THE PRESENT-DAY CLOUD CLASSIFICATION SYSTEM

The classification system for clouds is based on what we see above us. At about the same time at the turn of the 19th century, the process of classifying objectively the many shapes and sizes of something as ephemeral as a cloud was first accomplished by an English chemist, Luke Howard, in 1803, and a French naturalist, Jean Baptiste Lamarck, in 1802 (Hamblyn, 2001). Both published systems of cloud classifications. However, because Howard used Latin descriptors of the type that scientists were already using in other fields, his descriptions appeared to resemble much of what people saw, and because he published his results in a relatively well-read journal, *Tilloch's Philosophical Magazine*, Howard's system became accepted and was reproduced in books and encyclopedias soon afterward (Howard, 1803).

Howard observed, as had Lamarck before him, that there were three basic cloud regimes. There were fibrous and wispy clouds, which he called *cirrus* (Latin for hair), sheet-like laminar clouds that covered much or all of the sky, which he referred to as *stratus* (meaning flat), and clouds that were less pervasive but had a strong vertical architecture, which he called *cumulus* (meaning heaped up). Howard used an additional Latin term, *nimbus* (Latin for cloud), meaning in this case, a cloud or system of clouds from which precipitation fell. Today, *nimbus* itself is not a cloud, but rather a prefix or suffix to denote the two main precipitating clouds, *nimbostratus* and *cumulonimbus*. The question over clouds and their types generated such enthusiasm among naturalists in the 19th century that an ardent observer and member of the British Royal Meteorological Society, Ralph Abercromby, took two voyages around the world to make sure that no cloud type had been overlooked!

The emerging idea that clouds preferred just two or three levels in the atmosphere was supported by measurements using theodolites and photogrammetry to measure cloud height at Uppsala, Sweden, as well as at sites in Germany and in the United States in the 1880s. These measurements eventually led H. Hildebrandsson, Director of the Uppsala Observatory, and Abercromby to place the "low," "middle," and "high" cloud groupings of Howard more systematically in their own 1887 cloud classification. At this time, *cumulus* and *cumulonimbus* clouds were placed in a fourth distinct category representing clouds with appreciable updrafts and vertical development.

Howard's modified classification system was re-examined at the International Meteorological Conference at Munich in 1891 followed by the publication of a color cloud atlas in 1896 (Hildebrandsson et al., 1896). At this point, the definitions of clouds were close to their modern forms. Additional international committees made minor modifications to this system in 1926 that were realized with the publication of the 1932 International Cloud Atlas. Little change has been made since that time.

The most comprehensive version of the classification system was published in two volumes (International Cloud Atlas) by the World Meteorological Organization in 1956 (WMO, 1956). Volume I contained the cloud morphology and Volume II consisted of photographs. An abridged Atlas published in 1969 consisted of combined morphology and photographs. The descriptions of clouds and their classifications (Volume I) were published again in 1975 by the WMO (WMO, 1975). In 1987, a revised Volume II of photographs (WMO, 1975) was published that included photographs of clouds from more disparate places than in the previous volumes.

4 THE CLASSIFICATION OF CLOUDS

There are ten main categories or genera into which clouds are classified for official observations (e.g., British Meteorological Office, 1982; Houze, 1993): *cirrus*, *cirrostratus*, *cirrocumulus*, *altostratus*, *altocumulus*, *nimbostratus*, *stratocumulus*, *stratus*, *cumulus*, and *cumulonimbus*. Table 1 is a partial list of the nomenclature used to describe the most commonly seen species and varieties of these clouds. Figures 1 to 13 illustrate these main forms and their most common varieties or species.

Within these ten categories are three cloud base altitude regimes: "high" clouds, those with bases generally above 7 km above ground level (AGL); "middle-level" clouds, those with bases between 2 and about 7 km AGL; and "low" clouds, those with bases at or below 2 km AGL. The word "about" is used because clouds with certain visual attributes that make them, for example, a middle-level cloud, may actually have a base that is above 7 km. Similarly, in winter or in the Arctic, high clouds with cirriform attributes (fibrous and wispy) may be found at heights below 7 km. Also, some clouds that are still considered low clouds (e.g., *cumulus* clouds) can have bases that are a km or more above the general low cloud upper base limit of

TABLE 1 The Ten Cloud Types and Their Most Common Species and Varieties.

Genera	Species	Varieties
Cirrus (Ci)	Uncinus, fibratus, spissatus, castellanus	Intortus, radiatus, vertebratus
Cirrostratus (Cs)	Nebulosus, fibratus	
Cirrocumulus (Cc)	Castellanus, floccus lenticularis	Undulatus
Altocumulus (Ac)	Castellanus, floccus, lenticularis	Translucidus, opacus, undulatus, perlucidus
Altostratus (As)	None	Translucidus, opacus
Nimbostratus (Ns)	None	None
Stratocumulus (Sc)	Castellanus, lenticularis	Perlucidus, translucidus opacus
Stratus (St)	Fractus, nebulosus	
Cumulonimbus (Cb)	Calvus, capillatus	
Cumulus (Cu)	Fractus, humilis, mediocris, congestus	

Letters in parentheses denote accepted abbreviations.
Source: From World Meteorological Organization, 1975.

2 km AGL. Therefore, these cloud base height boundaries should be considered somewhat flexible. Note, too, that what is classified as an *altocumulus* layer when seen from sea level will be termed a *stratocumulus* layer when the same cloud is seen by an observer at the top of a high mountain because the apparent size of the cloud elements, part of the definition of these clouds, becomes larger the nearer one is to the cloud layer. The definitions are made on the basis of the distance of the observer from the cloud.

The classification of clouds is also dependent on their composition. This is because the composition of a cloud, all liquid, all ice, or a mixture of both, determines many of its visual attributes on which the classifications are founded (e.g., luminance, texture, color, opacity, and the level of detail of the cloud elements). For example, an *altocumulus* cloud cannot contain too many ice crystals and still be recognizable as an *altocumulus* cloud. It must always be composed largely of water drops to retain its sharp-edged compact appearance. Thus, it cannot be too high and cold. On the other hand, wispy trails of ice crystals comprising *cirrus* clouds cannot be too low (and thus, too warm). Therefore, having the ability to assess the composition of clouds (i.e., ice vs. liquid water) visually can help in the determination of a cloud's height.

Other important attributes for identifying a cloud are: How much of the sky does it cover? Does it obscure the sun's disk? If the sun's position is visible, is its disk sharply defined or diffuse? Does the cloud display a particular pattern such as small cloud elements, rows, billows, or undulations? Is rain or snow falling from it? If so, is the rain or snow falling from it concentrated in a narrow shaft, suggesting heaped cloud tops above, or is the precipitation widespread with little gradation, a characteristic that suggests uniform cloud tops? Answering these questions will allow the best categorization of clouds into their ten basic types.

Figure 1 Cirrus (*fibratus*) with a small patch of *cirrus spissatus* left between the trees.

4.1 High Clouds

Cirrus, *cirrostratus*, and *cirrocumulus* clouds (Figs. 1, 2, and 3, respectively) comprise high clouds. By WMO definition, they are not dense enough to produce shading except when the sun is near the horizon, with the single exception of a thick patchy *cirrus* species called *cirrus spissatus* in which gray shading is allowable.[1] *Cirrus* and *cirrostratus* clouds are composed of ice crystals with, perhaps, a few momentary exceptions just after forming and when the temperature is higher than −40°C. In this case, droplets may be briefly present at the instant of formation. The bases of *cirrus* and *cirrostratus* clouds, composed of generally low concentrations of ice crystals that are about to evaporate, are usually colder than −20°C. The coldest cirriform clouds (i.e., *cirrus* and *cirrostratus*) can be −80° C or lower in deep storms with high cloud tops (> 15 km above sea level) such as in thinning anvils associated with high-level outflow of thunderstorms.

Because of their icy composition, *cirrus* and *cirrostratus* clouds are fibrous, wispy, and diffuse. This wispy and diffuse attribute is because the ice crystals that comprise them are overall in much lower concentrations (often an order of magnitude or more) than are the droplet concentrations in liquid water clouds. In contrast, droplet clouds look hard and sharp-edged with the details of the tiniest elements clearly visible. The long filaments that often comprise cirriform clouds are caused by

[1]Many users of satellite data refer to *cirrus* or *cirriform* those clouds with cold tops in the upper troposphere without regard to whether they produce shading as seen from below. However, many such clouds so described would actually be classified as *altostratus* clouds by ground observers. This is because such clouds are usually thick enough to produce gray shading and cannot, therefore, be technically classified as a form of cirrus clouds from the ground.

Figure 2 Cirrostratus (*nebulosus*), a relatively featureless ice cloud that may merely turn the sky from blue to white.

the growth of larger ice crystals that fall out in narrow, sloping shafts in the presence of changing wind speeds and directions below the parent cloud. As a result of the flow settling of ice crystals soon after they form, mature *cirrus* and *cirrostratus* clouds are, surprisingly, often 1 km or more thick, although the sun may not be appreciably dimmed (e.g., Planck et al., 1955; Heymsfield, 1993).

Cirrus and *cirrostratus* clouds often produce haloes when viewed from the ground whereas thicker, mainly ice clouds, such as altostratus clouds (see below) cannot. This is because the cirriform clouds usually consist of small, hexagonal prism crystals such as thick plates, short solid columns—simple crystals that refract the sun's light as it passes through them. On the other hand, *altostratus* clouds are much deeper and are therefore composed of much larger, complicated ice crystals and snowflakes that do not permit simple refraction of the sun's light, or, if a halo exists in its upper regions, it cannot be seen due to the opacity of the *altostratus* cloud. The appearance of a widespread sheet of cirrostratus clouds in wintertime in the middle and northern latitudes has long been identified as a precursor to steady rain or snow (e.g., Brooks, 1951).

Cirrocumulus clouds are patchy, finely granulated clouds. Owing to a definition that allows no shading, they are very thin (less than 200 m thick), and usually very short-lived, often appearing and disappearing in minutes. The largest of the visible cloud elements can be no larger than the width of a finger held skyward when observed from the ground; otherwise it is classified as an *altocumulus* or, if lower, a *stratocumulus* cloud.

Figure 3 Cirrocumulus: the tiniest granules of this cloud can offer the illusion of looking at the cloud streets on a satellite photo.

Cirrocumulus clouds are composed mostly or completely of water droplets. The liquid phase of these clouds can usually be deduced when they are near the sun; a corona or irisation (also called iridescence) is produced due to the diffraction of sunlight by the cloud's tiny (<10-μm diameter) droplets. However, many *cirrocumulus* clouds that form at low temperatures ($<-30°C$) migrate to fibrous *cirriform* clouds within a few minutes, causing the granulated appearance to disappear as the droplets evaporate or freeze to become longer-lived ice crystals that fall out and can spread away from the original tiny cloudlet.

4.2 Middle-Level Clouds

Altocumulus, *altostratus*, and *nimbostratus* clouds (Figs. 4, 5, 6, and 7, respectively) are considered middle-level clouds because their bases are located between about 2 and 7 km AGL (see discussion concerning the variable bases of *nimbostratus* clouds below). These clouds are the product of slow updrafts (centimeters per second) often taking place in the middle troposphere over an area of thousands of square kilometers or more. Gray shading is expected in *altostratus*, and is generally present in *altocumulus* clouds. *Nimbostratus* clouds by definition are dark gray and the sun cannot be seen through them. It is this property of shading that differentiates these clouds from high clouds, which as a rule can have no shading.

Figure 4 Altocumulus translucidus is racing toward the observed with an altocumulus lenticularis cloud in the distance above the horizon. Cirrostratus nebulosus is also present above the altocumulus clouds.

Figure 5 Altocumulus lenticularis: these may hover over the same location for hours at a time or for just a few minutes. They expand and shrink as the grade of moisture in the lifted air stream waxes and wanes.

Figure 6 Altostratus translucidus overspreads the sky as a warm-frontal system approaches Seattle. The darker regions show where the snow falling from this layer (virga) has reached levels that are somewhat lower than from regions nearby.

Figure 7 Nimbostratus on a day with freezing level at the ground. Snowfall is beginning to obscure the mountain peaks nearby and snow reached the lower valley in the foreground only minutes later.

Altostratus and *altocumulus* are different from one another in the same way that *cirrus* and *cirrostratus* clouds are different from *cirrocumulus* clouds; in *altocumulus* clouds, droplets predominate, giving them a crisp, sharper-edged look. With *altostratus* clouds, ice crystals and snowflakes dominate or comprise the entire cloud, giving it a diffuse, fibrous look. The observation of *altocumulus* and *altostratus* clouds in the sky has long been a marker for deteriorating weather in the hours ahead.

In spite of its name, *altocumulus* clouds are generally rather flat clouds that strongly resemble *stratocumulus* clouds. An exception to this overall laminar architecture is in those species of *altocumulus* called *castellanus* and *floccus*. In these forms, *altocumulus* clouds resemble miniature, lofted *cumulus* clouds that usually occur in rows or patches rather than in widespread layers.

Altocumulus clouds are distinguished from *cirrocumulus* because they are lower and the cloud elements in *altocumulus* are, or appear to be from the ground, several times larger than those in *cirrocumulus* clouds. For example, the elements of an *altocumulus* cloud are typically the width of three fingers held skyward at the ground. Also, shading toward the center of the thicker elements is usually present in *altocumulus* clouds, a property that is not allowed in the classification of *cirrocumulus* clouds. *Altocumulus* clouds are distinguished from *stratocumulus* because they are higher above ground level than *stratocumulus* (at least 2 km) and because the individual cloud elements in *altocumulus* are, or appear to be from the ground, smaller than those in *stratocumulus*.

In spite of the gray shading that may be present, *altocumulus* clouds are rarely more than 1 km thick. This is because the concentrations of drops in them are relatively high (typically 50,000 or more per liter) compared with fibrous ice clouds whose particle concentrations may only be only tens to a few thousand per liter. This density of particles produces an optical depth in which the sun's disk can be obscured (optical depth of 4 or more) by an *altocumulus* cloud layer only 300–500 m thick.

Altocumulus clouds sometimes sport patchy "virga." *Virga* is light precipitation that falls from a cloud but does not reach the ground because it evaporates. Because virga is almost always caused by falling snow, it appears fibrous, often with striations or long filaments that often far surpass the depth of the cloud from which it is falling. Virga, because it is comprised of falling snow, can appear to be quite dense. *Altocumulus* clouds with virga are predominantly those clouds whose temperatures are lower than −10° C (Heymsfield, 1993). However, at the same time, they are rarely colder than about −30° C. This is because at very low temperatures they are likely to take on the attributes of ice clouds such as *cirrus* or its thicker brethren, *altostratus*.

The species of *altocumulus* clouds called *altocumulus castellanus* has always had a special significance in meteorology because these clouds reveal a conditionally unstable lapse rate in the middle troposphere. Instability of this sort has been viewed as a marker for likely releases of deeper convection in the hours ahead. Occasionally, *castellanus* clouds group and enlarge into higher based *cumulus* and *cumulonimbus* clouds.

When the winds are relatively strong aloft (greater than about $20 \mathrm{m\,s^{-1}}$) and moderately moist, but stable lapse rate conditions are present, a species of *altocumulus* called *lenticularis* (lens or almond-shaped) clouds (Fig. 5) may form over or downwind of mountains. *Altocumulus lenticularis* clouds can hover over the same location for minutes to hours while expanding and shrinking in response to fluctuations in the relative humidity of the air mass being lifted over the terrain. Because the conditions under which these clouds form are most often associated with advancing short wave troughs in the middle and upper atmosphere and their accompanying regions of low pressure, *lenticularis* clouds are usually precursors to deteriorating weather.

With *altostratus* clouds (Fig. 6), the dominance of ice causes a diffuse, amorphous appearance with striations or fallstreaks (virga) on the bottom; an observer is usually viewing relatively low concentrations of precipitation particles rather than a cloud per se. *Altostratus* clouds are rarely less than 2 km thick and often have tops at the same heights as *cirrus* and *cirrostratus* clouds. Because of this great altitude range, they are considerably colder and span a much greater temperature range than do *altocumulus* clouds. They also, by definition, cover the entire sky or at least wide portions of it; they are not patchy clouds. Precipitation is usually imminent when *altostratus* clouds are moving in because they are a sign that a widespread slab of air is being lifted, usually that due to an approaching cyclone and its frontal system (Brooks, 1951).

The relatively low concentrations of large particles in some *altostratus* clouds (often tens per liter) can allow the sun's position to be seen as though looking through ground or fogged glass: the sun's position is apparent, but the outline of its disk is not. This may be the case (*translucidus* variety) even when the cloud layer is two or more kilometers thick.

Somewhat surprisingly, when the top of an *altostratus* cloud layer is warmer than about -30 to $-35°C$, it is not uncommon to find that the top is comprised of a thin droplet cloud virtually identical to an *altocumulus* cloud layer that is producing virga. The survival, growth, aggregation, and breakup of ice crystals spawned by these cold, liquid clouds over a great depth usually obscures the ice-producing droplet cloud top in these cases. These kinds of situations were dubbed "the upside-down storm" when first noticed in the mid-1950s because the coldest part of the cloud (the top) was liquid and the warmer regions below were comprised of ice (Cunningham, 1957).

Optical phenomena seen from the ground with *altostratus* clouds are limited to *parhelia* ("sun dogs"). They are usually observed in thin portions of *altostratus* when the sun is low in the sky. Parhelia are bright, colored highlights, which sometimes rival the brightness of the sun, that are located 22° from the sun's position. They are most noticeable in the morning or before sunset.

However, since the composition of the uppermost regions of the deepest *altostratus* clouds are virtually identical to *cirriform* ice clouds with simpler, smaller ice crystals, haloes are often observed in the uppermost portions of deep *altostratus* clouds.

Nimbostratus clouds (Fig. 7) are virtually identical to *altostratus* clouds in their composition except that their bases are usually perceived from the ground as lower than in *altostratus* (from which it has usually derived due to a downward thickening). Therefore, they often appear somewhat darker than *altostratus* clouds and, by definition, do not allow the sun to be seen through them. The perceived base of *nimbostratus* is caused by snowflakes that are melting into raindrops. This apparent base of the cloud is a result of the greater opacity of snow particles giving the impression of a bottom or sharp increase in thickness of the cloud at the melting level. Thus, the base of an amorphous, precipitating *nimbostratus* cloud might be perceived at mid-levels on a day when the freezing level is high (above 2 km) such as in southern latitudes or the tropics, or be perceived as low when the freezing level is low as in northern latitudes in the winter. Generally, the bottom of *nimbostratus* clouds is obscured by lower detached clouds such as *stratus fractus*, or *stratocumulus*.

Nimbostratus clouds produce relatively steady precipitation that often continues for hours at a time. They are not clouds responsible for passing showers with periods of sun in between. The tops of dark, steadily precipitating *nimbostratus* clouds can be as shallow as 2–3 km and even be above freezing in temperature, or they may reach into the upper regions of the troposphere (to *cirriform* cloud levels) and be as cold as −80°C. At elevations above the freezing level *nimbostratus* is largely composed of ice crystals and snowflakes, though embedded thin (supercoooled) droplet cloud layers similar to *altocumulus* clouds are relatively common. Also, similar to *altostratus* clouds, when the temperature at the top of *nimbostratus* clouds is above about −30° to −35°C, a thin droplet cloud layer may be found in which the first ice crystals form.

A broken-to-overcast layer of shallow *stratus* or *stratocumulus* clouds often resides at the bottom of *nimbostratus* clouds. However, while usually not precipitating themselves, these lower cloud layers are important in enhancing the amount of rain or snow that falls from *nimbostratus* clouds. This enhancement occurs because of the accretion or riming of the relatively small cloud drops in the lower clouds by the rapidly falling precipitation-sized particles. This enhancement is especially evident in hilly or mountainous regions. However, just the existence of lower clouds even with drops too small to be collected by the faster falling precipitation indicates enhanced precipitation at those locations compared with cloud-free locations. This is because where there are no lower clouds, the precipitation is likely to be subject to a degree of evaporation, and the drops or snowflakes are slightly smaller in comparison to those locations where clouds exist below the *nimbostratus* and there is no evaporation through the depth of the cloud.

Cumulonimbus clouds (see below) may also be embedded in *nimbostratus* clouds. The presence of such clouds within *nimbostratus* is evident by sudden gushes of much heavier rain and sometimes lightning within a context of relatively steady rain.

4.3 Low Stratiform Clouds

Stratocumulus and *stratus* clouds (Figs. 8 and 9, respectively) are low-based, shallow stratiform clouds, almost always less than 1 km thick. They are composed of droplets unless the cloud top is cooler than about -5 to $-10°$ C in which case ice crystals may form (Hobbs and Rangno, 1985). *Stratocumulus* clouds differ from *stratus* clouds because they have a rather lumpy appearance at cloud base with darker and lighter regions due to embedded weak convection. Also, their bases tend to be higher, and more irregular in height than those of *stratus* clouds. *Stratus* presents a smoother, lower, more uniform sky than do *stratocumulus* clouds because the internal convective overturning that produces lighter and darker regions is slight. Drizzle (precipitation comprised of drops < 500-μm diameter that nearly float in the air) is likely to form when the cloud droplet concentrations are lower than about $100\,\text{cm}^{-3}$. Therefore, drizzle is common from both *stratus* and *stratocumulus* clouds at sea and along and inland of coastlines in onshore flow because in such situations the air is clean and there are relatively few cloud condensation nuclei on which droplets can form. However, recent measurements have also shown that drizzle and light rain producing shallow clouds with low droplet concentrations are much more common at inland locations in winter than was once believed (e.g. Huffman and Norman, 1988). Since such clouds are often supercooled in winter, they pose a severe potential for aircraft icing and freezing rain or drizzle at the ground.

Figure 8 Stratocumulus: note the uneven bases of this cloud layer. The darker regions show where there is enhanced modest convection and higher liquid water content.

Figure 9 Stratus: the clouds intercept hills only a few hundred meters above sea level. Note here the uniformity of this cloud layer compared with *stratocumulus* in Fig. 8.

4.4 Convective Clouds

Cumulus and *cumulonimbus* clouds (Figs. 10, 11, 12, and 13, respectively) are convective clouds created when the temperature decreases rather rapidly with increasing height. Differential heating and converging air currents in this circumstance can therefore send plumes of warmer air skyward with relative ease. Convective clouds are limited in coverage compared with stratiform clouds and, except for the anvil portions of *cumulonimbus* clouds, rarely cover the entire sky or do so only for short periods. This coverage characteristic differentiates, for example, *stratocumulus* clouds with their linked cloud bases covering large portions of the sky, and similar-sized *cumulus* clouds that by definition must be relatively scattered into isolated clouds or small clusters with large sky openings.

Cumulus clouds have a size spectrum of their own that ranges from *cumulus fractus*, those first cloud shreds that appear at the top of the convective boundary layer, to *congestus* size (more than about 2 km deep). Between these sizes are *cumulus humilis* and *cumulus mediocris* clouds, clouds that range between about 1 and 2 km in depth, respectively. The tops of these larger clouds are marked by sprouting portions called turrets that represent the growing and usually warmer parts of the cloud. Individual turrets are generally one to a few kilometers wide, although in strong storms individual turrets may coalesce into groups of many turrets to form

Figure 10 In this busy sky, *cumulus mediocris* is at the center, with *cumulus humilis* in the distance, and *cumulus congestus* on the horizon at left. Also present in this photo is *altocumulus translucidus* (left center), *cirrocumulus* (top center), and *cirrus uncinus* (center above *cumulus* and right center).

Figure 11 Cumulonimbus calvus: this rapidly changing and expanding *cumulonimbus* cloud is in the short-lived calvus stage where fibrousness of the top is just beginning to be apparent. No strong rainshaft has yet appeared below base, though the careful eye can see that considerable ice has already formed aloft (right half of cloud). An intense rainshaft emerged below cloud base a few minutes after this photograph was taken.

Figure 12 Cumulus congestus, *cumulonimbus calvus*, and *cumulonimbus capillatus*. In this line of convection north of Seattle, WA, (nonprecipitating) *cumulus congestus* clouds comprise the left side of the line, while in the center a taller single turret has emerged and reached the *cumulonimbus calvus* stage. The right third of the photograph shows a *cumulonimbus capillatus*, the stage that follows the calvus stage. In the capillatus stage, in this instance consisting of a conglomerate of turrets, the tops have lost their compact cumuliform look and are clearly fibrous and icy. Strands of precipitation are falling below the bases of both types of *cumulonimbus* clouds.

a large, tightly packed, and hard-appearing cauliflower mass that roils upward with little turret differentiation.

Prior to reaching the *cumulonimbus* stage, *cumulus* clouds are therefore composed of droplets and contain very few precipitation-sized particles. Precipitation, however, usually begins to develop in *cumulus congestus* clouds if they are more than about 3 km thick over land and about 1.5 to 2 km thick over the oceans (Ludlum, 1980; Wallace and Hobbs, 1977). The precipitation that falls may be caused by collisions with coalescence of cloud drops in the upper portions of the cloud or it may be a result of the formation of ice particles in clouds with cooler bases. However, in the winter, even small *cumulus* clouds with tops colder than about -10 to -15° C can produce virga, snow flurries, or even accumulating amounts of snow. These kinds of small, cold-based, and precipitating *cumulus* clouds are found in winter in such locations as the Great Lakes of the United States, off the east coasts of the continents, and over high mountains or desert regions (Rangno and Hobbs, 1994).

Figure 13 Cumulonimbus capillatus incus: this well-known species of *cumulonimbus* cloud also has a fibrous and icy top, but it is marked by a noticeable flattening there. Incus translates to "anvil." These types of *cumulonimbus* clouds suggest that the convection has spanned the entire troposphere.

If significant precipitation begins to develop in deep *cumulus* clouds, they quickly take on the visual attributes of *cumulonimbus* clouds (Figs. 11–13)—a strong precipitation shaft is seen below cloud base with a cloud top that is soft, fibrous, fraying, or wispy. The visual transition to a softer, fibrous appearance in the upper portion of *cumulus* clouds is caused by the lowering of the concentrations of the particles from hundreds of thousands per liter of relatively small cloud droplets (< 50-μm diameter), to only tens to hundreds per liter of much larger (millimeter-sized) precipitation-sized particles (rain drops or ice particles). These larger particles tend to fall in filaments and help produce a striated appearance.

In the period while this transformation is taking place and the fibrousness is just becoming visually apparent in the upper portions of a *cumulus congestus* cloud, the cloud is entering a short-lived period of its lifecycle when it is referred to as a *cumulonimbus calvus* ("bald") cloud. At this same time, a concentrated precipitation shaft may not be present yet or is just emerging below the cloud base (Fig. 11).

When the fibrousness of the upper portion of the cloud is fully apparent, the *cumulonimbus* cloud has transitioned to a *cumulonimbus capillatus* (hair) in which most or all of its upper portion consists of ice crystals and snowflakes (Figs. 12 and 13). In the tropics or in warm humid air masses, this visual transformation also occurs but can be due solely to the evaporation of the smaller drops leaving drizzle and raindrops rather than ice and snow in smaller *cumulonimbus* clouds. *Cumulo-*

nimbus capillatus clouds span a wide range of depths, from miniature versions only about 2 km deep in polar air masses over the oceans, to as much as 20 km in the most severe thunderstorms in equatorial regions, eastern China in the summer, and the plains and southeast regions of the United States. If a pronounced flattening of the top develops into a spreading anvil then the cloud has achieved the status of a *cumulonimbus capillatus incus* (incus meaning "anvil").

Hail or graupel (soft hail) are usually found, if not at the ground, then aloft in virtually all *cumulonimbus* clouds that reach above freezing level. Updrafts may reach tens of meters per second in *cumulus* and *cumulonimbus* clouds, particularly in warm air masses. These updrafts lead to large amounts of condensation and liquid water content. Depending on how warm the cloud base is, the middle and upper building portions of deep *cumulus* clouds might contain 1 to 5 $g\,m^{-3}$ of condensed water in the form of cloud droplets and raindrops. Supercooled water concentrations of these magnitudes are sufficient to cause a buildup of about 1 cm or more of ice buildup on an airframe for every one to two minutes in cloud. Therefore, they are avoided by aircraft. *Cumulonimbus* clouds are the only clouds, by definition, that produce lightning. If lightning is observed, the cloud type producing it is automatically designated a *cumulonimbus*.

REFERENCES

British Meteorological Office (1982). *Cloud Types for Observers*, Met. O.716. London: Her Majesty's Stationery Office, 38 pp.

Brooks, C. F. (1951). The use of clouds in forecasting. In *Compendium of Meteorology*, Boston: American Meteorological Society, 1167–1178.

Cunningham, R. M. (1957). A discussion of generating cell observations with respect to the existence of freezing or sublimation nuclei. In *Artificial Stimulation of Rain*, Ed. H. Weickmann, New York: Pergamon Press.

Hamblyn, R. (2001). *The Invention of Clouds*, New York: Farrar, Straus and Giroux, 244 pp.

Heymsfield, A. J. (1993). Microphysical structures of stratiform and cirrus clouds. In *Aerosol–Cloud–Climate Interactions*, New York: Academic Press, 97–121.

Hildebrandsson, H., Riggenbach, A., and L. Teeisserenc de Bort, (1896). *Atlas International des Nuages*, Comité Météorologique International.

Hobbs, P. V., and A. L. Rangno, (1985). Ice particle concentrations in clouds, *J. Atmos. Sci.*, **42**, 2523–2549.

Houze, R. A., Jr. (1993). *Cloud Dynamics*, New York: Academic Press, 573 pp.

Howard, L. (1803). On the modifications of clouds, and on the principles of their production, suspension, and destruction. *Phil. Mag.*, **16**, 97–107; 344–357.

Howell, W. E. (1951). The classification of cloud forms. In *Compendium of Meteorology*, Boston: American Meteorological Society, 1161–1166.

Huffman, G. J., and G. A. Norman, Jr., 1988. The supercooled warm rain process and the specification of freezing precipitation. Monthly Wea. Rev., *116*, 2172–2182.

Ludlum, F. H. (1980). *Clouds and Storms: The Behavior and Effect of Water in the Atmosphere*. The Pennsylvania State University Press, University Park, 123–128.

Plank, V. G., Atlas, D., and W. H. Paulsen, (1955). The nature and detectability of clouds and precipitation by 1.25 cm radar. *J. Meteor.*, **12**, 358–378.

Rangno, A. L., and P. V. Hobbs, (1994). Ice particle concentrations and precipitation development in small continental cumuliform clouds. *Quart. J. Roy. Meteor. Soc.*, **120**, 573–601.

Wallace, J. M., and P. V. Hobbs, (1977). *Atmospheric Science: An Introductory Survey*, New York: Academic Press, 216–218.

Warren, S. G., Hahn, C. J., and J. London, (1991). *Analysis of cloud information from surface weather reports*. World Climate Research Programme Report on Clouds, Radiative Transfer, and the Hydrological Cycle, World Meteorological Organization, 61–68.

World Meteorological Organization (1956). *International Cloud Atlas (Complete Atlas)*, Vol. I, Geneva: World Meteorological Organization, 175 pp.

World Meteorological Organization (1956). *International Cloud Atlas (Complete Atlas)*. Vol. II, Geneva: World Meteorological Organization, 124 pp.

World Meteorological Organization (1975). *International Cloud Atlas*, Vol. I, *Manual on the Observations of Clouds and Other Meteors*, Geneva: World Meteorological Organization, 155 pp.

World Meteorological Organization (1987). *International Cloud Atlas*. Vol. II, Geneva: World Meteorological Organization, 212 pp.

CHAPTER 22

ATMOSPHERIC ELECTRICITY AND LIGHTNING

WALTER A. LYONS AND EARLE R. WILLIAMS

1 GLOBAL ELECTRICAL CIRCUIT

The atmosphere is the thin veneer of gas surrounding our planet and is one of the most highly electrically insulating regions in the universe. The electrical conductivity of air near Earth's surface is some 12 orders of magnitude smaller than both Earth's crust and the conductive ionosphere. In such a medium, electric charges may be physically separated by both air and particle motions to create large electric fields. The laws of electrostatics apply in this situation, and, though the charges involved are seldom static, the application of these laws in the atmosphere (in fair weather regions, in electrified clouds which may produce lightning, and in the upper atmosphere where sprites are found) is the science of atmospheric electricity.

Earth itself, whose surface is composed of seawater and water-impregnated crustal rock, is a good electrical conductor. This conducting sphere is known to carry a net negative charge of half a million coulombs. The corresponding downward-directed electric field at the conductor's surface in fair weather regions is of the order of 100 V/m. The conductivity of highly insulating air near Earth's surface is of the order of 10^{-14} mho/m and, according to Ohm's law, invoking a linear relationship between electric field and current density, a downward-directed current is flowing with a current density of order 1 pA/m^2. Integrated over the globe, this amounts to a total current of about 1 kA.

The origin of Earth's persistent fair weather electrification remained a mystery for more than 100 years. In the 1920s, C.T.R. Wilson proposed that the electrification manifest in fair weather regions was caused by the action of electrified clouds and

Handbook of Weather, Climate, and Water: Dynamics, Climate, Physical Meteorology, Weather Systems, and Measurements, Edited by Thomas D. Potter and Bradley R. Colman.
ISBN 0-471-21490-6

thunderstorms worldwide. The fair weather region provided the return current path for electrified weather and is the collective battery for the global circuit. Nominally, 1000 storms are simultaneously active worldwide, each supplying a current of the order of 1 A, which flows upward from the cloud top into the conductive upper atmosphere where the current spreads out laterally and then turns downward in fair weather regions to the conductive earth to complete the circuit.

The finite conductivity of the lower atmosphere is primarily the result of ionization of neutral nitrogen and oxygen molecules by energetic cosmic radiation from space. The atmosphere attenuates the incoming ionizing radiation, and this causes the conductivity to increase exponentially with altitude toward the highly conductive ionosphere, with a scale height of about 5 km. Since the fair weather return current is uniform with altitude, Ohm's law guarantees that the electric field will decline exponentially with altitude from its maximum nominal value of 100 V/m at the surface, with the same scale height. The integral of the fair weather electric field with altitude from the conductive earth to the conductive upper atmosphere is referred to as the ionospheric potential, and amounts to some 250,000 V. This quantity is a global invariant and the standard measure of the global electrical circuit. The ionospheric potential and the net charge on Earth are not constant but vary by approximately $\pm 15\%$ over the course of every day in response to the systematic modulation of electrified weather by solar heating, centered in the tropics. Three major zones of deep convection—the Maritime Continent (including the Indonesian archipelago, Southeast Asia, and Northern Australia), Africa, and the Americas—exhibit strong diurnal variations, and the integrated effect lends a systematic variation to the electrification of Earth and the ionospheric potential in universal time. The consistent phase behavior between the action of the three tropical "chimney" regions and Earth's general electrification is generally regarded as key evidence in support of C.T.R. Wilson's hypothesis.

According to this general idea, the negative charge on Earth is maintained by the action of electrified clouds. One direct agent is cloud-to-ground lightning, which, as explained below, is most frequently negative in polarity. The less frequent but more energetic positive ground flashes, which are linked with sprites in the mesosphere should actually deplete the negative charge in Earth. Possibly more important (though more difficult to evaluate) agents for maintaining the negatively charged Earth are point discharge currents that flow from plants, trees, and other elevated objects in the kilovolt-per-meter level electric fields beneath electrified clouds, and precipitation currents carried to Earth by rain and graupel particles charged by microphysical processes within storms.

The global electrical circuit outlined here provides a natural framework for monitoring electrified weather on a worldwide basis. The expectation that lightning activity is temperature-dependent has spurred interest in the use of the global circuit as a monitor for global temperature change and for upper atmospheric water vapor. A second manifestation of the global circuit, sometimes referred to as the "AC" global circuit, is provided for by the same insulating air layer between two conductors. This configuration is an electromagnetic waveguide for a phenomenon known as Schumann resonances and maintained continuously by worldwide lightning activity. The wavelength for the fundamental resonant mode at 8 Hz is equal to the circumference of Earth. One advantage afforded by Schumann resonances over ionospheric potential as a measure of the global circuit is the capability for continuous monitoring from ground level. The signatures of the three major tropical zones referred to earlier are distinguishable in daily recordings from a single measurement station. Furthermore, variations in Schumann resonance intensity on longer time scale (semiannual, annual, interannual), for which global temperature variations are recognized, have also been documented.

The evidence that giant positive lightning can simultaneously create sprites in the mesosphere and "ring" Earth's Schumann resonances to levels higher than the contribution of all other lightning combined has also spurred interest in this aspect of the global circuit.

2 WHAT IS LIGHTNING?

Lightning is a transient, high-current atmospheric discharge whose path length is measured in kilometers (Uman and Krider, 1989). It is, in effect, a big spark. Benjamin Franklin's legendary (and incredibly risky) kite flying experiment in 1752 documented that, indeed, Thor's bolts were "merely" electricity. The number of scientists killed or injured in subsequent years attempting to duplicate Franklin's experiment attests to the threat to life and property inherent in a lightning discharge.

Lightning represents an electrical breakdown of the air within a cloud between separated pockets of positive and negative electrical space charge. What actually causes the formation and separation of the electrical charges within the thunderstorm? While considerable progress is being made in this research area, a concise theory has yet to emerge (Williams, 1988; MacGorman and Rust, 1998). Suffice it to say that in almost all cases of natural lightning, charge separation results from the interactions between ice particles, graupel, and supercooled droplets within thunderstorm clouds (Fig. 1). The complex updrafts and downdrafts separate the positive and negative charge pools allowing the electric field to attain ever-higher values until dielectric breakdown occurs. Storm updrafts penetrate well above the freezing (0°C) level. It is generally thought that significant updrafts ($>\sim 8$ m/s) in the layers where supercooled water droplets and ice particles coexist (0° to −40°C) are necessary though not sufficient for most lightning-generating scenarios. Lightning discharges also occur within the stratiform precipitation regions of large mesoscale convective systems. Though the sources of the charge centers have been thought caused by advection from the convectively active parts of the storm, there are also strong arguments favoring in situ charge generation. Occasional reports of lightning occurring in warm clouds (entirely above 0°C) have never been convincingly substantiated.

Several terminologies are employed to describe lightning. Discharges occurring within the cloud are called intracloud (IC) events. A cloud-to-ground event is called a CG. In most storms ICs greatly outnumber CGs. The typical ratio may be 4:1, but the ratio may approach 100:1 or more in some especially intense storms. The IC:CG

Figure 1 Cumulonimbus cloud over the U.S. High Plains unleashes a thunderstorm. See ftp site for color image.

ratio also shows geographical variations. It has been maintained that the ratio is lowest at more northerly latitudes. Recent satellite measurements have shown that, within the United States, certain regions, notably the High Plains and West Coast have much higher IC:CG ratios than other parts of the country such as Florida.

Cloud-to-ground events typically lower negative charge to earth (the negative CG, or −CG). CG discharges are sometimes initiated by leaders that are positively charged, and the resulting return stroke essentially lowers positive charge to ground (the +CG). Once thought to be very rare, the +CG is now thought to constitute about 10% of all CGs. For reasons not well understood, many severe local storms tend to have higher percentages of +CG flashes. Negative CGs typically lower on the order of 5 to 20 C of charge to ground, while +CGs can often be associated with much larger amounts of charge transfer.

When viewed with the naked eye, a CG flash often appears to flicker. This is a consequence of the fact that CG flashes are often composed of a sequence of multiple events called strokes. While flashes can consist of a single stroke (which is the norm for +CGs), the average stroke multiplicity for −CGs is around 4. Flashes with greater than 10 strokes are not uncommon, and there have been up to 47 strokes documented within a single flash. Strokes within the same flash tend to attach themselves to the same point on the surface, but this is not necessarily always the case. One Florida study (Rakov and Uman, 1990) found that about 50% of the flashes showed multiple terminations per flash to the ground. The multiple attach points within single flashes were separated by 0.3 to 7.3 km, with a mean of 1.3 km (Thottappillil et al., 1992). A very small percentage of flashes are initiated from the tops of tall towers, buildings, or mountains. Unlike conventional CG discharges, their channels branch upwards and outwards.

The CG flash is part of a complex series of events (Golde, 1977; Holle and Lopez, 1993; Uman, 1987; MacGorman and Rust, 1998). A typical flash begins with one or more negatively charged channel(s), called stepped leaders, originating deep within the cumulonimbus cloud and may emerge from the base of the cloud. Depending on the electrical charge distribution between cloud and ground, the leader proceeds erratically downward in a series of luminous steps traveling tens of meters in around a microsecond. A pause of about 50 μs occurs between each step. As the stepped leader approaches the ground, it is associated with electrical potentials on the order of 100 million volts. The intense electric field that develops between the front of the leader and the ground causes upward-moving discharges called streamers from one or more objects attached to the ground. When one of these streamers connects with the leader, usually about 100 m above the ground, the circuit is closed, and a current wave propagates up the newly completed channel and charge is transferred to the ground. This last process is called the return stroke. This is the brilliant flash seen by the naked eye, even though it lasts only tens to perhaps a few hundred microseconds. The peak current, which is typically on the order of 30 kA, is attained within about 1 μs. After the current ceases to flow through the leader-created channel, there is a pause ranging from 10 to 150 ms. Another type of leader, called a dart leader, can propagate down the same ionized channel, followed by a subsequent return stroke. The entire lightning discharge process can last, in extreme cases, for over 3 s, with the series of individual strokes comprising the CG flash sequence sometimes extending over a second. Certain phases of the lightning discharge, particularly the return stroke, proceed at speeds of more than one half the speed of light, while other discharge processes travel through the clouds up to two orders of magnitude more slowly.

While many CG return strokes are very brief (sub-100 μs), additionally many are followed by a long lasting (tens to many hundreds of milliseconds) continuing current. This behavior is particularly true of +CGs. This, along with the high initial peak currents often found in some +CGs, explains why such flashes are thought to ignite a substantial number of forest fires and cause other property damage (Fuquay et al., 1972). Since the temperature within the lightning channel has been estimated to reach 30,000 K, extending the duration of the current flow allows for greater transfer of heat energy and thus combustion.

There is great variability in the amount of peak current from stroke to stroke. While a typical peak current is in the 25 to 30 kA range, much smaller and larger currents can occur. Recent data suggest that peak currents of less than 10 kA may be more common than once believed. One survey of 60 million lightning flashes found 2.3% had peak currents of >75 kA; the largest positive CG reaching 580 kA and the largest negative CG 960 kA (Lyons et al., 1998). It has long been assumed that +CGs on the average had larger peak currents than −CGs, but again recent studies in the United States suggest that, while there are certainly many large peak current

+CGs, that for all classes between 75 and 400 kA, negative CGs outnumbered the positives (Lyons et al., 1998). It has generally been assumed that the first stroke in a flash contains the highest peak current. While on the average this is true (typically the first stroke averages several times that of subsequent strokes), recent research has found that subsequent strokes with higher peak currents are not that unusual (Thottappillil et al., 1992).

There are many reports of lightning strikes out of an apparently clear sky overhead. This is the so-called bolt from the blue. Powerful lightning discharges can leap for distances of 10 km (and sometimes more) beyond the physical cloud boundary. Under hazy conditions or with the horizon obscured by trees or buildings, the perception of a bolt from the blue is entirely understandable. Lightning tracking equipment at NASA's Kennedy Space Center has documented one discharge that came to ground more than 50 km from its point of origin within a local thunderstorm.

Lightning has also been associated with smoke plumes from massive forest fires, likely from the thunderstorm clouds induced by the intense local heat source (Fuquay et al., 1972). Plumes from volcanic eruptions have also produced lightning discharges (Uman, 1987). Lightning has also been associated with large underwater explosions as well as atmospheric thermonuclear detonations. Lightning is not confined to Earth's atmosphere. Space probes have detected lightning within the clouds of Jupiter, and possibly in Saturn and Uranus. Considerable additional information on the physics and chemistry of lightning can be found in Uman (1987), Golde (1977), Uman and Krider (1989), the National Research Council (1986), and Williams (1988).

3 HUMAN COST OF LIGHTNING

Lightning is a major cause of severe-weather-related deaths in the United States. According to the U.S. Department of Commerce's "Storm Data," during the period 1940–1981, lightning killed more people in the United States (7583) than hurricanes and tornadoes combined (7071). Between 1959 and 1995, the toll was 3322 reported deaths and 10,346 injuries. Thus while there was a downward trend in lightning casualties during the later part of the twentieth century in the United States, in many years lightning deaths still continue to outnumber those associated with tornadoes and hurricanes. Floods, however, are the number one killer. Between 1959 and 1987, the most lightning casualties were recorded in Florida, the nation's "lightning capital." The states of Michigan, North Carolina, Pennsylvania, Ohio, and New York followed in the rankings. Though having lower lightning frequencies, the large populations of these states translated into large numbers of casualties. Lightning casualties are less well documented in other parts of the world. It has been estimated by Andrews et al. (1992) that the annual worldwide lightning casualty figures are about 1000 fatalities and 2500 injuries.

Many of the published U.S. and worldwide losses from lightning may be significant underestimates (Lopez et al., 1993). In Colorado, a region with an efficient emergency response and reporting system, detailed examinations of hospital records found that U.S. government figures still underestimated lightning deaths by at least 28% and injuries by 40%. It appears many deaths ultimately caused by lightning (by heart attacks or in lightning-triggered fires) are not included within the casualty totals. Cooper (1995) suggests actual U.S. lightning deaths are probably 50% higher than reported, with as many as 500 injuries, many of them unreported or erroneously classified. Thus the lightning threat tends to be underestimated by public safety officials. This may be in part also due to the tendency of lightning to kill or injure in small numbers. It rarely receives the press notice accorded more "spectacular" weather disasters such as hurricanes, tornadoes, and flash floods. In recent years, however, there have been increasing reports of "mass casualties." In July 1991, at least 22 people were injured when lightning struck a crowded beach in Potterville, Michigan. In 1980 a high school football team practicing in Wickliffe, Ohio, was struck by lightning. All the players were knocked over and one was injured. In the Congo, 11 members of one soccer team were killed as a flash struck the playing field. In Nigeria, 12 were killed as a storm struck an outdoor funeral service. In the Colorado Rockies, a herd of 56 elk were found dead within a radius of 100 yards, apparently the result of a lightning strike.

4 ECONOMIC COSTS OF LIGHTNING

While some summaries of lightning property damages in the United States have listed relatively modest annual losses (under $100 million) recent research suggests the damages are many times greater. One study of insurance records (Holle et al., 1996) estimated about 1 out of every 55 Colorado CG flashes resulted in a damage claim. A major property insurer, covering about 25% of all U.S. homes, reports paying out 304,000 lightning-related claims annually. The average insurance lightning damage claim was $916. This suggests that U.S. homeowner losses alone may exceed $1 billion annually. Lightning has been responsible for a number of major disasters, including a substantial fraction of the world's wildfires in forests and grasslands. During the 1990s in the United States, some 100,000 lightning-caused forest fires consumed millions of acres of forests. Lightning is the leading cause of electric utility outages and equipment damage throughout the world, estimated to account for 30 to 50% of all power outages. On July 13, 1977, a lightning strike to a power line in upstate New York triggered a series of chain-reaction power outages, ultimately blacking out parts of New York City for up to 24 h. The resulting civil disturbances resulted in over $1 billion in property losses. Though the actual total of lightning's economic effects are unknown, the National Lightning Safety Institute estimates it could be as high as $4 billion to $5 billion per year in the United States alone. Given the substantial noninsured and poorly documented losses occurring in forestry, aviation, data processing, and telecommunications plus the lost wages due to power disruptions and stoppages of outdoor economic activities, this estimate seems plausible.

On December 8, 1963, a lightning strike ignited fuel vapors in the reserve tank of a Pan American airliner while the plane was circling in a holding pattern during a thunderstorm. The Boeing 707 crashed in Elkton, Maryland, with a loss of 81 lives. Engineering improvements make such deadly airliner incidents much less likely today. Spacecraft launches, however, still must contend with the lightning hazard. The 1969 flight of Apollo 12 almost ended in tragedy rather than the second moon landing when the vehicle "triggered" two lightning discharges during the initial moments of ascent. The event upset much of the instrumentation and caused some equipment damage, but the crew fortunately was able to recover and proceed with the mission. On March 26, 1987, the U.S. Air Force launched an Atlas/Centaur rocket from the Kennedy Space Center. Forty-eight seconds after launch the vehicle was "struck" by lightning—apparently triggered by the ionized exhaust plume trailing behind the rocket (Uman and Krider, 1989). The uninsured cost to U.S. taxpayers was $162 million. Ground facilities can fare poorly as well. On July 10, 1926, lightning struck a U.S. Navy ammunition depot. The resulting explosions and fires killed 19 people and caused $81 million in property losses. A lightning strike to a Denver warehouse in 1997 resulted in a $50 million fire.

Blaming lightning for equipment failure has become commonplace, and perhaps too much so. One estimate suggests that nearly one third of all insurance lightning damage claims are inaccurate or clearly fraudulent. Damage claim verification using data from lightning detection networks is now becoming commonplace in the United States.

5 CLIMATOLOGY OF LIGHTNING

An average of 2000 thunderstorm cells are estimated present on Earth at any one time (Uman, 1987). Over the past several decades, the worldwide lightning flash rate has been variously estimated at between 25 and 400 times per second. Based upon more recent satellite monitoring, the consensus is emerging that global lightning frequency probably averages somewhat less than 50 IC and CG discharges per second. Lightning is strongly biased to continental regions, with up to 10 times more lightning found over or in the immediate vicinity of land areas.

Before the advent of the National Lightning Detection Network (NLDN), various techniques were used to estimate CG flash density (expressed in flashes/km^2 year unless otherwise stated). One measure frequently employed is the isokeraunic level, derived from an analysis of the number of thunderstorm days per year at a point. A thunderstorm day occurs when a trained observer at a weather station reports hearing at least one peal of thunder (typically audible over a range of 5 to 10 km). Changery (1981) published a map of thunder days across the United States using subjective reports of thunder from weather observers at 450 stations over the period 1948–1977. A very rough rule of thumb proposed that for each 10 thunderstorm days there were between 1 and 2 flashes/km^2 year.

Needed, however, was a more accurate determination of the CG flash density, for a variety of purposes, including the design of electrical transmission line protection systems. During the 1980s, as discussed below, lightning detection networks capable of locating CG events began operating in the United States. By the mid-1990s, a reasonably stable annual pattern had begun to emerge (Orville and Silver, 1997). Approximately 20 to 25 million flashes per year strike the lower 48 states. The expected flash density maximum was found in central Florida (~10 to 15 flashes/km^2 year). Other regional maxima include values around 6 flashes/km^2 year along the Gulf Coast and around 5 flashes/km^2 year in the Midwest and Ohio River region. More than half the United States has a flash density of 4 flashes/km^2 year or greater. In any given year the highest flash densities may not be found in central Florida. For instance, the highest flash densities during 1993 were found in the Mississippi and Ohio valleys, in association with the frequent and massive thunderstorms leading to the great floods of that summer. Two annual maxima occurred in 1995, in southern Louisiana and near the Kentucky–Illinois border. In the intermountain west, values around 1.0 flashes/km^2 year are common with West Coast densities being less than 0.5 flashes/km^2 year.

Different thunderstorm types can produce lightning at very different rates. A small, isolated air mass shower may produce a dozen IC discharges and just one or two CGs. At the other extreme, the largest convective system, aside from the tropical cyclone, is the mesoscale convective complex (MCC). These frequent the central United States during the warm season as well as many other parts of the world including portions of South America, Africa, and Asia. The MCC, which can cover 100,000 km^2 and last for 12 h or more, has been known to generate CG strikes at rates exceeding 10,000 per hour. Goodman and MacGorman (1986) have noted that the passage of the active portion of a single MCC can result in 25% of a region's annual CG total.

There are distinct regional differences in such parameters as the percentage of flashes that are of positive polarity and also of peak current. The U.S. region with the largest number of positive polarity CGs (>12.5% of the total) is concentrated in a broad belt in the interior of the United States, stretching from Texas north to Minnesota. A study of the climatology of CGs with large peak currents (defined as >75 kA) found powerful +CGs were similarly clustered in a band stretching from eastern New Mexico and Colorado northeastward into Minnesota (Lyons et al., 1998). By contrast, large peak current −CGs were largely confined to the overocean regions of the northern Gulf of Mexico and along the southeastern U.S. coastline and the adjacent Atlantic ocean. The implications of these regional differences in flashes with large peak current upon facility design and protection as well as hazards to human safety are yet to be explored.

The geographic distribution of lightning flash density on a global basis is becoming better known. A combination of proliferating land-based CG detection networks and satellite observations should allow a more robust global climatology to emerge over the next decade or so. For now we must rely on observations of total lightning (IC + CG events) made by polar orbiting satellites such as NASA's Optical Transient Detector (OTD). Figure 2 shows the elevated lightning densities over parts of North America, the Amazon, central Africa, and the Maritime Continent of Southeast Asia.

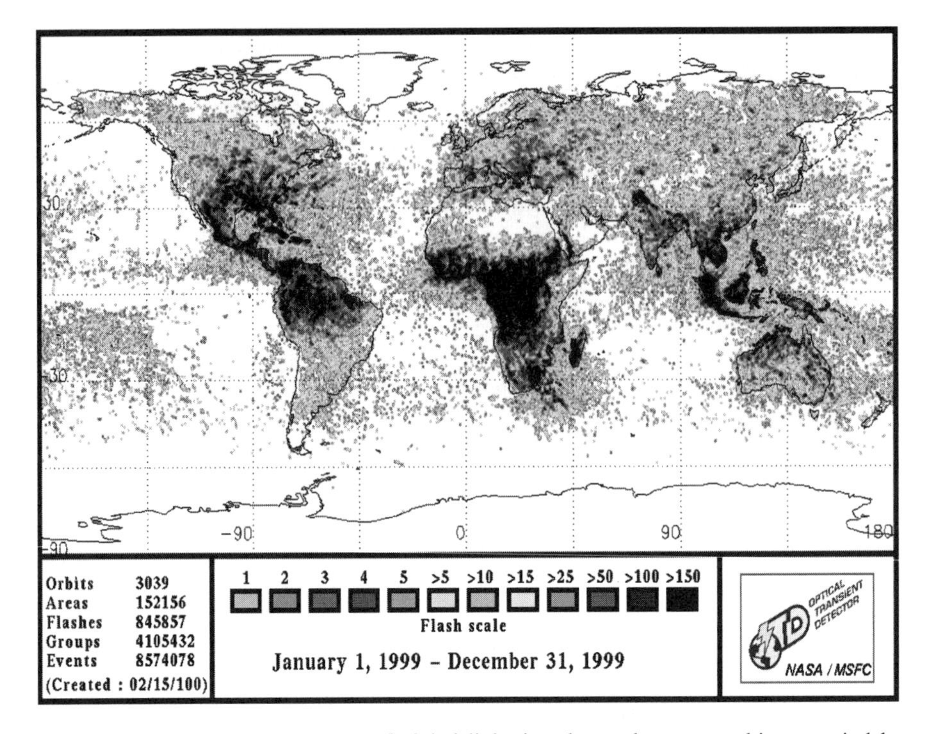

Figure 2 (see color insert) Map of global lightning detected over a multiyear period by NASA's Optical Transient Detector flying in polar orbit. See ftp site for color image.

6 LIGHTNING DETECTION

During the 1980s, a major breakthrough in lightning detection occurred with the development of lightning detection networks. Two distinct approaches were developed to detect and locate CGs. One was based on magnetic direction finding (MDF) and the second on time-of-arrival (TOA) approaches. Krider et al. (1980) described one of the first applications of an MDF-based system, the location of CG flashes in remote parts of Alaska, which might start forest fires. The TOA technique was applied to research and operations in the United States in the late 1980s (Lyons et al., 1985, 1989). As various MDF and TOA networks proliferated, it soon was evident that a merger of the two techniques was both technologically and economically desirable. The current U.S. National Lightning Detection Network (NLDN) (Cummins et al., 1998) consists of a hybrid system using the best of both approaches.

The NLDN uses more than 100 sensors communicating via satellite links to a central control facility in Tucson, Arizona, to provide real-time information on CG events over the United States. While employing several different configurations, it has been operating continuously on a nationwide basis since 1989. Currently, the mean flash location accuracy is approaching 500 m and the flash detection efficiency ranges between 80 and 90%, varying slightly by region (Cummins et al., 1998). It can provide coverage for several hundred kilometers beyond the U.S. coastline, although the detection efficiency drops off with range. The networks are designed to detect individual CG strokes, which can then be combined to flashes. The NLDN provides data on the time (nearest millisecond), location (latitude and longitude), polarity, and peak current of each stroke in a flash, along with estimates of the locational accuracy. National and regional networks using similar technology are gradually evolving in many areas, including Japan, Europe, Asia, Australia, and Brazil. The Canadian and U.S. networks have effectively been merged into a North American network (NALDN).

In the United States, lightning data are available by commercial subscription in a variety of formats. Users can obtain real-time flash data via satellite. Summaries of ongoing lightning-producing storms are distributed through a variety of websites. Pagers are being employed to alert lightning-sensitive facilities, such as golf courses, to the approach of CGs. Historical flash and stroke data can also be retrieved from archives (see www.LightningStorm.com). These are used for a wide variety of scientific research programs, electrical system design studies, fault finding, and insurance claim adjusting.

Future detection developments may include three-dimensional lightning mapping arrays (LMAs). These systems, consisting of a network of very high frequency (VHF) radio receivers, can locate the myriad of electrical emissions produced by a discharge as it wends its way through the atmosphere. Combined with the existing CG mapping of the NLDN, total lightning will be recorded and displayed in real time. Such detection systems are now being tested by several private firms, universities, and government agencies.

A new long-range lightning detection technique employs emissions produced by CG lightning events in the extremely low frequency (ELF) range of the spectrum. In the so-called Schumann resonance bands (8, 14, 20, 26, . . . Hz) occasional very large transients occur that stand out against the background "hum" of all the world's lightning events. These transients are apparently the result of atypical lightning flashes that transfer massive amounts of charge (>100 C) to ground. It is suspected many of these flashes also produce mesospheric sprites (see below) (Huang et al., 1999). Since the ELF signatures travel very long distances within Earth–ionosphere waveguide, just a few ELF receivers are potentially capable of detecting and locating the unusual events on a global basis.

7 LIGHTNING PROTECTION

Continued improvement in the detection of lightning using surface and space-based systems is a long-term goal of the atmospheric sciences. This must also be accompanied by improving alerts of people and facilities in harm's way, more effective medical treatment for those struck by lightning, and improved engineering practice in the hardening of physical facilities to withstand a lightning strike.

The U.S. National Weather Service (NWS) does not issue warnings specifically for lightning, although recent policy is to include it within special weather statements and warnings for other hazards (tornado, hail, high winds). Thus members of the public are often left to their own discretion as to what safety measures to take. When should personnel begin to take avoidance procedures? The "flash-to-bang method" as it is sometimes called is based on the fact that while the optical signature of lightning travels at the speed of light, thunder travels at 1.6 km every 5 s. By counting the time interval between seeing the flash and hearing the thunder, one can estimate the distance rather accurately. The question is then how far away might the next CG flash jump from its predecessor's location? Within small Florida air mass thunderstorms, the average value of the distance between successive CG strikes is 3.5 km (Uman and Krider, 1989). However, recently statistics from larger midwestern storms show the mean distance between successive flashes can often be 10 km or more (~30 s flash-to-bang or more). According to Holle et al. (1992), lightning from receding storms can be as deadly as that from approaching ones. People are struck more often near the end of storms than during the height of the storm (when the highest flash densities and maximum property impacts occur). It appears that people take shelter when a storm approaches and remain inside while rainfall is most intense. They fail to appreciate, however, that the lightning threat can continue after the rain has diminished or even ceased, for up to a half hour. Also, the later stages of many storms are characterized by especially powerful and deadly +CG flashes. Thus, has emerged the 30:30 rule (Holle et al., 1999). If the flash-to-bang duration is less than 30 s, shelter should be sought. Based upon recent research in lightning casualties, persons should ideally remain sheltered for about 30 min after the cessation of audible thunder at the end of the storm.

Where are people struck by lightning? According to one survey, 45% were in open fields, 23% were under or near tall or isolated trees, 14% were in or near water, 7% were on golf courses, and 5% were using open-cab vehicles. Another survey noted that up to 4% of injuries occurred with persons talking on (noncordless) telephones or using radio transmitters. In Colorado some 40% of lightning deaths occur while people are hiking or mountain climbing during storms. The greatest risks occur to those who are among the highest points in an open field or in a boat, standing near tall or isolated trees or similar objects, or in contact with conducting objects such as plumbing or wires connected to outside conductors. The safest location is inside a substantial enclosed building. The cab of a metal vehicle with the windows closed is also relatively safe. If struck, enclosed structures tend to shield the occupants in the manner of a Faraday cage. Only rarely are people killed directly by lightning while inside buildings. These include persons in contact with conductors such as a plumber leaning on a pipe, a broadcaster talking into a microphone, persons on the telephone, or an electrician working on a power panel.

A lightning strike is not necessarily fatal. According to Cooper (1995) about 3 to 10% of those persons struck by lightning are fatally injured. Of those struck, fully 25% of the survivors suffered serious long-term after effects including memory loss, sleep disturbance, attention deficits, dizziness, numbness/paralysis, and depression. The medical profession has been rather slow to recognize the frequency of lightning-related injuries and to develop treatment strategies. There has, fortunately, been an increasing interest paid to this topic over the past two decades (Cooper, 1983, 1995; Andrews et al., 1988, 1992). Persons struck by lightning who appear "dead" can often be revived by the prompt application of cardiopulmonary resuscitation (CPR). The lack of external physical injury does not preclude the possibility of severe internal injuries.

Lightning can and does strike the same place more than once. It is common for the same tall object to be struck many times in a year, with New York City's Empire State Building being struck 23 times per year on average (Uman, 1987). From a risk management point of view, it should be assumed that lightning cannot be "stopped" or prevented. Its effects, however, can be greatly minimized by diversion of the current (providing a controlled path for the current to follow to ground) and by shielding.

According to the U.S. Department of Energy (1996), first-level protection of structures is provided by the lightning grounding system. The lightning leader is not influenced by the object that is about to be struck until it is only tens of meters away. The lightning rod (or air terminal) developed by Benjamin Franklin remains a key component for the protection of structures. Its key function is to initiate an upward-connecting streamer discharge when the stepped leader approaches within striking distance. Air terminals do not attract significantly more strikes to the structure than the structure would receive in their absence. They do create, however, a localized preferred strike point that then channels the current to lightning conductors and safely down into the ground. To be effective, a lightning protection system must be properly designed, installed, and maintained. The grounding must be adequate. Sharp bends (less than a 20-cm radius) can defeat the conductor's function. The bonding of the components should be thermal, not mechanical, whenever possible. Recent findings suggest that air terminals with rounded points may be more effective than those with sharp points.

Sensitive electronic and electrical equipment should be shielded. Switching to auxiliary generators, surge protectors, and uninterruptible power supplies (UPSs) can help minimize damage from direct lightning strikes or power surges propagating down utility lines from distant strikes. In some cases, the most practical action is to simply disconnect valuable assets from line power until the storm has passed.

There is still much to be learned about atmospheric electrical phenomena and their impacts. The characteristics of lightning wave forms and the depth such discharges can penetrate into the ground to affect buried cables are still under investigation. Ongoing tests using rocket-triggered lightning are being conducted by the University of Florida at its Camp Blanding facility in order to test the hardening of electrical systems for the Electric Power Research Institute.

The notion that lightning is an "act of God" against which there is little if any defense is beginning to fade. Not only can we forecast conditions conducive to lightning strikes and monitor their progress, but we are beginning to understand how to prevent loss of life, treat the injuries of those harmed by a strike, and decrease the vulnerability of physical assets to lightning.

8 IMPACTS OF LIGHTNING UPON THE MIDDLE ATMOSPHERE

Science often advances at a deliberate and cautious pace. Over 100 years passed before persistent reports of luminous events in the stratosphere and mesosphere, associated with tropospheric lightning, were accepted by the scientific community. Since 1886, dozens of eyewitness accounts, mostly published in obscure meteorological publications, have been accompanied by articles describing meteorological esoterica such as half-meter wide snow flakes and toads falling during rain showers. The phenomena were variously described using terms such "cloud-to-space lightning" and "rocket lightning." A typical description might read, "In its most typical form it consists of flames appearing to shoot up from the top of the cloud or, if the cloud is out of sight, the flames seem to rise from the horizon." Such eyewitness reports were largely ignored by the nascent atmospheric electricity community—even when they were posted by a Nobel Prize winning physicist such as C.T.R. Wilson (1956). During the last three decades, several compendia of similar subjective reports from credible witnesses worldwide were prepared by Otha H. Vaughan (NASA Marshall) and the late Bernard Vonnegut (The State University of New York—Albany). The events were widely dispersed geographically from equatorial regions to above 50° latitude. About 75% of the observations were made over land. The eyewitness descriptions shared one common characteristic—they were perceived as highly atypical of "normal" lightning (Lyons and Williams, 1993). The reaction of the atmospheric science community could be summarized as casual interest at best. Then, as so often happens in science, serendipity intervened.

The air of mystery began to dissipate in July 1989. Scientists from the University of Minnesota, led by Prof. John R. Winckler, were testing a low-light camera system (LLTV) for an upcoming rocket flight at an observatory in central Minnesota (Franz et al., 1990). The resulting tape contained, quite by accident, two fields of video that provided the first hard evidence for what are now called sprites. The twin pillars of light were assumed to originate with a thunderstorm system some 250 km to the north along the Canadian border. The storm system, while not especially intense, did contain a larger than average number of positive polarity cloud-to-ground (+CG) lightning flashes. From this single observation emanated a decade of fruitful research into the electrodynamics of the middle atmosphere.

Spurred by this initial discovery, in the early 1990s NASA scientists searched videotapes from the Space Shuttle's LLTV camera archives and confirmed at least 17 apparent events above storm clouds occurring worldwide (Boeck et al., 1998). By 1993, NASA's Shuttle Safety Office had developed concerns that this newly discovered "cloud-to-space lightning" might be fairly common and thus pose a potential threat to Space Shuttle missions especially during launch or recovery. Based upon the available evidence, the author's hunt for these elusive events was directed above the stratiform regions of large mesoscale convective systems (MCSs), known to generate relatively few but often very energetic lightning discharges. On the night of July 7, 1993, an LLTV was deployed for the first time at the Yucca Ridge Field Station (YRFS), on high terrain about 20 km east of Fort Collins, Colorado. Exploiting an uninterrupted view of the skies above the High Plains to the east, the LLTV was trained above a large nocturnal MCS in Kansas, some 400 km distant (Lyons, 1994). Once again, good fortune intervened as 248 sprites were imaged over the next 4 h. Analyses revealed that almost all the sprites were associated with +CG flashes, and assumed an amazing variety of shapes (Fig. 3). Within 24 h, in a totally independent research effort, sprites were imaged by a University of Alaska team onboard the NASA DC8 aircraft over Iowa (Sentman and Wescott, 1993). The following summer, the University of Alaska's flights provided the first color videos detailing the red sprite body with bluish, downward extending tendrils. The same series of flights documented the unexpected and very strange blue jets (Wescott et al., 1995).

By 1994 it had become apparent that there was a rapidly developing problem with the nomenclature being used to describe the various findings in the scientific literature. The name sprite was selected to avoid employing a term that might presume more about the physics of the phenomena than our knowledge warranted. Sprite replaced terms such as "cloud-to-space" lightning and "cloud-to-ionosphere discharge" and similar appellations that were initially used. Today a host of phenomena have been named: sprites, blue jets, blue starters, elves, sprite halos, and trolls, with perhaps others remaining to be discovered. Collectively they have been termed transient luminous events (TLEs).

Since the first sprite observations in 1989, the scientific community's misperception of the middle atmosphere above thunderstorms as "uninteresting" has completely changed. Much has been learned about the morphology of TLEs in recent years. Sprites can extend vertically from less than 30 km to about 95 km. While telescopic investigations reveal that individual tendril elements may be of the order of 10 m across, the envelope of the illuminated volume can exceed $10^4\,km^3$. Sprites are

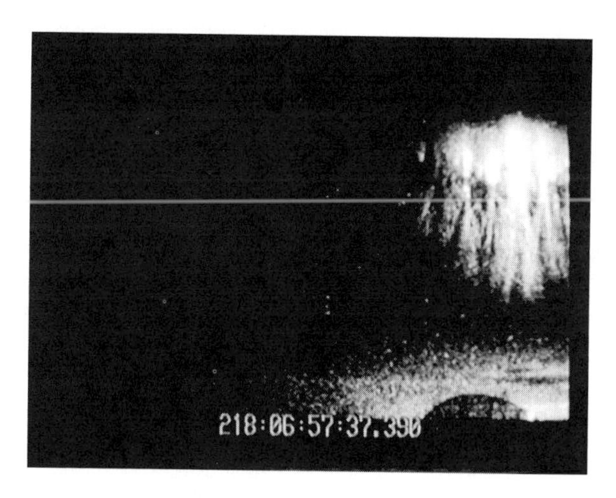

Figure 3 One of the many shapes assumed by sprites in the mesosphere. Top may be near 95 km altitude with tendril-like streamers extending downward toward 40 km or lower. Horizon is illuminated by the flash of the parent lightning discharge. Image obtained using a low-light camera system at the Yucca Ridge Field Station near Ft. Collins, CO. Sprite is some 400 km away.

almost always preceded by +CG flashes, with time lags of less than 1 to over 100 ms. To date, there are only two documented cases of sprites associated with negative polarity CGs. The sprite parent +CG peak currents range widely, from under 10 kA to over 200 kA, though on average the sprite +CG peak current is 50% higher than other +CGs in the same storm. High-speed video images (1000 fps) suggest that many sprites usually initiate around 70 to 75 km from a small point, and first extend downward and then upward, with development at speeds around 10^7 m/s. Sprite luminosity on typical LLTV videos can endure for tens of milliseconds. Photometry suggests, however, that the brightest elements usually persist for a few milliseconds, though occasionally small, bright "hot spots" linger for tens of milliseconds. By 1995, sprite spectral measurements by Hampton et al. (1996) and Mende et al. (1995) confirmed the presence of the N_2 first positive emission lines. In 1996, photometry provided clear evidence of ionization in some sprites associated with blue emissions within the tendrils and sometimes the sprite body (Armstrong et al., 2000). Peak brightness within sprites is on the order of 1000 to 35,000 kR. In 7 years of observations at Yucca Ridge, sprites were typically associated with larger storms ($>10^4$ km^2 radar echo), especially those exhibiting substantial regions of stratiform precipitation (Lyons, 1996). The TLE-generating phase of High Plains storms averages about 3 h. The probability of optical detection of TLEs from the ground in Colorado is highest between 0400 and 0700 UTC. It is suspected that sprite activity maximizes around local midnight for many storms around the world. The TLE counts observed from single storm systems has ranged from 1 to 776, with 48 as an average count. Sustained rates as high as once every 12 s have been noted, but more typical intervals are on the order of 2 to 5 min.

Can sprites be detected with the naked eye? The answer is a qualified yes. Most sprites do not surpass the threshold of detection of the dark-adapted human eye, but some indeed do. Naked-eye observations require a dark (usually rural) location, no moon, very clean air (such as visibilities typical of the western United States) and a dark-adapted eye (5 min or more). A storm located 100 to 300 km distant is ideal if it contains a large stratiform precipitation area with +CGs. The observer should stare at the region located some 3 to 5 times the height of the storm cloud. It is best to shield the eye from the lightning flashing within the parent storm. Often sprites are best seen out of the corner of the eye. The event is so transient that often observers cannot be sure of what they may have seen. The perceived color may not always appear "salmon red" to any given individual. Given the human eye's limitations in discerning color at very low light levels, some report seeing sprites in their natural color, but others see them as white or even green.

While most TLE discoveries came as a surprise, one was predicted in advance from theoretical arguments. In the early 1990s, Stanford University researchers proposed that the electromagnetic pulse (EMP) from CG flashes could induce a transient glow at the lower ledge of the ionosphere between 80 and 100 km altitudes (Inan et al., 1997). Evidence for this was first noted in 1994 using LLTVs at Yucca Ridge. Elves were confirmed the following year by photometric arrays deployed at Yucca Ridge by Tohoku University (Fukunishi et al., 1996). Elves, as they are now called, are believed to be expanding quasi-torroidal structures that attain an integrated width of several hundred kilometers. (The singular is elve, rather than elf, in order to avoid confusion with ELF radio waves, which are used intensively in TLE studies.) Photometric measurements suggest the elve's intrinsic color is red due to strong N_2 first positive emissions. While relatively bright (1000 kR), their duration is $<500\,\mu s$. These are usually followed by $\sim 300\,\mu s$ very high peak current (often >100 kA) CGs, most of which are positive in polarity. Stanford University researchers, using sensitive photometric arrays, documented the outward and downward expansion of the elve's disk. They also suggest many more dim elves occur than are detected with conventional LLTVs. It has been suggested that these fainter elves are more evenly distributed between positive and negative polarity CGs.

Recently it has been determined that some sprites are preceded by a diffuse disk-shaped glow that lasts several milliseconds and superficially resembles elves. These structures, now called "halos," are less than 100 km wide, and propagate downward from about 85 to 70 km altitude. Sprite elements sometimes emerge from the lower portion of the sprite halo's concave disk.

Blue jets are rarely observed from ground-based observatories, in part due to atmospheric scattering of the shorter wavelengths. LLTV video from aircraft missions revealed blue jets emerging from the tops of electrically active thunderstorms. The jets propagate upwards at speeds of ~ 100 km/s reaching terminal altitudes around 40 km. Their estimated brightness is on the order of 1000 kR. Blue jets do not appear to be associated with specific CG flashes. Some blue jets appear not to extend very far above the clouds, only propagating as bright channels for a few kilometers above the storm tops. These nascent blue jets have been termed blue starters. During the 2000 observational campaign at Yucca Ridge, the first blue starters ever imaged from the ground were noted. They were accompanied by very bright, short-lived (~ 20 ms) "dots" of light at the top of MCS anvil clouds. A NASA ER2 pilot flying over the Dominican Republic high above Hurricane Georges in 1998 described seeing luminous structures that resembled blue jets.

The troll is the most recent addition to the TLE family. In LLTV videos, trolls superficially resemble blue jets, yet they clearly contain significant red emissions. Moreover, they occur after an especially bright sprite in which tendrils have extended downward to near cloud tops. The trolls exhibit a luminous head leading a faint trail moving upwards initially around 150 km/s, then gradually decelerating and disappearing by 50 km. It is still not known whether the preceding sprite tendrils actually extend to the physical cloud tops or if the trolls emerge from the storm cloud per se.

Worldwide, a variety of storm types have been associated with TLEs. These include the larger midlatitude MCSs, tornadic squall lines, tropical deep convection, tropical cyclones, and winter snow squalls over the Sea of Japan. It appears that the central United States may be home to some of the most prolific TLE producers, even though only a minority of High Plains thunderstorms produce TLEs. Some convective regimes, such as supercells, have yet to be observed producing many TLEs and the few TLEs are mostly confined to any stratiform precipitation region that may develop during the late mature and decaying stages. Furthermore, the vast majority of +CGs, even many with peak currents above 50 kA, produce neither sprites nor elves, which are detectable using standard LLTV systems. While large peak current

+CGs populate both MCSs and supercells, only certain +CGs possess characteristics that generate sprites or elves.

Monitoring ELF radio emissions in the Schumann resonance bands (8 to 120 Hz) has provided a clue to what differentiates the TLE parent CG from "normal" flashes. The background Schumann resonance signal is produced from the multitude of lightning flashes occurring worldwide. It is generally a slowly varying signal, but occasionally brief amplitude spikes, called Q-bursts, are noted. Their origin was a matter of conjecture for several decades. In 1994, visual sprite observations at Yucca Ridge were coordinated in real time with ELF transients (Q-bursts) detected at a Rhode Island receiver station operated by the Massachusetts Institute of Technology (Boccippio et al., 1995). This experiment, repeated many times since, clearly demonstrated that Q-bursts are companions to the +CG flashes generating both sprites and elves. ELF measurements have shown that sprite parent +CGs are associated with exceptionally large charge moments (300 to >2000 C-km). The sprite +CG ELF waveform spectral color is "red," that is, peaked toward the fundamental Schumann resonance mode at 8 Hz. Lightning charge transfers of hundreds of coulombs may be required for consistency with theories for sprite optical intensity and to account for the ELF Q-burst intensity. Lightning causal to elves has a much flatter ("white") ELF spectrum, and though associated with the very highest peak current +CGs (often >150 kA), exhibits much smaller charge moments (<300 C-km) (Huang et al., 1999).

Recent studies of High Plains MCSs confirm that their electrical and lightning characteristics are radically different from the textbook "dipole" thunderstorm model, derived largely from studies of rather small convective storms. Several horizontal laminae of positive charge are found, one often near the 0°C layer, and these structures persist for several hours over spatial scales of ~ 100 km. With positive charge densities of 1 to 3 nC/m^3, even relatively shallow layers (order 500 m) covering 10^4 to $10^5\,km^2$ can contain thousands of coulombs. Some 75 years ago, C.T.R. Wilson (1956) postulated that large charge transfers and particularly large charge moments from CG lightning appear to be a necessary condition for conventional breakdown that could produce middle atmospheric optical emissions. Sprites occur most readily above MCS stratiform precipitation regions with radar echo areas larger than $\sim 10^4\,km^2$. It is not uncommon to observe rapid-fire sequences of sprites propagating above storm tops, apparently in synchrony with a large underlying horizontal lightning discharge. One such "dancer" included a succession of eight individual sprites within 700 ms along a 200-km-long corridor. This suggests a propagation speed of the underlying "forcing function" of $\sim 3 \times 10^5$ m/s. This is consistent with the propagation speed of "spider" lightning—vast horizontal dendritic channels tapping extensive charge pools once a +CG channel with a long continuing current becomes established (Williams, 1998).

It is suspected that only the larger MCS, which contain large stratiform precipitation regions, give rise to the +CGs associated with the spider lightning networks able to lower the necessary charge to ground. The majority of sprite parent +CGs are concentrated in the trailing MCS stratiform regions. The radar reflectivities associated with the parent +CGs are relatively modest, 30 to 40 dBZ or less. Only a

TABLE 1 Current Ideas on TLE Storm/Lightning Parameters in Selected Storm Types

	Core of Supercells[a]	MCS Stratiform Region	"Ordinary" Convection
+CG peak currents	$>\sim 40$ kA	$>\sim 60$ kA	$\sim$30 kA
Storm dimension	10–20 km	10–500 km	<20–100 km
Spider discharges	Few/small	Many/large	Some
Continuing current	Short if any	Longest, strongest	Low intensity
+CG channel height	10–15 km (?)	5 km (?)	10 km (?)
Sprites occur?	No (except at end)	Many	Rare (?)
Elves occur?	No (except at end)	Many	Rare (?)
Blue jets occur?	Yes (?)	Rare (?)	Rare (?)

[a]Some supercells may generate a few sprites during their final phase when/if extensive stratiform develops.

small subregion of trailing stratiform area produces sprite and elves. It would appear that this portion of the MCS possesses, for several hours, the requisite dynamical and microphysical processes favorable for the unique electrical discharges, which drive TLEs. Tables 1 and 2 summarize the relationships between lightning and the major TLE types

9 THEORIES ON TRANSIENT LUMINOUS EVENTS

Transient luminous events have captured the interest of many theoreticians (Rowland, 1998; Pasko et al., 1995). Several basic mechanisms have been postulated to explain the observed luminous structures. These include sprite excitation by a quasi-electrostatic (QE) mechanism. Sprite production by runaway electrons in the strong electric field above storms has been suggested. The formation of elves from lightning electromagnetic pulses (EMP) is now generally accepted. More than one mechanism may be operating, but on different temporal and spatial scales, which in turn produce the bewildering variety of TLE shapes and sizes. Absent from almost all theoretical modeling efforts are specific data on key parameters characterizing lightning flashes that actually produce TLEs. Many modelers refer to standard reference texts, which, in turn, tend to compile data taken in storm types and locales that are not representative of the nocturnal High Plains. Specifically, many invoke the conventional view that the positive charge reservoir for the lightning is found in the upper portion of the cloud at altitudes of $\sim$10 km. But the positive dipole (or tripole) storm model has been found wanting in many midcontinental storms. We surveyed the range of lightning parameters used in over a dozen theoretical modeling studies. While the proposed heights of the vertical +CG channel ranges from 4 to 20 km, there is a clear preference for 10 km and above. The amount of charge lowered varies over three orders of magnitude, as does the time scale over which the charge transfer occurs. Only a few studies consider the possible role of horizontal

TABLE 2 Current Speculations as to Characteristics of TLEs and Their Parent Lightning (None for Blue Jets)

	Sprite	Elve	Blue Jet
Color of emission	Red top/blue base	Red?	Deep blue
Polarity of parent CG	Positive (almost all)	Positive (mostly)	None
+CG peak current	$> \sim 50$ kA	$> \sim 100$ kA	None
Charge transferred (C)	Largest	Large	N/A
Charge moment (C-km)	Largest (>300)	Large (<300)	N/A
Parent CG location	Stratiform area	Stratiform area (?)	N/A
Parent CG vertical channel height	5–7 km (?)	10 km (?)	N/A
di/dt value	Moderate	Very large (?)	N/A
Total flash duration	Very long	Short (?)	N/A
Spider involved	Yes (?)	No (?)	N/A
Continuing current duration	Very long (?)	Short in any (?)	N/A
VLF/ELF slow tail	Distinct	Yes	N/A
ELF spectral color	Red	White	None
VLF audio character	Low freq.	Higher freq.	None
Duration of TLE	1–150 ms	0.5 ms	100–200 ms
Altitude range of TLE	25–95 km	75–105 km	Cloud-40 km
Onset delay after CG	1–100 ms	~0.3 ms	N/A
Brightness	50–35,000 kR	1000 kR	1000 kR
Horizontal size of emission	100 m–100 km	100–400 km	~1–2 km

components of the parent discharge. The charge moment (in C-km), not the peak current as measured by the NLDN, is the key parameter in the basic QE conventional breakdown mechanism first proposed by Wilson. The key physics of the problem appear to involve the altitude and magnitude of the removed charge and the time scale on which this occurs—parameters about which little agreement exists. Many theorists note that even with an assumed tall +CG channel (~10 km) this still requires extremely large (~100 C) charge transfers, typically 10 times larger than in "conventional" lightning. Some models yield a 1000-fold enhancement in optical intensity at 75 km for a doubling of the altitude of lightning charge removal from 5 to 10 km. The use of shorter channels to ground, say 5 km, would imply truly large charge transfers. Yet evidence is accumulating that indeed such may be the case.

While the various models simulate optical emissions bearing some (though in many cases rather minimal) resemblance to the observations, such wide ranges in the lightning source term parameters do not appear physically realistic. If, in fact, such a range of lightning characteristics could produce sprites, why does only a very small subset of +CGs (<1:20 even in active storms) actually produce observable TLEs (with current sensors)? It appears that most models have made assumptions about the lightning to produce something resembling a TLE—rather than starting with hard physical constraints on the source term. The reason, of course, is that there are very little data on the actual CGs that generate specific TLE occurrences.

During the summer of 2000, an extensive field research effort was conducted over the High Plains of Colorado, Kansas, and Nebraska. Several radars, research aircraft, and mobile storm monitor teams were deployed, along with a three-dimensional lightning-mapping array. As sprite-bearing storms passed over the LMA, LLTV systems at Yucca Ridge (some 300 km to the northwest) was able to document the sprites and other TLEs coincident with a variety of detailed measurements of the parent lightning discharges. Data analyses have just begun, but initial findings suggest that +CGs with very large charge moments (>300 C-km) are indeed associated with sprite events above their parent storms.

10 WHY STUDY TLEs?

Aside from their intrinsic scientific interest, there may be some rather practical reasons to explore TLEs in more depth. It has been suggested that there may be significant production of NO_x in the middle atmosphere by sprites. This becomes even more interesting in light of recent observations that regional smoke palls from biomass burns radically enhance the percentage of +CGs within storms, and thus increase sprite counts (and middle atmosphere NO_x production?). Currently, no global chemical model accounts for any potential effects of TLEs (Lyons and Armstrong, 1997). Once a better estimate of NO_x production per sprite is obtained, it will be necessary to know the global frequency and distribution of sprites. It has been demonstrated that several Schumann resonance monitoring sites working in tandem are capable of obtaining a worldwide TLE census.

There is growing interest in determining the sources of unusual infrasound emissions detected above sprite-capable MCSs as determined by the National Oceanic and Atmospheric Administration's (NOAA's) Environmental Technology Laboratory near Boulder, Colorado. TLEs thus produce optical, radio frequency (RF), and acoustic emissions that have the potential of mimicking or masking signatures from clandestine nuclear tests. Such findings may have important implications for global monitoring efforts supporting the Comprehensive Test Ban Treaty.

Transient luminous events may contribute in ways not yet understood to the maintenance of the global electrical circuit. To quantify the impacts of TLEs, we require information on their global frequency (now roughly estimated between 1 and 10 per minute) and their geographic distribution. It has been proposed (Williams, 1992) that the Schumann resonance can be used in the manner of worldwide tropical thermometer on the assumption that as global warming occurs, the amount of lightning may rise rapidly. The implications for sprite production of any global warming are uncertain.

ACKNOWLEDGMENTS

This material is based upon work supported in part with funds from the National Aeronautics and Space Administration (Office of Space Sciences), the U.S. Department of Energy, and the U.S. National Science Foundation (Contract number ATM-0000569).

11 SOME WEATHER AND LIGHTNING-RELATED WEBSITES

National Lightning Safety Institute: *http://www.lightningsafety.com*

Global Atmospherics, Inc.—National Lightning Detection Network: *http://www.LightningStorm.com*

Sprites and Elves: *http://www.FMA-Research.com*

Lightning Injury Research: *http://tigger.uic.edu/~macooper/cindex.htm*

Kennedy Space Center/Patrick Air Force Base: *www.patrick.af.mil/45og/45ws/ws1.htm*

The National Severe Storms Laboratory—Norman, OK: *http://www.nssl.noaa.gov*

U.S. National Oceanic and Atmospheric Administration: *http://weather.noaa.gov* and *http://www.srh.noaa.gov/ftproot/ssd/html/lightnin.htm*

The World Meteorological Organization: *http://www.wmo.ch*

National Center for Atmospheric Research: *http://www.rap.ucar.edu/weather*

Federal Emergency Management Agency: *http://www.fema.gov*

NASA Marshall Space Flight Center: *http://thunder.msfc.nasa.gov*

REFERENCES

Andrews, C. J., M. Darveniza, and D. Makerras (1988). A review of medical aspects of lightning injuries, Proceedings, Intl Aerospace and Ground Conference on Lightning and Static Electricity, Oklahoma City, NOAA Special Report, OK 231–250.

Andrews, C. J., M. A. Cooper, M. Darveniza, and D. Mackerras (1992). *Lightning Injuries: Electrical, Medical and Legal Aspects*, Boca Raton, FL, CRC Press.

Armstrong, R. A., D. M. Suszcynsky, W. A. Lyons, and T. E. Nelson (2000). Multi-color photometric measurements of ionization and energies in sprites, *Geophys. Res. Lett.* **27**, 653–656.

Boccippio, D. J., E. R. Williams, S. J. Heckman, W. A. Lyons, I. T. Baker, and R. Boldi (1995). Sprites, ELF transients, and positive ground strokes, *Science* **269**, 1088–1091.

Boeck, W. L., O. H. Vaughan, Jr., R. Blakeslee, B. Vonnegut, and M. Brook (1998). The role of the space shuttle videotapes in the discovery of sprites, jets and elves, *J. Atmos. Solar-Terr. Phys.* **60**, 669–677.

Changery, M. J. (1981). National thunderstorm frequencies for the contiguous United States. U.S. Nuclear Regulatory Commission, NUREG/CR-22452, Washington, DC.

Cooper, M. A. (1983). Lightning injuries, *Em. Med. Clin. N. Am.* **3**, 639.

Cooper, M. A., (1995). Myths, miracles, and mirages, *Seminars in Neurology* **15**, 358–361.

Cummins, K. L., M. J. Murphy, E. A. Bardo, W. L. Hiscox, R. B. Pyle, and A. E. Pifer (1998). A combined TOA/MDF technology upgrade of the U.S. National Lightning Detection Network, *J. Geophys. Res.* **103**, D8, 9035–9044.

Franz, R. C., R. J. Nemzek, and J. R. Winckler (1990). Television image of a large upward electrical discharge above a thunderstorm system, *Science* **249**, 48–51.

Fukunishi, H., Y. Takahashi, M. Kubota, K. Sakanoi, U. S. Inan, and W. A. Lyons (1996). Elves: Lightning-induced transient luminous events in the lower ionosphere, *Geophys. Res. Lett.* **23**, 2157–2160.

Fuquay, D. M., A. R. Taylor, R. G. Hawe, and C. W. Schmid, Jr. (1972). Lightning discharges that caused forest fires, *J. Geophys. Res.* **77**, 2156–2158.

Golde, R. H. (1977). *Lightning, Vol. 1, Physics of Lightning*, London, Academic.

Goodman, S. J., and D. R. MacGorman (1986). Cloud-to-ground lightning activity in mesoscale convective complexes, *Mon. Wea. Rev.* **114**, 2320–2328.

Hampton, D. L., M. J. Heavner, E. M. Wescott, and D. D. Sentman (1996). Optical spectra characteristics of sprites. *Geophys. Res. Lett.* **23**, 89–92.

Holle, R. H., and R. E. Lopez (1993). Overview of Real-time lightning detection systems and their meteorological uses, NOAA Technical Memorandum ERL NSSL-102, National Severe Storms Laboratory.

Holle, R. L., R. E. Lopez, R. Ortiz, A. I. Watson, D. L. Smith, D. M. Decker, and C. H. Paxton (1992). Cloud-to-ground lightning related to deaths, injuries and property damage in central Florida. Proceedings, Intl. Conf. on Lightning and Static Electricity, Atlantic City, NJ, FAA Report DOT/FAA/CT-92/20,66-1-66-12.

Holle, R. L., R. E. Lopez, L. J. Arnold, and J. Endres (1996). Insured lightning-caused property damage in three western states, *J. Appl. Meteor.* **35**, 1344–1351.

Holle, R. L., R. E. Lopez, and C. Zimmermann (1999). Updated recommendations for lightning safety—1998, *Bull. Am. Meteor. Soc.* **80**, 2035–2041.

Huang, E., E. Williams, R. Boldi, S. Heckman, W. Lyons, M. Taylor, T. Nelson, and C. Wong (1999). Criteria for sprites and elves based on Schumann resonance observations, *J. Geophys. Res.* **104**, 16943–16964.

Inan, U. S., C. Barrington-Lee, S. Hansen, V. S. Glukhov, T. F. Bell, and R. Rairden (1997). Rapid lateral expansion of optical luminosity in lightning-induced ionospheric flashes referred to as "elves," *Geophys. Res. Lett.* **24**, 583–586.

Krider, E. P., R. C. Noggle, A. E. Pifer, and D. L. Vance, 1980: Lightning direction-finding systems for forest fire detection, *Bull. Am. Meteor. Soc.* **61**, 980–986.

Lopez, R. E., R. L. Holle, T. A. Heitkamp, M. Boyson, M. Cherington, and K. Langford (1993). The underreporting of lightning injuries and deaths, Preprints, 17th Conf. on Severe Local Storms, Conf. on Atmospheric Electricity, St. Louis, American Meteorological Society, pp 775–778.

Lyons, W. A. (1994). Characteristics of luminous structures in the stratosphere above thunderstorms as imaged by low-light video, *Geophys. Res. Letts.* **21**, 875–878.

Lyons, W. A. (1996). Sprite observations above the U.S. High Plains in relation to their parent thunderstorm systems, *J. Geophys. Res.* **101**, 29641–29652.

Lyons, W. A., and R. A. Armstrong (1997). NO_x Production within and above Thunderstorms: The contribution of lightning and sprites, Preprints, 3rd Conf. on Atmospheric Chemistry, Long Beach, American Meteorological Society.

Lyons, W. A., and E. R. Williams (1993). Preliminary investigations of the phenomenology of cloud-to-stratosphere lightning discharges. Preprints, Conference on Atmospheric Electricity, St. Louis, American Meteorological Society, pp 725–732.

Lyons, W. A., K. G. Bauer, R. B. Bent, and W. H. Highlands (1985). Wide area real-time thunderstorm mapping using LPATS—the Lightning Position and Tracking System, Preprints, Second Conf. on the Aviation Weather System, Montreal American Meteorological Society.

Lyons, W. A., K. G. Bauer, A. C. Eustis, D. A. Moon, N. J. Petit, and J. A. Schuh (1989). R·Scan's National Lightning Detection Network: The first year progress report. Preprints, Fifth Intl. Conf. on Interactive Information and Processing Systems for Meteorology, Oceanography and Hydrology, Anaheim, American Meteorological Society.

Lyons, W. A., M. Uliasz, and T. E. Nelson (1998). Climatology of large peak current cloud-to-ground lightning flashes in the contiguous United States, *Mon. Wea. Rev.* **126**, 2217–2233.

MacGorman, D. R., and W. D. Rust (1998). *The Electrical Nature of Storms*, New York, Oxford University Press.

Mende, S. N., R. L. Rairden, G. R. Swenson, and W. A. Lyons (1995). Sprite spectra: N2 first positive band identification, *Geophys. Res. Lett.* **22**, 2633–2636.

National Research Council (1986). *The Earth's Electrical Environment*, Studies in Geophysics, Washington, DC, National Academy Press.

Orville, R. E., and A. C. Silver (1997). Annual Summary: Lightning ground flash density in the contiguous United States: 1992–95, *Mon. Wea. Rev.* **125**, 631–638.

Pasko, V. P., U. S. Inan, Y. N. Taranenko, and T. F. Bell (1995). Heating, ionization and upward discharges in the mesosphere due to intense quasi-static thundercloud fields, *Geophys. Res. Lett.* **22**, 365–368.

Rakov, V. A., and M. A. Uman (1990). Some properties of negative cloud-to-ground lightning flashes versus stroke order, *J. Geophys. Res.* **95**, 5447–5453.

Rowland, H. L. (1998). Theories and simulations of elves, sprites and blue jets, *J. Atmos. and Solar-Terrestrial Phys.* **60**, 831–844.

Sentman, D. D., and E. M. Wescott (1993). Observations of upper atmospheric optical flashes recorded from an aircraft, *Geophys. Res. Lett.* **20**, 2857–2860.

Thottappillil, R., V. A. Rakov, M. A. Uman, W. H. Beasley, M. J. Master, and D. V. Sheluykhin (1992). Lightning subsequent-stroke electric field peak greater than the first stroke peak and multiple ground terminations, *J. Geophys. Res.* **97**, 7503–7509.

Uman, M. A. (1987). *The Lightning Discharge*, International Geophysics Series, Vol. 39, Orlando, Academic.

Uman, M. A., and E. P. Krider (1989). Natural and artificially initiated lightning, *Science* **246**, 457–464.

U.S. Dept. of Energy (1996). *Lightning Safety*, Office of Nuclear and Facility Safety, USDOE, DOE/EH-0530, Washington, DC.

Wescott, E. M., D. Sentman, D. Osborne, D. Hampton, and M. Heavner (1995). Preliminary results from the Sprites94 aircraft campaign: 2. Blue jets, *Geophys. Res. Lett.* **22**, 1209–1212.

Williams, E. R. (1988). The electrification of thunderstorms, *Sci. Am.* **259**, 88–99.

Williams, E. R. (1992). The Schumann resonance: A global tropical thermometer, *Science* **256**, 1184–1186.

Williams, E. R. (1998). The positive charge reservoir for sprite-producing lightning, *J. Atmos. Solar-Terr. Phys.* **60**, 689–692.

Wilson, C. T. R. (1956). A theory of thundercloud electricity, *Proc., Royal Met. Soc., London* **236**, 297–317.

CHAPTER 23

WEATHER MODIFICATION

HAROLD D. ORVILLE

1 INTRODUCTION

Although much remains to be learned about the seeding of clouds, scientists and engineers have accomplished much in the 55 or so years since the discovery that dry ice and silver iodide are efficient ice nucleants, able to influence the natural precipitation processes. In addition, within the past 10 years renewed efforts using hygroscopic seeding materials have yielded very promising results in producing more rain from convective clouds. This chapter will discuss the basic physics and chemistry of the precipitation and hail processes and explain the primary methods for changing the various atmospheric precipitation components. Some of the key advances in weather modification in the past 25 years will be highlighted. The primary source material comes from a workshop report to the Board on Atmospheric Sciences and Climate (BASC) of the National Research Council (NRC) published in December 2000 (BASC, 2000). In addition, the most recent statement by the American Meteorological Society (AMS, 1998a) and the scientific background for the policy statement (AMS, 1998b) has much good information concerning the current status of weather modification.

Basic Physics and Chemistry of the Precipitation Processes

Only a brief review of these complex processes will be given here. Detailed treatments can be found in the textbooks or scientific studies listed in the references at the end of this chapter (e.g., Dennis, 1980; Rogers and Yau, 1989; Young, 1993). Concerning rain and snow, the basic problem of precipitation physics is: How do one million cloud droplets combine to form one raindrop or large snowflake in a

Handbook of Weather, Climate, and Water: Dynamics, Climate, Physical Meteorology, Weather Systems, and Measurements, Edited by Thomas D. Potter and Bradley R. Colman.
ISBN 0-471-21490-6

period of 10 to 20 min? Two primary processes are involved after the clouds have formed.

Atmospheric aerosols are important for the formation of clouds, the precursors of rain. The chemical content and distribution of the aerosol plus the cloud updraft speed determine the initial distribution and number concentration of the droplets. Normally there are ample suitable hygroscopic aerosols, called cloud condensation nuclei (CCN), for clouds to form at relative humidities very near 100% (at supersaturations of a few tenths of a percent only). The relative humidity with respect to a liquid or ice surface is defined as the ratio of the environmental vapor pressure to the saturation vapor pressure at the liquid or ice surface, respectively. Maritime and continental regions have very different aerosol populations, leading to characteristically different cloud droplet distributions.

For clouds with all liquid particles (certainly all those clouds warmer than 0°C) a process known as collection (a combination of collision and coalescence)—the *warm rain process*—is thought to be operative. This process depends on the coexistence of cloud droplets of different sizes and, consequently, of different fall velocities. Sizes range from a few micrometers in diameter to 30 or 40 μm. The larger droplets fall through a population of smaller droplets, collecting them as they fall. The larger particles become raindrop embryos. If conditions are right (depth of cloud, breadth of droplet distribution, cloud updraft, etc.), the embryos can form raindrops by this process in the necessary time period. Raindrops range in size from 200 μm to 5 mm diameter (the upper limit occurs because of raindrop breakup).

One might think that with billions and billions of cloud droplets in a cloud that it would be relatively easy for the droplets to collide, coalesce, and grow to raindrop size. However, even though very numerous, the droplets occupy less than one millionth of the volume of the cloud. Consequently, the droplets are relatively far apart and require special conditions to interact and grow. Indeed, even large thunderstorms and massive raining nimbostratus clouds are more than 99.99% clear space (dry air and vapor), but by the time precipitation particles have formed in such clouds the collection process is well developed and relatively efficient.

The second precipitation process is called the *cold rain process* and depends on the formation of ice in the cloud. This process also depends on nuclei (ice nuclei, very different particles from the CCN mentioned above), but in this case there is a shortage of ice-forming nuclei in the atmosphere so that liquid water does not freeze at 0°C. The water is then called *supercooled*. To understand this cold rain process, one other important fact is needed: For the same subzero temperature, the relative humidity will be much above 100% for the ice particle while staying at 100% over the liquid surface, due to the fact that the saturation vapor pressure is higher over the liquid surface than over the ice surface. The growth of the particles depends on the vapor pressure difference between the cloudy air and the particles' surface. The liquid droplets are generally much more numerous than ice crystals in a cloud and control the relative humidity (the vapor pressure in the cloudy air). The stage is then set for rapid growth of an ice particle if it is introduced into a cloud of supercooled liquid droplets. The ice crystal grows, depleting the water vapor in the cloudy environment, which is then replenished by water vapor from evaporating droplets trying to maintain 100% relative humidity with respect to liquid. The crystal rapidly increases in mass and falls through the cloud droplets, collecting them and further increasing the ice particle's mass. Snow or rain will appear at the ground depending on the temperature conditions of the lower atmosphere. Bergeron and Findeisen first described this process in the 1930s.

The formation of the ice particle by the ice nucleus can occur in at least four ways: (1) Deposition of water vapor molecules may occur directly on the nucleus. (2) Condensation may occur on a portion of the nucleus, followed by freezing of the liquid. (3) A drop or droplet may come in contact with a nucleus and freeze. (4) And finally a drop or droplet may have formed including a nucleus in its bulk water content, which then freezes the water when a low enough temperature is reached. How low? All of the processes mentioned above require a supercooling of from 15 to 20°C, that is, temperatures of -15 to -20°C in continental-type clouds (those clouds with a narrow drop size distribution and large number concentrations). Clouds with extremely strong updrafts may delay their primary ice formation to temperatures near -40°C (Rosenfeld and Woodley, 2000). There is some evidence accumulating that the ice nucleating temperature may depend on the type of cloud; maritime clouds with large and fewer droplets and a broad droplet distribution produce ice particles at temperatures as warm as -10°C. The more frequent collisions among large and small droplets may contribute to the early formation of ice in those clouds.

As mentioned above, in the absence of any ice nuclei, the water will remain in the liquid state until a temperature of -40°C is reached. The most effective nuclei are insoluble soil particles, kaolinites, and montmorillonites, but industrial and natural combustion products may also contribute to the ice-forming characteristics of clouds.

These ice processes refer to primary ice formation in a cloud. Secondary processes may also contribute (Mossop, 1985). Ice crystals may fracture or graupel, and droplets may interact to form additional ice particles, or ice particles may be transported from one cloud cell to another. A stormy, turbulent cloud situation has many ways to create and distribute ice among the clouds.

Basic Hail Processes

Hailstorms are normally large thunderstorms that produce hail. Hailstones are irregular ice particles larger than 5 mm in diameter (about one quarter of an inch). Large hailstones may grow to several centimeters in diameter and fall at many tens of meters per second. Hailstorms are characterized by large updrafts, usually greater than 15 m/s (30 mph) and sometimes as great as 50 m/s (100 mph). The large updrafts in a storm are closely associated with the largest hailstones in a storm. Such storms produce more than a billion dollars of crop damage and a like amount of property damage per year in the United States.

In addition to large updrafts to suspend the growing hailstones, a two-stage process is thought needed to produce hailstones. The first stage requires the formation of hailstone embryos. Either frozen raindrops or graupel particles may serve as the embryos. They range in size from 1 to 4 or 5 mm. The second stage involves the growth of the embryo to hailstone size. This requires the presence of ample supercooled water at sufficiently cold temperatures to grow the hailstone. The most efficient growth occurs between the temperatures of -10 to -30°C, at elevations of 5 to 8 or 9 km in the atmosphere. Unfortunately, the source regions of the embryos are not well known and can come from several locations in a storm. This makes it difficult to modify the hailstone growth process. This two-stage scenario indicates that the microphysical and dynamical processes in a storm must be matched in a special way to produce hail.

2 MODIFICATION METHODS

Rain and Snow

Knowledge and understanding of the precipitation processes lead to finding ways of modifying the processes. The two techniques used for rain and snow increases for cold clouds are called *microphysical* (or *static*) and *dynamic* seeding methods. The first method rests on the assumption that clouds lack the proper number of rain or ice embryos for the maximum precipitation to occur. Hence, great emphasis is placed on increasing the precipitation efficiency of the cloud, making sure that as much as possible of the cloud water is converted to precipitation. To affect the ice process, artificial nuclei, such as silver iodide, can be added or powdered dry ice (as cold as -80 to -100°C) can be dropped through the cloud, forming ice crystals by both homogeneous and heterogeneous nucleation. (Homogeneous freezing refers to the change of phase of water from vapor to ice without the assistance of a nucleus. Heterogeneous freezing occurs with the aid of a nucleus, which allows freezing at warmer temperatures than in homogeneous freezing.) The most striking effect of this microphysical seeding mode is the production of precipitation from marginal-type clouds, those on the brink of producing precipitation. However, the most effective precipitation increase may come from treating more vigorous but inefficiently raining or snowing clouds.

The concept of the second process, dynamic seeding, is that seeding a supercooled cloud with large enough quantities of artificial ice nuclei or a coolant such as solid carbon dioxide (dry ice) will cause rapid glaciation of the cloud. The resultant latent heat release from the freezing of supercooled drops (modified by the adjustment of saturation with respect to ice instead of with respect to liquid) will then increase the buoyancy of a cloud. The increased vigor of the cloud may result in a taller cloud, stronger updraft, more vapor and water flux, broader and longer lasting rain area, stronger downdrafts, and greater interaction with neighboring clouds, resulting in more cloud mergers and, hopefully, more rain. The reasoning sounds convincing, but there are many junctures where the process may go astray, and much research is needed to verify the hypothesis. Recent field studies of this seeding process in convective cloud systems in Texas (Rosenfeld and Woodley, 1993) have shown nearly twice as many mergers in the seeded complexes compared with the unseeded cloud complexes.

Theoretical results have indicated that dynamic seeding can also affect relatively dry stratus-type clouds as well as wet cumulus clouds (Orville et al., 1984, 1987). The action of the seeding is to cause embedded convection in the stratus clouds. One of the challenges with both of these seeding methods is to identify those clouds and cloud environments that will respond positively to the seeding.

The amount of seeding agent to use depends on the type of seeding agent, the seeding method, and the vigor of the cloud. One gram of silver iodide can produce about 10^{14} nuclei when completely activated. The activation starts at -4°C and becomes more efficient as the atmosphere cools, about one order of magnitude more activated nuclei for each 4° drop in temperature.

Normally the goal is to supply about one ice nucleus per liter of cloudy air to affect the precipitation process through the microphysical seeding method. Up to 100 times this number may be needed for the dynamic seeding method to be effective. The total amount of seeding material needed depends on the volume of cloud to be seeded, which will depend on the strength of the updraft. In general, a few tens of grams of silver iodide are needed for microphysical seeding and about 1 kg or more for dynamic seeding. Dry ice produces about 10^{12} ice crystals per gram of sublimed CO_2 in the supercooled portion of the cloud. Normally about 100 g of dry ice are used per kilometer of aircraft flight path for the microphysical seeding method, with only about 10 to 20% of the dry ice being sublimed before falling below the 0°C level in the cloud.

There are other determinants for the growth of precipitation-sized particles that can, in at least some cases, be artificially influenced. In the case of the collision–coalescence mechanism, opportunities sometimes exist for providing large hygroscopic nuclei to promote initial droplet growth or much larger salt particles to form raindrop embryos directly. The purpose of the hygroscopic seeding is thus to produce precipitation particles either directly or by enhancing the collision–coalescence mechanism. Two salt seeding methods are currently in use. One method applies hundreds of kilograms of salt particles (dry sizes are 10 to 30 μm in diameter) near cloud base to produce drizzle-size drops very soon after the salt particles enter the cloud. The second method uses salt flares to disperse 1 μm or smaller size particles into cloud updrafts, a method currently receiving renewed interest in cloud seeding efforts (Mather et al., 1996, 1997; Cooper et al., 1997; Bigg, 1997; Tzivion et al., 1994; Orville et al., 1998). The salt material is released from kilogram-size flares carried by aircraft; several flares are released per cloud cell. The salt particles change the size distribution of CCN in the updraft, creating a more maritime-type cloud. Coalescence is enhanced; rain forms in the seeded volume, eventually spreading throughout the cloud. This seeding thus accelerates the warm rain process. In addition, if the updraft lifts the rain to high enough altitudes, then the ice processes are also enhanced because of the larger drops and droplets, making it more likely that freezing of some drops will occur. Graupel and snow are more easily formed, increasing the precipitation efficiency of the cell. These hygroscopic seeding methods are thought to work only on continental-type clouds.

In the past, large water drops were added to clouds to accelerate the rain process, but this is not an economically viable way to increase warm rain.

Hail

The suppression of hail from a hailstorm is a much more complex matter than the increase of rain or snow from more benign cloud types. Two concepts appear to offer the most hope for the suppression of hail. They are called *beneficial competition* and *early rainout*. To suppress hail according to the beneficial competition theory, many more hail embryos must be introduced into the cloud cell than would occur naturally. (Cloud seeding with silver iodide is one way to provide the additional embryos.) According to this hypothesis, the sharing of the available supercooled water among a larger number of hailstones (i.e., their "competition" for the available water) reduces their size. If enough embryos are present, it should be possible to reduce the local supercooled liquid water content and hence the hailstone growth rates so that no particle grows large enough to survive without melting during fallout to the ground. Thus, less hail and more rain would be produced by the storm.

The consensus of many scientists is that this concept is most promising in storms containing large supercooled drops, since these drops when frozen (caused by the seeding) are large enough to act as efficient collectors and to provide the necessary competition. In storms containing only supercooled cloud droplets, this possibility is absent, and there is greater difficulty in creating graupel embryos in the correct place, time, and amount. Also, note that some storms may be naturally inefficient for the production of hail and the addition of potential hail embryos may actually increase hail production.

Creating early rainout from the feeder cells of an incipient hailstorm also appears to be an attractive method to suppress hail. Cloud seeding starts the precipitation process earlier than it would naturally. The weak updrafts in the feeder cells cannot support the rain particles and they fall out without participating in the hail formation process. This premature rainout removes liquid water and is accompanied by a reduction of the updraft strength due to the downward force components caused by both water loading in the lower part of the cloud and negative buoyancy caused by cooling resulting from melting of ice particles and the evaporation of precipitation beneath the cloud. These processes are thought to inhibit the hail generation process. Some operational projects use this method with encouraging results.

3 SOME SCIENTIFIC AND TECHNOLOGICAL ADVANCES IN THE PAST 25 YEARS

Weather modification as a scientific endeavor was last reviewed on a national basis in the late 1970s. At that time the Weather Modification Advisory Board (WMAB), established by an act of Congress in 1976, produced a report and a two-volume supporting document, detailing the status of weather modification in both research and operations (WMAB, 1978). Nearly $20 million per year was being spent on federal research in weather modification at the time of the Congressional act. An ambitious program was proposed to continue development of the technology, but little action was taken. Weather modification research funding dwindled to nearly zero in the ensuing years. However, operational weather modification continued and in recent years has shown signs of invigoration. Much of the operational work is being done in large areas of the midwestern and western United States (see Fig. 1).

Most of the following material comes from the BASC workshop summary (BASC, 2000).

Hygroscopic Seeding of Convective Clouds

An exciting breakthrough in the development of rain augmentation technology was made within the past decade. Three randomized experiments in three different parts of the world showed that hygroscopic seeding increased rainfall from convective clouds. Statistically significant increases in radar-estimated rainfall were achieved by hygroscopic flare seeding of cold convective clouds in South Africa, its replication on cold convective clouds in Mexico, and by hygroscopic particle seeding of warm convective clouds in Thailand. These efforts included physical studies as well as statistical experiments and resulted in strong evidence indicating that the increases in rainfall were due to seeded clouds lasting longer and producing rain over a larger area.

The common elements of the three randomized experiments were: (1) seeding with hygroscopic particles, (2) evaluation using a time-resolved estimate of storm rainfall based on radar measurements in conjunction with an objective software package for tracking individual storms (different software was used for each experiment), (3) statistically significant increases in radar-estimated rainfall, and (4) the necessity to invoke the occurrence of seeding-induced dynamic effects to explain the results.

The great significance of the Mexican experiment was that it replicated the results of the South African experiment in another area of the world. Several past programs in the world failed when attempts were made to replicate previously successful programs. Figure 2 displays a comparison of the quartile values of radar-derived rain mass in seeded (solid lines) and nonseeded (dashed lines) cases for the three different quartiles as a function of time after "decision time" (the time at which the treatment decision was made) for the South African (dark lines) and Mexican (gray lines) experiments.

The Mexican data were averaged over 5-min periods, while the South African data were averaged over 10 min. To be consistent, the combined results are plotted in Figure 2 as radar-derived "rain mass accumulated per minute." The indicated differences in rain mass for both the South African and Mexican experiments were statistically significant after 20 to 30 min and remained significant for the remainder of the period. It is clear that the results from both experiments are in good agreement. The main difference is that the storms in South Africa tended to last somewhat longer than those in Mexico.

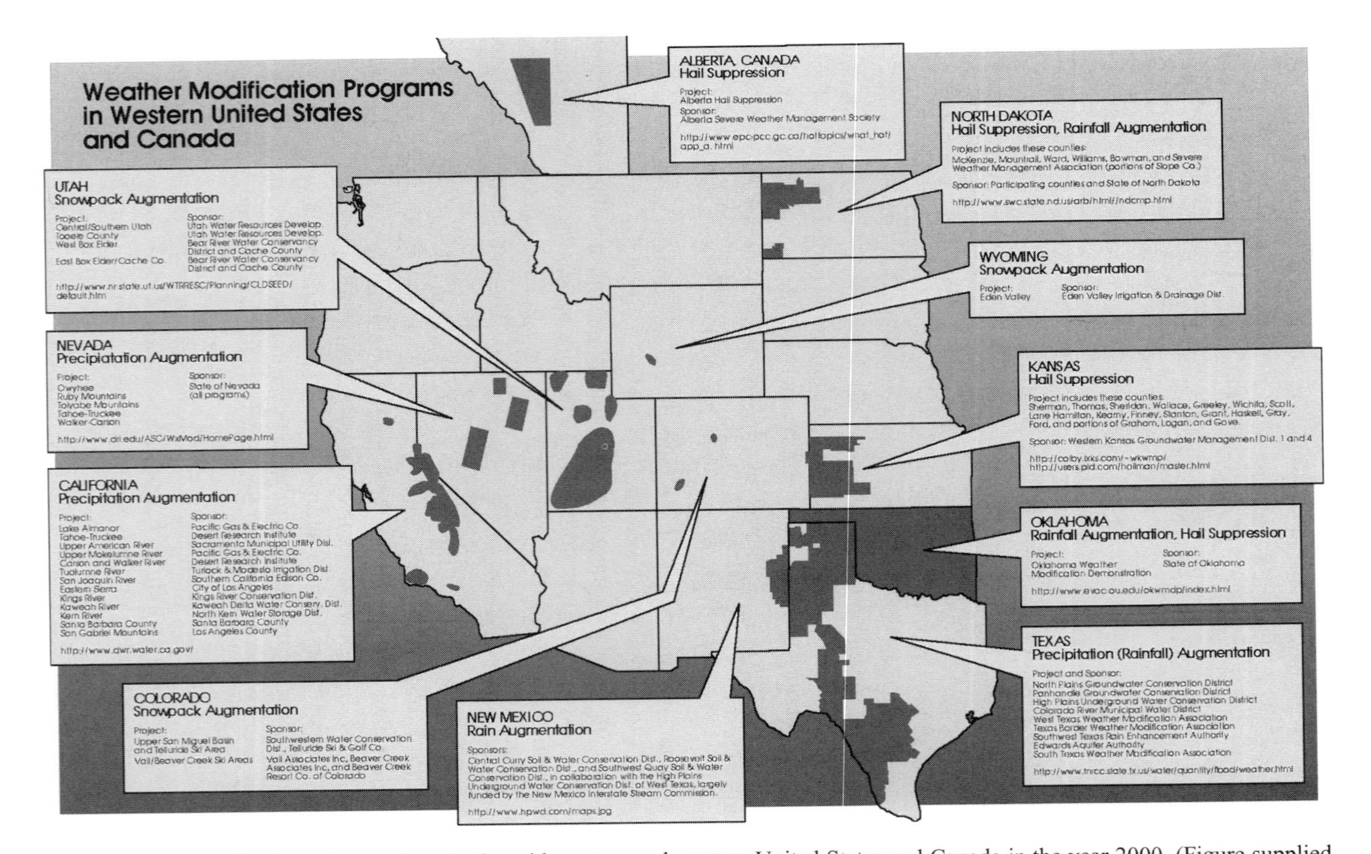

Figure 1 Active cloud seeding projects in the mid-western and western United States and Canada in the year 2000. (Figure supplied by Bruce Boe, Weather Modification, Inc.). See ftp site for color image.

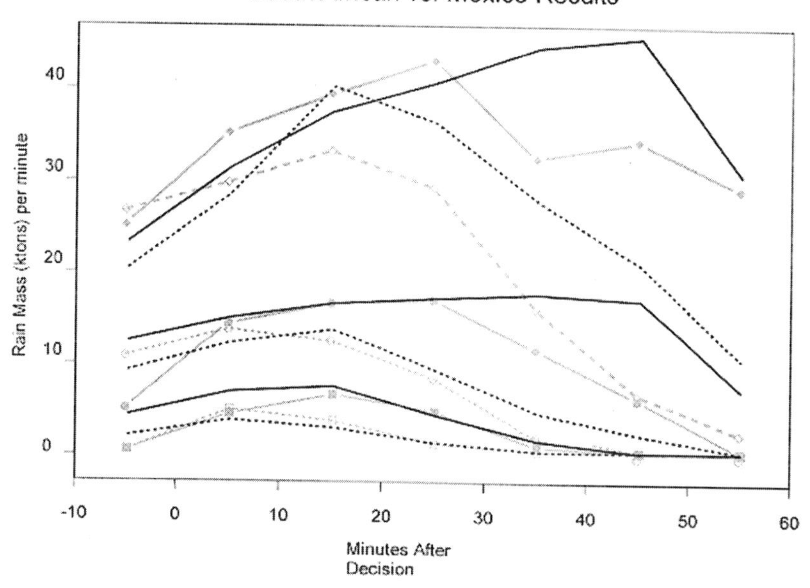

Figure 2 Quartile values of radar-derived rain mass (precipitation flux integrated over 1 min) versus time after "decision time" for seeded (solid lines) and nonseeded (dashed lines) cases for the South African (dark lines) and Coahuila (gray lines) experiments. First quartile is the value of rain mass that is larger than the value for 25% of the storms; second quartile value exceeds the value for 50% of the other storms; and third quartile value exceeds the value for 75% of the storms. (Figure supplied by Roelof Bruintjes, National Center for Atmospheric Research).

Glaciogenic Seeding Effects in Convective Clouds

The seeding results on these clouds are more controversial and more difficult to document (Rangno and Hobbs, 1995, 1997; Silverman, 2001; but also note Nicholls, 2001). Strong glaciogenic seeding signatures, which appear 2 to 3 min after initial seeding, have been documented in treated clouds by aircraft (Woodley and Rosenfeld, 2000) and, more recently, from space. Glaciation times after seeding are halved in both maritime and continental seeded clouds. These are consistent with the conceptual model of dynamic seeding guiding the glaciogenic seeding experiments. In addition, there is now limited field evidence and some numerical modeling results that indicate ice-phase seeding invigorates the internal cloud circulation concurrent with the growth of graupel to precipitation size and the depletion of the cloud water. These too are consistent with expectations from the conceptual model.

Radar estimation of the properties of treated and nontreated convective cells shows indications that under certain conditions the seeded cells produce more rainfall as shown by the increasing maximum radar reflectivities, maximum areas, maximum rain-volume rates, duration, clustering and merging of cells, and inferred maximum rainfall rates. The indicated effects on rainfall quantities require further validation by direct surface measurements. Earlier experimentation had also indicated that increases in the top height of the treated clouds accompanied the apparent rainfall increases. In recent years, however, this apparent height signal has been much weaker, due probably to a change in the method of estimating cloud top. In the early years, cloud tops were measured directly with a jet aircraft, while radar has been used in later years to estimate cloud top. Because radar may underestimate the tops of seeded clouds relative to nonseeded clouds as a consequence of the seeding-induced glaciation (glaciated clouds are less radar reflective than unglaciated clouds in the absence of hail), satellite measurements of cloud tops must be used to recheck this link in the conceptual chain.

Snowpack Augmentation

A significant accomplishment in recent snowpack augmentation research is the establishment of the direct link between the seeding activity and the water reaching the ground in the form of snow. The increases in precipitation rate caused by silver iodide seeding have been documented several times in the reviewed scientific literature (Reynolds, 1988; Super and Holroyd, 1997). The link has been established by physical and chemical techniques. The snow falling at particular targeted sites is connected directly to the seeding material (see Fig. 3).

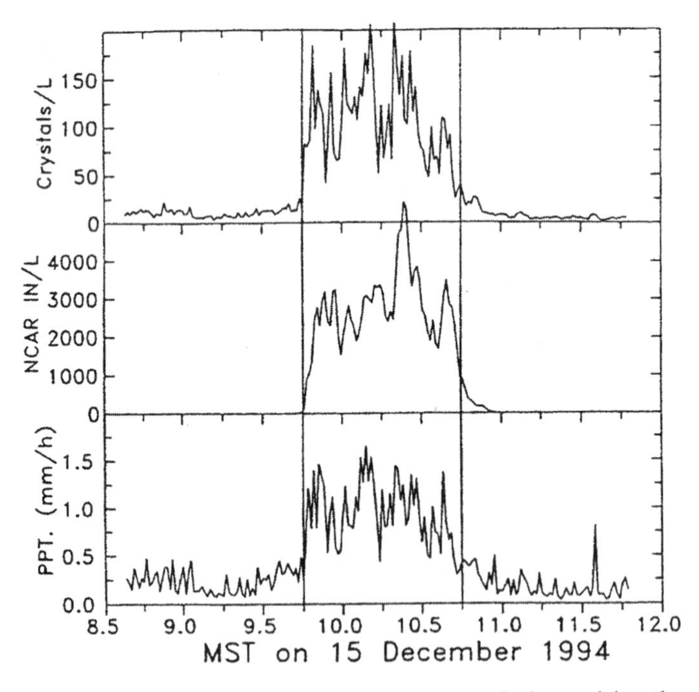

Figure 3 Observed concentrations of precipitating ice crystals, ice nuclei, and precipitation rate during one hour of AgI seeding between 0945 and 1045, December 15, 1994 in Utah (Super and Holroyd, 1997).

The methodologies used to establish this direct chemical linkage have been described by Warburton et al. (1985, 1994, 1995a, b), Super and Heimbach (1992), Chai et al. (1993), Stone and Huggins (1996), Super and Holroyd (1997), and McGurty (1999).

One big advantage of snowpack work is that the scientists are dealing with solid-state precipitation that can be sampled in fixed and targeted areas both during and after storm events and stored in the frozen state until analyzed.

Hail Suppression

In recent years, crop hail damage in the United States has typically been around $2.3 billion annually (Changnon, 1998). Susceptibility to damage depends upon the crop type, the stage of development, the size of the hail, and also the magnitude of any wind accompanying the hailfall.

Property damage from hail in recent years has been on the same order as crop hail damage, usually topping the $2 billion mark, sometimes more. A recent report by the Institute for Business and Home Safety (IBHS, 1998) indicated that losses from wind storms involving hail, from June 1, 1994 through June 30, 1997, totaled $13.2 billion. While some of this damage resulted from wind, hail certainly accounted for a significant fraction of the total damage.

Some of the recent high-dollar hailstorms include: Denver, 1990, $300 million; Calgary, 1991, $350 million; Dallas–Fort Worth, 1992, $750 million; Bismarck, ND, 1993, $40 million; Dallas–Fort Worth, 1995, $1 billion; and Calgary, 1996, $170 million. Wichita (Kansas), Orlando (Florida), and northern (Arlington) Virginia are just a few of the other U.S. locales that have recently been hard hit by hailstorms.

Results from North American hail suppression programs vary. The North Dakota Cloud Modification Project (NDCMP) reports reductions in crop hail damage on the order of 45% (Smith et al., 1997), while the Western Kansas Weather Modification Program (WKWMP) reports reduced crop hail losses of 27% (Eklund et al., 1999). Neither program reports any statistically significant changes in rainfall.

Both of these projects are operational, nonrandomized programs, and the evaluations are based upon analyses of crop hail insurance data. Projects elsewhere in the world (e.g., Dessens, 1986) have generally reported similar reductions in damage.

Considerable success has been achieved using numerical cloud models to simulate hailstorms and hail development (Farley et al., 1996; Orville, 1996). This has contributed significantly to the development of the contemporary conceptual models for hail suppression. Contemporary numerical models contain microphysical components, as well as cloud dynamics.

If a cloud model, after programming with actual atmospheric conditions, can successfully reproduce a cloud or storm like that actually observed in those same atmospheric conditions, the model has reinforced the physical concepts employed therein. Such cloud models can be used to test concepts for hail suppression. If the model employing a certain concept gets it right, this strengthens the confidence in the concept.

It is impossible to find two identical clouds, seed one, and leave the other untreated as a control, since no two clouds are exactly the same. However, the effects of seeding can be examined by modeling a cloud beginning with natural conditions and simulating seeding in a second run. Any differences in behavior can then be attributed to the seeding. In fact, this option is very attractive, as the timing (relative to the life cycle of the subject cloud), locations, and amounts of seeding can be varied in a succession of model runs to better understand the importance and effects of targeting.

Figure 4 gives an example of numerical model output, showing a decrease in hail in the large sizes and an increase in graupel size particles due to silver iodide cloud seeding. Total hailfall was decreased by 44%, hail impact energy by 58% in the seeded case as compared with the nonseeded case.

Advances in Technology

Instrumentation Unusual opportunities for determining, *by direct measurement*, the physical processes and water budgets of cloud systems and their changes due to purposeful and inadvertent cloud modification reside with new technologies, particularly remote-sensing technologies. Many other instruments and techniques have been developed that are applicable, including satellite systems and in situ systems including aircraft platforms. However, the main point to be noted is that *none* of these technologies were available during the period of about 1965–1985 when significant funding was directed toward weather modification research. There has been a paradigm shift in observational capabilities, and cloud modification can now be revisited with the new and emerging tools.

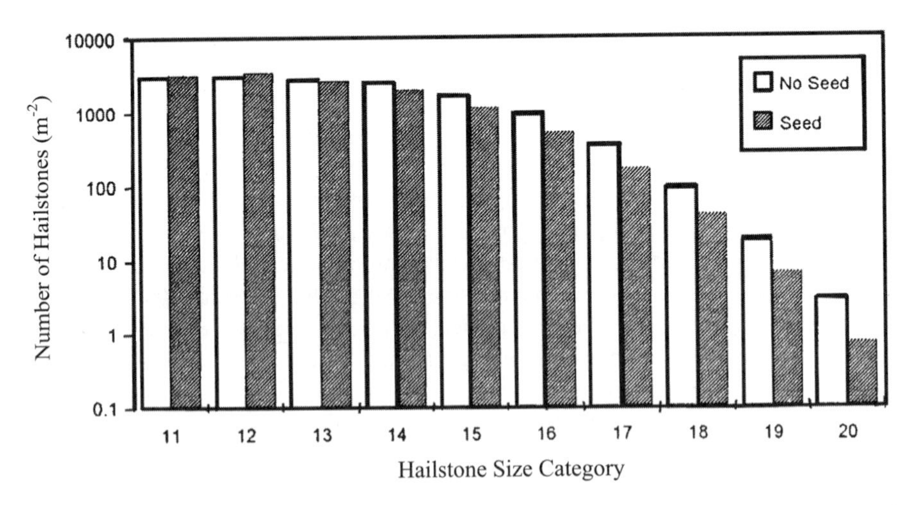

Figure 4 Size distributions of graupel and hail particles striking the ground (at the region of maximum hailfall from computer-simulated hailstorm). Categories 11, 14, 17, 20 represent 2-, 6-, 16-, 42-mm hailstones, respectively (Farley et al., 1996).

An abbreviated list of significant tools that have advanced our ability to observe cloud systems is summarized in Table 1. This table indicates that the microphysical and dynamic history of seeded and nonseeded clouds can now be documented and results compared between the two types of clouds.

Cloud Seeding Agents Practical capabilities have been established over the past 50 years, and especially since about 1980, for generating highly effective ice nucleating aerosols with well-characterized behaviors for both modeling and detecting their atmospheric fate after release for cloud modification. In the same period, our understanding of natural ice nucleation and our prospects for further elucidating ice formation processes have improved. Improvements in understanding how hygroscopic aerosols interact with clouds and how to use them to increase precipitation have also occurred in recent years.

By 1980, a clear need was recognized to account for the complex mechanisms of ice formation by specific ice nuclei used in weather modification field programs. In particular, the chemical and physical properties of aerosols were established to be very important in determining ice formation rates as well as efficiency. This recognition led to the development of new, highly efficient silver chloro-iodide ice nuclei (DeMott et al., 1983). These same nuclei can be generated with a soluble component to enhance the action of a fast condensation-freezing ice nucleation mechanism (Feng and Finnegan, 1989). The development of similar, fast-acting and highly efficient ice nuclei from pyrotechnic generation methods followed by the early 1990s (Fig. 5). These new nucleating agents represent substantial improvements

TABLE 1 Some Remote Sensors for Hydrometeorology*

Winds, air motions in clouds, clear air	Doppler radar, wind profiling radar, Doppler lidar
Temperature profiles	Radio acoustic sounding system (RASS)
Water vapor, cloud liquid water	Microwave radiometer
Cloud boundaries	Cloud (mm-wave) radar
Water vapor and liquid fluxes and profiles	(Combinations of the above)
Thermodynamic and microphysical cloud structures	"Retrievals" from Doppler radar data
Cloud phase, ice hydrometeor type and evolution	Dual-polarization cloud and precipitation radars, microwave radiometer, polarization microwave radiometer, polarization multi-field-of-view lidar
Transport and dispersion, mixing in clouds	Chaff wind tracking dual-circular-polarization radar, gaseous tracers
Precipitation trajectories	Chaff wind tracking and Doppler radar
Snowflake size, snowfall areal rate	Dual-wavelength radar
Rainfall rate, hail differentiation	Dual-polarization radar

*Table provided by Roger Reinking, National Oceanic and Atmospheric Administration.

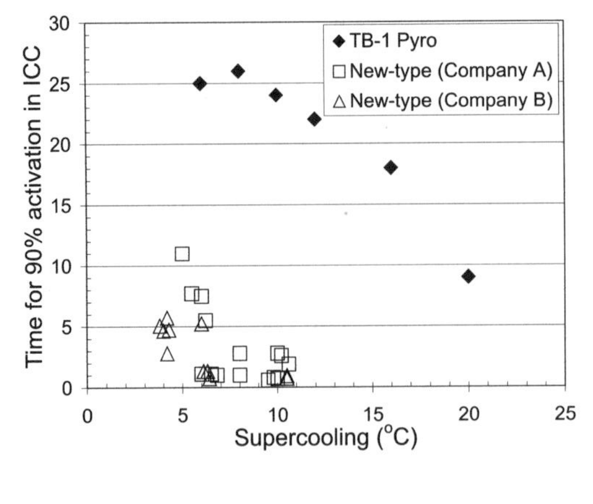

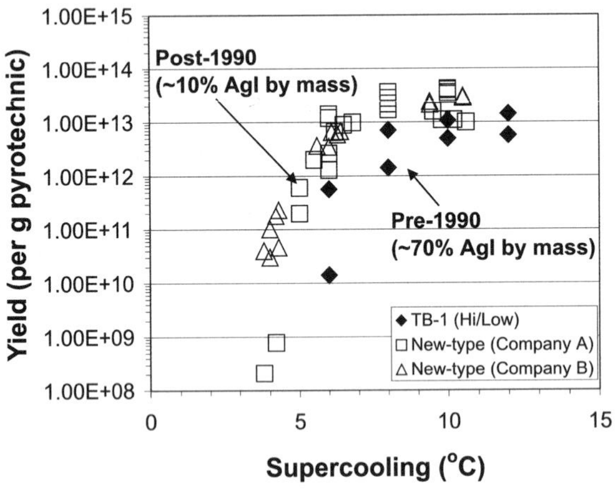

Figure 5 Yield (top panel) and rates (bottom panel) of ice formation by state-of-the-science pyrotechnic glaciogenic seeding generators, prior to (TB-1) and since about 1990. New-type pyrotechnics are more efficient in producing ice nuclei on a compositional basis, require less AgI (as $AgIO_3$), and "react" much faster in a water-saturated cloud. The new type pyrotechnics were developed in the former Yugoslavia but are now manufactured in North America. Results are from records of the Colorado State University isothermal cloud chamber (ICC) facility (Figure supplied by Paul DeMott, Colorado State University).

over prior nuclei generation capabilities and offer possibilities for engineering nuclei with specific and desired properties.

Numerical Modeling Capabilities

Simulations of Seeding Supercooled Orographic Clouds It is now possible to perform two- and three-dimensional simulations of clouds over individual mountains and entire mountain ranges with grid spacing on the order of a kilometer. For smaller mountain ridges, mixed-phase bin-resolving microphysics models can be used, while for larger mountain ranges, bulk microphysics is still needed. Such models can explicitly represent the transport and dispersion of seeding material, the primary and secondary ice nucleation of natural clouds, as well as nucleation of seeded material using laboratory-derived activity spectra.

An example of a numerical simulation of silver iodide transport over the Black Hills of South Dakota is given in Figure 6. Five generators were located in the simulation in the Northern Hills. The north winds spread the seeding material over the slopes and into the simulated orographic clouds. The AgI is activated in the clouds and forms snow at an earlier stage than the simulated natural nuclei (Farley et al., 1997).

Simulations of Seeding Cumulonimbi and Severe Convective Storms It is now quite common to perform fully three-dimensional time-dependent simulations of cumulonimbi and mesoscale convective systems with grid spacing of about 1 km or even 0.5 km, with mixed-phase bulk microphysics models. For hailstorms, a hybrid bulk microphysics/bin-microphysics model with continuous accretion approximations has been implemented in several storm models (Farley et al., 1996; Johnson et al., 1993). With access to advanced, state-of-the-art computers, stochastic bin-resolving microphysics models such as Reisin et al. (1996) could be applied to the simulation of natural and seeded hailstorms. Such a model could be used to examine in some detail beneficial competition, trajectory lowering (including hygroscopic seeding), early rainout, and glaciation hail suppression concepts.

Other Advances and General Comments

Research in the past 10 years has shown the close association of the cloud dynamics with the cloud microphysics and precipitation processes. The total precipitation from a cloud system is often the product of both the precipitation processes and the dynamic airflow regime. Consequently, reasoning about seeding effects on precipitation should include the effects of the dynamics of the cloud and cloud system. Cloud and mesoscale numerical models run on supercomputer systems are needed to keep track of all of the possible interactions. And efficient field projects are needed to test the concepts and validate the models. These are tasks that should be taken more seriously so that the technology can be appropriately developed and applied.

A comment regarding the magnitude of the seeding effect is warranted here. Cloud seeding for weather modification does not cause floods nor does it create droughts. Floods are caused by natural conditions much more vigorous than those caused by seeding. The rain processes are fully developed and highly efficient in flood situations. Likewise, droughts are caused by large-scale weather conditions not influenced by cloud seeding. The use of cloud seeding to produce more precipitation is most helpful to store water in reservoirs before droughts occur or to cause as much precipitation as possible as weather conditions improve. Cloud seeding will not break a drought, but it may make the drought less severe by extracting as much of the atmospheric water supply as possible from the existing clouds. This aspect of

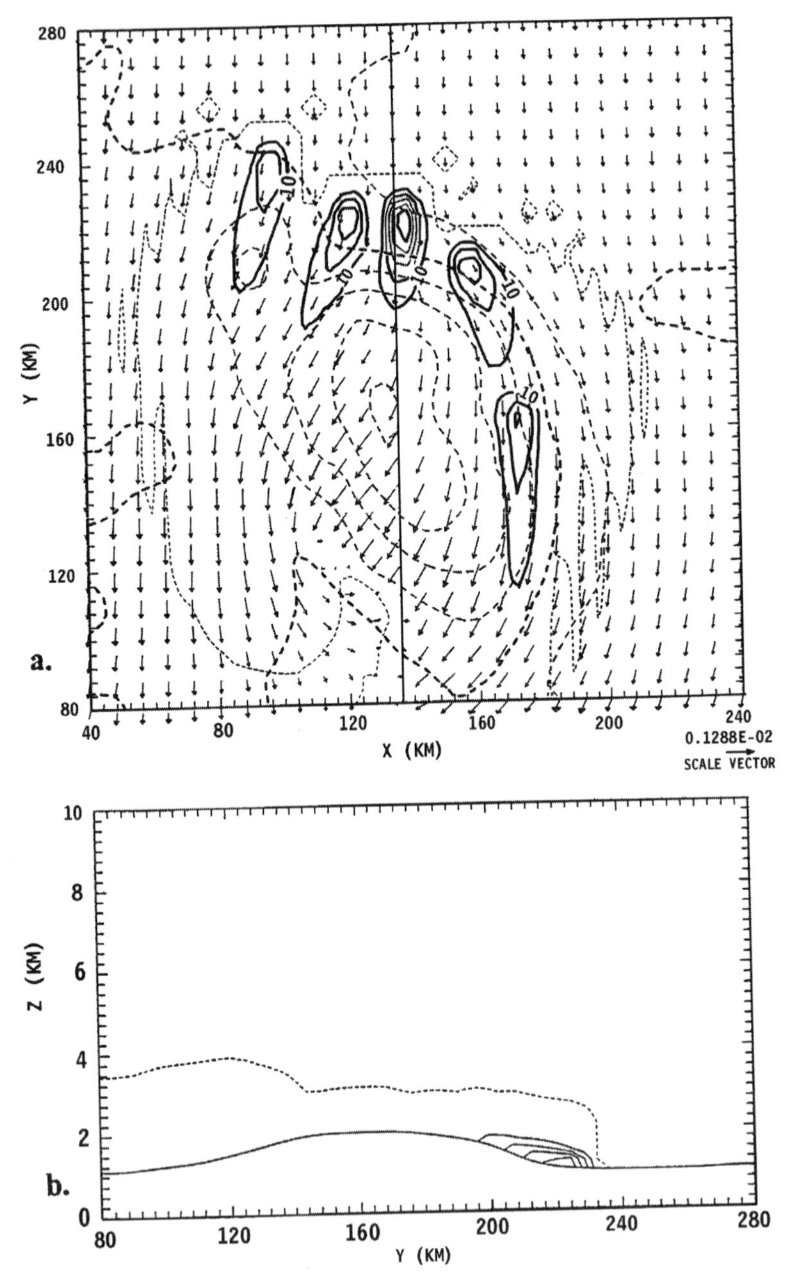

Figure 6 Seed case. (*a*) Surface locations where the inert tracer and AgI were released for day 3. (*b*) *Y–Z* cross section of the AgI field at $X = 136$ km. The contour interval is 0.1 g/kg for (*a*) and (*b*) (from Farley et al., 1997).

cloud seeding is still very much in the research stage. The research requires good cloud climatological data and well-conducted cloud seeding operations in drought conditions.

A common misconception follows from the fact that the desired result of cloud seeding for precipitation augmentation is to cause more rain or snow than might have occurred naturally. If the natural rainfall is far below normal, then the seeded rainfall will also be below normal, but not by as much, so the cloud seeding should not be blamed (but often is) for the below normal rainfall. This is another reason why more cloud seeding experiments are needed in various weather conditions, in the field and in computer models, to make more quantitative the effects of the cloud seeding and to develop methods that will distinguish the seeded rainfall from the natural component.

Nothing in this review has been said about fog suppression, the suppression of severe storms, and the reality of inadvertent weather modification, but the AMS statements give further information on these important topics (AMS, 1998a,b). Cold fog suppression is being practiced routinely in several airports around the world. The suppression of severe storms lacks critical hypotheses and should be tested in computer models before mounting field efforts. Large-scale land-use changes, widespread burning of forested lands, increased air pollution through industrialization, and extensive contrails caused by aircraft, have all contributed to inadvertent effects on the weather.

4 CONCLUDING REMARKS

Of great importance for proof of the efficacy of cloud seeding have been the significant strides made in the development of new or improved equipment, the use of more powerful and convenient computer power, and the development of more powerful statistical methods. Aircraft instrumentation, dual channel microwave radiometers, Doppler and multiparameter radars, satellite remote sensing, wind profilers, automated rain gauge networks, and mesoscale network stations have all contributed or have the potential to contribute to the detection of cloud seeding effects or to the identification of cloud seeding opportunities.

The increased computer power has made possible the greater use of cloud models in weather modification. Models have been developed that simulate explicitly the dispensing, transport, and nucleation effects of a seeding agent in convective and orographic clouds. Prediction and understanding of the seeding effects are products of the simulations. The results can also be used in evaluation and design of field projects.

The critical need for clean water in many parts of the world now, and even more so in the future, makes it imperative that the scientific basis of weather modification be strengthened. The conditions for increase, decrease, or redistribution of precipitation need to be determined. Stronger scientific and engineering projects in weather modification by cloud seeding give promise of more rapid progress in this scientifically important and socially relevant topic.

REFERENCES

AMS (1998a). Policy Statement: Planned and Inadvertent Weather Modification, *Bull. Am. Meteor. Soc.* **79**, 2771–2772.

AMS (1998b). Scientific background for the AMS policy statement on planned and inadvertent weather modification, *Bull. Am. Meteor. Soc.* **79**, 2773–2778.

BASC (2000). *New Opportunities in Weather Research, Focusing on Reducing Severe Weather Hazards and Providing Sustainable Water Resources*, Summary of the National Academy of Sciences Workshop for Assessing the Current State of Weather Modification Science as a Basis for Future Environmental Sustainability and Policy Development, prepared for the National Research Council Board on Atmospheirc Sciences and Climate (available at *ccrandal@taz.sdsmt.edu*).

Bigg, E. K. (1997). An independent evaluation of a South African hygroscopic cloud seeding experiment, 1991–1995, *Atm. Res.* **43**, 111–127.

Chai, S. K., W. G. Finnegan, and R. L. Pitter (1993). An interpretation of the mechanisms of ice crystal formation operative in the Lake Almanor cloud seeding program, *J. Appl. Meteor.* **32**, 1726–1732.

Changnon, S. A. (1998). In *Natural Hazards of North America*, National Geographic Maps, National Geographic Magazine, supplement, July, 1998, Washington, D.C.

Cooper, W. A., R. T. Bruintjes, and G. K. Mather (1997). Calculations pertaining to hygroscopic seeding with flares, *J. Appl. Meteor.* **36**, 1449–1469.

DeMott, P. J., W. C. Finnegan, and L. O. Grant (1983). An application of chemical kinetic theory and methodology to characterize the ice nucleating properties of aerosols used for weather modification, *J. Climate Appl. Meteor.* **22**, 1190–1203.

Dennis, A. S. (1980). *Weather Modification by Cloud Seeding*, New York, Academic.

Dessens, J. (1986). Hail in Southwestern France. II: Results of a 30-year hail prevention project with silver iodide seeding from the ground, *J. Climate Appl. Meteor.* **25**, 48–58.

Eklund, D. L., D. S. Jawa, and T. K. Rajala (1999). Evaluation of the Western Kansas weather modification program, *J. Wea. Mod.* **31**, 91–101.

Farley, R. D., H. Chen, H. D. Orville, and M. R. Hjelmfelt (1996). The numerical simulation of the effects of cloud seeding on hailstorms, Preprints, AMS 13th Conference on Planned and Inadvertent Weather Modification, Atlanta, GA, pp. 23–30.

Farley, R. D., D. L. Hjermstad, and H. D. Orville (1997). Numerical simulation of cloud seeding effects during a four-day storm period, *J. Wea. Mod.* **24**, 49–55.

Feng, D., and W. G. Finnegan (1989). An efficient, fast functioning nucleating agent—AgI•AgCl–4NaCl, *J. Wea. Mod.* **21**, 41–45.

IBHS (1998). *The Insured Cost of Natural Disasters: A Report on the IBHS Paid Loss Data Base*, Institute for Business and Home Safety, Tampa, FL (formerly Boston, MA).

Johnson, D. E., P. K. Wong, and J. M. Straka (1993). Numerical simulations of the 2 August 1981 CCOPE supercell storm with and without ice microphysics, *J. Appl. Meteor.* **32**, 745–759.

Mather, G. K., M. J. Dixon, and J. M. de Jager (1996). Assessing the potential for rain augmentation—the Nelspruit randomised convective cloud seeding experiment, *J. Appl. Meteor.* **35**, 1465–1482.

Mather, G. K., D. E. Terblanche, F. E. Steffens, and L. E. Fletcher (1997). Results of the South African cloud-seeding experiments using hygroscopic flares, *J. Appl. Meteor.* **36**, 1433–1447.

McGurty, B. M. (1999). Turning silver to gold: Measuring the benefits of cloud seeding, *Hydro. Rev.* April issue, 2–6.

Mossop , S. C. (1985). The origin and concentration of ice crystals in clouds, *Bull. Am. Meteor. Soc.* **66**, 264–273.

Nicholls, N. (2001). The insignificance of significance testing, *Bull. Am. Meteor. Soc.* **82**, 981–986.

Orville, H. D. (1996). A review of cloud modeling in weather modification, *Bull. Am. Meteor. Soc.* **77**, 1535–1555.

Orville, H. D., R. D. Farley, and J. H. Hirsch (1984). Some surprising results from simulated seeding of stratiform-type clouds, *J. Climate Appl. Meteor.* **23**, 1585–1600.

Orville, H. D., J. H. Hirsch, and R. D. Farley (1987). Further results on numerical cloud seeding simulations of stratiform-type clouds, *J. Wea. Modif.* **19**, 57–61.

Orville, H. D., C. Wang, and F. J. Kopp (1998). A simplified concept of hygroscopic seeding, *J. Wea. Mod.* **30**, 7–21.

Rangno, A. L. and P. V. Hobbs (1995). A new look at the Israeli cloud seeding experiments, *J. Appl. Meteor.* **34**, 1169–1193.

Rangno, A. L., and P. V. Hobbs (1997). Reply, *J. Appl. Meteor.* **36**, 272–276.

Reisin, T., S. Tzivion, and Z. Levin (1996). Seeding convective clouds with ice nuclei or hygroscopic particles: A numerical study using a model with detailed microphysics, *J. Appl. Meteor.* **35**, 1416–1434.

Reynolds, D. W. (1988). A report on winter snowpack augmentation, *Bull. Am. Meteor. Soc.* **69**, 1290–1300.

Rogers, R. R. and M. K. Yau (1989). *A Short Course in Cloud Physics*, 3rd ed., International Series in Natural Philosophy, Vol. 113, D. Ter Haar (Ed.), Butterworth-Heinemann, Woburn, MA.

Rosenfeld, D., and W. L. Woodley (1993). Effects of cloud seeding in west Texas: Additional results and new insights, *J. Appl. Meteor.* **32**, 1848–1866.

Rosenfeld, D. and W. L. Woodley (2000). Convective clouds with sustained highly supercooled liquid water down to −37.5°C, *Nature* **405**, 440–442.

Silverman, B. A. (2001). A critical assessment of glaciogenic seeding of convective clouds for rainfall enhancement, *Bull. Am. Meteor. Soc.* **82**, 903–923.

Smith, P. L., L. R. Johnson, D. L. Priegnitz, B. A. Boe, and P. J. Mielke, Jr. (1997). An exploratory analysis of crop hail insurance data for evidence of cloud seeding effects in North Dakota, *J. Appl. Meteor.* **36**, 463–473.

Stone, R. H., and A. W. Huggins (1996). The use of trace chemistry in conjunction with ice crystal measurements to assess wintertime cloud seeding experiments, 13th Conf. on Planned and Inadvertent Weather Modification, Atlanta, GA., Amer. Meteor. Soc., pp. 136–141.

Super, A. B., and J. A. Heimbach (1992). Investigations of the targeting of ground released silver iodide in Utah. Part I: Ground observations of silver in snow and ice nuclei, *J. Wea. Modif.* **24**, 19–34.

Super A. B., and E. W. Holroyd (1997). Some physical evidence of silver iodide and liquid propane seeding effects on Utah's Wasatch Plateau, *J. Wea. Modif.* **29**, 8–32.

Tzivion, S., T. Reisin, and Z. Levin (1994). Numerical simulation of hygroscopic seeding in a convective cloud, *J. Appl. Meteor.* **33**, 252–267.

Warburton, J. A., L. G. Young, M. S. Owens, and R. H. Stone (1985). The capture of ice nucleating and non ice-nucleating aerosols by ice-phase precipitationm, *J. Rech. Atmos*, **19**, (2–3), 249–255.

Warburton, J. A., L. G. Young, and R. H. Stone (1994). Assessment of seeding effects in snowpack augmentation programs: Ice nucleation and scavenging of seeding aerosols, *J. Appl. Meteor.* **33**, 121–130.

Warburton, J. A., R. H. Stone, and B. L. Marler (1995a). How the transport and dispersion of AgI aerosols may affect detectability of seeding effects by statistical methods, *J. Appl. Meteor.* **34**, 1930–1941.

Warburton, J. A., S. K. Chai, and L. G. Young (1995b). A new concept for assessing silver iodide cloud seeding effects in snow by physical and chemical methods, *Atmos. Res.* **36**, 171–176.

WMAB (1978). *The Management of Weather Resources.* Vol. I, Washington, DC, Dept. of Commerce.

Woodley, W. L., and D. Rosenfeld (2000). Evidence for changes in microphysical structure and cloud drafts following AgI Seeding, *J. Wea. Modif.* **32**, 53–67.

Young, K. C. (1993). *Microphysical Processes in Clouds*, New York, Oxford University Press.

CHAPTER 24

ATMOSPHERIC OPTICS

CRAIG F. BOHREN

1 INTRODUCTION

Atmospheric optics is nearly synonymous with light scattering, the only restrictions being that the scatterers inhabit the atmosphere and the primary source of their illumination is the sun. Essentially all light we see is scattered light, even that directly from the sun. When we say that such light is unscattered we really mean that it is scattered in the forward direction, hence it is *as if* it were unscattered. Scattered light is radiation from matter excited by an external source. When the source vanishes, so does the scattered light, as distinguished from light emitted by matter, which persists in the absence of external sources.

Atmospheric scatterers are either molecules or particles. A particle is an aggregation of sufficiently many molecules that it can be ascribed macroscopic properties such as temperature and refractive index. There is no canonical number of molecules that must unite to form a bona fide particle. Two molecules clearly do not a quorum make, but what about 10, 100, or 1000? The particle size corresponding to the largest of these numbers is about 10^{-3} μm. Particles this small of water substance would evaporate so rapidly that they could not exist long under conditions normally found in the atmosphere. As a practical matter, therefore, we need not worry unduly about hermaphrodite scatterers in the shadow region between molecule and particle.

A property of great relevance to scattering problems is *coherence*, both of the array of scatterers and of the incident light. At visible wavelengths air is an array of incoherent scatterers: The radiant power scattered by N molecules is N times that scattered by one (except in the forward direction). But when water vapor in air condenses, an incoherent array is transformed into a coherent array: Uncorrelated

Handbook of Weather, Climate, and Water: Dynamics, Climate, Physical Meteorology, Weather Systems, and Measurements, Edited by Thomas D. Potter and Bradley R. Colman.
ISBN 0-471-21490-6

water molecules become part of a single entity. Although a single droplet is a coherent array, a cloud of droplets taken together is incoherent.

Sunlight is incoherent but not in an absolute sense. Its lateral coherence length is tens of micrometers, which is why we can observe what are essentially interference patterns (e.g., coronas and glories) resulting from illumination of cloud droplets by sunlight.

This chapter begins with the color and brightness of a purely molecular atmosphere, including their variation across the vault of the sky. This naturally leads to the state of polarization of skylight. Because the atmosphere is rarely, if ever, entirely free of particles, the general characteristics of scattering by particles follow, setting the stage for atmospheric visibility.

Atmospheric refraction usually sits by itself, unjustly isolated from all those atmospheric phenomena embraced by the term *scattering.* Yet refraction is yet another manifestation of scattering, although in the forward direction, for which scattering is coherent.

Scattering by single water droplets and ice crystals, each discussed in turn, yields feasts for the eye as well as the mind. The curtain closes on the optical properties of clouds.

2 COLOR AND BRIGHTNESS OF MOLECULAR ATMOSPHERE

Brief History

Edward Nichols began his 1908 presidential address to the New York meeting of the American Physical Society as follows: "In asking your attention to-day, even briefly, to the consideration of the present state of our knowledge concerning the color of the sky it may be truly said that I am inviting you to leave the thronged thoroughfares of our science for some quiet side street where little is going on and you may even suspect that I am coaxing you into some blind alley, the inhabitants of which belong to the dead past."

Despite this depreciatory statement, hoary with age, correct and complete explanations of the color of the sky still are hard to find. Indeed, all the faulty explanations lead active lives: the blue sky is the reflection of the blue sea; it is caused by water, either vapor or droplets or both; it is caused by dust. The true cause of the blue sky is not difficult to understand, requiring only a bit of critical thought stimulated by belief in the inherent fascination of all natural phenomena, even those made familiar by everyday occurrence.

Our contemplative prehistoric ancestors no doubt speculated on the origin of the blue sky, their musings having vanished into it. Yet it is curious that Aristotle, the most prolific speculator of early recorded history, makes no mention of it in his *Meteorologica* even though he delivered pronouncements on rainbows, halos, and mock suns and realized that "the sun looks red when seen through mist or smoke." Historical discussions of the blue sky sometimes cite Leonardo da Vinci as the first to comment intelligently on the blue of the sky, although this reflects a European bias.

If history were to be written by a supremely disinterested observer, Arab philosophers would likely be given more credit for having had profound insights into the workings of nature many centuries before their European counterparts descended from the trees. Indeed, Möller (1972) begins his brief history of the blue sky with Jakub Ibn Ishak Al Kindi (800–870), who explained it as "a mixture of the darkness of the night with the light of the dust and haze particles in the air illuminated by the sun."

Leonardo was a keen observer of light in nature even if his explanations sometimes fell short of the mark. Yet his hypothesis that "the blueness we see in the atmosphere is not intrinsic colour, but is caused by warm vapor evaporated in minute and insensible atoms on which the solar rays fall, rendering them luminous against the infinite darkness of the fiery sphere which lies beyond and includes it" would, with minor changes, stand critical scrutiny today. If we set aside Leonardo as *sui generis*, scientific attempts to unravel the origins of the blue sky may be said to have begun with Newton, that towering pioneer of optics, who, in time-honored fashion, reduced it to what he already had considered: interference colors in thin films. Almost two centuries elapsed before more pieces in the puzzle were contributed by the experimental investigations of von Brücke and Tyndall on light scattering by suspensions of particles. Around the same time Clausius added his bit in the form of a theory that scattering by minute bubbles causes the blueness of the sky. A better theory was not long in coming. It is associated with a man known to the world as Lord Rayleigh even though he was born John William Strutt.

Rayleigh's paper of 1871 marks the beginning of a satisfactory explanation of the blue sky. His scattering law, the key to the blue sky, is perhaps the most famous result ever obtained by dimensional analysis. Rayleigh argued that the field $\mathbf{E}_s$ scattered by a particle small compared with the light illuminating it is proportional to its volume V and to the incident field $\mathbf{E}_i$. Radiant energy conservation requires that the scattered field diminish inversely as the distance r from the particle so that the scattered power diminishes as the square of r. To make this proportionality dimensionally homogeneous requires the inverse square of a quantity with the dimensions of length. The only plausible physical variable at hand is the wavelength of the incident light, which leads to

$$\mathbf{E}_s \propto \mathbf{E}_i \frac{V}{r\lambda^2} \tag{1}$$

When the field is squared to obtain the scattered power, the result is Rayleigh's inverse fourth-power law. This law is really only an often—but not always—very good approximation. Missing from it are dimensionless properties of the particle such as its refractive index, which itself depends on wavelength. Because of this *dispersion*, therefore, nothing scatters exactly as the inverse fourth power.

Rayleigh's 1871 paper did not give the complete explanation of the color and polarization of skylight. What he did that was not done by his predecessors was to give a law of scattering, which could be used to quantitatively test the hypothesis that selective scattering by atmospheric particles could transform white sunlight into blue skylight. But as far as giving the agent responsible for the blue sky, Rayleigh did not go essentially beyond Newton and Tyndall, who invoked particles. Rayleigh was circumspect about the nature of these particles, settling on salt as the most likely candidate. It was not until 1899 that he published the capstone to his work on skylight, arguing that air molecules themselves were the source of the blue sky. Tyndall cannot be given the credit for this because he considered air to be *optically empty*: When purged of all particles it scatters no light. This erroneous conclusion was a result of the small scale of his laboratory experiments. On the scale of the atmosphere, sufficient light is scattered by air molecules to be readily observable.

Molecular Scattering and the Blue of the Sky

Our illustrious predecessors all gave explanations of the blue sky requiring the presence of water in the atmosphere: Leonardo's "evaporated warm vapor," Newton's "globules of water," Clausius's bubbles. Small wonder, then, that water still is invoked as the cause of the blue sky. Yet a cause of something is that without which it would not occur, and the sky would be no less blue if the atmosphere were free of water.

A possible physical reason for attributing the blue sky to water vapor is that, because of selective *absorption*, liquid water (and ice) is blue upon transmission of white light over distances of order meters. Yet if all the water in the atmosphere at any instant were to be compressed into a liquid, the result would be a layer about 1 cm thick, which is not sufficient to transform white light into blue by selective absorption.

Water vapor does not compensate for its hundredfold lower abundance than nitrogen and oxygen by greater scattering per molecule. Indeed, scattering of visible light by a water molecule is slightly *less* than that by either nitrogen or oxygen.

Scattering by atmospheric molecules does not obey Rayleigh's inverse fourth-power law exactly. A least-squares fit over the visible spectrum from 400 to 700 nm of the *molecular scattering coefficient* of sea level air tabulated by Penndorf (1957) yields an inverse 4.089 scattering law.

The molecular scattering coefficient β, which plays important roles in following sections, may be written

$$\beta = N\sigma_s \tag{2}$$

where N is the number of molecules per unit volume and σ_s, the scattering cross section (an average because air is a mixture) per molecule, approximately obeys Rayleigh's law. The form of this expression betrays the incoherence of scattering by atmospheric molecules. The inverse of β is interpreted as the scattering *mean free path*, the average distance a photon must travel before being scattered.

To say that the sky is blue because of Rayleigh scattering, as is sometimes done, is to confuse an agent with a law. Moreover, as Young (1982) pointed out, the term *Rayleigh scattering* has many meanings. Particles small compared with the wavelength scatter according to the same law as do molecules. Both can be said to be

Rayleigh scatterers, but only molecules are necessary for the blue sky. Particles, even small ones, generally diminish the vividness of the blue sky.

Fluctuations are sometimes trumpeted as the "real" cause of the blue sky. Presumably, this stems from the fluctuation theory of light scattering by media in which the scatterers are separated by distances small compared with the wavelength. Using this theory, which is associated with Einstein and Smoluchowski, matter is taken to be continuous but characterized by a refractive index that is a random function of position. Einstein (1910) stated that "it is remarkable that our theory does not make *direct* use of the assumption of a discrete distribution of matter." That is, he circumvented a difficulty but realized it could have been met head on, as Zimm (1945) did years later.

The blue sky is really caused by scattering by molecules. To be more precise, scattering by bound electrons: free electrons do not scatter selectively. Because air molecules are separated by distances small compared with the wavelengths of visible light, it is not obvious that the power scattered by such molecules can be added. Yet if they are completely uncorrelated, as in an ideal gas (to good approximation the atmosphere is an ideal gas), scattering by N molecules is N times scattering by one. This is the only sense in which the blue sky can be attributed to scattering by fluctuations. Perfectly homogeneous matter does not exist. As stated pithily by Planck: "a chemically pure substance may be spoken of as a vacuum made turbid by the presence of molecules."

Spectrum and Color of Skylight

What is the spectrum of skylight? What is its color? These are two different questions. Answering the first answers the second but not the reverse. Knowing the color of skylight we cannot uniquely determine its spectrum because of *metamerism*: A given perceived color can in general be obtained in an indefinite number of ways.

Skylight is not blue (itself an imprecise term) in an absolute sense. When the visible spectrum of sunlight outside Earth's atmosphere is modulated by Rayleigh's scattering law, the result is a spectrum of scattered light that is neither solely blue nor even peaked in the blue (Fig. 1). Although blue does not predominate spectrally, it does predominate perceptually. We perceive the sky to be blue even though skylight contains light of all wavelengths.

Any source of light may be looked upon as a mixture of white light and light of a single wavelength called the *dominant wavelength.* The *purity* of the source is the relative amount of the monochromatic component in the mixture. The dominant wavelength of sunlight scattered according to Rayleigh's law is about 475 nm, which lies solidly in the blue if we take this to mean light with wavelengths between 450 and 490 nm. The purity of this scattered light, about 42%, is the upper limit for skylight. Blues of real skies are less pure.

Another way of conveying the color of a source of light is by its *color temperature*, the temperature of a blackbody having the same perceived color as the source. Since blackbodies do not span the entire gamut of colors, all sources of light cannot be assigned color temperatures. But many natural sources of light can. The color

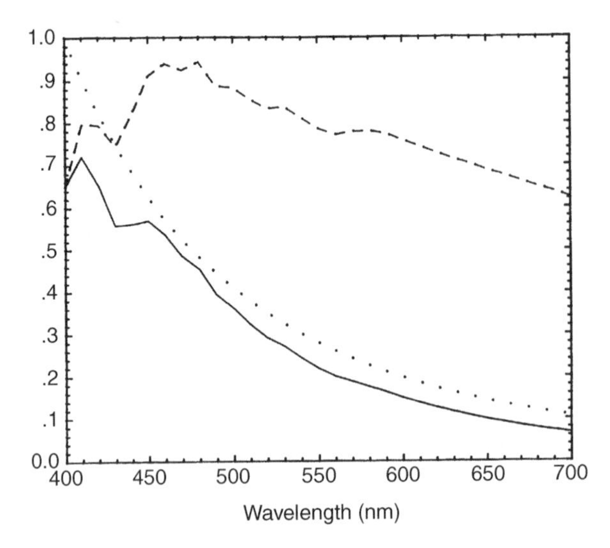

Figure 1 Rayleigh's scattering law (dots), the spectrum of sunlight outside the earth's atmosphere (dashes), and the product of the two (solid). The solar spectrum is taken from Thekaekara and Drummond (1971).

temperature of light scattered according to Rayleigh's law is infinite. This follows from Planck's spectral emission function $e_{b\lambda}$ in the limit of high temperature

$$e_{b\lambda} \simeq \frac{2\pi ckT}{\lambda^4} \qquad \frac{hc}{\lambda} \ll kT \tag{3}$$

where h is Planck's constant, k is Boltzmann's constant, c is the speed of light in vacuo, and T is absolute temperature. Thus the emission spectrum of a blackbody with an infinite temperature has the same functional form as Rayleigh's scattering law.

Variation of Sky Color and Brightness

Not only is skylight not pure blue, its color and brightness vary across the vault of the sky, with the best blues at zenith. Near the astronomical horizon the sky is brighter than overhead but of considerably lower purity. That this variation can be observed from an airplane flying at 10 km, well above most particles, suggests that the sky is inherently nonuniform in color and brightness (Fig. 2). To understand why requires invoking multiple scattering.

Multiple scattering gives rise to observable phenomena that cannot be explained solely by single-scattering arguments. This is easily demonstrated. Fill a blackened pan with clean water, then add a few drops of milk. The resulting dilute suspension illuminated by sunlight has a bluish cast. But when more milk is added, the suspen-

Figure 2 Even at an altitude of 10 km, well above most particles, the sky brightness increases markedly from the zenith to the astronomical horizon.

sion turns white. Yet the properties of the scatterers (fat globules) have not changed, only their *optical thickness*: the blue suspension being optically thin, the white being optically thick:

$$\tau = \int_1^2 \beta \, ds \tag{4}$$

Optical thickness is physical thickness in units of scattering mean free path, hence is dimensionless. The optical thickness τ between any two points connected by an arbitrary path in a medium populated by (incoherent) scatterers is an integral over the path: The *normal optical thickness* τ_n of the atmosphere is that along a radial path extending from the surface of Earth to infinity. Figure 3 shows τ_n over the visible spectrum for a purely molecular atmosphere. Because τ_n is generally small compared with unity, a photon from the sun traversing a radial path in the atmosphere is unlikely to be scattered more than once. But along a tangential path, the optical thickness is about 35 times greater (Fig. 4), which leads to several observable phenomena.

Even an intrinsically black object is luminous to an observer because of *airlight*, light scattered by all the molecules and particles along the line of sight from observer to object. Provided that this is uniformly illuminated by sunlight and that ground reflection is negligible, the airlight radiance L is approximately

$$L = GL_0(1 - e^{-\tau}) \tag{5}$$

where L_0 is the radiance of sunlight along the line of sight and τ is its optical thickness; G accounts for geometric reduction of radiance because of scattering of nearly monodirectional sunlight in all directions. If the line of sight is uniform in composition, $\tau = \beta d$, where β is the scattering coefficient and d is the physical distance to the black object.

If τ is small ($\ll 1$), $L \approx GL_0\tau$. In a purely molecular atmosphere, τ varies with wavelength according to Rayleigh's law; hence the distant black object in such an atmosphere is perceived to be bluish. As τ increases so does L but not proportionally.

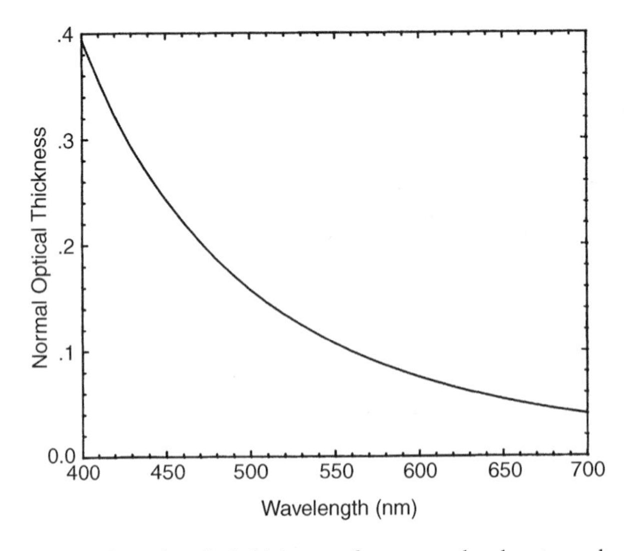

Figure 3 Normal optical thickness of a pure molecular atmosphere.

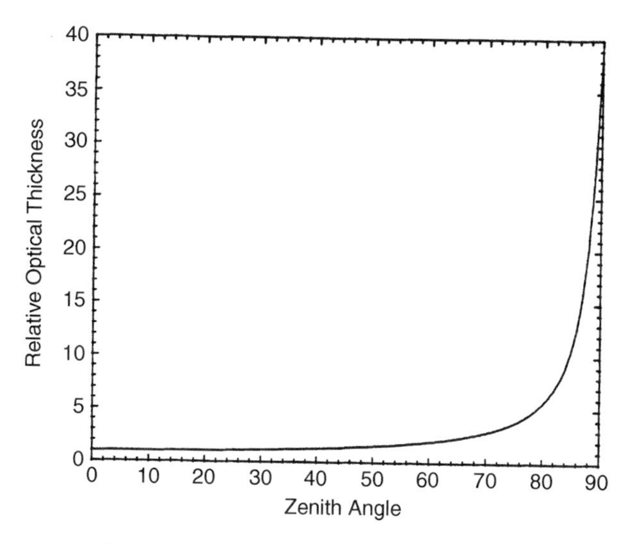

Figure 4 Optical thickness (relative to the normal optical thickness) of a molecular atmosphere along various paths with zenith angles between 0° (normal) and 90° (tangential).

Its limit is GL_0: The airlight radiance spectrum is that of the source of illumination. Only in the limit $d=0$ is $L=0$ and the black object is truly black.

Variation of the brightness and color of dark objects with distance was called *aerial perspective* by Leonardo. By means of it we estimate the distance of objects of unknown size such as mountains.

Aerial perspective belongs to the same family as the variation of color and brightness of the sky with zenith angle. Although the optical thickness along a path tangent to Earth is not infinite, it is sufficiently large (Figs. 3 and 4) that GL_0 is a good approximation for the radiance of the horizon sky. For isotropic scattering (a condition almost satisfied by molecules), G is around 10^{-5}, the ratio of the solid angle subtended by the sun to the solid angle of all directions (4π). Thus the horizon sky is not nearly so bright as direct sunlight.

Unlike in the milk experiment, what is observed when looking at the horizon sky is not multiply scattered light. Both have their origins in multiple scattering but manifested in different ways. Milk is white because it is weakly absorbing and optically thick, hence all components of incident white light are multiply scattered to the observer even though the blue component traverses a shorter average path in the suspension than the red component. White horizon light has escaped being multiply scattered, although multiple scattering is why this light is white (strictly, has the spectrum of the source). More light at the short wavelength end of the spectrum is scattered *toward* the observer than at the long wavelength end. But long wavelength light has the greater likelihood of being transmitted to the observer without being scattered *out of* the line of sight. For a long optical path, these two processes compensate, resulting in a horizon radiance spectrum of the source.

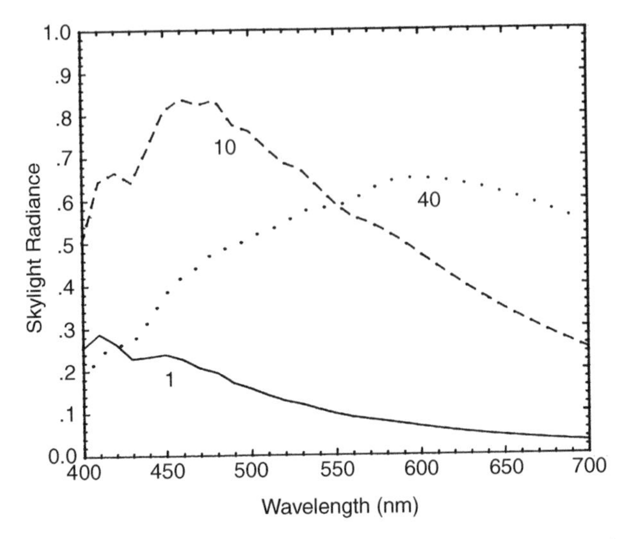

Figure 5 Spectrum of overhead skylight for the present molecular atmosphere (solid), as well as for hypothetical atmospheres 10 (dashes) and 40 (dots) times thicker.

Selective scattering by molecules is not sufficient for a blue sky. The atmosphere also must be optically thin, at least for most zenith angles (Fig. 4) (the blackness of space as a backdrop is taken for granted but also is necessary, which Leonardo recognized). A corollary of this is that the blue sky is not inevitable: An atmosphere composed entirely of nonabsorbing, selectively scattering molecules overlying a nonselectively reflecting Earth need not be blue. Figure 5 shows calculated spectra of the zenith sky over black ground for a molecular atmosphere with the present normal optical thickness as well as for hypothetical atmospheres 10 and 40 times thicker. What we take to be inevitable is accidental: If our atmosphere were much thicker, but identical in composition, the color of the sky would be quite different from what it is now.

Sunrise and Sunset

If short wavelength light is preferentially scattered out of direct sunlight, long wavelength light is preferentially transmitted in the direction of sunlight. Transmission is described by an exponential law (if light multiply scattered back into the direction of the sunlight is negligible):

$$L = L_o e^{-\tau} \tag{6}$$

where L is the radiance at the observer in the direction of the sun, L_o is the radiance of sunlight outside the atmosphere, and τ is the optical thickness along this path.

If the wavelength dependence of τ is given by Rayleigh's law, sunlight is *reddened* upon transmission: The spectrum of the transmitted light is comparatively richer than the incident spectrum in light at the long wavelength end of the visible spectrum. But to say that transmitted sunlight is reddened is not the same as saying it is red. The perceived color can be yellow, orange, or red, depending on the magnitude of the optical thickness. In a molecular atmosphere, the optical thickness along a path from the sun, even on or below the horizon, is not sufficient to give red light upon transmission. Although selective scattering by molecules yields a blue sky, reds are not possible in a molecular atmosphere, only yellows and oranges. This can be observed on clear days, when the horizon sky at sunset becomes successively tinged with yellow, then orange, but not red. Equation (6) applies to the radiance only in the direction of the sun. Oranges and reds can be seen in other directions because reddened sunlight illuminates scatterers not lying along the line of sight to the sun. A striking example of this is a horizon sky tinged with oranges and pinks in the direction *opposite* the sun.

The color and brightness of the sun changes as it arcs across the sky because the optical thickness along the line of sight changes with solar zenith angle Θ. If Earth were flat (as some still aver), the transmitted solar radiance would be

$$L = L_o \exp - \frac{\tau_n}{\cos \Theta} \tag{7}$$

This equation is a good approximation except near the horizon. On a flat Earth, the optical thickness is infinite for horizon paths. On a spherical Earth, optical thicknesses are finite although much larger for horizon than for radial paths.

The normal optical thickness of an atmosphere in which the number density of scatterers decreases exponentially with height z above the surface, $\exp(-z/H)$, is the same as that for a uniform atmosphere of finite thickness:

$$\tau_n = \int_0^\infty \beta \, dz = \beta_0 H \tag{8}$$

where H is the *scale height* and β_0 is the scattering coefficient at sea level. This equivalence yields a good approximation even for the tangential optical thickness. For any zenith angle, the optical thickness is given approximately by

$$\frac{\tau}{\tau_n} = \sqrt{\frac{R_e^2}{H^2}\cos^2 \Theta + \frac{2R_e}{H} + 1} - \frac{R_e}{H}\cos \Theta \tag{9}$$

where R_e is the radius of Earth. A flat Earth is one for which R_e is infinite, in which instance Eq. (9) yields the expected relation

$$\lim_{R_e \to \infty} \frac{\tau}{\tau_n} = \frac{1}{\cos \Theta} \tag{10}$$

For Earth's atmosphere, the molecular scale height is about 8 km. According to the approximate relation Eq. (9), therefore, the horizon optical thickness is about 39 times greater than the normal optical thickness. Taking the exponential decrease of molecular number density into account yields a value about 10% lower.

Variations on the theme of reds and oranges at sunrise and sunset can be seen even when the sun is overhead. The radiance at an observer an optical distance τ from a (horizon) cloud is the sum of cloudlight transmitted to the observer and airlight:

$$L = L_0 G(1 - e^{-\tau}) + L_0 G_c e^{-\tau} \tag{11}$$

where G_c is a geometrical factor that accounts for scattering of nearly monodirectional sunlight into a hemisphere of directions by the cloud. If it is approximated as an isotropic reflector with reflectance R and illuminated at an angle Φ, the geometrical factor G_c is $\Omega_s R \cos \Phi/\pi$. If $G_c > G$, the observed radiance is redder (i.e., enriched in light of longer wavelengths) than the incident radiance. If $G_c < G$, the observed radiance is bluer than the incident radiance. Thus distant horizon clouds can be reddish if they are bright or bluish if they are dark.

Underlying Eq. (11) is the implicit assumption that the line of sight is uniformly illuminated by sunlight. The first term in this equation is airlight, the second is transmitted cloudlight. Suppose, however, that the line of sight is shadowed from direct sunlight by clouds (that do not, of course, occlude the distant cloud of interest). This may reduce the first term in Eq. (11) so that the second term dominates. Thus under a partly overcast sky, distant horizon clouds may be reddish even when the sun is high in the sky.

The zenith sky at sunset and twilight is the exception to the general rule that molecular scattering is sufficient to account for the color of the sky. In the absence of molecular absorption, the spectrum of the zenith sky would be essentially that of the zenith sun (although greatly reduced in radiance), hence would not be the blue that is observed. This was pointed out by Hulburt (1953), who showed that absorption by ozone profoundly affects the color of the zenith sky when the sun is near the horizon. The Chappuis band of ozone extends from about 450 to 700 nm and peaks at around 600 nm. Preferential absorption of sunlight by ozone over long horizon paths gives the zenith sky its blueness when the sun is near the horizon. With the sun more than about 10° above the horizon, however, ozone has little effect on the color of the sky.

3 POLARIZATION OF LIGHT IN MOLECULAR ATMOSPHERE

Nature of Polarized Light

Unlike sound, light is a vector wave, an electromagnetic field lying in a plane normal to the propagation direction. The polarization state of such a wave is determined by the degree of correlation of any two orthogonal components into which its electric (or magnetic) field is resolved. Completely polarized light corresponds to complete

correlation; completely unpolarized light corresponds to no correlation; partially polarized light corresponds to partial correlation.

If an electromagnetic wave is completely polarized, the tip of its oscillating electric field traces out a definite elliptical curve, the *vibration ellipse.* Lines and circles are special ellipses, the light being said to be linearly or circularly polarized, respectively. The general state of polarization is elliptical.

Any beam of light can be considered an incoherent superposition of two collinear beams, one unpolarized, the other completely polarized. The radiance of the polarized component relative to the total is defined as the *degree of polarization* (often multiplied by 100 and expressed as a percent). This can be measured for a source of light (e.g., light from different sky directions) by rotating a (linear) polarizing filter and noting the minimum and maximum radiances transmitted by it. The degree of (linear) polarization is defined as the difference between these two radiances divided by their sum.

Polarization by Molecular Scattering

Unpolarized light can be transformed into partially polarized light upon interaction with matter because of different changes in amplitude of the two orthogonal field components. An example of this is the partial polarization of sunlight upon scattering by atmospheric molecules, which can be detected by looking at the sky through a polarizing filter (e.g., polarizing sunglasses) while rotating it. Waxing and waning of the observed brightness indicates some degree of partial polarization.

In the analysis of any scattering problem a plane of reference is required. This is usually the *scattering plane*, determined by the directions of the incident and scattered waves, the angle between them being the *scattering angle.* Light polarized perpendicular (parallel) to the scattering plane is sometimes said to be vertically (horizontally) polarized. Vertical and horizontal in this context, however, are arbitrary terms indicating orthogonality and bear no relation, except by accident, to the direction of gravity.

The degree of polarization P of light scattered by a tiny sphere illuminated by unpolarized light is (Fig. 6)

$$P = \frac{1 - \cos^2\theta}{1 + \cos^2\theta} \tag{12}$$

where the scattering angle θ ranges from 0° (forward direction) to 180° (backward direction); the scattered light is partially linearly polarized perpendicular to the scattering plane. Although this equation is a first step toward understanding polarization of skylight, more often than not it also has been a false step, having led countless authors to assert that skylight is completely polarized at 90° from the sun. Although $P = 1$ at $\theta = 90°$ according to Eq. (12), skylight is never 100% polarized at this or any other angle, and for several reasons.

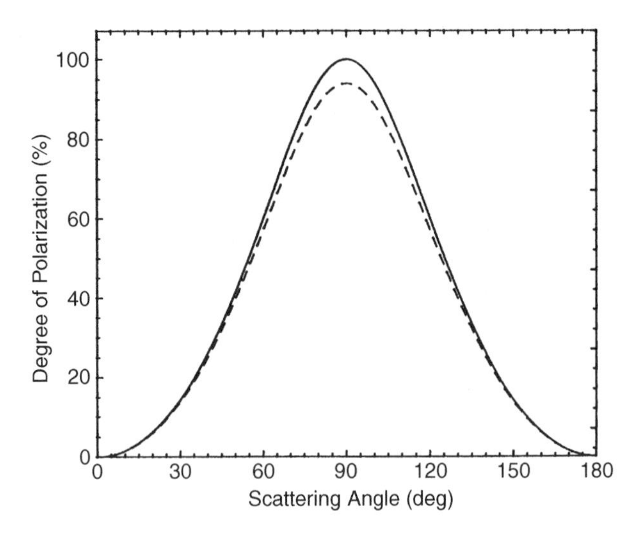

Figure 6 Degree of polarization of the light scattered by a small (compared with the wavelength) sphere for incident unpolarized light (solid). The dashed curve is for a small spheroid chosen such that the degree of polarization at 90° is that for air.

Although air molecules are very small compared with the wavelengths of visible light, a requirement underlying Eq. (12), the dominant constituents of air are not spherically symmetric.

The simplest model of an asymmetric molecule is a small spheroid. Although it is indeed possible to find a direction in which the light scattered by such a spheroid is 100% polarized, this direction depends on the spheroid's orientation. In an ensemble of randomly oriented spheroids each contributes its mite to the total radiance in a given direction, but each contribution is partially polarized to varying degrees between 0 and 100%. It is impossible for beams of light to be incoherently superposed in such a way that the degree of polarization of the resultant is greater than the degree of polarization of the most highly polarized beam. Because air is an ensemble of randomly oriented asymmetric molecules, sunlight scattered by air never is 100% polarized. The intrinsic departure from perfection is about 6%. ure 6 also includes a curve for light scattered by randomly oriented spheroids chosen to yield 94% polarization at 90°. This angle is so often singled out that it may deflect attention from nearby scattering angles. Yet the degree of polarization is greater than 50% for a range of scattering angles 70° wide centered about 90°.

Equation (12) applies to air, not to the atmosphere, the distinction being that in the atmosphere, as opposed to the laboratory, multiple scattering is not negligible. Also, atmospheric air is almost never free of particles and is illuminated by light reflected by the ground. We must take the atmosphere as it is, whereas in the laboratory we often can eliminate everything we consider extraneous.

Even if scattering by particles were negligible, because of both multiple scattering and ground reflection, light from any direction in the sky is not, in general, made up solely of light scattered in a single direction relative to the incident sunlight but is a superposition of beams with different scattering histories, hence different degrees of polarization. As a consequence, even if air molecules were perfect spheres and the atmosphere were completely free of particles, skylight would not be 100% polarized at 90° to the sun or at any other angle.

Reduction of the maximum degree of polarization is not the only consequence of multiple scattering. According to Figure 6, there should be two *neutral points* in the sky, directions in which skylight is unpolarized: directly toward and away from the sun. Because of multiple scattering, however, there are three such points. When the sun is higher than about 20° above the horizon there are neutral points within 20° of the sun, the *Babinet point* above it, the *Brewster point* below. They coincide when the sun is directly overhead and move apart as the sun descends. When the sun is lower than 20°, the *Arago point* is about 20° above the antisolar point, in the direction opposite the sun.

One consequence of the partial polarization of skylight is that the colors of distant objects may change when viewed through a rotated polarizing filter. If the sun is high in the sky, horizontal airlight will have a fairly high degree of polarization. According to the previous section, airlight is bluish. But if it also is partially polarized, its radiance can be diminished with a polarizing filter. Transmitted cloudlight, however, is unpolarized. Because the radiance of airlight can be reduced more than that of cloudlight, distant clouds may change from white to yellow to orange when viewed through a rotated polarizing filter.

4 SCATTERING BY PARTICLES

Up to this point we have considered only an atmosphere free of particles, an idealized state rarely achieved in nature. Particles still would inhabit the atmosphere even if the human race were to vanish from Earth. They are not simply byproducts of the "dark satanic mills" of civilization.

All molecules of the same substance are essentially identical. This is not true of particles: They vary in shape, size, and may be composed of one or more homogeneous regions.

Salient Differences Between Particles and Molecules

Magnitude of Scattering The distinction between scattering by molecules when widely separated and when packed together into a droplet is that between scattering by incoherent and coherent arrays. Isolated molecules are excited primarily by incident (external) light whereas the same molecules forming a droplet are excited by incident light and by each other's scattered fields. The total power scattered by an incoherent array of molecules is the sum of their scattered powers. The total power scattered by a coherent array is the square of the total scattered field, which in turn is the sum of all the fields scattered by the individual molecules. For an incoherent array we *may* ignore the wave nature of light, whereas for a coherent array we *must* take it into account.

Water vapor is a good example to ponder because it is a constituent of air and can condense to form cloud droplets. The difference between a sky containing water vapor and the same sky with the same amount of water but in the form of a cloud of droplets is dramatic.

According to Rayleigh's law, scattering by a particle small compared with the wavelength increases as the sixth power of its size (volume squared). A droplet of diameter 0.03 μm, for example, scatters about 10^{12} times more light than does one of its constituent molecules. Such a droplet contains about 10^7 molecules. Thus scattering per molecule as a consequence of condensation of water vapor into a coherent water droplet increases by about 10^5. Cloud droplets are much larger than 0.03 μm, a typical diameter being about 10 μm. Scattering per molecule in such a droplet is much greater than scattering by an isolated molecule, but not to the extent given by Rayleigh's law. Scattering increases as the sixth power of droplet diameter only when the molecules scatter coherently in phase. If a droplet is sufficiently small compared with the wavelength, each of its molecules is excited by essentially the same field and all the waves scattered by them interfere constructively. But when a droplet is comparable to or larger than the wavelength, interference can be constructive, destructive, and everything in between; hence scattering does not increase as rapidly with droplet size as predicted by Rayleigh's law.

The figure of merit for comparing scatterers of different size is their scattering cross section per unit volume, which, except for a multiplicative factor, is the scattering cross section per molecule. A scattering cross section may be looked upon as an effective area for removing radiant energy from a beam: The scattering cross section times the beam irradiance is the radiant power scattered in all directions.

The scattering cross section per unit volume for water droplets illuminated by visible light and varying in size from molecules (10^{-4} μm) to raindrops (10^3 μm) is shown in Figure 7. Scattering by a molecule that belongs to a cloud droplet is about 10^9 times greater than scattering by an isolated molecule, a striking example of the virtue of cooperation. Yet in molecular as in human societies there are limits beyond which cooperation becomes dysfunctional: Scattering by a molecule that belongs to a raindrop is about 100 times less than scattering by a molecule that belongs to a cloud droplet. This tremendous variation of scattering by water molecules depending on their state of aggregation has profound observational consequences. A cloud is optically so much different from the water vapor out of which it was born that the offspring bears no resemblance to its parents. We can see through tens of kilometers of air laden with water vapor, whereas a cloud a few tens of meters thick is enough to occult the sun. Yet a rain shaft born out of a cloud is considerably more translucent than its parent.

Wavelength Dependence of Scattering Regardless of their size and composition, particles scatter approximately as the inverse fourth power of wavelength if they are small compared with the wavelength and absorption is negligible, two important caveats. Failure to recognize them has led to errors, such as that yellow

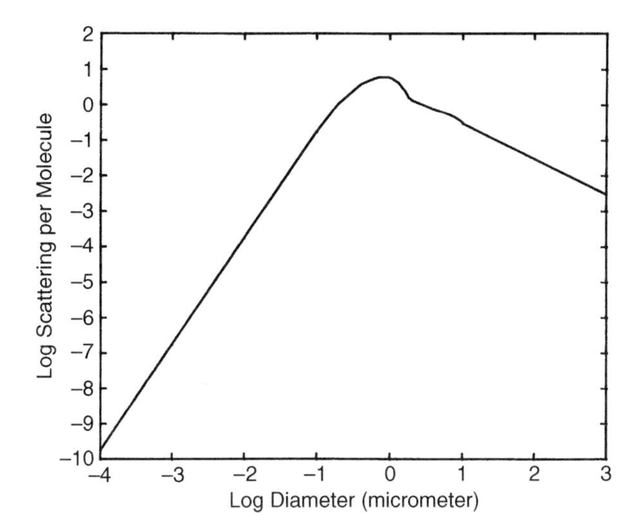

Figure 7 Scattering (per molecule) of visible light (arbitrary units) by water droplets varying in size from a single molecule to a raindrop.

light penetrates fog better because it is not scattered as much as light of shorter wavelengths. Although there may be perfectly sound reasons for choosing yellow instead of blue or green as the color of fog lights, greater transmission through fog is not one of them: Scattering by fog droplets is essentially independent of wavelength over the visible spectrum.

Small particles are selective scatterers, large particles are not. Particles neither small nor large give the reverse of what we have come to expect as normal. Figure 8 shows scattering of visible light by oil droplets with diameters 0.1, 0.8, and 10 μm. The smaller droplets scatter according to Rayleigh's law; the larger droplets (typical cloud droplet size) are nonselective. Between these two extremes are droplets (0.8 μm) that scatter long wavelength light more than short wavelength. Sunlight or moonlight seen through a thin cloud of these intermediate droplets would be bluish or greenish. This requires droplets of just the right size; hence it is a rare event, so rare that it occurs once in a blue moon. Astronomers, for unfathomable reasons, refer to the second full moon in a month as a blue moon, but if such a moon were blue it only would be by coincidence. The last reliably reported outbreak of blue and green suns and moons occurred in 1950 and was attributed to an oily smoke produced by Canadian forest fires.

Angular Dependence of Scattering The angular distribution of scattered light changes dramatically with the size of the scatterer. Molecules and particles that are small compared with the wavelength are nearly isotropic scatterers of unpolarized light, the ratio of maximum (at 0° and 180°) to minimum (at 90°) scattered radiance being only 2 for spheres, and slightly less for other spheroids. Although small particles scatter the same in the forward and backward hemispheres, scattering

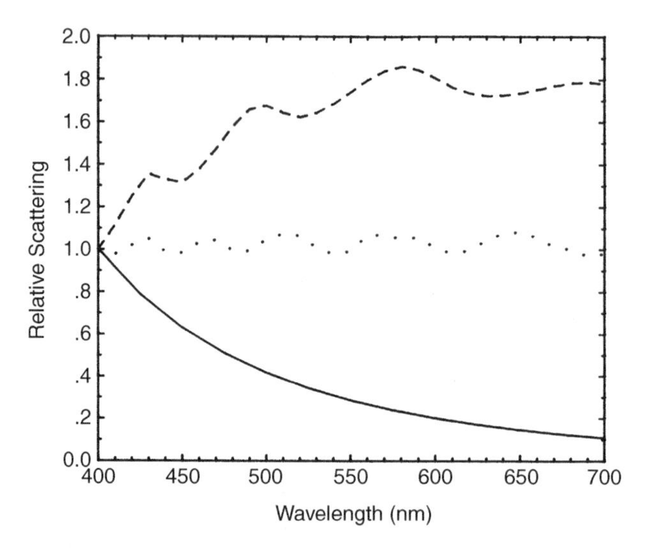

Figure 8 Scattering of visible light by oil droplets of diameter 0.1 μm (solid), 0.8 μm (dashes), and 10 μm (dots).

becomes markedly asymmetric for particles comparable to or larger than the wavelength. For example, forward scattering by a water droplet as small as 0.5 μm is about 100 times greater than backward scattering, and the ratio of forward to backward scattering increases more or less monotonically with size (Fig. 9).

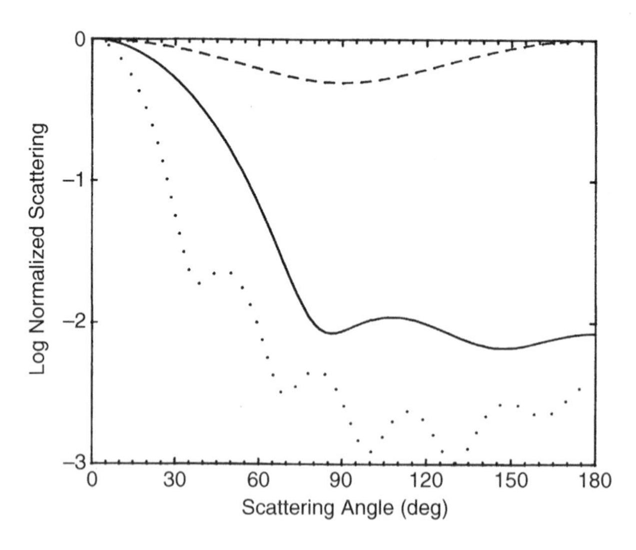

Figure 9 Angular dependence of scattering of visible light (0.55 μm) by water droplets small compared with the wavelength (dashes), diameter 0.5 μm (solid), and diameter 1.0 μm (dots).

The reason for this asymmetry is found in the singularity of the forward direction. In this direction, waves scattered by two or more scatterers excited solely by incident light (ignoring mutual excitation) are always in phase regardless of the wavelength and the separation of the scatterers. If we imagine a particle to be made up of N small subunits, scattering in the forward direction increases as N^2, the only direction for which this is always true. For other directions, the wavelets scattered by the subunits will not necessarily all be in phase. As a consequence, scattering in the forward direction increases with size (i.e., N) more rapidly than in any other direction.

Many common observable phenomena depend on this forward–backward asymmetry. Viewed toward the illuminating sun, glistening fog droplets on a spider's web warn us of its presence. But when we view the web with our backs to the sun, the web mysteriously disappears. A pattern of dew illuminated by the rising sun on a cold morning seems etched on a window pane. But if we go outside to look at the window, the pattern vanishes. Thin clouds sometimes hover over warm, moist heaps of dung, but may go unnoticed unless they lie between us and the source of illumination. These are but a few examples of the consequences of strongly asymmetric scattering by single particles comparable to or larger than the wavelength.

Degree of Polarization of Scattered Light All the simple rules about polarization upon scattering are broken when we turn from molecules and small particles to particles comparable to the wavelength. For example, the degree of polarization of light scattered by small particles is a simple function of scattering angle. But simplicity gives way to complexity as particles grow (Fig. 10), the scattered light being partially polarized parallel to the scattering plane for some scattering angles, perpendicular for others.

The degree of polarization of light scattered by molecules or by small particles is essentially independent of wavelength. But this is not true for particles comparable to or larger than the wavelength. Scattering by such particles exhibits *dispersion of polarization*: The degree of polarization at, say, 90° may vary considerably over the visible spectrum (Fig. 11).

In general, particles can act as polarizers or retarders or both. A polarizer transforms unpolarized light into partially polarized light. A retarder transforms polarized light of one form into that of another (e.g., linear into elliptical). Molecules and small particles, however, are restricted to roles as polarizers. If the atmosphere were inhabited solely by such scatterers, skylight could never be other than partially linearly polarized. Yet particles comparable to or larger than the wavelength often are present, hence skylight can acquire a degree of ellipticity upon multiple scattering: Incident unpolarized light is partially linearly polarized in the first scattering event, then transformed into partially elliptically polarized light in subsequent events.

Bees can navigate by polarized skylight. This statement, intended to evoke great awe for the photopolimetric powers of bees, is rarely accompanied by an important caveat: The sky must be clear. Figures 10 and 11 show two reasons—there are others—that bees, remarkable though they may be, cannot do the impossible. The simple wavelength-independent relation between the position of the sun and

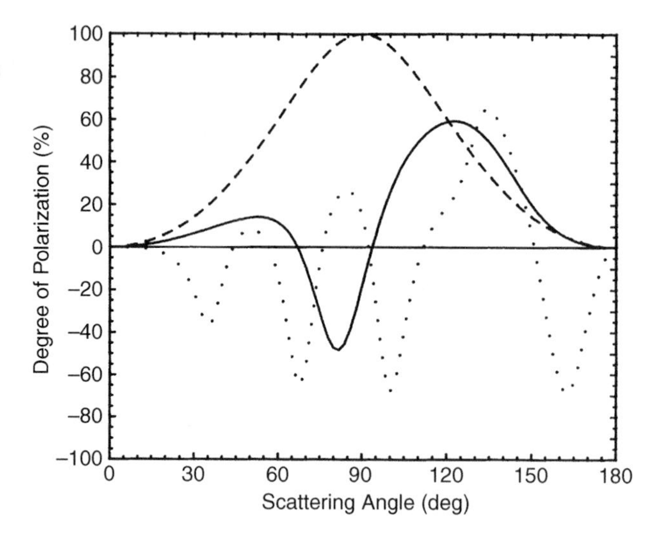

Figure 10 Degree of polarization of light scattered by water droplets illuminated by unpolarized visible light (0.55 μm). The dashed curve is for a droplet small compared with the wavelength; the solid curve is for a droplet of diameter 0.5 μm; the dotted curve is for a droplet of diameter 1.0 μm. Negative degrees of polarization indicate that the scattered light is partially polarized parallel to the scattering plane.

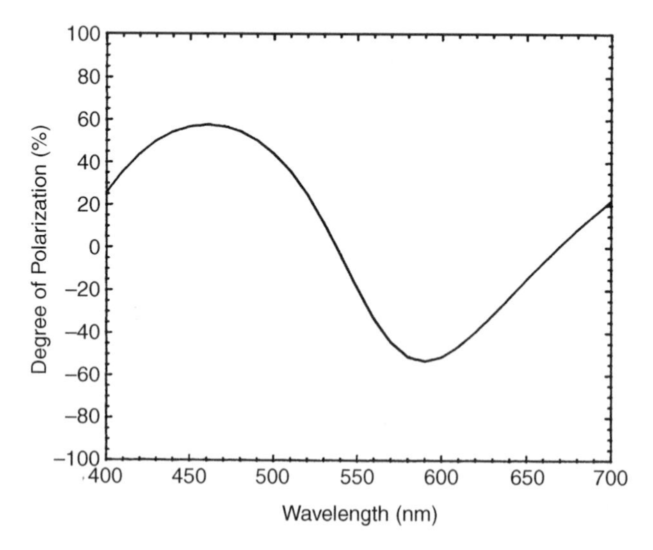

Figure 11 Degree of polarization at a scattering angle of 90° of light scattered by a water droplet of diameter 0.5 μm illuminated by unpolarized light.

the direction in which skylight is most highly polarized, an underlying necessity for navigating by means of polarized skylight, is obliterated when clouds cover the sky. This was recognized by the decoder of bee dances himself (von Frisch, 1971): "Sometimes a cloud would pass across the area of sky visible through the tube; when this happened the dances became disoriented, and the bees were unable to indicate the direction to the feeding place. Whatever phenomenon in the blue sky served to orient the dances, this experiment showed that it was seriously disturbed if the blue sky was covered by a cloud." But von Frisch's words often have been forgotten by disciples eager to spread the story about bee magic to those just as eager to believe what is charming even though untrue.

Vertical Distributions The scattering properties of particles are not only quite different, in general, from those of molecules, the different vertical distributions of particles and molecules by itself affects what is observed. The number density of molecules decreases more or less exponentially with height z above the surface: $\exp(-z/H_m)$, where the molecular scale height H_m is around 8 km. Although the decrease in number density of particles with height is also approximately exponential, the scale height for particles H_p is about 1 to 2 km. As a consequence, particles contribute disproportionately to optical thicknesses along near-horizon paths. Subject to the approximations underlying Eq. (9), the ratio of the tangential (horizon) optical thickness for particles τ_{tp} to that for molecules τ_{tm} is

$$\frac{\tau_{tp}}{\tau_{tm}} = \frac{\tau_{np}}{\tau_{nm}} \sqrt{\frac{H_m}{H_p}} \tag{13}$$

where the subscript t indicates a tangential path and n indicates a normal (radial) path. Because of the incoherence of scattering by atmospheric molecules and particles, scattering coefficients are additive, hence so are optical thicknesses. For equal normal optical thicknesses, the tangential optical thickness for particles is at least twice that for molecules. Molecules by themselves cannot give red sunrises and sunsets; molecules need the help of particles. For a fixed τ_{np}, the tangential optical thickness for particles is greater the more they are concentrated near the ground.

At the horizon the relative rate of change of transmission T of sunlight with zenith angle is

$$\frac{1}{T}\frac{dT}{d\Theta} = \tau_n \frac{R_e}{H} \tag{14}$$

where the scale height and normal optical thickness may be those for molecules or particles. Particles, being more concentrated near the surface, not only give disproportionate attenuation of sunlight on the horizon, they magnify the angular gradient of attenuation there. A perceptible change in color across the sun's disk (which subtends about 0.5°) on the horizon also requires the help of particles.

5 ATMOSPHERIC VISIBILITY

On a clear day can we really see forever? If not, how far can we see? To answer this question requires qualifying it by restricting viewing to more or less horizontal paths during daylight. Stars at staggering distances can be seen at night, partly because there is no skylight to reduce contrast, partly because stars overhead are seen in directions for which attenuation by the atmosphere is least.

The radiance in the direction of a black object is not zero because of light scattered along the line of sight (see discussion on variation of sky color and brightness under Section 2). At sufficiently large distances, this airlight is indistinguishable from the horizon sky. An example is a phalanx of parallel dark ridges, each ridge less distinct than those in front of it (Fig. 12). The farthest ridges blend into the horizon sky. Beyond some distance we cannot see ridges because of insufficient contrast.

Equation (5) gives the airlight radiance, a radiometric quantity that describes radiant power without taking into account the portion of it that stimulates the human eye or by what relative amount it does so at each wavelength. Luminance

Figure 12 Because of scattering by molecules and particles along the line of sight, each successive ridge is brighter than the ones in front of it even though all of them are covered with the same dark vegetation.

(also sometimes called brightness) is the corresponding photometric quantity. Luminance and radiance are related by an integral over the visible spectrum:

$$B = \int K(\lambda)L(\lambda)\, d\lambda \tag{15}$$

where the luminous efficiency of the human eye K peaks at about 550 nm and vanishes outside the range 385 to 760 nm.

The *contrast* C between any object and the horizon sky is

$$C = \frac{B - B_\infty}{B_\infty} \tag{16}$$

where B_∞ is the luminance for an infinite horizon optical thickness. For a uniformly illuminated line of sight of length d, uniform in its scattering properties, and with a black backdrop, the contrast is

$$C = -\frac{\int KGL_0 \exp(-\beta d)\, d\lambda}{\int KGL_0\, d\lambda} \tag{17}$$

The ratio of integrals in this equation defines an average optical thickness:

$$C = -\exp(-\bar{\tau}) \tag{18}$$

This expression for contrast reduction with (optical) distance is mathematically, but not physically, identical to Eq. (6), which perhaps has engendered the misconception that atmospheric visibility is reduced because of attenuation. Yet, since there is no light from a black object to be attenuated, its finite visual range cannot be a consequence of attenuation.

The distance beyond which a dark object cannot be distinguished from the horizon sky is determined by the *contrast threshold*: the smallest contrast detectable by the human observer. Although this depends on the particular observer, the angular size of the object observed, the presence of nearby objects, and the absolute luminance, a contrast threshold of 0.02 is often taken as an average. This value in Eq. (18) gives

$$-\ln|C| = 3.9 = \bar{\tau} = \overline{\beta d} \tag{19}$$

To convert an optical distance into a physical distance requires the scattering coefficient. Because K is peaked at around 550 nm, we can obtain an approximate value of d from the scattering coefficient at this wavelength in Eq. (19). At sea level, the molecular scattering coefficient in the middle of the visible spectrum corresponds to about 330 km for "forever": the greatest distance a black object can be seen against the horizon sky assuming a contrast threshold of 0.02 and ignoring the curvature of Earth.

We also observe contrast between elements of the same scene, a hillside mottled with stands of trees and forest clearings, for example. The extent to which we can resolve details in such a scene depends on sun angle as well as distance.

The airlight radiance for a nonreflecting object is Eq. (5) with $G = p(\Theta)\Omega_s$, where $p(\Theta)$ is the probability (per unit solid angle) that light is scattered in a direction making an angle Θ with the incident sunlight and Ω_s is the solid angle subtended by the sun. When the sun is overhead, $\Theta = 90°$; with the sun at the observer's back, $\Theta = 180°$; for an observer looking directly into the sun $\Theta = 0°$.

The radiance of an object with a finite reflectance R is given by Eq. (11). Equations (5) and (11) can be combined to obtain the contrast between reflecting and nonreflecting objects:

$$C = \frac{Fe^{-\tau}}{1 + (F - 1)e^{-\tau}}, \qquad F = \frac{R \cos \Phi}{np(\Theta)} \tag{20}$$

All else being equal, therefore, contrast decreases as $p(\Theta)$ increases. As shown in Figure 9, $p(\Theta)$ is more sharply peaked in the forward direction the larger the scatterer. Thus we expect the details of a distant scene to be less distinct when looking toward the sun than away from it if the optical thickness of the line of sight has an appreciable component contributed by particles comparable to or larger than the wavelength.

On humid, hazy days, visibility is often depressingly poor. Haze, however, is not water vapor but rather water that has ceased to be vapor. At high relative humidities, but still well below 100%, small soluble particles in the atmosphere accrete liquid water to become solution droplets (haze). Although these droplets are much smaller than cloud droplets, they markedly diminish visual range because of the sharp increase in scattering with particle size (Fig. 7). The same number of water molecules when aggregated in haze scatter vastly more than when apart.

6 ATMOSPHERIC REFRACTION

Atmospheric refraction is a consequence of molecular scattering, which is rarely stated given the historical accident that before light and matter were well understood refraction and scattering were locked in separate compartments and subsequently have been sequestered more rigidly than monks and nuns in neighboring cloisters. The connection between (lateral) scattering and refraction (forward scattering) can be divined from the expressions for the refractive index n of a gas and the scattering cross section σ_s of a gas molecule:

$$n = 1 + \frac{1}{2}\alpha N \tag{21}$$

$$\sigma_s = \frac{k^4}{6\pi}|\alpha|^2 \tag{22}$$

where N is the number density (not mass density) of gas molecules, $k = 2\pi/\lambda$ is the wavenumber of the incident light, and α is the polarizability of a molecule (induced dipole moment per unit inducing electric field). The appearance of the polarizability in Eq. (21) but its square in Eq. (22) is the clue that refraction is associated with electric fields whereas lateral scattering is associated with electric fields squared (powers). Scattering, without qualification, usually means scattering in all directions. Refraction, in a nutshell, is scattering in the forward direction. In this special direction incident and scattered fields superpose coherently to form the transmitted field, which is shifted in phase from that of the incident field by an amount determined by the polarizability and number density of scatterers.

Terrestrial Mirages

Mirages are not illusions, no more so than are reflections in a pond. Reflections of plants growing at its edge are not interpreted as plants growing into the water. If the water is ruffled by wind, the reflected images may be so distorted that they are no longer recognizable as those of plants. Yet we still would not call such distorted images illusions. And so is it with mirages. They are images noticeably different from what they would be in the absence of atmospheric refraction, creations of the atmosphere, not of the mind.

Mirages are vastly more common than is realized. Look and you shall see them. Contrary to popular opinion, they are not unique to deserts. Mirages can be seen frequently over ice-covered landscapes and highways flanked by deep snowbanks. Temperature per se is not what gives mirages but rather temperature gradients.

Because air is a mixture of gases, the polarizability for air in Eq. (21) is an average over all its molecular constituents, although their individual polarizabilities are about the same (at visible wavelengths). The vertical refractive index gradient can be written so as to show its dependence on pressure p and (absolute) temperature T:

$$\frac{d}{dz}\ln(n-1) = \frac{1}{p}\frac{dp}{dz} - \frac{1}{T}\frac{dT}{dz} \tag{23}$$

Pressure decreases approximately exponentially with height, where the scale height is around 8 km. Thus the first term on the right side of Eq. (23) is around 0.1/km. Temperature usually decreases with height in the atmosphere. An average lapse rate of temperature (i.e., its decrease with height) is around 6°C/km. The average temperature in the troposphere (within about 15 km of the surface) is around

280 K. Thus the magnitude of the second term in Eq. (23) is around 0.02/km. On average, therefore, the refractive index gradient is dominated by the vertical pressure gradient. But within a few meters of the surface, conditions are far from average. On a sun-baked highway your feet may be touching asphalt at 50°C while your nose is breathing air at 35°C, which corresponds to a lapse rate a thousand times the average. Moreover, near the surface, temperature can increase with height. In shallow surface layers, in which the pressure is nearly constant, the temperature gradient determines the refractive index gradient. It is in such shallow layers that mirages, which are caused by refractive index gradients, are seen.

Cartoonists by their fertile imaginations unfettered by science, and textbook writers by their carelessness have engendered the notion that atmospheric refraction can work wonders, lifting images of ships, for example, from the sea high into the sky. A back-of-the-envelope calculation dispels such notions. The refractive index of air at sea level is about 1.0003 (Fig. 13). Light from empty space incident at glancing incidence onto a uniform slab with this refractive index is displaced in angular position from where it would have been in the absence of refraction by

$$\delta = \sqrt{2(n-1)} \tag{24}$$

This yields an angular displacement of about 1.4°, which as we shall see is a rough upper limit.

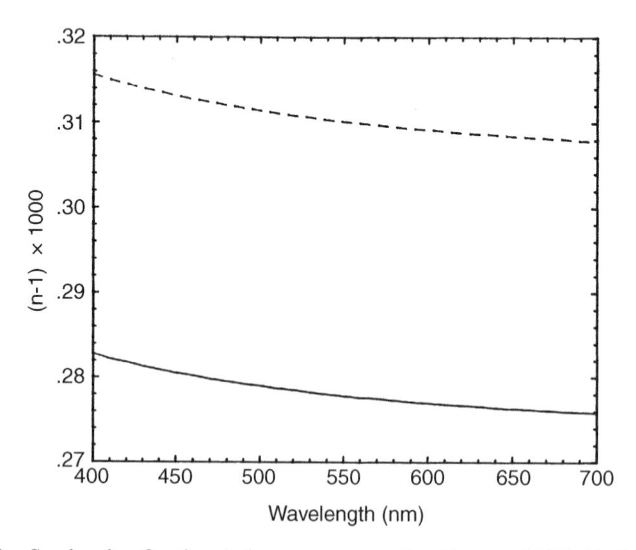

Figure 13 Sea-level refractive index versus wavelength at −15°C (dashes), and 15°C (solid). Data from Penndorf (1957).

Trajectories of light rays in nonuniform media can be expressed in different ways. According to Fermat's principle of least time (which ought to be extreme time), the actual path taken by a ray between two points is such that the path integral

$$\int_{1}^{1} n \, ds \tag{25}$$

is an extremum over all possible paths. This principle has inspired piffle about the alleged efficiency of nature, which directs light over routes that minimize travel time, presumably freeing it to tend to important business at its destination.

The scale of mirages is such that in analyzing them we may pretend that Earth is flat. On such an Earth, with an atmosphere in which the refractive index varies only in the vertical, Fermat's principle yields a generalization of Snell's law

$$n \sin \theta = \text{constant} \tag{26}$$

where θ is the angle between the ray and the vertical direction. We could, of course, have bypassed Fermat's principle to obtain this result.

Under the assumption that θ is small compared to 1, Eq. (26) yields the following differential equation satisfied by a ray:

$$\frac{d^2 z}{dy^2} = \frac{dn}{dz} \tag{27}$$

where y and z are its horizontal and vertical coordinates, respectively. For a constant refractive index gradient, which to good approximation occurs for a constant temperature gradient, Eq. (27) yields parabolas for ray trajectories. One such parabola for a constant temperature gradient about 100 times the average is shown in Figure 14. Note the vastly different horizontal and vertical scales. The image is displaced downward from what it would be in the absence of atmospheric refraction, hence the designation *inferior* mirage. This is the familiar highway mirage, seen over highways warmer than the air above them. The downward angular displacement is

$$\delta = \frac{1}{2} s \frac{dn}{dz} \tag{28}$$

where s is the horizontal distance between object and observer (image). Even for a temperature gradient 1000 times the tropospheric average, displacements of mirages are less than a degree at distances of a few kilometers.

If temperature increases with height, as it does, for example, in air over a cold sea, the resulting mirage is called a *superior* mirage. Inferior and superior are not designations of lower and higher caste but rather of displacements downward and upward.

For a constant temperature gradient, one and only one parabolic ray trajectory connects an object point to an image point. Multiple images therefore are not possible. But temperature gradients close to the ground are rarely linear. The

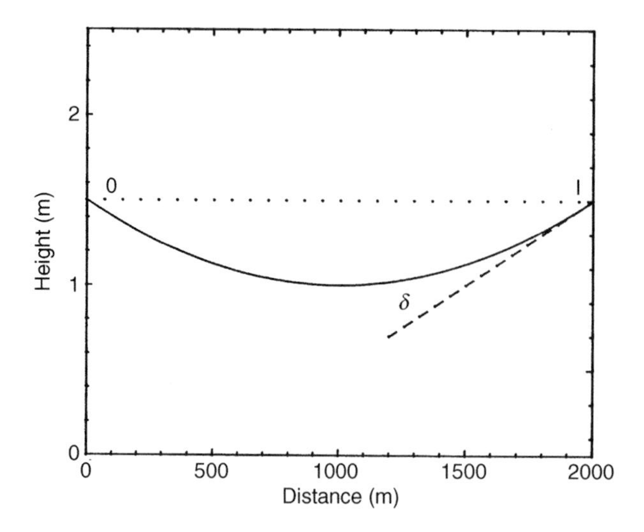

Figure 14 Parabolic ray paths in an atmosphere with a constant refractive index gradient (inferior mirage). Note the vastly different horizontal and vertical scales.

upward transport of energy from a hot surface occurs by molecular conduction through a stagnant boundary layer of air. Somewhat above the surface, however, energy is transported by air in motion. As a consequence, the temperature gradient steepens toward the ground if the energy flux is constant. This variable gradient can lead to two observable consequences: magnification and multiple images.

According to Eq. (28), all image points at a given horizontal distance are displaced downward by an amount proportional to the (constant) refractive index gradient. A corollary is that the closer an object point is to a surface where the temperature gradient is greatest, the greater the downward displacement of the corresponding image point. Thus nonlinear vertical temperature profiles may magnify images.

Multiple images are seen frequently on highways. What often appears to be water on the highway ahead but evaporates before it is reached is the inverted secondary image of either the horizon sky or of horizon objects lighter than dark asphalt.

Extraterrestrial Mirages

When we turn from mirages of terrestrial objects to those of extraterrestrial bodies, most notably the sun and moon, we can no longer pretend that Earth is flat. But we can pretend that the atmosphere is uniform and bounded. The total phase shift of a vertical ray from the surface to infinity is the same in an atmosphere with an exponentially decreasing molecular number density as in a hypothetical atmosphere with a uniform number density equal to the surface value up to height H.

A ray refracted along a horizon path by this hypothetical atmosphere and originating from outside it had to have been incident on it from an angle δ below the horizon:

$$\delta = \sqrt{\frac{2H}{R}} - \sqrt{\frac{2H}{R} - 2(n-1)} \tag{29}$$

where R is the radius of Earth. Thus, when the sun (or moon) is seen to be on the horizon, it is actually more than halfway below it, δ being about 0.36° whereas the angular width of the sun (and moon) is about 0.5°.

Extraterrestrial bodies seen near the horizon also are vertically compressed. The simplest way to estimate the amount of compression is from the rate of change of angle of refraction θ_r with angle of incidence θ_i for a uniform slab

$$\frac{d\theta_r}{d\theta_i} = \frac{\cos\theta_i}{\sqrt{n^2 - \sin^2\theta_i}} \tag{30}$$

where the angle of incidence is that for a curved but uniform atmosphere such that the refracted ray is horizontal. The result is

$$\frac{d\theta_r}{d\theta_i} = \sqrt{1 - \frac{R}{H}(n-1)} \tag{31}$$

according to which the sun near the horizon is distorted into an ellipse with aspect ratio about 0.87. We are unlikely to notice this distortion, however, because we expect the sun and moon to be circular, hence we see them that way.

The previous conclusions about the downward displacement and distortion of the sun were based on a refractive index profile determined mostly by the vertical pressure gradient. Near the ground, however, the temperature gradient is the prime determinant of the refractive index gradient, as a consequence of which the sun on the horizon can take on shapes more striking than a mere ellipse. For example, Figure 15 shows a nearly triangular sun with serrated edges. Assigning a cause to these serrations provides a lesson in the perils of jumping to conclusions. Obviously, the serrations are the result of sharp changes in the temperature gradient—or so one might think. Setting aside how such changes could be produced and maintained in a real atmosphere, a theorem of Fraser (1975) gives pause for thought. According to this theorem "in a horizontally (spherically) homogeneous atmosphere it is impossible for more than one image of an extraterrestrial object (sun) to be seen above the astronomical horizon." The serrations on the sun in Figure 15 are multiple images. But if the refractive index varies only vertically (i.e., along a radius), no matter how sharply, multiple images are not possible. Thus the serrations must owe their existence to horizontal variations of the refractive index, a consequence of gravity waves propagating along a temperature inversion.

Figure 15 A nearly triangular sun on the horizon. The serrations are a consequence of horizontal variations in refractive index.

The Green Flash

Compared to the rainbow, the green flash is not a rare phenomenon. Before you dismiss this assertion as the ravings of a lunatic, consider that rainbows require raindrops as well as sunlight to illuminate them, whereas rainclouds often completely obscure the sun. Moreover, the sun must be below about 42°. As a consequence of these conditions, rainbows are not seen often, but often enough that they are taken as the paragon of color variation. Yet tinges of green on the upper rim of the sun can be seen every day at sunrise and sunset given a sufficiently low horizon and a cloudless sky. Thus the conditions for seeing a green flash are more easily met than those for seeing a rainbow. Why then is the green flash considered to be so rare? The distinction here is between a rarely observed phenomenon (the green flash) and a rarely observable one (the rainbow).

The sun may be considered to be a collection of disks, one for each visible wavelength. When the sun is overhead, each disk coincides and we see the sun as white. But as it descends in the sky, atmospheric refraction displaces the disks by slightly different amounts, the red less than the violet (see Fig. 13). Most of each disk overlaps all the others except for the disks at the extremes of the visible spectrum. As a consequence, the upper rim of the sun is violet or blue, its lower rim red, whereas its interior, the region in which all disks overlap, is still white.

This is what would happen in the absence of lateral scattering of sunlight. But refraction and lateral scattering go hand in hand, even in an atmosphere free of particles. Selective scattering by atmospheric molecules and particles causes the color of the sun to change. In particular, the violet-bluish upper rim of the low sun can be transformed to green.

According to Eq. (29) and Figure 13, the angular width of the green upper rim of the low sun is about 0.01°, too narrow to be resolved with the naked eye or even to be seen against its bright backdrop. But, depending on the temperature profile, the atmosphere itself can magnify the upper rim and yield a second image of it, thereby enabling it to be seen without the aid of a telescope or binoculars. Green rims, which require artificial magnification, can be seen more frequently than green flashes, which require natural magnification. Yet both can be seen often by those who know what to look for and are willing to look.

7 SCATTERING BY SINGLE WATER DROPLETS

All the colored atmospheric displays that result when water droplets (or ice crystals) are illuminated by sunlight have the same underlying cause: Light is scattered in different amounts in different directions by particles larger than the wavelength, and the directions in which scattering is greatest depends on wavelength. Thus, when particles are illuminated by white light, the result can be angular separation of colors even if scattering integrated over all directions is independent of wavelength (as it essentially is for cloud droplets and ice crystals). This description, although correct, is too general to be completely satisfying. We need something more specific, more quantitative, which requires theories of scattering.

Because superficially different theories have been used to describe different optical phenomena, the notion has become widespread that they are caused by these theories. For example, coronas are said to be caused by diffraction and rainbows by refraction. Yet both the corona and the rainbow can be described quantitatively to high accuracy with a theory (the Mie theory for scattering by a sphere) in which diffraction and refraction do not explicitly appear. No fundamentally impenetrable barrier separates scattering from (specular) reflection, refraction, and diffraction. Because these terms came into general use and were entombed in textbooks before the nature of light and matter was well understood, we are stuck with them. But if we insist that diffraction, for example, is somehow different from scattering, we do so at the expense of shattering the unity of the seemingly disparate observable phenomena that result when light interacts with matter. What is observed depends on the composition and disposition of the matter, not on which approximate theory in a hierarchy is used for quantitative description.

Atmospheric optical phenomena are best classified by the direction in which they are seen and by the agents responsible for them. Accordingly, the following sections are arranged in order of scattering direction, from forward to backward.

When a single water droplet is illuminated by white light and the scattered light projected onto a screen, the result is a set of colored rings. But in the atmosphere we see a mosaic to which individual droplets contribute. The scattering pattern of a single droplet is the same as the mosaic provided that multiple scattering is negligible.

Coronas and Iridescent Clouds

A cloud of droplets narrowly distributed in size and thinly veiling the sun (or moon) can yield a spectacular series of colored concentric rings around it. This corona is most easily described quantitatively by the Fraunhofer diffraction theory, a simple approximation valid for particles large compared with the wavelength and for scattering angles near the forward direction. According to this approximation, the differential scattering cross section (cross section for scattering into a unit solid angle) of a spherical droplet of radius a illuminated by light of wavenumber k is

$$\frac{|S|^2}{k^2} \tag{32}$$

where the scattering amplitude is

$$S = x^2 \frac{1 + \cos\theta}{2} \frac{J_1(x \sin\theta)}{x \sin\theta} \tag{33}$$

where J_1 is the Bessel function of first order and the size parameter $x = ka$. The quantity $(1 + \cos\theta)/2$ is usually approximated by 1 since only near-forward scattering angles θ are of interest.

The differential scattering cross section, which determines the angular distribution of the scattered light, has maxima for $x \sin\theta = 5.137, 8.417, 11.62, \ldots$ Thus the dispersion in the position of the first maximum is

$$\frac{d\theta}{d\lambda} \approx \frac{0.817}{a} \tag{34}$$

and is greater for higher-order maxima. This dispersion determines the upper limit on drop size such that a corona can be observed. For the total angular dispersion over the visible spectrum to be greater than the angular width of the sun (0.5°), the droplets cannot be larger than about 60 μm in diameter. Drops in rain, even in drizzle, are appreciably larger than this, which is why coronas are not seen through rain shafts. Scattering by a droplet of diameter 10 μm (Fig. 16), a typical cloud droplet size, gives sufficient dispersion to yield colored coronas.

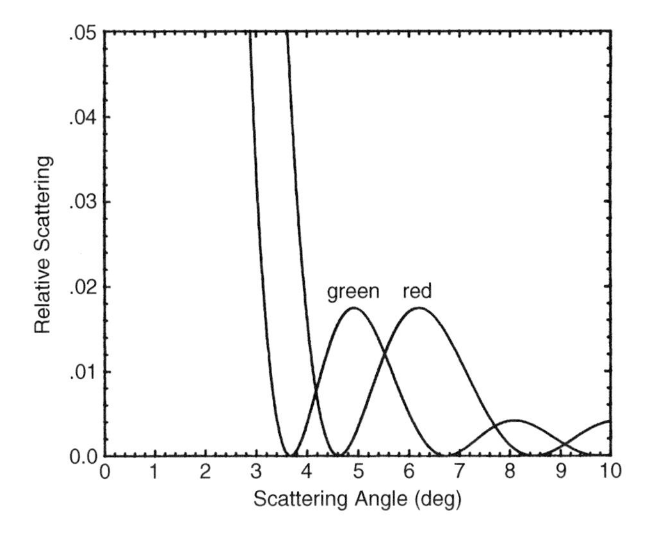

Figure 16 Scattering of light near the forward direction (according to Fraunhofer theory) by a sphere of diameter 10 μm illuminated by red and green light.

Suppose that the first angular maximum for blue light (0.47 μm) occurs for a droplet of radius a. For red light (0.66 μm) a maximum is obtained at the same angle for a droplet of radius $a + \Delta a$. That is, the two maxima, one for each wavelength, coincide. From this we conclude that coronas require narrow size distributions: If cloud droplets are distributed in radius with a relative variance $\Delta a/a$ greater than about 0.4, color separation is not possible.

Because of the stringent requirements for the occurrence of coronas, they are not observed often. Of greater occurrence are the corona's cousins, iridescent clouds, which display colors but usually not arranged in any obviously regular geometrical pattern. Iridescent patches in clouds can be seen even at the edges of thick clouds that occult the sun.

Coronas are not the unique signatures of spherical scatterers. Randomly oriented ice columns and plates give similar patterns according to Fraunhofer theory (Takano and Asano, 1983). As a practical matter, however, most coronas probably are caused by droplets. Many clouds at temperatures well below freezing contain subcooled water droplets. Only if a corona were seen in a cloud at a temperature lower than $-40\,^{\circ}\mathrm{C}$ could one assert with confidence that it must be an ice-crystal corona.

Rainbows

In contrast with coronas, which are seen looking toward the sun, rainbows are seen looking away from it and are caused by water drops much larger than those that give coronas. To treat the rainbow quantitatively we may pretend that light incident on a transparent sphere is composed of individual rays, each of which suffers a different fate determined only by the laws of specular reflection and refraction. Theoretical justification for this is provided by van de Hulst's (1957, p. 208) *localization principle*, according to which terms in the exact solution for scattering by a transparent sphere correspond to more or less localized rays.

Each incident ray splinters into an infinite number of scattered rays: externally reflected, transmitted without internal reflection, transmitted after one, two, and so on internal reflections. At any scattering angle θ, each splinter contributes to the scattered light. Accordingly, the differential scattering cross section is an infinite series with terms of the form

$$\frac{b(\theta)}{\sin\theta}\frac{db}{d\theta} \tag{35}$$

The *impact parameter* b is $a \sin \Theta_i$, where Θ_i is the angle between an incident ray and the normal to the sphere. Each term in the series corresponds to one of the splinters of an incident ray. A *rainbow angle* is a singularity (or *caustic*) of the differential scattering cross section at which the conditions

$$\frac{d\theta}{db} = 0 \qquad \frac{b}{\sin\theta} \neq 0 \tag{36}$$

are satisfied. Missing from Eq. (35) are various reflection and transmission coefficients (Fresnel coefficients), which display no singularities and hence do not determine rainbow angles.

A rainbow is not associated with rays externally reflected or transmitted without internal reflection. The succession of rainbow angles associated with one, two, three, ... internal reflections are called primary, secondary, tertiary,... rainbows. Aristotle recognized that "three or more rainbows are never seen, because even the second is dimmer than the first, and so the third reflection is altogether too feeble to reach the sun" (Aristotle's view was that light streams outward from the eye). Although he intuitively grasped that each successive ray is associated with ever-diminishing energy, his statement about the nonexistence of tertiary rainbows in nature is not quite true. Although reliable reports of such rainbows are rare (unreliable reports are as common as dirt), at least one observer who can be believed has seen one (Pledgley, 1986).

An incident ray undergoes a total angular deviation as a consequence of transmission into the drop, one or more internal reflections, and transmission out of the drop. Rainbow angles are angles of minimum deviation.

For a rainbow of any order to exist

$$\cos\Theta_i = \sqrt{\frac{n^2 - 1}{p(p+1)}} \tag{37}$$

must lie between 0 and 1, where Θ_i is the angle of incidence of a ray that gives a rainbow after p internal reflections and n is the refractive index of the drop. A

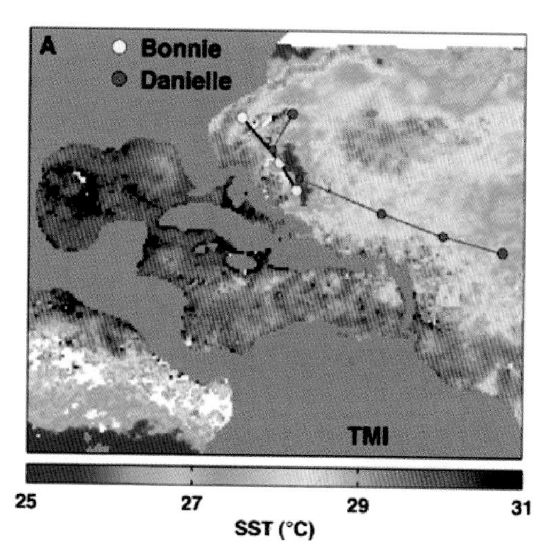

Figure 20 (Chapter 31) Cold wake produced by Hurricane Bonnie for August 24–26, 1998, as seen by the NASA TRMM satellite *Microwave Imager* (TMI). Small white patches are areas of persistent rain over the 3-day period. White dots show Hurricane Bonnie's daily position from August 24 to 26. Gray dots show the later passage of Hurricane Danielle from August 27 to September 1. Danielle crossed Bonnie's wake on August 29 and its intensity drops. [From F. J. Wentz, C. Gentemann, D. Smith, and D. Chelton, *Science* **288**, 847–850 (2000). Copyright owned by the American Geophysical Union. (*http:/www.sciencemag.org*)]

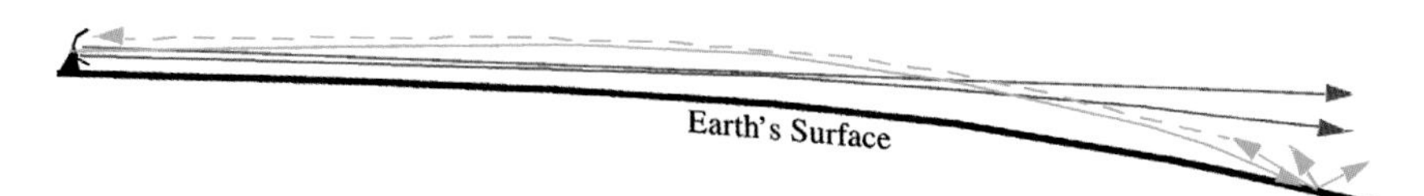

Figure 6 (Chapter 48) Beam paths assuming straight propagation (red), typical atmospheric density gradient bending beam partially back toward Earth (blue), strong density gradient, possibly temperature inversion, bending beam back into Earth (green) where scattering off surface sends energy back toward radar.

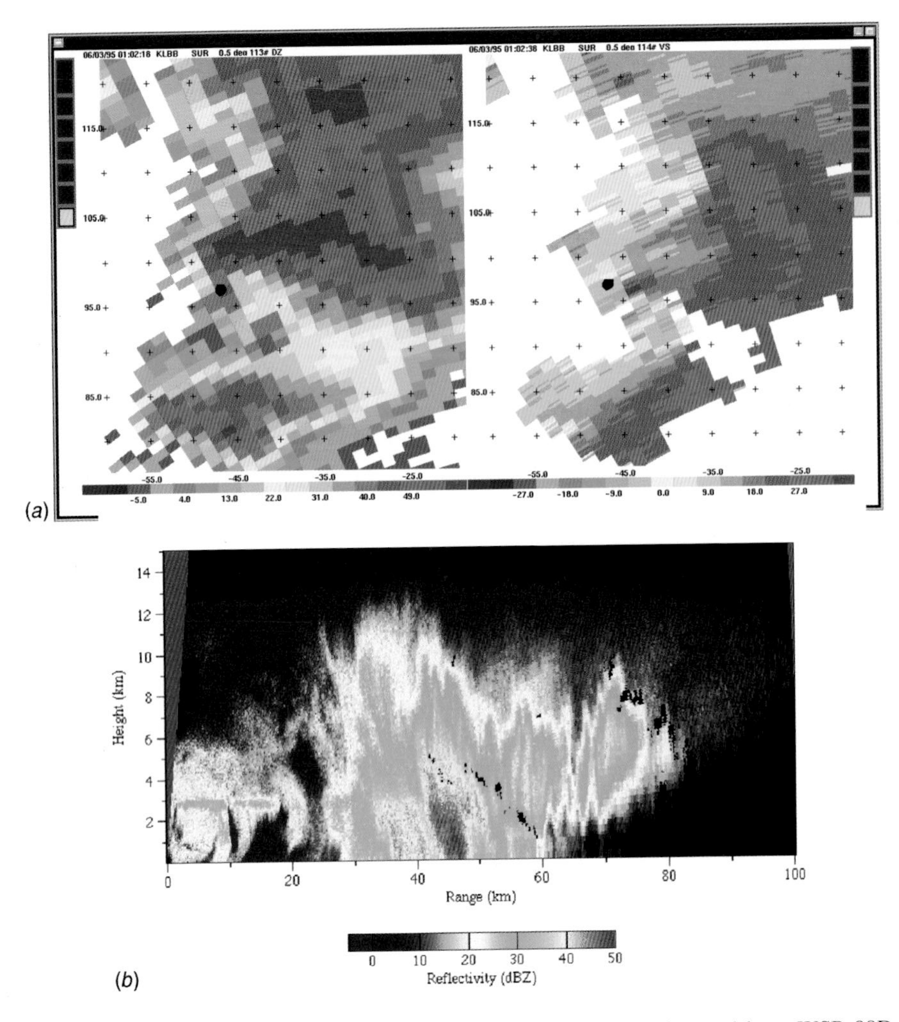

Figure 10 (Chapter 48) (*a*) A tornadic supercell thunderstorm observed by a WSR-88D operational radar. Reflectivity (*left*) and Doppler velocity (*right*) are shown. Classic hook echo extends from the western side of the supercell. An intense circulation, suggested by the strong away and toward velocities near the hook, is the mesocyclone associated with a tornado that was occurring. 5(*b, c*) Reflectivity and Doppler velocity in a vertical cross section (RHI) through a portion of a squall line. The high reflectivity core and lower reflectivity extending to 12km are visible. Strong toward and away Doppler velocities are associated with the up and down drafts of the cell, as indicated by arrows. Data from the MIT radar in Albuquerque, New Mexico. (Courtesy of MIT Wea. Rad. Lab. D. Boccippio.) (*d*) Reflectivity (*lower*) and Doppler velocity (*upper*) in a winter storm. Reflectivity is somewhat amorphous but is enhanced in a ring corresponding to the melting layer. In the melting layer, large, wet slow-moving particles cause high reflectivity. The velocity pattern provides a vertical sounding of the atmosphere. Winds are from the NNW at low levels (near the radar), but from the southwest aloft (away from the radar as the beams diverge from Earth's surface). Cold advection is implied. Data from the MIT radar in Cambridge, Massachusetts. (Courtesy of MIT Wea. Rad. Lab. D. Boccippio.)

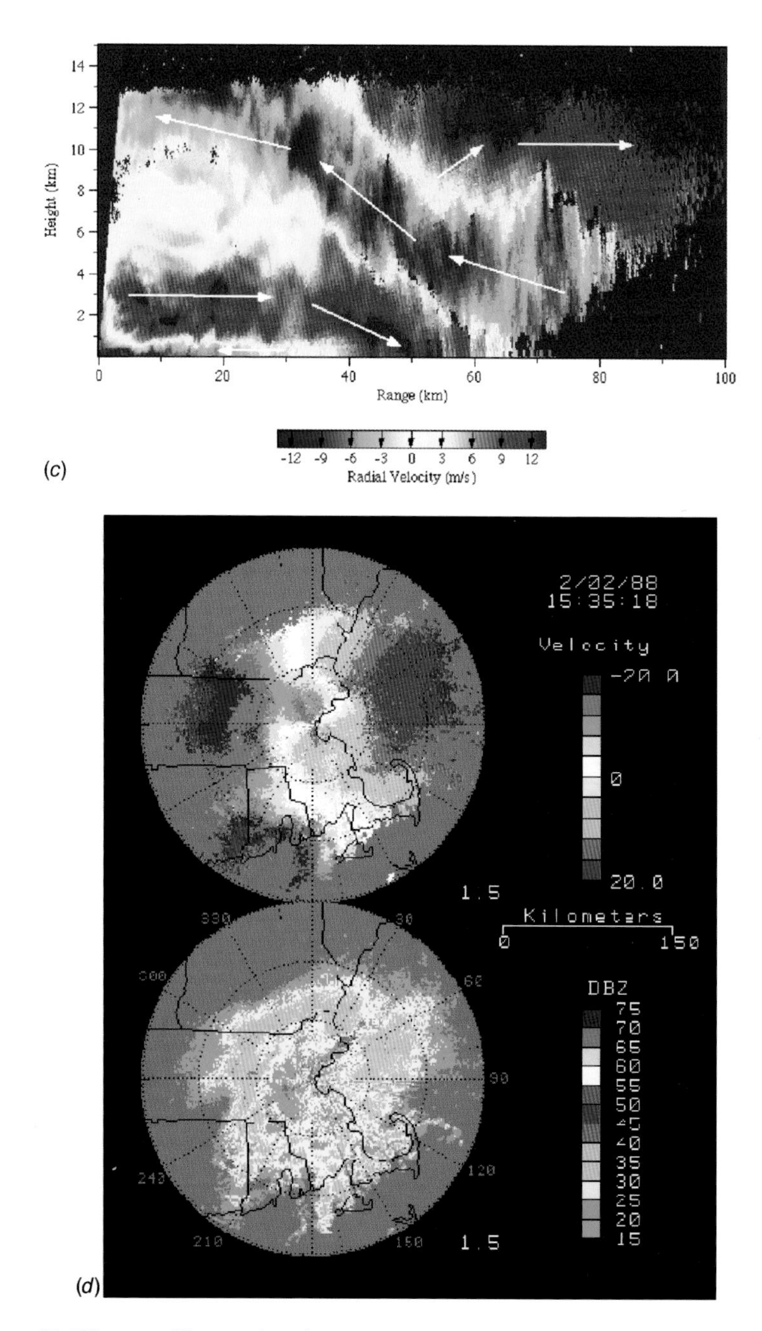

Figure 10 (Chapter 48) continued

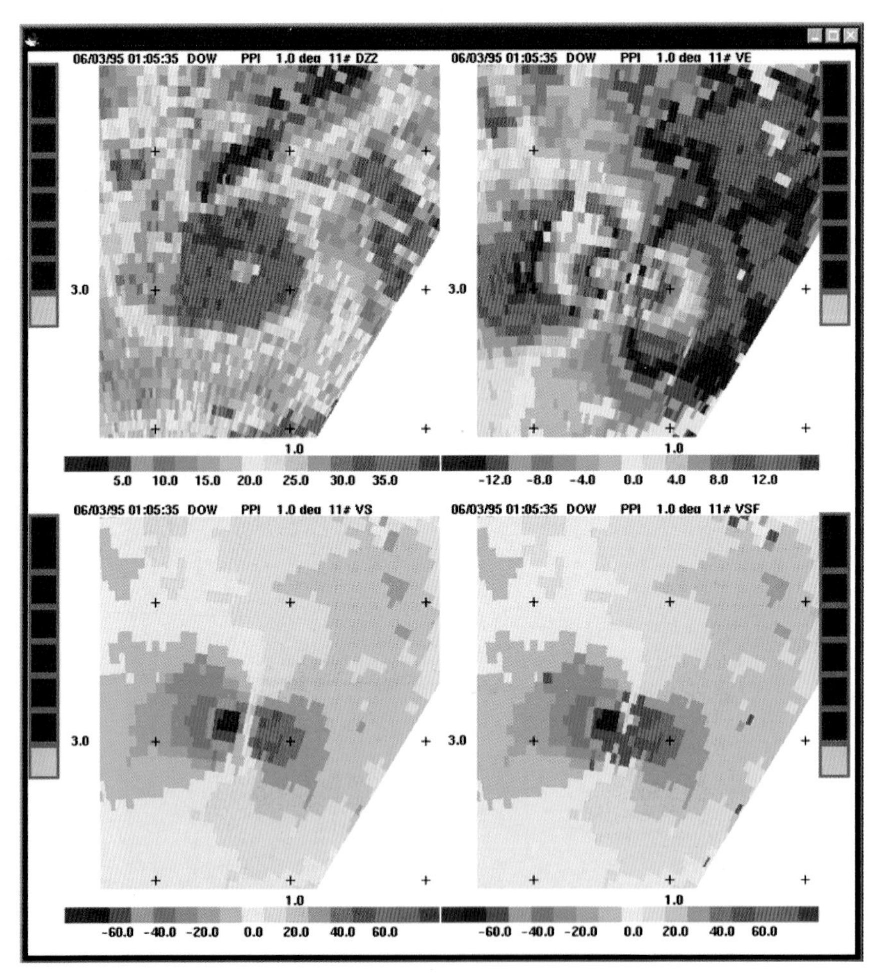

Figure 9 (Chapter 48) Illustration of dealiasing and cleaning of radar data. (*Top left*) Reflectivity in tornado showing ring debris. (*Top right*) Raw Doppler velocity with aliasing. (*Bottom left*) Velocity after dealiasing. Strong away and strong toward velocities adjacent to each other imply rotation, in this case over 70m/s. (*Bottom right*) Velocity after values with high spectral width or contaminated by echoes from the ground (ground clutter) have been removed. Data is from DOW mobile radar in the Dimmitt, Texas, Tornado on June 2, 1995, from a range of 3km.

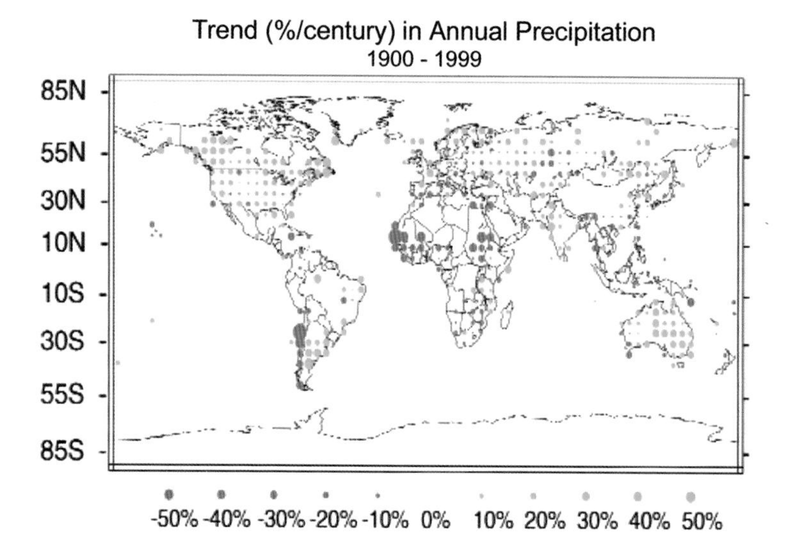

Figure 3 (Chapter 12) Estimates of globally and annually averaged radiative forcing (in W/m^{-2}) for a number of agents due to changes in concentrations of greenhouse gases and aerosols and natural changes in solar output from 1750 to the present day. Error bars are depicted for all forcings (from IPCC, 2001).

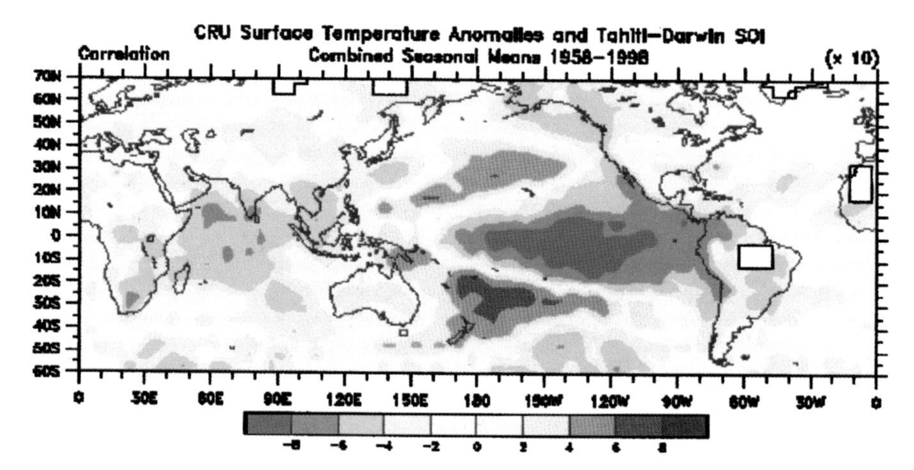

Figure 1 (Chapter 14) Correlation coefficients of the SOI with sea surface temperature seasonal anomalies for January 1958 to December 1998. It can be interpreted as the sea surface temperature patterns that accompany a La Niña event, or as an El Niño event with signs reversed. Values in the central tropical Pacific correspond to anomalies of about 1.5°C [From Trenberth and Caron, 2000].

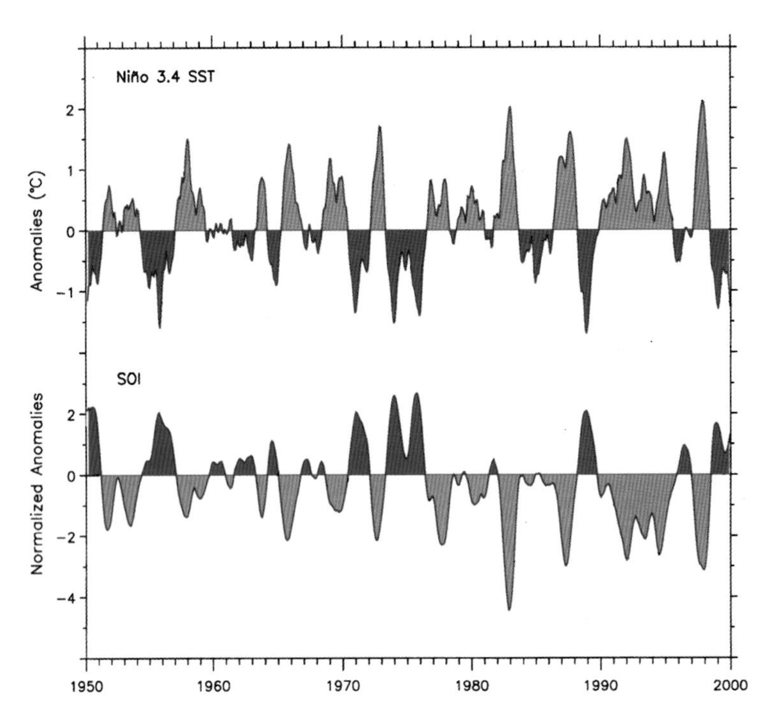

Figure 2 (Chapter 14) Time series of areas of sea surface temperature anomalies from 1950 through 2000 relative to the means of 1950–79 for the region most involved in ENSO 5° N–5° S, 170° E –120° W (top) and for the Southern Oscillation Index. El Niño events are in grey and La Niña events in black.

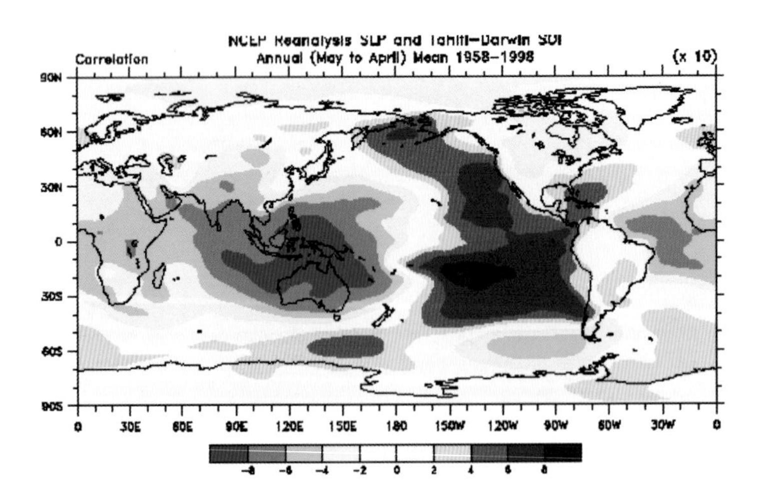

Figure 3 (Chapter 14) Map of correlation coefficients (×10) of the annual mean sea level pressures (based on the year May to April) with the SOI showing the Southern Oscillation pattern in the phase corresponding to La Niña events. During El Niño events the sign is reversed [From Trenberth and Caron, 2000].

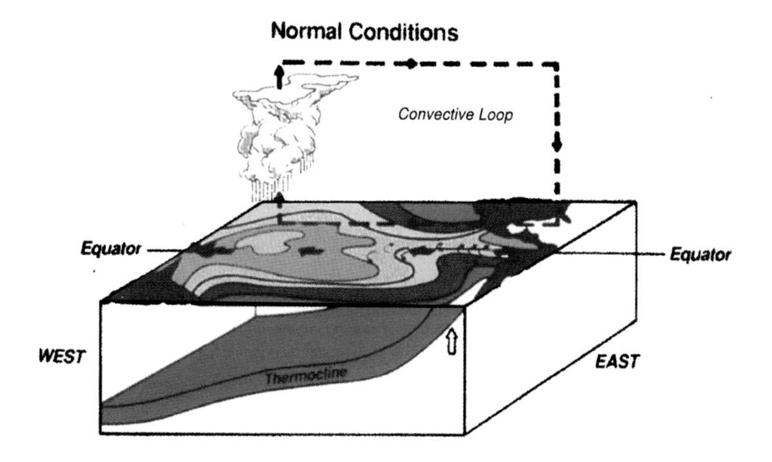

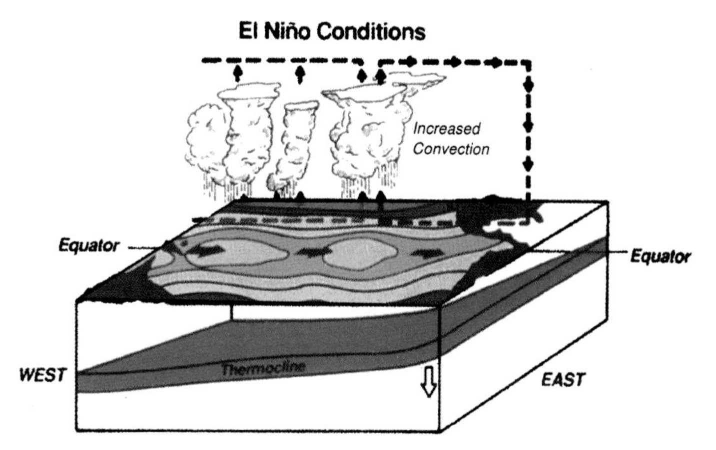

Figure 5 (Chapter 14) Schematic cross section of the Pacific Basin with Australia at lower left and the Americas at right depicting normal and El Niño conditions. Total sea surface temperatures exceeding 29°C are in gold and the colors change every 1°C. Regions of convection and overturning in the atmosphere are indicated. The thermocline in the ocean is shown in blue. Changes in ocean currents are shown by the black arrows. (Copyright University Corporation for Atmospheric Research. Reprinted by permission).

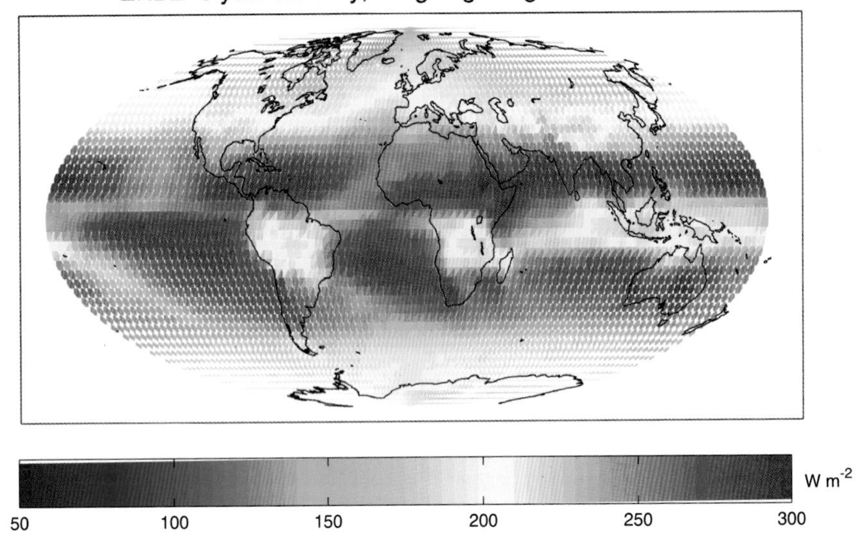

Figure 21 (Chapter 20) January mean OLR in $\mathrm{W/m^{-2}}$ determined from ERBE observations.

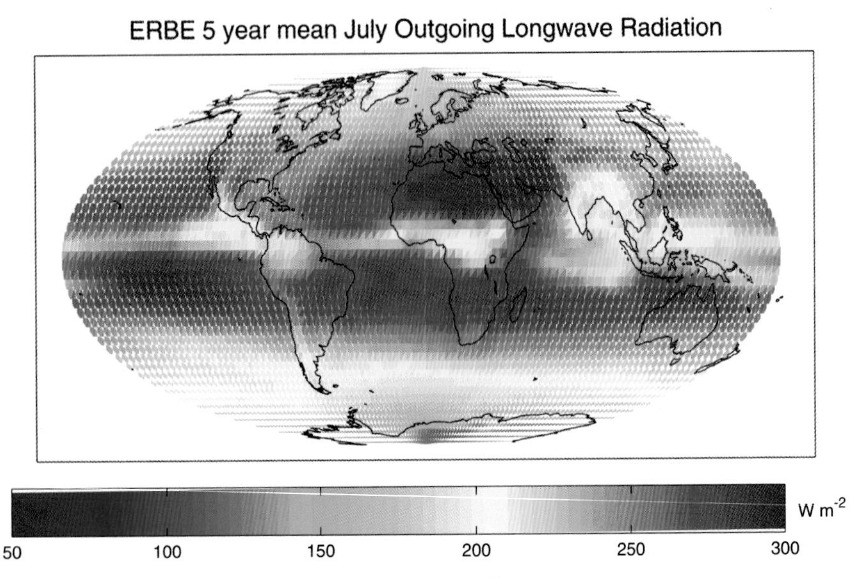

Figure 22 (Chapter 20) July mean OLR in $\mathrm{W/m^{-2}}$ determined from ERBE observations.

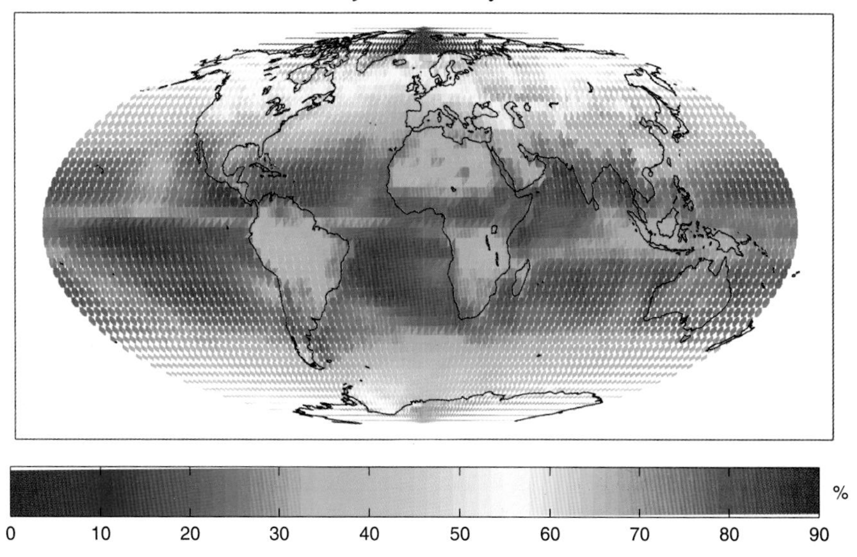

Figure 23 (Chapter 20) January mean albedo in percent determined from ERBE observations.

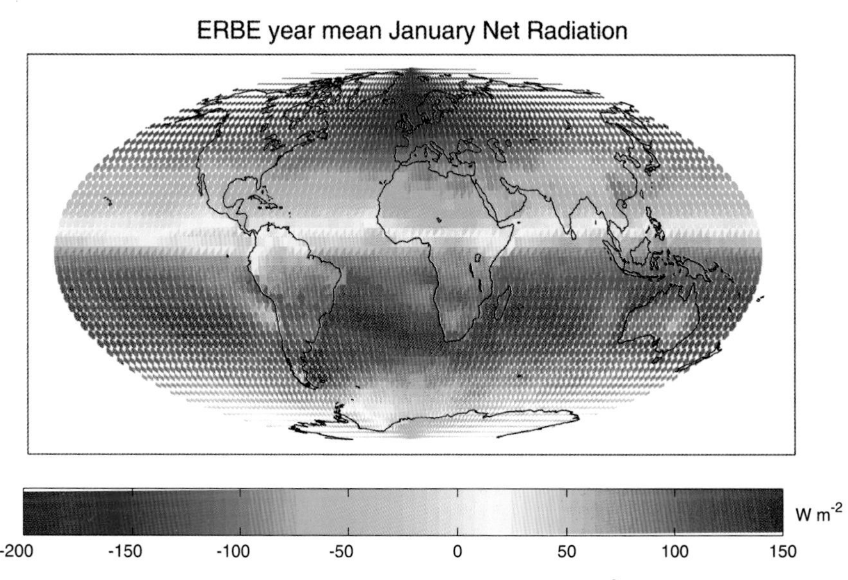

Figure 25 (Chapter 20) January mean net radiation in W/m^{-2} determined from ERBE observations.

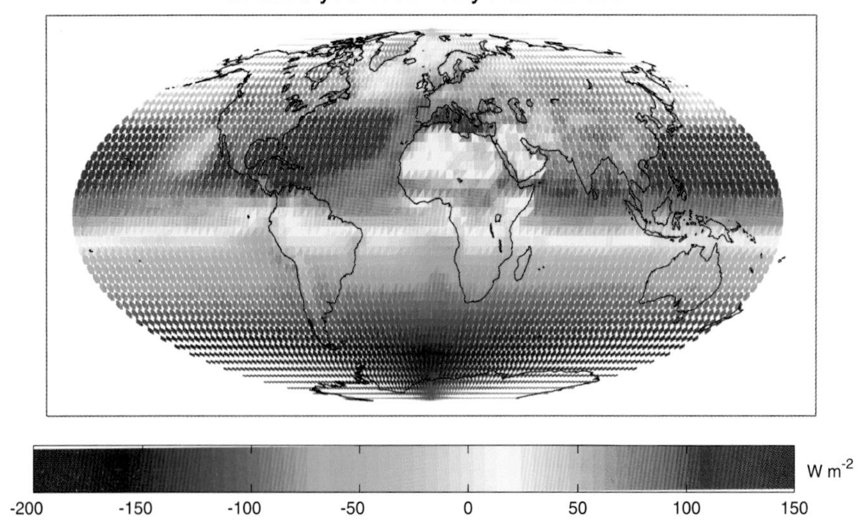

Figure 26 (Chapter 20) July mean net radiation in $\mathrm{W/m^{-2}}$ determined from ERBE observations.

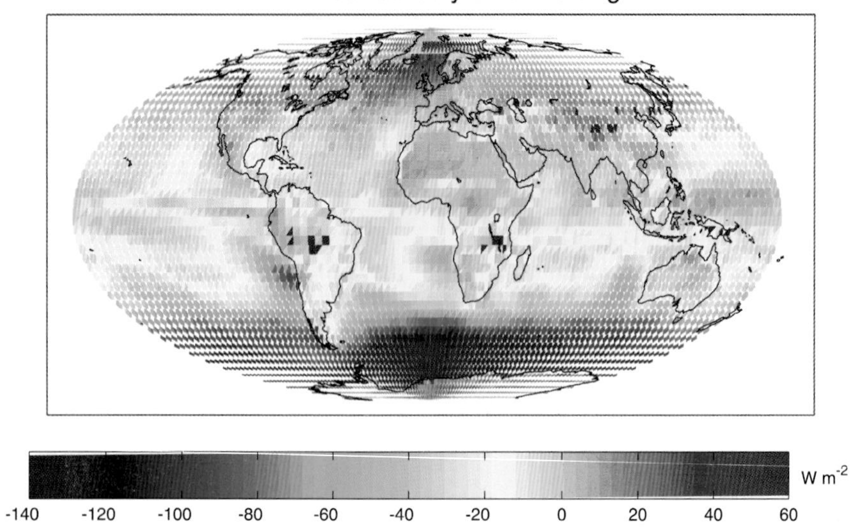

Figure 27 (Chapter 20) January cloud radiative forcing in $\mathrm{W/m^{-2}}$.

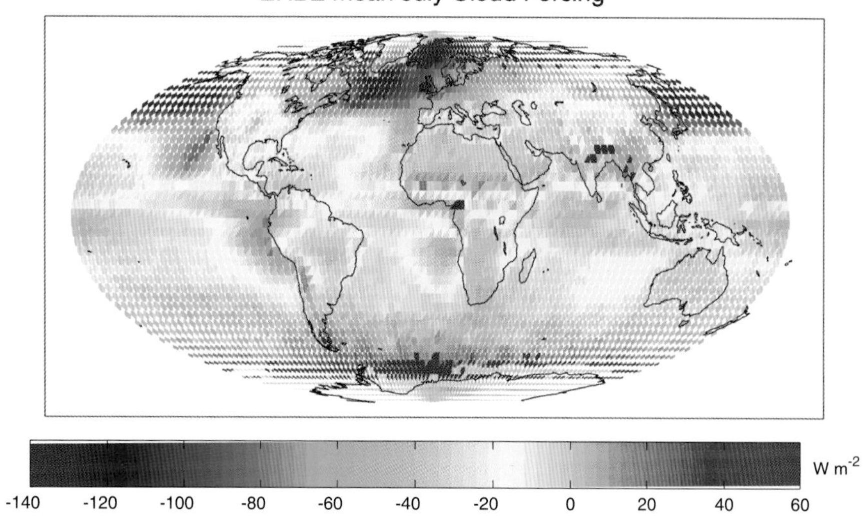

Figure 28 (Chapter 20) July cloud radiative forcing in W/m^{-2}.

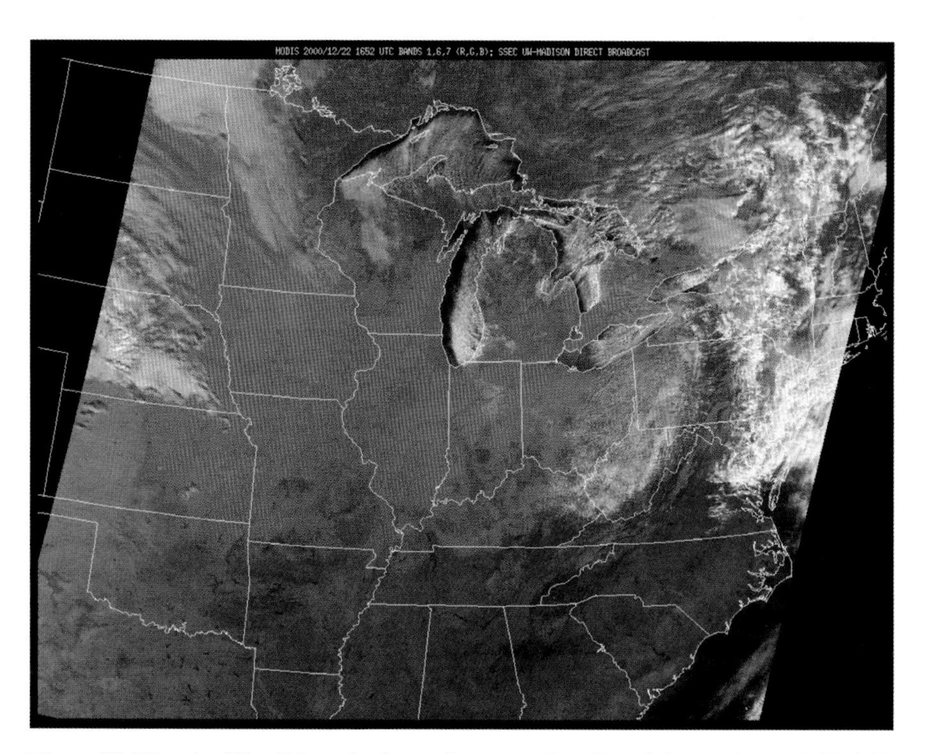

Figure 30 (Chapter 20) False color image from a combination of observations at 0.66, 1.64, and 2.14μm are used to detect snow. In this false color images, land surfaces are green, water surfaces are black, snow cover is red, and clouds are white.

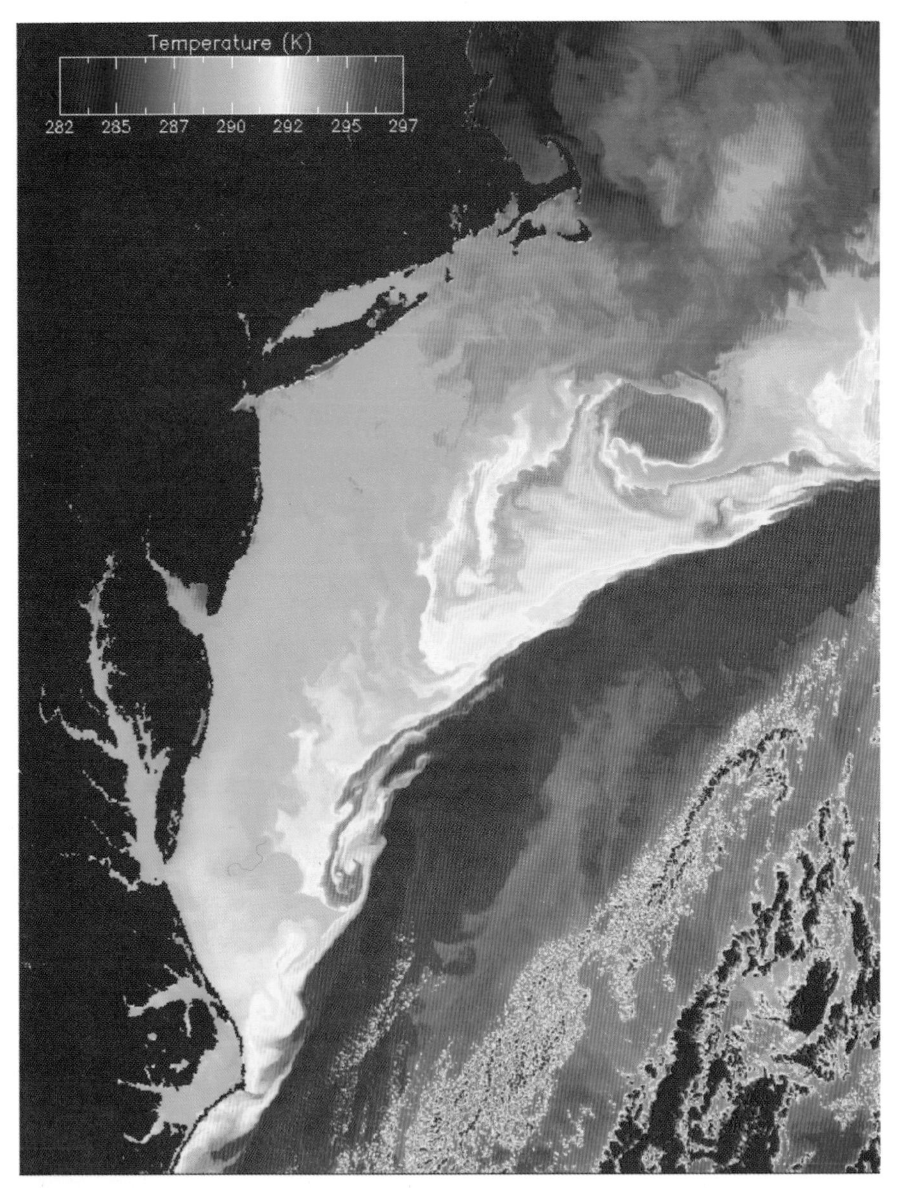

Figure 31 (Chapter 20) An 11-μm image from MODIS of Atlantic Ocean off east coast of North America.

Evolution of stability as seen in GOES LI DPI

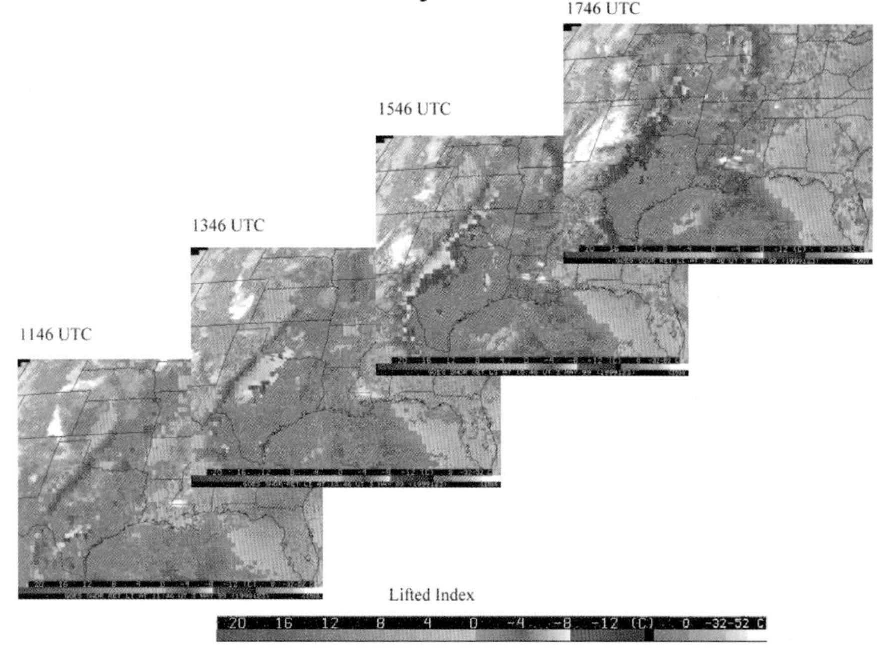

Figure 33 (Chapter 20) Sequence of GOES-derived lifted index on May 3, 1999. Red regions indicate the potential areas of thunderstorm development. Strong tornadoes developed in southwest Oklahoma before 22 UTC. The lifted index versus color code is given in the legend.

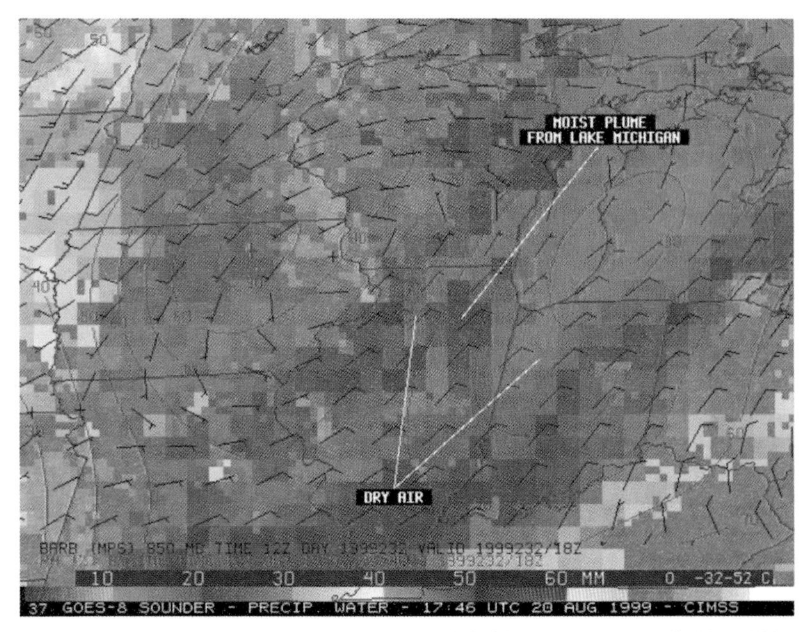

Figure 34 (Chapter 20) GOES-8-derived precipitable water on April 20, 1999. Blue regions indicate regions of low PW and green relative high amounts of PW. The satellite analysis clearly shows dry region south of Great Lakes.

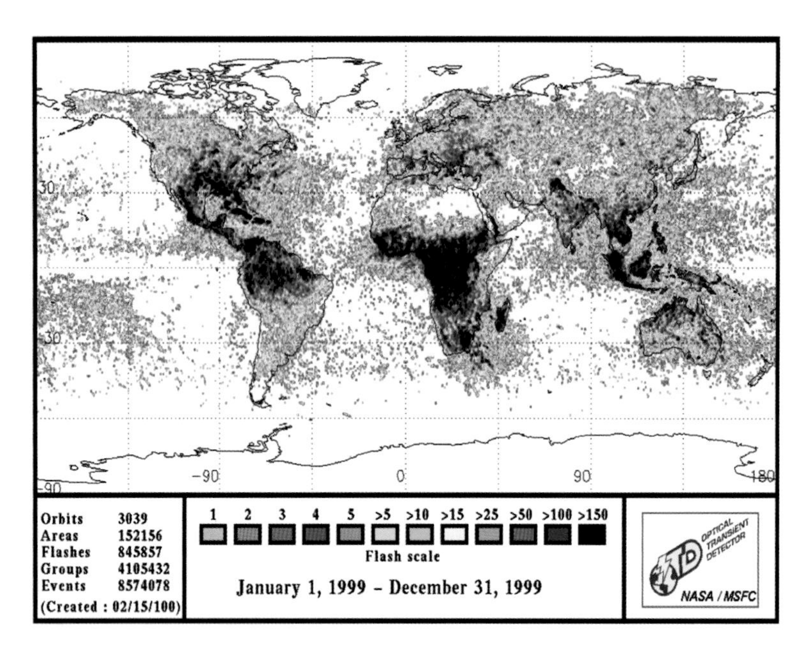

Figure 2 (Chapter 22) Map of global lightning detected over a multiyear period by NASA's Optical Transient Detector flying in polar orbit.

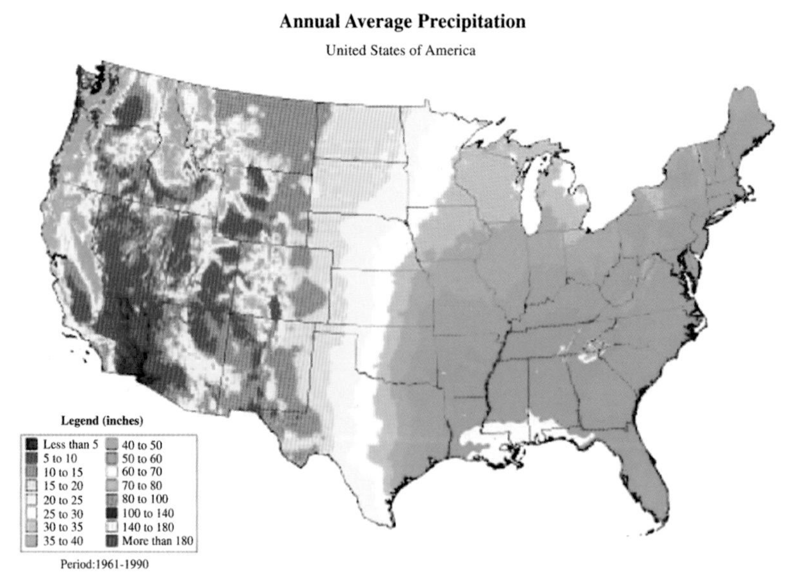

Figure 1 (Chapter 28) Average annual precipitation in inches, as determined from the Parameter-elevation Regression on Independent Slopes (PRISM) model by Chris Daly, based on 1961–1990 normals from National Oceanic and Atmospheric Administration (NOAA) cooperative stations and Natural Resources Conservation Service (NRCS) SNOwpack TELemetry (SNOTEL) sites. Modeling sponsored by USDA-NRCS Water and Climate Center, Portland, Oregon. Available from George Taylor, Oregon State Climatologist, Oregon Climate Service.

Figure 4 (Chapter 28) Cumulus clouds, indicative of shallow instability, above the Wasatch Mountains in northern Utah are capped by lenticular wave clouds, indicative of stable air flowing across the mountain barrier (photo by J. Horel).

Figure 5 (Chapter 28) Fog channelled through a gap in the coastal mountains of northern California (photo by J. Horel).

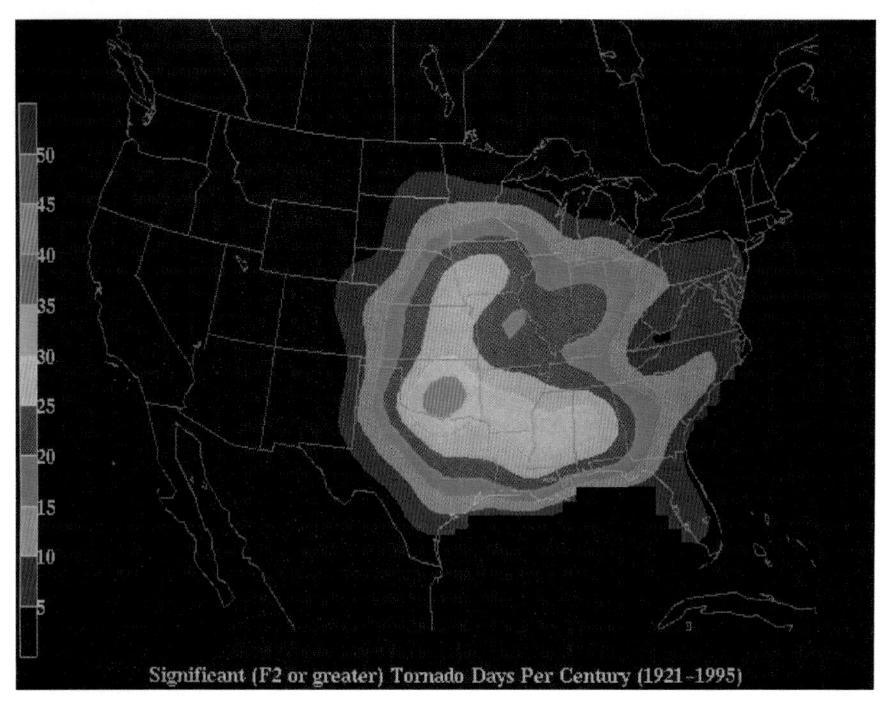

Figure 1 (Chapter 29) Number of days per century an F2 or greater strength tornado might occur within 25 miles of a point. For example, in southcentral Oklahoma one would expect an F2 tornado within 25 miles about once every 3 years. Calculations based on data from 1921 to 1995.

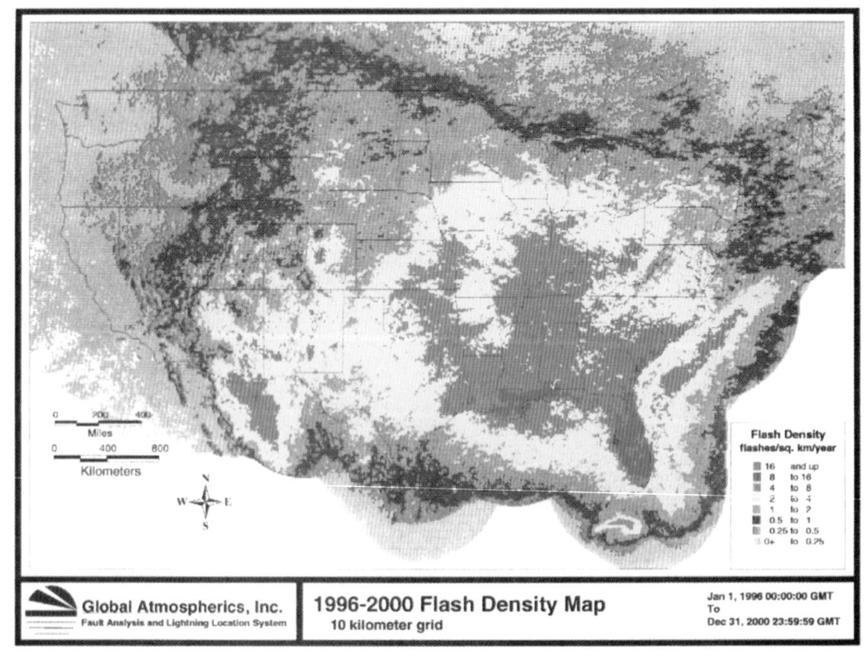

Figure 3 (Chapter 29) Number of cloud-to-ground lightning flashes per year for the United States based on data from 1996 to 2000. (Courtesy Global Atmospherics, Inc.)

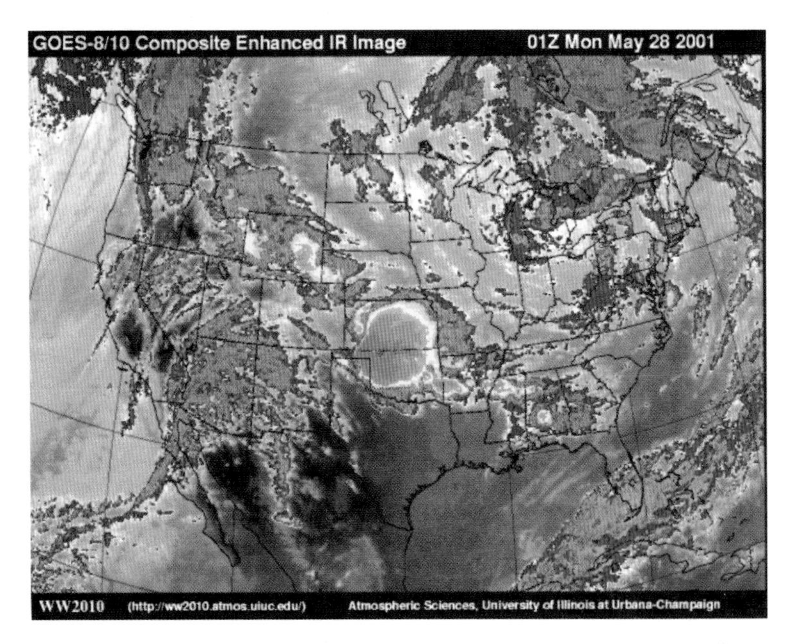

Figure 6 (Chapter 29) Infrared satellite picture of a developing MCS over south-central Kansas and northern Oklahoma at 0100 UTC May 28, 2001. A severe squall line was developing underneath the cloud shield and subsequently moved southeasterly for the next 12h to the Texas coast. [Image courtesy of University of Illinois WW2010 project (http://ww2010.atmos.uiuc.edu).]

Figure 16 (Chapter 29) High-resolution measurements of the reflectivity factor (an indicator of precipitation size and number concentration) by the Doppler on Wheels mobile radar in a tornadic storm with a hook echo near Rolla, Kansas, on May 31, 1996. The thin blue rings indicate range from the radar at intervals of 5km.

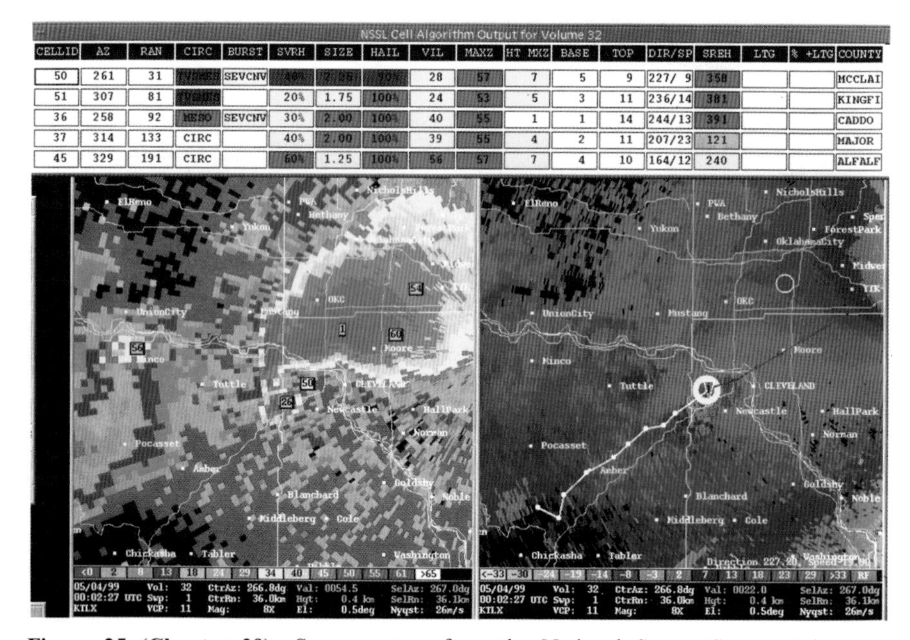

NSSL Cell Algorithm Output for Volume 32

CELLID	AZ	RAN	CIRC	BURST	SVRH	SIZE	HAIL	VIL	MAXZ	HT MXZ	BASE	TOP	DIR/SP	SREH	LTG	% +LTG	COUNTY
50	261	31	[illegible]	SEVCNV	40%	2.25	90%	28	57	7	5	9	227/ 9	358			MCCLAI
51	307	81	[illegible]		20%	1.75	100%	24	53	5	3	11	236/14	381			KINGFI
36	258	92	MESO	SEVCNV	30%	2.00	100%	40	55	1	1	14	244/13	391			CADDO
37	314	133	CIRC		40%	2.00	100%	39	55	4	2	11	207/23	121			MAJOR
45	329	191	CIRC		60%	1.25	100%	56	57	7	4	10	164/12	240			ALFALF

Figure 25 (Chapter 29) Screen output from the National Severe Storms Laboratory Warnings Decision Support System of the May 3, 1999, Oklahoma City tornado. The table at the top shows the system is tracking 5 different storms with a variety of algorithm-determined characteristics such as direction and speed of motion, presence of hail and its maximum size, whether a circulation or mesocyclone is present, and its echo top. Left panel is radar reflectivity with cities and county boundaries as background. Boxes with numbers correspond to locations of tracked storms. Right panel is radial velocity. White line is storm #50 track with future positions shown as the purple line. Yellow circle shows the presence of the tornado vortex signature. (Courtesy of Greg Stumpf of the National Severe Storms Laboratory.)

Figure 1 (Chapter 31) NOAA-14 AVHRR multispectral false color image of Hurricane Floyd at 2041 UTC, September 13 about 800km east of south Florida. (Photo courtesy of NOAA Operationally Significant Event Imagery website: *http://www.osei.noaa.gov/*.)

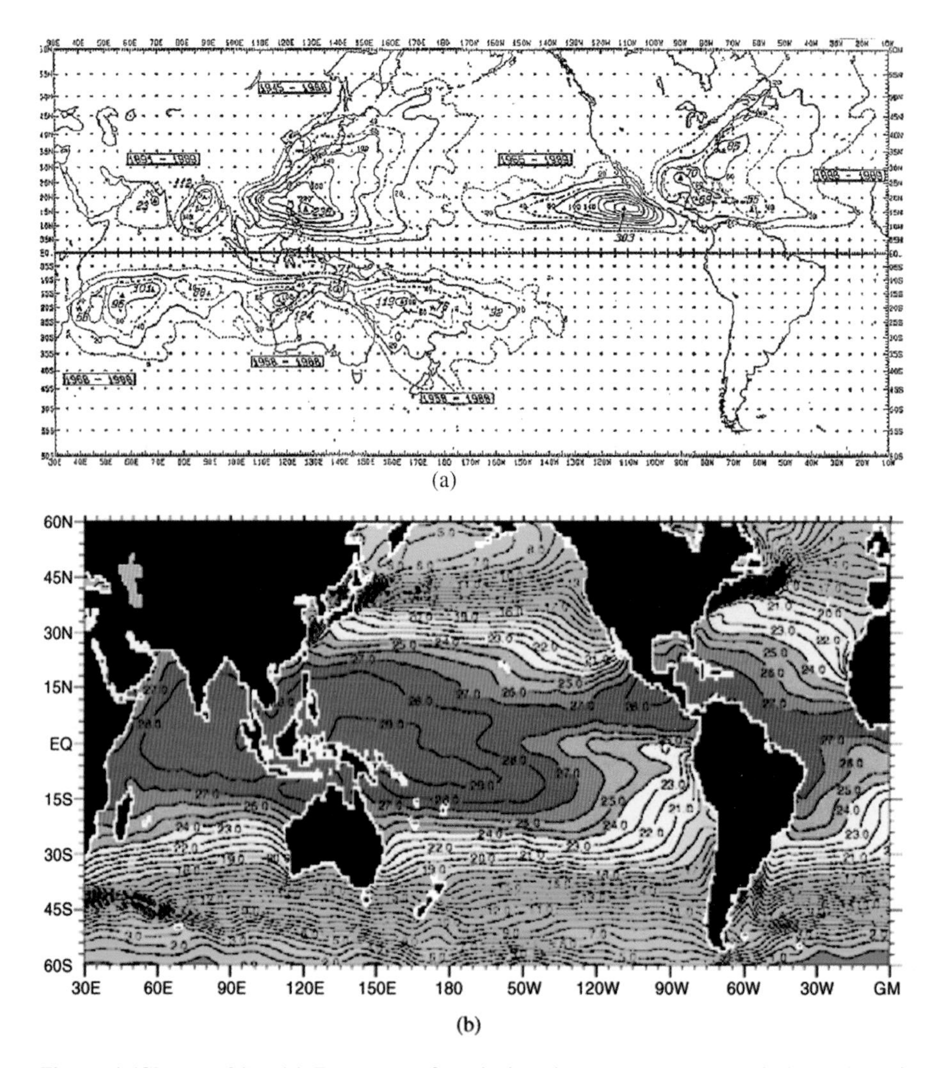

Figure 4 (Chapter 31) (*a*) Frequency of tropical cyclones per 100 years within 140km of any point. Solid triangles indicate maxima, with values shown. Period of record is shown in boxes for each basin. (*b*) Annual SST distribution (°C).

primary bow therefore requires drops with refractive index less than 2; a secondary bow requires drops with refractive index less than 3. If raindrops were composed of titanium dioxide ($n \approx 3$), a commonly used opacifier for paints, primary rainbows would be absent from the sky and we would have to be content with only secondary bows.

If we take the refractive index of water to be 1.33, the scattering angle for the primary rainbow is about 138°. This is measured from the forward direction (solar point). Measured from the antisolar point (the direction toward which one must look in order to see rainbows in nature), this scattering angle corresponds to 42°, the basis for a previous assertion that rainbows (strictly, primary rainbows) cannot be seen when the sun is above 42°. The secondary rainbow is seen at about 51° from the antisolar point. Between these two rainbows is *Alexander's dark band*, a region into which no light is scattered according to geometrical optics.

The colors of rainbows are a consequence of sufficient dispersion of the refractive index over the visible spectrum to give a spread of rainbow angles that appreciably exceeds the width of the sun. The width of the primary bow from violet to red is about 1.7°; that of the secondary bow is about 3.1°.

Because of its band of colors arcing across the sky, the rainbow has become the paragon of color, the standard against which all other colors are compared. Lee and Fraser (1990) (see also Lee, 1991), however, challenged this status of the rainbow, pointing out that even the most vivid rainbows are colorimetrically far from pure.

Rainbows are almost invariably discussed as if they occurred literally in a vacuum. But real rainbows, as opposed to the pencil-and-paper variety, are necessarily observed in an atmosphere the molecules and particles of which scatter sunlight that adds to the light from the rainbow but subtracts from its purity of color.

Although geometrical optics yields the positions, widths, and color separation of rainbows, it yields little else. For example, geometrical optics is blind to *supernumerary bows*, a series of narrow bands sometimes seen below the primary bow. These bows are a consequence of interference, hence fall outside the province of geometrical optics. Since supernumerary bows are an interference phenomenon, they, unlike primary and secondary bows (according to geometrical optics), depend on drop size. This poses the question of how supernumerary bows can be seen in rain showers, the drops in which are widely distributed in size. In a nice piece of detective work, Fraser (1983) answered this question.

Raindrops falling in a vacuum are spherical. Those falling in air are distorted by aerodynamic forces, not, despite the depictions of countless artists, into tear drops but rather into nearly oblate spheroids with their axes more or less vertical. Fraser argued that supernumerary bows are caused by drops with a diameter of about 0.5 mm, at which diameter the angular position of the first (and second) supernumerary bow has a minimum: Interference causes the position of the supernumerary bow to increase with decreasing size whereas drop distortion causes it to increase with increasing size. Supernumerary patterns contributed by drops on either side of the minimum cancel leaving only the contribution from drops at the minimum. This cancellation occurs only near the tops of rainbow, where supernumerary bows are seen. In the vertical parts of a rainbow, a horizontal slice through a distorted drop is more or less circular; hence these drops do not exhibit a minimum supernumerary angle.

According to geometrical optics, all spherical drops, regardless of size, yield the same rainbow. But it is not necessary for a drop to be spherical for it to yield rainbows independent of its size. This merely requires that the plane defined by the incident and scattered rays intersect the drop in a circle. Even distorted drops satisfy this condition in the vertical part of a bow. As a consequence, the absence of supernumerary bows there is compensated for by more vivid colors of the primary and secondary bows (Fraser, 1972). Smaller drops are more likely to be spherical, but the smaller a drop, the less light it scatters. Thus the dominant contribution to the luminance of rainbows is from the larger drops. At the top of a bow, the plane defined by the incident and scattered rays intersects the large, distorted drops in an ellipse, yielding a range of rainbow angles varying with the amount of distortion, hence a pastel rainbow. To the knowledgeable observer, rainbows are no more uniform in color and brightness than is the sky.

Although geometrical optics predicts that all rainbows are equal (neglecting background light), real rainbows do not slavishly follow the dictates of this approximate theory. Rainbows in nature range from nearly colorless fog bows (or cloud bows) to the vividly colorful vertical portions of rainbows likely to have inspired myths about pots of gold.

The Glory

Continuing our sweep of scattering directions, from forward to backward, we arrive at the end of our journey: *the glory.* Because it is most easily seen from airplanes it sometimes is called the *pilot's bow.* Another name is *anticorona*, which signals that it is a corona around the antisolar point. Although glories and coronas share some common characteristics, there are differences between them other than direction of observation. Unlike coronas, which may be caused by nonspherical ice crystals, glories require spherical cloud droplets. And a greater number of colored rings may be seen in glories than in coronas because the decrease in luminance away from the backward direction is not as steep as that away from the forward direction. To see a glory from an airplane, look for colored rings around its shadow cast on clouds below. This shadow is not an essential part of the glory; it merely directs you to the antisolar point.

Like the rainbow, the glory may be looked upon as a singularity in the differential scattering cross section of Eq. (35). Equation (36) gives one set of conditions for a singularity; the second set is

$$\sin\theta = 0 \qquad b(\theta) \neq 0 \tag{38}$$

That is, the differential scattering cross section is infinite for nonzero impact parameters (corresponding to incident rays that do not intersect the center of the sphere) that give forward (0°) or backward (180°) scattering. The forward direction is excluded because this is the direction of intense scattering accounted for by the Fraunhofer theory.

For one internal reflection, Eq. (38) leads to the condition

$$\sin \Theta_i = \frac{n}{2}\sqrt{4 - n^2} \tag{39}$$

which is satisfied only for refractive indices between 1.414 and 2, the lower refractive index corresponding to a grazing incidence ray. The refractive index of water lies outside this range. Although a condition similar to Eq. (39) is satisfied for rays undergoing four or more internal reflections, insufficient energy is associated with such rays. Thus it seems that we have reached an impasse: The theoretical condition for a glory cannot be met by water droplets. Not so, says van de Hulst (1947) in a seminal work. He argues that 1.414 is close enough to 1.33 given that geometrical optics is, after all, an approximation. Cloud droplets are large compared with the wavelength, but not so large that geometrical optics is an infallible guide to their optical behavior. Support for the van de Hulstian interpretation of glories was provided by Bryant and Cox (1966), who showed that the dominant contribution to the glory is from the last terms in the exact series for scattering by a sphere. Each successive term in this series is associated with ever larger impact parameters. Thus the terms that give the glory are indeed those corresponding to grazing rays. Further unraveling of the glory and vindication of van de Hulst's conjectures about the glory were provided by Nussenzveig (1979).

It sometimes is asserted that geometrical optics is incapable of treating the glory. Yet the same can be said for the rainbow. Geometrical optics explains rainbows only in the sense that it predicts singularities for scattering in certain directions (rainbow angles). But it can predict only the angles of intense scattering not the amount. Indeed, the error is infinite. Geometrical optics also predicts a singularity in the backward direction. Again, this simple theory is powerless to predict more. Results from geometrical optics for both rainbows and glories are not the end but rather the beginning, an invitation to take a closer look with more powerful magnifying glasses.

8 SCATTERING BY SINGLE ICE CRYSTALS

Scattering by spherical water drops in the atmosphere gives rise to three distinct displays in the sky: coronas, rainbows, and glories. Ice particles (crystals) also can inhabit the atmosphere, and they introduce two new variables in addition to size: shape and orientation, the second a consequence of the first. Given this increase in the number of degrees of freedom, it is hardly cause for wonder that ice crystals are the source of a greater variety of displays than are water drops. As with rainbows, the gross features of ice-crystal phenomena can be described simply with geometrical optics, various phenomena arising from the various fates of rays incident on crystals. Colorless displays (e.g., sun pillars) are generally associated with reflected rays, colored displays (e.g., sun dogs and halos) with refracted rays. Because of the wealth of ice-crystal displays, it is not possible to treat all of them here, but one example should point the way toward understanding many of them.

Sun Dogs and Halos

Because of the hexagonal crystalline structure of ice, it can form as hexagonal plates in the atmosphere. The stable position of a plate falling in air is with the normal to its face more or less vertical, which is easy to demonstrate with an ordinary business card. When the card is dropped with its edge facing downward (the supposedly aerodynamic position that many people instinctively choose), the card somersaults in a helter-skelter path to the ground. But when the card is dropped with its face parallel to the ground, it rocks back and forth gently in descent.

A hexagonal ice plate falling through air and illuminated by a low sun is like a 60° prism illuminated normally to its sides (Fig. 17). Because there is no mechanism for orienting a plate within the horizontal plane, all plate orientations in this plane are equally probable. Stated another way, all angles of incidence for a fixed plate are equally probable. Yet all scattering angles (deviation angles) of rays refracted into and out of the plate are not equally probable.

Figure 18 shows the range of scattering angles corresponding to a range of rays incident on a 60° ice prism that is part of a hexagonal plate. For angles of incidence less than about 13° the transmitted ray is totally internally reflected in the prism. For angles of incidence greater than about 70°, the transmittance plunges. Thus the only rays of consequence are those incident between about 13° and 70°.

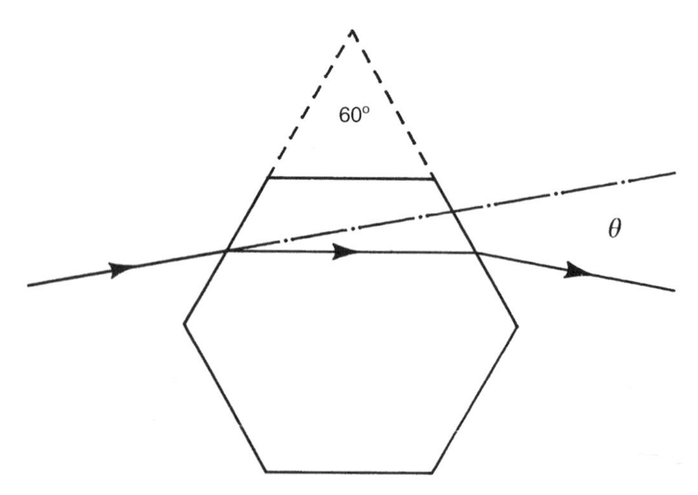

Figure 17 Scattering by a hexagonal ice plate illuminated by light parallel to its basal plane. The particular scattering angle θ shown is an angle of minimum deviation. The scattered light is that associated with two refractions by the plate.

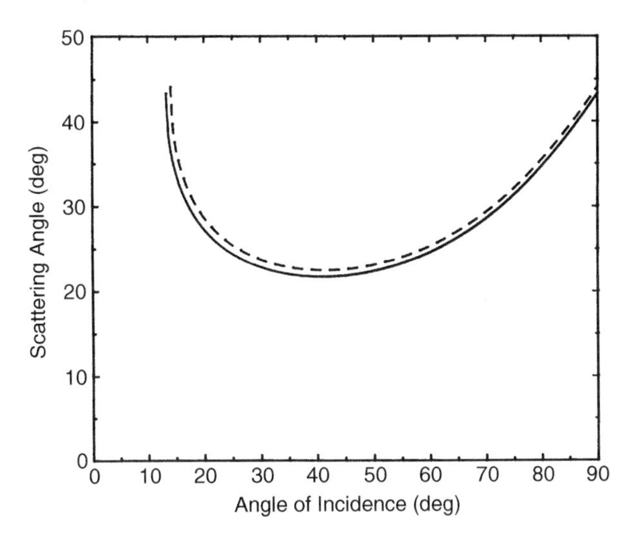

Figure 18 Scattering by a hexagonal ice plate (see Fig. 17) in various orientations (angles of incidence). The solid curve is for red light, the dashed curve is for blue light.

All scattering angles are not equally probable. The (uniform) probability distribution $p(\theta_i)$ of incidence angles θ_i is related to the probability distribution $P(\theta)$ of scattering angles θ by

$$P(\theta) = \frac{p(\theta_i)}{d\theta/d\theta_i} \tag{40}$$

At the incidence angle for which $d\theta/d\theta_i = 0$, $P(\theta)$ is infinite and scattered rays are intensely concentrated near the corresponding angle of minimum deviation.

The physical manifestation of this singularity (or caustic) at the angle of minimum deviation for a 60° hexagonal ice plate is a bright spot about 22° from either or both sides of a sun low in the sky. These bright spots are called *sun dogs* (because they accompany the sun) or *parhelia* or *mock suns.*

The angle of minimum deviation θ_m, hence the angular position of sun dogs, depends on the prism angle Δ (60° for the plates considered) and refractive index:

$$\theta_m = 2\sin^{-1}\left(n \sin\frac{\Delta}{2}\right) - \Delta \tag{41}$$

Because ice is dispersive, the separation between the angles of minimum deviation for red and blue light is about 0.7° (Fig. 18), somewhat greater than the angular width of the sun. As a consequence, sun dogs may be tinged with color, most noticeably toward the sun. Because the refractive index of ice is least at the red end of the spectrum, the red component of a sun dog is closest to the sun. Moreover, light of any two wavelengths has the same scattering angle for different angles of incidence if one of the wavelengths does not correspond to red. Thus red is the purest color seen in a sun dog. Away from its red inner edge a sun dog fades into whiteness.

With increasing solar elevation, sun dogs move away from the sun. A falling ice plate is roughly equivalent to a prism, the prism angle of which increases with solar elevation. From Eq. (41) it follows that the angle of minimum deviation, hence the sun dog position, also increases.

At this point you may be wondering why only the 60° prism portion of a hexagonal plate was singled out for attention. As evident from Figure 17, a hexagonal plate could be considered to be made up of 120° prisms. For a ray to be refracted twice, its angle of incidence at the second interface must be less than the critical angle. This imposes limitations on the prism angle. For a refractive index 1.31, all incident rays are totally internally reflected by prisms with angles greater than about 99.5°.

A close relative of the sun dog is the 22° halo, a ring of light approximately 22° from the sun (Fig. 19). Lunar halos are also possible and are observed frequently (although less frequently than solar halos); even moon dogs are possible. Until Fraser (1979) analyzed halos in detail, the conventional wisdom had been that they obviously were the result of randomly oriented crystals, yet another example of jumping to conclusions. By combining optics and aerodynamics, Fraser showed that if ice crystals are small enough to be randomly oriented by Brownian motion, they are too small to yield sharp scattering patterns.

But completely randomly oriented plates are not necessary to give halos, especially ones of nonuniform brightness. Each part of a halo is contributed to by plates with a different tip angle (angle between the normal to the plate and the vertical). The transition from oriented plates (zero tip angle) to randomly oriented plates occurs over a narrow range of sizes. In the transition region plates can be small enough to be partially oriented yet large enough to give a distinct contribution to the halo. Moreover, the mapping between tip angles and azimuthal angles on the halo depends on solar elevation. When the sun is near the horizon, plates can give a distinct halo over much of its azimuth.

When the sun is high in the sky, hexagonal plates cannot give a sharp halo but hexagonal columns—another possible form of atmospheric ice particles—can. The stable position of a falling column is with its long axis horizontal. When the sun is directly overhead, such columns can give a uniform halo even if they all lie in the horizontal plane. When the sun is not overhead but well above the horizon, columns also can give halos.

A corollary of Fraser's analysis is that halos are caused by crystals with a range of sizes between about 12 and 40 μm. Larger crystals are oriented; smaller particles are too small to yield distinct scattering patterns.

More or less uniformly bright halos with the sun neither high nor low in the sky could be caused by mixtures of hexagonal plates and columns or by clusters of bullets (rosettes). Fraser opines that the latter is more likely.

One of the byproducts of his analysis is an understanding of the relative rarity of the 46° halo. As we have seen, the angle of minimum deviation depends on the

Figure 19 A 22° solar halo. The hand is not for artistic effect but rather to occlude the bright sun.

prism angle. Light can be incident on a hexagonal column such that the prism angle is 60° for rays incident on its side or 90° for rays incident on its end. For $n = 1.31$, Eq. (41) yields a minimum deviation angle of about 46° for $\Delta = 90°$. Yet, although 46° halos are possible, they are seen much less frequently than 22° halos. Plates cannot give distinct 46° halos although columns can. Yet they must be solid and most columns have hollow ends. Moreover, the range of sun elevations is restricted.

Like the green flash, ice-crystal phenomena are not intrinsically rare. Halos and sun dogs can be seen frequently—once you know what to look for. Neuberger (1951) reports that halos were observed in State College, Pennsylvania, an average of 74 days a year over a 16-year period, with extremes of 29 and 152 halos a year. Although the 22° halo was by far the most frequently seen display, ice-crystal displays of all kinds were seen, on average, more often than once every 4 days at a location not especially blessed with clear skies. Although thin clouds are necessary for ice-crystal displays, clouds thick enough to obscure the sun are their bane.

9 CLOUDS

Although scattering by isolated particles can be studied in the laboratory, particles in the atmosphere occur in crowds (sometimes called clouds). Implicit in the previous two sections is the assumption that each particle is illuminated solely by incident sunlight; the particles do not illuminate each other to an appreciable degree. That is, clouds of water droplets or ice grains were assumed to be optically thin, hence multiple scattering was negligible. Yet the term *cloud* evokes fluffy white objects in the sky or perhaps an overcast sky on a gloomy day. For such clouds, multiple scattering is not negligible, it is the major determinant of their appearance. And the quantity that determines the degree of multiple scattering is optical thickness (see earlier discussion in Section 2).

Cloud Optical Thickness

Despite their sometimes solid appearance, clouds are so flimsy as to be almost nonexistent—except optically. The fraction of the total cloud volume occupied by water substance (liquid or solid) is about 10^{-6} or less. Yet although the mass density of clouds is that of air to within a small fraction of a percent, their optical thickness (per unit physical thickness) is much greater. The number density of air molecules is vastly greater than that of water droplets in clouds, but scattering per molecule of a cloud droplet is also much greater than scattering per air molecule (see Fig. 7).

Because a typical cloud droplet is much larger than the wavelengths of visible light, its scattering cross section is to good approximation proportional to the square of its diameter. As a consequence, the scattering coefficient [see Eq. (2)] of a cloud having a volume fraction f of droplets is approximately

$$\beta = 3f\frac{\langle d^2 \rangle}{\langle d^3 \rangle} \tag{42}$$

where the brackets indicate an average over the distribution of droplet diameters d. Unlike molecules, cloud droplets are distributed in size. Although cloud particles can be ice particles as well as water droplets, none of the results in this and the following section hinge on the assumption of spherical particles.

The optical thickness along a cloud path of physical thickness h is βh for a cloud with uniform properties. The ratio $\langle d^3\rangle/\langle d^2\rangle$ defines a mean droplet diameter, a typical value for which is 10 μm. For this diameter and $f = 10^{-6}$, the optical thickness per unit meter of physical thickness is about the same as the normal optical thickness of the atmosphere in the middle of the visible spectrum (see Fig. 3). Thus a cloud only 1 m thick is equivalent optically to the entire gaseous atmosphere.

A cloud with (normal) optical thickness about 10 (i.e., a physical thickness of about 100 m) is sufficient to obscure the disk of the sun. But even the thickest cloud does not transform day into night. Clouds are usually translucent, not transparent, yet not completely opaque.

The scattering coefficient of cloud droplets, in contrast with that of air molecules, is more or less independent of wavelength. This is often invoked as the cause of the colorlessness of clouds. Yet wavelength independence of scattering by a single particle is only sufficient, not necessary, for wavelength independence of scattering by a cloud of particles (see discussion in Section 2). Any cloud that is optically thick and composed of particles for which absorption is negligible is white upon illumination by white light. Although absorption by water (liquid and solid) is not identically zero at visible wavelengths, and selective absorption by water can lead to observable consequences (e.g., colors of the sea and glaciers), the appearance of all but the thickest clouds is not determined by this selective absorption.

Equation (42) is the key to the vastly different optical characteristics of clouds and of the rain for which they are the progenitors. For a fixed amount of water (as specified by the quantity f), optical thickness is inversely proportional to mean diameter. Raindrops are about 100 times larger on average than cloud droplets; hence optical thicknesses of rain shafts are correspondingly smaller. We often can see through many kilometers of intense rain whereas a small patch of fog on a well-traveled highway can result in carnage.

Givers and Takers of Light

Scattering of visible light by a single water droplet is vastly greater in the forward ($\theta < 90°$) hemisphere than in the backward ($\theta > 90°$) hemisphere (Fig. 9). But water droplets in a thick cloud illuminated by sunlight collectively scatter much more in the backward hemisphere (reflected light) than in the forward hemisphere (transmitted light). In each scattering event, incident photons are deviated, on average, only slightly, but in many scattering events most photons are deviated enough to escape from the upper boundary of the cloud. Here is an example in which the properties of an ensemble are different from those of its individual members.

Clouds seen by passengers in an airplane can be dazzling; but, if the airplane were to descend through the cloud, these same passengers might describe the cloudy sky overhead as gloomy. Clouds are both givers and takers of light. This dual role is exemplified in Figure 20, which shows the calculated diffuse downward irradiance below clouds of varying optical thickness. On an airless planet the sky would be black in all directions (except directly toward the sun). But if the sky were to be filled from horizon to horizon with a thin cloud, the brightness overhead would markedly increase. This can be observed in a partly overcast sky, where gaps between clouds

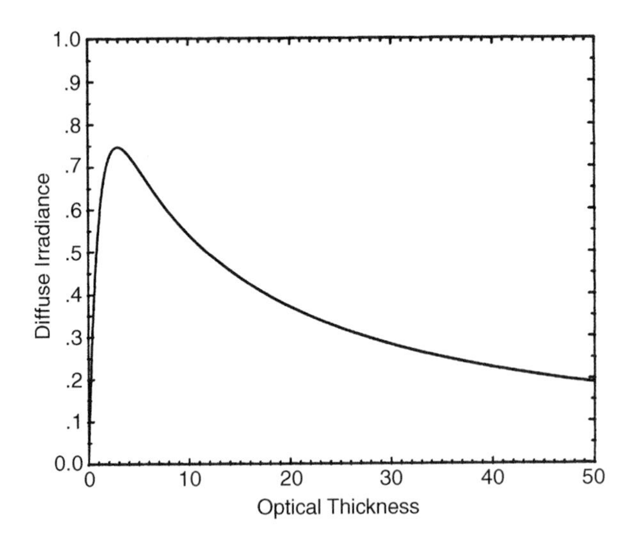

Figure 20 Computed diffuse downward irradiance below a cloud relative to the incident solar irradiance as a function of cloud optical thickness.

(blue sky) often are noticeably darker than their surroundings. As so often happens, more is not always better. Beyond a certain cloud optical thickness, the diffuse irradiance decreases. For a sufficiently thick cloud, the sky overhead can be darker than the clear sky.

Why are clouds bright? Why are they dark? No inclusive one-line answers can be given to these questions. Better to ask, Why is that particular cloud bright? Why is that particular cloud dark? Each observation must be treated individually, generalizations are risky. Moreover, we must keep in mind the difference between brightness and radiance when addressing the queries of human observers. Brightness is a sensation that is a property not only of the object observed but of its surroundings as well. If the luminance of an object is appreciably greater than that of its surroundings, we call the object bright. If the luminance is appreciably less, we call the object dark. But these are relative rather than absolute terms.

Two clouds, identical in all respects, including illumination, may still appear different because they are seen against different backgrounds, a cloud against the horizon sky appearing darker than when seen against the zenith sky.

Of two clouds under identical illumination, the smaller (optically) will be less bright. If an even larger cloud were to hove into view, the cloud that formerly had been described as white might be demoted to gray.

With the sun below the horizon, two identical clouds at markedly different elevations might appear quite different in brightness, the lower cloud being shadowed from direct illumination by sunlight.

A striking example of dark clouds can sometimes be seen well after the sun has set. Low-lying clouds that are not illuminated by direct sunlight but are seen against the faint twilight sky may be relatively so dark as to seem like ink blotches.

Because dark objects of our everyday lives usually owe their darkness to absorption, nonsense about dark clouds is rife: They are caused by pollution or soot. Yet of all the reasons that clouds are sometimes seen to be dark or even black, absorption is not among them.

GLOSSARY

Airlight: Light resulting from scattering by all atmospheric molecules and particles along a line of sight.

Antisolar Point: Direction opposite the sun.

Astronomical Horizon: Horizontal direction determined by a bubble level.

Brightness: Attribute of sensation by which an observer is aware of differences of luminace (definition recommended by the 1922 Optical Society of America Committee on Colorimetry).

Contrast Threshold: Minimum relative luminance difference that can be perceived by the human observer.

Inferior Mirage: Mirage in which images are displaced downward.

Irradiance: Radiant power crossing unit area in a hemisphere of directions.

Lapse Rate: Rate at which a physical property of the atmosphere (usually temperature) decreases with height.

Luminance: Radiance integrated over the visible spectrum and weighted by the spectral response of the human observer. Also sometimes called *photometric brightness*.

Mirage: Image appreciably different from what it would be in the absence of atmospheric refraction.

Neutral Point: Direction in the sky for which the light is unpolarized.

Normal Optical Thickness: Optical thickness along a radial path from the surface of Earth to infinity.

Optical Thickness: The thickness of a scattering medium measured in units of photon mean free paths. Optical thicknesses are dimensionless.

Radiance: Radiant power crossing a unit area and confined to a unit solid angle about a particular direction.

Scale Height: Vertical distance over which a physical property of the atmosphere is reduced to $1/e$ of its value.

Scattering Angle: Angle between incident and scattered waves.

Scattering Coefficient: Product of scattering cross section and number density of scatterers.

Scattering Cross Section: Effective area of a scatterer for removal of light from a beam by scattering.

Scattering Plane: Plane determined by incident and scattered waves.

Solar Point: Direction toward the sun.

Superior Mirage: Mirage in which images are displaced upward.

Tangential Optical Thickness: Optical thickness through the atmosphere along a horizon path.

REFERENCES

*Bryant, H. C., and A. J. Cox (1966). *J. Opt. Soc. Am.* **56**, 1529–1532.

Einstein, A. (1910). *Ann. der Physik* **33**, 175. English translation in J. Alexander (Ed.), *Colloid Chemistry*, Vol. I. New York, Chemical Catalog Company, 1926, p. 323.

Fraser, A. B. (1972). *J. Atmos. Sci.* **29**, 211–212.

*Fraser, A. B. (1975). *Atmosphere* **13**, 1–10.

*Fraser, A. B. (1979). *J. Opt. Soc. Am.* **69**, 1112–1118.

*Fraser, A. B. (1983). *J. Opt. Soc. Am.* **73**, 1626–1628.

Lee, R. (1991). *Appl. Opt.* **30**, 3401–3407.

Lee, R., and A. Fraser (1990). *New Scientist* **127** (1 September), 40–42.

*Hulburt, E. O. (1953). *J. Opt. Soc. Am.* **43**, 113–118.

Möller, F. (1972). "Radiation in the Atmosphere," in D. P. McIntyre (Ed.), *Meteorological Challenges: A History*, Ottawa, Information Canada, p. 43.

Neuberger, H. (1951). *Introduction to Physical Meteorology*, University Park, College of Mineral Industries, Pennsylvania State University.

*Nussenzveig, H. M. (1979). *J. Opt. Soc. Am.* **69**, 1068–1079.

*Penndorf, R. (1957). *J. Opt. Soc. Am.* **47**, 176–182.

Pledgley, E. (1986). *Weather* **41**, 401.

Takano, Y., and S. Asano (1983). *J. Meteor. Soc. Jpn.* **61**, 289–300.

Thekaekara, M. P., and A. J. Drummond (1971). *Nature Phys. Sci.* **229**, 6–9.

*van de Hulst, H. C. (1947). *J. Opt. Soc. Am.* **37**, 16–22.

van de Hulst, H. C. (1957). *Light Scattering by Small Particles*, New York, Wiley-Interscience.

von Frisch, K. (1971) *Bees: Their Vision, Chemical Senses, and Language*, Rev. Ed., Ithaca, New York, Cornell University Press, p. 116.

Young, A. T. (1982). *Physics Today*, January, 2–8.

Zimm, B. H. (1945). *J. Chem. Phys.* **13**, 141–145.9

BIBLIOGRAPHY

Many of the seminal studies in atmospheric optics, including those by Lord Rayleigh, are bound together in *Selected Papers on Scattering in the Atmosphere*, C. F. Bohren (Ed.), Bellingham, WA, SPIE Optical Engineering Press, 1989. References marked with an asterisk are in this collection.

M. Minnaert's *The Nature of Light and Colour in the Open Air*, New York, Dover, 1954, is the Bible for those interested in atmospheric optics. Like accounts of natural phenomena in the Bible, those in Minnaert's book are not always correct, despite which, again like the Bible, it has been and will continue to be a source of inspiration.

A history of light scattering, "From Leonardo to the Graser: Light Scattering in Historical Perspective," was published serially by J. D. Hey in *South African Journal of Science*, **79**, January 1983, 11–27; **79**, August 1983, 310–324; **81**, February 1985, 77–91; **81**, October 1985, 601–613; **82**, July 1986, 356–360. The history of the rainbow is recounted by C. B. Boyer, *The Rainbow*, Princeton, NJ, Princeton University Press, 1987.

Special issues of *Journal of the Optical Society of America* (August 1979 and December 1983) and *Applied Optics* (August 20, 1991) are devoted to atmospheric optics.

Several monographs on light scattering by particles are relevant to and contain examples drawn from atmospheric optics: H. C. van de Hulst, *Light Scattering by Small Particles*, New York, Wiley-Interscience, 1957 (reprinted by Dover, 1981); D. Deirmendjian, *Electromagnetic Scattering on Polydispersions*, New York, Elsevier, 1969; M. Kerker, *The Scattering of Light and Other Electromagnetic Radiation*, New York, Academic, 1969; C. F. Bohren and D. R. Huffman, *Light Scattering by Small Particles*, New York, Wiley-Interscience, 1983; H. M. Nussenzveig, *Diffraction Effects in Semiclassical Scattering*, Cambridge, Cambridge University Press, 1992.

The following books are devoted to a wide range of topics in atmospheric optics: R. A. R. Tricker, *Introduction to Meteorological Optics*, New York, Elsevier, 1970; E. J. McCartney, *Optics of the Atmosphere*, New York, Wiley, 1976; R. Greenler, *Rainbows, Halos, and Glories*, Cambridge, Cambridge University Press, 1980. Monographs of more limited scope are those by W. E. K. Middleton, *Vision Through the Atmosphere*, Toronto, University of Toronto Press, 1952; D. J. K. O'Connell, *The Green Flash and Other Low Sun Phenomena*, Amsterdam, North Holland, 1958; G. V. Rozenberg, *Twilight: A Study in Atmospheric Optics*, New York, Plenum, 1966; S. T. Henderson, *Daylight and Its Spectrum*, 2nd ed., New York, Wiley, 1977; R. A. R. Tricker, *Ice Crystal Haloes*, Washington, DC, Optical Society of America, 1979; G. P. Können, *Polarized Light in Nature*, Cambridge, Cambridge University Press, 1985.

Although not devoted exclusively to atmospheric optics, W. J. Humphreys, *Physics of the Air*, New York, Dover, 1964, contains a few relevant chapters. Two popular science books on simple experiments in atmospheric physics are heavily weighted toward atmospheric optics: C. F. Bohren, *Clouds in a Glass of Beer*, New York, Wiley, 1987; C. F. Bohren, *What Light Through Yonder Window Breaks?* New York, Wiley, 1991.

For an expository article on colors of the sky see C. F. Bohren and A. B. Fraser, *The Physics Teacher*, May, 267–272 (1985).

An elementary treatment of the coherence properties of light waves was given by A. T. Forrester, *Am. J. Phys.* **24**, 192–196 (1956). This journal also published an expository article on the observable consequences of multiple scattering of light: C. F. Bohren, *Am. J. Phys.*, **55**, 524–533 (1987).

Although a book devoted exclusively to atmospheric refraction has yet to be published, an elementary yet thorough treatment of mirages was given by A. B. Fraser and W. H. Mach in *Scientific American*, January, 102–111 (1976).

Colorimetry, the often (and unjustly) neglected component of atmospheric optics, is treated in, for example, *The Science of Color*, Washington, DC, Optical Society of America, 1963; F. W. Billmeyer and M. Saltzman, *Principles of Color Technology*, 2nd ed., New York, Wiley-Interscience, 1981; D. L. MacAdam, *Color Measurement*, 2nd ed., Berlin, Springer, 1985.

Understanding atmospheric optical phenomena is not possible without acquiring at least some knowledge of the properties of the particles responsible for them. To this end, the following are recommended: H. R. Pruppacher and J. D. Klett, *Microphysics of Clouds and Precipitation*, Dordrecht, Holland, Reidel, 1980; S. A. Twomey, *Atmospheric Aerosols*, New York, Elsevier, 1977.

SECTION 4

WEATHER SYSTEMS

Contributing Editor: John W. Nielsen-Gammon

CHAPTER 25

OVERVIEW OF WEATHER SYSTEMS

JOHN W. NIELSEN-GAMMON

1 INTRODUCTION

Meteorology is the only science in which it is common for a practitioner to appear every day on the evening news. Weather forecasting has immediate interest to people because it affects their day-to-day lives. The impact is great because weather itself changes from day to day. If weather were all good, or all bad, perhaps no one would notice—when's the last time you thought about the quality of the ground under your feet?

2 FOUNDATIONS OF WEATHER FORECASTING

The attempt to understand the weather is driven largely by the desire to be able to forecast the weather. However, the atmosphere is a complicated dynamical system, and an understanding of even the basic physical principles came only recently. Before that, scientists either attempted to understand the physical causes of weather events or recorded "weather signs," rules for weather forecasting. Because the physical understanding was incorrect, and most weather signs had no physical basis, neither approach was very successful. True progress in the study of the atmosphere that would lead to success in weather forecasting required a complete understanding of physical principles and a comprehensive set of observations of the atmosphere.

Aristotle's *Meteorologica*, written around 340 BC, was the leading book on meteorological principles in the Western world for two millenia. With his theories on the fundamental physical nature of the universe, Aristotle sought to explain a wide range of meteorological phenomena. For example, he argued that wind is

Handbook of Weather, Climate, and Water: Dynamics, Climate, Physical Meteorology, Weather Systems, and Measurements, Edited by Thomas D. Potter and Bradley R. Colman.
ISBN 0-471-21490-6

caused by the hot, dry exhalation of the earth when struck by sunlight. A leading book on weather signs, *De Signis Tempestatum*, written by Aristotle's pupil Theophrastus, focused on such maxims as "A dog rolling on the ground is a sign of violent storm" and "Reddish sky at sunrise foretells rain." Since some of these maxims are still in use today, it appears that Theophrastus's work has outlived that of Aristotle.

Astronomers during the Middle Ages often argued that weather could be predicted by careful examination of the sky, clouds, and stars. This method of weather forecasting is sometimes successful, but it more closely resembled astrology than modern-day meteorology. Astrological forecasting schemes grew more advanced and intricate with time, while the scientific study of weather systems made little or no progress.

Scientific progress required basic observational data for describing weather systems. Only with good observations could theories be tested and the correct nature of the atmosphere be determined. The primary tools for observation of weather conditions are the thermometer (for temperature), the barometer (for air pressure), and the hygrometer (for humidity). Prior to and during the Middle Ages, these basic weather instruments did not exist. These instruments were invented in the sixteenth and seventeenth centuries and refined throughout the eighteenth century. The resulting observations of the atmosphere eventually established the basis for meteorological breakthroughs of the nineteenth and twentieth centuries.

Meanwhile, Isaac Newton had laid the foundation for the development of physical laws governing the motion of objects. The great mathematician Leonhard Euler of Germany, in 1755, rewrote those laws in a form that applied to a continuous fluid such as air or water. But Euler's laws were incomplete. They described how air can be neither created nor destroyed and how air accelerates (and wind blows) in response to forces acting on it. Missing from Euler's equations were the crucial relationship between temperature and pressure and the consequences of evaporation and condensation.

With the new instruments, the study of air became an experimental science. Through numerous laboratory experiments, scientists gradually discovered the fundamental physical laws governing the behavior of gases. In particular, the missing key relationship, the first law of thermodynamics, became known from experiments during the first half of the nineteenth century.

Now that the basic physical laws were known, a comprehensive set of observations of the atmosphere was necessary for further progress. Weather observations from a single careful observer were insufficient, but dozens of observations taken simultaneously across Europe or the United States gave a much clearer picture of the distribution of weather elements within a winter storm. Such "synoptic" observations were also used directly in weather forecasting. As rapid long-distance communication became possible through the telegraph, individual weather systems could be tracked and therefore forecasted. The coordinated observations that have taken place since the middle of the nineteenth century form the foundation of modern weather forecasting and our understanding of weather systems of all scales.

3 DEVELOPING AN UNDERSTANDING OF WEATHER SYSTEMS

Progress in understanding meteorology in the nineteenth century was largely based on thermodynamics. James Pollard Espy of the United States conducted a variety of laboratory experiments involving the condensation of water vapor in air. In 1841, based on his experiments and those of others, Espy correctly described the basic principle of thunderstorm formation: Condensation of water vapor within ascending air causes the air to become warmer than its surroundings and thus to continue rising. Espy went on to assert that the resulting lower pressure near the ground accounts for the low pressure observed in large-scale storm systems.

Espy's theory lay dormant until later in the nineteenth century, when the complete first law of thermodynamics was established. With this development, scientists were able to put Espy's theory of convection on a solid mathematical footing. Karl Theodor Reye of Germany, for example, determined the specific conditions under which such ascending air would be unstable. Their ideas form our understanding of the fundamental nature of individual thunderstorms to this day.

Espy also suggested that upward motion and the release of latent heat by evaporation might be the driving force for large weather systems. The widespread development of synoptic observations finally established that surface winds spiral inward toward a large-scale low-pressure system. The strong temperature contrasts within low-pressure systems suggested that the ascent would be driven, at least initially, by a current of warm air impinging on a current of cold air and rising. The rising air, and the resulting latent heating from condensation, would cause surface pressures to fall. This theory for the cause of low-pressure systems gained rapid acceptance for both theoretical and observational reasons in the 1870s.

A leading proponent of this thermal theory of cyclones was the American William Ferrel. Earlier, Ferrel had completed the application of Euler's equations to the atmosphere by formally including the effect of Earth's rotation as the Coriolis force. Ferrel also saw a clear analog to the role of equatorial convection in the general circulation, which drives ascent in the tropics and descent in the subtropics.

Unfortunately for its proponents, the thermal theory was wrong. Coordinated mountaintop observations in Europe, analyzed by Julius von Hann, showed that above the ground the centers of low-pressure systems were actually cooler than the centers of high-pressure systems. This observation illustrates the self-limiting nature of moist convection: The downward motion outside the convective cell causes warming of that surrounding air, so that eventually the instability is eliminated.

4 WEATHER AND ENERGY

The above discussion leads to a key question for the understanding of weather systems: How do weather systems grow and maintain themselves? As noted above, the basic dynamics of ordinary moist convection were well established, both conceptually and mathematically, by the end of the nineteenth century.

Convection was also understood to be an important driver of the general circulation, or at least the tropical and subtropical circulation known as the Hadley cells.

By the turn of the century, the work of Hann and others showed that convection was not the energy source for extratropical cyclones, the midlatitude migratory low-pressure systems. Attention shifted from the role of latent heat release to the role of the large horizontal temperature gradients that were systematically observed within midlatitude low-pressure systems. In 1903, Max Margules, born in the Ukraine and working in Austria, calculated the amount of kinetic energy that could be obtained from the rising of hot air and the sinking of cold air in a low-pressure system. He found that the amount of energy that could be converted into kinetic energy was comparable to the actual amount of kinetic energy in a mature storm system. Margules had identified the correct energy source for extratropical cyclones. The process by which cyclones form and move was described by researchers working under Vilhelm Bjerknes in Bergen, Norway, shortly after World War I. Finally, a comprehensive theory by Jule Charney in 1947 explained that the structure and intensity of low-pressure systems was a consequence of the growth of unstable eddies on a large-scale horizontal temperature gradient.

Hurricanes did not have the prominent horizontal temperature variations found in midlatitude weather systems; indeed, many hurricanes seemed almost perfectly symmetrical. The prominence of convection throughout the hurricane, particularly within the eyewall, suggested that convection was fundamentally important in the development and maintenance of hurricanes. But how did the hurricane organize itself or maintain itself against large-scale subsidence within its environment? The currently accepted theory, published by Kerry Emanuel and Richard Rotunno in 1986 and 1987, relies on radiative cooling of the subsiding air at large distances from the hurricane. Emanuel also noted that as air spirals inward toward the eye it would become more unstable because it would be gaining heat and moisture from the sea surface at a progressively lower pressure.

Theories satisfactorily describing organized convection such as mesoscale convective systems and supercells also have been slow to develop. One reason for this is that no observing systems could accurately describe the structure of organized convection until the development of radar. Indeed, the term "mesoscale" was coined specifically to refer to in-between sizes (10 to 500 km) that were too large to be adequately observed at single locations and too small to be resolved by existing observing networks. The widespread use of weather radar in the 1950s helped fill the observation gap. The development of Doppler radar (for measuring winds within precipitating systems) for research purposes in the 1970s and as part of a national network in the 1990s helped even more. With comprehensive radar observations, it became clear that the long lifetime of organized convection was due to a storm keeping its updraft close to the leading edge of its cold, low-level outflow. The storm could then take advantage of the ascent caused by the cold air undercutting warm air without having its supply of warm air cut off completely. But even radar was not sufficient. The development of the first dynamical descriptions of supercells and squall lines in the 1980s, by Joseph Klemp, Richard Rotunno, Robert Wilhelmson, and Morris Weisman, relied upon numerical computer-generated simulations of the phenomena to provide an artificial data set unavailable from observations.

While the above discussion has focused on self-contained weather systems, certain "weather producers" may be thought of as byproducts of these weather systems. Many forms of severe weather, such as hail, lightning, and tornadoes, are essentially side effects of convection (organized or otherwise) but have little or no influence on the convection itself. Other mesoscale phenomena, such as sea breezes, upslope precipitation, and downslope windstorms, do not represent instabilities at all. Instead they are called "forced" weather phenomena because they are driven by such external features as solar heating gradients and topographic obstacles to the large-scale flow.

5 FORECASTING

The massive strides in weather forecasting during the past 50 years are due in large part to our growing understanding of the nature and dynamics of important weather systems. Forecasting of many phenomena has evolved from an exercise in extrapolation to specific predictions of the evolution of weather systems to computer simulations that accurately forecast the evolution of weather systems. Currently, the most skillful weather forecasts involve the prediction of large-scale extratropical weather systems.

One very important advance in our ability to forecast large-scale weather systems was the development of the omega equation, first presented in simplified form by Richard Sutcliffe in 1947. The omega equation is named after the Greek symbol that represents the change in pressure of an air parcel. In its original form, the omega equation is based upon a simplification of the equations of motion that assumes that all motions are large-scale and evolve slowly. When this approximation is made, it is possible to diagnose vertical motion entirely from large-scale pressure and temperature variations. Vertical motion is important for weather prediction because it is fundamentally related to clouds and precipitation, but large-scale vertical motion also can be used to infer the evolution of low-level and upper-level wind patterns. Vertical motion became the key to understanding the evolution of weather. In forecast offices, maps were designed and widely distributed that made it easy to look at the large-scale fields and diagnose vertical motion.

A second important advance was the development of numerical weather prediction, or NWP. NWP, discussed more extensively in the chapter by Kalnay in the dynamics part of this *Handbook*, has completely transformed the forecasting of large-scale weather systems. In the past, forecasters relied on their diagnosis of the current weather patterns to make forecasts. Nowadays, computers can make much more accurate forecasts than humans alone, and the best possible forecast is obtained by humans working with computer forecast output. The forecaster's task has become one of identifying likely errors in the model forecast, based on the forecaster's knowledge of systematic errors within the model, errors in the initial analysis, and computer forecast scenarios that run counter to experience.

In contrast to large-scale weather systems, individual convective storms are not directly simulated, nor accurately forecasted, by the numerical weather prediction models currently in use. Forecasting convective storms still relies heavily on extrapolation and a sound knowledge of what sorts of storms are likely to be produced by certain larger-scale weather conditions. The modern era of forecasting severe weather began in 1948, when two Air Force weather forecasters (Maj. Ernest Fawbush and Capt. Robert Miller) had responsibility for forecasting at Tinker Air Force Base near Oklahoma City. After a damaging tornado struck the base in 1948, the two meteorologists were given the task of identifying the days when tornadoes were likely to strike the base. We now know that forecasting the specific path of a tornado is essentially impossible, but when a similar low-pressure system evolved a few days later, the forecasters were compelled by the base commander to use a newly minted severe weather warning system to issue a forecast of a possible tornado at the base. Amazingly, the forecast came true, and preventive measures taken before the second tornado struck the base prevented considerable casualties to aircraft and personnel. This weather forecast, possibly the most serendipitous in the history of humankind, led directly to modern tools for diagnosing the likelihood of severe weather from large-scale conditions and laid the foundation for modern tornado forecasting.

6 CONCLUSION

This historical review has emphasized that progress in our understanding of the weather has required four ingredients: the need to understand the science of the atmosphere before one can hope to make accurate forecasts; a complete knowledge of the basic physical underpinnings of atmospheric behavior; observations adequate to test the theories and point the way toward new ones; and numerical models to provide simulations of the atmosphere more complete than any set of observations. The following chapters of this part of the *Handbook* discuss the current understanding of the weather phenomena discussed above, improvements in which have lead to vastly improved weather forecasts.

BIBLIOGRAPHY

Frisinger, H. H. (1983). *The History of Meteorology: To 1800*, Boston, American Meteorological Society.

Kutzbach, G. (1979). *The Thermal Theory of Cyclones: A History of Meteorological Thought in the Nineteenth Century*, Boston, American Meteorological Society.

Miller, R. C., and C. A. Crisp (1999). The First Operational Tornado Forecast—Twenty Million to One, *Weather and Forecasting*, **14**(4), 479–483.

CHAPTER 26

LARGE-SCALE ATMOSPHERIC SYSTEMS

JOHN W. NIELSEN-GAMMON

1 INTRODUCTION

The structure and evolution of large-scale extratropical weather systems is dominated by a fundamental contradiction: The airflow within such systems represents an almost exact balance among the forces affecting each air parcel, but the slight departures from balance are essential for vertical motion and the resulting clouds and precipitation, as well as changes in intensity of the systems. Furthermore, balanced flow could not be maintained without slight departures from balance. This section will explore the morphology and dynamics of large-scale weather systems and will never stray far from the concepts of balance and adjustment to balance.

The figures in this Chapter of the *Handbook* will typically use pressure as a vertical coordinate, in keeping with the standard practice of displaying all meteorological information above Earth's surface on levels of constant pressure. Constant-pressure surfaces are so nearly flat that they can be treated as horizontal for most purposes. Vertical motion in this coordinate system, represented as ω, is defined as dp/dt rather than dz/dt, and since pressure increases downward rather than upward, negative ω corresponds to upward motion.

2 HORIZONTAL BALANCE

In its simplest form, the balance of forces in the horizontal plane is geostrophic balance (see chapter by Salby in this *Handbook*) between the Coriolis force and the

Handbook of Weather, Climate, and Water: Dynamics, Climate, Physical Meteorology, Weather Systems, and Measurements, Edited by Thomas D. Potter and Bradley R. Colman.
ISBN 0-471-21490-6

horizontal pressure gradient force. Since the Coriolis force is proportional to the wind speed and directed to the right of the wind direction (to the left in the Southern Hemisphere), a balance of forces can only be attained if the horizontal pressure gradient force is directed to the left of the wind direction (right in the Southern Hemisphere), with sufficient speed that the magnitude of the Coriolis force equals the magnitude of the horizontal pressure gradient force. Thus, balanced flow is parallel to the isobars (in Cartesian coordinates) or height contours (in pressure coordinates) with a strength proportional to the horizontal pressure gradient. This balance (and the others described in this chapter) is generally stable, in the sense that air parcels initially out of balance will tend to approach balance, on a time scale of a few hours.

The full horizontal wind may be divided into geostrophic and ageostrophic components. The geostrophic wind is, by definition, that wind that would represent an exact balance between the horizontal pressure gradient force and the Coriolis force. For large-scale extratropical weather systems, the geostrophic wind very nearly equals the total wind. The ageostrophic wind is associated with an imbalance of forces and, therefore, acceleration. It may be thought of as that portion of the total wind whose associated Coriolis force is not balanced by a horizontal pressure gradient force. Thus, acceleration is proportional to the magnitude of the ageostrophic wind and is directed in the same direction as the Coriolis force, to the right of the ageostrophic wind (to the left in the Southern Hemisphere).

Other forms of balance may be defined that include nonzero ageostrophic winds. For example, circular flow represents a balance of three forces: the horizontal pressure gradient force, the Coriolis force, and the centrifugal force. In this circumstance, the Coriolis force associated with the geostrophic wind balances the horizontal pressure gradient force, and the Coriolis force associated with the ageostrophic wind balances the centrifugal force. The wind remains parallel to the isobars or height contours but is weaker (subgeostrophic) in a cyclonic vortex and stronger (supergeostrophic) in an anticyclonic vortex. Since ageostrophic winds can often be associated with balanced flow, it is often more useful to subdivide the wind as divergent and nondivergent rather than geostrophic and ageostrophic. The geostrophic wind is always nondivergent (except for effects due to the variation of the Coriolis parameter with latitude), and the divergent wind is directly related to vertical motion through the continuity equation.

3 VERTICAL BALANCE

The vertical balance of forces, known as hydrostratic balance, is between the vertical pressure gradient force and gravity. This balance may be assumed to be maintained exactly for extratropical weather systems. Indeed, the balance is so close that vertical accelerations must be deduced by indirect means, through diagnosis of the divergent wind.

As horizontal balance introduced a close connection between the instantaneous pressure and wind fields, vertical balance introduces a close connection between the instantaneous pressure and density/temperature fields. To balance gravity, the vertical pressure gradient force must be strong where air density is large and weak where density is small. Equivalently, the vertical separation between two given pressure surfaces (known as *thickness*) is small where the air is relatively cool and large where the air is relatively warm. Horizontal variations in temperature imply vertical differences in the horizontal variation of pressure; that is, the horizontal pressure gradient at one level will differ from the pressure gradient at another level if air of differing densities (temperatures) lies between the two levels. If one simultaneously assumes geostrophic (horizontal) and hydrostatic (vertical) balance, one obtains a relationship between temperature and wind known as the thermal wind law: If one represents the vertical derivative of the horizontal wind (vertical wind shear) by a vector, it will lie parallel to the isotherms with relatively warmer temperature to the right (to the left in the Southern Hemisphere). Thus, temperature, pressure, and wind are all constrained by each other in the limit of exact balance.

Vertical motion carries a strong influence on atmospheric temperature. Following an air parcel within which phase changes are not taking place and which is not exchanging heat with its surroundings, upward motion causes the parcel to cool at the dry adiabatic lapse rate, 9.8°C/km, and downward motion causes an identical amount of warming. In most circumstances away from the daytime boundary layer, the atmosphere is stably stratified, meaning that the instantaneous vertical derivative of temperature is greater than $-9.8\,\mathrm{K/km}$, so that, for example, an ascending air parcel replaces an air parcel that is warmer (and less dense) than itself. Consequently, at a given level, upward motion causes cooling at a rate proportional to the magnitude of upward motion and the degree of stratification. Similarly, downward motion causes warming. This stratification is called *stable* because an ascending air parcel, finding itself cooler and more dense than its surroundings, would tend to sink back down toward its original level, and a subsiding air parcel would be warmer than its surroundings and would tend to rise.

Extratropical large-scale weather systems can be directly affected by condensation of water vapor within ascending, cooling air parcels. This condensation releases latent heat, causing the air parcel to be warmer than it would have been without the release of latent heat. As a result, in areas of saturated ascent, the temperature at a given level does not fall as rapidly as it would in ascending dry air and may in fact rise if the atmosphere is unstable to moist convection. Condensation, and the generation of clouds and precipitation, represents the primary internal heat source for large-scale weather systems.

4 POTENTIAL VORTICITY

When the atmosphere is in approximate horizontal and vertical balance, the wind and mass fields are tightly interconnected. The distribution of a single mass or momentum variable may be used as a starting point to infer the distribution of all other such variables. We choose to focus on the variable known as potential vorticity (approximately equal to the vorticity times the stratification) for three reasons: (1) it tends to have a simple three-dimensional distribution; (2) it is conserved in the absence of diabatic and frictional processes, so it tends to have a simple evolution; (3) it is of direct relevance to the fundamental dynamics of atmospheric behavior.

Potential vorticity is related to the three-dimensional mass and wind fields through equations that involve three-dimensional inverse second-order operators whose exact form depends on the level of approximation. One consequence of the inverse Laplacian-like operator is that potential vorticity variations in one location are diagnostically related to variations in the height and wind fields at a distance. Indeed, the relationship between perturbation (deviation from some standard mean state) potential vorticity and perturbation height (or pressure) is mathematically and conceptually analogous to the relationship between electric charge and electric potential. Areas of locally high potential vorticity correspond to locally (and regionally) low heights, and therefore cyclones, while areas of locally low potential vorticity correspond to anticyclones.

While it is possible for isolated regions of high potential vorticity to form anywhere in the atmosphere, many potential vorticity anomalies are found at the tropopause. There, midlatitude potential vorticity changes suddenly from around $0.5 \times 10^{-6} m^2/s\ K\ kg$ (henceforth, 0.5 PVU) to around 5 PVU. Horizontal and vertical displacements in the position of the tropopause lead to sizable potential vorticity variations.

Figure 1 shows the wind and mass fields associated with a prototypical axially symmetric potential vorticity anomaly at the tropopause. This cyclonic anomaly (positive in the Northern Hemisphere) is associated with a cyclonic circulation that is strongest at the level of the anomaly and decreases downward and away from the anomaly. Temperatures are anomalously warm above the vortex (isentropic surfaces are deflected downward) and anomalously cold below the vortex. While consistent with the given simple potential vorticity distribution, the wind and mass fields are also in vertical and horizontal balance. The general characteristics of the

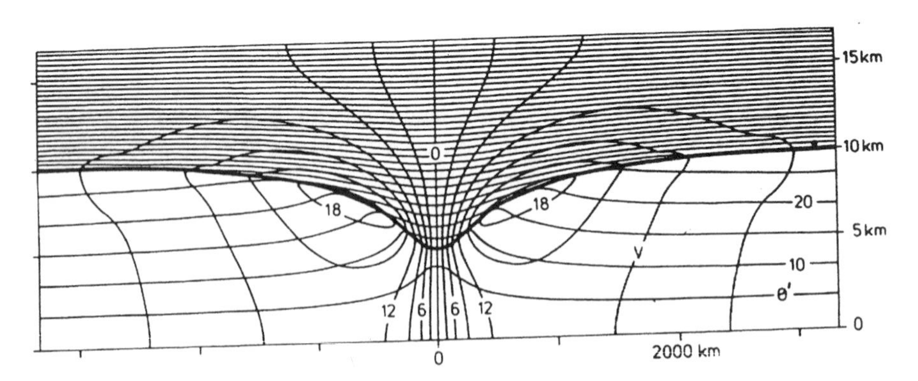

Figure 1 Wind (V) and potential temperature (θ') structure associated with a balanced axisymmetric vortex at the tropopause. Wind speeds are in m/s and are out of the page on the left and into the page on the right. Potential temperature (in K) is the departure from a uniform surface potential temperature. (From Thorpe, 1986).

wind and temperature distribution are common to cyclonic potential vorticity anomalies, with analogous atmospheric states for anticyclonic potential vorticity anomalies. Winds and temperatures associated with complex potential vorticity distributions may, to a first approximation, be recovered by treating the potential vorticity distribution as an assemblage of discrete anomalies, each with its own impact on the nearby wind and temperature fields.

5 FRONTS

Fronts are elongated zones of strong temperature gradient separating regions of relatively weak temperature gradient. Fronts can form as a consequence of deformation, convergence, and differential heating or cooling. Fronts produced by differential heating or cooling tend to be relatively small in scale, such as a gust front (produced by evaporative cooling) or a sea breeze front (produced by daytime heating over land). Such mesoscale fronts are discussed in Chapter 30 (Brooks et al). Here, we focus on synoptic-scale fronts, generally formed by a combination of deformation and convergence. Depending on their motion and structure, synoptic-scale fronts are called warm fronts, cold fronts, stationary fronts, or occluded fronts.

Characteristics of Fronts

While fronts are defined in terms of temperature, other atmospheric variables are also affected by the presence of a front. Often, the air on either side of a front originated from widely different locations, and a sharp humidity gradient will be present as well. Pressure and wind are also affected by the density contrast across a front. A pressure trough, wind shift, and vorticity maximum tend to be present along the warm side of the front. The temperature gradient also tends to be strongest at the warm edge of the front, and it is at this edge of the strong temperature gradient that the front itself is deemed to be located. The temperature gradient can be so strong that most of the temperature variation across the frontal zone takes place over a distance as small as 100 m. Under many circumstances, the leading edge of the frontal zone is marked by a discontinuity of low cloud or even a long, narrow band of low cloud apparent in visible satellite imagery and known as a rope cloud.

Cyclones (here used in the common meteorological sense to refer to synoptic-scale low-pressure systems) have a symbiotic relationship to fronts. Cyclones frequently form along preexisting surface frontal zones. If a preexisting frontal zone is not present, however, a cyclone is quite capable of generating its own surface fronts. A typical life cycle is shown in Figure 2. As the cyclone intensifies, both a warm and cold front are present, which may or may not intersect in the vicinity of the surface low. The warm front is marked by geostrophic wind from warm to cold air, and the cold front is marked by the opposite. As the cyclone matures, the cold front moves ahead of the low, and the former warm front between the cyclone and the

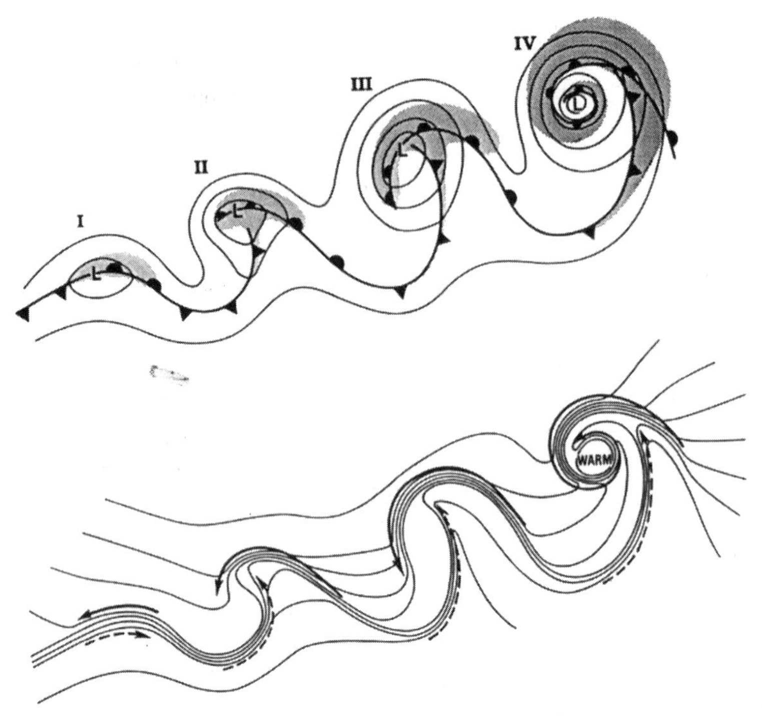

Figure 2 Typical life cycle of an intensifying offshore extratropical cyclone: (I) incipient frontal cyclone, (II) frontal fracture, (III) T-bone structure, and (IV) warm-core seclusion. *Top panel:* sea level pressure (solid), fronts (bold), and clouds (shaded). *Bottom panel:* temperature (solid), cold air currents (solid arrows), and warm air currents (dashed arrows). (From Shapiro and Keyser, 1990).

intersection of warm and cold fronts becomes the occluded front, marked with strong temperature gradients on both sides and the warmest air along the front.

Fronts are typically associated with a variety of clouds and precipitation. The precipitation may be ahead of the front, behind it, or both. Some distinguish between *anafronts*, with an updraft sloping upward toward the cold air and clouds and precipitation behind the front, and *katafronts*, with an updraft sloping upward toward the warm air and the bulk of the precipitation within the warm air. The distribution of clouds and precipitation associated with the surface fronts of a developing cyclone is shown in Figure 3.

Surface fronts tend to be strongest (i.e., possess the largest temperature gradient) near the ground and decrease in intensity upward. Typically, a surface front will lose its frontal characteristics by an altitude of 4 km or 625 mb (Fig. 4), although deeper fronts have been observed. Fronts are also favored at heights of 6 to 9 km. Such fronts are called upper-level fronts. An especially strong upper-level front is shown in Figure 5. On rare occasions, a surface and upper-level front may merge, yielding a single frontal zone stretching from the ground to the top of the troposphere (Fig. 6).

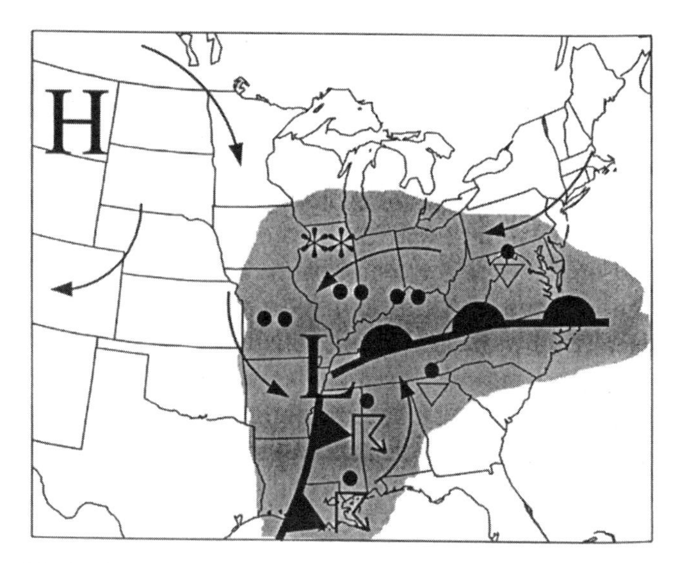

Figure 3 Distribution of clouds and fronts associated with a prototypical developing extratropical cyclone. Surface weather is represented by standard meteorological symbols.

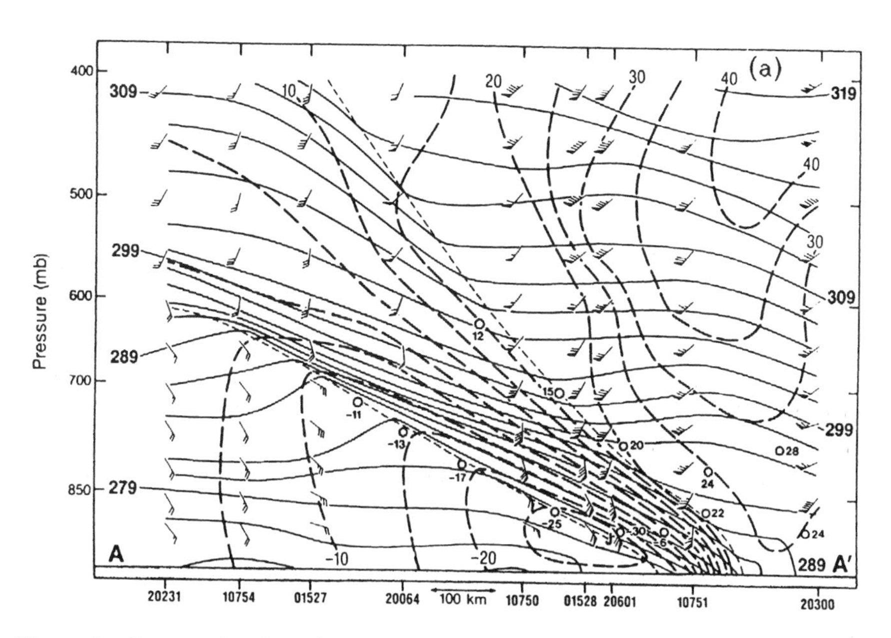

Figure 4 Cross section through strong front observed over the central Pacific on March 9, 1987. Data taken from dropwinsondes (potential temperature in K, solid lines; wind barbs, 1 long barb = 10 knots) and airborne Doppler radar (open circles with section-normal wind component in m/s beneath). Wind analyzed with bold contours in m/s; negative directed out of the page. (From Shapiro and Keyser, 1990).

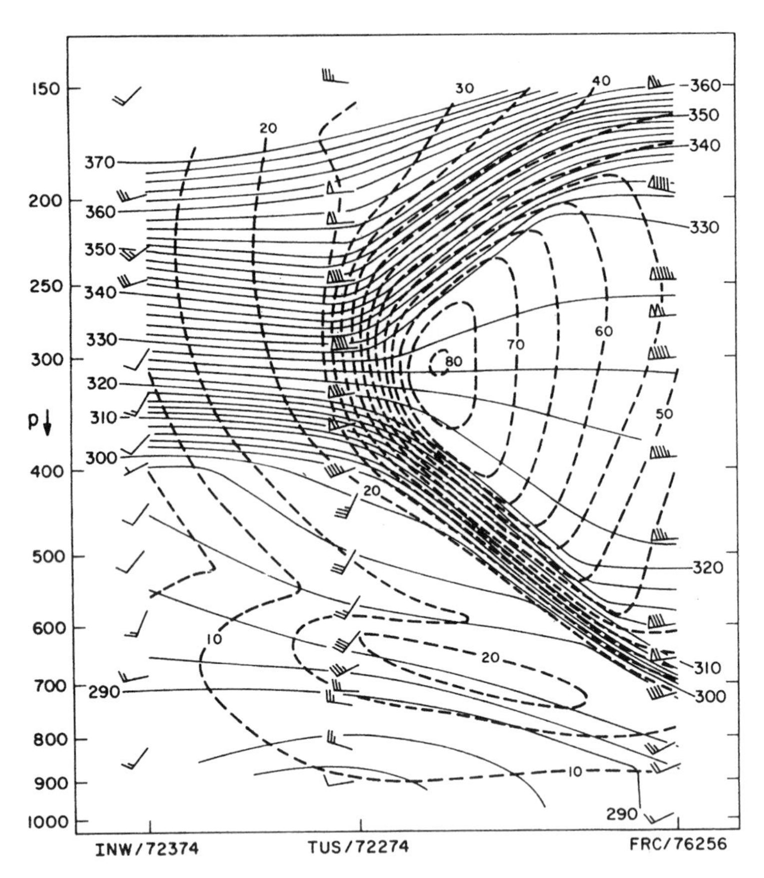

Figure 5 Cross section through a strong upper-level front, showing potential temperature (K, solid), observed wind (barbs, as in Fig. 4), and section-normal wind component (dashed, m/s, positive into the page). (From Shapiro, 1991).

In Figures 4 to 6, contours of potential temperature are shown rather than temperature. The fronts are indicated by the zones of sloping, tightly packed contours of potential temperature. At any given pressure level, potential temperature is proportional to temperature, so concentrated gradients of potential temperature imply concentrated gradients of temperature. One advantage of using potential temperature is that frontal zones consist of isentropes (lines of constant potential temperature) packed horizontally or vertically. A second advantage is that mixing can take place adiabatically along isentropes; cross-isentrope transport is only possible in conjunction with diabatic processes. Thus, the extent to which the front acts like an impenetrable wall is given by the extent to which the frontal zone is defined by a single isentrope.

In Figures 5 and 6, the frontal zone includes air from the stratosphere that has been drawn downward into the troposphere in a structure known as a tropopause

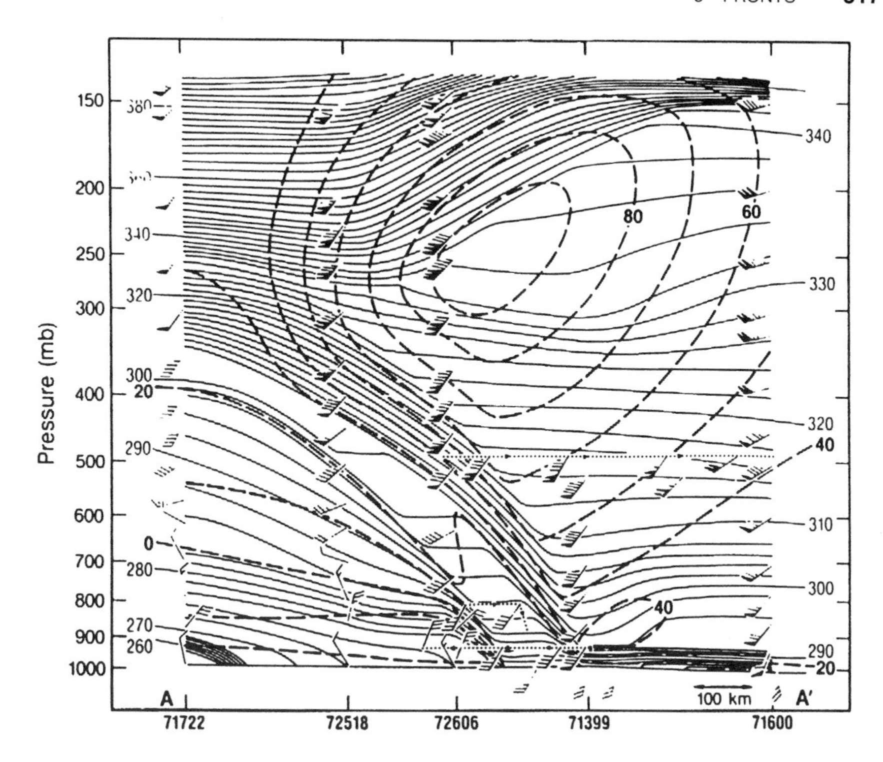

Figure 6 Cross section through a strong merged front, as in Fig. 5. Dashed line shows path of aircraft. (From Shapiro and Keyser, 1990).

fold. Tropopause folds are common to upper-level fronts; because of the large amounts of mixing that take place in the vicinity of upper-level fronts, tropopause folds are a primary mechanism for the exchange of air between the stratosphere and troposphere. The mixing associated with upper-level fronts is also one manifestation of clear-air turbulence. Extreme tropopause folding events can bring stratospheric air all the way to the ground. This is presently not thought to be dangerous, although it was a concern in the days of aboveground nuclear tests.

Dynamical Aspects of Fronts

Synoptic-scale surface fronts are often portrayed as zones of colliding air masses or war zones where battles are fought within thunderstorms. This portrayal is wrong, primarily because of the influence of the Coriolis force. Beneath a cold air mass will tend to be a region of high pressure, which in the absence of the Coriolis force would accelerate the cold air in the direction of the warm air and directly cause vertical motion. Instead, with the Coriolis force, the cold air comes into approximate balance and the wind is ultimately directed parallel to the isobars, with higher pressure to the right. Rather than two armies rushing forward toward each other, perhaps the appropriate image is of two armies nervously pacing sideways along a demilitarized zone.

In addition to the horizontal wind variations mentioned earlier, a front that is nearly in geostrophic and hydrostratic balance will be associated with significant vertical wind shear. Recall from the introduction to this section that because of the relationships between pressure and temperature on the one hand and pressure and wind on the other, a horizontal temperature gradient implies vertical shear of the horizontal wind. Since a front is a zone of concentrated temperature gradient, the vertical wind shear (or thermal wind) is large there too. The shear vector is oriented such that winds aloft will tend to blow parallel to the front with cold temperatures to the left and warm temperatures to the right. Because of the interrelationship between wind shear and temperature gradient, an upper-level jet stream is often located directly above a deep surface or upper-level front.

Because wind shear tends to improve the ability of midlatitude convection to organize itself into convective systems or supercells, the wind shear of a front is one reason severe weather tends to be located near fronts. A second reason for convection, and clouds in general, near fronts is the vertical motion that tends to be associated with fronts. This upward motion is not due to warm air being forced upward by cold air; it is a consequence of the atmosphere attempting to stay within balance.

In an intensifying synoptic-scale front, deformation and convergence is causing the magnitude of the temperature gradient to increase. At the same time, by means that will not be explained here, the deformation acts to reduce the vertical wind shear, even though the vertical wind shear would have to increase to remain within balance. The resulting imbalance between the pressure gradient force and Coriolis force produces accelerations: At low levels the air accelerates from the cold side of the front toward the warm side, and aloft the air accelerates from the warm side of the front toward the cold side. An ageostrophic circulation is thereby established, and mass continuity demands upward motion on the warm side of the front and downward motion on the cold side in order to complete the circulation cell. The horizontal ageostrophic flow, once it reaches finite magnitude, causes an acceleration to the right by the Coriolis force, eventually producing an increase in the vertical shear. Meanwhile, the vertical motion is acting to cool the air through ascent on the warm side of the front and warm it through descent on the cool side of the front, thereby acting to reduce the horizontal temperature gradient. This vertical circulation is called a *direct* circulation because relatively warm air rises and relatively cold air sinks; the opposite circulation, which is found when fronts are weakening, is called *indirect.*

Here, then, is the true "battle" within a front: The balanced flow is acting in one sense, and the unbalanced (ageostrophic) flow is acting in precisely the opposite sense! The net effect is that both the vertical wind shear and horizontal temperature gradient increase. By assuming that the ageostrophic circulation is precisely as strong as it has to be to maintain thermal wind balance, the horizontal ageostrophic and vertical motions can be diagnosed purely from the rate of frontogenesis (strengthening of the temperature gradient across an air parcel), using the so-called Sawyer–Eliassen equation. This constant adjustment, as the air attempts to keep up with thermal wind balance, is the primary cause of the vertical motion and resulting clouds and precipitation mainly on the warm side of active zones of strong temperature gradient.

The ageostrophic circulation also modifies the frontal structure. Since the vertical motions are constrained to be near zero at the ground and in the upper troposphere, the vertical motion serves to weaken the horizontal temperature gradient primarily in the middle troposphere, between 3 and 6 km, thus explaining the paucity of strong fronts at this level. Meanwhile, the low-level ageostrophic flow is directed from cold air toward warm air, where it ascends. The convergence on the warm air side of the front enhances the temperature gradient there, even as the gradient on the cold side of the front is being weakened by the same mechanism. This explains the tendency for surface fronts to have their strongest temperature gradient, or perhaps even a discontinuity, right at the warm-air edge of the frontal zone. By similar arguments, upper-level fronts ought to be strongest along the cold-air edge.

6 EXTRATROPICAL CYCLONES

Extratropical cyclones (also known as low-pressure systems) are large-scale circulations that develop spontaneously in midlatitudes. The sense of the circulation is cyclonic, or in the same direction as Earth's rotation, which is counterclockwise in the Northern Hemisphere and clockwise in the Southern Hemisphere. Extratropical cyclones have scales ranging from hundreds to thousands of kilometers and can be associated with widespread rain and snow and high winds. Extratropical anticyclones, or high-pressure systems, are just as common and form by similar means, but receive less attention since they tend to be associated with fair weather and light winds.

Observed Characteristics

The life cycle of an extratropical cyclone has been shown in Figure 2. The general evolution shown in Figure 2 is typical for extratropical cyclones forming over the ocean and associated with a single upper-tropospheric mobile trough. Over land, warm fronts tend to be weaker and the presence of mountain ranges and coastlines strongly alters the structure of extratropical cyclones and their associated precipitation.

Rather than simply discuss the typical distribution of weather about a developing extratropical cyclone, the wide range of cyclone characteristics will be illustrated here. Weather maps and infrared satellite images showing four different extratropical cyclones affecting North America are shown in Figures 7 to 10. The first cyclone (Fig. 7) is a mature extratropical cyclone striking the coast of the northwestern United States, the second (Fig. 8) is a typical cyclone in the central United States, the third cyclone (Fig. 9) is of a type known as an *Alberta clipper*, and the fourth (Fig. 10) is a cyclone redeveloping along the East Coast of the United States.

Wind Typically, the strongest winds and pressure gradients associated with an offshore extratropical cyclone are in the northwest quadrant (southwest in the

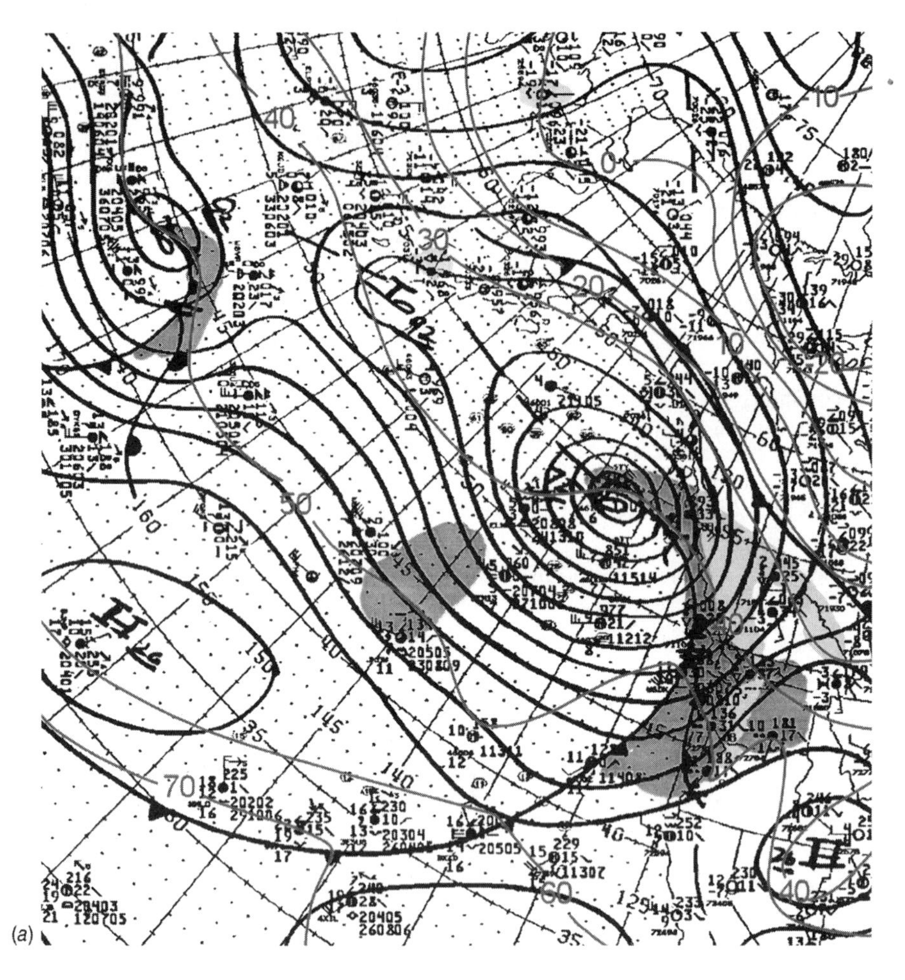

Figure 7 Mature extratropical cyclone in northwest United States, 0000 UTC Feb. 19, 1997. (*a*) Surface weather map with fronts (bold), sea level pressure (solid), temperature (°F, red), and areas of snow (light = light blue; moderate to heavy = dark blue), rain (light = light green; moderate to heavy = dark green), and sleet or freezing rain (light red). Surface data is plotted using conventional station model. Station temperatures and dewpoints are reported in Celsius in Fig. 7 and in Fahrenheit in Figs. 8 to 10. The temperature and precipitation analyses have been added by the author to the operational surface analyses generated by the National Centers for Environmental Prediction. See ftp site for color image.

Southern Hemisphere). Regional exceptions include the Pacific Northwest, where the strongest winds tend to be southerly ahead of the cyclone as it reaches the coast while high pressure remains in place inland (Fig. 7*a*), and the Northeast, where strong winds can often be found between the low-pressure center and the anticyclone in place ahead of it (Fig. 10*a*). Extratropical cyclones do not tend to have a tight

Figure 7 (*continued*) (*b*) Infrared satellite image, 2100 UTC Feb. 18, 1997, remapped to a polar stereographic projection similar to the surface weather map, with coldest temperatures (i.e., highest clouds) enhanced.

inner core like hurricanes, particularly over land, and the radius of maximum winds tends to be 100 to 500 km from the center.

Temperature The overall distribution of temperatures tends to be consistent with the locations of fronts. The warm sector, where the highest surface temperatures are found, is between the warm front and the cold front, generally the southeastern quadrant of the storm. Warmest temperatures are not in the center of the cyclone but tend to increase toward the southeast and south, with the warmest temperature at any given location typically occurring just prior to cold front passage. These patterns are modified by the history of air parcels within the extratropical cyclone. For example, cyclones in the northeastern Pacific tend to have small temperature variations at the surface (Fig. 7*a*), since the low-level air has been exchanging heat and moisture with the relatively uniform oceanic waters. Alberta clippers (Fig. 9*a*) and other cyclones close to the eastern edge of the Rocky Mountains often have their warmest temperatures to the southwest of the low in air that has descended over the mountains and warmed adiabatically. Along the East Coast (Fig. 10), relatively cold air ahead of the low often becomes trapped between the Appalachian Mountains to the west and the warm Atlantic Ocean to the east. The boundary between the cold air and the marine air is known as a coastal front and is analyzed as a trough in Figure 10.

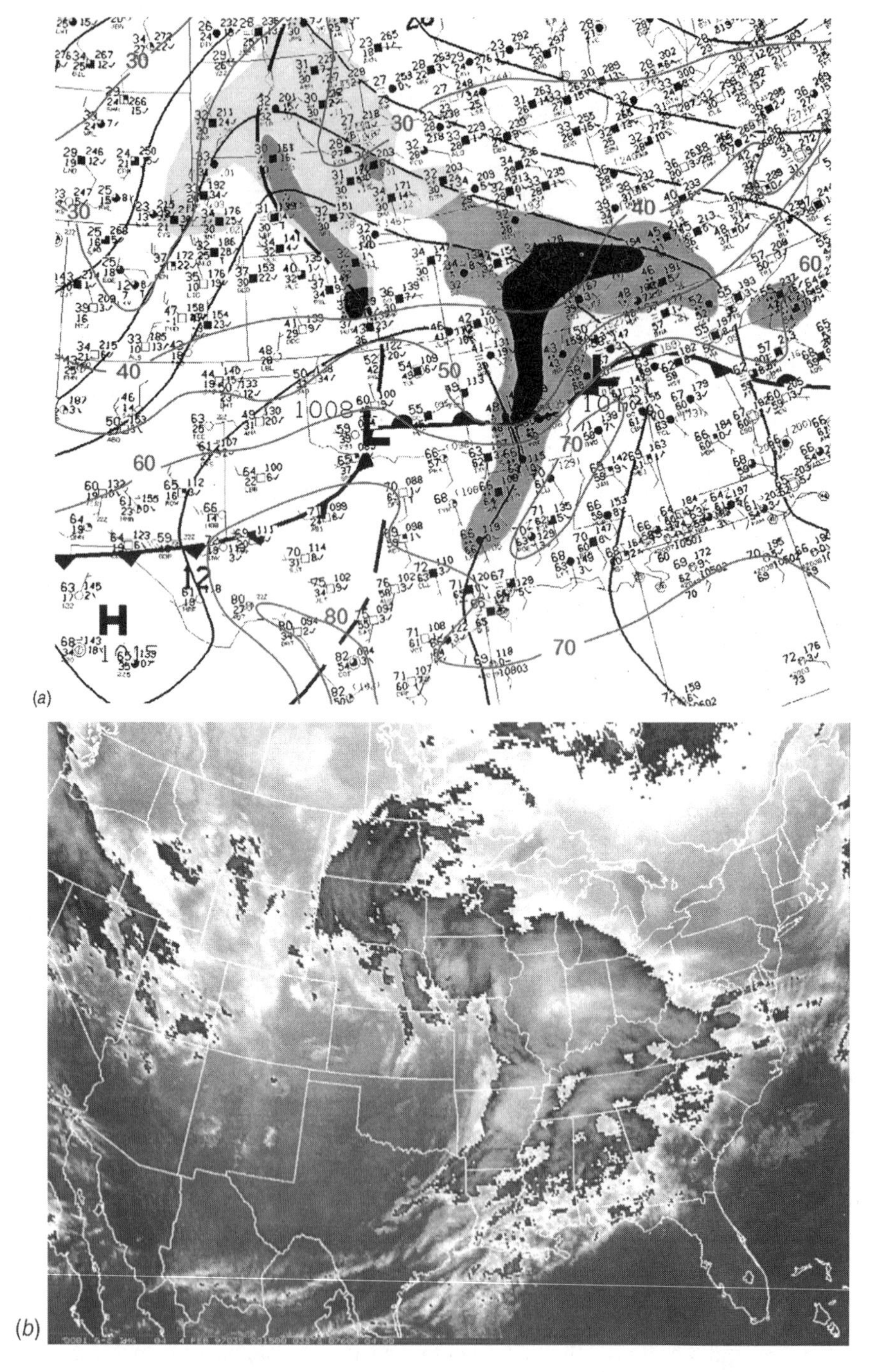

Figure 8 Typical cyclone in central United States, 0000 UTC Feb. 4, 1997. (*a*) Surface weather map. (*b*) Infrared satellite image, 0015 UTC Feb. 4, 1997. See Fig. 7 for details. See ftp site for color image.

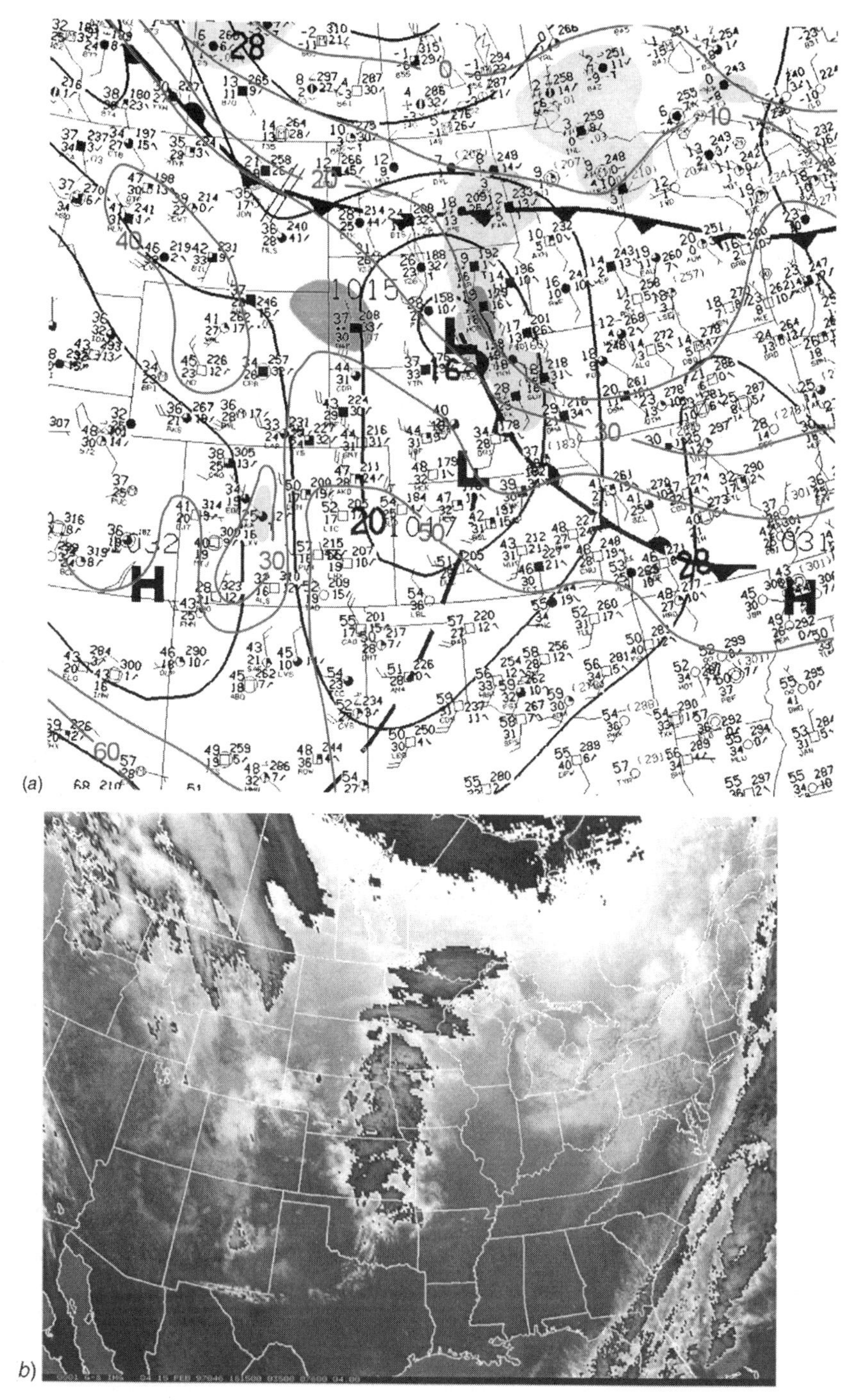

Figure 9 Alberta clipper cyclone, 1800 UTC Feb. 15, 1997. (*a*) Surface weather map. (*b*) Infrared satellite image, 1815 UTC Feb. 15, 1997. See ftp site for color image.

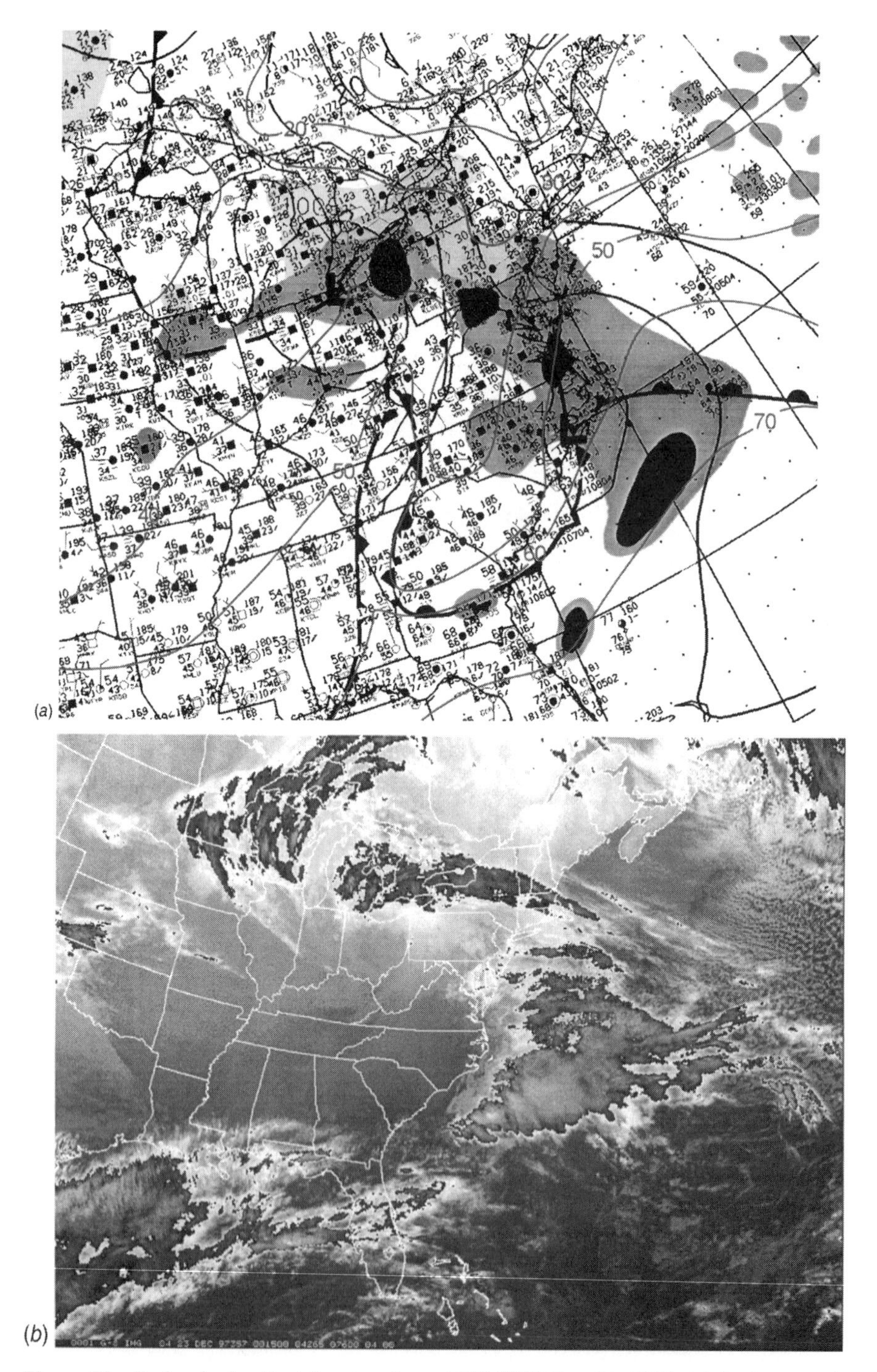

Figure 10 Redeveloping East Coast cyclone, 0000 UTC Dec. 23, 1997. (*a*) Surface weather map. (*b*) Infrared satellite image, 0015 UTC Dec. 23, 1997. See ftp site for color image.

Precipitation The distribution of precipitation is the most variable aspect of extratropical cyclone structure. The typical cyclone (Fig. 2) has extensive precipitation north of the warm front and in the vicinity of and behind the low, with the heaviest precipitation just ahead of the low center. Showers and rain bands tend to be present along the cold front, with scattered or widespread precipitation possible in the warm sector, especially near the low. The rain–snow line tends to be oriented roughly parallel to the track of the low, with a bulge northward near the low center, but its exact position depends on the temperature of the environment within which the extratropical cyclone is forming.

Precipitation requires both a moist air mass and a mechanism for lifting the air mass. The classical distribution of precipitation described above assumes a ready supply of moisture and a pattern of vertical motion determined by the dynamics of the cyclogenesis itself. The patterns can be strongly modified by both the air mass characteristics and the distribution of orography. Over water, where orography is nonexistent and moisture is readily available, precipitation tends to conform to the classical picture. However, the three cyclones over land in Figures 8 to 10 have markedly nonclassical precipitation patterns. The cyclone in Figure 8 possesses a rain band well ahead of the primary surface cold front, because air closer to the cold front has a history of descent over the Rocky Mountains and is relatively dry. Farther north of the cold front, another region of precipitation is being produced by upslope flow as air approaches the higher elevations of the Rocky Mountains and interacts with an inverted trough. Both of these areas of precipitation are common to cyclones in the central United States. The Alberta clipper in Figure 9 possesses no precipitation whatsoever within the warm sector or along the cold front because the entire air mass in the warm sector is dry after descending the mountains or originated as a cold air mass that has not passed over a large body of water and therefore remains dry. The storm along the East Coast in Figure 10 has an adequate supply of moisture from the adjacent Atlantic Ocean but is being affected by the Appalachian Mountains. Precipitation is relatively widespread on the eastern slope of the mountains where the air is warm and moist and is being forced to ascend over the mountains and over the cooler air trapped against them. Within the mountains, the trapped air often creates the necessary low-level temperature inversion for sleet or freezing rain. On the lee side of the mountains, precipitation is lighter.

The observations plotted in Figure 7 are too sparse to show the effects of the orography of the western United States on the distribution of precipitation. However, on scales of tens to thousands of kilometers, the orography modulates the precipitation by mechanically forcing ascent and descent and by altering the structure and tracks of weather systems. Precipitation tends to be high on the windward side of mountains and low on the leeward side. As air passes over successive mountain ranges, progressively less precipitation is produced.

The cloud patterns associated with extratropical cyclones are broadly similar to the precipitation patterns, but the determination of cloud top heights possible with infrared satellite imagery makes satellites a useful tool for diagnosing cyclone structure, particularly over oceans where other forms of data are scarce. Most common is a southwest to northeast oriented (in the Northern Hemisphere) band

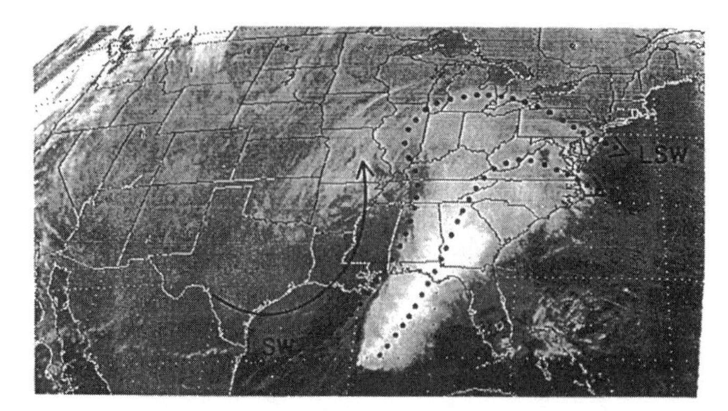

Figure 11 Infrared satellite images showing the location of the warm conveyor belt (dotted streamlines) and the dry airstream (solid streamline). "LSW" marks the leftmost streamline in the warm conveyor belt, which is known as the "limiting streamline" and typically is manifested as a sharp cloud boundary in satellite imagery. (From Carlson, 1991).

of high cloud with a sharp western edge. This cloud band represents the so-called warm conveyor belt, within which air from relatively warm latitudes is carried northward and upward through the storm roughly parallel to the cold front before curving anticyclonically beyond the warm front (Fig. 11). Precipitation beneath this conveyor belt may be convective, stratiform, or nonexistent. If the cyclone is sufficiently strong, the conveyor belt cloud band may curve cyclonically around the approximate position of the low center (as in Fig. 7*b*), or a separate cloud mass known as the cold conveyor belt may be present (as in Fig. 8*b* over the Dakotas and Fig. 10*b* over the Great Lakes and upper Midwest). This conveyor belt consists of air originating north of the warm front and ascending ahead of and on the poleward side of the cyclone before spreading northward or wrapping around the cyclone center. A third upper-tropospheric airstream, consisting of air descending behind the cyclone and curving cyclonically to the northeast, is known as the dry airstream and is indicated in Figure 11.

These cloud structures may be used to diagnose the location and intensity of the extratropical cyclone center, but they are most directly related to the horizontal and vertical wind fields in the upper troposphere and the distribution of precipitation. Because surface pressure features tend to be strongly modified by orography, the structure of the surface cyclone can be inferred reliably from cloud structures only over water. Notice the superficial similarity of Figures 7*b*, 8*b*, and 10*b* despite the wildly different low-level wind and pressure fields.

Statistical Climatology

The climatological distribution of extratropical cyclones reflects the influence of orography (Fig. 12). Cyclones are most common over the oceans between latitudes of 30°N and 60°N. In contrast, extratropical cyclones are rare over high orography.

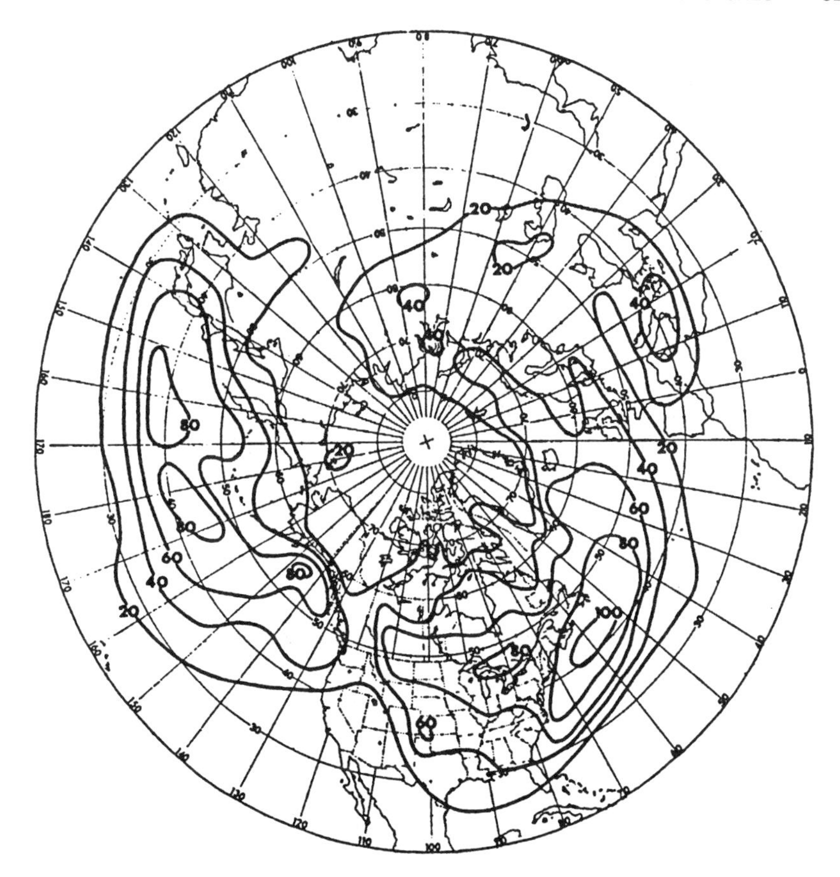

Figure 12 Distribution of total number of cyclones passing through 5×5 latitude–longitude boxes during 20 Januarys, 1958–1977. (From Whittaker and Horn, 1984).

Given the general tendency of cyclones to move toward the east or northeast, one can infer from Figure 12 the general distribution of cyclogenesis (formation and intensification of extratropical cyclones) and cyclolysis (weakening and dissipation of extratropical cyclones). Cyclones tend to form downstream of major mountain barriers such as the Rocky Mountains and the Alps, as well as over the midlatitude oceans. Semipermanent large-scale cyclones may be found in the extreme North Atlantic (the Icelandic low) and North Pacific (the Aleutian low). These lows are periodically reinvigorated as strong extratropical cyclones migrate northwestward across the jet and merge with them. At higher latitudes, small-scale intense extratropical cyclones known as *polar lows* sometimes form near the sea ice boundary. These polar lows are partly driven by deep convection as air masses are destabilized by the underlying warmer ocean surface, and share some of the characteristics of hurricanes.

The midlatitude extratropical cyclones that become particularly intense typically do so by deepening rapidly. Such explosively deepening cyclones, known as *bombs*,

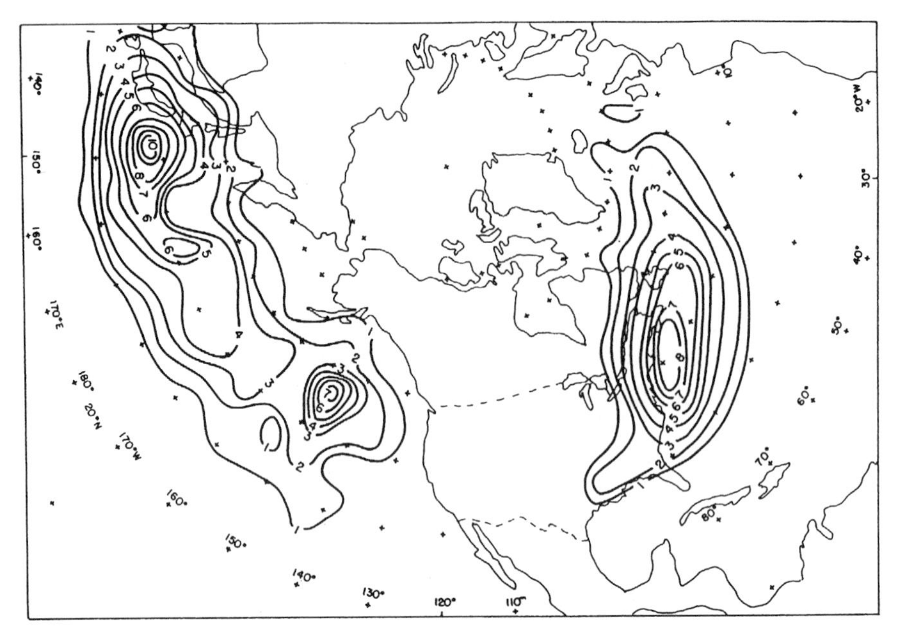

Figure 13 Distribution of the locations at which rapidly deepening extratropical cyclones undergo their most rapid intensification, expressed as total number within each 5 × 5 latitude–longitude box (normalized by latitude) for the period 1976–1982. (From Roebber, 1984).

develop almost exclusively over the oceans, preferentially near the eastern edge of continents (Fig. 13). Conditions favoring explosive deepening include a strong upper-level disturbance, a strong low-level horizontal temperature gradient, and a source of warm, moist air. The Kuroshio current (in the Pacific) and the Gulf Stream (in the Atlantic) help provide the latter two conditions by influencing the distribution of heat and moisture in the overlying atmosphere.

Vertical Structure and Dynamics of Extratropical Cyclones

For development, extratropical cyclones require vertical shear. This means that cyclones will typically develop beneath an upper-tropospheric jet stream. Since balanced vertical shear implies a horizontal temperature gradient, cyclones therefore develop within a large-scale temperature gradient or along a frontal zone. Discussing the vertical structure of extratropical cyclones requires the introduction of the terms upshear, meaning in the direction opposite the vertical shear vector (typically, the direction opposite the upper-tropospheric jet stream), and downshear, meaning in the same direction as the vertical shear vector (typically, downwind relative to the upper troposphere). These concepts are illustrated in Figure 14.

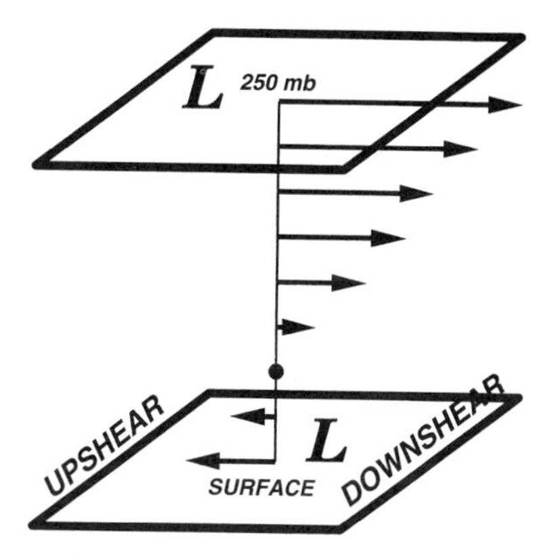

Figure 14 Schematic showing the definition of upshear and downshear for a simple, typical vertical distribution of horizontal wind. The low-pressure system in the figure tilts upshear with height.

An intensifying extratropical cyclone, as defined by the height and pressure distribution, typically tilts upshear with height. This means that the trough in the upper troposphere will be located upshear of the surface cyclone. The temperature distribution is tilted in the opposite sense, so that the warmest temperatures at the surface are just downshear of the surface cyclone position and the warmest temperatures in the upper troposphere are well downshear of the surface cyclone position. The largest variations in pressure and temperature are located in the upper troposphere and at the surface but are substantial at all levels in between. This pressure and temperature distribution is entirely consistent with thermal wind balance, so this particular pressure distribution implies this particular temperature distribution and vice versa. Winds, being in approximate geostrophic balance, vary in tandem with pressure.

The vertical motion is an integral part of the extratropical cyclone structure because without vertical motion cyclones would not intensify. Cyclone intensification can be thought of as an increase in the circulation about the cyclone center and, for large-scale frictionless flow, the fractional rate of increase of total circulation (including the circulation associated with Earth's rotation, which is essentially constant) is proportional to the horizontal convergence of the wind. Mass conservation implies upward motion in the middle troposphere wherever there is low-level convergence (excluding exotic effects associated with a sloping lower boundary), so vertical motion is not just a consequence of cyclogenesis; it is a necessary component of cyclogenesis. Since the stratosphere inhibits vertical motion, upper-tropospheric convergence (and intensification of the upper-tropospheric part of the cyclone) implies midtropospheric downward motion. In a typical intensifying cyclone, there is midtropospheric upward motion (with the associated clouds and precipitation) ahead of and over the surface cyclone, and downward motion beneath and behind the upper-tropospheric trough (Fig. 15).

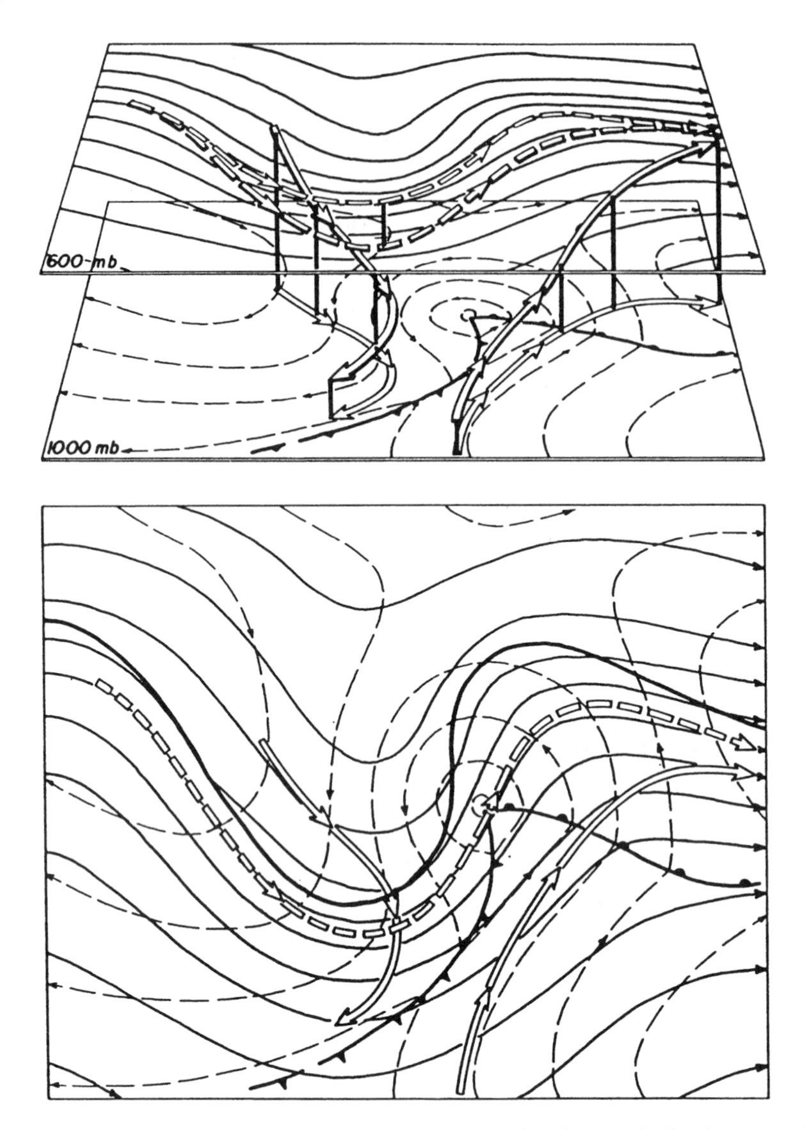

Figure 15 Distribution of upward and downward motion in an idealized extratropical cyclone. Depicted are a perspective view (top) and two-dimensional view (bottom) of height contours at 1000 mb (dashed) and 600 mb (thin solid), surface fronts, and ascending and descending airstreams (wide arrows) with their projections onto the 1000-mb or 600-mb surfaces. (From Palmen and Newton, 1969).

To diagnose and predict extratropical cyclogenesis, meteorologists use a relationship between the structure of the large-scale balanced winds and temperatures and the vertical motion known as the omega equation. The omega equation is a three-dimensional analog to the Sawyer–Eliassen equation and is based on the same dynamical principles of continuous adjustment to preserve nearly balanced flow. The *forcing*, the wind and temperature patterns that imply upward motion, is rather complicated when expressed mathematically, and several versions of the equation, some involving simplifying assumptions, are in use. The most common qualitative application roughly equates upward motion with the sum of two terms, one proportional to the vertical derivative of vorticity advection and the other proportional to the temperature advection. The first term tends to dominate in the upper troposphere and implies upward motion ahead of an upper-level trough and downward motion behind it. The second term tends to dominate in the lower troposphere and implies upward motion ahead of the surface cyclone (where warm advection is typically found) and downward motion behind it. Between the upper-level trough and the surface cyclone, it will be noted that the two terms are of opposite sign, leading to difficulty in inferring the vertical motion field there using this method. With the increasing use of computers, many of these simplifying assumptions are being discarded in favor of explicit mathematical calculation of the vertical motion forcing, but the qualitative interpretation is still essential for relating the vertical motion field to the large-scale features in the lower and upper troposphere.

In contrast to the simultaneous treatment of winds, temperatures, and vertical motion, cyclogenesis can also be understood from a dynamical point of view in terms of potential vorticity. Indeed, the basic theoretical paradigm for cyclogenesis, known as baroclinic instability, requires a potential vorticity gradient aloft opposite in orientation to the surface potential temperature gradient as a necessary condition for instability. Cyclogenesis itself, by any mechanism, requires an increase in the integrated perturbation potential vorticity and/or surface perturbation potential temperature in the vicinity of the surface cyclone. Baroclinic instability accomplishes this by having two waves exist simultaneously, one on the surface potential temperature gradient and the other on the upper-tropospheric potential vorticity gradient; if the two waves are of large enough horizontal scale and the upper one is upshear of the lower one, the circulation associated with each wave causes the other wave to intensify. Other mechanisms for growth that do not involve instability include rearrangement (specifically, compaction) of potential vorticity by horizontal shear, and a decrease in the vertical tilt between the upper and lower perturbations. Using the electrostatics analogy, the instability mechanism involves increasing the electric charge, while the other two mechanisms involve rearranging the electric charge into a compact area. The potential vorticity approach and the omega equation approach are mutually consistent and complementary.

7 UPPER-TROPOSPHERIC JETS AND TROUGHS

A meteorologist looks at the surface weather map to see the current weather but looks at an upper-tropospheric map to understand the current weather. The upper-tropospheric map, generally the 500-mb constant-pressure surface or above, shows the meteorologist the upper-level features associated with the current vertical motion field, the current motion of the upper-level features and related surface features, and the potential for intensification of surface cyclones and anticyclones. By examining the large-scale height field at or near jet stream level, the meteorologist can infer the likely distribution of surface temperatures and the likely path of any storms that might develop.

Jet Streams and Jet Streaks

The jet stream is often described as a band of strong winds in the upper troposphere (8 to 12 km above sea level) encircling the globe. Typically, however, the location of the jet stream is not so simple. There may be two or more jet streams at a given longitude, or the Northern Hemisphere jet stream may originate in the subtropical Atlantic, cross southern Asia, pass over the Pacific Ocean at about 35°N and the Atlantic at 45°N, and eventually weaken over northern Asia. Three examples of the wintertime Northern Hemisphere jet stream and one of the wintertime Southern Hemisphere jet stream are shown in Figure 16.

The first example (Fig. 16*a*) possesses many features common to the Northern Hemisphere wintertime circulation. The strongest jet is over the Pacific Ocean, although it is unusually far south in this example. Consistent with this departure from normal is an unusual northward displacement of the jet over North America; long-term departures from the average circulation tend to have wavelengths of about 8000 km. Over the Atlantic Ocean and over Asia there are at least two jets. The southern jet is called the subtropical jet and the northern jet is called the polar jet.

The second example (Fig. 16*b*) is unusual because the jet over the Atlantic is stronger than the jet over the Pacific. This unusually strong Atlantic jet corresponds to a location where the polar and subtropical jet streams appear to merge. Over the extreme northeastern Atlantic and northern Europe is a northward displacement of the jet stream known as a *block*, because weather disturbances are carried northward around the block rather than from west to east as would be normal. This particular block is known as an omega block because of the resemblance of the jet stream pattern to the Greek letter Ω.

The third example (Fig. 16*c*) occurred during an El Niño year and is also quite unusual. It features a subtropical jet that is essentially continuous around the globe, with a weaker polar jet that is nearly continuous except for its merger with the subtropical jet over the Pacific. Some resemblance may be found to the wintertime jet pattern over the Southern Hemisphere (Fig. 16*d*), which also features two nearly continuous jets. As a general rule, because the topographic and sea surface temperature variations in the Southern Hemisphere are comparatively small, the planetary-scale waves in the jet stream tend to be weaker there. Also (not shown), the jet streams actually tend to be stronger during the summer in the Southern Hemisphere than during the winter because the pole-to-subtropics temperature contrast increases during the summer season. In the Northern Hemisphere, temperatures are much more uniform during the summer and the upper-tropospheric jet speeds are correspondingly weaker.

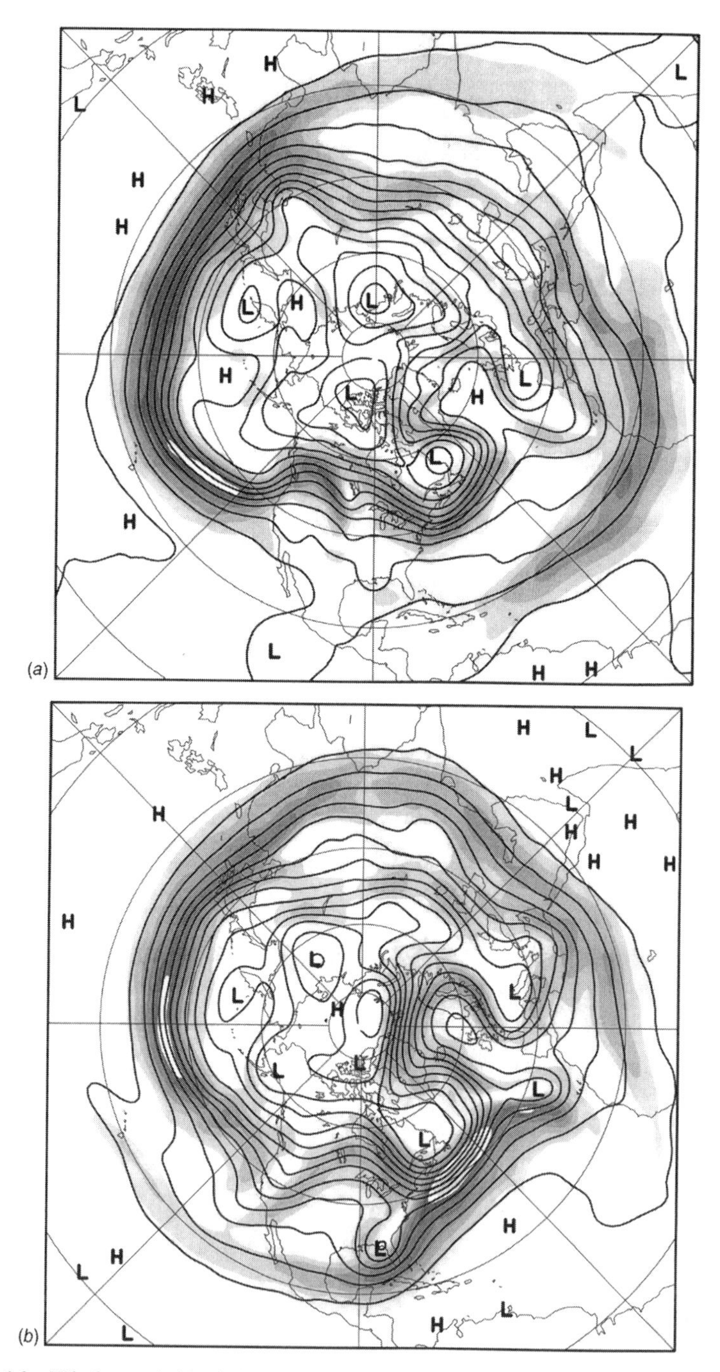

Figure 16 Wind speed (shaded, 10 m/s interval, beginning at 20 m/s) and height (solid, 120-m interval) at 250 mb for four wintertime hemispheres. (*a*) 0000 UTC Jan. 1, 1997. (*b*) 1200 UTC Dec. 15, 1997. (*c*) 0000 UTC Feb. 1, 1998. (*d*) 0000 UTC July 1, 1997.

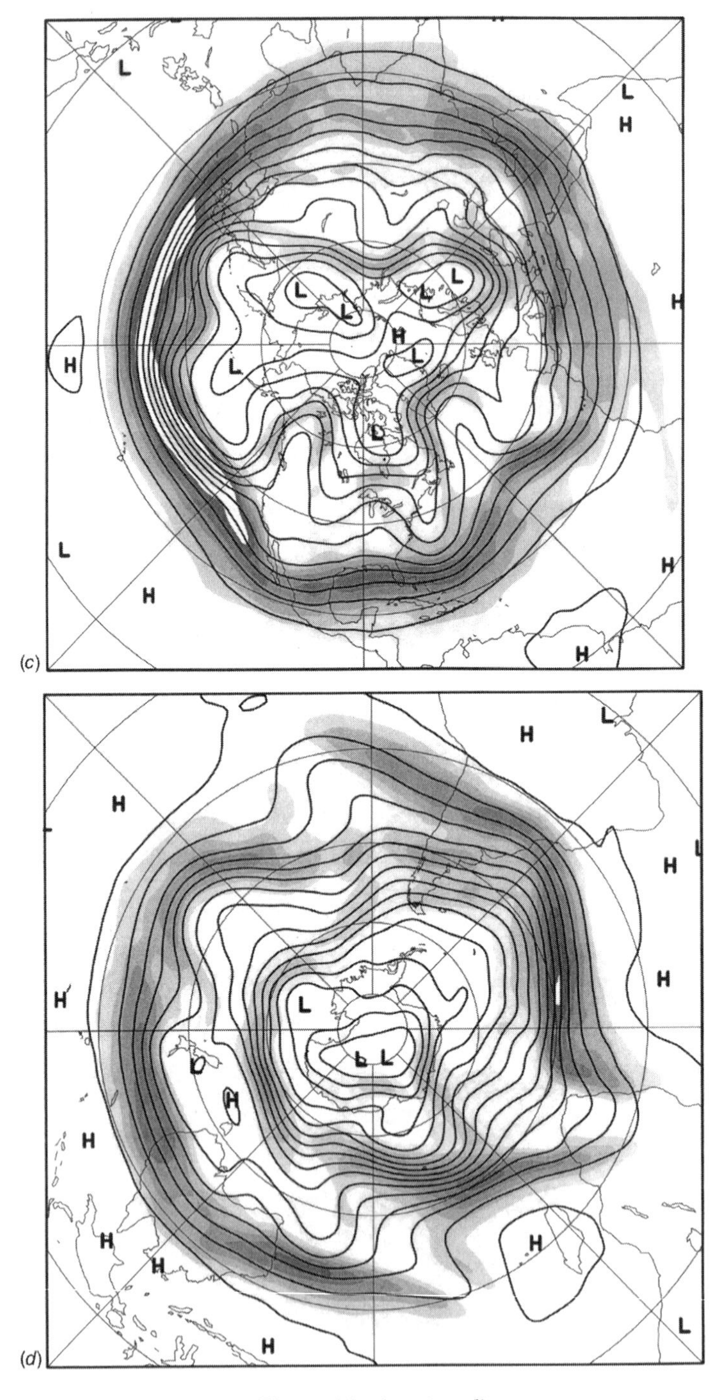

Figure 16 *(continued)*

In a cross section through a jet stream (Fig. 17), the close relationship between the horizontal derivative of temperature and the vertical derivative of wind speed is apparent. Beneath the jet is a strong horizontal temperature gradient, and the wind speed increases with height; above the jet the opposite is true. At jet steam level, the vertical derivative of wind speed is (by definition) zero and, to the extent that the airflow is balanced, the horizontal derivative of temperature is zero too. The height of the tropopause (marked by the transition between a rapid decrease of temperature with height and nearly uniform temperatures with height) also varies across the jet stream, being low on the poleward side and high on the equatorward side. Since potential vorticity is much higher in the stratosphere than the troposphere, a strong potential vorticity gradient would also be found in the vicinity of the jet.

A jet streak is essentially a local wind maximum within a jet stream. Being of smaller scale than a jet stream, air parcels frequently undergo rapid accelerations as they enter and exit a jet streak, implying strong ageostrophic winds and the likelihood of patterns of divergence and convergence. For example, as an air parcel enters a jet streak, it experiences a stronger pressure (or height) gradient than before, and accelerates to its left in the direction of lower heights. This downgradient-directed ageostrophic wind then implies a Coriolis force directed in the original direction of motion, causing the air parcel to speed up and ultimately approach balance with the height gradient. In terms of the ageostrophic wind, one expects a

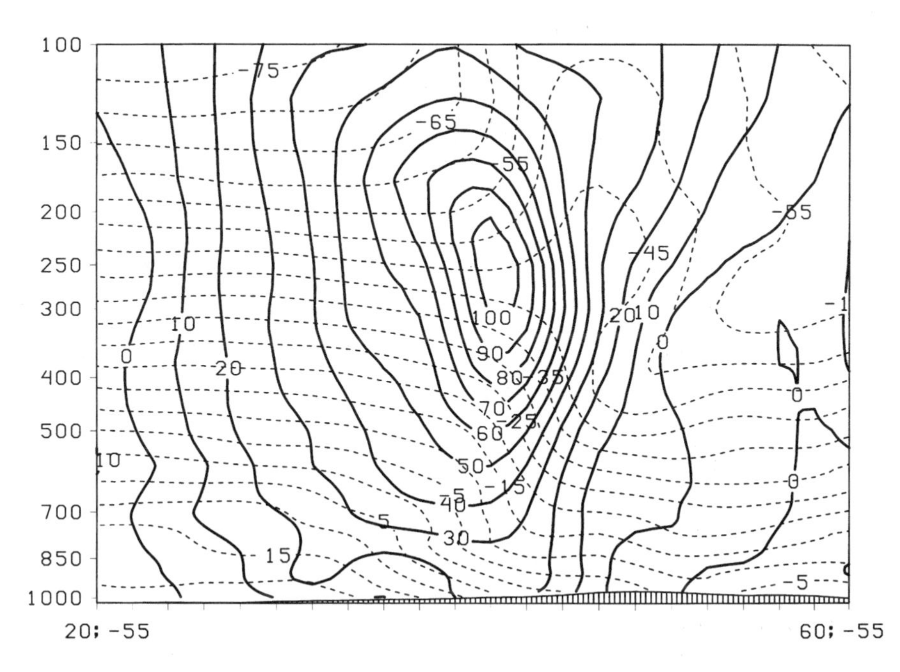

Figure 17 Vertical section taken through the strong Atlantic jet shown in Fig. 16*c* along 55°W. South is to the left. Solid contours show the component of wind into the page (m/s) and dashed contours show temperature (°C).

downgradient ageostrophic wind in the entrance region of a jet streak, and a corresponding upgradient ageostrophic wind in the exit region.

For an idealized straight jet streak (Fig. 18), this ageostrophic wind configuration produces characteristic patterns of convergence and divergence. Divergence is found in what is known as the right entrance region and the left exit region, with convergence in the left entrance region and the right exit region. The pattern is reversed in the Southern Hemisphere. And, since jet streaks are located near tropopause level and the overlying stratosphere inhibits vertical motion, the divergence and convergence imply rising and sinking motion beneath the jet streak in the middle troposphere. Consequently, clouds and precipitation are favored to the left or poleward side of an approaching jet streak and to the right or equatorward side of a receding jet streak.

An along-stream intensification of the height gradient, such as is found at the entrance region of a jet streak, implies confluence of the balanced geostrophic wind.

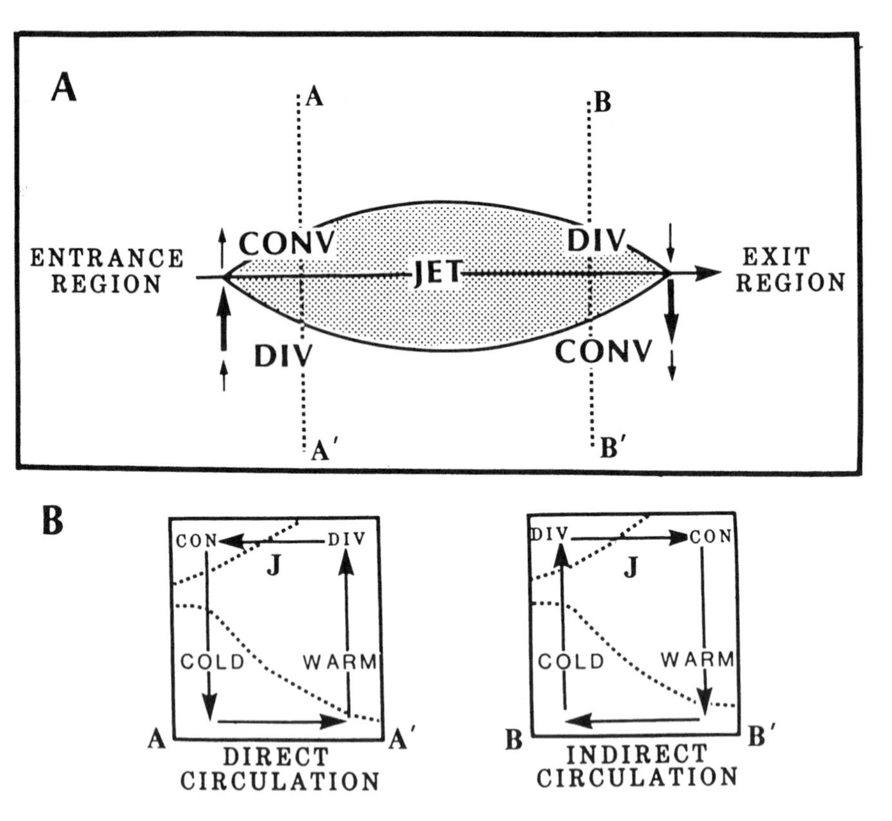

Figure 18 Idealized four-quadrant jet streak. (*a*) Patterns of divergence and convergence (DIV and CONV) and associated transverse ageostrophic winds (arrows) associated with a straight jet streak. (*b*) Vertical sections along A–A′ and B–B′ from (*a*) showing the vertical circulations in the entrance and exit regions. Also plotted in (*b*) are the jet position (J) and two representative isentropes (dotted lines). (From Kocin and Uccellini, 1990).

Beneath the jet streak is a horizontal temperature gradient, very often in the form of an upper-level front. The confluence implies frontogenesis, and the ageostrophic flow at jet streak level constitutes the upper branch of the direct circulation associated with frontogenesis.

Short and Long Waves

Just as straight, concentrated gradients of potential vorticity in the upper troposphere are associated with jets, waves in the potential vorticity gradient are associated with upper-tropospheric troughs and ridges. Indeed, the concentrated gradients serve as a medium along which the waves, known as Rossby waves on the synoptic scale and Rossby–Haurwitz waves on the planetary scale, propagate.

The propagation mechanism is fairly straightforward and can be applied also to the upper and lower waves in a baroclinic extratropical cyclone. Imagine that a straight jet has been perturbed into a periodic wave pattern (Fig. 19). Each equatorward excursion of the potential vorticity contours will be associated with a cyclonic circulation, by the principles outlined in the first part of this chapter, and each poleward excursion will be associated with an anticyclonic circulation. In a frame of reference moving with the large-scale westerlies, these circulations imply an alternating pattern of northerlies and southerlies, which act to redistribute the potential vorticity so as to cause the potential vorticity waves to propagate westward. The speed of propagation is proportional to the strength of the potential vorticity gradient and to the wavelength. As a general rule, the propagation speed of short waves (wavelengths less than 5000 km or so) tends to be much less than the speed of the westerlies in which they are embedded, so short waves (also known as mobile

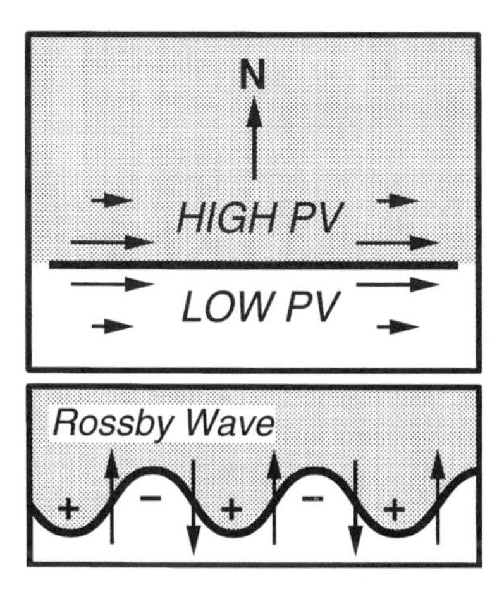

Figure 19 Schematic Rossby wave train. *Top panel:* undisturbed upper-level jet showing winds and PV variations. *Bottom panel:* Rossby waves along upper-level jet with associated north–south component of winds. The pluses and minuses represent cyclonic and anticyclonic potential vorticity anomalies, respectively.

troughs for this reason) tend to move in the same direction and slower speed as the upper-level winds in which they are embedded. Midlatitude long waves may move slowly, be stationary, or even move westward against the large-scale flow. Stationary ridges that persist for one or more weeks are known as blocks or blocking highs because extratropical cyclones follow the jet and migrate poleward around the ridge, and are blocked from entering the affected area.

Rossby waves are dispersive, and the group velocity is oriented opposite the phase velocity, so wave energy tends to propagate downstream at a speed faster than the jet stream winds. This phenomenon is illustrated schematically in Figure 20, which shows an isolated upper-tropospheric trough. The cyclonic winds associated with the trough act to move the trough westward, but at the same time they generate a ridge to the east of the trough. As time goes on, that ridge would then generate a trough farther downstream, and so on. Meanwhile, the original trough would weaken, unless it is involved in cyclogenesis or has some other energy source to allow it to maintain its strength. The process by which one wave triggers another wave downstream is known as downstream development. An example of downstream development in the atmosphere is shown in Figure 21.

Long waves may be formed by interaction between the jet and the underlying large-scale topography, by interaction between the jet and the embedded short waves, and by large-scale latent heating produced by convection. The latter mechanism is the primary means by which equatorial oceanic or atmospheric conditions affect the midlatitude and polar circulations. Equatorial tropospheric heating by convection and the associated upper-tropospheric divergence alters the potential vorticity pattern and initiates a wave train whose energy propagates poleward and eastward. Ideally, the waves follow a great circle path and ultimately return to the equator, but the exact path of the waves is influenced by the local wind and potential vorticity characteristics of the medium within which they are embedded. The wave train persists as long as the equatorial forcing persists.

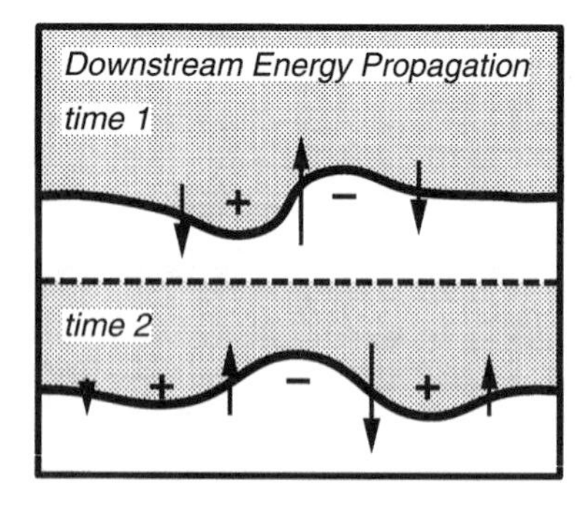

Figure 20 Schematic example of energy propagation associated with an isolated upper-level trough on a jet such as the one depicted in Fig. 19. As the pattern evolves, successive troughs and ridges appear downstream (to the east), while the original wave weakens. While individual crests and troughs propagate upstream, the wave packet as a whole propagates downstream.

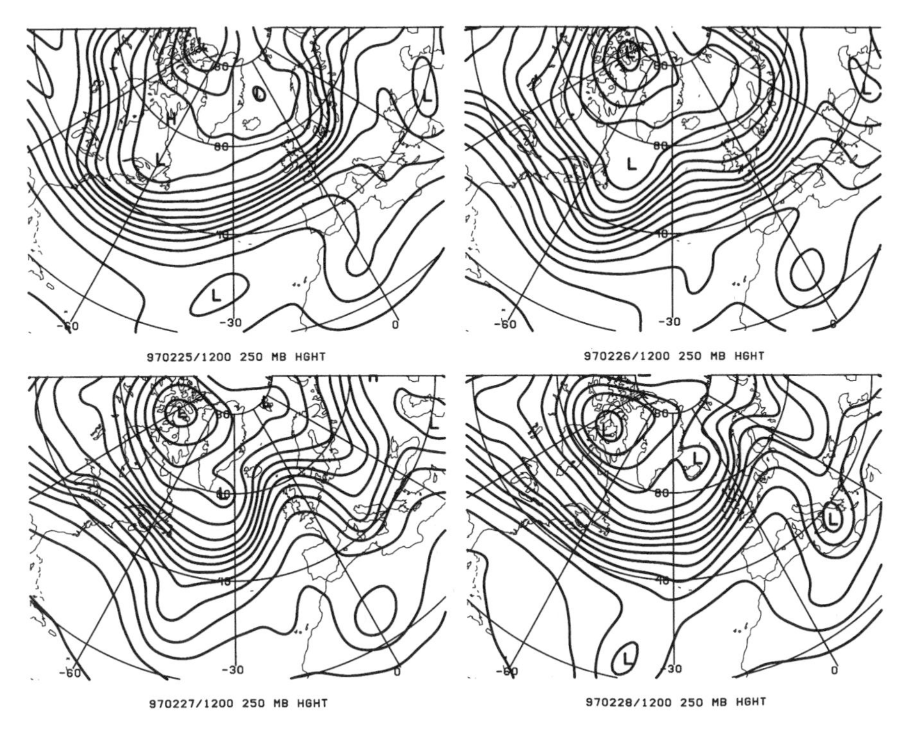

Figure 21 Example of downstream development in the atmosphere, Feb. 25–28, 1997. Contours represent 250-mb height at 120-m intervals. *First panel:* relatively straight jet across North Atlantic. *Second panel:* trough develops over west Atlantic associated with surface cyclogenesis (not shown); ridge begins forming in east Atlantic. *Third panel:* trough begins weakening over central Atlantic; ridge moves eastward and reaches maximum intensity over Great Britain; new trough forming over central Europe. *Fourth panel:* trough nearly gone over central Atlantic, ridge weakening over northern Europe, trough over central Europe attains maximum intensity.

8 WATER VAPOR IMAGERY

The primary satellite tool for detecting upper-tropospheric weather features is the water vapor image. This image is based on a near-infrared wavelength for which water vapor is a strong absorber and emitter, so that the radiation reaching the satellite typically comes from water vapor in the middle to upper troposphere. Where the upper troposphere is particularly dry, the radiation reaching the satellite originates from the lower troposphere, from water vapor at a higher temperature. The variation in intensity of radiation caused by the differing temperatures of the water vapor emitting the radiation is used to infer the distribution of moisture (or lack thereof) in the upper and middle troposphere.

A water vapor image can be used to pinpoint the locations of troughs and jets. Often, the jet stream is marked by a strong upper-level moisture gradient, with low moisture on the cyclonic side of the jet and higher moisture on the anticyclonic side of the jet. This pattern is partly related to the confluence of different air masses as air enters the jet, and partly related to the lower tropopause on the cyclonic side of the jet or in troughs, with comparatively low water vapor content in the stratosphere compared to the troposphere. However, similar features can be produced by ordinary advective processes and vertical motion, so loops of images or maps of upper-level height and wind must be consulted to properly interpret features appearing in water vapor images.

Figure 22 shows the water vapor satellite image for the time of the strong Atlantic jet shown in Figure 16*b*. The water vapor image features a band of moisture crossing Cuba and becoming brighter as it passes alongside the East Coast and moves south of Newfoundland. The sharp left edge of this band is associated with the axis of maximum winds at jet level. By contrast, the polar jet crossing into the Atlantic from Canada does not appear as a distinct feature in this water vapor image; as a general rule, southwesterly jets are easier to identify in water vapor imagery than northwesterly jets. The two upper-level vortices over Florida and off the coast of Spain are also prominent in the water vapor image. The dark area east of Newfoundland, north of the subtropical jet, indicates the position of an upper-level trough and is caused by the extremely low tropopause there. The trough appears to be initiating cyclogenesis, judging from the developing water vapor pattern just to its east (compare Fig. 22 to Fig. 10*b*). The much larger dark areas over the eastern Caribbean are associated

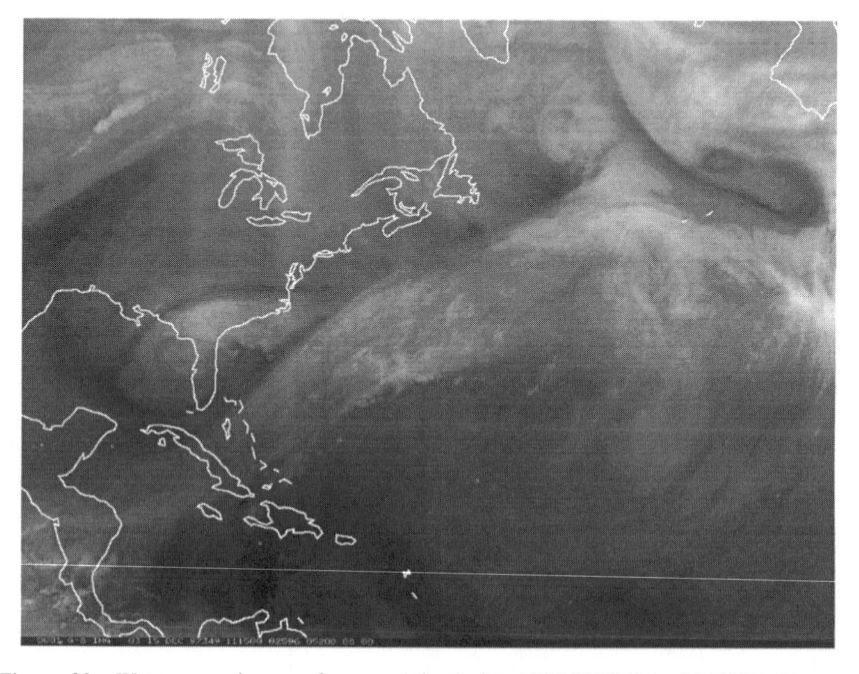

Figure 22 Water vapor image of strong Atlantic jet, 1115 UTC Dec. 15, 1997. Compare with Figs. 16*b* and 17.

with dry, subsiding air within the subtropics and are of no particular dynamical significance.

REFERENCES

Carlson, T. N. (1991). *Mid-Latitude Weather Systems*, London, HarperCollins.

Kocin, P. J., and L. W. Uccellini (1990). *Snowstorms Along the Northeastern Coast of the United States: 1955 to 1985*, *Meteor. Monogr.*, **22**, No. 44, Boston, American Meteorological Society, 280pp.

Nese, J. M., and L. M. Grenci, (1998). *A World of Weather: Fundamentals of Meteorology*, IA, Kendall/Hunt.

Roebber, P. J. (1984). Statistical analysis and updated climatology of explosive cyclones, *Mon. Wea. Rev.*, **112**, 1577–1489.

Shapiro, M. A. (1991). Frontogenesis and geostrophically forced secondary circulations in the vicinity of jet stream— frontal zone systems, *J. Atmos. Sci.*, **38**, 954–973.

Shapiro, M. A., and D. Keyser (1990). Fronts, jet streams, and the tropopause, *Extratropical Cyclones: The Erik Palmen Memorial Volume* (C. Newton and E. Holopainen, eds.), Boston, American Meterological Society, 167–191.

Thorpe, A. J. (1986). Synoptic scale disturbances with circular symmetry, *Mon. Wea. Rev.*, **114**, 1384–1389.

Whittaker, L. M., and L. H. Horn (1984). Northern hemisphere extratropical cyclone activity for four mid-season months, *J. Clim*, **4**, 297–310.

CHAPTER 27

WINTER WEATHER SYSTEMS

JOHN GYAKUM

1 INTRODUCTION

Winter weather in the extratropical latitudes is defined by the individual and collective contributions of cold air, wind, and precipitation. Each of these meteorological phenomena is determined by a combination of synoptic-scale weather systems and topography.

Economic impacts of winter weather can be substantial. Among the 48 weather disasters that cost at least \$1 billion in the United States during the period 1980–2000, 11 were winter weather events (NOAA, 2000). These included 4 flooding/heavy-rain events, three coastal cyclones, two freezes in Florida, and two ice storms.

This chapter focuses on the dynamics of winter weather systems and how topographic features affect these systems.

The weather maps used in this chapter show fields of sea level pressure (SLP), 500–1000 hPa thickness, and their anomalies from a climatological state. Each field is taken from the National Centers for Environmental Prediction (NCEP) global reanalysis of meteorological data (Kalnay et al., 1996). The thickness fields display the 500–1000 hPa layer mean virtual temperature. An incremental change of 60 m of 500–1000 hPa thickness represents an approximate temperature change of 3°C. The anomalies of this quantity are also displayed on these maps. We define an anomaly as the difference between the actual 500–1000 hPa thickness and the appropriate monthly climatological mean. This mean value is computed for the period 1963–1995. Therefore, a thickness anomaly field quantifies how much colder or warmer a particular region is, compared with its expected value for the season.

Handbook of Weather, Climate, and Water: Dynamics, Climate, Physical Meteorology, Weather Systems, and Measurements, Edited by Thomas D. Potter and Bradley R. Colman.
ISBN 0-471-21490-6

2 COLD AIR

Winter is characterized by relatively cold temperatures in extratropical latitudes. These colder temperatures are a manifestation of equatorward displacements of the zonal jet stream and cold-surface anticyclones. One especially interesting component of winter weather is the production of cold-air outbreaks. These outbreaks are associated with anomalously cold temperatures, where an anomaly is defined as the difference between the actual temperature and the long-term climatological mean.

For a cold-air outbreak to have a large impact, cold air must travel rapidly equatorward from its source region in the higher latitudes of continents. The rapid displacement is essential for a strong cold anomaly to occur, because cold air will warm at lower latitudes owing to the warmer ground or ocean surface and to stronger solar insolation. A favorable upper-tropospheric pattern includes an anomalously strong wave in the jet stream with the ridge and trough being amplified relative to the mean climatological state.

A severe cold-air outbreak occurred during December 1983 with damage to the Florida citrus crop exceeding \$2 billion. The structure of the highly amplified 1000–500 hPa thickness field (Fig. 1) included a strong ridge in Alaska with an anomaly

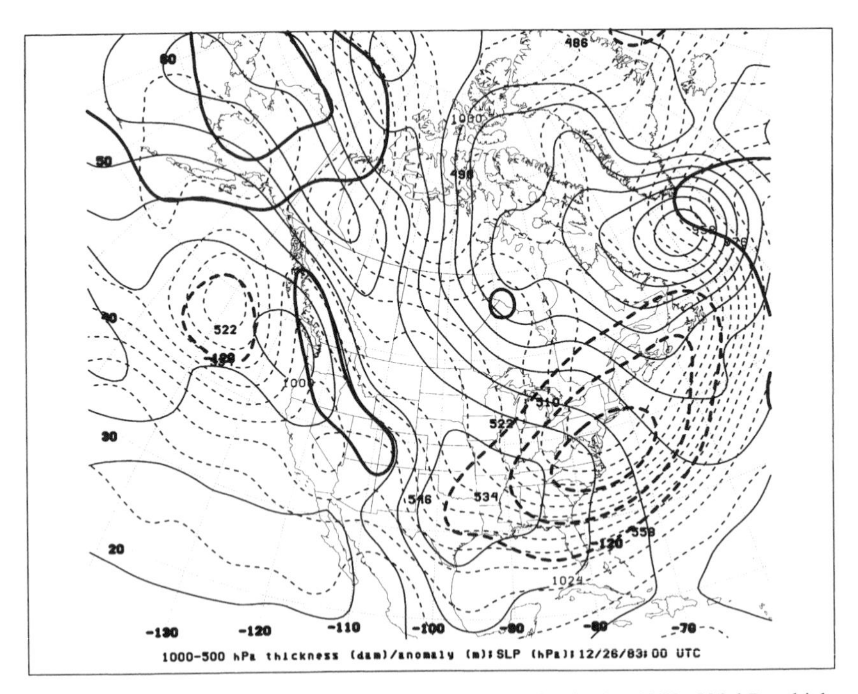

Figure 1 Sea-level pressure (light solid, interval of 8 hPa), 1000–500-hPa thickness (light dashed, interval of 6 dam), and 1000–500-hPa thickness anomaly (heavy solid/dashed dashed for positive/negative with interval of 120 m) for 0000 Universal Time Coordinated (UTC), December 26, 1983.

exceeding +24 dam and a cold anomaly (−44 dam) in the eastern states. The structure shown by Figure 1 is an extreme case of a tropo persisted for much of December 1983 (e.g., Fig. 8*b* of Quiroz, 19

The qualitative structure of Figure 1 illustrates the necessary anc tions for a significant cold outbreak. The tropospheric ridging in the thickness field develops as a result of strong surface cyclogenesis in the North Pacific, as poleward flow to its east advects warm air into Alaska. This strong ridging in western North America provides anomalously strong equatorward steering flow for cold-surface anticyclones. A cold-surface ridge axis extends southeastward from Alaska to the Mexican border east of the Rockies, which provides a natural barrier from the moderating influence of the North Pacific. The combination of the surface anticyclones and a deep cyclone centered south of Greenland provides a northerly low-level geostrophic flow extending from the Arctic to the Caribbean Sea.

Topographic features, such as the Rockies, are often important to the evolution of the cold-air outbreak. A relatively warm sea surface temperature (SST) in the vicinity of the tropospheric thickness ridge would enhance its amplitude. A snow cover over the eastern part of North America acts to amplify the thickness trough.

As in North America, cold-air outbreaks in other regions of the globe are characterized by an upper-level pattern that favors a cold continental surface anticyclone traveling into the affected region. Significant cold-air outbreaks in western Europe, for example, are characterized by southwestward-traveling anticyclones from the continental regions of Russia.

3 WIND

High winds during the winter are often associated with blizzard conditions. The official definition of a blizzard includes large amounts of falling *or* blowing snow, with wind speeds greater than 56 km/h *and* visibilities less than 0.4 km for a least 3 h. Blizzards are substantial threats to North American east coastal regions because intense surface cyclones and their accompanying strong horizontal pressure gradients occur preferentially in these regions. Coastal residents are therefore especially vulnerable to the effects of these deep surface lows, which may include hurricane-force winds and storm surges comparable to those associated with land-falling hurricanes. Other locations in North America that experience blizzard conditions include the high-plains region extending east of the Rocky Mountains. Blizzard conditions are more likely to occur in these areas in association with high winds and blowing (with small amounts of falling snow) over relatively smooth terrain.

North American east coast cyclones develop in response to a middle-upper tropospheric mobile trough that typically travels along a northwesterly upper-level flow. The initial development of the coastal surface low may be in a preexisting zone of coastal frontogenesis that separates warm, moist marine air from cold, dry air that is dammed east of the Appalachians. The interaction of this lower-tropospheric cyclonic potential vorticity anomaly with its upper-level mobile counterpart defines the cyclogenesis.

1200 UTC 12 MARCH 1993

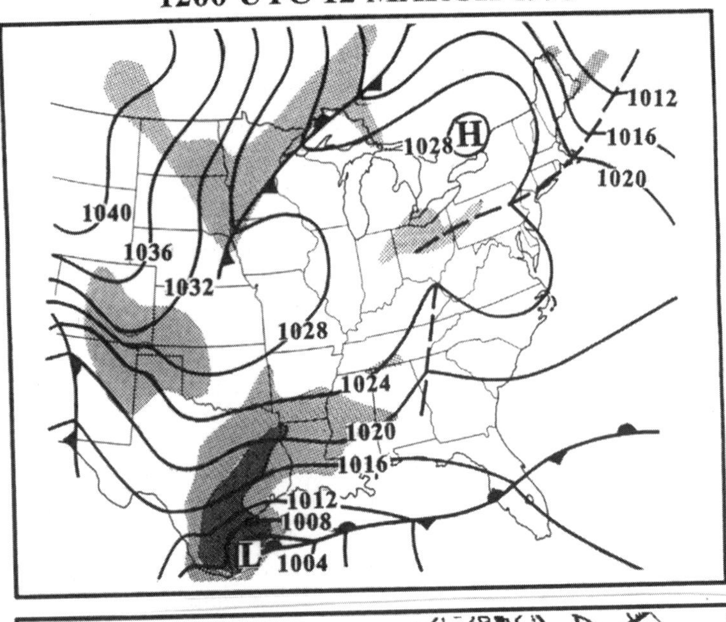

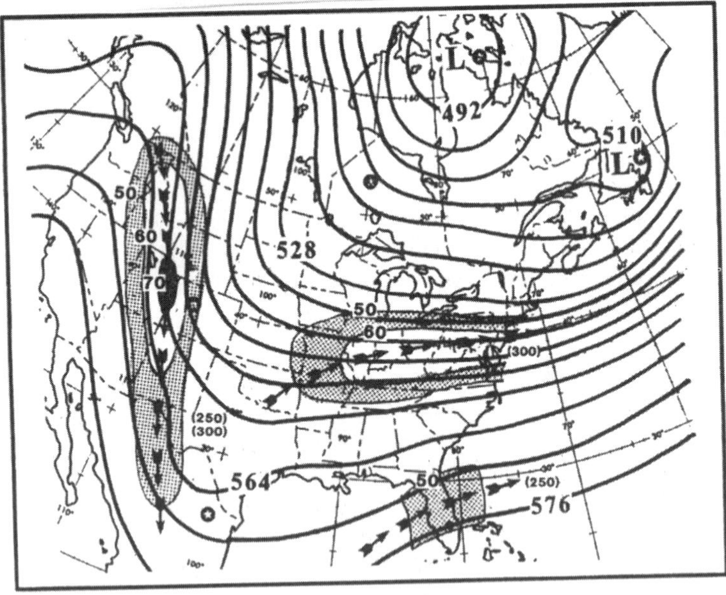

Figure 2 (*a*) Sea-level pressure (solid, interval of 4 hPa) for 1200 UTC, March 12, 1993. (*b*) Heights (solid, interval of 6 dam) for 1200 UTC, March 12, 1993. (From Kocin et al., 1995; reprinted by permission of AMS.)

An extreme example of North American cyclogenesis occurred in March 1993. The stage was set with a cold-air outbreak covering most of eastern North America southward into the Gulf of Mexico (Fig. 2*a*). Rapid cyclogenesis in the Gulf of Mexico occurred in response to multiple interactions with upper-tropospheric mobile troughs approaching from the west and northwest (Figs. 2*b* and 3). Additional intensification was associated with the latent heat release in embedded thunderstorms. Explosive intensification of 30 hPa during the next 24 h produced an unprecedented 968-hPa cyclone in central Georgia by 1200 Universal Time Coordinated (UTC), March 13 (Fig. 4*a*). At this time, the unusually large horizontal scale of the cyclonic circulation extended from the Caribbean Sea northward into New England and westward to the Mississippi River. The system continued to deepen to its minimum pressure of 962 hPa during the next 12 h, in a favorable location for interactions with the upper trough (Fig. 4*b*).

This Storm of the Century produced record low pressures at several locations along the North American coast, blizzard conditions to the north and west of its track, and a storm surge, comparable to that seen in hurricanes, along the Gulf of Mexico coast of 3 to 4 m. Its development was triggered by multiple troughs aloft, but the influence of topography was also substantial. First, the presence of the anomalously warm Gulf of Mexico in association with a cold-air outbreak destabilized the air. This destabilization enhanced the cyclogenesis by enhancing the interactions with upper-level troughs. Second, the warming and moistening of the air mass provided a favorable environment for the heavy thunderstorm activity occurring near the cyclone center. This enhanced thunderstorm activity aided the cyclogenesis through the addition of latent heat of condensation.

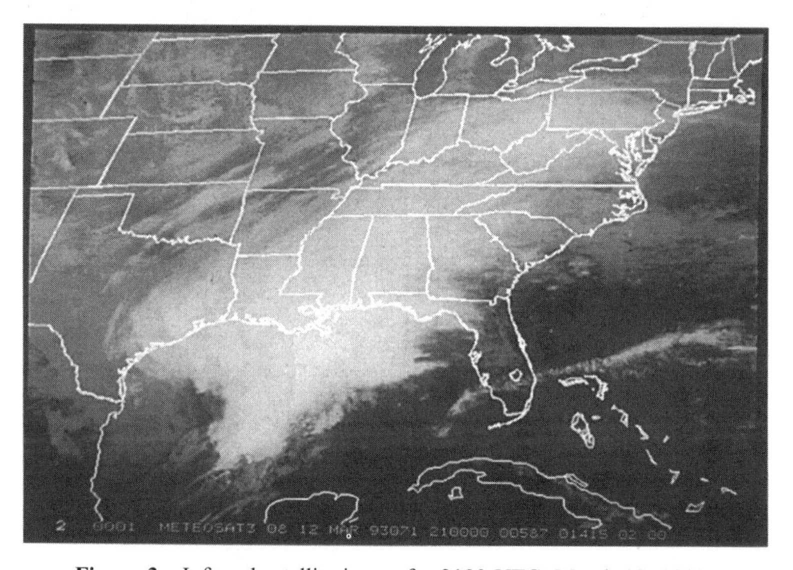

Figure 3 Infrared satellite image for 2100 UTC, March 12, 1993.

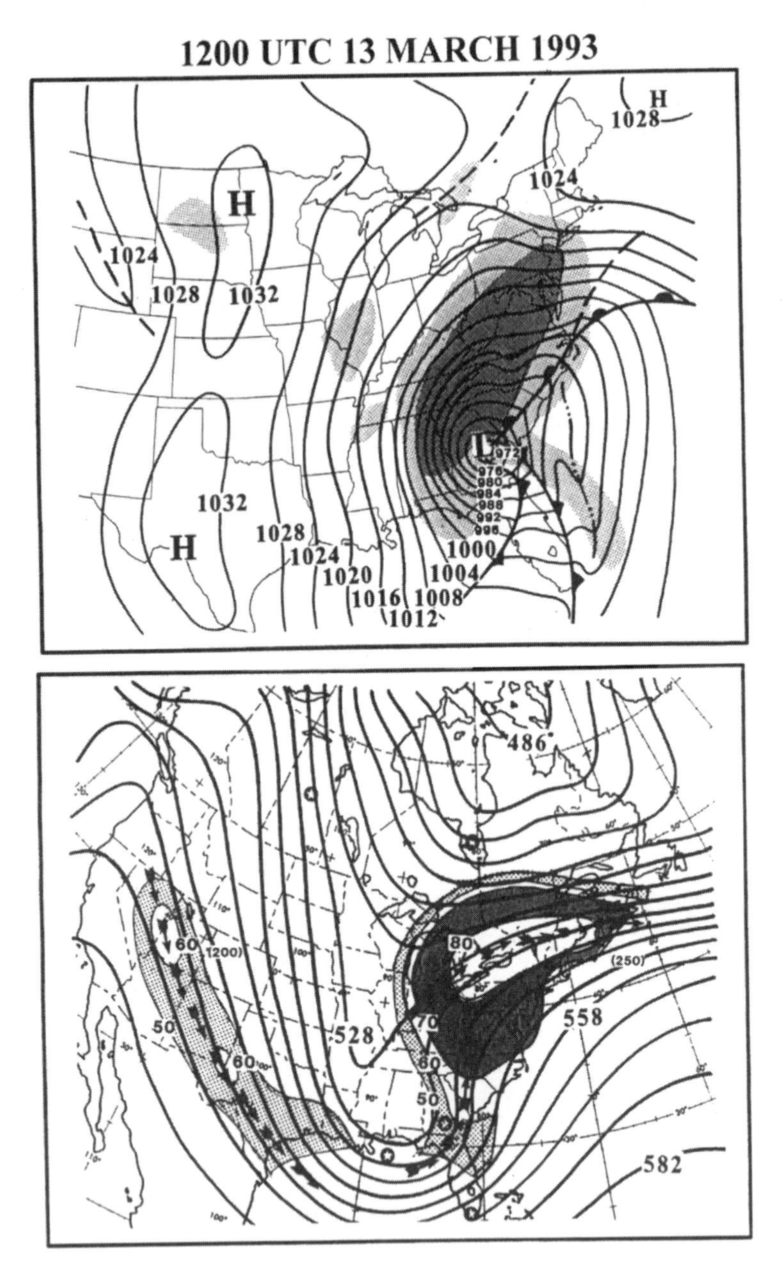

Figure 4 As for Fig. 2, except for 1200 UTC, March 13, 1993. (From Kocin et al., 1995; reprinted by permission of AMS).

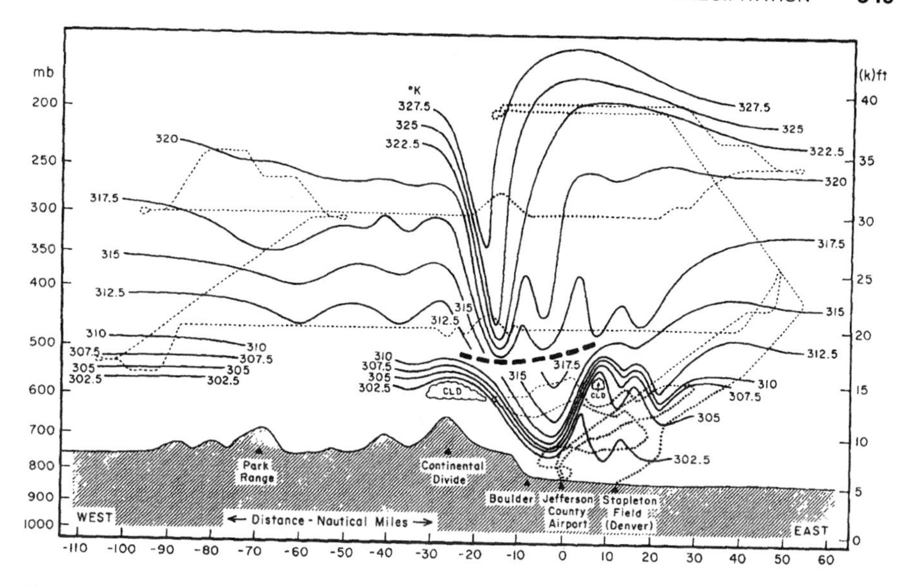

Figure 5 Cross section of potential temperature (K) along an east–west line through Boulder, as obtained from analysis of aircraft data on January 11, 1972. The isentropes are excellent indicators of streamlines for steady, adiabatic flow. Flight tracks are indicated by light dashed lines, with the heavy dashed line separating the tracks of the different research aircraft. (From Klemp and Lilly, 1975; reprinted by permission of AMS).

Strong winds also can occur during the cold season in the lee of mountain ranges. These wind storms, typically termed chinook winds, may contain hurricane-force wind speeds and produce extensive property damage. Klemp and Lilly (1975) studied a case that occurred in Boulder, Colorado, in which the phenomenon is found to be the surface signal of a standing gravity wave with a wavelength of ~60 km and an upwind temperature inversion approximately 60 hPa above the Continental Divide (Fig. 5). This extreme event was associated with surface wind gusts of 50 m/s and extensive property damage. The vertical cross section of wind speeds (Fig. 6) shows the core of extreme wind located 20 km to the east of the Divide. Generally, similar events with gusts in excess of hurricane force occur each winter. Such events occur when the westerlies are strong, the inversion above the mountains is also strong, and in the lee of where the westerlies arrive at the Divide unimpeded by upstream mountains.

4 PRECIPITATION

One of the more common synoptic-scale structures associated with cold-season precipitation along the west coast of North America is the so-called pineapple express. The synoptic-scale structure of this event is similar to that observed for most substantive cold-season rainfall events along the west coast of North America.

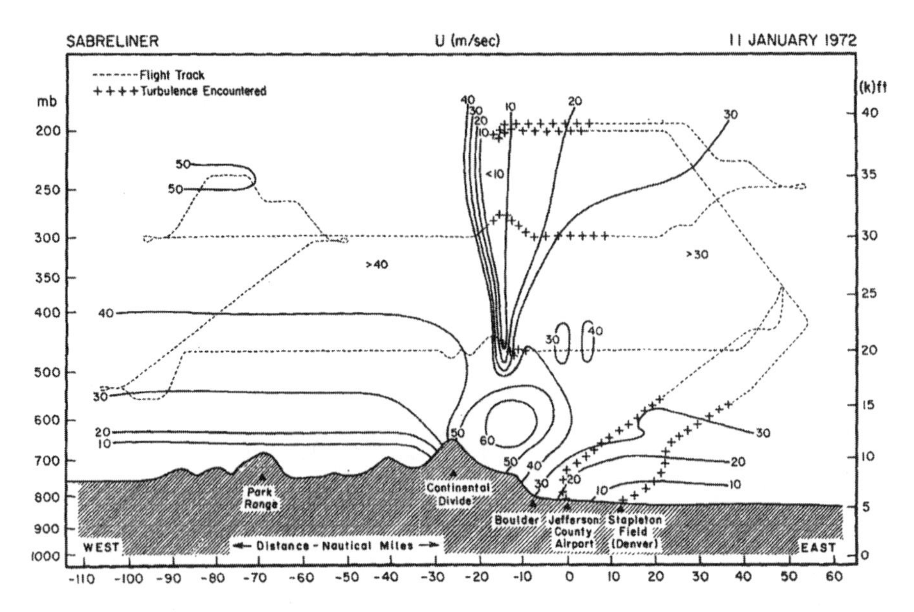

Figure 6 Horizontal velocity (m/s) contours along the cross section as in Fig. 5 and derived from research aircraft observations. (From Klemp and Lilly, 1975; reprinted by permission of AMS).

The dominant lifting mechanism is the effect of cyclonic thermal vorticity advection, in addition to anomalously large precipitable water amounts. Large amounts of water vapor are transported into the region from subtropical maritime regions near Hawaii by anomalously strong southwesterly tropospheric flow. These events can produce large amounts of precipitation, induce flooding and landslides, and generally produce great economic damage.

The meteorological effects of the pineapple express are amplified by the existence of the north–south oriented mountain ranges that exist along the west coastal regions of North America. The orographic lifting and cooling of the eastward-traveling moisture-laden subtropical air enhances the precipitation rates on the windward slopes of these mountains. The largest measured seasonal snowfalls in the world occur in these areas. Washington State's Paradise Ranger Station, at 1654 m on Mount Rainier in the Cascades, averages 1733 cm (682.4 in.) of snowfall each year. The Mount Baker ski resort, at 1286 m in the north Cascades of Washington, recorded a single-season total snowfall of 2896 cm (1140 in.) during 1998–1999, establishing a new world record [previously held by Paradise Ranger Station with 2850 cm (1122 in.) during 1971–1972].

The leeward regions to the east of the Cascades, in contrast, receive relatively small amounts of precipitation. This rain shadow effect is due to the compressional warming of the easterly traveling air as it subsides on the lee slopes of the mountains. A comparison of annual precipitation amounts illustrates this effect. Tacoma, Washington, at an elevation of 89 m, receives 939.8 mm (37.0 in.), whereas Coulee

Dam, to the east of the Cascades at an elevation of 518 m, receives only 274.3 mm (10.8 in.).

An example of the synoptic pattern relating to events in western Washington and Oregon is shown in Fig. 7 (Lackmann and Gyakum, 1999). The 46-case composite corresponds to days in which at least 12.5 mm of precipitation falls on each of the four stations shown in Fig. 7*d*, and in which the maximum temperature exceeded 10°C for each of the lowland stations and 5°C for Stampede Pass. The southwesterly geostrophic flow in the region of the precipitation exists from the surface to 500 hPa. The composite surface low in the Gulf of Alaska has an upper-level counterpart with a planetary-scale trough that extends throughout the Pacific Ocean. Figure 8 shows anomalously strong 500-hPa southwesterly flow prior, during, and after the event. This strong flow extends from the subtropical regions of the Pacific to Oregon, Washington, and British Columbia.

An extreme case of heavy rains along the western North American coast in late 1996 was part of a 2-week regime in which 250 to 1000 mm of rain fell, with economic damage of up to $3 billion. The large-scale SLP and 1000–500-hPa layer mean temperature fields for 1200 UTC, December 31, 1996 (Fig. 9), appear very similar to those shown in Figure 7, with a deep planetary-scale trough covering much of the south-central North Pacific Basin. The coldest tropospheric anomaly corresponds to the stratospheric intrusion of high-PV (potential vorticity) air seen as the dark region extending southwestward from the comma cloud of Figure 10. This long plume of large PV provides forcing for the 960-hPa surface cyclone seen in Figure 9. As pointed out by Lackmann and Gyakum (1999), these cyclonic disturbances are responsible for transporting the subtropical water vapor from marine areas near Hawaii directly toward the west coast of North America.

Snowfall may occur in ascending regions of surface cyclones and regions of frontogenesis. However, especially large snows are strongly affected by topography. One example of such a strong topographic influence on snowfall is that of the lake effect snowfalls. The most prominent region of North America that is affected by these snowfalls is that of the Great Lakes. However, any comparably large body of water in extratropical latitudes exerts a similar influence on the surrounding region. The Great Salt Lake in Utah and James Bay in Quebec are two other examples of North American lake effect regions.

Figure 11, derived from Eichenlaub's (1979) work and in turn printed in the review article by Niziol et al. (1995), shows the mean annual snowfall climatology for the Great Lakes. The amounts vary from less than 50 cm in the southern region to nearly 500 cm in regions east and south of Lake Superior. The existence of both the Great Lakes and orography controls the variability of these climatological snowfalls. The lake effect snowfall of December 17–19, 1985, which affected the Buffalo, New York, area, was typical of the associated large-scale conditions. The environment was characterized by a deep surface cyclone that traveled well to the north and east of the affected region (Fig. 12), in this case a 968-hPa low located between Greenland and Labrador. This low, combined with a 1040 surface anticyclone over the Dakotas, advected bitterly cold air directly over the Great Lakes. The 1000–500-hPa mean temperature over Lakes Superior, Michigan, and Huron was about 20°C colder

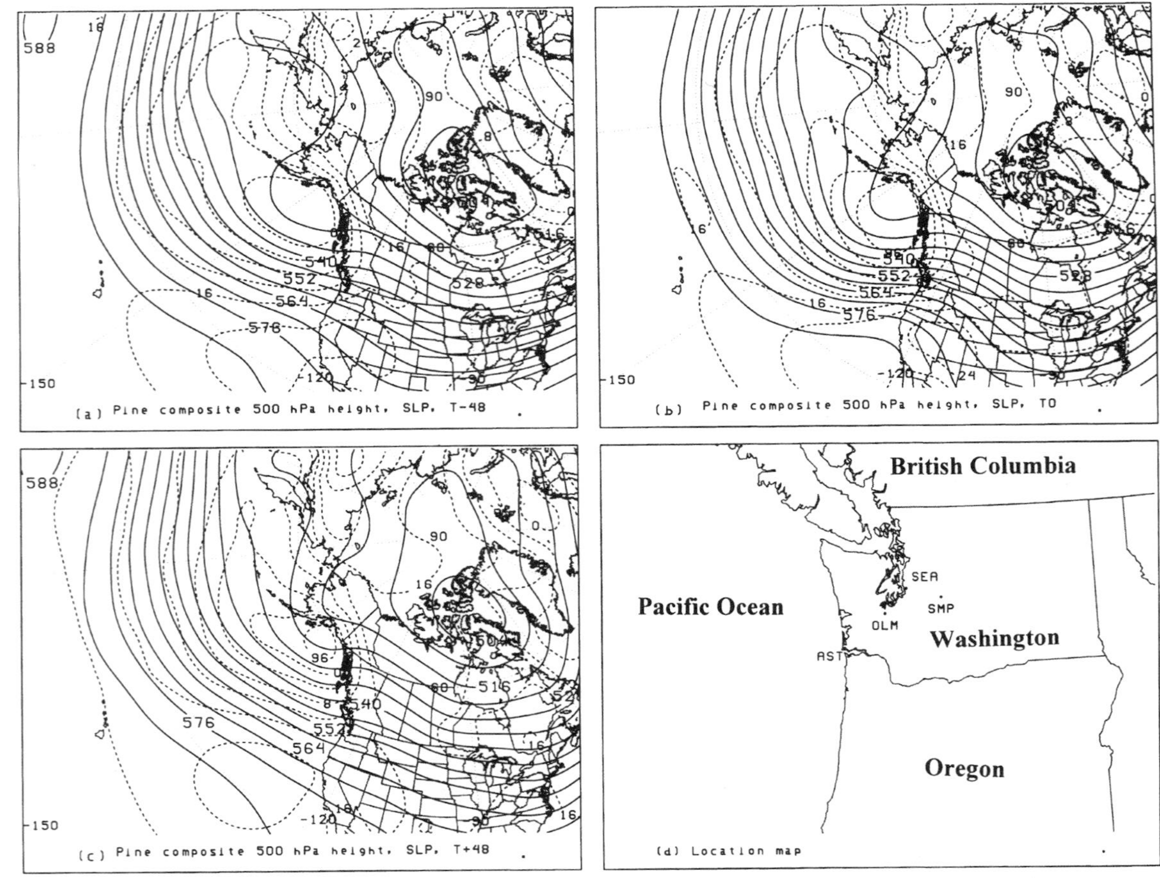

Figure 7 Pineapple Express composite of 500-hPa height (solid, interval of 6 dam) and sea-level pressure (dashed, interval of 4 hPa) of 46 cold-season cases for (*a*) 48 h prior, (*b*) the day of, and (*c*) 48 h after the event. Geographical references are displayed in (*d*), including Astoria, OR (AST); Olympia, WA (OLM); Seattle-Tacoma airport, WA (SEA); and Stampede Pass, WA (SMP). (From Lackmann and Gyakum, 1999; reprinted by permission of AMS.)

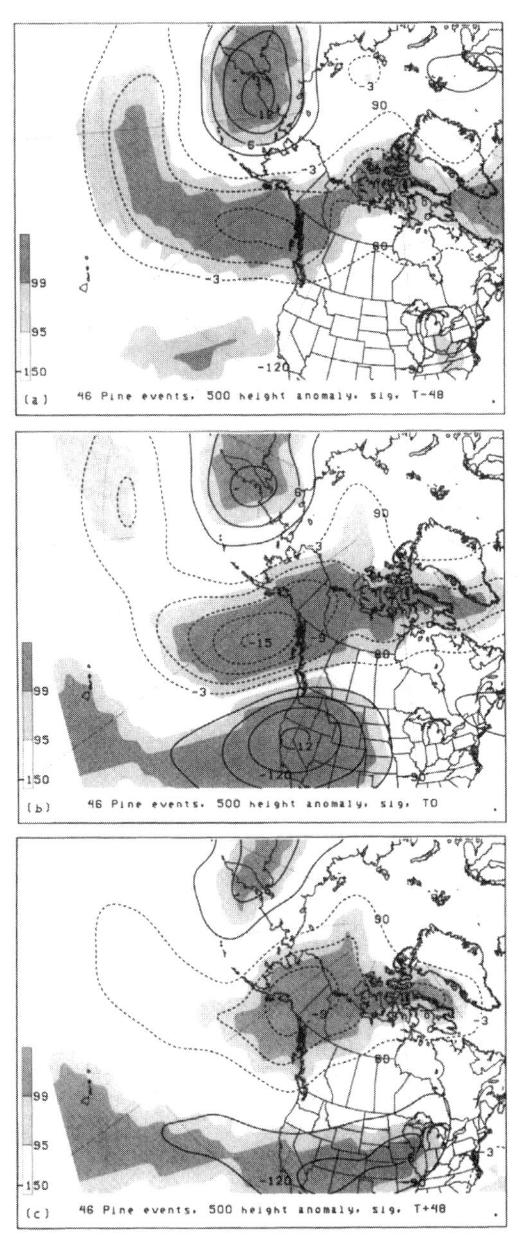

Figure 8 Composite 500-hPa geopotential height anomaly [contour interval 3 dam, positive (negative) values solid (dashed), zero contour omitted] and statistical significance determined from a two-sided Student's *t*-test (shading intervals correspond to 95 and 99% confidence limits as shown in legend at left of panels): (*a*) 48 h prior, (*b*) the day of, and (*c*) 48 h after the event. The light (dark) shading denotes regions where there exists a greater than 95% (99%) probability that the composite belongs to a population distinct from that of climatology. (From Lackmann and Gyakum, 1999; reprinted by permission of AMS).

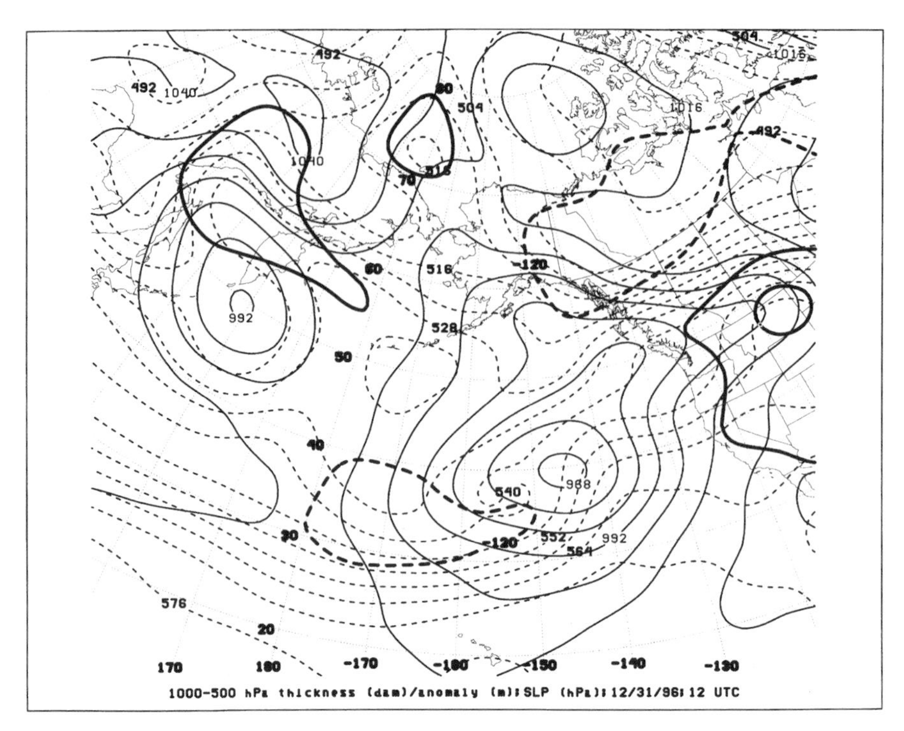

Figure 9 As for Fig. 1, except 1200 UTC, December 31, 1996.

Figure 10 Water vapor image for 1430 UTC, December 31, 1996. (Courtesy of NOAA).

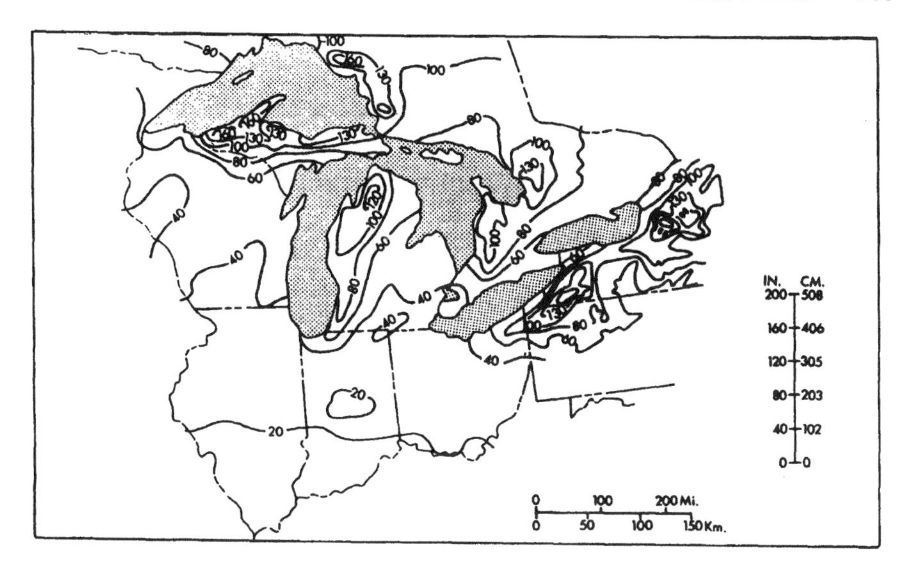

Figure 11 Mean annual snowfall (in.) for the Great Lakes region. (From Niziol et al., 1995; in turn from Eichenlaub, 1979; reprinted by permission of AMS).

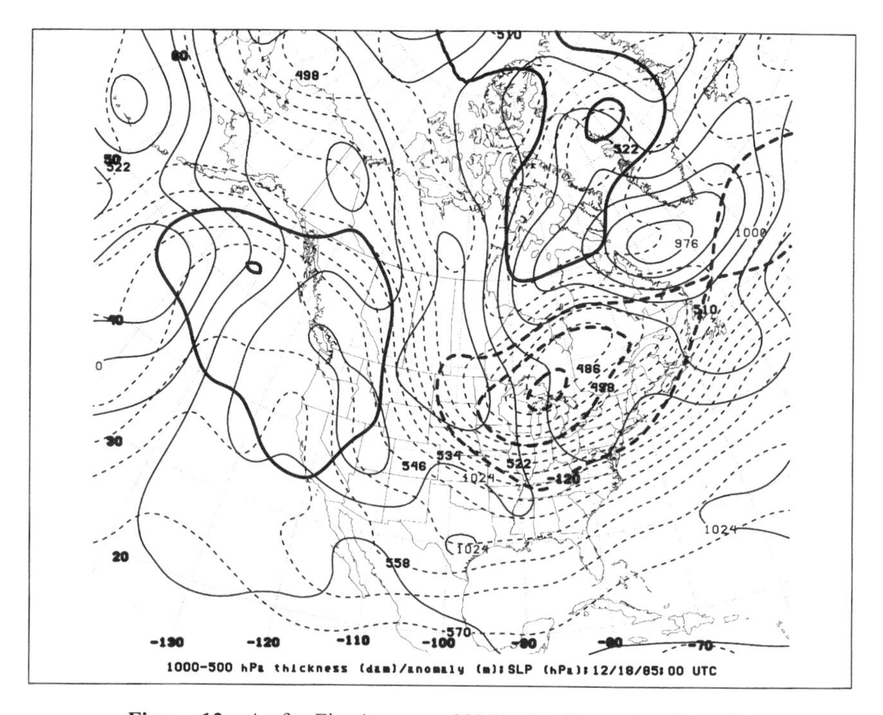

Figure 12 As for Fig. 1, except 0000 UTC, December 18, 1985.

than the climatological average. The tropospheric air was about 12°C colder than the mean in Buffalo. Despite the fact that the most substantial synoptic-scale ascent typically is located to the north and east of a developing surface low, owing to a favorable combination of warm advection and cyclonic vorticity effects, extremely large snowfall rates in excess of 5 cm/h were observed in the Buffalo area. Several factors contributed to such an extreme snowfall rate. These include large-scale ascent associated with a strong upper-level trough. Within this environment, several crucial mesoscale conditions amplify the response to this advection. First, the cold air traveling over the relatively warm waters will hydrostatically destabilize the air. Second, the evaporation of water vapor from the lakes will saturate the air, so that the effective hydrostatic stability will be reduced even further.

Additional physical processes are responsible for enhancing the snow. These include the development of a surface trough (Fig. 12) that provides the larger-scale ascent necessary to trigger moist convection in the presence of instability. The existence of sensible heat transfer from the unfrozen lakes to the atmosphere contributes to ascent. Additionally, the differential roughness between the lakes and the surrounding land will create low-level convergence areas with accompanying ascent. Often, as in the case of the December 17–19, 1985, event, mesoscale bands of heavy snow will develop (Fig. 13), and a key forecast problem is to predict the existence and movement of such band(s).

Freezing rain events can have devastating impacts on economic infrastructure. The synoptic environment of freezing rain is characterized typically by a surface extratropical cyclone advecting warm, moist air above relatively shallow, cold air masses. Areas that experience persistent shallow, cold air are therefore prone to freezing rain. Especially susceptible regions include larger valleys and basins.

Cortinas (2000) has documented the mean synoptic conditions for freezing rain events in the Great Lakes region. Figure 14 illustrates that freezing rain occurs to the northeast of a surface cyclone that advects warm, moist air poleward from either the Gulf of Mexico or the Atlantic Ocean. Typically, the air to the northeast of the low had been associated with a prior cold-air outbreak. The temperature stratification becomes very stable (Fig. 15), as the warm, moist air travels above the relatively cold dense near-surface air. As Figure 15 shows, the lowest layer during a freezing rain event is characterized by an inversion with near-surface air that is less than 0°C, with a deep layer of air aloft that is greater than 0°C. The strong inversion helps to prevent any turbulent mixing of the warm air aloft down to the surface. Furthermore, the stratification decreases markedly above the inversion, to the extent that nearly moist adiabatic conditions in the free atmosphere may exist. The implication for precipitation amounts is that substantial ascent, or even convection, may occur in these upper weakly stratified layers. The lowest-layer air is typically from the east or northeast, and this reinforces the inversion with cold-temperature advection from the down-shear polar air mass. The winds at the top of the inversion are typically warm and moist and blowing from the southwest. Therefore the vertical wind shear reinforces the temperature inversion.

An extreme example of such an event occurred during January 5–9, 1998, when an ice storm deposited greater than 100 mm of predominantly freezing rain in south-

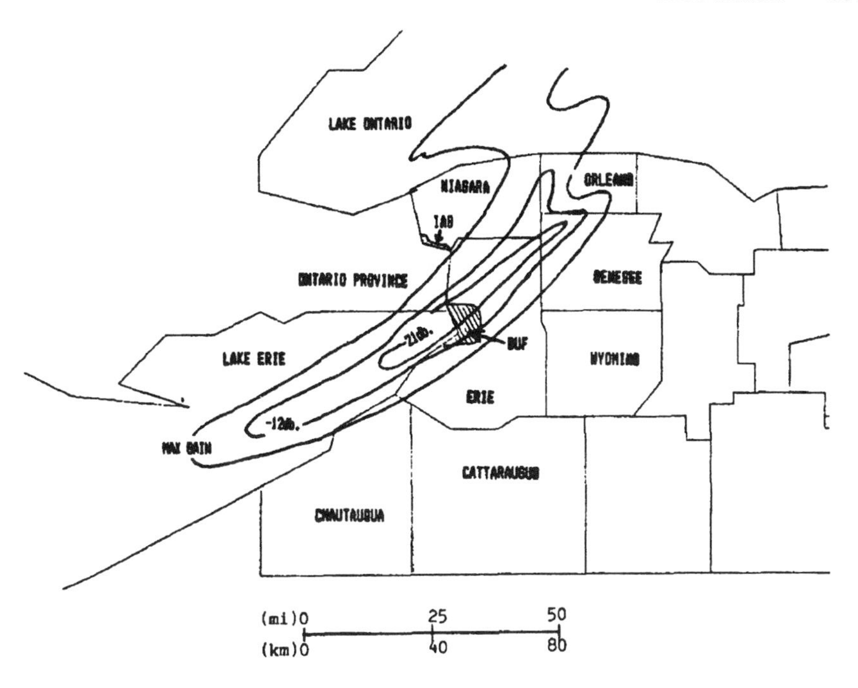

Figure 13 Precipitation echoes from the WSR-57 radar at Buffalo for the Lake Erie snowstorm on 2330 UTC, December 17, 1985. Snowfall intensities are contoured for local use, with linear attenuation, to depict the areas of heaviest snowfall more accurately. (From Niziol, 1987; reprinted by permission of AMS).

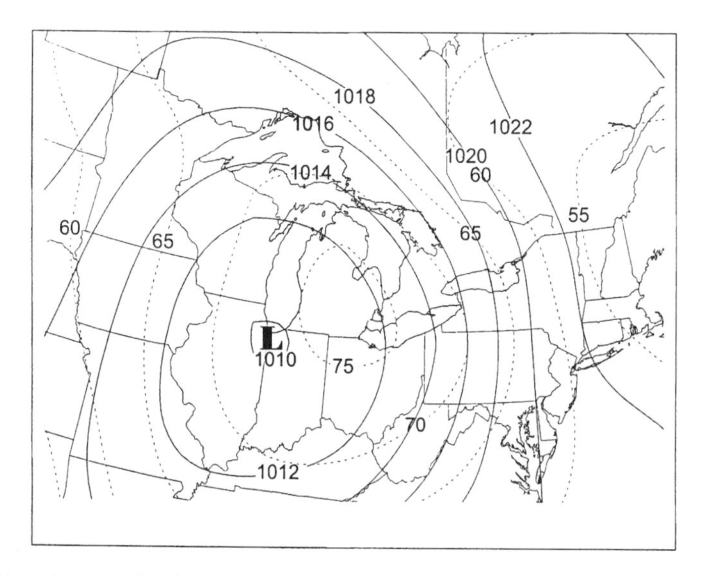

Figure 14 Mean sea-level pressure (hPa, solid line) and 1000–666-hPa relative humidity (%, dashed line) during freezing rain events over the central Great Lakes for the period 1976–1990. (From Cortinas, 2000; reprinted by permission of AMS).

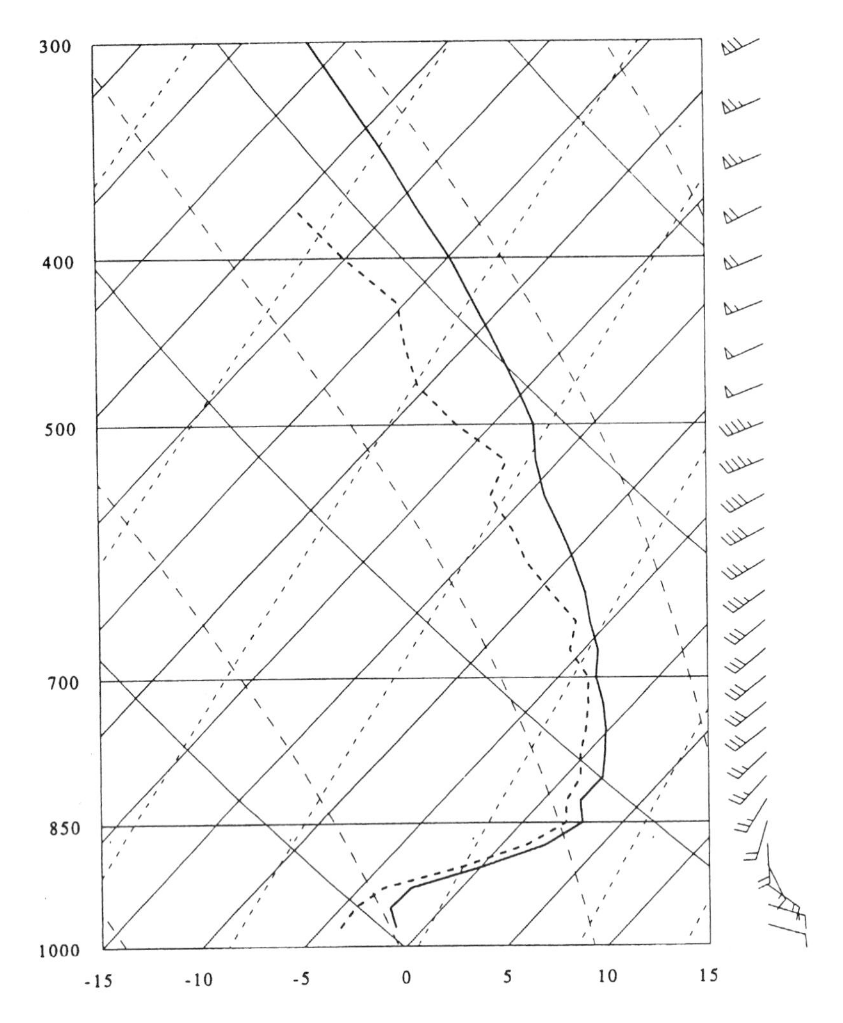

Figure 15 Median values of dry-bulb temperature (°C, dark solid line) and dewpoint temperature (°C, dark dashed line) from a distribution of freezing rain soundings at Flint, Michigan. Data are plotted on a skew T–log p diagram, with pressure (vertical axis) given in hPa and winds shown by full barb = 5 m/s and half barb = 2.5 m/s. (From Cortinas, 2000; reprinted by permission of AMS).

ern Quebec and upstate New York. Approximately 44 fatalities and damages in excess of $4 billion occurred (NOAA, 2000). The synoptic conditions at 1200 UTC, January, 8, 1998 (Fig. 16), were typical for the event. The freezing rain occurred in a strong geostrophic deformation zone that extended northeast of a surface low from upstate New York into New England. Surface temperatures during the event ranged from −5 to −1°C in extremely large 1000–500-hPa thicknesses of 552 dam or greater. The anomalously strong Bermuda anticyclone assisted

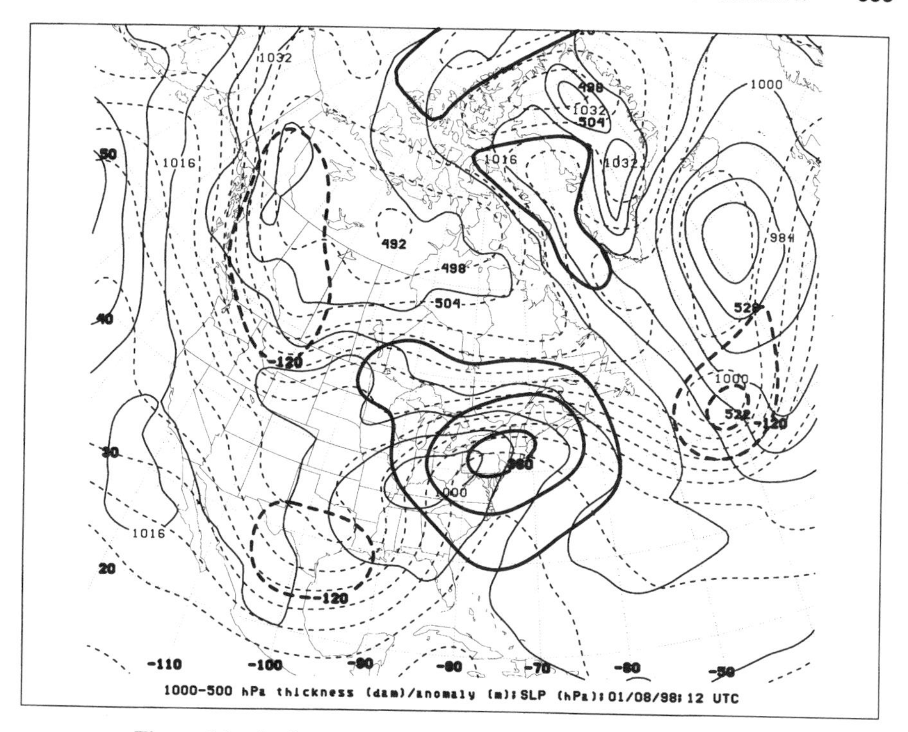

Figure 16 As for Fig. 1, except for 1200 UTC, January 8, 1998.

the tropospheric warming of the affected regions with northward transports of warm and moist air from the Caribbean Sea. Surface winds advected cold air from the northeast along the Saint Lawrence River Valley. These surface winds occurred along the southern flank of a bitterly cold 1044-hPa surface anticyclone centered in the Canadian Northwest Territories. While the instantaneous synoptic-scale features were unusually strong for a freezing rain event, the factor that produced such an extreme event was the persistence for 5 days of this freezing rain–producing pattern.

5 SUMMARY

We have discussed several meteorological phenomena associated with winter weather. These events, associated with extreme cold, wind, and precipitation, have the potential to produce substantial economic impact (NOAA, 2000). We have described these events in terms of the larger-scale meteorological patterns that are conducive to their existence. Additionally, we have seen how these events are produced by the modulation of larger-scale atmospheric patterns by interesting topographic features, such as mountains, lakes, and oceans. Although we have focused our attention on North American phenomena, similar events exist in other regions of the world, where similar interactions exist between the atmospheric flows and topography.

REFERENCES

Cortinas, J., Jr. (2000). A climatology of freezing rain in the Great Lakes region of North America, *Mon. Wea. Rev.* **128**, 3574–3588.

Eichenlaub, V. L. (1979). *Weather and Climate of the Great Lakes Region*, Notre Dame, IN, University of Notre Dame Press.

Kalnay, E., M. Kanamitsu, R. Kistler, W. Collins, D. Deaven, L. Gandin, M. Iredell, S. Saha, G. White, J. Woollen, Y. Zhu, A. Leetmaa, B. Reynolds, M. Chelliah, W. Ebisuzaki, W. Higgins, J. Janowiak, K. C. Mo, C. Ropelewski, J. Wang, R. Jenne, and D. Joseph (1996). The NCEP/NCAR 40-year reanalysis project, *Bull. Am. Meteor. Soc.* **77**, 437–471.

Klemp, J. B., and D. K. Lilly (1975). The dynamics of wave-induced downslope winds, *J. Atmos. Sci.* **32**, 320–339.

Kocin, P. J., P. N. Schumacher, R. F. Morales, Jr., and L. W. Uccellini (1995). Overview of the 12–14 March 1993 superstorm, *Bull. Am. Meteor. Soc.* **76**, 165–182.

Lackmann, G. M., and J. R. Gyakum (1999). Heavy cold-season precipitation in the northwestern United States: Synoptic climatology and an analysis of the flood of 17–18 January 1986, *Wea. Forecasting* **14**, 687–700.

NOAA (2000). Billion Dollar US. Weather Disasters, 1980–2000 (available on the World Wide Web at http://www.ncdc.noaa.gov/ol/reports/billionz.html).

Niziol, T. A. (1987). Operational forecasting of lake effect snowfall in western and central New York, *Wea. Forecasting* **2**, 310–321.

Niziol, T. A., W. R. Snyder, and J. S. Waldstreicher (1995). Winter weather forecasting throughout the Eastern United States. Part IV: Lake effect snow, *Wea. Forecasting* **10**, 61–77.

Quiroz, R. S. (1984). The climate of the 1983–84 winter—a season of strong blocking and severe cold in North America, *Mon. Wea. Rev.* **112**, 1894–1912.

CHAPTER 28

TERRAIN-FORCED MESOSCALE CIRCULATIONS

JOHN HOREL

1 INTRODUCTION

The characteristics of the Earth's surface affect climate and weather on all spatial scales (Barry, 1992). However, many weather phenomena that are influenced by surface inhomogeneities in elevation, moisture, temperature, snow cover, vegetation, or roughness are organized on the mesoscale, which spans the range from 2 to 200 km. According to Pielke (1984), weather systems on the mesoscale can be divided into two general categories: those that are forced primarily by instabilities in traveling large-scale disturbances (e.g., squall lines or mesoscale convective complexes) and those that are forced by surface inhomogeneities (e.g., mountain/valley circulations, sea breezes, or urban circulations).

The impact of mesoscale variations in the underlying land surface is evident in an estimate of the average annual precipitation over the United States (Fig. 1). As a general rule, precipitation increases locally as terrain height increases, for example over the Appalachians and the mountain ranges of the western United States. A further general rule is that precipitation is higher near the coasts, where moisture is more abundant, than further inland.

This section emphasizes mesoscale weather phenomena that are strongly modulated by the characteristics of the underlying surface. After discussion of thermally driven and mechanically driven flows, the impact of terrain upon precipitation processes will be presented. The influence of other variations in surface properties

Handbook of Weather, Climate, and Water: Dynamics, Climate, Physical Meteorology, Weather Systems, and Measurements, Edited by Thomas D. Potter and Bradley R. Colman.
ISBN 0-471-21490-6

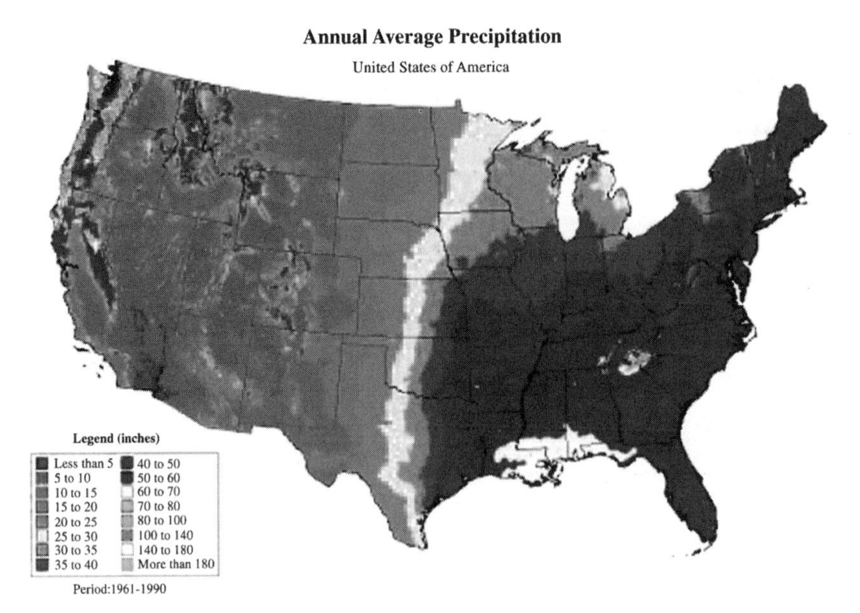

Figure 1 (see color insert) Average annual precipitation in inches, as determined from the Parameter-elevation Regression on Independent Slopes (PRISM) model by Chris Daly, based on 1961–1990 normals from National Oceanic and Atmospheric Administration (NOAA) cooperative stations and Natural Resources Conservation Service (NRCS) SNOwpack TELemetry (SNOTEL) sites. Modeling sponsored by USDA-NRCS Water and Climate Center, Portland, Oregon. Available from George Taylor, Oregon State Climatologist, Oregon Climate Service. See ftp site for color image.

upon mesoscale circulations is followed by discussion of the predictability of terrain-forced mesoscale circulations.

2 IMPACT OF TERRAIN UPON WIND

Thermally Driven Flows

Mesoscale wind circulations are produced frequently by temperature contrasts that develop as a result of terrain variations. As noted by Pielke and Segal (1986), the general nature of diurnal sea and land breezes was well understood in ancient times: "southward goes the wind, then turns to the north; it turns and turns again" (Ecclesiastes 1:6). Sea (or lake) and land breezes are driven by horizontal temperature contrasts that develop between water bodies and adjacent land surfaces (Whiteman, 2000). As shown in Figure 2 for the Great Salt Lake (the largest body of water in the continental United States to the west of the Great Lakes), differences in air temperature develop over the water and land surfaces as a result of the higher heat capacity of the water relative to that of the surrounding land. In the case of the Great Salt Lake during summer, the air over the lake is roughly 1°C warmer than the air over the surrounding land at night and 4°C cooler during the day. These temperature

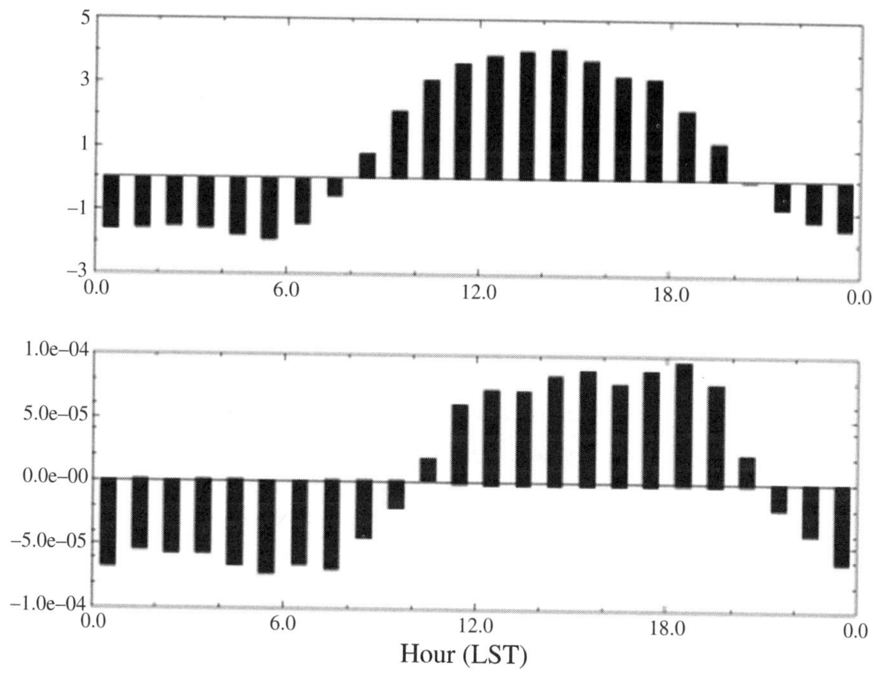

Figure 2 Top panel: Diurnal variation in the difference in air temperature (°C) over the land surrounding the Great Salt Lake, Utah, versus the air over the lake during August 1999. The air over the lake is warmer than the air over the land during the night. Bottom panel: Diurnal variation in surface wind divergence (s^{-1}) over the Great Salt Lake. The air converges over the lake at night and diverges away from the lake during the day (a similar diurnal variation in surface wind divergence is evident over Lake Ontario; Chen, 1977).

differences drive the air from over the water to over the land during the day (lake breeze) and drive the air from over the land to over the water at night (land breeze). The resulting convergence of the surface winds at night over the Great Salt Lake and divergence during the day are evident in Figure 2.

Mountain–valley circulations arise as a result of differential heating between the ground in regions of complex terrain and the free atmosphere at the same elevation. There are two broad categories for mountain–valley circulations: slope flows and along-valley winds (Whiteman, 1990, 2000). Typically, slope flows are driven by horizontal temperature contrasts between the air over the valley sidewalls and the air over the center of the valley. Since a larger diurnal temperature variation occurs near the ground, the higher terrain above the valley serves as a heat source during the day and a heat sink at night. These temperature differences between the slope and free air over the valley drive cold, dense air flowing down the slope at night and warm, light air surging up the slope during the day. Upslope flows tend to be deeper than the shallow nocturnal drainage flows as a result of strong turbulent mixing during the day.

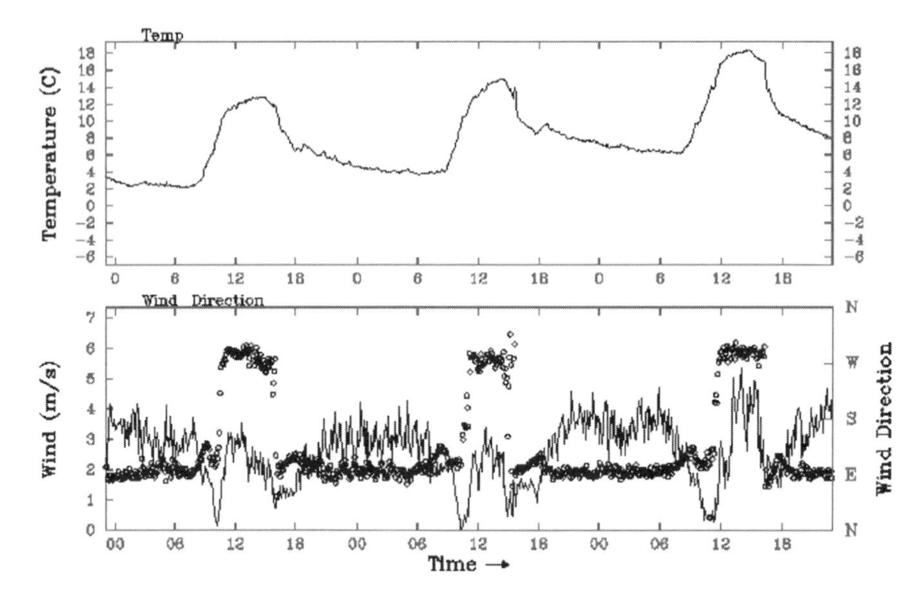

Figure 3 Diurnal mountain–valley circulation in Big Cottonwood canyon in the Wasatch Mountains of northern Utah. Top panel: temperature (°C) from midnight local standard time (LST) October 15 to midnight October 18, 2000. Bottom panel: wind speed (solid line in m/s) and wind direction (circles) for the same period. The weather station is located where the canyon is oriented east–west.

Along-valley winds develop as a result of variations in the strength of slope flows within the valley or differences in temperature between the air in the valley and that found over the adjacent lowlands. As an example of along-valley circulations, Figure 3 shows the evolution of wind and temperature at a station located roughly midway along the length of the Big Cottonwood Canyon in the Wasatch Mountains of northern Utah. During this 3-day period, a warming trend was underway and the skies remained clear; further, the station location within this narrow, steep canyon was shaded during the morning and late afternoon. The diurnal temperature swing of roughly 10°C drove a reversal in the wind from downvalley at night (from the east) to upvalley during the day (from the west). The sudden onsets of the wind reversals were particularly striking during these 3 days.

Dynamically Driven Flows

As summarized by Whiteman (2000), the degree to which a hill or mountain affects the air flow depends upon the characteristics of the terrain feature (e.g., height, width, roughness, orientation relative to flow direction) and the upstream speed and stability of the air. Since the speed and stability of the flow can vary significantly with height, the impact of the terrain upon the flow can change from one atmospheric layer to another (Fig. 4).

Figure 4 (see color insert) Cumulus clouds, indicative of shallow instability, above the Wasatch Mountains in northern Utah are capped by lenticular wave clouds, indicative of stable air flowing across the mountain barrier (photo by J. Horel). See ftp site for color image.

Whether an obstacle significantly obstructs the flow depends upon whether or not the approaching air has enough kinetic energy to lift the air over it. The non-dimensional Froude number (defined as $\mathrm{Fr} = u/Nh$, where u is the speed of the upstream flow, N is the Brunt–Vaisala frequency, a measure of stability, and h is the height of the terrain) provides a conceptual framework to assess the impact of terrain height and flow speed and stability. If the flow approaching a relatively low obstacle has strong winds and weak stability (Froude number greater than 1), then there is sufficient kinetic energy to cross the obstacle. On the other hand, if the flow approaches a relatively high barrier with weak winds and strong stability (Froude number less than 1), then there is not enough kinetic energy to force the air over the obstacle and the flow is either channeled through gaps in the terrain or forced to travel laterally around the obstacle. For example, Figure 5 shows fog spilling through a gap in the coastal mountains of northern California as a result of strong stability at crest level. The flow is more likely to be blocked and required to travel around obstacles if the barrier has a relatively short lateral extent or the flow impinging upon the mountain range is very shallow.

As stable air flows across a mountain barrier (with Froude number less than 1), mountain gravity waves are often created over or in the lee of the barrier (Durran, 1990). The structure of the mountain waves exhibited at any particular time depends upon the charateristics of the barrier (e.g., height, width, multiple ridges) as well as the stability, orientation of the flow relative to the ridge crest, and vertical change of wind with height. If sufficient moisture is present, the uppermost portion of the wave

Figure 5 (see color insert) Fog channelled through a gap in the coastal mountains of northern California (photo by J. Horel). See ftp site for color image.

may become visible (e.g., the lenticular clouds in Fig. 4). Viewed from space (Fig. 6), the wavelike nature of the gravity waves is readily apparent. The waves embedded within these flows result from the response to the forced ascent: After the air is carried aloft as high as possible, the lifted air is cooler (and heavier) than the surrounding air and displaced back toward its original level by gravity.

Figure 6 Mountain waves generated by southwesterly flow traversing the mountain ranges of the southwestern United States. This visible satellite image was taken at 2200 Universal Time Coordinated (UTC), February 9, 1999.

When the cross-barrier flow of stable air is sufficiently strong, damaging downslope wind storms in the lee of major mountain barriers are common during the winter season at many locales around the globe (Whiteman, 2000). Local residents refer to these wind storms by many different names (e.g., foehn in the European Alps, bora near the Adriatic Sea, chinook to the east of the Rockies, Santa Ana in southern California, canyon winds in Utah, zonda in Argentina, and oroshi in Japan). While local topography modulates the intensity of these storms, damaging downslope wind storms share one or more of the following common characteristics: pronounced mountain waves, precipitation on the upwind side of the mountain range coupled with adiabatic warming on the downwind side, strong cross-barrier pressure gradient (high pressure upstream, low pressure downstream), inversion (temperature increasing with height) in a layer above ridge crest, or wind reversal above the crest. The wind and temperature structure across the Rocky Mountains during a particularly damaging event is shown in Figures 5 and 6 in Chapter 27. Durran (1990) summarizes how the change in flow characteristics across the barrier (Froude number less than 1 upstream and greater than 1 downstream) may contribute to further acceleration of the wind in the lee of the barrier in a manner analogous to a hydraulic jump.

Barrier jets form when stable, low-level flow impinges upon a mountain range. As the flow piles up in front of the barrier, it is deflected to the left in the Northern Hemisphere as a result of a leftward-directed component of the pressure gradient force. Northward-directed barrier jets have been observed along the Pacific coast when the prevailing wind was from the west (Overland and Bond, 1995). Southward-deflected barrier jets have been observed on the east slopes of the Rockies and Appalachians during cold-air damming episodes (Dunn, 1987; Bell and Bosart, 1988).

3 INFLUENCE OF TERRAIN UPON PRECIPITATION

Orographic Precipitation

Local variations in precipitation as a result of the height, relief, and aspect (i.e., slope direction) of local terrain features can be striking: For example, annual precipitation increases over a distance of 35 km from 16.2 in. (41 cm) at Salt Lake City, Utah, to 58.5 in. (149 cm) at Alta in the nearby Wasatch Mountains. One of the most dramatic mesoscale variations in precipitation in the United States is evident in northwestern Washington where precipitation on the windward side of the Olympic Mountains is over 120 in. (305 cm) yet drops to less than 30 in. (76 cm) in the lee of those mountains (see Fig. 1).

Hills or mountains deflect air near the surface upward or downward, depending on the direction of the air flow relative to the slope of the topography. Banta (1990) notes that mountains have two major roles in forming clouds and precipitation: first, as obstacles to the flow and, second, as high-level heat sources during the day that cause the winds to converge toward the mountain. Clouds are likely to be generated if the ascent caused by either of these mechanisms is strong and there is sufficient moisture present in the air stream. If the ascent carries the cloud water and ice high enough such that the air becomes supersaturated, then the excess water and ice in the cloud may begin to fall and eventually may be deposited at the surface as precipitation.

Figure 7 summarizes the physical processes that control the development of orographic precipitation (Houze, 1993):

- *Upslope Condensation* Stable ascent of saturated air is forced by flow over mountains.
- *Orographic Convection* Lifting induced by terrain leads to convective release of instabilities present in the flow (e.g., Fig. 8). Orographic convection can be further subdivided into the following:
  - *Upslope and Upstream Triggering* Topographically induced motions and upstream blocking trigger convection leading to precipitation on the windward slope of the barrier.
  - *Thermal Triggering* Daytime heating produces an elevated heat source with local convergence near the top of the hill or mountain.
  - *Lee-Side Triggering* Low-Froude-number flow around a hill or isolated mountain leads to convergence in the lee of the obstacle.
  - *Lee-Side Enhancement of Deep Convection* Flow across a mountain range converges with low-level thermally induced upslope flow.

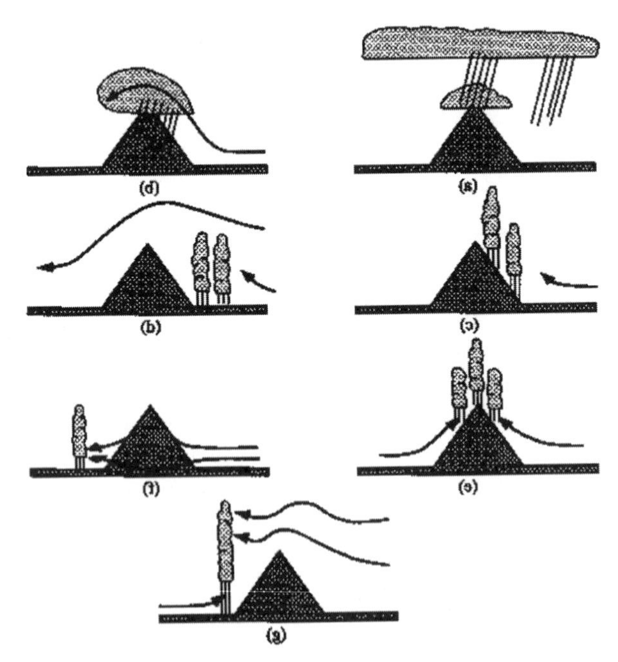

Figure 7 Mechanisms of orographic precipitation (adapted from Houze, 1993); (*a*) seeder-feeder; (*b*) upslope condensation; (*c*) upslope triggering; (*d*) upstream triggering; (*e*) thermal triggering; (*f*) lee-side triggering; (*g*) lee-side enhancement.

Figure 8 Convection developing over the Wasatch Mountains in the afternoon (photo by J. Horel). See ftp site for color image.

- *Seeder-feeder* Convective cells aloft produce cloud water or ice that fall into lower cloud decks; the falling cloud particles grow at the expense of the water content of the lower clouds.

Lake Effect Snow

During the cool season, enhanced precipitation is often observed in the lee of major water bodies such as the Great Lakes (Niziol et al., 1995). As shown in Figure 11 of Chapter 27, the greatest snowfall near the Great Lakes occurs where the prevailing winds blow across the longest fetch of a lake and is enhanced by local orography such as the Tug Hill plateau downwind of Lake Ontario (Niziol et al., 1995). A crippling snowstorm for the Buffalo, New York, area and many other locales downwind of Lakes Erie and Ontario occurred during November 20–23, 2000. Local total snowfall amounts were as large as 31 in. (79 cm) downwind of Lake Erie and 28 in. (71 cm) downwind of Lake Ontrario. Chapter 27 reviews the synoptic-scale and mesoscale weather associated with lake effect snowstorms in the Great Lakes area.

Residents far from the shores of the Great Lakes contend occasionally with narrow lake effect snow bands that extend a considerable distance inland (Niziol et al., 1995). However, the greatest impact of lake effect snowstorms tends to be closer to the shores of the Great Lakes. Ingredients that determine the intensity and structure of lake effect snowstorms include the synoptic setting, the upstream fetch (distance the air travels over water) and degree to which the water body is covered by snow or ice, instability in the boundary layer, wind shear, moisture availability, differences in surface roughness between the water and land surfaces, offshore-directed land breezes, and presence of upstream lakes or downstream orography.

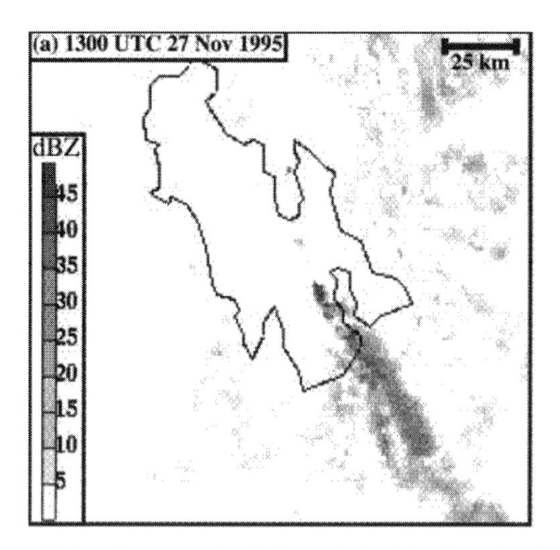

Figure 9 Example of lake effect snow band downwind of the Great Salt Lake, Utah. Lowest-elevation (0.5°) base-reflectivity analysis at 1300 UTC, November 27, 1995. Reflectivity shading based on scale at left. (From Steenburgh et al., 2000.)

Enhanced snowfall occurs as well downwind of smaller water bodies, such as the Great Salt Lake, that can have serious societal impacts (Steenburgh et al., 2000). For example, a wind-parallel band developed on November 27, 1995, over the Great Salt Lake and produced localized snow accumulations of 10 in. (25 cm) downstream (Fig. 9).

4 OTHER IMPACTS OF SURFACE INHOMOGENEITIES

Anthes (1984) and Zeng and Pielke (1995) demonstrate that variations in landscape characteristics can strongly affect the atmospheric boundary layer structure, formation of convective clouds, and fluxes of heat and moisture on the mesoscale. Mesoscale flows can be generated directly by differences in surface temperature and heat and moisture fluxes. In addition, biophysical processes that control moisture availability within the vegetative canopy can influence mesoscale circulations.

Pielke and Segal (1986) summarize the impact of horizontal contrasts in snow cover upon mesoscale circulations. Snow cover increases the albedo, reduces roughness, and alters surface fluxes of heat and moisture relative to nearby snow-free regions.

Urban areas affect the structure of the atmosphere and weather in a variety of ways (Dabberdt et al., 2000). Urban heat islands result from the combined effects of modified thermal and radiative properties of the surface and anthropogenic sources of sensible heat and moisture. The variability of surface roughness in urban areas affects the exchange of heat, mass, and momentum between the surface and the atmosphere. As a result of structures and pavement, hydrological processes are altered substantially.

Residents of urban areas are susceptible to the interactions of mesoscale and convective-scale weather phenomena, as illustrated in Figure 10, the F2 tornado that traversed downtown Salt Lake City on August 11, 1999 (Dunn and Vasiloff, 2001). It has been speculated (e.g., Landsberg, 1981; Bornstein and Lin, 1999) that large urban areas may affect the origin, strength, and movement of convective storms and other weather systems.

5 PREDICTABILITY OF TERRAIN-FORCED MESOSCALE CIRCULATIONS

As mentioned earlier, Pielke (1984) divided mesoscale atmospheric systems into two groups: terrain-induced mesoscale systems and synoptically induced mesoscale systems. He stated that the former are easier to simulate because they are forced by geographically fixed features in the underlying terrain. Paegle et al. (1990) also suggest that terrain-forced circulations may be inherently more predictable than synoptically induced flows, which are more sensitive to errors in the data used to specify the initial conditions of a numerical simulation. However, Paegle et al. (1990) note that the modeling of terrain-induced flows is difficult and susceptible to inaccurate numerical treatment of physical processes such as turbulent mixing,

Figure 10 Category F2 tornado of August 11, 1999, in downtown Salt Lake City, UT (photo courtesy of Department of Meteorology, University of Utah, Salt Lake City, UT).

radiative heating, cloud processes, and soil transfer and the numerical errors arising from steep terrain slopes.

From ALPEX (Kuettner, 1981) to CASES (LeMone et al., 2000) and MAP (Bougeault et al., 2001), field programs have been critical to improve the numerical simulation and prediction of mesoscale circulations forced by variations in the underlying surface. Comprehensive datasets are required to provide validation of the model simulations and lead to improved treatment of relevant physical processes in high-resolution numerical weather prediction models.

Considerable debate remains in the mesoscale modeling community regarding the inherent predictability of terrain-induced circulations. As noted by Mass and Kuo (1998), mesoscale predictability in regions of complex terrain may be enhanced due to the relatively deterministic interactions between the synoptic-scale flow and the underlying terrain. Nonetheless, while models are increasingly capable of simulating physically realistic responses to flow over terrain, a corresponding increase in forecast skill has not always been evident (e.g., Colle et al., 2000).

REFERENCES

Anthes, R. A. (1984). Enhancement of convective precipitation by mesoscale variations in vegetative covering in semi-arid regions, *J. Appl. Meteor.* **23**, 541–554.

Banta, R. M. (1990). The role of mountain flows in making clouds, in W. Blumen (Ed.), *Atmospheric Processes over Complex Terrain, Meteor. Monogr.* **23**(45), 229–282.

Barry, R. G. (1992). *Mountain Weather and Climate*, 2nd ed., London, Routledge.

Bell, G. D., and L. F. Bosart (1988). Appalachian cold-air damming, *Mon. Wea. Rev.* **116**, 137–162.

Bornstein, R., and Q. Lin (1999). Urban heat islands and summertime convective thunderstorms in Atlanta, *Atmos. Environ.* **34**, 507–516.

Bougeault, P., P. Binder, A. Buzzi, R. Dirks, R. Houze, J. Kuettner, R. B. Smith, R. Steinacker, and H. Volker (2001). The MAP special observing period, *Bull. Am. Meteor. Soc.* **82**, 433–462.

Chen, W. Y. (1977). Analysis of vorticity and divergence fields and other meteorological parameters over Lake Ontario during IFYGL, *Mon. Wea. Rev.* **105**, 1298–1309.

Colle, B. A., C. F. Mass, and K. J. Westrick (2000). MM5 precipitation verification over the Pacific Northwest during the 1997–99 cool seasons, *Wea. Forecasting* **15**, 730–744.

Dabberdt, W. F., J. Hales, S. Zubrick, A. Crook, W. Krajewski, J. C. Doran, C. Mueller, C. King, R. N. Keener, R. Bornstein, D. Rodenhuis, P. Kocin, M. A. Rossetti, F. Sharrocks, and E. M. Stanley (2000). Forecasting issues in the urban zone: report of the 10th prospectus development team of the U.S. Weather Research Program, *Bull. Am. Meteor. Soc.* **81**, 2047–2064.

Dunn, L. (1987). Cold air damming by the Front Range of the Colorado Rockies and its relationship to locally heavy snows, *Wea. Forecasting* **2**, 177–189.

Dunn, L., and S. Vasiloff (2001). Tornadogenesis and operational considerations of the 11 August 1999 Salt Lake City tornado as seen from two different doppler radars, *Wea. Forecasting* **16**, 377–398.

Durran, D. R. (1990). Mountain waves and downslope winds, in W. Blumen (Ed.), *Atmospheric Processes over Complex Terrain, Meteor. Monogr.* **23**(45), 59–82.

Houze, R. A., Jr. (1993). *Cloud Dynamics*, San Diego, CA, Academic.

Kuettner, J. P. (1981). ALPEX: the GARP mountain subprogram, *Bull. Am. Meteor. Soc.* **62**, 793–805.

Landsberg, H. E. (1981). *Urban Climate*, New York, Academic.

LeMone, M., R. Grossman, R. Coulter, M. Wesley, G. Klazura, G. Poulos, W. Blumen, J. Lundquist, R. Cuenca, S. Kelly, E. Brandes, S. Oncley, R. McMillen, and B. Hicks (2000). Land-atmosphere interaction research, early results, and opportunities in the Walnut River Watershed in southeast Kansas: CASES and ABLE, *Bull. Am. Meteor. Soc.* **81**, 757–779.

Mass, C. F., and Y.-H. Kuo (1998). Regional real-time numerical weather prediction: Current status and future potential, *Bull. Am. Meteor. Soc.* **79**, 253–263.

Niziol, T. A., W. R. Snyder, and J. S. Waldstreicher (1995). Winter weather forecasting throughout the eastern United States. Part IV: Lake effect snow, *Wea. Forecasting* **10**, 61–77.

Overland, J. E., and N. A. Bond (1995). Observations and scale analysis of coastal wind jets, *Mon. Wea. Rev.* **123**, 2934–2941.

Paegle, J., R. Pielke, G. Dalu, W. Miller, J. Garratt, T. Vukicevic, G. Berri, and M. Nicolini (1990). Predictability of flows over complex terrain, in W. Blumen (Ed.), *Atmospheric Processes over Complex Terrain, Meteor. Monogr.* **23**(45), 285–299.

Pielke, R. A. (1984). *Mesoscale Meteorological Modeling*, San Diego, CA, Academic.

Pielke, R. A., and M. Segal (1986). Mesoscale circulations forced by differential terrain heating, in P. S. Ray (Ed.), *Mesoscale Meteorology and Forecasting*, Boston, American Meteorological Society, pp. 516–548.

Steenburgh, W. J., S. F. Halvorson, and D. J. Onton (2000). Climatology of lake-effect snowstorms of the Great Salt Lake, *Mon. Wea. Rev.* **128**, 709–727.

Whiteman, C. D. (1990). Observations of thermally developed wind systems in mountainous terrain, in W. Blumen (Ed.), *Atmospheric Processes over Complex Terrain, Meteor. Monogr.* **23**(45), 5–42.

Whiteman, C. D. (2000). *Mountain Meteorology: Fundamentals and Applications*, New York, Oxford University Press.

Zeng, X., and R. A. Pielke (1995). Landscape-induced atmospheric flow and its parameterization in large-scale numerical models, *J. Clim.* **8**, 1156–1177.

CHAPTER 29

SEVERE THUNDERSTORMS AND TORNADOES

H. BROOKS, C. DOSWELL III, D. DOWELL, R. HOLLE, B. JOHNS, D. JORGENSON, D. SCHULTZ, D. STENSRUD, S. WEISS, L. WICKER, AND D. ZARAS

1 INTRODUCTION

Severe thunderstorms and tornadoes are phenomena that can occur at almost any place on the planet. Unlike hurricanes and synoptic-scale cyclones, these local storms affect areas of 10 to 100 km^2 (e.g., the size of a typical U.S. city) and last a few minutes to several hours. Nevertheless, these storms can produce devastating damage that rivals any other atmospheric storm on earth. Tornadoes kill nearly three-dozen people in the United States per year, and recent tornadoes such as the May 3, 1999, Oklahoma City tornado can cause more than $1 billion in damage. However, the real killers from severe storms are actually flash floods and lightning, as the fatality rate from these events is more than 200 people every year. Therefore, timely forecasts and warnings of severe weather are crucial for mitigating damage and protecting the public.

2 CLIMATOLOGY OF SEVERE THUNDERSTORMS

Severe weather associated with thunderstorms affects almost all of the planet and represents a significant threat to life and property in many locations. The definition of what is considered "severe" depends on operational forecasting considerations that vary from country to country but typically includes phenomena

Handbook of Weather, Climate, and Water: Dynamics, Climate, Physical Meteorology, Weather Systems, and Measurements, Edited by Thomas D. Potter and Bradley R. Colman.
ISBN 0-471-21490-6

such as tornadoes, large hail (usually of diameter at least approximately 2 cm), strong convective wind gusts (usually approximately 90 km/h or more), and extremely heavy precipitation associated with flash floods (frequently 50 mm/h at a single location). Criteria associated with heavy precipitation are the most variable from country to country and, in some places, even within one country. For instance, in the United States, flash flooding is not considered a severe thunderstorm event, and in Canada the objective definition of heavy precipitation is different in different geographical regions.

Tornadoes

Tornadoes have been observed on every continent except Antarctica, although they are most common in North America, particularly the Great Plains of the United States. Increased efforts to collect information about tornadoes in North America have led to an increase in the number of reports, with an average of about 1200 tornadoes reported annually in the United States in recent years, compared to only 600 just 50 years ago. The increase has been particularly apparent in the number of weak tornadoes (classified F0 or F1 on the Fujita damage scale that goes from F0 to F5). Similar efforts in other countries have also led to large increases in the reported number of tornadoes there, such as in Germany where the average prior to 1950 was about 2 per decade, but in the 1990s was 7 per decade, with more than 20 reported in the year 2000 alone. Climatologies of tornado occurrence in the United States have identified the temporal and spatial structure of the threat. The strongest tornadoes (F2 to F5 on the Fujita scale) are most often found in the Great Plains of the United States (Fig. 1). This is a result of the frequent production of favorable environments with warm, moist air near the ground, dry, relatively cool air aloft, and strong vertical speed and directional shear of the horizontal winds. The Gulf of Mexico acts as a source region for warm, moist low-level air flowing north, and the Rocky Mountains act as a source region for the dry, relatively cool air aloft flowing eastward toward the Plains. The presence of these two fixed geographic features appears to be the dominant reason for the frequency of tornadoes in the Plains.

The rarity of tornadoes in other regions of the world does not mean that, when events occur there, the effects are small. Landfalling tropical cyclones often produce tornadoes. Historically, devastating tornadoes have struck Europe approximately once every 20 years. Since 1984, individual tornadoes with hundreds of fatalities have occurred in Russia, northeast of Moscow, and in Bangladesh. While it seems that strong and violent tornadoes are much less common in other parts of the world compared to the United States, it also appears likely that tornadoes are vastly underreported in the rest of the world. The prime evidence for this is that the majority of reported tornadoes in many parts of the world are fatality-producing events or are especially newsworthy (such as the 1998 tornado in Umtata, South Africa, while the South African president was visiting the town). This situation is similar to that in the United States in the middle part of nineteenth century, when only approximately 25 tornadoes per year were reported. Recent studies have indicated that probability of a

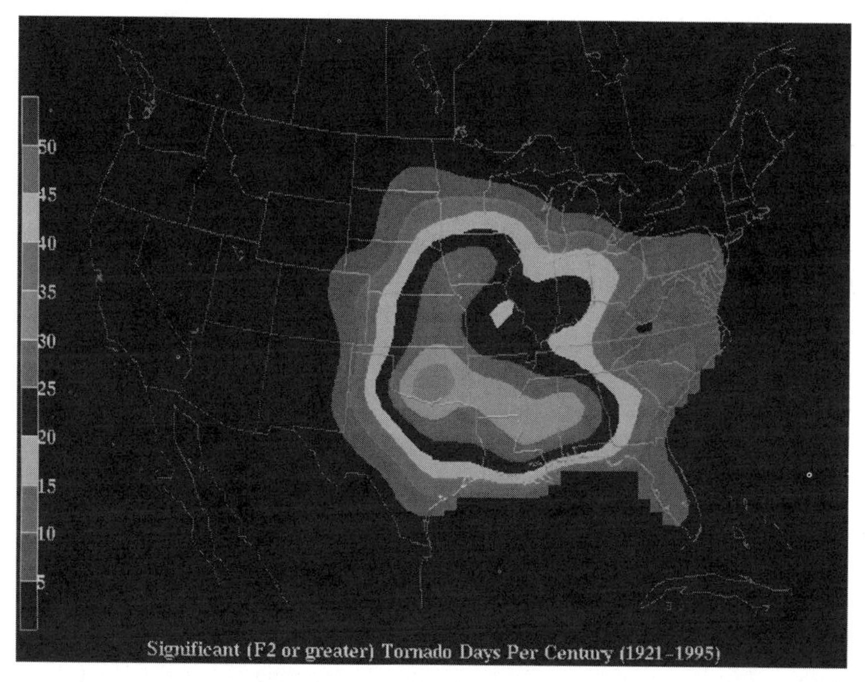

Figure 1 (see color insert) Number of days per century an F2 or greater strength tornado might occur within 25 miles of a point. For example, in southcentral Oklahoma one would expect an F2 tornado within 25 miles about once every 3 years. Calculations based on data from 1921 to 1995. See ftp site for color image.

particular reported tornado being strong or violent is approximately the same over most of the world (Fig. 2).

Hail

The definition of exactly what is severe hail is troublesome. For some agricultural interests during some times of the year, even 1 cm in diameter hail may be devastating. For urban areas, it may take much larger hail, say 4 cm in diameter, to cause problems. The distribution of regions prone to these levels of threat is very different. The smaller limit occurs in much of the temperate world during the warm season. Larger hail is typically limited to the central part of North America and regions near major mountain ranges in the rest of the world (e.g., the Himalayas and Alps). It has been suggested that extremely large hail is much more likely in supercell thunderstorms than in "ordinary" thunderstorms. This is consistent with the observed distribution of tornadoes, presumably associated with supercells, in the central part of the United States. The lack of a relationship when hail of any size is considered has been pointed out for China by showing that the frequency of hail is maximized in the high plateau regions of western China while tornadoes are more common in the eastern part of the country, particularly the Yangtze River valley.

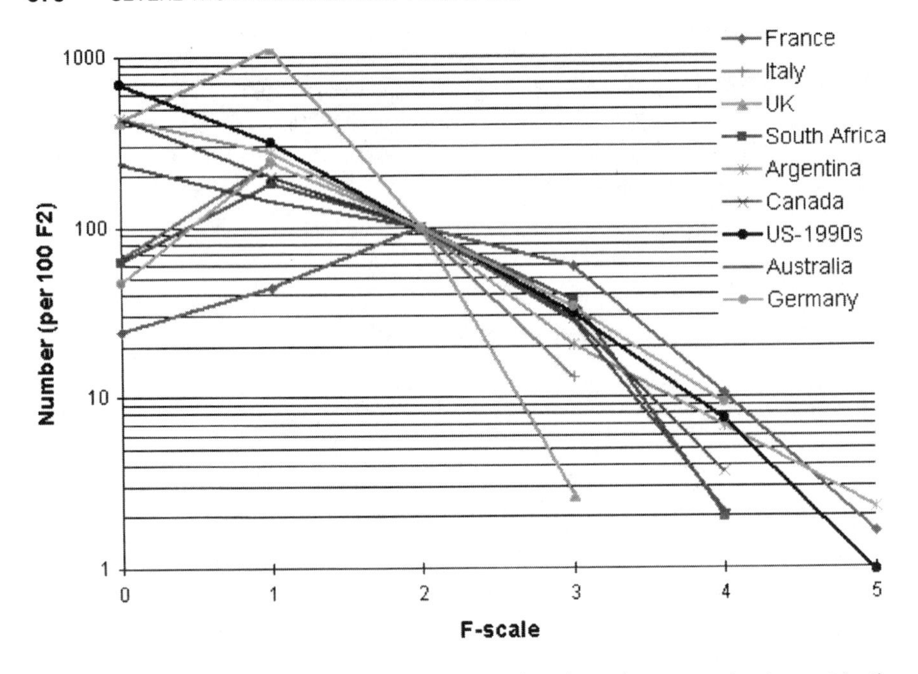

Figure 2 Distribution of F-scale rating for tornadoes in various countries is roughly the same, particularly for the violent (F4 to F5) tornado. See ftp site for color image.

The high plateaus and other regions downwind of mountains may produce large hail because of the steep tropospheric lapse rates that develop as air comes over mountains.

Unfortunately, observations of hail have not been consistent through the years. In the United States, the number of reports of severe hail (approximately 2 cm or larger) have increased by an order of magnitude in the past 30 years. Most of the increase has been in the smaller end of the severe range. As a result, attempts to develop a climatology based on the reports face significant challenges. Researchers are faced with the dilemma of having small sample sizes or an inhomogeneous record. Efforts to use insurance losses are complicated by the issue of what causes the losses (agricultural versus urban interests) and the temporal inhomogeneity of the insured base. Nevertheless, extremely large losses (greater than $500 million) have been associated with hailstorms in areas such as Munich (Germany), Denver, Colorado, and the Dallas–Fort Worth, Texas, region in the last 15 years.

Damaging Convective Wind Gusts

Strong winds associated with thunderstorms are a common feature. Little has been done to document their climatological occurrence until recently, although they are almost certainly the most common severe weather event. Damaging straight-line thunderstorm wind gusts are usually associated with cold air outflow as the downdraft of the storm reaches the ground, producing what has been termed a *downburst.* Factors that influence the generation of damaging wind gusts at the surface include: negative buoyancy enhanced by evaporative cooling within unsaturated air, precipitation loading within the downdraft, and downward transfer of horizontal momentum by the downdraft. Again, aspects of these processes are dependent on storm-scale microphysics including drop size distribution and liquid water content per unit volume, neither of which can be determined from standard observing systems.

Strong winds can occur in a variety of situations. They can be associated with small, short-lived downdrafts and even when they are relatively weak (say less than 25 m/s), they can be a significant hazard to aviation. Considerable effort has been expended in the last 20 years to decrease commercial aircraft accidents due to thunderstorm downdrafts. Radar detection and education of the aviation industry about the threats seems to have limited the number of accidents in the last decade, after several occurred from the early 1970s through the mid-1980s.

Larger areas can be affected by high winds when organized systems of thunderstorms occur. In the United States, widespread convective wind events are sometimes referred to as *derechos.* They occur in association with mesoscale convective systems, which are composed of a number of individual thunderstorms. Often, they are arranged as a squall line, producing a wide area of high winds with new convective cells initiated on the leading edge of the outflow from earlier cells. The system may maintain itself for many hours, provided sufficient low-level moisture and midlevel unstable air can be found as the system moves along.

Flash Floods

Flash floods are the most widespread severe local storm phenomena associated with large loss of life. They occur all over the world, especially in regions of complex terrain. They are the most difficult to forecast, in part because they involve both meteorological and hydrological aspects. Determining their effects is complicated further by interactions with people and buildings. If a flash flood occurs in a location where it does not impact life or property, it is unlikely to be reported. On the other hand, relatively minor precipitation events may produce significant flooding if antecedent conditions exacerbate the flooding as occurred in the Shadyside, Ohio, flood of 1990 with saturated soils or in the Buffalo Creek, Colorado, flood of 1996 when a forest fire cleared vegetation from the area a couple of months before the rain event.

Great loss of life has been associated with flash flooding, even in developed nations. Recently, a campground in Biescas in the Spanish Pyrenees was flooded with more than 80 deaths. In 1998, 11 hikers were killed by a flash flood in a "slot canyon" in northern Arizona when rain-generated runoff from a storm tens of kilometers away was funneled into the canyon. (This storm, by the way, was accurately located by NWS radar in southern Utah, and its potential impacts relayed to Park Service personnel. Unfortunately those who died chose to ignore the warning.) The three biggest convective-weather death toll single events in the United States (with the exception of aircraft crashes) since 1970 have all been flash floods. Death tolls in developing countries are frequently difficult to estimate.

Flash floods are distinguished from main-stem river floods by the extremely rapid rate of rise of water levels. While main-stem river floods may have water stages rising by tens of centimeters per day, flash floods are associated with water stages rising by tens of centimeters per hour or, in extreme cases, per minute. Small streams may carry 100 times their normal capacity and, often, it is very small basins that produce flash floods. (In operational practice, even in the United States, this can cause problems, since these small basins may not be mapped as well as larger basins, particularly for comparison to radar estimates of precipitation. As a result, forecasters may be unaware of the nature of the threat even if accurate estimates of rainfall are available.)

The threat from flash floods has increased in some regions because of the increased use of mountainous regions for recreation. Excellent examples include the Big Thompson (Colorado) and Biescas (Spain) floods. Public response to flash flood forecasts is frequently poorer than for other weather hazards forecasts, such as for tornadoes. Most people have experienced heavy rain events and fail to realize until it is too late that the particular event underway is more dangerous. Further, heavy rain often washes out roads and bridges, which makes escape and rescue difficult.

Lightning

The frequency and location of cloud-to-ground lightning in the United States are now known quite well. The deployment and operation during the last decade of automatic real-time lightning detection sensors have made this possible. An average of about 25 million cloud-to-ground flashes are detected by the National Lightning Detection Network (NLDN) in the United States every year.

The map of network-detected flashes in Figure 3 shows that much of peninsular Florida has the greatest frequency of flashes per area over a year. Flash density decreases to the north and west from there. Important local variations occur along the coast of the Gulf of Mexico, where sea breezes and urban areas enhance lightning frequency. Other maxima and minima are located in and around the western United States where there are mountains and large slopes in terrain.

Lightning is most common in summer (Fig. 4*a*); about two-thirds of the flashes occur in June, July, and August. In the southeastern states, lightning is more common throughout the year, since there often is a significant amount of moisture in the lower and middle levels of the atmosphere. Air needs to be lifted strongly to form lightning; coastlines and large mountains provide persistent updrafts that result in lightning-producing thunderstorms.

During the course of the day, lightning is most common in the afternoon (Fig. 4*b*). Nearly half of all lightning occurs from 1500 to 1800 Local Standard Time (LST). Flashes are most common in the afternoon because the updrafts needed for thunderstorm formation are strongest during the hours of the day when surface temperatures typically are the highest, which results in the greatest vertical instability.

The primary information for lightning deaths and injuries in the United States is the National Oceanic and Atmospheric Administration (NOAA) publication *Storm*

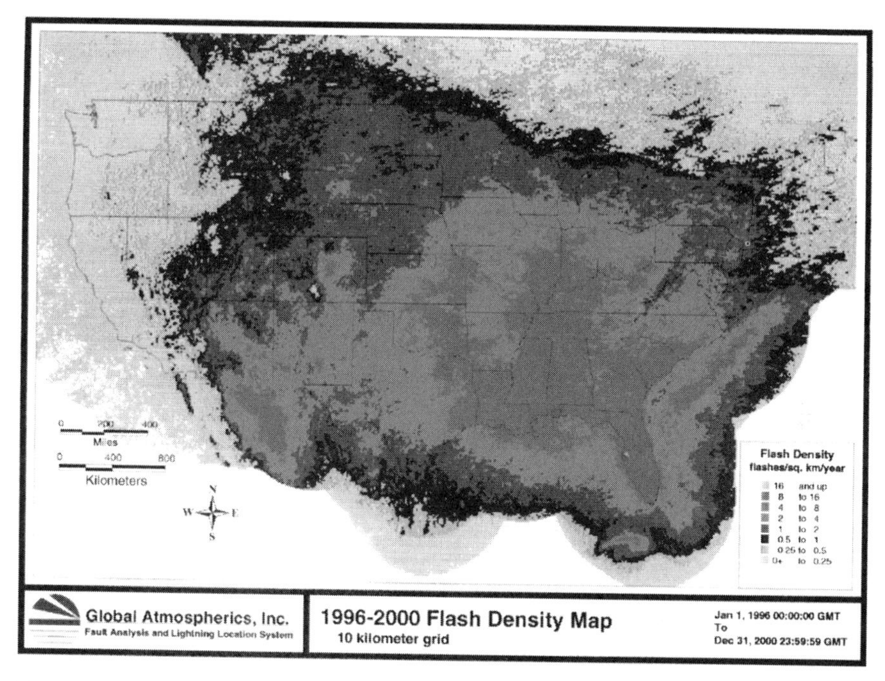

Figure 3 (see color insert) Number of cloud-to-ground lightning flashes per year for the United States based on data from 1996 to 2000. (Courtesy Global Atmospherics, Inc.) See ftp site for color image.

Data. Since most lightning casualties occur to one person at a time, and are more dispersed in time and space than other severe weather phenomena, lightning casualties are underreported.

The spatial distribution of deaths and injuries in Figure 5 shows the largest absolute number to be in Florida, whose total is twice that of any other state. Most of the other states with large numbers of casualties are among the most populous in the country. However, when population is taken into account, the highest rates of lightning casualties per million people are found in the Rocky Mountain states, Florida, and other states in the southeast. The time distributions of lightning deaths and injuries by month and day in Figure 4 show a close resemblance to the distribution of the actual lightning flashes in the same figure.

Over the last 100 years, the rate of lightning deaths has dropped significantly. The decrease parallels a major shift from rural to urban settings for much of the U.S. population and is apparent in similar records in Europe and Australia. The most common activity of lightning casualty victims has also changed during this period from agricultural to recreational.

Around the world, data from lightning sensors on satellites show that lightning occurs most often over land in the tropical and subtropical areas of Africa, South America, and Southeast Asia. The annual rates of lightning per area often exceed those in Figure 3 for Florida.

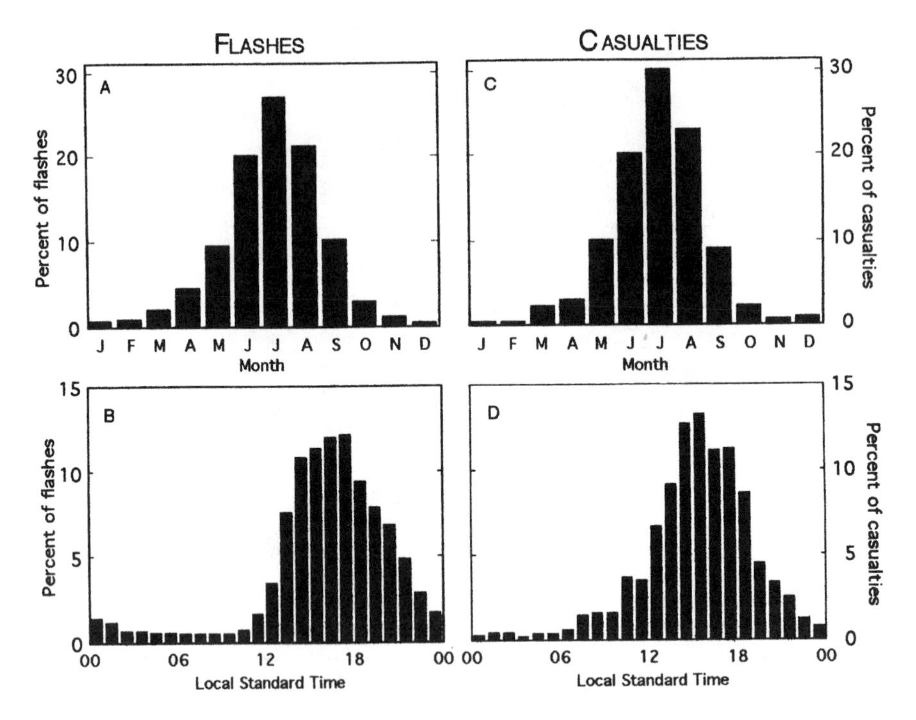

Figure 4 Lightning histograms showing flash rates by (*a*) month of year and (*b*) time of day. (Courtesy of National Severe Storms Laboratory.)

About 100 years ago, the annual U.S. death rate from lightning was 6 per million people, which is an order of magnitude higher than now. The United States then was a more agriculturally oriented society, living and working in ungrounded and less substantial buildings than are now common. This rate may continue to be appropriate for populous tropical and subtropical areas where lightning is frequent.

There are now about 100 lightning deaths per year in the United States, but this total could be around 1000 if the population was still rural, practiced labor-intensive agriculture, and lived in less substantial dwellings. So, the worldwide death total could be expected to be at least 10,000 per year, since many people continue to live in such situations. A ratio of 10 injuries per death gives a global total of 100,000 injuries a year from lightning.

3 CLOUD PROCESSES AND STORM MORPHOLOGY

Ingredients of Deep Moist Convection

Deep moist convection requires three ingredients to occur: instability, lift, and moisture. Interesting weather can occur in the absence of these ingredients, but it will not be deep moist convection. For example, in the absence of instability, forced

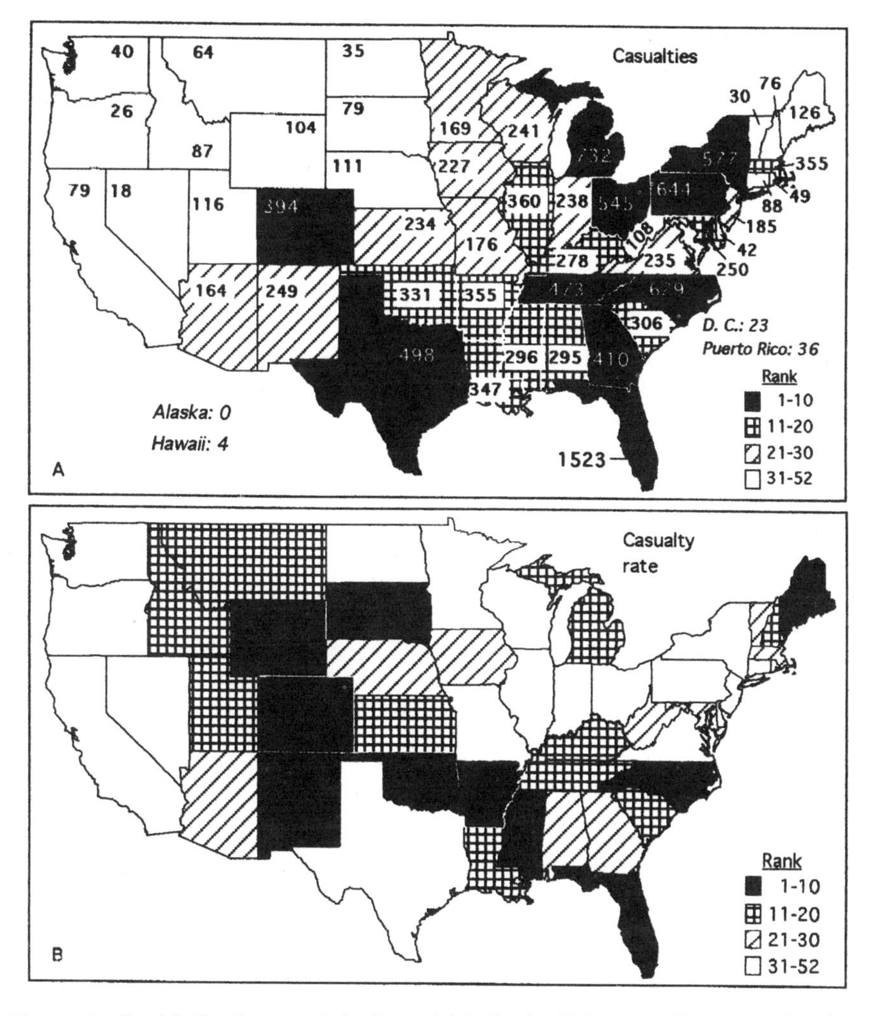

Figure 5 Spatial distribution of deaths and injuries by U.S. state. (Courtesy of National Severe Storms Laboratory.)

ascent of moist air over topography or over a frontal surface can produce heavy precipitation; but, without instability, it is not generally considered convection. Sometimes the term forced convection is applied to this situation. In this section, however, we will focus mainly on free convection, where all three ingredients are present.

Before discussing the ingredients of convection individually, it is necessary to introduce several important concepts. First is the concept of hydrostatic stability. In a hydrostatically stable atmosphere the downward gravitational force associated with the weight of the air is exactly balanced by the upward vertical pressure gradient force (the pressure near the surface of Earth is greater than the pressure aloft, resulting in a pressure gradient force that acts upward). Generally, the atmosphere is very close to hydrostatic stability at most times and locations. Another important concept is that of the lapse rate, a measure of how the temperature in a column of air changes with height. In the troposphere, the temperature almost always cools with increasing height, so a positive lapse rate indicates decreasing temperature with respect to height. As will be explained below, large lapse rates indicate rapidly decreasing temperatures with height, and this is a favorable environment for the development of convection.

Buoyant stability is a measure of the atmosphere's resistance to vertical motions. The primary way meteorologists measure stability is through parcel theory, a pedagogic tool in which idealized bubbles of air called air parcels are employed. An air parcel is a small bubble of air that does not exchange heat, moisture, or mass with its environment, i.e., the thermodynamic process is adiabatic. As air parcels sink in the atmosphere, the increasing pressure they encounter causes them to compress and warm. In fact, this rate of warming is a constant 9.8°C/km and is termed the dry adiabatic lapse rate. Likewise, when air parcels rise in the atmosphere, the decreasing pressure allows them to expand and cool at the dry adiabatic lapse rate. The dry adiabatic cooling rate occurs as long as the parcel's relative humidity is less than 100%. Saturated parcels (i.e., the parcel relative humidity is 100%) cool less as they rise because cooling causes condensation of water vapor. A process that releases latent heat and lessens the rate of cooling with height to a value ranging from 4 to 7°C/km, depending on the temperature and pressure of the parcel. The cooling rate for a saturated parcel is termed the moist adiabatic lapse rate.

Parcel buoyancy is defined by comparing the parcel's temperature to the surrounding environment's temperature. If an air parcel is warmer than its environment, it is said to be positively buoyant and the parcel will accelerate upward. If an air parcel is colder than its environment, it is said to be negatively buoyant and the parcel will accelerate downward. If an air parcel is the same temperature as its environment, it is said to be neutrally buoyant and there is no net force on the parcel. If the parcel buoyancy is large, the accelerations are significant and cause the atmosphere to deviate significantly from hydrostatic balance. These nonhydrostatic forces are often large and an important mechanism for the development and maintenance of severe thunderstorms.

Consider a moist, but unsaturated, air parcel near the surface of Earth. If the actual lapse rate of the atmosphere is between the moist and dry adiabatic lapse rates, then a rising air parcel will cool at the dry adiabatic lapse rate, a rate larger than the environmental lapse rate. Thus, this parcel will be negatively buoyant and will need to be forcibly lifted to continue to rise. As the air parcel cools and expands, it may eventually reach 100% relative humidity, or saturation. The height of this point is called the lifting condensation level (LCL). Further forced lifting will result in the air parcel cooling at the moist adiabatic lapse rate. Eventually, the temperature of the rising air parcel may become warmer than the environmental air at the same height, becoming positively buoyant. The height of this point is called the level of free convection (LFC). Typically, the positive buoyancy continues with height until the air parcel rises near the tropopause, where the stability becomes larger, to its equilibrium level, the point of neutral buoyancy. The integrated positive buoyancy from the LFC to the equilibrium level can be computed and is termed the convective available potential energy (CAPE), which is related to the maximum updraft possible under parcel theory. The integrated negative buoyancy needing to be overcome during forced parcel lifting near the surface is called the convective inhibition (CIN), and the layer over which the CIN occurs is called the cap or lid. When the environmental lapse rate lies between the moist and dry adiabatic lapse rates, conditional instability is said to exist. Conditional instability indicates that the atmosphere will form convection if enough air parcels are lifted to the LFC by some mechanism.

CAPE is an important measure of the integrated instability in the atmosphere. One way to generate a high CAPE environment is to get dry air with large lapse rates on top of warm, moist air. During the spring in the central United States, this so-called loaded-gun sounding frequently occurs when low-level warm moist air from the Gulf of Mexico is overrun by dry air at midlevels from the Mexican Plateau or the Rocky Mountains. This stratification can produce high-CAPE soundings, loaded for a classic Central Plains outbreak of severe weather. A certain degree of CIN is also required so that convection does not break out spontaneously and so that the CAPE can build to large values as the surface temperatures warm during the day.

The second ingredient for convection to occur is that sufficient lift for the air to reach its level of free convection is required. Often this arises due to surface airflow boundaries where convergence occurs, forcing low-level ascent. These surface boundaries can be fronts, drylines, sea-breeze convergence lines, horizontal convective rolls in the boundary layer, or even outflow boundaries from previous convection. These thunderstorms may occur due to buoyant thermals in the warm boundary layer air near the surface of Earth. Sufficiently strong thermals may have enough penetration out of the boundary layer that their LFC may be reached, thus initiating convection.

The final ingredient for convection is an adequate supply of moisture. For most environments, such as the loaded-gun sounding, an increase in low-level moisture results in an increase in CAPE. The dry air aloft is also important since it can quickly evaporate the warm moist thermals bubbling up from the boundary layer. Evidence suggests that it may take several generations of ever-deepening towering cumulus congestus to sufficiently moisten the lower part of the dry layer in the loaded-gun sounding. Without sustained vertical motion and moisture, deep moist convection may never develop, even from a well-primed loaded-gun sounding.

Although not essential for convection to occur, a factor that is important for controlling the type of convective system that may develop is the vertical change in the horizontal wind direction and speed. This is called vertical wind shear. (Readers may be familiar with the term wind shear as it relates to aircraft accidents. In that case, wind shear refers to the horizontal change in the horizontal wind speed and direction. It is important to distinguish between the two different types of wind shear.)

The vertical wind shear is important to deep moist convection for several reasons. Too much wind shear tears apart cloud elements, not allowing updrafts to develop deeply and nearly vertical. Too little wind shear results in the downdrafts suppressing the updrafts, shutting off the flow of warm moist air to the storm. The proper amount of wind shear is important for separating the updrafts from the downdrafts of a thunderstorm, thus permitting the storm to be long lived. More specifically, convection produces cold downdraft air that pools underneath the storm. A gust front, marking the boundary between the cool downdraft air and the warm environmental air, can lift the warm air, sparking new convective storms. In the absence of shear, the downdraft air will spread uniformly in all directions. New cells formed along the gust front will be quickly undercut by the expanding gust front from the parent cell, thus limiting further convective development. As the shear increases, new convective cells will move downshear away from the parent cell faster than the evolving gust front, allowing continual redevelopment of new cells. This effect is particularly important in long-lived multicells and squall lines.

Types of Convective Storms

Ultimately, the role of convection in the atmosphere is to take an unstably stratified sounding and make it more stable by lifting the warm, moist low-level air and bringing cooler, drier air down in the downdrafts. This removes the conditional instability present in the pre-convective atmosphere. It is the interplay between the release of this instability within the storm itself and the environment that produces the panoply of storm types that we see.

Although many of the factors that affect storm structure and evolution are understood, there is still much to be discovered about the relative roles of the large-scale environment and the internal dynamics of the storm itself. This is an important issue because it relates to the limits of predictability of convective storms. If the environment of the storm is a strong factor in storm evolution, then storms are potentially more predictable since the large-scale data in the storm environment is usually well observed. But if the internal dynamics of the storm are the most important factor, then it may be a long time, if ever, until we have measurements within the storm that could be useful for understanding, let alone predicting, the storm evolution. For the purposes of this section, we consider the effect of the environment on the type of convective storm.

At least three environmental factors affect the type of convective storm that forms: wind shear, instability, and synoptic setting (e.g., fronts, drylines, jet streams). The type and direction of wind shear that the convection initiates is important for the morphology (or mode) of convection that results. The amount of instability affects the strength of the convective updrafts. The flow pattern in which the storms develop may also play a role in the mode and strength of the resulting convection. In the following section, the types of convective storms and their characteristics are summarized. More detailed discussion of the individual storm types can be found in later sections of this chapter.

The individual cumulus clouds that constitute so-called *pulse* or *air-mass* thunderstorms are typically a few kilometers in diameter with updrafts on the order of 10 m/s or more. CAPE is usually less than 1000 J/kg and the deep-layer wind shear is weak (less than 10 m/s over 10 km). These storms typically last 30 to 50 min and produce short-lived localized showers with few, if any, reports of severe weather (tornadoes, hail, or damaging winds). Typically, large areas tend to erupt in convective cloud about the same time when the surface temperature reaches the convective temperature, the temperature required to eliminate any low-level CIN. While useful for these situations, the use of the convective temperature for other types of storms is not recommended.

With moderately strong low-level and deep-layer wind shear and moderate to high CAPE, *supercells* may form. The essence of a supercell thunderstorm is a single, nearly steady, rotating updraft. High winds, flooding rains, large hail, and potentially long-lived violent tornadoes can occur with supercells. If the wind shear is constant in direction with height (called a straight-line hodograph), maturing supercells will tend to split into left-moving and right-moving cells. If the wind shear curves counterclockwise with height (in the Northern Hemisphere), right-moving cells will tend to dominate with counterclockwise rotation. On the other hand, if the wind shear curves clockwise with height, left-moving cells will tend to dominate with clockwise rotation in the Northern Hemisphere. Right-moving cells tend to be more common. More will be said about supercells later.

Sometimes, individual convective clouds will join together as gust fronts caused by downdrafts from the parent cells combine to lift unstable environmental air to its LFC, forming secondary convection along the periphery of the parent cells. These storms are called *multicells* and range in organization from poorly organized clusters of individual convective elements to highly organized linear structures including bow echoes and larger-scale squall lines. Individual storms within a multicell usually move to the left (in the Northern Hemisphere) of the mean motion of the multicell itself. Because the warm moist air tends to be ingested on the right side (or equatorward side, in the Northern Hemisphere) of the multicell, new cells form preferentially on this side, forcing older cells toward the left of the multicell storm. Because multicells propagate away from the original convection, a certain amount of low-level shear is needed for gust-front propagation. If storm motion is very slow, local areas may be affected by prolonged periods of rain, making flooding a strong possibility.

Sometimes multicellular convection will organize in lines. Typically this occurs because linear features exist at the surface (such as a cold front) that organizes the convection. Because of the linear forcing, a common location for development of *squall lines* and bow echoes is along fronts. Sometimes these lines move ahead of the front that spawned them; at other times, forcing above the surface will produce squall lines in the warm sectors of extratropical cyclones. In some cases, often when the vertical wind shear is weak, mesoscale convective complexes form.

4 MESOSCALE CONVECTIVE SYSTEMS, MESOSCALE CONVECTIVE COMPLEXES, SQUALL LINES, AND BOW ECHOES

Mesoscale Convective Systems

While isolated thunderstorms are important producers of severe weather, it is just as common for thunderstorms to merge and interact, forming a more complex precipitation system. These more complex thunderstorm systems are called mesoscale convective systems (MCSs)—cloud and precipitation systems that are comprised of a group of thunderstorms that have a contiguous precipitation area of ~ 100 km in at least one direction (Fig. 6). MCSs are particularly important because they are observed worldwide, over both the Tropics and midlatitudes, and over land and water. MCSs also have been shown to produce approximately half of the warm season rainfall in the central United States, indicating that they are important components of the hydrologic cycle. Further, in the form of squall lines and bow echoes, such systems likely account for most, if not all, of the widespread convective windstorms that occur around the world.

Mesoscale Convective Complexes

Observations indicate that MCSs occur in a variety of shapes and sizes. The largest 1% of MCSs are called mesoscale convective complexes (MCCs). They are most commonly identified from infrared satellite imagery as large regions of cold, nearly circular, cloud tops. MCCs typically produce a contiguous cloud shield of 200,000 km^2 and last for over 15 h (Table 1). These cold cloud tops are the collective

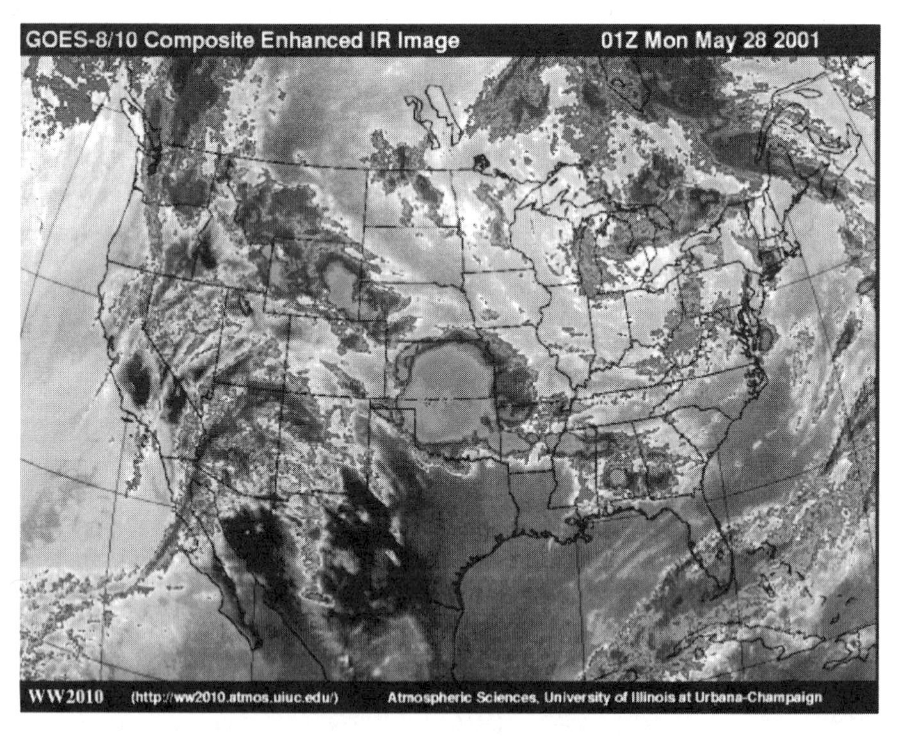

Figure 6 (see color insert) Infrared satellite picture of a developing MCS over south-central Kansas and northern Oklahoma at 0100 UTC May 28, 2001. A severe squall line was developing underneath the cloud shield and subsequently moved southeasterly for the next 12 h to the Texas coast. [Image courtesy of University of Illinois WW2010 project (http://ww2010.atmos.uiuc.edu).] See ftp site for color image.

TABLE 1 Mesoscale Convective Complex (MCC) Criteria

Size	A—Contiguous cold cloud shield with infrared (IR) temperatures ≤ 241 K with area $\geq 100{,}000\ \mathrm{km}^2$ B—Interior cold cloud region with IR temperatures ≤ 221 K with area $\geq 50{,}000\ \mathrm{km}^2$
Initiate	Size definitions A and B are first satisfied
Duration	Size definitions A and B must be met for a period of at least 6 h
Maximum extent	Contiguous cold cloud shield (IR temperatures ≤ 241 K) reaches maximum size
Shape	Eccentricity (minor axis/major axis) ≥ 0.7 at time of maximum extent
Terminate	Size definitions A and B no longer satisfied

anvils from interacting thunderstorm cells constituting the MCC. MCCs tend to initiate in the late afternoon, reach their maximum extent around midnight, and terminate early in the morning. They tend to form in situations with large-scale warm advection in flow regimes with strong anticyclonic shear or anticyclonic curvature at the jet-stream level. The processes by which individual cells join to form an MCC are poorly understood. Some MCCs last for several days and can influence large portions of a continent. In addition, when MCCs develop or move over warm ocean waters, they have been observed to evolve into tropical storms.

Squall Lines and Bow Echoes

While MCCs represent only a small portion of all MCSs, commonly observed subsets of MCSs include the squall-line and bow-echo complexes. Squall lines consist of a well-defined line of thunderstorms with an associated stratiform precipitation region and are often identified from radar data. (Fig. 7). The stratiform

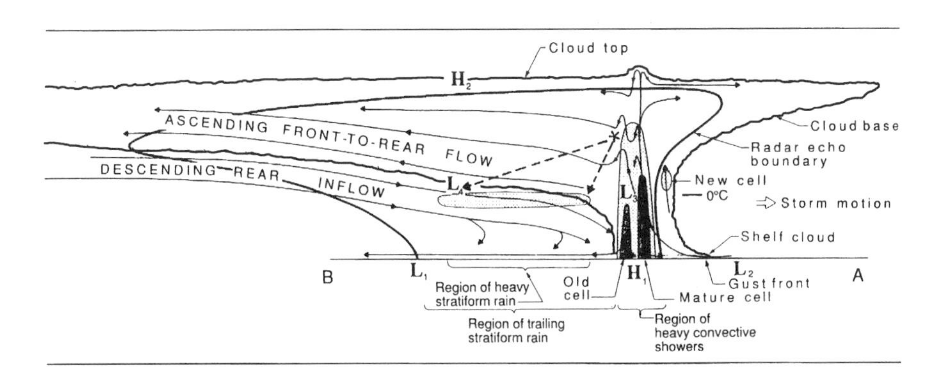

Figure 7 Conceptual model of a squall line with a trailing stratiform area view in a vertical cross section oriented perpendicular to the convective line (i.e., parallel to its motion). (Adapted from Houze et al., 1989, American Meteorological Society.)

precipitation region can be located either in front of, along, or behind the convective line, although it is most common for the stratiform region to be behind the convective line. The convective and stratiform regions often have a symmetric radar depiction early in their life cycle, evolving into a more asymmetric pattern with time (Fig. 8). The distinction between convective and stratiform precipitation is made by comparing the typical vertical air velocities with the terminal fall velocities of ice crystals and snow (~ 1 to $3\ \mathrm{m/s}$). If the vertical air velocities are larger in magnitude than the ice and snow fall velocities, then the precipitation is convective; otherwise the precipitation is stratiform. On radar, the convective regions have much larger values of reflectivity, indicating heavier precipitation or hail. While the typical vertical motions in stratiform precipitation regions are not large, these regions provide $\sim 40\%$ of the total precipitation reaching the ground for many squall lines owing to their large areal extent.

Four typical phases have been identified in the life cycles of MCSs (Fig. 9). The first phase is the formative stage in which the initial thunderstorms are developing and act independently from each other. This is often the time during which the most severe weather occurs. As the thunderstorms grow and merge, the MCS enters the intensifying stage. During this stage the MCS is seen first to have a contiguous precipitation region using radar. The mature stage occurs as a large stratiform rain region is produced, often from older convective cells that weaken and move to the rear of the convective line. This stratiform rain region often grows in size until it is

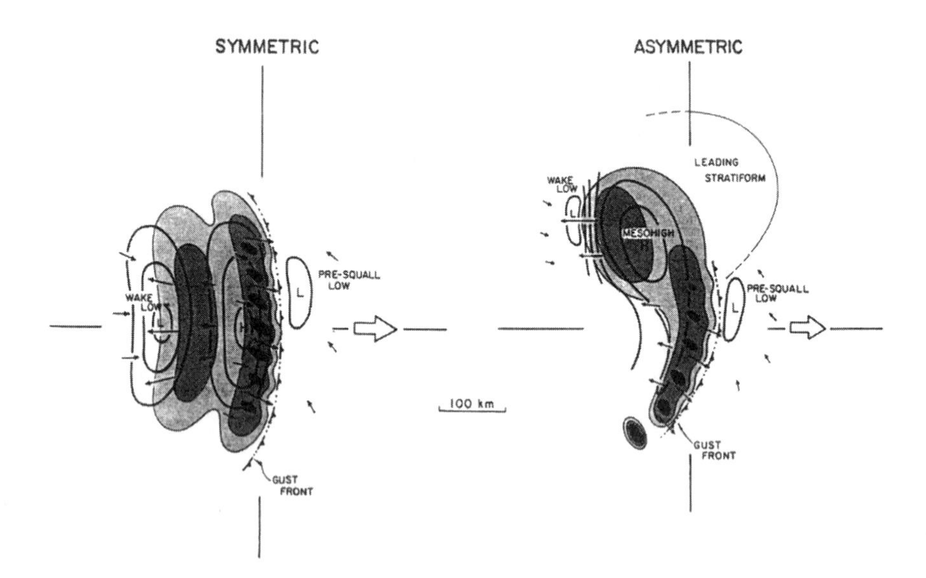

Figure 8 Conceptual model of a midlevel horizontal cross section through (*a*) an approximately two-dimensional squall line, and (*b*) a squall line with a well-defined mesoscale vortex in the stratiform region. In each case, the midlevel storm-relative flow is superimposed on the low-level radar reflectivity. The stippling indicates regions of higher reflectivity. (Adapted from Houze et al., 1989, American Meteorological Society.)

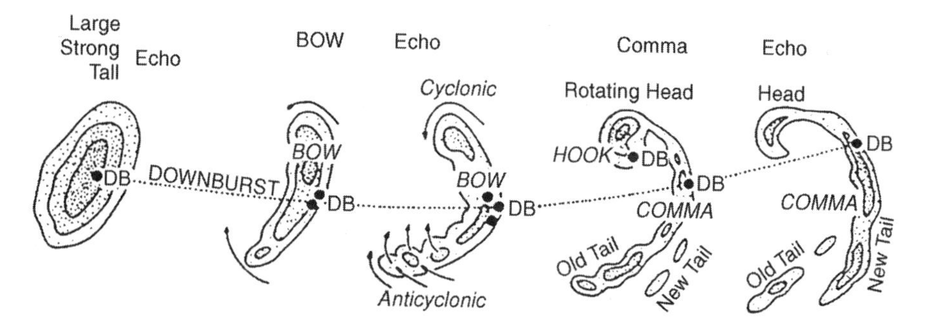

Figure 9 Idealized morphology of an isolated bow echo associated with strong and extensive downbursts. (after Fujita 1978).

an order of magnitude larger in area than the convective region. Significant lightning can occur within the stratiform precipitation region, along with a potential for heavy rainfall and flooding. Positive cloud-to-ground lightning strikes appear to be more common within the stratiform precipitation region than in the convective line. The final, dissipating stage is when the convective portion of the MCS weakens, with fewer and fewer convective cells developing along the convective line. Eventually, all that remains is a weakening region of stratiform precipitation.

One of the most important aspects of MCSs is that the combined effects of the individual thunderstorms and the stratiform precipitation region produce dynamical features that are unique to MCSs and influence both smaller and larger scales of motion. Pools of cooler air are produced near the ground by the evaporation of falling precipitation underneath the thunderstorm cells. In MCSs, these pools merge and influence the development of new convective cells, since warm air approaching these cold pools is lifted up and over the cooler surface air. Thus, MCSs often move in different directions than the individual thunderstorms that constitute the convective line. In the midlevels of the atmosphere, the latent heat released from the change of phase from water vapor to liquid water within the stratiform precipitation region often gives rise to the development of front-to-rear flow in the upper portion of the MCS (Fig. 7). A rear inflow jet also may develop below the front-to-rear flow region of the MCS and approaches the MCS from the rear. Other circulations can develop that produce vortices within the stratiform precipitation regions; these vortices can persist for several days after the initial parent MCS has dissipated. Finally, the upper-level outflows from MCSs alter the mass field near the tropopause and can produce upper-level jet streaks that move away from the MCS and can influence other weather systems downstream. On occasion smaller scale features known as bow echoes (20–200 km in length) form within squall lines or as individual MCSs (Fig. 9). In such situations the severe weather threat, particularly in the form of damaging wind gusts, is enhanced and may continue through much of the life cycle of the MCS. Widespread convection windstorms known as derechos may occur with the longer-lived bow echo dominated MCS events. In such cases it appears the cold pool is enhanced and often moving more rapidly than the mean tropospheric wind. This rapid movement enhances low level storm relative flow and the development of new cells along the gust front. As was mentioned in the last paragraph, predictions of MCS development is complicated. However, in those situations in which bow echo development is dominant, the environmental air above the boundary layer typically displays low relative humidity. This drier air likely enhances the development of the downdraft and cold pool strength as well as the risk of damaging winds at the surface. Owing to the myriad interactions that occur within MCSs, it is probably not surprising that the prediction of MCS development and evolution has proven to be a challenging problem.

5 SUPERCELLS

Definition

The idea that some severe thunderstorms have a markedly different character from other types of thunderstorms owes its origins to the development of radar as an observing tool. Radar gave meteorologists the ability to see the distribution and time evolution of precipitation in thunderstorms. Keith Browning and Frank Ludlam, in England, were pioneers in the interpretation of radar. They participated in field observation campaigns in the United States during the early 1960s after having observed a particularly severe and long-lived hailstorm that struck in and close to the town of Wokingham, England. After observing a number of severe storms with radars, Browning made use of the fact that the distribution of precipitation and its changes with time provides indirect evidence of a thunderstorm's up- and downdrafts. A careful examination of that radar-depicted structure and evolution gave meteorologists a detailed look at what was going on inside a thunderstorm.

Browning and Ludlam noticed that a few severe thunderstorms exhibited a particular set of characteristics: (1) they produced extreme severe weather events (tornadoes, giant hail, and violent wind gusts), (2) they tended to move to the right of the winds, whereas most storms moved more or less with the winds, (3) they exhibited a columnar region of reduced radar echo originally called a vault (later renamed the bounded weak echo region, or BWER), and (4) they had extended lifetimes in comparison to most thunderstorms, persisting for many hours on occasion. At first, these were referred to as severe, right-moving (or SR) storms. In an informal report published in 1962, Browning and Ludlam first used the term *supercell* to refer to these storms. As more and more storms were subjected to scrutiny by radar, the so-called hook echo became associated with supercells (Fig. 10).

The development in the early 1970s of Doppler radar, which can observe the motion of precipitation echoes and, therefore, can be used to infer the airflow, provided conclusive evidence of what Browning and his collaborators had surmised from non-Doppler radar observations: Supercells are characterized by rotation on the scale of the thunderstorm itself. Although from a formal viewpoint, the rotation can be either cyclonic or anticyclonic, the SR storms that Browning called supercells are

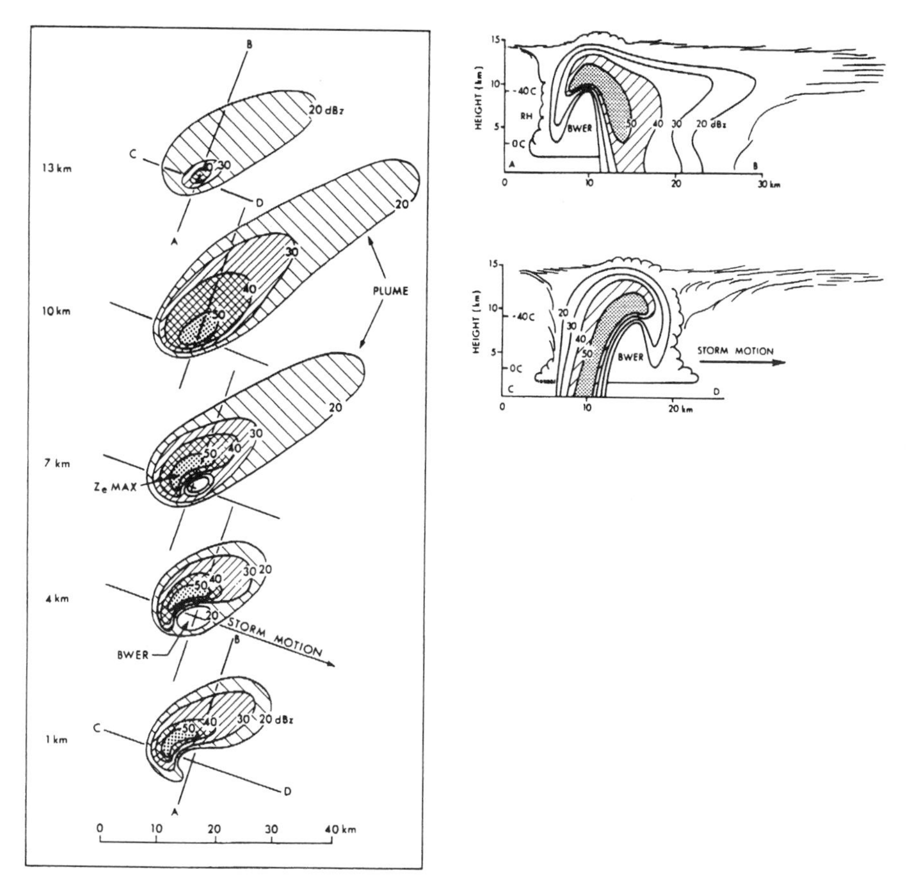

Figure 10 Plan view of supercell showing hook echo, bounded weak echo region (BWER), and forward flank precipitation regions. (Courtesy of National Severe Storms Laboratory.)

characterized by cyclonic rotation. The center of that cyclonic rotation came to be called the mesocyclone. It is the presence of this rotation that distinguishes supercells from other types of thunderstorms. Therefore, it is generally agreed that supercells are thunderstorms that have a deep, persistent mesocyclone. It is the mesocyclone that is deemed responsible for the high probability of severe weather and a supercell's characteristic features.

Supercell Structure and Evolution

Figure 10 illustrates the primary features of a supercell thunderstorm during its mature phase. The fact that supercells can persist for many hours suggests that their structure evolves relatively slowly. However, every storm must have a beginning and end, so no storm is ever truly steady. Some supercells are more nearly steady than others, but all exhibit a basic evolutionary pattern. Thunderstorms begin as pure updrafts, creating towers of cloud called cumulus congestus, which then develop precipitation aloft. The production of precipitation triggers the development of downdrafts, both from the drag effect of having precipitation particles and from the evaporation of some of those particles. Thus, whereas the thunderstorm is initially dominated by updrafts in its mature phase, the thunderstorm has both updrafts and downdrafts. In dissipation, a thunderstorm's updraft weakens and eventually ceases, leaving only the weakening downdraft and precipitation.

Supercells also follow this evolution although the mature phase is prolonged. During the early stages of a supercell, the updraft begins to rotate cyclonically, typically several kilometers above the surface. Thus, the rotation is initially on the same axis as the updraft. With the development of precipitation and the descent of downdrafts, the mesocyclone structure is transformed. Rather than being centered on the updraft, the cyclonic rotation extends into the downdraft, such that the mesocyclone comes to be centered near the interface between updraft and downdraft. The updraft changes its shape from being nearly circular as seen at a given horizontal level, to become elongated and crescent-shaped, with the crescent aligned with the low-level boundary between updraft and downdraft (the so-called gust front). With time, the downdraft and outflow at low levels goes through an evolution called occlusion, whereby the gust front undercuts the updraft, which then dissipates.

Most supercells exhibit a cyclic behavior during their extended mature phase, with new updrafts and mesocyclones forming on the leading edge of the outflow boundary, even as the previous updraft is in the process of dissipation. This process can go on many times, so the mature phase of supercells undergoes a fluctuation on a time scale of 20 to 30 min or so (although the time between "pulsations" is by no means constant). Eventually, of course, the storm can no longer be maintained and the last of updrafts and mesocyclones in the series finishes its life cycle and the storm dissipates.

Origins of Supercell Structure

Browning and his collaborators noticed right from the beginning that supercell storms formed in environments having enhanced vertical wind shear. That is, the winds typically changed both direction and speed with height. Poleward low-level winds became more westerly and increased in speed rapidly in the vertical. Vertical wind shear in the atmosphere is associated with regions of enhanced horizontal temperature differences (or fronts) and also means that the airflow possesses a property called vorticity or spin, which is created when regions of air move past each other at different speeds and/or different directions. When this occurs in association with vertical wind shear, this means the wind profile has horizontal vorticity. This is illustrated in Figure 11, where it can be seen that the change of wind in the vertical implies a potential for rotation about a horizontal axis.

When thunderstorm updrafts develop in regions of strong vertical wind shear, they cause the horizontal vorticity in their surroundings to be tilted upward into the vertical, and it is this uptilted vorticity that gives rise to the rotation of the updraft

Figure 11 Tilting of horizontal vorticity into the vertical via a thunderstorm updraft. (Courtesy of National Severe Storms Laboratory.)

about a vertical axis during its development. This has been demonstrated in computer simulation models and is clearly the source for mesocyclonic rotation in thunderstorms.

Available evidence indicates, however, that the creation of rotation near the surface is associated with a more complex process that involves a supercell's downdrafts. In some supercells, the mesocyclone aloft and that developing near the surface can interact strongly, creating a deep column of intense rotation. Such storms are the primary producers of long-lasting, strong tornadoes and often result in families of tornadoes, with each tornado being a reflection of another pulsation in a cyclic supercell. Other supercells fail to develop a strong interaction between the low-level mesocyclone and the mesocyclone aloft. Such storms can produce other forms of severe weather, notably giant hail, but tornadoes are relatively infrequent and tend to be brief and weak if they do form.

Hazardous Weather Associated with Supercells

It appears that the development of a mesocyclone has a strong influence on the storm, and that influence is such that the likelihood of severe weather in all forms increases if a storm becomes supercellular. Supercell storms constitute only a small fraction of the total number of severe thunderstorms, but they account for a disproportionate share of the most intense forms of hazardous weather.

Tornadoes are arguably the most hazardous weather event associated with convective storms. Although tornadoes are by no means limited to supercells, those tornadoes associated with supercells have a much greater likelihood to be intense and long-lived than those produced by nonsupercell thunderstorms. In turn, this means that such tornadoes have the highest potential for damage and casualties. This is exemplified by events on May 3, 1999, when an outbreak of 69 tornadoes across Oklahoma and Kansas was produced by only 10 supercell storms. One tornado from the first supercell of the day caused $1 billion in damage and 36 fatalities as it tracked first through rural areas southwest of Oklahoma City, Oklahoma, and then on into the metropolitan area.

As with tornadoes, the hail produced by supercells is much more likely to exceed 2 in. (5 cm) in diameter than in nonsupercell storms. There is strong scientific evidence to believe that updrafts are enhanced by the presence of a mesocyclone, and giant hail requires an intense updraft for its formation. Some supercells are prolific hail producers, creating long swaths of hail up to 10 cm in diameter. When such storms interact with populated areas, they can cause damage on the order of $100 million or more from broken glass, dented motor vehicles, roof destruction, and vegetation damage. Occasionally, people are seriously injured or even killed by being caught outdoors during a fall of giant hail.

Windstorms of a nontornadic nature in supercells can also reach extreme proportions. A few times per year in North America, supercells produce swaths of wind up to 20 km wide and more than 100 km long, within which winds can exceed 25 m/s for more than 30 min, with peak gusts approaching 50 m/s. In forested areas, such events produce vast blowdowns of trees. Interactions with populated areas are rare, but the potential for destruction is enormous. In such events, hail can accompany the strong winds, adding to their destructive potential.

Finally, a few supercells are responsible for prodigious rainfall production. It appears that most supercells are not very efficient at producing rainfall because they often are associated with processes promoting evaporation of precipitation. Moreover, supercells are mostly isolated storms that move over a given location relatively quickly. Nevertheless, their intense updrafts can process a lot of water vapor into precipitation, however inefficiently it is accomplished. Thus, supercells have produced bursts of precipitation exceeding 200 mm/h, even if only for a short time. Such intense rainfall rates can create flash floods, particularly in urban areas where runoff is so high owing to the lack of permeability of urban environments composed mostly of concrete, buildings, and other hard surfaces.

Variations on the Theme

Supercells are not all the same, but they have certain common features. The presence of a deep, persistent mesocyclone is the defining feature of a supercell, so it is possible to look only at storms meeting that criterion. Using this broad definition, we should not be surprised to learn that variations on the supercell theme exist.

The most widely used classification scheme for supercells is based loosely on the notion that mesocyclones vary in the amount of precipitation they contain. The prototypes are called the low-precipitation, classic, and high-precipitation supercell.

Low-precipitation (LP) supercells characteristically have little or no precipitation within their mesocyclones. The absence of heavy precipitation means they typically do not produce strong downdrafts and outflow, nor are they likely to be tornadic. However, they can produce falls of giant hail and, like other supercells, tend to do so in relatively long swaths. They typically are not very large storms and tend to be observed mostly in the transitional environments between arid and moist regions (e.g., the western Great Plains of North America). Figure 12 shows a schematic illustration of the appearance of an LP supercell, both visually and on radar.

Classic (CL) supercells most clearly resemble those described by Browning and his collaborators. They often exhibit most, if not all, of the traditional radar echo morphology associated with them in the literature. They produce all forms of severe weather, including the most extreme examples, although they are not likely to be flash flood producers. Figure 13 illustrates the archetypical appearance of a CL supercell.

Finally, high-precipitation (HP) supercells have mesocyclones deeply embedded in precipitation, although there may be a narrow corridor that, at low levels, remains free of precipitation (sometimes referred to as an inflow notch). Figure 14 shows the typical appearance of HP supercells. With heavy precipitation falling into the downdrafts on a supercell's rear flank, the potential for strong winds is quite high. Clearly, HP supercells also have a potential to be heavy rainfall producers, as well as giant hail. HP supercells stand somewhere between CL and LP supercells in their tornado potential. It is rare for a violent, long-track tornado to be associated with storms displaying a predominantly HP structure.

6 TORNADOES

Tornadoes are rotating columns of air that extend from the surface to the interior of a thunderstorm cloud (or, more correctly, a cloud associated with deep moist convection, with or without thunder). The distinction between tornadoes and other vortices within thunderstorms is not always clear, so tornadoes are sometimes defined as having wind speeds near the surface that have the potential to cause damage. Many tornadoes produce cylindrical or conical clouds of condensed water (Fig. 15*a*), but a funnel cloud is not always present. Some tornadoes are made visible only by lofted debris (Fig. 15*b*).

The number of reported tornadoes in the United States now typically exceeds 1000 per year. Annual totals of fatalities and property damage caused by tornadoes vary considerably from year to year. The death toll, as a fraction of the total population of the United States, has been decreasing on average since the mid-1920s. Improved forecasting, detection, and warning of tornadoes; improved communication of warnings; increased public awareness of safety precautions; urbanization; and changes in building standards may be some of the important factors in explaining

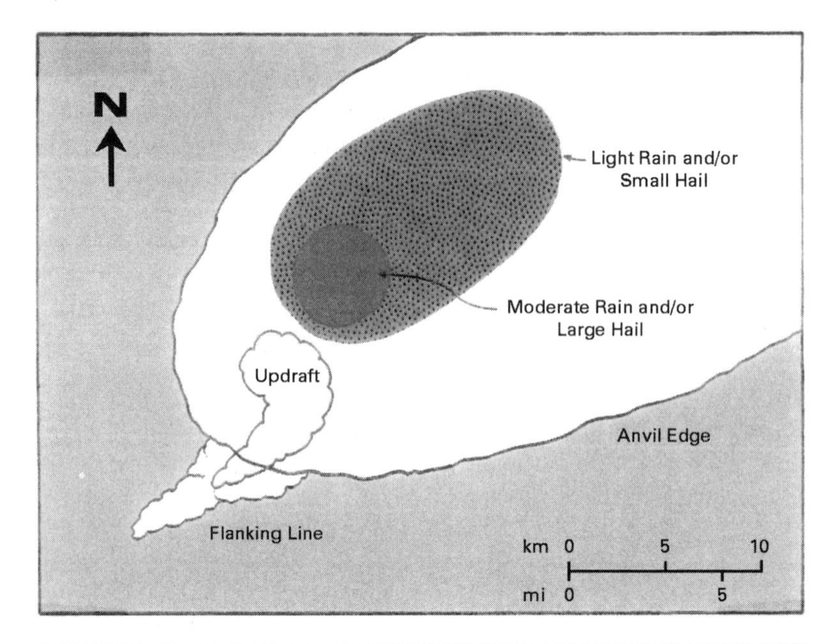

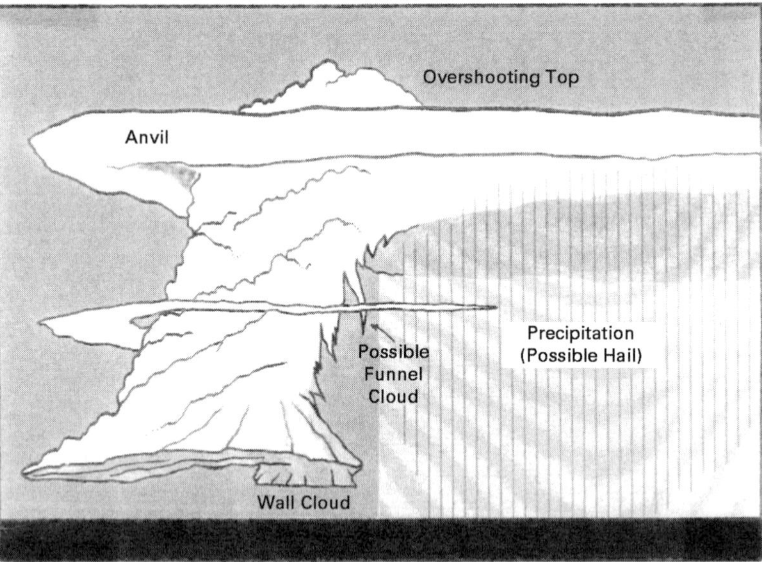

Figure 12 Schematic of a low-precipitation supercell. (Courtesy of National Severe Storms Laboratory.) See ftp site for color image.

this trend. Tornadoes typically last less than 10 min but have been known to persist for over an hour. Wind speeds may exceed 120 m/s (270 mph) in particularly strong tornadoes but are less than 60 m/s (135 mph) in most cases. In some tornadoes, the distribution of strong winds around the vortex core is relatively uniform; in others,

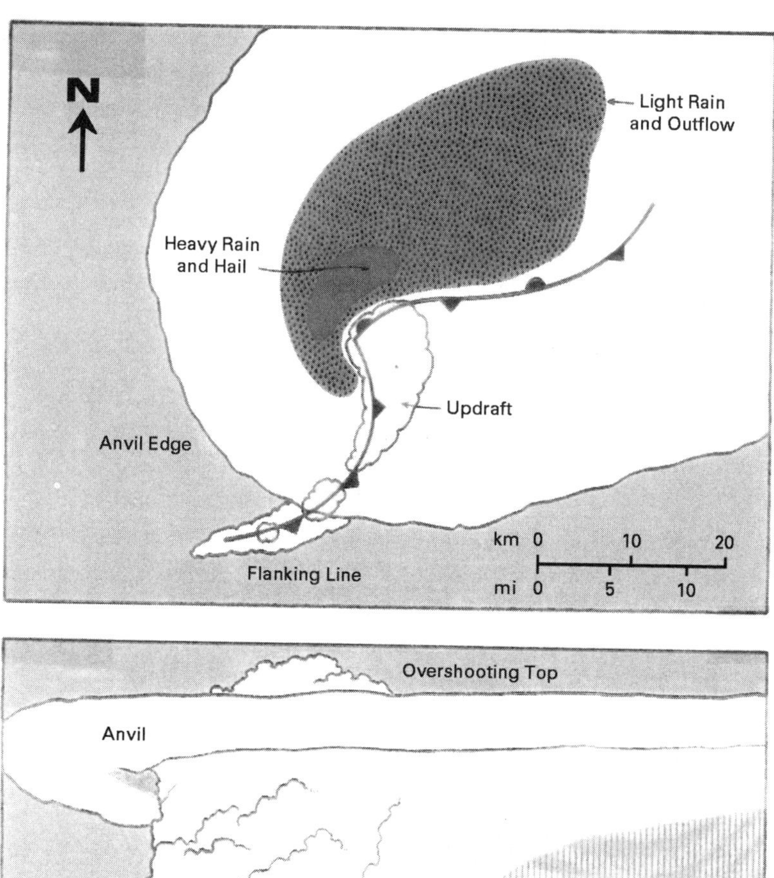

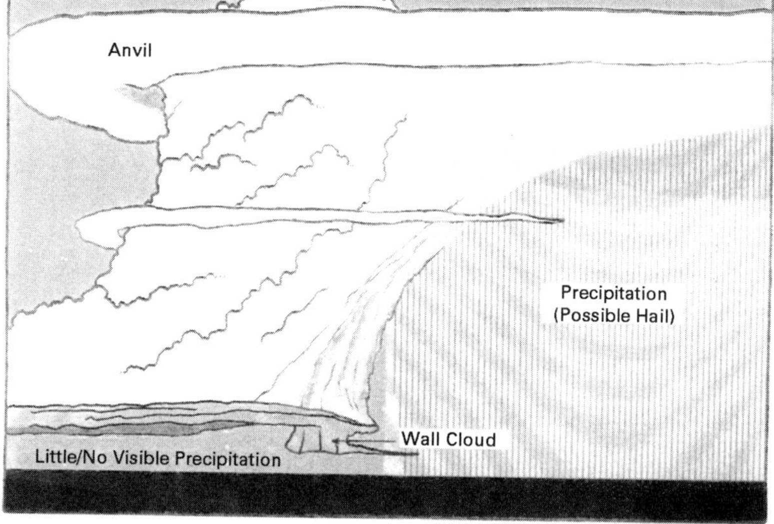

Figure 13 Schematic of a classic supercell. (Courtesy of National Severe Storms Laboratory.) See ftp site for color image.

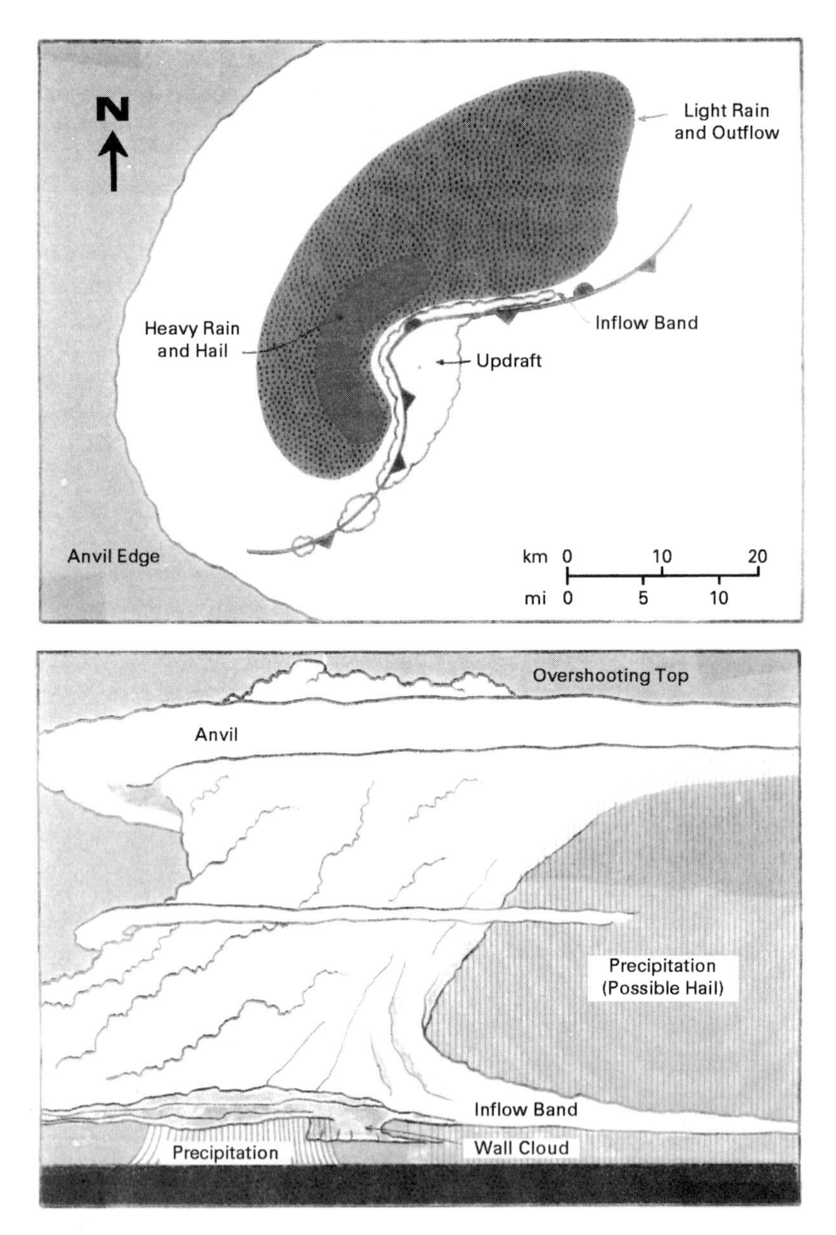

Figure 14 Schematic of a high-precipitation supercell. (Courtesy of National Severe Storms Laboratory.) See ftp site for color image.

(*a*) (*b*) (*c*)

Figure 15 Tornado photographs: (*a*) Tornado with a cone-shaped condesation funnel near Stockton, Kansas, on May 15, 1999 (copyright 1999 by David Dowell). (*b*) Tornado with no condensation funnel near Denver, Colorado, on July 2, 1987 (copyright 1987 by Bill Gallus). (*c*) Tornado with multiple subvortices near Elbert, Texas, on April 30, 2000 (copyright 2000 by David Dowell). See ftp site for color image.

the strongest winds may occur within relatively small subvortices (Fig. 15*c*). Although tornadoes may extend to over 10 km above ground within the cloud, the strongest horizontal winds in mature tornadoes occur within 100 m of the surface. Near the radius of strongest horizontal winds, air rises at speeds comparable to those of the horizontal winds; the extreme vertical velocities may loft debris to great heights. Near the center of the tornado, there may be descending motion. On rare occasions, the large-scale atmospheric pattern may be favorable for the formation of tornadic thunderstorms over a broad region covering multiple states. The "super outbreak" of April 3–4, 1974, is an extreme example. During this outbreak, at least 148 tornadoes occurred within 14 states from the Midwest to the Southeast. The majority of tornadoes, however, are isolated occurrences. Most thunderstorms form in environments that are not favorable for tornadoes, or when the environment would support the development of tornadoes, storms do not always form.

Scientists continue to be challenged to explain the details of how tornadoes form (and to explain why tornadoes do not always form on days when they are anticipated). Organized field programs to study tornadoes began in the early 1970s and continue today. Scientists use a number of mobile devices (Doppler radar, instrumented automobiles, weather balloons, etc.) to collect measurements of wind, temperature, pressure, humidity, and precipitation in and near tornadoes (e.g., Fig. 16). To assess how a tornado forms in a particular thunderstorm, scientists

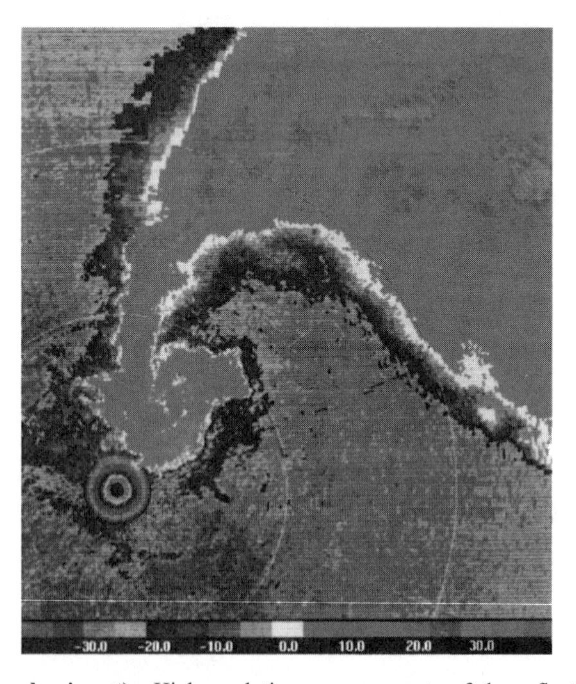

Figure 16 (see color insert) High-resolution measurements of the reflectivity factor (an indicator of precipitation size and number concentration) by the Doppler on Wheels mobile radar in a tornadic storm with a hook echo near Rolla, Kansas, on May 31, 1996. The thin blue rings indicate range from the radar at intervals of 5 km. See ftp site for color image.

require accurate measurements at high temporal and spatial resolution; the need for such complete observations continues to challenge observational capabilities.

Significant advances in our understanding of tornadoes and their parent storms have also come from numerical simulations with high-speed computers. In a numerical model, the thunderstorm and its environment are represented on a three-dimensional grid. During each time step of the simulation, the dynamical equations governing the changes of wind, temperature, pressure, humidity, liquid water content, and ice content are solved at each grid point. Numerical simulations (e.g., Fig. 17) have reproduced storm features analogous to those in observed storms.

The prevailing hypotheses for how tornadoes form all involve horizontal convergence of low-level air that has significant angular momentum. As air retains its angular momentum with respect to the center of the region of convergence while spiraling inward, the rate of rotation increases. The sequence of events leading up to the formation of a tornado is not identical in all cases. If the air near the surface in the environment of a thunderstorm already had significant rotation before the thunderstorm formed, then the mechanism of tornado formation could be relatively simple (Fig. 18). For example, a low-level boundary, along which there is a shift in the wind direction and speed, may be present in the environment. The wind shift may be organized into circulations that are initially a few hundred meters to a few kilometers wide (Fig. 18). If one of these circulations coincides with a growing

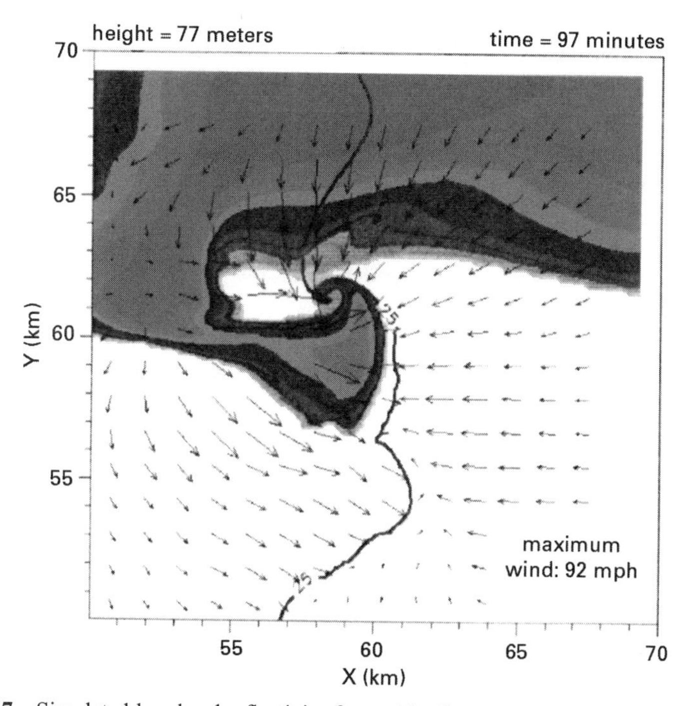

Figure 17 Simulated low-level reflectivity factor (shading) and horizontal wind (vectors) in a numerical simulation of a tornadic storm. (Courtesy of Matt Gilmore of Cooperative Institute of Mesoscale Meteorological Studies.) See ftp site for color image.

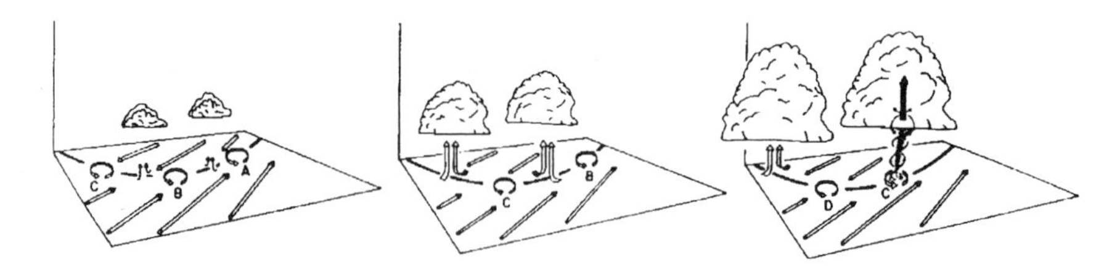

Figure 18 Schematic of nonsupercell tornado formation (by R. Wakimoto and J. Wilson; copyright 1989 by the American Meteorological Society). The arrows indicate the wind direction. Numbered loops represent circulations along a low-level boundary, which is indicated by a black line.

thunderstorm updraft, then a tornado may form as the circulating air at the base of the updraft converges to smaller and smaller radii. This mechanism involving the interaction of an updraft with a preexisting low-level circulation may apply to tornadoes that form within otherwise nonsevere deep moist convection over water ("waterspouts") but may also explain the formation of some tornadoes over land.

The formation of most of the tornadoes in supercell thunderstorms appears to be more complicated. Supercells develop in environments in which there is strong vertical shear (i.e., a change in direction and/or speed with height) of the horizontal wind; such shear is associated with rotation about a horizontal axis. In addition, rotation about a horizontal axis may develop within a mature thunderstorm when there are significant horizontal variations of air density in the region where rain and hail are falling. Updrafts and downdrafts within the supercell thunderstorm may tilt the orientation of the rotation such that a component of it becomes rotation about a vertical axis. If the air that is rotating about a vertical axis is drawn into the region of horizontal convergence at the base of the thunderstorm updraft, then a mesocyclone (a region of rotation a few kilometers wide) and perhaps tornado (a narrower, more intense vortex) may form. A major current research problem is to determine why tornadoes form in some supercell thunderstorms but not in others.

7 RADAR CHARACTERISTICS OF SEVERE STORMS

Use of Radar in Observing Severe Storms

Radar has been used to identify and track severe local storms since its invention in the 1940s. Weather radar detects hydrometeors within storms with a series of pulses of electromagnetic energy, directed by an antenna that is mechanically rotated in azimuth and elevation. It alternates transmitting and receiving those pulses and measures the distance to the target by the time delay between a pulse transmission and echo arrival. Figure 19 illustrates the typical radar beam geometry. Because of the hazards within severe storms to aircraft penetrations, radar has been the primary tool that has been used by meteorologists to probe the inner structure and circulations of storms to better understand the physics of their behavior.

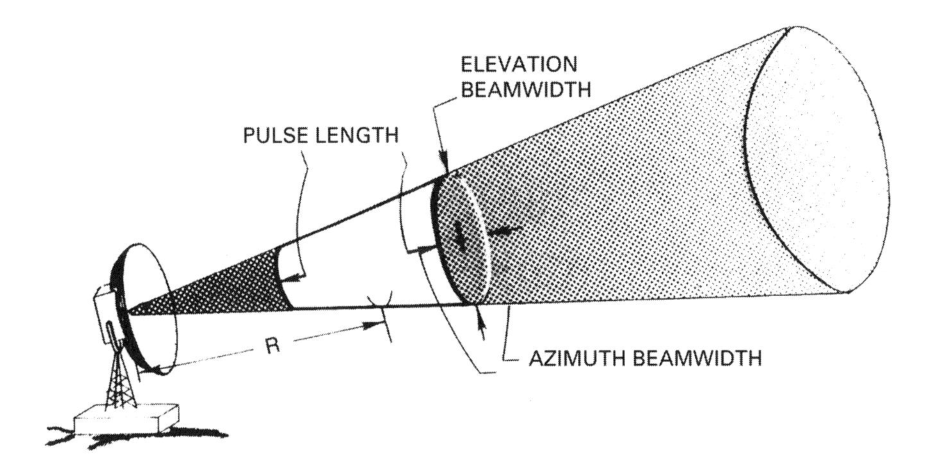

Figure 19 Radar beam geometry. The size of the beam is a function of antenna size and radar wavelength.

Early techniques utilized "incoherent" technology of simply displaying returned power normalized for range on a horizontal display device (termed a *plan position indicator*, or PPI). Severe storms, sometimes producing hazardous weather such as large hail, high winds, and tornadoes, very often exhibit a characteristic structure that could easily be tracked on a PPI display, and suitable warnings could be made for regions in the storm's path by simple extrapolation of storm motion. One of the key developments in radar technology that made radar valuable for distinguishing between tornado-producing storms and other less severe events such as hailstorms was the development of Doppler radar techniques. Doppler radars not only detect and measure the mean power received from a target but also its relative motion. Thus regions of a storm can be seen to approach or recede from a radar site and inferences about rotation within the storm (e.g., tornado mesocyclones that are the parent circulation of tornadoes) can easily be made. This motion detection capability greatly assists weather forecasters to better distinguish between storms that produce tornadoes from ones that do not.

The U.S. National Weather Service (NWS) was quick to recognize the value of radar in providing timely public warnings of severe weather. In the 1950s a network of incoherent radars was deployed (called the WSR-57). With the realization by the 1980s that Doppler technology could offer significant improvement in tornado warning lead time, a new generation of operational weather radars (termed at the time *NEXRAD*, now called the WSR-88D radar) was deployed to meet the warning missions of the NWS, U.S. Air Force, and the Federal Aviation Administration. The network consists of over 130 radars providing nearly complete coverage of the continental United States (Fig. 20). The WSR-88D has greatly increased the tornado warning lead time from nearly zero before the deployment of the WSR-88D network to nearly 10 min today.

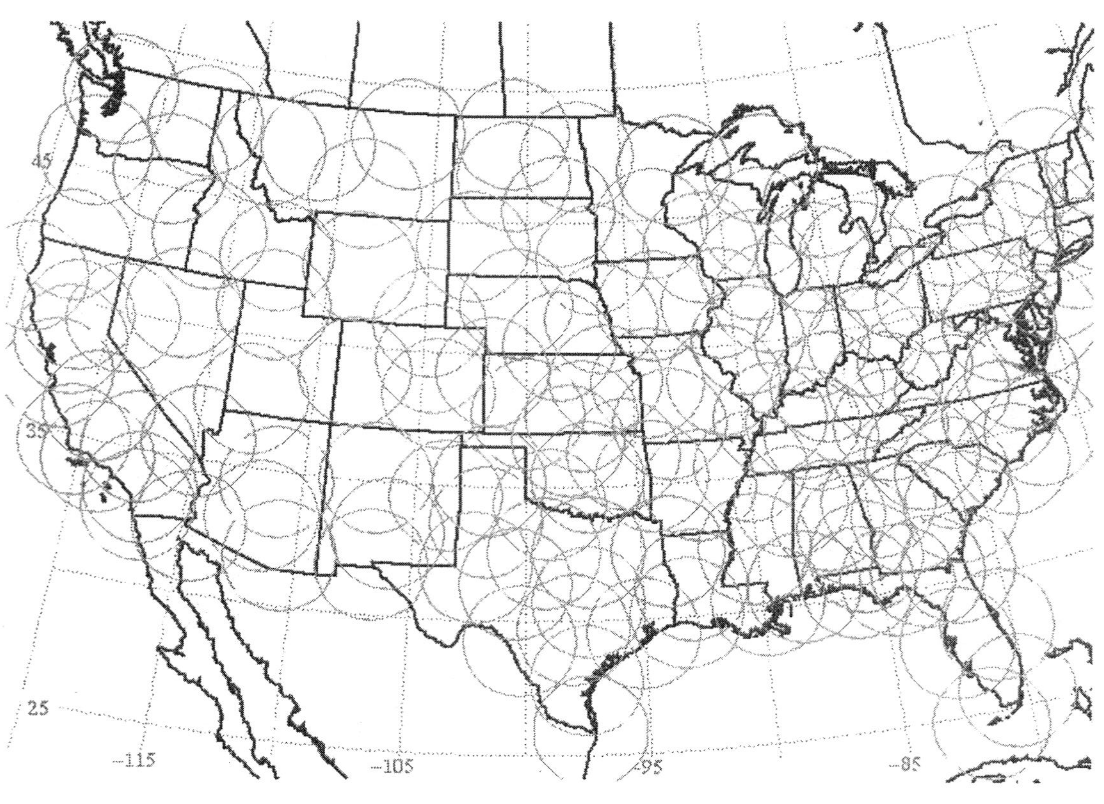

Figure 20 The 230-km range of each WSR-88D radar site.

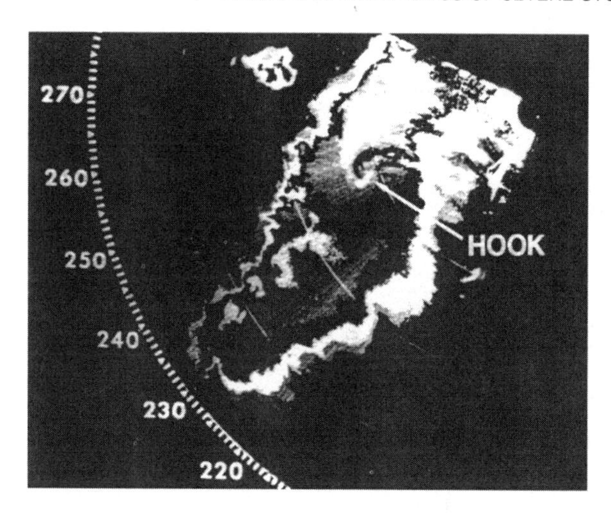

Figure 21 Plan Position Indicator (PPI) display of contoured echo power of the "Tabler" storm on June 6, 1974, in central Oklahoma (from Brandes, 1977). The region labeled "HOOK" is an indication of a likely region for a tornado. Radar location is in the upper right. Range marks (arcs) are spaced 20 km. Echoes are contoured in a gray–white–black pattern starting at 10 dBZ in 10 dBZ steps.

Storm Structure Revealed by Radar Observations

Precipitation echoes from mature thunderstorms usually are easily recognizable even by incoherent radars. On a PPI display they are characterized by high reflectivities (reflectivity is proportional to the sixth power of all the hydrometeor diameters within the pulse volume with units of mm^6/m^3 and is expressed in logarithmic units) and sharp gradients of intensity (Fig. 21). The presence of "hook" echoes, or echoes with notches, on their southern sides has been correlated with tornadic potential. The value of Doppler radar in identifying tornadoes was first shown in the early 1970s when velocities were measured from inside the Union City, Oklahoma, thunderstorm. The pattern of large wind shear (termed a *tornado vortex signature*) between adjacent beams on either side of the tornado location seen by ground observers persisted for over 40 min throughout a deep depth of the thunderstorm. A significant finding for possible tornado warning was the presence of this shear pattern aloft for about 20 min before the tornado was on the ground. Often, however, most Doppler radars do not actually "see" a tornado because of its small size relative to the beam size. It is the much larger parent circulation (termed a *mesocyclone*), often high up in the storm, which is first detected by the Doppler radar (Fig. 22).

While tornado detection by Doppler radar is an important component of the NWS warning responsibility, single Doppler radars also identify many other severe weather hazards. For example, strong downdrafts (often termed *downbursts*) that can affect the safety of airplanes can be identified by the pattern of Doppler velocity

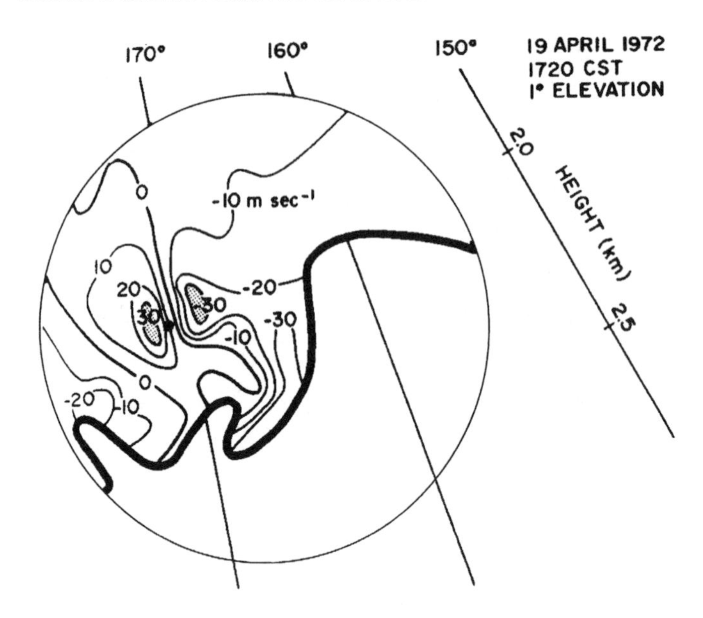

Figure 22 Single Doppler radar signature of a strong mesocyclone indicative of rotation. Dot indicates location of tornado. Negative velocities are approaching the radar. Straight lines labeled in degrees indicate azimuth radials from the radar (from Burgess and Lemon, 1990).

divergence near the ground. Storms that produce large hail are recognized by their elevated reflectivity structure aloft.

Although important for recognizing the presence of severe weather through patterns of velocity, single Doppler radars provide only limited information about the internal circulations of storms, particularly updraft and downdraft strengths that are critical for precipitation formation processes. A single Doppler radar can only provide information about the component of the flow directed either toward or away from the radar. To derive actual wind fields at least two Doppler radars are required. Thus, to understand why storms move the way they do and produce tornadoes, hail, and sometimes damaging straight-line winds, researchers have utilized multiple Doppler radars, and even Doppler radars mounted on airplanes, to provide simultaneous observations that can be combined to produce a three-dimensional description of the flow within severe storms. Figure 23 is an example of combining the radar data from two radars to produce an analysis of the updraft through a strong hailstorm observed on May 26, 1985. In this case the two radars were the Cimarron Doppler radar operated by the National Severe Storms Laboratory and a Doppler radar mounted on the WP-3D aircraft from the National Oceanic and Atmospheric Administration. By flying the aircraft close to the storm, the beams from the aircraft's radar, which was scanning in a vertical orientation normal to the aircraft's flight track, can be combined with the PPI scans from the Cimarron radar.

Progress toward understanding the complex dynamic and microphysical forces acting to control the behavior of severe storms has certainly been aided by multiple

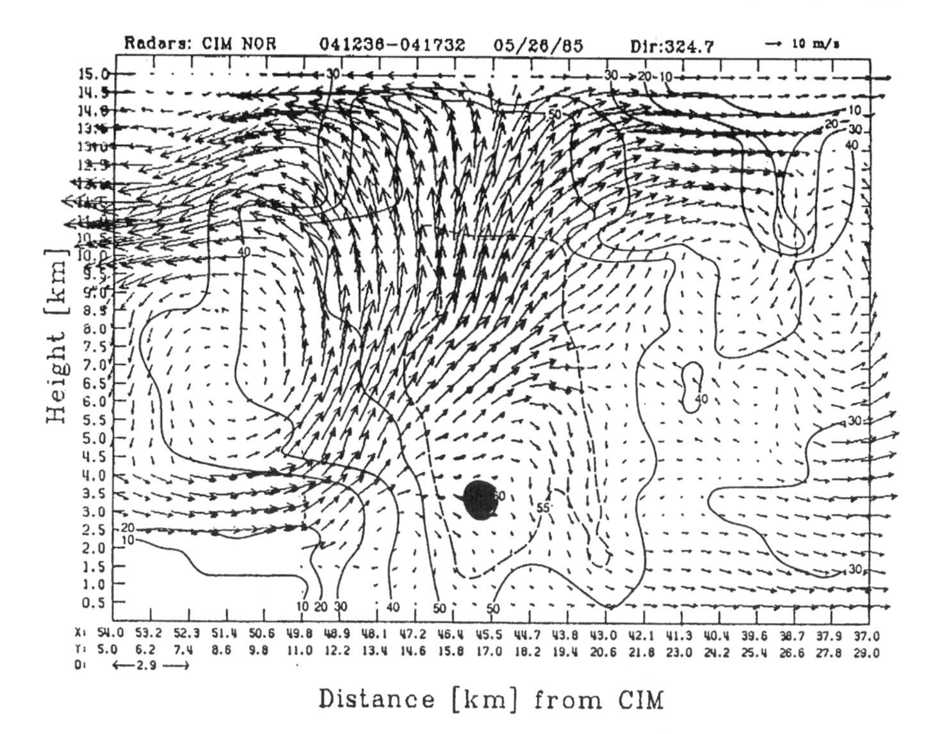

Figure 23 Vertical cross section through the core of a hailstorm simultaneously observed by the P-3 aircraft and the Cimarron Doppler radar on May 26, 1985. Solid lines are reflectivity (dBZ). The 10 m/s wind vector is shown in the upper right.

Doppler case studies. Doppler radar data, however, suffers from a number of limitations, the most severe of which is its restriction to areas where there are hydrometeors that scatter the electromagnetic radiation. Thus clear air regions are usually devoid of observations unless the storm is very close to the radar and the clear air contains some other particle scatterer such as insects. Even in those circumstances the amount of coverage is generally slight. One of the most powerful approaches has been to combine Doppler radar with numerical simulations of convective storms. The simulations use observed environmental conditions to initialize the grid domain and the Doppler observations to validate the computed solutions. Numerical simulations, even using state-of-the-art approaches, have significant limitations. Even so, useful insights have been derived from these simulations in the case of the strongest and longest-lived convective storm, the supercell thunderstorm, as evidenced by the closeness of simulations and Doppler observations where they overlap, such as the high reflectivity region of the storm core where tornadogenesis occurs.

The hypothesis that emerges of tornadogenesis from the analysis of Doppler winds and numerical simulations is that the vertical vorticity (or "spin") develops initially from the tilting of environmental vorticity set up by the low-level wind shear into the vertical by the principal storm updraft. Once initiated, the vorticity is

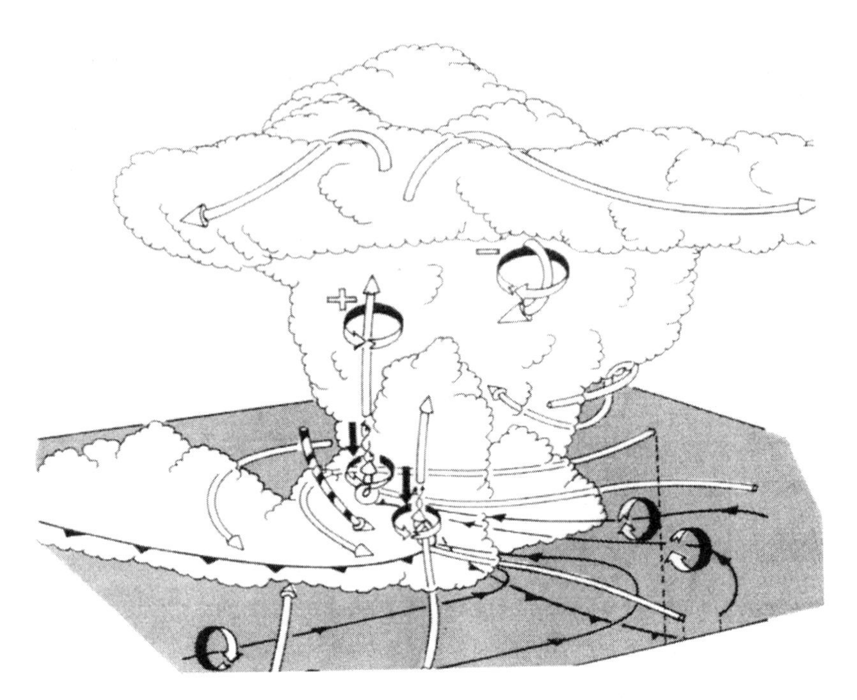

Figure 24 Three-dimensional schematic representation of the processes leading to tornado formation based on Doppler radar and numerical simulations. Cylindrical arrows depict flow in and around the storm. Thin lines show the low-level vortex lines, with vector direction by arrows along the lines and sense of direction also by circular-ribbon arrows. (Adapted by Klemp, 1987.)

increased by concentration of spin by convergence of air similar to the way in which a figure skater increase his or her spin by contracting their arms (called *conservation of angular momentum*). This process is illustrated schematically in Figure 24. Numerical simulations reveal the pressure gradient forces implied in the Doppler winds. Both observations and numerical models have shown that strong low-level rotation promotes a downdraft that spreads cold air at low levels around the storm and helps create convergence along the resultant "gust front." These features have also been seen visually by storm intercept teams.

Future Advances

Doppler radars have enabled considerable leaps in our understanding and improvements in warning lead-time of severe storms. The establishment of a national network of Doppler radars (Fig. 20) has provided nearly complete nationwide coverage. Some hazards, however, are not optimally detected by the WSR-88D network (e.g., aviation hazards such as downbursts and sudden near-surface wind shifts caused by gust frontal passage) and require specialized radars for each major airport called the Terminal Doppler Weather Radar (TDWR) used by the Federal Aviation

Administration (FAA). These specialized radars can also be used to augment the WSR-88D coverage and improve the NWS warning lead time.

Tied to the national network is the ability to rapidly process and interpret the single Doppler radar data for the detection of a variety of hazards through pattern recognition. This "marriage" of the computer and radar greatly accelerates the detection of patterns that can be very subtle and often are embedded in large areas of high reflectivity. WSR-88D radar data from the May 3, 1999, Oklahoma City tornado outbreak is shown integrated with the computerized warning decision process in Figure 25. The Warning Decision Support System (WDSS) can independently track many storm cells simultaneously and provide forecast guidance as to rain severity, hail size, tornadic potential, and probably accurate tornado location and tracking without requiring a dedicated radar scientist.

Other technological advances in development may help identify and warn of severe storm hazards. For example, improved hail detection and rainfall estimation may be possible with polarization of the radar beam. In this approach, two pulses are alternatively transmitted with orthogonal polarizations (e.g., horizontal and vertical). The polarization of each received pulse is measured and various products (such as the ratio of the horizontal-to-vertical polarizations) can be computed. According to electromagnetic scattering theory, if a particle were not circular, it would scatter preferentially in its long axis, i.e., a pancake-shaped raindrop would scatter more horizontally polarized radiation than vertical polarization. This information can be used to improve the accuracy of rainfall estimates, discriminate hail from heavy rain, and reduce uncertainties about the drop-size distribution that is being sampled. The application of polarimetric observation to weather forecasting is still an active area of research. Even so, the utility of polarimetric data in improving weather forecasting is already recognized and the NWS is already planning to upgrade the U.S. WSR-88D radar network to have this capability.

One of the most severe limitations to multiple Doppler radar analysis is the rather large uncertainty in vertical air motion estimates. If fundamental problems in convective dynamics are to be addressed, these uncertainties need to be reduced so that a more complete picture of the mass, momentum, pressure, vorticity, thermodynamic, electrical, and water substance interactions can be examined. One of the ways to reduce the uncertainties is to observe the phenomena at higher spatial and temporal density since it is known that convective elements possess large kinetic energy on spatial scales of 1 to 2 km and exhibit significant evolution over 1 to 3 min. Therefore "rapid-scan" radars that can sample a storm volume within 1 min at data densities of 200 to 300 m are needed. Such data would permit adjoint analysis methods with cloud resolving numerical models to be implemented. This type of radar is currently used in military applications and possibly could be adapted to examine severe weather.

8 SEVERE STORM FORECASTING

In the United States, the National Weather Service's Storm Prediction Center (SPC) in Norman, OK, is responsible for forecasting thunderstorm occurrence as well as

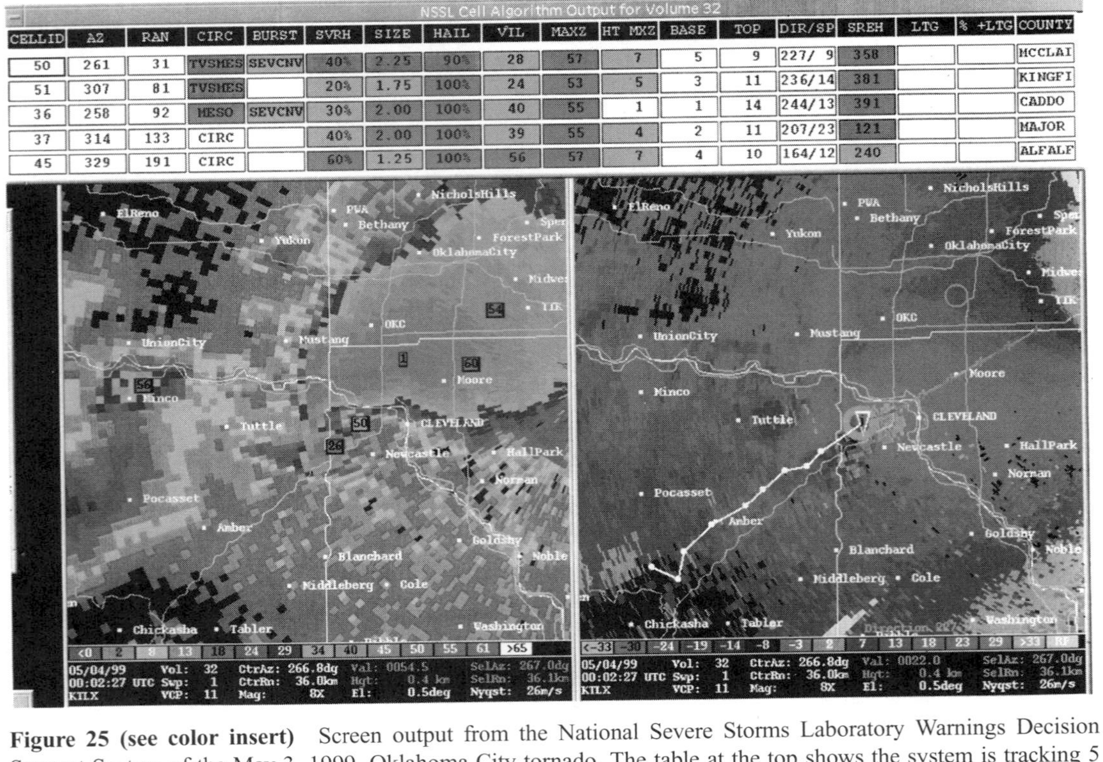

NSSL Cell Algorithm Output for Volume 32

CELLID	AZ	RAN	CIRC	BURST	SVRH	SIZE	HAIL	VIL	MAXZ	HT MXZ	BASE	TOP	DIR/SP	SREH	LTG	% +LTG	COUNTY
50	261	31	TVSMES	SEVCNV	40%	2.25	90%	28	57	7	5	9	227/ 9	358			MCCLAI
51	307	81	TVSMES		20%	1.75	100%	24	53	5	3	11	236/14	381			KINGFI
36	258	92	MESO	SEVCNV	30%	2.00	100%	40	55	1	1	14	244/13	391			CADDO
37	314	133	CIRC		40%	2.00	100%	39	55	4	2	11	207/23	121			MAJOR
45	329	191	CIRC		60%	1.25	100%	56	57	7	4	10	164/12	240			ALFALF

Figure 25 (see color insert) Screen output from the National Severe Storms Laboratory Warnings Decision Support System of the May 3, 1999, Oklahoma City tornado. The table at the top shows the system is tracking 5 different storms with a variety of algorithm-determined characteristics such as direction and speed of motion, presence of hail and its maximum size, whether a circulation or mesocyclone is present, and its echo top. Left panel is radar reflectivity with cities and county boundaries as background. Boxes with numbers correspond to locations of tracked storms. Right panel is radial velocity. White line is storm #50 track with future positions shown as the purple line. Yellow circle shows the presence of the tornado vortex signature. (Courtesy of Greg Stumpf of the National Severe Storms Laboratory.) See ftp site for color image.

most hazardous weather associated with thunderstorms. This includes tornadoes, damaging straight-line winds, large hail, and heavy rainfall that can result in flash floods. The SPC's primary focus is on the risk of tornadoes, damaging straight-line winds (58 mph or greater), and large hail ($\frac{3}{4}$ inch or greater in diameter). In this section we will explain how the SPC makes forecasts for thunderstorms and these three hazards.

Severe thunderstorm forecasting began in earnest in the mid-twentieth century. The U.S. Air Force began making rudimentary internal forecasts for severe weather in the late 1940s. By the early 1950s, the National Weather Service formed a national unit of specialists to forecast severe thunderstorms and tornadoes for the 48 contiguous states. Called the National Severe Storms Forecast Center, this unit was based in Kansas City, Missouri, for more than 40 years before relocating to Norman, Oklahoma, and being renamed the Storm Prediction Center in 1997 (Fig. 26). Currently, 20 specialized meteorologists work in teams of four to monitor weather conditions around the clock, 7 days a week, all year long. An "outlook" forecaster makes forecasts for severe weather out to 3 days ahead (Fig. 27). The other three specialists issue short-term forecasts (1 to 7 h ahead) that include tornado and severe thunderstorm watches as well as other products. Typically, watches are parallelogram in shape covering an area averaging about 25,000 square miles (about the size of the state of Iowa; Fig. 28). A severe thunderstorm watch is issued when there is a significant and concentrated threat of damaging straight-line winds and/or

Figure 26 Storm Prediction Center operations area. Forecaster Jeff Peters studies data on one of several high-speed workstations. See ftp site for color image.

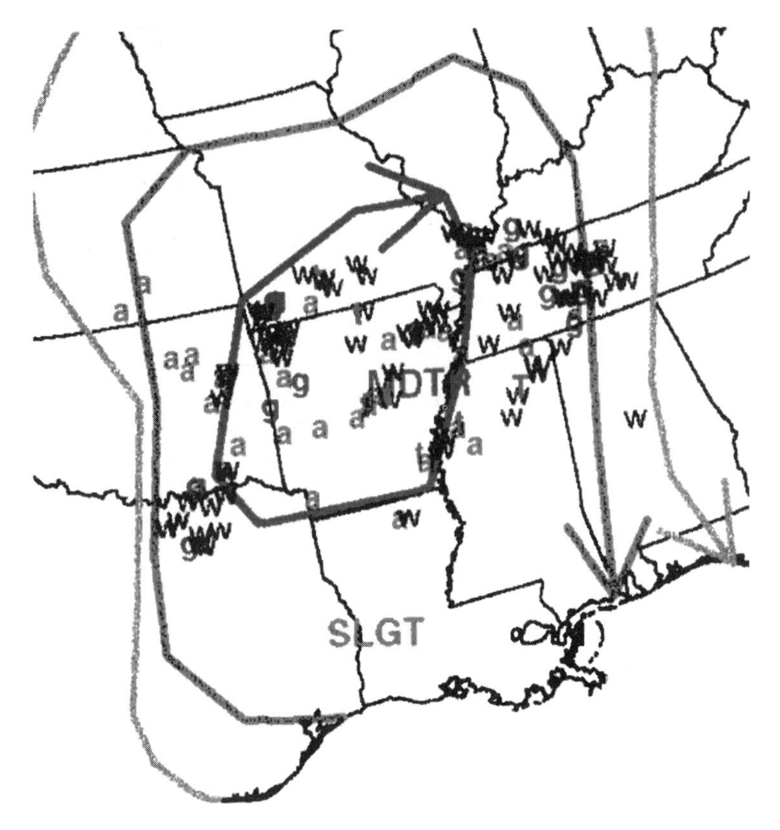

Figure 27 Example of SPC outlook and watch area. See ftp site for color image.

large hail. Tornado watches are issued when there is a threat of tornadoes. There may also be a threat for damaging winds and large hail within a tornado watch. Significant and/or concentrated severe weather typically results from "organized" severe thunderstorms, those that are of the supercell, bow echo, or strong multicell modes.

Generally, forecasting severe thunderstorms involves three concepts: climatology, pattern recognition, and parameter evaluation. Climatology is used by SPC forecasters to know what time of day, season, and area they should expect a higher likelihood of severe weather development. For example, in the Great Plains region of the United States, spring tornadoes are most likely to occur in the late afternoon and evening hours. Given a typical severe weather situation for the region, forecasters anticipate an enhanced risk at that time. In the winter, however, the highest risk of tornadoes in Florida and the coastal regions of the southeastern states is during late night and morning hours, nearly the diurnal opposite of the Plains states in spring. Therefore, given the typical severe weather situation in the southeastern United States in the winter, an SPC forecaster's anticipation of tornado development based on climatology is enhanced for the period from midnight until noon.

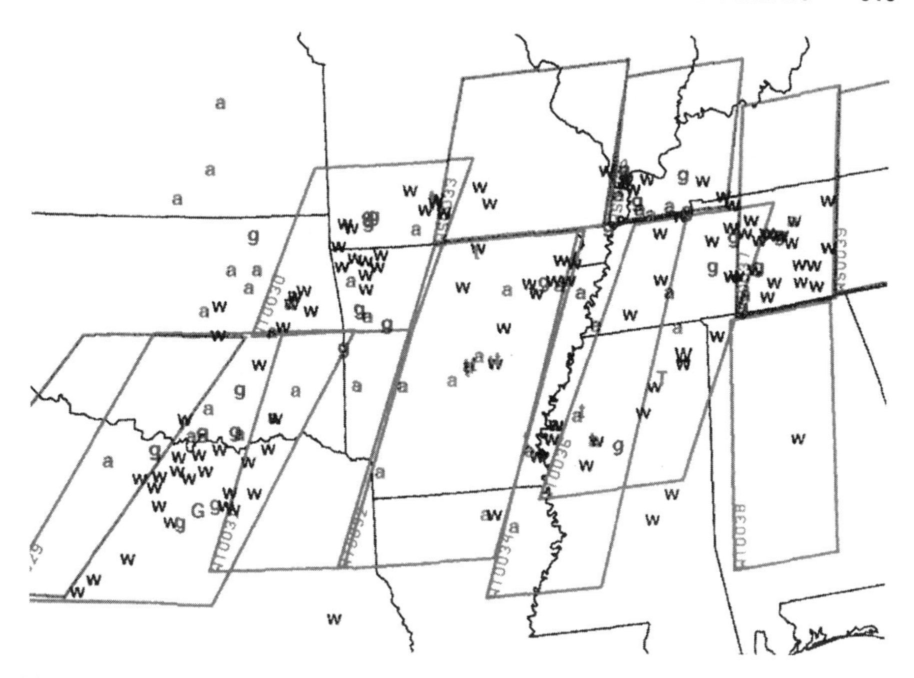

Figure 28 Composite forecast map used by SPC to overlay various surface and upper-level features to help poinpoint potential severe weather threats. See ftp site for color image.

Pattern recognition is used by SPC forecasters as a first approximation concerning severe thunderstorm development. Features such as the upper-level jet stream (20,000 ft or higher above ground level), the low-level jet stream (from 2000 to 5000 ft above ground level), fronts, thermal and moisture axes, etc. are examined. The intensity, orientation, juxtaposition, and movement of these features aid the forecaster in estimating the probability of occurrence, area affected, timing, and severity of potential severe weather episodes. As an example, most large tornado outbreaks are associated with a pattern that includes dual upper jet streams that diverge over the outbreak area and a strong southerly low-level jet stream that transports warm and moist air into the area. The strength and movement of these jet streams play a major role in the severity, area affected, and timing of the outbreak.

Finally, parameter evaluation (also called *ingredients-based forecasting*) has become increasingly important in forecasting severe thunderstorms in recent years. Atmospheric scientists have learned much about thunderstorm development and evolution over the five decades since the first severe weather forecasts were made. This knowledge is derived from observations, theoretical studies, and numerical modeling experiments. Application of this knowledge has resulted in forecast techniques that relate values of meteorological parameters to the type of storms that develops, their evolution, and the types of severe weather (large hail, damaging winds, and/or tornadoes) associated with them. As an example, "isolated" supercells can generally be associated with varying combinations of the amount of vertical wind shear and the degree of buoyancy for rising parcels of air in the troposphere. Given that thunderstorms will develop, SPC forecasters assess the values of these two parameters from both "real-time" observational data and operational numerical forecasts when deciding whether or not to forecast "isolated" supercells. If supercells are predicted, the forecaster then looks at other meteorological parameter values to assess the risk of tornadoes, damaging winds, and/or large hail with the expected supercells.

Making forecasts for severe thunderstorms requires detailed analysis of both real-time observations and operational numerical model forecast data and trends. Since the 1980s, computer workstations have been used to process and display ever-increasing quantities of meteorological data, and their use has helped forecasters gain a better understanding of processes important for severe storm development (Fig. 29). Outlook forecasters rely primarily on analysis of numerical model forecast data for forecasts that extend out as far as 3 days ahead. Composite forecasts of model data are typically constructed for key times within the forecast period (Fig. 30). Adjusting for known model biases and limitations, the forecaster then uses the patterns and positions of features on the composite forecast charts to determine timing and area covered by potential severe storms. Model forecasts of vertical profiles of wind, temperature, and moisture are then examined within the forecast area to help estimate storm mode, intensity, evolution, and severe weather type, if any.

Figure 29 Example of mesoscale analysis used by SPC to determine regions where the threat of severe weather exists. See ftp site for color image.

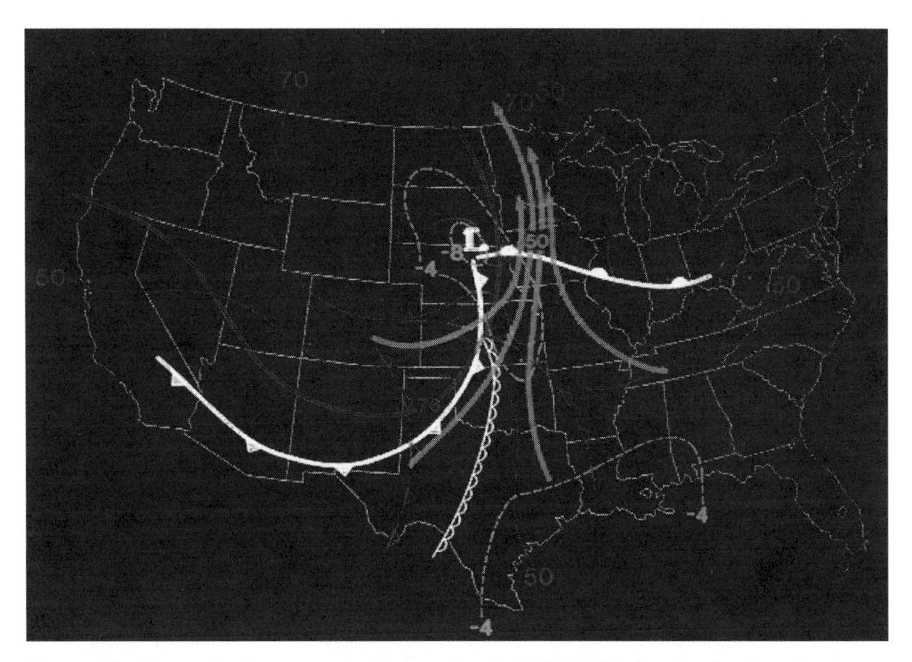

Figure 30 Example of a composite forecast for April 26, 1991. See ftp site for color image.

For short-term forecasts (1 to 7 h ahead), the most recent day 1 outlook forecast (current time to 24 h ahead) is used to focus attention on specific areas. Although short-term operational numerical model trends are noted, the primary emphasis for short-term forecasts is on analyzing real-time data and trends. Real-time data includes both observations (e.g., radar reflectivity and wind data, satellite imagery, lightning strike data, aircraft wind and temperature data, and upper air and surface observations of temperature, moisture, air pressure, and wind direction and speed.) and derived fields computed from this data (e.g., surface pressure changes over time). For timely and accurate short-term products, continuous attention to details and trends is very important because the atmosphere is constantly changing. For example, regional subjective analysis of surface observations (the densest operational data network available) is necessary to assess the short-term severe weather threat (Fig. 31). So, such analysis is typically done each hour. As significant small-scale patterns, parameter values, and trends are diagnosed, mesoscale discussion products consisting of one or two paragraphs are issued. These messages describe the situation and how storms are expected to develop and/or evolve. A severe thunderstorm or tornado watch is issued if the analysis reveals that there is a significant threat of "organized" severe thunderstorm development that will last 4 or more hours and affect more than a localized area (generally greater than 8000 square miles).

Although the ingredients-based approach to severe thunderstorm forecasting (parameter evaluation) has allowed more precision in severe weather forecasting

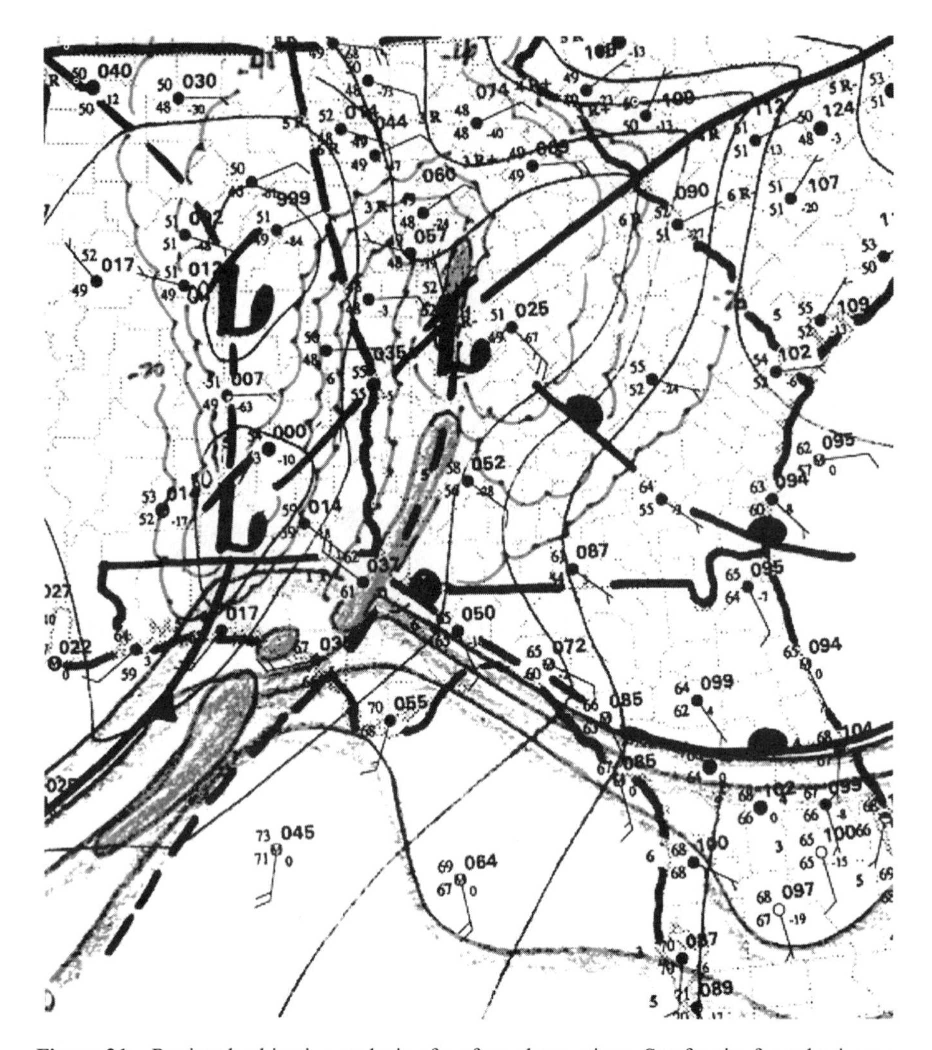

Figure 31 Regional subjective analysis of surface observations. See ftp site for color image.

during recent years, it appears that the climatological and pattern recognition aspects of forecasting will continue to be utilized in the foreseeable future. There has been a rapid increase in knowledge about storm processes in the past several years, but there is still much to learn. Further, despite the increase in the amount of real-time data available to forecasters, the operational data network is still not dense enough to diagnose all parameter values and other features in enough detail to take advantage of some newly understood storm-scale processes.

To complete the tornado warning process, local NWS Warning and Forecast Offices, which have responsibility for much smaller areas, issue very short-range warnings for events that are either occurring or imminent.

FOR FURTHER READING

Bluestein, H. B. (1999). *Tornado Alley. Monster Storms of the Great Plains*, New York, NY: Oxford University Press.

Grazulis, T. P. (2001). *The Tornado: Nature's Ultimate Wind Storm*, Norman, OK: University of Oklahoma Press.

Rinehart, R. E. (1997). *Radar for Meteorologists*, 3rd ed, Columbia, MO: Rinehart.

Vasquez, T. (2000). *Weather Forecasting Handbook*, Garland, TX: WeatherGraphics Technologies.

CHAPTER 30

TROPICAL PRECIPITATING SYSTEMS

EDWARD J. ZIPSER

Precipitation is influenced by phenomena on all scales of motion. In the tropics and subtropics, it is rare to find continuous precipitation on horizontal scales larger than mesoscale, whether related to a larger-scale disturbance or not. The reason is straightforward: The stratification of temperature and moisture is such that the equivalent potential temperature (θ_e) decreases with height in most of the low to midtroposphere. That is, the atmosphere is often both conditionally and convectively unstable, such that large-scale lifting will inevitably result not in slow steady ascent and light precipitation but in convective clouds and possibly heavy precipitation. Adjacent regions normally have subsidence without precipitation, *even within regions of large-scale ascent or disturbances*. This chapter surveys current knowledge of tropical convective and mesoscale precipitation and its organization. We focus first on the physical nature of the precipitation systems themselves, and only later examine the reasons for how those systems are forced, or organized. The organizing systems are then arranged in order of scale, beginning with small-scale orography and land–sea breezes, progressing to large and planetary-scale forcing.

1 DIFFERENCES BETWEEN TROPICAL AND MIDLATITUDE CONVECTION

There are varying perceptions about tropical convection, not always rooted in reality. It is important to examine the basis for statements about tropical phenomena. The old

Handbook of Weather, Climate, and Water: Dynamics, Climate, Physical Meteorology, Weather Systems, and Measurements, Edited by Thomas D. Potter and Bradley R. Colman.
ISBN 0-471-21490-6

climatological school of thought spoke of the "daily thunderstorm to which one could set one's clock." This belief has a grain of truth in some former British colonial outposts of Malaysia or Africa, some of the time. Knowledge of the oceanic tropics expanded during World War II, leading to descriptions not of boring unchanging climate but of synoptic-scale disturbances such as easterly and equatorial waves, which sometimes intensified into tropical cyclones. Synoptic models of these waves describe useful relationships between phases of the waves and weather. This represented an extension of synoptic meteorology thinking into the tropics, which was helpful in some regions, was irrelevant or misleading in regions where synoptic-scale systems do not control daily events, and mostly ignored mesoscale phenomena.

Knowledge depends upon observations *of appropriate scale*. Motivated by a series of devastating hurricanes in the 1950s, research aircraft penetrations of hurricanes provided such data, leading to major advances in description, understanding, and prediction of tropical cyclones (Marks, F. D. Jr., Chapter 32). In the meantime, quantitative radar and mesoscale data in the severe storm regions of the United States led to analogous knowledge of storms bearing hail and tornadoes (Brooks, H. et al., Chapter 30). It would be nearly 20 years before such tools would be used for "ordinary" tropical weather. The motivating factor was not weather forecasting but the increasing realization that global models of weather and climate required sound treatment of "subgrid-scale" phenomena in the tropics. The conceptual framework for the Global Atmospheric Research Program (GARP) and its Atlantic Tropical Experiment (GATE) was created in the 1960s, under the leadership of Jule Charney, Verner Suomi, and Joseph Smagorinsky. Following a series of smaller field experiments in the late 1960s, the GATE was carried out in the eastern Atlantic in 1974, ending forever any lingering thoughts that large-scale and small-scale phenomena can be treated independently.

The landmark "hot tower" study of Riehl and Malkus (1958) had already conditioned meteorologists to the belief that tropical cumulonimbus clouds were not just decorations, responsible for local showers and the heavy rains that constitute many tropical climates. These convective towers were shown to be a critical link in the general circulation, transporting heat, moisture, and moist static energy from the low to high troposphere for subsequent export to higher latitudes. They also have an essential role in hurricane formation and maintenance. Thus, it becomes easier to accept the truth that tropical convection must be parameterized if global models were to be successful. It perhaps was natural to believe that these hot towers of the deep tropics were also some of the biggest and most powerful storms on the planet, but observations demonstrated otherwise.

Research aircraft equipped to derive vertical velocity made thousands of penetrations of convective clouds, both isolated and embedded in mesoscale convective systems (MCSs, more about these below). Beginning in GATE, but over a 20-year period in other field programs, including the Bay of Bengal, offshore Borneo, offshore northern Australia, offshore Taiwan, the warm pool of the equatorial west Pacific, and tropical cyclones in several oceans, the results were remarkably similar.

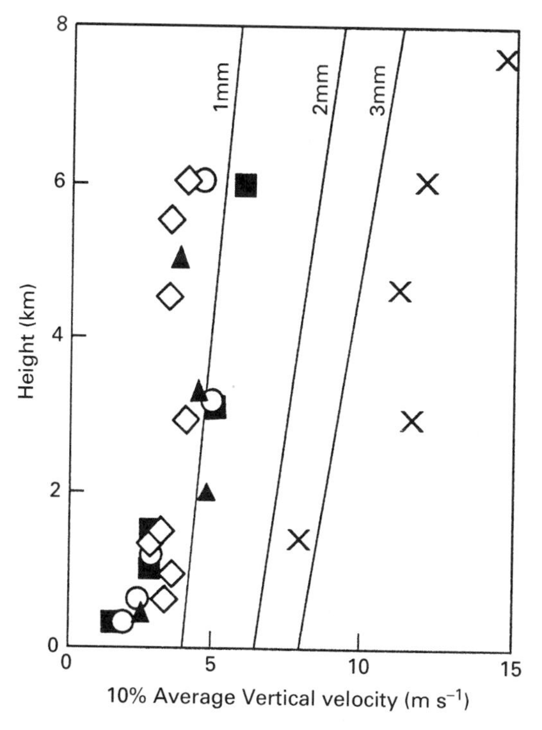

Figure 1 Average vertical velocity in the strongest 10% of updraft cores over tropical oceans from measurements in three different regions (triangles, circles, diamonds) and over land (crosses, Thunderstorm Project). The lines show terminal fall speeds of raindrops as a function of raindrops diameter and height. (after Zipser and Lutz, 1994).

Most tropical convective clouds were weak. Typical maximum updrafts were a factor of 2 to 3 lower than those in ordinary cumulonimbus clouds over land (Fig. 1). Updraft and downdraft velocity, diameter, and mass flux are approximately lognormally distributed.

But it would be wrong to assume that weak convection characterizes the entire tropics. The cited observations are entirely from the *oceanic* tropics. Few data exist from continental tropical regions, not coincidentally because local knowledge of intense convection over land would quickly discourage pilots from attempting penetrations. The true hot towers, therefore, are mainly confined to certain regions, e.g., India–Bangladesh, Argentina, southern United States, and Africa. Recent satellite data show the distribution of intense storms more quantitatively (see Section 7). The common oceanic cumulonimbi still perform their major role of energy transport and precipitation; the vertical speed is simply smaller than previously imagined.

2 MESOSCALE CONVECTIVE SYSTEMS (MCSs)

There is no precise definition of an MCS. The essence of an organized mesoscale convective system is that it is a *recognizable entity of a distinctly larger temporal and spatial scale than its constituent convective clouds*. That is, the MCS lives for several hours and covers a horizontal scale of the order of 100 km in at least one dimension, while the ordinary cumulonimbus cloud may live for one hour and cover the order of 10 km. Many MCSs are unmistakably well organized, such as the squall line or mesoscale convective complex (MCC; Section 4). Some are not.

Mesoscale convective systems are widely distributed throughout the world. The fundamental processes by which groups of cumulonimbus clouds become organized into MCSs, and which govern the structure and evolution of MCSs, are independent of latitude. (The sole exception is the Coriolis effect, which influences the asymmetry and inertial stability of particularly large, long-lived MCSs.) Given the important role of MCSs in the United States, there is some irony in the fact that so much of what we know about them comes from field programs over tropical oceans, transferred to midlatitudes later. The first field program explicitly targeting MCSs over the United States was carried out in Oklahoma and Kansas in 1985.

The literature on MCSs is dominated by studies of squall lines. The reason is the great simplification inherent in a quasi-two-dimensional system. Concentrating on squall lines results in little loss of generality, provided one remembers that more complex or disorganized systems are often more common.

3 SQUALL LINE AND MCS STRUCTURE

The best-known type of squall line has *a leading convective region and a trailing stratiform precipitation region*. It is fairly common in almost all regions, the easiest to study, and therefore the best-known MCS. The convective region is usually 10 to 30 km in width, followed by a region of stratiform precipitation that often exceeds 100 km in width. When such a system passes over a given location, the weather experienced includes a few heavy convective showers (often >25 mm/h) followed by several hours of steady precipitation (often 3 to 5 mm/h). Rainfall from the convective region depends upon details of individual cells, their intensity, and their relation to the observer; from the stratiform region it depends mainly on duration of the moderate rain.

Convective updrafts within MCSs are basically similar to those in more isolated storms but are concentrated in space and time. Updraft speeds rarely exceed 3 to 10 m/s over oceans and 10 to 25 m/s over land. Diameters rarely exceed 5 km except in supercells. The *mesoscale updrafts* occupy the upper half of the troposphere in the stratiform precipitation region, with updraft speeds in the 10 to 100 cm/s range over an area that may extend for 100 km.

The *convective downdrafts* in the convective region form a cold pool at low levels that quickly spreads out to cover a large area. This cold pool can be thought of as one of the exhaust products of the system, analogous to the more easily visible exhaust product of the convective updrafts: the anvil system, which spreads out near and below the level of neutral buoyancy (equilibrium level). The cold pool has the important function of helping to continuously generate new convective cells, usually along its advancing edge (e.g. Houze 1993, Chapters 8 and 9). The interior of the cold pool is usually sufficiently stabilized that new deep convective cell growth is impossible. Even over warm tropical oceans, while heat and moisture transfer can restore the boundary layer properties and make new convection potentially possible, another element of the MCS usually prevents it: the mesoscale downdraft.

The *mesoscale downdraft* is universally found below the melting level in the stratiform precipitation region, beneath the mesoscale updraft. The evaporation of precipitation fails to keep the mesoscale downdraft saturated, and relative humidity is often below 80% and can be as low as 50% in the 1 to 2 km levels. Therefore, in spite of thick anvil-type clouds above, steady rain, and occasionally lightning, there are usually no clouds at all below the melting layer. All precipitation in this region was initially frozen before falling through the melting layer. The mesoscale downdraft cannot penetrate through the shallow cold pool to the surface, so can be detected by aircraft or soundings (Fig. 2). The result is a characteristic onion-shaped sounding (Fig. 3). Under most circumstances, the combination of a shallow, capped cold pool and a warm, dry unsaturated downdraft dictates that the air in the stratiform region of an MCS cannot generate new deep convection for at least 12 to 24 h.

Mesoscale downdrafts have been attributed to three causal mechanisms, and all three mechanisms act together in most MCSs. The latent heat of fusion cools the air when ice melts. Although the magnitude is only 13% of the latent heat of evaporation, the cooling takes place in a concentrated layer only a few hundred meters thick, having the effect of lowering the 0°C isotherm on the mesoscale, generating or accelerating descent. The evaporation of falling precipitation is a powerful cooling process, assuming subsaturation of the ambient air. Once descent begins for any reason, however, evaporation can be effective. A final reason is that the cold pool near the surface always tends to diverge, so air above the cold pool must descend.

In both tropics and midlatitudes, some 40% of rainfall from MCSs falls in the stratiform precipitation region. The question arises: Which process is most important, the condensation/sublimation growth of ice particles in the strong ascent in the mesoscale updraft, or horizontal transfer of convective debris (ice) from the convective region? The answer is that both are required for substantial precipitation rates and that both are present in MCSs. While this result comes from simulation experiments, the remarkable fact is that *no documented case exists of an MCS cloud structure without prior existence of deep convection.* Some earlier descriptions of rainfall systems have mistakenly ascribed the existence of widespread light precipitation to midlevel convergence in subtropical cyclones or monsoon depressions. This is very doubtful; the misconception arises when synoptic-scale reasoning is applied to situations where mesoscale processes are responsible for the specific cloud and precipitation features.

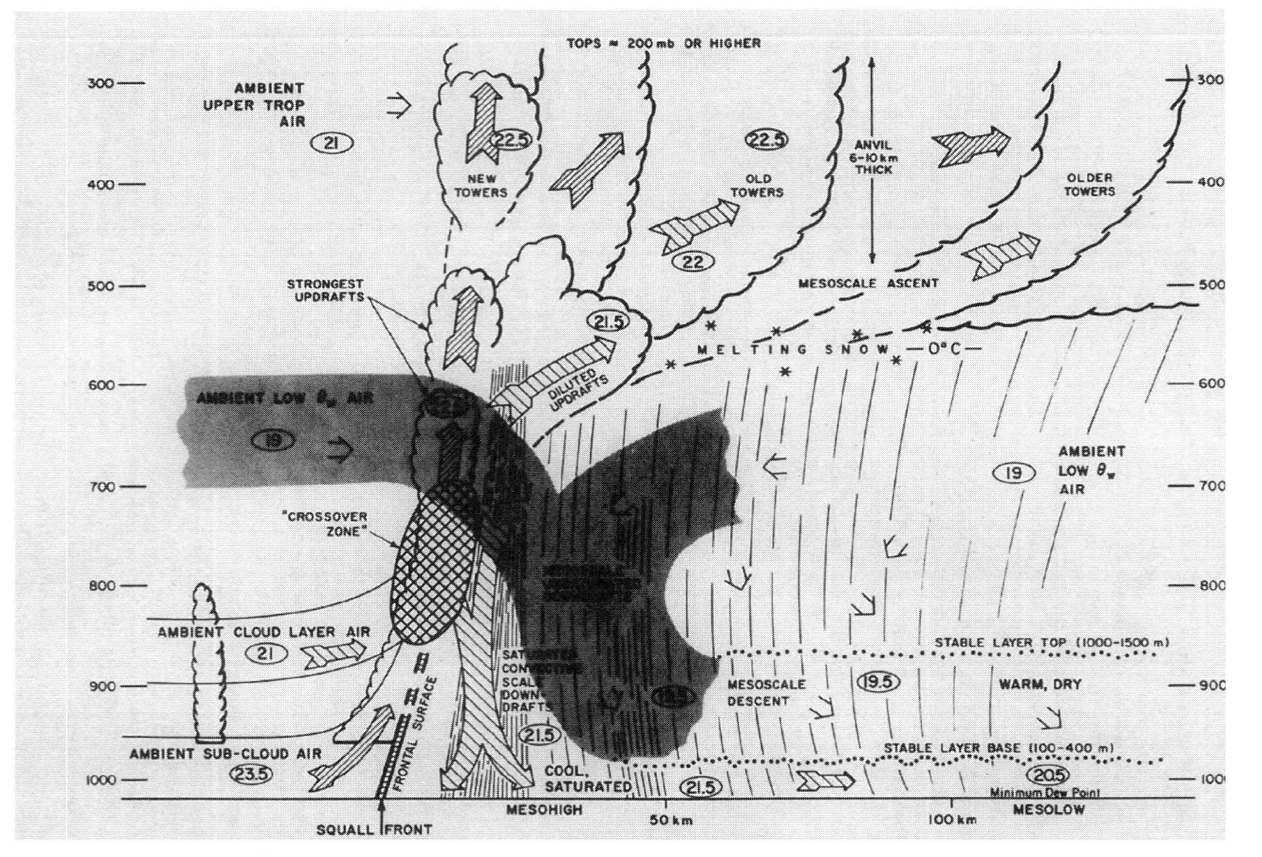

Figure 2 Schematic cross section through a typical squall line system. All flow is relative to the squall line, which is moving from right to left. Circled numbers are typical values of wet-bulb potential temperature, directly related to equivalent potential temperature (in °C). The convective region occupies the first 30 km behind the leading edge, and stratiform precipitation the next 100 km. The air descending in the mesoscale downdraft remains above the cooler air, which descended in the convective downdrafts and spread out in the lowest layers (after Zipser, 1977).

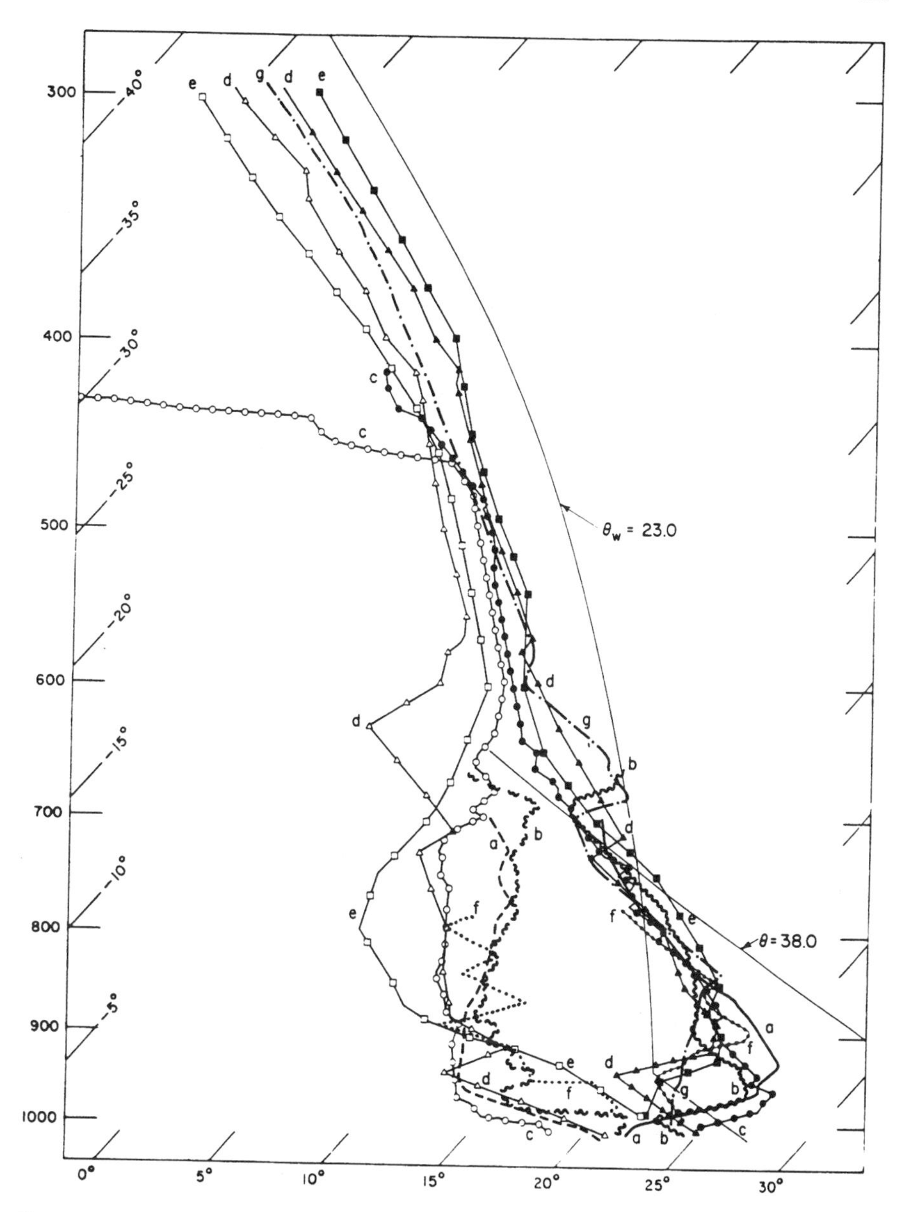

Figure 3 Soundings shown are taken behind squall lines, mostly within or toward the rear of the stratiform precipitation region. The temperature and dew-point curves are far apart in the mesoscale downdraft air below the melting level and near the surface, signifying low relative humidity even in the rain area (after Zipser, 1977).

4 SQUALL LINES, MCCs, OR DISORGANIZED MCSs

What are the environmental conditions that favor MCSs organizing in the form of squall lines? Empirically and theoretically (and confirmed in simulations) the determining factor is the low-level wind shear (LeMone et al., 1998). Substantial low-level wind shear usually results in convection rapidly organizing into lines perpendicular to the shear, moving downshear while leaning upshear and transporting line-perpendicular momentum opposite to the direction of motion (upshear).

What are the environmental conditions that favor MCCs (mesoscale convective complexes)? These are the largest and longest-lived of the MCS family. The question is what favors extremely large and persistent relative inflow of high θ_e air into the system at low levels? Local or purely orographic effects are unable to provide such inflow, which helps to explain why MCCs are often associated with a low-level jet stream. Also, there is often a region of forced ascent for the low-level jet, either a synoptic-scale front, orography, or large cold pool left behind by previous MCSs/MCCs (Laing and Fritsch, 2000). Squall lines and MCCs are not mutually exclusive. Only the criterion of approximately circular shape prevents many large squall lines from satisfying the other MCC threshold criteria.

5 EVOLUTION OF MCSs

All MCSs that have been documented in case studies evolve in much the same manner (Houze, 1993). A group of cumulonimbus clouds forms fairly close to one another for a variety of reasons. The convective rain volume often peaks rapidly but occasionally can persist at a high level for many hours (the typical MCC precursor). Stratiform precipitation usually rises in phase with the convective precipitation but displaced a few hours later in time. If the system is weak or short-lived, there may be little overlap in time between convective and stratiform regions—one may appear to evolve into the other. For strong, long-lived systems, including squall lines and MCCs, stratiform and convective regions may coexist in relative steady state for many hours. The end stages of MCSs inevitably consist of decaying stratiform-only regions.

6 CANDIDATE SYSTEMS THAT ORGANIZE RAINFALL AND MAIN CONTRIBUTORS TO RAINFALL IN SELECTED REGIONS

Organization of precipitation systems is governed by disturbances with a variety of horizontal length scales. The main requirement is that they include regions where ascending motion is organized. The *frequency and amount* of precipitation in a region is often determined by the nature of the disturbances, while the *character and intensity* of the rainfall system (e.g., strength of convection, MCS or not) is often determined by the local wind shear and thermodynamic profiles. There is too much variety to cover the entire range of tropical environments, so selected illustrative examples are given.

As a general rule, there tends to be an inverse correlation between total monthly or annual rainfall, and the intensity of the rainfall systems. For example, the heaviest monsoon rains over land, and the heavy rains of the oceanic intertropical convergence zones (ITCZ), often come without lightning and with poorly organized MCSs. Where are the most violent storms in the tropics? This may seem counterintuitive, but they are often on the fringes of deserts or during the premonsoon seasons or "break monsoon" periods. Some examples are given below.

Orographic Forcing (1) Strong flows of moist low-level air capped by warm dry air. The classic example is the Hawaiian Islands, where persistent, steady trade winds are forced to ascend the windward slopes. Annual rainfall in the adjacent oceans is probably <300 mm, yet parts of the windward slopes of most islands experience >300 mm per month. In the extreme case of Mt. Waileale on Kauai, annual rainfall exceeds 10 m. These orographic rains are not at all steady but consist of shallow convective clouds that produce periodic heavy showers. Other regions with similar orographic enhancement include Jamaica and Puerto Rico, January–May.

Orographic Forcing (2) Weak flow, with the moist layer weakly capped. This distinctly different kind of orographic enhancement is triggered by the heating of elevated terrain, which, in turn, forces a diurnal upslope flow. The organized mesoscale flow often triggers thunderstorms, which may be closely spaced and in turn organize into MCSs or MCCs, occasionally propagating off the mountains as organized squall lines. The Sierra Madre Occidental of western Mexico is the site of a classic example. These mesoscale rain events extend northward into southeastern Arizona during July and August, sometimes referred to as a monsoon. Rainfall can exceed 500 mm monthly in Mexico and 200 mm monthly in southeastern Arizona during this regime. Many other parts of the world experience similar regimes, including the tropical Andes, parts of east Africa, mountainous islands in the Indonesian region, and Puerto Rico in summer.

Land–Sea Breeze Circulation Systems Conditions for effective rain production from sea breeze circulations are similar to that for orographic forcing (2) above. Florida in summer is a classic example. Adjacent oceans have little rain except for tropical disturbances, while peninsular Florida generates some 200 mm rain and 20 to 25 thunderstorms per month, some of which can be locally severe, and/or organize into MCSs, including squall lines. It is obvious from satellite and surface data that these are generated by the ascent along sea-breeze convergence zones from either coastline, and numerous case studies of these storms have been undertaken.

Synoptic-Scale Waves The best known is the easterly wave. These may be the most regular synoptic waves on Earth, generated in Africa, coasting across the Atlantic into the eastern Pacific. They are less common and regular in the Pacific. They have a period of 4 to 5 days, a wavelength about 2500 km, propagating westward at about 6 to 8 m/s. Some 10% of these waves encounter favorable conditions for intensification into tropical cyclones. They modulate convection and MCSs strongly, with the wave troughs wetter than the wave ridges (Reed et al., 1977); in this way they are analogs of traveling waves in the higher-latitude westerlies.

The northern half of an easterly wave usually has a strong midlevel easterly jet that dictates squall line formation, while the southern half of the same waves may generate heavily raining but poorly organized MCSs in the light wind shear regime. Also of importance is the relative motion of the MCS and the system that may have generated it. The squall line moves at roughly twice the wave speed; therefore, it forms preferentially near the wave trough but tends to dissipate as it approaches the wave ridge. This resembles the case of MCSs in the United States, where it is common for the system to move eastward more rapidly than its larger-scale forcing mechanism. If it moves into an unfavorable area, it may dissipate rapidly. If it can regenerate convection on its right rear flank, it may not only live for a long time but also move slowly enough to be a serious flash flood threat (Chappell, 1986). Excessive rain events in the tropics are rarely attributable to a simple wave passage but to unusual interactions among systems of different scales.

Many other regions of the tropics are affected by westward traveling disturbances. Most of these do not fit the description of easterly waves but can be Rossby waves or Rossby-gravity waves. Their existence is well-documented by satellite data analyses, but case studies are rarely undertaken due to almost complete lack of appropriate in situ data.

Cyclones Other than tropical cyclones (Marks, F. D., Jr., Chapter 32), there are a variety of other cyclones that can organize tropical rainstorms. Once again, their existence is well known but definitive case studies are quite rare. They are often categorized by whether the maximum circulation (vorticity) is found in the upper, mid, or low troposphere.

The tropical upper tropospheric trough (TUTT) is a semipermanent feature of the summer hemisphere in the north Atlantic and north Pacific. It is a subtropical feature and generally exists over dry midtropospheric air masses. Upon occasion, some of the common low-pressure centers along this trough become strong enough to organize deep convection and MCSs, usually in the subtropical west Atlantic and west Pacific. It is possible that such activity requires the low to extend downward from its usual upper tropospheric location.

Midtropospheric cyclones may exist in the vicinity of Hawaii, where they are known as Kona lows, and they can generate heavy rainfall events, especially when they move slowly (Barnes, 2001). Similar cyclones have been documented in the Indian Ocean. It is not clear how often these cyclones propagate downward into the low levels to form monsoon depressions, as they are called, or whether the latter are generated independently in low levels. In any event, some of the heaviest rainfalls of the Indian subcontinent are produced in association with the passage of these depressions. They are most common in the Bay of Bengal and frequently move over northern India from there. They are not to be confused with tropical cyclones, which are unable to survive in the strong wind shear regime that dominates south Asia during the heart of the summer monsoon (strong low-level westerlies *and* strong upper-level easterlies).

Intertropical Convergence Zone: Oceans It has been known, literally for centuries, that the trade-wind flows of the Northern and Southern Hemispheres converge along rather narrow zones in the oceans, known as the ITCZ, along which synoptic-scale ascent must take place on average (not at all times). Rain systems are frequent along the oceanic ITCZ, although without great convective intensity. Typical rainfall totals are about 300 mm monthly; where the ITCZ stays over a given location for many months of the year, annual totals may exceed 3 m. There is a large-scale convergence zone extending southeastward from New Guinea toward Tahiti known as the South Pacific convergence zone, within which the rainfall may have similar properties to the ITCZ.

Intertropical Convergence Zone: Land, Monsoons The off-equatorial heating of the continents forces much stronger cross-equatorial flows well into the summer hemisphere. At very low levels, the ITCZ as defined by wind convergence often moves far from the equator, and the low-level flow of moist air attempts to follow suit. However, the deep convection and heavy rainfall does not usually reach the location of this ITCZ but is distributed over a large region.

A prime example is north Africa in August. The wind convergence marking the ITCZ at the surface appears to reach the central Sahara Desert, while the heaviest rainfall remains well to the south, along about 10°N. This is a region of strong temperature gradient, connected with the midlevel easterly jet and the generation of easterly waves. Very strong thunderstorms and squall lines form in the desert margins near 20°N, modulated by the waves, but rare in any one location. The Sahel zone near 15°N is in the strong rainfall gradient, with heavy rainfall during the July–September monsoon season when the large-scale flow brings in moist air from the southwest at low levels and the waves and squall lines are strongest (Zipser, 1994).

Short monsoon seasons are also experienced in western North America, noted above, and northern Australia in December to March. The latter has been the subject of several field programs near Darwin and the Tiwi Islands just to the north. The thunderstorms over these islands are so common in season that Darwinites call them by name (Hector). As in west Africa, the surface convergence zone extends well beyond the area of heaviest rain (near 10°S), into interior Australia. Near Darwin, research has established that the main modulation of these storms and rain systems is not by synoptic-scale waves but by longer period oscillations in the flow (also known as the MJO, see below). During monsoonal low-level westerly wind periods, rainfall is heavy, oceanic in character, and tropical cyclones may form. During the "break" periods, westerlies weaken or become easterlies, and rainfall decreases, but is concentrated into strong thunderstorms and squall lines moving from the continent (Rutledge et al., 1992).

The classic Asian monsoon affects billions of people and requires far more research before its rain systems have been properly described and understood, since few field programs have been undertaken with observations of appropriate scale. It is well known that the southwest monsoon flow is from ocean to continent during northern summer over India, China, and all of Southeast Asia. There is great variety of local and regional details, well beyond our scope here to describe. Over India, there is a characteristic alternation between rainy and break periods on a scale of weeks. Where persistent moist flow is forced to ascend rapidly, as over the hills of Assam in northeast India, world record rainfalls have been recorded.

Madden–Julian Oscillation (MJO) First discovered by Madden and Julian (1972, 1994), there is often a strong variability in tropical winds and rainfall on this intraseasonal time scale of 30 to 60 days. While many wave modes may come into play, the most significant one appears to be an equatorially trapped Kelvin wave (Meehl, G., Chapter 5). It is strongest within 10° of the equator (but its effects extend beyond), and from the western Indian Ocean eastward through the central Pacific Ocean (but its effects extend beyond). In these regions, it is considered normal for rainfall to be excessive for a few weeks and deficient for the next few weeks. The break period has anomalous easterlies at low levels and westerlies at high levels, while the rainy period has westerly wind anomalies at low levels and easterly anomalies at upper levels. During the rainy west wind periods near the equator, tropical cyclones may form on either side of the equator, sometimes on both sides (double vortex). Periods with especially strong west winds are known as westerly wind bursts. These phenomena have been extensively studied during a major field program in 1992–1993 (Godfrey et al., 1998).

7 DISTRIBUTION OF TROPICAL/SUBTROPICAL RAINFALL AND MESOSCALE RAIN SYSTEMS

Maps of annual rainfall totals in the tropics are often accuracy challenged but show the same major features. The "doldrums" or ITCZ regions near the meteorological equator are very rainy; the subtropical high-pressure regions and trade wind regions are generally dry except where orographic lifting occurs. Interestingly, the regions of heaviest rainfall, and the regions with numerous large, strong MCSs, are often quite different.

Total Rainfall Distribution

Figure 4 can be considered a "best estimate" of total rainfall in the tropics, using a combination of satellite and rain gauge data. The rainfall maps include the region covered by the *Tropical Rain Measuring Mission* (*TRMM*) satellite (36°N–36°S) and include one month from each season in 1998–1999. During the first 4 months of 1998 one of the strongest warm episodes of El Niño–Southern Oscillation (ENSO) occurred, and for the next 12 months one of the strongest cold episodes. The rainfall anomalies associated with these events are clearly seen in the expected areas: equatorial central and east Pacific (rainy in January and April 1998, dry in January and

April 1999); Indonesia (rainier in January and April 1999 than in the same months of 1998).

Despite these unusually strong anomalies, these rainfall maps clearly show the "normal" seasonal cycle of rainfall over the tropics and subtropics. Asia is dry in winter, wet in the summer monsoon, with similar extent in both years (despite different amounts in some regions). African rainfall migrates north and south of the equator with season. The oceanic ITCZ migrates very little in the tropical Atlantic and Pacific with the seasons.

MCC Distributions Using Satellite Infrared (IR) Measurements

Laing and Fritsch (1997) have systematically mapped out the regions subject to MCCs throughout the globe (Fig. 5). There are concentrations of MCCs in the central United States, Panama–Columbia, South America east of the Andes from 15° to 35°S, Africa between 0° and 15°N, northeast India–Bangladesh, and lesser concentrations in China and northern Australia. Laing and Fritsch (2000) show how the environments of MCCs in these regions vary, but the similarities are striking: a low-level jet bringing high θ_e air, and an approaching disturbance (which can be weak) providing ascent, triggering intense convection downwind of elevated terrain.

Some tropical regions are notable for high rainfall, but a near absence of MCCs. These include the entire oceanic ITCZ belt, equatorial South America, Southeast Asia, and the entire Maritime continent (Indonesian Archipelago). These regions are characterized by persistent deep and high cloudiness, and by low values of *average* outgoing longwave radiation (Laing and Fritsch, 1997). The contrast between the Amazon and Congo basins is especially striking. McCollum et al., (2000) have recently noted that the satellite estimates of rainfall are biased low in equatorial Africa but are approximately correct in the Amazon, for reasons yet to be determined, but which they speculate are related to differing properties of MCSs and their microphysics.

MCS Distributions Using Passive Microwave Measurements

Passive microwave radiances at 37 and 85 GHz can be used to identify, categorize, and map MCSs. The principle used to diagnose deep precipitating convection is that ice particles are efficient scatterers but poor absorbers and emitters of radiation at frequencies in this range. Therefore, regions with large depth of precipitation-sized ice particles, especially graupel, have low brightness temperatures compared with their surroundings. This approach has the advantage that precipitation particles are sensed directly, compared with IR techniques that sense only the cloud tops, even if they consist mainly of nonprecipitating cirrus. Data are available at microwave frequencies from a series of SSM/I Special Sensor Microwave Imager (SSM/I) instruments on DMSP satellites, since 1987, and on *TRMM* since 1997. The main disadvantage compared to IR techniques is that they are not yet available at high temporal frequencies from geosynchronous orbits.

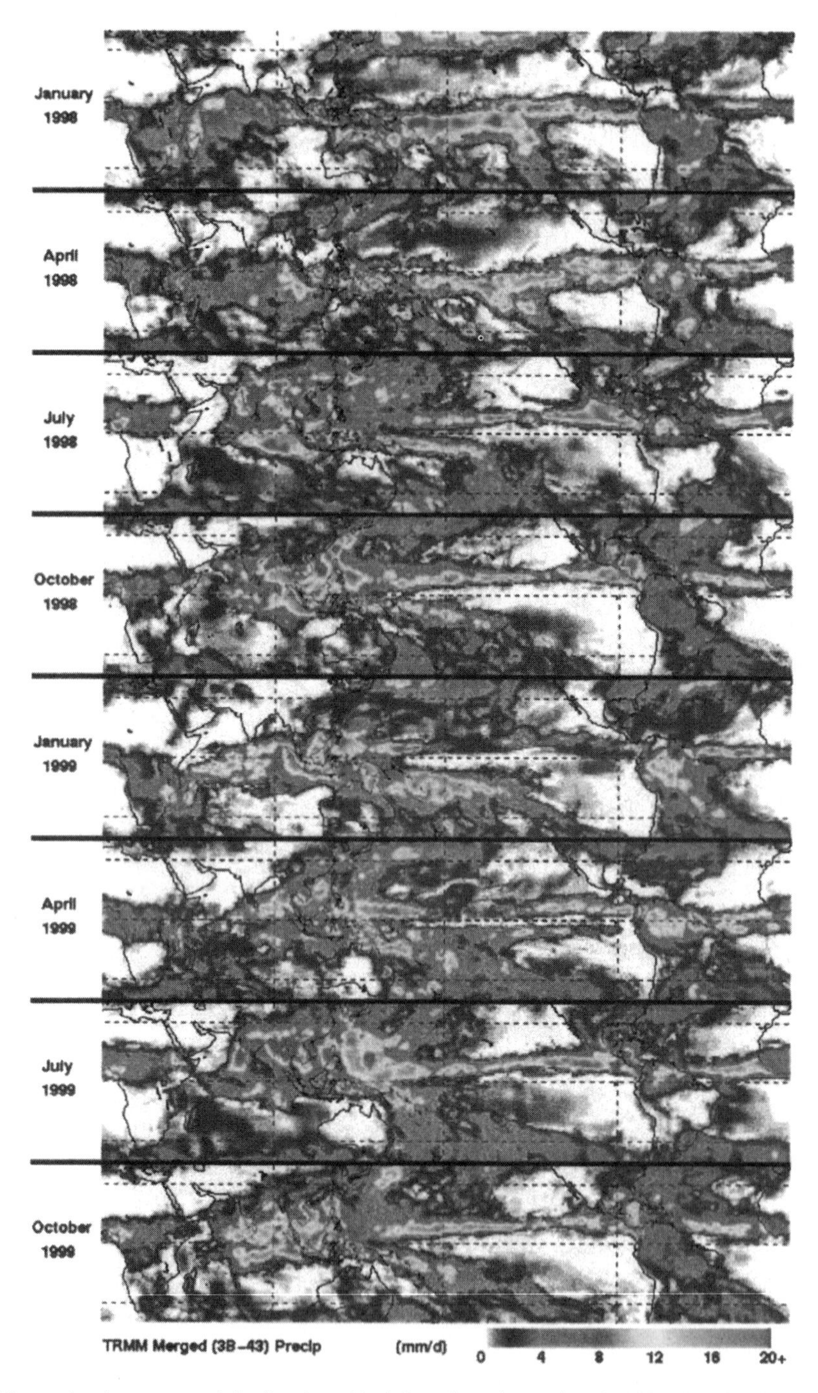

Figure 4 Average precipitation (mm/day) for selected months of 1998–1999 for the *TRMM*-merged analysis (Adler et al., 2000).

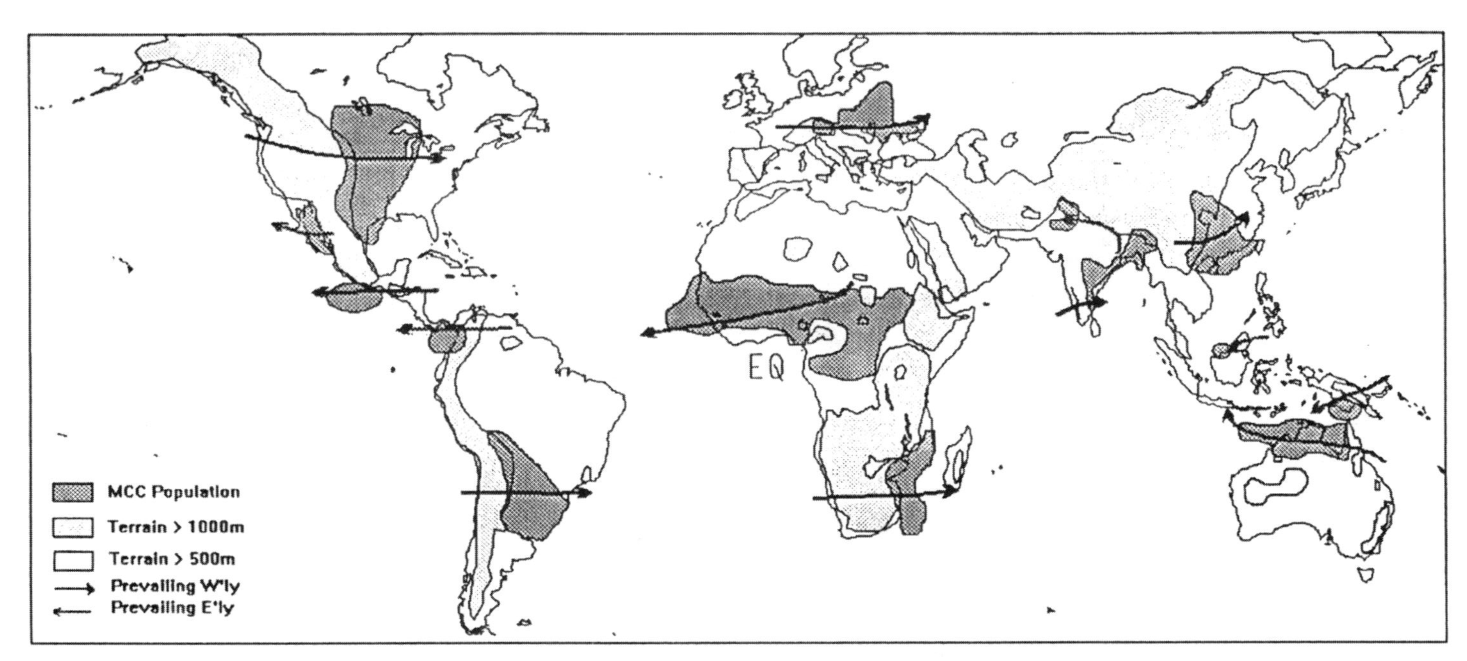

Figure 5 Relationship among mesoscale convective complex (MCC) population centers, elevated terrain, and prevailing midlevel flow (Laing and Fritsch, 1997).

Mohr and Zipser (1996*a*,*b*) mapped MCSs by size and by "intensity," defined by the minimum brightness temperature (i.e., by the pixel with largest optical depth of ice in a given MCS). They are mapped by season (Fig. 6) and by sunrise vs sunset observation time (not shown). The resulting maps give some additional insights into the properties of precipitating systems in different parts of the world. Unlike MCCs,

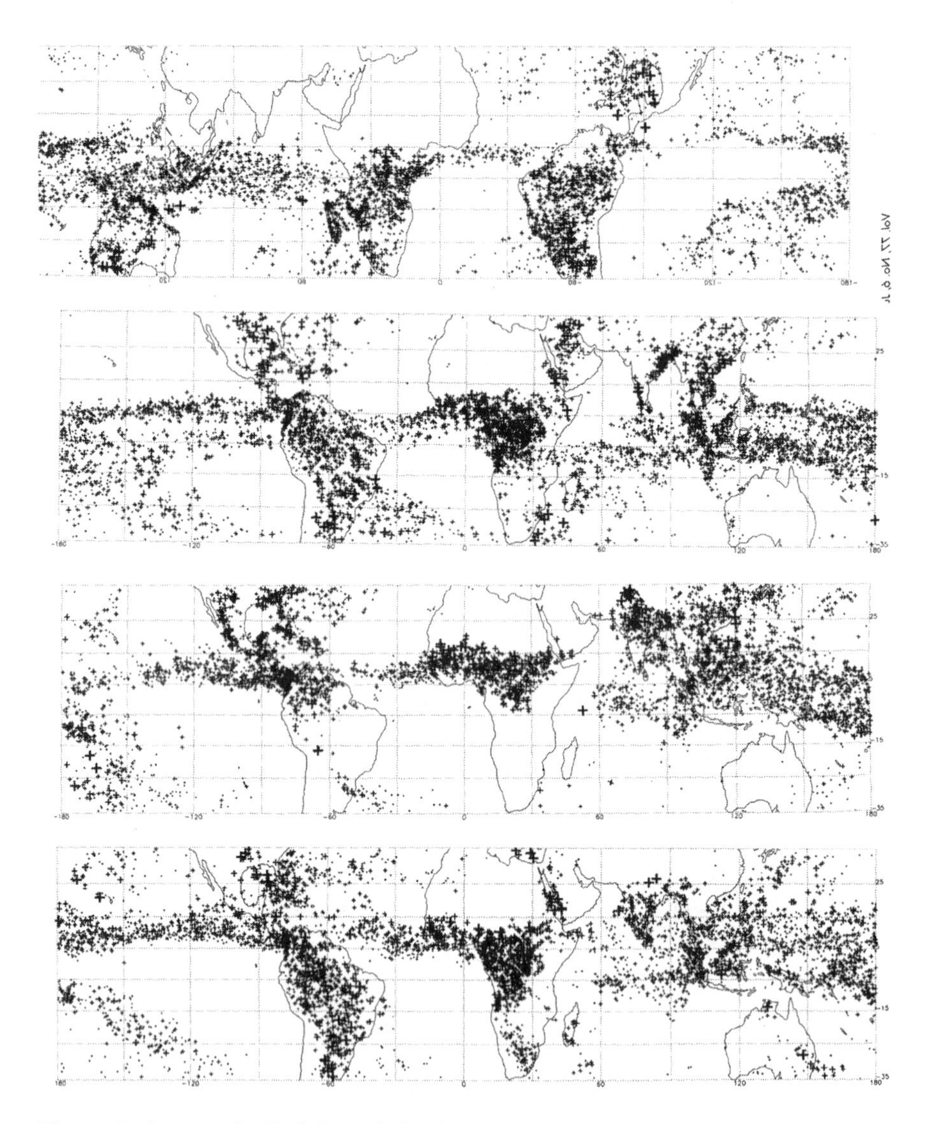

Figure 6 January, April, July, and October, 1993, top to bottom distribution of MCSs according to minimum (polarization-corrected) brightness temperatures. Symbols increase in size and thickness with decreasing temperature, which represents more intense convective pixels. The coldest range, <120 K, has the largest and thickest cross (after Mohr and Zipser, 1996a).

which are nearly absent in tropical oceans, the high rainfall regions of the oceans have abundant (ice-scattering) MCSs, including numerous MCSs with large areas. The MCSs with large (ice-scattering) intensity, however, favor land areas, especially at sunset, a time when growth of MCSs from intense thunderstorms is common. Oceanic regions have more MCSs at sunrise than at sunset.

Using Lightning Measurements

Satellite-derived maps of lightning discharges have been used increasingly in studies of global distributions of thunderstorms. Since 1995 such data have been available from a near-polar orbit, and since 1997 on *TRMM*. Research using these data is relatively new, and comparative studies of lightning and other databases only beginning. The distribution of lightning is even more skewed than the distribution of rainfall, with a relatively small number of storms producing a disproportionate fraction of total lightning. Lightning mapping clearly demonstrates a scarcity of lightning over tropical oceans, resembling the MCC distributions to some extent (Fig. 7). Comparing the lightning, MCS, and MCC distributions, one could hypothesize that the lightning distribution faithfully represents the distribution of strong MCSs and MCCs, which are common over continents and rare over oceans. Recent research that attempts to test this hypothesis indicates that it fails. That is, MCSs over land and ocean with similar brightness temperatures and

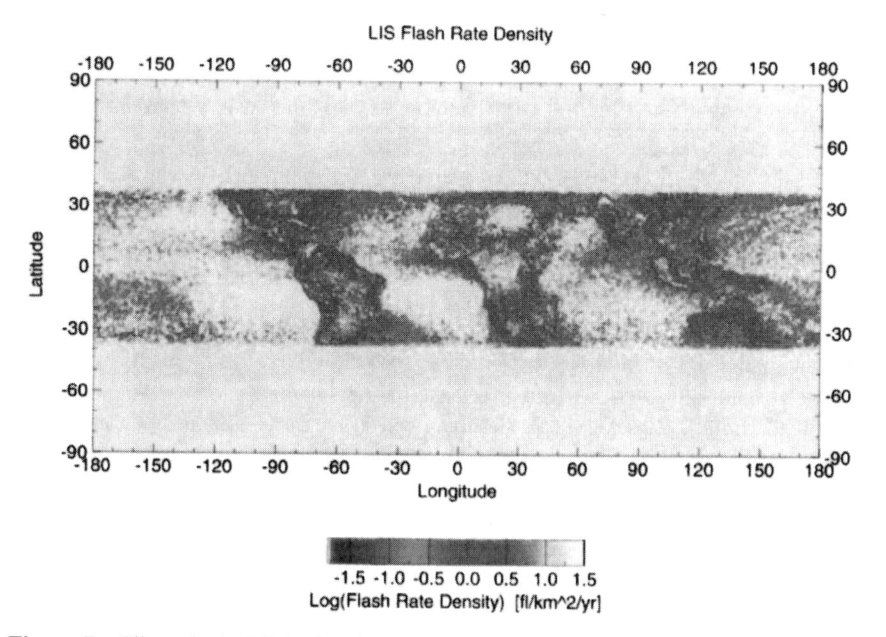

Figure 7 Climatological lightning imaging sensor (LIS instrument on *TRMM*) flash rate density, Dec. 1997–Nov. 1999. Data have been normalized by sensor view time and an assumed detection efficiency of 75% (after Boccippio et al., 2000).

similar radar echoes (from *TRMM*) still have higher lightning frequency over land, for reasons that are under investigation but must involve details of the microphysics.

REFERENCES

Adler, R. F., G. J. Huffman, D. V. Bolvin, S. Curtis, and E. J. Nelkin (2000). Tropical rainfall distributions determined using TRMM combined with other satellite and rain gauge information, *J. Appl. Meteor.* **39**, 2007–2023.

Barnes, G. M. (2001). Severe weather in the tropics, in *Severe Convective Storms*, *Meteor. Monographs*, **28**(50), C. Doswell (Ed.)., Am. Meteor. Soc., Boston MA.

Boccippio, D. J., S. J. Goodman, and S. Heckman (2000). Regional differences in tropical lightning distributions, *J. Appl. Meteor.* **39**, 2231–2248.

Chappell, C. F. (1986). Quasi-stationary convective events. Mesoscale analysis and forecasting, P. S. Ray (Ed.), *Am. Meteor. Soc.* 289–310.

Godfrey, J. S., R. A. Houze, Jr., R. H. Johnson, R. Lukas, J.-L. Redelsberger, A. Sumi, and R. Weller (1998). Coupled Ocean–Atmosphere Response Experiment: An interim report, *J. Geophys. Res.* **103**(C7), 14,395–14,450.

Houze, R. A., Jr. (1993). *Cloud Dynamics*, Academic Press, San Diego, CA.

Laing, A. G., and J. M. Fritsch (1997). The global population of mesoscale convective complexes, *Quart. J. Roy. Meteor. Soc.* **123**, 389–405.

Laing, A. G., and J. M. Fritsch (2000). The large-scale environments of the global populations of mesoscale convective complexes, *Mon. Wea. Rev.* **128**, 2756–2776.

LeMone, M. A., E. J. Zipser, and S. B. Trier (1998). The role of environmental shear and CAPE in determining the structure and evolution of mesoscale convective systems during TOGA COARE, *J. Atmos. Sci.* **55**, 3493–3518.

Madden, R. A., and P. R. Julian (1972). Description of global-scale circulation cells in the Tropics with a 40–50 day period, *J. Atmos. Sci.* **29**, 1109–1123.

Madden, R. A., and P. R. Julian (1994). Observations of the 40–50 day tropical oscillation—A review. *Mon. Wea. Rev.,* **122**, 814–837.

McCollum, J. R., A. Gruber, and M. B. Ba (2000). Discrepancy between gauges and satellite estimates of rainfall in equatorial Africa, *J. Appl. Meteor.* **39**, 666–679.

Mohr, K. I., and E. J. Zipser (1996*a*). Defining mesoscale convective systems by their ice scattering signature, *Bull. Am. Meteor. Soc.* **77**, 1179–1189.

Mohr, K. I., and E. J. Zipser (1996*b*). Mesoscale convective systems defined by their 85 GHz ice scattering signature: Size and intensity comparison over tropical oceans and continents, *Mon. Wea. Rev.* **124**, 2417–2437.

Reed, R. J., D. C. Norquist, and E. E. Recker (1977). The Structure and Properties of African Wave Disturbances as Observed During Phase III of GATE, *Mon. Wea. Rev.* **105**, 317–333.

Riehl, H., and J. S. Malkus (1958). On the heat balance in the equatorial trough zone, *Geophysica* **6**, 503–538.

Rutledge, S. A., E. R. Williams, and T. D. Keenan (1992). The Down Under Doppler and Electricity Experiment (DUNDEE): Overview and preliminary results, *Bull. Am. Meteor. Soc.* **73**, 3–16.

Zipser, E. J. (1977). Mesoscale and convective-scale downdrafts as distinct components of squall-line structure, *Mon. Wea. Rev.* **105**, 1568–1589.

Zipser, E. J. (1994). Deep cumulonimbus cloud systems in the tropics with and without lightning, *Mon. Wea. Rev.* **122**, 1837–1851.

Zipser, E. J., and K. Lutz (1994). The vertical profile of radar reflectivity of convective cells: A strong indicator of storm intensity and lightning probability? *Mon. Wea. Rev.* **122**, 1751–1759.

CHAPTER 31

HURRICANES

FRANK D. MARKS, Jr.

1 INTRODUCTION

The term *hurricane* is used in the Western Hemisphere for the general class of strong tropical cyclones, including western Pacific typhoons and similar systems, known simply as cyclones in the Indian and southern Pacific Oceans. A tropical cyclone is a low-pressure system that derives its energy primarily from evaporation from the sea in the presence of 1-min sustained surface wind speeds >17 m/s and the associated condensation in convective clouds concentrated near its center. In contrast, midlatitude storms (low-pressure systems with associated fronts) primarily get their energy from the horizontal temperature gradients that exist in the atmosphere. Structurally, the strongest winds in tropical cyclones are near Earth's surface (a consequence of being "warm core" in the troposphere), while the strongest winds in midlatitude storms are near the tropopause (a consequence of being "warm core" in the stratosphere and "cold core" in the troposphere). Warm core refers to being warmer than the environment at the same pressure surface.

A tropical cyclone with the highest sustained wind speeds between 17 and 32 m/s is referred to as a tropical storm, whereas a tropical cyclone with sustained wind speeds ≥ 33 m/s is referred to as a hurricane or typhoon. Once a tropical cyclone has sustained winds ≥ 50 m/s, it is referred to as a major hurricane or supertyphoon. In the Atlantic and eastern Pacific Oceans hurricanes are also classified by the damage they can cause using the Saffir–Simpson scale (Table 1).

The Saffir–Simpson scale categorizes hurricanes on a scale from 1 to 5, where category 1 hurricanes are the weakest, and category 5 the most intense. Major hurricanes correspond to categories 3 and higher. The reasons that some disturbances intensify to a hurricane, while others do not, are not well understood. Neither

Handbook of Weather, Climate, and Water: Dynamics, Climate, Physical Meteorology, Weather Systems, and Measurements, Edited by Thomas D. Potter and Bradley R. Colman.
ISBN 0-471-21490-6

TABLE 1 Saffir–Simpson Scale of Hurricane Intensity

Category	Pressure (hPa)	Wind (m/s)	Storm Surge (m)	Damage
1	>980	33–42	1.0–1.7	Minimal
2	979–965	43–49	1.8–2.6	Moderate
3	964–945	50–58	2.7–3.8	Extensive
4	944–920	59–69	3.9–5.6	Extreme
5	<920	≥ 70	≥ 5.7	Catastrophic

is it clear why some tropical cyclones become major hurricanes, while others do not. Major hurricanes produce 80 to 90% of the U.S. hurricane-caused damages despite accounting for only one-fifth of all landfalling tropical cyclones. Only three category 5 hurricanes have made landfall on the U.S. mainland (Florida Keys, 1935, Camille, 1969, and Andrew, 1992). Recent major hurricanes to make landfall on the United States were Hurricanes Bonnie and Georges in 1998, and Bret and Floyd in 1999.

As with large-scale extratropical weather systems, the structure and evolution of a tropical cyclone is dominated by the fundamental contradiction that while the airflow within a tropical cyclone represents an approximate balance among forces affecting each air parcel, slight departures from balance are essential for vertical motions and resulting clouds and precipitation, as well as changes in tropical cyclone intensity. As in extratropical weather systems, the basic vertical balance of forces in a tropical cyclone is hydrostatic except in the eyewall, where convection is superimposed on the hydrostatic motions. However, unlike in extratropical weather systems, the basic horizontal balance in a tropical cyclone above the boundary layer is between the Coriolis *force* (defined as the horizontal velocity, v, times the Coriolis parameter,* f), the centrifugal force (defined as v^2 divided by the radius from the center, r), and the horizontal pressure gradient force. This balance is referred to as gradient balance, where the Coriolis and centrifugal force are both proportional to the wind speed. Centrifugal force is an apparent force that pushes objects away from the center of a circle. The centrifugal force alters the original two-force geostrophic balance and creates a nongeostrophic gradient wind.

The inner region of the tropical storm, termed the cyclone *core*, contains the spiral bands of precipitation, the eyewall, and the eye that characterize tropical cyclones in radar and satellite imagery (Fig. 1). The primary circulation—the tangential or swirling wind—in the core becomes strongly axisymmetric as the cyclone matures. The strong winds in the core, which occupies only 1 to 4% of the cyclone's area, threaten human activities and make the cyclone's dynamics unique. In the core, the local Rossby number† is always >1 and may be as high

*f is the Coriolis parameter ($f = 2\Omega \sin \phi$), where Ω is the angular velocity of Earth ($7.292 \times 10^{-5}\ \mathrm{s}^{-1}$) and ϕ is latitude. The Coriolis parameter is zero at the equator and 2Ω at the pole.

†The Rossby number indicates the relative magnitude of centrifugal (v^2) and Coriolis (fv) accelerations, $\mathrm{Ro} = V/fr$, where V is the axial wind velocity, r the radius from the storm center, and f the Coriolis parameter. An approximate breakdown of regimes is: $\mathrm{Ro} < 1$ geostrophic flow; $\mathrm{Ro} > 1$ gradient flow; and $\mathrm{Ro} > 50$ cyclostrophic flow.

Figure 1 (see color insert) NOAA-14 AVHRR multispectral false color image of Hurricane Floyd at 2041 UTC, September 13 about 800 km east of south Florida. (Photo courtesy of NOAA Operationally Significant Event Imagery website: *http://www.osei.noaa.gov/*.) See ftp site for color image.

as 100. When the Rossby number significantly exceeds unity, the balance in the core becomes more cyclostrophic, where the pressure gradient force is almost completely balanced by the centrifugal force. The time scales are such that air swirling around the center completes an orbit in much less than a pendulum day (defined as $1/f$).

When the atmosphere is in approximate horizontal and vertical balance, the wind and mass fields are tightly interconnected. The distribution of a single mass or momentum variable may be used as a starting point to infer the distribution of all other such variables. One such variable is potential vorticity (PV), approximately equal to the vorticity times the thermal stratification, which is related to the three-dimensional mass and momentum fields through an inverse second-order Laplacian-like operator. The benefit of such a relationship is that PV variations in a single location are diagnostically related to variations in mass and wind fields at a distance. Areas of high PV correspond locally to low mass, or cyclones, while areas of low PV correspond to anticyclones.

Typical extratropical weather systems contain high PV values around $0.5 \times 10^{-6} m^2/s$ K/kg (0.5 PVU) to 5 PVU, whereas, typical values in the tropical cyclone core are ≥ 10 PVU. Figure 2 shows the wind and mass fields associated with an idealized axially symmetric tropical cyclone PV anomaly with the PV concentrated near the surface rather than in a vertical column. The cyclonic anomaly (positive in the Northern Hemisphere) is associated with a cyclonic circulation that is strongest at the level of the PV anomaly near the surface, and decreases upward. Temperatures are anomalously warm above the PV anomaly (isentropic surfaces are deflected downward). While consistent with the simple PV distribution, the wind and mass fields are also in horizontal and vertical balance. The tropical cyclone being a warm-core vortex, the PV inversion dictates that the winds that swirl about the center decrease with increasing height, but they typically fill the depth of

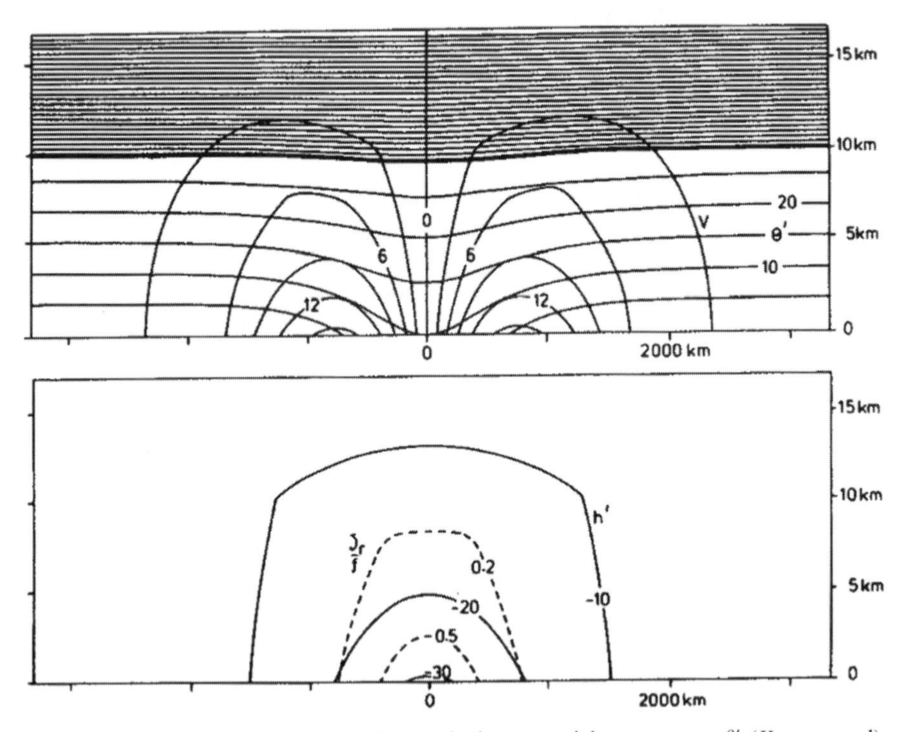

Figure 2 Gradient wind v (m/s) and perturbation potential temperature θ' (K, top panel); and geopotential/height perturbation h' (dm) and relative vorticity scale by Coriolis parameter ζ/f (bottom panel) for a warm core, lower cyclone. The tropopause location is denoted by the bold solid line, and the label 0 on the horizontal axis indicates the core (and axis of symmetry) of the disturbance. The equivalent pressure deviation at the surface in the center of the vortex is -31 hPa. [From A. J. Thorpe, *Mon. Wea. Rev.* **114**, 1384–1389 (1986). Copyright owned by the American Meteorological Society.]

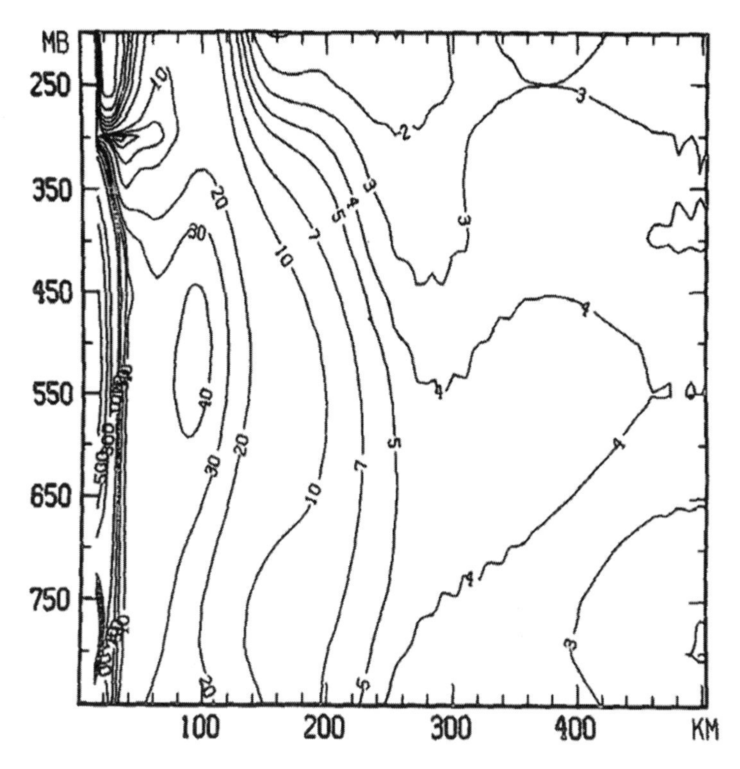

Figure 3 Radial-height cross section of symmetric PV for Hurricane Gloria, September 24, 1985. Contours are 0.1 PVU. Values in data sparse region, within 13 km of vortex center, are not displayed. [From L. J. Shapiro, and J. L. Franklin, *Mon. Wea. Rev.* **123**, 1465–1475, (1995). Copyright owned by American Meteorological Society.]

the troposphere. If the PV reaches values $\geq$10 PVU, the inner region winds can become intense as in Hurricane Gloria in Figure 3. Gloria had PV values exceeding 60 PVU just inside the radius of maximum winds of 15 km where the axisymmetric mean tangential winds exceeded 65 m/s.

Many features in the core, however, persist with little change for (pendulum) days (mean life span of a tropical cyclone is about 5 to 10 days). Because these long lifetimes represent tens or hundreds of orbital periods ($\sim$1 h), the flow is nearly balanced. Moreover, at winds >35 m/s, the local Rossby radius of deformation* is reduced from its normal $\sim 10^3$ km to a value comparable with the eye radius. In very intense tropical cyclones, the eye radius may approach the depth of the troposphere (15 km), making the aspect ratio unity. Thus, the dynamics near the center of a tropical cyclone are so exotic that conditions in the core differ from Earth's day-to-day weather as much as the atmosphere of another planet does.

*The Rossby radius of deformation is the ratio of speed of the relevant gravity wave mode and the local vorticity, or, equivalently, the ratios of the Brunt–Vaisala and inertial frequencies. This scale indicates the amount of energy that goes into gravity waves compared to inertial acceleration of the wind.

2 CLIMATOLOGY

There are 80 to 90 tropical cyclones worldwide per year, with the Northern Hemisphere having more tropical cyclones than the Southern Hemisphere. Table 2 shows that of the 80 to 90 tropical cyclones, 45 to 50 reach hurricane or typhoon strength and 20 reach major hurricane or super typhoon strength. The western North Pacific (27 tropical cyclones), eastern North Pacific (17 tropical cyclones), Southwest Indian Ocean (10 tropical cyclones), Australia/southwest Pacific (10 tropical cyclones), and North Atlantic (10 tropical cyclones) are the major tropical cyclone regions. There are also regional differences in the tropical cyclone activity by month with the majority of the activity in the summer season for each basin. Hence, in the Pacific, Atlantic, and North Indian Ocean, the maximum numbers of tropical cyclones occur in August through October, while in the South Pacific and Australia regions, the maxima are in February and March. In the South Indian Ocean, the peak activity occurs in June. In the western North Pacific, Bay of Bengal, and South Indian Ocean regions tropical cyclones may occur in any month, while in the other regions at least one tropical cyclone-free month occurs per year. For example, in the North Atlantic, there has never been tropical cyclone activity in January.

Some general conclusions can be drawn from the global distribution of tropical cyclone locations (Fig. 4*a*). Tropical cyclone formation is confined to a region approximately 30°N and 30°S, with 87% of them located within 20° of the equator. There is a lack of tropical cyclones near the equator, as well as in the eastern South Pacific and South Atlantic basins. From these observations there appears to be at least five necessary conditions for tropical cyclone development.

- Warm sea surface temperature (SST) and large mixed layer depth: Numerous studies suggest a minimum SST criterion of 26°C for development The warm water must also have sufficient depth (i.e., 50 m). Comparison of Figure 4*a* and 4*b* the annual mean global SST, shows the strong correlation between regions with SST > 26°C and annual tropical cyclone activity. SST > 26°C is sufficient but not necessary for tropical cyclone activity, evidenced by the regions with tropical cyclone activity when SST < 26°C. Some of the discrepancy exists because storms that form over regions where SST > 26°C are advected poleward during their life cycle. However, tropical cyclones are observed to originate over regions where SST < 26°C. These occurrences are not many, but the fact that they exist suggests that other factors are important.
- Background earth vorticity: Tropical cyclones do not form within 3° of the equator. The Coriolis parameter vanishes at the equator and increases to extremes at the poles. Hence, a threshold value of Earth vorticity (f) must exist for a tropical cyclone to form. However, the likelihood of formation does not increase with increasing f. Thus, nonzero Earth vorticity is necessary but not sufficient to produce tropical cyclones.
- Low vertical shear of the horizontal wind: In order for tropical cyclones to develop, the latent heat generated by the convection must be kept near the

TABLE 2 Mean Annual Frequency, Standard Deviation (σ), and Percent of Global Total of Number of Tropical Storms (Winds $\geq$17 m/s), Hurricane-Force Tropical Cyclone (Winds $\geq$33 m/s), and Major Hurricane-Force Tropical Cyclone (winds $\geq$50 m/s).

Tropical Cyclone Basin[a]	Tropical Storm Annual Frequency σ	Percent of Total	Hurricane Annual Frequency (σ)	Percent of Total	Major Hurricane Annual Frequency (σ)	Percent of Total
Atlantic (1944–2000)	**9.8** (3.0)	11.4	**5.7** (2.2)	12.1	**2.2** (1.5)	10.9
Northeast Pacific (1970–2000)	**17.0** (4.4)	19.7	**9.8** (3.1)	20.7	**4.6** (2.5)	22.9
Northwest Pacific (1970–2000)	**26.9** (4.1)	32.1	**16.8** (3.6)	35.5	**8.3** (3.2)	41.3
North Indian (1970–2000)	**5.4** (2.2)	6.3	**2.2** (1.8)	4.6	**0.3** (0.5)	1.5
Southwest Indian (30–100°E) (1969–2000)	**10.3** (2.9)	12.0	**4.9** (2.4)	10.4	**1.8** (1.9)	9.0
Australian/S.E. Indian (100–142°E) (1969–2000)	**6.5** (2.6)	7.5	**3.3** (1.9)	7.0	**1.2** (1.4)	6.0
Australian/S.W. Pacific (142°E) (1969–2000)	**10.2** (3.1)	11.8	**4.6** (2.4)	9.7	**1.7** (1.9)	8.5
Global (1970–2000)	**86.1** (8.0)		**47.3** (6.5)		**20.1** (5.7)	

[a]Dates in parentheses provide the nominal years for which accurate records are currently available.

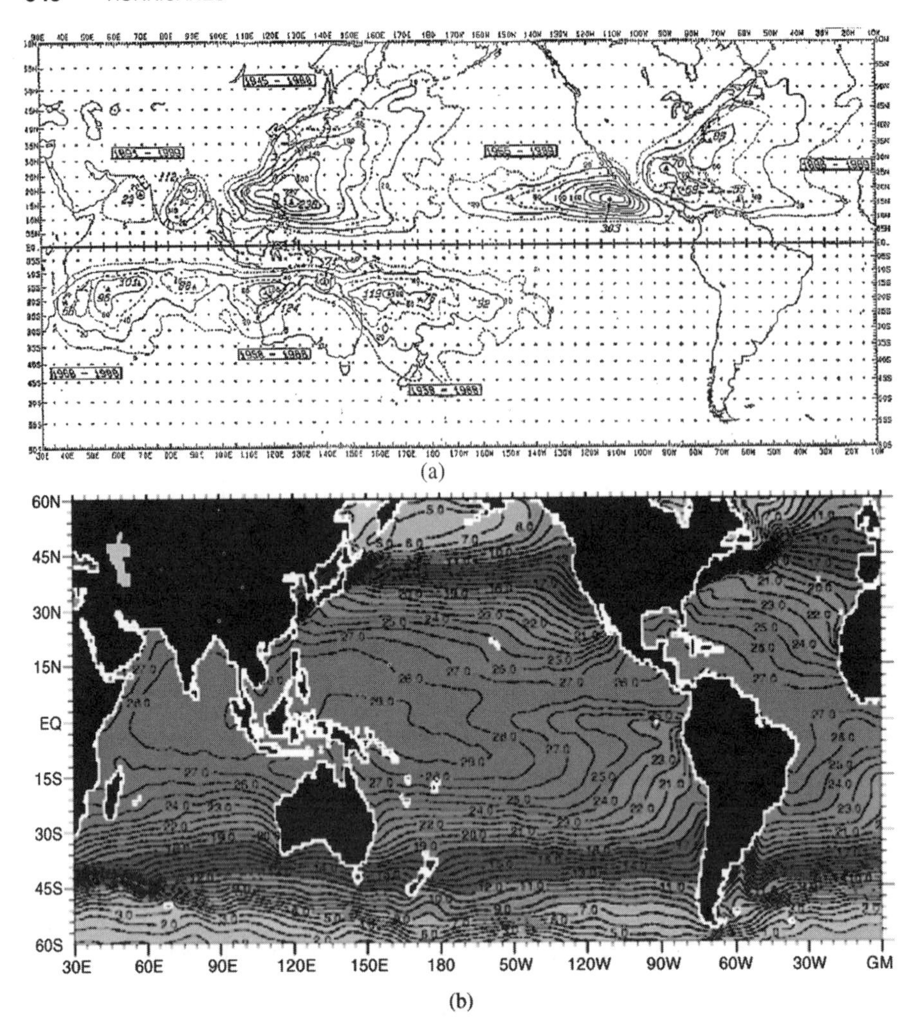

Figure 4 (see color insert) (*a*) Frequency of tropical cyclones per 100 years within 140 km of any point. Solid triangles indicate maxima, with values shown. Period of record is shown in boxes for each basin. (*b*) Annual SST distribution (°C). See ftp site for color image.

center of the storm. Historically, shear was thought to "ventilate" the core of the cyclone by advecting the warm anomaly away. The ventilation argument suggests that if the storm travels at nearly the same speed as the environmental flow in which it is embedded, its heating remains over the disturbance center. However, if it is moving slower than the mean wind at upper levels, the heating in the upper troposphere is carried away by the mean flow. Recent analysis suggests that the effect of shear is to force the convection into an asymmetric pattern such that the convective latent heat release forces flow asymmetry and irregular motion rather than intensification of the symmetric vortex. Thus, if the vertical shear is too strong (>16 m/s) existing tropical cyclones are ripped apart and new ones cannot form.

- Low atmospheric static stability: The troposphere must be potentially unstable to sustain convection for an extended period of time. Typically measured as the difference between the equivalent potential temperature (θ_e) at the surface and 500 hPa, instability must typically be >10 K for convection to occur. This value is usually satisfied over tropical oceans.
- Tropospheric humidity: The higher the midlevel humidity, the longer a parcel of air can remain saturated as it entrains the surrounding air during its ascent. Vigorous convection occurs if the parcel remains saturated throughout its ascent. A relative humidity of 50 to 60% at lower to midlevels (700 to 500 hPa) is often sufficient to keep a parcel saturated during ascent. This condition is regularly evident over tropical oceans.

These conditions are usually satisfied in the summer and fall seasons for each tropical cyclone basin. However, even when all of the above conditions are favorable, tropical cyclones do not necessarily form. In fact, there is growing evidence for significant interannual variability in tropical cyclone activity, where numerous tropical cyclones form in a given basin over a week to 10-day period, followed by 2 to 3 weeks with little or no tropical cyclone activity. Figure 5 shows just such an active period in the Atlantic basin in mid-September 1999, where two hurricanes (Floyd and Gert), both major, and an unnamed tropical depression formed within a few days of each other. During these active phases, almost every disturbance makes at least tropical storm strength, whereas in the inactive phase, practically none of the distur-

Figure 5 (*a*) *GOES* multispectral false color image of Hurricanes Floyd and Gert and an unnamed tropical depression at 1935 UTC, September 13. 1999. (Photo courtesy of NOAA Operationally Significant Event Imagery website: *http://www.osei.noaa.gov/*.) See ftp site for color image.

bances intensify. The two hurricanes and unnamed depression in Figure 5 represented the second 10-day active period during the summer of 1999. An earlier period in mid-August also resulted in the development of three hurricanes (Brett, Cindy, and Dennis), two of which were major, as well as a tropical storm (Emily). There is speculation that the variability is related to the propagation of a global wave. Because the SST, static stability, and Earth vorticity do not vary that much during the season, the interannual variability is most likely related to variations in tropospheric relative humidity and vertical wind shear.

It has long been recognized that the number of tropical cyclones in a given region varies from year to year. The exact causes of this remain largely speculative. The large-scale global variations in atmospheric phenomena such as the El Niño and Southern Oscillation (ENSO) and the Quasi-Biennial Oscillation (QBO) appear to be related to annual changes in the frequency of tropical cyclone formation, particularly in the Atlantic Ocean. The ENSO phenomenon is characterized by warmer SSTs in the eastern South Pacific and anomalous winds over much of the equatorial Pacific. It influences tropical cyclone formation in the western North Pacific, South Pacific, and even the North Atlantic.

During the peak phase of the ENSO, often referred to as El Niño (usually occurs during the months of July to October), anomalous westerly winds near the equator extend to the date line in the western North Pacific acting to enhance the intertropical convergence zone (ITCZ) in this area, making it more favorable for formation of tropical cyclones. Another effect of the El Niño circulation is warmer SST in the eastern South Pacific. During such years, tropical cyclones form closer to the equator and farther east. Regions, such as French Polynesia, which are typically unfavorable for tropical cyclones due to a strong upper-level trough, experienced numerous tropical cyclones. The eastern North Pacific is also affected by the El Niño through a displacement of the ITCZ south to near 5°N. Additionally, the warm ocean anomaly of El Niño extends to near 20°N, which enhances the possibility of tropical cyclone formation. The result is an average increase of two tropical cyclones during El Niño years. Cyclones also develop closer to the equator and farther west than during a normal year.

The QBO is a roughly 2-year oscillation of the equatorial stratosphere (30 to 50 hPa) winds from easterly to westerly and back. The phase and magnitude of QBO are associated with the frequency of tropical cyclones in the Atlantic. Hurricane activity is more frequent when the 30-hPa stratospheric winds are westerly. The exact mechanism by which the QBO affects tropical cyclones in the troposphere is not clear; however, there are more North Atlantic tropical cyclones when the QBO is in the westerly phase than in the easterly phase.

3 TROPICAL CYCLOGENESIS

Tropical cyclones form because of thermodynamic disequilibrium between the warm near-surface waters of the tropical ocean and the tropospheric column. If one adjusts an air column to equilibrium with the SST of the summertime tropical ocean, the resulting change in surface pressure is commensurate with that found in tropical cyclones. Thus, much of the tropical oceans contain enough moist enthalpy to support a major hurricane.

Throughout most of the trade-wind regions, gradual subsidence causes an inversion that traps water vapor in the lowest kilometer. Sporadic convection (often in squall lines) that breaks through the inversion exhausts the moist enthalpy stored in the near-surface boundary layer quickly, leaving a wake of cool, relatively dry air. This air comes from just above the inversion and is brought to the surface by downdrafts driven by the weight of hydrometeors and cooling due to their evaporation. The squall line has to keep moving or it quickly runs out of energy. A day, or even several days, may pass before normal fair-weather evaporation can restore the preexistent moist enthalpy behind the squall. The situation is like a poorly designed control loop that oscillates around its equilibrium point. The reasons why squall line convection generally fails to produce hurricanes lie in the limited amount of enthalpy that can be stored in the subinversion layer and the slow rate of evaporation under normal wind speeds in the trades.

To make a tropical cyclone, one needs to speed up evaporation and raise the equilibrium enthalpy at the sea surface temperature by lowering the surface pressure. Tropical cyclones are thus finite-amplitude phenomena. They do not grow by some linear process from infinitesimal amplitude. The normal paradigm of searching for the most rapidly growing unstable linear mode used to study midlatitude cyclogenesis through baroclinic instability fails here. The surface wind has to exceed roughly 20 m/s before evaporation can prevail against downdraft cooling.

How then do tropical cyclones reach the required finite amplitude? The answer seems to lie in the structure of tropical convection. As explained previously, behind a squall line the lower troposphere (below the 0°C isotherm at ~5 km) is dominated by precipitation-driven downdrafts which lie under the "anvil" of nimbostratus and cirrostratus that spreads behind the active convection. Above 5 km, condensation in the anvil forces rising motion. This updrafts-over-downdrafts arrangement requires horizontal convergence centered near 5 km altitude to maintain mass continuity. The important kinematic consequence is formation of patchy shallow vortices near the altitude of the 0°C isotherm. The typical horizontal scales of these "mesovortices" are tens to hundreds of kilometers. If they were at the surface or if their influence could be extended downward to the surface, they would be the means to get the system to the required finite amplitude.

The foregoing reasoning defines the important unanswered questions: (1) How do the midlevel mesovortices extend their influence to the surface? (2) What are the detailed thermodynamics at the air–sea interface during this process? Leading hypotheses for (1) are related to processes that can increase the surface vorticity through changes in static stability and momentum mixing, both horizontally and vertically. However, the answers to these questions await new measurements that are just becoming available through improved observational tools.

4 BASIC STRUCTURE

Primary and Secondary Circulations

Inner-core dynamics received a lot of attention over the last 40 years through aircraft observations of the inner-core structure. These observations show that the tropical cyclone inner-core dynamics are dominated by interactions between "primary" (horizontal axisymmetric), "secondary" (radial and vertical) circulations, and a wave number one asymmetry caused by the storm motion. The primary circulation is so strong in the cyclone core that it is possible to consider axisymmetric motions separately, if account is taken of forcing by the asymmetric motions. The primary circulation is in near-gradient balance and evolves when heat and angular momentum sources (often due to asymmetric motions) force secondary circulations, which in turn redistribute heat and angular momentum.

Figure 6 shows that the primary circulation is sustained by the secondary circulation, which consists of frictional inflow that loses angular momentum* to the sea as it gains moist enthalpy. The inflow picks up latent heat through evaporation and exchanges sensible heat with the underlying ocean, as it spirals into lower levels of the storm under influence of friction. The evaporation of sea spray adds moisture to the air, while at the same time cooling it. This process is important in determining the intensity of a tropical cyclone. Near the vortex center, the inflow turns upward and brings the latent heat it acquires in the boundary layer into the free atmosphere. Across the top of the boundary layer, turbulent eddies causes significant downward

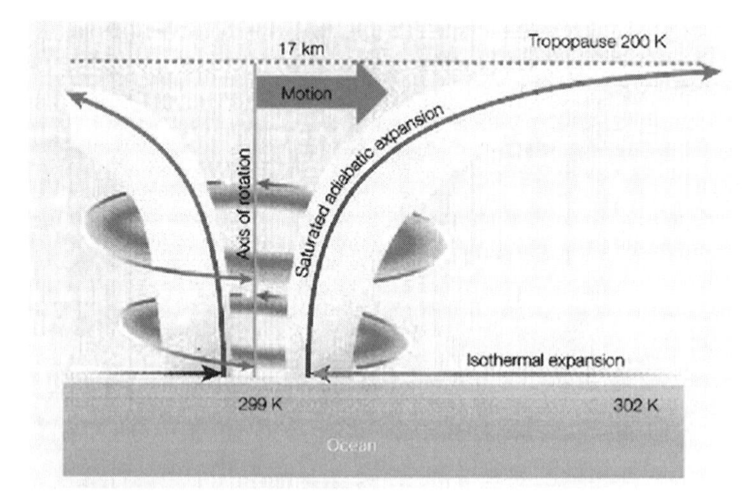

Figure 6 Schematic of the secondary circulation thermodynamics. [From H. E. Willoughby, *Nature* **401**, 649–650 (1999). Copyright owned by Macmillan Magazines, Ltd.] See ftp site for color image.

*Angular momentum, $M = Vr + fr^2/2$, where V is the tangential wind velocity, f is the Coriolis parameter, and r the radius from the storm center.

flux of sensible heat from the free atmosphere to the boundary layer. The energy source for the turbulent eddies is mechanical mixing caused by the strong winds. The eddies are also responsible for downward mixing of angular momentum. Hence, these turbulent eddy fluxes fuel the storm.

As the air converges toward the eye and is lifted in convective clouds that surround the clear eye, it ascends to the tropopause (the top of the troposphere, where temperature stops decreasing with height). As shown in Figure 6 the convective updrafts in the eyewall turn the latent heat into sensible heat through the latent heat of condensation to provide the buoyancy needed to loft air from the surface to tropopause level. The updraft entrains midlevel air promoting mass and angular momentum convergence into the core. It is the midlevel inflow that supplies the excess angular momentum needed to spin up the vortex. The net energy realized in the whole process is proportional to the difference in temperature between the ocean ($\sim$300 K) and the upper troposphere ($\sim$200 K). Storm-induced upwelling of cooler water reduces ocean SST by a few degrees, which has a considerable effect on the storm's intensity.

As shown in Figure 7 the secondary circulation also controls the distribution of hydrometeors and radar reflectivity. It is much weaker than the primary circulation except in the anticyclonic outflow, where the vortex is also much more asymmetric. Precipitation-driven convective updrafts form as hydrometeors fall from the outward sloping updraft. Condensation in the anvil causes a mesoscale updraft above the 0°C isotherm and precipitation loading by snow falling from the overhanging anvil causes a mesoscale downdraft below 0°C isotherm. The melting level itself marks the height of maximum mass convergence. Inside the eye, dynamically driven descent and momentum mixing leads to substantial pressure falls.

In order for the primary circulation to intensify, the flow cannot be in exact balance. Vertical gradients of angular momentum due to vertical shears of the primary circulation causes updrafts to pass through the convective heat sources because the path of least resistance for the warmed air lies along constant angular momentum surfaces. Similarly, horizontal temperature gradients due to vertical shears cause the horizontal flow to pass through momentum sources because the path of least resistance lies along isentropes (potential temperature or θ surfaces). Although the flow lies generally along the angular momentum or isentropic surfaces, it has a small component across them. The advection by this component not the direct forcing, is the mechanism by which the primary circulation evolves.

Some of the most intense tropical cyclones exhibit "concentric" eyewalls, two or more eyewall structures centered at the circulation center of the storm. In much the same way as the inner eyewall forms, convection surrounding the eyewall can become organized into distinct rings. Eventually, the inner eye begins to feel the effects of the subsidence resulting from the outer eyewall, and the inner eyewall weakens, to be replaced by the outer eyewall. The pressure rises due to the destruction of the inner eyewall are usually more rapid than the pressure falls due to the intensification of the outer eyewall, and the cyclone itself weakens for a short period of time. This mechanism, referred to as the *eyewall replacement* cycle, often accom-

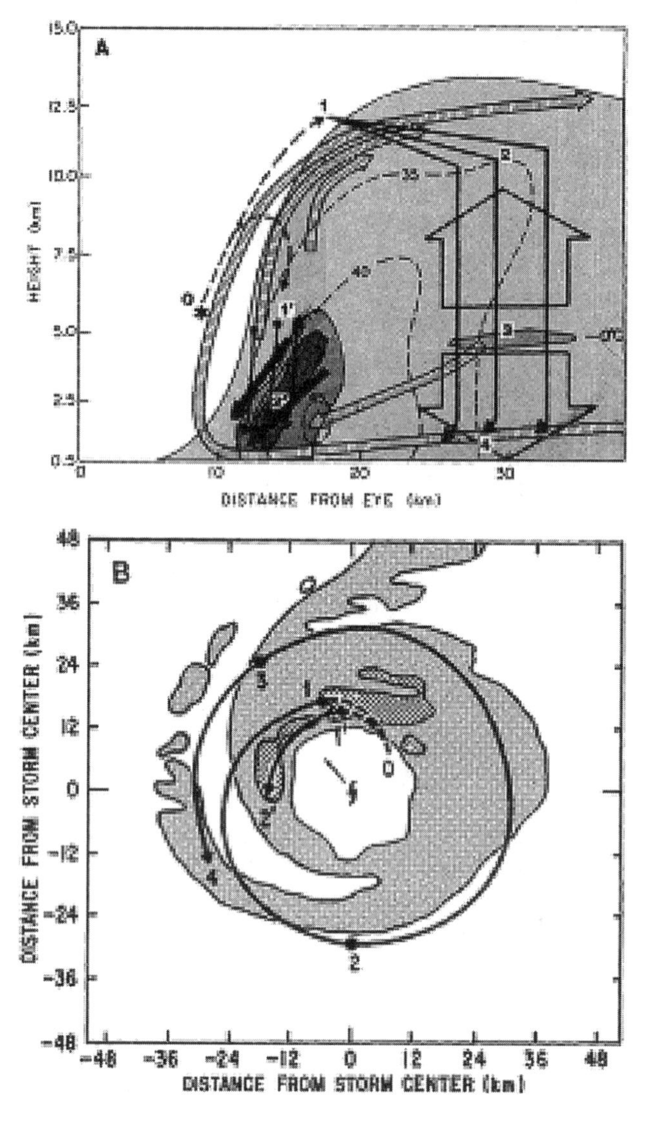

Figure 7 (*a*) Schematic of the radius–height circulation of the inner core of Hurricane Alicia. Shading depicts the reflectivity field, with contours of 5, 30, and 35 dBZ. The primary circulation (m/s) is depicted by dashed lines and the secondary circulation by the wide hatched streamlines. The convective downdrafts are denoted by the thick solid arrows, while the mesoscale up- and downdrafts are shown by the broad arrows. (*b*) Schematic plan view of the low-level reflectivity field in the inner core of Hurricane Alicia superimposed with the middle of the three hydrometeor trajectories in (*a*). Reflectivity contours in (*b*) are 20 and 35 dBZ. The storm center and direction are also shown. In (*a*) and (*b*) the hydrometeor trajectories are denoted by dashed and solid lines labeled 0–1–2–3–4 and 0′–1′–2′ [From F. D. Marks, and R. A. Houze, *J. Atmos. Sci.* **44,** 1296–1317 (1987). Copyright owned by American Meteorological Society.]

panies dramatic changes in storm intensity. The intensity changes are often associated with the development of secondary wind maxima outside the storm core.

A good example of contracting rings of convection effecting the intensification of a hurricane is shown in Figure 8 for Hurricane Gilbert on September 14, 1988. On September 14, two convective rings, denoted by intense radar reflectivity, are evident in Figure 8*a*. The outer ring is located near 80 to 90 km radius and the inner one at 10 to 12 km radius. Figure 8*b* shows that both are associated with maxima in tangential wind and vorticity. Figure 9 shows that in the ensuing 12 to 24 h the storm filled dramatically. However, it is not clear how much of the filling was caused by the storm moving over land and how much was due to the contracting outer ring and decaying inner ring of convective activity.

A process has been proposed whereby (i) nonlinear balanced adjustment of the vortex to the eddy heat and angular momentum sources associated with an upper trough in the storm's periphery produces an enhanced secondary circulation, (ii) a secondary wind maximum develops in response when the forcing reaches the inner radii, and (iii) the wind maximum contracts as a result of differential adiabatic warming associated with the convective diabatic heating in the presence of a inward radial gradient of inertial stability. Under these circumstances, understanding the intensification of the tropical cyclone reduces to determining how the secondary wind maximum develops.

Inner Core—Eyewall and Eye

The most recognizable feature found within a hurricane is the "eye" (Fig. 10). It is found at the center and is typically between 20 and 50 km in diameter. The eye is the focus of the hurricane, the point about which the primary circulation rotates and where the lowest surface pressures are found in the storm. The eye is a roughly circular area of comparatively light winds and fair weather found at the center of strong tropical cyclones. Although the winds are calm at the axis of rotation, strong winds may extend well into the eye. As seen in Figure 10 there is little or no precipitation and sometimes blue sky or stars can be seen. The eye is the region of warmest temperatures aloft—the eye temperature may be $\geq 10^{\circ}$C warmer at an altitude of 12 km than the surrounding environment, but only 0 to 2°C warmer at the surface.

The eye is surrounded by the eyewall, the roughly circular area of deep convection that is associated with the up-branch of the secondary circulation and the highest surface winds. The eye is composed of air that is slowly sinking, and the eyewall has a net upward flow because of many moderate—occasionally strong—updrafts and downdrafts. The eye's warm temperatures are due to warming by compression of the subsiding air. Most soundings taken within the eye are similar to that for Hurricane Hugo in Figure 11. They show a low-level layer that is relatively moist, with an inversion above, suggesting that the sinking in the eye typically does not reach the ocean surface, but instead only gets within 1 to 3 km of the surface. An eye is usually only present in hurricane-strength tropical cyclones.

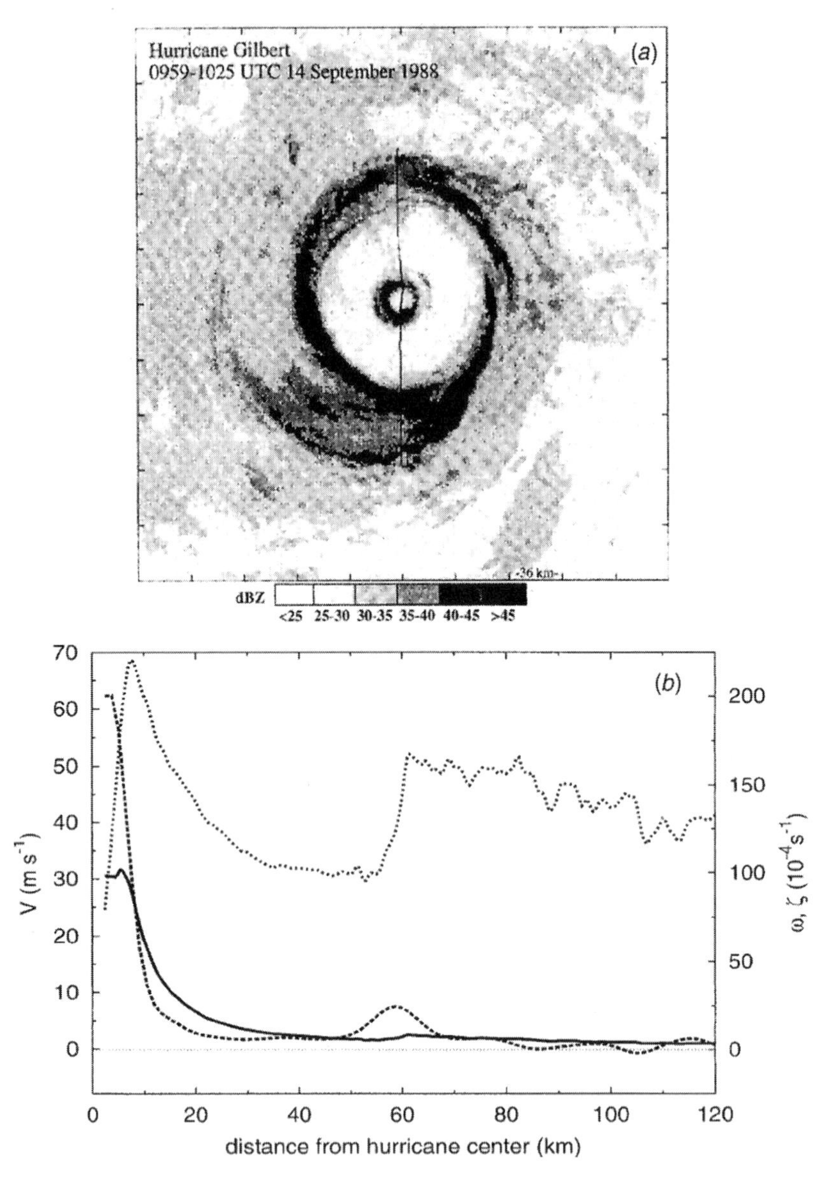

Figure 8 (*a*) Composite horizontal radar reflectivity of Hurricane Gilbert for 0959–1025 UTC September 14, 1988; the domain is 360 × 360 km, with tick marks every 36 km. The line through the center is the WP-3D aircraft flight track. (*b*) Profiles of flight-level angular velocity (solid), tangential wind (short dash), and smoothed relative vorticity (long dash) along the southern leg of the flight track shown in (*a*). [From J. P. Kossin, W. H. Schubert, and M. T. Montgomery, *Atmos. Sci.* **57**, 3893–3917 (2000). Copyright owned by American Meteorological Society.]

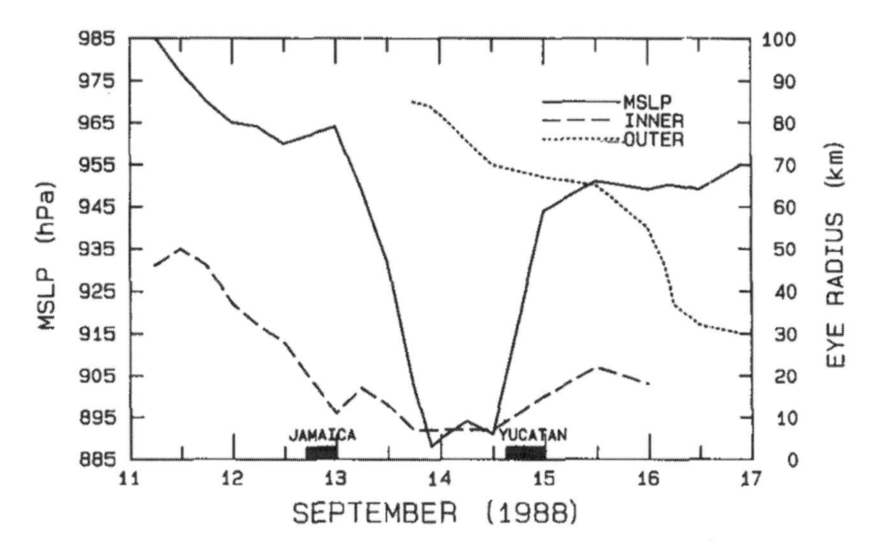

Figure 9 Hurricane Gilbert's minimum sea level pressure and radii of the inner and outer eyewalls as a function of time, September 1988. Solid blocks at bottom indicate times over land. [From M. L. Black, and H. E. Willoughby, *Mon. Wea. Rev.* **120**, 947–957 (1992). Copyright owned by American Meteorological Society.]

The general mechanisms by which the eye and eyewall are formed are not fully understood, although observations shed some light on the problem. The calm eye of the tropical cyclone shares many qualitative characteristics with other vortical systems such as tornadoes, waterspouts, dust devils, and whirlpools. Given that

Figure 10 Eyewall of Hurricane Georges 1945 UTC, September 19, 1998. (Photo courtesy of M. Black, NOAA/OAR/AOML Hurricane Research Division.) See ftp site for color image.

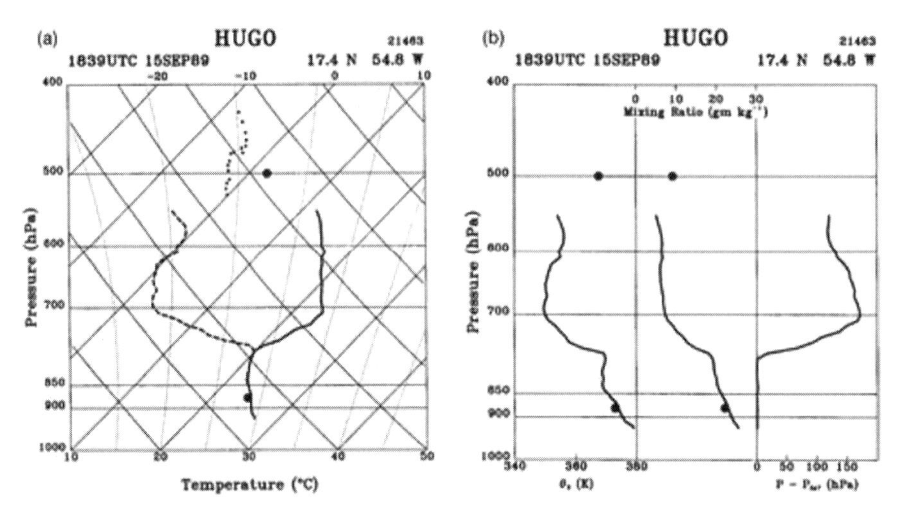

Figure 11 (a) Skew $T \log p$ diagram of the eye sounding in Hurricane Hugo at 1839 UTC on September 15, 1989. Isotherms slope upward to the right; dry adiabats slope upward to the left; moist adiabats are nearly vertical curving to the left. Solid and dashed curves denote temperature and dew point, respectively. The smaller dots denote saturation points computed for the dry air above the inversion, and the two larger dots temperature observed at the innermost saturated point as the aircraft passed through the eyewall. (*b*) θ_e, water vapor mixing ratio, and saturation pressure difference, $P–P_{SAT}$, as functions of pressure at 2123 UTC. [From H. E. Willoughby, *Mon. Wea. Rev.* **126**, 3189–3211 (1998). Copyright owned by American Meteorological society.]

many of these lack a change of phase of water (i.e., no clouds and diabatic heating involved), it may be that the eye feature is a fundamental component to all rotating fluids. It has been hypothesized that supergradient wind flow (i.e., swirling winds generate stronger centrifugal force than the local pressure gradient can support) present near the radius of maximum winds causes air to be centrifuged out of the eye into the eyewall, thus accounting for the subsidence in the eye. However, others found that the swirling winds within several tropical cyclones were within 1 to 4% of gradient balance. It may be though that the amount of supergradient flow needed to cause such centrifuging of air is only on the order of a couple percent and thus difficult to measure.

Another feature of tropical cyclones that probably plays a role in forming and maintaining the eye is the eyewall convection. As shown in Figure 12, convection in developing tropical cyclones is organized into long, narrow rain bands that are oriented in the same direction as the horizontal wind. Because these bands seem to spiral into the center of a tropical cyclone, they are sometimes called spiral bands. The earliest radar observations of tropical cyclones detected these bands, which are typically 5 to 50 km wide and 100 to 300 km long. Along these bands, low-level convergence is a maximum, and therefore, upper-level divergence is most pronounced. A direct circulation develops in which warm, moist air converges at the surface, ascends through these bands, diverges aloft, and descends on both sides

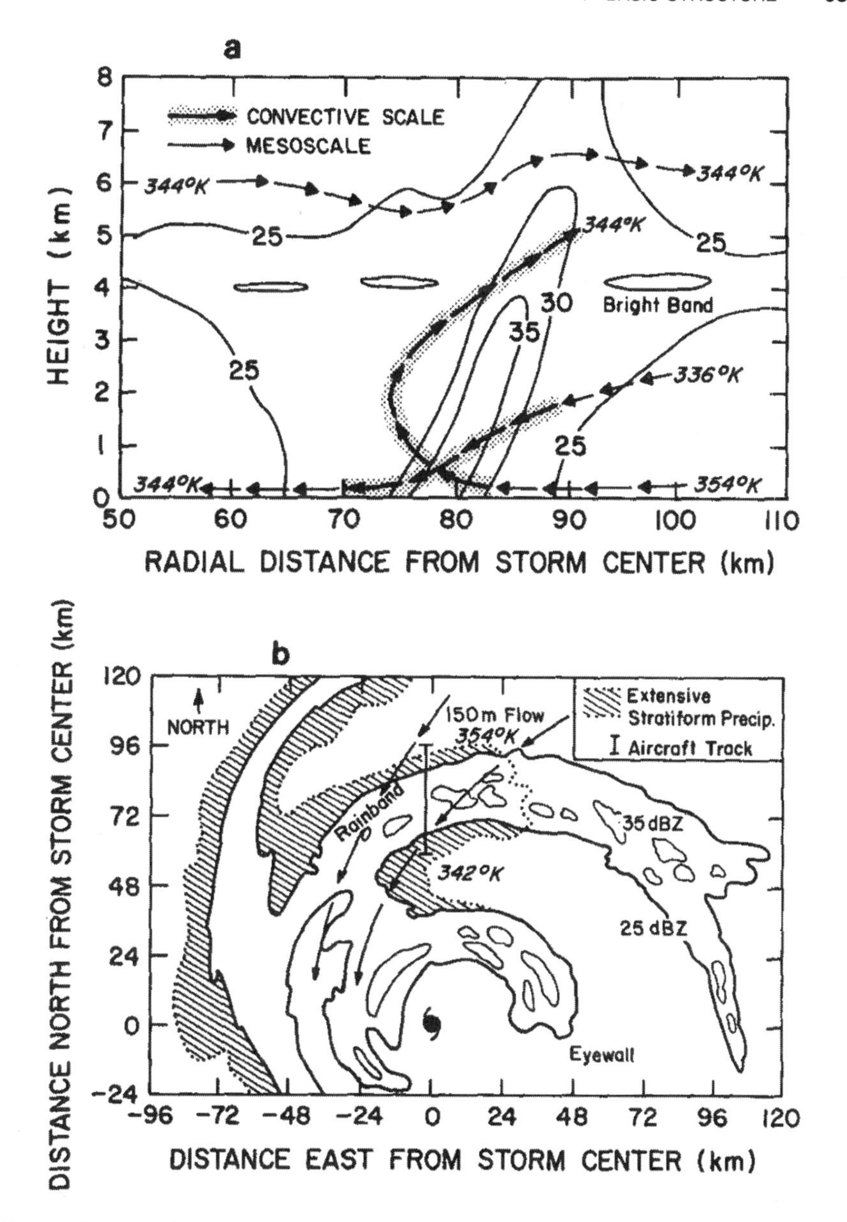

Figure 12 (*a*) Schematic of the rain band in radius–height coordinates. Reflectivity, θ_e, mesoscale (arrows), and convective-scale motions are shown. (*b*) Plan view. Aircraft track, reflectivities, cells, stratiform precipitation, 150-m flow, and θ_e values are shown. [From G. M. Barnes, E. J. Zipser, D. P. Jorgensen, and F. D. Marks, *J. Atmos. Sci.* **40**, 2125–2137 (1983). Copyright owned by American Meteorological Society.]

of the bands. Subsidence is distributed over a wide area outside of the rain band but is concentrated in the small inside area. As the air subsides, adiabatic warming takes place, and the air dries. Because subsidence is often concentrated on the inside of the band, the adiabatic warming is stronger inward from the band causing a sharp contrast in pressure falls across the band since warm air is lighter than cold air. Because of the pressure falls on the inside, the tangential winds around the tropical cyclone increase due to increased pressure gradient. Eventually, the band moves toward the center and encircles it and the eye and eyewall form.

The circulation in the eye is comparatively weak and, at least in the mature stage, thermally indirect (warm air descending), so it cannot play a *direct* role in the storm energy production. On the other hand, the temperature in the eye of many hurricanes exceeds that which can be attained by any conceivable moist adiabatic ascent from the sea surface, even accounting for the additional entropy (positive potential temperature, θ, anomaly) owing to the low surface pressure in the eye (the lower the pressure, the higher the θ at a given altitude and temperature). Thus, the observed low central pressure of the storm is not consistent with that calculated hydrostatically from the temperature distribution created when a sample of air is lifted from a state of saturation at sea surface temperature and pressure. The thermal wind balance restricts the amount of warming that can take place. In essence, the rotation of the eye at each level is imparted by the eyewall, and the pressure drop from the outer to the inner edge of the eye is simply that required by gradient balance.

Because the eyewall azimuthal velocity decreases with height, the radial pressure drop decreases with altitude, requiring, through the hydrostatic equation, a temperature maximum at the storm center. Thus, given the swirling velocity of the eyewall, the steady-state eye structure is largely determined. The central pressure, which is estimated by integrating the gradient balance equation inward from the radius of maximum winds, depends on the assumed radial profile of azimuthal wind in the eye.

In contrast, the eyewall is a region of rapid variation of thermodynamic variables. As shown in Figure 13, the transition from the eyewall cloud to the nearly cloud-free eye is often so abrupt that it has been described as a form of atmospheric front. Early studies were the first to recognize that the flow under the eyewall cloud is inherently frontogenetic. The eyewall is the upward branch of the secondary circulation and a region of rapid ascent that, together with slantwise convection, leads to the congruence of angular momentum and moist entropy (θ_e) surfaces. Hence, the three-dimensional vorticity vectors lie on θ_e surfaces, so that the moist PV vanishes. As the air is saturated, this in turn implies, through the invertibility principle applied to flow in gradient and hydrostatic balance, that the entire primary circulation may be deduced from the radial distribution of θ_e in the boundary layer and the distribution of vorticity at the tropopause.

In the classic semigeostrophic theory of deformation-induced frontogenesis, the background geostrophic deformation flow provides the advection of temperature across surfaces of absolute momentum that drives the frontogenesis whereas, in the hurricane eyewall, surface friction provides the radial advection of entropy across angular momentum surfaces. Also note that the hurricane eyewall is not

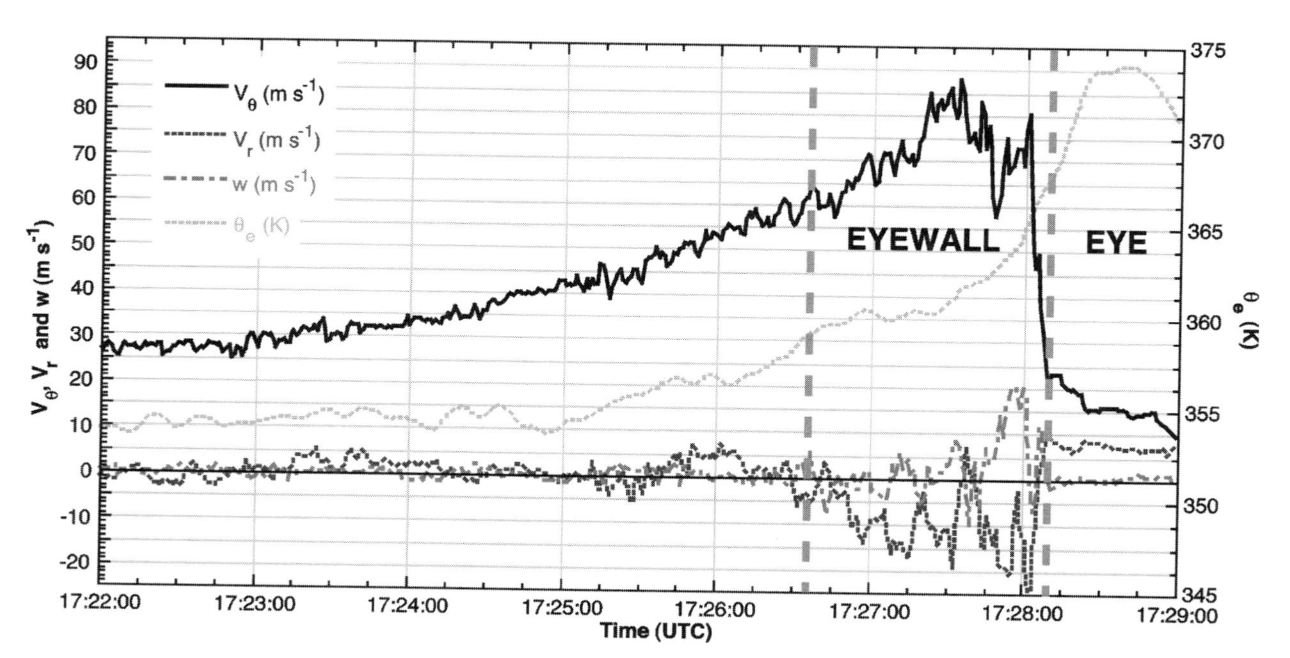

Figure 13 Time series plots of tangential wind (V_θ), radial wind (V_r), vertical velocity (w), and θ_e in Hurricane Hugo for 1721–1730 UTC, September 15, 1989. The aircraft flight track was at 450 m. Thick dashed vertical lines denote the width of the eyewall reflectivity maximum at low levels. See ftp site for color image.

necessarily a front in surface temperature, but instead involves the θ_e distribution, which is directly related to density in saturated air.

There is likely a two-stage process to eye formation. The amplification of the primary circulation is strongly frontogenetic and results, in a comparatively short time, in frontal collapse at the inner edge of the eyewall. The frontal collapse leads to a dramatic transition in the storm dynamics. While the tropical cyclone inner core is dominated by axisymmetric motions, hydrodynamic instabilities are potential sources of asymmetric motions within the core. In intense tropical cyclones the wind profile inside the eye is often "U-shaped" in the sense that the wind increases outward more rapidly than linearly with radius (Fig. 13). The strong cyclonic shear just inside the eyewall may result in a local maximum of absolute vorticity or angular momentum, so that the profile may actually become barotropically unstable. This instability leads to frontal collapse as a result of radial diffusion of momentum into the eye and also may explain the "polygonal eyewalls" where the eyewall appear on radar to be made up of a series of line segments rather than as a circle. It may also explain intense mesoscale vortices observed in the eyewalls of hurricanes Hugo of 1989 and Andrew of 1992.

Once the radial turbulent diffusion of momentum driven by the instability of the primary circulation becomes important, it results in a mechanically induced, thermally indirect (warm air sinking) component of the secondary circulation in the eye and eyewall. Such a circulation raises the vertically averaged temperature of the eye beyond its value in the eyewall and allows for an amplification of the entropy distribution. Feedbacks with the surface fluxes then allow the boundary layer entropy to increase and result in a more rapid intensification of the swirling wind. Thus, the frontal collapse of the eyewall is an essential process in the evolution of tropical cyclones. Without it, amplification of the temperature distribution relies on external influences, and intensification of the wind field is slow. Once it has taken place, the mechanical spinup of the eye allows the temperature distribution to amplify without external influences and, through positive feedback with surface fluxes, allows the entropy field to amplify and the swirling velocity to increase somewhat more rapidly.

Outer Structure and Rain Bands

The axisymmetric core is characteristically surrounded by a less symmetric outer vortex that diminishes into the synoptic "environment." In the lower troposphere, the cyclonic circulation may extend more than 1000 km from the center. As evident in Figure 14 the boundary between cyclonic and anticyclonic circulation slopes inward with increasing height, so that the circulation in the upper troposphere is primarily anticyclonic, except near the core. In the outer vortex, there are no scale separations between the primary and the secondary circulations, the asymmetric motions, or the vortex translation as they are all of the same rough magnitude. The asymmetric flows in this region control the vortex motion and sustain an eddy convergence of angular momentum and moisture toward the center. Interactions between the symmetric motions of the inner core with the more asymmetric

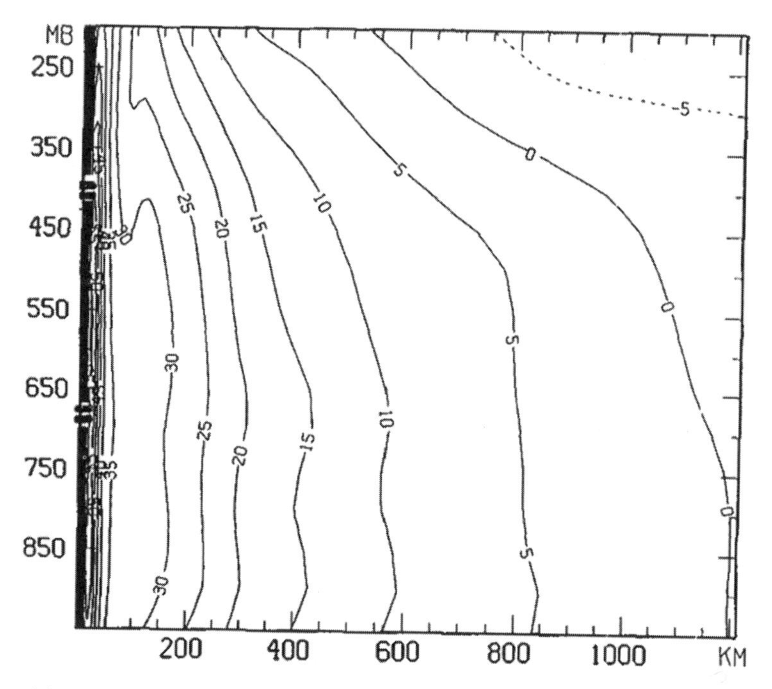

Figure 14 Vertical cross section of the azimuthal mean tangential wind for Hurricane Gloria on September 24, 1985. Anticyclonic contours are dashed. [From J. L. Franklin, S. J. Lord, S. E. Feuer, and F. D. Marks, *Mon. Wea. Rev.* **121**, 2433–2451 (1985). Copyright owned by American Meteorological Society.]

motions in the outer portion of the storm are the key to improved forecasts of tropical cyclone track and intensity.

Spiral bands of precipitation characterize radar and satellite images in this region of the storm (Figs. 1 and 15). As seen in Figures 8, 12, and 15, radar reflectivity patterns in tropical cyclones provide a good means for flow visualization although they represent precipitation, not winds. Descending motion occupies precipitation-free areas, such as the eye. The axis of the cyclone's rotation lies near the center of the eye. The eyewall surrounds the eye. In intense hurricanes, it may contain reflectivities as high as 50 dB(Z),* equivalent to rainfall rates of 74 mm/h. Less extreme reflectivities, 40 dB(Z) (13 mm/h), characterize most convective rainfall in the eyewall and spiral bands. The vertical velocities (both up and downdrafts) in convection with highest reflectivity may reach 25 m/s, but typical vertical velocities are <5 m/s. Such intense convection occupies $<10\%$ of the tropical cyclone's area. Outside convection, reflectivities are still weaker, 30 dB(Z), equivalent to a 2.4 mm/h rain rate. This "stratiform rain," denoted by a distinct reflectivity maximum or "bright band" at the altitude of the 0°C isotherm, falls out of the anvil cloud

*$10 \log_{10}(Z)$, where Z is equivalent radar reflectivity factor (mm^6/m^3).

that grows from the convection. The spiral bands tend to lie along the friction layer wind that spirals inward toward the eyewall (Fig. 12).

Many aspects of rain band formation, dynamics, and interaction with the symmetric vortex are still unresolved. The trailing-spiral shape of bands and lanes arises because the angular velocity of the vortex increases inward and deforms them into equiangular spirals. In the vortex core, air remains in the circulation for many orbits of the center; while outside the core, the air passes through the circulation in less than the time required for a single orbit. As the tropical cyclone becomes more intense, the inward ends of the bands approach the center less steeply approximating arcs of circles. Some bands appear to move outward, while others maintain a fixed location relative to the translating center.

As shown in Figure 15, motion of the vortex through its surroundings may cause one stationary band, called the “principal” band, to lay along a convergent streamline asymptote that spirals into the core. A tropical cyclone advected by middle-level steering with westerly shear moves eastward through surrounding air at low

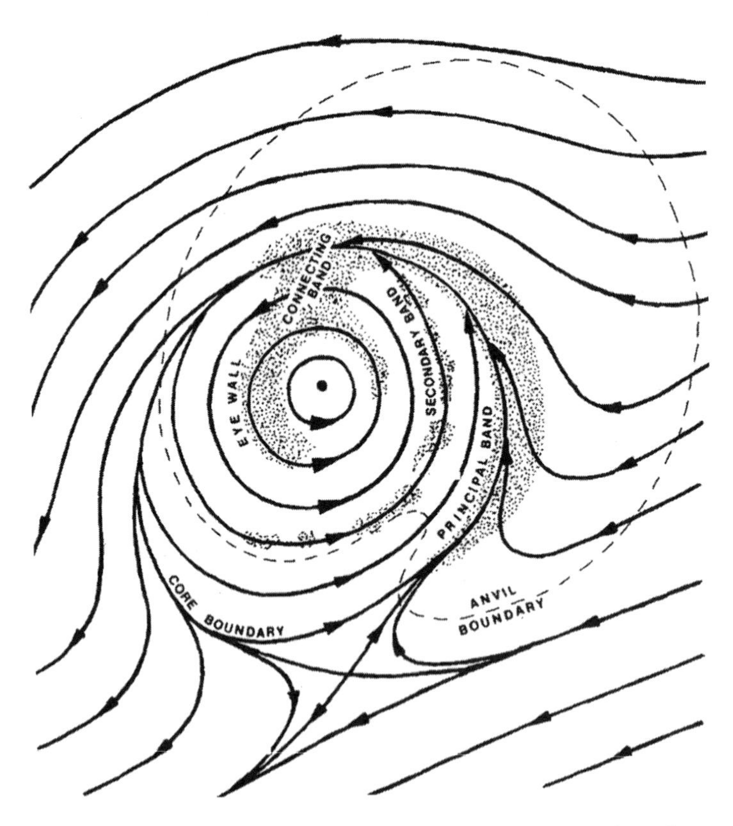

Figure 15 Schematic representation of the stationary band complex, the entities that compose it, and the flow in which it is embedded. [From H. E. Willoughby, F. D. Marks, and R. J. Feinberg, *J. Atmos. Sci.* **41**, 3189–3211 (1984). Copyright owned by American Meteorological Society.]

levels. Thus, the principal band may be a "bow wave" due to displacement of the environmental air on the eastern side of the vortex. Its predominant azimuthal wave number is one.

Moving bands, and other convective features, are frequently associated with cycloidal motion of the tropical cyclone center, and intense asymmetric outbursts of convection are observed to displace the tropical cyclone center by tens of kilometers. The bands observed by radar are often considered manifestations of internal gravity waves, but these waves can exist only in a band of Doppler-shifted frequencies between the local inertia frequency (defined as the sum of the vertical component of Earth's inertial frequency f and the local angular velocity of the circulation, V/r) and the Brunt–Vaisala frequency.* Only two classes of trailing-spiral, gravity wave solutions lie within this frequency band: (i) waves with any tangential wave number that move faster than the swirling wind, and (ii) waves with tangential wave number ≥ 2 that move slower than the swirling wind.

Bands moving faster than the swirling wind with outward phase propagation are observed by radar. They are more like squall lines than linear gravity waves. Waves moving slower than the swirling wind propagate wave energy and anticyclonic angular momentum inward, grow at the expense of the mean-flow kinetic energy, and reach appreciable amplitude if they are excited at the periphery of the tropical cyclone. Alternate explanations for these inward-propagating bands involve filamentation of vorticity from the tropical cyclone environment, asymmetries in the radially shearing flow of the vortex, and high-order vortex Rossby waves. Detailed observations of the vortex-scale rain band structure and wind field are necessary to determine which mechanisms play a role in rain band development and maintenance.

While the evolution of the inner core is dominated by interactions between the primary, secondary, and track-induced wave number one circulation, there is some indication that the local convective circulations in the rain bands may impact on intensity change. Although precipitation in some bands is largely stratiform, condensation in most bands tends to be concentrated in convective cells rather than spread over wide mesoscale areas. As shown in Figure 12, convective elements form, move through the bands, and dissipate as they move downwind. Doppler radar observations indicate that the roots of the updrafts lay in convergence between the low-level radial inflow and gust fronts that are produced by convective downdrafts. This convergence may occur on either side of the band. A 20 K decrease in low-level θ_e was observed in a rain band downdraft and suggested that the draft acts as a barrier to inflow. This reduction in boundary layer energy may be advected near the center, inhibit convection, and thereby alter storm intensity.

5 MOTION

Tropical cyclone motion is the result of a complex interaction between a number of internal and external influences. Environmental steering is typically the most prominent external influence on a tropical cyclone, accounting for as much as 70 to 90%

*Natural gravity wave frequency, the square root of the static stability defined as $(g/\theta)\,\partial\theta/\partial z$.

of the motion. Theoretical studies show that in the absence of environmental steering, tropical cyclones move poleward and westward due to internal influences.

Accurate determination of tropical cyclone motion requires accurate representation of interactions that occur throughout the depth of the troposphere on a variety of scales. Observations spurred improved understanding of how tropical cyclones move using simple barotropic and more complex baroclinic models. To first order, the storm moves with some layer average of the lower tropospheric environmental flow: The translation of the vortex is roughly equal to the speed and direction of the basic "steering" current. However, the observations show that tropical cyclone tracks deviate from this simple steering flow concept in a subtle and important manner. Several physical processes may cause such deviations. The approach in theoretical and modeling of tropical cyclones has been to isolate each process in a systematic manner to understand the magnitude and direction of the track deviation caused by each effect. The β effect,* due to the differential advection of Earth's vorticity (f), alone can produce asymmetric circulations and propagation. Models that are more complete describe not only the movement of the vortex but also the accompanying wave number one asymmetries. It was also discovered that the role of meridional and zonal gradients of the environmental flow could greatly add to the complexity even in the barotropic evolution of a vortex. Hence, the evolution of the movement depends not only on the relative vorticity gradient and on shear of the environment but also the structure of the vortex itself.

Generally, the propagation vector of these model baroclinic vortices is very close to that expected from a barotropic model initialized with the vertically integrated environmental wind. An essential feature in baroclinic systems is the relative vorticity advection through the storm center where the vertical structure of the tropical cyclone produces a tendency for the low-level vortex to move slower than the simple propagation of the vortex due to β. Vertical shear plays an important factor in determining the relative flow; however, there is no unique relation between the shear and storm motion. Diabatic heating effects also alter this flow and change the propagation velocity. Thus, tropical cyclone motion is primarily governed by the dynamics of the low-level cyclonic circulation, however, the addition of observations of the upper-level structure may alter this finding.

6 INTERACTION WITH ATMOSPHERIC ENVIRONMENT

A consensus exists that small vertical shear of the environmental wind and lateral eddy imports of angular momentum are favorable to tropical cyclone intensification. The inhibiting effect of vertical shear in the environment on tropical cyclone intensification is well known from climatology and forecasting experience. The appearance of the low-level circulation outside the tropical cyclones central dense overcast in satellite imagery is universally recognized as a symptom of shear and as an

*Asymmetric vorticity advection around the vortex caused by the latitudinal gradient of f, $\beta = 2\Omega \cos \phi$. β has a maximum value at the equator (i.e., $2.289 \times 10^{-1}\text{s}^{-1}$) and becomes zero at the pole.

indication of nonintensification or weakening. Nevertheless, the detailed dynamics of a vortex in shear has been the topic of surprisingly little study, probably because, while the effect is a reliable basis for practical forecasting, it is difficult to measure and model.

In contrast, the positive effect of eddy momentum imports at upper levels has received extensive study. Modeling studies with composite initial conditions show that eddy momentum fluxes can intensify a tropical cyclone even when other conditions are neutral or unfavorable. It has been shown theoretically that momentum imports can form a tropical cyclone in an atmosphere with no buoyancy. Statistical analysis of tropical cyclones reveals a clear relationship between angular momentum convergence and intensification, but only after the effects of shear and SST variations are accounted for. Such interactions occur frequently (35% of the time, defined by eddy angular momentum flux convergence exceeding 10 m/s day), and likely represent the more common upper boundary interaction for tropical cyclones. Frequently, they are accompanied by eyewall cycles and dramatic intensity changes.

The environmental flows that favor intensification, and presumably inward eddy momentum fluxes, usually involve interaction with a synoptic-scale cyclonic feature, such as a midlatitude upper-level trough or PV anomaly. Given the interaction of an upper-level trough and the tropical cyclone, the exact mechanism for intensification is still uncertain. The secondary circulation response to momentum and heat sources is very different. Upper-tropospheric momentum sources can influence the core directly. As can be seen in Figure 16, large inertial stability* in the lower troposphere protects the mature tropical cyclone core from direct influence by momentum sources; however, the inertial stability in the upper troposphere is smaller and a momentum source can induce an outflow jet with large radial extent just below the tropopause. If the eyewall updraft links to the direct circulation at the entrance region of the jet, as shown in Figure 17*c*, and 17*d*, the exhaust outflow is unrestricted. The important difference between heat and momentum sources is that the roots of the diabatically induced updraft must be in the inertially stiff lower troposphere, but the outflow jet due to a momentum flux convergence can be confined to the inertially *labile* upper troposphere. Momentum forcing does not spin the vortex up directly. It makes the exhaust flow stronger and reduces local compensating subsidence in the core, thus cooling the upper troposphere and destabilizing the sounding. The cooler upper troposphere leads to less thermal wind shear and a weaker upper anticyclone.

A two-dimensional balanced approach provides reasonable insight into the nature of the tropical cyclone intensification as a trough approaches. Isentropic PV analysis (Fig. 17), which express the problem in terms of a quasi-conserved variable in three dimensions, are used to describe various processes in idealized tropical cyclones with considerable success. The eddy heat and angular momentum fluxes are related to changes in the isentropic PV through their contribution to the eddy flux of PV.

*A measure of the resistance to horizontal displacements, based on the conservation of angular momentum for a vortex in gradient balance, and is defined as $(f + \zeta)(f + V^2/r)$, where ζ is the relative vorticity, V is the axial wind velocity, f is the Coriolis parameter, and r radius from the storm center.

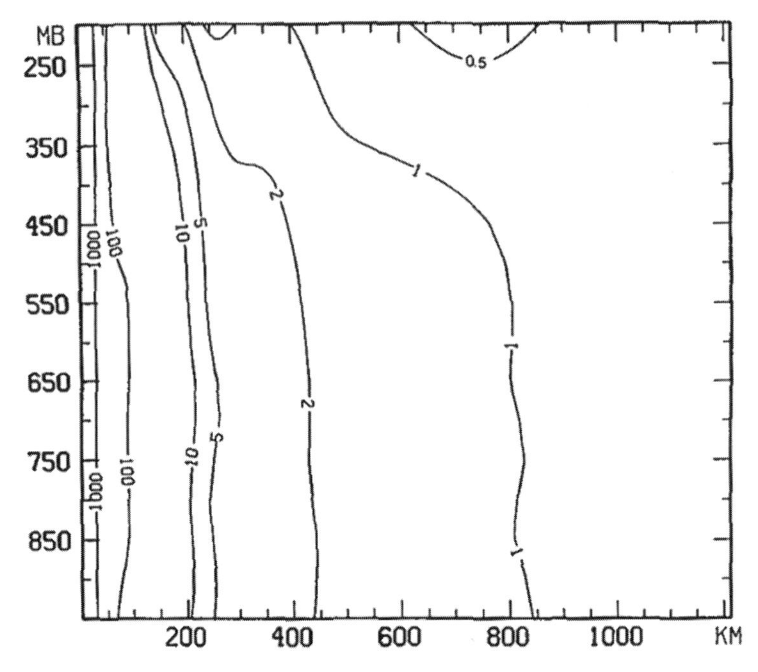

Figure 16 Axisymmetric mean inertial stability for Hurricane Gloria on September 24, 1985. Contours are shown as multiples of f^2 at the latitude of Gloria's center. [From J. L. Franklin, S. J. Lord, S. B. Feuer, and F. D. Marks, *Mon. Wea. Rev.* **121**, 2433–2451 (1993). Copyright owned by American Meteorological Society.]

It has been suggested that outflow-layer asymmetries, as in Figure 17, and their associated circulations could create a mid- or lower-tropospheric PV maximum outside the storm core, either by creating breaking PV waves on the midtropospheric radial PV gradient (Fig. 3) or by diabatic heating. It has been shown that filamentation of any such PV maximum in the "surf zone" outside the tropical cyclone core (the sharp radial PV gradient near 100 km radius) produces a feature much like a secondary wind maximum, which was apparent in the PV fields of hurricane Gloria in Figure 3. These studies thus provide mechanisms by which outflow-layer asymmetries could bring about a secondary wind maximum.

An alternative argument has been proposed for storm reintensification as a "constructive interference without phase locking," as shown in Figure 18. As the PV anomalies come within the Rossby radius of deformation, the pressure and wind perturbations associated with the combined anomalies are greater than when the anomalies are apart, even though the PV magnitudes are unchanged. The perturbation energy comes from the basic-state shear that brought the anomalies together. However, constructive interference without some additional diabatic component cannot account for intensification. It is possible that intensification represents a baroclinic initiation of a wind-induced surface heat exchange. By this mechanism,

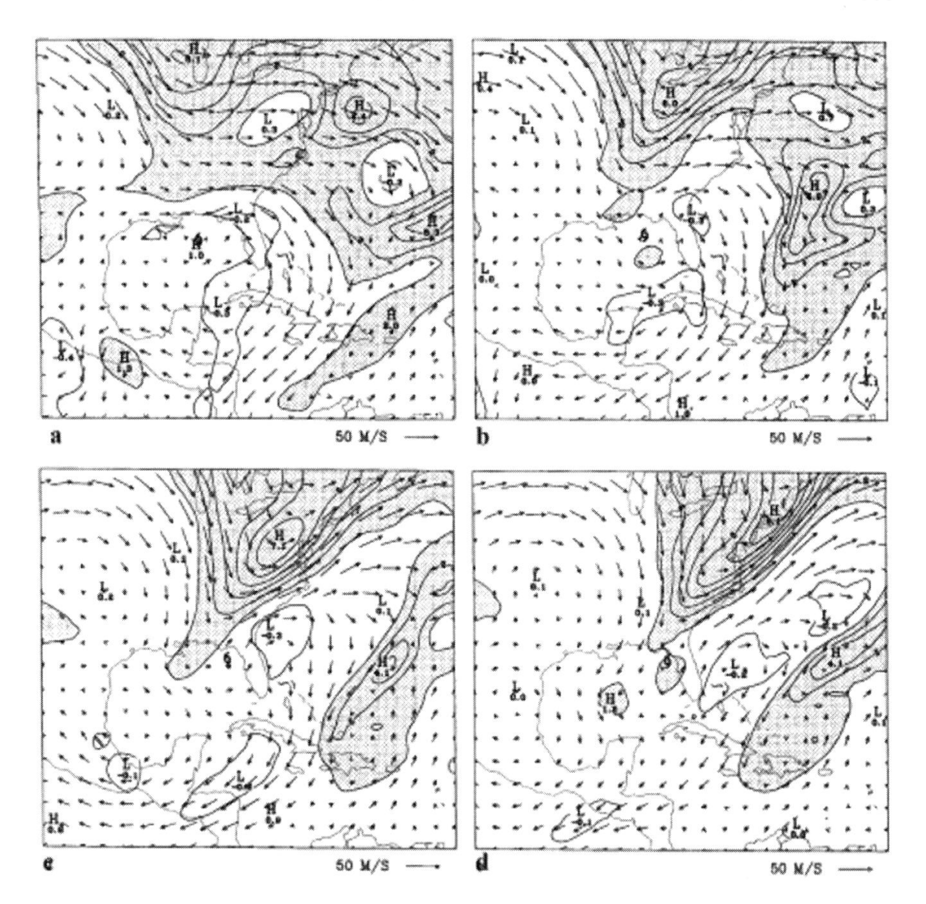

Figure 17 Wind vectors and PV on the 345 K isentropic surface at (*a*) 1200 UTC August 30; (*b*) 0000 UTC August 31; (*c*) 1200 UTC August 31, and (*d*) 0000 UTC September 1, 1985. PV increments are 1 PVU and values >1 PVU are shaded. Wind vectors are plotted at 2.25° intervals. The 345 K surface is approximately 200 hPa in the hurricane environment and ranges from 240 to 280 hPa at the storm center. The hurricane symbol denotes the location of Hurricane Elena. [From J. Molinari, S. Skubis, and D. Vollaro, *J. Atmos. Sci.* **52**, 3593–3606 (1995). Copyright owned by American Meteorological Society.]

the constructive interference induces stronger surface wind anomalies, which produce larger surface moisture fluxes and thus higher surface moist enthalpy. This feeds back through the associated convective heating to produce a stronger secondary circulation and thus stronger surface winds. The small effective static stability in the saturated, nearly moist neutral storm core ensures a deep response so that even a rather narrow upper trough can initiate this feedback process. The key to this mechanism is the direct influence of the constructive interference on the surface wind field as that controls the surface flux of moist enthalpy.

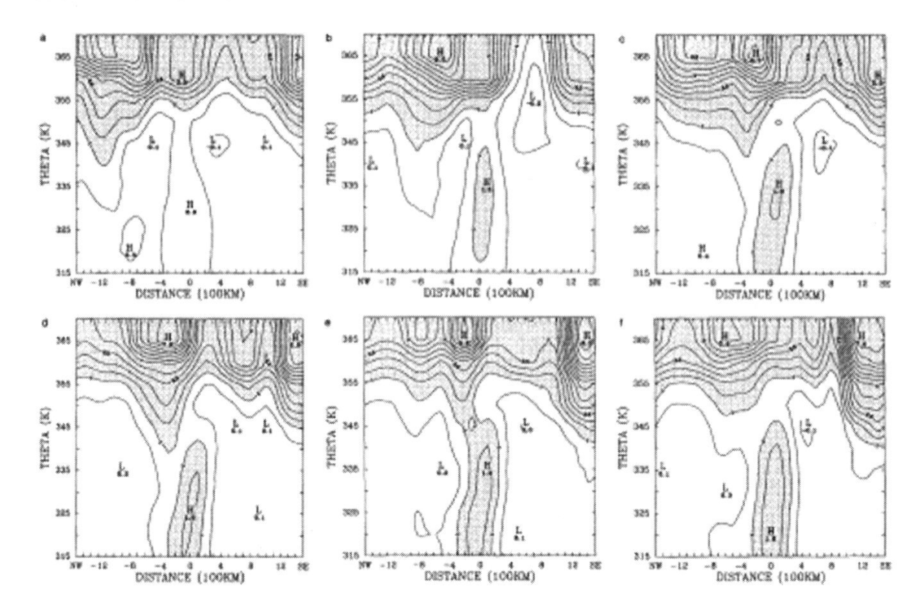

Figure 18 Cross sections of PV from northwest (left) to southeast (right) through the observed as center of Hurricane Elena for the same times as in Fig. 17, plus (*a*) 0000 UTC August 30 and (*f*) 1200 UTC September 1. Increment is 0.5 PVU and shading above 1 PVU. [From J. Molinari, S. Skubis, and D. Vollaro, *J. Atmos. Sci.* **52**, 3593–3606 (1995). Copyright owned by American Meteorological Society.]

7 INTERACTION WITH THE OCEAN

As pointed out in Section 2, preexisting SSTs >26°C are a necessary but insufficient condition for tropical cyclogenesis. Once the tropical cyclone develops and translates over the tropical oceans, statistical models suggest that warm SSTs describe a large fraction of the variance (40 to 70%) associated with the intensification phase of the storm. However, these predictive models do not account either for the oceanic mixed layers having temperatures of 0.5 to 1°C cooler than the relatively thin SST over the upper meter of the ocean or horizontal advective tendencies by basic-state ocean currents such as the Gulf Stream and warm core eddies. Thin layers of warm SST are well mixed with the underlying cooler mixed layer water well in advance of the storm where winds are a few meters per second, reducing SST as the storm approaches. However, strong oceanic baroclinic features advecting deep, warm oceanic mixed layers represent moving reservoirs of high-heat content water available for the continued development and intensification phases of the tropical cyclone. Beyond a first-order description of the lower boundary providing the heat and moisture fluxes derived from low-level convergence, little is known about the complex boundary layer interactions between the two geophysical fluids.

One of the more apparent aspects of the atmospheric-oceanic interactions during tropical cyclone passage is the upper ocean cooling as manifested in the SST (and

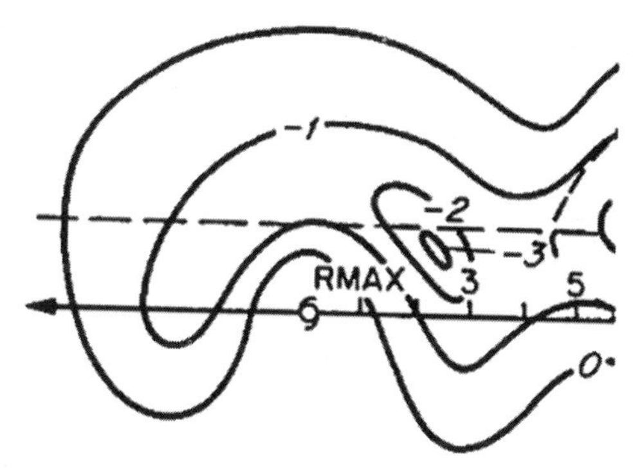

Figure 19 Schematic SST change (°C) induced by a hurricane. The distance scale is indicated in multiples of the radius of maximum wind. Storm motion is to the left. Horizontal dashed line is at 1.5 times the radius of maximum wind. [From P. G. Black, R. L. Elsberry, and L. K. Shay, *Adv. Underwater Tech., Ocean Sci. Offshore Eng.* **16**, 51–58 (1998). Copyright owned by Society for Underwater Technology (Graham & Trotman).]

mixed layer temperature) decrease starting just in back of the eye. As seen in Figure 19, ocean mixed layer temperature profiles acquired during the passage of several tropical cyclones revealed a crescent-shaped pattern of upper ocean cooling and mixed layer depth changes, which indicated a rightward bias in the mixed layer temperature response with cooling by 1 to 5°C extending from the right-rear quadrant of the storm into the wake regime. These SST decreases are observed through satellite-derived SST images, such as the one shown in Figure 20 of the post–Hurricane Bonnie SST, which are indicative of mixed layer depth changes due to stress-induced turbulent mixing in the front of the storm and shear-induced mixing between the mixed layer and thermocline in the rear half of the storm. The mixed layer cooling represents the thermodynamic and dynamic response to the strong wind that typically accounts for 75 to 85% of the ocean heat loss, compared to the 15 to 25% caused by surface latent and sensible heat fluxes from the ocean to the atmosphere. Thus, the upper ocean's heat content for tropical cyclones is not governed solely by SST, rather it is the mixed layer depths and temperatures that are significantly affected along the lower boundary by the basic state and transient currents.

Recent observational data has shown that the horizontal advection of temperature gradients by basic-state currents in a warm core ring affected the mixed layer heat and mass balance, suggesting the importance of these warm oceanic baroclinic features. In addition to enhanced air–sea fluxes, warm temperatures (>26°C) may extend to 80 to 100 m in warm core rings, significantly impacting the mixed layer heat and momentum balance. That is, strong current regimes (1 to 2 m/s) advecting deep, warm upper ocean layers not only represent deep reservoirs of high heat content water with an upward heat flux but transport heat from the tropics to the subtropical and polar regions as part of the annual cycle. Thus, the basic state of the mixed layer and the subsequent response represents an evolving three-dimensional

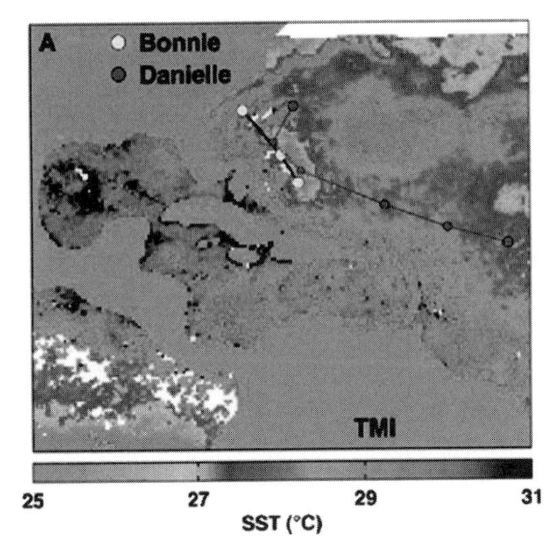

Figure 20 (see color insert) Cold wake produced by Hurricane Bonnie for August 24–26, 1998, as seen by the NASA TRMM satellite *Microwave Imager* (TMI). Small white patches are areas of persistent rain over the 3-day period. White dots show Hurricane Bonnie's daily position from August 24 to 26. Gray dots show the later passage of Hurricane Danielle from August 27 to September 1. Danielle crossed Bonnie's wake on August 29 and its intensity drops. [From F. J. Wentz, C. Gentemann, D. Smith, and D. Chelton, *Science* **288**, 847–850 (2000). Copyright owned by the American Geophysical Union. (*http:/www.sciencemag.org*)]. See ftp site for color image.

process with surface fluxes, vertical shear across the entrainment zone and horizontal advection. Simultaneous observations in both fluids are lacking over these baroclinic features prior, during, and subsequent to tropical cyclone passage and are crucially needed to improve our understanding of the role of lower boundary in intensity and structural changes to intensity change.

In addition, wave height measurements and current profiles revealed the highest waves and largest fetches to the right side of the storm where the maximum mixed layer changes occurred. Mean wave-induced currents were in the same direction as the steady mixed layer currents, modulating vertical current shears and mixed layer turbulence. These processes feed back to the atmospheric boundary layer by altering the surface roughness and hence the drag coefficient. However, little is known about the role of strong surface waves on the mixed layer dynamics, and their feedback to the atmospheric boundary layer under tropical cyclone force winds by altering the drag coefficient.

8 TROPICAL CYCLONE RAINFALL

Precipitation in tropical cyclones can be separated into either convective or stratiform regimes. Convective precipitation occurs primarily in the eyewall and rain bands, producing rains >25 mm/h over small areas. However, observations suggest that only 10% of the total rain area is comprised of these convective rain cores. The average core is 4 km in radius (area of 50 km^2) and has a relatively short lifetime, with only 10% lasting longer than 8 min (roughly the time a 1-mm-diameter raindrop takes to fall from the mean height of the 0°C isotherm at terminal velocity). The short life cycle of the cores and the strong horizontal advection produce a well-mixed and less asymmetric precipitation pattern in time and space. Thus, over 24 h the inner core of a tropical cyclone as a whole produces 1 to 2 cm of precipitation over a relatively large area and 10 to 20 cm in the core. After landfall, orographic forcing can anchor heavy precipitation to a local area for an extended time. Additionally, midlatitude interaction with a front or upper level trough can enhance precipitation, producing a distortion of the typical azimuthally uniform precipitation distribution.

9 ENERGETICS

Energetically, a tropical cyclone can be thought of, to a first approximation, as a heat engine; obtaining its heat input from the warm, humid air over the tropical ocean, and releasing this heat through the condensation of water vapor into water droplets in deep thunderstorms of the eyewall and rain bands, then giving off a cold exhaust in the upper levels of the troposphere (~12 km up). One can look at the energetics of a tropical cyclone in two ways: (1) the total amount of energy released by the condensation of water droplets or (2) the amount of kinetic energy generated to maintain the strong swirling winds of the hurricane. It turns out that the vast majority of the heat released in the condensation process is used to cause rising motions in the convection and only a small portion drives the storm's horizontal winds.

Using the first approach we assume an average tropical cyclone produces 1.5 cm/day of rain inside a circle of radius 665 km. Converting this to a volume of rain gives 2.1×10^{16} cm/day (a cm^3 of rain weighs 1 g). The energy released through the latent heat of condensation to produce this amount of rain is 5.2×10^{19} J/day or 6.0×10^{14} W, which is equivalent to 200 times the worldwide electrical generating capacity.

Under the second approach we assume that for a mature hurricane, the amount of kinetic energy generated is equal to that being dissipated due to friction. The dissipation rate per unit area is air density times the drag coefficient times the wind speed cubed. Assuming an average wind speed for the inner core of the hurricane of 40 m/s winds over a 60 km radius, the wind dissipation rate (wind generation rate) would be 1.5×10^{12} W. This is equivalent to about half the worldwide electrical generating capacity.

Either method suggests hurricanes generate an enormous amount of energy. However, they also imply that only about 2.5% of the energy released in a hurricane by latent heat released in clouds actually goes to maintaining the hurricane's spiraling winds.

10 TROPICAL CYCLONE–RELATED HAZARDS

In the coastal zone, extensive damage and loss of life are caused by the storm surge (a rapid, local rise in sea level associated with storm landfall), heavy rains, strong winds, and tropical cyclone-spawned severe weather (e.g., tornadoes). The continental United States currently averages nearly $5 billion (in 1998 dollars) annually in tropical cyclone–caused damage, and this is increasing, owing to growing population and wealth in the vulnerable coastal zones.

Before 1970, large loss of life stemmed mostly from storm surges. The height of storm surges varies from 1 to 2 m in weak systems to more than 6 m in major hurricanes that strike coastlines with shallow water offshore. The storm surge associated with Hurricane Andrew (1992) reached a height of about 5 m, the highest level recorded in southeast Florida. Hurricane Hugo's (1989) surge reached a peak height of nearly 6 m about 20 miles northeast of Charleston, South Carolina, and exceeded 3 m over a length of nearly 180 km of coastline. In recent decades, large loss of life due to storm surges in the United States has become less frequent because of improved forecasts, fast and reliable communications, timely evacuations, a better educated public, and a close working relationship between the National Hurricane Center (NHC), local weather forecast offices, emergency managers, and the media. Luck has also played a role, as there have been comparatively few landfalls of intense storms in populous regions in the last few decades. The rapid growth of coastal populations and the complexity of evacuation raises concerns that another large storm surge disaster might occur along the eastern or Gulf Coast shores of the United States.

In regions with effectively enforced building codes designed for hurricane conditions, wind damage is typically not so lethal as the storm surge, but it affects a much larger area and can lead to large economic loss. For instance, Hurricane Andrew's winds produced over $25 billion in damage over southern Florida and Louisiana. Tornadoes, although they occur in many hurricanes that strike the United States, generally account for only a small part of the total storm damage.

While tropical cyclones are most hazardous in coastal regions, the weakening, moisture-laden circulation can produce extensive, damaging floods hundreds of miles inland long after the winds have subsided below hurricane strength. In recent decades, many more fatalities in North America have occurred from tropical cyclone–induced inland flash flooding than from the combination of storm surge and wind. For example, although the deaths from storm surge and wind along the Florida coast from hurricane Agnes in 1972 were minimal, inland flash flooding caused more than 100 deaths over the northeastern United States. More recently, rains from Hurricane Mitch (1998) killed at least 10,000 people in Central America, the majority after the storm had weakened to tropical storm strength. An essential difference in the threat from flooding rains, compared to that from wind and surge, is that the rain amount is not tied to the strength of the storm's winds. Hence, any tropical disturbance, from depression to major hurricane, is a major rain threat.

11 SUMMARY

The eye of the storm is a metaphor for calm within chaos. The core of a tropical cyclone, encompassing the eye and the inner 100 to 200 km of the cyclone's 1000 to 1500 km radial extent, is hardly tranquil. However, the rotational inertia of the swirling wind makes it a region of orderly, but intense, motion. It is dominated by a cyclonic primary circulation in balance with a nearly axisymmetric, warm core low-pressure anomaly. Superimposed on the primary circulation are weaker asymmetric motions and an axisymmetric secondary circulation. The asymmetries modulate precipitation and cloud into trailing spirals. Because of their semibalanced dynamics, the primary and secondary circulations are relatively simple and well understood. These dynamics are not valid in the upper troposphere where the outflow is comparable to the swirling flow, nor do they apply to the asymmetric motions. Since the synoptic-scale environment appears to interact with the vortex core in the upper troposphere by means of the asymmetric motions, future research should emphasize this aspect of the tropical cyclone dynamics and their influence on the track and intensity of the storm.

Improved track forecasts, particularly the location and time when a tropical cyclone crosses the coast are achievable with more accurate specification of the initial conditions of the large-scale environment and the tropical cyclone wind fields. Unfortunately, observations are sparse in the upper troposphere, atmospheric boundary layer, and upper ocean, limiting knowledge of environmental interactions, angular momentum imports, boundary layer stress, and air–sea interactions. In addition to the track, an accurate forecast of the storm intensity is needed since it is the primary determinant of localized wind damage, severe weather, storm surge, ocean wave runup, and even precipitation during landfall. A successful intensity forecast requires knowledge of the mechanisms that modulate tropical cyclone intensity through the relative impact and interactions of three major components: (1) the structure of the upper ocean circulations that control the mixed layer heat content, (2) the storm's inner core dynamics, and (3) the structure of the synoptic-scale upper-tropospheric environment. Even if we could make a good forecast of the landfall position and intensity, our knowledge of tropical cyclone structure changes as it makes landfall is in its infancy because little hard data survives the harsh conditions. To improve forecasts, developments to improve our understanding through observations, theory, and modeling need to be advanced together.

SUGGESTED READING

Emanuel, K. A. (1986). An air–sea interaction theory for tropical cyclones. Part I: Steady-state maintenance, *J. Atmos. Sci.* **43**, 585–604.

Ooyama, K. V. (1982). Conceptual evolution of the theory and modeling of the tropical cyclone, *J. Meteor. Soc. Jpn.* **60**, 369–380.

WMO (1995). *Global Perspectives of Tropical Cyclones*, Russell Elsberry (Ed.), World Meteorological Organization Report TCP-38, Geneva, Switzerland.

CHAPTER 32

MODERN WEATHER FORECASTING

LAWRENCE B. DUNN

1 INTRODUCTION

It has often been said that weather forecasting is part science and part art. Essentially, weather forecasting is the application of science to a very practical problem that affects the lives and livelihoods of people and nations. How the science is applied varies somewhat from forecaster to forecaster, and this subjectivity is the aspect of weather forecasting that is still a bit of an art form. Many meteorologists and atmospheric scientists are uncomfortable with the subjectivity of weather forecasting since it inevitably leads to inconsistency, and there is a never-ending attempt to provide forecasters with tools to make the process as objective as possible. Given the preceding discussion, it should be clear that, like forecasting itself, any description of modern weather forecasting will vary, at least slightly, according to the experiences of the author.

Modern weather forecasting is applied to a broad spectrum of scales and applications. Forecasting is done by government and private meteorologists for a variety of users. Forecasts of general conditions for each of the next 7 days are widely distributed and are familiar to virtually everyone. In the United States, forecasts and warnings to protect life and property are provided by government meteorologists, although private-sector forecasters will often add value to these products, especially through dissemination of the products. Specific forecasts for commercial interests are done by private-sector meteorological firms. Forecast applications are numerous. Examples include weather predictions for aviation, the marine community, management of wildfires and prescribed burns, flood control, reservoir management, utility power demand, road surface conditions, agriculture, snowmaking at ski resorts, the retail industry, commodity traders and brokers, the film industry, special outdoor

Handbook of Weather, Climate, and Water: Dynamics, Climate, Physical Meteorology, Weather Systems, and Measurements, Edited by Thomas D. Potter and Bradley R. Colman.
ISBN 0-471-21490-6

sporting events, and many others. The increased skill of modern weather forecasts has resulted in more and more decision makers in all fields basing high resource commitments on weather predictions.

Early in the twentieth century weather forecasting was based on the collection of surface observations of wind, temperature, pressure, dew-point temperature, clouds, and precipitation. The observations were plotted and analyzed by hand. By the 1930s rawinsonde observations of the upper atmosphere began to supplement the surface observation network that was the backbone of operational meteorology. By mid-century the frontal theory of the time, often referred to as the Norwegian Frontal Model because of its origins, was being applied as a unifying concept to interpret the observations and to guide analysis. Numerous empirical rules existed that related clouds, precipitation, and frontal evolution to the observations. To a great extent these rules described how to prognosticate weather systems through extrapolation.

The nonlinear nature of atmospheric processes limits the utility of extrapolation techniques, and the problem of cyclone development and decay were addressed in the 1940s and 1950s by application of theoretical "development" equations that related upper-level vorticity and thermal advection to divergence and ultimately to changes in surface pressure. The quasi-geostrophic (QG) approximation, basically stating that the wind velocity remains in approximate balance with the pressure gradient, was invoked as part of these development equations. To apply these concepts to weather forecasting, vorticity charts were constructed via manual graphical methods that were quite laborious. The goal of all these efforts was to predict the future position of cyclone centers and the attendant fronts as accurately as possible and then to derive the sensible weather that was characteristically associated with each weather system.

By the late 1950s and 1960s numerical weather prediction using recently developed computers became skillful enough that humans were no longer preparing prognostic charts by hand, and forecasting instead focused on interpreting the output from the various computer models of the atmosphere. This continues to this day and will be discussed in detail later in this chapter.

The 1950s and 1960s were also the time when remote sensing of the atmosphere via radar and satellite began to have an impact on weather forecasting. In particular, these types of observations became primary tools along with the conventional surface observations to make forecasts and warnings of weather phenomena on a short temporal scale (0 to 6 h) and a spatial scale of approximately towns and counties. Remote-sensing technology continued to evolve through the remainder of the twentieth century, and short-term warnings and forecasts of severe and significant weather have become one of the primary activities of operational weather forecasters.

Modern weather forecasting has evolved as a result of changes in observational, computational, and communications technology. Increases in computational resources have made a significant impact on the complexity of numerical weather prediction, but, as importantly, also on display capabilities where forecasters view model output and observational data. Perhaps more important to the evolution of weather forecasting than the changes in technology are the application of new knowledge from research results that examine the fundamental structures and processes of atmospheric phenomena on all scales. Research has led to the development and application of numerous conceptual models of atmospheric phenomena that provide a framework for forecasters as they attempt to interpret the large volume of model output and observational data now available.

The rest of this chapter will be divided into three parts. A discussion of the forecast process will be followed by sections on long-term forecasting and short-term forecasting. The distinction between short-term and long-term will be made based on the time scale when numerical weather prediction begins to become the dominant tool used by the forecaster. At the time of this writing, this transition takes place between 6 and 12 h. For the purposes of this chapter, long-term forecasting covers the period from 12 h through 7 days. Forecasting at ranges longer than 7 days is beyond the scope of this chapter.

2 FORECAST PROCESS

A prediction of the future state of the atmosphere requires that a forecaster must understand what is taking place in the present. This fundamental first step is required regardless of the spatial or temporal scale of the prediction. This step is required even if the basis for the prediction is output from a numerical weather prediction model for reasons that will be described in the long-term section below.

As part of the process of understanding the present state of the atmosphere, a forecaster must pose and then answer a series of questions. Examples of some of the more obvious questions might be: Where is it raining? Where is it not raining? Why is it raining in one location but not another? How are the clouds distributed horizontally and vertically? How has this been changing with time? Where are the centers of high and low pressure? Where are the fronts? How are the troughs and ridges aloft distributed and how have they been evolving? Where are the temperature gradients at the surface and aloft and how are they changing? A quick look at an animated sequence of satellite or radar images will bring these and dozens of other questions quickly to the forefront of the problem to be solved by the forecaster.

This part of the forecast process is called analysis. During analysis, a forecaster will examine both direct and remotely sensed observations and attempt to answer the many questions that, in the aggregate, lead to an understanding of the present state of the atmosphere. Analysis of directly observed parameters is often done objectively by software, with graphical output displayed on a computer. Analysis is also done by hand, with a forecaster carefully scrutinizing a plot of observations with a pencil in hand, drawing contours and making annotations of significant features. Skillful hand analysis can provide a forecaster with insight that is not likely to be brought out by objective analysis routines, but the utility of hand analysis is dependent on the skill of the analyst.

Analysis of remotely sensed data is also performed to ascertain the current state of the atmosphere. The most common example of this type of analysis is the examination of satellite imagery. The three most commonly available satellite data types are infrared (IR), visible (VIS), and water vapor (WV) imagery. Other types of satellite imagery exist that detect precipitable water, rainfall rates, ozone, and other parameters, and these are seeing increased usage in forecasting applications. Animation of WV imagery quickly shows a forecaster the positions of mid and upper-level circulation features such as troughs, ridges, closed cyclones, and anticyclones, as well as areas of ascent and subsidence aloft. Animation of IR imagery shows development and decay of cloud systems by changes in cloud top temperature with time. Visible imagery, which is available only during sunlit hours of the day, shows regions of low and mid-level clouds, as well as regions of fog and snow cover. Visible imagery can also be used to identify areas of convection, especially at times of low sun angle, when convective clouds cast shadows. Forecasters use animation of the different types of satellite imagery to see movement and evolution of features.

Similar analysis techniques of radar data can quickly show a forecaster the location and intensity of both stratiform and convective precipitation. The evolution and vertical structure of individual convective cells can be discerned from careful analysis of radar data. Doppler radar offers forecasters the opportunity to remotely sense the wind speed and direction on a continuous basis at a variety of levels in the vertical if there are adequate reflectors. Wind profilers, which are a type of vertically pointing Doppler radar, also provide remotely sensed information about the three-dimensional wind field.

The various directly and remotely sensed observations described above, along with other parameters not included in this discussion, are often best analyzed in combination. Computational resources are commonly available that allow for the combination of the various observed data sets. To a great extent, this was not possible prior to the 1990s, and many of these data were often analyzed separately from each other. A graphical overlay of rawinsonde and surface observations on a satellite image can make it very easy to answer a fundamental question such as "where is the front at the surface and aloft," but this was not always the case and is still quite difficult over oceanic regions.

To make sense of the observations, forecasters will often attempt to fit a conceptual model to the observed data. A conceptual model is a mental picture of an atmospheric structure or process. They are usually the result of research where intensive observations have been made on a temporal and spatial scale that is far greater in resolution than the operational observational network. The Norwegian Frontal Model is one of the best-known examples of a conceptual model. The structure and evolution of a so-called supercell thunderstorm is another example of a conceptual model. There are hundreds, and perhaps thousands, of different conceptual models in use to describe virtually every type of phenomena on virtually every spatial and temporal scale. Conceptual models are convenient structures for organizing the huge volume of data with which forecasters are faced; however, they are inevitably simplifications of actual atmospheric phenomena and processes.

Research results lead to constant refinement of existing conceptual models, proposals of new models, and even the abandonment of some conceptual models. Models of phenomena are often applicable in some geographic regions, but not others. The Norwegian Frontal Model is a very old paradigm, and it has been revised many times, and its complete abandonment has been recommended on numerous occasions. However, a satisfactory replacement for this conceptual model has not been found, so at least parts of this model continue to be used by forecasters. Most conceptual models have similar life cycles.

Another method employed by forecasters to organize the forecast process is to start their analysis of the data on the largest scale and then move progressively downscale. This method has been called the "forecast funnel" in that a forecaster starts at the top, or wide portion, of the funnel, which corresponds to the planetary scale, and then descends into the synoptic-scale in the middle of the funnel, and finally the narrow portion of the funnel focuses on the forecast problem of a specific place and time, which is determined by conditions on the mesoscale. The forecast funnel is itself a conceptual model in that it is based on the idea that larger scale atmospheric conditions drive smaller scale phenomena, and that it will be difficult or impossible to accurately predict conditions on the smallest scales without taking into account conditions on the next larger scale, and so on.

A simple example of scale interaction and the application of the forecast funnel might be to consider the potential for a snowstorm along the mid-Atlantic states. On the planetary scale, a longwave trough position must be present over eastern North America for there to be a chance of cyclogenesis, and the track of the storm will, to some extent, be determined by the planetary-scale conditions. A migratory short-wave trough(s) on the synoptic-scale is required to initiate cyclogenesis. The interaction of the synoptic-scale troughs and ridges with the local topography will determine whether cold air remains trapped on the east side of the Appalachian Mountains. This synoptic-scale interaction with the terrain drives the mesoscale conditions and modulates the precipitation type and intensity.

Like all conceptual models, the forecast funnel is a simplification. In the example above, one could also argue that the preexisting conditions on the mesoscale are just as important, or more so, for the creation of a snowstorm. Where is the cold air already present? Are there any preexisting temperature gradients or old frontal boundaries? Is the air in the warm sector of the storm unstable enough to produce widespread precipitation and convection, which could lead to further intensification of the cyclone. Forecasters generally find that they must also ascend the forecast funnel at times to take into account "upscale" interactions.

3 FORECASTING 12 h TO 7 DAYS

Modern weather forecasting of conditions beyond 12 h in the future is primarily based on forecaster interpretation of output from numerical weather prediction models. Forecasters have access to many different models. Through the Internet, forecasters can access models from the national centers of various countries, military national centers, and universities around the world. Depending on their employer's computational resources, forecasters may even have access to models run locally. At the time of this writing, model output is typically used in a deterministic sense. This means that given realistic initial conditions and a model with adequate resolution and accurate representations of the pertinent physical process, the model prediction of the future state of the atmosphere can be used to explicitly derive the sensible weather elements at Earth's surface (or aloft for aviation forecasting). By the end of the twentieth century considerable effort was being expended on so-called ensemble forecasting, where output from an ensemble of models is used either by a forecaster or directly through objective techniques to produce probabilistic forecasts of sensible weather, with a goal of including information about uncertainty with each forecast element. A very brief discussion of ensemble techniques is included at the end of this section, but most of what follows is based on a deterministic approach, which is still the most widely used method for most forecasting tasks.

Before using any model, forecasters will go through the process of analysis in an attempt to understand current conditions. This must be done on all spatial scales. The analysis step is the cornerstone of proper use of deterministic model output. It is important for at least two reasons. First, the forecaster must decide if the model initial conditions and the early hours of the forecast capture the essential detail of what is happening in the real atmosphere. Second, through analysis of current conditions, the forecaster must grasp the physical processes that are important. Different models represent the physical processes of the atmosphere in different ways, and if the forecaster expects certain processes to be particularly important, then this information will be a factor in how the forecaster uses the output from the various models.

A critique of the initial conditions of models has become much easier with the use of modern graphic/image workstations. It is a simple task to overlay various model fields with animating satellite imagery. Comparison of model initial fields with plotted observations and satellite imagery allows the forecaster to quickly determine if troughs, ridges, vorticity centers, baroclinic zones, jet stream axes, moisture plumes, and other features are reasonably well depicted in the model's initial conditions. This critique is particularly important over data-sparse regions such as the oceans but is also important over relatively data-rich areas such as North America since there are times when relatively small errors in initial conditions can lead to significant errors in the forecast.

When the initial conditions of one model are superior to all the others, a forecaster may decide to give the solution from this model greater weight as the forecast is developed. Often the situation is not clear as to which model is best initialized. Differences are often subtle, and even with current satellite capabilities, there is considerable uncertainty about the exact locations in the horizontal and vertical of specific gradients and features over oceanic regions. There are times when all of the available models have a poor grasp of the real atmosphere. In these situations, a forecaster will have to fall back to some of the empirical rules of the pre-NWP (Numerical Weather Prediction) era to mentally modify model output in the direction most likely to minimize the error. Most often in this situation the forecaster will have low confidence in what the ultimate conditions will be, and this uncertainty will be expressed quantitatively and qualitatively within the allowable format of the forecast product.

Given a reasonable outcome from the initial condition critique, the forecaster will often next move onto consideration of the different model characteristics. Typically, resolution is greater in limited domain models, while lower resolution models have larger and often global domains. High-resolution models will have more realistic representations of Earth's topography, including mountains, valleys, lakes, and coastlines. Larger domain models will often perform best in regimes with strong westerly flow and fast-moving storm systems.

Similar differences typically exist between models with respect to vertical resolution, particularly in the boundary layer. Greater vertical resolution generally leads to more accurate representation of temperature inversions, unstable layers, and the vertical distribution of moisture. There can be significant differences between models in their treatment of moist processes such as convection and stratiform precipitation due to differences in their vertical resolution. It also can drastically affect how a model predicts temperature and winds near the surface. Models also differ in their vertical coordinate system and whether the basic equations assume hydrostatic balance or not. Some vertical coordinate systems are terrain-following; others are not. Some dynamically place the greatest resolution in regions of large potential temperature gradient; others are fixed regardless of the situation. The vertical coordinate can lead to different model characteristics for phenomena such as mountain-induced windstorms or cyclogenesis that takes place in the lee of orographic barriers.

Different methods of representing physical processes are used in the various models available to forecasters. Processes that cannot be resolved explicitly within the model's formulation of the primitive equations are represented through what are called parameterization schemes. These are basically subroutines that are invoked throughout the model run that use "state" variables that are explicitly predicted, such as wind, temperature, pressure, and some form of moisture, to handle such phenomena and conditions as soil moisture, moist and dry convection, shortwave and longwave radiation, evapotranspiration from vegetation, turbulence, and so forth. The parameterization schemes are designed to feedback into the model and alter the state variables. Although many of these schemes are quite complex, an understanding of some of the simpler schemes can be used effectively by forecasters in their subjective use of model output.

Model output is now widely available in gridded form. Prior to the 1990s, only a very limited set of graphics at specific levels were typically available. These were, to a great extent, the same set of graphics that were produced via manual methods prior to the NWP era. Gridded model output allows forecasters to examine the model solution in great detail, and in particular it allows forecasters to gain insight into the three-dimensional structures of the model's simulated atmosphere. Software and computational resources exist that allow the output to be viewed via three-dimensional renderings, but at the time of this writing these resources are not widely in use yet, and most model output is still viewed via two-dimensional graphics, such as plan views, cross sections, time–height sections, and forecast soundings.

Forecasters will apply their knowledge of the important physical processes gleaned from the analysis step to examination of the model output. Knowledge of local conditions plays a role in the forecaster's decision about which processes are most important in each situation, and this becomes a significant factor since the strengths and weaknesses of different models are considered. If flow interaction with terrain is a key consideration for a given event, such as in cases of orographic precipitation, then a model with high horizontal resolution might be favored. If lake effect clouds and precipitation are significant forecast problems, then a model that correctly initializes the areal extent of the lake and its water temperature might be favored. If two models are forecasting a trough to move through a mean longwave ridge position with the same timing, but with different depths, the larger domain model that has a better representation of the planetary-scale waves might be preferred. If the forecaster knows that precipitation has recently occurred and near-surface conditions are moist, then the model that has most correctly predicted precipitation in recent model runs might be expected to have the best representation of soil moisture and boundary layer conditions. Each situation must be evaluated and forecasters must apply their knowledge of initial conditions and model characteristics to decide, when possible, which model solution offers the greatest skill and utility.

The final step in the deterministic use of model output for forecasting is the derivation of sensible weather. This includes parameters such as precipitation amount, onset, end, type, and areal distribution, potential for severe convection, lightning activity, sustained wind speed and gusts at the surface, wind shift timing, cloud cover, cloud layers, visibility, areas of turbulence, icing, blowing snow, blowing dust, temperature, wind chill, heat index, wave and swell height, sea-ice accumulation on ships, and so forth, depending on the type of forecasting that is being done. Statistical postprocessing of model output exists that attempts to derive sensible weather elements directly from model output. In the United States, some of these statistical methods are called model output statistics (MOS). The MOS regression equations are derived by comparing past model forecasts with subsequent observations at individual locations. This method has the advantage that it can correct for model biases. It has the disadvantage that in the modern era operational models are undergoing nearly constant change, and the developmental data set for MOS techniques may not be representative of the current incarnation of the model. There are other objective algorithms that convert model output into sensible weather elements and more are always being developed, but a complete discussion of this topic is beyond the scope of this chapter. A popular display of this type of algorithm output is known as a meteogram, where a group of time series of derived sensible weather is displayed graphically in the same way that a graphic of observations might look. In general, experienced forecasters are quite expert at deriving the sensible weather from model output, once they have applied the knowledge and techniques described above to determine which model output is most appropriate for a given situation.

In deterministic forecasting, a forecaster will consider a number of different model solutions during the forecast process. In essence the forecaster is examining a small ensemble of forecasts. The number is typically five or fewer. The forecaster may be looking for model-to-model consistency and/or run-to-run consistency from a given model as an indication of the uncertainty associated with a given forecast. Ensemble forecasting techniques are based on the idea that there is inherent uncertainty in the initial conditions and perhaps similar uncertainty about which is the optimal model configuration for a given situation. The goal of ensemble forecast methods is to quantify the uncertainty by running many versions of NWP models with either perturbed initial conditions and/or versions of models with different parameterization schemes, different basic configurations, or different analysis schemes. The method of perturbing the initial conditions, or the basis for deciding the makeup of the ensemble members, is beyond the scope of this discussion, but two key premises are that each member has an equal probability of being correct and that the more ensemble members included the more skillful the probability distribution of the output.

Ensemble techniques are being developed, both as an alternative and as an adjunct to the deterministic forecast method. Although experienced forecasters are very skillful at deriving sensible weather from model output, the task of determining which model output is best in a given situation, and estimating the magnitude and shape of the model's most likely error is quite difficult. Some would argue that this task is essentially impossible. This last statement is a topic of active debate. Another reason for ensemble techniques is that an increasing number of users of forecast information, especially those who must make large resource commitments based on the weather forecast, require a quantification of the uncertainty associated with the forecast. While deterministic forecasts often include uncertainty, such as probability of precipitation statements, and narrative wording that indicates alternative scenarios, the method does not lend itself to quantification of uncertainty of all parameters in time and space, and certainly not with consistency from forecaster to forecaster. A strength of ensemble techniques is that probabilistic output falls out quite naturally, offering consistent quantitative assessment of uncertainty. Of course, skillful algorithms that convert model output to sensible weather elements must exist, and this is still problematic in many situations.

Ensemble forecasting is a rapidly emerging technique. It could become the primary method of weather forecasting in the future. A description of modern weather forecasting in 2010 may look back on deterministic forecasting methods as something from a bygone era, much as today we look back on the pre-NWP era with nostalgia. More likely, some forecast applications will lend themselves to ensemble techniques, especially where a sophisticated user community exists, while other applications will be best accomplished by deterministic methods for the foreseeable future.

4 FORECASTING 0 TO 12h

Output from numerical prediction models is of limited use in short-term forecasting for a number of reasons. The most obvious reason is that the collection of observations, quality control of the observations, objective analysis, the execution of the numerical model, and the distribution of the model output takes time. Usually these tasks take on the order of 1 to 3 h. Forecasts for this time frame will have to be made without the benefit of new model guidance, although output from a previous model run may still be applicable. Another reason NWP is of limited use in the short-term is that the initial conditions of numerical models do not yet include ongoing precipitation processes. This is an area of research and development and progress is being made, but, presently, there is a so-called spin-up period, typically 3 to 6 h, before a modern numerical model can produce realistic precipitation. Perhaps the most important reason that NWP is of limited use for short-term forecasting is that the atmospheric phenomena of interest are often not explicitly resolved by current NWP models. Predictions of whether an individual thunderstorm will produce large hail or not and where such a storm is likely to track in the next hour are examples of short-term forecasting for which current NWP offers little or no guidance, although research efforts are ongoing in this area.

The short-term forecast process begins much like the longer-term forecast process, the forecaster must understand the current state of the atmosphere. Analysis is even more important for short-term, small-scale forecasting. Objective and manual analyses are useful, and particular emphasis must be placed on careful examination of individual observations. Because specific observations may be smoothed by objective analysis schemes, or even eliminated by quality control algorithms, manual analysis is usually considered a routine part of short-term forecasting, particularly during the warm season when gradients are weak and important features may be subtle.

Most short-term forecasting deals with phenomena on the mesoscale. However, as part of the analysis step, forecasters must consider scale interaction. Although it is debatable whether planetary-scale considerations are important for short-term forecasting, there is no doubt that synoptic-scale conditions are critical to the evolution of mesoscale phenomena. The location and evolution of synoptic-scale features such as frontal boundaries, troughs and ridges aloft, regions of warm and cold temperature advection, high- and low-level jets, and the distribution of moisture in the horizontal and vertical is necessary knowledge for any forecaster attempting to predict mesoscale phenomena.

Because short-term forecasting does not have the benefit of primitive equation model output, it is absolutely critical that the forecaster understand the current atmospheric processes and phenomena. A prediction of conditions 1 to 3 h into the future will generally fail if the current conditions are misdiagnosed. The forecaster must decide whether conditions are evolving linearly or via nonlinear processes. An example of a linear forecast would be extrapolation. A line of rain showers is moving east at 10 mph and is expected to continue east with no change in speed, direction, or intensity. Again, modern graphic/image workstations allow tracking and extrapolation of features on radar and satellite imagery to make even this simple extrapolation forecast more accurate. However, if the forecaster fails to note that this line of rain showers is moving into an environment of greater instability and favorable vertical wind shear, then the nonlinear transformation of the line of showers into a severe squall line could be completely missed in the short-term forecast.

Conceptual models play a very important role in short-term forecasting, perhaps even more so than in longer-term situations because of the lack of NWP output. A good example of applying a conceptual model to a short-term forecast problem can be illustrated for the prediction of supercell thunderstorms. Once thunderstorms have started to develop over an area, the short-term forecast for that area can be very different depending on the character of the convection. Ordinary thunderstorms will have a life cycle of only an hour or so, they will generally not produce large hail or tornadoes, and they will tend to move in a straight line. In contrast, supercell thunderstorm life cycles can be many hours, they are likely to produce severe weather, including large hail and possible tornadoes, and their movement will likely be to the right or left of the movement of the ordinary cells. Clearly, the short-term forecast will be quite different if supercell thunderstorms are expected as opposed to ordinary thunderstorms.

Considerable research has shown that supercell thunderstorms develop in regions with at least 20 m/s of vertical wind shear in the 0 to 6 km depth above ground, and with adequate instability to support deep convection. The basic conceptual model of supercells is that the combination of buoyancy and shear allow a thunderstorm structure to become organized such that the updraft and downdraft do not interfere with each other, as is typically the case in ordinary thunderstorms. The result is a long-lived storm with very strong updrafts and downdrafts, and pressure perturbations that lead to rotation within the storm and propagation rather than just advection of the thunderstorm.

A forecaster will first examine the environment to determine if supercell thunderstorms are possible. Output from NWP can provide guidance about buoyancy and shear. However, detailed analysis of direct and remotely sensed observations is required to identify pertinent features such as old frontal zones, outflow boundaries, gradients in surface characteristics, and diabatic heating gradients due to cloud cover that may modulate the shear and instability on the mesoscale. Upon completion of this analysis the forecaster will know whether supercell thunderstorms are possible and also if some areas are more favorable than others for their development. The second step in this process is to use Doppler radar data to carefully examine the three-dimensional structure of thunderstorms as they develop. The forecaster will look for features such as tilted reflectivity cores, weak-echo regions, bounded weak-echo regions, rotation in the velocity data in the region of the storm updraft, and anomalous storm movement. These are structures that have been documented in research results and are the basis of current conceptual models of supercells. Once identified, appropriate forecasts and warnings can be issued for these storms.

A similar forecast process is used for virtually all short-term forecasting that attempts to predict the evolution of mesoscale phenomena. A precipitation band in a cool-season situation can be caused by many different conditions. A short-term forecast of the behavior of the band can only be made if the forecaster understands the structure and processes responsible for its existence. The short-term forecast will be very different depending upon whether the band is due to conditional symmetric instability, or an internal gravity wave, or postfrontal lake effect, or orographic lift, or low-level convergence.

Ensemble techniques may eventually play a role in the 0- to 12-h forecast process. Often, the differences between environmental conditions that can support certain mesoscale processes and those that cannot are quite subtle. An ensemble of models may be able to offer forecasters insight into which situations are borderline and which are not. Again, the value of the ensemble technique is the quantification of uncertainty.

SECTION 5

MEASUREMENTS

Contributing Editor: Thomas J. Lockhart

CHAPTER 33

ATMOSPHERIC MEASUREMENTS

THOMAS J. LOCKHART

The technology applied to sensing of atmospheric conditions and content improves constantly and at an ever-increasing rate as newly available technologies are applied to old, familiar questions. The increasing capability and decreasing price of personal computers is both a case in point and a cause of expanding technological applications. The best venues for describing and discussing advanced instrumentation are the technical conferences and peer-reviewed journals. The thrust of this section will be toward common applications and common concerns with respect to the representativeness and uncertainty of measurements. The contributors to the measurements section have practical experience dealing with such questions. The goal of this chapter is to describe the problems, compromises, and pitfalls with measurement programs of many kinds. This section will also include some examples of program failures, which seldom reach the readers of technical journals.

In centuries past, instrument design, siting, and data application began with individuals. Curious individuals used measurements to document their observations and to help find explanations for what was happening. This led to a need to duplicate instruments and introduced a commercial opportunity. As communication technology developed, the ability to assemble synchronous observations became possible. Synoptic meteorology generating forecasts and warnings followed. National weather services were born and skills in measurement technology with the necessary set of specifications became a specialty.

It is a fundamental principle that the application of data rests on an understanding of the measurement process and the size of the uncertainty or error bars. It is therefore incumbent on the organization taking the data to retain expertise in these disciplines. Those with such expertise usually are thinking about improvements and investigating new technology. This has led to the various national

Handbook of Weather, Climate, and Water: Dynamics, Climate, Physical Meteorology, Weather Systems, and Measurements, Edited by Thomas D. Potter and Bradley R. Colman.
ISBN 0-471-21490-6

laboratories and research organizations formed to pursue technological and application progress.

The atomic age introduced a need for radioactive fallout predictions. In recent decades (since the 1960s) federal regulations have driven many of the applications of special measurement programs in the atmospheric surface layer. First, in response to the U.S. Department of Defense (DOD) Army chemical warfare research programs, a need for better understanding of dispersion or diffusion processes was recognized. The Nuclear Regulatory Commission (NRC) required accurate measurements and a quality program to assure compliance with the requirements stated in the NRC Safety Guide 23. A group of volunteers working within the auspices of the American Nuclear Society (ANS) composed a standard (ANS 2.5) that defined the details necessary to meet the ANS requirements for measurement accuracy and the methods for quality assurance.

The U.S. Environmental Protection Agency (EPA) provided similar guidelines in response to the needs of the 1967 Clean Air Act. The EPA requirements were quite similar to those of the NRC. Clearly, a vacuum of measurement standards existed. No one could cite consensus standards to justify claims of data compliance. These two lists of requirements, with the force of the federal government behind them, brought the measurement community to action. It was immediately clear that standard definitions and test methods were necessary to determine if the requirements were being met by the measurement systems used.

The American Meteorological Society's (AMS) Committee on Atmospheric Measurements (CAM) considered the need to write standard methods to determine the various characteristics that were being specified. After due consideration and conversations between the AMS executive secretary and his counterpart in the American Society for Testing and Materials (ASTM), it was decided to form a subcommittee within ASTM D22, Sampling and Analysis of Atmospheres. Subcommittee D22.11, Meteorology, was formed in 1972. The work of volunteer members of D22.11 provided some of the early standards, but the process is open ended. ASTM has a requirement of a 5-year review and reapproval process. New standards are generated as the need, individual willingness to contribute, and expertise come together. This national program is now contributing to a similar international program through ISO (International Standards Organization) TC 146 (Technical Committee 146, Air Quality) SC 5 (Subcommittee 5, Meteorology).

It is generally recognized that approaching perfection in the definitions and performance of instruments is reducing the least important contributor to measurement uncertainty. Take wind speed as an example. A perfect anemometer will only suggest what the wind speed would be if the calibration facility were a true analog to the turbulent environment. Its performance has been tested in the wind tunnel. Its response to a speed change (sensitivity to inertia in a mechanical design) is defined in terms of its distance constant, defined by a standard test using a speed step from zero to the wind tunnel speed. The time constant, $1 - 1/e$, at that speed is measured between the 30 and 74% points (of maximum value) during its recovery to avoid the stall condition at release. The test is run at two or more speeds, which documents that the response is constant with distance rather than time.

It has been recognized that the distance constant is different for
from a higher speed to a lower speed than for a step change from lc
speeds. The ratio of the two distance constants will predict the "overs
cup anemometer. This overspeeding is the inertial effect that causes the anemometer to speed up faster than it slows down, creating an average that is slightly larger in turbulent flow than in laminar flow. The word *overspeeding* has also been used to describe the difference between the scalar average and the resultant vector average of samples taken over a period of time.

There is further confusion when the effect of off-axis or nonhorizontal flow is considered. Cup anemometers will usually turn a little faster when the angle of the flow is not quite horizontal. This is an aerodynamic or lift-drag effect that can be simulated in the wind tunnel. When the distance constant model, based on the wind tunnel test, is used to estimate the underreporting of average speed in turbulent flow, it fails. The step change in speed does not describe the speed changes in atmospheric flow. While the model might predict a 20% error, the measurements in the atmosphere show little or no effect.

There are effects, however, that do bias the measurements. Mounting hardware and/or supporting structures may influence the speed. Nearby trees or buildings will alter the flow. While the anemometer may faithfully report the speed where it is mounted, it may not represent an area as large as the application might require. The overall subject of representativeness will usually contain much more uncertainty than the uncertainty in the transfer function (relationship between wind tunnel speed and measured rate of rotation) of the anemometer. However, there is a general rule of thumb that suggests one should not ignore those parts of the uncertainty analysis that can be understood simply because there are other parts that are larger and more difficult to define.

There are a broad variety of atmospheric variables that can be measured and an equally broad variety of applications that need these measurements, each with their own unique requirements. The contributors to this section will touch on many of these. It remains the responsibility of the person who uses atmospheric measurements to understand, to whatever degree possible or practical, the process by which the measurements were made. This knowledge should range from the engineering specifications of the instrument to the siting biases that might contribute to representativeness of samples.

CHAPTER 34

CHALLENGES OF MEASUREMENTS

THOMAS J. LOCKHART

1 ETHICS

If there is one absolute in the ethics of measurement, it is that the data must speak for themselves. But what can numbers say? If one applies standard statistical processes to a series of numbers, the answer is reproducible. Of course, there is some control in selecting a subset of a series of numbers (see "All That Is Labeled Data Is Not Gold" below). Even if a time series is continuous, the beginning point and the ending point can influence the outcome. There is nothing unethical in presenting a continuous subset of data and listing their statistical parameters, if the fact that it was a selected subset is disclosed. When the parameters are used to make a political (or societal) point, the ice is thinner.

A classic example of mixing true science with "political" science comes from comments made by Stephen Schneider (1987), a proponent of the theory that cholro-fluorocarbons (CFCs) are depleting the ozone layer. He said: "We have to offer up scary scenarios, make simplified, dramatic statements, and make little mention of any doubts we may have. Each of us has to decide what the right balance is between being effective and being honest." Those of us who subscribe to the ethics of measurement will have no problem with that decision. Science, including its subset measurement, requires honesty. There is no type-A or type-B honesty. If any consideration influences how numbers are gathered or generalized, that consid-eration must be defendable without subjectivity.

The measurement community often expresses a genuine skepticism when consider-ing the results of model simulations. The assumptions used for the model simplifica-tions and for the "data" inputs deserve scrutiny. When simulations go beyond the data in time, they become a forecast. There is real value in general circulation models, and

Handbook of Weather, Climate, and Water: Dynamics, Climate, Physical Meteorology, Weather Systems, and Measurements, Edited by Thomas D. Potter and Bradley R. Colman.
ISBN 0-471-21490-6

there has been real improvement in most aspects of weather forecasting as a result of model outputs. However, it will be some time, if ever, before models will contain the intelligence exhibited by local meteorologists (or local farmers) in local short-term forecasts.

When assumptions go beyond the data in range, it is a guess. If anemometers are calibrated to 50 m/s and report speeds of 70 m/s, it is impossible to know the uncertainty of the measurement (unless some postanalysis is conducted). Extrapolation beyond experience is often the only course available, but a footnote is required to warn the user of that fact. Extrapolation between measurement points may not agree with new measurements specifically located to test the extrapolation. The measurement deserves the presumption of accuracy. There are instrument tests that can confirm the instrument performance. The smoothed model value should not be presumed to be correct.

It is true that caveats may adversely affect the literary quality of pronouncements. Perhaps there should be a sharp division between the statements of "scientists" on political issues and scientists reporting the results of their fundamental research. The public is not capable of sorting the "results" from a reputable scientist that are biased for advocacy from the results from a reputable scientist that contain all the dull but critical caveats that qualify the data supporting the results. It used to be that ethics took care of the problem. If ethics no longer apply, a method needs to be found to warn the public of the advocacy role of "political" scientists.

All That Is Labeled Data Is Not Gold

Most of us are quite comfortable with the idea of evolution in the animal kingdom or in climate, by which small departures from the norm can spread and eventually change the entire picture. We are not as ready to accept a similar evolution in classical data sets, the gold standard of climatology.

Recently, I have been using a small subset of the data generated by the GEOSECS (Geochemical Ocean Sections) program (1973–1974), in particular, the ocean profiles of oxygen-18 and deuterium measured by Harmon Craig at Scripps Institution of Oceanography in La Jolla, Calif. Until recently, I was secure in the knowledge that the data I was using, gathered from the National Climate Data Center repository, were complete and unadulterated. Earlier this year, while presenting my results and using these data for comparison, a member of the audience asked why I was not using the complete data set. Confused, I defended myself vigorously. Later, it appeared that although I had correctly presented the published data, there was an underground version of unpublished data that was known only to a small circle of initiates. My hosts were kind enough to include me among the cognoscenti and it was at this point that things started to become interesting.

Examining the new version, I started to find many inconsistencies in the two data sets: values different in one than the other, stations appearing in one or the other but not both, different measurement depths, and even different positions for the stations. Puzzled, I went further afield in search of information that could resolve the conflicts.

Looking at the published tables, I worryingly found that the mistakes in station positions seemed to be due to typographical errors.

Soon, though, I found someone else who had a subset of the unpublished data. This surely would clear things up. Unfortunately, it made things even worse. Where this data set purported to give the same data, it was different from both previous versions in seemingly random ways.

In despair, I tried to track down the source of the unpublished data. Finally, I found someone who had an old photocopy of an old-style computer printout whose provenance was claimed to be Harmon Craig's original measurements (including repeats). At last, truth!

It became clear very quickly that this printout was the original version of the unpublished data, and it was equally clear that the data had at some point been typed in by hand. The errors were of the sort caused by inadvertent line skipping, missing decimal points and minus signs, and incorrect averaging of repeat measurements. However, it was also evident that this was the source for the published tables. The published data though, had been subjected to some quality control; outliers had been thrown out, repeat samples with too much dispersion ignored, and possibly, further repeats had been performed. The measurement depths had clearly been refined using more accurate information.

The origin of the third data set remained mysterious until I was informed that it had been traced from a graph showing the data in an old article and then correlated to the original stations. That explained the small but random differences. Only with all this scientific detective work and the discovery of the ancestral data was I now able to amalgamate the published and unpublished results into a consistent data set.

These data are only 25 or so years old and yet there are at least three (maybe more) different and mutually inconsistent versions floating around. The lesson we should take from this is that, if unchecked, small mutations will occur during transcriptions. Unless we are vigilant, "data sets" will evolve and data that have been so painstakingly collected and saved will be distorted and twisted beyond all usefulness. Our care in using data should be a force analogous to natural selection, keeping data sets fit for the purposes intended. If we do not take care, the gold standards of climatology may in time turn to lead.

Gavin Schmidt, NASA Goddard Institute for Space Studies, New York; E-mail: gschmidt@giss.nasa.gov.

2 GENERATION AND IMPLEMENTATION OF A CONSENSUS ON STANDARDS FOR SURFACE WIND OBSERVATIONS

Generation

Robert L. Carnahan, the retired federal coordinator from the Office of the Federal Coordinator for Meteorology (OFCM), contacted the author in 1992 regarding a wind measurement question. Contact was arranged with David Rodenhuis, chairman of the Ad Hoc Group and the Interdepartmental Committee for Meteorological Services and Supporting Research (ICMSSR) on wind observing standards. The author was asked to assist Rodenhuis in organizing a workshop and presiding at the workshop; a plan was devised.

The purpose of the workshop was to seek a consensus standard method of characterizing surface wind measurements that would serve all applications of wind measurement. Two preparatory tasks were essential. First, a group of knowledgeable attendees to represent all applications had to be identified. Second, a syllabus needed to be prepared to set the background for the two-day workshop.

A list of 79 experts was assembled, and, of those, 40 attended; the 39 who could not attend were sent all the materials and reports and had the opportunity to participate by mail. The group included specialists from such applications as agriculture, aviation, climatology, forecasting and warnings—National Weather Service (NWS), manufacturers of instruments and systems, measurements, military, oceanography, buoy operation, and hurricane research, insurance, transport and diffusion, air quality, network operations, and quality control, and wind energy. The names, affiliations, and specialties of all participants are available from the OFCM.

The two-day workshop produced a group consensus, of which the most important elements are listed below.

- The metadata (information about the measurement system and the siting of the sensors) must be available where data are archived. Included will be the following: station name and identification number, station location in longitude and latitude or equivalent, sensor type, first day of continuous operation, sensor height, surface roughness analysis by sector and date, site photographs with 5-year updates, tower size and distance of sensors from centerline, size and bearing to nearby obstructions to flow, measurement system description with model and serial numbers, date and results of calibrations and audits, date and description of repairs and upgrades, data flowchart with sampling rates and averaging methods, statement of exceptions to standard requirements, and software documentation of all generated statistics.
- The basic period of time for surface wind observations should be 10 min. This period provides reasonable time resolution for continuous process analysis (air quality, wind energy, and research) while providing the building blocks to assemble longer averaging periods. Many current boundary layer models require hourly data. However, there is a growing world consensus for 10-min data periods. The periods must be synchronous with the Universal Time Clock (UTC) and labeled with the ending time. If a different time label is used, it must be stated and attached to the data.
- The operating range is either sensitive or ruggedized. The threshold and maximum speed of the sensitive range is 0.5 and 50 m/s. The threshold and maximum speed of the ruggedized range is 1.0 and 90 m/s.
- The dynamic-response characteristics for either type are as follows. The anemometer will have a distance constant of less than 5 m. The wind vane

will have a damping ratio greater than 0.3 and a damped natural wavelength of less than 10 m.

- The measurement uncertainty for wind speed is ±0.5 m/s below 10 m/s and 5% of the reading above 10 m/s. For wind direction, the accuracy to true north is ±5°.
- The measurement resolution is 0.1 m/s for both average speed and the standard deviation of the wind speed. Resolution for average direction is 1° and for the standard deviation of wind direction is 0.1°.
- The sampling rate will be from 1 to 3 s. Sampling for wind speed must be capable of achieving a credible 3-s average.
- The standard data output for archives will include the following:

  Ten-minute scalar-averaged wind speed
  Ten-minute unit vector or scalar-averaged wind direction
  Fastest 3-s gust during the 10-min period
  Time of the fastest 3-s gust
  Fastest 1-min scalar-averaged wind speed during the period
  Average wind direction for the fastest 1-min speed
  Standard deviation of the wind speed samples about the 10-min mean speed
  Standard deviation of the wind direction samples about the 10-min mean direction.

- Special guidance is offered for survivability to assure the support and power will not fail before the sensor itself fails. A suggestion is offered for preserving high-resolution data when the wind is above 20 m/s.
- Quality assurance is required with documentation in the site log.
- A specific method, based on Wieringa (1992), was described.

The consensus was reached with caveats from some members, generally preferring tighter requirements. However, this method was designed as a minimum that all applications could use. Wind energy, for example, needs greater accuracy. There is no reason why special networks or stations could not meet this specification and still provide greater accuracy. Uncertainty is largely a result of calibration. The hope of all the attendees was that the NWS would adopt the consensus and implement it in the new ASOS (Automated Surface Observing System) deployment. Implementation of new technology provides opportunities to improve the value of measurements. Digital data systems can preprocess samples. There is value in the potential richness of data that this consensus could provide as compared to manual observations. Technology has also provided capability to archive the richer data. The National Climatic Data Center (NCDC) representative joined the consensus and agreed that the data could be archived. All that remained was implementation.

Implementation

There was immediate implementation by some members of the workshop. The Oklahoma Mesonet used a version of the consensus in its new (at the time) statewide network of automatic stations. Campbell Scientific Inc. designed an instruction for its data loggers that would generate the required outputs each 10 min. The American Society of Civil Engineers (ASCE) adopted the 3-s average speed at 10 m as its basic wind speed (since the fastest mile has been discontinued). It has been published in its Minimum Design Loads for Buildings and Other Structures ANSI/ASCE 7-95.

Since there now was no expectation that the consensus would be used by the NWS, its essence was introduced to the D22.11, Meteorology, subcommittee of the American Society for Testing and Materials (ASTM). After the normal open process of consideration, modification, and balloting, the Standard Practice for Characterizing Surface Wind Using a Wind Vane and Rotating Anemometer D 5741-96 was adopted. As of 2001, this standard had not been adopted by the OFCM.

One element of the consensus is the definition of peak speed or gust. The ASOS provides a peak 5-s average. This clock-driven averaging period provides a smaller speed than the F420 anemometer attached to a dial or galvanometer recorder has provided during the past several decades. This human observation was considered "instantaneous" since there was no intentional averaging between the generator in the sensor and the dial or recorder in the office. The dial had to be seen at the moment of maximum value. The strip chart records the speed, but the galvanometer does have a frequency response. The damping by the recorder has been estimated as approximately equivalent with a 1- to 2-s average.

It became clear that, from a climate continuity standpoint, the change from the conventional peak speed to the ASOS peak speed decreased the value reported. This could have ramifications for storm or wind warnings. It is also clear that changing from a clock 5-s average to a running 3-s average would move the peak speed value closer to the conventional measurements. The most compelling reason for adopting the running 3-s average for peak speed is the fact that, without standards, the concept of peak speed has no meaning.

In the past, technology was such that whatever one could get was acceptable. Any measurement was better than no measurement. The totalizing anemometer had to be read where the anemometer was mounted. Gears turned a scale not unlike an electric meter. An interim step was the fastest-mile anemometer, which could drive an event pen on a strip chart in the office, thereby recording the time of each mile (so many revolutions of the cup) of air that passed. Time per mile could be converted to miles per hour by using an overlay with different line spacings labeled as speed or by counting the miles that passed in an hour for an hourly average.

With the advent of the cup anemometer driving an electric generator, the measurement could be observed in real time in the office, first on a dial and then on a dial and a chart recorder. When automatic weather stations came into use, the questions of digital sampling and averaging had to be answered. The Federal Aviation Administration (FAA) funded the National Oceanic and Atmospheric Administration (NOAA) and the NWS to make recommendations appropriate for aircraft operations for the ASOS design. The 5-s average for peak speed came from this study. Many other countries also considered the same questions. In the World Meteorological Organization (WMO), the Commission for Instruments and Methods of Observation (CIMO) adopted the 3-s average as the standard for peak wind speed.

It is hoped that our national network of surface weather stations will eventually adopt the consensus standard. In the interim, a campaign is underway to change the peak speed from the clock 5-s average to a running 3-s average on the basis of climate continuity. The American Association of State Climatologists (AASC) voted in their August, 1998, meeting to adopt the running 3-s average for climatological purposes.

3 UNCERTAINTY, ACCURACY, PRECISION, AND BIAS

A critical component of a chapter of this kind is a careful consideration of the meaning of the language used. There are two levels of language in this situation. One is the precise and unambiguous words used by the professionals in the standards field. The other is the routine communication used in the application of technology in a regulated society. A convergence of these two is the goal of this chapter.

Accuracy is a word commonly used in the specification of regulations, procurement documents, and manufacturer's data sheets. According to ISO (1993), measurement accuracy is simply the closeness of the agreement between the result of a measurement and a true value of the measurand. Accuracy is therefore a qualitative concept since the true value must remain indeterminate.

The concept of uncertainty, as carefully defined by standards professionals, provides a method to quantitatively characterize all doubt about the validity of the value of a measurement. There are two types of evaluation methods. Type A is an evaluation of a statistical analysis of a series of observations of the measurand. Type B is an evaluation of other information about the measurement. These two types of evaluations provide what is called the "combined standard uncertainty."

This chapter includes an introduction to these concepts by listing a number of definitions found in the 1993 report by the International Organization for Standardization (ISO). Points that always need reinforcement are also discussed. These include the practices of rounding and the display of significant figures. All of these subjects will be discussed with examples drawn from the measurement of air temperature at 1.5 m above the ground surface.

Background

An effort to find an international consensus on the expression of uncertainty in measurements began in 1978. The world's highest authority in metrology (measurement science, not atmospheric science), the Comité International des Poids et Mesures (CIPM) asked the Bureau International des Poids et Mesures (BIPM) to make recommendations. After consulting the national standards laboratories of 21 countries, a working group was established. This working group assigned the responsibility to the ISO Technical Advisory Group on Metrology (TAG 4). It, in turn, established Working Group 3 (ISO/TAG 4/WG 3) composed of experts nominated by BIPM IEC (International Electrotechnical Commission), ISO, OIML (International Organization of Legal Metrology), and appointed by the chairman of TAG 4.

Working Group 3 was assigned the following terms of reference:

> To develop a guidance document based upon the recommendations of the BIPM Working Group on the Statement of Uncertainties which provide rules on the expression of measurement uncertainty for use within standardization, calibration, laboratory accreditation, and metrology services:
>
> The purpose of such guidance is
>
> - to promote full information on how uncertainty statements are arrived at;
> - to provide a basis for the international comparison of measurement results.

ISO (1993) is the current guide from WG 3.

Definitions

The first paragraph of the introduction to the *Guide* (ISO, 1993) is reproduced to explain the purpose of the *Guide.*

> When reporting the result of a measurement of a physical quantity, it is obligatory that some quantitative indication of the quality of the result be given so that those who use it can assess its reliability. Without such an indication, measurement results cannot be compared, either among themselves or with reference values given in a specification or standard. It is therefore necessary that there be readily implemented, easily understood, and generally accepted procedures for characterizing the quality of a result of a measurement, that is, for evaluating and expressing its *uncertainty.*

In the scope of the *Guide* (ISO, 1993) is the following:

> This document, hereafter called the *Guide*, establishes general rules for evaluating and expressing uncertainty in physical measurement that can be followed at various levels of accuracy and in many fields—from the shop floor to fundamental research. Therefore, the principles of this *Guide* are intended to be applicable to a broad spectrum of measurements, including those required for:

- maintaining quality control and quality assurance in production;
- complying with and enforcing laws and regulations;
- conducting basic research, and applied research and development, in science and engineering;
- calibrating standards and instruments and performing tests throughout a national measurement system in order to achieve traceability to national standards; and

- developing, maintaining, and comparing international and national physical reference standards, including reference materials.

Annex B of the *Guide* (ISO, 1993) contains definitions of general metrological terms that will be repeated here, with air temperature examples where appropriate.

B.1 (measurable) **quantity**
attribute of a phenomenon, body or substance that may be distinguished qualitatively and determined quantitatively

EXAMPLE: the motion of air molecules interpreted as temperature.

B.2 **value** (of a quantity)
magnitude of a specific quantity generally expressed as a unit of measurement multiplied by a number

EXAMPLE: the temperature of the boiling point of water (at 1 atmosphere) is 100 °C.

B.3 **true value** (of a quantity)
value perfectly consistent with the definition of a given specific quantity

EXAMPLE: the theoretical absolute zero of 0 Kelvins.

B.4 **conventional true value** (of a quantity)

EXAMPLE: the temperature of air when there is no molecular motion is -273.15 ± 0.01°C.

B.5 **measurement**
set of operations having the object of determining a value of a quantity

EXAMPLE: the function of a temperature measurement system or an observer with a thermometer taking a reading.

B.6 **principle of measurement**
scientific basis of a method of measurement

EXAMPLE: the thermoelectric effect applied to the measurement of temperature.

B.7 **method of measurement**
logical sequence of operations, in generic terms, used in the performance of measurements according to a given principle

EXAMPLE: the resistance of a wire, exposed in an aspirated radiation shield, determined by a substitution method.

B.8 **measurement procedure**
set of operations, in specific terms, used in the performance of particular measurements according to a given method

EXAMPLE: a standard operating procedure used by an operator to obtain a measurement or manage an automatic measurement system.

B.9 **measurement process**
all the information, equipment and operations relevant to a given measurement.

(NOTE—This concept embraces all aspects relating to the performance and quality of the measurement; it includes for example the principle, method, procedure, values of the influence quantities, and the measurement standards.)

B.10 **measurand**
specific quantity subject to measurement

EXAMPLE: The temperature of the air at a specific location, often called a variable.

B.11 **influence quantity**
quantity that is not included in the specification of the measurand but that nonetheless affects the result of the measurement

EXAMPLE: the effect of solar or long-wave radiation, wind speed, water (in any phase), insects, environmental influence on measurement circuits, and the condition of any moving parts and connections.

B.12 **result of a measurement**
value attributed to a measurand, obtained by measurement

(NOTES—1. When the term "result of a measurement" is used, it should be made clear whether it refers to: the indication; the uncorrected result; the corrected result; and whether several values are averaged. 2. A complete statement of the result of a measurement includes information about the uncertainty of measurement.)

B.13 **uncorrected result**
result of a measurement before correction for the assumed systematic error

B.14 **corrected result**
result of a measurement after correction for the assumed systematic error

B.15 **accuracy of measurement**
closeness of the agreement between the result of a measurement and a true value of the measurand

(NOTES—1. "Accuracy" is a qualitative concept. 2. The term "precision" should not be used for "accuracy.")

B.16 **repeatability** (of results of measurement)
closeness of the agreement between the results of successive measurements of the same measurand carried out subject to all the following conditions:

- the same measurement procedure
- the same observer
- the same measuring instrument, used under the same conditions
- the same location
- repetition over a short period of time

B.17 **reproducibility** (of results of measurement)
closeness of the agreement between the results of successive measurements of the same measurand, where the measurements are carried out under changed conditions such as:

- principle or method of measurement
- observer
- measuring instrument
- location
- conditions of use
- time

B.18 **uncertainty** (of measurement)
a parameter, associated with the result of a measurement, that characterizes the dispersion of the values that could reasonably be attributed to the measurand

> (NOTES—1. The parameter may be, for example, a standard deviation (or a given multiple of it), or the half-width of an interval having a stated level of confidence. 2. Uncertainty of measurement comprises, in general, many components. Some of these components may be evaluated from the statistical distribution of the results of series of measurements and can be characterized by experimental standard deviations. The other components, which can also be characterized by standard deviations, are evaluated from assumed probability distributions based on experience or other information. 3. It is understood that all components of uncertainty, including those arising from systematic effects, such as components associated with corrections and reference standards, contribute to the dispersion.)

B.19 **error** (of measurement)
result of a measurement minus a true value of the measurand

B.20 **relative error** (of measurement)
error of measurement divided by a true value of the measurand

B.21 **random error**
result of a measurement minus the mean result of a large number of repeated measurement of the same measurand

B.22 **systematic error**
mean result of a large number of repeated measurements of the same measurand minus a true value of the measurand.

B.23 **correction**
value that, added algebraically to the uncorrected result of a measurement, compensates for an assumed systematic error

B.24 **correction factor**
numerical factor by which the uncorrected result of a measurement is multiplied to compensate for an assumed systematic error

The annex to the *Guide* (ISO, 1993) provides additional examples and comments. Annex C (not discussed here) provides the basic statistical terms and concepts for making a type A evaluation of uncertainty.

Discussion

The traditional use of the term *accuracy*, among applied meteorologists, is understood to include both a (constant) bias term and a (variable) precision term. These can only be estimated from comparisons with other measurements, calibration tests at a laboratory with accepted authority, or reliance on the manufacturer's data sheet claims. These estimations do NOT include comparisons with the true value nor do they include the influence quantity.

Using the 1.5-m air temperature example, the bias term might be +0.2 °C, based on either a calibration made by the National Institute of Standards and Technology

(NIST) or a manufacturer's calibration sheet that came with the sensor. More commonly, however, the sensor calibration is a variable with calibration temperature points. It should be considered a conditional bias with the bias correction dependent upon the temperature measured.

The precision term might be quite small, perhaps $\pm 0.02\,^{\circ}\mathrm{C}$ based on the standard deviation of a series of measurements with the sensor at equilibrium in a constant temperature thermal mass. This is really the precision of the measurement of resistance with the resistance kept constant by the thermal mass. The size of this precision measurement may be an indicator of the quality of the thermal mass.

A much larger precision will be found if the test uses a collocated transfer standard in the atmosphere. Functional precision (ASTM, 1990) is the root mean square of the distribution of differences in measurements made with two identical systems. The precision will be larger because the two sensors are not in the same thermal mass. It will probably be on the order of a few tenths of a degree Celsius.

The uncertainty will contain the influence quantity. This is like saying that the accuracy estimate is not complete until all of the siting and exposure errors are considered. These are the large errors when deciding how good the measurements are with respect to a specific application.

The influence quantities for the example of air temperature at 1.5 m above the ground include the following:

- The heat gained by the sensor from solar radiation arriving directly or indirectly by conduction, convection or reflection.
- The heat lost by the sensor to the night sky by radiation or conduction.
- Wind speed and, to a lesser extent, wind direction, as a modifying condition to the heat loss of the shield. Possible effect on the forced aspiration rate.
- Cloud cover and time of day as a modifying condition to radiative heat transfer. Also as a contributor to reflected radiation.
- Water droplets in the air or on the shield as a modifier to the air sample being measured. The relative humidity is a secondary effect to evaporation rate.
- Surface condition, including snow cover, as a radiation reflector. Parallax error for manually read liquid-in-glass thermometers.

The list can go on and on. The influence quantities will be different for different measurands or variables.

Measurement instrument output values can be reported to any resolution with modern digital systems. A sample measurement value should not be reported with a resolution finer than the uncertainty of the system. A sample air temperature value should be reported to whole degrees. An average of many samples should be reported to a tenth of a degree. Digital systems should carry sufficient resolution to avoid rounding errors. Maximum or minimum values should be reported to a tenth of a degree in order to preserve relative information. The uncertainty of a measurement will be much smaller in the context of a relative inquiry as opposed to an absolute value. Local measurement networks will have a relative uncertainty of value to the network application.

A recent study (Lockhart, 1996) showed that modern automatic data systems still fail to consider the need to maintain resolution for output values. In Lockhart (1979) it was shown that fastest-mile measurements were taken in integer knots and reported for climatological applications in miles per hour. The unit conversion was made with an integer table where the value of 19 mph could not exist, along with many other values. The modern ASOS (Automated Surface Observing System) takes measurements in knots but supplies daily summary data in miles per hour. The unit conversion method does not have the resolution to allow every integer to appear. There is still no 19 mph.

4 VALUE OF COMMON SENSE

During the process of writing a revision to the *EPA Quality Assurance Handbook for Air Pollution Measurement Systems, Volume IV, Meteorological Measurements* (EPA600/4-90-003), several subjects arose that could be resolved only by appeal to common sense. In the more than 30 workshops that have followed the publication of Volume IV, one appeal that is always made is to use common sense. When all else fails, common sense is a good standard.

For example, one always finds the requirement that calibrations be "traceable" to National Bureau of Standards (NBS) [now National Institute of Standards and Technology (NIST)]. What does this mean and how does one document the compliance with the requirement? The common interpretation is that there needs to be some anemometer calibration at NIST that starts the transfer standard path. It might go to another wind tunnel where the NIST calibration is transferred to the new wind tunnel. A calibration in the new wind tunnel might be further transferred to a third wind tunnel. A series of documents could be in the file that records all these calibrations. The operator of the third wind tunnel might provide calibrations traceable to NIST.

A paper trail is not enough when there is a possibility of error in the process. Two cases come to mind. The calibration at NIST has some uncertainty (NIST). Occasionally there is an outlier. The prudent summarization of a NIST calibration is a linear regression analysis, looking critically at differences at each point. Most mechanical anemometers are linear once the nonlinear starting speeds are passed. Outliers or problems with the calibration can usually be seen with this analysis. One wind tunnel operator transferred each point of the NIST calibration to a new anemometer, even the one obvious outlier in the NIST data. This was defendable on paper but failed the common sense test.

Technology is always improving. Cup anemometers block some of the flow in a wind tunnel. How much is not well known. The amount varies with the design of the anemometer but also with the size of the wind tunnel test section. Common sense says that a cup anemometer calibrated in the large test section at NIST ($3.25\ m^2$) will have a small blockage error, probably 1% or less. When the calibrated anemometer is used in a smaller wind tunnel with a test section of $0.4\,m^2$, there will be more blockage. If the calibration method involves transferring the NIST wind speeds to the wind tunnel fan rpm and then transferring the wind tunnel fan rpm to another identical anemometer, the blockage cancels out. Common sense says you can ignore the blockage effect. If, on the other hand, the wind tunnel fan rpm is used to "calibrate" a different kind of anemometer, the relative blockage of each anemometer in the small test section must be known. Since this effect is difficult to quantify, it is common sense to use NIST transfer standards for each type of anemometer of interest.

When, in the past, wet-bulb and dry-bulb temperatures were measured with a sling psychrometer, it was a new technology. Other methods for measuring dew-point temperature and relative humidity allowed for difference comparisons to be made. Then it became clear that siting bias was a big problem for the sling psychrometers. Even if the thermometer is moving rapidly, any incident solar radiation will heat the thermometer. If shade is found, there may still be a problem with reflected radiation. Body heat and humidity can bias the reading by modifying the air if the thermometer is down wind of the operator. Understanding the measurement process and the application of common sense will minimize these errors.

A very accurate air pressure transducer can be calibrated in the laboratory but when it is installed in the atmosphere the wind effect must be considered. Static ports are now available to minimize wind pressure effects, but two or more decades ago the exposure of the pressure transducer was not considered. The transducer would be mounted in a weatherproof box, but the inside box pressure was assumed to be the same as atmospheric pressure. When the wind was blowing, there was a bias, conditional on the speed and direction of the wind, which could be several times the calibration uncertainty. The common sense in this example must be uncommon until the knowledge of such effects reaches the operational meteorologist who must make decisions about the design of the instrument systems.

There is an "official" data archive for the United States at the National Climatic Data Center (NCDC) in Asheville, North Carolina. If one needs a copy of the "official" data for some purpose, often related to some lawsuit, one can get it from NCDC. It comes designated as official and, no doubt, judges and juries are duly impressed. It is true that the copy is certified to be correct by the head of NCDC, but this does not say anything about the accuracy of the data. The National Weather Service is responsible for the accuracy of the measurements it makes, while NCDC is responsible only for faithfully recording the NWS numbers and copying them when required (although NCDC does perform certain quality control checks for continuity, etc.)

EPA and National Research Council (NRC) list the performance specifications required for measurement systems used on projects under their authority. Performance audits are usually required on a periodic schedule to verify that the systems meet those specifications. The auditor may use auditing methods designed to document conformance. For wind speed there are two tests. One is starting threshold, measuring the starting torque of the shaft and bearings with the cup wheel or propeller removed. Bearings will degrade over time. The time is a function of the exposure environment. There is a starting torque that has been shown to be equivalent of 0.5 m/s. The audit will document the starting torque expressed as a speed. If the result is 0.6 m/s and the requirement is 0.5 m/s, are all of the speed data rejected? If the last audit was 6 months ago, perhaps only the last 30 days of data will be rejected since the bearings will degrade with time. The application of common sense by the auditor, operator, and regulator will result in keeping all the data. The bearings would be changed, but the performance of the anemometer was probably acceptable. The wind in the atmosphere does not start at steady slow speeds as the wind tunnel test requires. The simulation of starting speed by starting torque has some uncertainty.

If the audit showed a starting speed of 2 m/s, the answer becomes more difficult. The operator should examine the speed data for the 6 months or year since the last audit where the starting speed was shown to be 0.4 m/s. Perhaps the record will show a period where something happened to the anemometer. If the site is windy, there may not be many periods with winds less than 2 m/s, in which case a higher starting speed would not degrade the data. Common sense and critical examination will suggest an answer to which the regulator and operator will agree. The last answer to accept is the auditor rejecting all the speed data because of the starting torque test results.

The other auditing method for anemometers is imposing a series of known rates of rotation to the anemometer shaft. This challenges the ability of the measurement system to sense the rate of rotation of the shaft and express this rate in terms of wind speed at the output of the system. When the sensor output is a frequency and the data logger is digital, the test will always pass. The only thing being challenged is the ability of the system to count pulses and apply an algorithm to express frequency as speed. There is nothing in this test that confirms the algorithm of wind speed to frequency. This takes a wind tunnel test or the acceptance of the manufacturer's claim that the generic transfer function for the product is correct.

When dealing with regulators, operators, and consultants, a common sense discussion will usually result in an acceptable solution. What is even more important is that it will result in the exchange of information that leaves all parties more experienced and better prepared for the next question.

REFERENCES

ASTM (1990). Standard Practice for Determining the Operational Comparability of Meteorological Measurements, ASTM, West Conshohocken, PA.

ISO (1993). Guide to the Expression of Uncertainty in Measurement (International Organization for Standardization, Geneva, Switzerland), ISO/TAG 4 N 70 Rev. January.

Lockhart, T. J. (1979). Climate without 19 mph, *Bull. Am. Meteor. Soc.*, **60**, 660–661.

Lockhart, T. J. (1996). Wind climate data continuity study—II.

National Institute of Standards and Technology (1994). Guidelines for Evaluating and Expressing the Uncertainity of NIST Measurement Results, NIST Technical Note 1297,

Barry N. Taylor and Chris E. Kuyattt (Ed.), 12th International Conference on IIPS for Meteorology, Oceanography, and Hydrology, Atlanta, GA , Jan. 28–Feb. 2.

Schneider, S. (1987). *Discover* October, p. 47.

Wieringa, J. (1992). *J. Wind Engr. Ind. Aerodyn.* **41**, 357–368.

CHAPTER 35

MEASUREMENT IN THE ATMOSPHERE

JOHN HALLETT

The atmosphere is a turbulent medium. Any measurement is to be interpreted in the context of a set of measurements in a time and spatial dimension having a variance related to the scale considered. Such limits may be considered with respect to a fully developed turbulent field and limited by physical constraints—a solid boundary as at the surface, a variable boundary as with motions limited by the stability constraint as at an inversion top, or temporal as the diurnal or annual cycle. Any measurement is made with an instrument with a given time and spatial characteristic—whether it be a thermometer in sampling air flow or a satellite remotely measuring emissivity from a surface. The instrument design determines the lower limit of temporal and spatial measurement scales through the response time and the size of the instrument. Subsequent analysis determines how individual measurements are to be combined—as a mean and variance time series at a given location or a spatial field at a given time as a conventional synoptic map, with the lower scale necessarily limited through the initial instrument design. Such combinations of observations have been a central theme of understanding the different processes in the atmosphere and key in the initial development of meteorology (Fitz Roy, 1863; Brunt, 1917) and have been long realized (Middleton and Spilhaus, 1953); (*Handbook*, 1956).

We make measurements in the atmosphere for a variety of reasons. On a local scale, a measurement provides information for specific decisions, whether it be to wear a sweater, to begin harvesting a crop or to cancel an aircraft take off. On a broader scale, measurements are put together to provide information for a forecast for similar decisions at remote places. Such synthesis is required for initialization of a model of atmospheric processes whether it be to extrapolate frontal motion or to

Handbook of Weather, Climate, and Water: Dynamics, Climate, Physical Meteorology, Weather Systems, and Measurements, Edited by Thomas D. Potter and Bradley R. Colman.
ISBN 0-471-21490-6

solve numerically the equations of motion leading to such progression. Further synthesis may occur as such measurements are combined to give data to be used for climatology as a design for living, in terms of means and departures therefrom to estimate criteria for building design and for levels of investment in public utilities. Crudely, the local measurement requires a scale of meters and a time of minutes; the synoptic scale the dimensions of Earth and times of days; the climatological use requires a similar spatial scale but a time scale of a year to a decade or longer. Thus our instruments must be sufficiently small to be used locally, yet sufficiently robust and capable of calibration such that observations can be combined over extended scales of time and space.

Instrument response may be of first or second order depending on the nature of the design. Thus, a standard thermometer responds to an environmental change by approaching the new value almost exponentially and cannot overshoot. A wind vane, on the other hand, that approaches a new direction can be (and usually is) designed to overshoot so it can then approach the new direction more quickly than could be achieved if it were designed not to overshoot—by, for example, having a highly viscous damping system. Thus first-order instruments are to be designed having a time constant (lag) determined primarily by their size; a second-order instrument is designed to have a time constant determined by its size but also a period of oscillation determined separately by the damping characteristics, which also are influenced by size. The design must meanwhile ensure that any particular measurement responds solely to the quantity of interest; for example, rainfall should not influence the measurement of air temperature. The science and art of instrument design juggles these parameters to meet the needs of the way in which the data is to be used.

A further consideration lies in the spatial distribution of instruments and the frequency at which observations are made. While economic considerations may provide an upper boundary for the total number of instruments, the scale of the phenomenon of interest and its velocity of propagation should determine the optimum space distribution and frequency of measurement.

1 COMBINATION OF MEASUREMENTS

Many quantities of meteorological interest are to be derived by combination of independent measurements from separate instruments (Stankov, 1998) of the same region of interest or from regions which are to be expected on physical grounds to be highly correlated. Thus satellite measurements of ground emissivity at different wavelengths over the same area should be comparable, as should radiative properties of a given volume of particulates for identical viewing geometries. Under some circumstances instruments combine to give fluxes, as with a correlation between a temperature fluctuation (as measured by a thermocouple) and an air velocity fluctuation (as measured by a hot wire). Clearly the two instruments in the latter case are never quite collocated, and problems arise from fluctuations on the scale of the differences of location. Yet further complications arise should different instruments be of different order response (as with a wind vane and a thermocouple) and of different time lags. The question arises concerning the comparability of instruments located at a field site or on a moving platform as an aircraft. In the former case, distances of order meters may be involved, to be combined with differences in local airflow because of slightly different location geometry. In the case of aircraft, competition for space may lead to quite different locations for different instruments. Thus, air intakes for aerosol and gas measurements tend to be located at different sites on the fuselage, whereas particle measurement probes tend to be located on wing tip pylons; with different locations on the airplane, instruments are subject to different airflow geometries.

For measurement of properties of rare particles, in concentrations occurring, say, once per second in the sampling volume, it is clear that some hundred particles need to be sampled—or some hundred seconds of data need to be collected to obtain numbers meaningful, from a Poisson statistics viewpoint, at a 10% level. Thus a scale of 10 km is a practical limit of resolution. Relating this to meaningful measurements required to give insight into specific cloud processes—important, for example, for aircraft icing or the structure of cirrus—may require measurements on a scale well below 100 m (Liu et al., 2002). Hence the design of an instrument of adequate volume sample, of adequate time response and, for derived quantities, of appropriate collocation is no trivial undertaking. It is clear that instruments designed to measure *simultaneously* several properties of a given air particle sample will give more meaningful data to investigate the inhomogeneities shown by microwave studies (Mace et al., 1998).

2 PARTICLE MEASUREMENT

Particles in the atmosphere range from the smallest of aerosol some tens of nanometers in diameter in concentration in excess of $10^8\,cm^{-3}$ to large hailstones of 10 cm in diameter, in concentration less than one per $10\,m^3$. Such particles comprise precipitation, cloud, and the atmospheric aerosol itself. Measurement may characterize concentration, size distribution, and shape distribution. As an example, we examine ice crystal characteristics—spectra of habit, density, size, and concentration necessary for derivation of radiative fluxes in appropriate wavelengths and precipitation fluxes under differing dynamical conditions. A "point" measurement is idealized, but it is clear that a sample time, which differs for different size and other characteristics, needs to be specified. It is of interest to start with the pioneering work of cirrus crystal forms carried out from an open cockpit using a hand-held varnish-coated slide to give replicas for subsequent microscopic observation (Weickmann, 1947). This technique clearly showed the presence of three-dimensional bullet rosettes. Yet even in this work it is clear that an assessment of the *relative* occurrence of various crystal forms in the atmosphere—both as pristine and as complex crystal shapes—is a major undertaking because of the intrinsic variability of their nucleation and growth conditions. The early surface measurements (Bentley and Humphries, 1962; Nakaya, 1954) tended to select ideal forms as being of greater aesthetic value for sketching or photography, and later observations have similarly selected regions where crystal forms are relatively uniform. The occurrence of multiple habits and multipeaked size spectra at a point measurement is well documented in aircraft measurement and simplistically may be attributed initially to different nucleation and growth processes (Korolev et al., 2000; Bailey and Hallett, 2002), and subsequently to different fall speeds as well as the effects of mixing in lateral shear as in Kelvin–Helmholtz instability at an inversion top.

A more fundamental question needing to be addressed is the meaning of any data set of crystal shape, size, and habit distribution in a given measurement. The question of time and spatial scale of the sample is of major importance, and the detail of the averaging process is crucial as to how the data may be used. From a fundamental viewpoint, we may be interested in the nucleation and growth processes in a given volume of air that retains some coherence over growth times of interest—say some hundreds of seconds, with some hope of characterizing individual crystals over such a period. A Lagrangian observation strategy is therefore attractive, if not easily accomplished. From an applied viewpoint, crystals need to be characterized over, say, a volume of a lidar pulse some $1\,m^3$; the volume of a radar pulse some $10^6\,m^3$, the volume of a satellite footprint some $100\,km^3$. To assess precipitation from a frontal system over its precipitation history, a volume of air some $10^8\,km^3$ is more realistic; a precipitation of 1 cm over $1\,km^2$ requires some 10^{16} individual crystals. The realities of individual crystal measurement cannot compete, and the question of what is necessary for a meaningful sample arises. Surface collection and microscopy obviously gives a remarkably small sample and cannot provide a realistic sample for such a use. Electro-optical systems (PMS 2DC; PMS 2DP) give greater ease of data collection and are subject to some degree of automation, yet still provide a meager sample in relation to the above numbers. A similar consideration applies to more recent systems (Lawson et al., 1998). Some idea on variability in cirrus can be obtained on a broad scale from microwave radar (Mace et al., 1998), and it is clear that a cellular structure of order at least 100 m exists, as can readily be seen from a cursory visual inspection of any field of cirrus.

One may resort to a broader approach by assuming that in a sufficiently large volume of space, particle concentration, or indeed any other characteristic results from a combination of random events such that Shannon's maximum entropy principle applies. In this case, a Weibull distribution results (Liu and Hallett, 1998), implying that any spectrum measurement is but one of a family, and a sufficient data set may be specified to provide the "best" (most probable) distribution. It is necessary to specify a time or spatial boundary for such measurements; it may be convenient to do this on physical grounds. For example, limiting time by a well-determined effect (sea breeze/convection life time; Rossby wave transition time in Eulerian frame; a field of wave clouds, etc). Any individual measurement necessarily departs from this ideal. The reality of any particle measurement lies in the statistics of the numbers in each size bin. In general, there are fewer larger particles and an upper limit is set for realism by Poisson statistics for the large, rare particles. Recall that radar scattering relates to ΣNr^6, mass vertical flux to $\Sigma Nr^{5,4,3.5}$ depending on fall regime, mass to ΣNr^3, optical effects to ΣNr^2, particle diffusion growth rate to ΣNr, and nucleation processes to ΣN (Σ = sum over all particles). Uncertainties arise in derived quantities at different sizes depending where the size cut off in the measurements occurs for the selected sampling time and spatial average. A further consideration lies in direct measurement of properties of individual particles—such as impurity content and mean density, or density related to radius. The cloudscope class of instruments is a candidate for these measurements (Hallett et al., 1998).

From the viewpoint of an aircraft measurement, necessarily a long thin ribbon along its flight path, a longer path (space or time) to improve the statistics necessarily implies the likelihood of leaving the area (also defined in space or in time) where particles are occurring—meters or some tens of kilometers at most (arguably) for a cirrus regime or hundreds of kilometers for a field of convective storms (Fig. 1). Hence some formal definition of the *spatial and volume* scale and geometry of

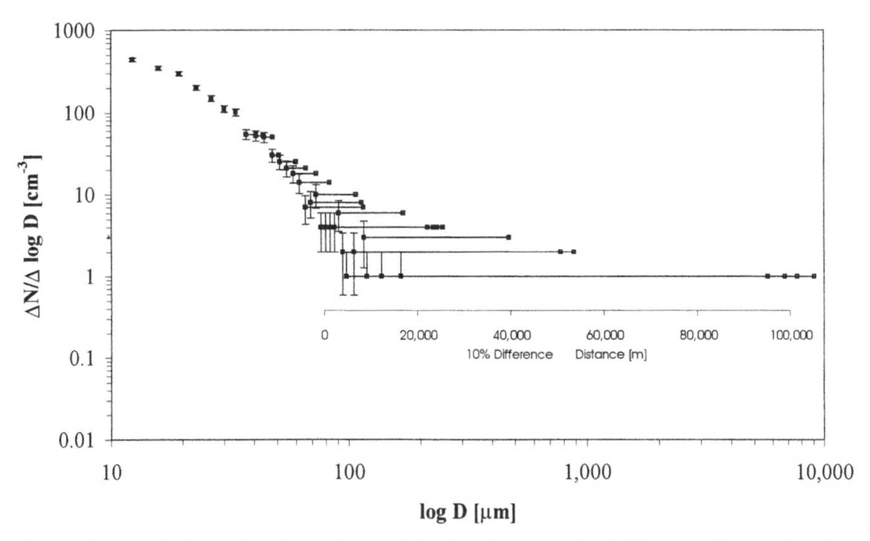

Figure 1 Analysis routine for an ice particle distribution collected by an airborne cloudscope (an instrument which images particles collected on a forward facing optical flat), sample volume about 5 liters per second. The protocol is set initially by the selection of the number of size bins on a logarithmic scale. It is further set by the level of uncertainty which can be tolerated—for example $\pm 10\%$. The uncertainty in the number of particles actually counted in each bin (N) is given by Poisson statistics as $N^{1/2}$, represented by the **vertical** error bars on each point. The uncertainty is obviously large for the small concentrations of large particles. The **horizontal** line from each point represents a flight distance necessary to sample 100 ± 10 particles (10% uncertainty) at the concentrations observed using the instrument of designated sample volume. The physical domain over which any set of measurements must be analyzed is then selectable. This could be (as this case) 10 km of a hurricane outflow for a specific size, but would be something like 1000 km for marine stratus or the whole Northern Pacific Ocean for frontal systems for larger rarer particles. In order to reduce the sampling uncertainty, it would be necessary to use an instrument with a greater sampling volume or sample for a longer time/distance in regions of interest delineated by other considerations—for example energy dissipation rate or radiance at a given wavelength. (Data collected on the NASA DC-8 in outflow of hurricane Earl approaching the Louisiana coast, September 2, 1998.)

measurement becomes imperative for synthesizing data from any set of observations, whether it be area of cloud cover or a particle distribution having the moments discussed above. It may be desirable under some circumstances to average particles by updraft location—updrafts in the upshear of convective clouds have quite different microphysical structure from the downshear, and combine the two averages for remote-sensing comparison. Other combinations may be required to achieve a sufficient approach to reality.

From the viewpoint of idealizing ice particle formation and linking the physics of the individual particle nucleation and growth with ambient conditions, specific situations may be identified. A mountain wave lenticular cloud, formed under conditions of high stability has the merit of having low turbulence levels with traceable air trajectories from conservative variables. There is a well-defined rate of change of relative humidity ahead of the cloud and rate of availability of water vapor in the cloud itself, enabling specification of growth conditions. This situation also occurs for orographic clouds in upslope flow, which can lead to continuous production of uniform crystal type and size under quasi-steady conditions as in a lenticular cloud. Conditions change along such trajectories over times of order several minutes as can be judged from observation of the location of the cloud leading edge. Shallower clouds formed in gravity waves on frontal surfaces give somewhat longer growth times. Of current interest is the ability to produce relatively low ice supersaturations that persist over a long time, giving low nucleation rates of Cloud Condensation Nuclei (CCN), which freeze homogeneously as they dilute at temperatures somewhat below -40°C. Small droplets freeze as single crystals and grow slowly under ambient conditions in the form of crystals with uniform flat facets to give well-defined and bright optical effects. The stability is very important since it maintains conditions so that a conceptional model may be put together (Hallett et al., 1999). Spectacular optical effects are reported extensively in polar regions and are known to be associated with such pristine crystals—both plates and columns. Less well reported are spectacular midlatitude summer displays at low temperatures and high levels. A way of producing a low supersaturation is a rapid overrunning of a cold layer by a warmer moister layer followed by a slow diffusion of properties from one to the other. Such effects can be idealized by a time-dependent solution of the heat/vapor diffusion equation in one dimension, with initial boundary conditions being uniform temperature and mixing ratio with an initial sharp discontinuity at the interface (Carslaw and Jaeger, 1959). Heat and vapor diffuse almost together (vapor is a little faster) and since the saturation vapor pressure of ice is near exponential with respect to $1/T$ (temperature in Kelvin), a pulse of supersaturation spreads into the cold air with time constant:

$$\frac{X^2}{D}, \quad \frac{X^2}{\kappa}$$

where X is the distance from the initial interface, D is water vapor diffusivity in air, and κ the thermal diffusivity of air. For distances of tens of centimeters, the times are of order 100 s; for meters of order a few hours. Shear can influence the boundary conditions and change these times, but maintaining stability is important to the process. This behavior results from the assumption of infinite lateral extent; in reality the discontinuity may be linear or circular, giving a decreasing supersaturation as the disturbance propagates. To maintain a crystal in growth conditions would require a special combination of progressive gravity wave having such a diffusion supersaturation field but moving with the growing crystal. This is necessarily an infrequent occurrence (the sky is not universally covered with spectacular optical events) but may be sufficiently frequent to explain the occasional occurrence of spectacular optical displays under both high cirrus and boundary layer diamond dust conditions, required for the local production of uniform crystal sizes and shapes. The detailed structure of inversions is far from well known and can be extremely sharp—with differences of 10 K sometimes extending over a meter or less. Such sharp discontinuities would not appear in any standard sounding analysis. The spatial distribution of supersaturation under these conditions is an interesting topic for further study.

The trail of defining and specifying cirrus cloud follows the process of nucleation, growth, and evaporation of individual crystals. It requires a knowledge of the dynamic and radiation conditions under which they evolve and ultimately specification on *physical grounds* of a spatial and temporal averaging domain. This domain is related to the time and spatial resolution of the instruments both for in situ and remote sensing. For multiple instruments, measurements of the *same* volume and geometry of air by the different techniques is crucial to a combination of observations. All need to be considered in any numerical comparison, a difficult but not impossible undertaking.

3 INSTRUMENTATION AND MODELING

Major advances in the atmospheric sciences often come from the development and use of specific instrumentation. The barometer and measurement of pressure led to the realization that weather systems existed. The development of telegraphy and radio communication enabled the construction of weather maps with only a few hours delay. The radiosonde, developed in the 1940s, opened our understanding of the upper troposphere, the basis for description of the weather on a global scale, and forecasting as a practical application. With the advent of major computing power (dependent on the production of fast processing chips and even smaller but more powerful memory devices by the semiconductor industry), opportunity to access very large databases has emerged, together with the ability to perform rapid calculations in the use of complex numerical models of a variety of atmospheric phenomena.

Current models of a variety of processes in Earth's atmosphere are an outcome of this technology. Such models are necessarily limited in application by our ability (or lack thereof) to provide measurements of initial and subsequent conditions over a sufficiently small grid length, both of the thermodynamic variables themselves (e.g., temperature, dew point, mixing ratio, radiative flux) and of what may be called structure-sensitive variables (cloud droplet nuclei, CCN, ice nuclei, trace NH_3, particle defect structure). The implication in the use of such models is that justification of their prediction comes from a scheme that compares their output with some set of measurements. This is called *verification* or perhaps *validation* (Hallett, 1996).

This sounds very grand, but when we look at things more closely, we find that it may be nothing of the sort. What it *is* is a check for consistency of the model against a set of measurements. It should be realized that any agreement may be coincidental and requires justification as the progenitor of the model assesses the sensitivity to both the physical assumptions together with the variability and rational for any parameterization of the measurement. This is the case for models on all scales in the atmosphere and also for the weather forecast itself.

The atmosphere is not a controlled laboratory; it is a turbulent, highly variable environment where parameterizations, sometimes overgenerous in their believability (e.g., size distributions of aerosol or precipitation for the reasons discussed above) may be no more than a figment of the modeler's imagination. Hopefully, it represents a space and time average that smoothes local effects—this is the nature of statistics. But how much larger a scale—how long a time? The conclusion is that physical and chemical concepts (still models in the generic terminology but based on our knowledge of the real world) need to guide closely what is real and what is not real in the numerical model. This may lead to quite specific measurements—not to validate the model (impossible) but to give some justification for the physical/chemical ideas on which it is based. The concept of parameterization of any atmospheric process needs to be offset against the reality of an appropriate space–time scale, whether it be mixing at the top of a cumulus or transfer of some property (heat, momentum) in a boundary layer or gravity wave.

The availability of suitable instruments is a necessary prerequisite to progress, here defined as insight into useful ideas that can be used in a predictive sense. In a completely unrealizable utopia, measurement ultimately provides a one-to-one comparison between a model and a measured reality. From a theoretical basis, self-consistency and logic necessarily have their place, but the reality is so complex that the assumptions and approximations dominate. The ideas of a theme song of an early radar conference "measure everything, everywhere, all the time" provide as much of a hindrance to progress as does the model in the absence of a prescribed set of measurements for its possible justification.

This brings us to the development of instruments for a specific purpose. A wide variety of instruments can be bought "off the shelf" for investigation of atmospheric processes—providing a sufficient market has existed to encourage their development. Thus, instrumentation is readily available, maybe for attacking yesterday's problem. But, what of today's problems? What of tomorrow's problems? This is where we need to be perceptive in development of instrumentation to optimize the future development of our subject. We need to train students not only in instrument design, construction, and use but in critical evaluation of such measurements. From a practical consideration we have a challenge to develop instruments and techniques that give us an edge for the evaluation of specific models and to enable a judgment to be made on their utility. There is an important caveat: The models themselves must be designed to be amenable to such an approach through the predication of specific instrumentation.

Thus, there is an important aspect of this concept in instrumentation evolution. If a class of instrument is to be developed with a specific output in mind, it requires input from the user whose ultimate aim is to justify an approach to a problem through physicochemical insight; it requires input from the engineer who makes the final design; and it requires input from the design of the numerical model such that there are very tight criteria for arguing that the model output makes sense in the overall scheme of things. This concept applies to the choice of sensors and their combination (as for turbulent flux measurement), it applies to the choice of wavelengths for assessing radiative flux budget both in solar and thermal infrared, and it applies to characterizing shapes and sizes of nonspherical particles for interaction with radiation fields of different wavelengths. The measurements made are to be justified in terms of the whole approach, as also is the model that is being tested for consistency.

Thus, success in new sensor technology lies in elegance of design for specific purposes by combining outputs to give meaningful quantities directly and measured *simultaneously* in the same volume of air. This is critical and predicates an integrated approach to instrumentation and model design at an early stage of any project in investigation of specific atmospheric problems. An observed consistency between model and measurement can then be used to give a better rationale both for the combination of observations and in the predictive use of the model output.

4 COMPLEXITY AND DIVERSITY

Diversity of biological species may result from the spread of ecological niches associated from a spread of microclimates. Measurements at widely separated locations miss such phenomena; remote sensing may miss such phenomena because of lack of resolution. The complexity of natural phenomena as they occur in the atmosphere (Rind, 1999) may remain hidden if it is not realized that extremes of environment may be quite unrealistically represented; similar considerations may apply elsewhere in our society (Arthur, 1999). Tails of distributions of such phenomena may be quite unsuitable for extrapolation, assuming standard distributions, as has been argued for years by the statistical community. This is certainly true for many physical processes—extremes of temperature may lead to species elimination both hot and cold. A few large drops lead to rainfall by coalescence and may or may not be present depending on the presence or absence of distinctive physical processes. Measurement of such phenomena must therefore proceed from a prior knowledge of such mechanisms, necessary to suggest the nature of the instruments to be built and used to verify (in its true sense) the physical processes responsible.

REFERENCES

Arthur, W. B. (1999). Complexity and the economy, *Science*, **284**, 107–109.

Bailey, M. and J. Hallett (2002). Nucleation effects on the habit of vapour grown ice crystals from -18 to $-42°C$, *Q. J. R. Meteorol. Soc.* **128**, 1461–1483.

Bentley, W. A., and W. J. Humphreys (1962). *Snow Crystals*, New York Dover (reprint of 1932 edition).

Brunt, D. (1917). *The Combination of Observations*, Cambridge, England, Cambridge University Press.

Carslaw, H. S., and J. C. Jaeger (1959). *Conduction of Heat in Solids*, 2nd ed., Clarendon Press, Oxford, UK.

Fitz Roy (Rear Admiral) (1863). *The Weather Book: A Manual of Practical Meteorology*, London, Longman, Green, Longman, Roberts, & Green.

Hallett, J. (1996). Instrumentation and modeling in atmospheric science, *Bull. Am. Meteorol. Soc.* **77**, 567–568.

Hallett, J., W. P. Arnott, R. Purcell, and C. Schmidt (1998). *A Technique for Characterizing Aerosol and Cloud Particles by Real Time Processing*, PM2.5: A Fine Particle Standard Proceedings of an International Specialty Conference, Sponsored by EPA Air & Waste Management Association, J. Chow, and P. Koutrakis, Eds., Vol. 1, pp. 318–325.

Hallett, J., M. P. Bailey, W. P. Arnott, and J. T. Hallett (2002). *Ice Crystals in Cirrus*, Ch. 3, pp. 41–77, New York, Cirrus, Oxford University Press.

Handbook (1956). *Handbook of Meteorological Instruments, Part 1, Instruments for Surface Observations*, Meteorological Office, Air Ministry, London, Her Majesty's Stationery Office, M.O. 577.

Korolev, A., G. A. Isaac, and J. Hallett (2000). Ice particle habits in stratiform clouds, *Q. J. R. Meteorol. Soc.* **126**, 2873–2902.

Lawson, P. R., A. V. Korolev, S. G. Cober, T. Huang, J. W. Strapp, and G. A. Isaac (1998). Improved measurements of the droplet size distribution of a freezing drizzle event, *Atmos. Res.* **47**, 181–191.

Liu, Y., and J. Hallett (1998). On size distributions of cloud droplets growing by condensation: A new conceptual model, *J. Atmos. Sci.* **55**, 527–536.

Liu, Y., P. H. Daum, and J. Hallett (2002). A generalized systems theory for the effect of varying fluctuations on cloud droplet size distributions, *Am. Meteorol. Soc.* **??**, 2279–.

Mace, G. G., K. Sassen, S. Kinne, and T. P. Ackerman (1998). An examination of cirrus cloud characteristics using data from millimeter wave radar and lidar: The 24 April SUCCESS case study, *G. Res. Lett.* **25**, 1133–1136.

Middleton, K. W. E., and A. F. Spilhaus (1953). *Meteorological Instruments*, 3rd ed., Toronto, University of Toronto Press.

Nakaya, U. (1954). *Snow Crystals, Natural and Artificial*, Cambridge, Harvard University Press.

Rind, D. (1999). Complexity and climate, *Science* **284**, 105–107.

Stankov, B. B. (1998). Multisensor retrieval of atmospheric properties, *Bull. Am. Meteorol. Soc.* **79**, 1835–1839.

Weickmann, H. K. (1947). Die Eisphase in der Atmosphare, Reports and translations, Ministry of Supply (A) Vbkenrode (H. M. Stationery Office, London, 1947, Volkrende R&T).

CHAPTER 36

INSTRUMENT DEVELOPMENT IN THE NATIONAL WEATHER SERVICE

JOSEPH W. SCHIESL

1 INTRODUCTION

Meteorological instrumentation is a very challenging arena. The first difficulty in the design of these instruments is finding a common set of requirements upon which the various users can agree. For example, hydrologists are generally happy with a precipitation resolution of 2.5 mm (0.1 in.) while meteorological modelers prefer a resolution of 0.25 mm (0.01 in.) for forecast verification.

Instruments must survive the very elements that they are measuring. There are the temperature and humidity extremes, high winds, heavy rains, ice accretion, sand and dust, solar insolation, ultraviolet radiation, salt spray, and altitude variations. Also, there are the nonmeteorological elements like corrosion, fungus, insects, birds, radio frequency interference, power line surges, pollen, and jet fuel. Many laboratory measurement technologies do not survive the passage to the outdoors with its diverse conditions.

There are many hundreds of instruments that have been used by the National Weather Service (NWS) and its predecessor organizations since 1870. Some were short lived while others that existed before the NWS are still in use. The instruments that emerged with the more interesting histories are those that measure the meteorological parameters in which the public is most interested, i.e., temperature, precipitation, and wind. The public wants to know if they have to dress for hot or cold, do they need an umbrella or boots, or will the wind blow their hats off or tip the trashcans?

Not all the instruments will be addressed, but those discussed will capture the flavor of the challenge of the times. There are many books available for those

Handbook of Weather, Climate, and Water: Dynamics, Climate, Physical Meteorology, Weather Systems, and Measurements, Edited by Thomas D. Potter and Bradley R. Colman.
ISBN 0-471-21490-6

wishing to know a complete history or the detailed theory of operation of meteorological instruments.

2 TEMPERATURE AND HUMIDITY

Air temperature readings have and still are made by liquid-in-glass thermometers. These simple devices are based on the principle that materials expand or contract with temperature changes. The liquids are either mercury or alcohol, depending on the range of temperatures to be measured. Mercury can only be used to −38.9°C where it reaches its freezing point.

Regardless of the sensor used, the temperature of the ambient air cannot be measured properly if the sensor is not adequately exposed. The sensor must be protected from direct and indirect solar radiation while still being exposed to a free flow of ambient air. To provide this protection, shelters or screens are constructed of wood and painted white with louvered sides for ventilation. The roof is double layered to provide air insulation. The door to the shelter faces north, in the Northern Hemisphere, to minimize the possibility of direct radiation when the door is opened. See Figures 1 and 2. Normally, temperatures measured in these shelters are representative of the ambient air. However, when the solar radiation is high and the winds are below 5 knots, the shelter temperature has been shown to be up to 3°C (5.4°F) higher than the ambient temperature. To overcome most of this bias, some shelters are aspirated by a fan rather than depending on natural aspiration.

Liquid-in-glass thermometers do not provide continuous temperature records, so the possibility of capturing the maximum and minimum temperature is very remote. A maximum thermometer is made by placing a constriction in the base of the bore

Figure 1 Cotton Region temperature shelter. See ftp site for color image.

Figure 2 Cotton Region temperature shelter with door open. See ftp site for color image.

just above the bulb. A temperature increase causes the liquid to rise, but the force of gravity is insufficient to allow the liquid to subside. After the maximum temperature is read, the mercury is shaken down below the constriction. See Figure 3.

In the minimum thermometer, a glass barbell-shaped object is placed in the alcohol column. The thermometer is placed in a near horizontal position with the bulb side down. When the alcohol contracts with decreasing temperatures, the meniscus pushes the barbell down. The top of the barbell stays at the minimum temperature position when the alcohol expands. After the minimum temperature is read, the barbell is tipped back to the meniscus. The minimum thermometer can also

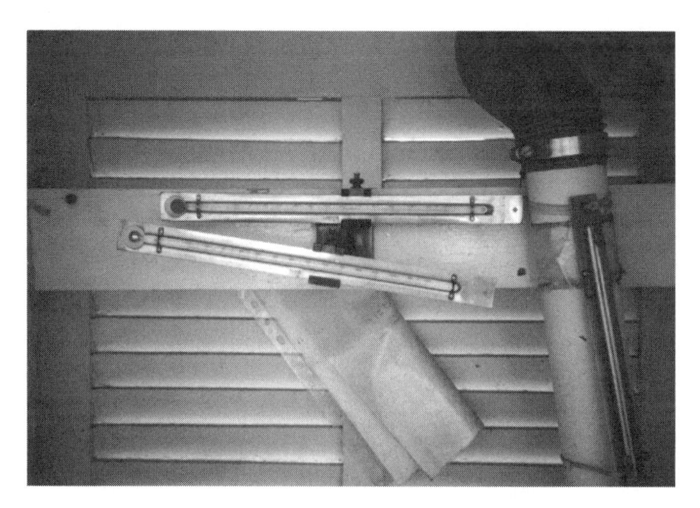

Figure 3 Maximum (bottom) and minimum (top) thermometers in Townsend support. See ftp site for color image.

be used to obtain the current temperature without disturbing its primary function. See Figure 3.

Other methods were used to record temperatures. One consisted of two bonded strips of metal with differing coefficients of expansion. Temperature changes caused the bonded strip to deform because of the different expansion coefficients. This motion was transformed by levers to make an ink trace on a chart called a thermograph.

In the nineteenth century, scientists were aware that the resistance of a material was a function of its temperature. This relationship was used in reverse to determine temperature. Electrical thermometers in which the resistance of the element increases with the temperature are known as resistance temperature devices. Those in which the resistance of the element decreases with increasing temperature are called thermistors. The small size of the electrical thermometer elements, usually less than an inch in length, allows them to have quick response times.

Measuring the amount of water vapor in the air accurately, over a wide range of temperatures outdoors, is one of the most challenging of meteorological parameters. Some terms used to express this measurement are wet-bulb temperature, dew-point temperature, vapor pressure, specific humidity, and relative humidity. The instruments used to measure water-vapor content are called hygrometers.

Many materials with which the public is familiar are coarse versions of hygrometers, although the specific values of moisture content are not apparent with them. Many materials expand with increasing humidity, and this is exhibited by sticking doors and windows. Certain fibers expand significantly with humidity, which caused problems in the non-air-conditioned buildings of the garment industry at the turn of the century. As the sewing threads expanded in high humidity, a point was reached where the thread could not pass through the sewing machine needle eye easily, and the thread snagged and broke. Humidity readings were taken to determine the proper needle eye size for that condition. This expansion property was used in one of the first hygrometers to measure humidity. The material used was human hair. As strands of hair changed in length with humidity changes, their movement was exaggerated with levers causing an inked pen arm to move. This record of humidity on a clock-driven drum was called a hygrograph. The problem with this instrument was that each hygrograph had to be calibrated individually and often, depending on the type of hair. Corrections also had to be made to separate the expansion caused by temperature versus that caused by the humidity.

One of the more common manual instruments used to measure the amount of water vapor in the air is the sling psychrometer. It consists of two liquid-in-glass thermometers mounted on a metal strip. One of the thermometer bulbs is wrapped in muslin. The muslin is dipped in distilled water and both thermometers on the strip are whirled, which evaporates water from the wetted bulb. The temperature decreases to a reading that will remain constant, even with further evaporation. This temperature is called the wet-bulb temperature. The amount of heat given up by the bulb is a function of the water vapor in the ambient air. If the dry-bulb and the wet-bulb temperature are the same after whirling, the air is saturated and the humidity is 100%. The difference in temperatures is called the wet-bulb depression. With psychrometric tables, the dry-bulb and wet-bulb temperatures can be used to determine the dew-point temperature, the humidity, and the amount of water vapor in the air per unit volume. Variations of this instrument were used, where the aspiration was provided by a motor-driven fan rather than by hand slinging.

The NWS, in addition to using the sling psychrometer to determine temperature and wet bulb, uses another instrument to measure both the temperature and the dew point. It is called the hygrothermometer and is in use at hundreds of airports across the United States. See Figure 4. The data are used in support of aircraft operations at airports and by meteorologists for forecasting. These systems were first deployed in the 1960s.

The dry-bulb temperature sensor is a resistance temperature device. The dew point is determined by a resistance temperature device imbedded in a mirror. A light-emitting diode provides a light source, and the dew point is determined by the ratio of the light received by two phototransistors. One measures the direct reflectivity from the source light, while the other measures the indirect. The heating and cooling of the mirror is effected by a Peltier device. See Figure 5. As the mirror is cooled, the formation of dew/frost on the mirror will cause an increase in the indirect sensor level. Cooling continues until the indirect sensor level is about 80% of the direct sensor level and the servo-loop equilibrium is achieved. This is the dew-point temperature. The device electronically compensates for a loss of mirror reflectivity as contaminates accumulate on the mirror surface. Another feature of this system is a circuit used to detect the reduction of flow through the instrument. This reduction can be caused by blockage (leaves and insects) or complete or partial failure of the fan. Specific mirror-cleaning procedures are critical to its proper operation. These procedures involve the use of isopropyl alcohol, lacquer thinner, and distilled water. The criticality of following specified procedures was emphasized, when a nonprescribed brand of cotton-tipped swabs was used to apply these

Figure 4 Hygrothermometer enclosure. See ftp site for color image.

Figure 5 Hygrothermometer interior showing intake air screen. Finned object on bottom is Peltier device. See ftp site for color image.

chemicals. The adhesive used in the substitute brand was dissolved by the lacquer thinner, which in turn contaminated the mirror.

The hygrothermometer sensors are housed in an enclosure to protect them from elements like precipitation, solar radiation, dirt, and insects. Any of these could cause false readings. The shelter also includes a fan that ensures a free flow of air over the sensor. Even with these precautions, the shelter is subject to small variations in temperature caused by solar radiation that varies from day to day and from place to place. Varying wind speeds also affect the amount of airflow through the shelter, which can cause temperature changes. In addition, the electronic components can cause a small calibration drift. Because of these factors affecting the temperature readings, the hygrothermometers are checked weekly using liquid-in-glass thermometers.

Even though these instruments at airports are not meant to meet climatological requirements (there are separate networks for those), these data are used as an index of change for the local area. In addition, airports are notoriously bad for climatological measurements because of the heating effects of runways, roads, buildings, and moving aircraft.

3 TEMPERATURE SITING STANDARDS

Temperature sensors should be located over terrain (earth or sod) that is typical of the area around the station. Unfortunately, some thermometers are located on rooftops for security purposes. Rooftops are not desirable locations for measuring air temperature. The siting problem is the main source of the bank thermometer phenomenon, whereby surface air temperatures are frequently reported too warm.

Because the sensing element of a thermometer absorbs more radiation when exposed to the sun, than the air itself, it must be shielded. The shielding also protects the sensing element from precipitation, dirt, and insects, any of which could cause a false reading.

The preferred temperature sensor height is about 2 m (5 ft) above the ground (eye level). This height needs to be adjusted in areas that accumulate deep snow cover. Sensors located too close to the ground will read too low for minimum temperatures and, conversely, maximum temperatures will read too high during the day. Sensors mounted on towers may be unrepresentative because temperatures can vary greatly in the lowest levels of the atmosphere near the ground. They can be especially unrepresentative during clear, calm mornings with inversions, where temperatures can vary 8°C (14.4°F) or more in the first 60 m (200 ft) above the ground.

Temperature sensors should be installed in a position to ensure that measurements are representative of the free air circulating in the locality and not influenced by artificial conditions such as large buildings, cooling towers, or expanses of concrete and tarmac.

4 PRECIPITATION

Many centuries ago, the people of the Middle East measured precipitation with buckets and measuring sticks. The data were used to levy taxes since precipitation was associated with agricultural output. Although the number of measuring techniques has increased, the simple bucket and measuring stick prevails in numbers over any other methodology used throughout the world.

The NWS uses the 20.3-cm (8-in.) (size of the orifice) precipitation gauge for manual observations. See Figure 6. The precipitation falling through the orifice is

Figure 6 Manual precipitation gauge, 20.3-cm diameter. See ftp site for color image.

Figure 7 Tipping bucket rain gauge. See ftp site for color image.

funneled into an inner tube whose cross-sectional area is one tenth that of the outer can. This ratio provides the magnification of the measuring resolution by a factor of 10. If the precipitation exceeds the total capacity of the measuring tube (2 in.), it spills over into the overflow can and must be poured into the receiving tube to be measured. For solid precipitation, the collector funnel and receiving tube are removed and the precipitation is caught in the overflow can. The catch is then melted to determine the water equivalent of the precipitation. This gauge also serves as the standard to which other NWS gauges are compared.

Unfortunately, such simplicity could not serve all of the NWS's requirements. In the latter part of the nineteenth century, the tipping bucket gauge was developed to record rainfall rates and amounts. See Figure 7. Rainfall was funneled from a 12-in.-diameter orifice through a small spout to a tipping bucket mechanism. This mechanism consisted of two buckets. The particular bucket under the spout filled to capacity, lost its balance, and tipped. As it tipped, it activated a mechanical counter to record the total rainfall. See Figure 8. In later years, the tip caused the closure of a mercury switch, which sent an electronic impulse to an electronic counter. With the concern over the potential hazard of mercury, the mercury switch is being replaced with a reed switch. The number of tips recorded over a period of time determined the rainfall rate. The rainfall amount was recorded in a collector tube beneath the gauge. The need to separate the means of measurement of rate and total amount is caused by splashing at the tipping mechanism during higher rainfall rates.

Another limitation of the tipping bucket gauge is that it does not work accurately with solid precipitation. A heated version was tried, but the heat caused further inaccuracies by causing evaporation and, in some cases, a thermal plume that prevented all the snowfall from falling into the gauge.

To catch all types of precipitation, the first of a number of weighing gauges was developed. See Figures 9 and 10. This "universal" gauge caught precipitation falling

Figure 8 Tipping bucket mechanism. See ftp site for color image.

through a 20.3-cm (8-in.) orifice and was directed into a galvanized bucket on the platform of a spring-scale weighing mechanism. The weight of the precipitation depressed a scale and through a mechanical linkage deflected a pen arm across a clock-driven rotating-paper chart. The most commonly used version of this gauge has a capacity of 12 in. of liquid precipitation and a 24-h recording period. For increased resolution, not accuracy, the recording is expanded across a dual traverse of the pen arm. During the winter season when snow, ice, and freezing temperatures are likely, the gauge is winterized with an antifreezing solution to prevent damage to the measuring bucket. Other precipitation gauges have been developed that measure

Figure 9 Universal precipitation gauge with wind shield. See ftp site for color image.

Figure 10 Weighing bucket in universal gauge. See ftp site for color image.

the amount of precipitation collected in a container. The amount is determined by technologies such as strain gauges, shaft encoders, and load cells.

5 PRECIPITATION STANDARDS

The gauge should be mounted so the orifice is in a horizontal plane. The height of the orifice should be as close to the ground as practicable. In determining the height of the orifice, consideration must be given to keeping the orifice above accumulated/drifting snow and minimizing the potential for splashing into the orifice. The immediate surrounding ground can be covered with short grass or be of gravel composition, but a hard flat surface, such as concrete, gives rise to splashing and should be avoided.

The catch of a precipitation gauge should represent the precipitation falling at that point. This is not always attainable because of the effect of wind, and thus care should be exercised in the selection of a precipitation gauge site to minimize the wind effects. Towers and rooftops are poor locations for precipitation gauges.

Precipitation gauges should be located on level ground at a distance from any object of a minimum of two, and preferably four, times the height of the object above the top of the gauge. An object is considered an obstruction if the included lateral angle from the sensor to the ends of the object is 10° or more. Beyond this range, objects that individually, or in groups, reduce the prevailing wind speed, the turbulence, and eddy currents in the vicinity of the gauge may provide a more accurate catch. Thus, the best exposures are often found in orchards, openings in groves of trees, bushes and shrubbery, or where fences and other objects together form effective windbreaks. Rain gauges should never be installed on roofs or towers because of increasing winds with height and the presence of eddy currents.

In order to reduce losses caused by wind, a windshield is recommended to be installed on gauges where 20% or more of the annual average precipitation (water equivalent) falls as snow.

6 WIND

From earliest times, attempts have been made to measure the effect of the speed of the wind. The wind force was referenced to the effect it had on objects. Without realizing it, people of old used a form of the Beaufort wind scale without the Beaufort number or the speed of the wind. The wind direction was something more obvious even without instruments. References to wind direction can be found in the early books of the Bible.

Somewhere in time, an early scientist must have tried to associate the revolutions of objects like a windmill, in a given amount of time, as a measure of wind speed. In the seventeenth century, seamen estimated the wind by the angular movement, from the vertical, of a flat plate that rotated on a horizontal bar. This was followed by the mounting of objects on rotating wheels to catch the wind. Some of the wind catchers were made of sail cloth, and different shaped wooden and metal disks.

By the time the federal government got into the weather business, the rotating cup anemometer was the most prominent method to measure wind speed. For many decades, meteorologists experimented with various anemometers. It was then known that the rotational speed of the cups was a function of wind speed and the density, viscosity, and turbulence of the air. What was also important was the diameter of the cups, cup shape, the number of cups, arm length, the moment of inertia of the system, and the effect of precipitation. By the late 1920s the four-cup, 10-cm (4-in.) diameter, hemispherical-shaped cups were replaced by the three-cup, 12.5-cm (5-in.) diameter, hemispherical-shaped cups. A few years later, the hemispherical-shaped cups were replaced by semiconical cups. See Figure 11. The semiconical cups did not overestimate gusty winds as much as the hemispherical. Throughout these transitions, the NWS made correction tables available so the speeds of the different systems could be compared. Other than for the methods of recording wind speed and direction, this system has remained essentially the same.

Early wind recorders were composed of worm and toothed gears that indicated the passage of a mile of wind. Subsequent to this, electrical contacts were used to measure miles of wind. These miles of wind were indicated by buzzers, blinking lights, or marks on a chart. Wind speed could then be determined by the amount of time it took between consecutive contacts. To improve the resolution of wind speed measurement, contacts were eventually made for every 1/60th of a mile.

This system was eventually replaced by a direct reading anemometer. See Figure 12. It contained a magneto or small electric generator using a permanent magnet. A spindle is connected to the armature of the magneto. The revolutions per minute of the spindle, caused by the rotating cups, determine the amount of electrical current generated by the magneto. See Figure 13.

Figure 11 Rotating cup anemometer and wind direction sensor atop 10-m tower. See ftp site for color image.

In the 1980s, the internal electronics were changed. Light from a light-emitting diode was pulsed by slits in a disk that rotated around the spindle in the anemometer. This "light chopper" replaced the magneto and provided improvements. One such improvement was reducing the starting torque with the removal of the magneto. Another improvement was the elimination of questions regarding time constants with the generator system. Time constants with the former system had to be estimated because the dial indicator and gust recorder have time constants determined by inertia. The output of the light chopper is simply the number of light counts per second.

Figure 12 Wind speed and wind direction indicators. See ftp site for color image.

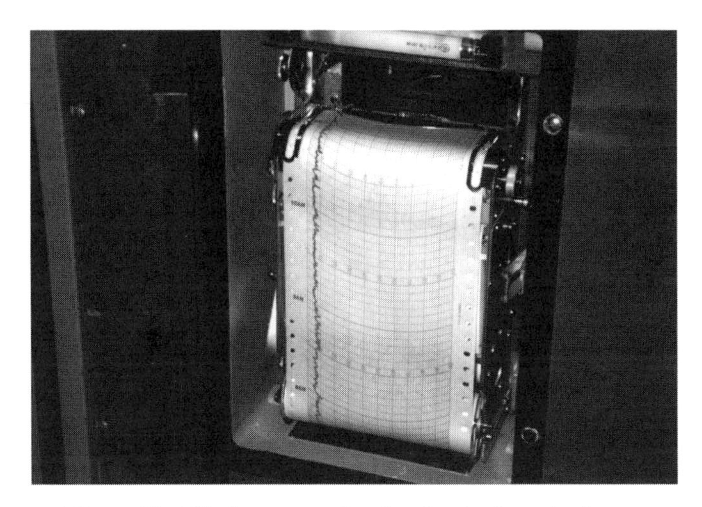

Figure 13 Wind gust recorder. See ftp site for color image.

Wind sensors should be oriented to true north. The site should be relatively level, but small gradual slopes are acceptable. The sensors should be mounted on a freely exposed tower, 10 m (30 to 33 ft) above the average ground height within a radius of 150 m (500 ft).

7 PRESSURE

In the seventeenth century, Torricelli balanced the pressure exerted by the atmosphere against the weight of a column of mercury with a vacuum above the mercury column. The height of this column was approximately 760 mm (30 in.). This is basically the same technology still being used today. These terms of height are still being substituted for pressure units in certain applications, such as aviation. Synoptic pressure analyses use the hectopascal, which is a recent standard change from the millibar.

The NWS has been phasing out the use of the Fortin tube to determine atmospheric pressure. In this device, a level of mercury in a cistern is raised by turning a thumbscrew to a zero scale. The zero scale is an ivory tip. Just before the top of the mercury in the cistern touches the ivory point, the image of the tip can be seen in the surface of the mercury. When the tip and its image just touch, or the mercury is "dimpled," the proper level has been reached. The height of the mercury is then measured by a vernier scale. Corrections are made for gravity and the expansion of glass and mercury with temperature.

With the advent of aviation, pressure readings were required more often. The reading of the mercury column was not something that could be done quickly, so the aneroid barometer was used. With the aneroid, the pressure of the atmosphere is balanced against the force of springs in an evacuated chamber made of metal. Later

Figure 14 Altimeter setting indicators. See ftp site for color image.

on, the construction of the metal bellows chamber itself provided the balancing force. The motion of these chambers were exaggerated by levers to a dial. See Figure 14.

Permanent graphic records of the pressure values were provided by using a pen arm that made a trace on a clock-driven drum. This recording device was called a barograph. Aneroids had to be calibrated against the mercury barometer periodically and differences were posted on the aneroid as a correction factor.

With the concern about the dangers of exposure to mercury from a broken mercurial barometer, these barometers are being replaced by pressure transducers. In these instruments, the atmospheric pressure is still balanced against a bellows, but the pressure is measured by the force on a resonator. The frequency of oscillation of a crystal quartz resonator varies with the atmospheric pressure. Corrections are made for pressure changes caused by temperature. The NWS has found that these instruments are very stable and are being used to replace the former aneroids and the mercury reference barometers.

Pressure-sensor-derived values are of critical importance to aviation safety and operations. This is one of the parameters that cannot be sensed by humans other than by the rapid rise or fall of pressure. This is usually experienced by "ear popping" or sinus pain.

Care should be taken to ensure that pressure sensor siting is suitable and accurate. The elevations of the sensors shall be determined to the nearest foot. Pressure sensors are usually located indoors. Sensors should not be exposed to direct sunlight or in drafts from heating and cooling. With pressure tight buildings, each pressure sensor should be individually vented to the outside to avoid pressure variations due to "pumping." These pumping or pressure variations occur with the cycling of heating and air-conditioning systems.

8 CLOUD HEIGHT EQUIPMENT

The height of clouds is most important for aviation. Earliest measurement methods were by the use of ceiling balloons. Balloons of various weights, typically 10 g with a nozzle lift of 45 g or a 30-g balloon with a nozzle lift of 139 g, were inflated with helium. A watch was used to time the interval between the release of the balloon and its entry into the base of the clouds. The point of entry for layers aloft was considered as midway between the time the balloon began to fade until the time the balloon completely disappeared. With surface-based clouds, the time interval ended when the balloon completely disappeared. During the day, red balloons were used with thin clouds and black balloons were used with thicker clouds. At night, a battery-powered light was attached.

In the 1930s, the beam from a ceiling light was projected at a 45° elevation angle into the sky. The projector was rotated about the vertical axis until the light beam hit the lowest cloud. The observer then paced off the distance from the projector to where he was directly under the cloud hit. With the geometry of this scheme, the paced distance equaled the height of the cloud. It soon became obvious that it would be less time consuming to project the light vertically into the air. See Figure 15. At a specified baseline away, a clinometer was used to measure the angle between the ground and the spot of light on the cloud. Knowing the baseline length and elevation angle in this right-triangle situation made it easy to determine the height with a lookup table. Clinometers were used to determine elevation angles. This instrument had wire cross hairs at one end and a narrowed neck at the siting end. This shape gave it the name "beer bottle."

The human detector end of this scheme was later automated by the use of a photocell that scanned the vertical path until the spot of light on the cloud was observed. The projector light was modulated so the photocell would only pick up

Figure 15 Fixed-beam ceilometer. See ftp site for color image.

Figure 16 Fixed-beam ceilometer (cylindrical shaped) and rotating-beam ceilometer detector on right. See ftp site for color image.

this type of light during the daytime. The angle of the cloud hits were displayed automatically at the observers console.

The next version of cloud height indicators was the rotating beam ceilometer. As the name implies, the beam of light rotated, and the vertically looking detector measured any cloud hits directly overhead. See Figure 16. The angle of the cloud hits was displayed either on a scope or on a recorder chart. Height measurements were limited to heights no greater than 10 times the baseline. Above this ratio, the value of the tangent function increased too quickly to ensure the accuracy of a measurement.

The latest cloud height indicator is one that was put into use in 1985. This sensor sends laser pulses vertically into the atmosphere. The pulse rate varies with the temperature to maintain a constant power output. The time interval between the pulse transmission and the reflected reception determines the height of the clouds. The reporting limit of this instrument had been 3800 m (12,000 ft). Indicators are now available with a range of 7600 m (25,000 ft). See Figures 17 and 18.

9 VISIBILITY

In the 1940s, attempts were made to automate visibility observations for aircraft carriers. Because of motion problems, these attempts were not fully successful. Work continued in this area and, in 1952 a system became operational at the Newark, New Jersey, airport. This airport was located in the meadowlands adjacent to Newark Bay and experienced dense fogs. An interesting aside was the effect of the fog on the adjoining New Jersey Turnpike. Large propellers were installed at the top of poles by local authorities along the turnpike to disperse the fog, but the attempts were unsuccessful.

Figure 17 Laser beam ceilometer. See ftp site for color image.

Runway visibility measurements were important to aviation for landings and takeoffs. A fixed intensity visible beam of light was transmitted horizontally toward a receiver along a baseline of 500 ft. The height of the sensors was at 5 m (14 ft) above the centerline of the runway, the average height of an aircraft cockpit of that time period. The particulates in the air attenuated the light beam. The photons arriving at the receiver photodiode generated a small current, and an extinction coefficient was determined. This system was called a transmissometer. See Figure 19. The transmissometer readings were converted to meteorological visibility at this time. In 1955, Runway Visual Range (RVR) was implemented. In addition to the transmissometer reading, this system took into account the light setting or

Figure 18 Laser beam ceilometer showing transmitter and receptor windows. See ftp site for color image.

Figure 19 Transmissometer. See ftp site for color image.

intensity of the runway edge lights and whether it was day or night by the use of a photometer. Daytime readings were based on the attenuation of contrast, while nighttime readings were based on the attenuation of flux density. The derived visibility was expressed in increments of 200 ft below 4000 ft and increments of 500 ft above 4000 ft. The initial range of values was from 1000 ft to 6000+ ft. Transmissometer baselines were lowered to 250 ft at certain airports to permit RVR readings down to 600 ft.

A forward scatter visibility sensor was developed in the 1970s. A beam of light is emitted from a projector and is scattered forward by any particulates in the air to a receiver. See Figure 20. The axis of the projector and the receiver are placed a few

Figure 20 Forward scatter visibility sensor showing projector and receiver. See ftp site for color image.

degrees apart to prevent any direct detection of the beam of light. The light source is visible xenon light pulsed twice per second. The sampling volume is 0.75 ft^3. An extinction coefficient is determined and a sensor equivalent visibility is determined with day/night input from a photometer. Experiments were conducted using infrared light, but it was found to be inferior to visible light, especially with certain atmospheric particles such as haze.

10 CALIBRATION, MAINTENANCE, AND QUALITY CONTROL OF INSTRUMENTS

Obtaining quality meteorological sensors and locating them according to siting and monitoring standards are only the first steps in establishing a quality data acquisition network. Provisions need to be made for maintenance, calibration, and quality control.

Sensors are usually calibrated initially at the factory but sometimes need to be recalibrated at the site because of shipping and site peculiarities. The sensors also need to be recalibrated from time to time because of mechanical and electrical drifting. Recalibration intervals vary with each type of sensor. Data network managers need to know what initial calibration was done, when the sensors need periodic calibration, and have procedures in place to ensure proper calibration.

Sensors fail and maintenance philosophies vary. They range from on-site repair to replacing the failed sensor, to abandonment. Another consideration is the availability and location of maintenance, the required response time and the cost.

Sensors produce data, but procedures need to be in place to determine if the data produced are within the sensor specifications. Users will usually detect grossly bad data. However, a lack of knowledge about local climatology or subtle changes occurring with time or particular weather events, where a number of different weather conditions are possible, may result in bad data going undetected. A monitor(s) needs to quality control the data from a network to ensure its quality. If the equipment is malfunctioning, this person needs to notify the appropriate maintenance personnel.

When a network is new, the above issues may not seem important, but with the passage of time, they become the most important issues threatening data quality, continuity, and availability.

11 TELEMETRY AND AUTOMATION

The sensors described previously are part of the surface and hydrologic data networks. The most familiar parts of this network are located at airports throughout the country. Observations from these networks were usually transmitted by teletypewriter. These observations from a large-scale network were essential for forecasting and making specific point observations. They fell short, however, in providing sufficient data for the hydrologic, marine, agriculture, climatic, and fire weather programs. In the early years of the NWS, data from these off-airport networks were taken manually and relayed to a weather office either by telephone or by radio. Sometimes data took hours to reach a forecast center. There was the time interval between the observation and the phone call. More time delays were caused by busy phones at a collection center and the time needed to relay the report to a forecast center by phone, radio, or teletypewriter. In every step, there was a chance for human error.

In 1945, a three-station experimental network was deployed south of Miami to report pressure, wind speed, and direction every 3 h. The data were transmitted by a radio powered by batteries that were charged by a wind generator. Data were deciphered by the position of an information pulse between two reference pulses.

During these early years of automation, the emphasis in systems development was on telemetering hydrometeorological parameters. Systems were designed that would permit data to be transmitted from a remote sensor to a collecting office by hardwire, telephone line, or radio. See Figure 21. Sensors were designed with cams, coded discs, weighing mechanisms, potentiometers, shaft encoders, and other apparatus that would transform analog data into an electrical signal. In general, these signals had to be manually decoded at a collection center by counting beeps.

In the 1960s, the transistor era, plans were made to automate at least part of this network for three reasons. The first was to make it possible to obtain data from sites where observers were unavailable. Second, data were needed to be available at any time to make forecasts and warnings timelier. The third reason was to reduce the workload at weather offices. During weather emergencies, many hours were required to manually collect data by telephone.

An automated hydrometeorological observing system was developed using solid-state electronics. Data were collected via telephone from remote instruments connected to a collection device and powered by battery and solar cells. See Figure 22. The remote system transmitted a fixed-format, variable-length message coded in American Standard Code for Information Interchange (ASCII) format. In this same time period, the Geostationary Operational Environmental Satellites (GOES) with communications relay facilities became available. Now data could be made accessible from areas where no or unreliable telephone service was available.

With time, further improvements were made. Automatic event reporting sensors replaced sensors reporting only on a fixed schedule to improve sampling for small areas and short events. These data are also being used to provide ground truth for satellite imagery and Doppler radar. This was a big step from when the beam from the old WSR-57 radars was stopped above a rain gauge equipped with a transponder. The amount of precipitation in the gauge was displayed as blips on the radar screen.

Almost any analog or digital sensor can now be interfaced to a microprocessor-based data collection system. The first level of processing converts raw sensor data to engineering units. The second level is used to determine such things as maximums, minimums, means, variances, standard deviations, wind roses, histograms, and hydrographs. Simple or sophisticated algorithms allow transmissions to be sent following an out-of-limits reading on a specific sensor.

Figure 21 Early remote automated meteorological observing system (RAMOS), late 1960s. See ftp site for color image.

Systems such as the GOES Data Collection System also afford effective data sharing among agencies rather than what agencies would encounter using their own independent systems. The data from the approximately 14,000 remote sites in this coordinated network, not only provide data for forecasting and warning needs but are also cost-effective to the taxpayer when interagency duplication is avoided. See Figure 23.

12 IMPORTANCE OF STANDARDIZATION AND CONTINUITY

Users of meteorological data can only validly compare those data that have been collected by standard methods. This conformity makes it possible to determine meteorological patterns and how they are changing with time. Good standards

Figure 22 Automated hydrologic observing system (AHOS) with GOES radio transmitter and antenna in background. See ftp site for color image.

evolve from a process that involves many of the users with their various interests. When the needs of participants such as academia, government, and the private sector are considered, the results will be a consensus and consequently accepted and used.

Changes in instrument types, modifications, and relocations are inevitable because of economic and scientific reasons or for changes in individual user missions. These changes need to be documented or data discontinuities will occur. Users need to know the resolution, accuracy, and the temporal and spatial specifics of each data site. This is becoming more important with the many studies being conducted on the state of the environment and determining changes in it. There is increasing cooperation among data providers within certain disciplines, but generally there is a lack of understanding across these disciplines with regard to data standards, quality, and continuity. See Figures 24 and 25. Without this cooperation, we

Figure 23 Interagency automatic remote collector (ARC) using satellite communications on Macaroni Ridge in Antarctica. Weather data are used in penguin breeding studies. Hundreds of penguins in lower right background. See ftp site for color image.

Figure 24 Sensors located too close to a jet takeoff area. Windscreen around the rain gauge is damaged from jet blast. Blast also causes temperature to rise and wind gusts to increase. See ftp site for color image.

Figure 25 Sensors poorly exposed in a parking lot. See ftp site for color image.

will not be able to determine whether we are measuring changes in the environment or just variations in the data system.

BIBLIOGRAPHY

Cornick, J. C., and T. B. McKee (1993). A Comparison of Ceiling and Visibility Observations for NWS Manned Observation Sites and ASOS Sites, Atmospheric Science Paper #529, Colorado State University, Fort Collins, CO.

Federal Aviation Agency and Department of Commerce (1965). Aviation Weather for Pilots and Flight Operations Personnel, Washington, DC, U.S. Government Printing Office.

Flanders, A. F., and J. W. Schiesl (1972). Hydrologic data collection via geostationary satellite, *IEEE Trans. Geosci. Electro.* **GE-10**(1).

Kettering, C. A. (1965). Automatic Meteorological Observing System, Weather Bureau Technical Note 12-METENG-16.

National Weather Service (1995). *Federal Meteorological Handbook No. 1, Surface Observations*, FCMH-1, Washington, DC, U.S. Government Printing Office.

Schiesl, J. W. (1958). Measurement of Dynamic Strain, Needle Research Report, Singer Manufacturing Company, Elizabethport, NJ.

Schiesl, J. W. (1976). Automatic Hydrologic Observing System, Paper presented at the International Seminar on Organization and Operation of Hydrologic Services, July 1976. Ottawa.

U.S. Department of Commerce, National Oceanic and Atmospheric Administration (1994). Federal Standard for Siting Meteorological Sensors at Airports, FCM-S4-1994.

U.S. Department of Commerce, National Oceanic and Atmospheric Administration (1997). National Geostationary Operational Environmental Satellite (GOES) Data Collection System (DCS) Operations Plan, FCM-P28-1997.

U.S. Department of Commerce, Weather Bureau (1963a). *Manual of Barometry*, Washington, DC, U.S. Government Printing Office.

U.S. Department of Commerce, Weather Bureau (1963b). *History of Weather Bureau Wind Measurements*, Washington, DC, U.S. Government Printing Office.

CHAPTER 37

CONSEQUENCE OF INSTRUMENT AND SITING CHANGES

JOSEPH W. SCHIESL AND THOMAS B. McKEE

1 INTRODUCTION

In 1984, the National Weather Service (NWS) began the deployment of the H083 as the standard operational hygrothermometer in use at approximately 400 NWS and Federal Aviation Administration (FAA) field stations across the United States. The data are used in support of aircraft operations at airports and by NWS meteorologists for forecasting.

The purpose of the H083 was to measure temperatures at airports where the required accuracy was $\pm 1^{\circ}$C (1.8°F) at temperatures between -50 and 50°C (-58 and 122°F). This was also the aviation standard of the World Meteorological Organization (WMO). Even though these instruments at airports were not meant to meet climatological requirements (there are separate networks for those), some people use these data as an index of change for the local area. In addition, airports are notoriously poor for climatological measurements because of the solar absorption characteristics of runways and roads plus the heating effects of buildings and moving aircraft. See Figure 1. It is important and even critical to determine these heating factors since they are essential in computing aircraft takeoff distance.

2 EARLY MODIFICATIONS

Improvements were made to the H083 since its initial deployment to improve its accuracy, reliability, maintainability, and to correct several deficiencies that were identified after long-term exposure in a field environment.

Handbook of Weather, Climate, and Water: Dynamics, Climate, Physical Systems, and Measurements, Edited by Thomas D. Potter and Bradley R. Colman.
ISBN 0-471-21489-2

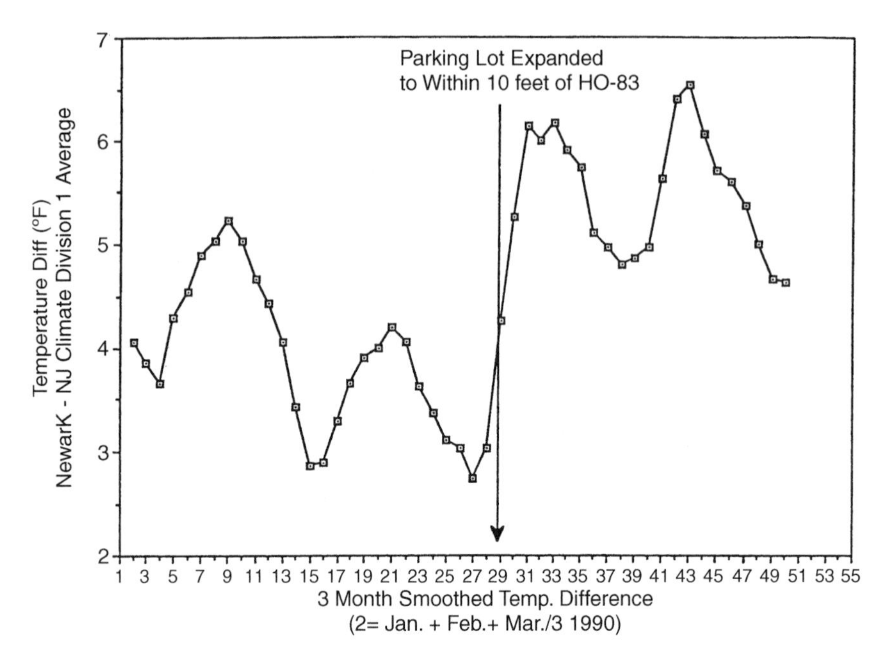

Figure 1 Temperature difference between Newark, New Jersey, airport, and surrounding climate stations. Note the change coincident with the parking lot expansion.

The improvements were concentrated in four areas: (1) an optical block, (2) board/connector corrosion resistance, (3) a new mirror, and (4) aspirator housing and aspirator improvements. See Figures 2 and 3.

The sensor assembly was constructed with an optical block, which was devised to simplify factory alignment and eliminate field misalignment of the electro-optical components during mirror cleaning. The electro-optical components of the dew-point sensor were previously soldered to a circuit board and were aligned at the factory by bending the component leads into position. These components are now rigidly mounted in an optical block assembly, which ensures precise component alignment relative to the reflective mirror surface. However, it is important that the optical loop calibration procedures be followed correctly. Electronic readings must be allowed to settle. This is easy to say but difficult to do in subzero temperatures or other inclement weather conditions.

The redesign of the new sensor assembly and the use of new fabrication techniques eliminated the formation of corrosion on the sensor board. Corrosion on the sensor assembly, especially in the vicinity of the connector, can result in electrical leakage paths on the printed circuit boards in high-humidity conditions. This leakage can lead to erroneous calibration and simulated temperature diagnostics, incorrect temperature readings, and, eventually, complete failure of the sensor. The new design featured a hand-wired, potted connector whose pigtails are soldered to plated-through holes on the sensor board. Printed wiring circuits on the boards were

Figure 2 Hygrothermometer sensor board. Pyramid shaped object is the optical block. A round mirror is beneath the optical block. Peltier device is below the mirror on the opposite side of the sensor block (See Fig. 3).

Figure 3 Peltier device on back side of sensor board. The aspirator would be located at the end of the sensor board toward the upper left in the figure.

virtually eliminated and the board was manufactured with a protective solder mask coating.

The new dew-point mirror was designed to provide proper operation with the optical block and to extend the life of the mirror surface. The mirror life was extended by improving fabrication techniques, including thorough cleaning prior to plating and using improved precision plating technology.

Consistency is critical for the components in the optical loop. Variations in the mirror, light-emitting diodes, and phototransistors cause calibration errors especially at high and low temperatures. Combinations of inconsistency, corrosion, excessive contaminants, improper cleaning, and calibration result in mirror icing.

The H083 aspirator/housing modifications were designed to minimize the effects of solar heating of the aspirator. The first modification was a "top hat" for the hemispherical part of the aspirator, which provided for an inch of expanded polystyrene insulation and additional shading of the stem. The second modification provided increased airflow through the aspirator by decreasing the airflow restrictions on the temperature side of the aspirator.

3 NEW CONCERNS

While plans were being made to do validation testing of the redesigned sensor and aspirator before deployment in the field, another concern surfaced. In July, 1989, meteorologists at the University of Arizona expressed concern about anomalous high-temperature readings at Tucson Airport. The Tucson Airport was breaking numerous records. Scientists at the National Climatic Data Center confirmed a warm bias at Tucson but also stated that they did not see this bias in other parts of the country.

At this point, NWS management established a Hygrothermometer Working Group (HWG) to investigate this problem and come up with an appropriate solution. The HWG was composed of meteorologists and engineers from the government, academia, and the manufacturer. To determine the extent of the problem, a survey of the weekly temperature comparisons around the country was conducted. The results showed that only Tucson was out of specifications, but there was definitely a warm bias in the Southwest on days with low winds and high solar radiation.

Tests conducted in the spring and summer of 1990 by NWS's Equipment Test and Evaluation Branch in Sterling, Virginia, indicated that the H083 was susceptible to solar-radiation-induced errors that could approach 1.7°C (3.0°F) under certain low wind speed conditions. While errors of this magnitude were acceptable in the H083 temperature specification as it existed at that time, steps were taken to improve the accuracy of the hygrothermometer.

The next task of the HWG was to clarify the H083 temperature specification. The original H083 requirements were for an accuracy of 0.5°C (0.9°F) root-mean-squared error (rmse) based on monthly computations, a maximum error of ±1.0°C (1.8°F), and 95% of the monthly data should be less than the specified errors. The revised H083 temperature specification was also adopted for the model 1088 hygrothermometer, also known as the Automated Surface Observing System (ASOS) hygrothermometer. The new specification, requires an accuracy of 0.5°C (0.9°F) rmse (computations based on 7 days of data) and an absolute maximum error of ±1.0°C (1.8°F). These requirements were for temperatures between −50 and 50°C (−58 and 122°F). The rmse and absolute errors were doubled for temperatures outside this range down to −62°C (−80°F) and up to 54°C (130°F). The 95% requirement was deleted, which tightened the accuracy requirements significantly.

Through the preliminary testing, all hygrothermometer testing had been done at Sterling. The HWG decided that concurrent testing should be conducted at additional field sites to determine whether the solar-radiation-induced errors seen at Sterling were representative of those that would be expected under worst-case conditions. It was decided to conduct the initial field testing at Tucson, Arizona. First, there was an interest in the hygrothermometer readings at Tucson. Second, Tucson is in a climatological area where we expected meteorological conditions to be most conducive to maximum solar-radiation-induced errors—high solar radiation with low winds that occur during portions of spring and again in late summer.

Besides Tucson, the group decided to test the modifications in the north central (Sault Ste. Marie, Michigan) and the southeastern (Daytona Beach, Florida) part of the country for the extremes of temperature, humidity, and solar radiation, both direct and indirect. Soon after this, another test site was established at the Ocean City, Maryland, airport because of its proximity to the Atlantic Ocean. This site proved valuable for accelerated corrosion testing.

The H083s that incorporated the early modifications were field tested at Sterling in late 1990 and early 1991 and were found to meet the improved H083 accuracy requirements. Based on these data, the HWG decided to validate the modifications in Tucson. A field test bed was installed in April, 1991, but it was soon evident that solar loading/wind conditions in Tucson contributed to errors that exceeded the accuracy specification.

The group then undertook the task of working to reduce the amount of solar-induced temperature measurement error. Several system modifications were made to insulate the temperature probe from solar heating and improve aspiration through the sensor assembly. Preliminary results showed a dramatic improvement in sensor performance.

4 INITIAL TUCSON TEST RESULTS

A prototype of a modified H083 sensor (mod 1) was installed in the Tucson, Arizona, test site on June 24, 1991. Mod 1 had a larger fan with approximately three times the airflow capacity of the earlier version. In addition, the flow direction was reversed. The reversal was made to minimize the heating of the ambient air by the sensor housing itself. Data, during days of high solar loading, showed the average solar-induced temperature rise to be approximately 0.6°C (1.0°F) versus the 1.1 to 1.7°C (2.0 to 3.0°F) for the early double domed insulated hat, and 1.7

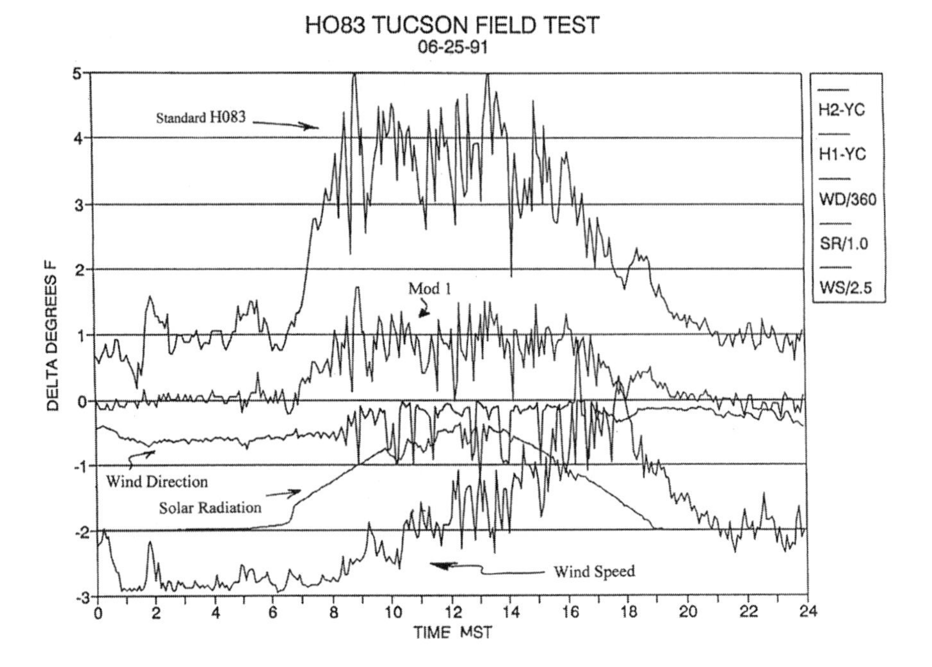

Figure 4 Comparison between standard H083 and Mod 1 at Tucson, Arizona. Mod 1 has a larger fan with approximately three times the air flow of the standard H083 and the flow direction has been reversed. Temperature deltas (°F) are in comparison to the Young reference sensor.

to 2.8°C (3.0 to 5.0°F) for the standard H083. See Figure 4. These deltas are all in comparison to the Young aspirated precision reference thermometer.

Even though good progress was made in solving the solar loading problem in the Southwest, the HWG was not in a position to firmly state that the problem was solved. The final modification had to be tested through the four seasons and in the different climatological regimes that were chosen. Since the temperature sensor was now located at the bottom of the instrument at the aspirator inlet, the group had to be certain the sensor was protected from any albedo effect either off snow or a natural reflective soil typified at Tucson, Arizona. Also, the HWG had to make sure the minimum temperatures were not affected or more importantly the dew point. Tucson, Arizona, with its temperature–dew-point depressions reaching 47°C (85°F), represents only one end of the humidity spectrum.

5 OVERALL TEST RESULTS

While the HWG was evaluating the solar loading modifications, it became apparent that there were a number of other problems with the H083. During temperature bath calibration tests, calibration errors were found that varied from −0.8 at 54°C to 1.2 at −51°C (−1.5 at 130°F to 2.1 at −60°F). It was also found during calibration stability tests that the sensor calibration drifted excessively with temperature, which would affect the actual sensor accuracy in the operational mode. These performance characteristics add a large degree of uncertainty to interpretation of H083 field performance test data. The problems were serious enough to suspend field testing during November and December of 1991.

Based on testing conducted in late December, 1991, and early January, 1992, the HWG determined that the latest modified sensor now had more stable electronics and deployed these modifications at the field test sites.

Both the standard and modified systems were installed at Tucson, Sault Ste. Marie, Daytona Beach, Ocean City, and Sterling. The HWG was able to measure not only the instrumental biases at these diverse climatological sites but also could determine why the biases occurred. This was possible because pyrheliometers, wind sensors, and reference temperature sensors were installed and carefully calibrated at each site. The effects of a partial solar eclipse were captured on both systems, which clearly demonstrated the solar loading problem. See Figure 5. These test beds were set up to determine the temperature biases between the different sensors under different wind and solar loading conditions. The resolution of the data was in minutes, rather than by the hour or the daily maximums and minimums presented in previous studies. Having a calibrated standard at each site also allowed the determination of how

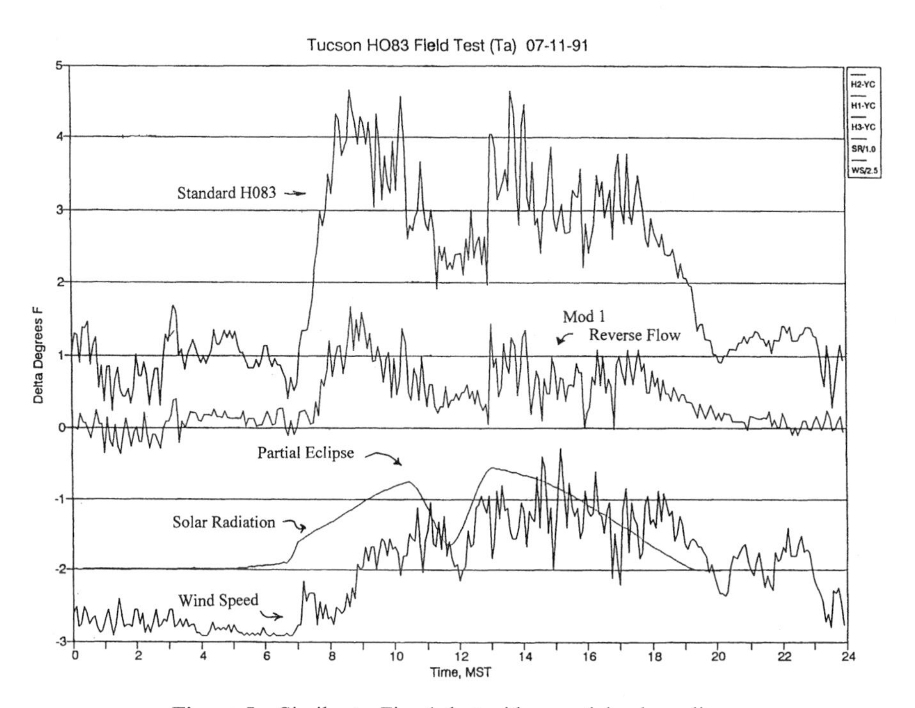

Figure 5 Similar to Fig. 4, but with a partial solar eclipse.

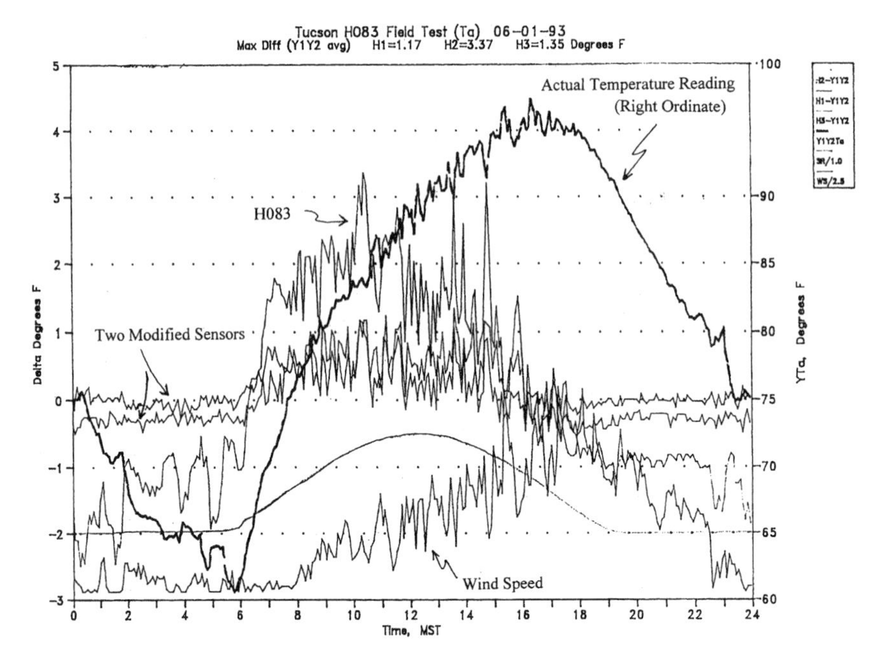

Figure 6 Similar to Fig. 4, but with two modified sensors.

much each sensor differed from an established reference, rather than just stating the difference between the standard and modified systems. See Figure 6.

6 AN OVERCONSERVATIVE DATA QUALITY CHECK

Of all the ASOS sensors, the hygrothermometer generated the most data quality failures. These failures generally indicate that the data coming from the sensor are suspicious. This suspicion can be caused by an internal self-test or the data quality algorithms. One of the processes initiated by this failure is the setting of the maintenance character ($) in all future observations that can only be cleared via the technician interface. In the case of the hygrothermometer, the data quality algorithm is the main cause of quality failures.

The ASOS hygrothermometer continually measures the ambient temperature and provides sample values about six times per minute. Processing algorithms in the hygrothermometer use these samples to determine a 1-min average temperature valid for a 60-s period ending at M+00 (minute + 00 seconds). Once each minute the 5-min average temperature is calculated from the 1-min average observations. These 5-min averages are rounded to the nearest degree Fahrenheit, converted to the nearest 0.1 degree Celsius, and reported once each minute as the current 5-min average temperature.

A number of data quality checks are performed by an acquisition control unit, including a rate of change check. Originally, if the current 1-min temperature differed from the last respective, nonmissing, 1-min reading within the previous 2 min by more than 3.3°C (6.0°F), it is marked as missing. If there are less than four valid 1-min average temperatures within the past 5 min, then the current 5-min average temperature is not computed. In this case, ASOS will use the most recent 5-min average calculated temperature within the last 15 min. If no valid 5-min average temperature is available within the last 15 min, a sensor failure is indicated. The 15-min delay allows for a once-a-day calibration heat cycle to occur without causing a data quality flag.

This initial rate turned out to be too conservative for the way the atmosphere was behaving. Many meteorologists were naturally skeptical of large temperature changes over short time periods for a number of reasons. Forty to 50 years ago, with the limited data available, it was envisioned that temperature fluctuations of more than two or three degrees Fahrenheit per minute were very rare events. High-resolution temperature data in both space and time from mesonetworks were not generally available. Thermographs were available, but their limitations will be discussed later.

Experience reinforced this conservative view of temperature changes. Most times, forecasters only saw the hourly weather reports where larger temperature rates were smoothed out. Observations in between the hourlies (specials) that were generated by significant changes or the onset or cessation of other meteorological elements did not even require the recording of temperature and dew point.

Those who examined thermograph traces saw more drastic changes over time, but these too were muted. In our efforts to protect the sensor from the elements as well as the sun or other radiating surfaces, we used thermometer screens. These lengthened the response time.

This conservative view was in contrast to some records that were available. For example, the following extraordinary events are noteworthy. There was the temporal change in the surface temperature on January 22, 1943, at Spearfish, South Dakota, from −20 to 7°C (−4 to 45°F) in 2 min, which was caused primarily by a Chinook wind. Then there were the spatial differences caused by cold air collecting in the hollows on low-wind nights. This was dramatized by the car ride by Middleton and Millar through Toronto in 1936 that showed differences of 14°C (26°F) over a mile.

Then there was the way thermometers were read. Temperatures and wet bulbs at airport stations were read to tenths of degrees primarily for the computation of the dew point and humidity. At low temperatures, a difference of only a few tenths in wet-bulb depressions may mean differences of relative humidity of about 10%. Temperatures can be measured electronically to a resolution of a thousandth of a degree; however, in meteorological applications this is meaningless.

In the late 1960s, with the advent of automated and telemetered data systems like the Automatic Hydrologic Observing System (AHOS), data were taken at 1-min intervals. More abrupt changes in hydrometeorological data became evident. Subsequent development of automated systems that also transmit data based on rate and threshold algorithms, such as the Automatic Remote Collector (ARC) in

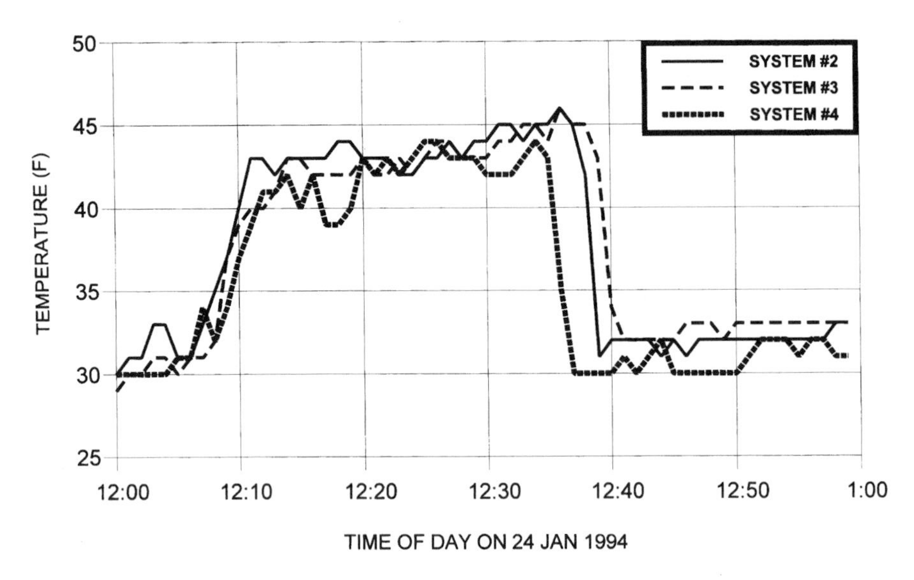

Figure 7 One-minute temperature data showing cold front passage from three noncollocated ASOS hygrothermometers at Sterling, Virginia.

the late 1970s, brought these changes to light even more. Today, these systems and ASOS provide us with very detailed data.

Figure 7 shows the passage of a cold front as recorded by three noncollocated hygrothermometers at Sterling, Virginia. Note the drastic changes in temperature over a very short period of time. All three systems logged temperature data quality errors and the sensors were flagged inoperational as a result of this temperature drop.

Based on temperature rate data from the HWG's studies over 3 years, the HWG recommended that the temperature and dew-point data quality algorithm be changed. The change was implemented in a subsequent firmware load in ASOS. Now, the quality failure does not occur until the last reading from the temperature or dew-point sensor varies from the previous reading by more than 5.6°C (10°F).

7 RELOCATION BIAS

In addition to the instrumental biases discussed so far, another factor was to affect temperature readings at most ASOS sites. ASOS sensors are typically located near the touchdown zone of the primary designated instrument runway and on occasion at a center field location. These exposures for temperatures at airports were generally better where the H083s were not already remotely located from the airport office. Some earlier station temperatures were taken on roofs or near parking lots. See Figure 8. Even where the H083 was already remotely located, the ASOS sensor was sometimes located a mile or two away from the previous location. There were

Figure 8 Sensors poorly exposed in a parking lot. See ftp site for color image.

sometimes significant exposure differences over these distances. Runways are flat but the areas around them are not always so. Suitable exposures for meteorological instruments are not always available because of the proximity to taxiways and instrument intrusion into safety regulated airspace. See Figure 9. As a result, sensors are sometimes located in swales where the temperatures are subject to cold air drainage. Regardless of the reason for the change in location, the differences needed to be accounted for.

Figure 9 Sensors located too close to a jet takeoff area. Windscreen around the rain gauge is damaged from jet blast. Blast also causes temperature to rise and wind gusts to increase. See ftp site for color image.

8 DATA CONTINUITY STUDIES

The basic dilemma was that even if the 1088 were perfect, it was now known that the H083 was not. Thus, at all sites, an account must be made for that bias given by the uncertain mixture of H083 bias and removal factors. This could only be done by leaving the H083 in place and comparing it to the modified and much improved 1088. It was obvious that the simple sling psychrometer would not suffice. If the sling is perfect and the 1088 is perfect and they are collocated, nothing is proved. For true data continuity, there was a need to develop for each station its unique bias signature.

Unfortunately, because of resource limitation and schedules, only a subset of the ASOS sites could be studied. The H083s had to be returned to the manufacturer and modified so they would be the sensor equivalent of the improved 1088. Only a limited number of the original H083s could be left in place with the ASOS 1088 version. With this restriction, the sites selected for a climate data continuity project took into account as many different climatological regimes as practical. This research was conducted by the Department of Atmospheric Science at Colorado State University in Fort Collins, Colorado, and was reported by McKee et al. (1996, 1997). Studies of data continuity began with the 15 sites given in Table 1 for data collected from June, 1994, through August, 1995. All sites had received all of the modifications described above by June, 1994. These 15 locations retained the H083 during this period so the maximum and minimum daily temperatures were available.

TABLE 1 Climate Data Continuity Study (CDCP) Comparison Sites for Daily Maximum and Minimum Temperatures

Number	Site ID	Station Name
1	AMA	Amarillo, TX
2	AST	Astoria, OR
3	BRO	Brownsville, TX
4	BTR	Baton Rouge, LA
5	COS	Colorado Springs, CO
6	DDC	Dodge City, KS
7	GLD	Goodland, KS
8	GRI	Grand Island, NE
9	ICT	Wichita, KS
10	LNK	Lincoln, NE
11	OKC	Oklahoma City, OK
12	PWM[a]	Portland, ME
13	SYR	Syracuse, NY
14	TOP	Topeka, KS
15	TUL	Tulsa, OK

[a]Station commissioned in August, 1994.

The goal of the analysis was to understand the physical causes of the observed temperature difference between the two hygrothermometers. An initial step was to confirm the absolute accuracy of the ASOS hygrothermometer. The NWS examined several of the ASOS instruments at Sterling, Virginia, and found a range of $\pm 0.20°$F. Three additional comparisons were made in the field at Colorado Springs, Colorado (COS), Oklahoma City, Oklahoma (OKC), and Tulsa, Oklahoma (TUL), which showed a range of $\pm 0.30°$F relative to a field standard (RM Young), which had been calibrated against a secondary standard at Sterling, Virginia. Thus, ASOS does not have a temperature bias. The model used to assess the temperature differences had the analytic form:

$$\Delta T = \Delta T_i + \Delta T_\lambda + \Delta T_s \qquad (1)$$

where ΔT is the observed temperature difference of ASOS–H083, and the subscripts are i (instruments), λ (local effect of location), and s (solar heating effect). All of the separations between the hygrothermometers were less than 1 mile and the hygrothermometers were collocated at four sites, which would have no local effect, by definition. Local effects due to site location could be different for day and night, various weather systems, and seasons. Solar heating was included since it was known that the H083 had a solar heating problem.

Results of the analyses are given in Table 2 for the 15-month period. Observed differences in the maximum and minimum temperatures (labeled M_x and M_n) show ASOS to be cooler by 1.17°F (M_x) and 0.86°F (M_n) averaged over all sites. Two methods were used to determine the instrument bias of the H083 (ΔT_i) assuming ASOS had no bias. The first was to examine comparisons when wind speeds were high enough to reduce local effects resulting in a narrow frequency distribution of ΔT. A speed of nearly 15 miles per hour was required, but there were not enough observations remaining to be meaningful at all sites. The second approach was to have an ASOS reported overcast sky at night, which meant cloud base was 12,000 ft or lower. Downward infrared radiation from an overcast low cloud would reduce horizontal temperature differences. In fact the frequency distribution of observed temperature differences was very narrow, which indicated the local effects were reduced. The instrument bias determined from this analysis is given in Table 2 in the column marked ΔT_i. The average ΔT_i was 0.57°F, and the range was quite large from 0.16 to 1.06°F. The ASOS at Lincoln, Nebraska, was moved midway in our data collection period. Two LNK sites are thus included. LNK-1was the first location and the ASOS instrument was moved to LNK-2, which was essentially collocated with the H083. The cloud analysis yielded the same estimate of ΔT_i from the two locations. Next the H083 bias was subtracted from the M_x and M_n observations, and the remainder for the M_n is ΔT_λ at night and for the M_x is a combination of ΔT_λ in the daytime plus the solar effect. Notice the resulting ΔT_λ from the M_x has a wide range of 0.56 to $-1.10°$F. This shows that local effects can be quite important for distances less than 1 mile. For the daytime local effect plus solar effects, several of the sites showed a magnitude of 1°F or larger indicating a likely strong solar effect, which was previously established as a problem for the H083. As a consequence of the

TABLE 2 ASOS–CONV (°F) June, 1994 Through August, 1995

Station	M_x	M_n	ΔT_i	Bias Removed M_x $(\Delta T_s + \Delta T_\lambda)$	Bias Removed M_n (ΔT_λ)	ABOS–CONV Diurnal Range
AMA	−0.76	−0.59	−0.35	−0.41	−0.24	−0.17
AST	−0.59	0.03	−0.28	−0.31	0.31	−0.62
BRO	−0.93	−0.32	−0.45	−0.48	0.13	−0.61
BTR	−1.96	−1.19	−0.89	−1.07	−0.30	−0.77
COS[a]	−1.38	−0.41	−0.16	−1.22	−0.25	−0.97
DDC	−0.48	−0.91	−0.61	0.13	−0.30	0.43
GLD	−1.16	−1.19	−0.21	−0.95	−0.98	0.03
GRI	−1.25	−0.69	−0.67	−0.58	−0.02	−0.56
ICT[a]	−0.79	−0.27	−0.29	−0.50	0.02	−0.52
LNK-1	−2.14	−2.00	−0.96	−1.18	−1.04	−0.14
LNK-2[a]	−2.38	−1.03	−0.96	−1.42	−0.07	−1.35
OKC	−0.64	−2.03	−0.93	0.29	−1.10	1.39
PWM	−0.78	−1.18	−0.47	−0.31	−0.71	0.40
SYR[a]	−0.80	−0.42	−0.29	−0.51	−0.13	−0.38
TOP	−0.48	0.06	−0.50	0.02	0.56	−0.54
TUL	−2.17	−1.55	−1.06	−1.11	−0.49	−0.62
Average	−1.17	−0.86	−0.57[b]	−0.60	−0.29	−0.31

[a]Co-located sites.
[b]Value has LNK twice.

differences for M_x and M_n, the ASOS observed diurnal range also decreased in comparison to the H083 observations. A summary of this data continuity analysis is that ASOS is a better instrument than the H083 with much less bias, less solar influence, and a new recognition that local effects are important.

9 INTERFACE WITH OUTSIDE GROUPS

The ongoing work on temperature measurement was done in coordination with groups such as the National Academy of Science, the National Center for Atmospheric Research, a few universities, and a number of state climatologists. What became apparent was that the measurement of temperature is more complex than originally thought. For instance, one reason electronic thermometers report higher temperatures is because of their quicker response time. Liquid-in-glass thermometers miss quick temperature fluctuations because of their slower response. Cotton Region temperature shelters have an even larger bias than the H083 under the low-wind, high-solar-load conditions. The climatological community is looking at these problems, and the NWS will be working with them on documenting the biases among the various sensors, climatological regimes, sensor locations, and by certain weather parameters.

Without this cooperation, unexplainable temperature shifts would be a murky blend of a number of biases caused by different factors, and certain data would go down in the books with the comment: “Data quality/continuity uncertain.”

REFERENCES

Federal Aviation Administration and Department of Commerce (1965). *Aviation Weather for Pilots and Flight Operations Personnel*, Washington, DC, U.S. Government Printing Office.

Flanders, A. F., and J. W. Schiesl (1972). Hydrologic data collection via geostationary satellite, *IEEE Trans. Geosci. Electron.* **GE-10**(1).

McKee, T. B., N. J. Doesken, and J. Kleist (1996). Climate Data Continuity with ASOS (Report for the period September 1994–March 1996), Climatology Report No. 96-1, Colorado Climate Center, Atmospheric Science Department, Fort Collins, CO, Colorado State University, March.

McKee, T. B., N. J. Doesken, J. Kleist, and N. L. Canfield (1997). Climate Data Continuity with ASOS—Temperature, Preprints, 13th International Conference on IIPS, AMS, 2–7 February, Long Beach, CA, pp. 70–73.

National Weather Service (1992). ASOS (Automated Surface Observing System) User's Guide. National Oceanic and Atmospheric Administration.

Schiesl, J. W. (1976). Automatic Hydrologic Observing System. Paper presented at the International Seminar on Organization and Operation of Hydrologic Services, July, 1976. Ottawa.

Schrumpf, A. D., and T. B. McKee (1996). Temperature Data Continuity with the Automated Surface Observing System, Atmospheric Science Paper No. 616, Climatology Report No. 96-2, Colorado State University, Fort Collins, CO.

U.S. Army Quartermaster Research and Development Command (1957). Environmental Protection Research Division, Weather Extremes around the World, Natick, MA, May.

U.S. Department of Commerce, National Oceanic and Atmospheric Administration (1994). Federal Standard for Siting Meteorological Sensors at Airports, FCM-S4-1994.

CHAPTER 38

COMMERCIAL RESPONSE TO MEASUREMENT NEEDS: DEVELOPMENT OF THE WIND MONITOR SERIES OF WIND SENSORS

ROBERT YOUNG

The Wind Monitor wind sensor is an example of a commercial product development in response to customer measurement needs. Requirements of a single government agency led to the eventual development of an entire series of wind speed and direction sensors. The current models are the result of input from a multitude of agencies and end users whose application requirements vary dramatically—from sensitivity and responsiveness to survival in high winds, snow, and ice as well as desert heat and blowing sand—from ease of mounting and maintenance to durability and long-term performance—from simple analog output signals to polled digital serial data streams.

1 BACKGROUND

The Wind Monitor was created in 1979 as a small development project to try to satisfy the requirements of the National Data Buoy Center (NDBC) at Stennis Space Center, Mississippi. Now, approximately 23 years later, seven standard models plus several additional special models of the Wind Monitor are being manufactured to serve different customer requirements. Over 50,000 units have been produced (as of February, 2002) and are in worldwide service in more than 85 countries as well as

Handbook of Weather, Climate, and Water: Dynamics, Climate, Physical Meteorology, Weather Systems, and Measurements, Edited by Thomas D. Potter and Bradley R. Colman.
ISBN 0-471-21490-6

the Arctic and Antarctic regions and ships and buoys at sea. New applications are still being discovered.

From our company's early days we were advocates of the use of helicoid propellers for measurement of wind speed. Several products, which originally were categorized as "sensitive wind instruments" combined a helicoid propeller with a wind vane to provide measurements of both wind speed and wind direction in a single instrument. We manufactured propellers from very lightweight polystyrene foam for sensitivity and also from injection-molded polypropylene for greater durability. The typical wind sensor matched one of these propellers with a wind vane of comparable sensitivity. Typical fabrication materials were aluminum castings, machined aluminum and stainless steel, standard metal and plastic component parts, with stainless steel fasteners used for assembly. The wind speed transducer was typically a miniature tachometer generator, which required slip rings and brushes, and the typical wind direction transducer was a wirewound or conductive plastic potentiometer.

2 DEVELOPMENT

In the fall of 1979 we entered into a development contract with the National Data Buoy Center (NDBC; then referred to as the National Data Buoy Office) to design a propeller-vane-type sensor that would address its future needs for deployment on offshore buoys and remote stations. At the time the operational wind sensors for these applications were the J-Tec, which measured wind speed by means of a vortex shedding technique (utilizing a detector that counted the number of vortices shed from an obstruction in the tail assembly), and the Bendix Aerovane, a rugged sensor that utilized a propeller and vane combination. The main problem encountered with these sensors was the difficulty in mounting at sea due to their size and weight plus the need to install mounting bolts between the base and the mounting fixture on the buoy. Many of the buoys were a long distance offshore, which sometimes meant servicing the sensors in difficult weather conditions.

The design goals of the development contract were: small size (maximum 36 cm overall height; 46 cm overall length), light weight (2.5 kg maximum; 1.5 kg desired), simple mounting allowing one-hand replacement, corrosion resistant, no electrical slip rings, three- or four-blade helicoid propeller (maximum diameter 20 cm), working range 0 to 60 m/s and 0 to 360°, cost reduction, and reliability (18-month MTBF). The contract called for delivery of three prototype sensors for testing and evaluation.

The first prototype, which was delivered to NDBC in November, 1980, was machined aluminum with a detachable tail assembly (Fig. 1). It was intentionally designed for testing flexibility. Many different size and shape tail fins were tested to determine the optimum trade-off between overall size and dynamic performance. The wind speed sensor was an injection-molded polypropylene four-blade propeller, 18-cm diameter by 30-cm pitch, which had been previously developed. The wind speed transducer was a magnetically operated Hall effect switch, actuated by a two-

Figure 1 First prototype wind sensor, November, 1980.

pole magnet on the propeller shaft, which provided an output of one pulse per propeller revolution. The wind direction transducer was a conductive plastic 1 kΩ, 352° function angle, potentiometer which was mounted concentrically in the sensor housing and coupled to the movable vane above. Replaceable ball bearings and sleeve bearings were provided to conduct torque and speed tests as well as to assess long-term reliability. Early wind tunnel testing showed that sleeve bearings would overheat at the higher wind speeds and would not be suitable. The most difficult part of the design was actuation of the wind speed transducer that, to eliminate the need for slip rings and brushes, had to be mounted on the nonrotating inner part of the main body. Since the potentiometer was also located concentrically in the same area and being actuated by a coupling from the vane assembly, it was necessary to mount the Hall switch slightly off-center to clear the coupling. To accomplish this, the Hall switch was sandwiched between two soft iron circular plates. The circular magnet attached to the propeller shaft would then always be the same distance from the edge of the plates and the magnetic flux would be directed through the Hall switch as the magnet rotated.

3 PLASTIC FABRICATION

Two identical units of a second prototype design were delivered in March, 1981 (Fig. 2). The main purpose of these prototypes was to develop a streamlined and economical fabrication method since cost reduction was one of the primary design goals. The main housing, nose cone, and tail assembly of these units were constructed of thermoformed black vinyl plastic. Wood molds were fabricated for forming the main housing and tail in two identical halves, which were then cemented together. These molds could be easily modified to incorporate any desired changes. The tail assembly was filled with polyurethane foam. To achieve the desired combination of rigidity and low weight several different thickness and color vinyl materials were tried as well as different foam densities. The wind speed and wind direction transducers were the same as those used in the first prototype. A total of six sensors of this design were fabricated for evaluation by a couple of different organizations.

Work on a third prototype design was already underway while prototype II units were being field tested. We acquired a small injection molding machine (previous injection-molded propellers were made by an outside supplier) and began to redesign the nose cone assembly and other internal parts for injection-molding. To minimize outdoor ultraviolet (UV) deterioration, we had selected a black vinyl material for

Figure 2 Second prototype wind sensor, March, 1981.

thermoforming the main housing and tail assembly. The wind speed and wind direction transducers remained the same. A machined aluminum mounting post with set screws provided for mounting on standard 1-in. schedule 40 pipe. This design was designated as Model 05101 Wind Monitor. Approximately 25 units of this design were produced in January, 1982. Field testing of these and the previous prototype units revealed a problem with the black vinyl softening in the hot sun and residual expansion of the internal foam causing irregular bumps in the surface of the tail assembly.

We changed to a superior material for thermoforming the main housing and tail assembly. The new material was a white ABS (Acrylonitrile-Butadiene-Styrene) plastic with UV inhibitors, which had been developed and extensively tested specifically for long-term outdoor exposure. Some changes were made in the shape of the main housing, and additional mating parts were designed for injection molding. An aluminum orientation ring was also supplied. The orientation ring, which was installed directly below the mounting post, had an index pin that allowed the sensor to be removed and replaced for service while maintaining its original orientation on the pipe support. These changes, which were included in the next production lot of 60 units produced in April, 1982 (still designated Model 05101), made a dramatic improvement in the performance of the sensor, especially in tropical conditions (Fig. 3).

Figure 3 Model 05101 Wind Monitor, April, 1982.

4 CUSTOMER INPUT

Meanwhile a number of changes in design continued to be made in response to customer requests. The most significant change was the wind speed transducer. The Hall switch was replaced with a coil that was wound on a bobbin with a central hole, which allowed space for the potentiometer coupling to pass through. The wind speed signal then became an alternating current (ac) sine wave (in place of the 5-V square wave from the Hall switch) with the frequency directly proportional to wind speed. The main advantages of the coil were greater reliability and the elimination of the magnetic attraction of the soft iron flux plates required for the Hall switch that caused a noticeable "jogging" of the propeller at threshold wind speeds. A separate sensor interface circuit was developed to convert the wind speed signal to a square-wave pulse output (similar to the Hall switch) as well as a calibrated analog voltage output (0 to 1 V = 0 to 100 mph). A small metal junction box with a terminal strip was added to the sensor mounting post to provide an easy cable connection point. Designated Model 05102 Wind Monitor, approximately 340 units of this design were produced between October, 1982, and July, 1984.

Early in the fall of 1983 we had begun to incorporate additional changes desired by NDBC and other customers. The potentiometer was changed from 1 kΩ, 352° function angle to 10 kΩ with a 355° function angle. NDBC also wanted a reduction in the number of external fasteners, which were difficult to deal with in a marine environment. Meanwhile Climatronics Corporation in Bohemia, New York, had been awarded a contract by the Federal Aviation Administration (FAA) to instrument 51 major U.S. airports with Low Level Wind Shear Alert Systems (LLWAS). The Climatronics proposal included the Model 05102 Wind Monitor as the system wind sensor. This contract resulted in a blanket order for 450 Wind Monitors that, added to the requirements of other organizations, was cause for a major redesign effort to incorporate numerous improvements that had been suggested by several different customers. One of the most significant changes was the elimination of nearly all external screws and set screws. The main mounting post and orientation ring were changed from machined aluminum, fastened with set screws, to injection molded plastic with stainless-steel band clamps for tightening on the standard 1-in. pipe. The entire nose cone assembly and the main body of the sensor were now being injection molded while the tail assembly continued to be fabricated by vacuum forming the white ABS plastic and filling with foam. A simple internal spring latch was designed to secure the main housing to the rotor of the mounting post assembly. The latch was accessed by removing the nose cone assembly that was now threaded with an O-ring seal. Previously, the nose cone was secured to the main housing with four screws. An injection-molded junction box with a slide cover was added to the mounting post. Special packaging was designed with die-cut foam fitted to a custom carton to provide optimum protection during shipment. This latest design was introduced in August, 1984, and designated Model 05103 Wind Monitor (Fig. 6). A sensor interface circuit and a 4- to 20-mA line driver circuit both with a choice of wind speed scaling in meters per second, miles per hour, or knots and a choice of 360° or 540° azimuth range were also made available to provide calibrated outputs for direct input into most recorders and data loggers, and also for long cable runs.

5 ICING PROBLEMS

The FAA was very concerned about the performance and survivability of the LLWAS Wind Monitor in icing conditions. A number of different ideas were tried for heating the sensor, and several test programs were carried out both in a controlled laboratory environment and in field conditions. A variety of commercially available wind sensors were tested simultaneously. It was apparent that wet snow, clear ice, and rime ice affected the performance and calibration of all mechanical wind sensors to varying degrees. Snow had a greater effect on cup anemometers than on propeller anemometers while both clear ice and rime ice caused degradation of performance of both types. The physically larger anemometers were able to operate longer in icing conditions before performance was badly deteriorated. Heating of rotating joints allowed some sensors to operate longer before complete freeze-up; however, because of loss of calibration with ice build up, the benefit of heating only rotating joints was doubtful. Climatronics developed a heating scheme that utilized a heat element on the sensor mounting pipe and an aluminum mounting post (in place of plastic) to conduct some of the heat to the sensor bearings. It was felt that this provided some benefit in helping the sensor to recover from freeze-up and was therefore adopted for the LLWAS program. Our conclusion was that only heating the entire sensor assembly with an array of heat lamps was sufficiently effective to warrant the cost. The heat lamp array worked satisfactorily for low wind freezing rain events when temperatures did not drop very far below freezing but was not useful for situations where higher wind speeds and colder temperatures made the heating ineffective.

During the winter of 1983–1984 a Wind Monitor was installed on Mt. Washington in New Hampshire to check on the effects of severe rime icing. During a prolonged rime ice event with sustained winds, the sensor continued to operate reasonably well due to the high rotation rate of the propeller, which prevented the rime ice from adhering, and the continuous action of the vane, which prevented the rime ice from bridging the gap between the vane skirt and the mounting post. After several hours of high winds and rime ice buildup, the sensor failed when it was impacted by an ice chunk that had dislodged from a structure up wind. From this test we concluded that the Wind Monitor could operate effectively in light rime ice situations with continuous wind but would probably not be suitable for heavy rime ice conditions.

6 WIDESPREAD ACCEPTANCE

Several organizations had been evaluating the Wind Monitor for some time. In 1984 Campbell Scientific in Logan, Utah, added it to its standard line of wind sensors and began to place multiple orders. In 1985 The U.S. Army Atmospheric Sciences Laboratory (ASL) at White Sands Missile Range began operational use of the Wind Monitor and began purchasing multiple units through New Mexico State University. Also in 1985 the National Oceanic and Atmospheric Administration (NOAA)/Pacific Marine Environmental Laboratory (PMEL) began operational use of the Wind Monitor for marine applications. Late in 1985 the National Data Buoy

Center began operational use of a special model of the Wind Monitor that was modified for its application. About 2 years later further modifications, including a machined aluminum mounting post, were made to provide NDBC a "ruggedized" version that could better withstand the rigors of buoy deployment in rough seas. Units previously shipped to NDBC were also retrofitted with these modifications.

In the fall of 1985 a new model of the Wind Monitor was developed to meet the requirements of the U.S. Environmental Protection Agency and the Nuclear Regulatory Commission for air pollution applications. This model utilized an expanded polystyrene foam tail for greater sensitivity (improved damping ratio and lower threshold). The polypropylene four-blade propeller was normally supplied; however, an optional larger and lighter propeller was offered at additional cost. This optional propeller was hand fabricated of carbon fiber and proved to be very popular. We imported these from a supplier in France; however, the cost of hand fabrication plus transportation and duty made them disproportionately expensive. The calibration of these propellers was also more variable than desired. Eventually we developed a much less expensive and more uniform injection-molded carbon fiber thermoplastic propeller, which we manufacture in our own plant and supply as standard. This second standard version of the instrument is designated Model 05305 Wind Monitor-AQ (air quality model) (Fig. 4).

Figure 4 (*Top*) Model 05305 Wind Monitor-AQ (air quality model), February 1986. (*Bottom*) Model 05701 Wind Monitor-RE (research model), July, 1987. See ftp site for color image.

7 SPECIALIZED MODELS

In early 1985 we received another development contract from the National Data Buoy Center to design a special model of the Wind Monitor with a tachometer generator for the wind speed transducer. This model required the use of slip rings. We calibrated the output of the tachometer generator so that it exactly matched the output of the Bendix/Belfort sensors, which were the operational sensors at many NDBC sites. With the addition of a special cable connector and special mounting adapter fabricated at the NDBC facility, this special model Wind Monitor could directly replace the other sensors in the field without any changes to the data collection package. When we began making deliveries in December, 1985, we designated this special-purpose Wind Monitor as Model 05203-1. It was only manufactured for NDBC plus a very few other customers for special applications. At a later date we developed a low-power circuit that converted the frequency output of the standard Wind Monitor into an analog voltage with the same calibration as the tachometer generator version. The combination of the special circuit box plus a standard Wind Monitor achieved the same result and was more reliable than the special Model 05203-1.

8 DESIGN IMPROVEMENTS

The magnet and coil wind speed transducer had proved to be very reliable; however, the signal level was very low when rotating slowly near threshold speeds providing a marginal signal-to-noise ratio. In 1986 we changed the wire size in the coil from 26AWG to 4OAWG, allowing more than twice as many turns and a corresponding increase in signal level. We were concerned about the durability of the much finer wire, but field testing indicated that it was quite satisfactory. We invested in a special coil winding machine that was able to properly control the tension and layering of the finer wire on the special injection-molded bobbins. Also in 1986 we changed the wind direction transducer to a custom-designed very high quality conductive plastic potentiometer, custom manufactured to our own specifications. Previously we had used potentiometers made by several different well-known manufacturers. These were all standard-type conductive plastic potentiometers with modifications to meet our requirements. While they all worked fairly well, the failure rate seemed to be unacceptably high with many warranty replacements. The new custom potentiometer, while much higher in initial cost, proved to be far more reliable and therefore less costly due to reduced repair costs and greatly improved customer satisfaction.

A fairly large number of Wind Monitors were deployed by Pacific Gas and Electric Company (PG&E) in a study of the dynamic thermal rating of power transmission lines. In 1986 PG&E reported a problem with the wind speed signal from an instrument located near a substation. The electromagnetic pickup from the 60-Hz power lines was causing a wind speed reading of about 6 mph when the actual wind speed was at zero. This problem was solved by adding an RC low-pass filter circuit that attenuated the 60-Hz signal below signal conditioning trigger levels. The circuit was designed to pass the normal ac wind speed signal at the very low frequencies near threshold. It also passes the 60-Hz real wind speed signal, which has a higher amplitude than the induced power line signal. This circuit has been incorporated on all subsequent models.

In July, 1987, we introduced a third standard model of the Wind Monitor, which was designed for maximum sensitivity for research and special applications requiring measurement of very low wind speeds. The tail assembly is similar in construction to the air quality model but with a much lower foam density, resulting in an optimum damping ratio of 0.65. The very sensitive helicoid propeller is made from the same expanded polystyrene foam as the tail, resulting in a very low threshold and a short distance constant. This sensor, designated Model 05701 Wind Monitor-RE (research model), is sold only in limited quantities (Fig. 4).

At the same time the different versions of the Wind Monitor (as well as other products) were being developed, we were gaining valuable experience in injection molding. We had acquired new CNC (Computer Numerical Control) machinery for mold making and also new and larger injection-molding machines with better controls and greater capability. In mid-1987 we completed a fairly sophisticated mold to injection mold the entire tail assembly of the standard Wind Monitor. This resulted in a much more uniform and durable tail assembly, which we could produce at a significantly reduced cost.

By now many Wind Monitors, especially the Wind Monitor-AQ, were being used in critical applications that required wind tunnel certification and field auditing. From the first production runs all units were serial numbered; however, since the propellers were interchangeable between sensors, in March, 1988, we began to serial number all three types of production propellers. Previously, serial numbers were added only to propellers that had been individually tested in our wind tunnel.

9 JUNIOR MODEL

In April, 1988, the National Data Buoy Center asked us to design a new reduced size model of the Wind Monitor. Its buoys were becoming progressively smaller and it was felt that a smaller sensor would be required in the future. All NDBC buoys use duplicate wind sensors for redundancy. A smaller sensor would allow the two sensors to be mounted far enough apart to avoid mutual interference while keeping the sensors close enough to avoid damage during buoy deployment. The new smaller Wind Monitor turned out to be about two-thirds the size of the standard model. Prototypes were delivered in October, 1989, for field testing; however, this model has not yet been deployed operationally by NDBC. This fourth standard model, which was designated Model 04101 Wind Monitor-JR (junior model) (Fig. 5), was not advertised until November, 1993. Eventually it proved useful for several special applications with a number of different customers and several hundred have been produced to date.

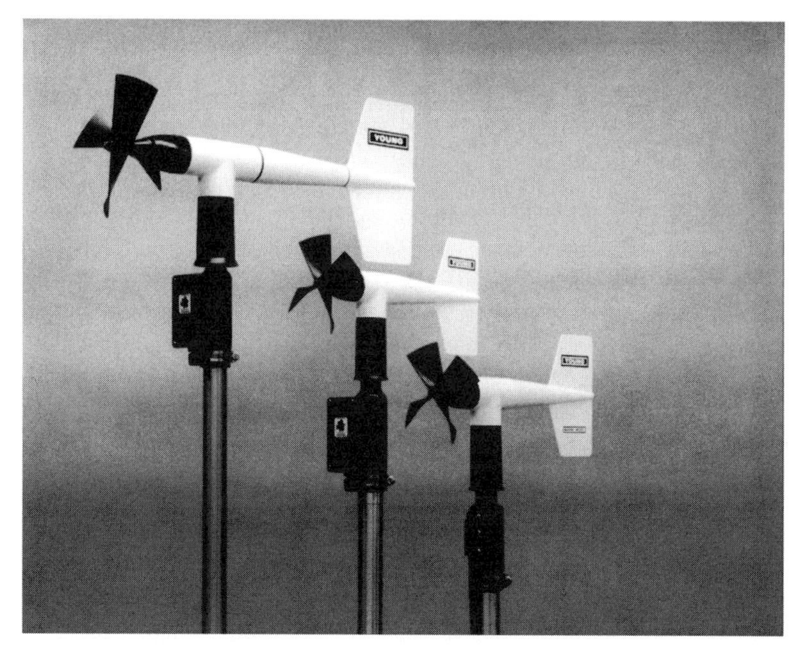

Figure 5 (*Top*) Model 05103 Wind Monitor (standard model, shown for relative size). (*Middle*) Model 04101 Wind Monitor-JR (junior model), November, 1993. (*Bottom*) Model 04106 Wind Monitor-JR/MA (junior marine model), October, 1994. See ftp site for color image.

10 CONTINUING DESIGN MODIFICATIONS

Until the spring of 1988 all Model 05103 Wind Monitors used ball bearings that had light-contacting plastic seals and were lubricated with instrument oil. The Model 05305 Wind Monitor-AQ and Model 05701 Wind Monitor-RE use the same size ball bearings but with noncontacting metal shields and instrument oil lubrication. This combination provides lower threshold as required by agency specifications, but the bearings are more easily contaminated and require routine maintenance. To achieve a longer field life, the bearing lubrication in Model 05103 was changed from oil to a special, wide-temperature-range grease. This resulted in a slight sacrifice in threshold but provided a dramatic increase in service life, which was extremely valuable to those customers with remote installations.

We had tested these sensors in our own wind tunnel and in the NDBC wind tunnel and had established a gust survival specification of 80 m/s. We had also done several sustained high-speed tests with the sensor mounted on a vehicle on an extended highway trip. We were beginning to get inquiries regarding survival at even higher wind speeds, especially for monitoring hurricane- and typhoon-prone areas. In August, 1988, we were able to make arrangements to test a Model 05103 Wind Monitor in the NASA/AMES Research Center wind tunnel at Moffett Field,

California. In this test the sensor was installed in their 7 × 10 ft Wind Tunnel No. 1 and subjected to wind speeds that were stepped from 0 to 250 mph (110 m/s) and back to 0 (in 12 steps). The wind tunnel was held for 2 min at each speed. The Wind Monitor performed well throughout the entire speed range. With these data we increased our gust survival specification to 100 m/s, and we began to recommend the sensor for applications that required higher wind exposure.

Woods Hole Oceanographic Institution (WHOI) had tested and deployed many Wind Monitors beginning with the early models. Beginning in March, 1989, WHOI began to incorporate the Model 05103 Wind Monitor into its IMET (Improved Meteorological Measurements from Buoys and Ships) "intelligent" sensor program. To satisfy its requirements, we had to adapt the upper portion of the sensor to a specially designed base assembly that WHOI supplied to us. The potentiometer, normally supplied for the wind direction transducer, was omitted and an elongated coupling was provided for connection to an optical encoder located in the special base assembly along with the signal conditioning and data logger circuits. Several of these units have been built and successfully field tested on buoys and research vessels by WHOI in an ongoing program. This same type of sensor package is now being proposed for the new NOAA/Voluntary Observing Ship (VOS) program, which may eventually include 300 to 400 vessels.

During the next several months additional design changes were made to improve performance and to overcome some reported field problems. The junction box, which contained the terminal strip for cable connections, was enlarged to accommodate a printed circuit board with transient protection devices. The circuit board also mounted an improved device for cable connections. There had been a number of potentiometer failures in the field that occurred during thunderstorm activity and also similar failures associated with extremely dry and windy sites such as the Antarctic. After extensive testing in our engineering department, we discovered that the terminations of the conductive plastic element in the potentiometer were susceptible to failure when a static charge would build up on the housing. Two changes were made to combat this problem. A ground wire was added to the housing of the potentiometer and connected directly to the earth ground terminal in the junction box. In addition the molded plastic parts surrounding both the coil and the potentiometer were changed to a conductive-type plastic. The mounting post was also changed to conductive plastic. With these changes and a properly grounded installation the static discharge failures were almost completely eliminated.

11 TESTING & CERTIFICATION

The World Meteorological Organization (WMO) had scheduled an intercomparison study of wind instruments to take place between July, 1992, and October, 1993. The Swiss Meteorological Institute (SMI) was already familiar with the Wind Monitor and recommended that we include it in the study. We submitted an application through the U.S. National Weather Service (NWS) to include the Model 05103 Wind Monitor, which was accepted. The WMO Wind Instrument Intercomparison was conducted jointly by METEO-FRANCE and the Swiss Meteorological Institute. The test site was the Mont Aigoual Observatory in the Cevennes Mountains in southern France. The observatory is at an altitude of 1567 m and normally experiences strong winds and heavy rime ice conditions during winter. The Wind Monitor was installed and operated satisfactorily without maintenance during the entire study, including winter, except during several periods of heavy rime ice. During the ice events the sensor became frozen and ceased to function; however, upon thawing it recovered and again functioned within specifications. The final report was released by METEO-FRANCE in October, 1997. We were surprised and pleased to learn that, after careful data analysis, the Wind Monitor had been selected as the reference sensor for the study. As the results of the intercomparison became known, there was a marked increase in acceptance of the Wind Monitor in Europe.

Late in 1992, to update our wind tunnel transfer standards, we arranged to have a Model 05103 Wind Monitor (with a four-blade polypropylene propeller) and a Model 05305 Wind Monitor-AQ (with a four-blade carbon fiber thermoplastic propeller) tested at the National Institute of Standards and Technology (NIST) in Gaithersburg, Maryland. Both models were run in the NIST 1.5 × 2.1 m (5 × 7 ft) Dual Test Section Wind Tunnel (DTSWT) from 0 to 46 m/s (103 mph) as well as the NIST Low Velocity Facility (LVF) from 0 to 10 m/s (22 mph). Based upon the data from these tests we found that the calibration of the four-blade polypropylene propeller differed from our published calibration by approximately 1%. The calibration of this propeller can be shifted slightly during manufacture by altering injection-molding temperature and pressure. Therefore we were able to correct the slight discrepancy in calibration. Unfortunately, the four-blade carbon fiber thermoplastic propeller used on the Model 05305 Wind Monitor-AQ exhibited a calibration discrepancy of 4.5%. The material used in the manufacture of this propeller does not allow any adjustment during the molding process. It was therefore necessary to publish a new calibration for this model and advise our customers of the revision. Fortunately many customers were able to apply a correction factor to their data to improve its quality.

12 MARINE MODEL

The number of Wind Monitors being used in marine applications was steadily increasing and in April, 1993, we introduced a standard model that incorporated many of the special modifications that we had done for the National Data Buoy Center. The sealed ball bearings are lubricated with a special waterproof grease. The junction box, normally attached to the mounting post, is omitted and a 3-m (10 ft) heavy-duty cable is supplied with a waterproof cable gland. The surge suppression components are contained in a potted module with the cable connections inside the mounting post. This fifth standard sensor is designated Model 05106 Wind Monitor-MA (marine model) (Fig. 6). The following year we introduced a marine version of the Wind Monitor-JR that incorporated the same special modifications. The Model

04106 Wind Monitor-JR/MA is the sixth standard model in the Wind Monitor line (Fig. 5).

Late in 1995 we had to do some extensive testing of all products, including all models of the Wind Monitor and associated signal conditioning and display modules, to determine conformance with the European requirements regarding electromagnetic emissions and susceptibility to electromagnetic interference known as the "EMC Directive" (Electro-Magnetic-Compatability). Most products met the requirements without modification, but a few circuits required some modification, and in most installations the use of shielded cable is required. After it was determined that the sensors met these requirements, we were required to affix a "CE" label (from the French 'Conformité Européan') to the sensor and also add a "Declaration of Conformity" to the instruction manual for each shipment destined for Europe.

13 SERIAL OUTPUT

After many months of development yet another standard model of the Wind Monitor was added to the product line. This new model incorporates an absolute optical encoder in place of the potentiometer for the wind direction transducer. Because of the very limited space available within the body of the Wind Monitor, it was necessary to design and fabricate a custom encoder that utilizes a unique scanning technique to determine azimuth angle to better than one degree. The operation of the encoder and the signal conditioning are controlled by a microprocessor in the junction box. There are three selectable output protocols, which are changed by moving a small jumper on the microprocessor circuit card. One output is specially developed by the National Center for Atmospheric Research (NCAR) in Boulder, Colorado, for its portable weather stations. A second output is a standard used in the marine industry developed by the National Marine Electronics Association (NMEA). The third output is our own, which we refer to as the "RMY" (R. M. Young) protocol. The outputs are a serial data stream, which can either be polled or continuous. In addition there are calibrated analog voltage outputs. This seventh standard model was introduced in December, 1995, as Model 09101 Wind Monitor-SE (serial output) (Fig. 6).

Icing continues to be a major limitation on Wind Monitor operation. The Wind Monitor is currently being deployed in many remote areas where there are occasional heavy rime ice conditions and limited power for any effective form of heating. We believe that the most promising potential solution is some form of passive surface modification. We have been investigating a couple of different surface treatments that we are hopeful will reduce the ability of clear and rime ice to adhere to the propeller and other outer surfaces of the sensor. Several researchers have reported varying degrees of success with a spray-on material called Vellox, which was developed specifically to reduce ice adhesion. This material seems to work quite well for a period of time following its application, but it rubs off easily and requires reapplication at regular intervals to be fully effective. Another form of permanent hydropho-

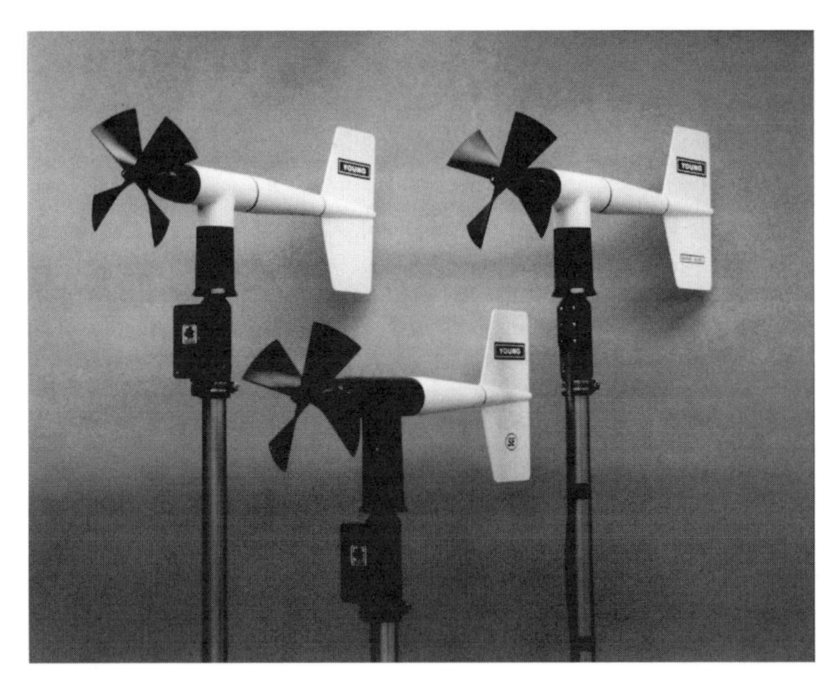

Figure 6 (*Top left*) Model 05103 Wind Monitor (standard model), August, 1984. (*Bottom middle*) Model 09101 Wind Monitor-SE (serial output), December, 1995. (*Top right*) Model 05106 Wind Monitor-MA (marine model), April, 1993. See ftp site for color image.

bic surface treatment is currently being investigated but results have so far been inconclusive and further testing is still needed. We are also investigating a new epoxy paint that is specially formulated to prohibit ice adhesion. If we are able to find a surface treatment that is practical and effective, it may result in the introduction of yet another model of the Wind Monitor.

The continually evolving needs of customers have dictated many changes as well as the ongoing development of a series of similar but quite different instruments. By this process products are continuously being improved and adapted to new demands.

BIBLIOGRAPHY

Anemometer Comparison Project (1986). A Report to the Atmospheric Environment Service of Canada, RVI/50.09, Resource Ventures Incorporated, Charlottetown, Prince Edward Island C1A 3W2.

Crawford, K. C., F. V. Brock, R. L. Elliott, G. W. Cuperous, S. J. Stadler, H. L. Johnson, and C. A. Doswell, The Oklahoma Mesonetwork—A 2 Pt Century Project, Preprints, Eighth International Conference on Interactive Information Processing Systems for Meteorology, Oceanography, and Hydrology, Atlanta, American Meteorological Society, pp. 27–33.

Frietag, H. P., M. J. McPhaden, and A. J. Shepard (1989). Comparison of equitorial winds as measured by cup and propeller anemometers, *J. Atmos. Oceanic Technol.* **6**, 327–332.

Hayes, S. P., L. J. Magnum, J. Picaut, A. Sumi, and K. Takeuchi (1991). TOGA-TAO: A moored array for real-time measurements in the tropical Pacific Ocean, *Bull. Am. Meteor. Soc.*, **72**(3), 339–347.

Jones, D. W., D. B. Hatton, D. A. Jenkins, and A. P. Scott (1992). A Field Comparison of Some Wind Sensors, Report No. 49, WMO Technical Conference on Instruments and Methods of Observation (TECO-92), Vienna, pp. 347–350.

Michelena, E., and J. Holmes (1983). A Rugged, Sensitive, and Lightweight Anemometer Used by NDBC for Marine Meteorology, Preprints, Fifth Symposium on Meteorological Observations and Instrumentation, Toronto, American Meteorological Society.

Michelena, E., and J. Holmes, The Meteorological and Oceanographic Sensors Used by the National Data Buoy Center, Proceedings MDS 9186 Marine Data Systems International Symposium, New Orleans, Marine Technology Society, pp. 596–601.

Payne, R. E. (1988). The MR, a meteorological sensing, recording and telemetering package for use on moored buoys, *J. Atmos. Oceanic Technol.* **5**, 286–297.

Phillips, C., D. Bumham, L. Jacobs, and D. Hazen (1992). 1991 LLWAS Anemometer Test Program, Final Report DOT/FAA/W/92-1, DOT-VNTSC-FAA-92-6, Research and Special Programs Administration, John A. Volpe National Transportation Systems Center, Cambridge, MA 02142-1093, September. (Document available through N.T.I.S., Springfield, VA 22161).

Qualid, G., P. Gregoire, M. Gilet, B. Hoegger, and A. Heimo (1993). WMO Wind Instrument Intercomparison, Preprints, Eighth Symposium on Meteorological Observations and Instrumentation, Anaheim, American Meteorological Society, pp. 274–278.

Weller, R. A., and D. S. Hosom (1989). Improved Meteorological Measurements from Buoys and Ships for the World Ocean Circulation Experiment, Proceedings Oceans 1989, Seattle, IEEE, pp. 1410–1415.

Weller, R. A., D. L. Rudnick, R. E. Payne, J. P. Dean, N. J. Pennington, and R. P. Trask (1990). Measuring near-surface meteorology over the ocean from an array of surface moorings in the subtropical convergence zone, *J. Atmos. Oceanic Technol.* **7**, 85–103.

Weller, R. A., M. A. Donelan, M. G. Briscoe, and N. E. Huang (1991). Riding the crest: A tale of two wave experiments, *Bull. Am. Meteor. Soc.* **72**(2), 163–183.

WMO (1997). Wind Instrument Intercomparison 1992–1993, Meteo-France/Swiss Meteorological Institute, World Meteorological Organization, Commission for Instruments and Methods of Observation, Final Report, October.

CHAPTER 39

COMMERCIAL RESPONSE TO MEASUREMENT SYSTEM DESIGN

ALAN L. HINCKLEY

1 INTRODUCTION

The commercial opportunity to produce multiple measurement systems began when measurement instrumentation was first used to document observations. While the following contains examples specific to Campbell Scientific's experience in measurement system development, the information presented reflects the direction taken by the measurement industry as a whole.

As a commercial producer of more than 10 different measurement systems, a great deal of practical experience has been gained in their design and in the representativeness and uncertainty of their measurements. Over 75,000 systems have been sold since the first battery-powered datalogger, the CR5, was shipped in August, 1975. Although this commercial success has given growth and economic stability, the greatest gain has been what we and our customers have learned about improving unattended environmental measurements.

Experience has taught us that measurement system design is optimized when it is based on theory and observation. The degree of confidence one can have in a system design is the degree to which the theory and observation agree. A measurement system may be designed and operational, but without some theoretical basis sources of uncertainty cannot be investigated (Campbell, 1991).

New designs should include a method by which the machine can detect and respond to its most probable failures—by observing and recording the errors and failures of current systems, better ones may be produced. Experience with similar designs that have been produced and supported in quantity is a most valuable asset in the design of new systems. Two principles that lead to good system design are (1)

Handbook of Weather, Climate, and Water: Dynamics, Climate, Physical Meteorology, Weather Systems, and Measurements, Edited by Thomas D. Potter and Bradley R. Colman.
ISBN 0-471-21490-6

making sure that the needs of the customer are met and (2) recognizing that there is an optimum high level of quality at which the total costs of designing, producing, selling, and servicing a product are minimized.

From manually measured mercury thermometers and staff gages, to chart recorders with spring-powered clocks, to microprocessor-based dataloggers, the tools for making good measurements have evolved. These new tools are more accurate, operate longer in harsher environments, and provide many more data retrieval options.

While the modern measurement system eliminates many of the uncertainties associated with manual measurements, others remain and new ones have been introduced. Using the measurement of temperature as an example, the modern systems have solved, within budget limits, the problem of insufficient human resources to measure the temperature in remote areas over extended periods of time. Most errors associated with manual data transcription and data transcription delays have been eliminated. Remaining are siting and exposure errors associated with the new temperature sensors and radiation shields. Inaccuracies in the liquid-in-glass thermometers are replaced by inaccuracies in the electronic temperature sensor (thermistor, resistance temperature detector (RTD), or thermocouple). The eye-sight-induced parallax error associated with manual measurement of a liquid-in-glass thermometer has been replaced by the datalogger's measurement error. Introduced are the effects of temperature changes on the measurement system's performance. While they do occur in the human observer, memory failure, run-down power supplies, and lightning damage are generally considered new uncertainties. Also introduced but often unresolved are a new set of "phantom" events like the rain caused by geckos (a tropical lizard) playing "teeter-totter" on the tipping bucket of a rain gage or the instantaneous half-meter of snow caused by an elk that lies down under the acoustic snow depth sensor.

The rest of this chapter will discuss some important features designed into the modern measurement system that improve its performance and reduce the uncertainties associated with unattended measurements. Where possible, the uncertainties will be defined and quantified. This information will facilitate the selection of good equipment and provided needed metadata (characterizing information) about the new measurement systems. Where they are helpful, anecdotes illustrate problems that have been overcome and failures that have led to greater understanding.

2 MEASUREMENT SYSTEM

Today's microprocessor-based measurement systems range from dedicated units with fixed sensor configurations, measurement rates, and reporting intervals to programmable systems with flexible measurement electronics. Dedicated systems have advantages in certain applications but, in general, the ability to accommodate different sensor types without additional signal conditioning as well as the ability to perform on-site calculations offers advantages for all but the most routine applications. Only electronic digital systems that are capable of operation without alternating current (ac) power are considered in this chapter.

All digital measurement systems must include (1) the measurement electronics to convert sensor signals to digital values, (2) either electronic storage media to collect the data on-site or telecommunications hardware to transmit the digital values, or both, (3) sensors that provide an electronic signal, and (4) mounting hardware, protective enclosures for the electronics, and power supplies (Tanner, 1990). In many of today's systems, a stand-alone environmental datalogger performs the measurement, processing, and on-site data storage functions as shown in Figure 1.

The datalogger, the cornerstone of a data acquisition system, has specific design requirements determined by the application. Our first fully digital datalogger, the CR21, was designed to meet the weather station needs of the agricultural and environmental studies disciplines (Schimmelpfennig et al., 1979). The datalogger design features required by these applications include:

- *Sensor Compatibility* Direct connection and measurement of sensor signals without external signal conditioning circuitry reduces cost, complexity, measurement error, and power requirements. The ability to resolve signals to the required precision affects the choice of sensor.
- *On-Site Processing* Field processing reduces data storage requirements, scales sensor signals to engineering units, and provides logic decisions for control applications.
- *Field Observation of Measurements* Field verification of sensors requires the ability to continuously observe instantaneous measurements, in engineering units, from a display or printer.
- *Input Transient Protection* Environmental datalogging is vulnerable to major hardware damage caused by large lightning-induced transients entering the system on sensor leads. Protection hardware such as transorbs and spark gaps, and proper grounding procedures are required to minimize damage.
- *Hardware Microprocessor Reset* Unattended, processor-based instrumentation should reset the processor, restoring normal execution in the event it is altered due to input transients or intermittent component failure. User-entered programs should exist in write-protected memory, minimizing the possibility that the processor can overwrite the program should an abnormal execution state occur.

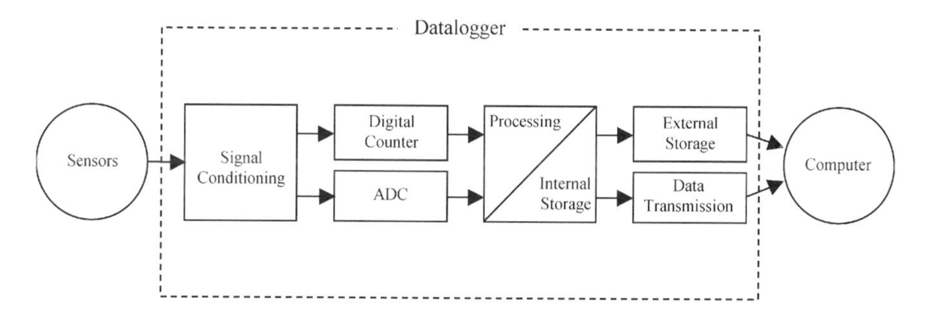

Figure 1 Generalized data acquisition sequence.

- *Low Power Consumption* Operation from direct current (dc) power supplies (batteries) and low current drain are required. An average power consumption of 50 mW (12 V at 4 mA) allows 2.5 months of operation on eight, alkaline D cells (7.5 Ah).
- *Operation in Adverse Environmental Conditions* Operation at high temperature and humidity are the main concerns in agricultural applications (at least until snow algae becomes the next food craze), but environmental dataloggers must operate and maintain measurement specifications over a minimum range of −20°C to +55°C. Solar heating can raise datalogger enclosure temperatures 20°C above air temperatures. Tight enclosures and desiccant provides the simplest and most cost-effective means of preventing water from condensing on the electronics.
- *Remote Communication Capability* Telephone was the only remote communication method available for the CR21. Since that time, data telemetry has become increasingly important. Models requiring climatic data are run daily to predict water use, crop development, disease and pest growth, etc. Synoptic models utilizing 15-min data are run several times a day. These applications require the timely transfer of data via radio, telephone, or satellite.
- *Logic and Control* The ability to compare values or time, with programmable limits and make decisions, provides powerful control capability. Samplers can be switched on or off, sensors with high current drain can be powered only during the measurement.

Good system design requires that the sensors be properly sited and mounted and that the electronics be protected and grounded. The accuracy and response times of the sensor needs to match those of the datalogger and the needs of the investigator. The data exchange between the measurement electronics and the data retrieval/storage hardware needs to be easy, reliable, and accurate.

Sales of our seventh measurement system, the Basic Data Recorder (BDR), did not meet expectations. Only 2000 BDRs were sold compared to sales of 35,000 of the datalogger built just before it. The BDR met the specific design requirements spelled out in a government bid, but it lacked the measurement versatility, telecommunication options, and software support required by our "typical" customer. Had the datalogger been built to fill both requirements, it would have been more successful. In spite of the failure, valuable lessons were learned. A new "table-based" data storage method was developed. Tighter government radio frequency (RF) noise requirements led to improved RF testing, shielding, and eventually to European market "CE Compliance." Both have been invaluable in the development of today's dataloggers.

3 DATALOGGER

A datalogger is made of pieces and processes that must fit and function together to make accurate measurements on low power under extreme environmental conditions. Seemingly simple design details like the style of a connector or the circuit board cleaning process are critical to the long-term performance of the unit. For example, the switch from a rosin-core solder flux to a water-soluble flux midway in the life of the first datalogger flooded the service department with repairs, threatening the existence of the new company. The dataloggers were failing because the flux remaining on the boards would adsorb water under humid conditions, become conductive, and cause shorts. Water-soluble fluxes were not as good as they are today and our procedure for completely cleaning it from the circuit boards had not yet been perfected.

Most modern datalogger uncertainties can be separated into two categories, failures or routine inaccuracies. Most failures result in the loss of part or all of the digital data. Only a few compromise the quality of the recorded data. Routine inaccuracies are the day-to-day errors in the datalogger's measurements that affect the quality of the recorded digital data.

Measurement system failures can be caused by:

- Failure of one of the components or circuit boards within the datalogger.
- Environmental conditions such as lightning, humidity, heat/cold, dust, floods, fire, wind, hail, ice, ultraviolet (UV) degradation, humans, animals, birds, insects, mold, etc.
- Human error in the wiring or programming of the datalogger.
- Failure of one of the power system components (battery, regulator, solar panel, etc.). Sometimes a datalogger component fails in a mode where it draws excessive current running down the battery prematurely.
- Failure of a sensor or sensor cable.
- Failure of the data storage/telemetry system.
- Failure of the protective enclosure or mounting hardware.

Inaccuracies of the measurement system that are routine in nature are caused by:

- Inaccuracies inherent in the voltage measurement, pulse measurement, or clock circuits over the temperature range of the datalogger. These uncertainties are quantified by the accuracy specification given by the manufacturer and should include the temperature range over which it is valid.
- Inaccuracies in a sensor's measured parameter induced by another environmental variable (e.g., temperature-induced inaccuracies in a relative humidity measurement).
- Changes over time of components in the voltage measurement, pulse measurement, and clock circuits–recalibration issue.
- Changes over time in the sensor's response to the parameter being measured—recalibration issue.

Fortunately, well-designed dataloggers rarely fail and their routine inaccuracies are relatively small and measurable. Many of the above-mentioned failures and inaccuracies are prevented or minimized by good datalogger design and manufacturing, system maintenance, and user training. Some are minimized by paying more for units with greater accuracy or reliability. Other failures, polar bears, for example, are simply Murphy's law at work. While it is not comprehensive, the following discussion details some of the more important features designed into a good environmental datalogger.

Low-Power Microprocessors

The microprocessor and quartz crystal clock are the brain and the heart of the datalogger. The processor does the computations and controls the operational sequences while the clock controls the timing and to some extent the speed. As evidenced by the cooling fans built on top of the Pentium processors in today's personal computers (PCs), most processors use a large amount of power. In a datalogger, which needs to run for 6 months on the power in eight alkaline D cells, it is important to use processors that require very little power.

In the early 1970s, a new logic circuit technology called CMOS (complementary metal-oxide semiconductor) was introduced. The first CMOS chips released were simple AND, OR, and NOR gates. About 2 years later, RCA introduced the 1802, the first CMOS-based microprocessor. The power consumed by a CMOS processor is low while changing states (processing), but it drops to less than 1 mA when it is inactive or "quiescent." Low-power system design requires that power consumption be minimized by placing CMOS or equivalent processors in the quiescent mode in between events.

The opportunities presented by this new microprocessor technology were recognized with great excitement by Eric Campbell, one of the founders of the company. His enthusiasm was such that he and his brother Evan built their younger brother Wayne a Christmas present, a CMOS-based "alarm clock" with speakers and light-emitting diodes (LEDs) that went off at 6 a.m. Christmas morning. The CMOS logic chips were first utilized in the CA9 path averaging laser anemometer and in the first datalogger, the CR5. Shortly after its release, the 1802 was added to the CR5 datalogger making it the first, or one of the first, microprocessor-based dataloggers. Three subsequent dataloggers—the CR21, CR7, and 21X—also utilized the low-power processing of the 1802. Quick recognition of the benefits of the 1802 processor placed the company right in the middle of an adventurous revolution in measurement system design.

Customer requirements and technology are constantly changing. Requirements for increased measurement and computational speed led to the use of two different versions of Hitachi's 6303 CMOS microprocessor in the next four dataloggers. The increased capability of this processor enabled subsecond measurement rates on a small number of channels or one-second rates on all channels. This speed is sufficient to meet or exceed most measurement needs in the weather, climate, and water fields. New microprocessors are introduced into new or upgraded dataloggers to meet changing customer requirements or because evolving technology has made the older parts unavailable. However, each microprocessor has its own quirks. Testing on the 6303 processor showed that its internal "overrun timer" was not reliable, so other methods had to be developed to work around the bug. Good system design requires extensive testing of a new processor to become familiar with its strengths and weaknesses.

"Watch Dogs"

Microprocessors specifically and digital circuits in general are quite susceptible to static electricity. Quite often, however, the static will "bomb" the processor or latch up a chip or change an address without causing permanent damage. Because lightning and other forms of static electricity are a very real part of the datalogger's environment, good system design requires that a "watch-dog" circuit be added to the system to reboot it when it crashes. A watch dog is a hardware count-down timer circuit that reboots the processor when the processor bombs for more than a preset time. It is like an automated "Ctrl–Alt–Del." The watch dog may also include a set of software checks that reboot the system if its system parameters are not within set limits. Once a watch-dog reset occurs, the processor reboots and special software is executed that attempts to restore the system to proper operation without the loss of data. The watch dog also assists in the development of datalogger software because it detects when a software bug causes the system to crash. The watch dog is a critical design requirement for all environmental dataloggers.

Clock

The clock needs to be inexpensive and low power but accurate enough that it can be left unattended for up to a year in places like Antarctica without a significant shift in time. The datalogger's clock is a quartz crystal that ideally oscillates at a constant frequency. In low-power less expensive crystals the frequency varies slightly as a function of temperature. The raw quartz crystal's accuracy of ± 4 min per month is improved in the datalogger to ± 1 min per month by measuring the clock's temperature and correcting the clock based on an algorithm permanently stored in the datalogger. While satellite applications require even greater clock accuracy, clocks that stay accurate to within 1 min per month meet most requirements.

Self-Contained Measurement Circuitry

Most of the sensors used in the fields of weather, climate, and water output a voltage or change resistance in response to the parameter they are sensing. The rest either output pulses or digital data via a serial protocol. Well-designed environmental dataloggers measure multiple voltage, pulse, and serial signals without the need for additional external signal conditioning, thus reducing the cost, complexity, errors, and power requirements associated with the extra circuitry.

Voltage Measurements

While environmental dataloggers can have analog inputs capable of measuring either voltage or current signals directly, Campbell Scientific dataloggers have always measured voltages and use an external shunt resistor to measure current signals. Voltage measurements are made by switching the signal from one of the analog channels into the analog-to-digital (A/D) circuitry. After waiting for the signal to stabilize at its correct value, the signal voltage is allowed to charge up a capacitor for a fixed amount of "integration" time. The voltage on the capacitor is then held for the A/D conversion. The A/D conversion is made using a successive approximation technique. For example, to measure a signal of +1000 mV on a ± 2730-mV range the microprocessor first compares the signal to 0.0 mV to determine the sign (+ or −) of the signal. Since +1000 mV is greater than 0.0 mV, a value would be stored indicating a positive signal and the 12-bit digital-to-analog converter (DAC) would then output +1365 mV, half of the remaining range. The signal is now less than the DAC output, so a 0 would be stored for the first bit and the DAC would next output 682.5 mV. The process repeats until all the bits have been set, each time closing in on the sensor's voltage by half of the remaining range.

A datalogger typically has one DAC that outputs voltages across the largest full-scale range (FSR). For a datalogger with a 12-bit DAC and a ±2500-mV FSR to measure a thermocouple with a 2.5-mV full-scale output, it must measure the small signal by switching in a gain circuit that electronically multiplies the voltage by a factor of 1000. Thus a 2.5-mV thermocouple signal is really measured on the same 2500-mV range with the same 12-bit resolution DAC as the largest full-scale voltage. Having multiple FSRs with voltage multipliers preserves the resolution and accuracy of the measurements. Good system design requires that the datalogger have multiple FSRs that have been carefully chosen to cover the voltages produced by the sensors it is to measure.

The resolution of a voltage measurement is the smallest incremental change detectable in the signal and therefore represents the fineness of the measurement. The greater the number of bits for a given full-scale voltage range, the finer the resolution of the measurement. An uncertainty of 1 bit must be assumed in the measurement. Continuing the previous voltage measurement example out to 12 bits would yield a signal of +1000.089 with a resolution of ±0.666 mV. A 13-bit and a 16-bit measurement would yield +999.925 ± 0.333 mV and +999.985 ± 0.042 mV, respectively. The number of bits a measurement is divided into is determined by the number of bits in the DAC plus a couple of tricks. A 12-bit DAC ($2^{12} - 1 = 4095$) is one that can change its output voltage in increments of 1/4095 of its full-scale output. An extra bit of resolution is gained by determining if the signal is positive or negative before the successive approximation starts. A second extra bit is gained when a measurement is the average of two integrations (e.g., a differential measurement includes two integrations). To convert the resolution of a measurement range in millivolts into engineering units, multiply the resolution of the appropriate FSR by the ratio of the engineering unit range to the signal range. For example, given a resolution of ±0.333 mV on the ±2500 mV range, and a −40 to $+60^\circ$C temperature signal represented by a 0- to 1000-mV signal, the resolution would be

$$\frac{\pm 0.333\,\text{mV} \times [60 - (-40)^\circ\text{C}]}{(1000 - 0\,\text{mV})} = \pm 0.0333^\circ\text{C}$$

When the resolution approaches the magnitude of the input noise, the latter determines the measurement uncertainty. Input noise is statistical and is specified in terms of the root-mean-square (rms) value. Numerical averaging of N measurements reduces the input noise by a factor of $N^{-1/2}$. One method of reducing the effect of noise on the measurement of a signal is to integrate the signal over a longer time period. Integration time is the time over which the signal is being electronically averaged. The integration time can be long for greater averaging, which smoothes the noise-induced peaks and valleys, short if more frequent measurements are required, or it can be broken into two integrations, one half of an ac cycle apart to remove either 60- or 50-Hz noise. The use of shielded sensor leads with the shield connected to system ground helps reduce high-frequency external noise.

Accuracy is often expressed in terms of the FSR, e.g., 0.1% FSR, and should be stated for a specific temperature range because of the temperature dependency of most errors. The accuracy of voltage measurements depends upon the accuracy of an internal voltage reference diode and the self-calibration ability of the datalogger over time and temperature. Input offset voltages also affect accuracy, but a good microprocessor-based system corrects for this error by shorting the input to ground and measuring the input offset voltage as part of the signal measurement sequence. The offset error is then removed from the result numerically.

The accuracy of each datalogger is carefully measured over several hot/cold temperature cycles in automated precision environmental test chambers using regularly calibrated National Institute of Standards and Technology (NIST) traceable voltage sources. The dataloggers are tested at the environmental extremes quoted in the specifications and then tested at points 10°C beyond those extremes to ensure their reliability and accuracy. Even in the “startup” days of the company when money for test equipment was limited, environmental testing was so important that dataloggers were manually calibrated while still cold from “soaking” in a chest freezer with dry ice and then again after soaking in a “hot box” made from an old refrigerator and a heat lamp.

In the field, the accuracy of the internal voltage reference is maintained over time and temperature by an automated self-calibration procedure that utilizes a precision reference diode to calibrate the rest of the system. To avoid interfering with the routine measurements, the calibration is done in the background in small segments throughout the 3-min self-calibration cycle. The calibration results are then averaged across a 15-min period, which keeps the logger calibrated across most naturally occurring environmental temperature changes.

“Single-ended” and “differential” amplifiers are commonly used for voltage measurements. With single-ended measurements the signal is measured with respect to instrumentation ground. Input connections therefore have one active side and one ground connection. With differential measurements, one side of the signal is measured with respect to the other side, requiring that both inputs be active. Often these connections are labeled as "high" and "low" or + and −. In some designs either the high or low side of the differential channel can be used to make two single-ended measurements. Most sensors require that the voltage channels have high impedance (use very little current).

Some signals can be measured either single-ended or differentially. Others, such as resistive full bridges, must be made differentially because the signal is referenced to one side of the bridge and not to ground. Single-ended measurements are adequate for high-level voltages, thermistors, potentiometers and similar signals. Differential measurements provide better rejection of noise common to both sensor leads, and for this reason they should be used for sensors having low-level signals (e.g., thermocouples). In addition, the accuracy of a differential measurement is improved by internally reversing the inputs and making a second measurement. The second measurement with the inputs reversed cancels out thermal or noise-induced offset errors in the following manner: Suppose a small change in the datalogger's temperature since the last self-calibration is making the high input read 1 mV high so with a 10-mV signal, the high input reads 11 mV, the low reads 0 mV, and the first differential voltage measured is 11 mV. When the inputs are reversed, the high input reads 1 mV, the low input reads 10 mV, so the second differential voltage measured is −9 mV. When the sign of the second is reversed and the two measurements are averaged, the 1-mV offset cancels yielding a differential voltage of 10 mV. Good system design requires careful attention to even the smallest measurement details.

Thermocouples

One of the common sensors used to measure temperature is the thermocouple. In addition to the need for an accurate measurement of a voltage signal usually less than 5 mV, thermocouple measurements require a reference junction temperature and polynomial equations to convert the voltage to temperature.

Precision Excitation

Resistive sensors, those whose resistance change in response to the parameter they are sensing, require an excitation voltage to output a voltage that corresponds to the changing parameter. The excitation voltage needs to be selectable so the output voltage fills one of the full-scale ranges. It also needs to have accuracy and resolution characteristics that match those of the input voltage range. Power consumption is reduced by shutting off the excitation after the sensors have been measured.

Most resistive sensors can be measured ratiometrically, which means that the measured voltage is divided by the excitation voltage. Because both the measurement of the signal voltage and the generation of the excitation voltage utilize the same voltage reference, dividing one by the other cancels out errors in the voltage reference. As a result, the accuracy of the measurement increases from 0.1% FSR across the −25 to +50°C range to 0.02% across the same temperature range.

Some of the sensors included in this group are thermistors, RTDs, potentiometers as used in wind vanes, weighing rain/snow gages, pressure transducers, and load cells.

Pulse Measurements

Several important sensors output low-level ac signals, square-wave pulses, or even just a switch closure. The datalogger needs a voltage supply and a bounce elimination circuit to measure switch closures. Amplifiers are required for the low-level ac signals. Dedicated counters that accumulate while the logger is sleeping in between measurement intervals keep the current drain low. The processor must be able to detect occasions when it is too busy to measure and reset the pulse accumulators at the designated time. The pulses from the excessively long interval must be discarded if the signal is a rate or accumulated if it is not. Vibrating wire transducers, commonly used for long-term monitoring of water pressure or deformation, require special circuitry and software to determine the frequency of their limited-duration signal.

Serial Inputs and Digital I/O Ports

In an effort to increase their accuracy or to even be able to make some of the more difficult measurements, several sensors are now microprocessor based. These sensors make the required measurement, digitize the value, and then transmit the value in digital form to a datalogger via a serial protocol such as SDI-12, RS232, RS485, etc. In the case of the SDI-12 protocol, the datalogger tells SDI-12-based sensors when to make a measurement and then requests the data after the wait time specified by the sensor. Some smart sensors simply transmit their data after the measurements are made on their programmed time interval. Most serial signals are transmitted and received using the digital input/output (I/O) port.

Digital ports are needed to input status signals or to output control signals. This capability allows the datalogger to measure, control, or alarm based on time or measured values.

Programming Versatility

A well-built datalogger is not much good if it cannot be easily told what needs to be done. A less versatile datalogger programming language works fine in markets where there is a lot of repetition. A more versatile programming language is needed in diverse markets, especially where research is the primary objective.

A powerful datalogger language consisting of canned measurement instructions and canned mathematical processing instructions take the work out of doing good complicated measurements. If the versatile programming language is backed up with good PC support software to facilitate the program development, routine programs become easy and complex programs become possible.

On-Site Data Processing

The advantage of processing measured values to obtain more efficient data storage has been mentioned. Data compression is particularly important in remote applications where site visitations are costly. Processed results such as averages, standard deviations, extremes, and values recorded conditionally all reduce data storage and handling logistics.

Linear calibration constants are entered into the datalogger to convert measurements into engineering units immediately. The datalogger displays the sensor signal in engineering units that enable the user to verify the correctness of the signal and its conversion. Field calibration of sensors is possible. User-entered polynomial coefficients are used to linearize nonlinear measurements. Linearization and the scaling of sensor outputs into engineering units make it possible to correct sensor readings on-site (e.g., correcting a piezometer reading for barometric pressure yields pore pressure). Nonlinear signals converted to engineering units can then be averaged, but in their nonlinear form they cannot. The ability to convert sensor signals into correct engineering units is extremely useful when verifying the performance of the sensors and the datalogger in the field.

The ability to compare values or time with programmable limits and make decisions provides useful control functions such as sampling at a faster rate, measuring a selected sensor, or initiating a radio or phone communication for an alarm message.

Internal Data Storage

The last 25 years of computer hardware innovation has greatly increased the amount of memory stored per chip. On-board data storage that used to be in increments of 2 kilobytes now comes in increments of 1 and 4 megabytes. To keep up with the rapid advances in memory storage technology, dataloggers are designed and built with the ability to add the next generation of memory chips as soon as they became available or cheap enough to justify their use.

The limited on-board memory of the past required one or more of the following: compression of the data into hourly or daily summaries, immediate data transfer to an external data storage device and even those had limited capability, or immediate data transfer to a computer via one of the telecommunication methods. Today's extensive memory often eliminates the need for on-site external data storage devices, allows for more frequent recording of the data, and permits data retrieval on less frequent intervals.

Today's on-board data storage is not only more extensive but it is also more reliable. Part of the data, the system program, and the clock are maintained by an internal lithium battery that takes over the instant the system voltage drops below the operational level of 9.6 V. The extended memory is stored in FLASH, a memory chip that only erases or records data when power is present.

Reliability

At first glance, reliability and low cost work against each other. For example, military-grade (mil-spec) parts have higher reliability than industrial-grade components but they cost much more. Only when the full costs of a datalogger, including warranty repair, support, and marketing (ever try to sell a lemon?) are understood, can one properly balance cost and reliability. For example, there are ways to significantly reduce the cost of a datalogger without sacrificing reliability. Industrial-grade parts can be tested upon arrival. Where possible, specific parts that have proven reliable in previous dataloggers can be designed into new dataloggers. Rigorous environmental testing of the completed units result in mil-spec or better reliability at a fraction of the cost.

Production and repair personnel keep careful records of the parts that fail, hoping to detect a bad batch of parts before many of the dataloggers ship. This method catches most of the bad parts but, unfortunately, there are those rare occasions when a part only fails after time or after exposure to humidity or temperature variations in the field. This happened to the very popular CR10 datalogger. In the spring of 1998, a sharp increase in capacitor failures in CR10s returned for warranty repair was noted. Examination of the carefully kept internal records finally linked the cap and datalogger failures to a change in the cap manufacturing process early in 1997. The cap manufacturing change caused an internal crack in the capacitor that eventually led to its failure. Although nearly 2500 dataloggers were shipped with the faulty parts, the carefully kept records helped resolve the problem. New capacitors from a new vendor were installed in units that began to ship in the fall of 1998. A recall was issued in the fall of 1999 when the rising failure rate indicated the seriousness of the problem.

Good system design requires good personnel in the test and repair groups. Feedback from these groups on the problems and parts that fail must make its way back to the system design group so that problems can be solved and better parts used in revisions of current dataloggers or in the design of new dataloggers. A monthly meeting of a quality control group made up of production, repair, testing, and engineering personnel helps maintain high quality. Reliability is a direct result of quality. High reliability requires that quality be an issue of continuous concern.

In the field, dataloggers are frequently exposed to voltage surges and RF noise. Dataloggers are best protected from lightning by providing it a low-resistance path to a *single* good earth ground. Datalogger lightning protection design has improved with each new datalogger. The most recent dataloggers are built with separate grounds for power while the rest of the circuit board is covered with a ground plane, which serves as the ground for analog signals and sensor shields. Up to 2 A of current can be run through the ground plane without inducing voltage gradients, which affect analog measurements. A good beefy connection from the ground plane to an earth ground gives lightning a good low-resistance path to travel. Spark gaps protect the analog input channels.

The datalogger both creates and receives RF noise. The amount of RF noise is greatly reduced by the metal package surrounding the electronics. Tighter RF standards and better test equipment have lead to much better RF shielding.

Inexpensive rugged packaging that offers portability, protection from humidity, dust, and temperature extremes helps increase system reliability. The ability to power the datalogger from a variety of power supplies facilitates its use in many environments.

In the competitive datalogger market, where a company's livelihood is earned by small sales to private and publicly funded researchers, most on limited budgets, success requires good technical support, word-of-mouth advertising, a reasonable price, and great reliability.

4 DATA RETRIEVAL

The reliability and convenience with which measurements are transferred from the field to the computer are important design goals in any measurement system. Most of the datalogger design requirements, especially low power and environmental ruggedness, apply to the data retrieval components. Whether the data are stored on-site and retrieved during site visits or retrieved remotely using various telecommunication options, the manufacturer should provide the necessary hardware and software tools to accomplish this task. The challenge has been to keep hardware and software current given the rapid changes in the fields of electronics, telecommunications, and personal computers. The following sections briefly describe most of the data retrieval options currently available. Examples of some specific benefits resulting from new or enhanced data retrieval methods are included.

On-Site Data Storage

An important part of almost all data retrieval systems is the storage of data on-site. On-site storage holds data between site visits or data transmissions. Should the telecommunication system fail, data may be retransmitted once the system is repaired. Printer and magnetic tape methods of on-site data storage have been replaced by data storage in memory either inside the datalogger or in an external memory module.

Our first fully digital datalogger, the CR21, stored data internally in random-access memory (RAM) for telecommunication and externally on either thermal printer paper or a magnetic cassette tape for data retrieval during a site visit. The cost of memory limited internal data storage to only 600 processed data values. The printer paper held 35,000 data values, but it had to be manually transcribed into the computer. Cassette tapes held 180,000 values on one side of a 60-min tape. A tape reader transferred the data to a computer. While the printer stuck in high humidity and the tape recorder failed below $-5^{\circ}C$, they were the only cost-effective on-site data storage options available at the time.

In 1983 Sequoia National Park requested a solid-state data storage module so it could gather winter precipitation data at high elevations in the Sierra Nevada. The desire to meet this customer's need combined with the falling cost of RAM led to the development of a RAM module powered by a small lithium battery. Due to its portability and reliability in cold and hot conditions, the memory module and its successors have become the preferred method of external on-site data storage and site visit data retrieval. Memory module development led to convenient data storage under cold winter conditions.

The increased portability and ruggedness of laptop computers and personal digital assistants (PDAs) allows them to be used to retrieve internally or externally stored data. Expansion of the dataloggers internal memory has contributed to this trend by reducing the frequency of site visits.

Data Retrieval via Telecommunication

Data can be transferred from remote field stations to a computer through wire, radio waves, or a combination of both. Variations on these options have greatly increased in number and quality in recent years. The new methods have increased data transfer speed and reliability. Cost, reliability, operating distances, and data rates are important in determining the usefulness of a telecommunication system for a particular application. Remote data retrieval provides timely reporting, early detection of equipment malfunctions, and, for more isolated sites, may be the only practical means of data recovery. The expense of manual data collection, particularly in larger networks, often justifies automated techniques even when near-real-time reporting is not required. Telecommunication improvements have changed data transmission methods from teletype or voice into a fully automatic digital transfer from the datalogger in the field to the computer running the synoptic forecast model or generating the reports.

Where only one-way communication is supported, such as with satellites or certain RF-based designs, redundant transmissions are often used to obtain high data capture rates. Interrogated networks based on two-way communication allow error checking and retransmission of improperly received data. Two-way communication also allows the user to remotely change the datalogger's program, clock, and control ports. The following information details advantages and disadvantages of most currently available telecommunication options.

Dedicated Cable

Short-haul or multidrop modems operate up to distances of several kilometers, the expense of the cable and its installation being the more practical distance limitation. Short-haul modems require a PC COM port and an independent cable (two twisted pairs) for each remote site accessed by the base station. A local area network (LAN) links multiple remote sites to a single PC COM port by a single two-conductor cable (typically COAX). Individual stations within the network are accessed through addresses set in the LAN modems. Direct cable links are useful in more permanent operations and for combining several local sites into a network (e.g., on a watershed), which can be accessed remotely by a single telephone or RF link. The communication rate is dependent upon both the cable characteristics and its length. Model SRM-5A short-haul modem manufactured by RAD Data Communications, Inc. specifies operation at 1200 and 9600 bit per second up to distances of 10.5 and 8 km, respectively, for 19 AWG wire. A network of 10 of Campbell Scientific's LAN modems (Model MD9) operates at 9600-bit per second up to a distance of 2 km assuming the loss characteristic of the coaxial cable is 0.02 dB/m.

Switched Telephone

Where available, standard voice-grade telephone lines provide a simple, reliable method for retrieving data. A typical PC and modem calls and retrieves data from each remote site. System support software facilitates the scheduling and setup of the call parameters. The modem at the measurement site needs to run on minimal dc power and function under hot and cold field conditions. Communication rates of 300, 1200, 4800, and 9600 bits per second are standard. Standard phone company procedures for transient suppression should be followed at the remote site. Suppression devices included in the modem may be augmented by external suppression devices.

Campbell Scientific's (CSI's) first rugged, low-power 300-bit per second phone modem (DC103A) combined nicely with the CR21-based automatic weather station (AWS). These telephone-linked weather stations were quickly utilized to form AWS networks in New Mexico, Nebraska, and California. In the CIMIS project in the Sacramento Valley of California, a computer polled each of 40 sites daily, providing data for evapo-transpiration (ET) models. The introduction of telephone telemetry to the AWS benefited farmers with daily crop water usage information.

Cellular Telephone

The cellular phone systems established in many countries provide vast networks of cellular repeaters in most urban areas and along major highways in rural areas. A cellular phone provides the radio link from a remote site to the nearest cellular repeater where the message is converted to telephone signals and passed down a standard phone line to the PC. Stationary sites able to utilize a cellular yagi antenna can communicate across distances of 20 km if there is line-of-sight to the cellular repeater site. The cellular link eliminates the installation costs of a standard phone line. In addition, the monthly "air time" fee for a cellular site accessed once a day is typically less than the monthly service fee for a standard phone line. The current used when transmitting is typically 1.8 A. The stand-by current drain of 170 mA requires either a larger power supply or that the cellular phone be turned on only a few hours during the day. Data throughput is limited by the bandwidth of the frequency to 4800-bit per second. Though the first cell phone systems were analog, overlapping digital networks are rapidly being installed.

Internet

Where Internet-grade T1 phone lines are available, an Internet modem is used to access data from one datalogger or a radio-linked network of dataloggers.

Single-Frequency RF

Single-frequency ultrahigh frequency (UHF) or very high frequency (VHF) networks with individual stations accessed by unique addresses are in general less expensive and logistically simpler than older style, multifrequency networks. Designs based on low-power transceivers (2 to 5 W) are available. Operating distances are limited to line-of-sight, typically 25 to 35 km, but this range is extended where individual stations may be used as repeaters to access more distant stations. Access of an RF network through telephone lines further extends the data collection range. Average power consumption of approximately 80 mA dc (1 A when transmitting) is typical for many low-powered radios. In standard UHF and VHF applications, a Federal Communications Commission (FCC) license is required. The bandwidth is regulated, limiting the RF bit per second rate. Data "through-put" rates depend upon the communication protocol, the overhead times required to establish the link, and the length of the transmitted data block. For example, data through-put rates of around 30 data values per second are reasonable.

At many sites, the cost of installing and maintaining a phone line are prohibitive. To obtain data from remote mountain tops economically, a radio modem and VHF/UHF radios were added to meet the needs of customers such as the military who were doing tests across the 8300 km^2 White Sands Missile Range. Ski resorts doing avalanche control with remote, mountain-top data also benefited from the radio links.

In 1992 a faster polling scheme was developed for radios. The increased speed of the new method allows data to be retrieved every 5 min from a 35-station network monitoring meteorological conditions across the 1400-km^2 research complex at Idaho National Engineering and Environmental Laboratory (INEEL). Kennecott Utah Copper smelter used the same setup to obtain air quality data from pollution monitors at 20 ambient sites and 3 stack sites every 2 min.

The same technology was taken one step further in the Oklahoma Mesonet Project (Brock et al., 1995). In that project, 108 weather stations uniformly distributed across the 180,000 km^2 of the state measure air temperature, humidity, barometric pressure, wind speed and direction, rainfall, solar radiation, and soil temperatures. Each station is polled every 15 min for its 5-min data. From a typical site, the data is transmitted several kilometers by VHF radio to the nearest police station or sheriff's office. From there it is relayed by the Oklahoma Law Enforcement Telecommunications System (OLETS) on a 9600-bit per second synchronous dedicated line to the central computer at the Oklahoma Climatological Survey office in Norman, Oklahoma. The central site ingests the data, runs quality assurance tests, archives the data, and disseminates weather information and warnings in real time to a broad community of users, primarily through a computerized bulletin board system.

In 2001, St. John's water management district in south Florida installed an even faster radio telemetry network of 85 sites with multiple repeaters. The 85 sites are polled every 15 min. The data are used to monitor levels and control gates and pumps thus controlling surface water levels within an acceptable range. The benefit of increased telemetry speed is real-time monitoring and control.

Spread-Spectrum RF

Spread-spectrum radio modems look and act like a wireless RS232 port. Constrained by the FCC to less than 1 W effective radiated power, an FCC license is not required to operate them. Transmitting and receiving on multiple frequencies in one of several bands (902 to 928 MHz was the first), they work well in industrial areas where line-of-sight is not available but distances are less than 5 km. With line-of-sight and yagi antennas, distances up to 30 km are possible. Spread-spectrum radios utilize one of two methods to transmit their data: frequency hopping or direct sequencing. The frequency hopping method is preferred for stationary sites. It also allows the frequency to be shared by more sites. Direct sequencing is preferred for mobile sites. Current spread-spectrum radios require 100 mA in the receive mode and 600 mA in the transmit mode. A spread-spectrum radio under design at CSI should require less than 75 mA in the transmit mode.

Satellite

Satellite transmitters provide data transfer from very remote sites and networks with wide spatial coverage. Most satellite links only provide one-way data transfer. Data transfer via satellite is generally more costly due to the required infrastructure. At the end of 1998, approximately 1300 satellites orbited Earth. Estimates put their number at 2400 by the year 2008. Satellites are grouped into three main classes based on height and type of orbit. They are: geosynchronous or geostationary Earth orbiting satellites (GEOs), middle Earth orbit satellites (MEOs), and Low Earth orbiting satellites (LEOs). The height of their orbits also determines the maximum geographical area they can cover.

Geosynchronous or Geostationary Earth Orbiting Satellites GEOs, positioned 22,300 miles (36,000 km) above the equator, appear to be stationary in the sky as they turn synchronously with Earth (hence the name geosynchronous or geostationary). This allows a ground station antenna to point at one place in the sky to send and receive signals. GEOs are the farthest from Earth and cover the largest surface area. Three GEOs are needed to cover the entire planet below 70° latitude.

The Geostationary Operational Environmental Satellite (GOES) system was installed and is maintained by the U.S. government. GOES channels (frequency and time slice) are reserved for use by federal, state, and local U.S. governments and some foreign governments. The recently introduced high data rate (HDR) GOES data throughput is limited to:

- 300- or 1200-bit per second data rate
- 20-s transmission window
- Transmissions intervals of 1, 3, or 4 h

The highest elevation weather station in the Americas was installed at 6542 m on the summit of Nevado Sajama in Bolivia in 1996 (Hardy et al., 1998). Installed to better understand tropical ice core variations and global climatic change, the station utilizes a GOES transmitter for near-real-time data and memory modules for on-site storage and manual data retrieval. The remote, high-elevation location precludes unscheduled service or repair visits. Snow and ice buildup on the antenna reduced near-real-time data recovery to 80% during the wettest month, but 100% of the on-site data was recovered during the annual site visit. In addition to permitting the start of the data analysis, the near-real-time data indicated that the failure of the lower sensors midyear was due to their burial by an El Niño induced larger than expected snowfall.

Middle Earth Orbit Satellites MEOs, positioned between GEOs and LEOs, orbit from 1000 to 22,300 miles. Depending on the altitude, as many as 10 or 12 are needed to cover the entire planet. In addition to voice and data transmission, MEOs are often used by surface navigation systems such as the Global Positioning System (GPS).

Many environmental measurement applications benefit from being able to record GPS position data. In addition, the GOES system now requires that transmitters use GPS clock data to keep their transmissions accurately timed.

Low Earth Orbiting Satellites LEOs orbit at altitudes from 100 to 1000 miles. Approximately 46 to 66 LEOs are needed to cover the entire planet. A LEO orbiting satellite is above the local horizon for less than 20 min. Messages must be short or LEO systems must support satellite-to-satellite hand-off to maintain communications.

One of the LEO systems called ARGOS, utilize the National Oceanic and Atmospheric Administration's (NOAA's) polar orbiting satellites. These are frequently used either in oceanography because station position can be tracked, or at latitudes greater than 70° where GOES satellites cannot be seen. The ARGOS platform provides good data transfer at the poles because the satellites cross the poles 24 times per day, which is three times more often than at the equator. ARGOS data throughput is limited by:

- The number of times per day the satellites pass overhead
- One satellite pass overhead lasts from 10 to 14 min

- Redundant 32-byte messages are transmitted every 90 or 200 s to ensure data integrity
- 400-bit per second data rate

Satellite communication is changing rapidly. Several projects currently underway will be launched in the near future. The satellite business is not easy due to the numerous governments involved and the fight for frequencies. Many projects have merged together so that the merged company has a better chance of seeing its project to completion. The following is an incomplete list of websites that provide links to companies that offer data transmission services via one of the GEO, MEO, or LEO satellites:

GEO Systems	Website
GOES	http://NOAA.WFF.NASA.GOV
Inmarsat-C	http://217.204.152.210/index.cfm
Qualcomm	http://www.qualcomm.com

MEO Systems	Website
ICO	http://www.ico.com
ORBLINK	http://www.orbital.com

LEO Systems	Website
Iridium	http://www.iridium.com
Orbcomm	http://www.orbcomm.com
Globalstar	http://www.globalstar.com
Argos	http://www.argosinc.com

5 SUMMARY

Commercial measurement system design requires careful attention to all details. Only the most important details were presented here. The experience gained through the design of similar measurement systems that have been produced and supported in quantity is valuable in the design of a new system. It is important to quickly utilize the technological advances in electrical components, sensors, and telecommunication methods. There is an optimum high level of quality at which the total costs of designing, producing, selling, and servicing a product are minimized. Careful, high-quality records help maintain high-quality standards. Low turnover of key personnel is important to keep expertise and continuity. Finally, the personal joy and satisfaction that comes from service to customers with legitimate needs is the basis for our existence as an organization (Campbell, 1987).

REFERENCES

Brock, F. V., and K. C. Crawford (1995). The Oklahoma mesonet: A technical overview, *J. Atmos. Oceanic Tech.* **12**, 5–19.

Campbell, E. C. (1987). Overview. "Good Morning, This is Campbell Scientific . . ." (in-house newsletter), March, 1.

Campbell, E. C. (1991). Overview, Independent Verification. The Campbell Update, A Newsletter for the Customers of Campbell Scientific, Inc., **2**(1), 2.

Hardy, D. R., M. Vuille, C. Braun, F. Keimig, and R. S. Bradley (1998). Annual and daily meteorological cycles at high altitude on a tropical mountain, *Bull. Am. Meteorolog. Soc.* **79**, 1899–1913.

Schimmelpfennig, H. G., B. D. Tanner, and E. C. Campbell (1979). Applications of a Minature, Low Power, Computing Datalogger in Environmental Investigations. *Fourteenth Conference on Agriculture & Forest Meteorology and Fourth Conference on Biometeorology, April 2–6, 1979, Minneapolis, MN*, American Meteorological Society, Preprint Volume, pp 154–155.

Tanner, B. D. (1990). Automated weather stations, *Remote Sensing Rev.* **5**(1), 73–98.

CHAPTER 40

DESIGN, CALIBRATION, AND QUALITY ASSURANCE NEEDS OF NETWORKS

SCOTT J. RICHARDSON AND FRED V. BROCK

1 INTRODUCTION

The calibration and quality assurance needs of meteorological networks vary significantly depending on factors such as end-user needs, data accuracy and resolution requirements, site maintenance schedule, climatology, and, of course, the budget of the project. This chapter describes general principles that can be used to ensure quality data in meteorological networks.

The design of a meteorological observation network depends on many factors, perhaps most importantly what the system is to measure and with what accuracy. However, design of a measurement system is also powerfully affected by other considerations such as choice of sensor and data logger. Selection of the measurement platform, data communication system, and type of power system has a profound affect on overall system design. Communication system limitations may dictate the location of remote sites forcing compromises in site location. Power limitations may prohibit the use of certain types of sensors.

A very important and sometimes overlooked aspect of system design is future data requirements and network upgrades. A system designed to meet today's requirements will not necessarily satisfy tomorrow's needs and/or may not work well with new technology. A data collection system should be designed with this in mind so that upgrades can be made without the need to replace the entire infrastructure of the network. Note that once a data collection system and/or vendor is chosen, it may be necessary to use this same system for the duration of the project including upgrades or modifications. This is because, for example, once a data logger is chosen it may

Handbook of Weather, Climate, and Water: Dynamics, Climate, Physical Meteorology, Weather Systems, and Measurements, Edited by Thomas D. Potter and Bradley R. Colman.
ISBN 0-471-21490-6

not be possible or feasible to switch to a different vendor or system because of compatibility problems. This is not necessarily bad but should be considered when choosing a manufacturer. For example, one should consider if the manufacturer will be in business when network upgrades are required and, if not, how much of the system will require replacement.

The text draws from the experiences of the Oklahoma Mesonet (Brock et al., 1995) as well as the Atmospheric Radiation Measurement (ARM) Program (Stokes and Schwartz, 1994; DOE, 1990); a brief description of both will be given to put the subsequent writing in context. More details on network design and instrument performance can be found in *Meteorological Measurement Systems* by Brock and Richardson (1991). *An Introduction to Meteorological Instrumentation and Measurement* by T. P. DeFelice (1998) is also a good reference.

The Oklahoma Mesonet

The Oklahoma Mesonet is a system of 115 remote surface observing stations across the state of Oklahoma and is a joint effort between Oklahoma State University and the University of Oklahoma. Parameters measured at all 115 sites include pressure, wind speed and direction at 10 m above ground level (agl), air temperature and relative humidity at 1.5 m agl, rainfall, global solar radiation, and ground temperature at 10 cm under bare soil and native ground cover. Additional parameters measured at about half the sites include wind speed at 2 m agl, air temperature at 9 m agl, soil moisture at 5, 25, 60, and 75 cm below ground, and additional soil temperature measurements down to 30 cm. Surface flux measurement capabilities are being added to the Oklahoma Mesonet through the OASIS (Oklahoma Atmospheric Surface-layer Instrumentation System) Project (Brotzge et al., 1999). Once completed in 2000, Oklahoma Mesonet sites will measure net radiation, ground heat flux, and sensible flux, and latent heat flux using a profile technique (at 90 sites) and eddy correlation techniques (at 9 sites).

Five-minute averaged data are collected from all 115 stations every 15 min and transmitted through the Oklahoma Law Enforcement Telecommunication System (OLETS) to an archival system at the University of Oklahoma. All stations are solar-powered with battery backup; site maintenance is typically performed 3 to 4 times per year. The system was commissioned in 1994 and each year more than 99.9% of possible observations (approximately 15,537,000 of a possible 15,547,000 observations) are archived. The use of OLETS was important to the success of the Oklahoma Mesonet, since it and allowed data collection and two-way communication between the base station and all remote sites without the need for expensive phone lines.

Atmospheric Radiation Measurement Program

The ARM Program is a multilaboratory, interagency program that was created in 1989 with funding from the U.S. Department of Energy (DOE). The ARM Program is part of the DOE effort to resolve scientific uncertainties about global climate change with a specific focus on improving the performance of general circulation models (GCMs) used for climate research and prediction. These improved models will help scientists better understand the influences of human activities on Earth's climate.

In pursuit of its goal, the ARM Program establishes and operates field research sites, called Cloud and Radiation Testbeds (CARTs), in several climatically significant locations (the north slope of Alaska, the tropical western Pacific, and the U. S. Southern Great Plains). Data are collected over extended periods of time (years) from large arrays of instruments (both state-of-the-science and conventional instrumentation are used) to study the effects and interactions of sunlight, radiant energy, and clouds on temperatures, weather, and climate. Specifically, ARM focuses on cloud–radiation interactions.

The ARM Program has taken advantage of the advances in communications and the World Wide Web. Data are ingested and available in near real time via the Web, trouble reports are submitted via the Web, complete site and instrument history information is archived and available via the Web, etc.

This chapter will emphasize design, calibration, and quality assurance needs of smaller networks such as the Oklahoma Mesonet or smaller.

2 SYSTEM DESIGN

Sensors are typically mounted on a stationary platform (a simple mast or tall tower) or on a moving platform (balloons, planes, ships, etc.). Ideally, data are communicated in real time from the measurement site or platform to a central archiving facility. In some cases, real-time communication is not possible but, instead, data are manually collected at periodic intervals, usually in some electronic form. Availability of electrical power, or the lack of it, may seriously affect the system design.

Instrument Platforms

It is not surprising that virtually every type of instrument platform is used in meteorology because the atmosphere is so extensive and because most of it is quite inaccessible. These platforms include masts, instrument shelters, tall towers, balloons, kites, cars, ships, buoys, airplanes, rockets, and satellites. Synoptic data platforms include balloons and satellites supplemented by buoys and ships over the ocean. In addition, aircraft are used for hurricane observation and some data are collected from commercial flights to fill in gaps in the observation networks. Aircraft are extensively used for research investigations around thunderstorms or wherever high-density upper air data are needed.

When selecting a platform, consideration should be given to where the measurement is to be made and whether the platform can be permanently fixed or is moving, cost, and exposure. To some extent, any platform, even a simple tower for surface measurements, interacts with the atmosphere and affects instrument exposure. A simple 10-m tower, shown in Figure 1, has a wind sensor at 10 m and temperature and a relative humidity (T&RH) sensor at 1.5 m in addition to a radio antenna for data transmission, a solar panel and battery for power, a barometer, and a data logger. These sensors must be mounted with due consideration for exposure to prevailing winds to minimize tower effects.

Communication Systems

A communications network is a vital part of almost every meteorological measurement system at all scales. Historically, meteorological communications have relied primarily on land-line and radio links. More recently, polar orbiting and geostationary satellites are used for data communications in macroscale or synoptic measurement systems and even in many mesoscale systems. Commercial satellites are used to broadcast data from central points, with sophisticated uplinks, to users equipped with fairly simple antennas and receivers (inexpensive downlinks).

The ideal communications system would reliably transmit data from the remote instrument platform to a central facility and in the reverse direction with little or no time delay, without limiting the volume of data to be transmitted. Communication both to and from the remote site is used to synchronize local clocks in the data loggers, to load operating programs into the data loggers, and to make special data requests, to name a few. Two-way communications is not essential but highly desirable.

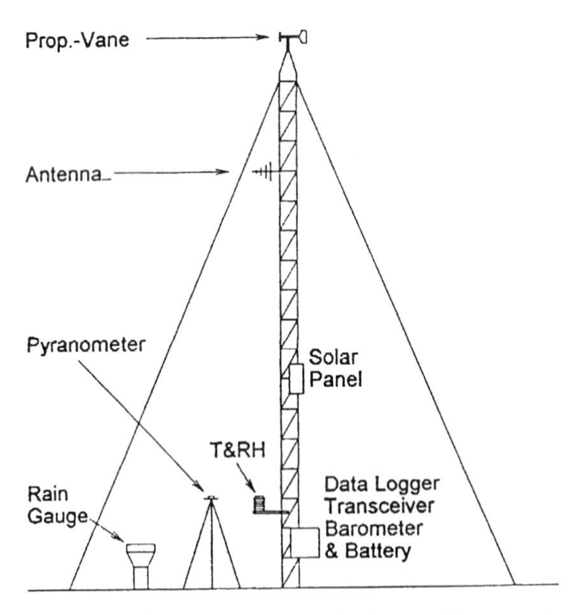

Figure 1 A 10-m tower for surface measurements. At the top of the tower is a propeller-vane anemometer that measures wind speed and direction. Below the anemometer is an antenna for two-way data communications, and power is provided by a solar panel. The pyranometer is mounted to the south where tower and guy-wire shadows will not affect the data.

Telephone Commercial telephone systems provide adequate signal bandwidth, are generally reliable, and cover most land areas. The cost is prohibitive if one must pay for running lines to each station, especially for a short-term project. Even for long-term projects like the Oklahoma Mesonet and the ARM Program, phone lines were either avoided entirely or used very sparingly due to the expense involved.

Recent advances in cellular telephone technology coupled with decreasing airtime charges means cellular data communications has become a viable alternative to traditional phone lines.

Direct Radio Direct radio links from the remote stations to a central base station are desirable because they offer flexibility but Earth curvature limits line-of-sight links. Figure 2 shows the maximum line-of-sight distance between two stations if the remote station antenna is at a height of 10 m. For a base station or repeater antenna height of 200 m, the line-of-sight link is only a little more than 60 km. Direct radio, even when augmented with repeaters, severely limits the size of a network and causes immense difficulties in complex terrain. For example, if the path of the signal from a remote station to the repeater or to the base station is too close to the ground, the signal could be trapped in an inversion layer and ducted away from the intended destination, thereby losing the connection between the sensor and the base station.

Satellite The first communications satellite that permitted an inexpensive uplink (low-power transmitter and simple antenna) was the *Geostationary Operational Environmental Satellite* (*GOES*). An inexpensive uplink is essential when a large

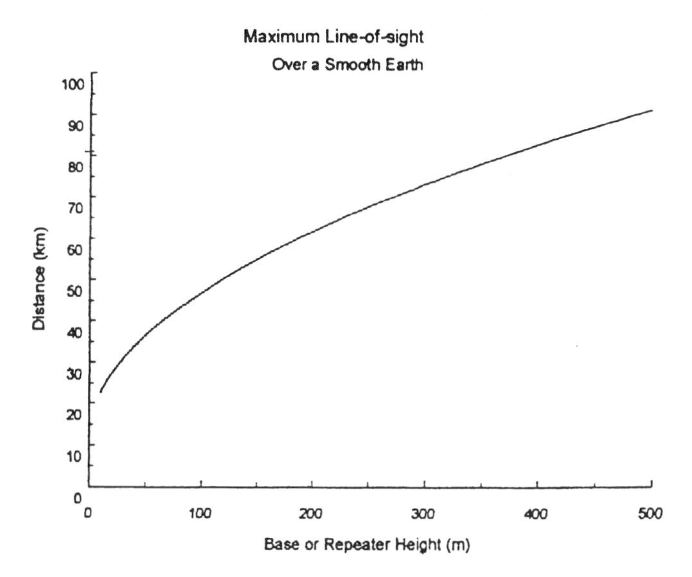

Figure 2 Line-of-sight distance over a smooth earth as a function of height of one end of the link when the other end is fixed at 10 m.

number of remote stations are involved. In addition, stations may be powered by batteries and solar panels, thereby requiring low-power radio transmitters. As satellite communications technology evolves, communication restraints will be eased, which will lead to vastly improved meteorological networks, especially for the mesoscale.

Power Source Electrical power consumption of a measurement system is often a vital consideration; the primary concern is cost. Where commercial power is available, cost is not usually a problem. However, many systems are required to be portable or to operate in locations where commercial power is not available. In these cases, the power source is usually batteries, perhaps supplemented with solar panels. Such systems must operate on a severely limited power budget and that constraint affects the selection of components and the overall system design. Battery-powered systems are constrained to select sensors with low power consumption and/or to switch the sensors on only as needed to conserve power. Heaters generally cannot be used and local computational capability may be severely limited. Therefore, all components must be rated for operation over the expected temperature range.

Unique methods are sometimes used to power remote meteorological stations and are often used out of necessity. The following is an example of what can be done when no traditional power source is available.

The year-long SHEBA (Surface Heat Budget of the Arctic Ocean) field experiment used PAMII ~1 (the third-generation Portable Automated Mesonet) remote meteorological stations and was located well above the Arctic Circle. PAMIII stations require 1.5 to 2.5 A at 12 V dc for continuous operation and, for most deployments, the power system consists of photovoltaically charged deep-cycle storage batteries with nominal capacity of 3 to 6 days of operation in the absence of sunshine. Obviously, modifications were required for SHEBA due to the lack of sunshine during the winter months and the extreme cold. Three separate power sources were eventually used for differing reasons and at different times of the year.

The primary source of power for each station was a small propane-fueled 12 V dc electro-thermal generator (TEG) and a single battery. To mitigate the effects of extreme cold, the generator, 2 propane bottles, the battery, and PAIII electronics were all placed inside an insulated container mounted on a sled for transportation between sites. Two 33-lb propane bottles allowed for 15 to 20 days of unattended operation. Heat from the TEG kept the electronics and more importantly the propane and battery warm when outside temperatures dropped below $-15\,^{\circ}$C. In the spring two photovoltaic panels were installed at each station and paralleled with the TEG to augment system power. In addition, two small wind generators were added in spring to the least accessible remote stations. The wind generators operated at 12 V dc that allowed parallel operation with both the TEG and photovoltaics through the PAIvIIIiI battery charging and monitoring system.

Some problems were encountered with the TEG, caused by poor fuel quality. However, the system described above was in general successful and was a unique solution to a difficult situation.

3 CALIBRATION

The calibration needs of a network depend on, among other factors, the desired accuracy of the field measurements. Sufficiently accurate laboratory standards are readily available for most meteorological sensors, but acquiring this same level of accuracy in the field is often very difficult if not impossible. For example, it is relatively easy to calibrate temperature sensors in the lab to an accuracy of $\pm 0.1^{\circ}$C or even $\pm 0.01^{\circ}$C. However, measuring atmospheric air temperature with this same accuracy requires great care and knowledge of the measurement system. Therefore, extremely accurate laboratory calibrations may not be required in some situations. Nevertheless, the total error associated with an instrument in the field is the square root of the sum squared errors (assuming the errors are independent) and, therefore, the total measurement error can be reduced if a careful laboratory calibration is performed.

Calibration standards are maintained by standards laboratories in each country such as the National Institute of Science and Technology (NIST) in the United States. Standards for temperature, humidity, pressure, wind speed, and for many other variables are maintained. The accuracy of these standards is more than sufficient for meteorological purposes. Every organization attempting to maintain one or more measurement stations should have some facilities for laboratory calibration including transfer standards. These are standards used for local calibrations that can be sent to a national laboratory for comparison with the primary standards. This is what is meant by traceability of sensor calibration to NIST standards. Ideally the calibration of all sensors can be traced back to such a standards laboratory.

Initial calibration of all instruments should include a full range test if at all possible. Even when instrumentation is new from the manufacturer, *all instruments should be tested prior to use in the field.* Testing each instrument can be time consuming and expensive but to not do so is a mistake that will inevitably cost many times more. Even seemingly simple instruments require a laboratory check before field use. For example, a tipping bucket rain guage has an output that is a function of rain rate and not the same for all instruments; therefore, a different calibration curve may have to be developed for each sensor. In addition, slight variations in bucket volumes can result in differences in measured rainfall and can be accounted for. As each rain guage is "calibrated," sensor deficiencies (e.g., faulty bearings or closure switches) may be detected prior to field deployment where such problems could go undetected.

The importance of checking, if not calibrating, each sensor prior to field use cannot be overemphasized. Instrument manufacturers, no matter how reputable, may not be able to maintain the desired standards. Even if sensor specifications meet or exceed the required accuracy, instruments can drift or be damaged during shipment and storage or electronic components can fail. Therefore, to maintain the highest level of confidence in a network's data, sensor performance should be verified prior to use in the field.

Fortunately, in many cases it is not necessary to purchase expensive calibration chambers to achieve the level of accuracy required for meteorological measurements. Simple yet effective "calibration" methods that require a minimal investment can be developed for many meteorological instruments (calibration is put in quotes to indicate that true calibrations, which require a sensor with an order of magnitude more accuracy, may not be occurring). For example, rain guage calibration can be performed with a reasonably accurate electronic scale with RS-232 or analog output, a simple water reservoir, a simple flow valve, and a data collection system like that used in the network. The scale output is connected to a computer (or datalogger) and the rain guage is connected to the data collection system. The water reservoir is placed on the scale and water is siphoned off into the rain guage. The scale output (weight of water remaining in the reservoir) is easily related to the water input to the rain guage and the siphon rate can be adjusted to simulate different rain rates. This calibration does not account for many factors that can affect the rain guage accuracy but does provide a simple yet effective check of the instrument performance.

An equally simple setup can be developed to test temperature sensors (Richardson, 1995). Required equipment includes a temperature transfer standard with errors less than about 0.030°C, a temperature chamber (e.g., a small freezer), a bath, and a stirrer. The use of a temperature standard means it is not necessary to maintain precise chamber temperatures nor is it necessary to hold the chamber temperature constant; it is only necessary to control the rate of change of temperature. The rate must be low enough that errors induced by spatial gradients between the temperature standard and the test sensors and the errors induced by the time response of the sensors are small compared to the acceptable calibration error.

Dynamic errors are fairly easy to detect if the test sensors lag the reference sensor during increasing temperature, and lead it when the temperature is falling, the rate of change is too high. The response of the sensors will be a function of the kind of bath (air or liquid) and the amount of stirring. The bath also affects spatial gradients. These can be detected by correlating errors with sensor position.

4 QUALITY ASSURANCE

Quality assurance methods and/or final data format may depend on the primary end user of the data, e.g., will they be used for research or by the general public? For example, relative humidity (RH) sensor inaccuracy specifications allow for the RH reported by the sensor to be in excess of 100% (as high as 103%). A sensor reporting an RH of 102% may be operating correctly but could cause problems for under-informed users. For example, modelers ingesting RH may need to be aware that values greater than 100% are possible and do not necessarily indicate supersaturation conditions. In addition, those unaware of instrument inaccuracy specifications may think the data is incorrect if RH is above 100%.

A more complete data quality assurance program can be developed for a network of stations than for a single measurement station, so this discussion will address the issue of quality assurance for a network. A single-station assurance program would be a subset of this. In designing a measurement system, it is useful to consider the impact of automation on data quality.

Schwartz and Doswell (1991) claim that automation sometimes has been accompanied by a decrease in data quality but that this decrease is not inevitable. When a measurement system is automated, some sources of human error are eliminated such as personal bias and transcription mistakes incurred during reading an instrument, to name only two. However, another, more serious form of error is introduced: the isolation of the observer from the measurement process.

One tends to let computers handle the data under the mistaken assumption that errors are not being made or that they are being controlled. Unless programs are specifically designed to test the data, the computer will process, store, transmit, and display erroneous data just as efficiently as valid data. Automatic transmission of data tends to isolate the end user from the people who understand the instrumentation system. Utilization of computers in a measurement system allows data to be collected with finer time and space resolution. Even if the system is designed to let observers monitor the data, they can be overwhelmed by the sheer volume and unable to effectively determine data quality.

Automation, or the use of computers in a measurement system, can have a beneficial impact on data quality if the system is properly designed. Inclusion of data monitoring programs that run in real time with effective graphical displays allow the observer to focus on suspect data and to direct the attention of technicians.

The objective of the data quality assurance (QA) system is to maintain the highest possible data quality in the network. To achieve this goal, data faults must be detected rapidly, corrective action must be initiated in a timely manner, and questionable data must be flagged. The data archive must be designed to include provision for status bits associated with each datum. The QA system should never alter the data but only set status bits to indicate the probable data quality. It is probably inevitable that a QA system will flag data that are actually valid but represent unusual or unexpected meteorological conditions. Flagged data are available to qualified users but may not be available for routine operational use.

The major components of a QA program are the measurement system design, laboratory calibrations, field intercomparisons, real-time monitoring of the data, documentation, independent reviews, and publication of data quality assessment. Laboratory calibrations are required to screen sensors as they are purchased and to evaluate and recalibrate sensors returned from the field at routine intervals or when a problem has been detected. Field intercomparisons are used to verify performance of sensors in the field since laboratory calibrations, while essential, are not always reliable predictors of field performance. QA software is used to monitor data in real time to detect data faults and unusual events. A variety of tests can be used from simple range and rate tests to long-term comparisons of data from adjacent stations. Documentation of site characteristics and sensor calibration coefficients and repair history is needed to answer questions about possible data problems.

Independent reviews and periodic publication of data quality are needed since people close to the project tend to develop selective awareness. These aspects of the overall QA program must be established early and enforced by project leaders. The whole QA program should, ideally, be designed before the project starts to collect data.

Laboratory Calibrations

As discussed previously, laboratory calibration facilities are required to verify instrument calibration and to obtain a new calibration for instruments that have drifted out of calibration or have been repaired. However, a laboratory calibration is not necessarily a good predictor of an instrument's performance in the field. This is because laboratory calibrations never replicate all field conditions. For example, a laboratory calibration of a temperature sensor does not include all the effects of solar and Earth radiation nor is it likely to simulate poor coupling with the atmosphere due to low wind speeds. This type of error is called "exposure error" and is a result of inadequate coupling between the sensor and the environment. Exposure errors can be very large (e.g., radiative heating of temperature sensors can result in air temperature errors in excess of 5°C; Gill, 1983) and are not accounted for in lab calibrations. In addition, the manufacturer does not specify this type of error because it is not under its control. Therefore, knowledge of the measurement system and instrument design is crucial to minimize measurement error.

Field Intercomparisons

Two types of field intercomparisons should be performed to help maintain data quality. First, a field intercomparison station should be established and, second, when technicians visit a station, they should carry portable transfer standards and make routine comparison checks.

The field intercomparison station should be comprised of operational sensors and a set of reference sensors (higher quality sensors). Both should report data to the base station but the reference station data should be permanently flagged and not made available for operational use.

Portable transfer sensors can be used to make reference measurements each time a technician visits a station. This method can detect drift or other sensor failures that could otherwise go undetected. These sensors can include a barometer and an Assmann psychrometer. In addition, technicians can carry a lap-top computer to read current data, make adjustments to calibration coefficients when sensors are changed, set the data logger clock (*if* it cannot be set remotely), and reload the data logger program after a power interruption.

Data Monitoring

Neither laboratory calibration nor routine field intercomparisons will provide a real-time indicator of problems in the field. In a system that collects and reports data in real time, bad data will be transmitted to users until detected and flagged. The volume of data flow will likely be far too great to allow human observers to effectively monitor the data quality. Therefore, a real-time, automatic monitoring system is required.

The monitor program should have two major components: scanning algorithms and diagnostic algorithms. The function of the scanning algorithms is to detect outliers while the diagnostic algorithms are used to infer probable cause. The monitor program can analyze the incoming data stream using statistical techniques adapted from exploratory data analysis, knowledge of the atmosphere, knowledge of the measurement system, and using objective analysis of groups of stations during suitable atmospheric conditions.

Exploratory data analysis techniques are resistant to outliers and are robust, i.e., insensitive to the shape of the data's probability density function. Knowledge of the atmosphere allows us to place constraints on the range of some variables such as relative humidity that would be flagged if it were reported greater than approximately 103% (due to sensor inaccuracy specifications). Knowledge of the measurement system places absolute bounds on the range of each variable. If a variable exceeds these limits, a sensor hardware failure may have occurred.

The monitor program must be tailored for the system and should be developed incrementally. Initially, it could employ simple range tests while more sophisticated tests can be added as they are developed. The QA monitoring program will never be perfect; it will fail to detect some faults and it will label some valid data as potentially faulty.

Therefore, *the monitor program must not delete or in any way change data* but set a flag associated with each datum to indicate probable quality. The monitor program should have a mechanism to alert an operator whenever it detects a probable failure. Some of these alerts will be false alarms, i.e., not resulting from hardware failure, but may in fact indicate interesting meteorological events.

Documentation

There are several kinds of documentation needed: Of individual station characteristics, a station descriptor file, and a sensor database.

Station characteristics can be documented by providing an article describing the station and its instrumentation. In addition, there should be a file of panoramic photographs showing the fetch in all directions and the nature of the land. Aerial photographs can also provide valuable information about a site, as can high-resolution topographical maps.

As part of the system database, a system descriptor file must include the location and elevation of each station and the station type (e.g., standard meteorological station, special-purpose agricultural station, or sensor research station).

It is necessary to maintain a central database of sensors and other major components of the system including component serial number, current location, and status. Some sensors have individual calibration coefficients so there must be a method of accounting for sensors to ensure that the correct calibration coefficients have been entered into the appropriate data logger. It is also necessary to keep records of how long a component has been in service and where it was used so that components that suffer frequent failures can be identified. This would help to determine if the component was seriously flawed or if the defect was characteristic of the component design.

This kind of accounting cannot be left to chance; if it is, sensors will inevitably be matched with the wrong calibration coefficients or put back into service without having been repaired or recalibrated. Some sensors require periodic recalibration, but it is not feasible to recalibrate them all at once. Therefore a formal database system should be set up. All technicians should be required to report maintenance activity including swapping components, and this information should be entered into the database. The database system should be able to generate reports indicating the serial number of every component at a station, the number of components awaiting repair at any given time, the number of spares available, the history of any given sensor or sensor type, etc.

Independent Review

For much the same reason that scientific proposals and papers are reviewed, periodic, independent reviews of a network's performance should be invited. It is always possible for people in constant close proximity to a project to become blind to problems; an independent review would help alleviate this at the root cause.

Publication of Data Quality Assessment

There will be frequent data faults in any network even with the data quality assurance program outlined above. To assist critics in making a realistic assessment, it is desirable to publish, periodically, an honest appraisal of the network performance including all data faults, causes when known, and action taken.

REFERENCES

Brock, F. V., K. C. Crawford, R. L. Elliott, G. W. Cuperus, S. J. Stdaler, H. L. Johnson, and M. D. Eihs (1995). The Oklahoma Mesonet: A technical overview. *J. Atmos. Oceanic Technol.* **12**, 5–19.

Brock, F. V., and S. J. Richardson (2001). Meteorological Measurement Systems, Oxford University Press, Inc. New York.

DeFelice, T. P. (1998). *An Introduction to Meteorological Instrumentation and Measurement*, Prentice Hall, New Jersey.

DOE (1990). U.S. Department of Energy, Atmospheric Radiation Measurement Program Plan, DOE/ER 0441, National Technical Information Service, Springfield VA.

Gill, G. C. (1983). Comparison Testing of Selected Naturally Ventilated Solar Radiation Shields, Report submitted to the NOAA Data Buoy Office, Louis, MS.

Richardson, S. J. (1995). Automated temperature and relative humidity calibrations for the Oklahoma Mesonetwork, *J. Atmos. Oceanic Tech.* **12**, 951–959.

Schwartz, B. E., and C. A. Doswell III (1991). North American rawinsonde observations: Problems, concerns and a call to action, *Bull. AMS* **72**, 1885–1896.

Stokes, G. M., and S. E. Schwartz (1994). The Atmospheric Radiation Measurement (ARM) Program: Programmatic background and design of the cloud and radiation test bed, *Bull. Am. Meteor. Soc.* **75**, 1201–1221.

CHAPTER 41

DATA VALIDITY IN THE NATIONAL ARCHIVE

GRANT W. GOODGE

1 INTRODUCTION

As we enter a new millennium the United States and indeed the whole world is becoming increasingly sensitive to various meteorological and climatological extremes. Given the potential for economic and political disruption as a result of these anomalies, there has been an increased interest in their prediction. Such predictions include frequency, magnitude, areal extent, and location. As of this time one of our best methods of understanding the potential for future climatic anomalies is to look at the past climatic record.

The integrity of any scientific research depends a great deal on the quality of the data used as well as the user's understanding of the data and all relevant information associated with its collection and processing (metadata).

These metadata include, but are not restricted to, instrument type, exposure, automatic, manual, maintenance/calibration, and the nature of data transfer from the point of measurement to archive. The final archive for much of the meteorological data for the United States and its territories resides at the National Climatic Data Center (NCDC) in Asheville, North Carolina. A considerable volume of similar data from other countries and the world's oceans are also archived at NCDC. The NCDC is one of several environmental and geophysical data centers in the United States and is administered by the National Environmental Satellite Data and Information Service (NESDIS), which is one of the major components of the National Oceanic and Atmospheric Administration (NOAA).

Handbook of Weather, Climate, and Water: Dynamics, Climate, Physical Meteorology, Weather Systems, and Measurements, Edited by Thomas D. Potter and Bradley R. Colman.
ISBN 0-471-21490-6

The NCDC is not directly involved with the observation of meteorological variables, sensor maintenance, or observer training. That portion of data collection is managed by the National Weather Service (NWS), which is also a major component of NOAA. The NCDC does, however, actively engage in providing relevant feedback to NWS managers concerning sensor and/or observer problems that are discovered during the quality assurance processes at NCDC.

Unfortunately, errors in the meteorological databases may be introduced anywhere between the point of observation and its final storage in the digital archive or websites. Several examples are observer transcription errors, radio frequency (RF) interference or lightning damage to electronically driven sensors, transmission signal interruption, keying errors of manually observed data during data entry and/or update during quality assurance, and computer program transfer of those files to the digital archives or websites. Also an increasingly problematic issue in today's electronic world is that of missing data. This can be of particular significance with the transmission of data from the Automated Surface Observing System (ASOS) where data are stored on site for only 12 h. Obviously, any interruption of the link between the sites and NCDC that exceeds this period will result in the permanent loss of data from affected sites until the link is restored. We will return to a more detailed discussion of some of these issues later in this chapter.

Before proceeding further, this writer would like to note that within the constraints of limited budgets and personnel, NCDC has always endeavored to produce high-quality databases and publications. Any discussion of the weaknesses in the observational systems and the quality assurance of the resulting observations are mentioned only to alert the research community to known data problems and encourage a better understanding of the data prior to use in any study. It is *not* intended as an indictment or criticism of those responsible for the collection and archive of these environmental observations.

2 TEMPERATURE

In the recent era there has been much public and scientific debate of the issue of "global climate change," and many of those debates center on the parameter of temperature. Certainly the fact that the magnitude of atmospheric model predicted centennial mean global temperature rises are on the same order (1°F) as the required accuracy (±1°F) of NOAA-sponsored instruments indicate the necessity of quality data. Prior to the 1960s official Weather Bureau thermometers were all compared against a National Bureau of Standard's "certified" thermometer, and any mercurial thermometer having a greater error than 0.3°F (above 0°F) from the certified standard was not accepted for observational use. During the 1960s most primary Weather Bureau Stations transitioned from the use of mercurial thermometers to an HO-60 series temperature sensor that employed the use of a fan-aspirated thermistor exposed in a metal shield with the sensed temperature electrically transmitted to an electromechanically driven dial in the weather office. Unfortunately, some of these instruments did not meet the more relaxed standard of ±1°F as compared to the ±0.3°F standard of the mercurial thermometers they replaced.

In one documented case the mean temperatures, sensed by the HO-62 hygrothermometer, at the Asheville, North Carolina, NWS Airport station, migrated upward more than 4°F (as compared to three surrounding cooperative stations) in less than 4 years (1965–1969). Since that time there have been other migrations and/or shifts of ±2°F. The most recent of these have been as a result of sensor changes and/or relocations at the airport. Certainly this station's record would not be a good candidate for climate trend determinations. Unfortunately, sensor-driven data errors and discontinuities for this station and others like it remain unadjusted in the official databases and archives and in some cases undocumented.

There are several reasons why NCDC does not adjust or correct such errors during the quality assurance (QA) process. The primary reason is that NWS stations (i.e., Chicago, Charlotte, Chattanooga, etc.) are quality assured on a single station basis and as a result are unable to identify instrument "drift" unless that drift were to exceed station monthly climatic extremes or cause conflict with another reported value such as ambient temperature versus dew-point temperature. Second, there are seldom any coincident data that would allow comparison or correction of the station in question. Lastly, the historic digital data storage at NCDC has been sequential in nature, which makes the correction of identified, postprocessing errors very expensive.

It should be noted that a change in a temperature sensor's exposure can make as much difference in the observed readings as a change in the type of sensor. Relocation of meteorological instruments at primary weather stations has been a problem for climatologists for many years. Some have been moved from rooftops of post offices in a downtown (urban) environment to rooftops of terminal buildings at airports and finally from the terminal buildings to airport field level exposures. After moving to the field level, sensors have continued to be moved due to airport expansions or alterations. Even if the sensor has not moved, increased extent of paved surfaces may alter the radiative flux of the area surrounding the sensor.

These relocations of sensors highlight the importance and need for timely and accurate station history information called metadata. Without it, a user or researcher is left to guess as to what may have caused the sudden change in the data. Accurate metadata is just as important in the documentation of observations from the cooperative observer network. Even though these stations only record data on a daily basis, it is still essential that any retrospective user know the type of instruments used, the exposure, and the time of day when the readings are made. Unfortunately, NCDC's receipt of the B-44 (metadata) forms lagged several months behind the date of actual station move or instrument change, and thus the publication and databases improperly indicated the location, instrument type, time of observation, or other critical station information.

In about 1984 the NWS began replacing the liquid-in-glass thermometers at cooperative observing sites with an electronic resistance thermometer (hereafter referred to as the Maximum Minimum Temperature System (MMTS)). The sensing accuracy of these MMTS units are generally compatible with the liquid-in-glass thermometer they replaced, but the introduction of the MMTS brought several new undesirable results. First is its susceptibility to electrical interference whether that be generated by radio frequency (RF) from nearby radio transmitters or induction surges from nearby lightning strikes. Usually, the display inside the observer's home or the thermistor outside will be burned out by a nearby lightning strike and result in data loss until the field manager either makes a site visit to replace the unit or ships the affected part(s) to the observer for their own replacement. Other causes of a data loss from the MMTS resulted from the connecting cable being cut by nearby utility construction or in the western states by various energetic burrowing animals. Electrical power interruptions to the units either by thunderstorm or winter snow and ice have also caused data loss over the years. These data losses can be significant when the power outages are widespread and extended, such as was the case in the New England ice storm of January 1998.

As undesirable as the above-mentioned data losses are, they are at least apparent and recognizable to the user. This, however, is not true when RF, lightning, or the salt air of the marine environment cause small changes in the resistance in the electrical loop between the thermistor and the readout in the observer's home or business. Such a resistance change caused abnormally cool temperatures, some of which were below freezing, to be recorded at a low elevation site in Hawaii. In another case, a lightning strike caused a resistance change equal to about 4.5°F that went undetected by the observer, NWS, and NCDC for 6 months. In these cases the errors were not detected by NCDC's automated QA program but rather by quality assurance specialists.

3 PRECIPITATION

In terms of identifying global climate changes, anthropogenic or natural, quality precipitation data is equally as important as temperature; however, because of the noncontinuous temporal nature and spatial variability of precipitation, its quality assurance is more difficult. As was the case with temperature data, errors in precipitation data can be introduced at any stage from the original observation to the final quality assurance and archive.

Before continuing with the discussion of precipitation data, it should be noted that there are many more uses of meteorological data than just long-term climate analysis. In fact there are thousands of users that depend on high-quality data, and since many of those users are interested in the condition at a single point and time, they cannot afford the perceived luxury expressed by some climatologists that errors in the data sets will likely be random and thus average out over the long run. These "single point and time" uses of the data range from the meteorological conditions at the site of an auto accident, aircraft mishap, or the structural failure of a bridge or building. Even though the original reasons for the collection of meteorological data did not envision such applications, they are none the less vital to our understanding of the effects of extreme meteorological conditions on human activities and their supporting infrastructure.

As in the case of temperature measurements, one of the best assurances of quality precipitation data at the observation site is to have some element of redundancy or reality check. The high costs of even the basic manual precipitation gages employed by the government-sponsored networks proved to be too expensive to install more than one identical gage at most cooperative stations. There are, however, about 850 of those stations that are equipped with two gages, one manual and the other automatic. We will discuss this subject later in the chapter. There also was more than one precipitation gage located at most primary NWS sites that were staffed by trained government or contract weather observers.

Observational procedures at these sites, however, did not require a comparison of the measured amounts of each gage. The high quality of precipitation data from these sites came primarily through the aspect of their reporting of hourly, 6 hourly, and 24 hourly amounts, which provided for cross reference at the station as well as at NCDC when processing the data for archive and publication.

Unfortunately this cross-referencing process was significantly compromised with the advent of the Automated Surface Observing System (ASOS) in the mid 1990s. ASOS is equipped with a heated tipping bucket precipitation sensor that does not always dependably and accurately measure precipitation. Its greatest difficulty occurs during periods of frozen or freezing precipitation. In the years prior to ASOS, human observers monitored conditions so they could resolve such problems through the use of a backup gage or make an estimation of precipitation amounts; however, now with no observers at most ASOS sites to augment the automated system, the weather indicator Light Emitting Diode Weather Identifier (LEDWI) may report snow while the precipitation gage does not report any corresponding meltwater equivalent and, if temperatures warm to near or above freezing, the snow will melt and produce a false time and intensity for the precipitation event.

Even in those cases where observers are in place to augment the ASOS observations, they do not have the ability to alter the one-minute data file from which the hourly precipitation files and publications are produced, the net effect being that no longer does the hourly precipitation data from the ASOS sites always agree with the 6-hourly, daily, or monthly totals. Needless to say such discrepancies seriously undermine the integrity and use of the data, particularly in litigation or insurance adjustment cases.

As was earlier mentioned about 850 of the cooperative weather stations are equipped with both manual and automatic precipitation gages. One of the gages is the "standard 8," which has an 8-in. diameter receiving funnel that directs the precipitation into a measuring tube that is one-tenth the area of the receiving funnel and thus expands the true depth of the rainfall by a factor of 10 and provides for an accurate measurement of the precipitation to one hundredth (0.01) of an inch. Today the vast majority of the 2250 automatic gages are Fischer–Porter type of gages, which also have an 8-in. diameter receiving chute that directs the precipitation into a large basin where the precipitation is weighed. That weighed amount is recorded onto a binary punched paper tape every 15 min to a maximum precipitation resolution of one tenth (0.10) of an inch.

The use of a weighing-type recording precipitation gage in the cooperative network began in the mid-1930s and continued until the mid-1960s when the Fischer–Porter type began to replace it. This earlier weighing gage was known as the Universal and was not only used at about 3000 cooperative stations but also became the primary gage for the primary NWS stations from the early 1960s until the mid-1990s when they were phased out with the commissioning of ASOS.

There were two major advantages of the Universal-type gage over the Fischer–Porter. First the Universal recorded data to the hundredth of an inch (0.01), and secondly it made a constant trace of the precipitation event on a chart that allowed a time resolution of as little as one minute. The accepted world record 1 min rainfall of 1.23 in. (3.12 cm) was extracted from one of the Universal autographic charts.

For quality assurance purposes at NCDC it has been unfortunate that for many years the data from the co-located standard and automated gages at the cooperative stations could not be compared during quality assurance and publication because the receipt and reduction of the punched paper tapes and or autographic charts lagged about a month or more behind that of the manually reported daily data forms.

The resulting inability to compare the data from these co-located gages prevented NCDC from effecting a higher level of QA for those sites and in some cases allowed significantly different precipitation amounts to be published for the same location. The worst known difference was published for a site in Puerto Rico. A heavy tropical rain event moved into the south central part of the island in October, 1985. The heavy rains washed out a major highway bridge during the night hours and allowed 29 motorists to drive to their death, one vehicle at a time.

A $65 million lawsuit was later initiated. Obviously, the true amount of rainfall was critical to the case as well as the knowledge of the true return period rainfall potential for the area. Unfortunately, the official published daily total from the manual 8-in. gage was 8.80 in. greater than the sum of the hourly values later published for the co-located automated Fischer–Porter gage.

In this case the automated gage was equipped with a self-siphoning overflow feature that not only drained the collecting bucket when it reached its capacity, but because of the extremely heavy rain, continued to siphon until the heavy rain ended, and thus the gage never accumulated or recorded the balance of the precipitation during the storm.

Fortunately, a more timely receipt and processing of the Fischer–Porter punched paper tapes now allows NCDC to do a comparison of the co-located station's precipitation records. Results of this QA process have also aided NWS cooperative network field managers to identify and correct gage or observer problems in a more timely fashion.

During the twentieth century various automated precipitation gages have been employed not only to measure and report precipitation in real time but also in a retrospective mode to help identify the frequency and nature of extreme precipitation events throughout the United States and its territories. Any user of these data files should pay particular attention to the significant changes that have occurred in the measurement and data extraction over the years.

The most significant of these changes occurred in 1973 when the derived extreme values changed from that of "excessive precipitation" to "maximum short duration precipitation." Excessive precipitation data began to be summarized in 1896 and continued (with several procedural changes) until 1973 when it was terminated in favor of the simpler maximum short duration precipitation (MSDP), which continues to this day. Both the excessive and MSDP systems used the same time periods (5, 10, 15, 20, 30, 45, 60, 80, 100, 120, 150, and 180 min) for which amounts were determined; however, there were two fundamental differences in their determination.

The first was that in determining values for excessive events there were certain precipitation intensity criteria that had to be met before any values were computed. There were some variations in the criteria over the early years, but most centered around that of $.01 \times t + 0.20$ in. (where $t =$ time in minutes) except in the southeastern third of the United States where $.02 \times t + 0.20$ in. was used until 1949 when these southeastern states reverted back to $.01 \times t + 0.20$, which placed the whole nation on the same criteria. Anytime a precipitation event met the above criteria at a given station, excessive precipitation values were determined for that event, and, if there were 10 excessive events during a station-month, then there would be 10 events documented.

This brings us to the second primary difference between the excessive and MSDP systems. Under MSDP there is no intensity criteria that must be met before computation. Therefore, if the maximum intensity for an event were only 0.01 in. in a station-month, it would be summarized and become the extreme for the month. That is because under MSDP procedures there is one extreme summarized, recorded, and published for each station for each month. As was earlier indicated, the excessive method might have any number (10, 12, 15, or none) excessive events in a station-month, while under MSDP if there were multiple excessive events only the most extreme precipitation of each time period (i.e., 5, 10, 15, ..., 180 min) would be summarized, recorded, and published for that station. Anyone who might combine the data from these two files would truly be mixing meteorological apples and oranges. NCDC has a digital file (TD 9656) of excessive precipitation data for the period of 1962–1972, and maximum short duration precipitation (TD 9649) for the period 1973 to the present. Prior to 1962 all excessive precipitation is available only in manuscript form. The MSDP data is also available in manuscript form in the monthly *Local Climatological Data* (LCD) publications from January 1982 through present.

There are also some similar extreme precipitation values that are published in the back pages of the NCDC monthly publication *Hourly Precipitation Data* (HPD). These data are primarily derived from the automated Fischer–Porter gages in the cooperative observing network but also include data from the primary NWS stations as well. Since the maximum time resolution for the automated gages in the cooperative network is 15 min, the "precipitation maxima" summary pages only report precipitation values for 15, 30, and 45 min as well as 1, 2, 3, 6, 12, and 24 h for each station-month.

Once again caution should be exercised when comparing these values with those recorded/published in either the excessive or MSDP systems. The reason being that all values presented in the precipitation maxima tables of the HPD publication are based on data derived from automated gages that are confined to the fixed 15-min clock times while the excessive and MSDP systems are free to move minute by minute along the duration of an extreme event(s) searching for the most extreme precipitation intensity. A close examination of the precipitation maxima table will also show that there are no 15, 30, or 45-min values published for the primary NWS stations. These time periods were omitted in an effort to avoid confusion with the values presented for the same time periods presented in the MSDP tables in the LCD publication for the same station. There are 1-, 2-, 3-, 6-, 12-, and 24-h values published for primary NWS stations in the precipitation maxima table of the HPD; however, for continuity purposes, the maxima table precipitation values for the primary stations are fixed "clock" hour times like the cooperative stations that constitute the majority of stations in the HPD publication. Therefore, these 1-, 2-, and 3-h values will seldom agree with the 60-, 120-, and 180-min values published in the MSDP table of the LCD for that same primary NWS station.

4 SNOW

Given the numerous problems in measuring snow accurately and uniformly (Doesken and Judson, *The Snow Booklet*, 1997) it is no surprise that the quality of snow data in the climate databases are not always what we would like to see. This is true not only of data from the cooperative network stations but unfortunately also from some of the primary NWS stations where, prior to the installation of ASOS (which does not measure snowfall or snow depth) in the mid-1990s, there were professionally trained personnel making the observations.

Obviously, no after-the-fact assurance review of snow data, or indeed any meteorological data, can know exactly what occurred at each reporting site; however, with other accompanying meteorological data or information from that site one can make some reasonable assessments of the validity of the snow reports. The scope of this chapter does not allow a full discussion of all the observational problems encountered in the measurement and quality assurance of snow, but some of the major difficulties encountered will be listed. These will be grouped by (1) observations from the cooperative network observers and (2) those from the NWS primary station network.

Snow observations from cooperative stations presented the greatest quality assurance challenges not only because of a wide variety in the knowledge and training of each observer, but also because of limited supporting meteorological data from the same station. For example, at least 2000 of the average 6800 published cooperative stations report precipitation only (no temperatures), and if an observer at any one of those 2000 sites reports liquid precipitation but no snowfall or snow depth, it is obviously difficult to know whether to assign a "missing" or a "0" to the database and publication.

Such determinations are not too difficult in North Dakota in January or even in Alabama during the blizzard of March, 1993. Unfortunately, these two cases are not representative of most stations during the transitional months of the snow season or for stations located at the geographic margins of a snowstorm.

For those cooperative stations that do report temperatures in addition to precipitation, a "snow–no snow" determination is greatly improved. Automated checks can easily find precipitation reports with both maximum and minimum temperatures below freezing and flag any that do not also include frozen precipitation.

Once the validity of a frozen precipitation event has been established, then the next challenge is to confirm the magnitude of the reported amount. This thought may come as a surprise to some, when one realizes that a significant portion of the U.S. population at large (which includes weather observers) does not clearly understand the proper placement of decimals in writing a bank check (thus the required worded expression of the numerical value) or reporting precipitation on a weather form. Liquid precipitation and/or melted snow is reported in the United States to inches and hundredths (0.00); snowfall to inches and tenths (0.0), and snow depth to whole inches (0).

A survey (by this writer) of precipitation (including snowfall and snow depth) reports during the March, 1993, snowstorm showed that only about 25% of the reports were correct as received by NCDC. Some had only liquid (meltwater) amounts reported, while most reported snowfall or snow depth with no meltwater. Combine the above reports with snow depth written in the snowfall column or snowfall (inches and tenths) written in the snow depth column (whole inches) and a 24.5-in. snowfall would become a 245-in. snow depth. Several reports had the snowfall reported in feet, which uncorrected would have a 3-ft (3′) snowfall recorded as 0.3 in. in the database and publication. Despite these numerous observational problems, NCDC's quality assurance specialists are quite adept in extracting the correct information from the manuscript reports. Much of that ability came from a knowledge of how each station entered their data in the past as well as an areal station-to-station check for magnitude errors.

The automated checks of cooperative precipitation and snow observations at NCDC is very intensive, and the major reason this could be accomplished was because the observation of each precipitation element (i.e., meltwater, snowfall, and snow depth) is made at the same time each day at any given station. Unfortunately, the same cannot be said for the primary NWS stations because their observing practices require a midnight-to-midnight [local standard time (LST)] period for precipitation/meltwater and snowfall, but a 1200 UTC time (7 a.m. EST, 6 a.m. CST, 5 a.m. MST, and 4 a.m. PST, etc.) for the snow depth observation, and if the snow depth on the ground was ≥ 2 in. at 1800 UTC time for the measurement of the meltwater equivalent of the snow on the ground. These observational time differences caused automated comparison checks to falsely flag more good data than bad, and, as a result, the automated checks were turned off. As one might expect, data errors passed on through to the archives and publications (Schmidlin, 1990).

The presence of these errors (mostly meltwater equivalent of snow on the ground) in the database was not totally due to NCDC oversight. There has been a long-standing verbal understanding between NCDC quality assurance specialists and NWS observers that NCDC would not change any observed values without agreement from the observing site in question. Even when presented with significant errors, some observers refused to agree to any correction. One observer from a northern state admitted to this writer that his station only made meltwater equivalent observations once a week instead of the required once daily time period. Following are two other examples of obvious water equivalent errors that were recorded at two primary NWS stations during the March, 1993, blizzard and still exist in the database and publications to this day: A station in the southeast United States recorded 1.85 in. of liquid precipitation for the 18.2 in. of snowfall that fell during the storm; however, the station also reported 3.8 in. of liquid meltwater equivalent for the 18 in. of snow on the ground. Note there was no snow on the ground prior to the storm. The second station was in the northeast and reported 6.3 in. of liquid meltwater, equivalent for 23.2 in. of new snow depth when most other stations in the area reported 1.5 to 2.0 in. of liquid equivalent for similar depths of new snow. Clearly, these two cases were in error.

5 WIND

Throughout the history of the U.S. Weather Bureau and the National Weather Service, the basic instrumentation used for the determination of wind speed has been a rotating cup anemometer (History of Weather Bureau Wind Measurements 1963). Despite this common design, there have been and continue to be many other factors that influence the values of wind speed that are recorded in the publications and databases. These factors include cup design, time constants/averaging periods, units of measure, and sensor exposure.

The first real cup anemometer (History of Weather Bureau Wind Measurements 1963) was introduced by Robinson in 1846, but it was not until the mid-1870s that the U.S. Army Signal Service (predecessor to the U.S. Weather Bureau, which began in 1890 under the U.S. Department of Agriculture) began using these cup anemometers to record wind speed data. These early anemometers were constructed with four cups that were hemispherical in shape (i.e., shaped like half of a ball) and had smooth edges. Over the early decades of the twentieth century scientists learned that three cups shaped like cones and possessing beaded (rounded) edges on the windward face of the cups dramatically improved the accuracy of cup-derived wind measurements. Just to give the reader some sense of the inaccuracy of the four-cup anemometer speeds, they were about 15% too high at 20 mph, 20% too high at 50 mph, and 25% at 100 mph. In comparison the three-cup speeds were <1% too high at 20 mph, 2% too high at 50 mph, and <5% at 100 mph.

The above information becomes more important when one learns that the vast majority of the wind speeds from the four-cup and early three-cup anemometers were *not* corrected prior to entry on the official observation forms until January 1, 1932. Beginning on that date speeds were corrected prior to entry on the forms. These corrections continued until the 1950s when the accuracy of the official three-cup anemometer (Model F-420C) became sufficiently accurate that corrections were no longer needed.

The three-cup, conical-shaped, beaded-edge design of the F-420C was carried through to the advent of the Automated Surface Observing System (ASOS) in the 1990s. However, the method of measuring the rotational speed of the cups was changed from an analog method to a digital. We will revisit this important issue later.

Despite the fact that electricity was used as far back as the 1870s to assist in the recording of wind speed and direction, the technology of the time still did not provide the instantaneous measurement of wind speed. At that time wind speeds were determined by counting (manually and/or graphically) the number of miles or tenths of miles per unit of time of the anemometer as indicated by the number of marks drawn on an autographic chart or the number of times an indicator light was illuminated and counted by the weather observer. This type of measurement is essentially the same as reading the miles or tenths of miles off the odometer of an automobile. If one measures the time required to travel a mile, he or she will then be able to compute the average speed for that mile. However, there is no way to know what the maximum speed was during the mile. Therefore, there was no way to obtain a maximum "instantaneous speed" without a continuous direct read out of the speed (i.e., a speedometer in the above analogy). It was not until the 1950s that instantaneous "peak gust" wind values began to be recorded by the military and not until the 1960s to 1970s for the U.S. Weather Bureau. These direct reading anemometers used a small electric generator (also known as a magneto) that produced an electric current that is linearly related to the rotational speed of the cups of the anemometer and thus the wind speed. The generated electrical current was displayed on a wind dial or strip chart recorder that was located in the weather office.

Operational requirements of the aviation community were influential in bringing about the measurement of these shorter duration wind speeds, and the resulting recorded values were of a great value to the engineering community for determining building and construction standards as well as the analysis of post-storm-related damage or structural failure. Prior to the advent and use of these direct reading (speedometer) type of anemometers the shortest duration (odometer) wind speed value that was recorded and published was the "fastest mile." The observer usually derived these values at the end of the observational day by finding the two closest mile marks on the single or triple register autographic recorder and then enter that value on the observation form. It has probably already occurred to the reader that a "mile" is a distance and not a duration. Obviously, the time that it takes the wind or a car (to continue our analogy above) to go 1 mile will vary greatly with the speed. (For an average speed of 30 mph, 1 mile will take 2 min; at 60 mph 1 min; and at 120 mph only 30 s.) The extreme wind databases contain not only these fastest mile values but also "fastest one-minute" values. These one-minute values were introduced randomly in time during the 1970s and 1980s at the National Weather Service stations as the old triple-register/multiregister recorders broke down. Whenever these instruments ceased operating, there was no longer any standard method by which the fastest mile could be observed or recorded, therefore the NWS instructed the weather observers to observe the wind dial (speedometer) for 1 min and record the value both for the regular hourly observation as well as the "fastest observed one-minute" for the day. Assuming an observer could accurately integrate the movement of the wind speed dial for a

1-min period, the derived values would NOT be comparable (for climatic or engineering use) to the earlier discussed fastest mile values. Remember that 1 min is a unit of time while 1 mile is a unit of distance and the only speed where the two values are equal is 60 mph. Let's go back to our earlier example of 30, 60, and 120 mph. A true "one-minute" wind of 30 mph is one-half the time of a fastest mile of 30 mph, an equal time at 60 mph, and twice the time of a fastest mile of 120 mph. One other reason why the fastest observed one-minute and fastest mile values are not comparable without statistical adjustment is the fact that the fastest mile value was taken from a recorder chart that would include every mile of wind that passed the anemometer during the day while the fastest observed one-minute value only represented the observed wind speed for one minute out of each hour. Obviously, the true peak one-minute wind would likely not be recorded under these circumstances.

During the 1970s and 1980s the NWS began installing strip chart "gust recorders" that constantly recorded the wind speed and allowed the observer to see the near instantaneous speed of the wind. This made it easier for the observer to make a more accurate estimation of a one-minute wind speed but was still difficult during periods of strong or gusty winds.

However, these gust recorders did provide the accurate measurement of wind "gusts," i.e., short duration wind speeds of 1 to 3 s depending upon the speed and gustiness of the wind. As was previously mentioned, these short duration values were important for aviation, structural design, and forensic use. For instance, the maximum dynamic pressure of the wind upon an average size home can occur in as little as 2 to 3 s. Smaller structures or objects can be affected by even shorter duration gusts.

Starting in 1992 the NWS and later the Federal Aviation Administration (FAA) and U.S. Military Services began installing automated weather observing platforms (ASOS/AWOS). Even though most of the platforms use a three-cup anemometer that is nearly identical to the older F-420 series used by the NWS, the internal workings of the anemometer are different in that a "light-chopper" device sends digital pulses to the computer, which samples the wind speed every second and builds a temporary storage of five (1 s) wind speed values and then computes a 5-s average that is stored for the observational day to be transmitted in real-time reports as well as the now "peak 5-s" wind for the day that is passed on to the climate databases and publications.

A 5-s gust differs significantly from the 1- to 3-s gusts that were recorded on the old analog strip chart gust recorders when you realize that the 5-s digital value is truncated. That is because the 5-s sum and resulting computer average is not recomputed with each new 1-s value. Rather the ASOS wind algorithm ingests five values and computes an average and then repeats the process 5 s later. The net effect of this logic dampens the magnitude of a true 5-s wind gust if computed with a running 1-s update.

Unfortunately, for data users the difference between the old analog gust values and the newer digital ASOS records were further complicated by the discovery that the first 3 or 4 years of ASOS-derived 5-s gust values were actually generated by the manufacturer's "test" algorithm that produced wind speed values that were lower than the correct running 5-s means. The erroneously determined values remain in the official databases and publications without adjustment or documentation as to when the correct algorithm was installed.

In addition to the previously discussed changes in wind speed observations, there is one other significant change that coincided with the implementation of ASOS and that was a doubling of the averaging period from 1 to 2 min for the regular hourly wind speed observations.

The final issue to be discussed here in relation to wind speed observations is not as significant as those previously discussed but nevertheless may show up in the reader's research or analysis. This issue deals with the units of measure.

Historically, wind speed measurements in the United States have been in miles per hour (mph); however, with the increasing influence of the military services on civilian weather observations during and after World War II, the knot (nautical mile of 6076 ft or 1.1508 statute miles) was eventually adopted by the U.S. Weather Bureau as the standard unit for wind speed in the mid-1950s. However, for public consumption, miles per hour were still used both in terms of nonmarine forecasts as well as historic publications like the *Local Climatological Data* (LCD) published by the NCDC. The NCDC digital databases and publications contain wind speed values in units of mph and/or knots.

ASOS wind observations are also recorded and disseminated to the aviation community in knots, but disseminated to the general public and published by NCDC in the daily summaries of the LCD in mph.

The reason for raising this issue is that in the arithmetic conversion (1.1508 × knots) of a whole knot value to a whole mph value causes certain mph values to be omitted. For example, 3 knots × 1.1508 = 3.45 mph when rounded becomes 3 mph while 4 knots × 1.1508 = 4.60 mph rounded becomes 5 mph and thus the value of 4 mph never shows up.

The same is true for the mph values of 11, 19, 27, 34, 42, etc., thus undermining wind speed frequency distributions of 1 mph classes.

6 WEATHER

Changes in equipment and observational practices are not restricted to those discussed above; rather, this is probably just the tip of the proverbial iceberg. The transition from human to automated determination of sky cover and visibility has brought its own set of discontinuities. In terms of sky cover the human observer viewed the whole sky while the automated (ASOS/AWOS) systems use a fixed laser beam and depend on the sky/clouds to move past them to be able to integrate a sky condition. This integration process works well most of the time, but still is limited to a ceiling height of 12,000 ft above ground level (agl) and thus does not indicate the presence of any high-level altocumulus, altostratus, or cirrus clouds. As for visibility, it is determined from an approximately 3-ft baseline at one location and is limited to a maximum reported distance of 10 miles. Therefore, any historic visibility studies that involved values greater than 10 miles would end at the time ASOS was commissioned at each observation site.

One of the other data losses that resulted from ASOS was that of the duration and intensity of thunderstorms. Work continues on the improvement and installation of a lightning sensor from which proximate thunder may be inferred.

This writer would be remiss if he did not mention one other major discontinuity to climatological studies and other retrospective uses of meteorological data. That is the conversion (on July 1, 1996) from the long-standing "airways" observational codes to a modified version of the European "METAR" code. This transition brought about numerous changes, but the scope of this chapter does not allow a full treatment of all these changes. Hopefully, the listing of a few will cause the reader to dig further before using any related data.

1. Ambient and dew-point temperatures are now recorded in Celsius (degrees and tenths) instead of whole degrees Fahrenheit.
2. Sky cover is now reported in oktas (eighths) instead of tenths.
3. Reported weather types of which there used to be about 28 have been reduced in number and altered in their two-letter abbreviations. Following are a few examples of the better known ones:

Weather Type	Airways	Metar
Hail	A	GR
Rain	R	RA
Ice pellets	IP	PE
Snow	S	SN
Drizzle	L	DZ
Fog	F	FG
Thunderstorm	T	TS

4. Rounding of negative temperatures changed from arithmetically absolute (i.e., -1.5° became -2°) to temperature absolute (i.e., -1.5° now becomes -1.0°). This change most often affects the computation of heating degree days (HDD) when the sum of the maximum and minimum temperature are odd and, when divided by 2 to obtain the mean temperature for the day, leave a product of whole units and tenths.

For example, Max Temp of -2° + Min Temp of $-9^\circ = -11^\circ \div 2 = -5.5^\circ$, which used to become -6° but now becomes -5°

7 SUMMARY

It is the strong recommendation of the author that anyone using meteorological and climatological data from any source investigate its background prior to analysis for research, publication, or use in the applied and forensic fields. Not to do so is an abrogation of scientific integrity. Hopefully, the discussion of this chapter will cause the reader to pause before blindly accepting the validity of any data set or value.

BIBLIOGRAPHY

Doesken, N. J., and A. Judson, (1997). *The Snow Booklet*, Colorado State University, 87 pp.

Engelbrecht, H. H., and G. N. Brancato (1959). World record one minute rainfall at Unionville, MD, *Mon. Wea. Rev.*, **87**, (8), 303–306.

Fujita, T. T. (1992). U.S. Department of Commerce, National Oceanic and Atmospheric Administration, NCDC, Asheville, NC, *Storm Data*, September 1992, Vol. 34, No. 9, p. 27.

Karl, T. R., and R. G. Quayle (1988). Climate change in fact and in theory; Are we collecting the facts? *Climate Change* **13**, 15–17.

Karl, T. R. (1995). *Long-Term Climate Monitoring by the Global Climate Observing System*. Kluwer Academic Publishers, pp. 51–56.

Schmidlin, T. W. (1990). A critique of the climate record of water equivalent of snow on the ground in the United States. *J. Appl., Meteor.* **29**, 1136–1141.

Schmidlin, T. W. et al. (1992). Design ground snow loads for Ohio. *J. Appl. Meteor.* **31**, 622–627.

Sherlock, R. H. (1947). Gust factors for the design of buildings, International Assoc. For Bridge and Structural Engineering, Vol. 8, pp. 207–235.

Sissenwine, et al. (1973). Extreme wind speeds, gustiness, and variations with height for MIL-STD 210B. U.S. Air Force, Cambridge Research Labs. TR-73-0560, p. 12.

U.S. Department of Commerce, National Oceanic and Atmospheric Administration (1985). Storm Data; October 1985, Vol. 27, No. 10. National Climatic Data Center, Asheville, NC, 40 pp.

U.S. Department of Commerce, National Oceanic and Atmospheric Administration (1992). ASOS Users Guide; National Weather Service, Silver Spring, MD, 98 pp.

U.S. Department of Commerce, National Oceanic and Atmospheric Administration (1993). Climatological Data, New York; March 1993, Vol. 105, No. 3, National Climatic Data Center, Asheville, NC, 40 pp.

U.S. Department of Commerce, National Oceanic and Atmospheric Administration (1993). Climatological Data, North Carolina; March 1993, Vol. 98, No. 3, National Climatic Data Center, Asheville, NC, 36 pp.

U.S. Department of Commerce, National Oceanic and Atmospheric Administration (1994). ASOS Progress Report—ASOS Wind Sensor, Problems Identified and Solved. National Weather Service, Silver Spring, MD, 8 pp.

U.S. Department of Commerce, National Oceanic and Atmospheric Administration (1994). Natural Disaster Survey Report, "Superstorm of March 1993." National Weather Service, Silver Spring, MD, 152 pp.

U.S. Department of Commerce, National Oceanic and Atmospheric Administration (1996). National Weather Service, Observing Handbook No. 7—Surface Weather Observations and Reports, July 1996. National Weather Service, Silver Spring, MD, 461 pp.

U.S. Department of Commerce, Weather Bureau (1957). History of Observational Instructions as Applied to Temperature Recordings. U.S. Government Printing Office, Washington, D.C., 7 pp.

U.S. Department of Commerce, Weather Bureau (1958). Excessive Precipitation Techniques. U.S. Government Printing Office, Washington, D.C., 12 pp.

U.S. Department of Commerce, Weather Bureau (1963). History of Weather Bureau Precipitation Measurements. U.S. Government Printing Office, Washington, D.C., 19 pp.

U.S. Department of Commerce, Weather Bureau (1963). History of Weather Bureau Wind Measurements. U.S. Government Printing Office, Washington, D.C., 68 pp.

U.S. Department of Commerce, Weather Bureau (1968). Final Report—Test and Evaluation of the Fischer and Porter Precipitation Gage. U.S. Government Printing Office, Washington, D.C., 28 pp.

U.S. Department of Commerce, Weather Bureau (1969). Specification No. 450. 1016 for Liquid in Glass Thermometers. U.S. Government Printing Office, Washington, D.C., 9 pp.

U.S. Department of Commerce, National Oceanic and Atmospheric Administration (1985). Climatological Data, Puerto Rico and Virgin Islands; October 1985, Vol. 31, No. 10. National Climatic Data Center, Asheville, NC, 23 pp.

U.S. Department of Commerce, National Oceanic and Atmospheric Administration (1985). Hourly Precipitation Data, Puerto Rico and Virgin Islands; October 1985, Vol. 15, No. 10. National Climatic Data Center, Asheville, NC, 15 pp.

U.S. Department of Commerce, National Oceanic and Atmospheric Administration, National Climatic Data Center, Asheville, NC continues to publish monthly and annual Local Climatological Data volumes that began in 1948. They are published for about 270 cities in the United States and some of its territories.

CHAPTER 42

DEMANDS OF FORENSIC METEOROLOGY

W. H. HAGGARD

1 SEARCH FOR THE TRUTH

The role of the forensic meteorologist is to assist the court in the search for the truth by presenting the most accurate description possible of the meteorological events pertinent to the case in litigation. To fulfill this role, the forensic meteorologist must utilize the best possible data set in his or her retrospective reconstruction of the weather. The quality of the analysis is dependent on the availability of representative measurements of meteorological parameters.

2 DATA NEEDED

The weather event reconstructions frequently require data and information beyond the standardized meteorological observations collected by governmental agencies. These frequently include eyewitness statements, field investigations, interviews, and data from other fields (e.g., recordings of radar-deduced tracks of aircraft, ship log books, air traffic control voice transcripts, and the like).

Handbook of Weather, Climate, and Water: Dynamics, Climate, Physical Meteorology, Weather Systems, and Measurements, Edited by Thomas D. Potter and Bradley R. Colman.
ISBN 0-471-21490-6

Reliability

Legal tests of the courtroom admissibility of expert testimony have been devised at various times. In 1923 the *Frye test* required that scientific evidence be of a type that is "generally accepted" in the relevant scientific community. In the late 1990s the *Daubert test* to eliminate "junk science" in the courtroom requires the proffered scientific evidence to be "reliable and based upon scientific methodology."

Hindsight

Several factors impact on the reliability of the results of a retrospective meteorological analysis. There must be data available upon which to rely. The data should be representative of the conditions of the atmosphere at the time and place of measurement. The instruments, sensors, observers, radars, satellites, etc. should have the capability of accurately sensing the requisite data. They should be properly calibrated, and their clocks should be set correctly. Since forensic meteorological analyses for litigation involve retrospective reconstructions, the data utilized come from the archives, and high quality of archived data is vital to proper reconstruction.

The forensic meteorologist enjoys the luxury of hindsight—usually believed to be "better" (or at least easier) to apply correctly than foresight. Like the forecaster, however, he or she is limited in analytical capability by the limitations in availability, representativeness, and quality of the available measurements pertinent to the circumstances.

Reconstructing the bases of the cloud above a 2900-ft mountain ridge between Nome and Koyuk, Alaska,[1] when there were no cloud base data within 100 miles of the crash site is more difficult than reconstructing the cloud bases at Flagstaff, Arizona[2] (where a continuously recording laser beam ceilometer record was available) (see Fig. 1).

Expert Testimony

Meteorology comes into the courtroom through the testimony of expert witnesses who are there because they possess an accumulation of education, knowledge, and experience that qualifies them to provide weather-related testimony beyond the normal capability of the court to obtain by other means. The qualified practitioner of forensic meteorology is a formally educated atmospheric scientist, often with specialized training in applied climatology, hydrometeorology, micrometeorology, aviation meteorology, marine meteorology, and/or disaster preparedness who is relied upon to assist legal counsel in clarifying scientific or technical issues that relate to the weather relevant to specific cases being litigated (Falconer and Haggard, 1990).

Forensic meteorologists do more than a "simple" weather event reconstruction (Haggard, 1980a,c, 1983, 1985, 1989; Haggard and Smutz, 1994). They may:

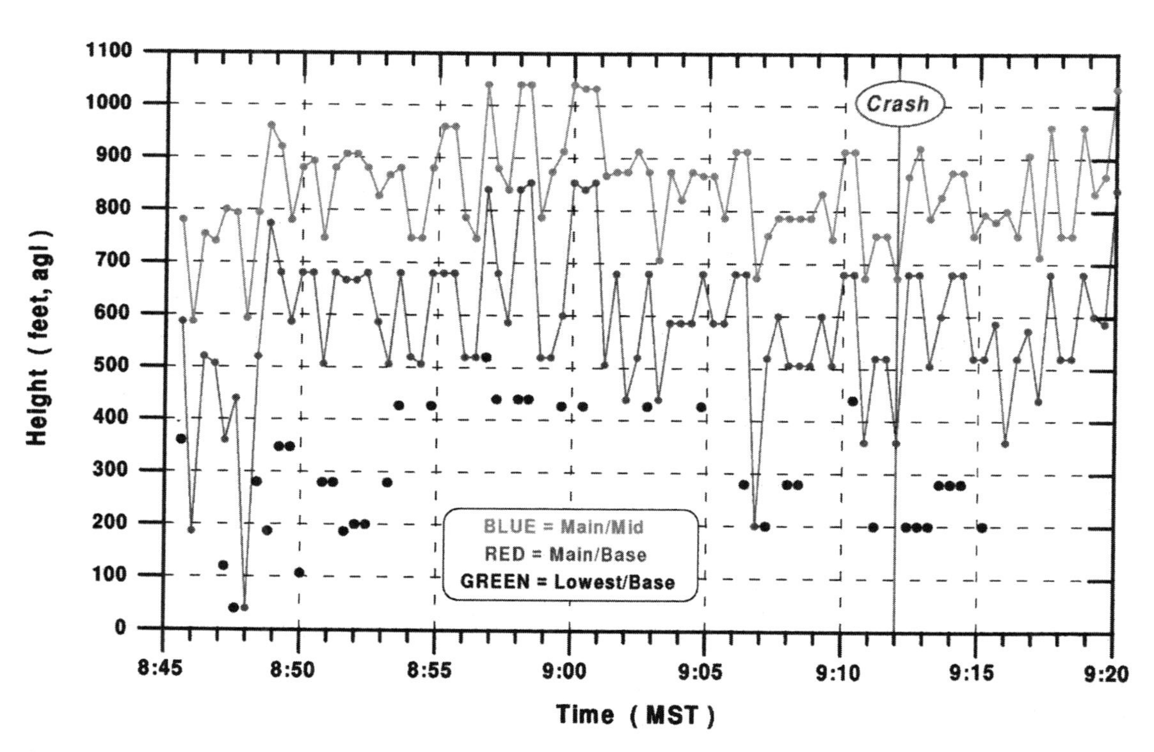

Figure 1 Decoded laser beam ceilometer data, Flagstaff, Arizona, 8:45–9:20 a.m. MST, December 20, 1991, showing values converted to heights above ground (agl) of lowest points of reflectivity (green) above ground; height of the base of main reflectivity (red); and midpoint of recorded main reflectivity (blue). See ftp site for color image.

- Acquire and interpret basic weather records (i.e., observations, operational charts and analyses, archived forecasts, or advisories, etc.).
- Assess the validity, representativeness, pertinence, and adequacy of the available data.
- Advise engaging counsel on the meaning and applicability of these basic weather data to issues pending before the court.
- Perform specialized meteorological analyses based on weather and other records (e.g., topographic surveys, eyewitness reports, damage photos, etc.).
- Prepare a comprehensive report.
- Provide credible expert testimony as to the weather pertinent to the matters under litigation.

High-quality records of measurements are essential to the accomplishment of these functions.

3 CASE EXAMPLES

Many retrospective reconstructions are for a site- and time-specific weather event (e.g., the collision of the *Summit Venture* with the Skyway Bridge, which spans the entrance to Tampa Bay, on May 9, 1980,[3] or the crash of a "TriStar" airliner on approach to Runway 17L at Dallas–Ft. Worth Airport at 6:04 pm CDT on August 2, 1985.[4] Others may involve a moderate span of time and an area (e.g., devastating floods over 16 counties of western North Carolina on November 5–6, 1977,[5] affecting thousands of square miles and several days). A few will be quite dependent on remotely sensed data (e.g., a microburst at Mayaguez, Puerto Rico[6]).

Some may involve a climatological study of decades of data (e.g., a wind study relating to the legality of constructing a N–S runway at Anchorage International Airport as a "crosswind runway"—a study involving detailed analysis of 23 years of hourly wind data[7]). The availability, adequateness, representativeness, pertinence, and quality control of the data impacted the credibility of the analytical results in these exemplar cases.

Visibility

Key questions in the litigation over the *Summit Venture*/Skyway Bridge collision (Haggard, 1985) related to the visibility in the ship channel and speed of the vessel, just prior to the collision with the bridge, far removed from any official weather-reporting site. Rain was so intense the bow of the ship could not be seen from the bridge.

Probably the most pertinent objective weather measurements relative to determination of the visibility were the rainfall rates at the ship as it traversed the channel. The National Weather Service Weather Surveillance Radar (WSR-57) located at Ruskin, Florida, 14 miles from the collision was in a favorable location to provide such data.

The collision occurred at 7:34 a.m. EDT. The radar at Ruskin was struck by lightning at 7:13 a.m. EDT, became inoperable, and did not provide further measurements until 7:43 a.m. EDT. The most potentially valuable and pertinent measurements were not available for 21 min before and 9 min after the event.

The meteorological system producing the intense rain—which reduced visibility to tens of feet—was sufficiently time consistent that is was possible to interpolate the "probable radar image" at the collision time by "bridging the gap" from the series of images prior to and subsequent to the radar outage. The resultant hypothetical image showed the ship and bridge channel spans to be within the area of the strongest reconstructed radar reflectivity, consistent with nonobjective lay eyewitness reports.

Microbursts

The principal litigation resulting from the crash of Delta Air Lines Flight 191 involved a 14-month trial in Federal District Court in Fort Worth Texas. Dr. T. Theodore Fujita (1986) performed an intensive retrospective weather analysis that relied on the combined meteorological measurements made by standard instrumentation, the 7-second LLWAS [low-level wind (shear)–alert system] data at sensors near and on the airport, weather satellite and radar imagery, the parametric data measured by the digital flight data recorder on the accident aircraft, and on-site physical evidence. This artful combination of all available pertinent data permitted a reconstruction of the interaction of aircraft and the atmosphere and led to a detailed reconstruction of the dynamics of the microburst that the aircraft penetrated.

Another microburst-related accident occurred at Mayaguez, on the western coast of Puerto Rico, on June 7, 1992, killing crew and passengers of a commuter plane attempting to land. The detailed weather reconstruction (Fujita et al., 1992) was largely dependent on a sequential analysis of satellite and radar measurements made remotely. The growth and collapse of thunderstorms augmented by the topography and sea breezes of the island were remotely sensed. The results of analyses of those measurements were integrated with multiple witness statements from both airborne and ground-based persons closer to the event. Finally the dynamics of the aircraft and its track were fitted into the meteorological analysis to locate the point of impact of the microburst relative to the trajectory of the plane immediately prior to impact (see Fig. 2). The coordinated use of remotely sensed meteorological measurements, their augmentation by anecdotal local witness statements, and their combination with tracked and timed aircraft positions were essential to the reconstruction of the weather scenario.

Winds

Measurement validity was an element of dispute in the litigation over construction of the "crosswind" north–south runway at Anchorage International Airport (Haggard, 1980b). The Federal Aviation Administration (FAA) conducted a study of winds relative to the existing east–west runway. Though 23 years of wind data existed, they

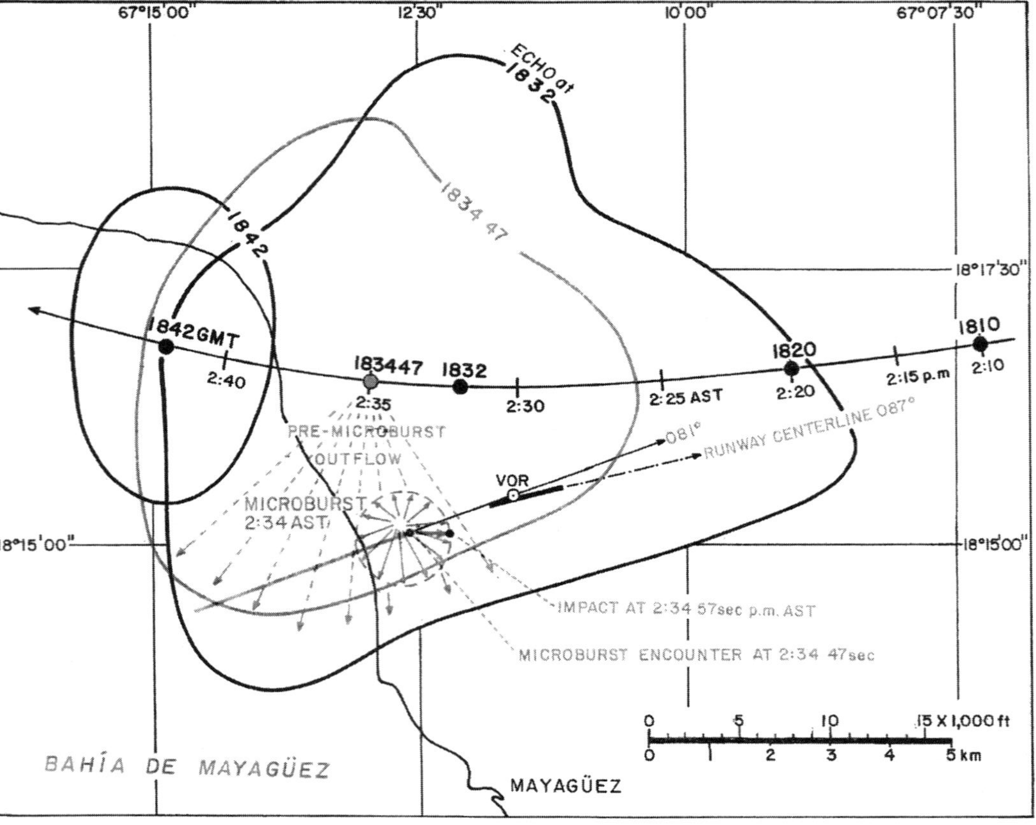

Figure 2 Motion and collapse of San Juan observed radar echo over Mayaguez, Puerto Rico, on June 7, 1992. Curved black line shows center of reflectivity at 1810, 1820, 1832, and 1842 UTC; echo outlines at 1832, and 1842 are shown in black; interpolated echo outline at 1834:47 is shown in blue; aircraft track is shown in red; and probable microburst encounter is depicted in blue (times are shown in Atlantic Standard Time = AST; AST = UTC minus 4 h, such that 1820 UTC = 1420 AST = 2:20 pm AST). See ftp site for color image.

were not compatible. Some were observed at 16 compass points, others at 36 compass directions; some were recorded in miles per hour, others in knots; some were digitized at hourly intervals, others only at 3-h times; the "official wind observations" were made at 12 different consecutive sites over the 23 years of available "on airport" measurements. The life span of the various sites ranged from 11 months to 10 years, averaging 5.2 years.

In its design study, the FAA used the latest 7 years of compatible 36 point, 3-h wind data, which yielded a qualifying (for cross-wind runway construction) cross wind of 15 knots or greater 5.88% of the time (5.0% is required for a cross-wind runway).

An affected property owner had a 17-year wind study made and determined the qualifying cross winds were apparently present less than 5% of the time. A federal judge stopped the $33 million construction project and ordered a wind study utilizing "all available data."

In the undertaking of this complex task, utilizing separate analyses for the differing periods of noncompatible data units, it was discovered that all data from one anemometer location (3/21/61 to 1/17/64) yielded strikingly low cross-wind components compared to prior and subsequent data. Wind speeds were compatible with earlier and later records, but the direction appeared anomalous. A statistical (F) test showed the data from that nearly 3-year sample to be statistically from a different population than the remaining data.

Adjusting the wind data of this sample by 30° (the difference between true and magnetic north at Anchorage) removed the statistical "bias" and made the sample compatible with the prior and subsequent data, suggesting the wind vane directional calibration might have been based on magnetic rather than true north during that sampling period.

Clouds

In aviation meteorology, the *sky condition* is a key element, determining whether aircraft operations may be conducted under visual flight rules (VFR) or instrument flight rules (IFR). While most scheduled commercial aviation flights are conducted under IFR, regardless of the observed conditions, a large percentage of "general aviation" flights are conducted under VFR conditions, and many pilots are not "IFR qualified and equipped." To them the accuracy of the "measurement" of sky condition is vital for a determination of whether they can legally operate their aircraft under visual flight rules.

The measurement of sky condition has historically been made by human observers viewing the sky and recording the percentage (in oktas or tenths) of the sky obscured by clouds. Recently, this task has been increasingly taken over by the Automated Surface Observing System (ASOS), vertically scanning laser ceilometer beams that substitute percentage of time clouds are sensed at various heights in the atmosphere below 12,000 ft above the ground level for the areal estimates formerly made by human observers.

Recent litigation[8] resulting from a six-death crash of a private plane attempting a nighttime VFR landing at a Florida airport hinged on whether the sky condition "measured" by a human observer was "measured 600 foot broken clouds" (covering 6/10ths of the sky with opaque clouds), technically IFR conditions, or whether these clouds had moved away, allowing the field to become VFR.

The case was complicated by a number of conflicting eyewitness postaccident statements regarding the sky condition at the airport at the time of the accident. Two testifying forensic meteorologists offered very differing opinions in testimony before the court during trial. Statements by two IFR-rated pilots who "lost sight of the airport on arrival" and "encountered an unreported cloud layer on departure" near the accident time were significant in the retrospective weather reconstruction.

Conflicting Views

Another aviation litigation case in which the testimony of two eminently qualified forensic meteorologists offered dramatically different opinions on the atmospheric conditions involved a tragic in-flight breakup of a twin-engine aircraft over Anthony, Kansas[9] (Haggard, 1983).

Far removed from any direct measurements, the question was whether:

1. The plane was in level flight between cloud layers (perhaps in light snow) with no wind shear, no turbulence, and no icing when both engines and various control surfaces broke off the aircraft; or
2. The plane flew through a strong wind shear zone with a dramatic weather change into freezing rain and severe turbulence, resulting in an upset and overstresses, which led to the in-flight breakup.

Both experts utilized the same archived weather measurements. One relied upon a computer analysis that indicated strong rotation but minimal shear in the wind field at flight altitude. The other relied on a "hand analysis" (not dependent on computer-programmed analysis procedures) that showed less rotation but a very strong shear zone in the wind field.

Changing Environment

The validity of weather measurements may be degraded by changing environmental circumstances. A recent tragic example was the failure of the LLWAS at Charlotte, North Carolina, to detect and warn of the small but strong microburst associated with the crash of USAir Flight 1016 on the evening of July 2, 1994 (Fujita and Haggard, 1995; Haggard and Fujita, 1994).

Installed in 1981, the six sensors of the system were mounted on poles at heights ranging from 20 to 68 ft on and near the Charlotte, North Carolina, Douglas International Airport in rolling terrain largely covered by early-growth pine. Between 1981 and 1994 many of the pines grew to heights equal to or greater than the heights

Figure 3 LLWAS anemometers at Charlotte, North Carolina, in 1994. Upper photo is of northwest sensor; lower photo is of southeast sensor; each shows sensors in relation to forest growth. See ftp site for color image.

of the wind sensors (see Fig. 3), degrading their wind measurements (and their ability to sense and warn of significant wind shear).

4 WITNESS CREDIBILITY

In these (and other) examples of weather-related litigation, the court (judge and/or jury) must consider the "testimony and demeanor" of each witness to determine their credibility. Credibility is enhanced by demonstrated reliance on valid meteorological measurements and sound scientific procedures.

The adversarial system of justice in the United States should not impact on the expert witness. The parties and the attorneys are adversaries and advocates. The expert witnesses should be objective scientists utilizing valid measurements and objective techniques. Procedures for bringing scientific facts forth in the courtroom are very different than in a scientific symposium (Bradley, 1989).

Decisions by various courts suggest (Haggard, 1989) that it is essential that the testifying expert:

- Reduce his or her opinion to the simplest terms possible.
- State (and visually illustrate) the opinion(s) clearly and concisely.
- Demonstrate (visually and orally) the factual basis for the opinion.
- Produce authenticated copies of all weather measurements relied upon.
- Demonstrate reliance upon the most accurately observed and adequately quality controlled meteorological measurements available.
- Ensure the bridge from demonstrable facts to the stated opinion is as short and solid as possible (i.e., "let the data speak for themselves").

The meteorological expert must avoid any appearance of advocacy (Saks, 1987) and adhere to the scientific analysis of the highest quality validated meteorological measurements available (Haggard, 1985).

CASE CITATIONS

1. *Kavairlook v. Ryan Air Service*, Superior Court for the State of Alaska (Nome-Koyuk 12/10/94).
2. *Kaufman v. Beech Aircraft*, Superior Court of the State of California for the County of Los Angeles (Flagstaff, AZ, 12/20/91).
3. *M/V Summit Venture*, United States District Court Middle District of Florida Tampa Division (Sunshine Skyway Bridge, Tampa, FL, 5/9/80).
4. *Kathleen Connors et al. v. USA*, United States District Court for the Northern District of Texas, Ft. Worth Division (DL 191, DFW, TX, 8/2/85).
5. *American Enka v. Southern Railroad*, United States District Court for the North Carolina District of Asheville, NC (Enka, NC, 11/7/7).
6. *Leslie et al. v. American Airlines et al.*, United States District Court for the District of Puerto Rico (Mayaguez, PR 6/7/92).
7. *John Overby v. USA*, U.S. District Court for the District of Alaska (N-S runway at Anchorage International Airport).
8. *McNair v. USA*, United States District Court Northern District of Florida Gainesville Division (Gainesville, FL, 6/7/95).
9. *Buzzard v. Piper*, District Court for Oklahoma County, Oklahoma City, OK (Anthony, KS, 2/11/77).

REFERENCES

Bradley, M. D. (1983). The Scientific and Engineer in Court, Am. Geophys. Union Water Resources Monograph 8, Washington, DC.

Falconer, P. D., and W. H. Haggard (1990): *Forensic Meteorology, Forensic Sciences*, New York, Matthew Bender, Chapter 35.

Fujita, T. T. (1986). DFW Microburst on August 2, 1985, SMRP Res. Paper No. 217; Chicago, University of Chicago Press.

Fujita, T. T., W. H. Haggard, and W. A. Bohan (1992). Puerto Rico's Weather of June 7, 1992 Related to the Crash of Executive Air Flight 5456 at Mayaguez, Puerto Rico, submission by Construcciones Aeronauticus, SA [CASA] to the National Transportation Safety Board, Washington, DC, November.

Fujita, T. T., and W. H. Haggard (1995). The Microburst at Charlotte, North Carolina in Relation to the Crash of USAir 1016 on July 2, 1994, A Report to The Air Line Pilots Association ALPA, January.

Haggard, W. H. (1980a). Radar Imagery for Past Analysis of Mountain Valley Flash Floods, Proc. 2nd Conf. on Flash Floods, Atlanta, GA, Boston, American Meteorological Society.

Haggard, W. H. (1980b). What's in a Wind Study? Proc. 2nd Joint Conf. on Industrial Meteorology, New Orleans, LA, Boston, American Meteorological Society.

Haggard, W. H. (1980c). Some Micro-Economic Aspect of Applied Climatology, Proc. Conf. on Climatic Aspects and Social Responses, Milwaukee, WIs, Boston, American Meteorological Society.

Haggard, W. H. (1983). Weather after the Event, Proc. 9th Conf. on Aerospace and Aeronautical Meteorology, Omaha, NE, Boston, American Meteorological Society.

Haggard, W. H. (1985). Meteorologists as Expert Witnesses, Proc. 15th Conf. of Broadcast Meteorology, Honolulu, HI, Boston, American Meteorological Society.

Haggard, W. H. (1989). Weather Testimony in Litigation, Proc. 3rd Int. Conf. on the Aviation Weather System, Anaheim, CA, Boston, American Meteorological Society.

Haggard, W. H., and T. T. Fujita (1994). The LLWAS at Charlotte, North Carolina in Relation to the Microburst of July 2, 1994, A Report to The Air Line Pilots Association ALPA, September.

Haggard, W. H., and S. W. Smutz (1994). Forensic Meteorology; Proc of 7th Aviation Law/Insurance Symposium, Daytona Beach, Fl, Embry Riddle Aeronautical University.

Saks, M. J. (1987). *MIT Tech. Rev.*, Aug/Sept.

CHAPTER 43

SURFACE LAYER IN SITU OR SHORT-PATH MEASUREMENTS FOR ELECTRIC UTILITY OPERATIONS

ROBERT N. SWANSON

1 INTRODUCTION

Electric utility operations, as interpreted in the following discussion, involves day-to-day operations, engineering requirements for planning, designing, and operating a system, applied research, as well as environmental concerns and/or requirements. Several aspects of measurements are discussed elsewhere in this *Handbook* and will not be repeated in this chapter. Meteorological data sets, collected as part of any operational or research program, require detailed knowledge of sensor and data retrieval system characteristics and employ proper quality assurance/quality control (QA/QC) procedures. It also involves sensor siting guidelines to assure maximum information is obtained from the measurement program. Since utilities measurement needs may extend throughout and above Earth's boundary layer, their programs are not necessarily restricted to the use of in situ or short-path sensors.

2 RECENT HISTORY OF METEOROLOGICAL REQUIREMENTS BY ELECTRIC UTILITIES

Until the early 1950s, meteorological information used by electric utilities, if any, were those collected and transmitted by the federal government. Simple adjustments were sometimes made to these data to compensate for local extreme conditions for purposes of power generation or system line loading. Occasionally systems were

Handbook of Weather, Climate, and Water: Dynamics, Climate, Physical Meteorology, Weather Systems, and Measurements, Edited by Thomas D. Potter and Bradley R. Colman.
ISBN 0-471-21490-6

operated at above rated levels, and the situations were not considered serious until they impacted operations in some manner. Reliability of power delivered to the customer was not always an overriding concern with power outages generally tolerated and/or expected by the user. In the 1950s, when nuclear power plants were first being built on a commercial basis, there was a developing interest in the transport and dispersion of potential releases of nuclear material. This interest or concern resulted in the development of meteorological monitoring programs at both nuclear and conventional power plants. These monitoring programs were developed to provide inputs to existing dispersion models. Early monitoring programs were quite simplistic in that they frequently collected very limited information of questionable quality. For example, at Humbolt Bay Nuclear Plant in northern California, wind measurements at this nuclear plant consisted of a single aerovane mounted on top of a tower near the plant. Calibration and/or maintenance of this sensor was not well established and possibly was never done.

As time passed, new and more complex dispersion and transport models were developed that, in turn, required additional meteorological information. Because of this requirement, as well as because of an increased concern by the public over safety, more sophisticated and better designed monitoring programs were developed. After the belief that nuclear power would be both inexpensive and plentiful was challenged, there was an increased interest in alternate energy development and in energy conservation. Both of these programs had additional meteorological requirements beyond those provided by the National Weather Service (NWS) with the largest uncertainties associated with planned alternate energy projects. This situation was readily apparent because NWS sites were generally located at airports with a lesser number of observation sites in urban areas. Neither of these sites would be representative of wind energy farms that were being evaluated. Until the mid-1970s wind energy and/or photovoltaic power sources were given little or no attention by electric utilities. This lack of interest was a result of many things, including cost, equipment availability, equipment reliability, and, of course, the already existing power-generating network.

In general each of the alternate energy programs, including wind, photovoltaic, hydroelectric, or geothermal power, have their own particular meteorological monitoring requirements. Also increasing concerns over public health issues and conservation of natural resources have created many changes in meteorological monitoring needs after the early 1950s.

3 MEASUREMENT REQUIREMENTS

Meteorological measurement requirements for electric utility operations can normally be met with existing off-the-shelf sensors and digital recording systems. Siting of these sensors to best address specific concerns or needs of a project is likely a more difficult task than is proper sensor selection. It is important to consider that measurements collected for one program may well be integrated into another, possibly unrelated, program. Therefore consideration should be given to this possibility when selecting and/or siting sensors and in data processing and archiving incoming data. This consideration should in no way detract from the original goal of collecting the best possible data set for the specific project to be evaluated. Additional information generated in data processing and archiving should always be considered since it represents only a minimal increase on the overall level of effort and may well provide valuable information to be used with another project.

Measurement programs should be designed to meet as many of the objectives as possible within budget constraints. These designs can become very difficult at times. For example, when monitoring for wind energy or photovoltaic farms in complex terrain, consideration must be given to many factors including terrain elevation and contours, cloud cover changes, temperature differences, and winds. If the site(s) locations are near large water masses, additional issues may exist. Frequently discussions with long-time residents in the area will provide valuable insight into specific conditions that would otherwise take an extended monitoring period to determine.

4 MEASUREMENT GUIDELINES

As mentioned above, most measurement program needs for electric utility operations can be easily satisfied with existing equipment. However, needs arise to examine the atmosphere above the lower boundary layer. When these needs exist in situ and/or short-path sensor measurements will have to be augmented with additional sensors such as long-path remote sensors or acoustic sounders. One example of such a program is cloud seeding for snow pack enhancement in an attempt to increase hydroelectric power. For this effort the atmospheric structure must be well defined through a very deep layer. Siting off the non-in-situ sensors for use in special programs is project specific and therefore cannot be given detailed guidelines except to follow rigid QA/QC policies. In designing a measurement program, the primary consideration is to determine the overall objective(s) of the project and then design a total system that will best address all concerns within the budgetary constraints that are levied on the effort.

Effective measurement programs require adhering to good QA/QC guidelines. Examples of such guidelines are ANSI/ANS-3.11 (ANS, 2000b) and EPA (1989). Both of these documents provide guidelines that will, if followed, assure a reasonably successful monitoring program, but, as with most efforts, deviations from the guidelines may have to occur. One typical deviation is in actual sensor siting where it is frequently impossible to meet all guidelines on distances of the sensor from obstructions. In such cases common sense must prevail and the sensors should be located where the smallest negative impacts will be found.

5 DATA ACQUISITION

Digital data acquisition systems can provide totally adequate data collection, storage, and transmission packages because of their speed, accuracy, and overall reliability.

Backup devices, such as strip-chart recorders, can be included in a monitoring program, but they add a significant level of effort and cost to the program while providing little redeeming information not already available with properly programmed digital systems. All maintenance and calibration efforts must be done using technically competent personnel. Data retrieval, processing, checking, editing and archiving should also be patterned after an accepted guideline, such as ANSI/ANS-3.11, in a manner similar to programs for field measurements. Statistical procedures should be established, using incoming data, to assist in assessing data quality and operating condition of measurement devices. These statistical programs should be established at or near the beginning of a measurement program to capture diurnal and seasonal variations. It is critical that any deterioration of sensor performance be detected and corrected as soon as possible to maintain a valid collection efficiency of 90% or better from all sensors on an annual basis. The definition of "valid" data capture must also be quantified to prevent all incoming values from becoming valid readings regardless of quality.

6 EXAMPLE OF METEOROLOGICAL REQUIREMENTS BY AN ELECTRIC UTILITY

A large electric utility in western United States, in the early 1990s, had a complete weather forecast office in addition to a significant sized group of personnel in a meteorological projects section. Many requests for information and/or studies done within the weather forecast office will not be discussed below as the following tasks represent only the major items requested of the projects section. This list is not intended to be all inclusive but is given only to illustrate the types of programs with which meteorology may become involved. Also it is not intended to reflect the meteorological research needs of other electric utilities since it is unique in size, type, and expanse of service territory and generation mix. Major study efforts by the projects section included:

1. *Electric load management and load research.* These are primarily regulatory issues because electric rates are determined by typical meteorological conditions within specific geographical areas. Routine measurements of temperature and humidity at about 25 locations throughout the utilities service area satisfied the requirements for this project.
2. *Transmission line loading (dynamic thermal rating).* This program consisted of wind and associated temperature measurements along major transmission lines. Critical conditions that may cause transmission line ratings to be exceeded occur with very low ventilation rates and high ambient temperatures. Wind speeds of interest, perpendicular to the power line, are in the 0.25 to 2.0 m/s range so the wind sensors must have a low starting threshold. Temperature measurements within about 1.0° of true are quite sufficient for this type of study. Since this program was implemented in the early 1990s,

new and probably better measurement systems for dynamic thermal rating have entered the market.

3. *Alternate energy system feasibility.* Studies in this program include those for photovoltaic, wind, and geothermal development. Photovoltaic and wind energy farm potentials are very site specific. Photovoltaic farm feasibility is dependent on distribution of cloud cover, evaporative cooling, atmospheric turbidity, and ambient temperature. Wind power is dependent on wind flow distribution in time and space, quality of wind such as wind shear and turbulence, time of day, and time of year. Geothermal power concerns are primarily transport and diffusion of released pollutants and of cooling tower siting and their operating efficiency. Each of these alternate energy programs had its own measurement requirements and each monitoring effort was unique. Wind tunnel modeling, covering a several square kilometer range, was completed for a potential wind energy farm site.
4. *Power line and transformer contamination.* This study centered around transmission lines, transformers, insulators, and substations where contaminant buildup can create equipment malfunction. Costs of washing these pieces of equipment are very high and a realistic washing schedule was desired. Conventional sensors for measuring winds, humidity, net radiation, and surface temperatures of transformers were used.
5. *Environmental impacts from released pollutants.* These studies were primarily done for planned and/or operational fossil fuel, nuclear, geothermal, and hydroelectric power plants. General concepts, techniques, and instrumentation needed for estimating pollutant transport and dispersion are discussed elsewhere in this *Handbook* and so will not be discussed here. However, most of the power plants are located in complex terrain where released pollutant transport and dispersion is modified by the plume's flow around terrain obstacles. Realistic estimates of environmental impacts from released material frequently required additional measurement sites along the plume's trajectory. Three-dimensional wind tunnel modeling of several of the geothermal sites were completed as part of a transport and dispersion study.
6. *Wind, ice, and snow loading on transmission towers and power lines.* Much of the information used in this study was derived from climatological data with a limited number of measurement sites in the field. Since this program was site or area specific, the measurement program was tailored for each area of concern. One major problem area in the implementation of this study was to employ sensors that would provide accurate data at times when icing or snowing conditions occurred. This problem was not adequately solved, but new systems such as use of line tension monitoring sensors, may be of considerable assistance in this type of study.
7. *Cloud seeding to increase snow pack for hydroelectric operations.* This effort involves three-dimensional wind speeds and directions, temperatures, and humidities from the seeding area to heights above clouds having seedable moisture. In situ measurements of wind and precipitation used in this study

must be capable of operating under adverse weather conditions where freezing rain, ice, and snow can be expected. Potential benefits of this study are very large in that, for a small expense, a considerable increase in water runoff from melting snow may occur. This water runoff is then used to generate hydroelectric power.

8. *Long-range transport of pollutants.* This study was initially designed to describe the fate of pollutants from one or more power plants as they were transported to downwind areas far removed from them. This study was co-sponsored by several governmental groups and had the mission of developing models for both air quality and acid deposition over distances of hundreds of kilometers. Meteorological sensors used in this project included many in situ and short-path sensors along with long-path sensors, upper air sensors, and a variety of aircraft measurements.

7 EXAMPLE OF DATA PROCESSING PROCEDURES

As part of the transmission line loading (dynamic thermal rating) effort, a simple prearchiving program was developed for in situ wind speed and direction measurements where inexpensive on-site digital recording systems were used. Data resulting from this program provide basic data sets that may be somewhat applicable to other project studies. In general the program was designed to collect, generate, store, and, on demand, transmit the following data for any desired time period:

1. Peak scalar speed
2. Time of peak speed
3. Unit vector wind direction at time of peak speed
4. Mean scalar wind speed over sampling period
5. Mean unit vector wind direction over sampling (period excludes observations of calm winds)
6. Mean resultant vector wind direction over sampling period
7. Standard deviation of wind speeds over sampling period
8. Standard deviation of unit vector wind directions over sampling period
9. Standard deviation of resultant vector wind directions over sampling period
10. Standard deviation of nonoverlapping 1-min mean speeds over sampling period
11. Standard deviation of nonoverlapping 1-min mean unit vector directions over sampling period

If the desired sampling period was longer than the basic time of 10 min, e.g., 1 h, then values for items 10 and 11 would be repeated for longer term means of 2, 5, or even 10 min. The purpose of generating the many standard deviations from averaged values was to detect the energy-containing frequencies within the lengths of record considered. More extensive or different analyses of incoming raw data can be easily programmed, but cost–benefit relationships as well as storage capacities and data QC must be considered. Again the example is intended only to be an illustration of a practical data processing approach and not a recommended guideline.

8 CONCLUSIONS

The above discussion of the recent history of meteorological support for an electric utility and of programs used within a selected utility are not complete but do serve as an indication of the importance of high-quality measurement programs as well as the need to match measurement requirements with study efforts. Hopefully, it also provides some insight into the need for making as complete as practical measurements for any given study so resulting data packages can be used in other studies. Other electric utilities in different areas of the country, or world, will undoubtedly have a different list of meteorologically related problems or concerns such as lightning protection and cooling water temperatures.

REFERENCES

ANS (2000a). American Nuclear Society, *Determining Meteorological Information at Nuclear Facilities*, La Grange Park, IL.

ANS (2000b). ANS/ANS-3.11-2000, American Nuclear Society, La Grange Park, IL.

U.S. EPA (1989). *Quality Assurance Handbook for Air Pollution Measurement Systems*. Vol. IV, *Meteorological Measurements*, prepared by Thomas J. Lockhart for the U.S. Environmental Protection Agency, Research Triangle Park, NC.

CHAPTER 44

INDEPENDENT AUDITING ASPECTS OF MEASUREMENT PROGRAMS

ROBERT A. BAXTER

There are a number of aspects that need consideration in the design and execution of measurement programs to assure the data collected are of documented quality and meet the program data quality goals. This chapter looks at the independent auditing aspects of monitoring as an integral tool to the overall data collection effort and provides examples of how independent audits contribute to the understanding of the quality of the data collected.

Prior to entering the role of audits, it is helpful to understand the details of monitoring programs and see where audits fit into the overall data collection scheme. The monitoring program and its goals are described in a monitoring plan.

1 MONITORING PLAN

A monitoring plan is a general description of the overall plan to collect data. A monitoring plan includes the monitoring program goals, methods of data collection, locations where data will be collected, internal and external checks of the measurement program, and the overall management and reporting structure. It is typically prepared well in advance of the collection of data so that management and/or regulatory review can identify and resolve any deficiencies in the measurement program.

Once the monitoring plan is agreed upon and ready to be implemented, a quality assurance project plan is prepared that addresses all of the details of the data collection program.

Handbook of Weather, Climate, and Water: Dynamics, Climate, Physical Meteorology, Weather Systems, and Measurements, Edited by Thomas D. Potter and Bradley R. Colman.
ISBN 0-471-21490-6

2 QUALITY ASSURANCE PROJECT PLAN (QAPP)

A quality assurance project plan is a formal document describing in comprehensive detail the necessary activities that must be implemented to ensure that the results of the work performed will satisfy the stated performance criteria.

The QAPP defines all details of the program including the methods of data collection, the specific instrumentation used for collection of each variable, methods of calibration, data storage, backup, validation, and reporting. Also included is the audit plan to verify the implementation of the methods and procedures defined in the QAPP. The U.S. Environmental Protection Agency (EPA) provides guidance for the preparation of QAPPs for environmental monitoring programs (U.S. EPA, 1998). After reading the guidance and recognizing the recommended level of detail in a QAPP, one may ask the question "Are all of the checks and documentation really necessary? After all, I have been collecting data using a variety of measurement methods for many years." The answer is a qualified yes. The qualification steps back to what the data will be used for and how defendable one wants to make it. Backyard weather forecasting requires little in oversight and documentation. Permit applications and compliance on the other hand requires well-documented, defensible data.

One learns through years of experience that even the most competent professional or scientist can make mistakes through the redundant process of setup, data collection, and data validation. These mistakes may be minor but could have a detrimental effect on the quality of the collected data. In some cases, quality and the proper way to operate a system may be compromised because of budget limitations or lack of physical resources. It is extremely important to identify the instances when data has been compromised and then quantify the impact it has on the quality of the data reported. This will provide end users with records and information needed to assess whether the collected data will meet their needs.

A properly designed program will have cross checks built into the overall measurement plan. These checks will be defined in the QAPP in the sections on quality control and quality assurance.

Quality Control (QC) The overall system of technical activities that measures the attributes and performance of a process, item, or service against defined standards to verify that they meet the stated requirements established by the client.

Quality Assurance (QA) An integrated system of management activities involving planning, implementation, assessment, reporting, and quality improvement to ensure that a process, item, or service is of the type and quality needed and expected by the client.

Quality control activities are designed to control the quality of a product so that it meets the user's needs. This includes the routine calibrations, data validation, preventive maintenance, equipment certification, etc. Quality assurance is the process whereby the implementation of the QC program and other activities is checked and verified. It encompasses the various activities needed to assure the QC program is being implemented and is working. QA includes QC as one of the activities needed to ensure that the product meets defined standards of quality. Details on quality assurance in meteorological measurements can be found in U.S. EPA (1995).

A key element in QA is the use of independent audits to review operations and make reports to management on the status of the measurement program. These audits are performed by an individual or group that is independent of those making the measurements. This independence allows the identification of potential problems without any conflict of interest the findings may have with either the technical merit of the data or financial resources needed to correct deficiencies.

Audit A systematic and independent examination to determine whether quality activities and related results comply with planned arrangements and whether these arrangements are implemented effectively and are suitable to achieve objectives.

In performing the audits, there are specific roles of the auditor and auditee.

Auditor A person that is qualified to perform audits.

Auditee Person or organization that is operating a measurement program and is the one being audited.

Ideally, the auditor will be an expert in the field of the measurements being performed. While not always true, the primary prerequisite is that the auditor understand the measurements being performed and be able to identify if the methods followed are consistent with the monitoring and quality assurance plans. The auditor is always a guest at the measurement site and should not perform any of the instrument removal or other activities that are part of the normal operation of the site. In past instances the auditor has helped in the removal of equipment and performance of duties for the operator and inevitably an instrument breaks or becomes inoperable that becomes attributable to the auditor.

3 CASE STUDIES

Given the above overview of the elements in the planning and oversight of a measurement program, it is useful to review some case studies to demonstrate the value in independent audits to support overall data collection and documentation. The case studies below describe field experiences in auditing some aspects of meteorological measurement programs.

In this first case study, a major measurement program was carried out in the western United States that included measurements made by federal, state, and local agencies as well as private contractors. The design of the program included an independent QA contractor responsible for system and performance auditing of the installation and operation of the surface and upper air meteorological systems. The surface measurements included wind measurements using conventional cup-

and-vane or propeller-vane anemometers. Some of the systems were part of existing measurement programs while the majority were installed specifically for this study.

The auditing plan for the study called for initial siting audits that assessed the appropriateness of the selected sites for the measurements. The results of these audits helped program management determine if changes were needed in the selected sites. Once final sites were selected and instruments installed, audits were performed on the surface meteorological systems. While there may have been several specific problems noted for each of the sites, there was a common equipment alignment problem noted with one of the measurement groups. This problem was present at most of their sites. The alignment and orientation of sensors have always presented challenges. While the most common method of orientation is to use the local magnetic declination to correct the alignment to true north, local anomalies in the magnetic field can create errors of 10° or more. With this particular measurements group (called group A), the number of sites set up and operated led them to streamline the alignment process using a hand-held "data scope." The scope allowed the direct entry of the local declination to the magnetic readings making the alignment checks quick and easy. A second group (called group B) used magnetic methods to determine the alignment but did not apply a specific declination. Instead, the magnetic readings were corrected to true north based on a measurement of the sun's azimuth angle and the calculated true angle of the sun for the site's latitude and longitude.

Figure 1 shows the alignment audit results from groups A and B for the surface sensors. In the overall program plan, the data quality objective for the wind direction

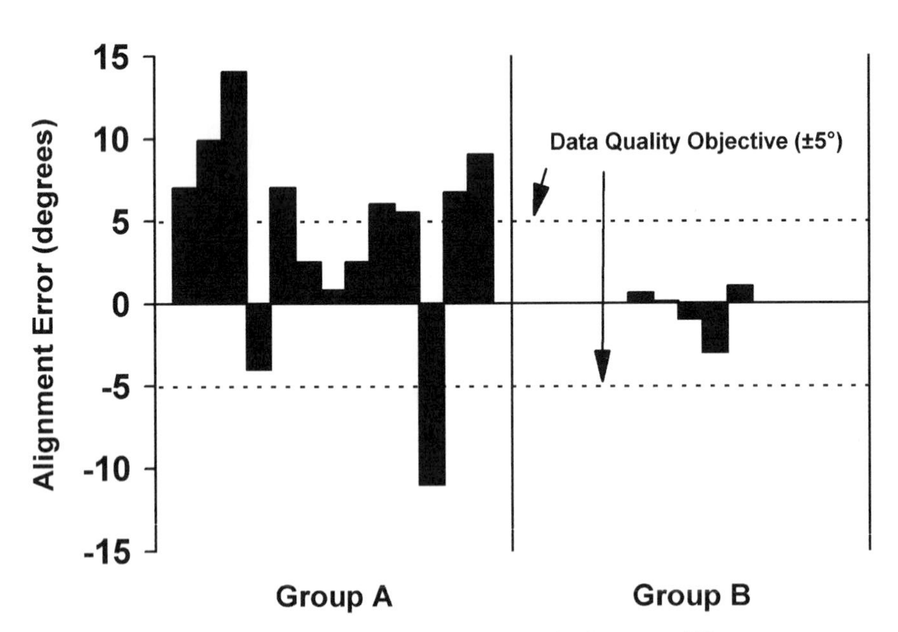

Figure 1 Surface wind direction sensor alignment results from two different measurement groups.

data was established as $\pm 5^\circ$, as indicated in the figure. The results of the alignment audits showed group A had much more scatter in the alignment accuracy, which was a direct result of the method used for alignment. The problem was systematic throughout the sites set up by group A. Agreement was reached between the auditee and auditor on appropriate methods for alignment of the systems. The procedures used by the auditee were modified to obtain the needed accuracy. The audit in this case served as a valuable training and teaching exercise for group A performing the measurements and corrected a long-term systematic problem in aligning sensors.

In a second example, a station was audited that collected both air quality and meteorological data. Meteorological sensors were located on a 10-m tower adjacent to the trailer. This particular station demonstrated an all-too-familiar example of how the meteorological sensors generally take second place to the air quality measurements. Due to the relative complexity of the air quality instrumentation, much of the maintenance effort focused on those instruments. The meteorological sensors were set up, operation verified, and then the sensors left alone to collect data without any further checks. This is a common scenario that arises from the ability to visually look at the sensors, see them rotating in the wind and aiming in appropriate directions. This gives the impression that the equipment is operating acceptably and no further checks need to be performed. To the contrary, bearings in the wind sensors may fail, cups or propellers may be damaged, and potentiometers that convert the vane direction to an electrical signal may wear. Other sensors such as those used for temperature may become corroded and produce erroneous signals. The audit in this example looked at wind and temperature measurements and included a system and siting audit to determine the appropriateness of the site for the intended measurements.

Results of the audit showed the following:

1. The entire meteorological system had not been serviced since installation with no calibrations performed.
2. The temperature sensor had corroded into the radiation shield and could not be removed to verify its proper calibration.
3. The wind direction sensor had corroded into the mounting and the connector failed when removed from the sensor. This hampered the performance evaluation of the sensor.
4. The wind speed bearings had corroded to the point where the starting threshold of the sensor was almost 2 m/s.
5. The alignment of the wind direction sensor was incorrect, resulting in a fixed offset in the measurements of nearly 10°.

While the items noted above are critical in the collection of valid data, the biggest problem was in the siting of the sensors. Figure 2 shows the location of the measurements relative to a nearby building and pollution sources. The building was about 8 to 10 m high and located about 20 m from the sensors. The problem was compounded by the proximity to the major air pollution source that was the focus

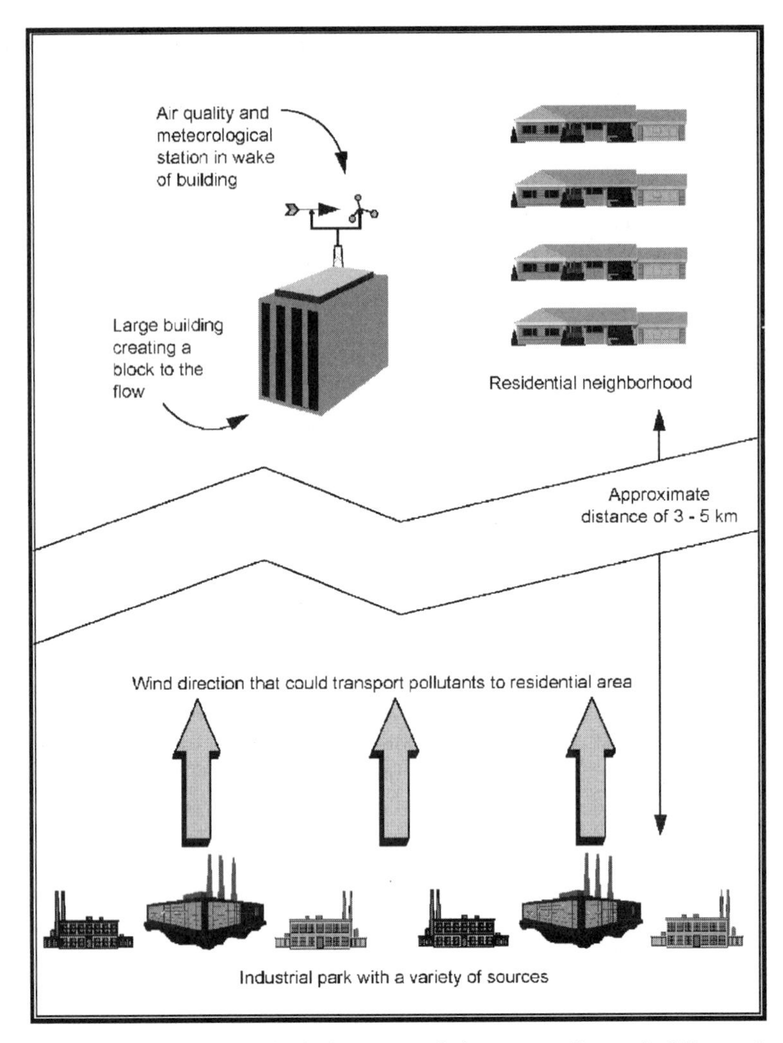

Figure 2 Location of meteorological sensors relative to an adjacent building and nearby pollution sources. See ftp site for color image.

of the overall measurement program. Located to the south was an industrial park with many sources of pollution. This air quality and meteorological station was intended to document the transport of pollutants from the industrial park to the station, which was located adjacent to residential housing. The data was then used in subsequent computer modeling of the sources to determine the impacts. With the building between the station and the source, winds from the south would be significantly altered and not be representative of the area meteorological conditions. The results of the independent audit noted these problems and made recommendations on where an acceptable site could be located and how the sensors should be maintained to collect accurate, defendable data. The five sensor-specific items noted above were highlighted and recommended servicing intervals provided in accordance with EPA guidance. The results of the audit will eventually lead to data that can be used for the intended modeling purposes.

The third case study draws more on the role of quality auditing in the planning and execution of a measurement program in the early stages of setup. As part of a large air quality and meteorological measurement program, the quality assurance contractor held a workshop for all who were making meteorological measurements as well as those who were performing the audits. The purpose of the workshop was to introduce the QA program to all measurement personnel and identify the needs and requirements of the program. This included the audit procedures, criteria used for passing or failing the audits, and the overall data quality objectives. In this manner all personnel making the measurements would understand the common goals of the program and be prepared for the auditing steps that would follow.

All participants in the program had many years of experience in the measurements being performed. One in particular had indicated objections to the workshop and the time it would take to go through the "orientation" process when it still had half of its 35 plus sites to install. This contractor had vast experience in setup of large networks of meteorological instrumentation on the east coast of the United States. Reluctantly, but fortunately, they agreed to participate. As part of the workshop exercise, one of the tasks was for all auditors and measurement contractors to measure the true direction of a distant landmark. A variety of methods were employed by the participants and the results were compared. For reference, the workshop director, who was also the technical director for the meteorological audit program, used his measurement as the "standard" by which all others were compared. For the most part all results were within $\pm 2^\circ$. One participant was heard mumbling and groaning, indicating his answer differed by more than 30°. After a brief review of the method used to determine the true pointing direction, it was learned that he had applied the local declination in the wrong direction. This happened to be the contractor with extensive east coast experience where the local declination is applied in the opposite direction. It was also the same contractor that had responsibility for setup of over 35 stations, half of which were already in operation with the improper alignment, an error that was eventually corrected.

The lesson of this story is that the independent audit program implemented at the start of the field study identified and corrected a major problem early. This allowed all participating in the project to understand the requirements, perform the measurements in a consistent manner, learn from each other's expertise, and ultimately collect data of known quality that are defensible.

As a final example, it is valuable to address the traditional challenge of calculating an "average" wind direction. This is not a trivial problem since the wind direction is a circular function that is not amenable to simple analog averages. For years there have been debates on the appropriate method to handle the average, whether it be dealt with on a vector basis or through some algorithms that deal with the north crossing through 360°. The EPA has provided guidance on procedures for calculating wind direction in regulatory driven and other monitoring programs (U.S. EPA, 2000) that have been incorporated in one form or another into commercially available data logging systems. The discussion below summarizes an experience with an audit of a program that identified a flaw in a widely accepted algorithm for calculating the scalar wind direction using a single pass method. More extensive details on the findings can be found in Baxter (1995).

Wind speed and direction data were collected as part of a dust monitoring program at a construction site with the data used to assess the contribution of the construction activities to the downwind particulate matter concentrations. The site was located in the middle of a densely populated urban area with a number of tall buildings surrounding it. This made it virtually impossible to meet the EPA siting criteria for exposure of wind sensors (U.S. EPA, 2000). While subject to building wake turbulence, the measurements were deemed adequate to assess the general wind direction and aid in the evaluation of the dust-producing activities.

As part of the overall program an audit was performed that identified unusual patterns in the collected data. The site was located in southern California, and the location was strongly influenced by the afternoon sea breezes. These late morning to early evening flow patterns produce a very consistent southwest wind. The data collected by the meteorological system, however, showed frequent interruptions of the afternoon southwest flow with winds from a variety of other directions, including those from the northeast. Further investigation into the data logging system identified the logger used a single pass scalar average algorithm generally accepted and described in the EPA guidelines. This algorithm corrects for the rotation through north, allowing the proper interpretation of wind direction when winds vary from northwest (270°–360°) to the northeast (0°–90°).

With the identification of the anomalies in the afternoon wind direction data, several tests were performed to determine whether the problems were in the physical instruments or whether it was in the calculation methodology. The first test involved programming the existing data logger with an additional unit vector algorithm and comparing the two sets of wind direction calculations. The unit vector algorithm is also described in the EPA guidelines. Figure 3 shows the comparison of 20 days of wind direction data when wind speeds were 1 m/s or greater. The comparison shows some agreement, but for a significant number of values there is no obvious relationship.

The second test placed an identical wind sensor near the first one and logged wind direction data on an independent data logger. This second system collected data using the unit vector algorithm. Figure 4 shows unit vector comparison data between the first and second systems when the wind speeds were 1 m/s or greater. It is clear there is good agreement between the two systems when the unit vector algorithm was used.

On the basis of the first two tests it was obvious there were questions about the calculation results of the single pass scalar algorithm. The third test performed used a model to generate test wind data and perform wind direction averaging calculations with a variety of methods. A simulated 1-s interval wind data set was generated and a simple rotation introduced in the middle of the 3600-point hourly data set. This data set was then evaluated using three averaging techniques: simple arithmetic, unit vector, and the single pass scalar algorithm. The data set and results are shown in

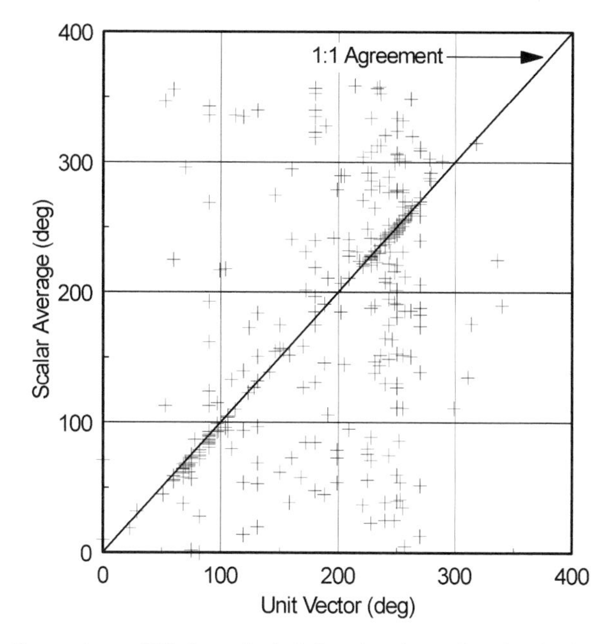

Figure 3 Comparison of 20 days of wind direction data using the single pass scalar and unit vector averaging algorithms.

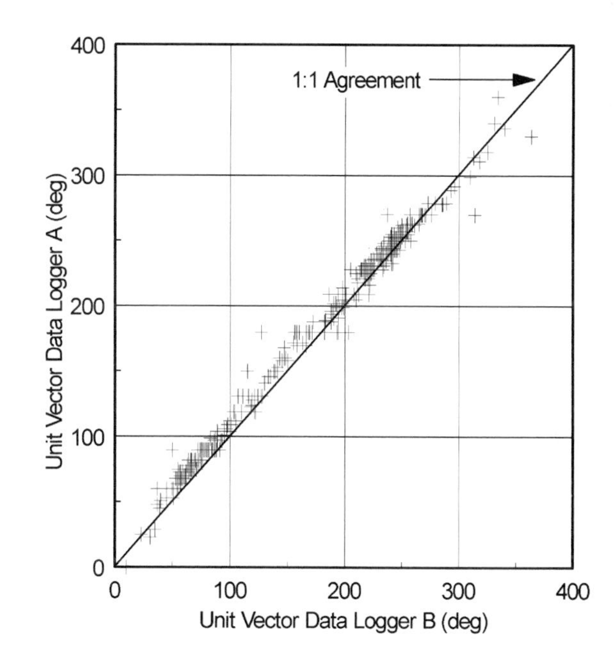

Figure 4 Comparison of 20 days of wind direction data using the unit vector averaging algorithm programmed into two different data loggers.

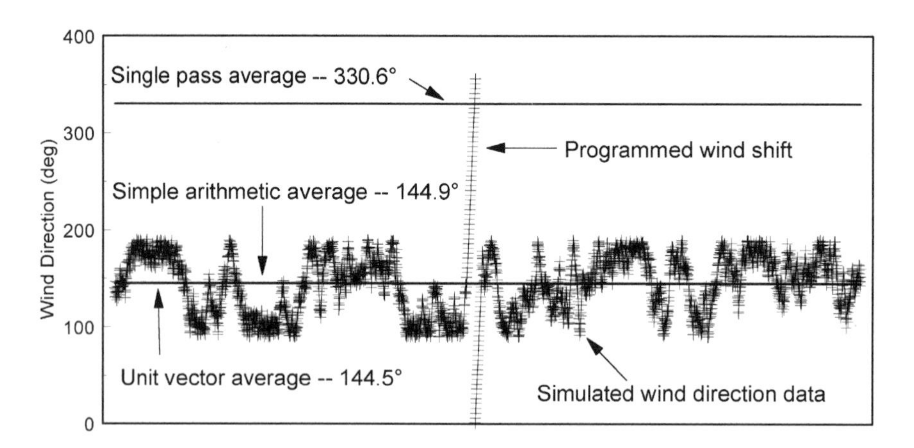

Figure 5 Simulated hourly wind direction data showing a 360° wind shift in the middle of the averaging period. The results of three different averaging techniques are plotted.

Figure 5. It was obvious from the results that there was something wrong with the single pass scalar method technique. Further investigation into the algorithm revealed that a complete circular rotation of the wind direction anytime during the hour would result in erroneous values with the magnitude of the error effectively being random. If there was not a complete rotation during the averaging period, then there was excellent agreement between the methods. However, introducing one or more complete 360° rotations during the averaging period resulted in an unrecoverable error in the reported average.

The results of this audit were extremely important in understanding potential errors in the collected data and resolving the ambiguous values obtained during what should have been very consistent wind directions. Further discussions with individuals who developed the scalar algorithm recognized the problem but also carried the purpose of the algorithm one step further. Part of the derivation of the algorithm was to develop a method that would correctly quantify the variability of the wind direction expressed as the standard deviation or sigma theta. That calculation is performed correctly using the scalar technique, but at the expense of the reported wind direction. Given these results, the optimum technique may be to use a vector method for the wind direction (either a unit vector for non-wind-speed weighted or straight vector for wind-speed weighted) and the scalar method for the standard deviation of the wind direction. Each of these methods are described in the EPA guidelines (U.S. EPA, 2000).

4 CONCLUSION

In summary, independent audits have become an integral part of many measurement programs. The extent of the audit scope and frequency of performance depends on the specific data needs and any applicable requirements. Not only do the audits aid in the improvement of the data quality, but they can in many instances identify potential problems that could invalidate collected data. Additionally, most regulatory driven programs such as Prevention of Significant Deterioration (PSD), National Air Monitoring and State and Local Air Monitoring Stations (NAMS and SLAMS), and Photochemical Assessment Monitoring Stations (PAMS) require independent audits as part of the normal measurement program. So not only are audits good for the data quality, they are also a regulatory requirement that must be fulfilled in order to use the data in modeling or analysis projects.

REFERENCES

Baxter, R. A. (1995). Evaluation of the Wind Data Collected Using Different USEPA Approved Calculation Algorithms. Paper presented at the Ninth Symposium on Meteorological Observations and Instrumentation, March 27–31, Charlotte, North Carolina.

U.S. EPA (2000). United States Environmental Protection Agency, Meteorological Monitoring Guidance for Regulatory Modeling Applications EPA-454/R-99-005, Office of Air Quality Planning and Standards.

U.S. EPA (1995). United States Environmental Protection Agency, *Quality Assurance Handbook for Air Pollution Measurement Systems, Vol. IV, Meteorological Measurements*, Document EPA/600/R-94/038d, Atmospheric Research and Exposure Assessment Laboratory.

U.S. EPA (1998). United States Environmental Protection Agency, EPA Guidance for Quality Assurance Project Plans EPA QA/G-5, Document EPA/600/R-98/018, Office of Research and Development.

CHAPTER 45

REGULATORY APPROACHES TO QUALITY ASSURANCE AND QUALITY CONTROL PROGRAMS

PAUL M. FRANSOLI

1 BACKGROUND

Meteorological data are often obtained to support environmental and engineering studies. Such data are evaluated by regulatory bodies for suitability in various risk, compliance, and design analyses. Information used in the regulatory arena is subject to intense scrutiny by all involved parties, including those attempting to challenge the credibility of work performed. To ensure that meteorological data are acceptably correct and complete, and that the regulations are being applied consistently to all applicants, some regulatory groups have promulgated quality assurance and quality control (QA/QC) requirements and guidelines. Meeting the monitoring requirements and following the guidelines helps to ensure acceptance of the data. It is usually preferable to agree on a monitoring plan with the regulatory group prior to proceeding with the monitoring program. It also helps to protect proposed projects from time-consuming demonstrations that the monitoring equipment and methods were satisfactory for the intended purpose.

Site characterization in engineering design or regulatory environmental studies differs from typical weather data collection programs. Site characterization programs seldom require real-time information flow, which is an important element in routine weather observations made for forecasting purposes. Instead, data from remotely operated stations can be stored on-site for long time intervals prior to collection. Some on-site processing is to obtain some statistical properties of the measurements, such as means, extremes, and standard deviations.

Handbook of Weather, Climate, and Water: Dynamics, Climate, Physical Meteorology, Weather Systems, and Measurements, Edited by Thomas D. Potter and Bradley R. Colman.
ISBN 0-471-21490-6

The primary quality factors appearing in most regulatory measurement programs are accuracy, precision, validity, completeness, and representativeness.

- **Accuracy and precision** are addressed in depth in another chapter in this part. The accuracy and precision requirements contained in regulatory guidance are based on both application needs and capability of the equipment used. Early atmospheric dispersion models had very simple data input requirements; typical synoptic weather observations were adequate model input. The use of site-specific (also known as on-site) data, that is, data intended to be representative of the source and/or receptor areas, helped reduce the uncertainty from the dispersion modeling portion of environmental studies. Leapfrog improvement advances between model and measurement capabilities continue to challenge both the measurement and the modeling fields.
- **Validity** of the data implies compliance with the accuracy and precision requirements and goes beyond to ensure that erroneous measurements have been removed from the validated data set. A data validation protocol often begins with removing data from known periods such as quality control checking and maintenance work that interrupts the measurements and continues with identifying periods of sensor or on-site data processing and recording equipment failures. While modern sensor and data system reliability has increased immensely, all equipment is subject to error. Some sensor failures can start as intermittent problems that may introduce a slight error that is difficult to distinguish from normal variations. Other problems can occur with sensors becoming temporarily incapacitated by external forces such as ice. Many measurement programs have found that a meteorologist knowledgeable of the measurement process and the specific program should be included in the data review team.
- **Completeness** means attaining an adequate amount of data for the intended purpose. Modeling programs intending to identify short-term phenomena that produce unacceptable hazards or risks need to be based on a sufficiently long monitoring program time period to identify the high-risk meteorological episodes. The most typical completeness requirement is 90% data recovery, though 80% applies to some remote measurement stations. Modern equipment reliability makes this requirement attainable, though frequent site checks by knowledgeable operators are important to ensure that sensors have not been damaged.
- **Representativeness** refers to measurements being made in a location that is representative of the area being characterized. The area could be the source itself, or the potential receptors of an airborne effluent, or significant points along a potential airflow pathway. Representativeness of the measurement location is addressed in siting guidelines. In addition to the measurement location being representative of an intended area, siting requirements also include instrument exposure. Wind measurements can be affected by obstacles, including the structure supporting the sensors. Temperature measurements can

be affected by nearby heat sources or sinks, such as parking lots and cooling towers.

2 STANDARDIZATION AND REGULATORY GUIDANCE

Regulatory monitoring requirements and guidance are based on the requirements set by the governing bodies, such as the U.S. Nuclear Regulatory Commission (NRC) and the U.S. Environmental Protection Agency (EPA). Others include the American National Standards Institute (ANSI) working in conjunction with the American Nuclear Society (ANS) or the American Society for Quality Control (ASQC). Of these, the EPA guidelines on prevention of significant deterioration (PSD) monitoring made the most significant recent advances in quality assurance and quality control applied to meteorological monitoring. The primary EPA monitoring guidance document is *Meteorological Monitoring Guidance for Regulatory Modeling Applications* (U.S. EPA, 2000). Volume IV (U.S. EPA, 1994) in a series of quality assurance handbooks for air pollution measurement systems contains some of the most detailed technical guidelines on monitoring techniques. Volume IV became the "how-to" book that explained many of the "why" questions behind the guidelines and guided technicians through the rigorous detail of correctly performing the equipment tests.

Early work in the nuclear power industry to site, license, and operate large nuclear-powered electric-generating stations triggered some guidelines such as *Meteorology and Atomic Energy* (Slade, 1968) and the Nuclear Regulatory Commission Safety Guide 23, which became Regulatory Guide 1.23. Technical guidance in these documents established the equipment and network design specifications for numerous nuclear-power-related projects. The 60-m tall meteorological tower with wind and temperature measurements at the 10- and 60-m levels above the ground became synonymous with RG 1.23 programs. Another important nuclear-related guidance document appeared in 1984 as ANSI/ANS-2.5: Standard for Determining Meteorological Information at Nuclear Power Sites (American Nuclear Society, 1984). This document helped to update the outdated RG 1.23 and was used for many safety plans at nuclear power plants. This standard became less useful by advances in monitoring technology. The replacement for this document was recently approved as ANSI/ANS-3.11: Determining Meteorological Information at Nuclear Facilities (American Nuclear Society, 2000).

Regulatory guidance is increasingly focused on using voluntary consensus standards, rather than having separate regulatory agencies promulgating independent requirements. The American Society for Testing and Materials (ASTM) subcommittee D22.11 (Meteorology) has produced consensus standards and practices related to meteorological monitoring equipment and application methods. The standards describe equipment design and testing considerations; the practices describe techniques to use the equipment effectively in operational programs. Many of the standards and practices address wind measurement, though atmospheric pressure, temperature, and humidity are also covered. These standards and practices are continually being reevaluated, and new material is being developed.

3 PRIMARY ELEMENTS

The basic elements of meteorological monitoring programs that achieve the data quality factors described above are summarized in this section. This material is intended to be a "primer" on the topic; compliance with current requirements and guidance should be based on the material effective during the time of a given program. The primary information needs are input to atmospheric dispersion models, with other applications of on-site measurements such as engineering and hydrology.

Siting

Siting criteria for meteorological monitoring programs include the number of stations, their locations, and the measurements to be made. Simple programs in flat terrain may only require a single station. The primary wind measurements are made at 10 m above ground level (agl), and temperature and atmospheric moisture at 2 m agl. Additional wind measurements may be necessary at higher levels to properly document airflow in complex terrain, near large manmade obstacles or bodies of water, particularly if the source is located much higher than 10 m agl. Such terrain or obstacles may also require more stations to properly document trajectories of airborne material or the locations of extreme wind or temperature conditions.

Siting criteria also include instrument exposure at the station itself. Wind, temperature, moisture, and precipitation sensors are easily influenced by supporting structures, such as towers and poles. Nearby natural and manmade obstacles can also adversely influence a measurement, making it unrepresentative of the area intended.

Equipment Selection and Procurement

Equipment selection should be driven by guidance specifications, availability of vendors to perform calibration services (which could include the manufacturers), product and service reliability, and cost. When initial cost is one of the few factors used in the selection process, the overall cost over the first few years of operation can far exceed initial differences. The vendor exhibits at trade shows and discussions with other program operators are valuable information resources.

Operator Training

Even the best equipment systems can produce unusable data if the station operators and data processing staff are inadequately trained and supervised. Some regulatory groups and private companies offer valuable training programs. As with most quality control and quality assurance activities, it is wise to document the training received.

Installation (Testing, Location Documentation)

Proper equipment installation involves adequate field testing to ensure that the system was not damaged in transit and has been properly installed. Equipment checking is addressed in the next section. Another essential station startup step all too easily overlooked is proper station location documentation. Seemingly obvious local landmarks can easily change in time, particularly when a station is the first step in a major development. The resolution and reference coordinate system should be compatible with the purposes of the data.

Calibrations, Checks, and Corrective Actions

Three important quality control activities are equipment calibrations, simple checks, and corresponding corrective actions. Calibrations imply formal comparisons of equipment responses with known conditions produced by a standard. Standards must have performance traceable to reliable sources, such as the National Institute of Standards and Technology (NIST). Less complex testing and checks can be accomplished in the field to ensure continued reliability of the measurement process. Such checks can be comparative measurements made by the monitoring equipment and a collocated standard, or by placing the equipment in a known operating condition, such as aiming a wind vane toward a known direction. Calibration and check results can either be made to demonstrate operation within the required tolerance limits or to allow for instrument adjustments that bring the monitoring equipment response within specified limits of the known condition.

Equally important to the calibration or check itself is the corresponding corrective action to be applied when the instrument response does not meet the desired tolerance limit. In addition to performing adjustment or maintenance to bring the response within the desired range, the nonconforming response of equipment operating on-line to collect data should be carefully documented and provided to the data validation staff to ensure that the out-of-tolerance data can be removed from a validated data set.

Routine Operations and Maintenance

Continuing the credibility of the data beyond the careful equipment selection, operator training, and checks requires vigilant routine operations and proper maintenance. Procedures specific to the given operating program should be available to (and used by) equipment operators and data validation staff. Site checks should be documented on checklist forms or in logbooks. Many meteorological sensors used in monitoring programs for regulatory purposes are designed for a balance of sensitivity to subtle airflow conditions and reliability of operation. Such equipment can be susceptible to damage from natural hazards, such as ice, blowing dust, birds, and other animals. It is difficult to tell from data acquired by telemetry if one of the anemometer cups is missing.

Maintenance includes preventive and corrective actions. A solid preventive maintenance program can preclude a large portion of equipment failure events. Corrective maintenance should be applied as promptly as possible to minimize downtime, provided that proper documentation is made of the failure symptoms or condition and that the replacement or repaired equipment is properly tested when installed.

Data Processing

Data processing begins during the measurement phase of the program; sensor responses are translated to a numerical value that can be averaged, totaled, stored as an extreme, or used to calculate a variance about the mean. Modern data recording equipment offers the user numerous options, as well as pitfalls. As with other activities, proper testing and documentation of the on-site processing routines is an essential first step in demonstrating that data processing is correctly applied. Most programs are enhanced by including data-identifier information with all data records, so that raw files can be uniquely identified by time period and station location. Once the data are collected from the field station, the data should be traceable throughout the data validation and editing processes.

Audits

An important step in demonstrating data credibility to outside groups is to document the results of independent verifications of compliance with established procedures and tolerance limits. The term *performance audit* implies tests made by knowledgeable staff independent of those performing the routine site operations and checks using independent testing equipment, which is also traceable to standards. The complementary function is a *system audit*, in which independent staff examine the work products and documentation to ensure that activities comply with the established procedures.

Reports

The final link in the chain of demonstrating data quality is to produce credible reports of the monitoring program results. Report content and format should fit the objectives of the program. It is wise to include summaries of the quality control and quality assurance activities in sufficient detail that the results can be accepted.

REFERENCES

American Nuclear Society (1984). ANSI/ANS-2.5: Standard for Determining Meteorological Information at Nuclear Power Sites, American Nuclear Society.

American Nuclear Society (2000). ANSI/ANS-3.11: Determining Meteorological Information at Nuclear Facilities, American Nuclear Society.

Slade, D. H. (Ed.) (1968). *Meteorology and Atomic Energy*, U.S. AEC, July.

U.S. EPA (1994). *Quality Assurance Handbook for Air Pollution Measurement Systems, Vol. IV: Meteorological Measurements*, U.S. EPA Office of Research and Development, Washington, DC, EPA/600/R-95-038d, April.

U.S. EPA (2000). *Meteorological Monitoring Guidance for Regulatory Modeling Applications*, U.S. EPA Office of Air Quality Planning and Standards, Research Triangle Park, NC, EPA-454/R-99-005, February.

CHAPTER 46

MEASURING GLOBAL TEMPERATURE

JOHN R. CHRISTY

Every part of the Earth system may be described by the variable of state we call temperature. Fundamentally, temperature is the magnitude of the average molecular kinetic energy of a substance—the higher the molecule's average speed, the greater the temperature. Today, temperature is estimated by several kinds of instruments using a variety of techniques.

In situ measurements require, in some way, direct contact between the instrument and the moving molecules of the targeted substance. The most familiar of these types of instruments is the liquid-in-glass thermometer whose bulb contains a liquid that expands or contracts, allowing the liquid to move through an opening to a narrow, graduated tube. The bulb is inserted into the medium of interest (e.g., air, water, ice, or earth) and the liquid responds according to the amount of molecular motion detected on contact. Another in situ device utilizes the fact that the electrical resistance of a conductor is proportional to its molecular kinetic energy so that the amount of electricity able to flow through the conductor indicates temperature. The velocity of sound through a substance (usually air or water) is also directly related to its temperature and thus is an indirect characteristic that can be used to monitor temperature in situ.

Remote methods, in contrast to in situ, have the advantage of measuring temperature from a distance. The most direct of these methods employs radiometers that measure the intensity of radiation emitted by a substance. The magnitude of the intensity is often proportional to the substance's temperature. Satellite radiometers fall into this category of devices as they monitor the upwelling radiation from the various components of Earth's system. Other relatively new remotely sensed methods estimate temperature by measuring (a) the speed and refraction of radio signals through a material (e.g., Global Positioning System satellites), (b) the physical

Handbook of Weather, Climate, and Water: Dynamics, Climate, Physical Meteorology, Weather Systems, and Measurements, Edited by Thomas D. Potter and Bradley R. Colman.
ISBN 0-471-21490-6

height of the sea surface (higher altitudes mean expanded or warmer water) and even the thermal "color" of the ocean affected by microorganisms that preferentially appear in waters of certain temperatures.

The distinction between in situ and remote may be rather blurred. One thinks of a bucket dropped over the side of a ship, filled with seawater, hoisted back up on deck into which a thermometer is inserted, and from which a temperature reading is determined after some time as in situ. However, one probably thinks of a satellite radiometer, which measures the actual, unaltered photons representing the exact character of the substance in question and traveling at the speed of light, as a remote measurement. One can see that mere proximity to the intended medium may not assure the most accurate estimate of temperature.

A unique and quasi-direct method is one that measures the temperature at various depths of a very stable borehole (e.g., in bedrock or ice cap) and then estimates what the surface temperature would have been in the past to produce the temperatures observed at each given depth. Simply put, the deeper the temperature reading, the longer in the past it was influenced by the surface temperature.

Temperatures from all of the above devices are referred to as being part of the "instrumental" record and in some sense may be called direct measurements. There are, however, several indirect methods available in which some organism or physical process preserves in its history the character of the environmental factors, including temperature, affecting it. In this category of "proxy" records are tree rings, ice core composition and thickness, isotope ratios in ice, sea floor, and sediment cores, pollen distributions in sediments, erosion rates, coral bands, sea level height, evidence of the extent of mountain glaciation, plant and animal fossil types and distributions, and many more. With these types of proxy and borehole records, some estimates of the climate prior to the instrumental record are possible.

The *global temperature* is a rather ambiguous term since every part of the global system has its own temperature. When used in the context of climate change, it usually means the temperature of Earth's atmosphere about a meter or two above the surface, often termed *near-surface air temperature*. However, one could speak of the temperature of the land itself, or of the sea water at various depths, or of the ice in the ice caps, or of the atmosphere at any of several altitudes. It is even possible to measure the temperature of the "cold space" in which Earth orbits and find a value of about 2.7 K. Because Earth is a system of many interactive components, the temperature of each is truly necessary to document global temperature.

Most data sets of global temperature are in fact not global in extent nor systematic in quantity measured. So, not only are there variations in "how" and "where" temperature is measured, one must be careful to know "what" aspect of the Earth system is being measured. Because we as humans experience weather and sustain our existence on the surface of this planet, the near-surface air temperature is usually the quantity of first importance to us.

For all of the temperature data sets above and to which the term global is applied, there are issues of uncertainty—spatial and temporal homogeneity, calibration (or lack thereof) of sensors and techniques, changes in instrumentation (type, method etc.), degradation of instruments over time, corruption of proxies over time, changes in local environment due to nonclimatic factors, poor information on observational practices, and many others. Each of these present potential problems and are usually difficult to assess so that the total level of uncertainty in any measurement is not completely quantifiable. In the attempt to understand precisely what the climate system is doing in terms of temperature, the unpleasant notion of potential error is never totally absent.

The various measures of global temperature have been described in publications too numerous to list. The four Intergovernmental Panel on Climate Change reports (IPCC, 1990, 1992, 1996, 2001) are probably the best sources of information that document the more prominent data sets.

1 PROXY RECORD OF SURFACE TEMPERATURES

The average of proxy-estimated temperatures of widely scattered sites prior to instrumental record gives a rough idea of what the temperature may have been in the past. One set of these time series is shown in Figure 1 (see also IPCC, 1990, Fig. 7.1). Even in these very low time-resolution diagrams, there is clear evidence of large variability in global temperatures. Fluctuations in volcanism, solar insolation, orbital parameters, atmospheric composition (including changes due to human activities), aerosols, natural chaos, etc. certainly play their roles, sometimes in isolation and at other times in interdependent ways, as probable explanations for the variations. Efforts have been attempted to quantify these causes, usually by a combination of empirical calibration and analytical theory (e.g., Mann et al., 1998; Tett et al., 1997; Wigley and Raper, 1990), but such discussion is not the focus of this chapter.

Figure 1 (top) shows two periods of interglacial (warm) temperatures over the past 150,000 years, the current being the Holocene that began over 10,000 years ago. The mid-Holocene (~6000 years ago) was relatively warm as was the period about 1000 years ago (Fig. 1, middle and lower). There is considerable evidence that the period between 1400 and 1850 was somewhat cool and is commonly called the Little Ice Age, though recent evidence suggests that it was a period of several types of climate fluctuations in various parts of the globe. Solar variability and significant volcanism are among those forcing parameters thought to play roles in causing these "recently" depressed temperatures.

2 INSTRUMENTAL RECORD OF SURFACE TEMPERATURES

For most of the instrumental period, the past 150 years or so, temperature has been determined from the familiar liquid-in-glass thermometers housed in various types of shelters, usually 1.5 m above the ground, around the world's land areas. Over the oceans, the sea surface temperature (or SST, i.e., the temperature of the seawater, not the air) is the preferred quantity because of the ocean's more spatially and temporally coherent temperature field. The manner by which the SSTs are measured has varied considerably through the years and requires careful adjustment factors to account for

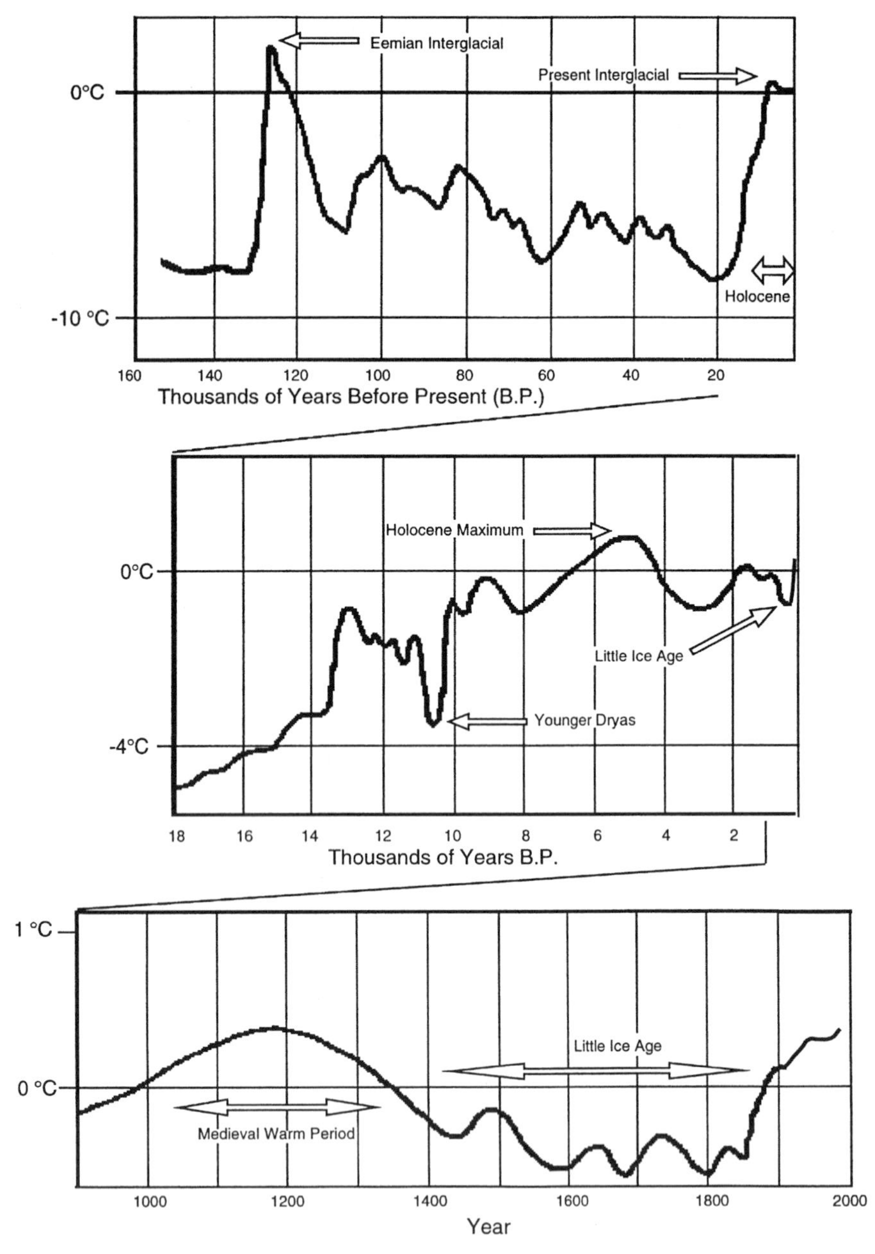

Figure 1 Estimates of global temperature variations over three long periods ending with the present day determined from proxy information (*EarthQuest*, Office of Interdisciplinary Earth Studies, Spring 1991, Vol. 5, p. 1).

biases introduced when, for example, buckets or ship-intake values are used (Folland and Parker, 1995). The "global" temperature most commonly reported is a combination of the near-surface air over land and SST reports over oceans, however geographically sparse they might be.

Only a handful of serious efforts have achieved the goal of collecting as many reports as are presently accessible to document the temperature in several places around the globe over the past 150 years. (Scattered records go back further, one to the midseventeenth century in central England; see Manley, 1974.) The IPCC usually focuses on the data sets produced by (1) the Climate Research Unit of the University of East Anglia (near-surface air temperature over land; Jones and Briffa, 1992), (2) UK Meteorological Office (SSTs; Parker et al., 1995), (3) Goddard Institute for Space Studies, NASA (near-surface air temperature over land; Hansen and Lebedeff, 1988), (4) Russian (near-surface air temperature over land, Vinnikov et al., 1990), and (5) NOAA satellite-based SSTs (Reynolds and Smith, 1995). Two other significant efforts are the Global Historical Climate Network (NOAA/NCDC; Vose et al., 1995) and a database of individual SST reports known as COADS (Woodruff et al., 1987). At present, some of these groups are combining the data sets of temperatures over land with SSTs to produce global coverage (e.g., IPCC, 1996; Hansen et al., 1996).

The time series of global, annual surface temperature anomalies for three of the groups is displayed in Figure 2. (All of the time series produced from the various groups are quite similar, and this is to be expected since the critical broad variations depend on basically the same primary source data for each data set.) These time

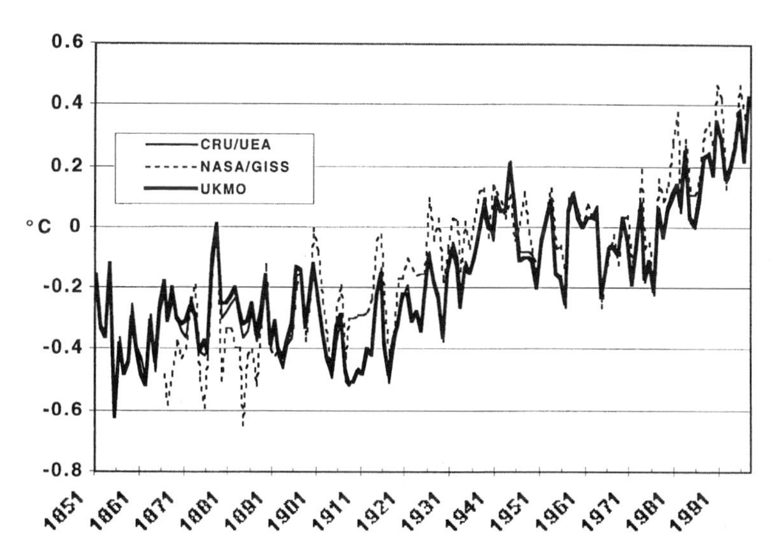

Figure 2 Annual, global anomalies of surface temperature 1851–1997 as a combination of near-surface air temperatures over land and SSTs over oceans (CRU/UEA, UKMO) and land-only (NASA/GISS).

series were described in the IPCC (1996) as showing a 0.3 to 0.6°C rise over the past century, the range being due to the uncertainty factors mentioned above.

The features of this time series are well-known; a fairly significant rise from about 1910 to 1940 (~0.4°C), then somewhat random variations to 1980 and a rise since that time (an additional ~0.2°C). A comparison of the most recent 30 years with those of 1870–1899 reveal an increase in temperature of about 0.4°C while a comparison of the most recent 10 years versus the earliest 10 (1860–1869) shows a rise of about 0.6°C.

3 BOREHOLE TEMPERATURES

An insulated, homogeneous column of material with one end exposed to temperature variations over a long period will reveal temperature variations throughout the column according to the speed at which the fluctuations propagate from the exposed end. An analysis of the temperatures throughout the column at a given point in time, then, may contain enough information to recover those external temperature variations. The general idea is that the greater the depth from the exposed end, the further back in time the temperature at that depth might represent. Of course, as time passes, the perturbations tend to smear, and it becomes difficult to extract meaningful information about what may have happened at the surface.

Certain types of stable, homogeneous bedrock and large ice plateaus lend themselves to temperature recovery through measurements taken in boreholes. The theory and complications regarding the inversion of the temperature profiles into time series are quite complex, but the general results show that many land regions have experienced warming in the past three centuries, though the results show variations among the individual sites (Deming, 1995). Also, evidence from boreholes on the Greenland Plateau suggest that the period around 1000 years ago was warmer than the present century (Cuffey et al., 1994; Dahl-Jensen et al., 1997). Both of these results are consistent with the temperature estimates in Figure 1.

4 UPPER AIR TEMPERATURES

Upper air measurements, which could be compiled into large-scale averages, became possible in the late 1950s due to the expansion of the network of radiosonde stations—sites that release balloon-borne instruments to measure temperature, wind, and humidity to altitudes up to and exceeding 10 km. Sources of large-scale averages of temperatures at various levels and for various layers have been provided by NOAA (Angell, 1988; 63 stations; Oort and Liu, 1993; 800+ stations), UK Meteorological Office (UKMO) (Parker et al., 1997; 350+ stations), and Russian IHMI-WDC (Sterin et al., 1997; 800+ stations).

Temperature time series constructed from individual radiosonde sites often display inhomogeneities most often related to changes in the instrumentation or the algorithms by which the raw data are processed into pressure-level data (Gaffen, 1994). These inhomogeneities can be quite prominent at the highest elevations because errors or changes tend to accumulate as the balloon ascends. Oort and Liu (1993) generate global maps by objective analysis of their world-wide radiosonde dataset. The UKMO product uses selected stations with better records of homogeneity and length and produces a quasi-global analyses with limited interpolation for filling in vacant grids. As with Oort and Liu (1993), the RIHMI-WDC produces global analyses by objective means but also applies a complex quality control algorithm to the data to remove obvious inconsistencies that violate hydrostatic constraints and those of spatial coherency.

Since 1979, nine NOAA polar orbiting satellites have carried an instrument, the microwave sounding unit, or MSU, designed to provide temperature information in both clear and cloudy areas where infrared (IR) methods are ineffective. The sensor measures the intensity of radiation near the 60-GHz oxygen absorption band, which is proportional to atmospheric temperature. Though not intended to provide long-term climate information, data from the nine MSUs, which have orbited since 1979, have been calibrated and merged into a single time series (Christy et al., 1995). With daily global coverage (over 30,000 observations per day) of a very robust quantity (a volume of air over 50,000 km^3 per observation), these data have some advantages over other types of data sets.

Two MSU temperature products are widely used: the lower troposphere (the average temperature of the surface to about 7 km or 1000 to 400 hPa) and the lower stratosphere (17 to 22 km layer or about 120 to 70 hPa). Version D of the data described in Christy et al. (2000) takes into account recently discovered influences on the satellites that affect the observations.

We have found in the several years of constructing the MSU data sets that the issues of greatest impact on the long-term record are (1) the intersatellite biases, (2) the orbital time drift, and (3) the orbit decay (see Wentz and Schabel, 1998). Because the MSU data sets are products we produce, I shall devote a relatively large amount of discussion to them.

Though calibrated to high precision on Earth in thermally controlled vacuum chambers, the MSU, once in the environment of space, may acquire or display unexpected characteristics. In one case (NOAA-12) an electronic gain change occurred after launch so that the anticipated calibration target temperature of cold space (2.7 K) was not correct, being measured at about 6 K. Corrections for this effect were generated by Mo (1995). Also, the MSU, as a cross-track scanner, monitors temperatures to the left and right of the track. In an unexpected result, temperature comparisons of these two sides produce average differences as great as 3°C and as little as 0.05°C in different instruments. This asymmetry is likely due to variations in the antenna beam patterns from instrument to instrument—i.e., the actual location of the "cone" through which the energy upwells to the sensor is not as precisely located as anticipated. This is a systematic effect (it will not change during the instrument's life) but will produce overall biases if not accounted for.

Continuing with nonclimatic effects, there are two spurious consequences of the slow east–west drift of a satellite. One is that the MSU observes Earth at later or earlier local times as it passes over a given region as it drifts. Thus, the natural diurnal cycle of Earth's temperature is aliased through time and can appear as an artificial trend in the data. In version D, we apply an independently determined diurnal trend correction to all satellites based on the differences of the individual footprints across a scan line.

In Christy et al. (1998) we noted a newly discovered effect in which the temperature of the MSU instrument itself influences the temperature of the observations. This effect is manifested as both intraannual and interannual variations, especially for the satellites in the 0200/1400 orbit time node. The temperature of the instrument fluctuates due to the east–west drift, which induces a changing solar shadowing effect on the instrument itself. The net effect of these east–west drifts is to introduce an artificial warming into the time series.

Wentz and Schabel (1998) have discovered yet another effect that for the lower troposphere product has an important influence. During periods of high solar activity, the upper atmosphere expands and exerts greater drag on the satellites, causing them to descend a few kilometers over the 2 or 3 years of increased solar activity from their average altitude of about 850 km. The lower tropospheric temperature utilizes a retrieval that is very sensitive to satellite altitude. The net effect of the "falling" satellites is to introduce an artificial cooling to the lower tropospheric time series.

To summarize, then, we know that the MSU data contain two artificial warming effects and one artificial cooling effect. These have been quantified and removed from the data set, and in fact they are almost exactly offsetting in their net effect. However, we must not presume that this set of issues, combined with what we have discovered previously, constitutes the complete body of knowledge regarding the long-term stability of the MSUs and their peculiarities. These instruments are in space, so we cannot examine them directly for anomalies we suspect may be developing. We are forced to diagnose problems based on the data they transmit, and this is a difficult enterprise. However, this effort has been a largely successful endeavor because we have the ability to examine the MSU data in light of completely independent data from radiosonde observations. The results, in our view, are very interesting.

In Figure 3 we show the annual anomalies of global tropospheric temperature from Angell, UK Met. Office, RIHMI-WDC, and the MSU D (see Christy, 1995). We must note that Angell studies the thickness temperature of the 850- to 300-hPa layer, RIHMI looks at the average temperature of the 850- to 300-hPa layer, and the UKMO at the raob-simulated MSU temperature. Geographical coverage of the radiosonde data sets is limited, while the MSU essentially observes the entire planet. However, the spatial coherence of the troposphere is such that fewer spatial degrees of freedom exist than at the surface; thus fewer sites are necessary to define the average global temperature.

For the period 1979–1997 all data sets indicate that the temperature of the troposphere has declined slightly. Before that period, all show a warming trend, principally a sharp jump during 1976–1980, though the RIHMI anomalies are much less in magnitude (the reasons are probably due to a stiff interpolation method). Many such

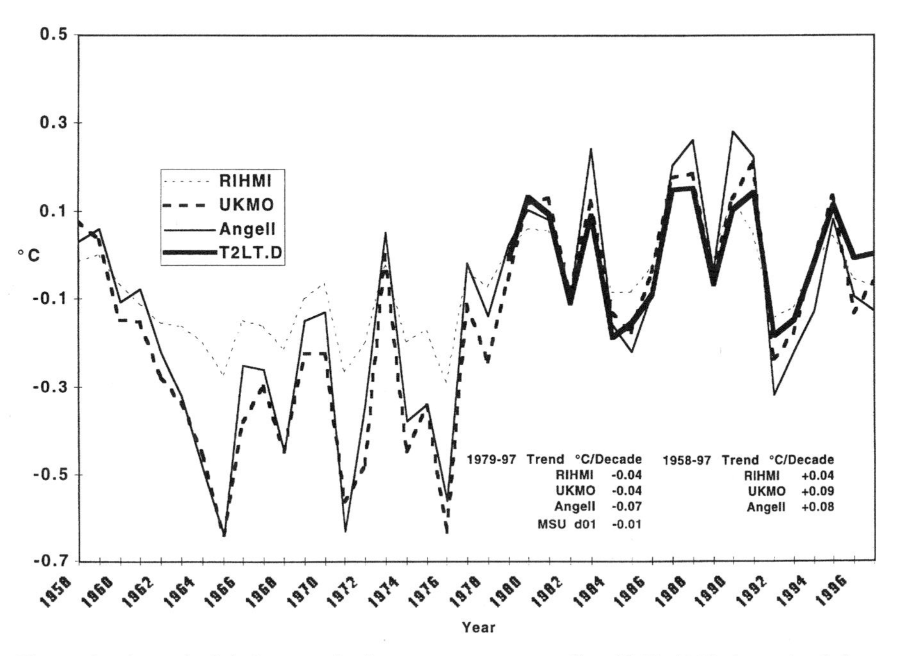

Figure 3 Annual global tropospheric temperature anomalies 1958–1997 determined from radiosonde data sets and the MSUs on polar orbiting satellites. The layer is roughly between 1 and 8 km altitude.

comparisons of multistation averages have been performed, and they indicate excellent agreement between the radiosondes and MSU temperatures.

Some concern has been raised regarding the lack of a tropical warming trend observed independently by radiosondes and the MSU since 1979. It has been proposed that the tropical troposphere should behave in concert with variations in tropical SSTs as indicated by climate model results, which are forced with observed SSTs. The observed atmospheric data seem to indicate the tropical troposphere is somewhat more decoupled from the boundary layer than presumed. The issue was put forth by Hurrell and Trenberth (1997) in which they suggested that when compared with tropically averaged SSTs, there was evidence for two "spurious" downward jumps in the MSU record, one of 0.25°C in late 1981 and of 0.1°C in late 1991. Since these events seemed to be near the time when new satellites were merged into the data set (NOAA-07 and NOAA-12, respectively), they suggested that improper biases might have been applied to those two satellites.

It is possible to test these claims by producing the anomalies of each of the satellites independently. A comparison of the anomalies indicate no such break occurred in 1981 or in 1991, i.e., the anomalies of the time series remained the same with or without the addition of the new satellites. In addition, comparisons with radiosonde anomalies across both of the boundaries of alleged jumps verified the MSU anomalies (Christy et al., 1997, 1998). Speculation cited by Hurrell and

Trenberth (1997) for the causes of the "jumps" (i.e., surface emissivity effects, bias errors, etc.) were investigated thoroughly and shown to be without foundation. Mentioned, but not emphasized, by Hurrell and Trenberth was the possibility that the SSTs and troposphere indeed do show evidence of slight independence in multi-decadal trends (a phenomenon already verified in other regions, Ross et al., 1996).

The apparent jumps between SSTs and the troposphere are an intriguing observation. We examined this effect further and found the real source of the shifts appear to be due to a relative warming of the SSTs in the tropical oceans from the Indian eastward through the central Pacific. We compared the tropical SST time series with that of the Night Marine Air Temperatures (NMATs), and in Figure 4 we see that, remarkably, there are shifts between the water and air temperatures at the suspected points in time (between 1981–1982 and 1991–1992). These shifts are evident without any appeal to tropospheric temperatures and shows that the NMATs produce a trend 0.11°C/decade cooler than the SSTs immediately underneath for this very large region (Christy et al., 1998). These shifts are not seen in the tropical Atlantic (Fig. 5). One possible factor being observed here is an apparent slight increase in the instability of the tropical atmosphere on the order of 0.2 to 0.3°C over a vertical extent of about 4 to 5 km (but mostly in the lowest 20 m) since 1979.

An additional possible factor for explaining the difference between surface and tropospheric trends is related to the change in the SST observing system of the tropical Indian and Pacific Oceans. Since the mid-1980s, moored and drifting buoys have supplied a massive increase in observations. In some months in the near-equatorial Pacific, over 90% of the observations now are derived from buoys. It is possible, though not proven as of yet, that the buoy temperatures may be warmer than traditional ship reports because their nominal depth of measurement is 1 m, a

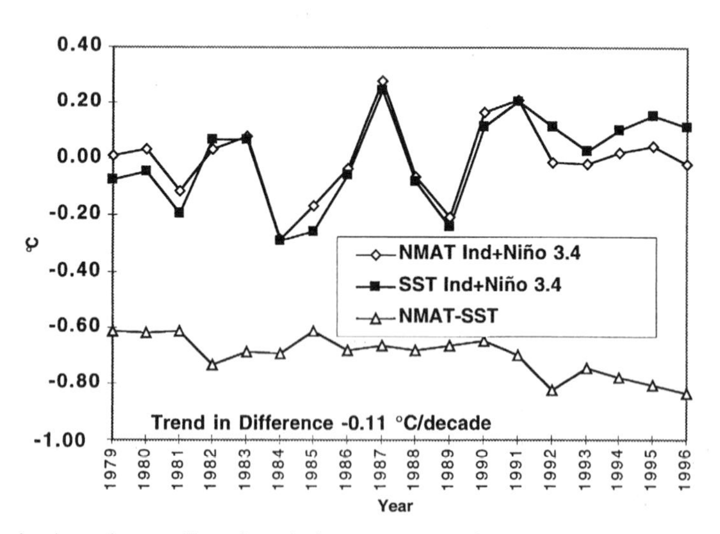

Figure 4 Annual anomalies of tropical temperatures of seawater temperatures (SSTs) and night marine air temperatures (NMATs) from the Indian eastward to the central Pacific oceans (from UK Meteorology Office).

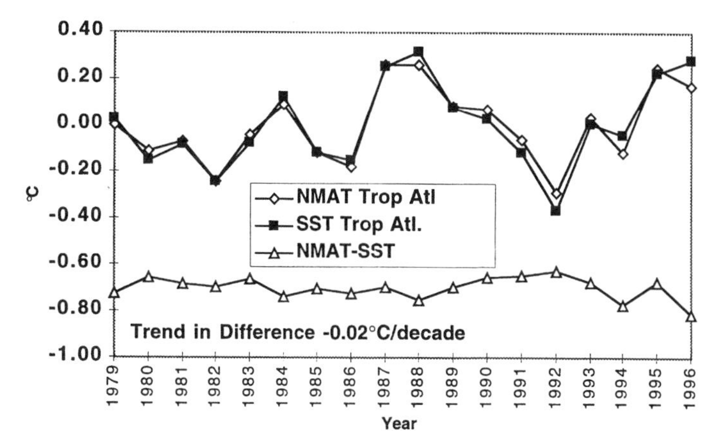

Figure 5 As in Fig. 4 but for the tropical Atlantic Ocean.

shallower and warmer depth than the typical ship-intake depth of 3 to 20 m. The increase in buoy reports may have introduced a slight warming bias since the mid-1980s. This is another issue that will be investigated to assure that the time series of SSTs is as homogeneous as possible (C. Folland, personal communication).

It appears that notions of vertically fixed linkages between the tropical SSTs and troposphere require some refinement in light of the comparisons shown above. In any case, the evidence presented here shows that over the past 20 years (a woefully brief period in terms of climate variations) the global troposphere has not experienced any significant warming or cooling. What may be of most interest in terms of climate is finding the explanation for the apparent difference in warming rates between the tropical surface and the free atmosphere.

The lower-stratospheric temperatures reveal considerable interannual variations (Fig. 6). Warming episodes related to eruptions of Agung (1963), El Chichon (1982), and Mt. Pinatubo (1991) stand out as the largest features. Underlying this punctuated time series is a general downward trend in the global stratospheric temperature, with an apparent acceleration in the last years. The location and degree of decline correlates well with similar decreases in the concentration of stratospheric ozone (Christy and Drouilhet, 1994; McCormack and Hood, 1994). With only a few decades available, however, it is difficult to know the character and extent of natural variations for this layer.

5 CONCLUSION

It should be apparent that all of the data sets mentioned above contain uncertainties in their ability to answer for us "what is the global temperature?" They provide information for different parts of the global system and thus should be viewed as pieces that imperfectly highlight specific components of the climate puzzle.

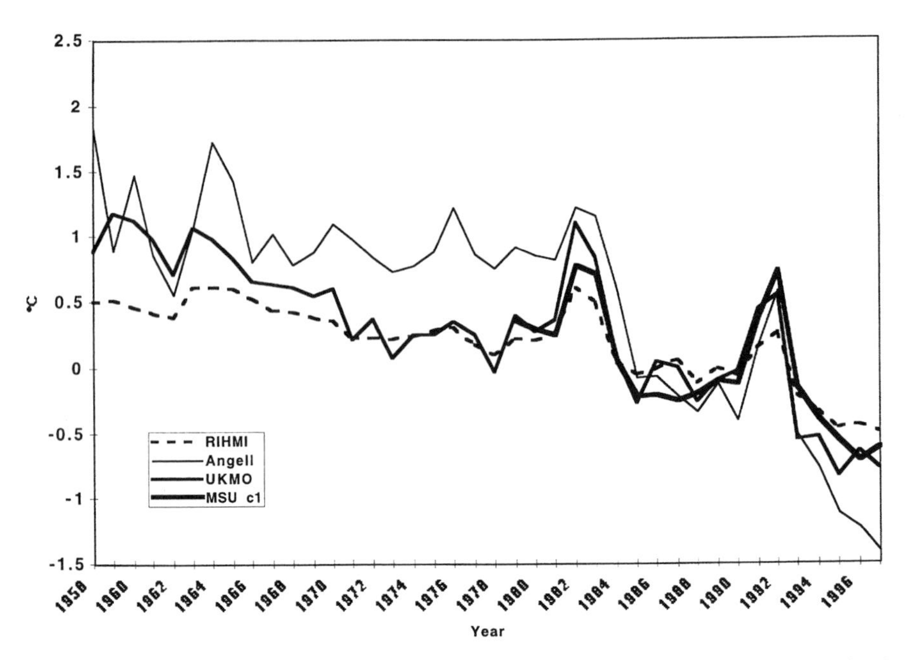

Figure 6 Annual global lower-stratospheric temperature anomalies 1958–1997 determined from radiosonde data sets and the MSUs on polar orbiting satellites. The layer is centered near 20 km altitude.

The global surface air temperature has surely risen in the past 150 years, while between 1400 and 1850 it appears to have been relatively level. In the centuries before 1400, the scant and murky evidence implies a period of relative warmth unrelated to any human factor. The troposphere has shown warming since 1958, which is due to a relatively rapid shift during only 5 years, 1976–1980. Without this shift, there would be no warming observed (which points to the limitations of discussing linear trends determined from a fluctuating time series). The global stratosphere clearly has experienced a decline in temperature, consistent with both ozone depletion and enhanced concentrations of greenhouse gases, though the character of natural variability is largely unknown for this layer.

The presence of so many types of uncertainty in our ability to determine the global temperature should draw attention to the critical needs we currently face. Our global observing system requires a more spatially and temporally complete set of both in situ and remote observations made with instruments that are precisely calibrated and consistent in quality. The current concerns of climate change have brought the present inadequacies to light and hopefully will spur action to increase our global network of observations so we will have the necessary information to make reasonable decisions about possible human impacts on climate.

Author's note: Much of the material contained in this chapter was developed for the Norwegian Academy of Technological Sciences and appears in *Do We Under-*

stand Global Climate Change? (an international seminar at Oslo at Holmen Fjordhotell, Asker, 11–12 June, 1998) Trondheim, June, 1998.

REFERENCES

Angell, J. K. (1988). Variations and trends in tropospheric and stratospheric global temperatures, 1958–87, *J. Climate* **1**, 1296–1313.

Christy, J. R., and S. J. Drouilhet (1994). Variability in daily, zonal mean lower-stratospheric temperatures, *J. Climate* **7**, 106–120.

Christy, J. R., R. W. Spencer, and R. T. McNider (1995). Reducing noise in the MSU daily lower-tropospheric global temperature dataset, *J. Climate* **8**, 888–896.

Christy, J. R. (1995). Temperature above the surface layer, *Clim. Change* **31**, 455–474.

Christy, J. R., R. W. Spencer, and W. D. Braswell (1997). How accurate are satellite "thermometers"? *Nature* **389**, 342–343.

Christy, J. R., R. W. Spencer, and W. D. Braswell (2000). MSU Tropospheric temperatures: Data set construction and radiosonde comparisons, *J. Atmos. Oceanic Tech.* **17**, 1153–1170.

Christy, J. R., R. W. Spencer, and E. S. Lobl (1998). Analysis of the merging procedure for the MSU daily temperature time series, *J. Climate*, **11**, 2016–2041.

Cuffey, K. M., R. B. Alley, P. M. Grootes, J. F. Bolzan, and S. Anandakrishnan (1994). Calibration of the $\delta^{18}O$ isotopic paleothermometer for central Greenland, using borehole temperatures, *J. Glaciology* **40**, 341–349.

Dahl-Jensen, D., N. S. Gundestrup, K. Mosegaard, and G. D. Clow (1997). Reconstruction of the past climate from the GRIP temperature profile by Monte Carlo inversion, *EOS Abstracts*, AGU 1997 Fall Meeting, San Francisco, CA, F6.

Deming, D. (1995). Climatic warming in North America: Analysis of borehole temperatures, *Science* **268**, 1576–1577.

Folland, C. K., and D. E. Parker (1995). Correction of instrumental biases in historical sea surface temperature data, *Q. J. R. Meteorol. Soc.* **121**, 319–367.

Gaffen, D. J. (1994). Temporal inhomogeneities in radiosonde temperature records, *J. Geophys. Res.* **99 D2**, 3667–3676.

Hansen, J., and S. Lebedeff (1988). Global surface temperatures: update through 1987, *Geophys. Res. Lett.* **21**, 2693–2696.

Hansen, J., R. Ruedy, and M. Sato (1996). Global surface air temperature in 1995: Return to pre-Pinatubo level, *Geophys. Res. Lett.* **23**, 1665–1668.

Hurrell, J. W., and K. E. Trenberth (1997). Spurious trends in MSU satellite temperatures due to merging of different satellite records, *Nature* **386**, 164–167.

IPCC, 1990: Climate Change, The IPCC Scientific Assessment, J. T. Houghton, G. J. Jenkins, and J. J. Ephraums, Eds., Cambridge University Press, Cambridge, UK, 365 pp.

IPCC, 1992: Climate Change, 1992: The Supplementary Report to the IPCC Scientific Assessment, J. T. Houghton, B. A. Callander, and S. K. Varney, Eds., Cambridge University Press, Cambridge, UK, 198 pp.

IPCC, 1996: Climate Change 1995, The Science of Climate Change, Contribution of Working Group I to the Second Assessment Report of the Intergovernmental Panel on Climate

Change, J. T. Houghton, L. G. Meira Filho, B. A. Callander, N. Harris, A. Kattenberg, and K. Maskell, Eds., Cambridge University Press, Cambridge, UK, 572 pp.

IPCC 2001: Climate Change 2001, The Scientific Basis, Contribution of Working Group I to the Third Assessment Report of the Intergovernmental Panel on Climate Change, J. T. Houghton et al. Eds., Cambridge University Press, UK, 881 pp.

Jones, P. D., and K. R. Briffa (1992). Global surface air temperature variations over the twentieth century, Part 1: Spatial, temporal and seasonal details, *Holocene* **2**, 165–179.

Jones, P. D., T. J. Osborn, and K. R. Briffa (1997). Estimating sampling errors in large-scale temperature averages, *J. Climate* **10**, 2548–2568.

Manley, G. (1974). Central England temperatures: Monthly means 1659 to 1973, *Q. J. R. Meteorol. Soc.* **100**, 389.

Mann, M. E., R. S. Bradley, and M. K. Hughes (1998). Global scale temperature patterns and climate forcing over the past six centuries, *Nature* **392**, 779–788.

McCormack, J. P., and L. L. Hood (1994). Relationship between ozone and temperature trends in the lower stratosphere: Latitude and seasonal dependencies, *Geophys. Res. Lett.* **21**, 1615–1618.

Mo, T. (1995). A study of the Microwave Sounding Unit on the NOAA-12 satellite, IEEE Trans. *Geosci. and Remote Sensing* **33**, 1141–1152.

Oort, A. H., and H. Liu (1993). Upper-air temperature trends over the globe, 1958–1989, *J. Climate* **6**, 292–307.

Parker, D. E., C. K. Folland, and M. Jackson (1995). Marine surface temperature observed variations and data requirements, *Clim. Change* **31**, 559–600.

Parker, D. E., M. Gordon, D. P. N. Cullum, D. M. H. Sexton, C. K. Folland, and N. Rayner (1997). A new global gridded radiosonde temperature data base and recent temperature trends, *Geophys. Res. Lett.* **24**, 1499–1502.

Reynolds, R. W., and T. M. Smith (1995). A high-resolution global sea surface temperature climatology, *J. Climate* **8**, 1571–1583.

Ross, R. J., J. Otterman, D. O'C. Starr, W. P. Elliot, J. K. Angell, and J. Susskind (1996). Regional trends of surface and tropospheric temperature and evening-morning temperature difference in northern latitudes: 1979–93, *Geophys. Res. Lett.* **23**, 3179–3182.

Sterin, A. M., V. A. Orshekhovskaya, and N. M. Mishina (1997). Comparison of upper-air temperature variations in the past and current decade, derived from the global radiosonde database and from the microwave sounding unit, *Proc. 22nd Ann. Climate Diagnostics and Prediction Workshop, Berkeley, CA, USA.* U.S. Dept. Commerce, Sills Bldg., 5285 Port Royal Road, Springfield VA 22161.

Tett, S. F. B., J. F. B. Mitchell, D. E. Parker, and M. R. Allen (1997). Human influence on the atmospheric vertical temperature structure: detection and observations, *Science* **247**, 1170–1173.

Vinnikov, K. Ya., P. Ya. Groisman, and K. M. Lugina (1990). Empirical data on contemporary global climate changes (temperature and precipitation), *J. Climate* **3**, 662–677.

Vose, R. S., T. C. Peterson, R. L. Schmoyer, and J. E. Eischeid (1995). The Global Historical Climatology Network, a preview of version 2. Ninth Conf. on Applied Climatology, Dallas TX, *Amer. Meteor. Soc.* 59–64.

Wentz, F. J., and M. Schabel (1998). Effects of satellite orbital decay on MSU lower tropospheric temperature trends, *Nature*, **394**, 361–364.

Wigley, T. M. L., and S. C. B. Raper (1990). Natural variability of the climate system and detection of the greenhouse effect, *Nature* **344**, 324–327.

Woodruff, S. D., R. J. Slutz, R. L. Jenne, and P. M. Steurer (1987). A Comprehensive Ocean-Atmosphere Dataset, *Bull. Amer. Meteor. Soc.* **68**, 1239–1250.

CHAPTER 47

SATELLITE VERSUS IN SITU MEASUREMENTS AT THE AIR–SEA INTERFACE

KRISTINA B. KATSAROS

In this chapter we explore the trade-offs in selecting surface in situ versus satellite platforms to measure properties near or at the air–sea interface. The most obvious difference between the two observing platforms is sampling coverage in time and space. A surface platform can obtain continuous measurements at a point, while a polar-orbiting satellite instrument samples, at most, twice per day depending on the swath width of the sensor. A geostationary satellite can sample the surface as frequently as every 15 min (once per hour is typical), but the high altitude (38,000 km) limits the resolution that is achievable for some sensors. To focus the discussion, we compare the following two variables commonly measured by both in situ and satellite systems: the sea surface temperature (SST) and surface wind speed, U, or wind vector, $\bar{U}$.

1 SEA SURFACE TEMPERATURE

SSTs were traditionally measured from ships by the insertion of a mercury thermometer into water samples obtained by buckets lowered from deck. Currently, the temperature of the water intake of ships is recorded and reported every 3 h via satellite to the international World Meteorological Organization (WMO) weather telecommunications network (Global Transmission System, or GTS) and distributed to all weather services globally. Moored and drifting buoys have expanded the in situ network to cover larger areas of the global ocean. In the case of moored buoys,

Handbook of Weather, Climate, and Water: Dynamics, Climate, Physical Meteorology, Weather Systems, and Measurements, Edited by Thomas D. Potter and Bradley R. Colman.
ISBN 0-471-21490-6

continuous time series are obtained. Examples of such networks are the U.S. National Data Buoy Center (NDBC) buoys located around the U.S. coastline, the Tropical Atmosphere Ocean (TAO) network in the tropical Pacific Ocean, operational since the mid-1980s (McPhaden et al., 1998), and the global drifter bouys (Swenson and Niiler, 1996; Bushnell, 1996).

Sea surface temperatures from satellites are obtained by measuring the radiance emitted from the sea surface at electromagnetic frequencies within spectral regions where Earth's atmosphere is only weakly absorbing, so-called atmospheric window regions. The main window for SST measurements is in the infrared spectrum between 8 and 12 μm. Clouds are opaque at infrared wavelengths, however, which is a serious limitation for observing SST in certain regions of the world. Certain wavelengths in the microwave spectrum do penetrate clouds, but the technology does not allow fine spatial resolution at these wavelengths. Table 1 presents the main atmospheric window regions used for SST measurements.

For SST, the in situ and satellite systems are complementary and are used in conjunction for the National Centers for Environmental Prediction (NCEP) SST product, which provides global SST data at 1° latitude by 1° longitude resolution, averaged over one week (Reynolds and Smith, 1994). The in situ SST are used to calibrate the satellite values inferred from infrared signals in the atmospheric window regions. Measurements are obtained at two wavelengths whose absorption values by the intervening atmosphere are different. By differencing the two measurements, the effect of atmospheric absorption can be measured and corrections applied. This technique provides the primary atmospheric corrections (McClain et al., 1985). The in situ SSTs from moored and drifting buoys are combined with the satellite data after this correction for atmospheric transmission in an optimal interpolation scheme (Barton, 1995).

TABLE 1 Spectral Regions Used for Determining SST Satellites[a]

	Frequency/Wavelength Region		
Satellite Type	3.6 μm	10–11 μm	5 cm (6.6 GHz)
Polar orbiting	*TIROS/NOAA* (1960s onward)		*Seasat*, (1978) *Nimbus 7* (1978–1985)
Resolution	4 km	4 km (1 km maximum)	100s of km
Low orbit in subtropical/tropical latitudes	—		*Tropical Rainfall Measuring Mission* (TRMM, 1997–)
Resolution	—	—	15 km
Geostationary	*GOES Series*		—
Resolution	4 km	4 km	—

[a]The satellites carrying the instruments and the typical surface resolution.

The basic measurement of temperature in situ is not a problem: Thermistor or resistance thermometer units are usually reliable and the calibration remains stable, although drift in the electronics due to seasonal changes in the mean environmental temperature must be accounted for. Similarly, satellite infrared technology, filters, and lenses have had a 35-year history or longer and are also very reliable. The most difficult problem for achieving stable results has been due to injections of aerosols from volcanos (Reynolds, 1993).

The measurement problems associated with SST determination include, for the in situ, heating of the platform by solar radiation on the buoys or ships and, in the latter case, also the ship's engine heating the cooling water and general heat diffusion from the bulk of the ship. For the satellite infrared observations, effects of atmospheric aerosols and clouds in the field of view are persisting problem areas, which have regional variations whose corrections have not been readily available. Table 2 compares in situ and satellite accuracies.

As the much relied-on SST product of NCEP is produced with a fit of the satellite information to a "calibration surface" generated with the available in situ data, inaccuracies develop when the sampling coverage of the in situ data is inadequate. This is the case in large regions of the Southern Hemisphere oceans, where the "fit" may become unreliable and can even *generate* errors. The presence of sea ice can also confuse the correspondence between in situ and satellite observations (Reynolds, 1999, personal communication).

Another difficulty with SST determination is that a satellite instrument senses the radiation from the top 0.5 mm or less of the sea surface (Katsaros, 1980; Robinson et al., 1984), while a buoy or ship measures the temperature from 0.5 to 5 m depth. Since the oceans lose heat to the atmosphere from the sea surface, there is typically a negative gradient in temperature toward the surface mainly confined to the nonturbulent region nearest the interface. We refer to the layer exhibiting the temperature difference, ΔT, across it as the "cool film". Many processes affect the magnitude of the ΔT, as illustrated in Figure 1. The ΔT values of the order of -0.3 to -0.7°C have been observed on the open ocean (Schlüssel et al., 1990; Wick et al., 1996).

TABLE 2 Sea Surface Temperature Uncertainties and Absolute Accuracies

	Intrinsic Errors	
	Random Error	Absolute Error
In situ sensors		
Thermistor	± 0.1	Of the order of ± 0.5
Resistance	± 0.1	
Satellites		
Polar orbiting	± 0.5	Of the order of ± 0.7
Geostationary	± 0.5	

However, on a calm sea during the day, the uppermost layer can be warmer than the subsurface layers by several degrees Celsius due to solar heating. This is not a common occurrence but has been observed, particularly in the tropics, and even from space (Katsaros et al., 1983). It is more likely to occur where fresh water added by heavy rain may stabilize the surface layer (Wijesekera et al., 1999).

Estimates of temperature drop across the "cool" film on the sea surface would be required to correctly relate the buoy in situ SST to the satellite observations. One proposed method would be to apply a simple model for the estimated ΔT across the cool film (Soloviev and Schlüssel, 1994).

The uncertainties in both the satellite and surface SST measurements are of the same order of magnitude as the typical temperature drop across the cool film, but since the cool film is present over most of the ocean most of the time, it could be argued that ignoring it introduces a bias or additional uncertainties. Most heat flux parameterizations and other uses for SST information have been based on the bulk value of SST provided by buoys and ships, so it is wise to consider that the satellite information represents the temperature at the interface.

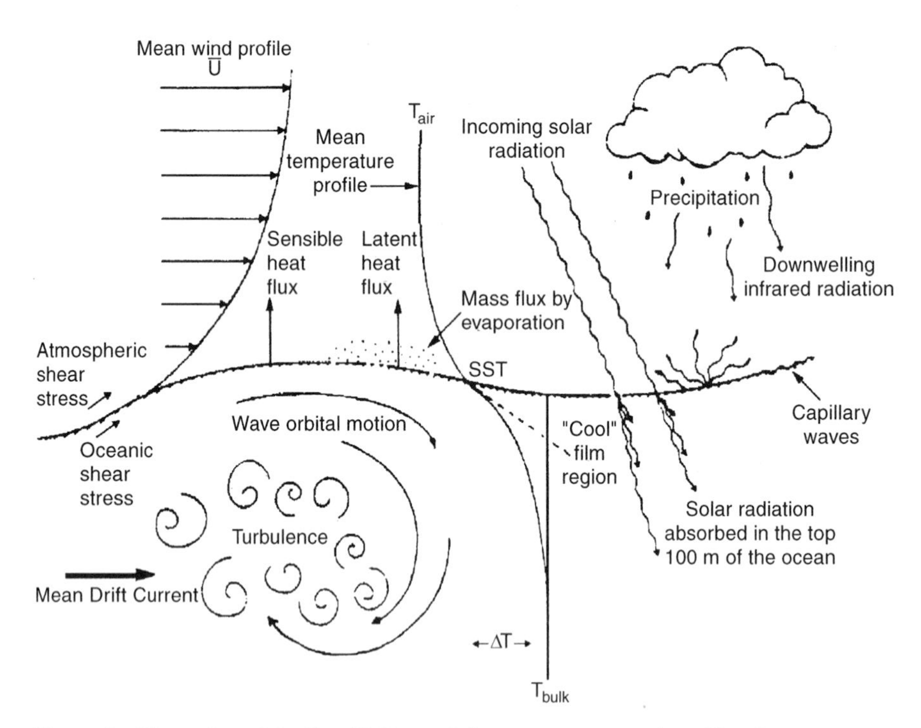

Figure 1 Illustration of the "cool" film and the many processes that affect the temperature difference across it. These processes include variable shear stress on the surface, radiative, sensible and latent heat fluxes, precipitation, wave breaking, and the presence of oil films and slicks (after Katsaros, 1980).

2 SURFACE WIND

Another sea surface variable observable by in situ sensors or from satellites is the surface wind. The classical sea surface wind speed observations were qualitative descriptions by mariners of the appearance of the sea surface as it was roughened by the wind. The observations were reported via the Beaufort scale (Weather Bureau, 1948; Shaw and Simpson, 1906). This scale has since been calibrated versus direct measurements by anemometers. Table 3 gives the Beaufort scale and conversions. The wind direction was determined from the directions in which the shorter gravity waves were traveling with respect to the ship's compass.

The replacement of these qualitative observations by anemometers on ever larger and larger ships has created much confusion and, perhaps, more erroneous data because to provide unaffected exposure of an anemometer on a bulky, moving ship is almost impossible due to severe distortion of the flow both in speed and direction (Yelland et al., 1998). Modern research ships carry two anemometers to provide alternate sensors to choose from, depending on the wind direction relative to the ship's heading.

Anemometers on buoys have the problem of being "pumped" by the rocking of the buoy in the wave field . The cup anemometers rectify the buoy motion, i.e., the anemometer rotation is increased, whether the buoy is in the "to" or "fro" cycle of its rocking, while propeller anemometers can turn in both directions and, therefore, average out the effect of the rocking. Propeller anemometers are, therefore, used almost exclusively on buoys (exceptions are stable buoys, such as the spar buoys, which do not follow the waves). A directional vane also suffers errors due to buoy motion.

The satellite method for measuring the fleeting wind is either by microwave backscatter from the sea surface by an instrument called a scatterometer (Ulaby

TABLE 3 Beaufort Scale for Wind Speed and Its Conversion to Knots and Meters per Second (m/s)

Beaufort Number	International Description	Wind Speed (knots)	Wind Speed (m/s)
1	Light air	1–3	0.5–1.5
2	Light breeze	4–6	2–3
3	Gentle breeze	7–10	4–5
4	Moderate breeze	11–16	6–8
5	Fresh breeze	17–21	9–10
6	Strong breeze	22–27	11–13
7	Moderate or near gale	28–33	14–17
8	Gale or fresh gale	34–40	18–20
9	Strong gale	41–47	21–24
10	Whole gale or storm	48–55	25–28
11	Violent storm	58–63	29–33
12	Hurricane	>64	>33

et al., 1982) or by variations in the emissivity (for microwave radiometers). Scatterometers measure the diffuse backscatter of a radar signal from the sea surface with two to three antennas, or by a rotating antenna, such that the same piece of ocean is observed several times from different look angles, e.g., the European Remote Sensing Satellite, ERS, scatterometer, the NASA scatterometer NSCAT, and QuikSCAT, so named for its short cycle from planning to launch. The difference in the backscatter power at the different look angles allows inference on the wind direction, as well as the wind speed. The *Seasat*, *Nimbus 7*, and SSM/I microwave radiometers only provide the wind speed as a consequence of the increased emissivity and presence of foam as the wind speed increases. These sensors were pioneered in the early satellite days, but saw their main debut with the *Seasat* satellite launched in 1978 (Jones et al., 1979; Lipes et al., 1979). The satellite wind measurement techniques have in common with the qualitative Beaufort scale that it is the effect of the wind on the sea that is being observed. The reason that this works for the satellite radars of wavelengths between a few to tens of centimeters is that the surface waves that interact with the electromagnetic radiation (via the Bragg scattering mechanism) are the capillary-gravity waves, which have a short life time and are closely linked to the wind speed (or more exactly the wind stress).

After a long hiatus of 13 years, a scatterometer was first launched again in 1991 by the European Space Agency and in 1996 by the National Aeronautics and Space Administration in cooperation with the Japan Space Agency. The *Seasat* instruments, as well as the instruments of the 1990s, were calibrated against buoy anemometer measurements (Bentamy, 1998; Graber et al., 1996), as well as against numerical weather prediction analyses with the buoy calibration having a somewhat more realistic wind speed range; atmospheric numerical models tend to smooth the wind variations and, therefore, lose the extremes. However, for very low wind speeds and for wind speeds $>25\,\mathrm{m/s}$, serious questions about the representativeness of the buoy observations are also emerging (Large et al., 1995; Bentamy et al., 1996).

The workhorse for satellite surface wind estimates from 1987 to the present has been the microwave radiometer, the Special Sensor Microwave Imager (SSM/I), partly for lack of a scatterometer in space. Microwave radiometry is less well adapted to satellite wind observations due to interference by heavy clouds and rain, exactly the condition of paramount meteorological interest. The active systems, scatterometers [and other radars such as altimeters and synthetic aperture radars (SARs)], have more signal power and are, therefore, better able to penetrate clouds. A new polarimetric microwave radiometer is expected to overcome this difficulty (but has not yet been flown on a satellite).

Again, sampling is the major problem for ship, buoy, and satellite systems. Reliance on one system alone is often unsatisfactory because of the large temporal variations of the wind. If a measurement at one specific place in the ocean is sought, then the data from an anemometer on a moored buoy will suffice, but if horizontal variability of the wind is desired, a combination of satellite and observations with an array of buoys may best serve the purpose.

Another important challenge of wind analysis lies in how to join various wind measurements together. The *surface wind* is typically defined to apply at 10 m height above the sea (this height is an arbitrary choice and others have been in use). Ships and buoys rarely have their anemometers at 10 m height. To make measurements intercomparable, we convert the wind speed to a common height by applying a correction to an individual measurement at height z to translate it vertically to the 10 m or other desired reference height. For that calculation, we must adopt an analytical vertical wind profile that depends on the atmospheric stratification, which is a function of the turbulence, which, in turn, depends on the wind stress and the buoyancy. For a neutrally stratified atmospheric boundary layer (no buoyancy fluxes or very strong wind shear), the profile is logarithmic. For stable and unstable stratification (negative or positive buoyancy fluxes) caused by sensible heat and/or water vapor flux, the profile differs from the logarithmic shape (Kraus and Businger, 1994).

Methods to provide the height correction have been established (Liu et al., 1979; Fairall et al., 1996), mostly derived from simultaneous flux and profile measurements over land. The application of the established profile shapes to the ocean have not yet been fully proven due to the difficulty in measuring profiles from moving or bulky platforms or the effects of the wave field on the lowest level atmospheric measurements for stationary towers.

The satellite measurement of the wind vector and wind speed depends on the roughening of the sea surface by the wind, particularly by the centimeter-scale gravity-capillary waves. These are forced by the wind stress exerted on the surface, and the stress is directly related to the wind profile. Thus, the fields of air–sea interaction, wave generation, and electromagnetic backscatter theory are involved in the remote measurement of surface wind. The enterprise is saved from the whole of this complexity under most circumstances by the footprint size of the radar data (12.5 to 50 km), which assures a certain averaging over the variability and, thereby, a randomization of the errors. However, in regions of large wind speed gradients and variable wave fields, this may not be true.

The measurements of SST and surface wind are now well developed and are only being refined by better sampling techniques, data recordings, and inclusion of auxiliary measurements to improve the corrections. An example is simultaneous measurements on recent (and future) satellites of the aerosol content in the atmosphere, which allows the elimination of the radiation error caused by the aerosol via radiative transfer models. Therefore, we can now embark on producing long, consistent time series of both SST and sea surface wind for climatological studies. The crux is in keeping the long time series consistent with proper intercalibration between sequential satellite sensors and when some inevitable design improvements are made.

REFERENCES

Barton, I. J. (1995). Satellite-derived sea surface temperatures: Current status, *J. Geophys. Res.* **100**, 8777–8790.

Bentamy, A., Y. Quilfen, F. Gohin, N. Grima, M. Lenaour, and J. Servain (1996). Determination and validation of average wind fields from ERS-1 scatterometer measurements, *Global Atmos. Ocean Syst.* **4**(1), 1–29.

Bentamy, A., N. Grima, and Y. Quilfen (1998). Validation of the gridded weekly and monthly wind fields calculated from ERS-1 scatterometer wind observations, *Global Atmos. Ocean Syst.*, **5**, 373–396.

Bushnell, M. H. (1996). Preliminary Results from Global Lagrangian Drifters Using GPS Receivers, WMO/DBCP Technical Doccument No. 10, pp. 23–26.

Fairall, C. W., E. F. Bradley, J. S. Godfrey, G. A. Wick, J. B. Edson, and G. S. Young (1996). Cool-skin and warm-layer effects on sea surface temperature, *J. Geophys. Res.* **101**(1), 1295–1308.

Graber, H. C., N. Ebutchi, and R. Vakkayil (1996). Evaluation of ERS-1 scatterometer winds with wind and wave ocean buoy observations, *Tech. Rep.*, *RSMAS 96-003*, Div. of Applied Marine Physics, Univ. of Miami, FL.

Jones, W. L., P. G. Black, D. M. Boggs, E. N. Bracalente, R. A. Brown, G. Dome, J. A. Ernst, I. N. Alberstan, J. E. Overland, F. Peteherych, W. J. Pierson, F. J. Wentz, P. M. Woicefiyn, and N. J. Wurtele (1979). SEASAT scatterometer: Results of the Gulf of Alaska workshop, *Science* **204**(4400), 1413–1415.

Katsaros, K. B. (1980). The aqueous thermal boundary layer, *Boundary-Layer Meteorol.* **18**, 107–127.

Katsaros, K. B., A. F. G. Fiuza, F. Sousa, and V. Amann (1983). Sea surface temperature patterns and air–sea fluxes in the German Bight during MARSEN 1979, Phase 1, *J. Geophys. Res.* **88**(C14), 9871–9882.

Kraus, E. B., and J. A. Businger (1994). *Atmosphere-Ocean Interaction*, Oxford University Press, New York.

Large, W. G., J. Morzel, and G. B. Crawford (1995). Accounting for surface wave distortion of the marine wind profile in low-level ocean storms wind measurements, *J. Phys. Oceanogr.* **25**(11), 2959–2971.

Lipes, R. G., R. L. Bernstein, V. J. Cardone, K. B. Katsaros, F. G. Njoku, A. L. Riley, D. B. Ross, C. T. Swift, and F. J. Wentz (1979). SEASAT scanning multichannel microwave radiometer: Results of the Gulf of Alaska workshop, *Science* **204**(4400), 1415–1417.

Liu, W. T., K. B. Katsaros, and J. A. Businger (1979). Bulk parameterization of air–sea exchanges of heat and water vapor including the molecular constraints at the interface, *J. Atmos. Sci.* **36**(9), 1722–1735.

McClain, E. P., W. G. Pichel, and C. C. Walton (1985). Comparative performance of AVHRR-based multichannel sea surface temperatures, *J. Geophys. Res.* **90**, 11,587–11,601.

McPhaden, M. J., A. J. Busalacchi, R. Cheney, J.-R. Donguy, K. S. Gage, D. Halpern, M. Ji, P. Julian, G. Meyers, G. T. Mitchum, P. P. Niiler, J. Picaut, R. W. Reynolds, N. Smith, and K. Takeuchi (1998). The Tropical Ocean-Global Atmosphere observing system: A decade of progress, *J. Geophys. Res.* **103**, 14,169–14,240.

Reynolds, R. W. (1993). Impact of Mount Pinatubo aerosols on satellite-derived sea surface temperatures, *J. Climate* **6**(4), 768–776.

Reynolds, R. W., and T. M. Smith (1994). Improved global sea surface temperature analyses using optimum interpolation, *J. Climate* **7**(6), 929–948.

Robinson, I. S., N. C. Wells, and H. Charnock (1984). The sea surface thermal boundary layer and its relevance to the measurement of sea surface temperature by airborne and spaceborne radiometers, *Int. J. Remote Sens.* **5**, 19–45.

Schlüssel, P., W. J. Emery, H. Grassl, and T. Mammen (1990). On the bulk-skin temperature difference and its impact on satellite remote sensing of sea surface temperature, *J. Geophys. Res.* **95**(C8), 13,341–13,356.

Shaw, W. N., and G. C. Simpson (1906). *The Beaufort Scale of Wind Force: Report of the Director of the Meteorological Office upon an Inquiry into the Relation between the Estimates of Wind-Force According to Admiral Beaufort's Scale and the Velocities Recorded by Anemometers Belonging to the Office, with a Report Upon Certain Points in Connection with the Inquiry*, Darling and Son, London.

Soloviev, A. V., and P. Schlüssel (1994). Parameterizations of the cool skin of the ocean and of the air-ocean gas transfer on the basis of modeling surface renewal, *J. Phys. Oceanogr.* **24**(6), 1339–1346.

Swenson, M. S., and P. P. Niiler (1996). Statistical analysis of the surface circulation of the California Current, *J. Geophys. Res.* **101**(C10), 22,631–22,645.

Ulaby, F. T., R. K. Moore, and A. K. Fung (1982). *Microwave Remote Sensing, Active and Passive, Vol. II; Radar Remote Sensing and Surface Scattering and Emission Theory*, Addison-Wesley, Reading, Mass.

Weather Bureau (1948). Beaufort Scale of Wind Force, WB Form 4042A, Washington, DC.

Wick, G. A., W. J. Emery, L. H. Kantha, and P. Schlüssel (1996). The behavior of the bulk-skin sea surface temperature difference under varying wind speed and heat flux, *J. Phys. Oceanogr.* **26**(10), 1969–1988.

Wijesekera, H. W., C. A. Paulson, and A. Huyer (1999). The effect of rainfall on the surface layer during a westerly wind burst in the western equatorial Pacific, *J. Phys. Oceanogr.* **29**(4), 612–632.

Yelland, M. J., B. I. Moat, P. K. Taylor, R. W. Pascal, J. Hutchings, and V. C. Cornell (1998). Wind stress measurements from the open ocean corrected for airflow distortion by the ship, *J. Phys. Oceanogr.* **28**(7), 1511–1526.

CHAPTER 48

RADAR TECHNOLOGIES IN SUPPORT OF FORECASTING AND RESEARCH

JOSHUA WURMAN

1 INTRODUCTION

Forecasters and researchers, and the computer models that support them, require detailed information concerning the three-dimensional state of the atmosphere and how it evolves with time. Some forecasting and research needs overlap; others are particular to those respective tasks. But whether the goal is weather prediction or scientific understanding, no tool can provide more wealth and diversity of information than existing and emerging technologies in weather radar (Fig. 1). Weather radars can sample large volumes of the atmosphere nearly continuously in space and time at fairly high resolution vertically, horizontally, and temporally. Radars can penetrate through clouds and precipitation to measure precipitation intensity, precipitation type, and wind motions in many weather conditions, throughout the depth of the atmosphere. They are a uniquely versatile tool.

A forecaster needs to know where it is raining or snowing, how hard, and whether snow or hail is occurring. He needs to know whether the precipitation is caused by convective systems or more stratiform uplift, whether it is moving toward the forecast area or away, whether it is strengthening or weakening. He needs to know if an airport runway will soon be affected by a gust front or microburst, or if a mesocyclone will move into his forecast area, whether a snow band will move onshore, or whether a sea breeze will initiate storms. For many forecasts of 6 h or less, radar is the crucial tool that provides the most up-to-date and comprehensive information to a forecaster. Mesoscale forecasting models

Handbook of Weather, Climate, and Water: Dynamics, Climate, Physical Meteorology, Weather Systems, and Measurements, Edited by Thomas D. Potter and Bradley R. Colman.
ISBN 0-471-21490-6

Figure 1 The MIT 5-cm (C-band) research radar deployed in Albuquerque, New Mexico. The nearly spherical radome protects the radar antenna and sits on top of a tower that places the antenna above blocking objects. The transmitter, receiver, and operating consoles are located in the buildings below the tower. (Photo courtesy of MIT Weather Radar Lab, D. Boccippio.) See ftp site for color image.

of the future will become more and more dependent on initializations and nudgings provided by the nation's radar network.

Research needs are similarly diverse. A scientist may want to study large-scale rainfall patterns, not only at weather stations, but in between, or to study much smaller scales, down to suction vortices in tornadoes or rolls in microbursts. He may want to understand the microphysical processes that cause hail, icing, or charging of particles and lightning, or to study the flow along a front or dry line, or the wind field of a hurricane rain band. Researchers use radar to probe the upper atmosphere and the boundary layer, intense and violent weather, and quiescent overturning on clear days. Almost all atmospheric phenomena with scales of 10 m to 100 km have been and are being intensively studied using radar.

2 BASIC RADAR OPERATION

This chapter is not a radar textbook, but it is valuable to briefly review some basics of the theory and operation of radars to understand the nature and quality of the data that they produce, the limitations of these data, and potential uses both now and in the future. The following discussion will not be exhaustive and will cover only some major technologies.

Most weather radars focus pulsed beams of microwaves on meteorological targets and listen for returned signals. (Not all radars operate in this fashion. But the exceptions are primarily fairly exotic research systems.) By measuring the strength, timing, and other parameters of these signals, information about the targets can be obtained. A pulse of focused radiation, typically about 1 to 2 μs in duration, leaves the radar (Fig. 2). At various points along its travel, the pulse encounters precipitation, insects, suspended particles, airplanes, mountains, or density discontinuities in the air and some of the radiation is scattered back toward the radar and elsewhere, while the remainder continues to travel outward. After the pulse leaves the radar, hardware and software listen for returned energy. The delay time between transmission and return uniquely defines the distance to the precipitation that caused the scattering, $R = c\ \Delta t/2$, where R is the distance to the precipitation, c is the speed of light, and Δt is the time between transmission and return of the energy. The returned signals are grouped or sampled at intervals, typically about 1 to 2 μs, resulting in distance resolution, using the above formula, of 150 to 300 m. After approximately 0.5 to 2 ms, another pulse is transmitted and the process is repeated. The strength of

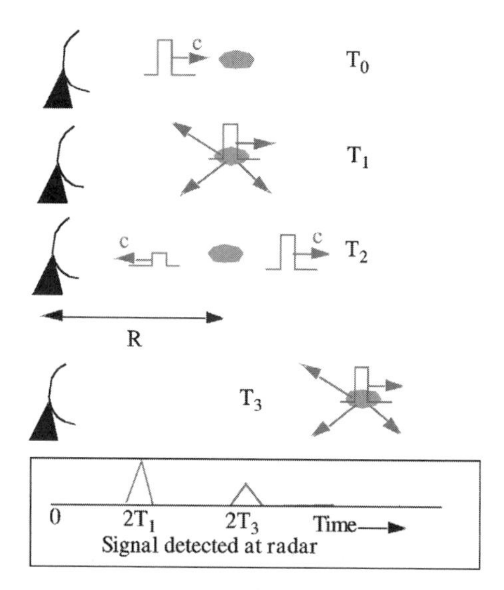

Figure 2 Pulse of microwaves travels from a radar toward a raindrop at T_0. At T_1, the pulse encounters the raindrop, which acts as an antenna and radiates in many directions. At T_2, the transmitted pulse continues outward while some of the microwaves emitted by the raindrop continue back toward the radar. The microwaves travel at the speed of light, c, from the radar to the drop and back again, so the distance to the drop can be calculated from the round-trip time, $2T_1$. Radiation emitted by drops encountered further along the transmitted pulse's travel T_3, will return to the radar at later times. See ftp site for color image.

the returned signals can be used to estimate the intensity of precipitation that causes the scattering, using what is known as the radar equation. The difference in phase of the returned radiation from subsequent pulses is used to calculate the motion of the precipitation. (Most radars do not, in fact, actually measure the Doppler shift of returned radiation.) The transmitting antenna rotates, usually horizontally, surveying a wide area, then inclines and repeats the rotation, surveying a similar area at greater elevation. Some of these processes will be discussed further below.

Choice of Wavelength: Attenuation Versus Antenna Size

A typical weather radar generates microwaves with a wavelength of 3 cm (X-band), 5 cm (C-band), or 10 cm (S-band). Other wavelengths are used for specialized research purposes as will be discussed later. (For comparison, a fluorescent light bulb generates energy at about 500 nm, an FM radio station 300 m, and a cordless phone at 40 cm.)

All other things being equal, it is usually best to use the longest practical wavelengths for a weather radar. This is because when a radar beam of microwaves passes through precipitation, it becomes attenuated due to scattering by the raindrops. Essentially, the beam is consumed by the process of scattering (back toward the radar and in other directions), and some is absorbed and converted to heat. (Not much heat, however; you cannot evaporate a cloud with a normal weather radar.) Shorter wavelength radiation suffers much more from attenuation; the effect is proportional to worse than the inverse square of wavelength $1/\lambda^2$. Precisely calculating how much a beam will be attenuated is difficult since complex scattering effects and absorption must be taken into account, but a rough approximation is possible. The attenuation rate is roughly proportional to the intensity of rain occurring in the cloud. A beam will lose a certain fraction of its intensity each kilometer: attenuation (dB/km) $= CR$ where R is in mm/h and $C = 0.01$ (3 cm), 0.02 (5 cm) or 0.003 (10 cm). So, if the rain rate is 10 mm/h, a 5-cm radar beam will lose $0.02 \times 10 = 0.2$ dB/km, a fairly significant amount since the effect is cumulative during the outward and return travel of the beam. After passing through 50 km of 10 mm/h rain and returning to the radar, the measured signal will be 20 dB weaker than otherwise expected. There are techniques for correcting for this attenuation, but they are prone to large errors. The velocity data from attenuated beams can still be good, but, in intense rain or hail, the beams can become mostly or completely extinguished, complicating or preventing the retrieval of useful data. In heavy rain all data from ranges beyond 20 km can be lost if a 3-cm radar is used. Additional attenuation can be caused by heavy rain forming opaque sheets of water on the radomes that protect antennas from the weather, sun, and wind.

But, all other things are not equal. Due to the physics of diffraction, the ability of an antenna to focus radiation into a narrow beam is proportional to its size. A large antenna can produce a much more focused beam than a small antenna. The focusing power of an antenna is also dependent, inversely, on the wavelength being transmitted, with longer wavelengths being more difficult to concentrate into a narrow beam. Thus, the beamwidth of the transmitted energy is roughly equal to $80\lambda/D$,

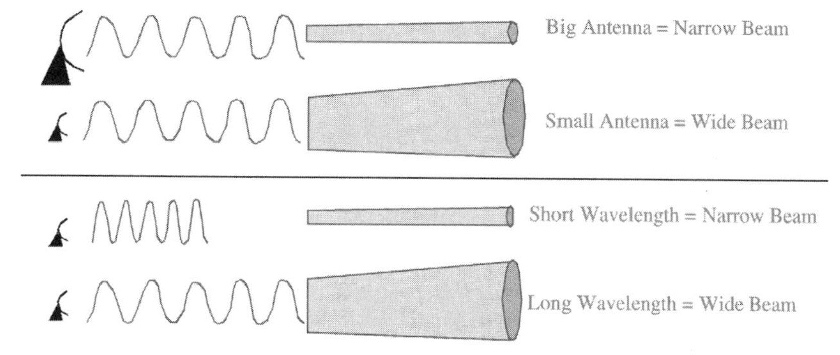

Figure 3 How beamwidth is affected by antenna size and wavelength. See ftp site for color image.

where λ is the radar wavelength and D the antenna diameter. As a result, it requires a 8-m diameter antenna to produce a 1° wide beam of 10-cm radiation, but only a 2.4-m antenna if 3-cm microwaves are used (Fig. 3). The area, weight, wind resistance to rotation, and the power required to turn an antenna are proportional to the square, or worse, of its diameter. The design trade-off is between penetration ability, better at 10 cm, versus low cost and logistical ease, better with smaller antennas transmitting at 3 cm. When the current operational weather radar network in the United States was constructed, a significant investment was required to produce radars with 1° beamwidths using 10-cm wavelengths. Each radar used a 8.5-m diameter antenna inside a 12-m radome, and cost approximately $5 million, requiring significant infrastructure, power, and maintenance. The benefit is that the network of WSR-88D radars can penetrate through many kilometers of intense rain and hail. Many other countries, researchers, and media have chosen lower cost, shorter wavelength options with the limitation of only moderate or poor penetration ability.

In certain specialized cases, other factors enter into the choice of wavelength, notably with mobile and airborne systems as will be discussed later.

Transmitter Type

Weather radars usually use one of two basic types of transmitter, magnetron or klystron, though some are designed with other mechanisms. Klystron transmitters are most expensive but produce very stable and coherent signals. Transmitted radiation varies very little in frequency and the phase of the radiation is synchronous from pulse to pulse. Until very recently, klystron radars were far superior in their ability to measure velocities since the phase of returned radiation is used in these calculations. Cheaper magnetron transmitters produce radiation that exhibits random phase from pulse to pulse, and the frequency drifts rapidly within a narrow band from second to second. In the past, many radars that did not attempt to measure precipitation motion (non-Doppler radars) used magnetron transmitters. Today, however, almost all

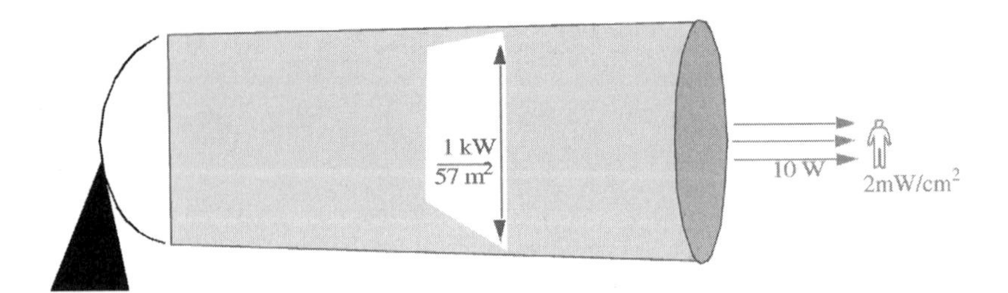

Figure 4 Intensity of microwaves emitted by typical weather radar. A person standing directly in front of a typical weather radar would absorb 10 W of microwave radiation with an exposure of $2\,\mathrm{mW/cm^2}$. See ftp site for color image.

weather radars in the United States, including magnetron systems, measure precipitation motion (Doppler) using modern hardware or software techniques.

A more intense transmitted beam results in more energy scattered back toward the radar, permitting the measurement of the properties of smaller or more tenuous weather targets. More intense beams also penetrate further into precipitation, despite attenuation. As with antenna and frequency choice, there are trade-offs involving transmitter strength. It costs more to construct high-intensity transmitters. It requires more power to transmit large amounts of energy. Also, intense electric fields can be hazardous and can complicate system design requiring pressurization of parts of the system with common or exotic gasses (SF_6) to prevent arcing. Transmitter strength is usually reported in terms of peak power, the power transmitted during the short pulses. Typical weather radars transmit pulses with 40 kW to 1 MW power. Some military systems use much more; some specialized research radars much less. Since the transmitter is off most of the time between pulses (e.g., on 1 μs, then off 999 μs, . . .), the average transmitted power is about 500 to 1000 times less, typically 100 W (like a light bulb) to 1 kW (like a hair drier). U.S. federal safety standards prohibit exposure to more than $10\,\mathrm{mW/cm^2}$ average power (other nations have similar standards, but with different levels of allowable exposure). If a person stood immediately in front of a WSR-88D 1000 W radar with a 8.5-m diameter antenna (area $57\,\mathrm{m^2}$), he or she would be exposed to approximately $2\,\mathrm{mW/cm^2}$, well below the safety limit (Fig. 4). The person (cross-sectional area about $0.5\,\mathrm{m^2}$) would intercept about 1% of the radar beam, or about 10 W total. The intensity at the center of the beam might be a couple of times higher. However, if one were to place a hand over the feedhorn (area $< 0.05\,\mathrm{m^2}$), much higher levels would be experienced.

Scattering

When the pulse of radiation impinges on raindrops in its path, scattering occurs. This is a complex electromagnetic process but, with some approximations, some simple statements can be made. When radiation interacts with a raindrop, it excites the water molecules in the drop. The molecules become electrically polarized. The radar beam causes this polarization to change orientation rapidly, at the radar frequency.

Thus, opposite sides of the drops become charged one way, then the next, billions of times per second. The drops become miniature antennas, analogous to the antenna of a walkie talkie, and radiate microwaves outward. Some of this radiation is radiated back toward the transmitting radar. (In drops that are similar in size to the transmitted radiation, e.g., hailstones, this process is considerably more complicated.)

The amount of radiation that a drop emits is proportional to D^6, the sixth power of the drop's diameter, so a 5-mm diameter drop radiates 15,000 times more than a 1-mm drop and 10^8 times more than a 50-μm cloud droplet. This complicates the interpretation of returned data since a single 5-mm drop returns as much energy as 15,000 1-mm drops, but the latter contain 125 times as much mass (Fig. 5). Therefore, it is difficult to know whether a particular strength of radar echo is due to a few large drops or a plethora of smaller ones. A particular amount of water mass scatters more microwaves if the water is contained in a few large drops. Raindrops are more efficient radiators than ice particles, scattering about 5 times as much than equivalent diameter ice particles. The D^6 approximation, called the Rayleigh approximation, applies only to drops that are much smaller than the radar wavelength. So, when short wavelength radars are used or when hail is present, more complicated formulations must be used.

Scattering can occur off other airborne objects, most notably insects and birds. Being mostly composed of water, these animals scatter in a fashion similar to raindrops of similar diameter. Importantly, however, they may not be passively moving with the wind. Bird echoes, moving at a different velocity than the air, can cause significant contamination to measured wind fields. Sometimes the military or researchers will release small strips of aluminized mylar or other objects into the air, called chaff. In the former case this is to confuse enemy radars, in the latter to provide passive scatterers carried by the wind in regions of little natural scattering. Scattering will occur where there is any contrast in the refractive index, usually caused by density changes, in the air. So, where turbulence is mixing cold and warm air parcels, or along precipitation-free dry lines or gust fronts, some energy is reflected back toward the radar by this process, called Bragg scattering. Researchers use these signals to observe the edges of clouds and other phenomena.

Scattering or reflections also occur when radar energy hits land, vegetation, or water surfaces. These echoes, called ground clutter or sea clutter, can be very intense and overwhelm the signals from the raindrops in the air near them. Often, clutter signals are filtered out by software that effectively blocks signals that have very low velocity and are presumed to originate from stationary objects. But this technique

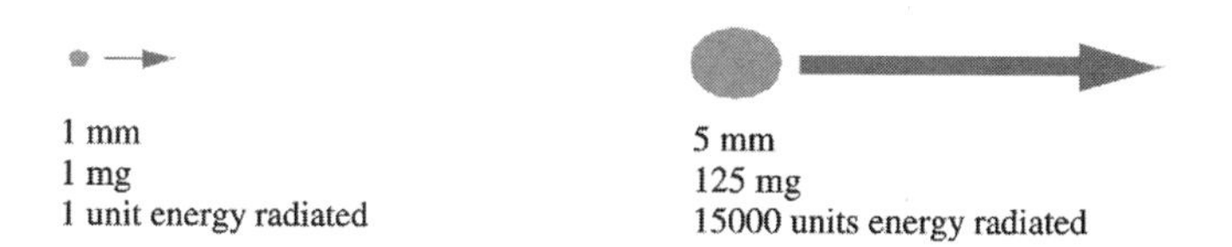

Figure 5 Relative diameter, mass, and scattered energy from small and large drops. Large drops scatter much more energy than small drops. See ftp site for color image.

does not work well with sea clutter contamination. Much of the clutter contamination arises from scattering from stray radiation that is not perfectly focused by the antenna into a narrow beam. This energy, called side-lobe radiation, hits objects in all directions and is scattered back to the radar. Since side-lobe energy scattered back by a very strong radar target, like a mountain or water tower, can be stronger than the weak signals scattered back by raindrops arriving back at the radar at the same time, this can cause significant contamination to weather radar data.

Propagation Paths

The narrow beam produced by a radar spreads with increasing distance from the radar. Even a 1° beam is 160 m wide at 10-km range and 1.6 km wide at 100-km range. At the maximum range measured by typical radars, 200 to 300 km, the beam may have spread to 3 to 5 km in extent. This critically affects the ability of the radar to detect weather, since objects smaller than the beamwidth cannot be resolved and only objects several times larger than a beamwidth can be accurately measured. Thus, microbursts and tornadoes are often difficult to detect at great range. Some very specialized research radars use extremely large antennas or short wavelengths to produce ultra-narrow beams, but this is not practical for most weather radar applications.

To a first approximation, radar beams travel in straight lines. Since Earth is curved, this means that a radar beam aimed at the horizon will soon become significantly raised from the surface and eventually depart into space (Fig. 6). This means that objects behind the horizon cannot be detected, preventing the resolution of near surface weather beyond a limited range. Fortunately, the atmosphere is more dense near the ground, resulting in an index of refraction gradient that bends radar beams partially back toward the curving surface of Earth, permitting some over-the-horizon visibility. The approximate height of the center of a beam aimed 0.5° above the ground, in "average" weather conditions, is 1.5 km at 100-km range and 3.5 km at 200-km range. The bottom of the beam would be approximately 400 m and 1.5 km above the ground at the same ranges.

Sometimes the gradient of atmospheric density is so high that it can bend the radar beams back into the earth. This can occur if very cold dense air lies near the surface. In this case, called anomalous propagation, energy will reflect off the ground or water surface, some back toward the radar.

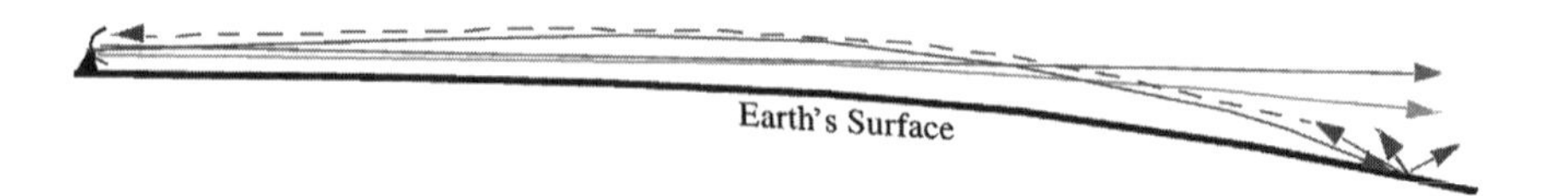

Figure 6 (see color insert) Beam paths assuming straight propagation (red), typical atmospheric density gradient bending beam partially back toward Earth (blue), strong density gradient, possibly temperature inversion, bending beam back into Earth (green) where scattering off surface sends energy back toward radar. See ftp site for color image.

Data Processing

Once the transmitter generates a pulse, it is focused by the antenna, interacts with objects in its path, and scattered energy returns to the radar. The returned signal must be converted into useful meteorological data. This is accomplished by the radar receiver and signal processing system. This is one of the most complex, varied, and rapidly evolving areas of radar technology. The basic concepts are relatively straightforward, however.

Radar Gates Most radars digitize (sample) the received signals. The digitalization rate determines the gate size and is one determiner of the resolution of a radar. If the returned signals are digitized at a rate of 1 MHz, or every 1 μs, the gate size will be 150 m ($c\ \Delta t/2$). Faster sampling will result in shorter gates. However, sampling intervals less than the duration of a pulse of the radar have diminishing added utility since the length of the transmitted pulse effectively blurs the returned signals and is another determining factor in true radar resolution. Frequently, the pulse length and gate length are matched.

Reflectivity The amount of power that returns to the radar from any scattering volume (defined by the beamwidth and sampling interval) is dependent on the amount of energy that impinges on the volume, the nature, number, size, shape, and arrangement of the scattering particles, radar wavelength, and distance to the weather target, attenuation, and other factors. These are related through the radar equation, which appears in many forms, but can be simplified to $P_r = CZ_e/R^2$, where P_r is the returned power, C is called the radar constant and contains all information about the transmitter, pulse length, antenna, wavelength, etc., R is the distance to the target, and Z_e is equivalent radar reflectivity factor, more commonly referred to as Z, or reflectivity. Z is a rather strange parameter; it has units of volume ($\mathrm{mm}^6/\mathrm{m}^3$) and it is usually expressed in terms of 10 times its base 10 logarithm, or $\mathrm{dB}Z = 10\ \log_{10} Z$. The amount of Z that would be measured from a raindrop is proportional to D^6, the sixth power of the drop diameter. The Z measured from a volume of drops is thus $\Sigma N_i D^6$, where N_i is the number of drops of each diameter in the volume. Because large drops are much more effective radiators, a certain value of Z can be due to a very small number of large particles or a large number of small particles; it is impossible to tell which by using Z alone.

It is difficult to precisely relate Z values to meteorologically useful quantities like liquid water content or rain rate. This is because it is dependent on the sum of the sixth power of raindrop sizes, not the sum of the masses of the raindrops. Numerous theoretical and empirical relationships, called Z–R relationships, exist to convert between Z, rain rate (R) and other quantities. Very roughly, 15 dBZ corresponds to light rain, 30 dBZ to moderate rain of several mm/h, 45 to 50 dBZ to 50 mm/h, 50 to 57 dBZ to 100 mm/h, and higher dBZ levels, 55 to 70, to hail or rain/hail mixes.

Typically Z is averaged over many pulses, 32 to 256, since it can vary greatly due to constructive and destructive interference from the radiation emitted from each drop in the illuminated volume. It is necessary to obtain several "independent" measurements to calculate an accurate value of Z. Independence means that the particles in the illuminated volume have reshuffled so that their arrangement is effectively decorrelated with their arrangement during the passage of the previous pulse. It can require a time spacing of several pulses before independence occurs, so the measurement of Z cannot take full advantage of all the 32 to 256 pulses mentioned above. The time to independence is shortest when there is high turbulence and/or short (i.e., X-band) transmissions. It is very difficult to calculate the radar constant, C, accurately, and the measurement of Z is prone to errors of approximately ± 2 dB. This can be very significant since small changes of Z can result in large differences in predicted R, particularly at high Z and R, where one cares the most.

The minimum power that a typical weather radar can measure is about 10^{-14} W, which corresponds to about -5 dBZ at a range of 50 km. This depends on the wavelength, antenna, quality of electronics, pulse length, number of pulses per average, etc. Though rarely an issue except in very close range research applications, there is a maximum power that can be detected before radar hardware/software saturates and is effectively blinded. This is seldom realized except in heavy rain within a few kilometers of a radar.

Doppler Velocity Most weather radars can measure the component of scatterer motion toward or away from the radar. While these radars are usually called Doppler, most do not directly use the Doppler effect to measure this motion. Typically the radar measures the path length to a raindrop (actually the sum of the path lengths to a volume of raindrops) during subsequent pulses (actually the remainder, noninteger portion) to calculate the motion (Fig. 7). The most common calculation technique is called pulse-pair processing, whereby the phase of the returned energy from each pulse is measured. Another technique called spectral processing, or Fourier processing, can be used also. While an acceptable velocity measurement can be made using just two pulses (in stark contrast to the several independent measurements needed to get an accurate Z measurement), typically many pulses are averaged to reduce error.

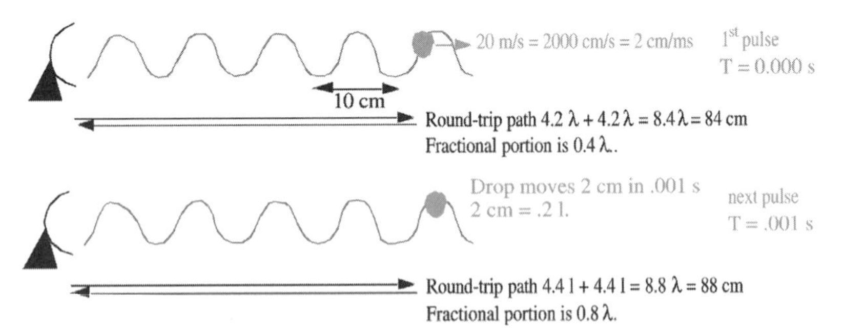

Figure 7 Illustration of pathlength changes used to calculate toward/away component of velocity. Signal processing is able to measure the fractional portion of the pathlength change. In reality, this calculation is performed on the energy scattered by many raindrops. See ftp site for color image.

The calculation is conceptually simple when the energy from just one raindrop is considered. However, typical radar volumes contain many raindrops, each moving with the wind, but with some random component, each radiating an amount of energy proportional to D^6, interfering constructively and destructively. If a radar beam is pointed horizontally, it is usually assumed that the drops are moving with the wind, $V_d = V_a$. But, if the radar beam is inclined, the terminal velocity of the drop will enter into the measurement: $V_d = V_a + V_t \sin\theta$. Estimation of V_t is difficult, and must take into account that V_d is the D^6 weighted average. Typically, it is assumed that V_t is a function of Z and atmospheric density, and is about 8 m/s in heavy rain near sea level.

A critical limitation of single radar "wind" measurements is that they can only detect the wind component toward and away from the radar (the radial wind) of the three-dimensional wind field. Even very strong cross-beam wind components cannot be detected with a single normal weather radar (see multiple Doppler and bistatic sections below).

Spectral Width In addition to the radial wind, averaged over many pulses, many radars also calculate the spectral width, which is just the standard deviation of the individual pulse-to-pulse wind measurements or frequency domain calculations. The drops in a radar volume can exhibit different motions for several reasons. They may have different terminal velocities as just discussed. They may be embedded in sub-resolution-volume-scale turbulence. The resolution volume may span a large-scale meteorological feature like a front or mesocyclone, so different portions of the beam illuminate different portions of the phenomena containing different characteristic velocities. There is also always some measurement error.

Range Ambiguity Once a pulse is transmitted from a radar, it will continue indefinitely until it is totally consumed by reflection, absorption, or scattering. Elevated radar beams quickly pass above the troposphere into regions where there are few scatterers other than the moon and planets. But, beams that are oriented almost horizontally can remain in the troposphere for hundreds of kilometers. However, the useful range of a radar is frequently limited by what is called the ambiguous range. The ambiguous range is determined by the maximum range to which a pulse can travel and return before the next pulse is sent. If the pulse repetition time (PRT) is 1 ms, then this range is 150 km. It takes 1 ms for the pulse to travel to 150 km, scatter off raindrops at that range, and return to the radar. Energy emitted by raindrops beyond that range will reach the radar after the next pulse is sent. The radar has no simple way of knowing whether this energy originated from raindrops illuminated by the first pulse beyond 150 km or by the second pulse just a short distance from the radar (Fig. 8). Since the energy returning from both pulses is superimposed, the data is contaminated. The amount of energy that is returned by the raindrop at great range is reduced significantly due to distance ($P_r \sim 1/R^2$), but if the distant weather system is intense, and the nearby weather weak, the data from the nearby weather can be obscured.

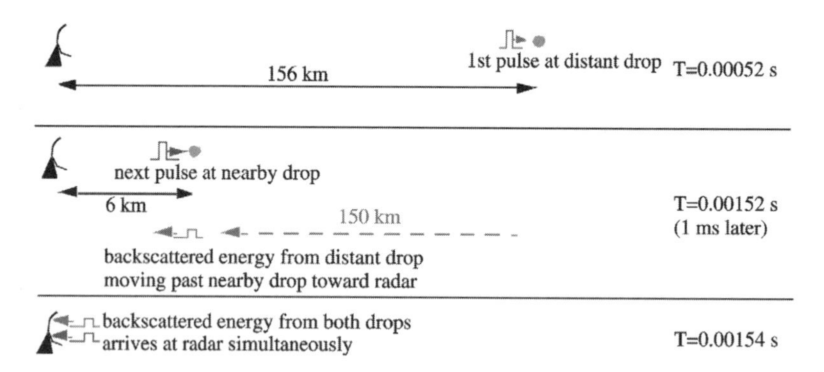

Figure 8 Illustration of range ambiguity phenomena. Energy scattered back from raindrops at 156-km range arrives at the radar simultaneously with energy from the next transmitted pulse scattered back from raindrops at 6-km range. See ftp site for color image.

The ambiguous range can be increased by slowing the PRT. This is frequently done to measure storms at great range. A PRT of 2 ms increases the range to 300 km. But this may complicate velocity processing as discussed in the next section. Newer techniques include the addition of phase offsets to the transmitted pulses so that the true range to the scatterers can be retrieved.

So-called second-trip echoes can be detected by trained observers since they have elongated and unrealistic shapes. In the case of random phase magnetron radars, the Doppler velocities in the second-trip echoes will be incoherent, not smoothly varying as in correctly ranged echoes.

Velocity Ambiguity: The Nyquist Interval For a given transmitted wavelength and PRT, there is a maximum velocity that can be ambiguously measured. This is because Doppler radars do not actually measure the Doppler shift of returned radiation. Referring back to Figure 7, the fractional portion of the path length from radar to target back to the radar is measured. It is usually assumed that this path length changes by less than one full wavelength during the PRT. But, this may not be so. Consider a 10-cm radar with a PRT of 1 ms. If a raindrop is embedded in a strong wind, say 40 m/s away from the radar, then it will move 4 cm during the PRT. This means that the round-trip path length will increase by 8 cm, say from 10^9 to $10^9 + 0.8$ wavelengths. Ambiguity arises because it is difficult to distinguish between a 8 cm increase in path length and a 2-cm decrease since each will result in the same fractional portion difference, namely 0.8 wavelengths [i.e., $\text{frac}(10^9 + 0.8) = \text{frac}(10^9 - 0.2) = 0.8$]. The Nyquist interval is the range of velocities that will produce path length changes between $\pm\frac{1}{2}$ wavelength and can be calculated as Nyquist Interval $= \pm\lambda/(4\ \text{PRT})$. Thus, in the above case, the Nyquist interval would be $0.1\ \text{m}/(4 \times 0.001) = \pm 25\ \text{m/s}$.

The Nyquist interval can be increased by decreasing the PRT, essentially giving the raindrops less time to change the round-trip path length. Note, however, that this would increase data contamination from storms beyond the now shortened ambiguous range. There is a trade-off between maximizing Nyquist interval and maximizing ambiguous range. Some research radars use techniques called dual PRT or staggered PRT whereby alternating (or other patterns) of long and short PRTs are used to gain a least-common-multiple effect and much larger effective Nyquist intervals.

Data from regions exhibiting velocities greater in magnitude than the ambiguous velocity are considered to be folded or aliased and need to be corrected, or unfolded, or dealiased, to produce correct values (Fig. 9). This can be a difficult process and can be conducted either automatically or manually.

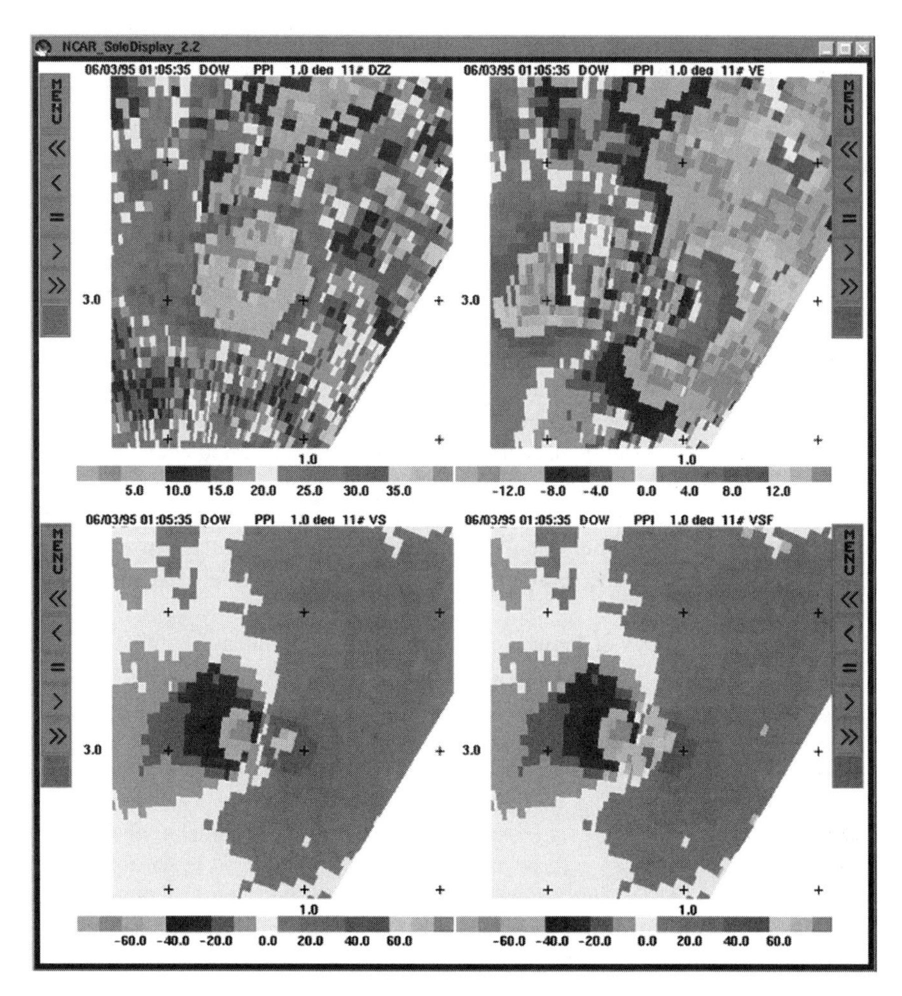

Figure 9 (see color insert) Illustration of dealiasing and cleaning of radar data. (*Top left*) Reflectivity in tornado showing ring debris. (*Top right*) Raw Doppler velocity with aliasing. (*Bottom left*) Velocity after dealiasing. Strong away and strong toward velocities adjacent to each other imply rotation, in this case over 70 m/s. (*Bottom right*) Velocity after values with high spectral width or contaminated by echoes from the ground (ground clutter) have been removed. Data is from DOW mobile radar in the Dimmitt, Texas, Tornado on June 2, 1995, from a range of 3 km. See ftp site for color image.

Scanning Techniques and Displays Most radars scan the sky in a very similar way most of the time. They point the antenna just above the horizon, say 0.5° in elevation, then scan horizontally (in azimuth), until 360° is covered. Then the radar is moved to a higher elevation, say 1.0° or 1.5°, and the process is repeated. This continues for several to many scans. Using this method, several coaxial cones of data are collected. These can be loaded (interpolated) onto three-dimensional grids or displayed as is to observe the low and mid/high levels of the atmosphere. There are infinite variations on this theme, including interleaving scans, fast and slow scans, repeated scans at different PRTs, scanning less than 360° sectors, etc. When a single scan is displayed, it is usually called a plan position indicator (PPI).

Sometimes, mostly during research applications, a radar will keep azimuthal angle constant and move in elevation, taking a vertical cross section through the atmosphere. These can be very useful when observing the vertical structure of thunderstorms, the melting layer, the boundary layer, and other phenomena. These types of scans are called range height indicators (RHI).

Since PPI scans have a polar, conical, geometry, lower near the radar and higher as the beams travel outward, it is often useful to use a computer to load the data from several scans into a Cartesian grid and display data from several different scans as they pass through a roughly constant altitude above Earth's surface, say 1 or 5 km above ground level (agl). These reconstructions are called constant altitude plan position indicators (CAPPIs). Since the data at a given altitude can originate from several scans, there can be ringlike interpolation artifacts in these displays.

Scanning strategy strongly influences the nature of the collected data. Operational radars typically scan fairly slowly through 360°, using many scans, requiring about 5 to 6 min for each rotation. This provides excellent overall coverage, but can miss the rapidly evolving weather such as tornadoes and microbursts. In specialized research applications, much more rapid scanning, through limited regions of the sky, is often employed.

New radar technology is being developed that may someday permit very rapid scanning of the entire sky as discussed below.

3 RADAR OBSERVATIONS OF SELECTED PHENOMENA

There are literally thousands of examples of weather phenomena observed by weather radars. Only a few will be illustrated here in Figure 10 to show some the range of phenomena that can be observed and the typical nature of the data. The interpretation of the data from weather radar is a complete study unto itself and could occupy far more space than is appropriate in this short overview.

4 GOAL: TRUE WIND VECTORS

Radial velocities provide much qualitative information about weather phenomena. However, the true wind field is a three-dimensional wind field comprised of three-

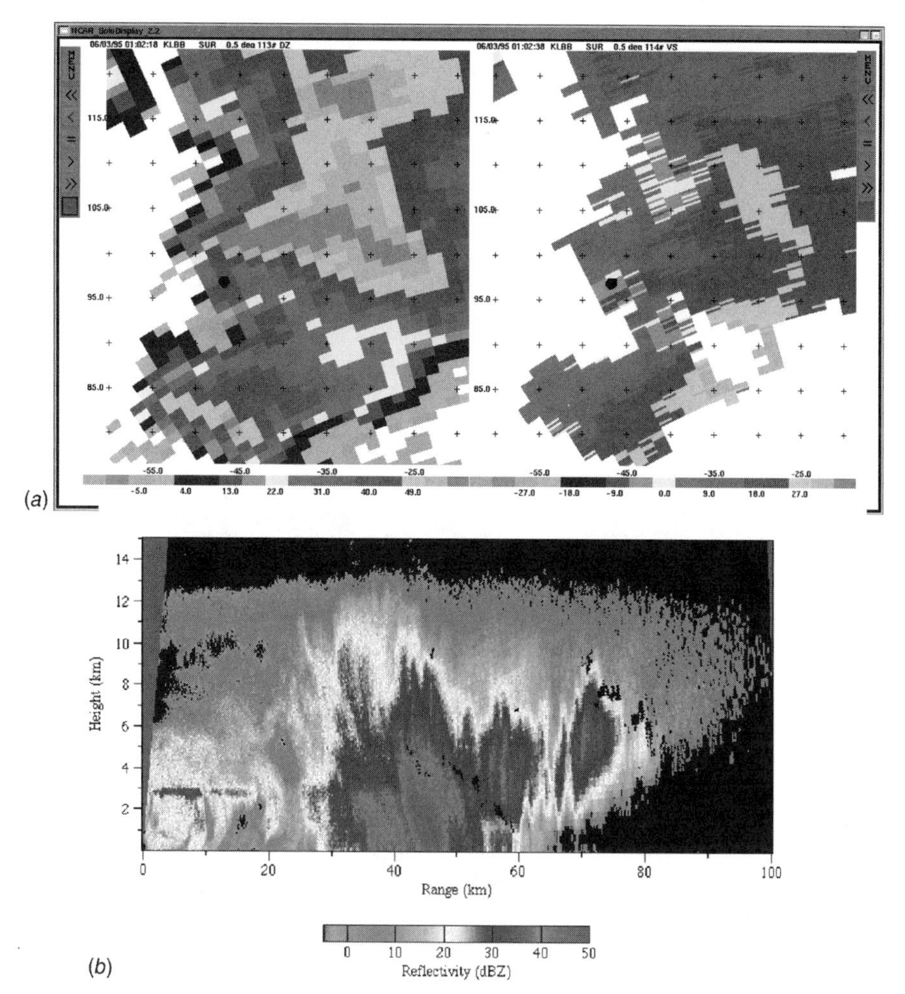

Figure 10 (see color insert) (*a*) A tornadic supercell thunderstorm observed by a WSR-88D operational radar. Reflectivity (*left*) and Doppler velocity (*right*) are shown. Classic hook echo extends from the western side of the supercell. An intense circulation, suggested by the strong away and toward velocities near the hook, is the mesocyclone associated with a tornado that was occurring. (*b, c*) Reflectivity and Doppler velocity in a vertical cross section (RHI) through a portion of a squall line. The high reflectivity core and lower reflectivity extending to 12 km are visible. Strong toward and away Doppler velocities are associated with the up and down drafts of the cell, as indicated by arrows. Data from the MIT radar in Albuquerque, New Mexico. (Courtesy of MIT Wea. Rad. Lab. D. Boccippio.) (*d*) Reflectivity (*lower*) and Doppler velocity (*upper*) in a winter storm. Reflectivity is somewhat amorphous but is enhanced in a ring corresponding to the melting layer. In the melting layer, large, wet slow-moving particles cause high reflectivity. The velocity pattern provides a vertical sounding of the atmosphere. Winds are from the NNW at low levels (near the radar), but from the southwest aloft (away from the radar as the beams diverge from Earth's surface). Cold advection is implied. Data from the MIT radar in Cambridge, Massachusetts. (Courtesy of MIT Wea. Rad. Lab. D. Boccippio.) See ftp site for color image.

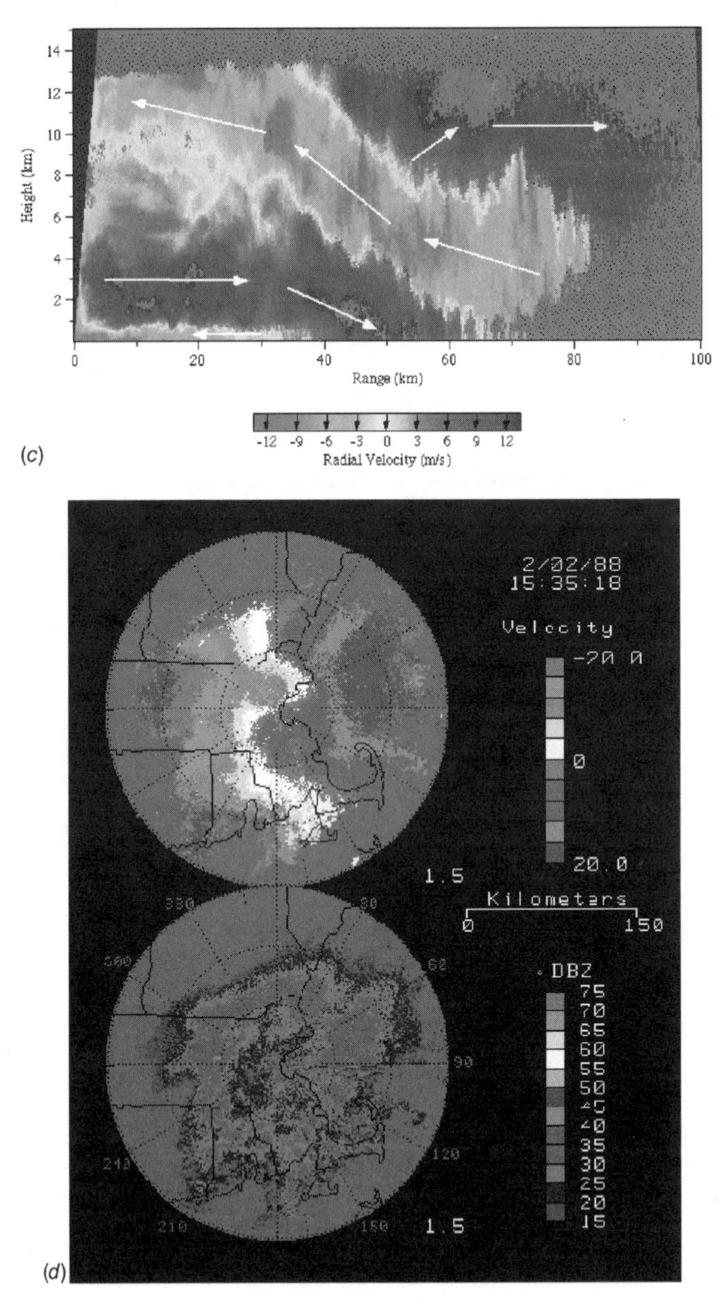

Figure 10 (*continued*)

dimensional vectors, evolving in time. Physical equations used in research and in forecasting models operate on these vector wind fields, which really contain the physics of the phenomena. So there is great value in estimating or measuring the full vector wind field and several techniques have been developed.

Single-Doppler Retrievals

One class of techniques for obtaining the vector wind field is called single-Doppler wind field retrievals. These techniques use various physical assumptions to convert data from a single radar into vector wind fields. One simple method assumes that the reflectivity field is a passive tracer that moves with the wind. In simple terms the Z field is examined at different times, and the wind field necessary to move the Z features from one place to another is calculated. Of course, evolving systems complicate these analyses. Another assumes that the wind field is composed of certain simple mathematical components. These are extensions of what radar meteorologists do visually when they look at the zero line of the Doppler velocity and assume that the wind is moving perpendicularly to it. Several sophisticated and combination methods are in development.

These techniques are very useful since they work with just one radar, but they are limited by the validity of the physical assumptions. Some also only work in cases of high reflectivity or velocity gradients, others only when precipitation covers much of the surveyed volume.

Currently, these techniques are being developed in the research environment, but it is hoped that they will be used to introduce wind fields into operational computer forecasting models in the future.

Dual and Multiple Doppler

In some research experiments, two or more radars can be deployed. Each radar can survey a target region from a different vantage point. A simple mathematical calculation can convert the two or more Doppler velocity measurements into a wind vector (Fig. 11). This technique is very powerful and has been a favorite of research meteorologists. It is not used frequently in operational forecasting because few permanent radars are close enough for dual-Doppler calculations to be useful. The large spacing of the U.S. WSR-88D network makes dual-Doppler calculations, while possible, not very useful due to the large beamwidths at the typical 200-km ranges to weather targets.

Typically, but not always, the horizontal components of the vector wind are calculated directly from the radar measurements. The vertical component is calculated by integrating the equation of mass conservation. Essentially, this says that if there is strong divergence in the boundary layer as in, say, a microburst, the air must have come from above, implying downward vertical motion (Fig. 12). Strong convergence near the ground implies an updraft. Similarly, strong divergence at storm top indicates an updraft from below, etc. Unfortunately, the vertical motions calculated in this manner result from the integration of quantities that are derivatives

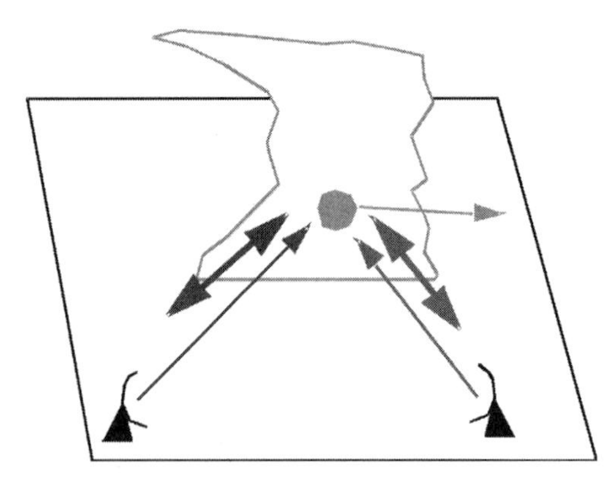

Figure 11 Dual-Doppler network with radars measuring motion of raindrops from different vantage points. The different Doppler radial winds (red and green) are combined mathematically to produce the horizontal projection of the true raindrop motion vector (blue). See ftp site for color image.

of actual measurements. Both the integration and differential processes are prone to errors. The resultant vertical wind estimates can be substantially incorrect. Accurate determination of vertical motions from radar data is probably the largest unsolved problem in radar meteorology today. An example of a dual-Doppler reconstruction of the vector wind field near and in a tornado is shown in Figure 13.

Rarely, three or more radars are in close enough proximity that the vertical component of the raindrop motion vector can be calculated directly. This method is called triple-Doppler and is exclusively a research technique.

There are two major limitations to multiple-Doppler techniques. The first is that radars are very expensive. Multiple-Doppler data is only affordable in a small frac-

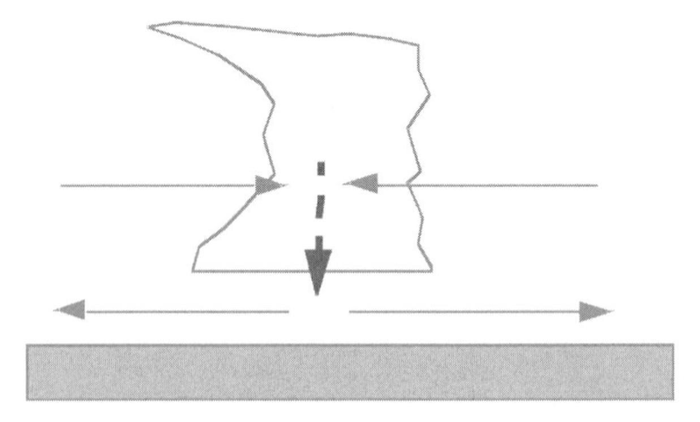

Figure 12 Since the vertical component of motion is rarely measured directly, the equation of mass continuity is usually used (sometimes in very complex formulations) to derive the vertical component of air motion. The physics of the method is very straightforward. If divergence is observed at low levels (*right*), then air must be coming from above to replace the departing air, implying a downdraft. See ftp site for color image.

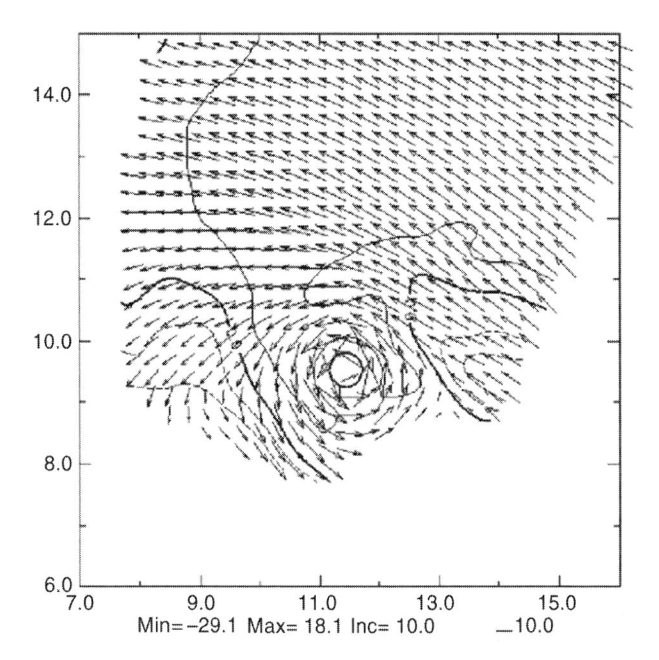

Figure 13 Dual-Doppler analysis showing horizontal component of the vector wind field in a tornado. Contours are Z with the small circle representing the low Z "eye" of the tornado. The axes are labeled in kilometers. Peak winds are about 60 m/s. (Courtesy Y. Richardson.)

tion of short-term research experiments and rarely in operational applications. The second is that the observations of weather targets by the different radars may occur at different times, sometimes a few minutes apart. Rapid evolution of some of the most interesting phenomena between these observations will contaminate the calculated vector wind fields.

A common expression among multiple-Doppler users is "you always get a vector," meaning that the technique always produces a result, but the result can be quite bogus. Multiple-Doppler and single-Doppler reconstructions should always be viewed with a sceptical eye.

Bistatic Radars

When the transmitted radar beam interacts with raindrops, only some of the reemitted energy travels back toward the transmitter. Most is scattered in other directions. The bistatic radar technique involves placing small passive radar receivers (Fig. 14) at various places to measure this stray radiation. The data from the bistatic receivers is combined with that from the transmitter using a variation on standard multiple-Doppler formulations.

The biggest advantage of the bistatic technique is the comparatively low cost of the bistatic receivers, less than 10% that of a WSR-88D. Several to many receivers

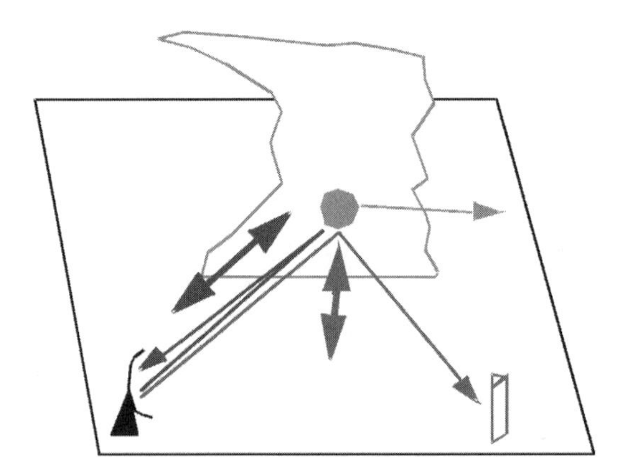

Figure 14 Bistatic dual-Doppler network with receive-only radar (red) measuring a different component (red arrow) of the raindrop motion vector (blue arrow) than the transmitter (green arrow). The components are combined mathematically in a fashion similar to that used in traditional dual-Doppler radar networks to calculate the horizontal projection of the raindrop motion vector (blue arrow). See ftp site for color image.

can result in more accurate data at a low cost. Another advantage is that the observations of individual weather targets are made simultaneously since there is only one source of radar illumination. Rapidly evolving weather can be well resolved. Thus two of the major limitations of traditional dual-Doppler networks are avoided.

Currently there are several research bistatic networks. Data from one are illustrated in Figure 15. It is anticipated that operational networks and operational computerized forecast models will use bistatic data in the future.

5 DATA ASSIMILATION INTO COMPUTER MODELS

Computerized weather forecasting models require accurate initializations to produce meaningful predictions. A major thrust of current modeling research involves how to best introduce radar data into these initializations, particularly into mesoscale simulations. Some models have successfully ingested both reflectivity and single-Doppler-retrieved wind fields, but none are yet used operationally.

6 NEW AND NONCONVENTIONAL TECHNOLOGIES

Dual and Multiple Polarization Radars

Most radars emit microwaves with just one polarization. This means that when they cause charge to move in raindrops as discussed above, the charge moves one direction then the opposite (say left, then right), but the charge distributions in other directions (say up and down) are largely unaffected. The intensity of microwaves

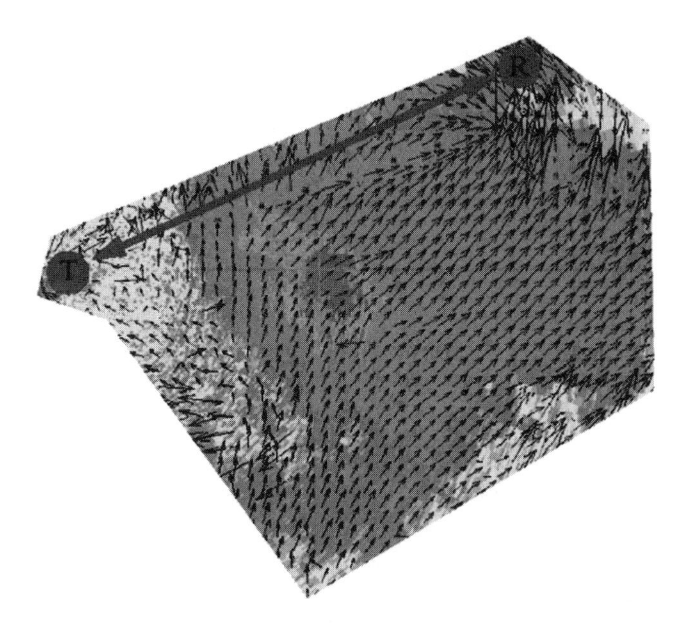

Figure 15 Bistatic dual-Doppler horizontal wind field. The transmitter (T) and passive, receive-only radar (R) are located 35 km distant. Z field is shaded. Data from NCAR CASES experiment in Kansas in 1997. See ftp site for color image.

emitted by a drop is proportional to D^6. But, large raindrops are hamburger bun shaped, ice particles have many shapes, and hail and insects can be very irregular. The intensity of the emitted microwaves is proportional to the D^6 in the direction of polarization of the radar.

It is possible to obtain information about the shape of the rain or ice particles by transmitting both horizontally and vertically polarized radiation. (There are other exotic techniques too, like using circularly polarized radiation, or radiation polarized at intermediate values between H and V.) If the beam hits a large hamburger-shaped drop, more horizontal energy will return than vertical energy (Fig. 16). The sixth power dependence means that small differences in D_h and D_v can cause large

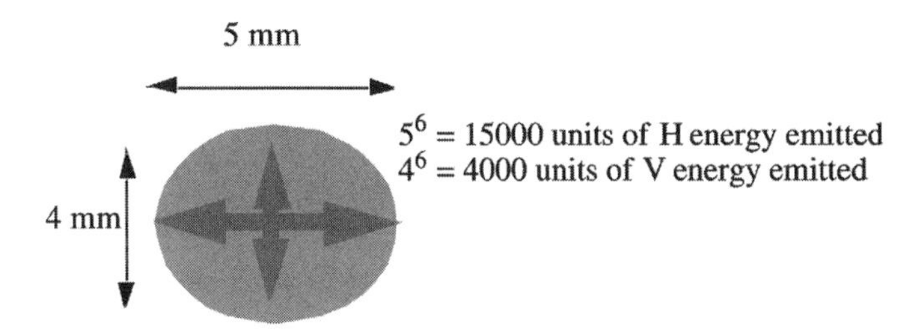

Figure 16 Oblate, hamburger-shaped raindrops scatter much more horizontally polarized energy than vertically polarized energy. See ftp site for color image.

differences in the emitted energy. This difference is called ZDR and can be as much as several decibels. Even though hailstones are often very irregular, they tumble randomly and the sum of the ZDR returns from thousands of hailstones is very close to zero (Fig. 17). Ice particles, however, can exhibit preferred orientations just like raindrops, and can produce ZDR. A good rule of thumb is that high Z with high ZDR = heavy rain while high Z with low ZDR = hail.

New techniques are in development to make use of other measurements possible with multiple polarization radars, such as the linear depolarization ratio (LDR), which measures how much radiation from the horizontal beam is reemitted from drops with vertical polarization, and specific differential phase, Φ_{dp}, which measures the differences in the phase of the horizontally and vertically polarized returned energy. Some of these hold promise to refine the identification of particle types (rain, hail, large hail, ice, etc.) and rain rate. Φ_{dp} holds particular promise since changes in Φ_{dp} are thought to be proportional to D^3, the volume of water in a resolution volume, and therefore more directly related to rain rate than Z. LDR is a very difficult measurement that requires a very precisely manufactured, therefore expensive, antenna.

Polarization radars are now primarily used in research but will probably be used operationally by forecasters in coming decades.

Mobile Radars

No matter which choices are made for radar wavelength and antenna size, the radar beam will always spread out with distance. Few radars have beams that are much narrower than 1°. This means that at 60-km range the beams are 1 km wide, and at 120 km they are 2 km wide, much too large to resolve small phenomena such as tornadoes, microbursts, etc. The best solution found to this problem so far has been

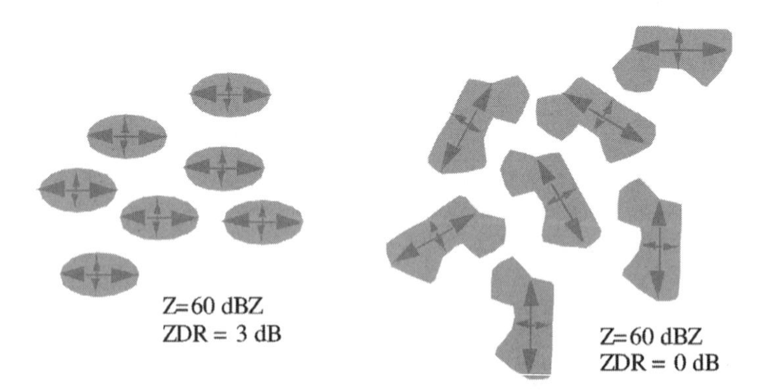

Figure 17 Large raindrops (*left*) are oriented similarly, so the ZDR effect from individual drops adds constructively to produce large ZDR values. Hailstones (*right*) tumble and are therefore oriented randomly so individual ZDR values tend to cancel out producing ZDR = 0. See ftp site for color image.

to put the radars on vehicles: ground based, in the air, and on the ocean. The vehicles are then deployed near the weather to be measured.

These systems tend to be very specialized and, with the exception of the hurricane hunter aircraft, are used mostly for research. Their mobile nature makes design and operation difficult.

Ground-Based Mobile Radars A leading example of the mobile radar concept is the Doppler On Wheels (DOW) research radars (Fig. 18). These radars have obtained unprecedented high-resolution three-dimensional data in tornadoes, hurricane boundary layers, etc., at scales as small as 3 to 60 m. Typically two or more DOWs are deployed in a mobile multiple-Doppler network to retrieve high-resolution vector wind fields. The DOWs use 3-cm radiation and 2.44-m antennas to produce 0.93° beam widths, which are comparable to the WSR-88Ds (but 3-cm radiation suffers more from attenuation in heavy precipitation). Fast scanning and very short pulses and gate lengths are combined for fine spatial and temporal scale observations. Other mobile radars using energy with wavelengths ranging from 3 mm to 10 cm have also been used by researchers to study similar phenomena.

None are currently in use for forecasting. But the idea of using DOW-type radars to augment the stationary radar network, part of a concept called adaptive observations, is being explored. In the future, forecasters who need information in specific areas, say a hurricane landfall, severe weather outbreak, flood, or the Olympics, may be able to request high-resolution tailored multiple-Doppler radar measurements.

Figure 18 DOW2 mobile radar. The DOW2 uses a 2.44-m antenna transmitting 3-cm radiation with a 0.93° beam. It is used to intercept tornadoes, hurricanes, and is deployed in mountain valleys or wherever an easily movable system is needed. See ftp site for color image.

Airborne and Ship-Based Radars Since many areas are inaccessible to DOW-type trucks, particularly oceanic areas that cover 70% of Earth including hurricane spawning grounds, and because trucks are limited to highways speeds of 30 m/s, limiting deployment ranges, researchers and operational meteorologists use radars mounted on aircraft (Fig. 19) and ships to get closer to the weather of interest. These aircraft regularly fly into hurricanes before they make landfall to aid in predictions. They have been used to study meteorology as diverse as tornadoes in the Midwest and tropical climates in the western Pacific.

Bistatic Radar Networks

These collect stray radiation emitted in many directions by raindrops to measure vector wind fields and were discussed above.

Rapid Scan Radars

Military radars have long been able to move their beams electronically rather than having to mechanically move an antenna. Since the beam can be moved almost instantaneously, these hold the promise of extremely rapid scanning of weather. Instead of requiring 6 min to survey the entire sky, it might be sampled in only 10 to 30 s. A type of antenna called a phased array is used. Very few exist outside military applications, in part due to the high cost of construction. There are efforts being made to adapt this military technology to meteorological use.

A new type of rapid scan radar is under development and holds promise as a research and operational tool primarily due to its low cost, compared to phased array systems. These radars transmit multiple frequencies from an unusual antenna designed to split the various frequencies into simultaneous multiple beams (Fig. 20).

Figure 19 ELDORA airborne radar can be deployed quickly to almost any point in the world, even over oceans. It has a sophisticated radar in the protuberance extending beyond the normal tail. Similar systems are used operationally to intercept hurricanes. (Courtesy NCAR/ATD.) See ftp site for color image.

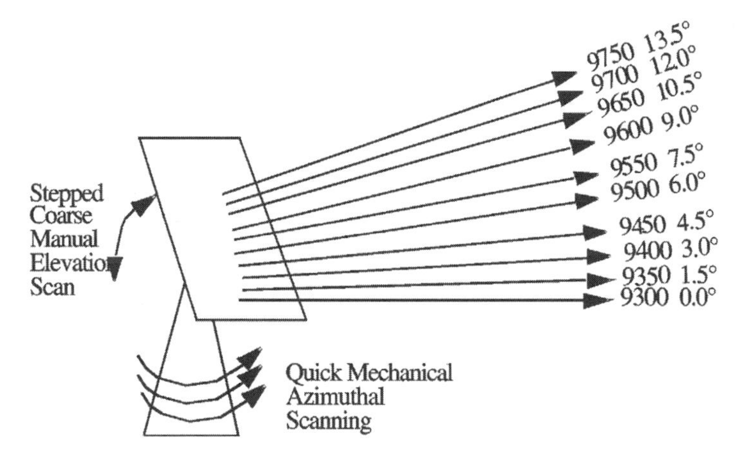

Figure 20 A flat panel, slotted waveguide array with 100 individual slotted waveguide antennas produces beams that emanate at different elevation angles depending on the frequency of the transmitted radiation. Therefore, nearly simultaneous transmissions at multiple frequencies, can be simultaneously received from several elevation angles at once. Ten or more elevation angles can be surveyed in $\sim$10 s, providing truly rapid scanning with a nonphased array system.

Figure 21 NCAR ISS in Kapingamarangi, Pohnpei State, Federated States of Micronesia. A wind profiler antenna is pointed vertically and shrouded by a clutter fence visible on top of the shelter. Aluminum cylinders shield a portion of a RASS radio acoustic sounding system. See ftp site for color image.

A single mechanical sweep of this antenna produces multiple beams, as many as 10 or more, permitting a typical volumetric scan to occur in just 2 sweeps, in as little as 10 s.

Wind Profilers

There is another class of radars called wind profilers that usually point vertically, do not scan, and sample vertical cross sections of wind and temperature. The principles of operation share some similarities to conventional weather radars, but there are substantial differences. They usually collect just one vertical sounding, averaged over about 30 min. Many can collect wind velocity data in clear air, in the absence of precipitation, through substantial depths of the lower troposphere. There is a network of profilers in the Midwest that augments the balloon sounding network by providing half hourly measurements at intermediate locations. Data from wind profilers is valuable for the forecasting of severe weather outbreaks in the Midwest. Deployable wind profilers are used by researchers to provide vertical soundings of wind between balloon launches (Fig. 21). An instrument called a RASS, or radio acoustic sounding system, uses an ingenious method whereby emitted sound waves disturb the atmosphere in a manner that can be measured by the profiler radar to measure the temperature of the atmosphere in the vertical column.

CHAPTER 49

BASIC RESEARCH FOR MILITARY APPLICATIONS

W. D. BACH, Jr.

> If you know the enemy and know yourself, your victory will never be endangered; if you know Heaven and know Earth, you may make your victory complete
>
> —Sun Tzu, The Art of War X, 31 circa 500 BC

The ancient Chinese general succinctly summarizes necessary ingredients for success in war. Understanding that "Heaven" represents the atmospheric environment, the need to know the weather is deeply rooted in military preparation and tactics. History is replete with examples of commanders using weather as an ally or suffering its bad effects. Even with today's modern technology, the dream of an all-weather military has not become a reality. Thus the military continues to seek ways to use the atmosphere as a "combat multiplier" in the order of battle.

1 MILITARY PERSPECTIVE

The military must understand the atmosphere in which it operates. That understanding will ultimately depend on scientific understanding of the atmospheric processes, an ability to use all of the information available, and an ability to synthesize and display that information in a quickly understandable fashion.

The military's needs for understanding the atmosphere's behavior has changed as a result of geopolitics and advancing technology. On the battlefield, future enemies are more apt to resort to chemical or biological attacks as a preferred method of mass destruction. The atmospheric boundary layer is the pathway of the attack. Increased

Handbook of Weather, Climate, and Water: Dynamics, Climate, Physical Systems, and Measurements, Edited by Thomas D. Potter and Bradley R. Colman.
ISBN 0-471-21490-6

reliance on "smart" weapons means that turbulent and turbid atmospheric effects will influence electromagnetic and acoustic signals propagating through the boundary layer. Intelligence preparation of the battlespace, a key to successful strategies, depends on full knowledge of the enemy, weather, and terrain. It requires an ability to estimate atmospheric details at specific locations and future times to maximize strategic advantages that weather presents a commander. It also avoids hazards and strategic disadvantages.

A convergent theme of basic research in the atmosphere is the interdependence of research progress in acoustic and electromagnetic propagation with the measurement and modeling of the dynamical boundary layer. The scattering of propagating energy occurs because of fine structure changes in the index of refraction in the turbulent atmosphere. The connection between the two—the refractive index structure function parameter, C_n^2 —is proportional to the dissipation rate of the turbulent kinetic energy.

Military operations are often in time (seconds to hours) and space (millimeters to tens of kilometers) scales that are not addressed by conventional weather forecasting techniques. Atmospheric information is required for meteorological data-denied areas. Furthermore the real conditions are inhomogeneous at various scales. Significant adverse effects may have short lifetimes at high resolution. Such breadths of scales make accurate quantification of important boundary layer processes very difficult. Carefully planned and executed experiments in the uncontrolled laboratory of the atmosphere are needed for almost every phase of the research. The combined range of atmospheric conditions and potential propagation types and frequencies that are of interest to the military is too broad to condense into a few nomograms. The propagation studies require intensive and appropriate meteorological data to understand the atmospheric effects. Understanding of heterogeneous atmospheric fields arising from inhomogeneous conditions requires measurements achievable only by remote sensors.

Chemical and biological agents constitute major threats to military field units as well as to civil populations. Electromagnetic propagation and scattering are the principal means of remote or in-place detection and identification of agents. Current models of atmospheric transport and diffusion of the agents over stable, neutral, and convective conditions are marginally useful and based on 40-year-old relationships. Significantly improved models of the transport and diffusion are needed to estimate concentrations and fluctuations at various time and space scales for simulated training, actual combat, and for environmental air quality.

2 RESEARCH ISSUES

Basic research for the military in the atmospheric sciences comes under two broad, interdependent scientific efforts:

Atmospheric wave propagation, which is concerned about the effects of the atmosphere, as a propagating medium, on the transmission of electromagnetic (EM) or acoustic energy from a source to a receptor, and

Atmospheric dynamics, which is concerned with quantifying the present and future state of the atmosphere.

Atmospheric Wave Propagation

Electromagnetic and acoustic propagation and scattering is a large subject that addresses problems in both characterizing the battlespace and in detecting targets for engagement and situational awareness. Although there has been much previous research in the general area of wave propagation and scattering, there remains a pressing need for research that specifically address issues related to propagation and scattering in realistic atmospheric environments.

Electro-optical and infrared propagation in the atmosphere is limited by turbulence and by molecular and aerosol extinction by both absorption and scattering. The relative importance of turbulence and extinction is dependent on the wavelength of the radiation and the atmospheric conditions. Modern, high-resolution imaging systems are often turbulence limited, rather than diffraction limited, impairing long-range, high-resolution target acquisition, recognition, and identification. Laser-based systems are limited in their ability to retain spatio-temporal coherence and effectively focus at tactical ranges. Like acoustic signals, electromagnetic signals are primarily useful for target detection or ranging and for remote sensing of the atmosphere itself. Although much research has been done in the area of electromagnetic extinction, and current models are generally good, applications of these models for remote-sensing purposes is an area of active research. A large body of research exists in the area of atmospheric turbulence effects for electromagnetic propagation, but with less conclusive results. Hence understanding of atmospheric turbulence for optical and infrared propagation is still an active area of research.

Understanding of the interaction between propagation and turbulence is rather straightforward in the theoretical sense. Practically, the time scales of the EM propagation interactions with the air are much shorter, $O(10^{-9}\,\mathrm{s})$ along a path, than the capability to locally sample the atmospheric turbulence or density fluctuations, $O(10^{-1}\,\mathrm{s})$ and to characterize the turbulence spectra, $O(10^{+3}\,\mathrm{s})$. Acoustic interactions also occur on short time scales. Second, the turbulent structure of the atmosphere between the source and detector is largely unknown, so the scale of the disturbance is undetermined. Furthermore, with models, grid volume samples are realized at $O(10^{0}\,\mathrm{s})$ in large eddy simulation models and at $O(10^{2}\,\mathrm{s})$ in fine mesoscale models. To make the connection between propagation effects, turbulence characteristics, and atmospheric models, new measurement techniques are needed. In careful field campaigns to relate propagation to atmospheric conditions, the biggest problem is independent ground truth.

Atmospheric Dynamics

The military emphasis on global mobility requires weather forecasting support at all scales from global to engagement scales. In recent years, the explosion in computer capacity enabled highly complex numerical weather prediction models on most of these scales. Model accuracy has improved even as complexities are added. Model results have become more accepted as a representation of the atmospheric environment. As computation power increases, finer and finer scales of motion are represented in smaller grid volumes. Lately, large eddy simulations represent dynamics at grid sizes of a few meters in domains of $5\,\text{km}^2$ by 1 km deep.

As the models go to finer scales, the variability imposed by large-scale synoptic and mesoscale influences the Atmospheric Boundary Layer (ABL) is modified by both the cyclical nature of solar radiation and metamorphosis due to the stochastic behavior of clouds and other natural processes as well as anthropogenic causes. The resulting (stable and unstable) boundary layers are sufficiently different that current models of one state do not adequately capture the essential physics of the other or the transition from one state to the other. Accurate predictions become more difficult, principally because the atmosphere is poorly represented. Data are lacking at the scales of the model resolutions. Parameterizations required to close the set of equations are inadequate. Observations are lacking to describe the four-dimensional fields of the forecast variables at the model resolution. The models are unlikely to improve until better theories of small-scale behavior are implemented and observations capable of testing the theories are available.

For the military, techniques to represent the inhomogeneous boundary layer in all environments—urban, forest, mountains, marine, desert—is absolutely essential for mission performance. In most cases, this must be done with limited meteorological data. Furthermore, the military is more frequently interested in the effects—visibility, trafficability (following rain), ceiling—than in the "weather" itself. Nevertheless, without high-quality, dependable models that represent the real atmosphere in time and space, the effects will not be representative.

Basic research attempts to improve modeling capability by increasing the knowledge base of the processes of the atmosphere. Until new measurement capabilities are developed and tested, our ability to characterize the turbulent environment—affecting propagation and dispersion of materials—will be severely limited.

3 PROTOTYPE MEASUREMENT SYSTEMS

Military basic research has participated in several new techniques to measure winds and turbulence effects in the atmospheric boundary layer. These techniques concentrate on sampling a volume of the atmosphere on the time scales of (at least) the significant energy containing eddies of the atmosphere.

The Turbulent Eddy Profiler, developed at the University of Massachusetts, Amherst, is a 90-element phased-array receiving antenna operating with a 915-MHz 25-kW transmitter. It is designed to measure clear air echoes from refractive index fluctuations. The received signal at each element is saved for postprocessing. These signals are combined to form 40 individual beams simultaneously pointing in different directions every 2 s. In each 30-m range gate of each beam, the intensity of the return, the radial wind speed is computed from the Doppler shift, and the spread of the Doppler spectrum is calculated. These data are displayed to show a four-dimensional evolution of refractive index structures in the boundary layer. Further processing gives estimates of the three components of the wind velocity. Combining the wind vectors with refractive index structures has shown significant horizontal convergence occurring with strong refractive index structures. At 500 m above the ground the volume represented is approximately a 30-m cube.

The Volume Imaging Lidar at the University of Wisconsin measures backscatter from atmospheric aerosols in three dimensions at 7.5-m range resolutions. The evolution of boundary layer structures can be seen in a variety of experiments. Capabilities to estimate horizontal winds at selected altitudes have been developed and demonstrated.

An eye-safe, scanning Doppler lidar operating at 2 μm has been jointly developed with NOAA/ERL/ETL to measure radial wind speeds at 30-m resolution. The error of the measurement is < 0.2 m/s. Various scanning approaches have shown several different evolutions of the morning transition and breakdown of a low-level jet.

Another lidar system has been developed at the University of Iowa to measure the horizontal wind in the boundary layer at about 5-m increments every minute. Teamed with existing FMCW radars, having comparable vertical resolution, should add to meteorological understanding of layers of high refractive index.

4 CURRENT MEASUREMENT CAPABILITIES

Reliable measures of atmospheric temperature and moisture at high resolution in space and time are still lacking. Some progress has been made but does not yet achieve the resolutions of the wind measurements.

Prototype and existing wind/wind field data do not yet measure the turbulence. Although the Doppler spectrum width is an indicator of the eddy dissipation rate, few researchers or equipment developers report the variable or its spatial variability. To date, none have done so operationally.

5 CONCLUSIONS

Some progress is being made by the military to develop necessary instrumentation to measured atmospheric fields at high time and spatial resolution. This is motivated by the need to improve the theory and subsequently the models of atmospheric motions at small scales to provide the armed services with reliable models on which to gain superiority over the adversary and successfully complete the mission.

CHAPTER 50

CHALLENGE OF SNOW MEASUREMENTS

NOLAN J. DOESKEN

1 INTRODUCTION—CHARACTERISTICS AND IMPORTANCE OF SNOW

Snow remains one of the truly incredible wonders of nature. Its delicate beauty, its pure whiteness, its endless variety and changeability delight children and adults. Its beauty is offset for some by the reality of the cold obstacle it presents to life's daily activities. The older we get, the greater an obstacle it becomes. How tiny and fragile crystals of ice totally transform a landscape even to the point of bringing temporary silence where the clamor of urban life usually prevails is indeed remarkable. The process of snow formation has gradually been explained by generations of ardent scientists (Bergeron, 1934; Nakaya, 1954; Mason, 1971; Hobbs, 1974; Takahashi et al., 1991). Still the reality of trillions of ice crystals forming and efficiently harvesting atmospheric water vapor and tiny cloud droplets as they fall through cloud layers on their path toward bringing moisture to Earth still seems miraculous to those that think and ponder.

The importance of snow in society today cannot be understated. While it is loved by some and hated by others, it greatly affects economic activities in the mid- and high-latitude nations of the world. Each year millions travel long distances to ski, snowboard, snow mobile, or participate in other winter recreation. Millions of others spend their hard-earned money escaping the cold and snow. Hundreds of millions of dollars are now spent annually in the United States alone clearing sidewalks, streets, highways, parking lots, and airport runways so that commerce and transportation are slowed as little as possible (Minsk, 1998). Despite these efforts, hundreds of lives are

Handbook of Weather, Climate, and Water: Dynamics, Climate, Physical Meteorology, Weather Systems, and Measurements, Edited by Thomas D. Potter and Bradley R. Colman.
ISBN 0-471-21490-6

lost each winter to snow and ice-related accidents. Thousands more are injured in a variety of types of accidents. The economic cost of closed highways, blocked businesses, canceled flights, and lost time is enormous.

Snow is much more than an impediment to commerce. Just think of all the forts, snowballs, and snowmen made each year. Snow is also a structural material providing practical temporary shelter and protection from extreme cold. When smoothed and compacted, it can make an effective temporary aircraft runway in cold, remote areas. Left uncompacted, snow is an excellent insulation material protecting what lies below it from the extreme cold that may exist in the air immediately above.

The weight of snow is a necessary consideration in the design and construction of buildings. Almost every year some buildings are damaged or destroyed following extreme storms or periods of great or prolonged snow accumulation.

The high albedo (ability to reflect light) of fresh snow has profound impacts on the surface energy budget of the globe, which, in turn, dramatically affects climate both locally and over larger areas. Snow-covered areas are significantly colder than adjacent land areas under most weather conditions. Interest in global snow accumulation patterns has risen greatly during the past two decades due to its great significance in the global climate system (Bamzai and Shukla, 1999; Walsh et al., 1982) and the extent to which snow affects and is affected by large-scale global climate change.

In some mountainous areas, the largest threats posed by snow are avalanches. Subtle changes in crystal structure within the snowpack are always occurring. Certain weather patterns, temperature changes, and snowfall sequences lead to layering within the snowpack where some layers are "weak" and do not adhere well to adjacent layers. With the addition of new snow, these unstable snowpacks can be very prone to avalanches that claim dozens of lives in North America and Europe each year. The study of avalanches is a field of its own, involving hundreds of scientists around the world and requiring some unique measurements of internal snowpack characteristics.

Great improvements have been made during recent decades in weather prediction on the scale of a few hours to a few days. Improvements are also evident in long-range forecasts. One of the biggest challenges in weather prediction today remains the quantitative prediction of precipitation. Predictions of snowstorms and the location of the rain/snow/ice boundaries remain difficult even just hours in advance. Almost every year there are examples of large snowstorms that catch us by surprise, or forecasted storms that never materialize. If forecasts are to continue to improve, adequate observations must be taken on the scale needed to track and model precipitation processes.

Perhaps most important of all, snow is water. The hydrologic consequences of snow are so great that it is imperative to carefully track snow accumulation and its water content. Accumulating snowfall becomes frozen reservoirs that later release water into the soil, into aquifers, into streams and rivers, and into reservoirs. This water source is critical to water availability and hydroelectric power generation throughout the year, especially in mountainous regions and where snow contributes a significant fraction of the year's precipitation. If snow melts quickly or if heavy rains fall on melting snow or downstream of melting snow, flooding also becomes a possible consequence.

2 MEASUREMENTS OF SNOW

Because of the importance of snow within our natural and socioeconomic environment, it is essential that we measure this remarkable substance and its salient features. We measure so that we can study, learn, explain, and teach others about snow—its properties and its impacts. We also measure so that we can describe and document what has occurred and note changes that occur over time. This allows us to compare, prepare, and predict so that our society can adapt as well as possible to the challenges and benefits derived from snow.

Much of what we know about snow, its spatial distributions and its contribution to the hydrologic cycle, we have learned from very simple observations taken at a large number of locations for a long period of time. Here is a list of common measurements. Details about instrumentation and methodologies are provided later in this chapter.

- *Precipitation Amount* This refers to the water content of snowfall plus any other liquid, freezing or frozen precipitation falling during the same period such as rain, freezing rain, or ice pellets (sleet). Measurements of precipitation amount are traditionally taken with a recording or nonrecording precipitation gage.
- *Snowfall* The accumulation of new snowfall or other forms of frozen precipitation that have fallen and accumulated in the past day or other specified time period. (Glaze from rain that freezes on contact is not included in this category.) The measurement of snowfall has traditionally been done manually by trained observers using a simple measurement stick and their own good judgment.
- *Snow Depth* This is simply the total depth of snow and ice, including both freshly fallen and older layers. For shallow snows, a ruler or longer measurement stick is all the equipment needed. For deeper snows, fixed snow stakes or special calibrated probing bars are used. Electronic methods for measuring snow depth have been developed in recent years and are used at some sites.
- *Snow Water Equivalent* This term, commonly abbreviated SWE, is the total water content expressed as an equivalent depth of the existing snowpack at the date and time of observation. Since snow does not accumulate or melt uniformly, the SWE is often the average of a set of representative measurements taken in the vicinity of the point of interest. The measurement of SWE is often taken by weighing a full core sample (snow surface down to ground surface) or averaging the weights from several samples. Other methods will also be described.
- *Density* The mass per unit volume is an important variable for describing the nature and potential impacts associated with snow (Judson and Doesken,

2000). A common but often incorrect assumption used in the United States is that 10 in. of new snow has a liquid water content of 1 in., which is a density of 0.10. This could also be expressed as a percentage—10%. Under ideal conditions, the density can be obtained by dividing the measured precipitation amount (gage catch) for the time period of interest, by the measured depth of new snowfall for that same period. However, a direct measurement of water content per carefully determined volume is often more accurate since gage measurements of snowfall often do not catch all the precipitation that actually falls, especially under windy conditions.

- *Precipitation Type and Intensity* For purposes of weather forecasting and verification as well as airport operations and other aspects of transportation, continuous monitoring of the type of precipitation (rain, freezing rain, ice pellets, snow pellets, snow, hail, etc.), its intensity (light, moderate, or heavy), rate of accumulation, and how much the horizontal visibility is restricted are very important. For many years, a simple definition of snowfall intensity has been used in the United States at airport weather stations based on the degree to which the snowfall reduces visibility (U.S. Dept. of Commerce, 1996). For example, unless the horizontal visibility was restricted to less than $\frac{3}{4}$ mile, the snowfall intensity could only be reported as "light." Precipitation type, intensity, and visibilities were all determined manually until the mid-1990s. Electronic sensors have been introduced at many airport weather stations in recent years.
- *Snowcover Extent* This is an assessment of how much of a specified land area is covered by sufficient snow to whiten the surface at any specified time. Until the use of aircraft and satellites devoted to observing snow cover, this was accomplished simply by mapping individual weather station snow depth observations and approximating the location of the edge and area of snow-covered regions.

Other types of snow measurements are taken for basic research, for special applications such as water quality assessments and avalanche prediction, and for military applications in cold climates.

- Albedo
- Crystal types and evolution
- Insulation
- Acidity (snow and the first flushes of snowmelt have been found to be among the most acidic forms of precipitation in some areas)
- Electrical conductivity
- Trafficablity/compactability
- Layer structure and stability
- Forest canopy snow accumulation and sublimation

By no means is this an exhaustive set of measurements. An excellent source of additional information about snow properties and measurements is the *Handbook of Snow* (Gray and Male, 1981).

3 PROPERTIES OF SNOW THAT MAKE BASIC MEASUREMENTS DIFFICULT

All environmental measurements have difficulties and limitations. For snow, its dynamic changing features provide challenges for observations. Snow melts, sublimates, settles, and drifts. Its crystal structure changes from storm to storm and from time to time within a storm. Once on the ground, the crystals change again in the presence of surrounding crystals, temperature gradients, and vapor density gradients. Snow is not deposited uniformly on the ground, and it melts even more unevenly depending on factors such as shading, slope, aspect, wind exposure, vegetation height, color, and amount. Traditional precipitation gages that work fine for measuring rain are often grossly inadequate for capturing and measuring the water content of snow since the feather-light crystals are easily deflected around the precipitation collector even by light to moderate winds so some of the snow does not fall into the gage. Furthermore, snow may cling to the side of collectors, effectively changing the collection diameter of the gage. Additionally, the compressibility of snow makes it difficult to gather the appropriate core samples. When all is said and done, measuring snow is easy. Measuring it accurately and consistently is the problem.

The following example taken from *Weatherwise* magazine (Doesken, 2000) demonstrates the challenge of measuring snow.

> Snowyville, USA
>
> For an example of some of the difficulties that snow observers face, let's imagine we are in Snowyville, USA, on a cold February morning. At 7:00 a.m., it begins to snow. For 12 hours it snows hard and steadily, until it ends abruptly at 7:00 p.m
>
> Five volunteer weather observers all live in lovely Snowyville and all receive the same amount of snow. Each observer has an excellent location for measuring snow in wind-protected locations, and all take measurements using identical rulers and snowboards properly placed on the surface of the snow. [Details on how and where to set up an accurate snow measurement station will be discussed later in the chapter.]
>
> The first observer is a retired engineer and is home all day watching and enjoying the snow. He diligently goes outside every hour on the hour throughout the storm, measures the depth of the new snow on his snowboard, and then clears it in preparation for his next measurement. Each hour there is exactly one inch of new snow on the board. When the storm ends, the observer adds up the snowfall amounts for each hour and comes up with a total of 12 inches of new snow. He writes it down and calls the NWS with his report.
>
> The second observer is also very diligent, but had to go to work in an office downtown. She knows it's snowing hard, so she takes a few minutes after lunch and comes home at 1:00 p.m. to take a measurement. There are five inches of fresh snow on the snowboard when the observer takes her measurement. She clears the surface of the snowboard and places it back on the top of the new snow. She then goes back to her office. When the snow ends at 7:00 p.m., she goes back outside and measures the new snow on the snowboard. There is another five inches. She clears her snowboard and goes inside. She adds up the two measurements, and calls the NWS with her report of ten inches of new snow.

The third observer is a school teacher. She keeps an eye on the snow all day and knows it's snowing hard. She hopes that school will be canceled, but in little Snowyville, people are used to heavy snow. Not even the after-school events are canceled. Between teaching and after-school parent-teacher conferences, she doesn't get home until 7:00 p.m., just as the snow ends. The observer goes straight to her snowboard and measures 8.4 inches. She promptly calls the NWS with her daily report.

The fourth observer normally works in a distant town but instead decides to take the day off due to the heavy snow. He looks out his window nearly constantly, thrilled by the millions of snowflakes. Every hour or so, he journeys out to his snowboard and measures the depth. He just lets it accumulate, though, and does not brush it away. The snow is one-inch deep by 8:00 a.m., five inches deep by 1:00 p.m., and reaches a maximum depth of 9.3 inches around sunset. When the snow ends at 7:00 p.m., it has settled back to 8.4 inches. The observer gets on the phone and calls in his daily snowfall total of 9.3 inches to the NWS.

Finally, the official cooperative observer for the town leaves that morning for a meeting in the nearest town. When he returns that evening, he goes straight to bed. The next morning, precisely at his scheduled reporting time of 7:00 a.m., he goes out to his snowboard, inserts the ruler, and measures 7.2 inches of snow. He clears the board and goes inside. Not realizing how much snow had settled overnight, he records this on his observation form and calls in a 7.2-inch snowfall total to the NWS.

So how much snow actually fell in Snowyville? 12 inches? 7.2 inches? Or something in between? Each observer received the same amount and measured very carefully. Yet their reports were quite different. This is due to the fact that snow melts, settles, and compacts after it falls to the ground. Climatologists and many data users would say that the 9.3-inch report was the correct value. Weather forecasters might argue that the 12-inch report was accurate, but if measuring every hour is good, who is to say that measuring every 30 minutes or even every 15 minutes isn't even better. Then, the sum might have been 14 or 15 inches. Meanwhile, engineers and water resource officials may not care at all about these debates, as long as the observers each accurately reported the same water content of the snow.

This one example shows just how complicated measuring snow really is. And although it does not include the problems of drifting and melting snow, it does demonstrate how important it is to follow similar procedures if measurements are meant to be compared from one location to another.

4 PROCEDURES FOR MEASURING SNOWFALL, SNOW DEPTH, AND WATER CONTENT

As with all other measurements of our environment, the key to success is finding and preserving a consistent and representative location for measurement and maintaining strict standards for instrumentation and observing practices. For comparing data from many locations, consistent procedures and representative measurement locations are essential (Doesken and Judson, 1997).

Precipitation

The measurement of precipitation amount is arguably the most basic and most useful. The most common device for measuring precipitation is a straight-sided cylinder of a sufficient diameter and depth to effectively catch rain, snow, and other forms of precipitation. The National Weather Service (NWS) standard precipitation gage has a diameter of 8 in. and is approximately 2 ft tall. In addition to the outer cylinder, it comes with a funnel, inner measuring tube, and calibrated measuring stick as shown in Figure 1. The area of the opening to the funnel and the outer cylinder (overflow can) is precisely 10 times greater than the area of the top of the inner measuring tube. This allows greater precision in the measurement of precipitation. In the United States, observers measure precipitation to the nearest 0.01 in. while much of the rest of the world measures to the nearest millimeter. For the measurement of rain, the inner cylinder and funnel are installed in and on top of

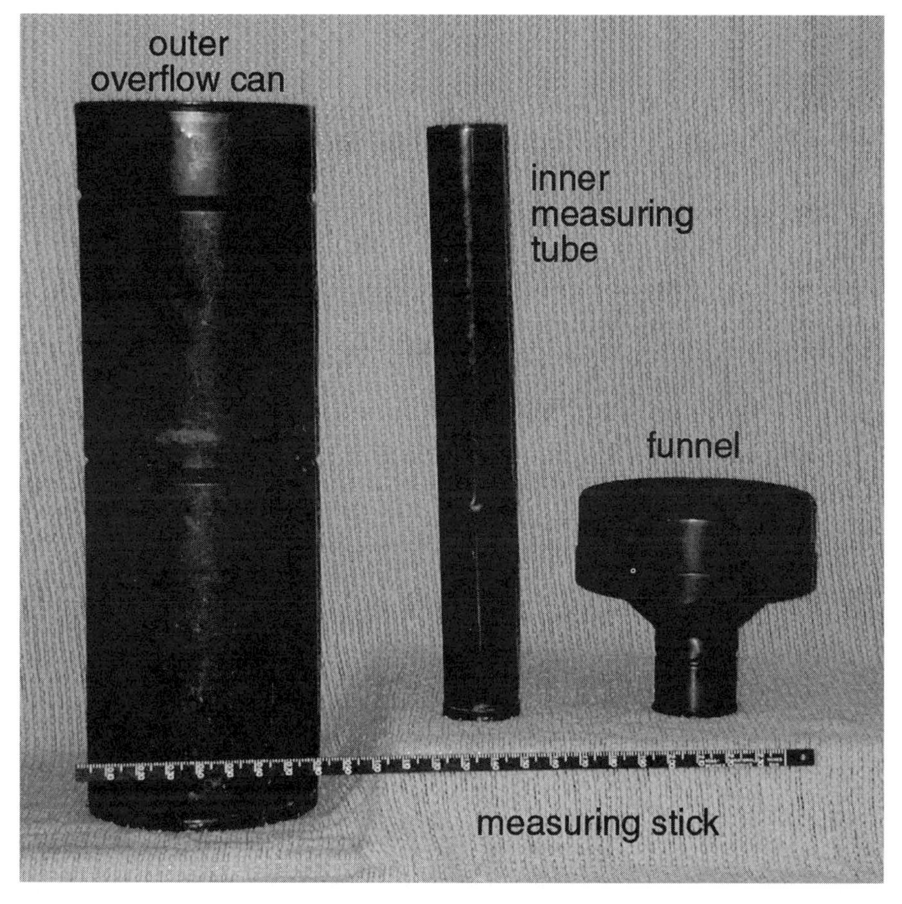

Figure 1 Standard precipitation gage consisting of funnel, inner measuring tube, outer overflow can, and calibrated measuring stick (photo by Clara Chaffin).

the large "overflow can," respectively. For capturing snow, the funnel and inner tube are removed.

Most manual weather stations read and record precipitation daily at a specified time of observation. A few manual stations may measure more frequently. When measuring liquid precipitation, the observer simply inserts the calibrated measurement stick straight down into the inner measurement tube and reads the depth of accumulated water. The observer then records that amount, empties the gage, and is ready for the next observation. Manual rain gages are very accurate for the measurement of liquid precipitation under most conditions. However, when winds are very strong during a rain event, the gage will not catch all of what falls from the clouds since wind movement over the top of the gage will deflect a portion of the raindrops.

For measuring the water content of snowfall only the large outer cylinder (overflow can) is left outdoors. Following a snow event, the standard observing procedure is to bring the gage inside at the scheduled time of observation. The snow sample must first be melted. This is accomplished either by setting the gage in a container of warm water until the snow and ice in the gage are melted or by adding a known amount of warm water directly to the contents of the gage to hasten its melt. Observers then pour the contents of the gage through the funnel into the inner cylinder for measurement, being careful to subtract any volume of water that was added to hasten the melt. In very snowy locations, some observers may be equipped with special calibrated scales so that the gage and its contents can be weighed to determine precipitation. This simplifies the observation considerably, especially for locations where warm water is not readily available.

A variety of other precipitation gages are also used. Weighing-type recording rain gages have been used for many years by the NWS for documenting the timing of precipitation (see, e.g., Fig. 2). For winter operation, an antifreeze solution is required. An oil film on the surface of the fluid reservoir is also recommended to suppress evaporation losses. The use of oil and antifreeze could be environmental hazards so care must be taken in the selection and use of these materials.

Storage precipitation gages, large gages with the capacity to hold several feet of snow water content, have been used for measuring total accumulated precipitation at remote locations. These gages also require oil and antifreeze. The volume of additives must be accurately measured since their density differs from water. Tipping bucket precipitation gages, a favorite gage because of its low cost, relative simplicity, and ease of use for automated applications, are not very effective for measuring the precipitation from snow (McKee et al., 1994). Heat must be applied to the surface of the funnel of these gages in order to melt the snow. Since most snow falls at rates of only a few hundredths of an inch per hour (1 or 2 mm per hour or less), even small amounts of added heat can lead to the sublimation or evaporation of much of the moisture before it reaches the tipping buckets. Furthermore, the addition of heat can create small convective updrafts above the surface of the gage, further reducing gage catch.

The National Weather Service continues to search for a satisfactory and affordable all-weather precipitation gage. As simple as it may seem, the measurement of precipitation amounts has yet to be perfected.

Figure 2 National Weather Service Fischer–Porter recording rain gage. Close to 3000 of these gages are in use in the United States for recording hourly and 15-min precipitation totals in increments of 0.10 in. (photo by Clara Chaffin).

Even if a perfect gage were available, there is still a major problem limiting the accuracy and introducing uncertainty into gage measurements of the water content of snow. *The problem is wind.* Gages protruding into the air present an effective obstacle to the wind resulting in a deflection. Lightweight snow crystals are easily deflected. The result is gage undercatch of precipitation. The degree of undercatch is a complex function of wind speed, snow crystal type, gage shape, and exposure (see Fig. 3). In the measurement of rain, gage undercatch is not significant until winds are very strong. However for snow, even a 5 mph breeze can have a very large effect on gage undercatch.

One approach to improving gage catch efficiency is the installation of a wind shield surrounding the gage to reduce the effects of wind-caused undercatches (see Fig. 4). The most common shield in use in the United States is called an Alter shield, named in honor of its designer. While the Alter shield clearly improves gage catch efficiency, it by no means solves the problem. The Nipher shield is a favorite in Canada, and the results above show why. Unfortunately, this shield does not perform well in heavy, wet snows common in many regions of the world and does not adapt to other types and sizes of precipitation gages, thus making it impractical for use with most precipitation gages in use in the United States. Currently, only a fraction of the U.S. precipitation gages are equipped with wind shields.

The World Meteorological Organization (WMO) has been diligently investigating the challenge of measuring solid precipitation. An extensive international study was completed during the 1990s thoroughly investigated the performance characteristics of a variety of gages and wind shields used in snowy regions of the world (Goodison and Metcalf, 1992). Many consider the Double Fence Intercomparison Reference (DFIR) to produce the most representative gage catch under a full range and snow-

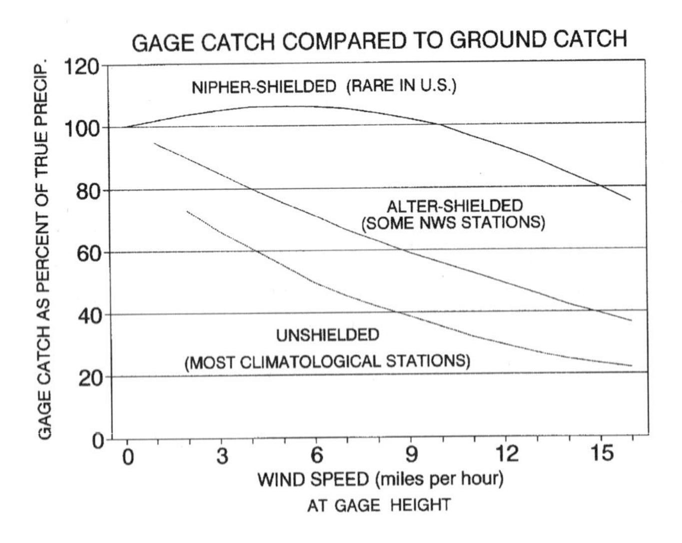

Figure 3 Gage catch efficiency in percent as a function of wind speed at the top of the gage for unshielded, Alter-shielded and Nipher-shielded gages (after Goodison, 1978).

Figure 4 Alter wind shield on a tower-mounted precipitation gage (from Doesken and Judson, 1997).

fall and wind conditions (Yang et al., 1993). Unfortunately, the size (12-m diameter) alone makes this wind shield too bulky and expensive for most common weather stations. No simple solution exists, and few countries show a willingness to change their long-standing practices. But there is no excuse to ignore the problem of gage undercatch since it is now well documented.

A practical approach to the gage catch problem is to be as wise as possible when first deploying weather stations in order to achieve the best exposure for gages. The ideal exposure for a precipitation gage is a delicate compromise between an open and unobstructed location and a protected site where the winds in the vicinity of the gage are as low as possible during precipitation events. The center of a small clearing in a forest or an open backyard in a suburban neighborhood are examples of good sites. The closer to the ground the gage is, the lower the winds will be—a plus for improving gage catch. But the gage must also be high enough to always be above the deepest snow. Rooftop exposures are not recommended because of the enhanced wind problems and the potential for building-induced updrafts that will further reduce gage catch.

When gage undercatch is apparent, observers are encouraged to take a secondary observation by finding a location where the accumulation of fresh snow approximates the area average. A core sample is then taken, and this core is melted or weighed in order to determine its water content. Unless an unrepresentative sample is taken or melting has reduced the water in the remaining layer of fresh snow, the water content of a core often exceeds that of the gage. In practice taking supplemental core samples is seldom done to check for gage undercatch, but it could improve data quality considerably.

Snowfall

The traditional measurement of snowfall requires only a measurement stick (ruler) and a bit of experience. While this may be the simplest meteorological measurement, in practice, it may also be the most inconsistent. The public interest in snowfall is always great, but the measurements are often more qualitative than people realize. The inconsistency is a direct result of the dynamic nature of fresh snow. It lands unevenly, melts unevenly, moves with the wind, and settles with time. The location where measurements are taken, the time of day, the time interval between measurements, the length of time since the snowfall ended, the temperature, the cloudiness, and even the humidity affect snow measurements. The imaginary but realistic example from Snowyville dramatized these points.

For climatological and business applications, for spatial mapping, and for station-to-station comparisons, the best functional definition of snowfall is "the greatest observed accumulation of fresh snow since the previous day prior to melting, settling, sublimation, or redistribution." The typical observer may only go out to measure snowfall once per day at a designated time. Depending on weather conditions and timing, that may or may not coincide with the time of greatest accumulation. If there had clearly been 4 in. of fresh snow on the ground early in the day but only 1 in. still remains at the scheduled observation, did it snow 4 in. or 1 in.? Four inches is obviously the better answer, but if the observer were not there to see it, he or she wouldn't know for sure. Ideally, the observer is available to continuously watch the snow accumulate, note the greatest accumulation, and then note the settling and melting that occurs later. But, in reality, this may not be the case, since much of the historic snowfall data in the United States has come from the National Weather Service Cooperative Observer Program, many of whom are volunteers who can take only one observation each day (National Research Council, 1998).

Because of the nature of fresh snow, perfectly consistent measurements may not be possible. However, there are several simple steps that lead to measurements that achieve a high degree of consistency.

- The location for taking measurements is critical. An unobstructed yet relatively protected location (such as a forest clearing or open backyard away from buildings, trees, and fences) is best since the goal is to measure where snow accumulation is as uniform and undrifted as possible and represents the average snow accumulation of the area.
- The use of a snowboard (square or rectangular flat, white surface positioned on the ground and repositioned daily on the top of the existing snow surface) for measuring the accumulation of new snowfall greatly helps to achieve consistency by providing a smooth, solid surface on which to measure and from which core samples can be taken (Doesken and Judson, 1997).
- When snow is falling, observers should periodically check the accumulation of new snow on the snowboard and note the greatest amount before settling or melting begin to reduce the depth of fresh snow. The greatest accumulation will typically occur just before the snowfall diminishes or changes to rain. Observers should only clear the new snow at the scheduled observation time and then reposition the snowboard on the top of the new snow surface.
- To account for blowing and drifting snow and the resulting uneven accumulation patterns, observers should assess the representativeness of the snowboard measurement by taking an average of several measurements in the surrounding environment, making sure to include only that snow that has fallen since the previous observation. If the snowboard measurement is found to not be representative, either because it has partially or totally blown clear, because drifts have formed in its immediate vicinity, or because snow melted on the snowboard but not on most ground surfaces, the reported snowfall should be the average of as many readings as are needed to obtain an appropriate average over an area including both moderate drifts and moderate clearings.

With the creation and expansion of airport weather stations from the 1930s through the 1950s, came a new emphasis on weather forecasting for air transportation and safety. Hourly weather observations were initiated that included manual measurements of precipitation type and intensity, visibility, and many other weather elements. These became the foundation of the surface airways weather observations (reference) and provided more details about snowfall characteristics than had been previously available. Every hour a complete report of weather conditions was used in a consistent manner from a large number of stations. Conditions were also monitored between hourly reports, and any significant changes were reported in the form of "special" observations. During snowfalls, depths were measured every hour, and special remarks were appended to observations whenever snow depth increased by an inch or more. These "SNOINCR" remarks always caught the attention of meteorologists since they signaled a significant storm in progress. But this also introduced a new complexity into the observation of snow. Instead of observing snowfall once daily, some weather stations reported more frequently. The instructions to airways observers stated that snowfall was to be measured and reported every 6 h. The daily snowfall was then the sum of four 6-h totals. Some weather stations then used the seemingly appropriate procedure to measure and clear their snowboards every hour and add these hourly increments into a 6-h and daily totals.

For some applications, short-interval measurements are extremely useful. However, for climatological applications, snowfall totals derived from short intervals are not the same as measurements taken once daily. For rainfall, a daily total can be obtained by summing short-interval measurements. However, for snowfall, the sum of accumulations for short increments often exceeds the observable accumulation for that period. To demonstrate this, volunteer snow observers were recruited from several parts of the country and measured snowfall for several winters at several locations in the United States. Each observer deployed several snowboards and measured snow accumulations on each board. One board was cleared each hour, one every 3 h, one every 6 h, and every 12 h, and finally one was only measured and cleared every 24 h (see Fig. 5). While results vary from storm to storm, it was very clear that snowfall totals are consistently and significantly higher based on short-interval measurements (Doesken and McKee, 2000). Based on 28 events where measurements taken every 6 h throughout the storms were summed and compared to measurements taken once every 24 h, the 6-h readings summed to 164.4 in., 19% greater than the 138.4-in. total from the once-daily observations. When hourly readings were summed and compared to once-daily measurements, the total was 30% greater. What this means to the user of snowfall data is that two stations, side by side, measuring the same snow amounts at different time intervals may report greatly different values.

In an effort to standardize procedures among different types of weather stations, the National Weather Service issued revised snow measurement guidelines in 1996 (U.S. Dept. of Commerce, 1996a). These guidelines stated that observers should measure snowfall at least once per day but could measure and clear their snowboards as often as but no more frequently than once every 6 h (consistent with long-standing airways instructions). These guidelines were promptly put to the test when an extreme lake-effect snowstorm in January, 1997, produced a reported 77 in. of snowfall in 24 h. Subsequent investigations by the U.S. Climate Extremes Committee (U.S. Dept. of Commerce, 1997) found that this total was the sum of six observations, some of which were for intervals less than 6 h. While the individual measurements were taken carefully, the summation did not conform to the national

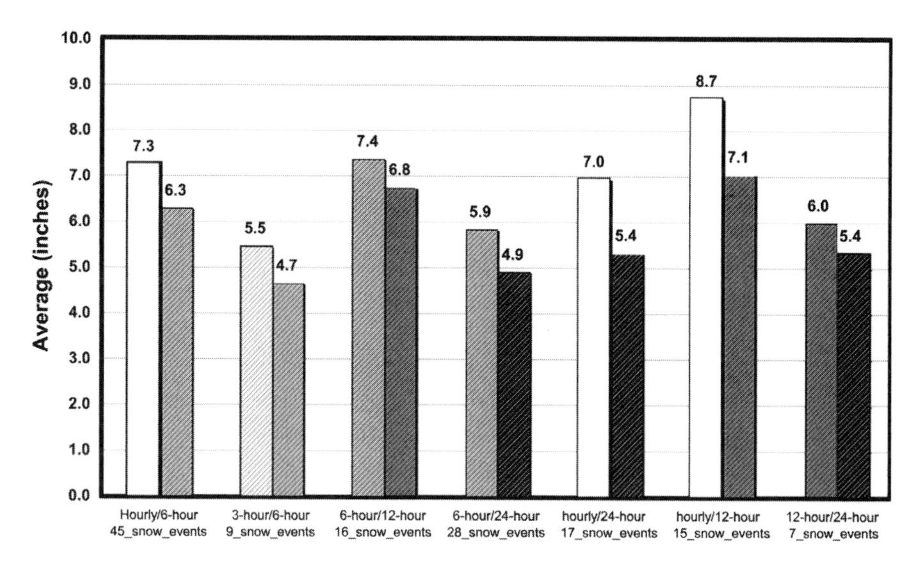

Figure 5 Snowfall comparisons for different measurement intervals. Values were normalized by dividing the total accumulated snowfall in each category by the number of events sampled (Doesken and McKee, 2000).

guidelines and hence could not be recognized as a new record 24-h snowfall for the United States.

The main conclusion here is that where station-to-station comparability and long-term data continuity are the goals, stations must measure in a consistent manner. There may be justification for different observation times and increments, but data from incompatible observation methods should not be interchanged or compared. Many of the data sets in common use today are not fully consistent, so end users may have the responsibility to determine the compatibility and comparability of various station data.

Snow Depth

The measurement of snow depth is not subject to the qualitative and definitional issues that plague the measurement of snowfall. The equipment is again very basic. For areas with shallow or intermittent snowpacks, a sturdy measurement stick is normally carried by an observer to the point(s) of observation and inserted vertically through the entire layer of new and old snow down to ground level. In areas of deep and more continuous snow cover, a fixed snow stake, round or square, is often used—clearly marked in whole inches or in centimeters and permanently installed in a representative location and read "remotely" by an observer standing at a convenient vantage point so that the snow near the snow stake is not disturbed by foot traffic.

The key to useful comparable measurements of snow depth is identifying and maintaining representative locations for taking measurements. Blowing and drifting are inevitable challenges. Uneven melting adds further complications. For uneven snow accumulation, measurements should be taken from several locations proportionally representing both the deeper and shallower areas. Since most traditional weather stations are long distances apart, it is imperative that each measurement represents the predominant conditions in the vicinity of each station.

Snow Water Equivalent (SWE)

This is an extremely important measurement for hydrologic applications. Both flood and overall water supply forecasting rely on accurate measurements of SWE from as many locations as possible to represent the spatial patterns of snow water content that will contribute to subsequent runoff.

The measurement of SWE is taken by only a fraction of National Weather Service stations and is usually accomplished by taking full core samples of total snow on the ground using the 8-in. diameter precipitation gage overflow can. Under deep snow conditions, the melting of snow cores requires considerable amounts of warm water and is tedious and time consuming for observers. Some stations are equipped with special scales that make it relatively easy to take a core sample and immediately estimate the SWE from the weight of the sample. When snow depths exceed 2 ft, the NWS overflow can is inadequate for effectively coring the snowpack.

The Natural Resources Conservation Service and other water resources organizations have a long history of measuring SWE in high snow accumulation areas. Records date back to the 1930s throughout the western mountains with even longer records from a few sites. A wealth of literature, much of it informal and non-peer-reviewed, exists detailing the results of years of experimentation and field testing of devices and techniques to measure snow water in the deep snow regions of North America. Proceedings of the Western Snow Conference and Eastern Snow Conference are great sources for information on the evolution of snow measurements and related research.

Over time, two devices have emerged as the standards for measuring snow water equivalent in deep snow accumulation regions. The Federal Snow Sampler, a portable set of tubes, handle, and a cutter to cleanly penetrate deep snow and ice layers, has been in use for many years (see Fig. 6). Core samples of the snowpack are taken with this instrument, and the sample is then weighed in situ with specially calibrated scales to determine the snow water equivalent of the core. To account for the nonuniform accumulation of snow, several cores are taken at each site. A measurement site is called a "snow course." The final SWE value for a snow course is the average of a set of measurements across the snow course. Each time a snow course is read, core measurements are taken at the same approximate set of individual points. Snow courses have been traditionally read once or twice a month beginning in midwinter and continuing into the spring until all annual snow has melted.

The second instrument, in wide use since the late 1970s, is called a snow pillow (see Fig. 7). This is essentially a scale built at ground level to measure the weight of

Figure 6 Federal snow sampler consists of a cutter, tubes, and scales for measuring snow water equivalent in moderate and deep snowpacks (from Doesken and Judson, 1997).

the snowpack as it accumulates and melts. Snow pillows were developed to provide remote measurements of SWE without requiring the huge expense of time and effort involved in sending teams of scientists and hydrologic technicians into the back country every month.

As with snowfall and snow depth, the utility and comparability of the observations are only as good as the representativeness of the measurement location and the averaging process. Snow pillow measurements have their own set of challenges. For example, "bridging" and other nonuniformities in load-bearing characteristics within the snowpack can compromise the accuracy of snow pillow measurements.

5 CONTRIBUTION OF TECHNOLOGY TO SNOW MEASUREMENTS

Many aspects of the measurement of snow continue to use only the simplest of instrumentation in the hands of trained and experienced observers. Increasingly,

Figure 7 U.S. Department of Agriculture Natural Resources Conservation Service SNOTEL (Snow Telemetry) site in the western United States including snow pillow (in foreground), a shielded standpipe precipitation storage gage (left background), and radio telemetry equipment (from Doesken and Judson, 1997).

scientists and practitioners turn to new technologies and new measurement techniques to acquire the data needed to better understand and apply our knowledge of snow to improve forecasts. Remote sensing is providing data sets showing greater areal coverage and finer resolution information on the spatial distribution of snow. This leads to improved models of snowmelt processes and water supplies. At the same time, improved physical models have pointed out the deficiencies in surface data motivating greater efforts to employ technology to gather more and better data about snow.

In these few pages we cannot do justice to all the technological innovations that are helping improve the measurements of snow and its properties. Rather, we will touch on a few technologies and instruments currently in use. Some of these ideas have been around for many years, but the actual implemention of operational measurement systems is made possible by the remarkable expansion in low-cost computing power in recent years.

Visual Satellite Imagery

From the time the first satellites orbited Earth, it was apparent that snow cover was easily detectable from space. Depth and water content are not easily estimated from reflected visible light, but the extent of snow cover can be readily evaluated from space. The primary limitation is cloud cover, which makes it impossible to see the snow cover below using only visible wavelengths.

Meteorological Radar

Radar refers to the remote-sensing technique of transmitting microwaves of a specified wavelength and receiving, processing, and displaying that portion of the transmitted energy reflected back to the transceiver. Rain and mixed-phase precipitation reflect microwave energy relatively effectively. Snow crystals can also be detected but, depending on crystal structure and temperature, are not detected as well as "wetter" forms of precipitation. The U.S. National Weather Service routinely uses radar to monitor the development, movement, and intensity of snowfall. Improvements in radar realized by the NWS WSR-88D allow estimates of snowfall intensity (Holyrod, 1999) and potential accumulation rates. However, ground truth data remain essential for radar calibration. Detection efficiency varies greatly with distance from the radar, cloud height, and other factors. Still, this technology affords wonderful opportunities for studying snowfall processes in action.

Passive Microwave Remote Sensing

Energy in the form of microwaves is continuously emitted from Earth's surface. Snow crystals within the snowpack scatter and attenuate these microwaves. Aircraft or satellites overhead can detect these emissions and through a series of approximations and algorithms can estimate regional snow cover, water content, and approximate depth. There are various limitations of this technology. For example, shallow or wet snows are more difficult to detect than deeper snows containing little or no liquid water. However, a clear advantage of this technology is that it is not greatly affected by cloud cover, so measurements can be taken day and night during both clear and cloudy conditions.

Gamma Radiation Remote Sensing

The soil near the surface of Earth constantly emits radiation to space in the form of gamma waves. Snow cover attenuates this radiation in proportion to the water content of the snow on the ground. This form of radiation is best detected over relatively narrow bands by receivers mounted on aircraft. Levels of background gamma emissions must be measured in the fall prior to snow accumulation and then along the identical flight path at different times throughout the winter. This method of mapping snow water equivalent is used operationally in several parts of the United States where large river flooding from snowmelt is a common problem.

Acoustic Snow Depth Sensing

Point measurements of snow depth can be taken continuously and remotely. Sound waves from an above-ground transmitter reflect off the snow surface. By measuring the time for the reflected wave to reach the receiver, the distance can be measured that corresponds to a snow depth. The depth of older snow is easiest to measure since the surface tends to become smoother and harder with time. But recent improvements in signal processing have led to accurate measurements of the depth of fresh snow as well. While heavy snow is falling, measurements may be compromised as a portion of the sound wave is reflected by ice crystals in the air.

Satellite Snow Depth Measurements

High-resolution mapping of the elevation of Earth's surface is leading to the opportunity to do similar mapping of snow depth by computing the difference between the elevation of a current observation with the known elevation of the ground from previous satellite measurements. The accuracy of this method requires extremely high-resolution background data and nearly perfect navigation of the data.

Portable Depth/Water Content Sensors

New sensors are being developed for use by ski areas and others concerned about detailed spatial patterns of snow depth and water content. These devices can be sled mounted and pulled by a skier. Using geographical positioning systems to automatically map the sensors location, detailed maps of snow depth/water content can be made.

6 SUMMARY OF SNOW DATA CONTINUITY

More than 100 years of measurements of snowfall, snow depth, and water content are available at hundreds of locations in the United States providing a remarkable resource for meteorological, hydrological, environmental, engineering, and societal applications. The data, however, are far from perfect. All the challenges to accurate observations that are described here have been handled from the very beginning of quantitative observations with varying degrees of success. Observational consistency has been hard to maintain due, in part, to the fact the much of the data have been gathered by volunteers who have received only modest training and who often can take only one observation per day. Even at the nation's primary weather stations, many changes have occurred over time affecting observational consistency. The large natural variability in snowfall sometimes hides the impact of observing changes. Yet, in historical perspective, seemingly small observational changes such as station exposure, time and frequency of observation, and observing procedures do have profound impacts on historical time series. The example in Figure 8 for Sault Ste. Marie, Michigan, is fairly typical of snowfall time series found in the United States. It can be very difficult to distinguish between actual climate variations and true data inhomogeneities.

Perhaps the greatest change in surface observations was the deployment of the Automated Surface Observing System (ASOS) by the National Weather Service during the 1990s as a part of a very extensive national modernization effort. A change in precipitation gages and the automation of the measurement of visibility

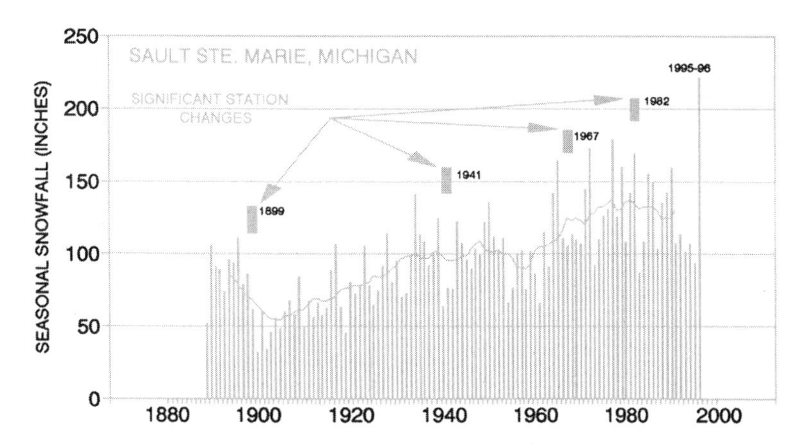

Figure 8 Historic seasonal snowfall totals at Sault Ste. Marie, Michigan. A combination of station moves, changes in exposure, and changes in observational procedures (some of which have not been documented are superimposed on what appears to be a very significant long-term increase in snowfall. Long-term snowfall time series for other parts of the country are plagued by similar problems. Improving the consistency in observations would make it easier to confidently identify important climate variations and trends (from Doesken and Judson, 1997).

and precipitation type have resulted in discontinuous records at several hundred stations. Furthermore, the measurement of snowfall was discontinued completely at many stations since it was not a requirement of the Federal Aviation Administration at that time and did not lend itself to automation.

7 SOURCES OF SPECIALIZED SNOW DATA

Incredible resources about snow, historic data, new measurement technologies, and applications are available, many via the World Wide Web. Some recommended sources include:

National Weather Service, National Operational Hydrologic Remote Sensing Center: *http://www.nohrsc.nws.gov*

National Snow and Ice Data Center (University of Colorado in collaboration with the National Oceanic and Atmospheric Administration: *http://nsidc.org*

National Oceanic and Atmospheric Administration, National Climatic Data Center: *http://lwf.ncdc.noaa.gov*

U.S. Department of Agriculture, Natural Resources Conservation Service, National Water and Climate Center: *http://www.wcc.nrcs.usda.gov*

U.S. Army Corps of Engineers Cold Regions Research and Engineering Laboratory: *http://www.crrel.usace.army.mil*

National Aeronautics and Space Administration: *http://www.nasa.gov*

Website addresses are as of autumn 2001.

REFERENCES

Bamzai, A. S., and J. Shukla (1999). Relation between Eurasian snow cover, snow depth, and the Indian summer monsoon: An observational study, *J. Clim.* **12**, 3117–3132.

Bergeron, T. (1935). On the Physics of Cloud and Precipitation. Proces Verbaux des Seances de l'Association de Metiorolgie, UGGI, 5th assemblee generale, Lisbon, Sept. 1933.

Doesken, N. J., and A. Judson (1997). *The Snow Booklet: A Guide to the Science, Climatology and Measurement of Snow in the United States*, Atmospheric Science Department, Colorado State University, Fort Collins.

Doesken, N. J., and R. J. Leffler (2000). Snow foolin', *Weatherwise* **53**(1), 30–37.

Doesken, N. J., and T. B. McKee (2000). Life after ASOS (Automated Surface Observing System)—Progress in National Weather Service Snow Measurement. Proceedings, 68th Annual Meeting, Western Snow Conference, April 18–20, Port Angeles, WA, pp. 69–75.

Goodison, B. E. (1978). Accuracy of Canadian snow gauge measurements, *J. Appl. Meteorol.* **27**, 1542–1548.

Goodison, B. E., and J. R. Metcalf (1992). The WMO Solid Precipitation Intercomparison: Canadian Assessment, WMO Technical Conference on Instruments and Method of Observation, WMO/TD No. 462, WMO, Geneva, pp. 221–225.

Gray, D. M., and D. H. Male (Eds.) (1981). *The Handbook of Snow: Principles, Processes, Management and Use*, Toronto, Pergamon.

Hobbs, P. V. (1974). *Ice Physics*, Bristol, Oxford University Press.

Holyrod, E. W. (1999). Snow Accumulation Algorithm for the WSR-88D Radar: Supplemental Report, Report #R-99-11, USDI, Bureau of Reclamation, Technical Service Center, Denver.

Judson, A., and N. J. Doesken (2000). Density of freshly fallen snow in the Central Rocky Mountains, *Bull. Am. Meteor. Soc.* **81**(7), 1577–1587.

Mason, B. J. (1971). *The Physics of Clouds*, 2nd ed., Oxford, Clarendon.

McKee, T. B., N. J. Doesken, and J. Kleist (1994). Climate Data Continuity with ASOS—1993 Annual Report for the Period September 1992–August 1993, Climatology Report 94-1, Atmospheric Science Department, Colorado State University, Fort Collins.

Minsk, L. D. (1998). *Snow and Ice Control Manual for Transportation Facilities*, New York, McGraw-Hill.

Nakaya, U. (1954). *Snow Crystals, Natural and Artificial*, Cambridge, MA, Harvard University Press.

National Research Council (1998). *Future of the National Weather Service Cooperative Observer Network*, Washington, DC, National Academy Press.

National Weather Service (2001). Snow Measurement Video, prepared by Department of Atmospheric Science, Colorado State University for National Weather Service, Silver Spring, MD, 30 min.

Takahashi, T., T. Endoh, G. Wakahama, and N. Fukuta (1991). Vapor diffusional growth of free-falling snow crystals, *J. Meteorol. Soc. Jpn.* **69**, 15.

U.S. Dept of Commerce (1996a). Snow Measurement Guidelines (rev. 10/28/96), NOAA, Silver Spring, MD, National Weather Service.

U.S. Dept. of Commerce (1996b). *Surface Weather Observations and Reports*, National Weather Service Handbook No. 7, NOAA, Silver Spring, MD, National Weather Service, Office of Systems Operations.

U.S. Dept. of Commerce (1997). Evaluation of the Reported January 11–12, 1997 Montague, New York, 77-inch 24-hour Lake-Effect Snowfall, NOAA, Silver Spring, MD, National Weather Service, Office of Meteorology.

Walsh, J. E., D. R. Tucek, and M. R. Peterson (1982). Seasonal snow cover and short-term climatic fluctuations of the United States, *Mon. Wea. Rev.* **110**, 1474–1485.

Yang, D., J. R. Metcalfe, B. E. Goodison, and E. Mekis (1993). An Evaluation of the Double Fence Intercomparison Reference Gauge, Proceedings, Eastern Snow Conference, 50th Meeting, Quebec City, Canada, pp. 105–111.

INDEX